BITSAT

Birla Institute of Technology & Science Admission Test

10 YEARS' SOLVED PAPERS (2021-2012)

5 MOCK TESTS

for 2022 Exam

BITSAT

Birla Institute of Technology & Science Admission Test

10 YEARS' SOLVED PAPERS (2021-2012)

5 MOCK TESTS

for 2022 Exam

arihant

ARIHANT PRAKASHAN (Series), MEERUT

arihant

ARIHANT PRAKASHAN (Series), MEERUT

卐 **Administrative & Production Offices**

Regd. Office

'Ramchhaya' 4577/15, Agarwal Road, Darya Ganj, New Delhi -110002
Tele: 011- 47630600, 43518550

卐 **Head Office**

Kalindi, TP Nagar, Meerut (UP) - 250002
Tel: 0121-7156203, 7156204

卐 **Sales & Support Offices**

Agra, Ahmedabad, Bengaluru, Bareilly, Chennai, Delhi, Guwahati, Hyderabad, Jaipur, Jhansi, Kolkata, Lucknow, Nagpur & Pune.

卐 **ISBN** 978-93-25799-84-4

PO No : TXT-XX-XXXXXXX-X-XX

Published by Arihant Publications (India) Ltd.

For further information about the books published by Arihant, log on to www.arihantbooks.com or e-mail at info@arihantbooks.com

Follow us on

PREFACE

Whenever a student decides to prepare for any examination his/her first and foremost curiosity is to know about the type of questions that are expected in the exam. This becomes more important in the context of competitive examinations where there is neck-to-neck competition.

We feel great pleasure in presenting before you this book containing Error Free Solutions of previous years' **BITSAT** Engineering Entrance Exams. Going through this book, you will get an exact idea of the questions generally asked in BITSAT.

We have made maximum efforts to provide correct solutions to the best of our knowledge and opinion. Solutions are not just sketchy–rather, have been drafted in a manner that the students will surely be able to solve some other related problems too. In this way, the students shall be able to judge the extent upto which they have been able to comprehend the methods.

Along with past exams solutions, the book also has **5 Mock Tests** framed exactly on latest pattern & trend of BITSAT. These Mock Tests will help you to assess your preparation level before the exam.

We hope, these sets of papers along with error free solutions would be highly beneficial to the students. We would be grateful if any discrepancy or mistake in the questions or answers is brought to our notice so that these could be rectified in subsequent editions.

Publisher

CONTENTS

BITSAT

Birla Institute of Technology & Science Admission Test

Solved Paper 2021*

Instructions

- There are 150 questions in all. The number of questions in each part is as given below.

		No. of Questions
Part I	Physics	1-40
Part II	Chemistry	41-80
Part III	a. English Proficiency	81-95
	b. Logical Reasoning	96-105
Part IV	Mathematics	106-150

- All questions are Multiple Choice Questions having four options out of which **only one** is correct.
- Each correct answer fetches 3 marks while incorrect answer has a penalty of 1 mark.
- Time allotted to complete this paper is 3 hrs.

PART I

Physics

1. An ideal monoatomic gas is taken round the cycle *ABCDA* as shown in the *p-V* diagram.

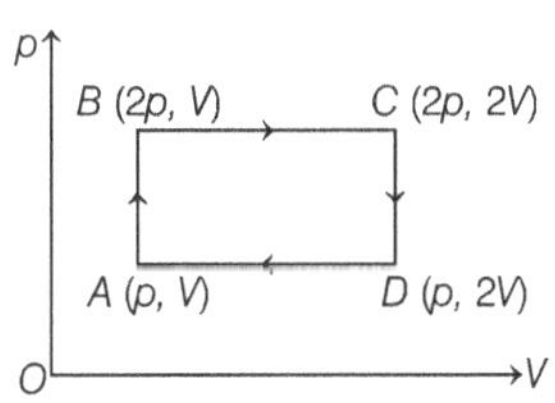

The work done during the cycle is

a. pV ***b.*** $2pV$ ***c.*** $pV/2$ ***d.*** zero

2. The initial speed of a body of mass 2.0 kg is 5 m/s. A force acts for 4 s in the direction of motion of the body, as shown in force-time graph. The impulse of force is

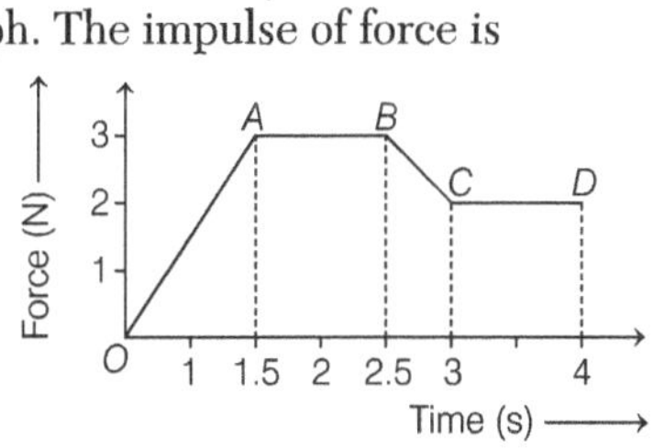

a. 8.50 Ns ***b.*** 8 Ns
c. 5.5 Ns ***d.*** 6 Ns

3. For a constant hydraulic stress on an object, the fractional change in the object's volume $\left(\frac{\Delta V}{V}\right)$ and its bulk modulus *B* are related as

a. $\frac{\Delta V}{V} \propto B$ ***b.*** $\frac{\Delta V}{V} \propto B^2$
c. $\frac{\Delta V}{V} \propto \frac{1}{B}$ ***d.*** $\frac{\Delta V}{V} \propto \frac{1}{B^2}$

4. A shunt is connected in parallel with a galvanometer. Why?

a. To prevent galvanometer from strong current
b. To convert galvanometer into ammeter
c. To increase the range of ammeter
d. All of the above

5. The phase difference between the V_{out} and V_{in} of *CE*-amplifier circuit is

a. 90° ***b.*** 180° ***c.*** 0° ***d.*** 270°

*Memory Based

6. If nth division of main scale coincides with $(n+1)$th divisions of vernier scale. The least count of the vernier is (Given, one main scale division is equal to a units)

a. $\frac{a+1}{n+1}$ **b.** $\frac{a}{n}$

c. $\frac{a}{n+1}$ **d.** $\frac{a+1}{n}$

7. A uniform rod AB of length $L = 1$ m is sliding along two mutually perpendicular surfaces OP and OQ as shown in figure.

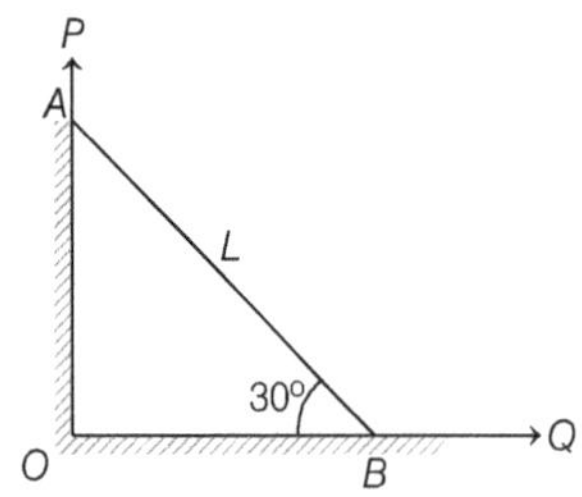

When the rod subtends an angle $\theta = 30°$ with OQ, then the end B has a velocity $\sqrt{3}$ m/s. The velocity of end A at that time is

a. 1 m/s **b.** 0.5 m/s

c. $\sqrt{3}$ m/s **d.** $\frac{1}{\sqrt{3}}$ m/s

8. A tray of mass (M) 12 kg is supported by a spring as shown in figure.

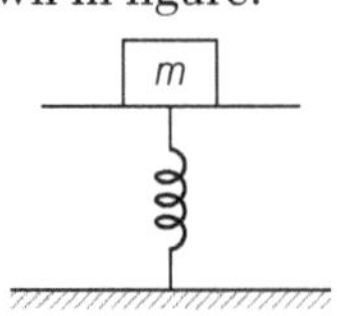

When the tray is pressed down and released, it executes SHM with a period of 1.5 s. When a block of mass m placed on the tray, then the period of SHM changes to 3.0 s. The mass of block is

a. 36 kg **b.** 48 kg **c.** 12 kg **d.** 24 kg

9. The phasor diagram of a load represents which circuit?

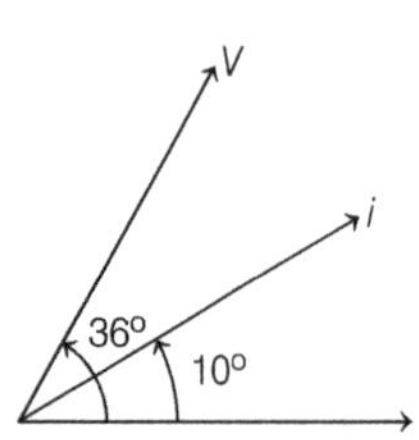

a. Purely capacitive

b. Purely inductive

c. R-L-C circuit with X_L more than X_C

d. R-L-C circuit with X_L less than X_C

10. The equation of progressive wave is given by $Y = \sin\left[x\left(\frac{t}{5} - \frac{x}{9}\right) + \frac{\pi}{6}\right]$ cm. Which one of the following is correct?

a. $v = 5$ cm/s **b.** $\lambda = 18$ cm

c. $A = 0.04$ cm **d.** $f = 50$ Hz

11. The property of light used in optical fibre cables is

a. total internal reflection

b. refraction

c. interference

d. polarisation

12. A man walks in a straight line for 5 min with a velocity of 45 m/s. What is the speed with which he has to move in order to comeback to its original position in 1.5 min?

a. 90 m/s **b.** 150 m/s

c. 135 m/s **d.** 115 m/s

13. From a solid sphere of mass M and radius R, a spherical portion of radius $\frac{R}{2}$ is removed as shown in the figure.

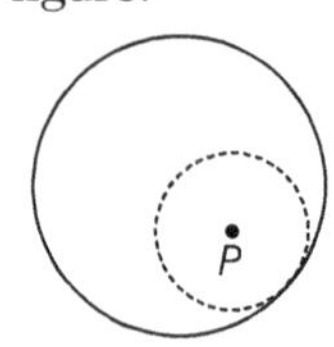

Taking gravitational potential $V = 0$ at $r = \infty$, the potential at the centre of the cavity thus formed is

a. $-\frac{GM}{2R}$ **b.** $-\frac{GM}{R}$

c. $-\frac{2GM}{3R}$ **d.** $-\frac{2GM}{R}$

14. Four charges equal to $+Q$ are placed at the four corners of a square and a charge $(-q)$ is at its centre. If the system is in equilibrium, then the value of $-q$ is

a. $-\frac{Q}{4}(1+2\sqrt{2})$ **b.** $\frac{Q}{2}(2\sqrt{2}+1)$

c. $\frac{Q}{4}(1+2\sqrt{2})$ **d.** $\frac{Q}{4}(1+\sqrt{2})$

15. A force $F = a + bx$ acts on a particle in the x-direction, where a and b are constants. The work done by this force during a displacement from $x = 0$ to $x = d$ is

a. $\left(a + \frac{bd}{2}\right)d$ **b.** $\left(ad + \frac{bd}{2}\right)$

c. $\left(\frac{ad+bd}{2}\right)$ **d.** $(ad+b)\frac{d}{2}$

16. A proton has kinetic energy $E = 100$ eV which is equal to that of a photon. The wavelength of photon is λ_2 and that of proton is λ_1. The ratio $\frac{\lambda_2}{\lambda_1}$ is proportional to

a. E^2 *b.* $E^{\frac{1}{2}}$ *c.* E^{-1} *d.* $E^{-\frac{1}{2}}$

17. A block of mass 10 kg rests on a rough inclined plane making an angle of 30° with the horizontal. The coefficient of static friction between the block and the plane is 0.1. The friction force on the block is

a. 4.9 N *b.* $49\sqrt{3}$ N
c. 49 N *d.* $0.1 \times 49\sqrt{3}$ N

18. Which of the following graphs show the correct relation between conductivity and temperature for a metallic conductor?

a.
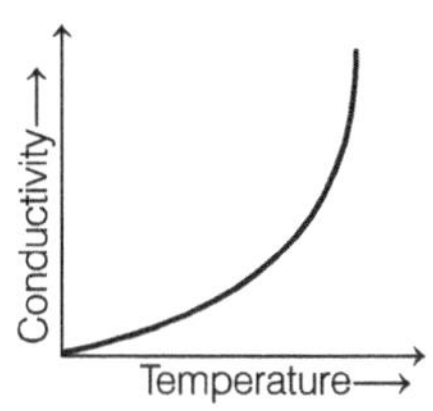

b.
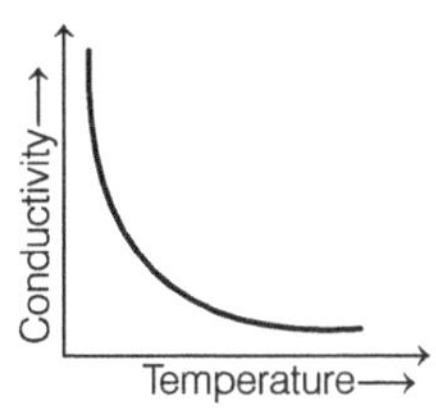

c.
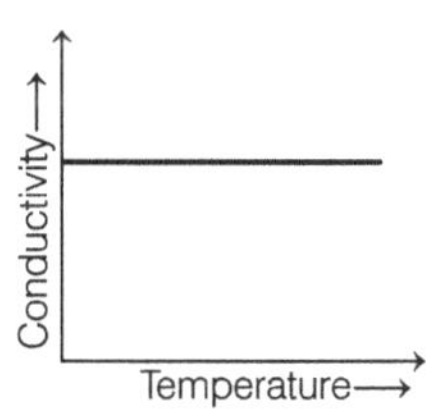

d.
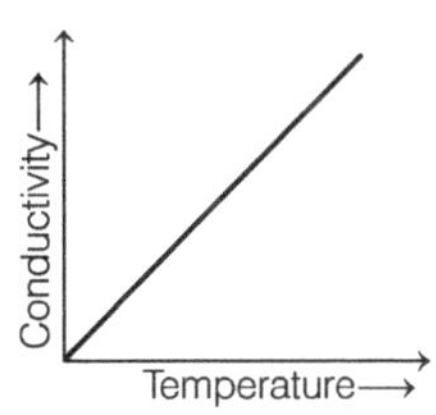

19. The radius of a muonic hydrogen atom is 2.5×10^{-13} m. The total atomic volume (in m^3) of a mole of such hydrogen atoms is
(Take, $\pi = 3.14$)

a. 3.94×10^{-14} *b.* 3.09×10^{-14}
c. 4×10^{-14} *d.* 3.9×10^{-14}

20. The angular momentum of a body placed at origin of mass 1 kg and having position vector $\mathbf{r} = 3t\,\hat{\mathbf{i}} + 4\hat{\mathbf{j}}$ is

a. time dependent *b.* $3\,\hat{\mathbf{k}}$ J-s
c. $-12\hat{\mathbf{k}}$ J-s *d.* 0

21. If the earth stops rotating about its axis, then what will be the change in the value of g at a place in the equatorial plane?
(Radius of earth = 6400 km)

a. 3.7 cm/s^2 *b.* 9.8 m/s^2
c. 0 *d.* 3.4 cm/s^2

22. Two cars approach a stationary observer from opposite sides as shown in the figure.

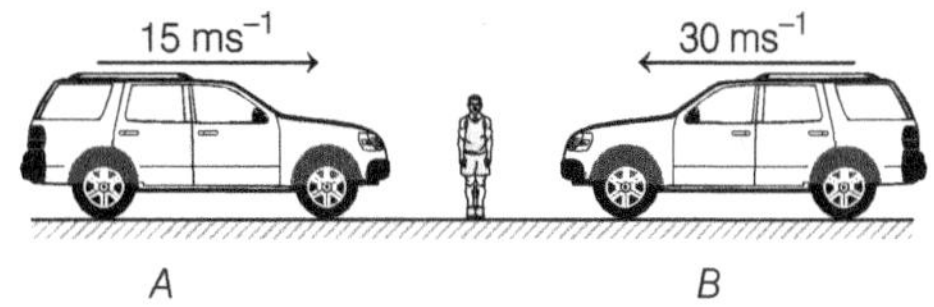

The observer hears no beats. If the frequency of the horn of the car B is 504 Hz, then the frequency of the horn of the car A will be

a. 529.2 Hz *b.* 440.5 Hz
c. 295.2 Hz *d.* None of these

23.

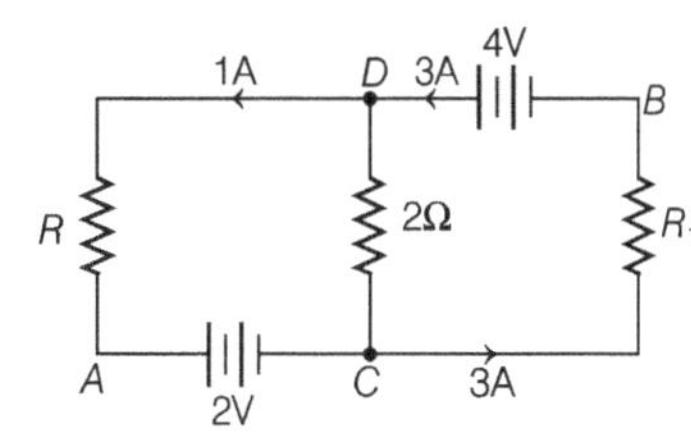

In the following circuit, assuming point A to be at zero potential, then what is the potential at point B?

a. 1 V *b.* 2 V *c.* 4 V *d.* 3 V

24. Two plane mirrors A and B are aligned parallel to each other as shown in the figure.

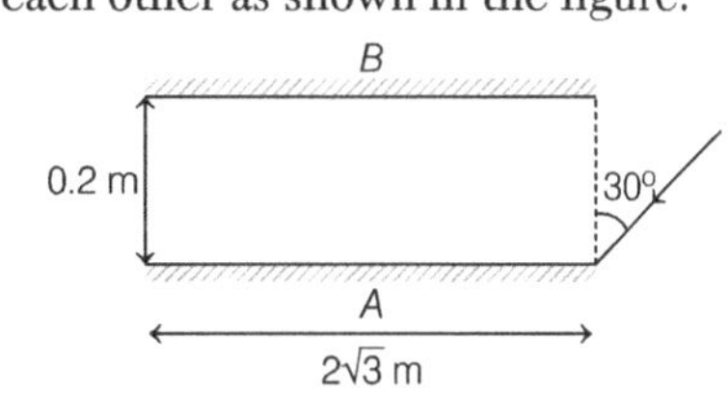

A ray of light is incident at an angle 30° at a point just inside one end of A. The plane of incidence coincides with the plane of the figure. The maximum number of times, the ray undergoes reflection (excluding the first one) before it emerges out is

a. 28 *b.* 34
c. 30 *d.* 29

25. A projectile is given an initial velocity of $(\hat{\mathbf{i}} + \hat{\mathbf{j}})$ m/s, where $\hat{\mathbf{i}}$ is along the ground and $\hat{\mathbf{j}}$ is along the vertical. If $g = 10$ m/s^2, then the equation of its trajectory is

a. $y = x + 5x^2$ *b.* $y = x - 5x^2$
c. $y = x^2 + 5x$ *d.* $y = x^2 - 5x$

26. Two drops of equal radius R coalesce to form a bigger drop. What is the ratio of surface energy of bigger drop to smaller one?

a. $2^{-1/3} : 1$ *b.* $2^{2/3} : 1$
c. $1 : 1$ *d.* $2^{1/2} : 1$

27. Two particles A and B of masses m_A and m_B respectively, are having same charge and moving on same plane. A uniform magnetic field exists perpendicular to this plane. The speeds of the particles are v_A and v_B respectively and the trajectories are as shown in figure. Then,

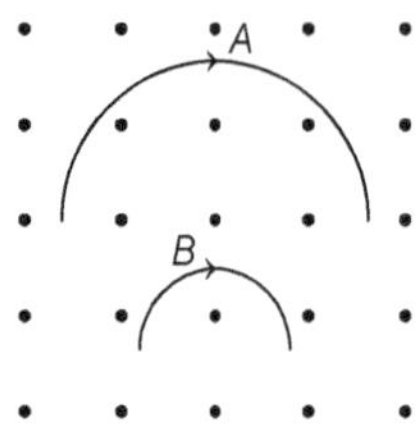

a. $m_A = m_B$ and $v_A = v_B$ **b.** $m_A v_A > m_B v_B$
c. $m_A < m_B$ and $v_A < v_B$ **d.** $m_A v_A < m_B v_B$

28. The potential difference applied to an X-ray tube is decreased. As a result, in the emitted radiation,

a. the intensity increases
b. the intensity decreases
c. the minimum wavelength increases
d. the minimum wavelength decreases

29. The ratio of the specific heats $\frac{C_V}{C_p} = \frac{1}{\gamma}$ in terms of degrees of freedom n is given by

a. $\left(1+\frac{n}{3}\right)$ **b.** $\left(1+\frac{2}{n}\right)$
c. $\left(1+\frac{n}{2}\right)$ **d.** $\left(1+\frac{1}{n}\right)$

30. A cyclist speeding at 6 m/s in a circle of 18 m radius makes an angle θ with the vertical. The minimum possible value of coefficient of friction between the tyres and the ground is

a. 12.041 **b.** 0.2041
c. 11.32 **d.** 10.020

31. Three particles each of mass m are kept at the vertices of an equilateral triangle of side b. Moment of inertia of the system about an axis passing through the centroid and perpendicular to its plane is

a. $3\,mb^2$ **b.** mb^2 **c.** $\frac{mb^2}{3}$ **d.** $\frac{2}{3}mb^2$

32. An electric dipole is placed at an angle of 30° in a non-uniform electric field. The dipole will experience

a. torque only
b. translational force only in the direction of the field
c. translational force only in a direction normal to the direction of the field
d. torque as well as a translational force

33. Two coils A and B have a mutual inductance 0.001 H. The current changes in the first coil according to the equation $i = i_0 \sin \omega t$, where $i_0 = 10\text{A}$ and $\omega = 10\pi$ rads^{-1}. The maximum value of emf in the second coil is

a. $0.01\,\pi$ V **b.** 1πV
c. $0.1\,\pi$ V **d.** $0.05\,\pi$ V

34. The current in the circuit will be

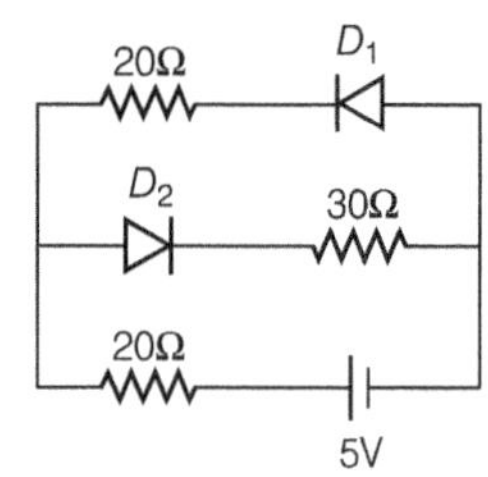

a. 0.125 A **b.** 0.1 A
c. 0.5 A **d.** 0.25 A

35. The bob of a pendulum of length 2m lies at P, when it reaches Q, it loses 10% of its total energy due to air resistance.

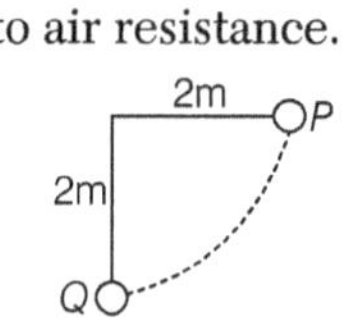

The velocity of bob at Q is

a. 6 m/s **b.** 1 m/s
c. 2 m/s **d.** 8 m/s

36. A particle executes SHM with a frequency f. The frequency with which its kinetic energy oscillates is

a. $\frac{f}{2}$ **b.** f **c.** $2f$ **d.** $4f$

37. A radioactive sample at any instant has its disintegration rate 5000 disintegrations per min. After 5 min, the rate is 1250 disintegrations per min. Then, the disintegration constant (per min) is

a. $0.4 \log_e 2$ **b.** $0.2 \log_e 2$
c. $0.1 \log_e 2$ **d.** $0.8 \log_e 2$

38. The separation between two parallel plates of capacitor is 1 mm. What is the electric potential generates between the plates of capacitor, when electric field of 2000 N/C is applied on it?

a. 2 V **b.** 2000 V
c. 0.2 V **d.** 200 V

39. Choose the correct statement.

a. The speed of light in the meta-material is $v = c|n|$.

b. The speed of light in the meta-material is $v = \frac{c}{|n|}$.

c. The speed of light in the meta-material is $v = c$.

d. The wavelength of the light in the meta-material (λ_m) is given by $\lambda_m = \lambda_{air}|n|$.

40. A T-shaped object with dimensions shown in figure, is lying on a smooth floor. A force **F** is applied at the point P parallel to AB, such that the object has only the translational motion without rotation. Find the location of P with respect to C.

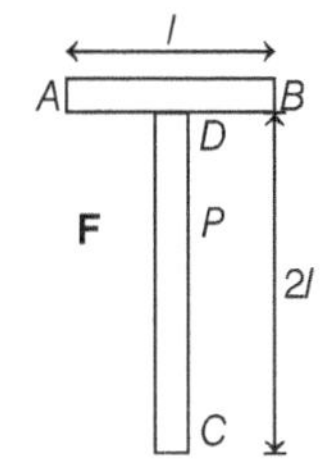

a. $\frac{2}{3}l$ *b.* $\frac{3}{4}l$ *c.* $\frac{3}{2}l$ *d.* $\frac{4}{3}l$

PART II

Chemistry

41. The most volatile compound among the given option is

a. *o*-nitrophenol *b.* *p*-nitrophenol
c. *m*-nitrophenol *d.* can't say

42. What is the magnetic moment of Ti^{2+}? (Given : Atomic number = 22)

a. $\sqrt{8}$ *b.* $\sqrt{10}$ *c.* $\sqrt{6}$ *d.* $\sqrt{9}$

43. Why only Xe can form compounds with fluorine among noble gases?

a. Large size
b. Low electronegativity
c. High ionisation energy
d. Low electron gain enthalpy

44. Which of the following inert gas is used as cryogenic agent?

a. He *b.* Ne *c.* Ar *d.* Kr

45. For the following reaction,

$$2A + 3B \longrightarrow 3C + 4D$$

expression for rate of reaction is

a. $\frac{d[A]}{dt}$ *b.* $\frac{1}{2}\frac{d[B]}{dt}$
c. $\frac{1}{3}\frac{d[C]}{dt}$ *d.* $\frac{1}{4}\frac{d[D]}{dt}$

46. Cyclohexanol on reacting with H_2SO_4 and then heating gives

a. cyclohexene *b.* cyclohexanone
c. cyclohexanol *d.* None of these

47. The major product obtained on reaction of 3-methylbutene with HCl is

a. 2-chloro-2-methylbutane
b. 3-chloro-2-methylbutane
c. 1-chloro-2-methylbutane
d. 3-chloro-3-methylbutane

48. If Rydberg constant is same for all elements, the angular momentum and energy of Li^{2+} of which orbital is equal to angular momentum and energy of 1*s*-orbital of hydrogen atom?

a. 3*s* *b.* 4*s* *c.* 2*p* *d.* 3*d*

49. Sodium iodide reacts with ammonia to give

a. $[Na(NH_3)_4]I$ *b.* $[Na(NH_3)_4]I_3$
c. $[Na(NH_3)I_3]I$ *d.* $[Na(NH_3)_3I]$

50. In photography, which compound is used as a fixing agent

a. sodium thiosulphate *b.* ammonium thiosulphate
c. sodium chloride *d.* Both (a) and (b)

51. The compound formed on reaction of epoxy ethane with NH_3 and H_2O is

a. mono-ethanol amines *b.* di-ethanol amines
c. tri-ethanol amines *d.* All of these

52. The process of removal of excess electrolyte from colloidal solution is

a. coagulation *b.* dialysis
c. ultra-filtration *d.* peptisation

53. Maltase converts maltose into

a. glucose *b.* sucrose
c. fructose *d.* starch

54. Starch is a polymer of

a. glucose *b.* fructose
c. Both (a) and (b) *d.* None of these

55. Methyl alcohol can be distinguished from ethyl alcohol using

a. Fehling solution
b. Schiff's reagent
c. Sodium hydroxide and iodine
d. Phthalein fusion test

56. Sodium stearate ($C_{17}H_{35}COO^{-}Na^{+}$) is

a. anionic soap *b.* cationic detergent
c. anionic detergent *d.* Non-ionic soap

57. Which is not an anti-fluorite structure

a. Rb_2S *b.* BaF_2
c. K_2O *d.* Li_2O

58. The structure of ClF_3 is

a. T-shape *b.* bent shape
c. linear *d.* trigonal planar

59. Which of the following is diamagnetic in nature?

a. O_2 *b.* O_2^{2-}
c. N_2^{2+} *d.* B_2

60. The correct order of bond order in SO_2, SO_3, SO_4^{2-}, SO_3^{2-} is

a. $SO_3^{2-} > SO_4^{2-} > SO_3 = SO_2$
b. $SO_4^{2-} > SO_3 > SO_2 > SO_3^{2-}$
c. $SO_2 = SO_3 > SO_4^{2-} > SO_3^{2-}$
d. Can't say

61. Ionisation energy for H^{+} ion is proportional to r^n, then value of n is

a. 1 *b.* –1 *c.* 2 *d.* –2

62. Role of BHA in food industries

a. prevent oxidation *b.* prevent reduction
c. add sweeteners *d.* None of these

63. Permanganate ion (MnO_4^{-}) is dark purple coloured though Mn is in + 7 oxidation state with d^0 configuration. This is due to

a. d - d transition
b. charge transfer from metal to ligand
c. charge transfer from ligand to metal
d. All of the above

64. Rutherford model could not explain

a. electronic structure of an atom
b. stability of an atom
c. Both (a) and (b)
d. None of the above

65. Which is true in case of $[Ni(CO)_4]$?

a. Hybridisation of Ni is sp^3
b. Hybridisation of Ni is dsp^2
c. Paramagnetic
d. Square planar

66. SI unit of Boltzmann's constant is

a. JK^{-1} *b.* eVK^{-1}
c. erg K^{-1} *d.* JK

67. Acetone does not undergo which type of reaction?

a. Substitution reaction
b. Polymerisation reaction
c. Condensation reaction
d. Addition reaction

68. Magnetic moment of Co in $[CoF_6]^{2-}$ of unpaired electron is

a. $\sqrt{35}$ *b.* $\sqrt{45}$
c. $\sqrt{40}$ *d.* $\sqrt{30}$

69. Some statements about heavy water are given below

1. Heavy water is used as a moderator in nuclear reactors.
2. Heavy water is more associated than ordinary water.
3. Heavy water is more effective solvent than ordinary water.

Which of the above statements are correct?

a. 1 and 2 *b.* 1, 2 and 3
c. 2 and 3 *d.* 1 and 3

70. $CH_3—CHCl—CH_2—CH_3$ has a chiral centre, which of the following represents its R configurations?

a. 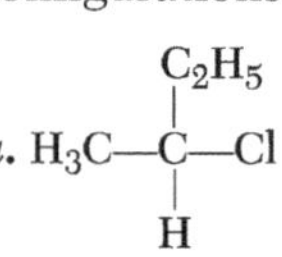
b. 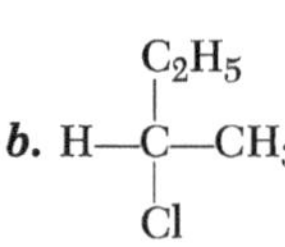
c. 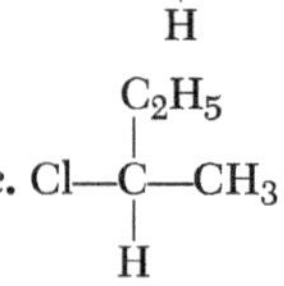
d. 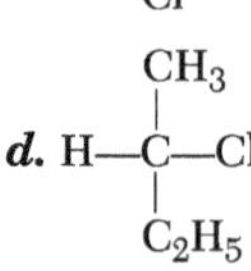

71. In Victor Meyer's method 0.2 g of an organic substance displaced 56 mL of air at STP the molecular weight of the compound is

a. 56 *b.* 112 *c.* 80 *d.* 28

72. For the complete combustion of ethanol, $C_2H_5OH(l) + 3O_2(g) \longrightarrow 2CO_2(g) + 3H_2O(l)$, the amount of heat produced as measured in bomb calorimeter is 1364.47 kJ mol^{-1} at 25°C. Assuming ideality the enthalpy of combustion, ΔH_C, for the reaction will be ($R = 8.314\, JK^{-1}\, mol^{-1}$)

a. – 1366.95 kJ mol^{-1}
b. – 1361.95 kJ mol^{-1}
c. – 1460.50 kJ mol^{-1}
d. – 1350.50 kJ mol^{-1}

73. The incorrect expression among the following is

a. $\frac{\Delta G_{system}}{\Delta S_{total}} = -T$

b. In isothermal process, $W_{reversible} = -nRT \ln \frac{V_f}{V_i}$

c. $\ln K = \frac{\Delta H^\circ - T\Delta S^\circ}{RT}$

d. $K = e^{-\Delta G^\circ/RT}$

74. The solubility of $Pb(OH)_2$ in water is 6.7×10^{-6} M. Its solubility in a buffer solution of pH = 8 would be

a. 1.2×10^{-2} *b.* 1.6×10^{-3}
c. 1.6×10^{-2} *d.* 1.2×10^{-3}

75. The density (in g mL^{-1}) of a 3.60 M sulphuric acid solution that is 29% H_2SO_4 (molar mass = 98 g mol^{-1}) by mass will be

a. 1.64 *b.* 1.88
c. 1.22 *d.* 1.45

76. The relative lowering of vapour pressure of a dilute aqueous solution containing non-volatile solute is 0.0125. The molality of the solution is about

a. 0.70 *b.* 0.50 *c.* 0.90 *d.* 0.80

77. Equal masses of methane and oxygen are mixed in an empty container at 25°C. The fraction of the total pressure exerted by oxygen is

a. $\frac{2}{3}$ *b.* $\frac{1}{3} \times \frac{273}{298}$ *c.* $\frac{1}{3}$ *d.* $\frac{1}{2}$

78. The molar conductivities of KCl, NaCl and KNO_3 are 152, 128 and 111 S cm^2 mol^{-1} respectively. What is the molar conductivity of $NaNO_3$?

a. 101 S cm^2 mol^{-1} *b.* 87 S cm^2 mol^{-1}
c. – 101 S cm^2 mol^{-1} *d.* – 391 S cm^2 mol^{-1}

79. The approximate time duration in hours to electroplate 30 g of calcium from molten calcium chloride using a current of 5 A is (Atomic mass of Ca = 40)

a. 80 *b.* 10 *c.* 16 *d.* 8

80. Given, the reduction potential of Na^+, Mg^{2+}, Al^{3+} and Ag^+ as $E^\circ_{Na^+/Na} = -2.17$ V; $E^\circ_{Mg^{2+}/Mg} = -2.37$ V; $E^\circ_{Ag^+/Ag} = -0.08$ V; $E^\circ_{Al^{3+}/Al} = -1.66$ V
The least stable oxide is

a. Ag_2O *b.* Al_2O_3 *c.* MgO *d.* Na_2O

PART III

a. English Proficiency

Directions (Q. Nos. 81-83) *Choose the word which best expresses the meaning of the underlined word in the sentence.*

81. Decay is an <u>immutable</u> factor of human life.

a. important *b.* unique
c. unchangeable *d.* awful

82. It was an <u>ignominious</u> defect for the team.

a. shameful *b.* admirable
c. unaccountable *d.* worthy

83. His <u>conjecture</u> was the better than mine.

a. guess *b.* fact *c.* surprise *d.* doubt

Directions (Q. Nos. 84-86) *Fill in the blanks.*

84. Freedom and equality are the .... rights of every human.

a. inalienable *b.* inscrutable
c. incalculable *d.* institutional

85. Pradeep's face spoke .... of the happiness he was feeling.

a. elegantly *b.* tons *c.* volumes *d.* much

86. His speech was disappointing : it .... all the major issues.

a. projected
b. revealed
c. skirted
d. analysed

Directions (Q. Nos. 87-89) *Choose the word which is closest to the opposite in meaning of the given italicised word.*

87. Hydra is biologically believed to be *immortal.*

a. undying *b.* perishable
c. ancient *d.* eternal

88. The Gupta rulers *patronised* all cultural activities and thus Gupta period was called the golden era in Indian History.

a. criticised *b.* rejected
c. opposed *d.* spurned

89. This is a *barbarous* act.

a. bad *b.* good
c. civilised *d.* exemplary

Directions (Q. Nos. 90-92) *In each of the following questions, out of the four alternatives, choose the one which can be substituted for the given words/sentence.*

90. A person who does not believe in any religion

a. Philatelist *b.* Rationalist
c. Atheist *d.* Pagan

91. A person who believes that pleasure is the chief good

a. Stoic *b.* Hedonist *c.* Epicure *d.* Sensual

92. One who loves mankind

a. Anthropologist *b.* Philanthropist
c. Seismologist *d.* Optometrist

Directions (Q. Nos. 93-95) *Choose the order of the sentences marked A, B, C, D and E to form a logical paragraph.*

93. A. Tasty and healthy food can help you bring out their best.

B. One minute they are toddlers and next you see them in their next adventure.

C. Your young ones seem to be growing so fast.

D. Being their loving custodians, you always want to see them doing well.

E. Their eyes sparkle with curiosity and endless questions on their tongues.

Codes

a. DBCEA *b.* CADEB
c. CBEDA *d.* ECABD

94. A. It is hoping that overseas friends will bring in big money and lift the morale of the people.

B. But a lot needs to be done to kick start industrial revival.

C. People had big hopes from the new government.

D. So far government has only given an incremental push to existing policies and programmes.

E. Government is to go for big time reforms, which it promised.

Codes

a. BCDAE *b.* EADCB
c. DABEC *d.* CDEAB

95. A. However, women hiring is catching up at a slow and steady rate in the recent times.

B. Gender ratio has been inclined more towards male employees.

C. As a result, recent reports have highlighted the rise in demand for women employees.

D. Women constitute a little over half of world's total population.

E. But, their contribution to measured economic activity is far below the potential.

Codes

a. DEBAC *b.* CDAEB
c. BCDEA *d.* AEDBC

b. Logical Reasoning

96. Choose the correct answer figure which will make a complete square on joining with the problem figure.

Problem figure

Answer figures

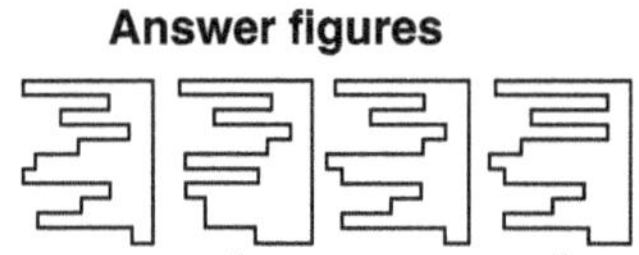

a. *b.* *c.* *d.*

97. In the following question, five figures are given. Out of them, find the three figures that can be joined to form a square.

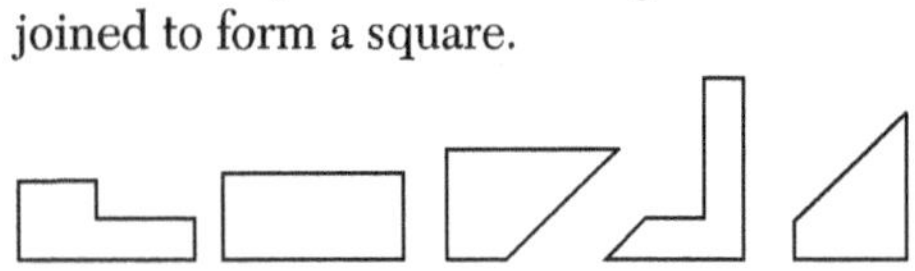

(A) (B) (C) (D) (E)

a. ABD *b.* BCD *c.* ACE *d.* CDE

98. Choose the answer figure which completes the problem figure matrix.

Problem Figure

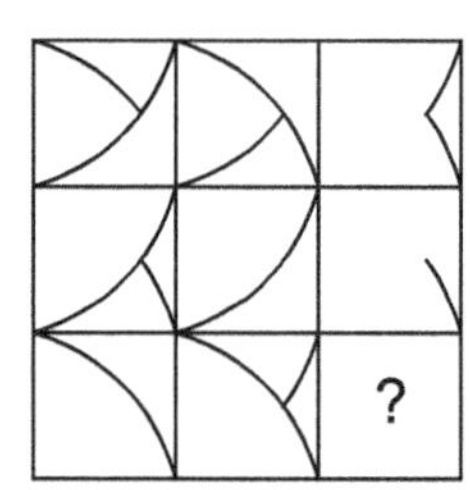

Answer Figures

a. *b.* *c.* *d.*

99. From the given four positions of a single dice, find the colour at the face opposite to the face having red colour.

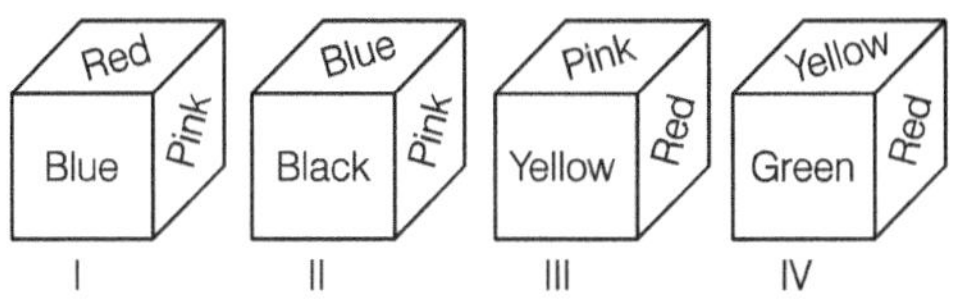

a. Yellow *b.* Pink *c.* Green *d.* Black

100. In the following questions, one or more dots are placed in the figure marked as (A). The figure is followed by four alternatives marked as (a), (b), (c) and (d). One out of these four options contains region(s) common to the circle, square, triangle, similar to that marked by the dot in figure (A).

Problem Figure

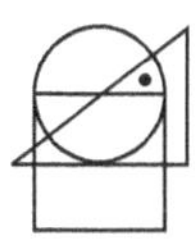

(A)

Answer Figures

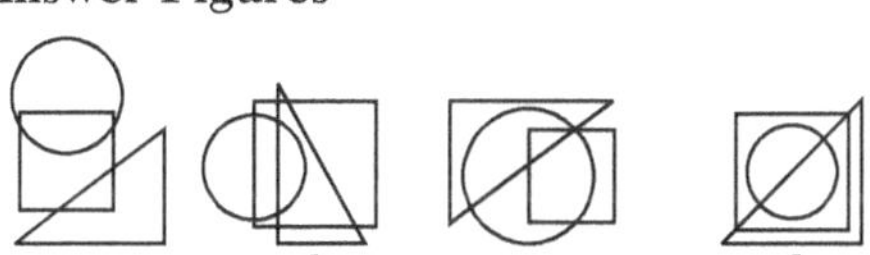

a. *b.* *c.* *d.*

101. Complete the series by replacing '?' mark.

G4T, J9R, M20P, P43N, S90L, ?

a. S90L *b.* V185J *c.* M20P *d.* P43N

102. Neeraj starts walking towards South. After walking 15 m, he turns towards North. After walking 20 m, he turns towards East and walks 10 m. He then turns towards South and walks 5 m. How far is he from his original position and in which direction?

a. 10 m, East *b.* 10 m, South-East
c. 10 m, West *d.* 10 m, North-East

103. The average age of 8 men is increased by 2 yrs. when one of them whose age is 20 yr is replaced by a new man. What is the age of the new man?

a. 28 yr *b.* 36 yr *c.* 34 yr *d.* 35 yr

104. Shikha is mother-in-law of Ekta who is sister-in-law of Ankit. Pankaj is father of Sanjay, the only brother of Ankit. How is Shikha related to Ankit?

a. Mother-in-law *b.* Aunt
c. Wife *d.* Mother

105. In a row of forty children, P is thirteenth from the left end and Q is ninth from the right end. How many children are there between P and R, if R is fourth to the left of Q?

a. 12 *b.* 13 *c.* 14 *d.* 15

PART IV

Mathematics

106. If $\text{Re}(z+2) = |z-2|$, then the locus of z is

a. parabola *b.* circle *c.* ellipse *d.* hyperbola

107. If $a \in R, b \in R$, then the equation $x^2 - abx - a^2 = 0$ has

a. one positive root and one negative root
b. Both positive roots
c. Both negative roots
d. Non-real roots

108. If $a + 2b + 3c = 12, (a, b, c \in R^+)$, then the maximum value of ab^2c^3 is

a. 2^3 *b.* 2^4 *c.* 2^6 *d.* 2^5

109. Sum of n terms of the infinite series

$1.3^2 + 2.5^2 + 3.7^2 + \dots \infty$ is

a. $\frac{n}{6}(n+1)(6n^2+14n+7)$

b. $\frac{n}{6}(n+1)(6n^2+14n+5)$

c. $\frac{n}{6}(n+1)(2n+1)(3n+1)$

d. $4n^3 + 4n^2 + n$

110. If $\log_7 5 - a$, $\log_5 3 - b$ and $\log_3 2 = c$, then the logarithm of the number 70 to the base 225 is

a. $\frac{1-a+abc}{2a(1+b)}$ *b.* $\frac{1-a-abc}{2a(1+b)}$

c. $\frac{1+a-abc}{2a(1+b)}$ *d.* $\frac{1+a+abc}{2a(1+b)}$

111. The maximum number of points of intersection of 10 circles is

a. 80 *b.* 90
c. 85 *d.* 95

112. $\frac{C_1}{C_0} + 2\frac{C_2}{C_1} + 3\frac{C_3}{C_2} + 4\frac{C_4}{C_3} + \dots 20\frac{C_{20}}{C_{19}} =$

a. 120 *b.* 260
c. 210 *d.* 180

113. If $p \neq q \neq r$ and $\begin{vmatrix} 0 & x-p & x-q \\ x+p & 0 & x-r \\ x+q & x-r & 0 \end{vmatrix} = 0$, then the value of x which satisfy the equation is

a. $x = p$ b. $x = q$ c. $x = r$ d. $x = 0$

114. Matrix $A = \begin{bmatrix} x & 3 & 2 \\ 1 & y & 4 \\ 2 & 2 & z \end{bmatrix}$, if $xyz = 60$ and $8x + 4y + 3z = 20$, then $A(adj\ A)$ is equal to

a. $\begin{bmatrix} 64 & 0 & 0 \\ 0 & 64 & 0 \\ 0 & 0 & 64 \end{bmatrix}$ b. $\begin{bmatrix} 88 & 0 & 0 \\ 0 & 88 & 0 \\ 0 & 0 & 88 \end{bmatrix}$

c. $\begin{bmatrix} 68 & 0 & 0 \\ 0 & 68 & 0 \\ 0 & 0 & 68 \end{bmatrix}$ d. $\begin{bmatrix} 34 & 0 & 0 \\ 0 & 34 & 0 \\ 0 & 0 & 34 \end{bmatrix}$

115. If $f(x) = 4x - x^2, x \in R$, and $f(a+1) - f(a-1) = 0$, then a is equal to

a. 0 b. 2 c. 1 d. 3

116. Which of the following is not an equivalence relation in z?

a. $aRb \Leftrightarrow a + b$ is an even integer
b. $aRb \Leftrightarrow a - b$ is an even integer
c. $aRb \Leftrightarrow a < b$
d. $aRb \Leftrightarrow a = b$

117. Which of the following is always true?

a. $(\sim p \vee \sim q) \equiv (p \wedge q)$
b. $(p \to q) \equiv (\sim q \to \sim p)$
c. $\sim(p \to \sim q) \equiv (p \wedge \sim q)$
d. $\sim(p \leftrightarrow q) \equiv (p \to q) \to (q \to p)$

118. The solution of the inequation $4^{-x+0.5} - 7.2^{-x} < 4,\ x \in R$ is

a. $(-2, \infty)$ b. $(2, \infty)$

c. $\left(2, \frac{7}{2}\right)$ d. None of these

119. If $\cos^3 x \cdot \sin 2x = \sum_{m=1}^{n} a_m \sin mx$ is identity in x, then

a. $a_3 = \frac{3}{8}, a_2 = 0$ b. $n = 6, a_1 = \frac{1}{2}$

c. $n = 5, a_1 = \frac{3}{4}$ d. $\Sigma a_m = \frac{1}{4}$

120. Total number of solutions of $|\cot x| = \cot x + \frac{1}{\sin x}, x \in [0, 3\pi]$ is equal to

a. 1 b. 2 c. 3 d. 0

121. The minimum value of $(\sin^{-1} x)^3 + (\cos^{-1} x)^3$ is equal to

a. $\frac{\pi^3}{32}$ b. $\frac{5\pi^3}{32}$ c. $\frac{9\pi^3}{32}$ d. $\frac{11\pi^3}{32}$

122. The origin is shifted to (1, 2). The equation $y^2 - 8x - 4y + 12 = 0$ changes to $y^2 = 4ax$, then a is equal to

a. 1 b. 2 c. -2 d. -1

123. The equations of the bisector of the angles between the straight lines $3x + 4y + 7 = 0$ and $12x + 5y - 8 = 0$ are

a. $7x + 9y + 17 = 0,\ 99x + 77y + 51 = 0$
b. $7x - 9y - 17 = 0,\ 99x + 77y - 51 = 0$
c. $7x - 9y + 17 = 0,\ 99x + 77y + 51 = 0$
d. None of the above

124. Equation of circle which passes through the points $(1, -2)$ and $(3, -4)$ and touch the X-axis is

a. $x^2 + y^2 + 6x + 2y + 9 = 0$
b. $x^2 + y^2 + 10x + 20y + 25 = 0$
c. $x^2 + y^2 + 6x + 4y + 9 = 0$
d. None of the above

125. If $x = 9$ is the chord of contact of the hyperbola $x^2 - y^2 = 9$, then the equation of the corresponding pair of tangent is

a. $9x^2 - 8y^2 + 18x - 9 = 0$
b. $9x^2 - 8y^2 - 18x + 9 = 0$
c. $9x^2 - 8y^2 - 18x - 9 = 0$
d. $9x^2 - 8y^2 + 18x + 9 = 0$

126. The points with position vectors $10\hat{i} + 3\hat{j}$, $12\hat{i} - 15\hat{j}$ and $a\hat{i} + 11\hat{j}$ are collinear, if a is

a. -8 b. 4

c. 2 d. $\frac{82}{9}$

127. Let **a**, **b**, **c** be vectors of lengths 3, 4, 5 respectively and **a** be perpendicular to $(\mathbf{b} + \mathbf{c})$, **b** to $(\mathbf{c} + \mathbf{a})$ and **c** to $(\mathbf{a} + \mathbf{b})$, then the value of $(\mathbf{a} + \mathbf{b} + \mathbf{c})$ is

a. $2\sqrt{5}$ b. $2\sqrt{2}$

c. $10\sqrt{5}$ d. $5\sqrt{2}$

128. For non-zero vectors **a**, **b**, **c**; $|(\mathbf{a} \times \mathbf{b}) \cdot \mathbf{c}| = |\mathbf{a}||\mathbf{b}||\mathbf{c}|$ holds if and only if

a. $\mathbf{a} \cdot \mathbf{b} = 0,\ \mathbf{b} \cdot \mathbf{c} = 0$ b. $\mathbf{b} \cdot \mathbf{c} = 0,\ \mathbf{c} \cdot \mathbf{a} = 0$

c. $\mathbf{c} \cdot \mathbf{a} = 0,\ \mathbf{a} \cdot \mathbf{b} = 0$ d. $\mathbf{a} \cdot \mathbf{b} = \mathbf{b} \cdot \mathbf{c} = \mathbf{c} \cdot \mathbf{a} = 0$

129. Angle between the diagonals of a cube is

a. $\pi/3$ b. $\pi/2$

c. $\cos^{-1}(1/3)$ d. $\cos^{-1}(1/\sqrt{3})$

130. Consider the two lines

$$L_1: \frac{x+1}{3} = \frac{y+2}{1} = \frac{z+1}{2}$$

and $L_2: \frac{x-2}{1} = \frac{y+2}{2} = \frac{z-3}{3}$

The unit vector perpendicular to both the lines L_1 and L_2 is

a. $\frac{-\hat{i}+7\hat{j}+7\hat{k}}{\sqrt{99}}$ b. $\frac{-\hat{i}-7\hat{j}+5\hat{k}}{5\sqrt{3}}$

c. $\frac{-\hat{i}+7\hat{j}+5\hat{k}}{5\sqrt{3}}$ d. $\frac{7\hat{i}-7\hat{j}+\hat{k}}{\sqrt{99}}$

131. The distance between the line $\mathbf{r} = 2\hat{i} - 2\hat{j} + 3\hat{k} + \lambda(\hat{i} - \hat{j} + 4\hat{k})$ and the plane $\mathbf{a}\cdot(\hat{i}+5\hat{j}+\hat{k}) = 5$ is

a. $\frac{10}{9}$ b. $\frac{10}{3\sqrt{3}}$ c. $\frac{10}{3}$ d. None of these

132. Two cards are drawn from a pack of 52 cards. What is the probability that either both are red or both are kings?

a. 7/13 b. 63/221 c. 55/221 d. 3/26

133. If A and B are two independent events such that $P(A) = \frac{1}{2}$ and $P(B) = \frac{1}{5}$, then which of the following is correct?

a. $P\left(\frac{A}{B}\right) = \frac{1}{2}$ b. $P\left(\frac{A}{A \cup B}\right) = \frac{5}{6}$

c. $P\left(\frac{A \cap B}{A' \cup B'}\right) = 0$ d. All of these

134. Box I contains 5 red and 2 blue balls, while box II contains 2 red and 6 blue balls. A fair coin is tossed. If it turns up head, a ball is drawn from box I, else a ball is drawn from box II. The probability ball drawn is from box I, if it is blue, is

a. 27/56 b. 8/29 c. 21/29 d. 29/56

135. For a random variable X, $E(X) = 3$ and $E(X^2) = 11$. The variable of X is

a. 8 b. 5 c. 2 d. 1

136. The sum of 10 items is 12 and the sum of their squares is 18, then the standard deviation will be

a. $-3/5$ b. 6/5 c. 4/5 d. 3/5

137. The height of the chimney when it is found that on walking towards it 50 m in the horizontal line through its base, the angle of elevation of its top changes from 30° to 60° is

a. 25 m b. $25\sqrt{2}$ m

c. $25\sqrt{3}$ m d. None of these

138. The value of $\lim_{x \to 0} \frac{\sqrt{1-\cos x^2}}{1-\cos x}$ is

a. 1/2 b. 2 c. $\sqrt{2}$ d. None of these

139. If $f(x) = \begin{cases} ax^2 + 1, & x \le 1 \\ x^2 + ax + b, & x > 1 \end{cases}$ is differentiable at $x = 1$, then

a. $a = 1, b = 1$ b. $a = 1, b = 0$

c. $a = 2, b = 0$ d. $a = 2, b = 1$

140. The slope of the tangent to the curve $x = t^2 + 3t - 8$, $y = 2t^2 - 2t - 5$ at the point $t = 2$ is

a. 7/6 b. 5/6 c. 6/7 d. 1

141. $\int \frac{1}{1-2\sin x} dx$ is equal to

a. $\frac{1}{2\sqrt{3}} \log \left| \frac{\tan\frac{x}{2} - 2 - \sqrt{3}}{\tan\frac{x}{2} - 2 + \sqrt{3}} \right| + c$

b. $\frac{\sqrt{3}}{2} \log \left| \frac{\tan\frac{x}{2} - 2 - \sqrt{3}}{\tan\frac{x}{2} - 2 + \sqrt{3}} \right| + c$

c. $\frac{1}{\sqrt{3}} \log \left| \frac{\tan\frac{x}{2} - 2 - \sqrt{3}}{\tan\frac{x}{2} - 2 + \sqrt{3}} \right| + c$

d. None of the above

142. $\int_0^1 \frac{\log(1+x)}{1+x^2} dx$ is equal to

a. $\frac{\pi}{8}\log 2$ b. $\frac{\pi}{8}\log\frac{1}{2}$

c. $\frac{\pi}{4}\log 2$ d. None of these

143. The area of one curvilinear triangle formed by curves $y = \sin x$, $y = \cos x$ and X-axis, is

a. 2 sq units b. $(2+\sqrt{2})$ sq units

c. $(2-\sqrt{2})$ sq units d. None of the above

144. Solution of $\left(\frac{x+y-1}{x+y-2}\right)\frac{dy}{dx} = \left(\frac{x+y+1}{x+y+2}\right)$, given that $y = 1$ when $x = 1$ is

a. $\ln\left|\frac{(x-y)^2 - 2}{2}\right| = 2(x+y)$

b. $\ln\left|\frac{(x+y)^2 - 2}{2}\right| = 2(x-y)$

c. $\ln\left|\frac{(x-y)^2+2}{2}\right| = 2(x+y)$

d. $\ln\left|\frac{(x+y)^2+2}{2}\right| = 2(x-y)$

145. If $y=\sin\left(2\tan^{-1}\sqrt{\frac{1-x}{1+x}}\right)$, then $\frac{dy}{dx}$ is

a. 1 **b.** -1 **c.** $\frac{x}{\sqrt{x^2-1}}$ **d.** $\frac{-x}{\sqrt{1-x^2}}$

146. The maximum value of the function $y=x(x-1)^2$, is

a. 0 **b.** 4/27 **c.** -4 **d.** None of these

147. The solution of $x^3\frac{dy}{dx}+4x^2\tan y=e^x\sec y$ satisfying $y(1)=0$, is

a. $\tan y=(x-2)e^x\log x$ **b.** $\sin y=e^x(x-1)x^{-4}$

c. $\tan y=(x-1)e^x x^{-3}$ **d.** $\sin y=e^x(x-1)x^{-3}$

148. The runs of two players for 10 innings each are as follows

A	58	59	60	54	65	66	52	75	69	52
B	94	26	92	65	96	78	14	34	98	13

The more consistent player is

a. player *A*

b. player *B*

c. both player *A* and *B*

d. None of the above

149. The linear programming problem minimise $z=3x+2y$ subject to constraints $x+y\ge 8$, $3x+5y\le 15$, $x\ge 0$ and $y\ge 0$, has

a. one solution **b.** no feasible solution

c. two solutions **d.** infinitely many solutions

150. Find the area enclosed by the loop in the curve $4y^2=4x^2-x^3$.

a. 128/5 **b.** 15/128 **c.** 130/17 **d.** 17/130

Answers

Physics

1. (a)	**2.** (a)	**3.** (c)	**4.** (d)	**5.** (b)	**6.** (c)	**7.** (a)	**8.** (a)	**9.** (c)	**10.** (b)
11. (a)	**12.** (b)	**13.** (b)	**14.** (c)	**15.** (a)	**16.** (d)	**17.** (c)	**18.** (b)	**19.** (a)	**20.** (c)
21. (d)	**22.** (a)	**23.** (b)	**24.** (d)	**25.** (b)	**26.** (a)	**27.** (b)	**28.** (c)	**29.** (b)	**30.** (b)
31. (b)	**32.** (d)	**33.** (c)	**34.** (b)	**35.** (a)	**36.** (c)	**37.** (a)	**38.** (a)	**39.** (b)	**40.** (d)

Chemistry

41. (a)	**42.** (a)	**43.** (a)	**44.** (a)	**45.** (c,d)	**46.** (a)	**47.** (a)	**48.** (a)	**49.** (a)	**50.** (d)
51. (d)	**52.** (b)	**53.** (a)	**54.** (a)	**55.** (c)	**56.** (a)	**57.** (b)	**58.** (a)	**59.** (b)	**60.** (c)
61. (b)	**62.** (a)	**63.** (c)	**64.** (c)	**65.** (a)	**66.** (a)	**67.** (a)	**68.** (a)	**69.** (a)	**70.** (c)
71. (c)	**72.** (a)	**73.** (c)	**74.** (d)	**75.** (c)	**76.** (a)	**77.** (c)	**78.** (b)	**79.** (d)	**80.** (a)

English Proficiency

81. (c)	**82.** (a)	**83.** (a)	**84.** (d)	**85.** (c)	**86.** (c)	**87.** (b)	**88.** (c)	**89.** (c)	**90.** (c)
91. (c)	**92.** (b)	**93.** (c)	**94.** (d)	**95.** (a)					

Logical Reasoning

96. (c)	**97.** (c)	**98.** (b)	**99.** (d)	**100.** (c)	**101.** (b)	**102.** (a)	**103.** (b)	**104.** (d)	**105.** (c)

Mathematics

106. (a)	**107.** (a)	**108.** (c)	**109.** (a)	**110.** (d)	**111.** (b)	**112.** (c)	**113.** (c)	**114.** (c)	**115.** (b)
116. (c)	**117.** (b)	**118.** (a)	**119.** (a)	**120.** (b)	**121.** (a)	**122.** (b)	**123.** (d)	**124.** (b)	**125.** (b)
126. (d)	**127.** (d)	**128.** (d)	**129.** (c)	**130.** (b)	**131.** (b)	**132.** (c)	**133.** (d)	**134.** (b)	**135.** (c)
136. (d)	**137.** (c)	**138.** (c)	**139.** (c)	**140.** (c)	**141.** (c)	**142.** (a)	**143.** (c)	**144.** (b)	**145.** (d)
146. (b)	**147.** (b)	**148.** (a)	**149.** (b)	**150.** (a)					

Hints & Solutions

Physics

1. ***(a)*** Work done in the cyclic process

$= \text{Area of the loop } ABCD$

$= (2p - p) \times (2V - V) = pV$

2. ***(a)*** Impulse of a force = Area between the force-time graph

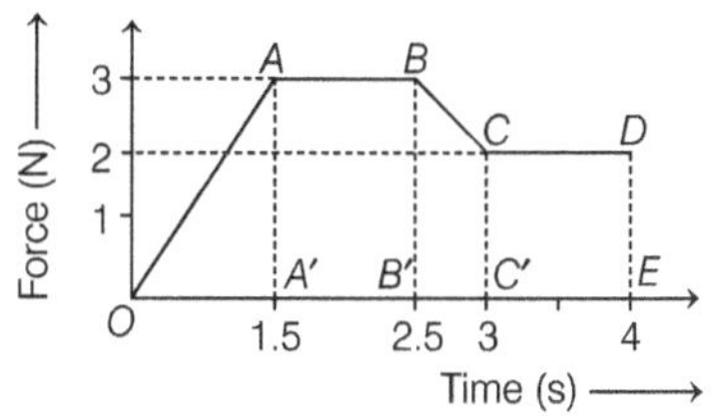

Hence, impulse of force = area of $\Delta OA'A$ + area of rectangle $A'B'BA$ + area of trapezium $BB'C'C$ + area of rectangle $CC'ED$

$$= \frac{1}{2} \times 1.5 \times 3 + 1 \times 3 + \frac{1}{2}(3+2)(3-2.5) + 2 \times 1$$

$$= 2.25 + 3 + 1.25 + 2$$

$$= 8.50 \text{ Ns}$$

3. ***(c)*** Bulk modulus $= \dfrac{\text{Hydraulic stress}}{\text{Volumetric strain}}$

$$\Rightarrow \quad B = \frac{\text{Hydraulic stress}}{\dfrac{\Delta V}{V}}$$

or $$\frac{\Delta V}{V} = \text{Hydraulic stress} \times \frac{1}{B}$$

$\therefore$ For constant hydraulic stress, $\dfrac{\Delta V}{V} \propto \dfrac{1}{B}$

4. ***(d)*** A shunt is a low resistance which is connected in parallel with a galvanometer (or ammeter) to protect it from large current and to increase the range of ammeter.

5. ***(b)*** There is a phase difference of $180°$ between the signal voltage and output voltage in a common emitter amplifier. It is also known as phase reversal.

6. ***(c)*** $(n+1)$th divisions of vernier scale = n division of main scale

$$\therefore \quad 1 \text{ VSD} = \frac{n}{n+1} \text{ MSD}$$

Least count = 1 MSD − 1 VSD

$$= 1 \text{ MSD} - \frac{n}{n+1} \text{ MSD} = \frac{1}{n+1} \text{ MSD}$$

$$= \frac{1}{n+1} \times a \text{ units} = \frac{a}{n+1} \text{ units}$$

[$\because$ given, 1MSD= a units]

7. ***(a)*** According to the question,

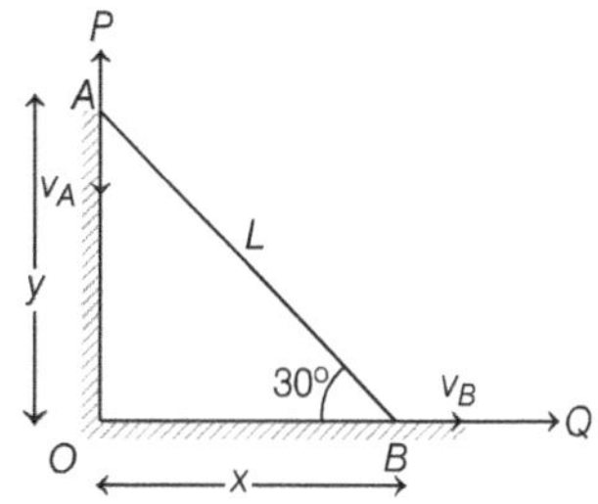

Let $OB = x$, $OA = y$ and $x = L\cos\theta$

So, by Pythagoras theorem,

$$\Rightarrow \quad x^2 + y^2 = L^2$$

Differentiating w.r.t. t, we get

$$2x\frac{dx}{dt} + 2y\frac{dy}{dt} = 0$$

$$2xv_B + 2yv_A = 0 \quad \left[\because \frac{dx}{dt} = v_B \text{ and } \frac{dy}{dt} = v_A\right]$$

$$\Rightarrow \quad v_A = -\frac{x}{y}v_B$$

$$\Rightarrow \quad |v_A| = \frac{x}{y}v_B = v_B\cot\theta \quad \left[\because \frac{x}{y} = \cot\theta\right]$$

$$= \sqrt{3} \times \cot 30°$$

$$= \sqrt{3} \times \frac{1}{\sqrt{3}} = 1 \text{ m/s}$$

8. ***(a)*** Let k be the force constant of spring.

$$\text{Time period, } T = 2\pi\sqrt{\frac{M}{k}} \quad \ldots\text{(i)}$$

When block of mass m is placed in the tray, the time period of oscillations becomes

$$T' = 2\pi\sqrt{\frac{M+m}{k}} \quad \ldots\text{(ii)}$$

On dividing Eq. (ii) by Eq. (i), we get

$$\Rightarrow \quad \frac{T'}{T} = \frac{\sqrt{M+m}}{\sqrt{M}}$$

$$\Rightarrow \quad \frac{3}{1.5} = \sqrt{\frac{12+m}{12}} \quad \left[\begin{array}{r}\text{where, } M = 12 \text{ kg,} \\ T = 1.5 \text{ s} \\ \text{and } T' = 3 \text{ s,}\end{array}\right]$$

$$\Rightarrow \quad \sqrt{\frac{12+m}{12}} = 2$$

$$\Rightarrow \quad 12 + m = 48$$

$$\Rightarrow \quad m = 36 \text{ kg}$$

9. ***(c)*** As shown in phasor diagram, voltage leads the current.

As in purely inductive circuit, current lags behind the voltage by an angle of 90°, here angle is not 90° but it is lesser than that.

So, this type of case arise in *R-L-C* circuit when *R-L-C* load with inductive reactance X_L is more than the capacitive reactance X_C as shown below.

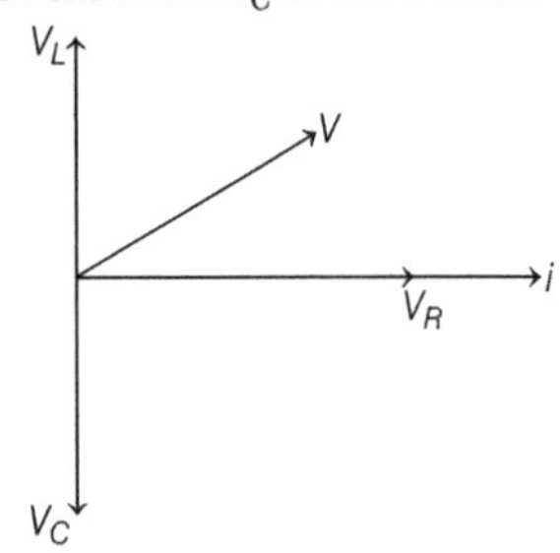

10. ***(b)*** Given,

$$Y = \sin\left[\pi\left(\frac{t}{5} - \frac{x}{9}\right) + \frac{\pi}{6}\right] \text{cm} \quad \ldots\text{(i)}$$

The general equation of progressive wave is

$$Y = A\sin\left[2\pi\left(\frac{t}{T} - \frac{x}{\lambda}\right) + \phi\right] \text{cm} \quad \ldots\text{(ii)}$$

On comparing Eq. (i) with Eq. (ii), we get

$$A = 1 \text{ cm}, f = \frac{1}{T} = \frac{1}{10} = 0.1 \text{ Hz}$$

and $\quad \dfrac{2\pi}{\lambda} = \dfrac{\pi}{9}$

$\Rightarrow \quad \lambda = 18 \text{ cm}$

$$v = f\lambda = 0.1 \times 18 = 1.8 \text{ m/s}$$

11. ***(a)*** In fibre optic communication, signals are transmitted through an optical fibre. The property of light used in transmission through optical fibre cables is total internal reflection.

12. ***(b)*** Distance moved by man in 5 min with velocity of 45 m/s is

$$\text{Distance} = \text{Speed} \times \text{Time}$$
$$= 45 \text{ m/s} \times (5 \times 60) \text{ s}$$
$$= 13500 \text{ m}$$

When he move back, it covers the same distance to come back to its original position.

Now, time taken = 1.5 min

$$= 1.5 \times 60 \text{ s} = 90 \text{ s}$$

Distance travelled = 13500 m

$$\therefore \text{Speed} = \frac{\text{Distance}}{\text{Time}} = \frac{13500}{90} = 150 \text{ m/s}$$

13. ***(b)*** Potential at point *P* due to the solid sphere,

$$V_1 = -\frac{GM}{2R^3}\left[3R^2 - \left(\frac{R}{2}\right)^2\right]$$

$$= -\frac{11\,GM}{8R}$$

Potential at point *P* due to the cavity part,

$$V_2 = -\frac{3}{2}\frac{G\left(\frac{M}{8}\right)}{\frac{R}{2}} = -\frac{3GM}{8R}$$

Potential due to the remaining part at point *P*,

$$V = V_1 - V_2 = \frac{-11GM}{8R} + \frac{3GM}{8R}$$

$$\Rightarrow \quad V = -\frac{GM}{R}$$

14. ***(c)*** Consider the equilibrium of charge *Q* at corner *A*

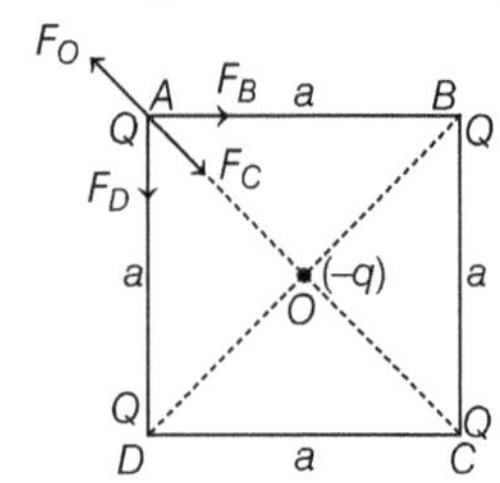

$$F_B = F_D = \frac{1}{4\pi\varepsilon_0}\frac{Q^2}{a^2} = \frac{kQ^2}{a^2} \quad \left[\because \frac{1}{4\pi\varepsilon_0} = k\right]$$

$$F_C = \frac{kQ^2}{(\sqrt{2}a)^2} \quad [\because AC = \sqrt{2}a]$$

$$= \frac{kQ^2}{2a^2}$$

$$F_O = \frac{k\,Qq}{\left(\frac{a}{\sqrt{2}}\right)^2} = \frac{k\,2Qq}{a^2} \quad \left[\because OC = \frac{a}{\sqrt{2}}\right]$$

For equilibrium of charge *Q* at corner *A*, the net force on this charge along *AC* must be zero.

$$\therefore \quad F_O = F_C + F_B\cos 45° + F_D\cos 45°$$

$$= F_C + 2F_B\cos 45° \quad [\because F_B = F_D]$$

$$= F_C + 2F_B \times \frac{1}{\sqrt{2}} \quad \left[\because \cos 45° = \frac{1}{\sqrt{2}}\right]$$

$$\Rightarrow \quad \frac{k\,2Qq}{a^2} = \frac{kQ^2}{2a^2} + \frac{2 \times kQ^2}{a^2} \times \frac{1}{\sqrt{2}}$$

$$\Rightarrow \quad 2Qq = \frac{Q^2}{2} + \frac{Q^2}{\sqrt{2}} \times 2$$

$$\Rightarrow \quad 2q = \frac{Q}{2} + Q\sqrt{2}$$

$$\Rightarrow \quad q = \frac{Q}{4}(1 + 2\sqrt{2})$$

15. ***(a)*** Work done by the particle in *x*-direction is given by

$$dW = F\,dx = (a + bx)\,dx$$

Total work done in displacement from
$x = 0$ to $x = d$ will be

$$W = \int dW = \int_0^d (a + bx)\, dx = \left[ax + \frac{bx^2}{2}\right]_0^d$$

$$\Rightarrow W = \left(a + \frac{bd}{2}\right)d$$

16. ***(d)*** Kinetic energy of a proton is given by

$$E = \frac{1}{2}m_p v_p^2$$

$$\Rightarrow \quad E = \frac{p^2}{2m_p} \quad [\because p = mv]$$

$$\Rightarrow \quad p = \sqrt{2m_p E}$$

$\therefore$ Wavelength of proton is, $\lambda_1 = h/p$

$$\Rightarrow \quad \lambda_1 = \frac{h}{\sqrt{2m_p E}} \quad \ldots(\text{i})$$

Now, energy of photon is given by

$$E = h\nu_2 \Rightarrow E = \frac{hc}{\lambda_2}$$

$$\Rightarrow \quad \lambda_2 = \frac{hc}{E} \quad \ldots(\text{ii})$$

where, λ_2 is wavelength of photon.
On dividing Eq. (ii) by Eq. (i), we get

$$\frac{\lambda_2}{\lambda_1} = \frac{hc}{E} \times \frac{\sqrt{2m_p E}}{h}$$

$$\Rightarrow \quad \frac{\lambda_2}{\lambda_1} = \frac{c}{\sqrt{E}}\sqrt{2m_p} \Rightarrow \frac{\lambda_2}{\lambda_1} \propto E^{-1/2}$$

17. ***(c)*** The FBD of block of mass 10 kg is

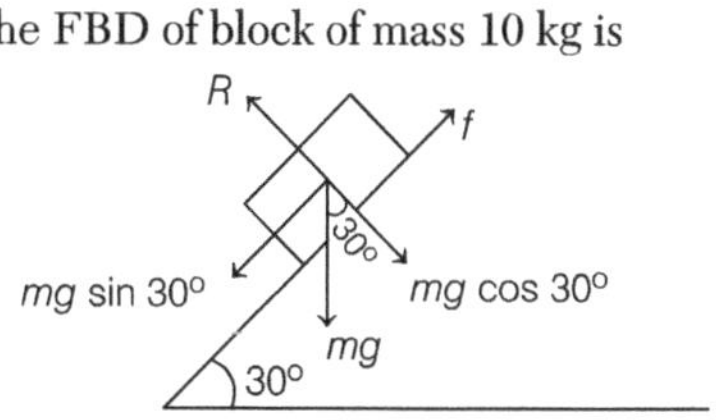

For the equilibrium of block, the frictional force must balanced the sine component of mg.

$$\therefore \quad f = mg \sin 30°$$

$$= 10 \times 9.8 \times \frac{1}{2} = 49 \text{ N}$$

18. ***(b)*** Conductivity (σ) of a metallic conductor is given by

$$\sigma = \frac{1}{\rho} = \frac{ne^2\tau}{m}$$

where, ρ = resistivity of conductor,
τ = relaxation time,
n = number of free electrons
and m = mass.

For metals, as the temperature increases, the relaxation time τ decreases because collisions become more frequent. Hence, the conductivity of a metallic conductor decreases with the increases of temperature.

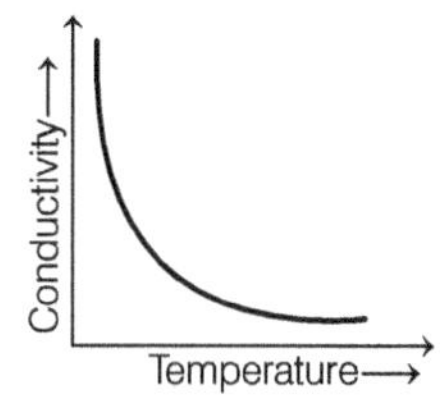

19. ***(a)*** Given, radius of hydrogen atom, $r = 2.5 \times 10^{-13}$ m

Volume of one atom, $V = \frac{4}{3}\pi r^3$

Number of atoms in 1 mole = 6.023×10^{23}

Volume of 1 mole of H-atoms = $N \times \frac{4}{3}\pi r^3$

$$= 6.023 \times 10^{23} \times \frac{4}{3} \times 3.14 \times (2.5 \times 10^{-13})^3$$

$$= 3.94 \times 10^{-14} \text{ m}^3$$

20. ***(c)*** Given, $m = 1$ kg, $\mathbf{r} = 3t\,\hat{\mathbf{i}} + 4\hat{\mathbf{j}}$

Velocity of body, $\mathbf{v} = \frac{d\mathbf{r}}{dt} = \frac{d}{dt}(3t\hat{\mathbf{i}} + 4\hat{\mathbf{j}})$

$$\Rightarrow \quad \mathbf{v} = 3\hat{\mathbf{i}}$$

Angular momentum, $\mathbf{L} = m(\mathbf{r} \times \mathbf{v})$

$$= 1 \times [(3t\hat{\mathbf{i}} + 4\hat{\mathbf{j}}) \times 3\hat{\mathbf{i}}]$$

$$= \begin{vmatrix} \hat{\mathbf{i}} & \hat{\mathbf{j}} & \hat{\mathbf{k}} \\ 3t & 4 & 0 \\ 3 & 0 & 0 \end{vmatrix}$$

$$= [3t \times 0 - 4 \times 3]\hat{\mathbf{k}}$$

$$\Rightarrow \quad \mathbf{L} = -12\hat{\mathbf{k}} \text{ J-s}$$

21. ***(d)*** The value of g at equator is expressed as

$$g_e = g - R\omega^2 \cos\phi = g - R\omega^2 \quad [\because \cos\phi = 1]$$

Change in value of g is $g - g_e$

$$= R\omega^2 = R\left(\frac{2\pi}{T}\right)^2$$

$$= 6.4 \times 10^6 \times \left[\frac{2\pi}{24 \times 60 \times 60}\right]^2$$

$$= 3.37 \times 10^{-2} \text{ m/s}^2 \simeq 3.4 \text{ cm/s}^2$$

22. ***(a)*** Apparent frequency of car A = Apparent frequency of car B

$$\Rightarrow \frac{v}{v - v_s} \times \nu = \frac{v}{v - v_s'} \times \nu'$$

$$\Rightarrow \quad \frac{\nu}{v - 15} = \frac{504}{v - 30}$$

$\Rightarrow \quad \nu = \frac{340-15}{340-30} \times 504$ [$\because$ velocity of sound in air is 340 m / s.]

$\Rightarrow \quad \nu = 529.2$ Hz

23. ***(b)*** Using KCL, $I_{DC} = 3 - 1 = ZA$

Using Kirchhoff's loop law, in loop $BDCR_1 B$, we get

$$2 \times 2 + 3R_1 = 4 \quad \text{or} \quad R_1 = 0$$

$\therefore$ Potential drop at B = Potential drop at C

$$\Rightarrow \quad V_B = V_C = 2 + V_A = 2 + 0 \quad [\because V_A = 0]$$

$$\Rightarrow \quad V_B = 2\text{V}$$

24. ***(d)*** Let N be the maximum number of reflection that ray undergoes.

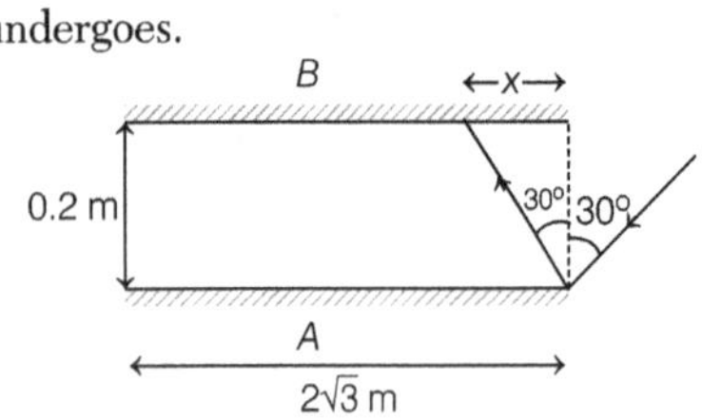

Suppose it covers x distance in one reflection, then

$$Nx = 2\sqrt{3}$$

$$\Rightarrow \quad N = \frac{2\sqrt{3}}{x} = \frac{2\sqrt{3}}{0.2 \tan 30^\circ} \quad [\because x = 0.2 \tan 30^\circ]$$

$$N = \frac{6}{0.2} = 30$$

Number of reflections ray undergoes before it emerges excluding the first one are $N - 1$

$$= 30 - 1 = 29$$

25. ***(b)*** According to equation of motion, in vector from,

$$\mathbf{v} = \mathbf{u} - gt\,\hat{\mathbf{j}}$$

$$\Rightarrow \quad \mathbf{v} = \hat{\mathbf{i}} + \hat{\mathbf{j}} - 10t\,\hat{\mathbf{j}}$$

Integrating w.r.t. t both sides, we get

$$\mathbf{r} = \int \mathbf{v}\,dt = \int (\hat{\mathbf{i}} + \hat{\mathbf{j}} - 10t\hat{\mathbf{j}})\,dt$$

or $\quad x\hat{\mathbf{i}} + y\hat{\mathbf{j}} = t\hat{\mathbf{i}} + t\hat{\mathbf{j}} - 5t^2\,\hat{\mathbf{j}} = t\hat{\mathbf{i}} + (t - 5t^2)\hat{\mathbf{j}}$

$\therefore$ On comparing, we get

$$x = t$$

and $\quad y = t - 5t^2$

$$\Rightarrow \quad y = x - 5x^2$$

This is the required equation of trajectory of particle.

26. ***(a)*** Volume of bigger drop = Volume of two smaller drops

$$\Rightarrow \quad \frac{4}{3}\pi R'^3 = 2 \times \frac{4}{3}\pi R^3$$

where, R' is the radius of bigger drop.

$$\Rightarrow \quad R'^3 = 2R^3$$

$$\Rightarrow \quad R' = 2^{1/3} R$$

Initial surface energy of two small drops,

$$U_1 = 8\pi R^2 \sigma \quad \ldots\text{(i)}$$

Final surface energy of big drop, $U_2 = 4\pi R'^2 \sigma$

$$= 4\pi \times 2^{2/3} R^2 \sigma \quad \ldots\text{(ii)}$$

On dividing Eq. (ii) by Eq. (i), we get

$$\frac{U_2}{U_1} = \frac{4\pi \times 2^{2/3} R^2 \sigma}{8\pi R^2 \sigma} = \frac{2^{2/3}}{2} = \frac{2^{-1/3}}{1}$$

Ratio of surface energy of bigger drop to smaller one is $2^{-1/3} : 1$.

27. ***(b)*** Force on charge particle in magnetic field is balanced by centripetal force, so $\frac{mv^2}{r} = qvB \sin 90^\circ$

[$\theta = 90^\circ$, therefore magnetic field is perpendicular]

$$\frac{mv^2}{r} = qvB \quad [\sin 90^\circ = 1]$$

$$\Rightarrow \quad r = \frac{mv}{qB}$$

For both particles, q and B are same.

Therefore, $\quad r \propto mv$

or $\quad \frac{r_A}{r_B} = \frac{m_A v_A}{m_B v_B}$

As, $\quad r_A > r_B \quad$ [given]

$$\Rightarrow \quad m_A v_A > m_B v_B$$

28. ***(c)*** Minimum wavelength for X- ray is given by

$$\lambda_{\min} = \frac{12375}{V} \text{Å}$$

As the accelerating voltage V is decreased, $\lambda_{\min}$ increases.

The intensity of emitted radiation is determined by the number of electrons bombarding the target.

$\therefore$ Accelerating voltage does not change the intensity of X-rays emitted.

29. ***(b)*** As we know, $C_V = \frac{n}{2}R \quad \ldots\text{(i)}$

$$C_p = \left(\frac{n}{2} + 1\right)R \quad \ldots\text{(ii)}$$

On dividing Eq. (ii) by Eq. (i), we get

$$\frac{C_p}{C_V} = \gamma = \frac{\left(\frac{n}{2} + 1\right)}{\frac{n}{2}} = 1 + \frac{2}{n}$$

$$\Rightarrow \quad \frac{C_p}{C_V} = \gamma = 1 + \frac{2}{n}$$

30. ***(b)*** Given, $v = 6$ m/s, $r = 18$ m, $g = 9.8$ m/s^2

Now, the velocity of cyclist is given by

$$v^2 = rg \tan\theta \quad \text{or} \quad \tan\theta = \frac{v^2}{rg} = \frac{6 \times 6}{18 \times 9.8}$$

$\Rightarrow \quad \tan\theta = 0.2041$

The minimum value of coefficient of friction,

$$\mu = \tan\theta$$

$\Rightarrow \quad \mu = 0.2041$

31. ***(b)*** Let us consider an equilateral ΔABC of side b.

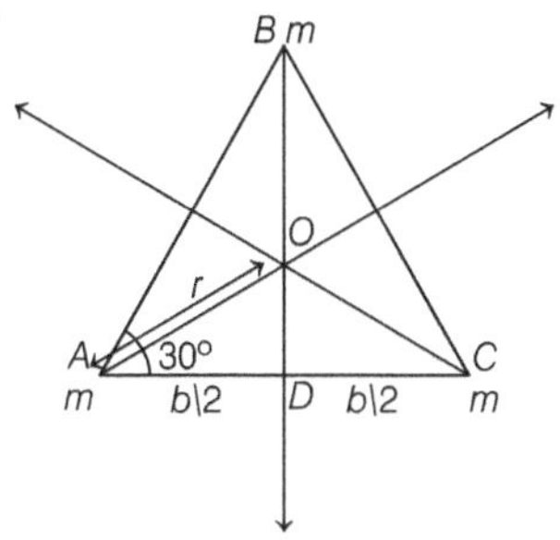

From figure,

In ΔAOD, $\cos 30° = \dfrac{AD}{AO}$

$\Rightarrow \quad \dfrac{\sqrt{3}}{2} = \dfrac{\frac{b}{2}}{r} \qquad \left[\because \cos 30° = \dfrac{\sqrt{3}}{2}\right]$

$\Rightarrow \quad r = \dfrac{b}{2} \times \dfrac{2}{\sqrt{3}} = \dfrac{b}{\sqrt{3}}$

Moment of inertia about line passing through centroid,

$I = 3 \times mr^2$ [due to 3 masses present in system]

$\Rightarrow \quad = 3 \times m \times \left(\dfrac{b}{\sqrt{3}}\right)^2$

$\Rightarrow \quad I = mb^2$

32. ***(d)*** When a dipole is placed in a non-uniform electric field, opposite charges of the dipole experiences a force due to **E**. It is as shown in the figure below

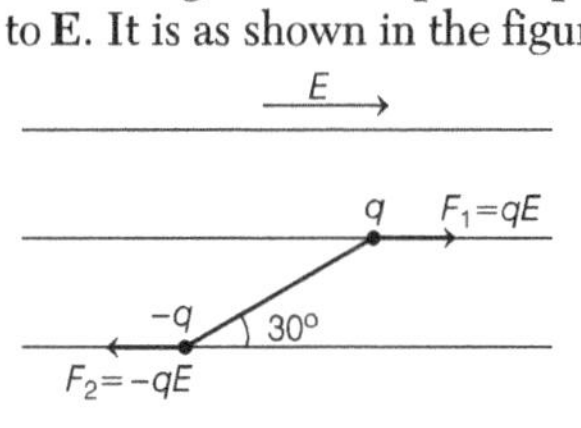

These forces will act in opposite direction. Now, since the electric field is not uniform, so the force experienced by the charges will be unequal.

Hence, their will be a net force acting on the dipole, as they do not cancel each other.

Also, forces on the charges are not linear as shown above. So, they will experience a not non-zero torque. Hence, the dipole will experience both translational force and torque in a non-uniform electric field.

33. ***(c)*** Induced emf, $\varepsilon = -M\dfrac{di}{dt}$

where, i is given as, $i = i_0 \sin\omega t$

$\Rightarrow \quad \varepsilon = -M\dfrac{d}{dt}[i_0 \sin\omega t]$

$$= M\, i_0\, \omega \cos\omega t$$

For maximum value of emf in second coil, $\cos\omega t$ is maximum, i.e. 1.

$E_{max} = M\, i_0\, \omega \qquad [\because \cos\omega t = 1]$

$\Rightarrow \quad = 0.001 \times 10 \times 10\pi$

$\Rightarrow \quad E_{max} = 0.1\,\pi V$

Hence, maximum emf value induced in second coil is $0.1\,\pi V$.

34. ***(b)*** Diode D_1 does not conduct because it is connected in reversed biased position.

Hence, only D_2 conducts as it is forward biased.

$\therefore$ Current in circuit is given by

$$i = \frac{V}{R}$$

$$= \frac{5}{30+20} = \frac{5}{50}\,A$$

$\Rightarrow \quad i = 0.1\,A$

35. ***(a)*** Kinetic energy at Q = 90% of potential energy at P

$\Rightarrow \quad \dfrac{1}{2}mv^2 = \dfrac{90}{100} \times mgh$

$\Rightarrow \quad v = \sqrt{1.8gh} \quad \Rightarrow \quad v = \sqrt{1.8 \times 10 \times 2}$

$\Rightarrow \quad v = 6\,m/s$

36. ***(c)*** Let displacement, $x = A\sin\omega t$

Velocity, $v = \dfrac{dx}{dt} = A\omega\cos\omega t$

Kinetic energy of particle in SHM,

$K = \dfrac{1}{2}mv^2 \quad \Rightarrow \quad K = \dfrac{1}{2}m\omega^2 A^2 \cos^2\omega t$

$= \dfrac{1}{2}m\omega^2 A^2 \left(\dfrac{1+\cos 2\omega t}{2}\right) \qquad \left[\because \cos^2\theta = \dfrac{1+\cos 2\theta}{2}\right]$

$= \dfrac{1}{4}m\,\omega^2 A^2(1+\cos 2\omega t)$

$\therefore$ New angular frequency, $\omega_K = 2\omega$

$\therefore$ Frequency of oscillation of kinetic energy

$= 2f \qquad \left[\because f = \dfrac{\omega}{2\pi}\right]$

37. ***(a)*** According to law of radioactivity, the rate of disintegration is given as

$$R = \lambda N = \lambda N_0\, e^{-\lambda t}$$

Given, $\quad R = 5000$

$\Rightarrow \quad 5000 = \lambda N_0 e^{-\lambda t} \qquad \ldots(i)$

After 5 min, $R = 1250$

$\therefore \quad 1250 = \lambda N_0 e^{-\lambda(t+5)} \qquad \ldots(ii)$

On dividing Eq. (i) by Eq. (ii), we get

$$\frac{5000}{1250}=\frac{e^{-\lambda t}}{e^{-\lambda(t+5)}} \Rightarrow 4=e^{5\lambda}$$

Taking log both sides, we get

$$\log_e 4=\log_e(e^{5\lambda})$$

$$\Rightarrow \log_e 4=5\lambda$$

$$\Rightarrow \lambda=\frac{1}{5}\log_e 2^2$$

$$\Rightarrow \lambda=0.4\log_e 2$$

38. ***(a)*** In a uniform electric field, potential difference across two parallel plates of capacitor, $V=Ed$

where, E is electric field and d is separation between plates.

$$\Rightarrow V=2000\times(1\times10^{-3})=2\text{V}$$

39. ***(b)*** When light travel in meta-material, physical characteristics remain unchanged.

$$\therefore \quad v=\frac{c}{|n|}$$

where, $|n|$ is the relative refractive index of meta-material.

As, $$|n|=\frac{c}{v}$$

$$\Rightarrow |n|=\frac{\lambda_{\text{air}}}{\lambda_{\text{med}}} \Rightarrow \lambda_{\text{med}}=\frac{\lambda_{\text{air}}}{|n|}$$

40. ***(d)*** Let O be the centre of mass of part AB and Q that of CD. Let M be the mass per unit length of the parts AB and CD.

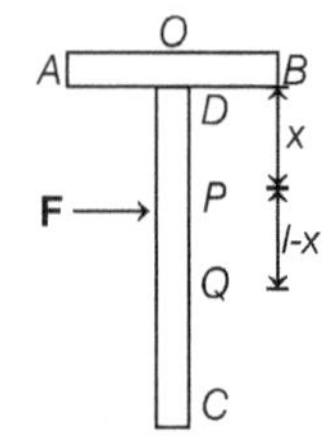

As, no rotation is set up about point P, so

moment of part AB about P = moment of part CD about P

or $(ml)\,x=(2\,ml)\,(l-x)$ $\left[\because m=\frac{M}{l}\right]$

or $x=2l-2x$

or $x=\frac{2l}{3}$

Distance of P from the end

$$C=2l-x=2l-2l/3=4l/3$$

Chemistry

41. ***(a)*** *o*-nitrophenol forms intramolecular H-bonding whereas molecules of *p*-nitrophenol get associated through intermolecular H-bond.

$\therefore$ *o*-nitrophenol is the most volatile compound.

42. ***(a)*** The magnetic moment is given by

$$\mu=\sqrt{n(n+2)}\text{ BM.}$$

(where, n = number of unpaired electron).

Electronic configuration of

$$Ti^{2+}(22): 1s^2\,2s^2\,2p^6\,3s^2\,3p^6\,3d^2$$

$\because$ Number of unpaired electrons = 2

$$\therefore \quad \mu=\sqrt{2(2+2)}=\sqrt{8}$$

43. ***(a)*** Only Xe can form compounds with fluorine among noble gases because, Xe is large in size and have high atomic mass. Due to having larger atomic radius the force of attraction between the outer electrons and protons in the nucleous is weaker. Hence, they are easily available to form compound.

44. ***(a)*** Helium is used as a cryogenic agent due to its very low boiling point.

45. ***(c,d)*** The rate of reaction for the given reaction is

$$r=-\frac{1}{2}\frac{d[A]}{dt}=-\frac{1}{3}\frac{d[B]}{dt}=\frac{1}{3}\frac{d[C]}{dt}=\frac{1}{4}\frac{d[D]}{dt}$$

46. ***(a)*** Cyclohexanol on reaction with H_2SO_4 and heating gives cyclohexene as follows

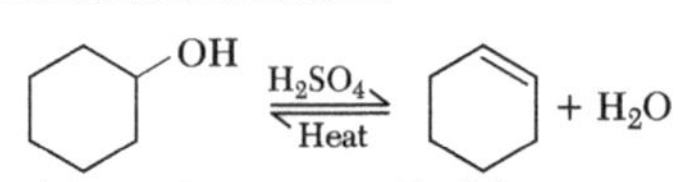

Cyclohexanol Cyclohexene

47. ***(a)*** Addition of HCl to 3-methylbutene gives 2-chloro-2-methylbutane as major product.

The mechanism reaction is as follows

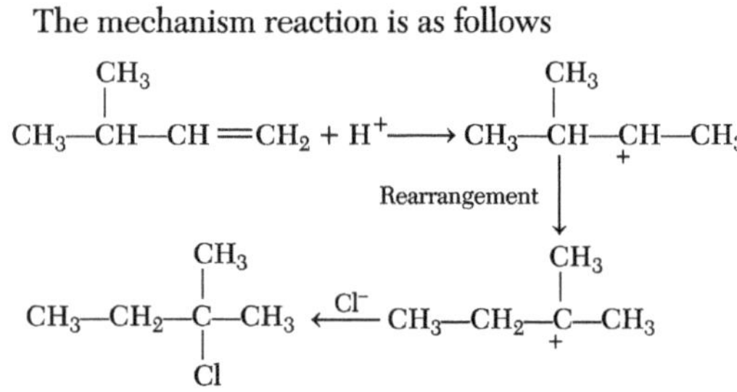

2-chloro-2-methylbutane

48. ***(a)*** Energy for atom, $E\propto\frac{Z^2}{n^2}$

For hydrogen, $Z=1, n=1$

For lithium $Z=3, n=?$

$\therefore$ For same energy,

$$E_H = E_{Li} \Rightarrow \frac{Z_H^2}{n_H^2} = \frac{Z_{Li}^2}{n_{Li}^2}$$

$$\therefore \quad n_{Li}^2 = \frac{Z_{Li}^2 \times n_H^2}{Z_H^2} \Rightarrow n_{Li}^2 = \frac{3^2 \times 1}{1}$$

$$n_{Li} = 3$$

While angular momentum $(nh/2\pi)$ is independent of atomic number.

$\therefore$ For 3*s*-orbital of lithium atom, energy and angular momentum is equal to that of 1*s* orbital of hydrogen atom.

49. ***(a)*** Sodium iodide reacts with ammonia to produce tetraamminesodium iodide. The reaction is conducted in liquid ammonia. The reaction is as follows

$$\underset{\text{Sodium iodide}}{NaI} + \underset{\text{Ammonia}}{4NH_3} \longrightarrow \underset{\text{Tetraammine-sodium iodide}}{[Na(NH_3)_4]\,I}$$

50. ***(d)*** Sodium or ammonium thiosulphate is used as a fixing agent in photography. They converts the silver halide into soluble, complex silver salts that dissolve in the fixer. During this process the film loses its original silver halide milkiness overlaying the image and becomes clear.

51. ***(d)*** Epoxy ethane reacts with ammonia forming a mixture of mono, di, tri-ethanol amines, in presence of small amount of water as follows.

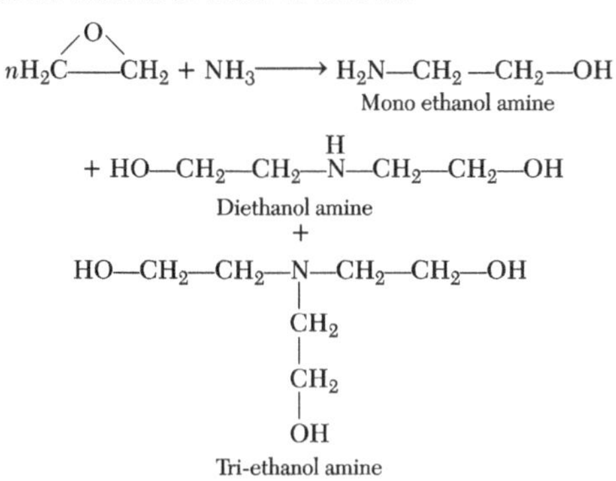

52. ***(b)*** Dialysis is a process of removing excess electrolyte from colloidal solution by means of diffusion through a suitable membrane.

53. ***(a)*** Maltase converts maltose into glucose by hydrolysis as follows

$$\text{Maltose} \xrightarrow[\text{Maltase}]{\text{Hydrolysis}} \text{glucose} + \text{glucose}$$

54. ***(a)*** Starch is a polymer of glucose in which glucopyranose units are bonded by α-linkage. It is made up mixture of amylose and amylopectin.

55. ***(c)*** Methyl alcohol can be distinguished from ethyl alcohol by iodoform test as follows by using sodium hydroxide and iodine.

$$C_2H_5OH \xrightarrow[\text{Iodoform test}]{NaOH/I_2} CHI_3 \text{ (yellow ppt.)}$$

$$CH_3OH \xrightarrow{NaOH/I_2} \text{No ppt.}$$

56. ***(a)*** Sodium stearate $(C_{17}H_{35}COO^-Na^+)$ is an anionic soap. Anionic part of sodium stearate is involved in cleansing action.

57. ***(b)*** BaF_2 is not an anti-fluorite structure. It is of AB_2 type i.e. fluorite (CaF_2) structure. While Rb_2S, K_2O and Li_2O are anti-fluorite structures.

58. ***(a)*** The shape of ClF_3 is as follows

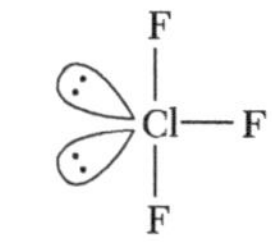

T-shape structure

It has two lone pairs at equatorial position.

59. ***(b)*** Molecular electronic configuration for given species are

(a) $O_2 = \sigma\, 1s^2\, \sigma^*1s^2\, \sigma 2s^2\, \sigma^*2s^2\, \sigma 2p_x^2\, \pi\, 2p_y^2\, \pi 2p_z^2$
$\pi^* 2p_y^1\, \pi^* 2p_z^1$

It has two unpaired electrons.

(b) O_2^{2-} : It has no unpaired electron.

(c) N_2^{2+} : It has one unpaired electron.

(d) B_2 : It has two unpaired electrons.

$\therefore$ O_2^{2-} is diamagnetic in nature.

60. ***(c)*** Bond order for the various compounds are as follows

For SO_2,

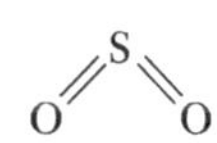

Number of bonds = 4

Number of (SO) groups = 2

$\therefore$ Bond order $= \frac{4}{2} = 2$

For SO_3,

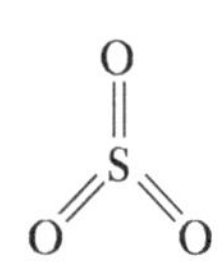

Number of bonds = 6

Number of (SO) groups = 3

$\therefore$ Bond order $= \frac{6}{3} = 2$

For SO_4^{2-},

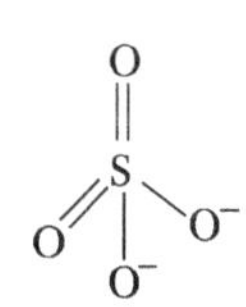

Number of bonds = 6

Number of (SO) groups = 4

BITSAT 2021

$\therefore$ Bond order $= \frac{6}{4} = 1.5$

For SO_3^{2-},

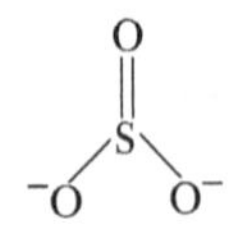

Number of bonds = 4

Number of (SO) groups = 3

$\therefore$ Bond order $= \frac{4}{3} = 1.33$

Hence, correct order is $SO_2 = SO_3 > SO_4^{2-} > SO_3^{2-}$.

61. ***(b)*** The energy required for ionisation H-atom is given by, $E = \frac{-kZe^2}{2r}$

where, Z = atomic number

e = charge on proton/electron

r = radius of that electron

$$\therefore \quad E \propto \frac{1}{r} \quad \text{or} \quad E \propto r^{-1}$$

$$\therefore \quad n = -1$$

62. ***(a)*** BHA is antioxidant (i.e. prevent oxidation). It help in food preservation by retarding the action of oxygen on food. As it is more reactive towards oxygen than the food material.

63. ***(c)*** In MnO_4^-,

$$x + (-2)4 = -1 \Rightarrow x = +7$$

$_{25}Mn^{+7}$ = [Ar], no unpaired electrons.

Thus, it will not show *d-d* transition. It is dark purple coloured due to charge transfer from ligand to metal.

64. ***(c)*** Rutherford model could not explain, the

(a) electronic structure of an atom.

(b) stability of an atom.

65. ***(a)*** The oxidation state of Ni in $[Ni(CO)_4]$ is 0. The electronic configuration of Ni is $3d^8\ 4s^2$.

Ni (ground state) = 3d: [⥮|⥮|⥮|↿|↿] 4s: [⥮] 4p: [| |]

As CO is a strong ligand, pairing of electrons occur.

$Ni(CO)_4$ = 3d: [⥮|⥮|⥮|⥮|⥮] 4s: [ ] 4p: [| |] (sp^3)

As, it has sp^3 hybridisation, so the geometry is tetrahedral. Since, it has no unpaired electrons. So, the complex is diamagnetic.

66. ***(a)*** The SI unit of Boltzmann's constant is JK^{-1}.

67. ***(a)*** Acetone does not undergo substitution reaction because it does not contain suitable leaving group.

68. ***(a)*** Oxidation number of Co in $[CoF_6]^{2-}$

$$x + (-1 \times 6) = -2$$

$$\therefore \quad x = 4$$

$\therefore$ Electronic configuration of Co^{4+},

$$Co^{4+} = [Ar]\,3d^5$$

Co^{4+} = $3d^5$: [↿|↿|↿|↿|↿]

$\therefore$ Number of unpaired electrons = 5

$\therefore$ Magnetic moment $= \sqrt{n(n+2)} = \sqrt{5 \times 7} = \sqrt{35}$ BM

69. ***(a)*** Heavy water is used as a moderator in nuclear reactors. It has higher boiling point compare to ordinary water, thus it is more associated as compared to ordinary water. Dielectric constant of $H_2O > D_2O$. Therefore H_2O is more effective solvent.

70. ***(c)*** According to sequence rule, the priority order is

$\underset{(1)}{Cl} > \underset{(2)}{C_2H_5} > \underset{(3)}{CH_3} > \underset{(4)}{H}$, so in *R* configuration.

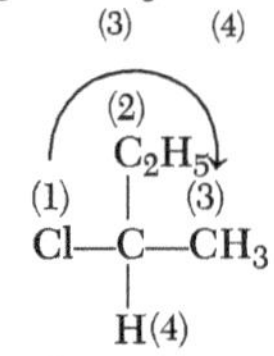

R-configuration

71. ***(c)*** Molecular mass

$$= \frac{\text{Weight of organic substance taken}}{\text{Air displaced at STP}} \times 22400$$

$$= \frac{0.2}{56} \times 22400 = 80$$

72. ***(a)*** $C_2H_5OH(l) + 3O_2(g) \longrightarrow 2CO_2(g) + 3H_2O(l)$

$$\Delta U = -1364.47 \text{ kJ/mol},$$

$$\Rightarrow \quad \Delta n_g = -1,\ T = 25°C = 298 \text{ K}$$

$$\therefore \quad \Delta H = \Delta U + \Delta n_g RT$$

$$\Delta H = -1364.47 + \frac{-1 \times 8.314 \times 298}{1000}$$

[Here, value of *R* in unit of J must be converted into kJ]

$$= -1364.47 - 2.4776 = -1366.95 \text{ kJ/mol}$$

73. ***(c)*** (a) $\Delta G = \Delta H - T\Delta S$

For a system, total entropy change is ΔS_{total},

$$\Delta H_{total} = 0$$

$$\therefore \quad \Delta G_{system} = -T\Delta S_{total}$$

$$\therefore \quad \frac{\Delta G_{system}}{\Delta S_{total}} = -T$$

Thus, (a) is true,

(b) For isothermal reversible process, $\Delta E = 0$.

By first law of thermodynamics,

$$\Delta E = q \times W$$

$$\therefore \quad W_{reversible} = -q = -\int_{V_i}^{V_f} p\, dV$$

$$W_{reversible} = -nRT \ln \frac{V_f}{V_i}$$

Thus, (b) is also true.

(c) $\Delta G^\circ = \Delta H^\circ - T\Delta S^\circ$

$\Delta G^\circ = -nRT \text{ In } K = -RT \text{ In } K$ (for $n = 1$)

$\therefore \quad -RT \text{ In } K = \Delta H^\circ - T\Delta S^\circ$

$\therefore \quad \text{In } K = -\left(\dfrac{\Delta H^\circ - T\Delta S^\circ}{RT}\right)$

Thus, (c) is false.

(d) $\Delta G^\circ = -RT \ln K$

$\therefore \quad \text{In } K = -\dfrac{\Delta G^\circ}{RT}$

$\therefore \quad K = e^{-\Delta G^\circ/RT}$

Thus, (d) is also true.

74. ***(d)*** $Pb(OH)_2 \rightleftharpoons Pb^{2+} + 2OH^-$

$K_{sp} = [Pb^{2+}][OH^-]^2 = S \times (2S)^2$

$K_{sp} = 4S^3$

$= 4 \times (6.7 \times 10^{-6})^3 = 1.20 \times 10^{-15}$

In a solution with pH = 8

$[H^+] = 10^{-8}$ and $[OH^-] = 10^{-6}$

$1.20 \times 10^{-15} = [Pb^{2+}][10^{-6}]^2$

$[Pb^{2+}] = \dfrac{1.2 \times 10^{-15}}{[10^{-6}]^2} = 1.2 \times 10^{-3}$ M

75. ***(c)*** Molarity

$= \dfrac{10 \times \text{density} \times \text{percentage weight of solute}}{\text{molecular weight of the solute}}$

$\Rightarrow \text{Density} = \dfrac{3.60 \times 98}{10 \times 29} = 1.216 = 1.22 \text{ g mol}^{-1}$

76. ***(a)*** According to Raoult's law, relative lowering of vapour pressure is equal to mole fraction of solute,

$\therefore \quad \dfrac{p - p_s}{p} = \dfrac{n}{n+N} = \dfrac{w/m}{\dfrac{w}{m} + \dfrac{W}{M}} \Rightarrow \dfrac{w}{m} <<< \dfrac{W}{M}$

$\therefore \quad \dfrac{p^\circ - p_s}{p^\circ} = \dfrac{w/m}{\dfrac{W}{M}} = \dfrac{w}{m} \times \dfrac{M}{W}$

$0.0125 = \dfrac{w \times 18}{m \times W} \Rightarrow \dfrac{w}{mW} = \dfrac{0.0125}{18}$

$m = \dfrac{\text{Weight of solute } (w) \times 1000}{\text{Molar mass of solute } (m) \times \text{Weight of } H_2O(W)}$

Now, molality, $m = \dfrac{0.0125}{18} \times 1000$

$= 0.69 \approx 0.7$

77. ***(c)*** Suppose the equal mass of methane and oxygen, $w = 1$ g

Mole fraction of oxygen $= \dfrac{\dfrac{w}{32}}{\dfrac{w}{32} + \dfrac{w}{16}} = \dfrac{\dfrac{1}{32}}{\dfrac{3}{32}} = \dfrac{1}{3}$

Let the total pressure = p

Pressure exerted by oxygen (partial pressure)

$= \chi_{O_2} \times p_{\text{total}} = p \times \dfrac{1}{3}$

$\therefore$ The fraction of total pressure exerted by oxygen is $\dfrac{1}{3}$.

78. ***(b)*** $NaCl + KNO_3 \longrightarrow NaNO_3 + KCl$

Sum of molar conductivity of reactant

= Sum of molar conductivity of products

$\Lambda^\circ(NaNO_3) = \Lambda^\circ_m(NaCl) + \Lambda^\circ_m(KNO_3) - \Lambda^\circ_m(KCl)$

$= 128 + 111 - 152$

$= 87 \text{ S cm}^2 \text{ mol}^{-1}$

79. ***(d)*** $Ca^{2+} + 2e^- \longrightarrow Ca$ (at cathode)

1 mole $CaCl_2 = 2F$

From $Q = it \Rightarrow n \cdot 2F = it$

$\Rightarrow \quad t = \dfrac{n \cdot 2F}{i} = \dfrac{30}{40} \times \dfrac{2 \times 96500}{5} \quad \left(\because n = \dfrac{w}{m} = \dfrac{30}{40}\right)$

$= 28950 \text{ s} = \dfrac{28950}{60 \times 60}$

$= 8.04 = 8 \text{h}$

80. ***(a)*** As the value of reduction potential of metal ion increases, the tendency of metal oxide to get reduce into metal increases.

Since, reduction potential of only Ag is positive among the given, thus Ag_2O readily gets reduced to Ag metal. In other words, it can be said that Ag_2O is the least stable oxide among the given.

BITSAT 2021

a. English Proficiency

81. ***(c)*** 'Immutable' means 'unchangeable'.

82. ***(a)*** 'Ignominious' means 'shameful'.

83. ***(a)*** 'Conjecture' means 'making a guess'.

84. ***(d)*** Institutional is the correct word to fill the blank. Institutional is the word which means relating to principles esp of law, so legally also every human has rights of freedom and equality.

85. ***(c)*** Volumes is the most appropriate option here. Other options do not match here.

86. ***(c)*** In the context of the sentence, the option 'skirted' is a appropriate word which means to avoid or evade.

87. ***(b)*** Immortal means living forever; never dying or decaying. So, among the given options, 'perishable' would be its correct opposite meaning word. Perishable means likely to decay easily.

88. ***(c)*** 'Opposed' is the correct option. Opposite meaning word to the given italicised word is 'patronised'.
Patronised means provide favour or support.
Opposed means disagreeing with someone/something.

89. ***(c)*** Barbarous' means extremely brutal, uncivilised. So, 'civilised' would be its correct opposite meaning word.

90. ***(c)*** 'Atheist' is the best alternative.

91. ***(c)*** 'Epicure' is the best alternative.

92. ***(b)*** 'Philanthropist' is the correct answer, other alternatives are not relevant.

93. ***(c)*** CBEDA **94.** (d) CDEAB **95.** ***(a)*** DEBAC

b. Logical Reasoning

96. ***(c)*** Figure shown in option (c) will make a complete square on joining with the problem figure.

97. ***(c)*** From figures (A), (C) and (E),

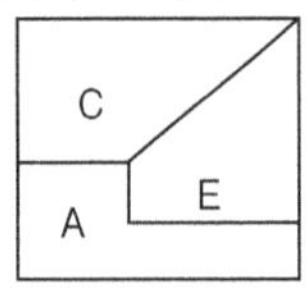

98. ***(b)*** The contents of the third figure in each row (and column) are determined by the contents of the first two figures. Lines are carried forward from the first two figures to the third one, except where two lines appear in the same position, in which they are cancelled out.

99. ***(d)*** From the given four positions of a single block,
Faces adjacent to face having red colour
= blue, pink, yellow, green
Clearly, black is opposite to red.

100. ***(c)*** In figure (A), the dot is placed in the region which is common to the circle and triangle. Now, we have to search similar common region in the four options. Only in figure (c), we find such a region which is common to the circle and triangle.

101. ***(b)*** The pattern is as following

$$G \xrightarrow{+3} J \xrightarrow{+3} M \xrightarrow{+3} P \xrightarrow{+3} S \xrightarrow{+3} \boxed{V}$$

$$4 \xrightarrow{\times 2+1} 9 \xrightarrow{\times 2+2} 20 \xrightarrow{\times 2+3} 43 \xrightarrow{\times 2+4} 90 \xrightarrow{\times 2+5} \boxed{185}$$

$$T \xrightarrow{-2} R \xrightarrow{-2} P \xrightarrow{-2} N \xrightarrow{-2} L \xrightarrow{-2} \boxed{J}$$

So, V185J will replace the question mark.

102. ***(a)*** According to the given information, the direction of Neeraj is as following

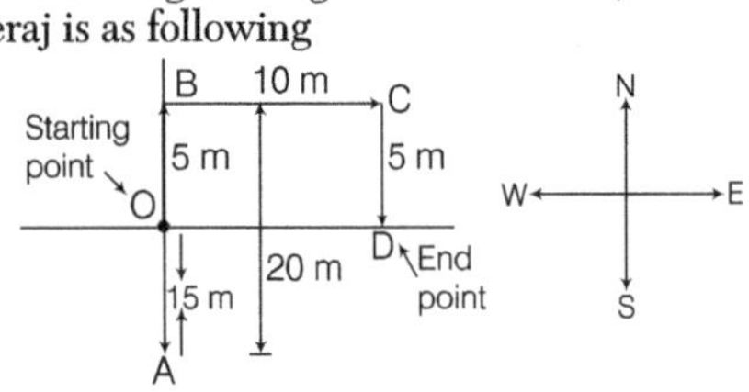

So, it is clearly shown that, Neeraj is 10 m for in East direction from his starting position.

103. ***(b)*** Let the average age of 8 men $= x$ yr
Total age of 8 man = Average × Total man
$= 8x$ yr
Now, new average age $= x + 2$ yr
Total age $= 8(x + 2)$ yr
Difference of ages $= 8(x+2) - 8x$
$= 8x + 16 - 8x$
$= 16$ yr
$\therefore$ Age of new man $= 20 + 16 = 36$ yr
So, the new man is 36 yr older to the man by whom the new man is replaced.

104. ***(d)*** The relation is as following

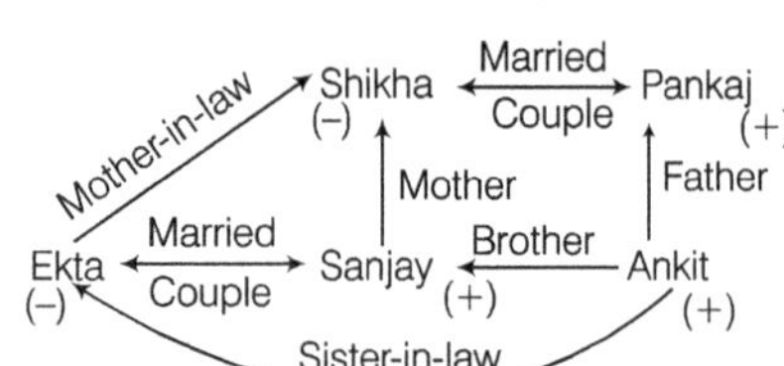

It is clearly shown that, Shikha is the mother of Ankit.

105. ***(c)*** According to the question

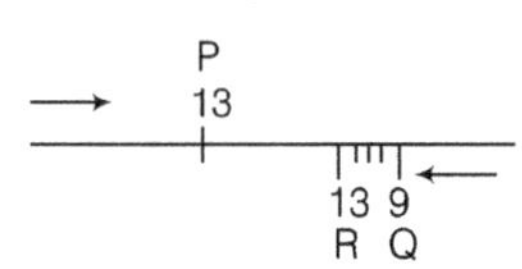

From left end, position of R = Total children − Position from right end +1
$= 40 - 13 + 1 = 28$th

Hence, number of children between P and R
$= (28 - 13) - 1 = 14$

Mathematics

106. *(a)* Let $z = x + iy$

$\text{Re}(z+2) = |z-2|$

$\text{Re}(x+iy+2) = |x+iy-2| \Rightarrow x+2 = \sqrt{(x-2)^2+y^2}$

Squaring on both sides,

$$(x+2)^2 = (x-2)^2 + y^2$$

$$x^2+4+4x = x^2+4-4x+y^2 \Rightarrow y^2 = 8x$$

The locus of z is a parabola.

107. *(a)* Let α, β be the roots of $x^2 - abx - a^2 = 0$ where, $\alpha+\beta = ab$ and $\alpha\beta = -a^2$, which shows product of roots < 0, i.e. one root must be negative and the other must be positive. Hence, equation has one positive root and one negative root.

108. *(c)* Given, $a + 2b + 3c = 12 \forall\ a, b, c \in R^+$

As, AM $\geq$ GM

$$\frac{a+b+b+c+c+c}{6} \geq \sqrt[6]{ab^2c^3}$$

$$\Rightarrow \quad \frac{12}{6} \geq \sqrt[6]{ab^2c^3} \quad \Rightarrow \quad ab^2c^3 \leq 2^6$$

Hence, the maximum value of ab^2c^3 is 2^6.

109. *(a)* a_n of the series $= n(2n+1)^2 \forall\ n \in N$

$$a_n = n[4n^2+4n+1] \Rightarrow a_n = 4n^3+4n^2+n$$

$$S_n = \Sigma a_n = 4\Sigma n^3 + 4\Sigma n^2 + \Sigma n$$

$$= 4\left(\frac{n(n+1)}{2}\right)^2 + \frac{4n(n+1)(2n+1)}{6} + \frac{n(n+1)}{2}$$

$$= \frac{n(n+1)}{2}\left[2n^2+2n+\frac{4(2n+1)}{3}+1\right]$$

$$= \frac{n(n+1)}{2}\left[\frac{6n^2+6n+(8n+4)+3}{3}\right]$$

$$= \frac{n(n+1)}{2}\left[\frac{6n^2+14n+7}{3}\right]$$

$$= \frac{n}{6}(n+1)(6n^2+14n+7)$$

110. *(d)* $\log_{225} 70 = \dfrac{\log_e 70}{\log_e 225} = \dfrac{\log_e 7 + \log_e 5 + \log_e 2}{\log_e 25 + \log_e 9}$

$$= \frac{1 + \frac{\log_e 5}{\log_e 7} + \frac{\log_e 2}{\log_e 7}}{2\frac{\log_e 5}{\log_e 7} + 2\frac{\log_e 3}{\log_e 7}}$$

$$= \frac{1 + \log_7 5 + \log_3 2 \times \log_5 3 \times \log_7 5}{2[\log_7 5 + \log_5 3 \times \log_7 5]}$$

$$= \frac{1+a+abc}{2[a+ab]} = \frac{1+a+abc}{2a(1+b)}$$

111. *(b)* Two circles intersect maximum at two distinct points.

Now, two circles can be selected in ${}^{10}C_2$ ways.

The total number of points of intersection are ${}^{10}C_2 \times 2$

$$= \frac{10\times 9}{1\times 2} \times 2 = 90$$

112. *(c)* $\dfrac{C_1}{C_0} = n, \dfrac{C_2}{C_1} = \dfrac{(n-1)}{2}, \dfrac{C_3}{C_2} = \dfrac{n-2}{3}$ and so on.

$$\frac{C_1}{C_0} + 2\frac{C_2}{C_1} + 3\frac{C_3}{C_2} + \ldots n\frac{C_n}{C_{n-1}} = \Sigma n = \frac{n(n+1)}{2}$$

On putting $n = 20$,

$$\frac{C_1}{C_0} + 2\frac{C_2}{C_1} + 3\frac{C_3}{C_2} + \ldots + 20\frac{C_{20}}{C_{19}} = \frac{20\times 21}{2} = 210$$

113. *(c)* Given, $\begin{vmatrix} 0 & x-p & x-q \\ x+p & 0 & x-r \\ x+q & x-r & 0 \end{vmatrix} = 0$

$$-(x-p)\begin{vmatrix} x+p & x-r \\ x+q & 0 \end{vmatrix} + (x-q)\begin{vmatrix} x+p & 0 \\ x+q & x-r \end{vmatrix} = 0$$

$$\Rightarrow (x-p)(x+q)(x-r) + (x-q)(x+p)(x-r) = 0$$

$$\Rightarrow (x-r)[(x-p)(x+q)+(x-q)(x+p)] = 0$$

$$\Rightarrow (x-r)[x^2 - px + qx - pq + x^2 - qx + px - pq] = 0$$

$$\Rightarrow (x-r)[2x^2 - 2pq] = 0 \Rightarrow x - r = 0 \text{ or } x^2 - pq = 0$$

either $x = r$ or pq

114. *(c)* $A \cdot \text{adj } A = |A| I$

$|A| = xyz - 8x - 3(z-8) + 2(2-2y)$

$|A| = xyz - (8x+3z+4y) + 28$

$\Rightarrow 60 - 20 + 28 = 68$

$\Rightarrow (\text{adj } A)^{-1}$ always exists whenever $(A)^{-1}$ exists.

$\therefore\ A \cdot \text{adj} A = |A| I$

$$= 68\begin{bmatrix} 1 & 0 & 0 \\ 0 & 1 & 0 \\ 0 & 0 & 1 \end{bmatrix} = \begin{bmatrix} 68 & 0 & 0 \\ 0 & 68 & 0 \\ 0 & 0 & 68 \end{bmatrix}$$

115. *(b)* Given, $f(x) = 4x - x^2$

$f(a+1) - f(a-1)$

$$= [4(a+1) - (a+1)^2] - [4(a-1) - (a-1)^2]$$

$$= [4a + 4 - a^2 - 1 - 2a] - [4a - 4 - a^2 - 1 + 2a]$$

$$\Rightarrow \quad 8 - 4a = 0$$

$$\therefore \quad a = 2$$

116. *(c)* Let $R = \{(a, b) : a + b \text{ is an even integer } a, b \in Z\}$

For $a \in Z$, $a + a = 2a$ is an even integer.

$\therefore (a, a) \in R\ \forall\ a \in Z$

$\therefore$ R is reflexive.

Let $(a, b) \in R \Rightarrow a + b$ is an even integer.

$\Rightarrow (b + a)$ is an even integer $\Rightarrow (b, a) \in R$

$\Rightarrow R$ is symmetric.

Let $(a, b), (b, c) \in R$

$\Rightarrow (a + b)$ and $(b + c)$ are even integers.

$\Rightarrow (a + b) + (b + c) = (a + c + 2b)$ is an even integer.

$\Rightarrow (a + c + 2b) - 2b = (a + c)$ is an even integer.

$\Rightarrow (a, c) \in R \Rightarrow R$ is transitive.

$\therefore$ R is an equivalence relation.

Let $R = \{(a, b) : (a - b)$ is an even integer, $a, b \in Z\}$.

For $a \in Z$, $a - a = 0$ is an even integer.

$\therefore (a, a) \in R \ \forall \ a \in Z$

$\therefore$ R is reflexive.

Let $(a, b) \in R \Rightarrow (a - b)$ is an even integer.

$\Rightarrow -(a - b)$ is an even integer.

$\Rightarrow (b - a)$ is an even integer $\Rightarrow (b, a) \in R$

$\Rightarrow R$ is symmetric.

Let $(a, b), (b, c) \in R$

$\Rightarrow (a - b), (b - c)$ are even integers.

$\Rightarrow (a - b) + (b - c)$ is an even integer.

$\Rightarrow (a - c)$ is an even integer.

$\Rightarrow (a, c) \in R$

$\Rightarrow R$ is transitive.

$\therefore$ R is an equivalence relation.

Let $R = \{(a, b) : a < b, a, b \in Z\}$

Let $a \in Z$. $a < a$ is false.

$\therefore$ R is not reflexive.

$\therefore$ R is not an equivalence relation.

Let $R = \{(a, b) : a = b, a, b \in Z\}$.

It is quite easy to check that R is an equivalence relation.

117. ***(b)*** Since, $\sim(p \vee q) \equiv (\sim p \wedge \sim q)$ and $\sim(p \wedge q) \equiv (\sim p \vee \sim q)$

So, options (a) and (d) are not true.

$\sim(p \to q) \equiv p \wedge \sim q$, so option (c) is not true.

Now, $p \to q \sim p \vee q$

$$\sim q \to \sim p \equiv [\sim(\sim q) \vee \sim p]$$

$$\equiv q \vee \sim p \equiv \sim p \vee q \ \Rightarrow \ p \to q \equiv (\sim q \to \sim p)$$

118. ***(a)*** $4^{-x + 0.5} - 7.2^{-x} < 4$

Let $2^{-x} = t$

The equation becomes

$$2t^2 - 7t - 4 < 0 \ \Rightarrow \ (t - 4)(2t + 1) < 0$$

$$\Rightarrow \ 2(t - 4)\left\{t - \left(-\frac{1}{2}\right)\right\} < 0 \Rightarrow -\frac{1}{2} < t < 4$$

Since, $t = 2^{-x} > 0 \ \forall \ x \in R$

$$\Rightarrow \quad 0 < t < 4 \Rightarrow 0 < 2^{-x} < 2^2$$

As, 2^x is an increasing function,

$$-x < 2 \text{ or } x > -2$$

Solution is $(-2, \infty)$.

119. ***(a)*** $\cos^3 x \cdot \sin 2x = \dfrac{\cos 3x + 3\cos x}{4} \times \sin 2x$

$$= \frac{1}{8}(\sin 5x - \sin x) + \frac{3}{8}(\sin 3x + \sin x)$$

$$= \frac{1}{4}\sin x + \frac{3}{8}\sin 3x + \frac{1}{8}\sin 5x$$

Here, $n = 5, a_1 = \dfrac{1}{4}, a_2 = 0, a_3 = \dfrac{3}{8},$

$a_4 = 0, a_5 = 1/8$

120. ***(b)*** $|\cot x| = \cot x + \dfrac{1}{\sin x}$

Let $\cot x > 0 \Rightarrow \cot x = \cot x + \dfrac{1}{\sin x} = 0$

$\Rightarrow \dfrac{1}{\sin x} = 0$ which is not possible.

Let $\cot x \le 0 \Rightarrow -\cot x = \cot x + \dfrac{1}{\sin x}$

$$\Rightarrow \ -2\cot x = \frac{1}{\sin x}$$

$$\Rightarrow \quad \cos x = -\frac{1}{2} \Rightarrow x = \frac{2\pi}{3}, \frac{8\pi}{3}$$

$\therefore$ The number of solutions is 2.

121. ***(a)*** Let $I = (\sin^{-1} x)^3 + (\cos^{-1} x)^3$

$$= (\sin^{-1} x + \cos^{-1} x)[(\sin^{-1} x)^2 + (\cos^{-1} x)^2 - (\sin^{-1} x)(\cos^{-1} x)]$$

$$= \frac{\pi}{2}\left[(\sin^{-1} x + \cos^{-1} x)^2 - 3\sin^{-1} x\left(\frac{\pi}{2} - \sin^{-1} x\right)\right]$$

$$= \frac{\pi}{2}\left(\frac{\pi^2}{4} - \frac{3\pi}{2}\sin^{-1} x + 3(\sin^{-1} x)^2\right)$$

$$= \frac{\pi}{2}\left[3\left((\sin^{-1} x)^2 - \frac{\pi}{2}(\sin^{1} x) + \frac{\pi^2}{16} - \frac{\pi^2}{16}\right) + \frac{\pi^2}{4}\right]$$

$$= \frac{\pi}{2}\left(\frac{\pi^2}{16} + 3\left(\sin^{-1} x - \frac{\pi}{4}\right)^2\right)$$

$$\left(\sin^{-1} x - \frac{\pi}{4}\right)^2 \ge 0$$

For minimum value put $\left(\sin^{-1} x - \dfrac{\pi}{4}\right)^2 = 0$

Minimum values $= \dfrac{\pi}{2}\left[\dfrac{\pi^2}{16} + 0\right] = \dfrac{\pi^3}{32}$

122. *(b)* Let $P(x, y)$ be the original position of the point w.r.t. the original axes. Let us move the origin at new position to (h, k).

Hence, the position of the same point P in the new system is

$$x' = x - h \Rightarrow y' = y - k$$

Here, $(h, k) = (1, 2)$

$\therefore \quad x' = (x - 1), y' = (y - 2)$

According to the question,

$y^2 - 8x - 4y + 12 = (y - 2)^2 - 4a(x - 1)$

$\Rightarrow y^2 - 8x - 4y + 12 = y^2 - 4y + 4 - 4ax + 4a$

On comparing respective coefficients, we get

$$4a = 8 \Rightarrow a = 2$$

123. *(d)* Equations of the bisectors of the angles between the given straight lines are given by

$$\frac{3x + 4y + 7}{\sqrt{9 + 16}} = \pm \frac{12x + 5y - 8}{\sqrt{144 + 25}}$$

$\Rightarrow 13(3x + 4y + 7) = \pm 5(12x + 5y - 8)$

$\Rightarrow 39x + 52y + 91 = \pm (60x + 25y - 40)$

Taking positive signs,

$$39x + 52y + 91 = 60x + 25y - 40$$

$\Rightarrow \; -21x + 27y + 131 = 0 \Rightarrow 21x - 27y - 131 = 0$

Taking negative signs,

$$(39x + 52y + 91) = -(60x + 25y - 40)$$

$\Rightarrow \quad 99x + 77y + 51 = 0$

124. *(b)* Since, the circle touches X-axis,

$$(x - h)^2 + (y - k)^2 = k^2 \qquad \dots (i)$$

Also, it passes through the points $(1, -2)$ and $(3, -4)$.

$$(1 - h)^2 + (-2 - k)^2 = k^2 \qquad \dots (ii)$$

and $$(3 - h)^2 + (-4 - k)^2 = k^2 \qquad \dots (iii)$$

Subtracting Eq. (iii) from Eq. (ii), we get

$$h = k + 5$$

On solving these equations, we get

$$k = -10, -2 \text{ and } h = -5, 3$$

By putting the values of $(h, k) = (-5, -10)$ or $(3, -2)$ in Eq. (i), we get

$$x^2 + y^2 + 10x + 20y + 25 = 0$$

or $$x^2 + y^2 - 6x + 4y + 9 = 0$$

125. *(b)* $x = 9$ meets the hyperbola at $(9, 6\sqrt{2})$ and $(9, -6\sqrt{2})$. Then, the equations of tangent at these points are $3x - 2\sqrt{2}y - 3 = 0$ and $3x + 2\sqrt{2}y - 3 = 0$.

The combined equation of these two tangent is $9x^2 - 8y^2 - 18x + 9 = 0$.

126. *(d)* Let the points be $A(10\hat{\mathbf{i}} + 3\hat{\mathbf{j}})$, $B(12\hat{\mathbf{i}} - 15\hat{\mathbf{j}})$ and $C(a\hat{\mathbf{i}} + 11\hat{\mathbf{j}})$.

$AB = |2\hat{\mathbf{i}} - 18\hat{\mathbf{j}}|$ and $AC = (a - 10)\,\hat{\mathbf{i}} + 8\hat{\mathbf{j}}$

Since, A, B and C are collinear, then

$$\frac{2}{a - 10} = -\frac{18}{8} \Rightarrow a = \frac{82}{9}$$

127. *(d)* We have, $|\mathbf{a}| = 3, |\mathbf{b}| = 4$ and $|\mathbf{c}| = 5$.

It is given that

$$\mathbf{a} \perp (\mathbf{b} + \mathbf{c}), \mathbf{b} \perp (\mathbf{c} + \mathbf{a}) \text{ and } \mathbf{c} \perp (\mathbf{a} + \mathbf{b})$$

$\Rightarrow \mathbf{a} \cdot (\mathbf{b} + \mathbf{c}) = 0, \mathbf{b} \cdot (\mathbf{c} + \mathbf{a}) = 0$ and $\mathbf{c} \cdot (\mathbf{a} + \mathbf{b}) = 0$

$\Rightarrow \mathbf{a} \cdot \mathbf{b} + \mathbf{a} \cdot \mathbf{c} = \mathbf{b} \cdot \mathbf{c} + \mathbf{b} \cdot \mathbf{a} = \mathbf{c} \cdot \mathbf{a} + \mathbf{c} \cdot \mathbf{b} = 0$

or $\mathbf{a} \cdot \mathbf{b} + \mathbf{b} \cdot \mathbf{c} + \mathbf{c} \cdot \mathbf{a} = 0$

(adding all the above equations)

Now, $|\mathbf{a} + \mathbf{b} + \mathbf{c}|^2 = |\mathbf{a}|^2 + |\mathbf{b}|^2 + |\mathbf{c}|^2 + 2(\mathbf{a} \cdot \mathbf{b} + \mathbf{b} \cdot \mathbf{c} + \mathbf{c} \cdot \mathbf{a})$

$$= 3^2 + 4^2 + 5^2 = 50$$

$\therefore \quad |\mathbf{a} + \mathbf{b} + \mathbf{c}| = 5\sqrt{2}$

128. *(d)* We have, $|(\mathbf{a} \times \mathbf{b}) \cdot \mathbf{c}| = |\mathbf{a}|\,|\mathbf{b}|\,|\mathbf{c}|$

$\Rightarrow ||\mathbf{a}|\,|\mathbf{b}|\,|\mathbf{c}| \sin\theta \cos\alpha| = |\mathbf{a}|\,|\mathbf{b}|\,|\mathbf{c}|$

$\Rightarrow |\sin\theta|\,|\cos\alpha| = 1 \Rightarrow \theta = \frac{\pi}{2}$ and $\alpha = 0$

$\Rightarrow \mathbf{a} \perp \mathbf{b}$ and $\mathbf{c} \parallel \hat{\mathbf{n}} \Rightarrow \mathbf{a} \perp \mathbf{b}$ and $\mathbf{c} \perp$ both $\mathbf{a}$ and $\mathbf{b}$.

$\Rightarrow \mathbf{a}, \mathbf{b}, \mathbf{c}$ are mutually perpendicular.

$\Rightarrow \mathbf{a} \cdot \mathbf{b} = \mathbf{b} \cdot \mathbf{c} = \mathbf{c} \cdot \mathbf{a} = 0$

129. *(c)* Let each edge of cube be a, then coordinates of the vertices of cube are

$O(0, 0, 0), A(a, 0, 0), B(0, a, 0), C(0, 0, a), N(a, a, 0), P(a, a, a), L(0, a, a), M(a, 0, a)$

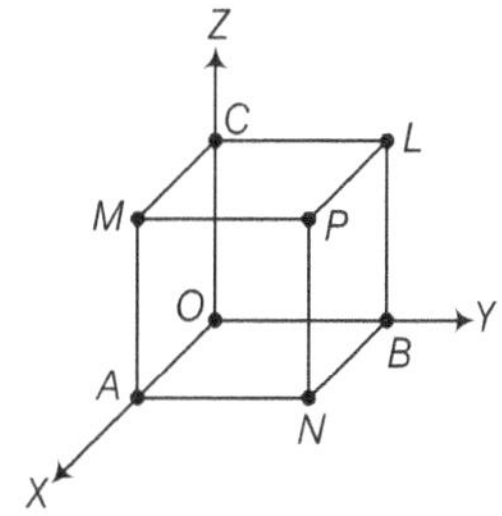

Direction ratios of the diagonals OP, AL, BM and CN are $(a, a, a), (-a, a, a), (a, -a, a)$ and $(a, a, -a)$.

Let θ be the acute angle between diagonals OP and AL.

$$\therefore \quad \cos\theta = \frac{a_1a_2 + b_1b_2 + c_1c_2}{\sqrt{a_1^2 + b_1^2 + c_1^2}\sqrt{a_2^2 + b_2^2 + c_2^2}}$$

$$= \frac{a \times (-a) + a \times a + a \times a}{\sqrt{a^2 + a^2 + a^2}\sqrt{(-a)^2 + a^2 + a^2}}$$

$$= \frac{-a^2 + a^2 + a^2}{\sqrt{3a^2}\,\sqrt{3a^2}} = \frac{a^2}{a\sqrt{3} \times a\sqrt{3}}$$

$$\Rightarrow \cos\theta = \frac{1}{3} \quad \Rightarrow \quad \theta = \cos^{-1}\left(\frac{1}{3}\right)$$

130. ***(b)*** Given lines are

$L_1 : \frac{x+1}{3} = \frac{y+2}{1} = \frac{z+1}{2}$ and $L_2 : \frac{x-2}{1} = \frac{y+2}{2} = \frac{z-3}{3}$

Now, convert into vector form

$L_1 : (-\hat{\mathbf{i}} - 2\hat{\mathbf{j}} - \hat{\mathbf{k}}) + \lambda(3\hat{\mathbf{i}} + \hat{\mathbf{j}} + 2\hat{\mathbf{k}})$

$L_2 : (2\hat{\mathbf{i}} - 2\hat{\mathbf{j}} + 3\hat{\mathbf{k}}) + \mu(\hat{\mathbf{i}} + 2\hat{\mathbf{j}} + 3\hat{\mathbf{k}})$

Line L_1 comparing with $\mathbf{a}_1 + \lambda \mathbf{b}_1$,
and L_2 comparing with $\mathbf{a}_2 + \mu \mathbf{b}_2$, then we have
$\mathbf{b}_1 = 3\hat{\mathbf{i}} + \hat{\mathbf{j}} + 2\hat{\mathbf{k}}$ and $\mathbf{b}_2 = \hat{\mathbf{i}} + 2\hat{\mathbf{j}} + 3\hat{\mathbf{k}}$

Perpendicular to both $\mathbf{b}_1$ and $\mathbf{b}_2 = \mathbf{b}_1 \times \mathbf{b}_2$

$$= \begin{vmatrix} \hat{\mathbf{i}} & \hat{\mathbf{j}} & \hat{\mathbf{k}} \\ 3 & 1 & 2 \\ 1 & 2 & 3 \end{vmatrix}$$

$$= \hat{\mathbf{i}}(3-4) - \hat{\mathbf{j}}(9-2) + \hat{\mathbf{k}}(6-1) = -\hat{\mathbf{i}} - 7\hat{\mathbf{j}} + 5\hat{\mathbf{k}}$$

$\therefore$ Required unit vector $= \frac{-\hat{\mathbf{i}} - 7\hat{\mathbf{j}} + 5\hat{\mathbf{k}}}{\sqrt{(-1)^2 + (-7)^2 + (5)^2}}$

$$= \frac{-\hat{\mathbf{i}} - 7\hat{\mathbf{j}} + 5\hat{\mathbf{k}}}{\sqrt{75}} = \frac{-\hat{\mathbf{i}} - 7\hat{\mathbf{j}} + 5\hat{\mathbf{k}}}{5\sqrt{3}}$$

131. ***(b)*** The given line is $\mathbf{r} = \mathbf{a} + t\mathbf{b}$, where
$\mathbf{a} = 2\hat{\mathbf{i}} - 2\hat{\mathbf{j}} + 3\hat{\mathbf{k}}$, $\mathbf{b} = \hat{\mathbf{i}} - \hat{\mathbf{j}} + 4\hat{\mathbf{k}}$ and given plane is
$\mathbf{r} \cdot \mathbf{n} = \mathbf{d}$, where $\mathbf{n} = \hat{\mathbf{i}} + 5\hat{\mathbf{j}} + \hat{\mathbf{k}}$, $\mathbf{d} = 5$.

Since, $\mathbf{b} \cdot \mathbf{n} = 1 - 5 + 4 = 0$

$\therefore$ Given line is parallel to given plane.

$\therefore$ The distance between the line and the plane is equal to length of the perpendicular from the point $\mathbf{a} = 2\hat{\mathbf{i}} - 2\hat{\mathbf{j}} + 3\hat{\mathbf{k}}$ on the line to the given plane.

$\therefore$ Required distance

$$= \left| \frac{(2\hat{\mathbf{i}} - 2\hat{\mathbf{j}} + 3\hat{\mathbf{k}}) \cdot (\hat{\mathbf{i}} + 5\hat{\mathbf{j}} + \hat{\mathbf{k}}) - 5}{\sqrt{1 + 25 + 1}} \right|$$

$$= \left| \frac{2 - 10 + 3 - 5}{\sqrt{27}} \right| = \frac{10}{3\sqrt{3}}$$

132. ***(c)*** Let E_1 = event of getting both red cards

E_2 = event of getting both kings

and $E_1 \cap E_2$ = event of getting 2 kings of red cards

$$\therefore P(E_1) = \frac{{}^{26}C_2}{{}^{52}C_2} = \frac{325}{1326},\ P(E_2) = \frac{{}^{4}C_2}{{}^{52}C_2} = \frac{6}{1326}$$

and $P(E_1 \cap E_2) = \frac{{}^{2}C_2}{{}^{52}C_2} = \frac{1}{1326}$

$\therefore$ P(both red or both kings) $= P(E_1 \cup E_2)$

$$= P(E_1) + P(E_2) - P(E_1 \cap E_2)$$

$$= \frac{325}{1326} + \frac{6}{1326} - \frac{1}{1326} = \frac{330}{1326} = \frac{55}{221}$$

133. ***(d)*** Since, A and B are independent events.

$$\therefore \quad P(A/B) = P(A) = \frac{1}{2}$$

$$\Rightarrow P\left(\frac{A}{A \cup B}\right) = \frac{P[A \cap (A \cup B)]}{P(A \cup B)}$$

$$= \frac{P(A)}{P(A \cup B)} = \frac{1/2}{\frac{1}{2} + \frac{1}{5} - \frac{1}{10}} = \frac{1/2}{\frac{1}{2} + \frac{1}{5} - \frac{1}{10}} = \frac{1/2}{\frac{6}{10}} = \frac{5}{6}$$

Similarly, $P\left(\frac{A \cap B}{A' \cup B'}\right) = P\left(\frac{A \cap B}{(A \cap B)'}\right) = 0$

134. ***(b)*** Let E_1 = coin shows head, E_2 = coin shows tail,
A = drawn ball is blue

$$P(E_1) = \frac{1}{2} = P(E_2)$$

$P(A/E_1)$ = Probability of drawing a blue ball from bag I = 2/7

$P(A/E_2)$ = Probability of drawing a blue ball from bag II = 6/8

By Baye's theorem, we have

$$P\left(\frac{E_1}{A}\right) = \frac{P(E_1)\, P\left(\frac{A}{E_1}\right)}{P(E_1)\, P\left(\frac{A}{E_1}\right) + P(E_2)\, P\left(\frac{A}{E_2}\right)}$$

$$= \frac{\frac{1}{2} \times \frac{2}{7}}{\frac{1}{2} \times \frac{2}{7} + \frac{1}{2} \times \frac{6}{8}} = \frac{\frac{2}{7}}{\frac{2}{7} + \frac{6}{8}} = \frac{2}{7} \times \frac{56}{58} = \frac{8}{29}$$

135. ***(c)*** Given, $E(X) = 3$ and $E(X^2) = 11$

Variable of $X = E(X^2) - [E(X)]^2 = 11 - 3^2 = 11 - 9 = 2$

136. ***(d)*** Given, $x_1 + x_2 + x_3 + \ldots + x_{10} = 12$
and $x_1^2 + x_2^2 + \ldots + x_{10}^2 = 18$

$$\sigma^2 = \frac{1}{n} \sum x^2 - \left(\frac{1}{n} \sum x\right)^2$$

$$= \frac{1}{10} \times 18 - \left(\frac{1}{10} \times 12\right)^2 = \frac{9}{5} - \frac{36}{25} = \frac{9}{25}$$

$\therefore$ Standard deviation = 3/5

137. ***(c)*** Let PQ be the chimney whose height is h metres.

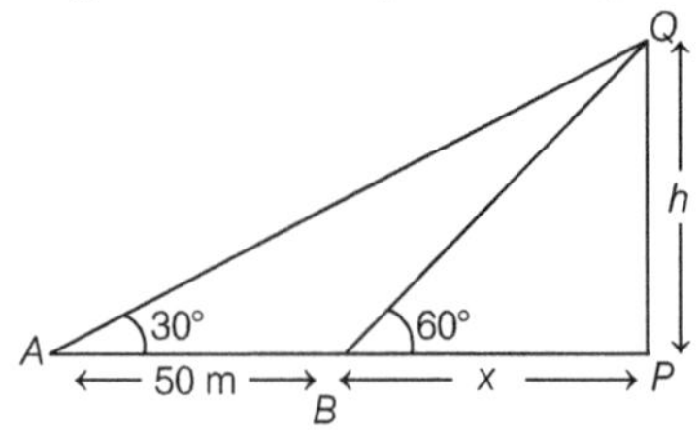

In ΔBPQ, $\tan 60° = \frac{PQ}{BP} \Rightarrow \sqrt{3} = \frac{h}{x}$

$$\Rightarrow \quad h = \sqrt{3}\,x \quad \ldots(i)$$

and in ΔAPQ, $\tan 30^\circ = \dfrac{PQ}{AP}$

$$\Rightarrow \quad \frac{1}{\sqrt{3}} = \frac{h}{50+x}$$

$$\Rightarrow \quad 50 + x = h\sqrt{3}$$

$$\Rightarrow \quad 50 + x = 3x \Rightarrow x = 25 \quad \text{[using Eq. (i)]}$$

$\therefore$ Height of the chimney $= 25\sqrt{3}$ m

138. ***(c)*** The required limit

$$= \lim_{x\to 0} \frac{\sqrt{(1-\cos x^2)(1+\cos x^2)}}{(1-\cos x)\sqrt{1+\cos x^2}}$$

$$= \lim_{x\to 0} \frac{\left(\dfrac{\sin x^2}{x^2}\right)}{2\left(\dfrac{\sin^2(x/2)}{x^2}\right)} \cdot \frac{1}{\sqrt{1+\cos x^2}}$$

$$= \frac{1}{2\times\frac{1}{4}\times\sqrt{2}} = \sqrt{2}$$

139. ***(c)*** $f'(1-0) = \lim_{h\to 0} \dfrac{f(1-h)-f(1)}{-h}$

$$= \lim_{h\to 0} \frac{a(1-h)^2+1-(1+a)}{-h}$$

$$= \lim_{h\to 0} \frac{a(h^2-2h)}{-h} = 2a$$

$$f'(1+0) = \lim_{h\to 0} \frac{f(1+h)-f(1)}{h}$$

$$= \lim_{h\to 0} \frac{(1+h)^2+a(1+h)+b-(1+a)}{h}$$

$$= \lim_{h\to 0} \frac{h^2+2h+ah+b}{h}$$

$$= 2+a, \text{ if } b = 0$$

Thus, $2a = 2 + a$, $b = 0$

$$\Rightarrow \quad a = 2, b = 0$$

140. ***(c)*** We have, $\dfrac{dx}{dt} = 2t+3$ and $\dfrac{dy}{dt} = 4t-2$

$$\therefore \quad \frac{dy}{dx} = \frac{\frac{dy}{dt}}{\frac{dx}{dt}} = \frac{4t-2}{2t+3}$$

Thus, slope of the tangent to the curve at the point $t = 2$ is

$$\left[\frac{dy}{dx}\right]_{t=2} = \frac{4(2)-2}{2(2)+3} = \frac{6}{7}$$

141. ***(c)*** Let $I = \int \dfrac{1}{1-2\sin x}\,dx$

On putting $\sin x = \dfrac{2\tan\frac{x}{2}}{1+\tan^2\frac{x}{2}}$, we get

$$I = \int \frac{1}{1 - \dfrac{4\tan\frac{x}{2}}{1+\tan^2\frac{x}{2}}}\,dx$$

$$= \int \frac{1+\tan^2\frac{x}{2}}{1+\tan^2\frac{x}{2} - 4\tan\frac{x}{2}}\,dx$$

$$= \int \frac{\sec^2\frac{x}{2}}{1+\tan^2\frac{x}{2} - 4\tan\frac{x}{2}}\,dx$$

$$= 2\int \frac{1}{\left(\tan\frac{x}{2} - 2\right)^2 - (\sqrt{3})^2}\,d\left(\tan\frac{x}{2}\right)$$

$$= 2\int \frac{dt}{(t-2)^2 - (\sqrt{3})^2}, \text{ where } t = \tan\frac{x}{2}$$

$$= 2\cdot\frac{1}{2\sqrt{3}}\log\left|\frac{t-2-\sqrt{3}}{t-2+\sqrt{3}}\right| + c$$

$$= \frac{1}{\sqrt{3}}\log\left|\frac{\tan\frac{x}{2} - 2 - \sqrt{3}}{\tan\frac{x}{2} - 2 + \sqrt{3}}\right| + c$$

142. ***(a)*** Let

$$l = \int_0^1 \frac{\log(1+x)}{1+x^2}\,dx = \int_0^{\pi/4} \frac{\log(1+\tan\theta)}{(1+\tan^2\theta)}\sec^2\theta\,d\theta$$

[Let $x = \tan\theta \Rightarrow dx = \sec^2\theta d\theta$]

$$= \int_0^{\pi/4} \log(1+\tan\theta)\,d\theta$$

$$= \int_0^{\pi/4} \log\left[1+\tan\left(\frac{\pi}{4}-\theta\right)\right]d\theta$$

$$= \int_0^{\pi/4} \log\left(1+\frac{1-\tan\theta}{1+\tan\theta}\right)d\theta$$

$$= \int_0^{\pi/4} \log\left(\frac{2}{1+\tan\theta}\right)d\theta$$

$$= \log 2\int_0^{\pi/4} 1\,d\theta - l$$

$$\Rightarrow \quad 2l = \frac{\pi}{4}\log 2 \Rightarrow l = \frac{\pi}{8}\log 2$$

143. *(c)* Required area

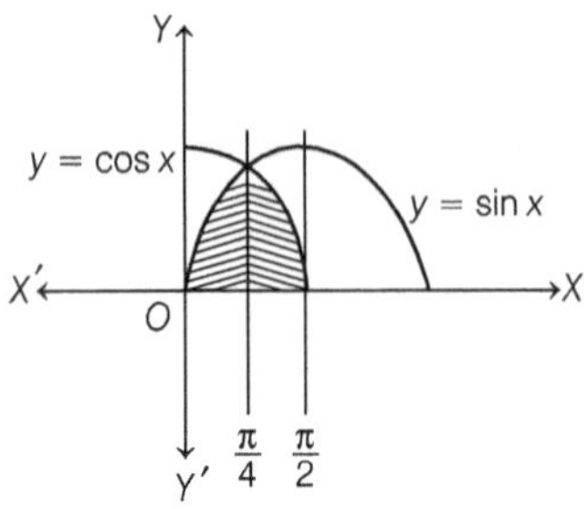

$$= \int_0^{\pi/4} \sin x\,dx + \int_{\pi/4}^{\pi/2} \cos x\,dx$$

$$= -[\cos x]_0^{\pi/4} + [\sin x]_{\pi/4}^{\pi/2}$$

$$= 2\left(1 - \frac{1}{\sqrt{2}}\right) = (2 - \sqrt{2}) \text{ sq units}$$

144. *(b)* $\left(\frac{x+y-1}{x+y-2}\right)\frac{dy}{dx} = \left(\frac{x+y+1}{x+y+2}\right)$

Put $x + y = v$

$\therefore\ 1 + \frac{dy}{dx} = \frac{dv}{dx}$ or $\left(\frac{v-1}{v-2}\right)\left(\frac{dv}{dx} - 1\right) = \left(\frac{v+1}{v+2}\right)$

$\Rightarrow\ \frac{dv}{dx} - 1 = \frac{(v+1)(v-2)}{(v-1)(v+2)} = \frac{v^2 - v - 2}{v^2 + v - 2}$

or $\quad \frac{dv}{dx} = \frac{2v^2 - 4}{(v^2 + v - 2)}$

$\Rightarrow\ \frac{(v^2 + v - 2)}{(v^2 - 2)}\,dv = 2dx$

$\Rightarrow\ \left(1 + \frac{v}{v^2 - 2}\right)dv = 2dx$

On integrating, we get

$$v + \frac{1}{2}\log|v^2 - 2| = 2x + c$$

or $\quad (y - x) + \frac{1}{2}\log\left|(x+y)^2 - 2\right| = c$

Given, $y = 1$ when $x = 1$

$\therefore\ 0 + \frac{1}{2}\log 2 = c$ or $(y - x) + \frac{1}{2}\log\left|\frac{(x+y)^2 - 2}{2}\right| = 0$

or $\quad \log\left|\frac{(x+y)^2 - 2}{2}\right| = 2(x - y)$

145. *(d)* Given, $y = \sin\left(2\tan^{-1}\sqrt{\frac{1-x}{1+x}}\right)$

Let $x = \cos 2\theta \Rightarrow y = \sin\left(2\tan^{-1}\sqrt{\frac{1 - \cos 2\theta}{1 + \cos 2\theta}}\right)$

$\Rightarrow\ y = \sin\left(2\tan^{-1}\sqrt{\frac{2\sin^2\theta}{2\cos^2\theta}}\right)$

$\Rightarrow\ y = \sin[2\tan^{-1}(\tan\theta)]$

$\Rightarrow\ y = \sin 2\theta \Rightarrow y = \sqrt{1 - \cos^2 2\theta}$

$\Rightarrow\ y = \sqrt{1 - x^2}$

$\therefore\ \frac{dy}{dx} = \frac{1}{2\sqrt{1 - x^2}}(-2x) = -\frac{x}{\sqrt{1 - x^2}}$

146. *(b)* We have, $y = x(x-1)^2$

$\therefore\ \frac{dy}{dx} = (x-1)^2 \cdot 1 + 2x(x-1)$

$$= 3x^2 - 4x + 1 = (x-1)(3x-1)$$

For maximum or minimum, $\frac{dy}{dx} = 0$

$\Rightarrow\ (x-1)(3x-1) = 0$

$\Rightarrow\ x = \frac{1}{3}, 1$

Now, $\frac{d^2y}{dx^2} = 6x - 4 \Rightarrow \left[\frac{d^2y}{dx^2}\right]_{x=\frac{1}{3}} = -2 < 0$

$\therefore$ y is maximum when $x = \frac{1}{3}$ and maximum value is

$$[y]_{x=\frac{1}{3}} = \frac{4}{27}$$

Also, $\quad \left[\frac{d^2y}{dx^2}\right]_{x=1} = 2 > 0$

$\therefore$ y is minimum when $x = 1$.

147. *(b)* We have, $\cos y\frac{dy}{dx} + \frac{4}{x}\sin y = \frac{e^x}{x^3}$

Let $\quad \sin y = t$

$\Rightarrow\ \cos y\frac{dy}{dx} = \frac{dt}{dx}$

$\therefore\ \frac{dt}{dx} + \frac{4}{x}t = \frac{e^x}{x^3}$

$$\text{IF} = e^{\int \frac{4}{x}dx} = e^{4\log x} = x^4$$

The solution is

$\therefore\ tx^4 = \int x^4 \cdot \frac{e^x}{x^3}dx = xe^x - e^x + c$

$\because\ \sin y\, x^4 = xe^x - e^x + c$

$\because x = 1,\ y = 0$

$\therefore\ \sin y = e^x(x-1)x^{-4}$ $\quad [\because\ c = 0]$

148. ***(a)*** Let us make the table from the given data

Player A			Player B		
x_i	$x_i - 61$	$(x_i - 61)^2$	y_i	$y_i - 61$	$(y_i - 61)^2$
58	− 3	9	94	33	1089
59	− 2	4	26	− 35	1225
60	− 1	1	92	31	961
54	− 7	49	65	4	16
65	4	16	96	35	1225
66	5	25	78	17	289
52	− 9	81	14	−47	2209
75	14	196	34	−27	729
69	8	64	98	37	1369
52	− 9	81	13	−48	2304
$\sum x_i = 610$		$\sum (x_i - 61)^2 = 526$	$\sum_{yi} = 610$		$\sum (y_i - 61)^2 = 11416$

For player *A*, $\text{Mean} = \dfrac{\sum x_i}{n} = \dfrac{610}{10} = 61$

$$\text{SD} = \sqrt{\frac{\sum (x_i - 61)^2}{N}}$$

$$= \sqrt{\frac{526}{10}} = 7.25$$

For player *B*, $\text{Mean} = \dfrac{\sum y_i}{n} = \dfrac{610}{10} = 61$

$$\text{SD} = \sqrt{\frac{\sum (y_i - 61)^2}{N}}$$

$$= \sqrt{\frac{11416}{10}}$$

$$= 33.79$$

Since, SD for player *A* is 7.25 < SD for player *B* is 33.79.

Hence, player *A* is more consistent player.

149. ***(b)*** Table for equation $x + y = 8$ is

x	0	8
$y = 8 - x$	8	0

Table for equation $3x + 5y = 15$ is

x	0	5
$y = \dfrac{15 - 3x}{5}$	3	0

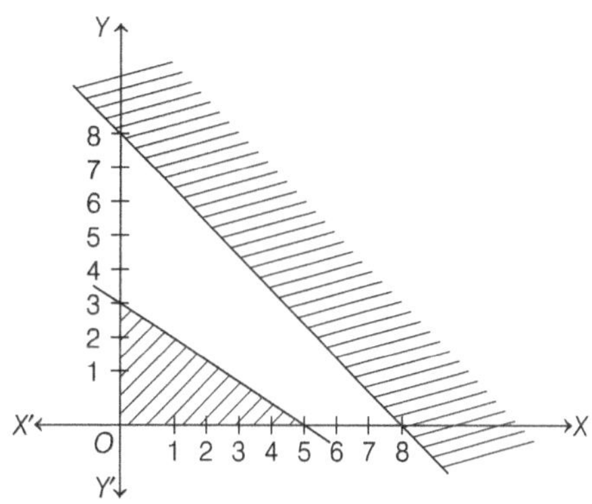

It can be concluded from the graph, that there is no point which can satisfy all the constraints simultaneously. Therefore, the problem has no feasible solution.

150. ***(a)*** Substitute 0 for y in the equation $4y^2 = 4x^2 - x^3$,

$$0 = x^2(4 - x) \Rightarrow x = 0, x = 4$$

It means curve makes the loop symmetric about *X*-axis between 0 and 4.

$$\text{Area} = 2\int_0^4 y\,dx = \frac{2}{2}\int_0^4 \sqrt{4x^2 - x^3}\,dx = \int_0^4 x\sqrt{4 - x}\,dx$$

Let $4 - x = t \Rightarrow -dx = dt$

$$\text{Area} = -\int_4^0 (4 - t)\sqrt{t}\,dt$$

$$= \int_0^4 (4\sqrt{t} - t\sqrt{t})dt = \left[4 \times \frac{2}{3}t^{3/2} - \frac{2}{5}t^{5/2}\right]_0^4$$

$$= \frac{8}{3} \times 8 - \frac{2}{5} \times 32 = \frac{64}{3} - \frac{64}{5} = 64 \times \frac{2}{15} = \frac{128}{15}$$

BITSAT

Birla Institute of Technology & Science Admission Test

Solved Paper 2020*

Instructions

- There are 150 questions in all. The number of questions in each part is as given below.

		No. of Questions
Part I	Physics	1-40
Part II	Chemistry	41-80
Part III	a. English Proficiency	81-95
	b. Logical Reasoning	96-105
Part IV	Mathematics	106-150

- All questions are Multiple Choice Questions having four options out of which **only one** is correct.
- Each correct answer fetches 3 marks while incorrect answer has a penalty of 1 mark.
- Time allotted to complete this paper is 3 hrs.

PART I

Physics

1. Three solenoid coils of same dimensions, same number of turns and same number of layers of windings are taken. Coil 1 has inductance L_1 wounded by Mn wire of resistance $6\,\Omega\text{m}^{-1}$, coil 2 with inductance L_2 wounded by similar wire but in reverse direction in each layer. Coil 3 with inductance L_3 wounded by a superconducting wire. The relation between their self inductances will be

a. $L_1 = L_2, L_3 = 0$ ***b.*** $L_1 = L_2 = L_3$
c. $L_1 = L_3, L_2 = 0$ ***d.*** $L_1 > L_2 > L_3$

2. Three capacitors C_1, C_2 and C_3 are connected as shown in the figure below. If capacitor C_3 breaks down electrically, then the change in total charge on the combination of capacitors, is

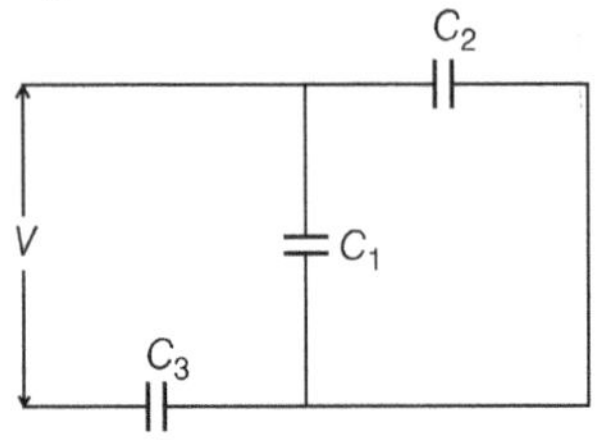

a. $(C_1+C_2)V\left[1-\left(\frac{C_3}{C_1+C_2+C_3}\right)\right]$

b. $(C_3+C_2)V\left[1-\left(\frac{C_1}{(C_1+C_2+C_3)}\right)\right]$

c. $(C_1+C_2)V\left[1-\left(\frac{C_2+C_1}{C_1+C_3}\right)\right]$

d. $(C_1+C_2)V\left[1-\left(\frac{C_2}{(C_1+C_3)}\right)\right]$

3. A black body radiates energy at the rate $E\ \text{Wm}^{-2}$ at high temperature T K. When the temperature is reduced to $(T/4)$K, then the new radiant energy is

a. $E/256$ ***b.*** $4E$ ***c.*** $E/4$ ***d.*** $E/16$

4. The length of the rectangle is $l = 15.2$ cm and breadth is $b = 2.9$ cm and the minimum possible measurement by scale = 0.1 cm. Then, the area of the rectangle is
(Taking, significant figures into consideration)

*Memory Based

a. 44.08 cm^2 *b.* 24.8 cm^2
c. 44 cm^2 *d.* 94.008 cm^2

5. In an adiabatic process, where pressure is decreased by $\frac{3}{4}\%$, if $\frac{C_p}{C_V} = \frac{4}{3}$, then the volume increases by

a. $\frac{3}{4}\%$ *b.* $\frac{9}{16}\%$
c. $\frac{16}{9}\%$ *d.* $\frac{4}{3}\%$

6. The vibrations of a string of length 60 cm fixed at both ends are represented by the equation

$$y = 4\sin\left(\frac{\pi x}{15}\right)\cos(96\pi t)$$

where, x and y are in cm and t is in second. Calculate the velocity of the particle at $x = 7.5$ cm and $t = 0.25$ s.

a. 4 m/s *b.* Zero
c. 16 m/s *d.* 9.8 m/s

7. A charge $+q$ is placed at the origin O of XY-axes as shown in the figure. The work done in taking a charge Q from A to B along the straight line AB is

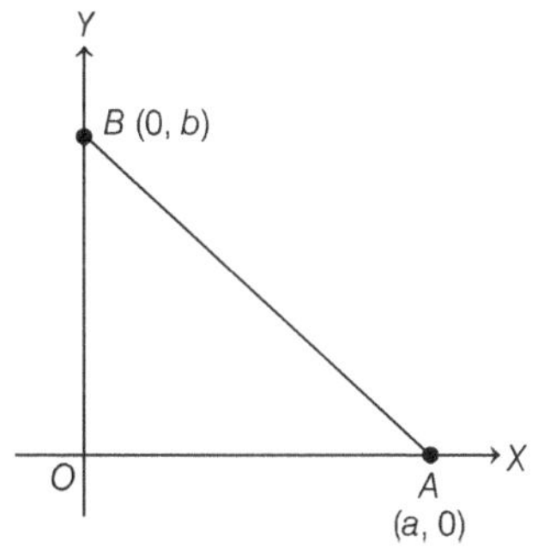

a. $\frac{qQ}{4\pi\varepsilon_0}\left(\frac{a-b}{ab}\right)$ *b.* $\frac{qQ}{4\pi\varepsilon_0}\left(\frac{b-a}{ab}\right)$
c. $\frac{qQ}{4\pi\varepsilon_0}\left(\frac{b}{a^2}-\frac{1}{b}\right)$ *d.* $\frac{qQ}{4\pi\varepsilon_0}\left(\frac{a}{b^2}-\frac{1}{b}\right)$

8. The hydrogen-like element that has a spectrum whose lines have wavelength four times shorter than those of atomic hydrogen is

a. lithium *b.* helium
c. berilliyum *d.* potassium

9. Two small conducting spheres of equal radius have charges $+20\mu$C and -40μC respectively and placed at a distance R from each other experience force F_1. If they are brought in contact and separated to the same distance, they experience force F_2. The ratio of F_1 to F_2 is

a. $1:4$ *b.* $8:1$
c. $-8:1$ *d.* $1:8$

10. A Carnot engine has the same efficiency between 600 K to 300 K and 1600 K to x K, then the value of x is

a. 1600 K *b.* 800 K
c. 819 K *d.* 900 K

11. A ball of mass 0.5 kg is thrown up with initial speed 16 ms^{-1} and reaches maximum height of 9 m. How much energy is dissipated by air drag acting on the ball during the ascent?

a. 199 J *b.* 19.9 J
c. 20.9 J *d.* 9.9 J

12. In the circuit shown, if the 10 Ω resistor is replaced by a resistor of 15 Ω, then what is the amount of current drawn from the battery?

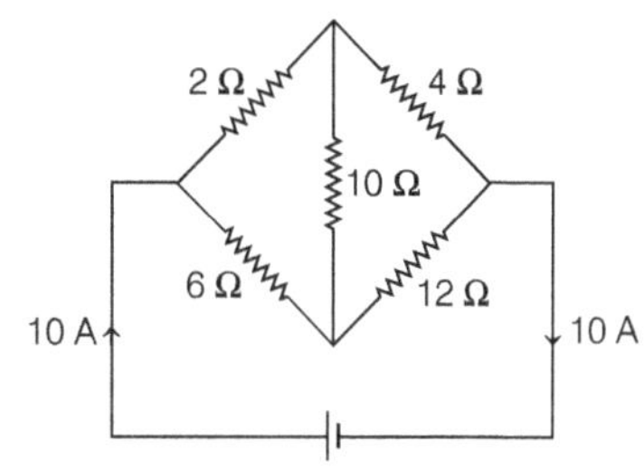

a. 100 A *b.* 10 A
c. 1 A *d.* 2.4 A

13. A time dependent force $F(=8t)$ acts on a particle of mass 2 kg. If the particle starts from rest, the work done by the force during the first 1s will be

a. 0.4 J *b.* 4 J
c. 19 J *d.* 4.5 J

14. If L, R, C and V represent inductance, resistance, capacitance and potential difference respectively, then dimensions of $\frac{L}{RCV}$ are the same as those of

a. current *b.* $\frac{1}{\text{current}}$
c. charge *d.* $\frac{1}{\text{charge}}$

15. The magnetic field of a beam emerging from a fitter facing a flood light is given by

$B = 10 \times 10^{-8} \sin(1 \times 10^7 z - 3.6 \times 10^{15} t)$T. The average intensity of the beam is

a. 1.82 W/m^2 *b.* 1.19 W/m^2
c. 1.18 W/m^2 *d.* 1.17 W/m^2

16. From the velocity-time graph of a body moving in a straight line, the distance travelled and the average velocity in the time interval $t = 0$ to $t = 20$ s are, respectively,

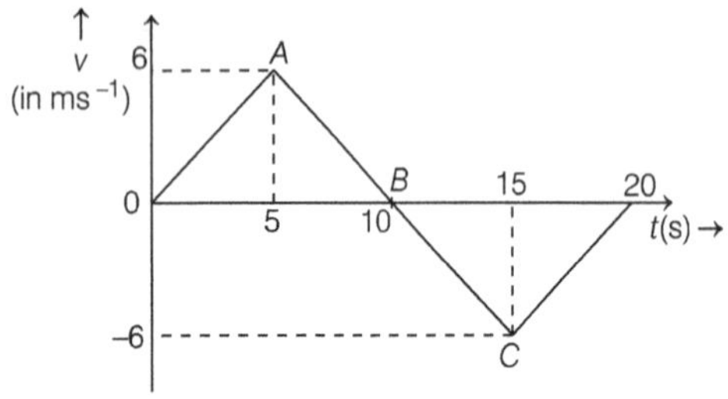

a. 0, 0 *b.* 120 m, 60 m
c. 60 m, 0 *d.* 0, 60 m

17. A thin equi-convex lens is made of glass of refractive index 1.5 and its focal length is 0.2 m. If it acts as a concave lens of focal length 0.5 m when dipped in a liquid, the refractive index of the liquid is

a. $\frac{17}{8}$ *b.* $\frac{15}{8}$ *c.* $\frac{13}{8}$ *d.* $\frac{9}{8}$

18. A moving coil galvanometer has a resistance of 60 Ω and it indicates full deflection on passing a current of 4.5 mA. A voltmeter is made using this galvanometer and a 4. 5 kΩ resistance. The maximum voltage, that can be measured using this voltmeter, will be close to

a. 21 V *b.* 20.5 V *c.* 20 V *d.* 19.5 V

19. The combination of the gates shown in following figure yields

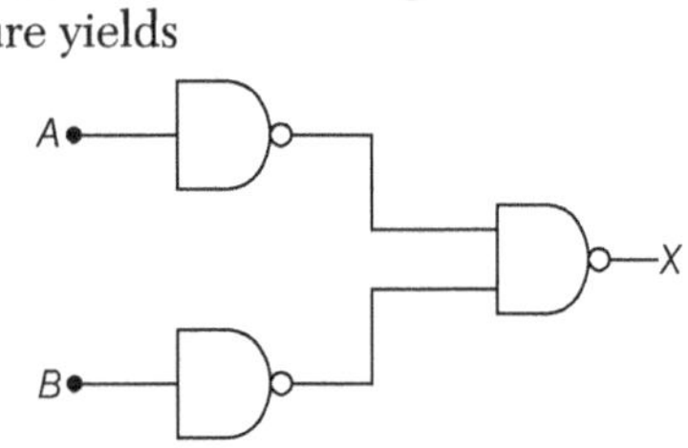

a. NAND gate *b.* OR gate
c. NOT gate *d.* XOR gate

20. A vessel contains one mole of O_2 gas (molar mass 32) at a temperature T. The pressure of the gas is p. An identical vessel containing one mole of He gas (molar mass 4) at a temperature $2T$ has a pressure of

a. $\frac{p}{8}$ *b.* p *c.* $2p$ *d.* $8p$

21. The acceleration of a particle in m/s^2 is given by $a = (3t^2 - 2t + 1)$, where t is in second. If the particle starts with a velocity $v = 1$ m/s at $t = 1$ s, then velocity of the particle at the end of 4s is

a. 40 m/s *b.* 52 m/s *c.* 48 m/s *d.* 84 m/s

22. The figure shows graphs of pressure (p) *versus* density (d) for an ideal gas at two temperatures T_1 and T_2, then

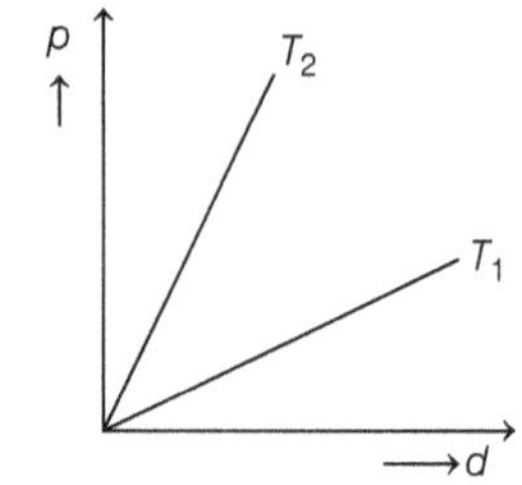

a. $T_1 > T_2$ *b.* $T_1 < T_2$
c. $T_1 = T_2$ *d.* None of these

23. Two spheres of the same material and same radii r are touching each other. The gravitational force between the spheres is proportional to

a. $\frac{1}{r^2}$ *b.* r^2 *c.* $\frac{1}{r^4}$ *d.* r^4

24. A spherical lens of power –4 D is placed at a distance of 15 cm from another spherical lens of power 5 D. A beam of parallel light falls on the first spherical lens. The final image formed is

a. real and at a distance of 40 cm from the lens of power 5 D
b. real and at a distance of 10 cm from the lens of power – 4 D.
c. virtual and at a distance of 40 cm from the lens of power – 5 D
d. None of the above

25. The wheel of a car, accelerated uniformly from rest, rotates through 5 rad during the first second. The angle (in rad) rotated during the next second is

a. 15 *b.* 7.5 *c.* 12.5 *d.* 20

26. Lights of two different frequencies whose photons have energies 1.5 eV and 2.5 eV respectively illuminate a metallic surface whose work function is 0.5 eV successively. Ratio of maximum speeds of emitted electrons will be

a. 3 : 2 *b.* 2 : 3
c. $\sqrt{3} : \sqrt{2}$ *d.* $\sqrt{2} : \sqrt{3}$

27. When a body is dropped from a height h, then it hits the ground with a momentum p. If the same body is dropped from a height which is three times more than previous height, the percentage change in momentum when it hits the ground is

a. 25% *b.* 50%
c. 75% *d.* 100%

28. The reading of the ammeter for a germanium diode in the given circuit is

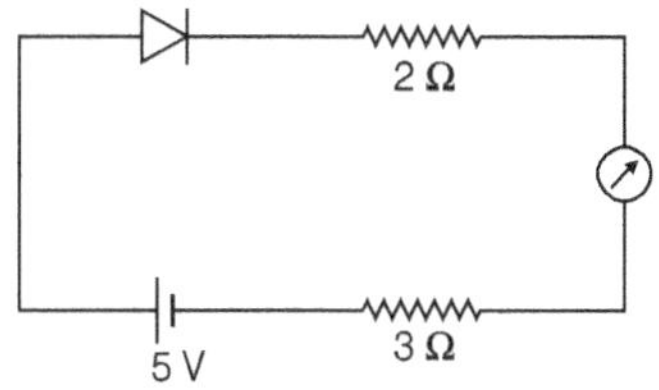

a. 0.94 A *b.* 0 *c.* 2.8 A *d.* 5 A

29. The decay constants of two radioactive substances X and Y are 4λ and λ respectively. At $t = 0$, a sample has the same number of two nuclei. The time taken for the ratio of number of nuclei to become $\frac{1}{e^3}$ will be

a. $\frac{1}{3\lambda}$ *b.* $\frac{1}{2\lambda}$ *c.* $\frac{2}{3\lambda}$ *d.* $\frac{3}{2\lambda}$

30. The magnitude of the force vector acting on a unit length of a thin wire carrying a current $I = 10$ A at a point O, if the wire is bent in the form of a semi-circle (shown below) with radius $R = 20\pi$ cm, is

a. 30 μN/m *b.* 40 μN/m
c. 50 μN/m *d.* 60 μN/m

31. A long cylindrical iron core of cross-sectional area 5 cm^2 is inserted into a long solenoid having 4000 turns/metre and carrying a current 5 A. The magnetic field inside the core is π T. Find the pole strength developed.

a. 1000 A-m *b.* 1240 A-m
c. 882 A-m *d.* 760 A-m

32. A plane requires for take off a speed of 72 kmh^{-1}, the run on the ground being 50 m. The mass of the plane is 10000 kg and the coefficient of friction between the plane and the ground is 0.2. Assume that the plane accelerates uniformly during take off. The minimum force required by the engine of the plane for take off is

a. 4.43×10^4 N *b.* 5.96×10^4 N
c. 2.25×10^4 N *d.* 3.45×10^4 N

33. In a fluorescent lamp choke, 120 V of reverse voltage is produced when the choke current changes uniformly 0.50 A to 0.20 A in a duration of 0.030 ms. The self inductance of the choke (in mH) is estimated to be

a. 12 H *b.* 12×10^{-3} mH
c. 12×10^{-3} H *d.* 0

34. A man grows into a giant such that his linear dimensions increase by a factor of 8. Assuming that his density remains same, the stress in his leg will change by a factor of

a. 1/8 *b.* 8 *c.* 36 *d.* 64

35. When 2 moles of a monoatomic gas are mixed with 3 moles of a diatomic gas, the value of adiabatic exponent for the mixture is

a. 15/16 *b.* 7/5 *c.* 31/21 *d.* 38/59

36. A simple pendulum is placed inside a lift, the lift is moving with a uniform acceleration. If the time periods of the pendulum, while the lift is moving upwards and downwards are in the ratio of 1 : 3, then the acceleration of the lift is [Take, acceleration due to gravity, $g = 10\ m/s^2$]

a. 4 m/s^2 *b.* 6 m/s^2 *c.* 8 m/s^2 *d.* 10 m/s^2

37. A satellite is moving around the earth with speed v in a circular orbit of radius r. If the orbit radius is decreased by 2%, its speed will increase by

a. 1% *b.* 2% *c.* 1.5% *d.* 1.414%

38. In the circuit shown in figure, the AC source has angular frequency $\omega = 2000\ rad\ s^{-1}$. The amplitude of the current will be nearest to

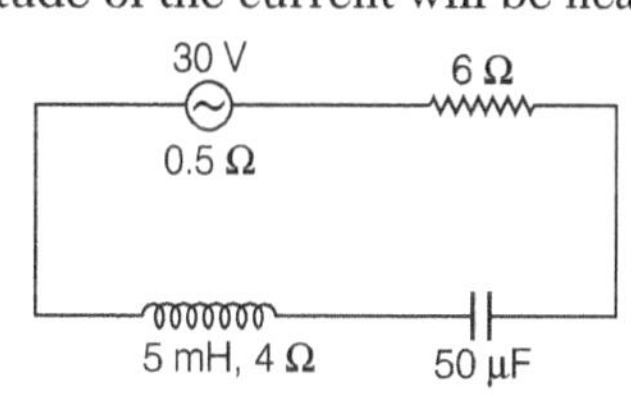

a. 2.85 A *b.* 3 A *c.* 0.5 A *d.* 3.625 A

39. The work done by a force acting on a body is as shown in the following graph. The total work done in covering an initial distance of 40 m is

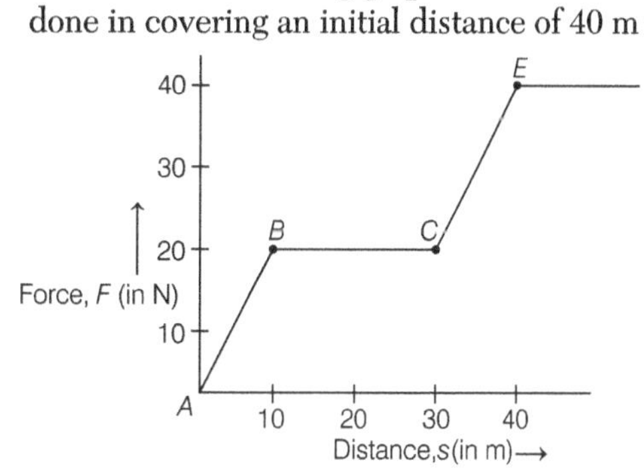

a. 500 J *b.* 600 J *c.* 400 J *d.* 800 J

40. A potentiometer wire, 10 m long, has a resistance of 40 Ω. It is connected in series with a resistance box and a 4 V storage cell. If the potential gradient along the wire is 0.4 $mVcm^{-1}$, the resistance unplugged in the box is

a. 220 Ω *b.* 360 Ω *c.* 760 Ω *d.* 848.3 Ω

PART II

Chemistry

41. In a set of reactions, *m*-bromobenzoic acid gives a product *D*. Identify the product *D*.

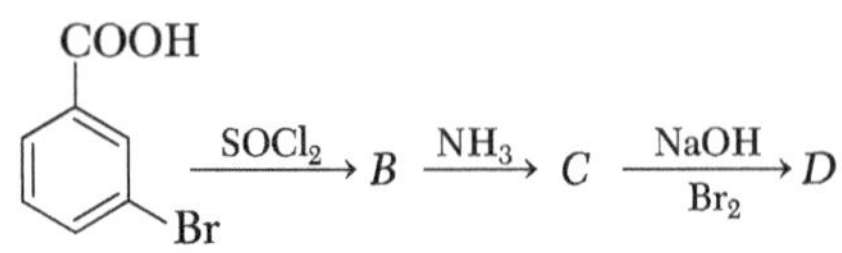

a. 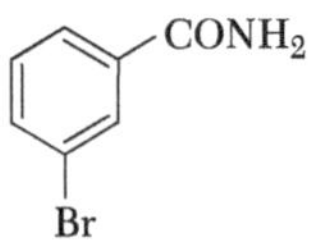 *b.* 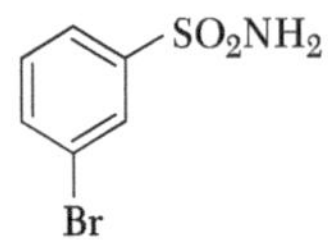

c. 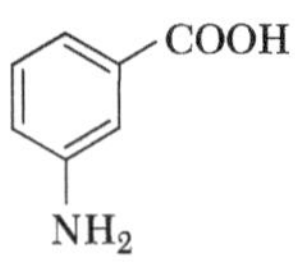 *d.* 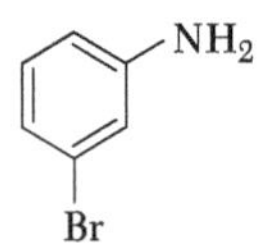

42. A volume of 50.00 mL of a weak acid of unknown concentration is titrated with 0.10 M solution of NaOH. The equivalence point is reached after 39.30 mL of NaOH solution has been added. At the half equivalence point (19.65 mL) the pH is 4.85. Thus, initial concentration of the acid and its pK_a values are

	[HA]	pK_a
a.	0.1 M	4.85
b.	0.079 M	4.85
c.	0.1 M	3.70
d.	0.097 M	2.93

43. 1-phenyl-1, 3-dibromopropane on treatment with alc. KOH gives diastereomeric mixture in which compound (*A*) is major product. (*A*) gives the following reaction

$$(A) \xrightarrow{AlCl_3/\Delta} (B) \xrightarrow{:CuBr} (C) + (D)$$

Compounds (*C*) and (*D*) are

a. 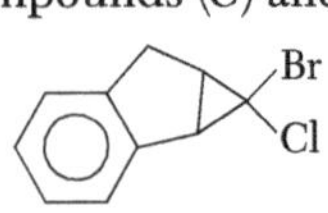 and 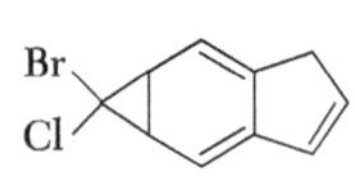

b. 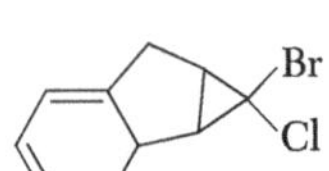 and 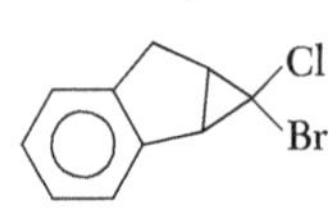

c. 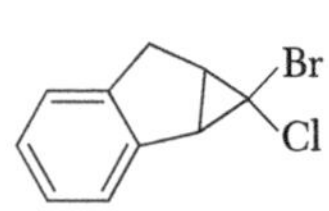 and 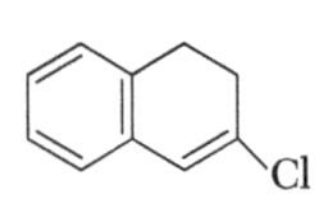

d. 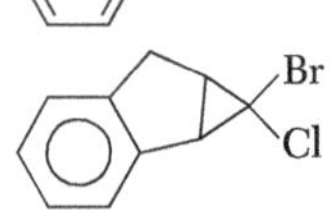 and 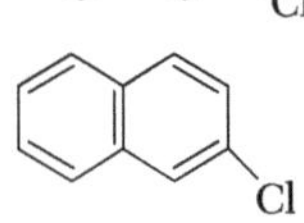

44. Which of the following statement is incorrect?

a. α-D-fructose and β-D-fructose are enantiomers of each other.

b. D-glyceraldehyde and L-glyceraldehyde are enantiomers of each other.

c. The reserve carbohydrate of animals is glycogen.

d. Aldohexoses which react with phenyl hydrazine to give identical osazones are C-2 epimers.

45. Which of the following is not a mixed pair of oxides?

a. Mn_3O_4 and Co_3O_4

b. Co_3O_4 and Pb_3O_4

c. Pb_3O_4 and Mn_3O_4

d. Fe_3O_4 and Fe_2O_3

46. A solution of copper sulphate is electrolysed between copper electrodes by a current of 10.0A passing for one hour. Which of the following statements is correct regarding the changes that occur at the electrodes and in the solution?

a. 11.84 g of copper will deposit on the cathode

b. 11.84 g of copper will deposit on the anode

c. 11.84 g of copper will deposit on the anode as well as on the cathode

d. copper will not deposit on any of the electrode

47. 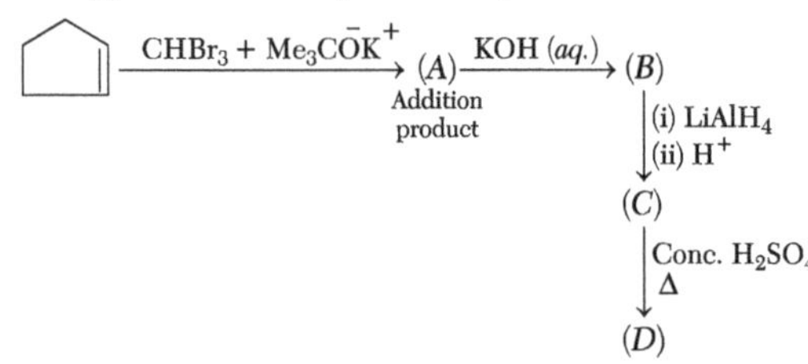

Identify structure of compound (*D*).

a. 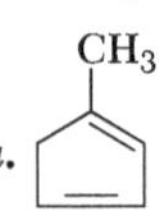 *b.* 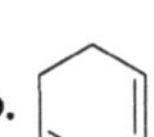 c. 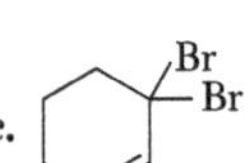 *d.* 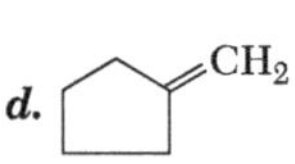

48. Which of the following statement is correct?

a. ΔS for $\frac{1}{2}Cl_2(g) \longrightarrow Cl(g)$ is positive.

b. $\Delta E < 0$ for combustion of $CH_4(g)$ in a sealed container with rigid adiabatic system.

c. ΔG is always zero for a reversible process in a closed system.

d. $\Delta G°$ for an ideal gas reaction is a function of pressure.

49. Which of the following statements on critical constants of gases are correct?

I. Larger the T_c/p_c value of a gas, larger would be the included volume.

II. Critical temperature (T_c) of a gas is greater than its Boyle temperature (T_B).

III. At the critical point in the van der Waals' gas isotherm, $\left(\frac{\partial p}{\partial V_m}\right)_{T_c} = 0$

Select the correct answer using the codes given below.

a. Both I and II *b.* Both I and III
c. Both II and III *d.* I, II and III

50. In the given reaction,

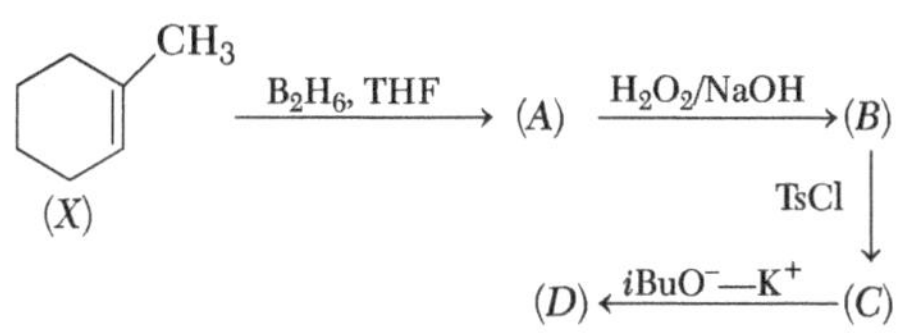

Product (D) is

a. a positional isomer of X
b. identical to X
c. chain isomer of X
d. an oxidation product of X

51. $Na_2B_4O_7 + conc.H_2SO_4 + H_2O$

$\downarrow$

(A)

$\downarrow$ (i) C_2H_2OH (ii) Ignite

(B)

(B) is identified by the characteristic colour of the flame. (A) and (B) are respectively

a. H_3BO_3 and $Na_2B_4O_7$
b. $B(OC_2H_5)_3$ and H_3BO_3
c. $NaBO_2$ and H_3BO_3
d. H_3BO_3 and $B(OC_2H_5)_3$

52. How many mL of perhydrol is required to produce sufficient oxygen which can be used to completely convert 2 L of SO_2 gas?

a. 10 mL *b.* 5 mL
c. 20 mL *d.* 30 mL

53. At the top of a mountain the thermometer reads 0°C and the barometer reads 710 mm Hg. At the bottom of the mountain the temperature is 30°C and the pressure is 760 mm Hg. The ratio of the density of air at the top to that of the bottom is

a. 1 : 1.04 *b.* 0.4 : 1
c. 1.04 : 1 *d.* 1 : 04

54. Match the following columns.

	Column I (Compound)		Column II (Structure)
A.	ClF_3	1.	Square planar
B.	PCl_5	2.	Tetrahedral
C.	IF_5	3.	Trigonal bipyramidal
D.	CCl_4	4.	Square pyramidal
E.	XeF_4	5.	T-shaped

Codes

	A	B	C	D	E
a.	5	4	3	2	1
b.	5	3	4	2	1
c.	5	3	4	1	2
d.	4	3	5	2	1

55. Most acidic hydrogen is present in

a. 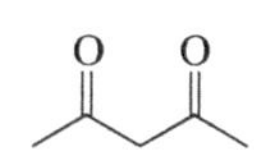 *b.* 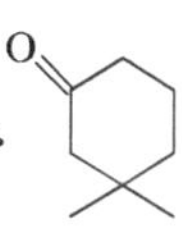

c. $(CH_3CO)_3CH$ *d.* $(CH_3)_3COH$

56. The process of 'eutrophication' is due to the

a. increase in concentration of insecticide in H_2O
b. increase in concentration of fluoride ion in H_2O
c. reduction in concentration of the dissolved oxygen in water due to phosphate pollution in water.
d. attack of younger leaves of a plant by peroxyacetyl nitrate

57. Ionisation energy of H-atom is 13.6 eV. The wavelengths of the spectral line emitted when an electron in Be^{3+} comes from 5th energy level to 2nd energy level is

a. 43.5 nm *b.* 4350 nm
c. 4.35 nm *d.* 435 nm

58. The enthalpies of combustion of carbon and carbon monoxide in excess of oxygen at 298 K and constant pressure are – 393.5 kJ/mol and –280.0 kJ/mol respectively. The heat of formation of carbon monoxide at constant volume is

a. + 111.7 kJ/mol *b.* –1111.7 kJ/mol
c. –111.7 kJ/mol *d.* –11.7 kJ/mol

59. If the quantum number l could have the value n also then, Sc(21) would have electronic configuration as (other rules strictly followed)

a. $1s^2\ 1p^6\ 2s^2\ 2p^6\ 2d^3\ 3s^2$

b. $1s^2\ 1p^6\ 2s^2\ 2p^6\ 3s^2\ 2d^3$

c. $1s^2\ 2s^2\ 2p^6\ 3s^2\ 3p^6\ 3d^1\ 4s^2$

d. $1s^2\ 2s^2\ 2p^6\ 3s^2\ 3p^6\ 3d^3$

60. 1.1 mole of A mixed with 2.2 moles of B and the mixture is kept in a 1 L flask and the equilibrium, $A + 2B \rightleftharpoons 2C + D$ is reached. If at equilibrium 0.2 mole of C is formed then the value of K_C will be

a. 0.1 *b.* 0.01 *c.* 0.001 *d.* 1

61. Choose the correct chemical reaction among the following:

a. $CaCN_2 + H_2O \longrightarrow Ca(OH)_2 + C_2H_2 + NH_3$

b. $2NH_3 + CaSO_4 + CO_2 + H_2O \longrightarrow CaCN_2 + (NH_4)_2SO_4$

c. $CaCl_2 + Na_2SO_4 \longrightarrow CaSO_4 + 2NaCl$

d. $CaC_2 + H_2O \longrightarrow Ca(OH)_2 + 2NaCl$

62. Which of the following reactions represents disproportionation?

a. $CrO_5 \longrightarrow Cr^{3+} + O_2$

b. $IO_3^- + I^- + H^+ \longrightarrow I_2$

c. $CrO_2Cl_2 + NaOH \longrightarrow Na_2CrO_4 + NaCl + H_2O$

d. $Na_2S_2O_3 + H_2SO_4 \longrightarrow Na_2SO_4 + SO_2 + S_8 + H_2O$

63. 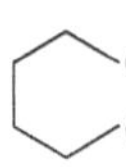 COOEt, COOEt $\xrightarrow{NaOEt}$ (B) $\xrightarrow[\text{(ii) }\Delta]{\text{(i) }H_3O^+}$ (C)

(A)

Compound C is

a. 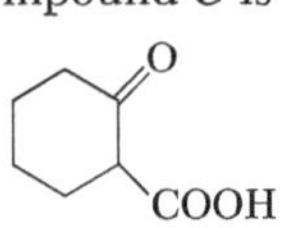

b. 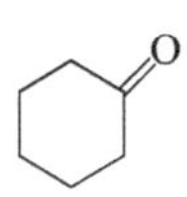

c. 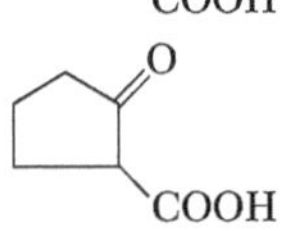

d. 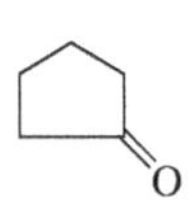

64. On the basis of Ellingham diagram, which of the following is not correct?

a. Entropy change for all metal oxides is roughly same.

b. Below the boiling point, $T\Delta S$ factor is same irrespective of metal.

c. Above $\Delta G = 0$ line, oxide decomposes into metal and oxygen.

d. If randomness increases the slope increases.

65. Which of the following is not explained by adsorption?

a. When acetic acid solution is shaken with charcoal the concentration of acid decreases.

b. The white ppt. of $Mg(OH)_2$ attains blue colour when precipitated in the presence of magneson reagent.

c. An aqueous solution of NaOH attains pink colour with a drop of phenolphthalein.

d. When animal charcoal is shaken with coloured methylene blue solution, the solution turns colourless.

66. Identify 'A' and 'B' in the following reaction

$B \xleftarrow[h\nu]{HBr}$ 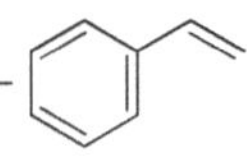 $\xrightarrow{HBr} A$

a. Both A and B are 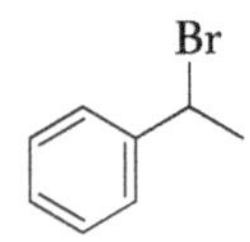

b. Both A and B are 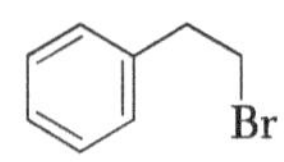

c. A is 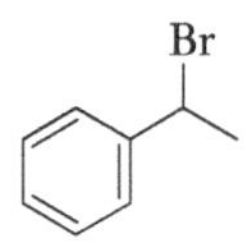 and B is 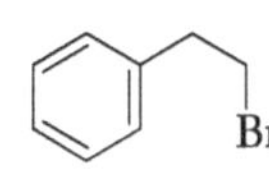

d. A is 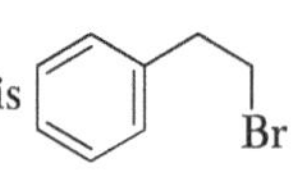 and B is 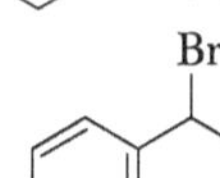

67. An element occurs in two crystalline forms α and β. The α-form has an fcc with $a = 3.68$ Å and β form has a bcc with $a = 2.92$ Å. Calculate the ratio of their densities.

a. 1 : 1 *b.* 1 : 2

c. 2 : 1 *d.* 2 : 3

68. The molar heat of vaporisation of water at 100°C is 40.585 kJ/mol. The temperature at which a solution containing 5.60 g of $Al_2(SO_4)_3$ per 1000 g of water boil is .... (Assuming the degree of ionisation of salt to be 1).

a. 1000.042° C *b.* 10.0042°C

c. 100.042°C *d.* 105°C

69. An organic base $C_8H_{11}N(X)$ reacts with nitrous acid at 0°C to give a clear solution. Heating the solution with KCN and cuprous cyanide followed by continued heating with conc. HCl gives a crystalline solid. Heating this solid with alkaline potassium permanganate gives a compound which dehydrates on heating to an anhydride $(C_8H_4O_3)$.

Compound X is

a. 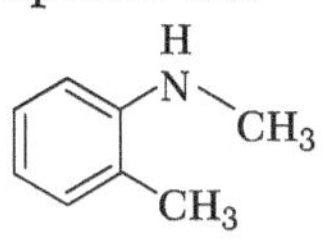

b. 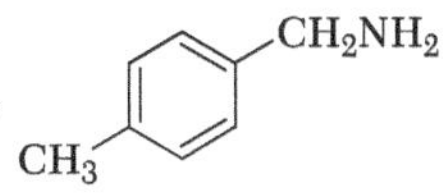

c. 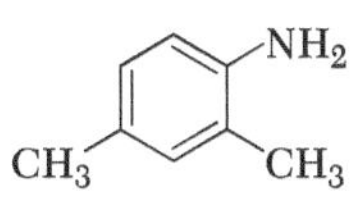

d. 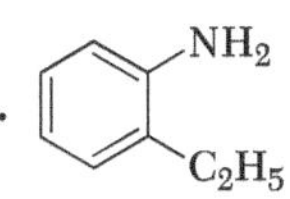

70. The final product of the following reaction is/are

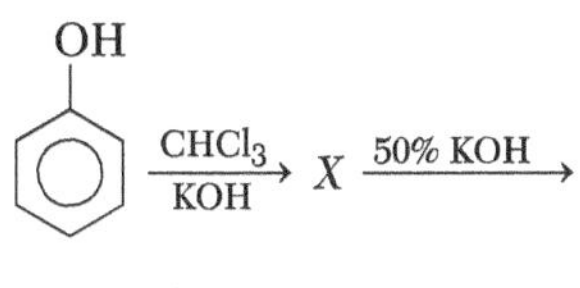

a. 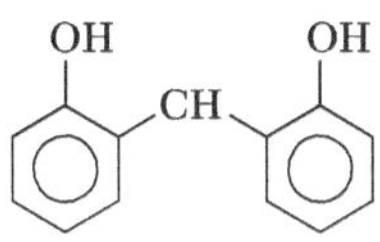

b. 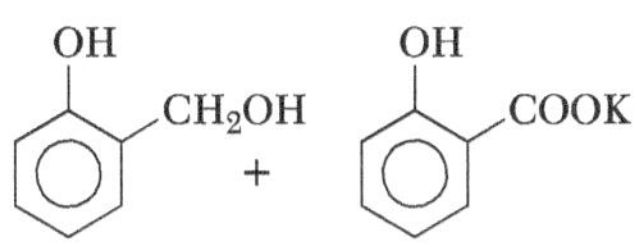

c. 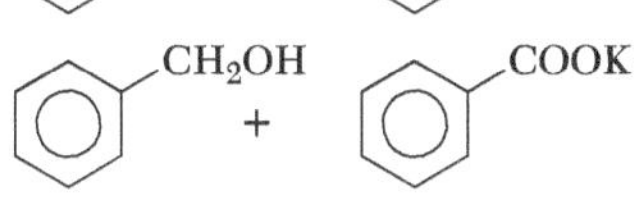

d. 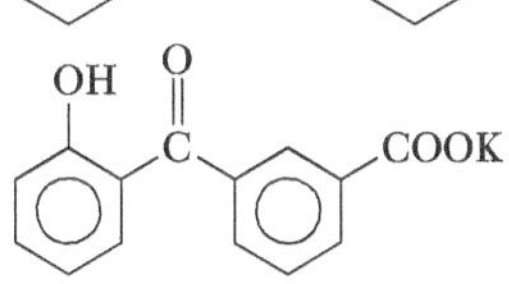

71. The correct order of pseudohalide, polyhalide and interhalogen is

a. BrI_2^-, OCN^-, IF_5 *b.* IF_5, BrI_2^-, OCN^-
c. OCN^-, IF_2, BrI_2^- *d.* OCN^-, BrI_2^-, IF_5

72. An inorganic halide (A) reacts with water to form two acids (B) and (C). (A) also reacts with NaOH to form two salts (D) and (E) which remain in solution. The solution gives white precipitate with both $AgNO_3$ and $BaCl_2$ solutions respectively. (A) is a useful organic reagent. The compound (A) is

a. $SOCl_2$ *b.* SO_2Cl_2 *c.* S_2Cl_2 *d.* SF_4

73. Following statements regarding the periodic trends of chemical reactivity to the alkali metals and the halogens are given. Which of these statements gives the correct picture?

a. The reactivity decreases in the alkali metals but increases in the halogens with increase in atomic number down the group.
b. In both the alkali metals and the halogens the chemical reactivity decreases with increase in atomic number down the group.
c. Chemical reactivity increases with increase in atomic number down the group in both the alkali metals and halogens.
d. In alkali metals the reactivity increases but in the halogens it decreases with increase in atomic number down the group.

74. According to IUPAC nomenclature sodium nitroprusside is named as

a. sodiumpentacyanonitrosylferrate (II)
b. sodiumpentacyanonitrosylferrate (III)
c. sodiumnitroferricyanide
d. sodiumnitroferrocyanide

75. $A(g) \longrightarrow P(g) + Q(g) + R(g)$,

Follow first order kinetics with a half-life of 69.3 s at 500°C. Starting from the gas 'A' an container at 500°C and at a pressure of 0.4 atm, the total pressure of the system after 230 s will be

a. 1.15 atm *b.* 1.32 atm
c. 1.22 atm *d.* 1.12 atm

76. Which of the following gives paracetamol on acetylation?

a. 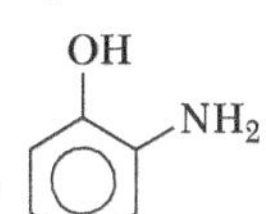

b. 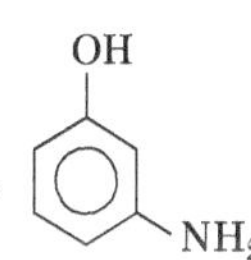

c. 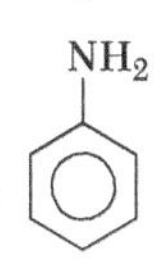

d. 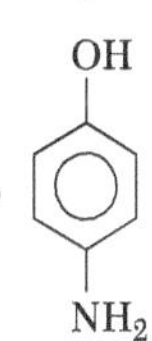

77. In qualitative analysis when H_2S is passed through an aqueous solution of salt acidified with dilute HCl, a black ppt. is obtained. On boiling the precipitate with dil. HNO_3, it forms a solution of blue colour. Addition of excess of aqueous solution of NH_3 to this solution gives

a. deep blue ppt. of $Cu(OH)_2$
b. deep blue solution of $[Cu(NH_3)_4]^{2+}$
c. deep blue solution of $Cu(NO_3)_2$
d. deep blue solution of $Cu(OH)_2 \cdot Cu(NO_3)_2$

78. Titration of 0.1467 g of primary standard $Na_2C_2O_4$ required 28.85 mL of $KMnO_4$ solution. Calculate the molar concentration of $KMnO_4$ solution.

a. 0.01518 M
b. 0.001518 M
c. 0.15180 M
d. 1.5180 M

79. Which of the following facts about the complex $[Cr(NH_3)_6]Cl_3$ is wrong?

a. The complex involves d^2sp^3 hybridisation and its octahedral in shape.
b. The complex is paramagnetic.
c. The complex is an outer orbital complex.
d. The complex gives white precipitate with silver nitrate solution.

80. Sulphuric acid is a dibasic acid. It ionises in two stages and hence, has two dissociation constants K_{a_1} and K_{a_2}. Which of the following is the correct observation regarding K_{a_1} and K_{a_2}?

a. $K_{a_1} > K_{a_2}$ *b.* $K_{a_1} < K_{a_2}$
c. $K_{a_1} = K_{a_2}$ *d.* $K_{a_1} = 1.2 \times 10^{-2}, K_{a_2} > 10$

PART III

a. English Proficiency

Directions (Q. Nos. 81-83) *In the following questions, find out which part of a sentence has an error. If there is no error, mark part (d) as your answer.*

81. The road **a.** / to famous monument **b.** / passes through a forest. **c.**/ No error **d**

82. The master did not know **a.**/ who of the servants **b.**/ broke the glass **c.**/ No error **d**

83. Had I come **a.**/ to know about his difficulties **b.**/ I would have certainly helped **c.**/ No error **d**

Directions (Q. Nos. 84 and 85) *Fill in the blanks with suitable preposition from the alternatives given under each sentence.*

84. Her trekking was met .... obstacles.

a. with *b.* from *c.* by *d.* of

85. She has not got .... the shock of losing her father.

a. over *b.* at *c.* from *d.* with

Directions (Q. Nos. 86-88) *Select the word or phrase which is closest to the opposite in meaning of the italicised word or phrase.*

86. Yuvraj Singh is suffering from a *BENIGN* cancer.

a. Unfriendly *b.* Friendly
c. Fatal *d.* Malignant

87. He is a *NOTED* figure of film industry.

a. Known *b.* Unknown
c. Famous *d.* Infamous

88. *SAGACIOUS* decisions taken at right time in one's career has long effects.

a. Foolish *b.* Intelligent
c. Thoughtful *d.* Intuitive

Directions (Q. Nos. 89 and 90) *Choose the word nearest in meaning to the italicised word.*

89. The actor got *PEEVISH* on asking personal questions.

a. Irritated *b.* Happy *c.* Shy *d.* Satisfied

90. The engineer *ROUGHED OUT* his ideas on a piece of paper while he talked.

a. Shaped soughly *b.* Rejected
c. Drew a quick plan *d.* Describe inaccurately

Directions (Q. Nos. 91-95) *Read the passage given below and answer the questions that follow.*

The Centre and the States must become partners in the planning process to determine national priorities together. The process of planning would undergo a change in view of the changes in domestic economic situation and momentous trends emerging in the world. The development of human resource and the building up of an institutional framework would have to receive priority attention. The role of the government would also have to be examined so as to fully involve the people in the process of nation-building. The main task would be to ensure that the real initiative is transferred to the people. The private sector which would register expansion hereafter should keep this objective firmly in view. The need for an effective population policy is an urgent necessity in the country's planning strategy. The family welfare programme should not be treated as the Centre's responsibility alone. The States should evolve a suitable mechanism for closer involvement of the Government agencies, Zilla Parishads and Panchayats for making the family welfare programme a success.

91. Which one of the following statements is correct?

a. Effective family welfare programme is Centre's responsibility alone.
b. Population policy and planning process are interlinked.
c. Family welfare programme should be left to the State Governments alone.
d. The State Government should use punitive measures to control population.

92. What should be given priority attention?

a. Role of the Government
b. Decentralisation of power
c. Involvement of people in labour welfare
d. Human resource and institutional framework

93. Which one of the following statements is not correct?

a. Role of the government in nation-building should be examined.

b. Real initiatives should be transferred to the people.

c. There should be no role for the government as far as planning is concerned.

d. The Centre and the States must become equal partners in the planning process.

94. What would force the planning process to undergo a change?

a. Free market forces

b. Domestic economic situation and world trends

c. Domestic compulsions

d. International pressures

95. Which one of the following is implied by the expression 'momentous trends'?

a. GDP growth of the country

b. Memorable historical events

c. Important changes in the international scene

d. Improvement of foreign exchange reserves

b. Logical Reasoning

96. Find out the wrong number.

2, 6, 12, 72, 865, 62208

a. 72 *b.* 12 *c.* 62208 *d.* 865

97. Each of P, Q, T, A and B has different heights. T is taller than P and B but shorter than A and Q. P is not the shortest, who is the tallest?

a. A *b.* Q *c.* A or Q *d.* P or B

98. Identify the missing part of the question figure and select it from given answer figures.

Question Figure

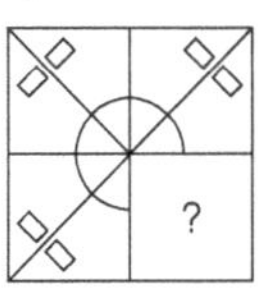

Answer Figures

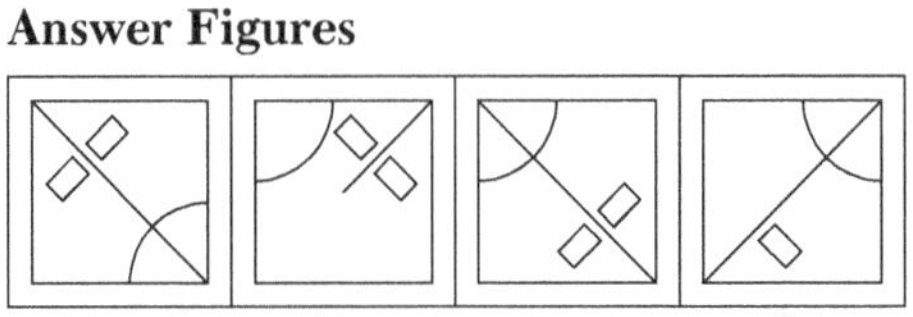

a. *b.* *c.* *d.*

99. Select the related word from the given alternatives.

Mechanic : Spanner : : Carpenter :?

a. tree *b.* wood *c.* furniture *d.* saw

100. How many rectangles are there in the following figure?

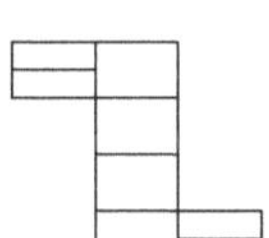

a. 8 *b.* 18 *c.* 17 *d.* 20

101. In the following question find the odd letters/group from the given alternatives.

a. BADC *b.* JILK

c. NMPO *d.* VUWX

102. Find out which of the answer figures (a), (b), (c) and (d) completes the figure matrix?

Question Figure

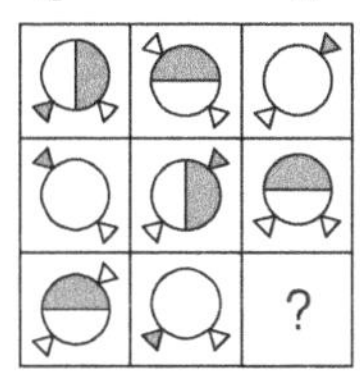

Answer Figures

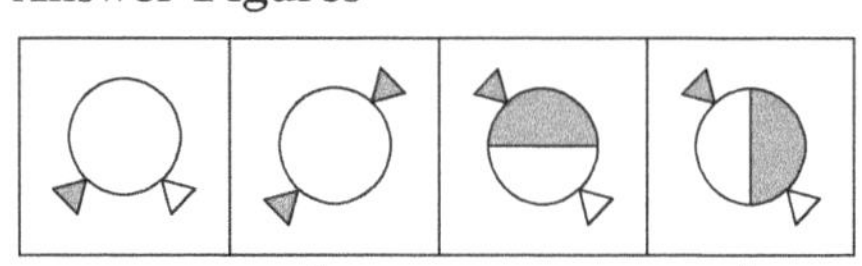

a. *b.* *c.* *d.*

103. Among the four answer figures, which one can be formed from the cut out pieces given below in the question figures?

Question Figure

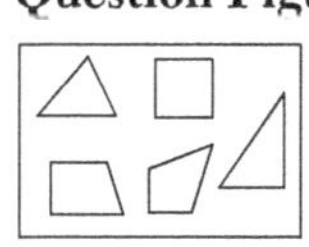

Answer Figures

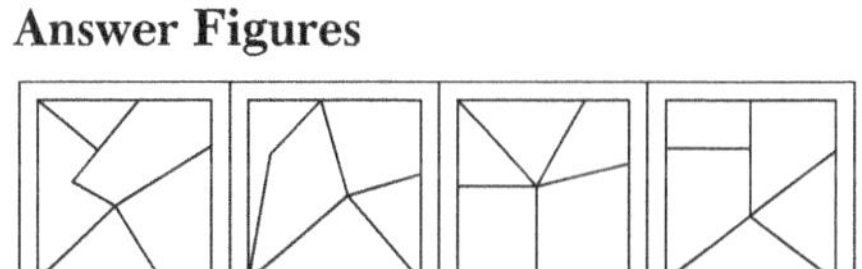

a. *b.* *c.* *d.*

104. A piece of paper is folded and cut as shown below in the question figures. From the given answer figures, indicates how it will appear when opened.

Question Figures

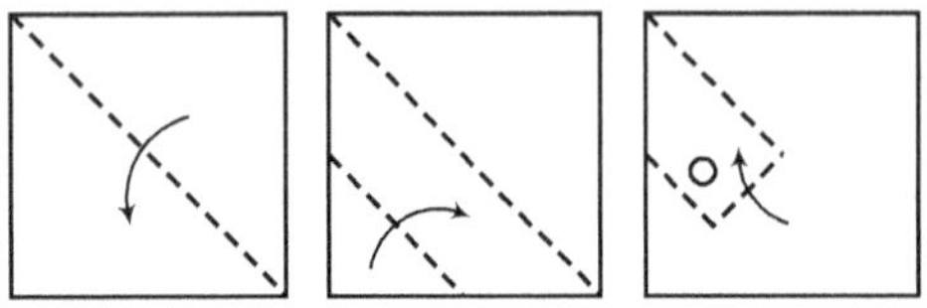

Answer Figures

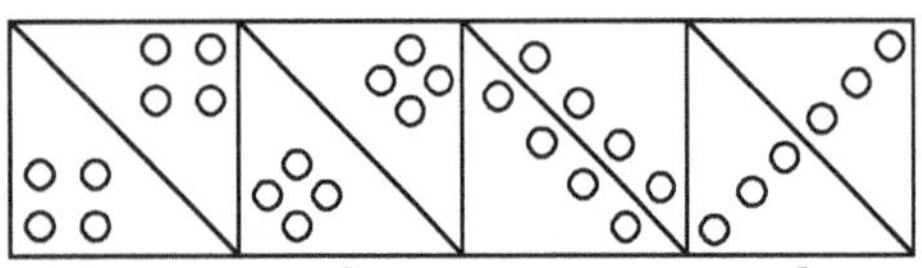

a. *b.* *c.* *d.*

105. In the following question three dots are placed in the figure marked as (A). The figure is followed by four alternatives marked as (a), (b), (c) and (d). One out of these four options contains region(s) common to the circle, square, triangle, similar to that marked by the dot in figure (A).

Question Figure

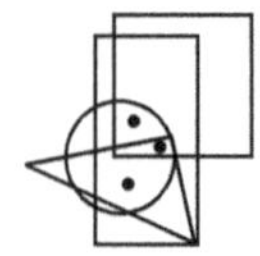

(A)

Answer Figure

a. 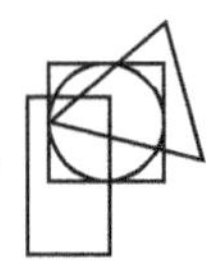

b. 

c. 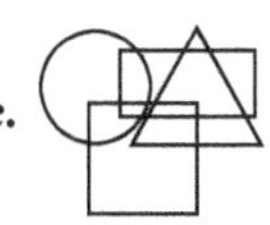

d. 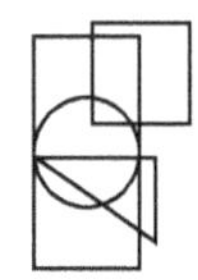

PART IV

Mathematics

106. If $A=\{x:x^2=1\}$ and $B=\{x:x^4=1\}$, then $A\,\Delta\,B$ is equal to

a. $\{-i, i\}$ *b.* $\{-1, i\}$
c. $\{-1, 1, -i, i\}$ *d.* None of these

107. If $2f(xy)=(f(x))^y+(f(y))^x$ for all $x, y\in R$ and $f(1)=a\,(\neq 1)$. Then $\sum\limits_{k=1}^{n} f(k)$ is equal to

a. $(a^n-1)/(a-1)$
b. $a(a^{n-1}-1)/(a-1)$
c. $a(a^n-1)/(a-1)$
d. $(a^n-1)/a+1$

108. Let $f(x)=x-3$ and $g(x)=4-x$. Then the set of values of x for which $|f(x)+g(x)|<|f(x)|+|g(x)|$ is true, is given by :

a. R *b.* $R-(3, 4)$
c. $R-[3, 4]$ *d.* None of these

109. If $a_1, a_2, a_3, \ldots, a_{20}$ are AM's between 13 and 67 then the maximum value of $a_1\cdot a_2\cdot a_3\ldots, a_{20}$ is equal to

a. $(20)^{20}$ *b.* $(40)^{20}$
c. $(60)^{20}$ *d.* $(80)^{20}$

110. If p, q, r are in AP and are positive, the roots of the quadratic equation $px^2+qx+r=0$ are all real for

a. $\left|\frac{r}{p}-7\right|\geq 4\sqrt{3}$ *b.* $\left|\frac{p}{r}-7\right|<4\sqrt{3}$
c. All p and r *d.* No p and r

111. The value of ${}^{47}C_4+\sum\limits_{r=1}^{5}{}^{52-r}C_3$ is equal to

a. ${}^{47}C_6$ *b.* ${}^{52}C_5$ *c.* ${}^{52}C_4$ *d.* None of these

112. The number of numbers divisible by 3 that can be formed by four different even digits is

a. 36 *b.* 18 *c.* 0 *d.* None of these

113. If $n(A)=1000$, $n(B)=500$, $n(A\cap B)\geq 1$ and $n(A\cup B)=P$, then

a. $500\leq P\leq 1000$ *b.* $1001\leq P\leq 1498$
c. $1000\leq P\leq 1498$ *d.* $1000\leq P\leq 1499$

114. $\left\{x\in R:\dfrac{2x-1}{x^3+4x^2+3x}\in R\right\}$ is equal to

a. $R-\{0\}$ *b.* $R-\{0, 1, 3\}$
c. $R-\{0, -1, -3\}$ *d.* $R-\left\{0, -1, -3, \frac{1}{2}\right\}$

115. Let $f(x)$ be a polynomial function of second degree. If $f(1) = f(-1)$ and a, b, c are in AP, then $f'(a), f'(b)$ and $f'(c)$ are in

a. AP
b. GP
c. Arithmetic-Geometric progression
d. None of the above

116. The value of $\lim_{x\to\infty} \frac{1}{n}\left\{\frac{1}{n+1}+\frac{2}{n+2}+\ldots+\frac{3n}{4n}\right\}$ is

a. $5-2\log 2$ *b.* $4-2\log 2$
c. $3-2\log 2$ *d.* $2-2\log 2$

117. The coefficient of x^8 in the polynomial $(x-1)(x-2)\ldots(x-10)$

a. 2640 *b.* 1320 *c.* 1370 *d.* 2740

118. If $z = \frac{7+i}{3+4i}$, then z^{14} is

a. 2^7 *b.* $2^7 i$ *c.* $(-2)^7$ *d.* $(-2)^7 i$

119. The solution of the equation

$$\frac{dy}{dx}+\frac{1}{x}\tan y = \frac{1}{x^2}\tan y \sin y \text{ is}$$

a. $2y = \sin y(1-2cx^2)$ *b.* $2x = \cot y(1+2cx^2)$
c. $2x = \sin y(1-2cx^2)$ *d.* $2x \sin y = 1-2cx^2$

120. The value of the definite integral

$$\int_0^{\pi/2} \frac{dx}{\tan x + \cot x + \operatorname{cosec} x + \sec x}$$

a. $1-\frac{\pi}{4}$ *b.* $1+\frac{\pi}{4}$
c. $\pi+\frac{1}{4}$ *d.* None of these

121. If **a** and **b** are two vectors such that $|\mathbf{a}| = 1, |\mathbf{b}| = 4$ and $\mathbf{a}\cdot\mathbf{b} = 2$, If $\mathbf{c} = (2\mathbf{a}\times\mathbf{b}) - 3\mathbf{b}$, then angle between **b** and **c**

a. $\frac{\pi}{6}$ *b.* $\frac{\pi}{3}$
c. $\frac{2\pi}{3}$ *d.* $\frac{5\pi}{6}$

122. Let x_1 and x_2 be the real roots of the equation $x^2-(k-2)x+(k^2+3k+5)=0$, then maximum value of $x_1^2+x_2^2$ is

a. 19 *b.* 22
c. 18 *d.* 17

123. Circle centered at origin and having radius π units is divided by the curve $y = \sin x$ in two parts. Then area of upper parts equals to

a. $\frac{\pi^2}{2}$ *b.* $\frac{\pi^3}{4}$ *c.* $\frac{\pi^3}{2}$ *d.* $\frac{\pi^3}{8}$

124. The root of the equation $2(1+i)x^2-4(2-i)x-5-3i=0$, where $i=\sqrt{-1}$, which has greater modulus, is

a. $\frac{3-5i}{2}$ *b.* $\frac{5-3i}{2}$ *c.* $\frac{3+i}{2}$ *d.* $\frac{3i+1}{2}$

125. The equation $(\cos\beta-1)x^2+(\cos\beta)x+\sin\beta=0$ in the variable x has real roots, then β lies in the interval

a. $(0, 2\pi)$ *b.* $(-\pi, 0)$ *c.* $\left(-\frac{\pi}{2}, \frac{\pi}{2}\right)$ *d.* $(0, \pi)$

126. An ordered pair (α, β) for which the system of linear $(1+\alpha)x+\beta y+z=2$, $\alpha x+(1+\beta)y+z=3$ and $\alpha x+\beta y+2z=2$ has a unique solution.

a. $(1, -3)$ *b.* $(-3, 1)$ *c.* $(2, 4)$ *d.* $(-4, 2)$

127. A bird is sitting on the top of a vertical pole 20 m high and its elevation from a point O on the ground is 45°. If flies off horizontally straight way from the point O. After one second, the elevation of the bird from O is reduced to 30°, then the speed (in m/s) of the bird is

a. $40(\sqrt{2}-1)$ *b.* $40(\sqrt{3}-\sqrt{2})$
c. $20\sqrt{2}$ *d.* $20(\sqrt{3}-1)$

128. If one GM, g and two AM's p and q are inserted between two numbers a and b, then $(2p-q)(p-2q)$ is equal to

a. g^2 *b.* $-g^2$ *c.* $2g$ *d.* $3g^2$

129. When x^{100} is divided by x^2-3x+2, the remainder is $(2^{k+1}-1)x-2(2^k-1)$, then k is

a. 97 *b.* 99 *c.* 100 *d.* 101

130. The mean of five observation is 5 and their variance is 9.20. If three of the given five observation are 1, 3 and 8, then a ratio of other two observations is

a. 4 : 9 *b.* 6 : 7
c. 5 : 8 *d.* 10 : 3

131. How many three digit number satisfy the property that the middle digit is arithmetic mean of the first and the last digit.

a. 41 *b.* 45 *c.* 43 *d.* 44

132. If $z = re^{i\theta}$, then $\arg(e^{iz})$ is

a. $-r\sin\theta$ *b.* $r\cos\theta$
c. $e^{-r\sin\theta}$ *d.* $-r\cos\theta$

133. If 4 dice are rolled, then the number of ways of getting the sum 10 is

a. 56 *b.* 64
c. 72 *d.* 80

134. Distance of point $A(1, 2)$ measured parallel to the line $3x - y = 10$ from the line $x + y + 5 = 0$, is

a. $2\sqrt{5}$ *b.* $2\sqrt{10}$ *c.* $4\sqrt{5}$ *d.* $4\sqrt{10}$

135. Let $f(x) = a_0 + a_1x^2 + a_2x^4 + a_3x^6 + \ldots + a_nx^{2n}$ be a polynomial in a real variable x with $0 < a_1 < a_2 < a_3 < \ldots < a_n$, the function $f(x)$ has

a. neither a maxima nor a minima
b. only one maxima
c. both maxima and minima
d. only one minima

136. If $f(x) = 2x^3 + x^4 + \log x$ and g is the inverse of f, then $g'(3)$ is equal to

a. 1/9 *b.* 1/7 *c.* 1/11 *d.* 1/8

137. A line passing through $P(3, 7, 1)$ and $R(2, 5, 7)$ meet the plane $3x + 2y + 11z - 9 = 0$ at Q. Then PQ is equal to

a. $\frac{5\sqrt{41}}{59}$ *b.* $\frac{\sqrt{41}}{59}$ *c.* $\frac{50\sqrt{41}}{59}$ *d.* $\frac{25\sqrt{41}}{59}$

138. $\int \frac{8x^{43} + 13x^{38}}{(x^{13} + x^5 + 1)^4} dx$ equals to

a. $\frac{x^{39}}{3(x^{13} + x^5 + 1)^3} + C$ *b.* $\frac{x^{39}}{(x^{13} + x^5 + 1)^3} + C$

c. $\frac{x^{39}}{5(x^{13} + x^5 + 1)^5} + C$ *d.* None of these

139. If $\mathbf{a} = -\hat{\mathbf{i}} + \hat{\mathbf{j}} + \hat{\mathbf{k}}$ and $\mathbf{b} = 2\hat{\mathbf{i}} + \hat{\mathbf{k}}$, then find z component of a vector $\mathbf{r}$, which is coplanar with $\mathbf{a}$ and $\mathbf{b}$, $\mathbf{r} \cdot \mathbf{b} = 0$ and $\mathbf{r} \cdot \mathbf{a} = 7$.

a. 0 *b.* 3 *c.* 6 *d.* 5/2

140. If x, y, z are three consecutive positive integers, $x - z + 2 = 0$, then

$$\frac{1}{2}\log_e x + \frac{1}{2}\log_e z + \frac{1}{2xz + 1} + \frac{1}{3}\left(\frac{1}{2xz + 1}\right)^3 + \ldots$$

is equal to

a. $\log_e x$ *b.* $\log_e y$
c. $\log_e z$ *d.* None of these

141. The solution of differential equation $(xy^5 + 2y)\,dx - xdy = 0$, is

a. $9x^8 + 4x^9y^4 = 9y^4C$ *b.* $9x^8 - 4x^9y^4 - 9y^4C = 0$
c. $x^8(9 + 4y^4) = 10y^4C$ *d.* None of these

142. The solution set of $\frac{|x - 2| - 1}{|x - 2| - 2} \le 0$ is

a. $[0, 1] \cup (3, 4)$ *b.* $[0, 1] \cup [3, 4]$
c. $[-1, 1] \cup (3, 4]$ *d.* None of these

143. Let $f(x) = \frac{x}{\sqrt{1 + x^2}}$, the $\underbrace{fofofo\ldots\ldots of(x)}_{n \text{ times}}$ is

a. $\frac{x}{\sqrt{1 + \left(\sum_{r=1}^{n} r\right)x^2}}$ *b.* $\frac{x}{\sqrt{1 + \left(\sum_{r=1}^{n} 1\right)x^2}}$

c. $\left(\frac{x}{\sqrt{1 + x^2}}\right)^x$ *d.* $\frac{nx}{\sqrt{1 + nx^2}}$

144. If $\log_5 \frac{(a + b)}{3} = \frac{\log_5 a + \log_5 b}{2}$, then $\frac{a^4 + b^4}{a^2b^2}$ is equal to

a. 50 *b.* 47 *c.* 44 *d.* 53

145. The number of distinct solutions of the equation $\frac{5}{4}\cos^2 2x + \cos^4 x + \sin^4 x + \cos^6 x + \sin^6 x = 2$ in the interval $[0, 2\pi]$ is

a. 8 *b.* 10 *c.* 6 *d.* 15

146. If the tangent at a point $\left(4\cos\phi, \frac{16}{\sqrt{11}}\sin\phi\right)$ to the ellipse $16x^2 + 11y^2 = 256$ is also a tangent to $x^2 + y^2 - 2x = 15$, then ϕ equals

a. $\frac{\pi}{3}$ *b.* $\frac{\pi}{6}$ *c.* $-\frac{\pi}{6}$ *d.* $\frac{\pi}{4}$

147. The distance of point of intersection of the tangents to the parabola $x = 4y - y^2$ drawn at the points where it is meet by Y-axis, from its focus is

a. 11/4 *b.* 17/4 *c.* 13/4 *d.* 3

148. The value of the sum $\sum_{k=1}^{\infty} \sum_{n=1}^{\infty} \frac{k}{2^{n+k}}$ is

a. 5 *b.* 4 *c.* 3 *d.* 2

149. A curve passes through (2, 0) and the slope of the tangent at $P(x, y)$ is equal to $\frac{(x + 1)^2 + y - 3}{x + 1}$ then the equation of the curve is

a. $y = x^2 - 2x$ *b.* $y = x^3 - 8$
c. $y^2 = x^2 + 2x$ *d.* $y^2 = 5x^2 - 6$

150. Consider matrix $A = \begin{bmatrix} 2 & 1 \\ 1 & 2 \end{bmatrix}$, if $A^{-1} = \alpha I + \beta A$, where $\alpha, \beta \notin R$, then $(\alpha + \beta)$ is equal to (where A^{-1} denotes the inverse of matrix A)

a. 1 *b.* $\frac{4}{3}$ *c.* $\frac{5}{3}$ *d.* $\frac{1}{3}$

BITSAT 2020

Answers

Physics

1. (c)	**2.** (a)	**3.** (a)	**4.** (c)	**5.** (b)	**6.** (b)	**7.** (a)	**8.** (b)	**9.** (c)	**10.** (b)
11. (b)	**12.** (b)	**13.** (b)	**14.** (b)	**15.** (b)	**16.** (c)	**17.** (b)	**18.** (b)	**19.** (b)	**20.** (c)
21. (b)	**22.** (b)	**23.** (d)	**24.** (a)	**25.** (a)	**26.** (d)	**27.** (d)	**28.** (a)	**29.** (b)	**30.** (c)
31. (b)	**32.** (b)	**33.** (c)	**34.** (b)	**35.** (c)	**36.** (c)	**37.** (a)	**38.** (a)	**39.** (d)	**40.** (b)

Chemistry

41. (d)	**42.** (b)	**43.** (d)	**44.** (a)	**45.** (d)	**46.** (a)	**47.** (b)	**48.** (a)	**49.** (b)	**50.** (a)
51. (d)	**52.** (a)	**53.** (c)	**54.** (b)	**55.** (c)	**56.** (c)	**57.** (d)	**58.** (c)	**59.** (b)	**60.** (c)
61. (c)	**62.** (d)	**63.** (d)	**64.** (a)	**65.** (c)	**66.** (c)	**67.** (a)	**68.** (c)	**69.** (d)	**70.** (b)
71. (d)	**72.** (b)	**73.** (d)	**74.** (b)	**75.** (d)	**76.** (d)	**77.** (b)	**78.** (a)	**79.** (c)	**80.** (a)

English Proficiency

81. (b)	**82.** (b)	**83.** (c)	**84.** (a)	**85.** (b)	**86.** (c)	**87.** (b)	**88.** (a)	**89.** (a)	**90.** (c)
91. (b)	**92.** (d)	**93.** (c)	**94.** (b)	**95.** (c)					

Logical Reasoning

96. (d)	**97.** (c)	**98.** (c)	**99.** (d)	**100.** (b)	**101.** (d)	**102.** (d)	**103.** (c)	**104.** (b)	**105.** (b)

Mathematics

106. (a)	**107.** (c)	**108.** (c)	**109.** (b)	**110.** (a)	**111.** (c)	**112.** (a)	**113.** (d)	**114.** (c)	**115.** (a)
116. (c)	**117.** (b)	**118.** (b)	**119.** (c)	**120.** (a)	**121.** (d)	**122.** (a)	**123.** (c)	**124.** (a)	**125.** (d)
126. (c)	**127.** (d)	**128.** (b)	**129.** (b)	**130.** (a)	**131.** (b)	**132.** (b)	**133.** (d)	**134.** (b)	**135.** (d)
136. (c)	**137.** (d)	**138.** (a)	**139.** (b)	**140.** (b)	**141.** (a)	**142.** (b)	**143.** (b)	**144.** (b)	**145.** (a)
146. (a)	**147.** (b)	**148.** (d)	**149.** (a)	**150.** (a)					

Hints & Solutions

Physics

1. ***(c)*** As coil 2 with inductance L_2 has been wounded by a similar wire as coil 1 but the direction of windings is reversed in each layer, so flux through L_2 will be zero.

As, $$L_2 \propto \phi \Rightarrow L_2 = 0$$

Also, since the dimensions, number of turns and number of layers of windings are equal in coil 1 and coil 3.

So, $$L_1 = L_3$$

2. ***(a)*** Since, capacitors C_1 and C_2 are parallel, so their equivalent capacitance will be $(C_1 + C_2)$. Now, $(C_1 + C_2)$ and C_3 are in series, so the net equivalent capacitance of circuit will be

$$\frac{1}{C} = \frac{1}{C_3} + \frac{1}{C_1 + C_2} = \frac{C_1 + C_2 + C_3}{(C_1 + C_2)C_3}$$

or $$C = \frac{(C_1 + C_2)C_3}{C_1 + C_2 + C_3}$$

Since, V is the voltage of the battery, so charge on this system,

$$q = CV \Rightarrow q = \frac{(C_1 + C_2)C_3V}{(C_1 + C_2 + C_3)}$$

If the capacitance C_3 breaks down, then total equivalent capacitance will become

$$C' = C_1 + C_2$$

$\therefore$ New charge stored,

$$q' = C'V$$

$$\Rightarrow \quad q' = (C_1 + C_2)V$$

Now, change in total charge,

$$\Delta q = q' - q \qquad [\because q' > q]$$

$$= (C_1 + C_2)V - \frac{(C_1 + C_2)C_3V}{C_1 + C_2 + C_3}$$

or $$\Delta q = (C_1 + C_2)V\left[1 - \left(\frac{C_3}{C_1 + C_2 + C_3}\right)\right]$$

3. ***(a)*** Let E' be the new radiant energy.

According to Stefan's law,

$$E \propto T^4$$

$$\Rightarrow \quad \frac{E_2}{E_1} = \left(\frac{T_2}{T_1}\right)^4$$

$$\Rightarrow \quad \frac{E_2}{E} = \left\{\frac{(T/4)}{T}\right\}^4 = \left(\frac{1}{4}\right)^4$$

or $$E_2 = \frac{E}{256}$$

4. ***(c)*** Area of rectangle, $A = lb = 15.2 \times 2.9 = 44.08\,\text{cm}^2$

Minimum possible measurement of scale = 0.1 cm

As we know, the product should have significant figures that are present in the measurement with least number of significant figures which is 2 in this case.

So, area measured by scale = 44 cm^2.

5. ***(b)*** In adiabatic process,

$$pV^{\gamma} = K$$

where, K is a constant.

Given that, $\gamma = \frac{4}{3} \Rightarrow pV^{4/3} = K$

Taking logarithm on both sides, we get

$$\log p + \frac{4}{3}\log V = \log K$$

$$\Rightarrow \quad \frac{\Delta p}{p} + \frac{4}{3}\frac{\Delta V}{V} = 0 \Rightarrow \frac{\Delta V}{V} = -\frac{3}{4}\frac{\Delta p}{p}$$

Now, % change in volume,

$$\frac{\Delta V}{V} \times 100 = -\frac{3}{4}\left(\frac{\Delta p}{p} \times 100\right)$$

It is given that, pressure decreases by $\frac{3}{4}$%,

i.e. $$\frac{\Delta p}{p} \times 100 = -\frac{3}{4}$$

$$\therefore \quad \frac{\Delta V}{V} \times 100 = -\frac{3}{4} \times \left(-\frac{3}{4}\right) = \frac{9}{16}\%$$

Hence, the volume increases by $\frac{9}{16}$%.

6. ***(b)*** The given equation is

$$y = 4\sin\left(\frac{\pi x}{15}\right)\cos(96\pi t) \qquad \ldots(\text{i})$$

Now, the velocity of the string at a point x at time t is obtained by differentiating Eq. (i) w.r.t.

$$\therefore v = \frac{dy}{dt} = -(4 \times 96\pi) \times \sin\left(\frac{\pi x}{15}\right)\sin(96\pi t)$$

At $x = 7.5$ cm $= 7.5 \times 10^{-2}$ m and $t = 0.25$ s,

$$v = -384\pi\sin\left(\pi \times \frac{7.5 \times 10^{-2}}{15}\right)\sin\left(96\pi \times \frac{25}{100}\right)$$

$$= -384\pi\sin\left(\frac{\pi}{200}\right)\sin(24\pi)$$

$$= 0 \qquad (\because \sin 24\pi = 0)$$

Hence, the velocity of the particle is zero.

7. ***(a)*** Potential at point A,

$$V_A = \frac{1}{4\pi\varepsilon_0} \cdot \frac{q}{a}$$

Potential at point B,

$$V_B = \frac{1}{4\pi\varepsilon_0} \cdot \frac{q}{b}$$

Work done in taking a charge Q from A to B,

$$W = Q(V_B - V_A) = \frac{Qq}{4\pi\varepsilon_0}\left[\frac{1}{b} - \frac{1}{a}\right] = \frac{Qq}{4\pi\varepsilon_0}\left[\frac{a-b}{ab}\right]$$

8. ***(b)*** For hydrogen-like atom,

$$\frac{1}{\lambda'} = RZ^2\left(\frac{1}{n_1^2} - \frac{1}{n_2^2}\right) \quad \ldots\text{(i)}$$

Similarly, for hydrogen atom,

$$\frac{1}{\lambda} = R\left(\frac{1}{n_1^2} - \frac{1}{n_2^2}\right) \quad \ldots\text{(ii)}$$

For the element to have hydrogen-like spectrum, the ratio of the wavelengths for transition of electron between any two levels n_1 and n_2 should be constant.

Dividing Eq. (ii) by Eq. (i), we get

$$\frac{\lambda'}{\lambda} = \frac{R\left(\frac{1}{n_1^2} - \frac{1}{n_2^2}\right)}{RZ^2\left(\frac{1}{n_1^2} - \frac{1}{n_2^2}\right)} \quad \text{or} \quad \frac{\lambda'}{\lambda} = \frac{1}{Z^2}$$

It is given that, wavelength of the hydrogen atom is four times the wavelength of the hydrogen-like atom, i.e.

$$\lambda = 4\lambda'$$

Hence, $Z = 2$

Now, the element with atomic number 2 is helium.

9. ***(c)*** Given, two small conducting spheres of equal radius R.

Charge on first sphere, $q_1 = +20\mu C$

Charge on second sphere, $q_2 = -40\mu C$

$$\therefore \quad F_1 = \frac{k(+20)(-40)}{R^2} = \frac{-k(800)}{R^2} \quad \ldots\text{(i)}$$

The spheres have equal radii, so their capacities will be same. Now, if they are brought in contact, then the net charge,

$$q_{\text{net}} = (+20 - 40)\mu C = -20\mu C$$

So, each sphere will have charge, $q = \frac{q_{\text{net}}}{2} = -10\mu C$

$$\text{Now,} \quad F_2 = \frac{k(q)(q)}{R^2} = \frac{k(-10)(-10)}{R^2} = \frac{k(100)}{R^2} \quad \ldots\text{(ii)}$$

$$\text{Ratio,} \quad \frac{F_1}{F_2} = -\frac{800}{100} = -8:1$$

Thus, ratio of F_1 to $F_2 = -8:1$

10. ***(b)*** Efficiency of Carnot engine is given by

$$\eta = 1 - \frac{T_2}{T_1}$$

Given, $T_1 = 600$ K, $T_2 = 300$K,

$T_1' = 1600$K and $T_2' = x$K

Carnot engine has same efficiency between 600 K to 300 K and 1600 K to xK, so

$$\left(1 - \frac{T_2}{T_1}\right) = \left(1 - \frac{T_2'}{T_1'}\right)$$

Substituting all the given values in above equation, we get

$$1 - \frac{300}{600} = 1 - \frac{x}{1600}$$

$$1 - \frac{1}{2} = 1 - \frac{x}{1600}$$

$$\frac{x}{1600} = 1 - \frac{1}{2} = \frac{1}{2}$$

$$\Rightarrow \quad x = \frac{1600}{2} = 800 \text{ K}$$

11. ***(b)*** Given, initial speed of ball, $v = 16 \text{ ms}^{-1}$

Maximum height, $h = 9$ m

Mass of ball, $m = 0.5$ kg

Energy supplied to ball while throwing up,

$$E_1 = \frac{1}{2}mv^2 = \frac{1}{2} \times 0.5 \times 16 \times 16 = 64 \text{ J}$$

Energy stored when ball reaches the maximum height,

$$E_2 = mgh = 0.5 \times 9.8 \times 9 = 44.1 \text{ J}$$

Energy dissipated by air drag acting on the ball during the ascent,

$$E = E_1 - E_2 = 64 - 44.1 = 19.9 \text{ J}$$

12. ***(b)*** From the following circuit diagram,

$\frac{R_1}{R_3} = \frac{R_2}{R_4}$, so the given Wheatstone bridge is balanced.

Hence, there is no current in the wire of 10 Ω resistance.

As,

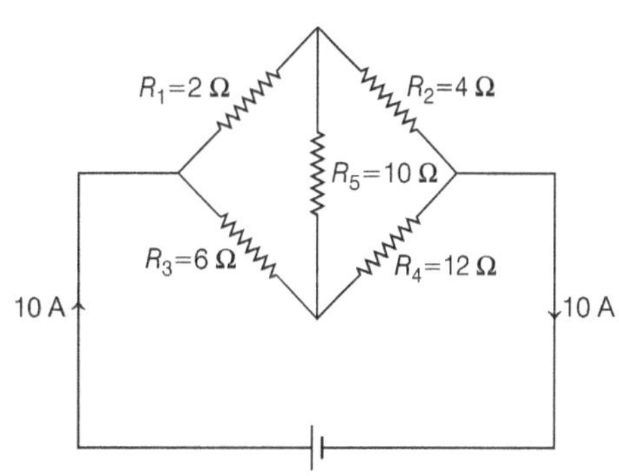

Now, if the 10 Ω wire is replaced with 15 Ω wire, the bridge still remains balanced. Hence, current drawn from battery will be 10 A.

13. ***(b)*** Given, force, $F = 8t$ N

Mass, $m = 2\text{kg}$

From Newton's second law,

Force = Rate of change of momentum

$$F = \frac{dp}{dt}$$

i.e. $dp = Fdt$

$\therefore$ Momentum, $p = \int dp = \int_0^1 F\cdot dt = \int_0^1 8t\cdot dt$

$$= 8\left[\frac{t^2}{2}\right]_0^1 = 8\left(\frac{1}{2} - 0\right)$$

$\Rightarrow$ $p = 4 \text{ kg ms}^{-1}$

Change in kinetic energy,

$$\Delta K = \frac{p^2}{2m} = \frac{(4)^2}{2\times 2} = 4\text{ J}$$

From work-energy theorem,

Work done = Change in kinetic energy

So, work done by the force during the first 1s,

$$W = \Delta K = 4\text{J}$$

14. ***(b)*** As, RC is the time constant, so it has the dimension of time and V has the dimensions of $[L]\left[\frac{di}{dt}\right]$.

$\therefore$ Dimensions of $\frac{L}{RCV}, \left[\frac{L}{RCV}\right] = \frac{[L]}{[RC]\,[V]}$

$$\frac{[L]}{[T][L]\left[\frac{A}{T}\right]} = \frac{[L][T]}{[T][L][A]} = \frac{1}{[A]}$$

So, $\left(\frac{L}{RCV}\right)$ has the dimensions $[A^{-1}]$, i.e. same as the dimensions of $\frac{1}{\text{current}}$.

15. ***(b)*** Given, magnetic field,

$$B = 10\times 10^{-8}\sin(1\times 10^7 z - 3.6\times 10^{15} t)\text{ T} \quad \ldots\text{(i)}$$

Standard equation of magnetic field,

$$B = B_0 \sin(kz - \omega t) \quad \ldots\text{(ii)}$$

On comparing Eqs. (i) and (ii), we get

$$B_0 = 10\times 10^{-8}\text{ T}$$

The average intensity of the beam,

$$I_{\text{av}} = \frac{B_0^2 c}{2\mu_0} = \frac{1}{2}\times\frac{(10\times 10^{-8})^2\times 3\times 10^8}{4\pi\times 10^{-7}}$$

$[\because c = 3\times 10^8\text{ ms}^{-1}]$

$$= 1.19\text{ W/m}^2$$

16. ***(c)*** Distance = area under speed-time graph as given in question.

$\therefore$ Distance travelled in 20 s = area of ΔOAB + area of ΔBCD

$$= \frac{1}{2}\times 6\times 10 + \frac{1}{2}\times 6\times 10 = 60\text{ m}$$

Displacement in 20 s = area of ΔOAB – area of ΔBCD

$$= \frac{1}{2}\times 6\times 10 - \frac{1}{2}\times 6\times 10 = 30 - 30 = 0$$

Hence, average velocity = $\frac{\text{displacement}}{\text{time}} = \frac{0}{t} = 0$

17. ***(b)*** The focal length of a convex lens of refractive index μ_g in air,

$$\frac{1}{f_{\text{air}}} = (\mu_g - 1)\left(\frac{1}{R_1} - \frac{1}{R_2}\right) \quad \ldots\text{(i)}$$

where, R_1, R_2 = radii of curvatures of its first and second surface.

When the lens is immersed in a liquid of refractive index μ_l, then refractive index of material of lens (glass) w.r.t. liquid is

$${}_l\mu_g = \frac{\mu_g}{\mu_l} \quad \ldots\text{(ii)}$$

Now, focal length of lens in liquid,

$$\frac{1}{f'} = ({}_l\mu_g - 1)\left(\frac{1}{R_1} - \frac{1}{R_2}\right) \quad \ldots\text{(iii)}$$

Dividing Eq. (i) by Eq (iii), we get

$$\frac{f'}{f_{\text{air}}} = \frac{\mu_g - 1}{({}_l\mu_g - 1)} = \frac{\mu_g - 1}{\left(\frac{\mu_g}{\mu_l} - 1\right)}$$

Given, $f' = -0.5$ m, $f_{\text{air}} = 0.2$ m, $\mu_g = 1.5$

Substituting all these values in above equation, we get

$$-\frac{0.5}{0.2} = \frac{1.5 - 1}{\left(\frac{1.5}{\mu_l} - 1\right)} = \frac{0.5}{\left(\frac{1.5}{\mu_l} - 1\right)}$$

$$\Rightarrow \quad \frac{1.5}{\mu_l} - 1 = -0.2$$

$$\Rightarrow \quad \frac{1.5}{\mu_l} = 1 - 0.2 = 0.8$$

$$\Rightarrow \quad \mu_l = \frac{15}{8}$$

$\therefore$ Refractive index of liquid = 15/8

18. ***(b)*** Given, resistance of galvanometer, $R_g = 60\,\Omega$

Current, $I_g = 4.5\text{ mA} = 4.5\times 10^{-3}$ A

Resistance used in converting a galvanometer into voltmeter, $R = 4.5\text{ k}\Omega = 4.5\times 10^3\,\Omega$

To make the voltmeter, the galvanometer and the high resistance (R) have to be connected in series combination as shown below.

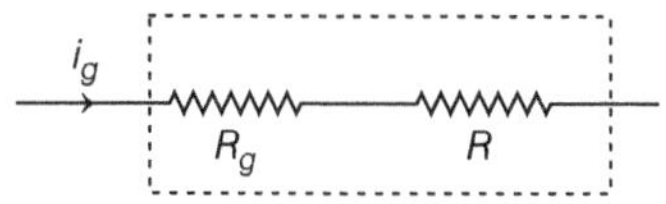

$\therefore$ Maximum current in galvanometer is

$$I_g = \frac{E}{R + R_g}$$

$$\Rightarrow \quad E = I_g(R + R_g)$$
$$= 4.5 \times 10^{-3} \times (4.5 \times 10^3 + 60)$$
$$= 20.5 \text{ V}$$

19. ***(b)*** The given combination of gates is shown below

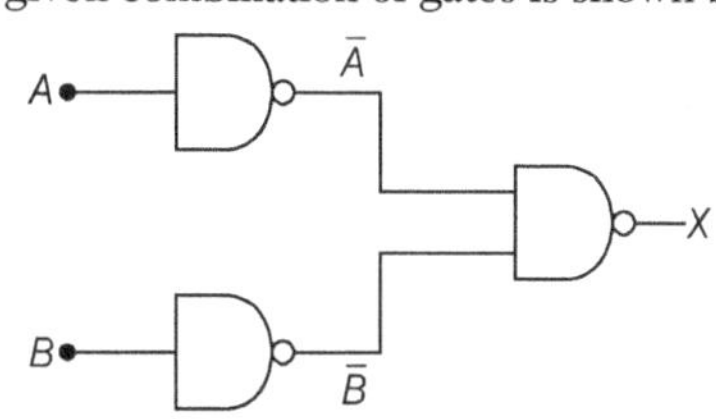

The output of this combination can be determined by Boolean algebra.

$\therefore$ Output, $X = \overline{\overline{A} \cdot \overline{B}} = \overline{\overline{A}} + \overline{\overline{B}}$

[Using de-Morgan's theorem]

$$\Rightarrow \quad X = A + B$$

This is the output of OR gate. Hence, the given combination of gates yield OR gate.

20. ***(c)*** From gas equation,

$$pV = nRT \Rightarrow p = \frac{nRT}{V}$$

Pressure of one mole of O_2 gas,

$$(p)_{O_2} = \frac{1 \times RT}{V} \quad \text{...(i)}$$

Pressure of one mole of He gas,

$$(p)_{\text{He}} = \frac{1 \times R(2T)}{V} \quad \text{...(ii)}$$

Dividing Eq. (ii) by Eq. (i), we get

$$\therefore \quad \frac{(p)_{\text{He}}}{(p)_{O_2}} = 2 \Rightarrow (p)_{\text{He}} = 2(p)_{O_2} = 2p$$

21. ***(b)*** Acceleration, $a = \frac{dv}{dt} \Rightarrow dv = adt$

Integrating both sides, $\int dv = \int adt$

Given, $a = (3t^2 - 2t + 1) \text{ ms}^{-2}$

Substituting this value in above equation, we get

$$v = \int (3t^2 - 2t + 1)dt$$
$$= t^3 - t^2 + t + C \quad \text{...(i)}$$

At $t = 1$ s, $v = 1$ m/s

Substituting these values in Eq. (i), we get

$$\because \quad 1 = 1 - 1 + 1 + C \quad \Rightarrow C = 0$$

Substituting the value of C in Eq. (i), we get

$$v = t^3 - t^2 + t$$

When $t = 4$ s, $v = (4)^3 - (4)^2 + 4 = 52$ m/s

22. ***(b)*** From ideal gas equation,

$$pV = nRT = \frac{mRT}{M}$$

where, $n = \frac{\text{Molar mass } (m)}{\text{Molecular weight } (M)}$

Here, n = number of moles,

R = gas constant

and V = volume.

$$\Rightarrow \quad \frac{pV}{m} = \frac{RT}{M}$$

$$\Rightarrow \quad \frac{p}{(m/V)} = \frac{RT}{M}$$

$$\frac{p}{d} = \frac{RT}{M} \quad \left[\because \frac{m}{V} = d\right]$$

$$\Rightarrow \quad \frac{p}{d} \propto T$$

$\Rightarrow$ Slope of p *versus* d graph $\propto T$

Since, slope is less for the gas at temperature T_1.

Therefore, $T_1 < T_2$

23. ***(d)*** The two spheres are touching each other as shown below.

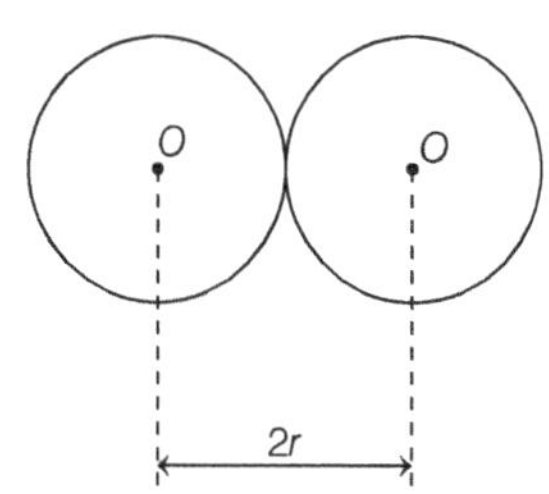

Let m be the mass of each sphere.

Gravitational force between the spheres,

$$F = G\frac{m \cdot m}{(2r)^2} = G.\frac{m^2}{4r^2}$$

$$= G.\frac{\left[\frac{4}{3}\pi r^3 \cdot \rho\right]^2}{4r^2} \quad [\text{Mass} = \text{volume} \times \text{density}]$$

$$= \frac{4G\pi^2 r^6 \rho^2}{9 \times 4r^2}$$

$$\Rightarrow \quad F = \left(\frac{4G\pi^2\rho^2}{9}\right)r^4 \Rightarrow F \propto r^4$$

24. ***(a)*** Power of first spherical lens, $P_1 = -4$ D

The power of this lens is negative, hence it is diverging lens.

Focal length of first spherical lens,

$$f_1 = \frac{1}{P_1} = \frac{1}{(-4\text{D})} = -0.25 \text{ m} = -25 \text{ cm}$$

Power of second spherical lens, $P_2 = 5$D

The power of this lens is positive, hence it is converging lens.

Focal length of second spherical lens,

$$f_2 = \frac{1}{P_2} = \frac{1}{5\text{D}} = 0.2 \text{ m} = 20 \text{ cm}$$

The given situation is shown in the following figure.

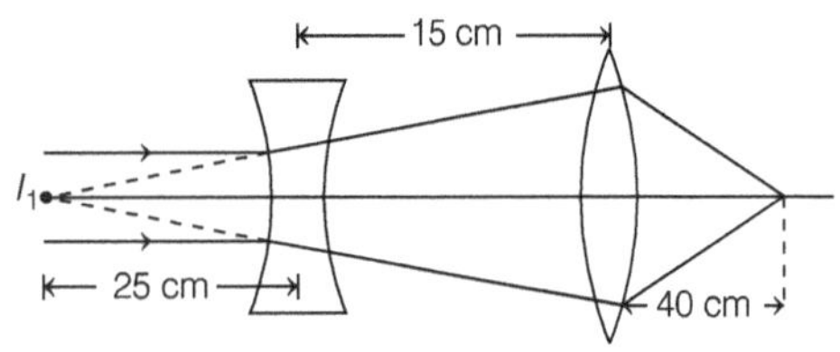

Now, the image I_1 will be considered as object for converging lens.

For converging lens, distance of object,

$$u = -(25 + 15) = -40 \text{ cm}$$

and $\quad f_2 = 20$ cm

$\therefore$ Using lens formula,

$$\frac{1}{f_2} = \frac{1}{v} - \frac{1}{u}$$

$$\Rightarrow \quad \frac{1}{20} = \frac{1}{v} - \frac{1}{-40}$$

$$\Rightarrow \quad v = 40 \text{ cm}$$

Since v is positive, so image formed will be real and will be formed at a distance of 40 cm from the lens of power 5 D (converging lens).

25. ***(a)*** From second equation of rotational motion,

$$\theta = \omega_0 t + \frac{1}{2}\alpha t^2$$

Given, $\theta = 5$ rad, $t = 1$ s, $\omega_0 = 0$

Substituting all these values in above equation, we get

$$5 = 0 + \frac{1}{2}\alpha(1)^2$$

$$\Rightarrow \quad \alpha = 0 \text{ rad/s}^2$$

The angle rotated by wheel of car during first two seconds,

$$\theta_2 = 0 + \frac{1}{2} + \alpha t^2 = \frac{1}{2} \times 10 \times 2^2 = 20 \text{ rad}$$

Hence, angle rotated during the 2nd second

$$= \theta_2 - \theta$$

$$= 20 - 5 = 15 \text{ rad}$$

26. ***(d)*** Kinetic energy in photoelectric effect can be given as

$$\text{KE} = W - W_0 \qquad \begin{bmatrix} W \to \text{incident energy} \\ W_0 \to \text{work function} \end{bmatrix}$$

Given, $(\text{KE})_1 = 1.5 - 0.5 = 1$ eV

$$(\text{KE})_2 = 2.5 - 0.5 = 1.5 \text{ eV}$$

$$\because \quad \frac{(\text{KE})_1}{(\text{KE})_2} = \frac{1}{1.5} = \frac{2}{3}$$

$$\Rightarrow \quad \frac{\frac{1}{2}mv_1^2}{\frac{1}{2}mv_2^2} = \frac{2}{3}$$

$$\Rightarrow \quad \frac{v_1}{v_2} = \sqrt{\frac{2}{3}}$$

$$\therefore \quad v_1 : v_2 = \sqrt{2} : \sqrt{3}$$

So, the ratio of maximum speeds of emitted electrons is $\sqrt{2}:\sqrt{3}$.

27. ***(d)*** Initial momentum, $p_i = mv$

$$p_i = m\sqrt{2gh} \qquad [\because v = \sqrt{2gh}]$$

New height of body, $h_1 = h + 3h = 4h$

Now, final momentum,

$$p_f = m\sqrt{2g.h_1} = m\sqrt{2g(4h)}$$

$$= m \cdot 2\sqrt{2gh} = 2p_i$$

$$\therefore \text{ \% change in momentum} = \frac{p_f - p_i}{p_i} \times 100$$

$$= \frac{2p_i - p_i}{p_i} \times 100 = 100\%$$

28. ***(a)*** In this case, the diode is forward biased. Potential drop in a germanium diode in forward bias is around 0.3 V.

Hence, current through the given circuit,

$$I = \frac{V_{\text{net}}}{R_{\text{eq}}} = \frac{5 - 0.3}{2 + 3} = \frac{4.7}{5}$$

$$= 0.94 \text{ A}$$

29. ***(b)*** Number of active nuclei after time t in a sample of radioactive substance is given by

$$N = N_0 e^{-\lambda t}$$

where, N_0 = initial number of nuclei at $t = 0$

and $\quad \lambda$ = decay constant.

Here, at $t = 0$, number of nuclei in sample X and Y are equal,

i.e., $(N_0)_X = (N_0)_Y = N_0$

Also, $\lambda_X = 4\lambda$ and $\lambda_Y = \lambda$

So, after time t, number of active nuclei in X and Y are

$N_X = N_0 e^{-4\lambda t}$ and $N_Y = N_0 e^{-\lambda t}$

$\therefore \quad \frac{N_X}{N_Y} = e^{-3\lambda t}$

$\Rightarrow \quad \frac{1}{e^3} = \frac{1}{e^{3\lambda t}} \quad \left[\text{Given, } \frac{N_X}{N_Y} = \frac{1}{e^3}\right]$

$\Rightarrow \quad 3\lambda t = 3 \Rightarrow t = \frac{1}{2\lambda}$

30. ***(c)*** Given, current in the wire, $I = 10\text{A}$

Radius of the wire, $R = 20\pi \text{ cm} = 20\pi \times 10^{-2}$ m

Magnetic field produced by semi-circular current carrying thin wire at centre O,

$$B = \frac{\mu_0 I}{4R}$$

Substituting all the given values in above equation, we get

$$B = \frac{4\pi \times 10^{-7} \times 10}{4 \times 20\pi \times 10^{-2}} = 5 \times 10^{-6}\text{ T}$$

Now, magnitude of force per unit length,

$$F = IB = 10 \times 5 \times 10^{-6} = 50 \times 10^{-6}\text{ N/m} = 50\,\mu\text{N/m}$$

31. ***(b)*** Given, area, $A = 5\text{ cm}^2 = 5 \times 10^{-4}\text{ m}^2$

Number of turns per unit length, $n = 4000$ turns/m

Magnetic field, $B = \pi T$

Current, $i = 5$ A

Now, magnetic intensity,

$H = ni = 4000 \times 5 = 2 \times 10^4\text{ Am}^{-1}$

As, $B = \mu_0 (H + I)$ [I = intensity of magnetisation]

Substituting all the values in above equation, we get

$$\pi = 4\pi \times 10^{-7} \{(2 \times 10^4) + I\}$$

$$\frac{1}{4} \times 10^7 - (2 \times 10^4) + I$$

$$250 \times 10^4 = 2 \times 10^4 + I$$

$$I = 250 \times 10^4 - 2 \times 10^4$$

$\Rightarrow \quad I = 248 \times 10^4$ A/m

Now, pole strength, $m = I \times A = 248 \times 10^4 \times 5 \times 10^{-4}$

$= 1240$ A-m

32. ***(b)*** Given, mass of plane, $M = 10000$ kg

Coefficient of friction, $\mu = 0.2$

Distance, $s = 50$ m

Initial speed, $u = 0$

Final speed, $v = 72\text{ kmh}^{-1} = 20\text{ ms}^{-1}$

Let a be the acceleration of the plane.

From third equation of motion,

$$v^2 - u^2 = 2as$$

$$(20)^2 - (0)^2 = 2 \times a \times 50$$

$\therefore \quad a = \frac{20 \times 20}{100} = 4\text{ ms}^{-2}$

Force required to provide necessary acceleration to the plane,

$$F_1 = ma = 10000 \times 4 = 4 \times 10^4\text{ N}$$

Force required to overcome friction by engine,

$$F_2 = \mu R = \mu mg = 0.2 \times 10000 \times 9.8 = 1.96 \times 10^4\text{ N}$$

Therefore, the minimum force required by the engine,

$$F_1 + F_2 = 4 \times 10^4 + 1.96 \times 10^4 = 5.96 \times 10^4\text{ N}$$

33. ***(c)*** Given, reverse voltage, $E = 120$ V

Change in current, $\Delta i = 0.50 - 0.20 = 0.30$A

Time, $\Delta t = 0.030\text{ ms} = 0.030 \times 10^{-3}$ s

In an inductor (choke),

Emf induced, $E_{\text{induced}} = L\frac{di}{dt}$

$\therefore$ Self inductance of choke, $L = \frac{E \times \Delta t}{\Delta i}$

$= \frac{120 \times 0.030 \times 10^{-3}}{0.30}$

$\Rightarrow \quad = 12 \times 10^{-3}$ H

34. ***(b)*** Given, linear dimensions increase by a factor of 8.

So, volume will become $(8)^3$ times and area of cross-section will become $(8)^2$ times.

Now, stress $= \frac{\text{weight}}{\text{area}}$

Weight = (volume × density × g) will also become $(8)^3$ times.

$\therefore$ New stress $= \frac{(8)^3 \times w_0}{(8)^2 \times A_0} = 8\left(\frac{w_0}{A_0}\right)$

Hence, the stress increases by a factor of 8.

35. ***(c)*** Let, γ be the adiabatic exponent for the mixture.

For monoatomic gas, $n_1 = 2$ and $\gamma_1 = 5/3$

For diatomic gas, $n_2 = 3$ and $\gamma_2 = 7/5$

For the mixture, $\frac{n_1 + n_2}{\gamma - 1} = \frac{n_1}{\gamma_1 - 1} + \frac{n_2}{\gamma_2 - 1}$

Substituting all the given values in above equation, we get

$$\frac{2+3}{\gamma - 1} = \frac{2}{\frac{5}{3} - 1} + \frac{3}{\frac{7}{5} - 1}$$

$\Rightarrow \quad \frac{5}{\gamma - 1} = \frac{21}{2} \Rightarrow \gamma = \frac{31}{21}$

36. *(c)* Let a be the acceleration of the lift.
Given, the ratio of time periods of pendulum, while the lift is moving upwards and downwards,

$$T_1 : T_2 = 1:3$$

When lift is moving upwards, then total time period,

$$T_1 = 2\pi\sqrt{\frac{l}{g+a}} \quad \text{...(i)}$$

When lift is moving downwards, then total time period,

$$T_2 = 2\pi\sqrt{\frac{l}{g-a}} \quad \text{...(ii)}$$

Dividing Eq. (i) by Eq. (ii), we get

$$\frac{T_1}{T_2} = \sqrt{\frac{g-a}{g+a}}$$

$$\Rightarrow \quad \frac{1}{3} = \sqrt{\frac{g-a}{g+a}}$$

or $$\frac{g-a}{g+a} = \left(\frac{1}{3}\right)^2$$

or $$\frac{g-a}{g+a} = \frac{1}{9}$$

$$\Rightarrow \quad 9g - 9a = g + a$$

$$\Rightarrow \quad 9g - g = 9a + a \Rightarrow 10a = 8g$$

$$\Rightarrow \quad a = \frac{8g}{10} = \frac{8 \times 10}{10} = 8 \text{ m/s}^2$$

37. *(a)* As, orbital velocity of a satellite, $v = \sqrt{\frac{GM}{r}}$

Hence, $$v \propto \frac{1}{\sqrt{r}}$$

% increase in speed $= \frac{1}{2}$(% decrease in radius)

$$= \frac{1}{2}(2\%) = 1\%$$

Thus, the speed of satellite will increase by 1%.

38. *(a)* Given, $\omega = 2000 \text{ rad s}^{-1}$

$$L = 5 \text{ mH} = 5 \times 10^{-3} \text{ H}$$

$$C = 50 \mu\text{F} = 50 \times 10^{-6} \text{ F}$$

$$V = 30 \text{ V}$$

Inductive reactance,

$$X_L = \omega L = 2000 \times 5 \times 10^{-3} = 10\,\Omega$$

and capactive reactance,

$$X_C = \frac{1}{\omega C} = \frac{1}{50 \times 10^{-6} \times 2000} = \frac{100}{10} = 10\,\Omega$$

$$\Rightarrow \quad X_L = X_C$$

The given circuit is resonant circuit.
Hence, impedance of circuit, $Z = R$

$$= 4 + 6 + 0.5 = 10.5\,\Omega$$

Amplitude of current, $I_0 = \frac{V}{Z} = \frac{30}{10.5} = 2.85 \text{ A}$

39. *(d)* Work done, W = area under F-s graph

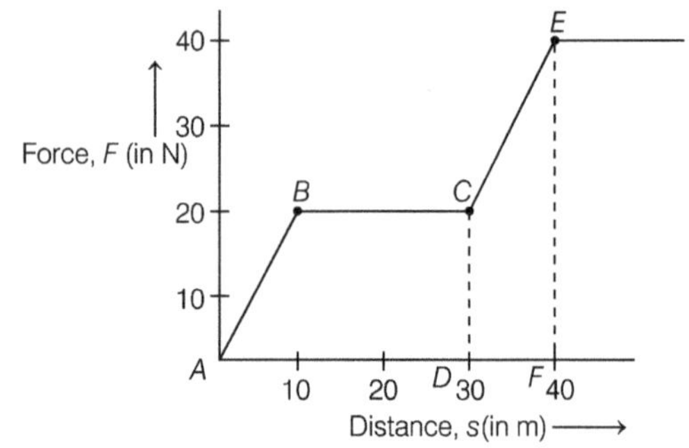

∴ Total work done is covering an initial distance of 40 m,

W = area of trapezium $ABCD$ + area of trapezium $CEFD$

$$= \frac{1}{2} \times 20(20+30) + \frac{1}{2} \times 10 \times (20+40) = 800 \text{ J}$$

[∵ Area of trapezium $= \frac{1}{2} \times$ height × sum of parallel sides]

$$\Rightarrow \quad W = 800 \text{ J}$$

40. *(b)* Given, potential gradient $= 0.4 \text{ mV cm}^{-1}$

$$= 0.4 \times 10^{-3} \text{ V cm}^{-1}$$

Length of wire, $l = 10 \text{ m} = 1000 \text{ cm}$
Electromagnetic force of cell, $E = 4$ V
Potential gradient along wire

$$= \frac{\text{Potential difference along wire}}{\text{Length of wire}}$$

$$\Rightarrow \quad 0.4 \times 10^{-3} = \frac{i \times 40}{1000}$$

$$\therefore \quad \text{Current in wire, } i = \frac{4}{400} \text{ A} \quad \text{...(i)}$$

Let R be the required resistance unplugged in the box.

$$\therefore \quad i = \frac{E}{R+R'} = \frac{4}{40+R} \quad \text{...(ii)}$$

Equating Eqs. (i) and (ii), we get

$$\frac{4}{40+R} = \frac{4}{400}$$

$$\Rightarrow \quad R = 400 - 40 = 360\,\Omega$$

Chemistry

41. ***(d)*** The given reactions take place as follows

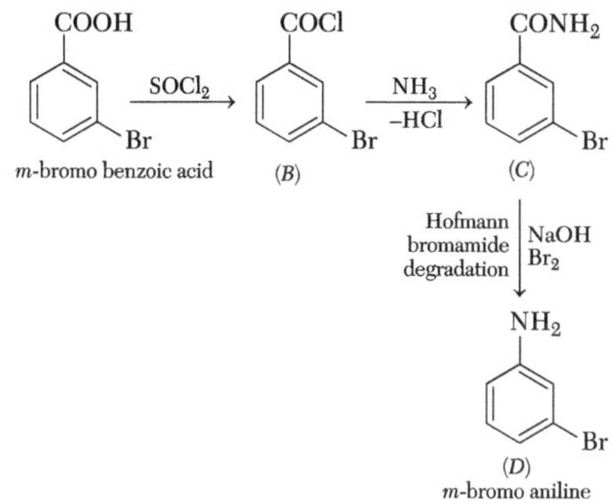

42. ***(b)*** From molarity equation, $\underset{\text{Acid}}{M_1V_1} = \underset{\text{Base}}{M_2V_2}$

$$M_1 \times 50.0 = 0.10 \times 39.30$$

$$\therefore \quad M_1 = \frac{0.10 \times 39.30}{50}$$

$$= 0.0786 \text{ M}$$

$$\approx 0.079 \text{ M}$$

When half equivalence point is reached,

$$[\text{HA}] = [A^-]$$

Using Handerson-Hasselbalch equation

$$\text{pH} = \text{p}K_a + \log\frac{[A^-]}{[\text{HA}]}$$

$$\therefore \quad \text{pH} = \text{p}K_a = 4.85$$

43. ***(d)*** The complete equation is given as follows:

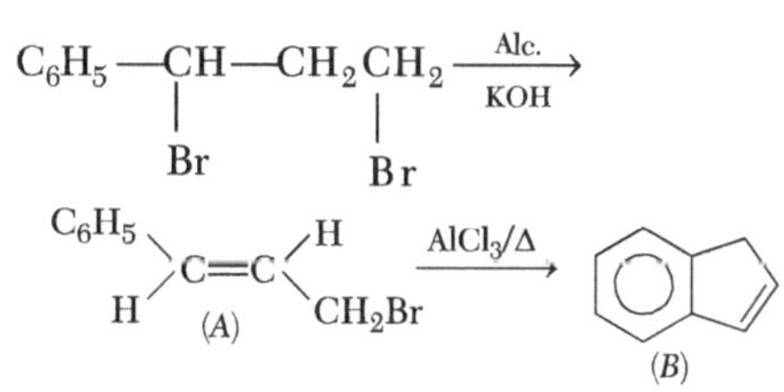

:CuBr compound gives :CClBr. This reaction is addition of carbene to double bond.

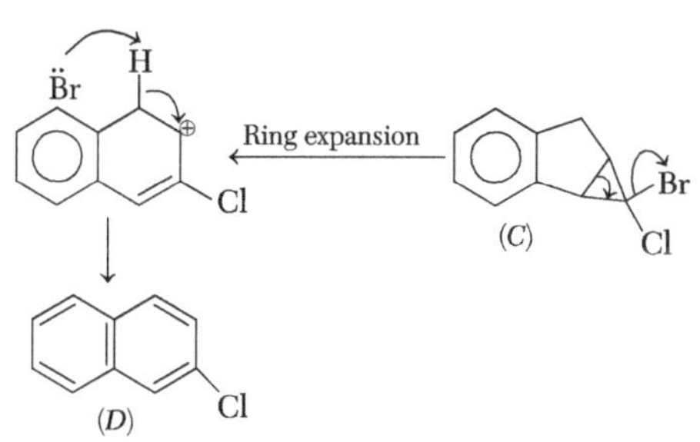

44. ***(a)*** Statement (a) is incorrect because α-D-fructose and β-D-fructose are anomers, diastereomers and geometrical isomers but not enantiomers.

Rest of the given statements are correct.

45. ***(d)*** Option (d) does not contain a pair of mixed oxides. Fe_2O_3 is not a mixed oxide.

While, Mn_3O_4 ($2MnO + MnO_2$); Co_3O_4 ($CoO + Co_2O_3$); Fe_3O_4 ($Fe_2O_3 + FeO$) and Pb_3O_4 ($2PbO + PbO_2$), all are mixed salts.

46. ***(a)*** According to Faraday's first law of electrolysis:

The reaction at cathode,

$$\underset{63.5\text{ g}}{Cu^{2+}} + 2e^- \longrightarrow \underset{2 \times 96500\text{ C}}{Cu}$$

The quantity of charge passed

$$= I \times t = (10 \text{ amp}) \times (60 \times 60\text{s}) = 36000 \text{ C}.$$

2×96500 C of charge deposit copper = 63.5 g

36000 C of charge deposit copper

$$= \frac{(63.5 \text{ g})}{(2 \times 96500 \text{ C})} \times (36000\text{C}) = 11.84 \text{ g}$$

Thus, 11.84 g of copper will dissolve from the anode and the same amount from the solution will get deposited on the cathode. The concentration of the solution will remain unchanged.

47. ***(b)*** The complete reaction is given as follows:

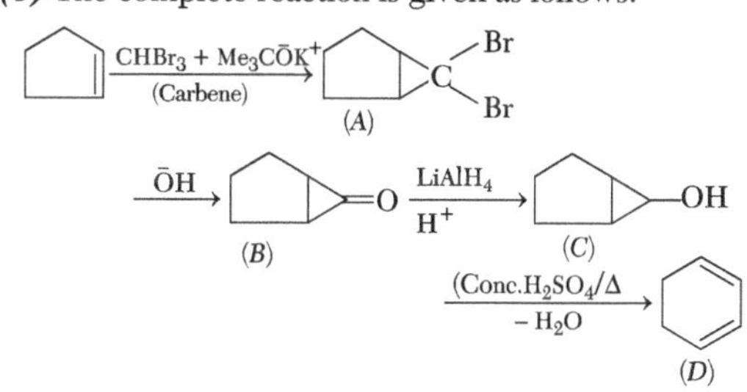

48. ***(a)*** Statement (a) is correct. Rest of the all statements are incorrect. These are explained as follows:

(a) $Cl_2(g) \longrightarrow 2Cl(g)$

As randomness is increasing. Therefore, ΔS is positive for this reaction.

∴ Statement is correct.

(b) In closed container, $\Delta V = 0$, hence work done is zero. There is no heat exchange as system is adiabatic. Hence, $\Delta E = q + W = 0$.

∴ Statement is incorrect.

(c) ΔG will be zero only when equilibrium is reached.

∴ Statement is incorrect.

(d) $\Delta G° = -RT \ln K_{eq}$, not a function of pressure. Thus, this statement is also incorrect.

49. ***(b)*** Statements I and III are correct and statement II is incorrect.

(I) $\dfrac{T_c}{p_c} = \dfrac{8a}{27Rb} \cdot \dfrac{27b^2}{a} = \dfrac{8b}{R}$

Thus, $\frac{T_c}{p_c} \propto b$

Thus, larger the T_c / p_c value, larger would be b i.e. included volume.

(II) $T_C = \frac{8a}{27Rb}$ and $T_B = \frac{a}{Rb}$

Thus, $T_B > T_C$

(III) When, T_C is attained.

$\therefore \quad (\partial p / \partial V)_{T_C} = 0$

50. ***(a)*** The reaction can be completed as follows:

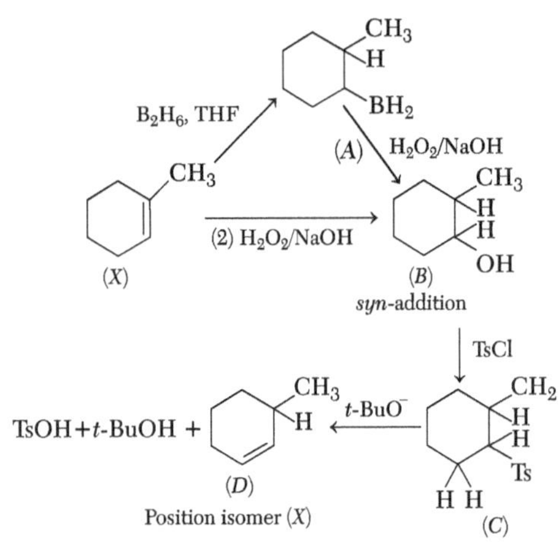

51. ***(d)*** The reaction can be completed as follows

$$Na_2B_4O_7 + conc.H_2SO_4 + 5H_2O \longrightarrow Na_2SO_4 + \underset{(A)}{4H_3BO_3}$$

$$H_3BO_3 + 3C_2H_5OH \longrightarrow \underset{(B)}{B(OC_2H_5)_3}$$

Volatile

(Burn with green edged flame)

52. ***(a)*** Perhydrol means 30% solution of H_2O_2.

H_2O_2 decomposes as :

$$2H_2O_2 \longrightarrow 2H_2O + O_2$$

Volume strength of 30% H_2O_2 solution is 100 that means 1 mL of this solution on decomposition gives 100 mL oxygen.

$$SO_2 + \frac{1}{2}O_2 \longrightarrow SO_3$$

1 L	1/2 L	1 L
2 L	1 L	2 L

Since, 100 mL of oxygen is obtained by 1 mL of H_2O_2.

$\therefore$ 1000 mL of oxygen will be obtained by

$$= \frac{1}{100} \times 1000 \text{ mL of } H_2O_2$$

$$= 10 \text{ mL of } H_2O_2$$

53. ***(c)*** At top of a mountain,

$T_1 = 0 + 273K = 273K, p_1 = \frac{710}{760}$ atm and density $= \rho_1$

At bottom of mountain,

$T_2 = 30 + 273 = 303$ K, $p_2 = 1$ atm and density $= \rho_2$

From ideal gas equation,

$\because \quad pV = \frac{w}{M}RT \qquad \left[p = \frac{w}{MV}RT; \rho = \frac{w}{V}\right]$

or $\quad p = \rho\frac{RT}{M}$

$\therefore \quad \frac{\rho_1}{\rho_2} = \frac{p_1}{T_1} \times \frac{T_2}{p_2}$

or $\quad \frac{\rho_1}{\rho_2} = \frac{710/760 \times 303}{273 \times 1} = \frac{1.04}{1}$

54. ***(b)*** The correct option is (b).

$A \to 5, B \to 3, C \to 4, D \to 2, E \to 1$.

The shapes of molecules are given below.

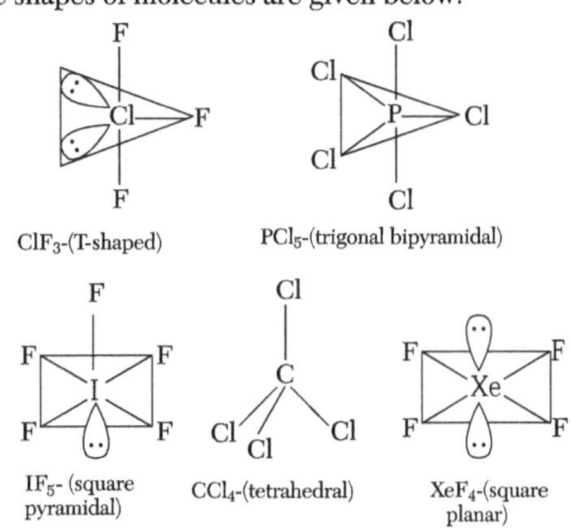

55. ***(c)*** In option (c), the hydrogen of —CH group is surrounded by three $-\underset{\underset{O}{\|}}{C}-CH_3$ groups as:

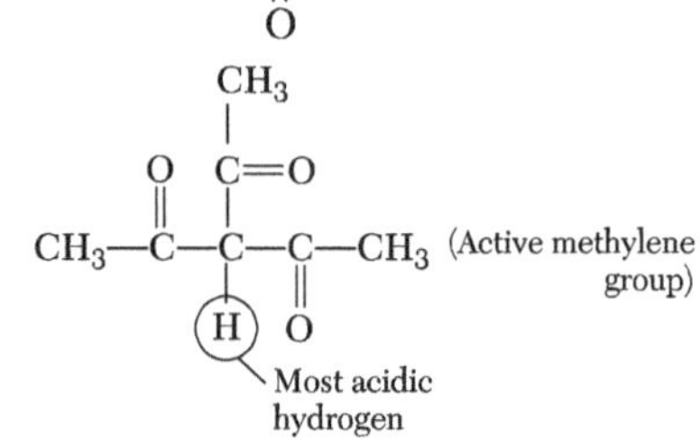

Here, O of CO attracts, the electron density towards itself.

Or

The anion formed after removal of acidic proton in

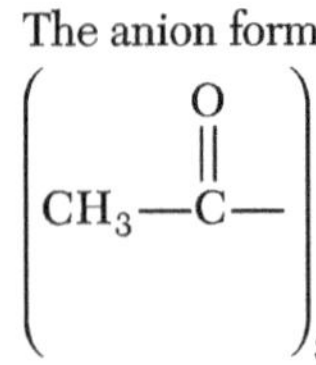 is stablised by $-R$-effect of three

$C=O$ groups.

Thus, (c) has most acidic hydrogen among given compounds.

56. ***(c)*** The process by which the aquatic life gets deprived oxygen and results in subsequent loss of biodiversity is known as eutrophication. It is because of reduction in concentration of the dissolved oxygen in water due to phosphate pollution in water.

57. ***(d)*** As we know,

$$\nu = \frac{1}{\lambda} = \frac{(\text{IE})_H Z^2}{hc} x - \left[\frac{1}{n_1^2} - \frac{1}{n_2^2}\right]$$

Given, $n_1 = 2, n_2 = 5$

$(\text{IE})_H = 13.6 \text{ eV} = 13.6 \times 1.6 \times 10^{-19}$ J

$$\nu = \frac{1}{\lambda} = \frac{13.6 \times 1.6 \times 10^{-19} \text{J}}{6.6 \times 10^{-34} \text{Js} \times 3 \times 10^8 \text{ ms}^{-1}}\left[\frac{1}{2^2} - \frac{1}{5^2}\right]$$

$= 2.2988 \times 10^6 \text{ m}^{-1}$

$\therefore \quad \lambda = \frac{1}{\nu} \Rightarrow \lambda = 4.35 \times 10^{-7} \text{m} = 435 \times 10^{-9}$ m

$= 435$ nm

58. ***(c)*** Heat change at constant pressure means enthalpy change ($\Delta_r H = q_p$).

Heat of formation of CO is written as :

$$C(s) + \frac{1}{2}O_2(g) \longrightarrow CO(g); \Delta_f H^\ominus = ?$$

Now we have,

I. $C(g) + O_2(g) \longrightarrow CO_2(g);$

$$\Delta_c H_1^\ominus = -393.5 \text{ kJ/mol}$$

II. $CO(g) + \frac{1}{2}O_2(g) \longrightarrow CO_2(g);$

$$\Delta_c H_2^\ominus = -283 \text{ kJ/mol}$$

Substracting appropriately (i.e., I–II), we get,

$$C(s) + \frac{1}{2}O_2(g) \longrightarrow CO(g); \ \Delta_f H^\ominus$$

From Hess's law, $\Delta_f H^\ominus = \Delta_c H_1^\ominus - \Delta_c H_2^\ominus$

$\Rightarrow \Delta_f H^\ominus = -393.5 - (-283) = 110.5$ kJ/mol

Now, calculation of the heat of formation at constant volume means that we have to calculate change in internal energy (i.e. $\Delta_r U$).

Using, $\Delta_r H = \Delta_r U + p\Delta V$ [For a chemical reaction]

$= \Delta_r U + \Delta n_g RT$ [As $pV = nRT$]

[Δn_g = gaseous moles of products – gaseous moles of reactant]

$\Rightarrow \quad \Delta_r U = \Delta_r H - \Delta n_g RT$

Now putting the values : $\Delta n_g = 1 - \frac{1}{2} = \frac{1}{2}, T = 298$ K,

$R = 8.314$ J/K mol and $\Delta_f H^\ominus = -110.5$ kJ/mol

$\Rightarrow \ \Delta_f U^\ominus = -110.5 - \frac{1}{2} \times 8.314 \times 2.98 \times 10^{-3}$

$= -111.7$ kJ/mol

59. ***(b)*** If $n = n$, then $l = 0, 1, 2 \ldots n$.

n	l	Sub-orbit	$(n + l)$
1	0	1s	1
	1	1p	2
	0	2s	2
2	1	2p	3
	2	2d	4
	0	3s	3
3	1	3p	4
	2	3d	5
	3	3f	6

Thus, energy order is,
$1s < 1p < 2s < 2p < 3s < 2d < 3p$

Sc (21) will have electronic configuration as :
$$1s^2\,1p^6\,2s^2 2p^6\,3s^2\,2d^3$$

60. ***(c)***

	A	+	$2B$	$\rightleftharpoons$	$2C$	+	D	
Initial	1.1		2.2		0		0	
At equili.	1.1–x		2.2 – 2x		2x		x	[2x = 0.2]
	1.1 – 0.1		2.2–0.2		0.2		0.1	[x = 0.1]
	1		2		0.2		0.1	

$$K_C = \frac{[C^2][D]}{[A][B^2]} = \frac{(0.2)^2(0.1)}{(1)(2)^2} = 0.001$$

61. ***(c)*** Chemical reaction given in option 'c' is correct. Rest of the given chemical reactions are incorrect. These can be corrected as:

(a) $CaCN_2 + 3H_2O \longrightarrow CaCO_3 + 2NH_3$

(b) $2NH_3 + CaSO_4 + CO_2 + H_2O \longrightarrow (NH_4)_2SO_4 + CaCO_3$

(d) $CaC_2 + H_2O \longrightarrow 2C_2H_2 + Ca(OH)_2$

62. ***(d)*** Disproportionation means a reaction in which a substance is oxidised as well as reduced simultaneously,

$Na_2S_2O_3 + H_2SO_4 \longrightarrow Na_2SO_4 + SO_2 + S_8 + H_2O$

Oxidation number of S in $Na_2S_2O_3 = +2$

Oxidation number of S in $H_2SO_4 = +6$

Oxidation number of S in $SO_2 = +4$

Oxidation number of S in $Na_2SO_4 = +6$

Oxidation number of S in $S_8 = 0$

So, sulphur can oxidise from + 2 to + 4 as well as reduce to 0 (S_8).

63. ***(d)*** Right answer is *d* because after heating (Δ), dcarboxylation occurs.

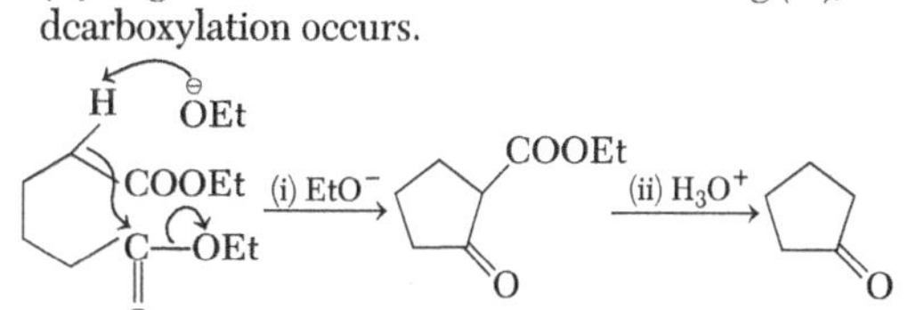

(A) (B) (C)

64. ***(a)*** $-\Delta S = \frac{\Delta G}{T}$, also $\frac{\Delta G}{T}$ is slope in Ellingham diagram. ΔS is different for different reactions as they have different slopes.

Rest of the statements (b), (c) and (d) are correct about Ellingham diagram.

65. ***(c)*** Phenolphthalein indicator changes into pink colour due to change in pH and not due to adsorption.

66. ***(c)*** The reaction take place as follows

$$HBr \longrightarrow H^+ + Br^-$$

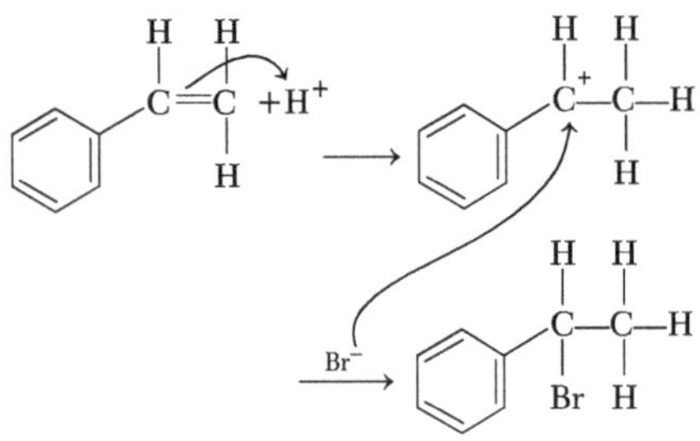

(A)

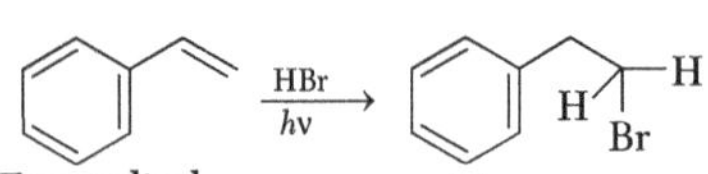

(Free radical reaction) (B)

67. ***(a)*** Z_{eff} for fcc = 4, Z_{eff} for bcc = 2

Atomic volume of α-form $= \frac{(3.68 \times 10^{-8})^3}{4} \times N_A$

Atomic volume of β-form $= \frac{(2.92 \times 10^{-8})^3}{2} \times N_A$

As atomic weight is same, element is same, so the density ratio is

$$\rho_\alpha : \rho_\beta = V_\beta : V_\alpha = \frac{(2.92)^3}{2} : \frac{(3.68)^2}{4}$$

$$= \frac{24.9}{2} : \frac{4}{49.8} = 1:1$$

68. ***(c)*** $\therefore \; K_b = \frac{RT_b^2 M_1}{\Delta H_{vap}}$

$$= \frac{8.314 \times (373)^2 \times 0.018}{40585 \text{ J/mol}}$$

$$= 0.513 \text{ K kg/mol}$$

Now, molecular weight of $Al_2(SO_4)_3 = 342$

$$\underset{1-\alpha}{Al_2(SO_4)_3} \rightleftharpoons \underset{2\alpha}{2Al^{3+}} + \underset{3\alpha}{3SO_4^{2-}}$$

$i = 1 - \alpha + 2\alpha + 3\alpha = 1 + 4\alpha \Rightarrow i = 5$ $\quad (\because \alpha = 1)$

$\Delta T_b = i \times K_b . m$

$$= 5 \times 0.513 \times \frac{5.6}{342 \times 1} = 0.042$$

Boiling point of solution = 100 + 0.042 = 100.042°C

69. ***(d)*** The reaction take place as follows

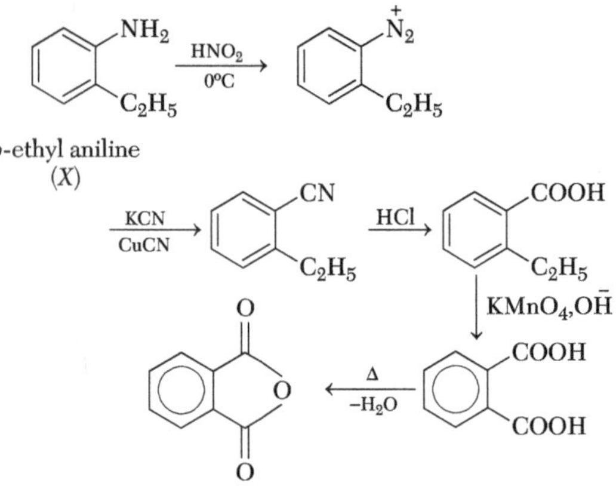

70. ***(b)*** Phenol on reaction with chloroform and KOH gives salicylaldehyde, which with 50% KOH solution undergoes Cannizzaro's reaction.

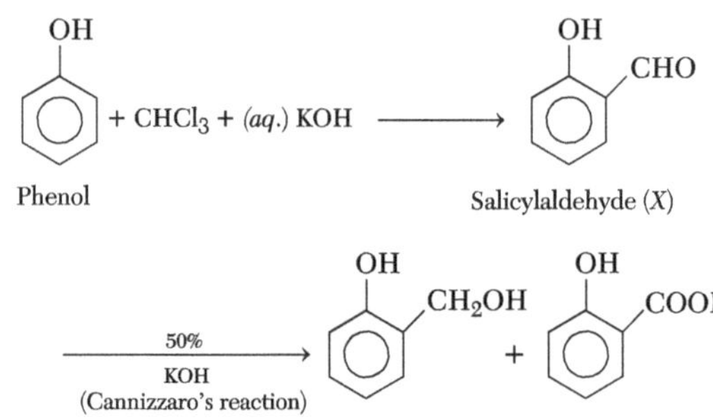

71. ***(d)*** Some complex ions of nitrogen show similarities with halide ions. These are called pseudohalide ions e.g., OCN^-, NCO^-, CN^- etc.

Halogens, among themselves, form complex ions which are called polyhalide ions. e.g., I_3^-, BrI_2^- etc.

Similarly, interhalogens are the compound of halogens in which one halogen is cation and other halogen is anion. e.g., IF_7, ICl_5, BrF_5, IF_5 etc.

$\therefore$ The correct order of pseudohalide polyhalide and interhalogen is

$$OCN^-, BrI_2^-, IF_5$$

72. ***(b)*** The inorganic halide (A) is SO_2Cl_2.

$$\underset{(A)}{SO_2Cl_2} + 2H_2O \longrightarrow \underset{(B)}{H_2SO_4} + \underset{(C)}{2HCl}$$

$$\underset{(A)}{SO_2Cl_2} + 4NaOH \longrightarrow \underset{(D)}{Na_2SO_4} + \underset{(E)}{2NaCl} + 2H_2O$$

$$\underset{(D)}{Na_2SO_4} + BaCl_2 \longrightarrow \underset{\text{White ppt.}}{BaSO_4} + 2NaCl$$

$$\underset{(E)}{NaCl} + AgNO_3 \longrightarrow \underset{\text{White ppt.}}{AgCl} + NaNO_3$$

73. ***(d)*** In alkali metals, reactivity increases down the group as electropositivity increases but for halogens reactivity decreases down the group as molecular stability of halogens increases.

74. ***(b)*** IUPAC name of sodiumnitroprusside $Na_2[Fe(CN)_5NO]$ is sodium pentacyanonitrosyl ferrate (III) sodium pentacyanonitrosyl ferrate (III) because in it NO is neutral ligand and the oxidation number of Fe is III.

75. ***(d)***

	A(g) $\longrightarrow$	$P(g)$ +	$Q(g)$ +	$R(g)$
At $t = 0$,	0.4 atm	0	0	0
At time $= t$,	$0.4 - x$	x	x	x

$$P_t = 0.4 - x + x + x + x = 0.4 + 2x \text{ or } x = \frac{P_t - 0.4}{2}$$

$$P_A = P_0 - x = P_0 - \frac{P_t - 0.4}{2} = \frac{2 \times 0.4 - P_t + 0.4}{2}$$

$$= \frac{1.2 - P_t}{2}$$

For a first order reaction,

$$k = \frac{2.303}{t} \log \frac{P_0}{P_A}$$

$$\frac{0.693}{t_{1/2}} = \frac{2.303}{230} \log\left(\frac{0.4 \times 2}{1.2 - P_t}\right)$$

$$\frac{0.693}{t_{1/2}} = \frac{2.303}{230} \log\left(\frac{0.8}{1.2 - P_t}\right)$$

$$\log \frac{0.8}{1.2 - P_t} = \frac{0.693 \times 230}{69.3 \times 2.303} = 0.9987$$

$$\frac{0.8}{1.2 - P_t} = 10^{0.9987}$$

$$0.8 = (10)(1.2 - P_t)$$

$$0.8 = 12 - 10P_t$$

$$10P_t = 12 - 0.8$$

$$10P_t = 11.2$$

$$P_t = \frac{11.2}{10} = 1.12 \text{ atm}$$

76. ***(d)*** The structure of an important antipyretic paracetamol is

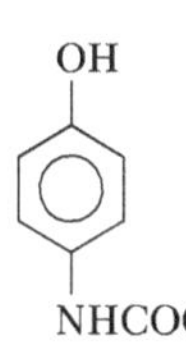

So, *p*-amino phenol gives paracetamol on actylation.

77. ***(b)*** On passing H_2S gas in the aqueous solution of salt in the presence of dilute HCl, black ppt. of CuS is formed.

On boiling CuS(ppt.) with dil. HNO_3, $Cu(NO_3)_2$ (blue colour) is formed. Which gives deep blue solution of $[Cu(NH_3)_4]^{2+}$.

$$3CuS + 8HNO_3 \longrightarrow 3Cu(NO_3)_2 + 2NO + 3S + 4H_2O$$

$$Cu^{2+} + 4NH_4OH \longrightarrow \underset{\text{Deep blue complex}}{[Cu(NH_3)_4]^{2+}} + 4H_2O$$

78. ***(a)*** $5C_2O_4^{2-} + 2MnO_4^- + 16H^+ \longrightarrow 2Mn^{2+} + 10CO_2 + 8H_2O$

Note In a reaction (redox/neutralisation) number of equivalents of different reactants are same.

Equivalents of $C_2O_4^{2-}$ = Equivalents of MnO_4^-

$$\text{Equivalent weight of } C_2O_4^{2-} = \frac{\text{Molecular weight}}{2}$$

$$\therefore \quad \text{Molarity} = 2 \times \text{Normality}$$

$$\text{Equivalent weight of } MnO_4^- = \frac{\text{Molecular weight}}{5}$$

$$\text{Molarity} = 5 \times \text{Normality}$$

$$\text{Number of equivalents of } MnO_4^- = \frac{28.85 \times 5\, M_1}{1000}$$

$$\text{Number of equivalents of } C_2O_4^{2-} = \frac{0.1467}{67}$$

$$\therefore \quad \frac{28.85 \times 5\, M_1}{1000} = \frac{0.1467}{67}$$

$$M_1 = 0.01518 \text{ M}$$

79. ***(c)*** $\underbrace{[Cr(NH_3)_6]}_{\text{Coordination sphere}} \underbrace{Cl_3}_{\text{Ionisable}} \rightleftharpoons [Cr(NH_3)_6]^{3+} + 3Cl^-$

$$[Cr(NH_3)_6]Cl_3 \xrightarrow{AgNO_3} \underset{\text{(White ppt.)}}{AgCl\downarrow}$$

Configuration of Cr (24) $= 3d^5\ 4s^1$

Configuration of $Cr^{3+} = 3d^3 4s^0$

$\therefore$ It is paramagnetic nature due to 3 unpaired electrons.

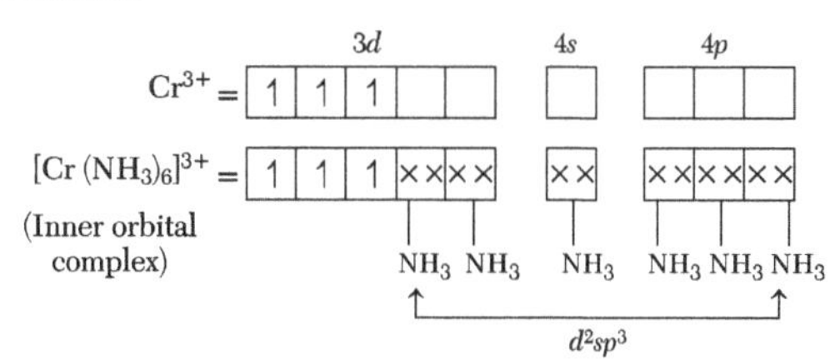

(a) d^2sp^3 hybridisation, octahedral. Thus, option (a) is correct.

(b) Here are three unpaired electrons, hence it is paramagnetic. Thus, option (b) is correct.

(c) d^2sp^3 inner orbital complex. Thus option (c) is incorrect.

(d) Due to ionisable Cl^- ions, white ppt. with $AgNO_3$. Thus, option (d) is correct.

So, the incorrect statement is (c).

80. ***(a)*** $H_2SO_4(aq) \longrightarrow H^+(aq) + HSO_4^-(aq)$;

$$H_2SO_4^-(aq) \longrightarrow H^+(aq) + SO_4^-(aq)$$

Here, $K_{a_2} < K_{a_1}$ because the negatively charged HSO_4^- ion has much less tendency to donate a proton to H_2O as compared to neutral H_2SO_4.

a. English Proficiency

81. ***(b)*** Article 'the' should be used before 'famous' as the sentence refers to a particular thing i.e. a famous monument.

82. ***(b)*** As the sentence refers to a choice among more than two persons (servants), 'which' will be used in place of 'who'.

83. ***(c)*** The sentence is incomplete as it does not answer the question 'helped whom?', So we add 'him' at the end of the sentence.

84. ***(a)*** Use of preposition 'with' is suitable to fill the given blank.

85. ***(b)*** Use of preposition 'at' is suitable to fill the given blank.

86. ***(c)*** 'Benign' means not likely to cause death. 'Fatal' would be its opposite.

87. ***(b)*** 'Noted' is known by many people because of particular qualities. Its opposite would be 'Unknown'.

88. ***(a)*** 'Sagacious' means having or showing understanding and the ability to make good judgements. So, 'Foolish' would be its correct antonym.

89. ***(a)*** 'Getting peevish' means to get irritated.

90. ***(c)*** 'Roughed out' is a phrasal verb which means to draw a 'rough draft'. So, 'drew a quick plan' is its correct synonym.

91. ***(b)*** The need for an effective population policy is an urgent necessity in the country's planning strategy. Hence, option (b) is the correct answer.

92. ***(d)*** The development of human resource and the building up of an institutional framework would have to receive priority attention.

93. ***(c)*** The Centre and the States government must become partners in the planning process to determine national priorities together. Hence, the statement given in option (c) is not correct.

94. ***(b)*** Domestic economic situation and world trends would force the planning process to undergo a change.

95. ***(c)*** Important changes in the international scene is implied by the expression 'momentous trends.'

b. Logical Reasoning

96. ***(d)*** The pattern is as follows

$$2\times 6 = 12$$
$$6\times 12 = 72$$
$$12\times 72 = 864 \neq \boxed{865}$$
$$72\times 864 = 62208$$

Hence, number 865 is wrong and should be replaced by 864.

97. ***(c)*** According to the question,

$$A/Q > T > P/B$$

But P is not the shortest

$$A/Q > T > P > B$$

So, tallest is either A or Q.

98. ***(c)*** Option figure (c) will complete the given pattern as

99. ***(d)*** 'Spanner' is used by 'mechanic' as a tool to loose or tight nut and bolt. In the same way, 'saw' is used by 'carpenter' as a tool to cut wood. Hence, option (d) is correct.

100. ***(b)***

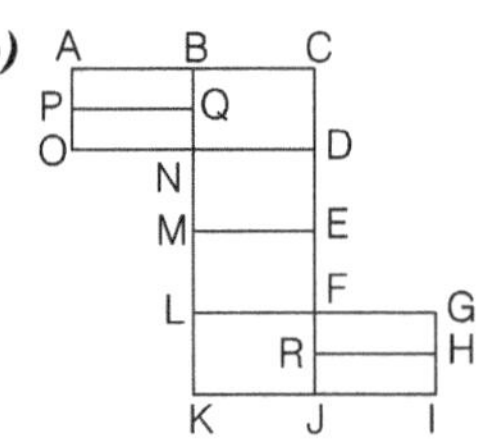

Total 18 rectangles are as follow

ABQP, PQNO, BCDN,
NDEM, MEFL, LFJK,
FGHR, RHIJ, ABNO,
BCEM, NDFL, MEJK,
FGIJ, ACDO, BCFL,
NDJK, LGIK and BCJK

101. ***(d)*** As, $B \xrightarrow{-1} A \xrightarrow{+3} D \xrightarrow{-1} C$

$$J \xrightarrow{-1} I \xrightarrow{+3} L \xrightarrow{-1} K$$
$$N \xrightarrow{-1} M \xrightarrow{+3} P \xrightarrow{-1} O$$

But, $\boxed{V \xrightarrow{-1} U \xrightarrow{+2} W \xrightarrow{+1} X}$

Except option (d), all others follow same pattern (− 1, + 3, − 1).

102. ***(d)*** Clearly, there are three types of shading of circles, one is unshaded another has its right half shaded and yet another has its upper half shaded. Also, two of the figures in each row have one triangle shaded.

103. ***(c)*** Option figure (c) can be formed from the cut out pieces given in question figure.

104. ***(b)*** After unfolding the folded and cut piece of paper it will look like as answer figure (b).

105. ***(b)***

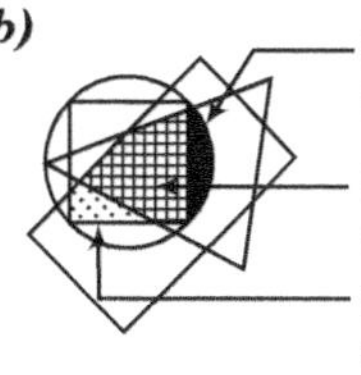

Mathematics

106. ***(a)*** Given that, $A = \{x : x^2 = 1\} = \{-1, 1\}$ and $B = \{x : x^4 = 1\} = \{-1, 1, -i, i\}$

Now, $A - B = \phi$ and $B - A = \{-i, i\}$

$\therefore \quad A \Delta B = (A - B) \cup (B - A) = \{-i, i\}$

107. ***(c)*** We have, $2f(xy) = (f(x))^y + (f(y))^x$, $x, y \in R$ and $f(1) = a$.

Putting $x = 1, y = 2$, we get

$2f(2) = (f(1))^2 + f(2) \Rightarrow f(2) = a^2$

Putting $x = 1, y = 3$, we get

$2f(3) = (f(1))^3 + f(3) \Rightarrow f(3) = a^3$, and so on.

$\therefore \quad f(n) = a^n$

$$\sum_{i=1}^{n} f(i) = a + a^2 + .. + a^n = \frac{a(a^n - 1)}{a - 1}.$$

108. ***(c)*** The given inequality is equivalent to

$$1 < |x - 3| + |4 - x|$$

i.e., $1 < |x - 3| + |x - 4|$ ...(i)

For $x < 3$, we have

$$3 - x + 4 - x = 7 - 2x > 1$$

i.e., $x < 3$, which is true in the domain.

For $3 \le x < 4$, Eq. (i) gives

$x - 3 + 4 - x = 1 \not> 1$ ...(ii)

For $4 \le x$, we have

$2x - 7 > 1 \Rightarrow x > 4$ ...(iii)

From Eqs. (i) and (iii), we have the solution of the inequality as $x < 3$ or $x > 4$.

i.e., $x \in]-\infty, 3) \cup (4, \infty[$ which is equal to $R - [3, 4]$.

109. ***(b)*** $\because$ 13, $a_1, a_2, ..., a_{20}$, 67 are in AP.

$$\therefore \; a_1 + a_2 + a_3 + \ldots + a_{20} = 20\left(\frac{13 + 67}{2}\right) = 800$$

Also, AM > GM

$$\Rightarrow \quad \frac{a_1 + a_2 + \ldots + a_{20}}{20} \ge (a_1 \cdot a_2 \cdot a_3 \cdot \ldots \cdot a_{20})^{\frac{1}{20}}$$

$$\Rightarrow \quad 40 \ge (a_1 \cdot a_2 \cdot a_3 \ldots a_{20})^{\frac{1}{20}}$$

Hence, maximum value of $a_1 \cdot a_2 \cdot a_3 \ldots a_{20}$ is $(40)^{20}$.

110. ***(a)*** Since, p, q and r are positive and are in AP.

$\therefore \quad q = \dfrac{p + r}{2}$ ...(i)

$\because$ The roots of the equation, $px^2 + qx + r = 0$ are real.

$\Rightarrow \quad q^2 \ge 4pr \Rightarrow \left[\dfrac{p + r}{2}\right]^2 \ge 4pr$ [from Eq. (i)]

$\Rightarrow \quad p^2 + r^2 - 14pr \ge 0$

$\Rightarrow \quad \left(\dfrac{r}{p}\right)^2 - 14\left(\dfrac{r}{p}\right) + 1 \ge 0 \; (\because p > 0)$

$\Rightarrow \quad \left(\dfrac{r}{p} - 7\right)^2 - 48 \ge 0$

$\Rightarrow \quad \left(\dfrac{r}{p} - 7\right)^2 - (4\sqrt{3})^2 \ge 0 \Rightarrow \left|\dfrac{r}{p} - 7\right| \ge 4\sqrt{3}$

111. ***(c)*** $^{47}C_4 + \sum_{r=1}^{5} {}^{52-r}C_3 = {}^{47}C_4 + {}^{51}C_3 + {}^{50}C_3 + {}^{49}C_3 + {}^{48}C_3 + {}^{47}C_3$

$= {}^{51}C_3 + {}^{50}C_3 + {}^{49}C_3 + {}^{48}C_3 + ({}^{47}C_3 + {}^{47}C_4) = {}^{52}C_4$

112. ***(a)*** Possible even digits are 2,4,6,8,0.

***Case* I** Number has digits 4,6,8,0.

(Here, sum of digits is 18, divisible by 3)

□ □ □ □

Number of arrangements $= 3 \times 3!$

[Ist place can be filled using 4, 6, 8]

$= 3 \times 6 = 18$

***Case* II** Number has digits 2, 4, 6, 0

(Here, sum of digits is 12, divisible by 3)

□ □ □ □

1st place cannot be filled by 0.

Number of arrangements $= 3 \times 3! = 18$

$\therefore$ Number of numbers $= 18 + 18 = 36$

113. ***(d)*** We know that,

$n(A \cup B) = n(A) + n(B) - n(A \cap B)$

$\therefore \quad P = 1500 - n(A \cap B)$

$\Rightarrow \quad n(A \cap B) = 1500 - P$

Clearly, $1 \le n(A \cap B) \le 500$

$\Rightarrow \quad 1 \le 1500 - P \le 500$

$\Rightarrow \quad -1499 \le -P \le -1000$

$\Rightarrow \quad 1000 \le P \le 1499$

114. *(c)* Here, $\frac{2x-1}{x^3+4x^2+3x} \in R$ only when

$x^3 + 4x^2 + 3x \ne 0$

$\because \quad x(x^2+4x+3) = 0$

$\Rightarrow \quad x(x+1)(x+3) = 0$

$\Rightarrow \quad x = 0, -1, -3$

$\therefore \quad \left\{x \in R : \frac{2x-1}{x^3+4x^2+3x} \in R\right\}$

$= R - \{0, -1, -3\}$

115. *(a)* Let $f(x) = Ax^2 + Bx + C$

$\therefore \quad f(1) = A + B + C$

and $\quad f(-1) = A - B + C$

$\because \quad f(1) = f(-1)$

$\Rightarrow \quad A + B + C = A - B + C$

$\Rightarrow \quad 2B = 0 \Rightarrow B = 0$

$\therefore \quad f(x) = Ax^2 + C \Rightarrow f'(x) = 2Ax$

$\Rightarrow \quad f'(a) = 2Aa$

$f'(b) = 2Ab$ and $f'(c) = 2Ac$

Also, a, b, c are in AP

So, $2Aa$, $2Ab$ and $2Ac$ are in AP.

$\Rightarrow f'(a), f'(b)$ and $f'(c)$ are also in AP.

116. *(c)* $\lim_{n \to \infty} \frac{1}{n}\left\{\frac{1}{n+1} + \frac{2}{n+2} + \ldots + \frac{3n}{n+3n}\right\}$

$= \lim_{n \to \infty} \sum_{r=1}^{3n} \frac{1}{n}\left(\frac{r}{n+r}\right) = \int_0^3 \frac{x}{1+x}\,dx$

$= \int_0^3 \left(1 - \frac{1}{1+x}\right) dx = [x - \log(1+x)]_0^3$

$= 3 - \log 4 = 3 - 2\log 2$

117. *(b)* We have,

$(x-1)(x-2)(x-3)\ldots(x-10)$

Coefficient of x^8 = Sum of the terms taken two at a time i.e., $\sum_{1 \le i < j \le 10} x_i \cdot x_j = \frac{1}{2}\left[\left(\sum_{i=1}^{10} x_i\right)^2 - \left(\sum_{i=1}^{10} x_i^2\right)\right]$

$= \frac{1}{2}[(1+2+3+..+10)^2 - (1^2+2^2+3^2+\ldots+10^2)]$

$= \frac{1}{2}\left[\left(\frac{10 \times 11}{2}\right)^2 - \left(\frac{10 \times 11 \times 21}{6}\right)\right]$

$= \frac{1}{2}((55)^2 - 385) = \frac{1}{2}(3025 - 385) = \frac{2640}{2} = 1320$

118. *(b)* We have, $Z = \frac{7+i}{3+4i}$

$\Rightarrow \quad Z = \left(\frac{7+i}{3+4i}\right)\left(\frac{3-4i}{3-4i}\right)$

$= \frac{21 + 4 - 28i + 3i}{9+16} = 1 - i$

$\therefore \quad Z^{14} = (1-i)^{14} = [(1-i)^2]^7 = (1 - 1 - 2i)^7$

$= (-2)^7 (i)^7 = 2^7 i$

119. *(c)* We have, $\frac{dy}{dx} + \frac{\tan y}{x} = \frac{\tan y \sin y}{x^2}$

Now, divide by $\tan y \sin y$ both sides, we get

$\cot y \operatorname{cosec} y \frac{dy}{dx} + \frac{\operatorname{cosec} y}{x} = \frac{1}{x^2}$ ...(i)

Put $\operatorname{cosec} y = z$

$\therefore \quad -\operatorname{cosec} y \cot y \frac{dy}{dx} = \frac{dz}{dx}$

$\therefore$ Eq. (i) reduces to

$\frac{dz}{dx} - \frac{z}{x} = -\frac{1}{x^2}$

$\text{IF} = e^{\int -\frac{1}{x}dx} = e^{-\log x} = 1/x$

Hence solution is

$\frac{z}{x} = \int -\frac{1}{x^3}\,dx = \frac{1}{2x^2} - c$

$\Rightarrow \quad \frac{1}{x \sin y} = \frac{1 - 2cx^2}{2x^2}$

$\Rightarrow \quad 2x = \sin y(1 - 2cx^2)$

120. *(a)* We have,

$I = \int_0^{\pi/2} \frac{dx}{\tan x + \cot x + \operatorname{cosec} x + \sec x}$

$\Rightarrow I = \int_0^{\pi/2} \frac{dx}{\frac{\sin x}{\cos x} + \frac{\cos x}{\sin x} + \frac{1}{\sin x} + \frac{1}{\cos x}}$

$\Rightarrow I = \int_0^{\pi/2} \frac{\sin x \cos x\,dx}{1 + \sin x + \cos x}$

$\Rightarrow I = \int_0^{\pi/2} \frac{2\sin\frac{x}{2}\cos\frac{x}{2}\left(\cos^2\frac{x}{2} - \sin^2\frac{x}{2}\right)}{2\cos^2\frac{x}{2} + 2\sin\frac{x}{2}\cos\frac{x}{2}}\,dx$

$\Rightarrow I = \int_0^{\pi/2} \sin\frac{x}{2}\left(\cos\frac{x}{2} - \sin\frac{x}{2}\right) dx$

$\Rightarrow I = \frac{1}{2}\int_0^{\pi/2} (\sin x + \cos x - 1)\,dx$

$\Rightarrow I = \frac{1}{2}[-\cos x + \sin x - x]_0^{\pi/2}$

$$\Rightarrow I = \frac{1}{2}\left[\left(-\cos\frac{\pi}{2} + \sin\frac{\pi}{2} - \frac{\pi}{2}\right) - (-\cos 0 + \sin 0 - 0)\right]$$

$$\Rightarrow I = \frac{1}{2}\left[0 + 1 - \frac{\pi}{2} + 1\right] = 1 - \frac{\pi}{4}$$

121. ***(d)*** We have given,

$$|\mathbf{a}| = 1, |\mathbf{b}| = 4 \text{ and } \mathbf{a}\cdot\mathbf{b} = 2$$

Now, $\mathbf{c} = (2\mathbf{a}\times\mathbf{b}) - 3\mathbf{b}$

Multiplying by $\mathbf{b}$ both sides, we get

$$\mathbf{b}\cdot\mathbf{c} = (2\mathbf{a}\times\mathbf{b})\cdot\mathbf{b} - 3|\mathbf{b}|^2$$

$$\Rightarrow \mathbf{b}\cdot\mathbf{c} = 0 - 3(4)^2 = -48$$

and $|\mathbf{c}|^2 = |(2\mathbf{a}\times\mathbf{b}) - 3\mathbf{b}|^2$

$$|\mathbf{c}|^2 = (2\mathbf{a}\times\mathbf{b})^2 + 9|\mathbf{b}|^2 = 4(\mathbf{a}\times\mathbf{b})^2 + 9(4)^2$$

$$\Rightarrow |\mathbf{c}|^2 = 4(|\mathbf{a}|^2\cdot|\mathbf{b}|^2 - (\mathbf{a}\cdot\mathbf{b})^2) + 144$$

$$= 4(16-4) + 144 = 48 + 144 = 192$$

$$\therefore \mathbf{b}\cdot\mathbf{c} = |\mathbf{b}||\mathbf{c}|\cos\theta$$

$$\Rightarrow \cos\theta = \frac{\mathbf{b}\cdot\mathbf{c}}{|\mathbf{b}||\mathbf{c}|} = \frac{-48}{4\times\sqrt{192}} = -\frac{\sqrt{3}}{2}$$

$$\theta = \pi - \frac{\pi}{6} = \frac{5\pi}{6}$$

122. ***(a)*** We have,

$$x^2 - (k-2)x + (k^2 + 3k + 5) = 0$$

x_1, x_2 are roots of equations

$$\therefore x_1 + x_2 = k - 2, x_1x_2 = k^2 + 3k + 5$$

$$x_1^2 + x_2^2 = (x_1 + x_2)^2 - 2x_1x_2$$

$$= (k-2)^2 - 2(k^2 + 3k + 5)$$

$$= k^2 - 4k + 4 - 2k^2 - 6k - 10$$

$$= -(k^2 + 10k + 6) = -[(k+5)^2 - 19]$$

$$= 19 - (k+5)^2 \le 19$$

$\therefore$ Maximum value of $x_1^2 + x_2^2 = 19$

123. ***(c)*** Graph of circle whose centre is origin and radius is π units and $y = \sin x$ is

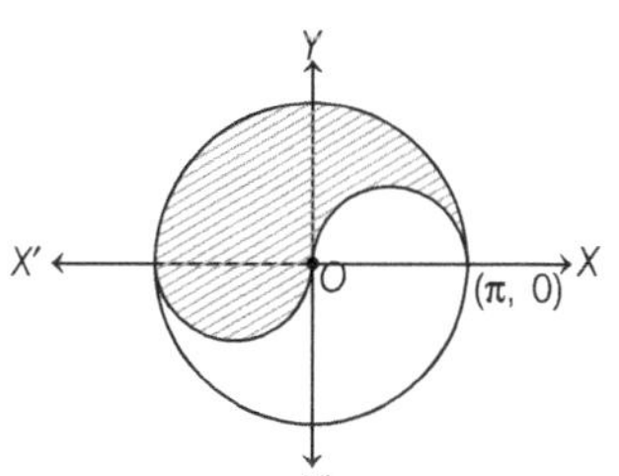

Area of shaded region $= \frac{1}{2}$ Area of circle

$$= \frac{1}{2}(\pi r^2) = \frac{1}{2}\pi(\pi^2) \quad [\because r = \pi]$$

$$= \pi^3/2$$

124. ***(a)*** The given equation is

$$2(1+i)x^2 - 4(2-i)x - 5 - 3i = 0$$

$$\Rightarrow x = \frac{4(2-i) \pm \sqrt{16(2-i)^2 + 8(1+i)(5+3i)}}{4(1+i)}$$

$$= -\frac{i}{1+i} \text{ or } \frac{4-i}{1+i} = \frac{-1-i}{2} \text{ or } \frac{3-5i}{2}$$

Now, $\left|\frac{-1-i}{2}\right| = \sqrt{\frac{1}{4} + \frac{1}{4}} = \sqrt{\frac{1}{2}}$

and $\left|\frac{3-5i}{2}\right| = \sqrt{\frac{9}{4} + \frac{25}{4}} = \sqrt{\frac{17}{2}}$

Also, $\sqrt{\frac{17}{2}} > \sqrt{\frac{1}{2}}$

Hence, required root is $\frac{3-5i}{2}$.

125. ***(d)*** For real roots, discriminant,

$$D = b^2 - 4ac \ge 0$$

$$= \cos^2\beta - 4(\cos\beta - 1)\sin\beta \ge 0$$

$$= \cos^2\beta + 4(1 - \cos\beta)\sin\beta \ge 0$$

So, $\sin\beta$ should be > 0. $[\because \cos^2\beta \ge 0, 1 - \cos\beta \ge 0]$

$$\Rightarrow \beta \in (0, \pi)$$

126. ***(c)*** We have,

$$(1+\alpha)x + \beta y + z = 2$$

$$\alpha x + (1+\beta)y + z = 3$$

$$\alpha x + \beta y + 2z = 2$$

For unique solution $\begin{vmatrix} 1+\alpha & \beta & 1 \\ \alpha & 1+\beta & 1 \\ \alpha & \beta & 2 \end{vmatrix} \ne 0$

Apply $R_1 \to R_1 - R_2$ and $R_2 \to R_2 - R_3$, we get

$$\begin{vmatrix} 1 & -1 & 0 \\ 0 & 1 & -1 \\ \alpha & \beta & 2 \end{vmatrix} \ne 0$$

Expanding along R_1, we get

$$1(2+\beta) + 1(0+\alpha) \ne 0 = \alpha + \beta + 2 \ne 0$$

Only option (c) satisfied above equation.

127. ***(d)*** Let $AB = h$ = Height of vertical pole

$OA = x$ = Horizontal distance from O to pole

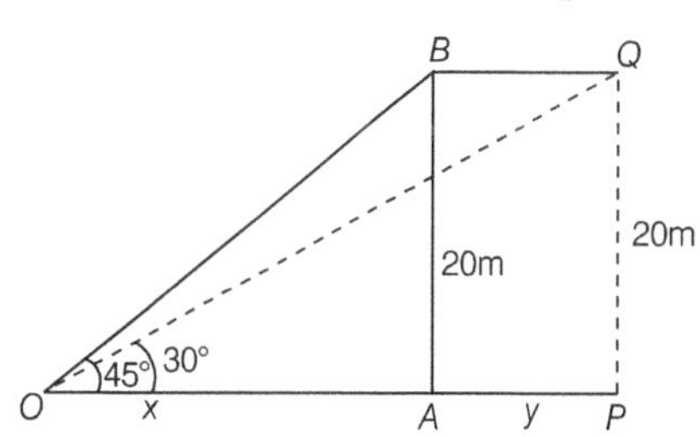

In ΔOAB

$$\tan 45° = \frac{AB}{OA} \Rightarrow 1 = \frac{20}{x}$$

$\Rightarrow \quad x = 20$ m

In ΔOPQ

$$\tan 30° = \frac{PQ}{OP} \Rightarrow \frac{1}{\sqrt{3}} = \frac{20}{x+y}$$

$\Rightarrow \quad x + y = 20\sqrt{3}$

$\Rightarrow \quad y = 20\sqrt{3} - x = 20\sqrt{3} - 20 = 20(\sqrt{3} - 1)$

$\therefore$ Speed of bird is $20(\sqrt{3} - 1)$ m/s.

128. ***(b)*** Since, $g = \sqrt{ab}$. Also, a, p, q and b are in AP.

So, common difference d is $\frac{b-a}{3}$

$$\therefore \quad p = a + d = a + \frac{b-a}{3} = \frac{2a+b}{3}$$

$$q = b - d = b - \frac{b-a}{3} = \frac{a+2b}{3}$$

Now, $(2p - q)(p - 2q)$

$$= \frac{(4a+2b-a-2b)}{3} \cdot \frac{(2a+b-2a-4b)}{3}$$

$$= -ab = -g^2$$

129. ***(b)*** Let $f(x) = x^{100}$

$g(x) = x^2 - 3x + 2$

$q(x)$ = quotient

$r(x)$ = remainder $= (ax + b)$

$\therefore \quad f(x) = g(x)\,q(x) + r(x)$

$x^{100} = (x^2 - 3x + 2)\,q(x) + ax + b$

Put $x = 1 \Rightarrow 1 = a + b$ ...(i)

Put $x = 2 \Rightarrow 2^{100} = 2a + b$ ...(ii)

From Eqs. (i) and (ii), we get

$a = 2^{100} - 1, b = 2 - 2^{100}$

Remainder $= (2^{100} - 1)x + (2 - 2^{100})$

$= (2^{99+1} - 1)x - 2(2^{99} - 1)$

$\therefore \quad k = 99$

130. ***(a)*** Let two observations be x and y.

$$\therefore \text{ Mean} = \frac{1+3+8+x+y}{5} = \frac{12+x+y}{5}$$

$$\Rightarrow \quad 5 = \frac{x+y+12}{5} \Rightarrow x + y = 13 \quad \text{...(i)}$$

Now, variance $= \frac{\Sigma x_i^2}{n} - (\bar{x})^2$

$$\Rightarrow \quad 9.20 = \frac{1^2 + 3^2 + 8^2 + x^2 + y^2}{5} - 25$$

$$\Rightarrow \quad 46 = 74 + x^2 + y^2 - 125$$

$$\Rightarrow \quad x^2 + y^2 = 97 \quad \text{...(ii)}$$

From Eqs. (i) and (ii), we get

$(x + y)^2 - 2xy = 97$

$\Rightarrow \quad (13)^2 - 2xy = 97$

$\Rightarrow \quad 2xy = 169 - 97 = 72$

$\Rightarrow \quad xy = 36$

$\therefore \quad x : y = 4 : 9$

131. ***(b)*** Let the three digit number be abc.

Given, middle digit is AM of first and last

$$\therefore \quad b = \frac{a+c}{2}$$

$\Rightarrow \quad 2b = a + c$

$2b$ is an even.

$\therefore$ a and c should be both even or both odd.

$\therefore$ Total number of ways $= 5C_1 \times 5C_1 + 4C_1 \times 5C_1$

$= 5 \times 5 + 4 \times 5 = 25 + 20 = 45$

132. ***(b)*** We have,

$Z = re^{i\theta}$

$Z = r(\cos\theta + i\sin\theta)$

$e^{iz} = e^{ir(\cos\theta + i\sin\theta)}$

$\Rightarrow \quad e^{iz} = e^{ir\cos\theta + i^2 r\sin\theta}$

$\Rightarrow \quad e^{iz} = e^{ir\cos\theta - r\sin\theta}$

$\Rightarrow \quad e^{iz} = e^{-r\sin\theta} \cdot e^{ir\cos\theta}$

$\therefore \quad \arg(e^{iz}) = r\cos\theta$

133. ***(d)*** With four 6-sided dics, there are possible combinations that can be rolled. Out of these, only 80 sum to 10 as the table shows,

$$6 \times 6 \times 6 \times 6 = 6^4 = 1296$$

Combinations	Count
$1-1-2-6$	$\frac{4!}{2!} = 12$
$1-1-3-5$	$\frac{4!}{2!} = 12$
$1-1-4-4$	$\frac{4!}{2!2!} = 6$
$1-2-2-5$	$\frac{4!}{2!} = 12$
$1-2-3-4$	$4! = 24$
$1-3-3-3$	$\frac{4!}{3!} = 4$
$2-2-2-4$	$\frac{4!}{3!} = 4$
$2-2-3-3$	$\frac{4!}{2!2!} = 6$
	Total Sum = 80

134. ***(b)*** A line passing through (1, 2) and parallel to the line $3x - y = 10$ is

$$3(x-1)-(y-2)=0$$

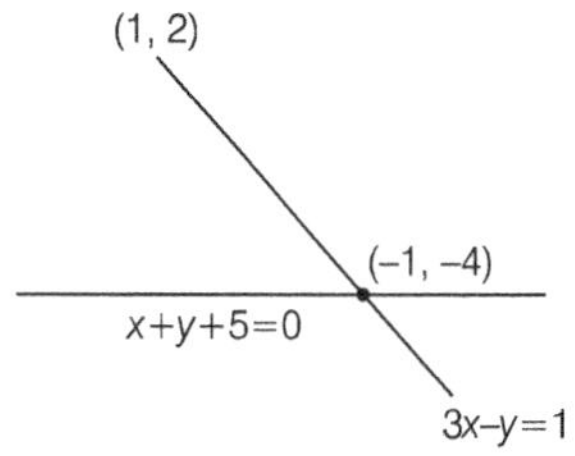

$\Rightarrow \quad 3x-3-y+2=0 \;\Rightarrow\; 3x-y-1=0$

The intersection point of line

$$x+y+5=0 \qquad \ldots\text{(i)}$$

and $\quad 3x-y-1=0 \qquad \ldots\text{(ii)}$

are $\quad (-1,-4)$

$\therefore$ Distance from the line $x+y+5$ to the point (1, 2) measured parallel to the line $3x-y=10$ is

$$\Rightarrow \quad \sqrt{(-1-1)^2+(-4-2)^2}$$

$$\Rightarrow \quad \sqrt{4+36}=\sqrt{40}=2\sqrt{10}$$

135. ***(d)*** We have,

$$f(x)=a_0+a_1x^2+a_2x^4+a_3x^6+\ldots+a_nx^{2n}$$

$$f'(x)=2a_1x+4a_2x^3+6a_3x^5+\ldots+2na_nx^{2n-1}$$

$$f'(x)=2x(a_1+2a_2x^2+3a_3x^4+\ldots+na_nx^{2n-2})$$

For maxima or minima

Put $f'(x)=0 \Rightarrow x=0$

$$f''(x)=2(a_1+6a_2x^2+15a_3x^4+\ldots+n(2n-1)a_nx^{2n-2})$$

$$(f''(x))_{x=0}=2a_1>0$$

$\therefore\; f(x)$ has only one minima at $x=0$.

136. ***(c)*** We have,

$$f(x)=2x^3+x^4+\log x$$

Put $\quad x=1$, we get

$\therefore \quad f(1)=3 \Rightarrow 1=f^{-1}(3)$

Now, given g is the inverse of f

$\therefore \quad g(x)=f^{-1}(x) \Rightarrow f(g(x))=x$

Differentiating with respect to x, we get

$$f'(g(x))\cdot g'(x)=1$$

$$g'(x)=\frac{1}{f'(g(x))}$$

$$g'(x)=\frac{1}{6(g(x))^2+4(g(x))^3+\dfrac{1}{g(x)}}$$

$$\left[\because f'(x)=6x^2+4x^3+\frac{1}{x}\right]$$

$$g'(3)=\frac{1}{6(g(3))^2+4(g(3))^3+\dfrac{1}{g(3)}}$$

$$g'(3)=\frac{1}{6+4+1} \qquad [\because g(3)=f^{-1}(3)=1]$$

$$g'(3)=\frac{1}{11}$$

137. ***(d)*** We have $P(3, 7, 1)$ and $R(2, 5, 7)$

DR's of $PR=(-1,-2,6)$

Angle between line PQ and plane is

$$\sin\theta=\frac{(-1)(3)+(-2)\times 2+6\times 11}{\sqrt{1+4+36}\,\sqrt{9+4+121}}$$

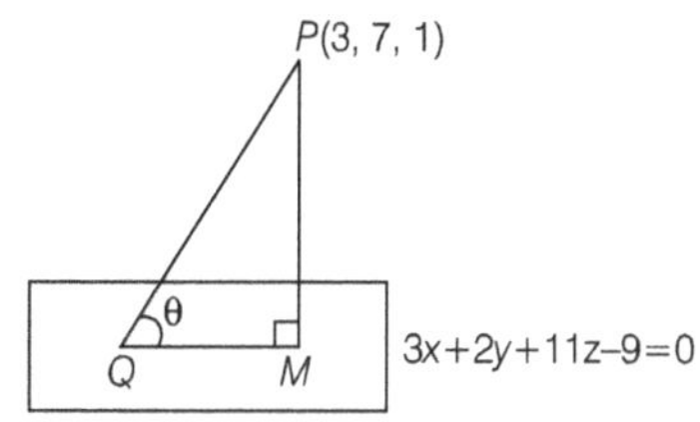

$$\sin\theta=\frac{59}{\sqrt{41}\,\sqrt{134}}$$

$$PM=\left|\frac{3(3)+2(7)+11(\times 1)-9}{\sqrt{9+4+121}}\right|$$

$$=\left|\frac{9+14+11-9}{\sqrt{134}}\right|=\frac{25}{\sqrt{134}}$$

$$\sin\theta=\frac{PM}{PQ}$$

$$\Rightarrow \quad PQ=\frac{PM}{\sin\theta}=\frac{25}{\sqrt{134}}\times\frac{\sqrt{41}\times\sqrt{134}}{59}=\frac{25\sqrt{41}}{59}$$

138. ***(a)*** We have,

$$I=\int\frac{8x^{43}+13x^{38}}{(x^{13}+x^5+1)^4}\,dx$$

$$=\int\frac{8x^{43}+13x^{38}}{x^{52}\left(1+\dfrac{1}{x^8}+\dfrac{1}{x^{13}}\right)^4}\,dx=\int\frac{8\left(\dfrac{1}{x^9}\right)+13\left(\dfrac{1}{x^{14}}\right)}{\left(1+\dfrac{1}{x^8}+\dfrac{1}{x^{13}}\right)^4}\,dx$$

Put, $\quad 1+\dfrac{1}{x^8}+\dfrac{1}{x^{13}}=t$

$$\Rightarrow \quad \left(\frac{-8}{x^9}-\frac{13}{x^{14}}\right)dx=dt$$

$$\therefore \quad I=-\int\frac{dt}{t^4}=\frac{1}{3t^3}+C$$

$$\Rightarrow I=\frac{1}{3\left(1+\dfrac{1}{x^8}+\dfrac{1}{x^{13}}\right)^3}+C=\frac{x^{39}}{3(x^{13}+x^5+1)^3}+C$$

BITSAT 2020

139. ***(b)*** We have,

Vector **r** is coplanar with vector **a** and **b**

$\therefore \quad \mathbf{r} = \lambda(\mathbf{b} \times (\mathbf{a} \times \mathbf{b}))$

$\Rightarrow \quad \mathbf{r} = \lambda((\mathbf{b}\cdot\mathbf{b})\mathbf{a} - (\mathbf{b}\cdot\mathbf{a})\mathbf{b})$

$\Rightarrow \quad \mathbf{r} = \lambda(|\mathbf{b}|^2\mathbf{a} - (\mathbf{a}\cdot\mathbf{b})\mathbf{b})$

$\Rightarrow \quad \mathbf{r} = \lambda[5\mathbf{a} + \mathbf{b}] \qquad [\because |\mathbf{b}| = \sqrt{2^2+1^2} = \sqrt{5}$

and $\quad \mathbf{a}\cdot\mathbf{b} = (-\hat{\mathbf{i}} + \hat{\mathbf{j}} + \hat{\mathbf{k}})\cdot(2\hat{\mathbf{i}} + \hat{\mathbf{k}}) = -1]$

Given, $\quad \mathbf{r}\cdot\mathbf{a} = 7$

$\therefore \quad \lambda(5\mathbf{a} + \mathbf{b})\cdot\mathbf{a} = 7$

$\Rightarrow \quad \lambda[5|\mathbf{a}|^2 + \mathbf{a}\cdot\mathbf{b}] = 7$

$\Rightarrow \quad \lambda[15 - 1] = 7 \quad [\because |a|^2 = 1+1+1 = 3]$

$\Rightarrow \quad \lambda = 1/2$

$\therefore \quad \mathbf{r} = \frac{1}{2}(5\mathbf{a} + \mathbf{b})$

$$\mathbf{r} = \frac{1}{2}(-5\hat{\mathbf{i}} + 5\hat{\mathbf{j}} + 5\hat{\mathbf{k}} + 2\hat{\mathbf{i}} + \hat{\mathbf{k}})$$

$$\mathbf{r} = \frac{-3}{2}\hat{\mathbf{i}} + \frac{5}{2}\hat{\mathbf{j}} + \frac{6}{2}\hat{\mathbf{k}}$$

z component of $\mathbf{r} = \frac{6}{2} = 3$

140. ***(b)*** Since x, y and z are three consecutive integers

$\therefore \quad 2y = x + z$

$\Rightarrow \quad (4y^2) = (x+z)^2$

$= (x-z)^2 + 4xz$

$= (-2)^2 + 4xz \qquad [\because x - z = -2]$

$\Rightarrow \quad 4y^2 = 4 + 4xz$

$\Rightarrow \quad y^2 = 1 + xz$

Now, $\frac{1}{2}\log_e x + \frac{1}{2}\log_e z + \frac{1}{2zx+1} + \frac{1}{3(2xz+1)^3} + \dots$

$$= \frac{1}{2}\left[\log_e(xz) + 2\left(\frac{1}{2zx+1} + \frac{1}{3}\cdot\frac{1}{(2zx+1)^3} + \dots\right)\right]$$

$$= \frac{1}{2}\left[\log_e(xz) + \log_e\left(\frac{1 + \frac{1}{2zx+1}}{1 - \frac{1}{2zx+1}}\right)\right]$$

$$= \frac{1}{2}\left[\log_e(xz) + \log_e\left(\frac{1+xz}{xz}\right)\right]$$

$$= \frac{1}{2}[\log_e(1+xz)] = \frac{1}{2}[\log_e(y)^2] = \log_e(y)$$

141. ***(a)*** We have, $\quad (xy^5 + 2y)dx = x\,dy$

$\Rightarrow \quad x\frac{dy}{dx} - 2y = xy^5$

$\Rightarrow \quad \frac{dy}{dx} - \frac{2y}{x} = y^5$

$\Rightarrow \quad y^{-5}\frac{dy}{dx} - \frac{2y^{-4}}{x} = 1 \qquad \dots(i)$

Put, $\quad y^{-4} = t$

$\Rightarrow \quad -4y^{-5}\frac{dy}{dx} = \frac{dt}{dx}$

$\Rightarrow \quad y^{-5}\frac{dy}{dx} = \frac{-1}{4}\frac{dt}{dx} \qquad \dots(ii)$

From Eqs. (i) and (ii), we get

$$-\frac{1}{4}\frac{dt}{dx} - \frac{2t}{x} = 1$$

$\Rightarrow \quad \frac{dt}{dx} + \frac{8t}{x} = -4$

Now, IF $= e^{\int \frac{8}{x}dx} = e^{8\log x} = x^8$

and the solution is $t\cdot x^8 = \int(-4)x^8 dx + C$

$\Rightarrow \quad \frac{x^8}{y^4} = -\frac{4\cdot x^9}{9} + C$

$\Rightarrow \quad 9x^8 + 4x^9\cdot y^4 = 9y^4 C$

142. ***(b)*** Given, $\frac{|x-2|-1}{|x-2|-2} \le 0$

Let $|x-2| = k$

Then, given equation,

$$\frac{k-1}{k-2} \le 0 \Rightarrow \frac{(k-1)(k-2)}{(k-2)^2} \le 0$$

$\Rightarrow \quad (k-1)(k-2) \le 0 \Rightarrow 1 \le k \le 2$

$\Rightarrow \quad 1 \le |x-2| \le 2$

Case I When $1 \le |x-2|$

$\Rightarrow \quad |x-2| \ge 1 \Rightarrow x - 2 \ge 1$ or $x - 2 \le -1$

$\Rightarrow \quad x \ge 3$ and $x \le 1 \qquad \dots(i)$

Case II When $|x-2| \le 2$

$\Rightarrow \quad -2 \le x - 2 \le 2 \Rightarrow -2+2 \le x \le 2+2$

$\Rightarrow \quad 0 \le x \le 4 \qquad \dots(ii)$

From Eqs. (i) and (ii), $x \in [0, 1] \cup [3, 4]$

143. ***(b)*** We have, $f(x) = \frac{x}{\sqrt{1+x^2}}$

$$\Rightarrow f(f(x)) = \frac{f(x)}{\sqrt{1+(f(x))^2}} = \frac{\frac{x}{\sqrt{1+x^2}}}{\sqrt{1+\frac{x^2}{1+x^2}}} = \frac{x}{\sqrt{1+2x^2}}$$

Similarly, $f(f(f(x))) = \frac{x}{\sqrt{1+3x^2}}$

$\vdots$

$$\underbrace{fofo\dots of}_{n \text{ times}}(x) = \frac{x}{\sqrt{1+nx^2}} = \frac{x}{\sqrt{1+\left(\sum_{r=1}^{n}1\right)x^2}}$$

144. ***(b)*** We have,

$$\log_5\left(\frac{a+b}{3}\right)=\frac{\log_5 a+\log_5 b}{2}$$

$$\Rightarrow \quad 2\log_5\left(\frac{a+b}{3}\right)=\log_5 ab$$

$$\Rightarrow \quad \log_5\left(\frac{a+b}{3}\right)^2=\log_5 ab$$

$$\Rightarrow \quad \frac{(a+b)^2}{9}=ab \Rightarrow (a+b)^2=9ab$$

$$\Rightarrow \quad a^2+b^2=7ab$$

$$\Rightarrow \quad \frac{a}{b}+\frac{b}{a}=7$$

$$\Rightarrow \quad \left(\frac{a}{b}+\frac{b}{a}\right)^2=49$$

$$\Rightarrow \quad \frac{a^2}{b^2}+\frac{b^2}{a^2}+2=49$$

$$\Rightarrow \quad \frac{a^4+b^4}{a^2b^2}=49-2=47$$

145. ***(a)*** We have,

$$\frac{5}{4}\cos^2 2x+\cos^4 x+\sin^4 x+\cos^6 x+\sin^6 x=2$$

$$\Rightarrow \frac{5}{4}\cos^2 2x+(\cos^2 x+\sin^2 x)^2-2\sin^2 x\cos^2 x+(\cos^2 x+\sin^2 x)^3-3\sin^2 x\cos^2 x(\sin^2 x+\cos^2 x)=2$$

$$\Rightarrow \frac{5}{4}\cos^2 2x+1-2\sin^2 x\cos^2 x+1-3\sin^2 x\cos^2 x=2$$

$$\Rightarrow \frac{5}{4}\cos^2 2x-\frac{1}{2}\sin^2 2x-\frac{3}{4}\sin^2 2x=0$$

$$\Rightarrow \frac{5}{4}(\cos^2 2x-\sin^2 2x)=0$$

$$\Rightarrow \cos 4x=0$$

$$\Rightarrow 4x=2n\pi\pm\frac{\pi}{2}, n\in Z$$

$$\Rightarrow x=(4n\pm 1)\frac{\pi}{8}, n\in Z$$

$$\Rightarrow x=\frac{\pi}{8},\frac{3\pi}{8},\frac{5\pi}{8},\frac{7\pi}{8},\frac{9\pi}{8},\frac{11\pi}{8},\frac{13\pi}{8},\frac{15\pi}{8}$$

So, number of solution = 8

146. ***(a)*** Given equation of ellipse,

$$16x^2+11y^2=256 \quad \Rightarrow \frac{x^2}{16}+\frac{11y^2}{256}=1$$

Equation of tangent of ellipse at $\left(4\cos\phi, \frac{16}{\sqrt{11}}\sin\phi\right)$ is

$$\frac{4x\cos\phi}{16}+\frac{11\times 16y}{256\sqrt{11}}\sin\phi=1$$

$$\Rightarrow \quad \frac{x\cos\phi}{4}+\frac{\sqrt{11}}{16}\sin\phi\, y=1$$

is also a tangent of circle $x^2+y^2-2x=15$

$$\Rightarrow \quad (x-1)^2+y^2=16$$

radius of circle = 4 and centre = (1, 0)

Now, $\frac{x\cos\phi}{4}+\frac{\sqrt{11}}{16}\sin\phi\, y=1$ is tangent of circle

$$\therefore \; 4=\left|\frac{\frac{\cos\phi}{4}-1}{\sqrt{\frac{\cos^2\phi}{16}+\frac{11}{256}\sin^2\phi}}\right| \Rightarrow \phi=\pm\frac{\pi}{3}$$

$$\therefore \quad \phi=\frac{\pi}{3}$$

147. ***(b)*** Given equation of parabola is

$$x=4y-y^2$$

$$\Rightarrow \quad y^2-4y+4=-x+4$$

$$\Rightarrow \quad (y-2)^2=-(x-4)$$

$$a=\frac{1}{4}$$

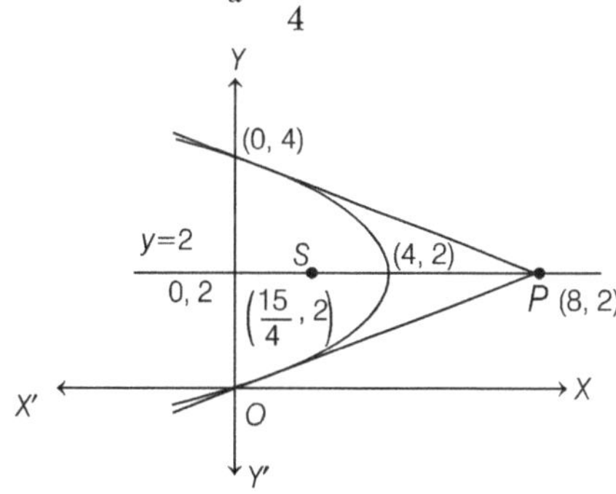

$$\text{Focus}=\left(\frac{15}{4}, 2\right)$$

Parabola cuts Y-axis at (0, 0) and (0, 4).

Now, equation of tangent at (0, 0) is

$$x=4y \qquad \text{...(i)}$$

and equation of tangent at (0, 4) is

$$x+4y=16 \qquad \text{...(ii)}$$

Solving Eqs. (i) and (ii), we get

$$x=8, y=2$$

∴ Intersection point of tangent is $P(8, 2)$.

Distance from point P to focus S

$$\text{Distance}=\sqrt{\left(8-\frac{15}{4}\right)^2+(2-2)^2}$$

$$=8-\frac{15}{4}=\frac{17}{4}$$

BITSAT 2020

148. ***(d)*** We have, $S_\infty = \sum_{k=1}^{\infty}\sum_{n=1}^{\infty}\frac{k}{2^{n+k}}$

$$S_\infty = \sum_{k=1}^{\infty}\frac{k}{2^k}\cdot\sum_{n=1}^{\infty}\frac{1}{2^n}$$

$$S_\infty = \sum_{k=1}^{\infty}\frac{k}{2^k}\left(\frac{1}{2}+\frac{1}{2^2}+\frac{1}{2^3}+..\right)$$

$$S_\infty = \sum_{k=1}^{\infty}\frac{k}{2^k}\left[\frac{1}{2}\left(1+\frac{1}{2}+\frac{1}{2^2}\cdots\right)\right]$$

$$S_\infty = \sum_{k=1}^{\infty}\frac{k}{2^k}\left(\frac{1}{2}\times\frac{1}{1-\frac{1}{2}}\right) = \sum_{k=1}^{\infty}\frac{k}{2^k}$$

$$S_\infty = \frac{1}{2}+\frac{2}{2^2}+\frac{3}{2^3}+\frac{4}{2^4}+\ldots \quad \ldots(i)$$

$$\frac{1}{2}S_\infty = \frac{1}{2^2}+\frac{2}{2^3}+\frac{3}{2^4}+\ldots \quad \ldots(ii)$$

From Eqs. (i) and (ii), we get

$$\frac{1}{2}S_\infty = \frac{1}{2}+\frac{1}{2^2}+\frac{1}{2^3}+\ldots$$

$$\frac{1}{2}S_\infty = \frac{1}{2}\left(\frac{1}{1-\frac{1}{2}}\right) = 1$$

$$S_\infty = 2$$

149. ***(a)*** We have,

Slope of tangent of the curve is $\frac{(x+1)^2+y-3}{x+1}$

$$\therefore \quad \frac{dy}{dx} = \frac{(x+1)^2+y-3}{x+1}$$

$$\Rightarrow \quad \frac{dy}{dx} = \frac{(x+1)^2-3}{x+1}+\frac{y}{x+1}$$

$$\Rightarrow \quad \frac{dy}{dx}-\frac{y}{x+1} = \frac{(x+1)^2-3}{x+1}$$

This is a form of linear differential equations

$$\therefore \quad \text{IF} = e^{\int\frac{-1}{x+1}dx} = e^{\log\frac{1}{1+x}} = \frac{1}{1+x}$$

Solution of given differential equation is

$$\frac{y}{x+1} = \int\frac{(x+1)^2-3}{(x+1)^2}dx$$

$$\frac{y}{x+1} = \int\left(1-\frac{3}{(x+1)^2}\right)dx$$

$$\frac{y}{x+1} = x+\frac{3}{x+1}+c$$

$\because$ Curve passes through (2, 0)

$$\therefore \quad 0 = 2+\frac{3}{3}+c$$

$$\Rightarrow \quad c = -3$$

Hence, equation of curve is

$$\frac{y}{x+1} = x+\frac{3}{x+1}-3$$

$$y = x^2+x+3-3x-3$$

$$y = x^2-2x$$

150. ***(a)*** We have,

$$A = \begin{bmatrix}2 & 1\\1 & 2\end{bmatrix}$$

We know that, $|A-\lambda I| = 0$

$$\left|\begin{bmatrix}2 & 1\\1 & 2\end{bmatrix}-\begin{bmatrix}\lambda & 0\\0 & \lambda\end{bmatrix}\right| = 0 = \begin{vmatrix}2-\lambda & 1\\1 & 2-\lambda\end{vmatrix} = 0$$

$$\Rightarrow \quad (2-\lambda)^2-1 = 0$$

$$\Rightarrow \quad 4-4\lambda+\lambda^2-1 = 0$$

$$\Rightarrow \quad \lambda^2-4\lambda+3 = 0$$

$$\therefore \quad A^2-4A+3I = 0$$

Multiply by A^{-1}, we get

$$A-4I+3A^{-1} = 0$$

$$\Rightarrow \quad A^{-1} = \frac{-1}{3}A+\frac{4}{3}I$$

$$\therefore \quad \alpha = -\frac{1}{3} \text{ and } \beta = \frac{4}{3}$$

$$\alpha+\beta = -\frac{1}{3}+\frac{4}{3} = \frac{3}{3} = 1$$

BITSAT

Birla Institute of Technology & Science Admission Test

Solved Paper 2019*

Instructions

- There are 150 questions in all. The number of questions in each part is as given below.

	No. of Questions
Part I Physics	1-40
Part II Chemistry	41-80
Part III a. English Proficiency	81-95
b. Logical Reasoning	96-105
Part IV Mathematics	106-150

- All questions are Multiple Choice Questions having four options out of which **only one** is correct.
- Each correct answer fetches 3 marks while incorrect answer has a penalty of 1 mark.
- Time allotted to complete this paper is 3 hrs.

PART I

Physics

1. A radioactive sample of half-life 23.1 days is disintegrating continuously. The percentage decay of it in 15th to 16th days will be [Take, $e^{0.03} = 1.03$]

a. 5% *b.* 1%
c. 2.9% *d.* 3.5%

2. Two seperate soap bubbles of radii 3×10^{-3} m and 2×10^{-3} m respectively, formed of same liquid (surface tension $= 6.5 \times 10^{-2}$ N/m) come together to form a double bubble. The radius of interface of double bubble is

a. 6×10^{-3} m *b.* 4×10^{-3} m
c. 1.5×10^{-3} m *d.* 0.66×10^{-3} m

3. An electron revolves in a circular orbit of radius r with angular speed ω. The magnetic field at the centre of electron orbit is

a. $\dfrac{\mu_0 e\omega}{\pi r}$ *b.* $\dfrac{\mu_0 e\omega}{4\pi r}$
c. $\dfrac{\mu_0 e^2\omega}{4\pi r}$ *d.* $\dfrac{\mu_0 e\omega}{2\pi r}$

4. A steel wire of cross-sectional area 4 cm^2 has elastic limit of 2.2×10^8 N/m^2. The maximum upward acceleration that can be given to a 1000 kg elevator supported by this steel wire if the stress is to exceed one-fourth of the elastic limit is [Take, $g = 10$ m/s^2]

a. 10 m/s^2 *b.* 9 m/s^2 *c.* 15 m/s^2 *d.* 12 m/s^2

5. An ideal monoatomic gas at 300 K expands adiabatically to twice its volume. The final temperature of gas is

a. $300\sqrt{2}$ *b.* $300\sqrt{3}$ *c.* $300\left(\dfrac{1}{2}\right)^{2/3}$ *d.* $300(2)^{2/3}$

6. Two sound producing sources A and B are moving towards and away from a stationary observer with same speed respectively. If frequency of sound produced by both sources are equal as 400 Hz, then speed of sources (approximately) when observer detects 4 beats per second, is [Given, speed of sound = 340 m/s]

a. 1.7 m/s *b.* 3.4 m/s *c.* 2.4 m/s *d.* 1 m/s

*Memory Based

7. Pressure *versus* temperature graph of an ideal gas is shown in the given figure. Density of gas at point A is ρ_0, then density of gas at point B will be

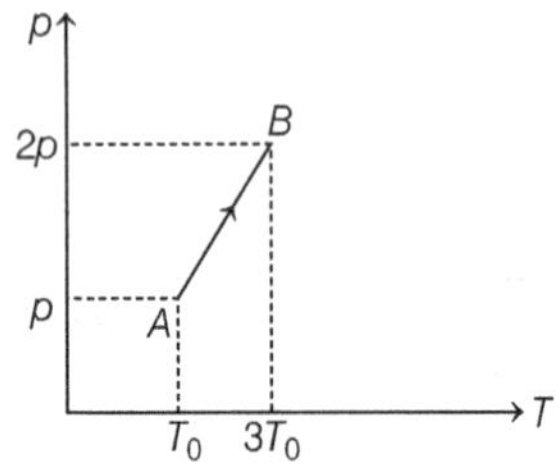

a. $2\rho_0$ *b.* $\frac{2}{3}\rho_0$ *c.* $\frac{3}{2}\rho_0$ *d.* $3\rho_0$

8. A rubber cord has a cross-sectional area 10^{-6} m^2 and total unstretched length 0.1 m. It is stretched to 0.125 m and then released to project a particle of mass 5.0 g. The velocity of projection is [Given, Young's modulus of rubber, $Y = 5 \times 10^8$ N/m^2]

a. 45 m/s *b.* 30 m/s
c. 25 m/s *d.* 15 m/s

9. A body is moving unidirectionally under the influence of a sources of constant power. The square of its displacement in time t is proportional to

a. t^3 *b.* t^2
c. t *d.* $\sqrt{t}$

10. Two electrons are moving with speed of 5×10^5 m/s parallel to each other, then the electrostatic and magnetic force between them is

a. 5×10^5 *b.* 2.5×10^4
c. 3.6×10^5 *d.* 4.4×10^3

11. In the following circuit diagram, the current through battery is

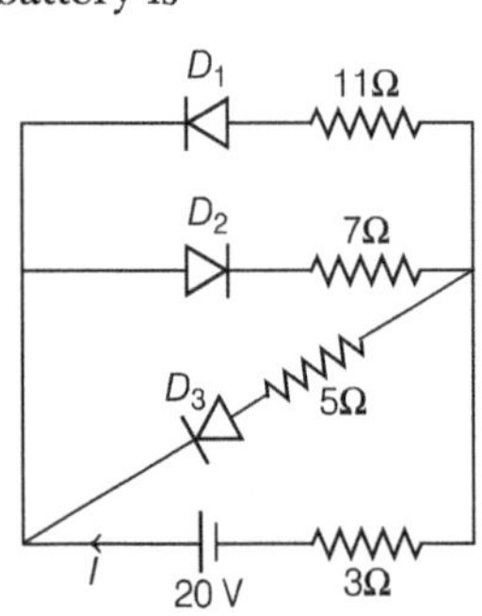

a. 4 A *b.* 3 A
c. 1 A *d.* 2 A

12. A semicircular disc of mass M and radius R is free to rotate about its diameter. The moment of inertia of semicircular disc about a line perpendicular to its plane through centre is

a. $\frac{3}{4}MR^2$ *b.* $\frac{MR^2}{2}$ *c.* $\frac{MR^2}{3}$ *d.* $\frac{MR^2}{4}$

13. 12 eV energy is given to electron in third orbit of H-atom, then final energy of electron when it ionise from third orbit, is

a. 5.25 eV *b.* 10.49 eV
c. 12 eV *d.* 11.15 eV

14. In Young's double slit experiment, the distance between slits and screen is 2 m and distance between slits is 0.25 mm. A light of wavelength 800 nm is used to find fringes on the screen. If screen moves with a speed of 5 m/s, then first maxima will move with a speed of

a. 16 mm/s *b.* 8 mm/s
c. 20 m/s *d.* 50 mm/s

15. The magnetic field at the point O in the following current carrying square loop at the centre will be

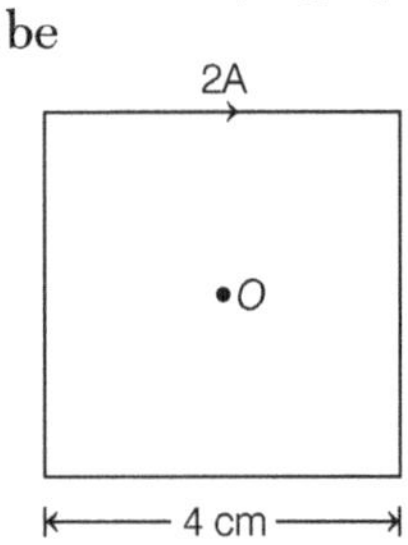

a. 12×10^{-5} T *b.* 4×10^{-5} T
c. $4\sqrt{2} \times 10^{-5}$ T *d.* $2\sqrt{2} \times 10^{-5}$ T

16. The value of current I as shown in the given circuit diagram is

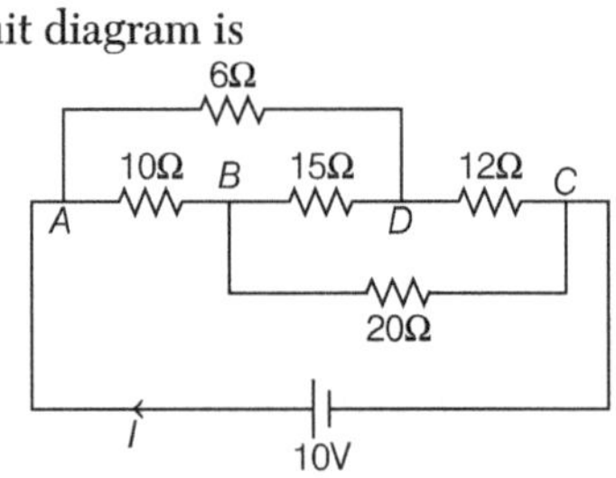

a. 2 A *b.* 1.5 A *c.* 0.9 A *d.* 0.2 A

17. A semicircular wire of radius r rotates in uniform magnetic field B about its diameter with angular velocity ω. If the total resistance of the circuit is R, then the mean power generated per period of rotation is

a. $\frac{(\pi Br^2\omega)^2}{8R}$ *b.* $\frac{\pi Br\omega^2}{8R}$ *c.* $\frac{(\pi Br\omega)^2}{8R}$ *d.* Zero

18. A convex lens of focal length 25 cm produces images of the same magnification 2, when an object is kept at two positions x_1 and x_2 $(x_1 > x_2)$ from the lens. The ratio of x_2 and x_1 is

a. 2 : 1 *b.* 1 : 3 *c.* 1 : 2 *d.* 3 : 1

19. A particle x of mass m and initial velocity u collide with another particle y of mass $\frac{3m}{4}$ which is at rest. The collision is head on and perfectly elastic. The ratio of de-Broglie wavelengths λ_y and λ_x after the collision is

a. 4 : 3 *b.* 2 : 32 *c.* 3 : 4 *d.* 3 : 32

20. The dimensional formula of $\sqrt{\frac{\mu_0}{\varepsilon_0}}$ is

a. $[ML^2T^{-3}A^2]$ *b.* $[M^0LT^{-1}A^0]$

c. $[ML^2T^{-3}A^{-2}]$ *d.* $[M^{-1}L^{-2}T^3A^2]$

21. A body of mass m is moving in a straight line with momentum p. Starting at time $t = 0$, a force $F = at$, acts in the same direction on the moving particle during time interval of T. So that its momentum changes from p to $2p$. The value of T is

a. $\sqrt{\frac{2p}{a}}$ *b.* $\sqrt{\frac{p}{a}}$

c. $2\sqrt{\frac{2p}{a}}$ *d.* $\frac{2p}{a}$

22. Unless a maximum force of 3 N is applied on a body kept on rough inclined plane of shown in the figure, then body remains stationary. Also when the maximum external force up the inclined plane that does not move the block is 12 N. The coefficient of static friction between the block and the plane is [Take, $g = 10\ m/s^2$]

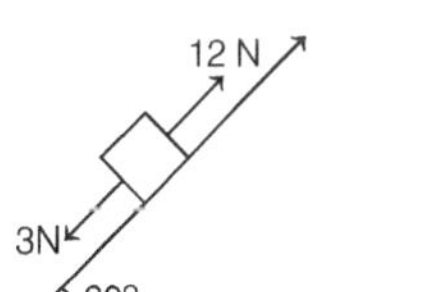

a. $\frac{\sqrt{3}}{5}$ *b.* $\frac{1}{\sqrt{3}}$ *c.* $\frac{3\sqrt{3}}{5}$ *d.* $\frac{5}{3\sqrt{3}}$

23. In the shown situation, if middle portion of the lens is painted black, then

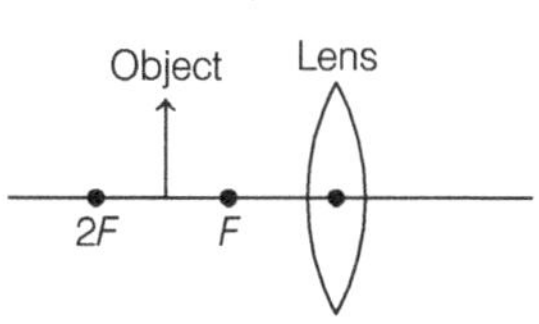

a. complete image of the object will form with low intensity

b. complete image of the object will form with high intensity

c. incomplete image of the object will form with low intensity

d. incomplete image of the object will form with high intensity

24. If sound travels in air with the speed of 340 m/s, then number of tones present in an open organ pipe of length 2 m and its maximum frequency 1200 Hz, are

a. 17 *b.* 11 *c.* 9 *d.* 14

25. A proton is revolving on a circular path of radius 2 mm with frequency 10 Hz. Magnetic dipole moment associated with proton is

a. 2×10^{-24} A-m^2 *b.* 4×10^{-24} A-m^2

c. 3×10^{-20} A-m^2 *d.* 6×10^{-20} A-m^2

26. When a capacitor is fully charged as shown in the following figure, then current drawn from the cell is

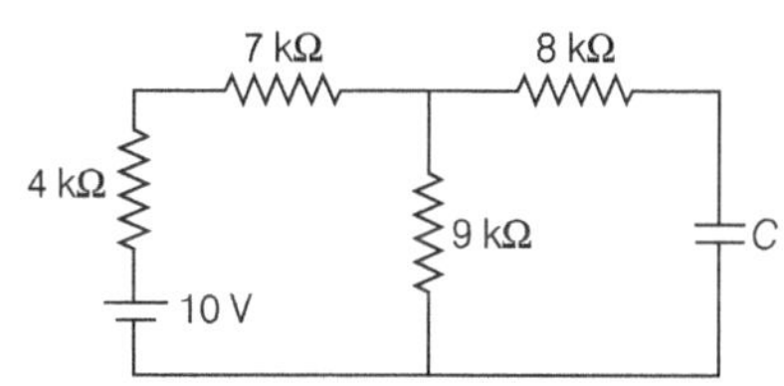

a. 1 mA *b.* 0.5 mA *c.* 2 mA *d.* 5 mA

27. A charged capacitor of capacitance C is discharging through a resistor of resistance R. At what time the charge on the capacitor is equal to one half of its initial value?

a. $\frac{RC}{2}$ *b.* $\frac{RC}{\log_e 2}$

c. $RC \log_e 2$ *d.* $\frac{1}{RC}\log_e 2$

28. The escape velocity of the earth surface is v_e. A body is projected with velocity $3v_e$. With what constant velocity will it move in the inter planetary space?

a. v_e *b.* $3v_e$ *c.* $\sqrt{2}\, v_e$ *d.* $2\sqrt{2}\, v_e$

29. The time period of a bob performing simple harmonic motion in water is 2s. If density of bob is $\frac{4}{3} \times 10^3$ kg/m^3, then time period of bob performing simple harmonic motion in air will be

a. 3s *b.* 4s *c.* 2s *d.* 1s

BITSAT 2019

30. A uniform solid cylindrical roller of mass m is being pulled on horizontal surface with force F parallel to the surface applied at its centre. If the acceleration of the cylinder is a and it is rolling without slipping, then the value of F is

a. $2ma$ **b.** $3\,ma$ **c.** $\frac{3ma}{2}$ **d.** $\frac{5\,ma}{2}$

31. A projectile is given an initial velocity of $(\hat{\mathbf{i}}+\sqrt{3}\,\hat{\mathbf{j}})$ m/s, where $\hat{\mathbf{i}}$ is along the ground and $\hat{\mathbf{j}}$ is along the vertical. Then, the equation of the path of projectile is
[Take, $g = 10\,\text{m/s}^2$]

a. $y=\sqrt{3}x-5x^2$ **b.** $y=\sqrt{3}\,x+5x^2$
c. $x=\sqrt{3}\,y+5x^2$ **d.** $x^2=y^2+\sqrt{3}$

32. The force on a body of mass 1 kg is $(20\hat{\mathbf{i}}+10\hat{\mathbf{j}})$ N. If is starts from rest, then the position of the body at time $t = 2$ s, is

a. $-20\hat{\mathbf{i}}-40\hat{\mathbf{j}}$ **b.** $20\hat{\mathbf{i}}-40\hat{\mathbf{j}}$
c. $40\hat{\mathbf{i}}-20\hat{\mathbf{j}}$ **d.** $40\hat{\mathbf{i}}+20\hat{\mathbf{j}}$

33. A cylinder of mass 2 kg is released from rest from the top of an inclined plane of inclination 30° and length 1m. If the cylinder rolls without slipping, then its speed when it reaches the bottom, is [Take, $g = 10\,\text{m/s}^2$]

a. $\frac{20}{3}$ m/s **b.** $\sqrt{\frac{20}{3}}$ m/s **c.** $\frac{10}{3}$ m/s **d.** $\sqrt{\frac{10}{3}}$ m/s

34. Frequency of oscillation of a body is 5 Hz when a force F_1 is applied and 12 Hz when another force F_2 is applied. If both forces F_1 and F_2 are applied together, then frequency of oscillation of the body will be

a. 13 Hz **b.** 169 Hz **c.** 62 Hz **d.** 52 Hz

35. The figure shows the graph of $\frac{pV}{T}$ *versus* p for 2×10^{-4} kg of hydrogen gas at two different temperatures, where p, V and T represents pressure, volume and temperature respectively. Then, the value of $\frac{pV}{T}$, where the curve meet on the vertical axis, is

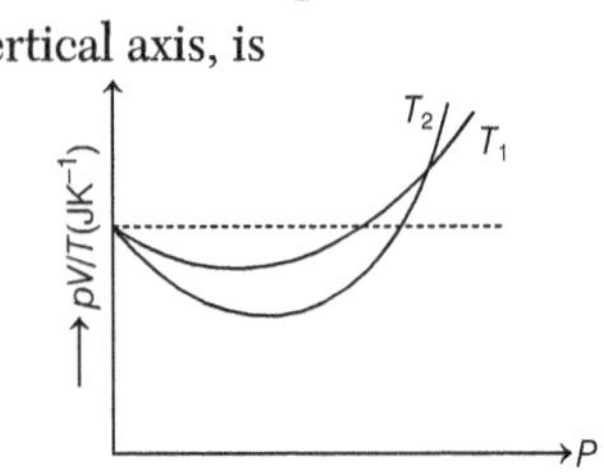

a. $\frac{R}{5}$ **b.** $\frac{2R}{3}$ **c.** $\frac{R}{10}$ **d.** $\frac{R}{15}$

36. An ammeter of resistance 1.5 Ω can measure currents upto 1 A. The value of shunt resistance to measure current upto 4 A is

a. 0.5 Ω **b.** 1 Ω
c. 1.5 Ω **d.** 0.2 Ω

37. In the following figures, which represents the variation of the electric field with distance from the centre of a uniformly charged non-conducting spheres of radius R?

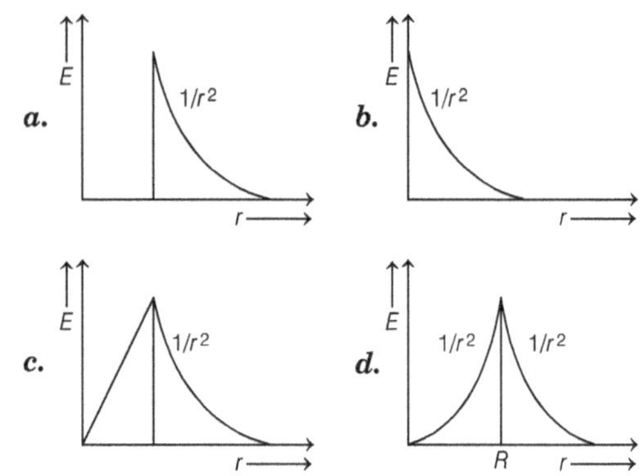

38. A carbon resistor of (56 ± 5.6) kΩ is to be marked with rings of different colours for its identification. The colour code sequence will be

a. green, blue, orange, silver
b. blue, yellow, orange, silver
c. green, red, orange, silver
d. red, green, blue, gold

39. Two monochromatic light waves are travelling with same frequency and constant phase different. If both waves interfere, then

a. there is a gain of energy
b. there is a loss of energy
c. the energy is redistributed and the distribution remains constant in time
d. the energy is redistributed and the distribution changes with time

40. In a coil of resistance 50 Ω, the induced current developed by changing magnetic flux through it, is shown in figure as a function of time. The magnitude of change in flux through the coil is

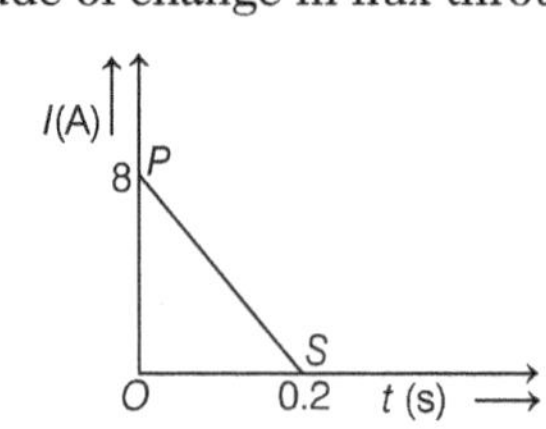

a. 10 Wb **b.** 20 Wb
c. 32 Wb **d.** 40 Wb

PART II

Chemistry

41. In a compound, atoms of elements Y form ccp lattice and those of element X occupy 2/3 rd of tetrahedral voids. The formula of the compound can be

a. X_4Y_3 **b.** X_2Y_3
c. X_2Y **d.** X_3Y_4

42. The density of a certain gas at 30°C and 768 torr is 1.35 kg/m^3, then density of the gas at STP is

a. 1.48 kg/m^3 **b.** 1.27 kg/m^3
c. 1.35 kg/m^3 **d.** 1.00 kg/m^3

43. For the octahedral complex, which of the following d-electron configuration will give maximum value of CFSE?

a. High spin, d^6 **b.** Low spin, d^4
c. Low spin, d^5 **d.** High spin, d^7

44. The wavelength corresponding to maximum energy for hydrogen is 91.2 nm. The corresponding wavelength for He^+ ion is

a. 2.28 nm **b.** 22.8 nm
c. 182.4 nm **d.** 364.8 nm

45. The heat of reaction for
$C_{18}H_8(s) + 12\,O_2(g) \longrightarrow 10CO_2(g) + 4H_2O(l)$
at constant volume is – 1228.2 kcal at 25°C. The heat of reaction at constant pressure and same temperature is

a. – 1228.2 kcal **b.** – 1229.3 kcal
c. – 1232.9 kcal **d.** –1242.6 kcal

46. Given the hypothetical reaction mechanism

$A \xrightarrow{\text{I}} B \xrightarrow{\text{II}} C \xrightarrow{\text{III}} D \xrightarrow{\text{IV}} E$ and the rate as

Species formed	Rate of its formation
B	0.002 mol/h per mole of A
C	0.030 mol/h per mole of B
D	0.011 mol/h per mole of C
E	0.420 mol/h per mole of D

The rate determining step is

a. step I **b.** step II **c.** step III **d.** step IV

47. For the reaction

$C_6H_5NH_2 \xrightarrow[180°\,C]{H_2SO_4} \underset{(para)}{H_2NC_6H_4(SO_3H)}$

the true statement about the product is

a. it does not exist as Zwitter ion
b. —NH_2 group displays a powerful basic character
c. it does not act as inner salt
d. —SO_3H, diminishes the basic character of —NH_2

48. The hybridisation of phosphorus in PO_4^{3-} is

a. sp **b.** sp^2
c. sp^3 **d.** sp^3d

49. Which of the following lanthanoid is diamagnetic?

[Atomic numbers are Sm = 62, Ce = 58, Eu = 63, Yb = 70]

a. Sm^{2+} **b.** Eu^{2+}
c. Yb^{2+} **d.** Ce^{2+}

50. Which of the following is/are aromatic alcohol?

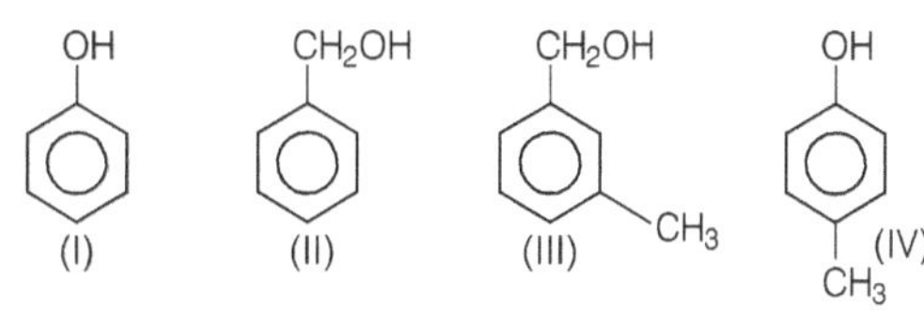

a. I,II,III,IV **b.** II and III
c. I and IV **d.** Only I

51. Among the following substituted silanes, the one which will give rise to cross-linked silicon polymer on hydrolysis?

a. R_4Si **b.** $RSiCl_3$
c. R_2SiCl **d.** R_3SiCl

52. For the chemical reaction,
$2O_3 \longrightarrow 3O_2$, the reaction proceeds as follows

$$O_3 \rightleftharpoons O_2 + [O]\ \text{(Fast)}$$
$$[O] + O_3 \longrightarrow 2O_2,\ \text{(Slow)}$$

the rate law expression should be given as

a. $r = k[O_3]^2[O_2]^{-1}$ **b.** $r = k[O_3]^2$
c. $r = k[O_3][O_2]$ **d.** unpridictable

53. An aqueous solution freezes at 272.4 K, while pure water freezes at 273 K, given $K_f = 1.86$ K kg mol^{-1}, and $K_b = 0.512$ K kg mol^{-1}; the molality of solution and boiling point of solution respectively will be

a. 0.322 and 373.16 K
b. 0.222 and 273.15 K
c. 0.413 and 400 K
d. 0.5 and 300.73 K

54. The order of acidic strength of boron trihalides

a. $BF_3 < BCl_3 < BBr_3 < BI_3$
b. $BI_3 < BBr_3 < BCl_3 < BF_3$
c. $BCl_3 < BBr_3 < BI_3 < BF_3$
d. $BBr_3 < BCl_3 < BF_3 < BI_3$

55. The correct order of mobility of alkali metal ions in aqueous solution is

a. $Li^+ > Na^+ > K^+ > Rb^+$
b. $Na^+ > K^+ > Rb^+ > Li^+$
c. $K^+ > Rb^+ > Na^+ > Li^+$
d. $Rb^+ > K^+ > Na^+ > Li^+$

56. Thermosetting polymer bakelite is formed by the reaction of phenol with

a. CH_3CH_2CHO *b.* CH_3CHO
c. HCHO *d.* HCOOH

57. The drug that is antipyretic as well as analgesic is

a. chloroquine
b. penicillin
c. paracetamol
d. chloropromazine hydrochloride

58. The correct statement about the following disaccharide is

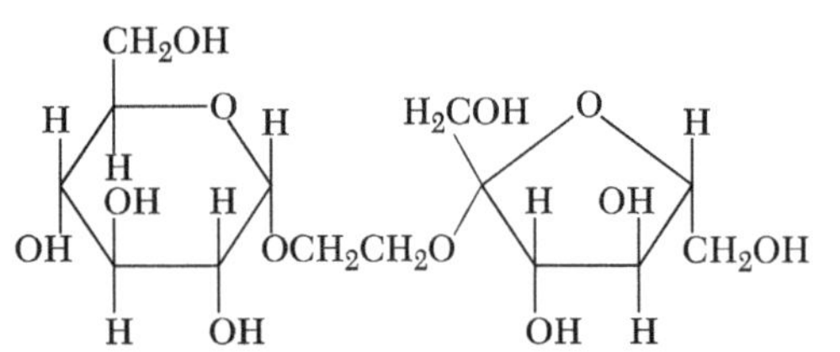

I II

a. Ring (I) is pyranose with α-glycosidic link
b. Ring (I) is furanose with α-glycosidic link
c. Ring (II) is furanose with α-glycosidic link
d. Ring (II) is pyranose with α-glycosidic link

59. If one strand of DNA has sequence ATCGTATG, the sequence in the complementary strand would be

a. TAGCTTAC *b.* TCACATAC
c. TAGCATAC *d.* TACGATAC

60. Which of the following salt would give SO_2 with hot and dil. H_2SO_4 and also decolourises Br_2 water?

a. Na_2SO_3 *b.* $NaHSO_4$ *c.* Na_2SO_4 *d.* Na_2S

61. In the reaction (*X*) and (*Y*) are respectively.

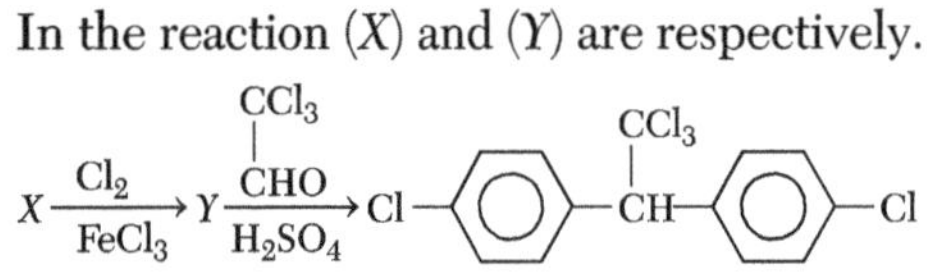

a. 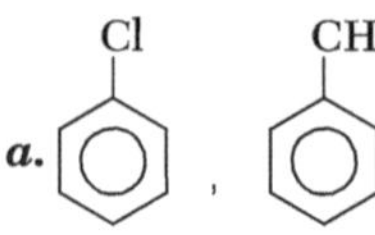
b. 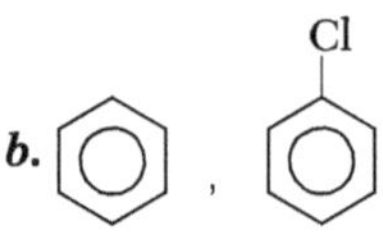
c. 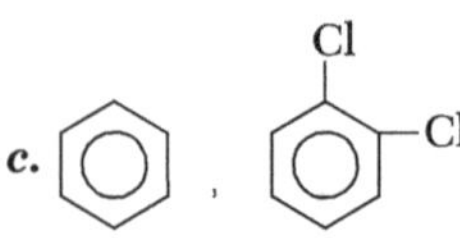
d. 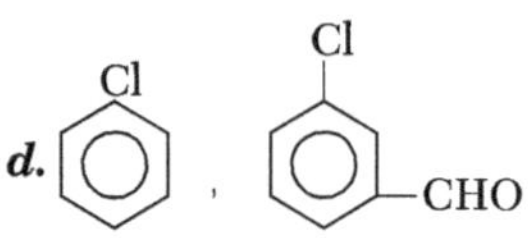

62. Which of the following reactions is an example of use of water gas in the synthesis of other compounds?

a. $CH_4(g) + H_2O(g) \xrightarrow[Ni]{1270\ K} CO(g) + H_2(g)$
b. $CO(g) + H_2O(g) \xrightarrow[Catalyst]{673\ K} CO_2(g) + H_2(g)$
c. $C_nH_{2n+2} + nH_2O(g) \xrightarrow[Ni]{1270\ K} nCO + (2n+1)H_2$
d. $CO(g) + 2H_2(g) \xrightarrow[Catalyst]{Cobalt} CH_3OH(l)$

63. Which of the following compounds, on reaction with NaOH and Na_2O_2 gives yellow colour?

a. $Zn(OH)_2$ *b.* $Al(OH)_3$
c. $Cr(OH)_3$ *d.* $CaCO_3$

64. For the reaction,

$$R{-}C(=O){-}Z + NCl^- \xrightarrow{H_2O} R{-}C(=O){-}NCl + Z^-$$

Rate of reaction is faster, when Z is

a. Cl *b.* NH_2
c. OC_2H_5 *d.* $OCOCH_3$

65. E_1, E_2 and E_3 are the emf of the following three galvanic cells respectively

(i) $Zn(s)|Zn^{2+}(0.1\ M)||Cu^{2+}(1M)|Cu(s)$
(ii) $Zn(s)|Zn^{2+}(1M)||Cu^{2+}(1M)|Cu(s)$
(iii) $Zn(s)|Zn^{2+}(1M)||Cu^{2+}(0.1M)|Cu(s)$

Which of the following is true?

a. $E_2 > E_1 > E_3$
b. $E_1 > E_2 > E_3$
c. $E_3 > E_1 > E_2$
d. $E_3 > E_2 > E_1$

66. Sodium nitroprusside, when added to an alkaline solution of sulphide ions, produce a

a. red colouration
b. blue colouration
c. purple colouration
d. brown colouration

67. For a reverse reaction, $A \rightleftharpoons B$, which one of the following statement is wrong from the given energy profile diagram?

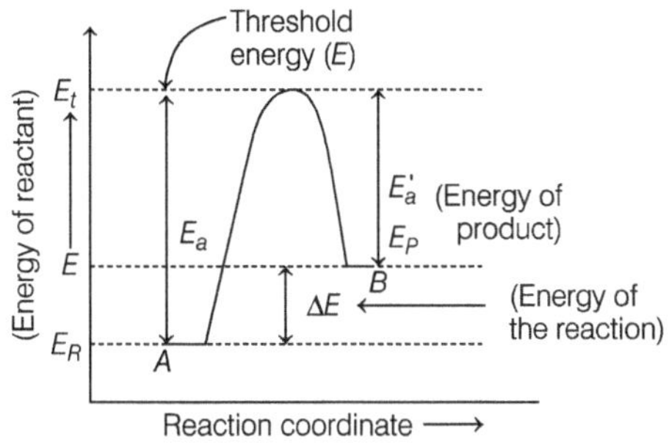

a. Activation energy of forward reaction is greater than backward reaction
b. The forward reaction is endothermic
c. The threshold energy is less than that of activation energy
d. The energy of activation of forward reaction is equal to the sum of heat of reaction and the energy of activation of backward reaction

68. Which of the following reactants is used for the preparation of ethyl benzene, where anhyd. $AlCl_3$ is a catalyst?

a. $CH_3CH_2OH + C_6H_6$
b. $CH_3—CH{=}CH_2 + C_6H_6$
c. $H_2C{=}CH_2 + C_6H_6$
d. $CH_3—CH_3 + C_6H_6$

69. Given pH of a solution 'A' is '3' and it is mixed with another solution 'B' having pH '2'. If both solutions are mixed, the resultant pH of the solution will be

a. 3.2 b. 1.9 c. 3.4 d. 3.5

70. In Freundlich adsorption isotherm, the value of $1/n$ is

a. between 0 and 1 in all cases
b. between 2 and 4 in all cases
c. always 1 in case of physical adsorption
d. always 1 in use of chemical adsorption

71. Spin only magnetic moment of Mn^{4+} ion is nearly

a. 3 BM b. 6 BM c. 4 BM d. 5 BM

72. 0.45 g acid of molecular mass 90 was neutralise by 20 mL of 0.5 N KOH. The basicity of the acid is

a. 2 b. 4 c. 1 d. 3

73. 4 moles of *A* are mixed with 4 moles of *B*. At equilibrium for the reaction $A + B \rightleftharpoons C + D$, 2 moles of *C* and *D* are formed, the equilibrium constant for the reaction will be,

a. 1 b. 1/2
c. 4 d. 1/4

74. What time (in s) is required for depositing all the silver present in 125 mL of 1 M $AgNO_3$ solution by passing a current of 241.25 A (Given 1F= 96500 C)

a. 10 b. 50
c. 100 d. 1000

75. In a homonuclear molecule, which of the following set of orbitals is degenerated?

a. $\sigma 1s$ and $\sigma 2s$
b. $\pi 2p_x$ and $\pi 2p_y$
c. $\pi^* 2p_x$ and $\sigma^* 2p_z$
d. $\sigma 2p_x$ and $\sigma^* 2p_z$

76. If the photon of wavelength 150 pm striks an atom and one of its inner bond electron is ejected out with a velocity of 1.5×10^7 m/s, what is the energy with which it is bond to the nucleus?

a. 1.2×10^2 eV
b. 2.15×10^3 eV
c. 7.6×10^3 eV
d. 8.12×10^3 eV

77. Identify compound (*C*) in the following reaction

$$C_7H_8 \xrightarrow{3Cl_2/\Delta} (A) \xrightarrow{Br_2/Fe} (B) \xrightarrow{Zn/HCl} (C)$$

a. *o*-bromotoluene
b. *m*-bromotoluene
c. *p*-bromotoluene
d. 3-bromo, 2, 4, 6-trichlorotoluene

78. In a compound C, H and N are present in 9 : 1 : 3.5 by weight. If molecular weight of compound is 108, the molecular formula of compound is

a. $C_2H_6N_2$ b. C_3H_4N
c. $C_6H_8N_2$ d. $C_9H_{12}N_3$

79. Which of the following compounds contain all the carbon atoms in the same hybridisation state

a. $H—C{\equiv}C—C{\equiv}C—H$
b. $CH_3—C{\equiv}C—CH_3$
c. $CH_2{=}C{=}CH_2$
d. $CH_2{=}CH—C{\equiv}CH$

80. The final step for extraction of copper from copper pyrite in bessemer converter involves the reaction

a. $Cu_2S + 2Cu_2O \longrightarrow 6Cu + SO_2$
b. $4Cu_2O + FeS \longrightarrow 8Cu + FeSO_4$
c. $2Cu_2O + FeS \longrightarrow 4Cu + Fe + SO_2$
d. $Cu_2S + 2FeO \longrightarrow 2Cu + Fe + SO_2$

BITSAT 2019

PART III

a. English Proficiency

Directions (Q. Nos 81-83) *In the following questions, the sentences may or may not be grammatically correct. Find out which part of a sentence has an error and mark that part. If there is no error mark part 'd' as your answer.*

81. Many a man **a.** / have come to India from Bangladesh **b.** / to live here permanently. **c.**/ No error **d.**

82. How long it takes **a.**/ to travel from Chennai to Trichy **b.** / by train? **c.**/ No error **d.**

83. One should respect **a.**/ the religions of others **b.**/ as much as his own. **c.**/ No error **d.**

Directions (Q. Nos. 84 and 85) *Fill in the blanks with suitable preposition from the alternatives given under each sentence.*

84. The master assured her .... success in the examination.
a. of *b.* in *c.* for *d.* with

85. Why do we always have to submit .... his authority?
a. under *b.* with *c.* for *d.* to

Directions (Q. Nos. 86-88) *Select the word or the phrase that is closest to the opposite in meaning of the italicised or phrase.*

86. The officer made an *audacious* remark.
a. Mute *b.* Visible *c.* Cultivated *d.* Timid

87. They put forward *cogent* arguments for their membership.
a. Logical *b.* Weighty *c.* Dissuasive *d.* Persuasive

88. His administration was *tainted* by scandal
a. Construct *b.* Clear *c.* Purify *d.* Repair

Directions (Q. Nos. 89 and 90) *Choose the word nearest in meaning to the underlined word.*

89. Her mother is vivacious and elegant.
a. Imaginary *b.* Lively *c.* Perceptible *d.* Languid

90. Sporadic fighting broke out in the office.
a. Timely *b.* Scattered *c.* Frequent *d.* Irrelevant

Directions (Q. Nos. 91-95) *Read the passage given below and answer the questions that follow.*

Radically changing monsoon patterns, reduction in the winter rice harvest and a quantum increase in respiratory diseases-all part of the environmental doomsday scenario which is reportedly playing out in South Asia. According to a United Nations Environment Programme report, a deadly 3-km deep blanket of pollution comprising a fearsome cocktail of ash, acids, aerosols and other particles has enveloped this region.

For India, already struggling to cope with a drought, the implications of this are devastating and further crop failure will amount to a life and death question for many Indians. The increase in premature deaths will have adverse social and economic consequences and a rise in **morbidities** will place an unbearable burden on our crumbling health system. And there is no one to blame but ourselves.

Both official and corporate India has always been **allergic** to any mention of clean technology. Most mechanical two wheelers roll off the assembly line without proper pollution control system.

Little effort is made for R & D on simple technologies, which could make a difference to people's lives and the environment. However, while there is no denying that South Asia must clean up its act, skeptics might question the timing of the haze report.

91. Both official and corporate India is allergic to
a. failure of monsoon
b. poverty and inequality
c. slowdown in industrial production
d. mention of clean technology

92. Which, according to the passage, is a life and death question to many Indians?
a. Increase in respiratory diseases
b. Use of clean technology
c. Thick blanket of pollution over the region
d. Failure in crops

93. Choose the word which is similar in meaning to the word 'profligacy' as used in the passage.
a. Wastefulness *b.* Conservation
c. Upliftment *d.* Criticalness

94. According to the passage, India cannot tolerate any further
a. crop failure
b. deterioration of healthcare system
c. increase in respiratory diseases
d. proliferation of nuclear devices

95. What could be the reason behind timing of the haze report just before the Johannesburg meet, as indicated in the passage?
a. United Nations is working hand-in-glove with US
b. Organises the forthcoming meet to teach a lesson to the US
c. Drawing attentions of the world towards devastating effects of environment degradation
d. US wants to use it as a handle against the developing countries in the forthcoming meet

b. Logical Reasoning

96. College' is related to 'Student' in the same way as 'Hospital' is related to

a. Doctor *b.* Nurse
c. Medicine *d.* Patient

97. Choose the odd one.

a. 27 *b.* 35
c. 18 *d.* 9

98. Complete the series by replacing question mark "?"

1, 2, 2, 4, 3, 8, 7, 10, ?

a. 9 *b.* 8
c. 11 *d.* 13

99. Mohan is taller than Shyam but shorter than Ramesh. Ramesh is taller than Rajat but shorter than Gautam. If Shyam is taller than Rajat, then who is the shortest among all?

a. Gautam
b. Rajat
c. Shyam
d. Ramesh

100. In the following question, complete the missing segment by selecting the appropriate figure from the given alternatives, (a), (b), (c) and (d).

Problem Figure

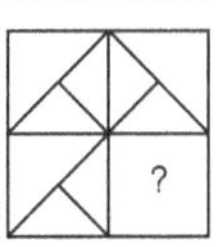

Answer Figures

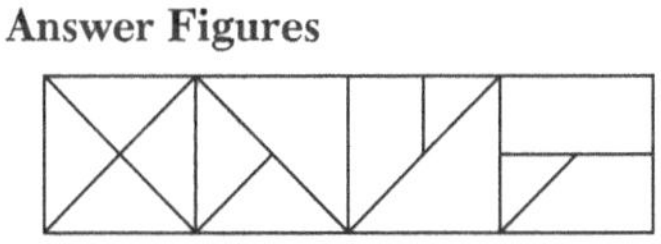

a. *b.* *c.* *d.*

101. In the question given below, find out which of the figures can be formed from the pieces given in the problem figure.

Problem Figure

Answer Figures

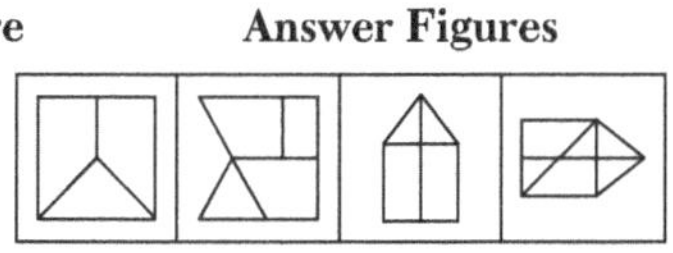

a. *b.* *c.* *d.*

102. In the following question, a set of three figures (X), (Y) and (Z) have been given, showing a sequence in which paper is folded and finally cut from a particular section. These figures are followed by a set of answer figures marked a., b., c. and d. showing the design which the paper actually acquires when it is unfolded. You need to select the answer figure which is closest to the unfolded piece of paper.

Problem Figures

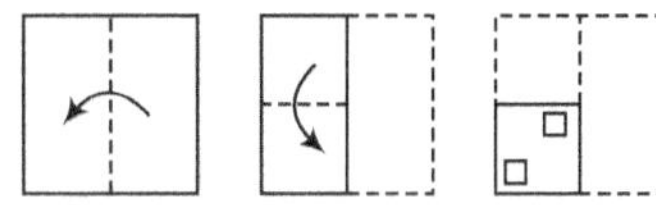

X Y Z

Answer Figures

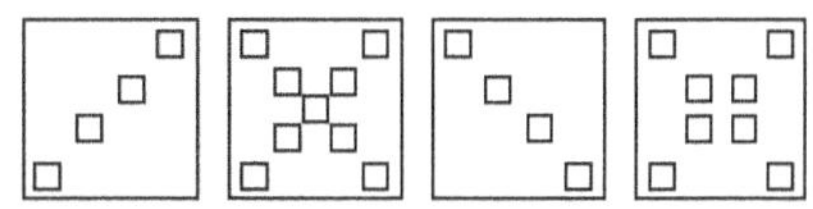

a. *b.* *c.* *d.*

103. Which of the answer figure a, b, c or d completes the figures matrix?

Problem Figure

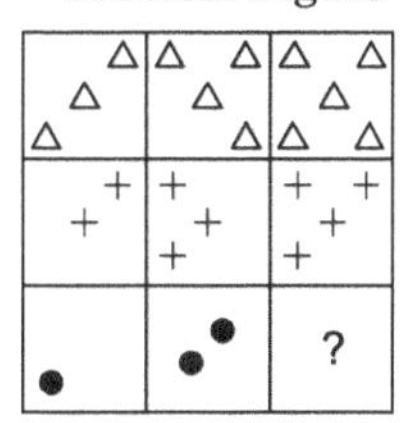

Answer Figures

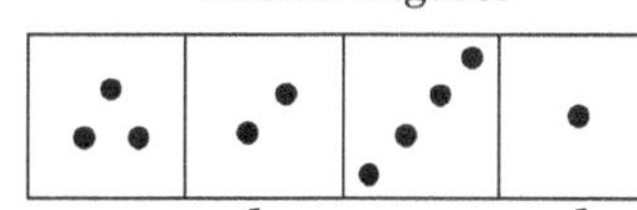

a. *b.* *c.* *d.*

104. In the following question, some dots are placed in the figure marked as (A). The figure is followed by four alternatives marked as (a), (b), (c) and (d). One out of these four options contains region(s) common to the circle, square, triangle, similar to that marked by the dot in figure (A). Choose the correct option.

Question Figure

(A)

Answer Figures

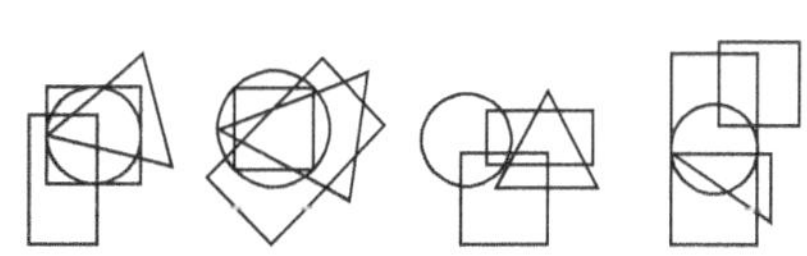

a. *b.* *c.* *d.*

105. How many triangles are there in the given figure?

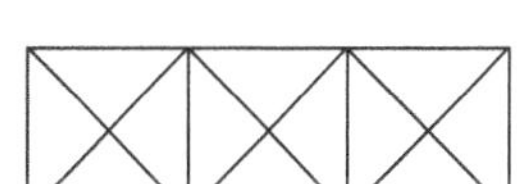

a. 28 *b.* 24
c. 25 *d.* 26

PART IV

Mathematics

106. If the length of the major axis of the ellipse $\frac{x^2}{a^2}+\frac{y^2}{b^2}=1$ is three times the length of minor axis, its eccentricity is

a. 1/3   b. $1/\sqrt{3}$ c. $\sqrt{2/3}$ d. $2\sqrt{2}/3$

107. $\sum_{m=1}^{n} \tan^{-1}\left(\frac{2m}{m^4+m^2+2}\right)$ is equal to

a. $\tan^{-1}\left(\frac{n^2+n}{n^2+n+2}\right)$ b. $\tan^{-1}\left(\frac{n^2-n}{n^2-n+2}\right)$

c. $\tan^{-1}\left(\frac{n^2+n+2}{n^2+n}\right)$ d. None of these

108. Let $(2x^2+3x+4)^{10}=\sum_{r=0}^{20} a_r x^r$, then the value of $\frac{a_8}{a_{12}}$, is

a. 2   b. 4   c. 8   d. 16

109. Let ABC be an acute-angled triangle and let D be the mid-point of BC. If $AB = AD$, then $\frac{\tan B}{\tan C}$ equal

a. $\sqrt{2}$ b. $\sqrt{3}$ c. 2   d. 3

110. Suppose the limit $L=\lim_{n\to\infty}\sqrt{n}\int_0^1\frac{1}{(1+x^2)^n}dx$ exists and is larger than $\frac{1}{2}$, then

a. $\frac{1}{2}<L<2$ b. $2<L<3$

c. $3<L<4$ d. $L\geq 4$

111. In the real number system, the equation $\sqrt{x+3-4\sqrt{x-1}}+\sqrt{x+8-6\sqrt{x-1}}=1$ has

a. No solution

b. Exactly two distinct solution

c. Exactly four distinct solution

d. Infinitely many solution

112. If system of equation $ax+y+z=a$, $x+by+z=b$ and $x+y+cz=c$ is inconsistent, then which of the following is correct?

a. $abc-a-b-c+2=0$

b. $abc-a-b-c+3=0,\ a=1$

c. $abc-a-b-c+3=0$

d. $abc-a-b-c+2=0,\ a\neq 1,\ b\neq 1, c\neq 1$

113. A die is thrown 7 times. What is the probability that an even number turns up at least 4 times?

a. $\frac{1}{2}$ b. $\frac{31}{64}$

c. $\frac{51}{128}$ d. $\frac{35}{128}$

114. The equation of plane containing line $x-y=1$, $z=1$ and parallel to $\frac{x}{2}-\frac{z}{3}=1$, $y=3$ is

a. $3x+3y-2z=1$ b. $3x-3y-2z=1$

c. $3x+3y+2z=1$ d. $3x+3y+2z=-1$

115. If $(1+x+x^2)^{20}=\sum_{r=0}^{40} a_r\cdot x^r$, then $\sum_{r=0}^{39}(-1)^r\cdot a_r\cdot a_{r+1}$ is equal to

a. 79   b. $2^{39}\cdot {}^{78}C_{39}$

c. $3^{39}\cdot {}^{78}C_{39}$ d. 0

116. The value of $\lim_{x\to\frac{3\pi}{4}}\frac{4\sin^2 x\cos x-\cos x+\sin x}{\sin x+\cos x}$ is equal to

a. -1 b. 0

c. 1   d. None of these

117. If **a**, **b** and **c** are non-coplaner vectors such that $\mathbf{b}\times\mathbf{c}=\mathbf{a}$; $\mathbf{c}\times\mathbf{a}=\mathbf{b}$ and $\mathbf{a}\times\mathbf{b}=\mathbf{c}$, then which of the following is not TRUE?

a. $|\mathbf{a}|-|\mathbf{b}|=0$ b. $|\mathbf{a}|=|\mathbf{b}|=|\mathbf{c}|=2$

c. $[\mathbf{a}\ \mathbf{b}\mathbf{c}]=1$ d. $|\mathbf{a}||\mathbf{b}||\mathbf{c}|=1$

118. The value of λ for which the loci $\arg z=\frac{\pi}{6}$ and $|z-2\sqrt{3}i|=\lambda$ on the argand plane touch each other is

a. 3   b. 4   c. 5   d. 6

119. The angle between the lines whose direction cosine satisfy the equations $l+m+n=0$ and $l^2=m^2+n^2$ is

a. $\frac{\pi}{3}$ b. $\frac{\pi}{4}$ c. $\frac{\pi}{6}$ d. $\frac{\pi}{2}$

120. The locus of the foot of perpendicular drawn from the centre of the ellipse $x^2+3y^2=6$ on any tangent to it is

a. $(x^2-y^2)^2=6x^2+2y^2$ b. $(x^2-y^2)^2=6x^2-2y^2$

c. $(x^2+y^2)^2=6x^2+2y^2$ d. $(x^2+y^2)^2=6x^2-2y^2$

121. If f and g are differentiable function in [0, 1] satisfying $f(0)=2=g(1)$, $g(0)=0$ and $f(1)=6$, then for some $c \in (0, 1)$

a. $2f'(c)=g(c)$ **b.** $2f'(c)=3g'(c)$
c. $f'(c)=g'(c)$ **d.** $f'(c)=2g'(c)$

122. The integral $\int \left(1+x-\frac{1}{x}\right)e^{x+\frac{1}{x}}dx$ is equal to

a. $(x-1)e^{x+\frac{1}{x}}+C$ **b.** $xe^{x+\frac{1}{x}}+C$
c. $(x+1)e^{x+\frac{1}{x}}+C$ **d.** $-xe^{x+\frac{1}{x}}+C$

123. If the coefficient of x^3 and x^4 in the expansion of $(1+ax+bx^2)(1-2x)^{18}$ in powers of x are both zero, then (a, b) is equal to

a. $\left(16, \frac{251}{3}\right)$ **b.** $\left(14, \frac{251}{3}\right)$
c. $\left(14, \frac{272}{3}\right)$ **d.** $\left(16, \frac{272}{3}\right)$

124. An equation of plane parallel to plane $x-2y+2z-5=0$ and at a unit distance from the origin is

a. $x-2y+2z=3$ **b.** $x-2y+2z+1=0$
c. $x-2y+2z-1=0$ **d.** $x-2y+2z+5=0$

125. Three numbers are choosen at random without replacement from $\{1, 2, 3, \ldots, 8\}$. The probability that their minimum is 3, given that their maximum is 6, is

a. $\frac{3}{8}$ **b.** $\frac{1}{5}$ **c.** $\frac{1}{4}$ **d.** $\frac{2}{5}$

126. The number of real numbers λ for which the equality $\frac{\sin(\lambda\alpha)}{\sin\alpha}-\frac{\cos(\lambda\alpha)}{\cos\alpha}=\lambda-1$, holds for all real α which are not integral multiple of $\pi/2$ is

a. 1 **b.** 2
c. 3 **d.** ∞

127. Suppose a parabola $y=ax^2+bx+c$ has two x intercepts, one positive and one negative, and its vertex is $(2, -2)$, then which of the following is true?

a. $ab>0$ **b.** $bc>0$
c. $ac>0$ **d.** $a+b+c>0$

128. The larger of two angles made with the X-axis of a straight line drawn through (1, 2) so that it intersects the line $x+y=4$ at a point distant $\sqrt{6}/3$ from the point (1, 2) is

a. 60° **b.** 75°
c. 105° **d.** None of these

129. The point $([P+1], [P])$ (where, $[x]$ is the greatest integer function) lying inside the region bounded by the circle $x^2+y^2-2x-15=0$ and $x^2+y^2-2x-7=0$, then

a. $P \in [-1, 2)-\{0,1\}$
b. $P \in [-1, 0) \cup (0, 1) \cup (1, 2]$
c. $P \in (-1, 2)$
d. None of the above

130. Solution of the equation $\frac{dy}{dx}=e^{x-y}(e^x-e^y)$ is

a. $e^y=e^x-1+ce^{-e^x}$ **b.** $e^{y-x}=-1+ce^{-e^x}$
c. $e^x+e^y=ce^{-e^x}$ **d.** None of these

131. Area bounded by two branches of the curve $(y-x)^2=x^3$ and $x=1$ is

a. $\frac{5}{4}$ sq unit **b.** $\frac{3}{2}$ sq unit
c. $\frac{1}{4}$ sq unit **d.** $\frac{4}{5}$ sq unit

132. The least value of the function

$f(x)=\int_0^x(3\sin x+4\cos x)dx$ on the interval $\left[\frac{5\pi}{4}, \frac{4\pi}{3}\right]$ is

a. $\frac{3}{2}-\frac{\sqrt{3}}{2}$ **b.** $\frac{5-4\sqrt{3}}{2}$
c. $\frac{7-4\sqrt{3}}{2}$ **d.** $\frac{9-4\sqrt{3}}{2}$

133. If z_1 and $\bar{z}_1$ represents adjacent vertices of a regular polygon of n sides and if $\frac{I_m(z_1)}{R_e(z_1)}=\sqrt{2}-1$, then n is equal to

a. 8 **b.** 16 **c.** 18 **d.** 24

134. In the expansion of $(1+x+x^3+x^4)^{10}$, the coefficient of x^4 is

a. 235 **b.** 310 **c.** 285 **d.** 325

135. A person writes letter to six friends and addresses the corresponding envelopes. Let x be the numbers of ways so that at least two of the letters are in wrong envelopes and y be the numbers of ways so that all the letters are in wrong envelopes. Then $x-y$ is equal to

a. 719 **b.** 265
c. 454 **d.** 720

136. If $x=\log_5 3+\log_7 5+\log_9 7$, then

a. $x>\frac{3}{2}$ **b.** $x>\frac{1}{\sqrt[3]{2}}$ **c.** $x>\frac{3}{\sqrt[3]{2}}$ **d.** $x>\sqrt[3]{2}$

137. Let p and q be roots of the equation $x^2 - 2x + A = 0$ and let r and s be the roots of the equation $x^2 - 18x + B = 0$, If $p < q < r < s$ are in AP, then A and B are

a. $-3, -77$ *b.* $3, -77$
c. $-3, 77$ *d.* $3, 77$

138. Let a, b and $c \in R^+$ and the system of equations $(1-a)x + y + z = 0$, $x + (1-b)y + z = 0$ and $x + y + (1-c)z = 0$ has infinitely many solutions, the minimum value of 'abc' is

a. $3\sqrt{3}$ *b.* 9
c. 27 *d.* 3

139. If A is an 3×3 non- singular matrix, such that $AA' = A'A$ and $B = A^{-1}A'$, then BB' equals to

a. $(B^{-1})'$ *b.* $I + B$
c. I *d.* B^{-1}

140. If $\sin^{-1} x + \tan^{-1} x = \pi/2$, then $2x^2 + 1 =$

a. $\sqrt{5}$ *b.* $\frac{\sqrt{5}-1}{2}$
c. 2 *d.* $\frac{\sqrt{5}+1}{2}$

141. Number of solution of the equation $|\cos x| = 2[x]$ are (where $|x|$, $[x]$ are modulus and greatest integer function respectively).

a. 0 *b.* 2
c. 1 *d.* infinitely many

142. Let $\cos(\alpha + \beta) = \frac{4}{5}$ and let $\sin(\alpha - \beta) = \frac{5}{13}$, where $0 \le \alpha, \beta \le \frac{\pi}{4}$, then $\tan 2\alpha =$

a. $\frac{20}{7}$ *b.* $\frac{25}{16}$
c. $\frac{56}{33}$ *d.* $\frac{19}{2}$

143. Cards are drawn one by one at random from a well shuffled pack of 52 cards until 2 aces are obtained for the first time. The probability that 18 draws are required for this is

a. 3/34 *b.* 17/455
c. $\frac{561}{15925}$ *d.* None of these

144. If $\mathbf{a} = \hat{\mathbf{i}} + 2\hat{\mathbf{j}} + 3\hat{\mathbf{k}}$, $\mathbf{b} = -\hat{\mathbf{i}} + 2\hat{\mathbf{j}} + \hat{\mathbf{k}}$ and $\mathbf{c} = 3\hat{\mathbf{i}} + \hat{\mathbf{j}}$ and $\mathbf{a} + p\mathbf{b}$ is normal to $\mathbf{c}$, then p is equal to

a. 0 *b.* 1
c. 5 *d.* 3

145. Let $f(x) = x[x]$, $x \notin Z$, ([·] denotes greatest integer function), then $f'(x)$ is equal to

a. $2x$ *b.* $[x]$
c. $2[x]$ *d.* 1

146. The difference of maximum and minimum values of $f(x) = x^2 e^{-x}$ is

a. e *b.* $1/e$
c. $1 - \frac{1}{e}$ *d.* $1 + \frac{1}{e}$

147. If $f'(x) = f(x) + \int_0^1 f(x)dx$, $f(0) = 1$, then $f(x) =$

a. $\frac{2e^x}{3-e} + \frac{1-e}{3-e}$ *b.* $\frac{e^x}{3-e} + \frac{1+e}{1-e}$
c. $\frac{3e^x}{2-e} + \frac{1+e}{1-e}$ *d.* $\frac{3e^x}{2-e} + \frac{1-e}{3+e}$

148. The distance between the origin and the normal to curve $y = e^{2x} + x^2$ at $x = 0$ is

a. $\frac{2}{\sqrt{3}}$ *b.* $\frac{2}{\sqrt{5}}$
c. $\frac{1}{\sqrt{3}}$ *d.* $\frac{1}{\sqrt{5}}$

149. Reflection of the line $\frac{x-1}{-1} = \frac{y-2}{3} = \frac{z-4}{1}$ in the plane $x + y + z = 7$ is

a. $\frac{x-1}{3} = \frac{y-2}{1} = \frac{z-4}{1}$
b. $\frac{x-1}{-3} = \frac{y-2}{-1} = \frac{z-4}{1}$
c. $\frac{x-1}{-3} = \frac{y-2}{1} = \frac{z-4}{-1}$
d. $\frac{x-1}{3} = \frac{y-2}{1} = \frac{z-4}{1}$

150. If $x, y, z \in R$, $x + y + z = 5$ and $x^2 + y^2 + z^2 = 9$, then length of interval in which x lies is

a. 8/3 *b.* 4/3
c. 2/3 *d.* 1/3

BITSAT 2019

Answers

Physics

1. (c)	**2.** (a)	**3.** (b)	**4.** (d)	**5.** (c)	**6.** (a)	**7.** (b)	**8.** (c)	**9.** (a)	**10.** (c)
11. (d)	**12.** (b)	**13.** (b)	**14.** (a)	**15.** (c)	**16.** (c)	**17.** (a)	**18.** (b)	**19.** (d)	**20.** (c)
21. (a)	**22.** (d)	**23.** (a)	**24.** (d)	**25.** (*)	**26.** (b)	**27.** (c)	**28.** (d)	**29.** (d)	**30.** (c)
31. (a)	**32.** (d)	**33.** (b)	**34.** (a)	**35.** (c)	**36.** (a)	**37.** (c)	**38.** (a)	**39.** (c)	**40.** (d)

Chemistry

41. (a)	**42.** (a)	**43.** (c)	**44.** (b)	**45.** (b)	**46.** (a)	**47.** (d)	**48.** (c)	**49.** (c)	**50.** (c)
51. (b)	**52.** (a)	**53.** (a)	**54.** (a)	**55.** (d)	**56.** (c)	**57.** (c)	**58.** (a)	**59.** (c)	**60.** (a)
61. (b)	**62.** (d)	**63.** (c)	**64.** (a)	**65.** (b)	**66.** (c)	**67.** (c)	**68.** (c)	**69.** (b)	**70.** (a)
71. (c)	**72.** (a)	**73.** (a)	**74.** (b)	**75.** (b)	**76.** (c)	**77.** (b)	**78.** (c)	**79.** (a)	**80.** (a)

English Proficiency

81. (b)	**82.** (a)	**83.** (c)	**84.** (a)	**85.** (d)	**86.** (d)	**87.** (c)	**88.** (b)	**89.** (b)	**90.** (b)
91. (d)	**92.** (d)	**93.** (a)	**94.** (a)	**95.** (d)					

Logical Reasoning

96. (d)	**97.** (b)	**98.** (c)	**99.** (b)	**100.** (b)	**101.** (b)	**102.** (d)	**103.** (a)	**104.** (b)	**105.** (a)

Mathematics

106. (d)	**107.** (a)	**108.** (b)	**109.** (d)	**110.** (a)	**111.** (d)	**112.** (d)	**113.** (a)	**114.** (b)	**115.** (d)
116. (a)	**117.** (b)	**118.** (a)	**119.** (a)	**120.** (c)	**121.** (d)	**122.** (b)	**123.** (d)	**124.** (a)	**125.** (b)
126. (b)	**127.** (b)	**128.** (b)	**129.** (d)	**130.** (a)	**131.** (d)	**132.** (d)	**133.** (a)	**134.** (b)	**135.** (c)
136. (c)	**137.** (c)	**138.** (c)	**139.** (c)	**140.** (a)	**141.** (a)	**142.** (c)	**143.** (c)	**144.** (c)	**145.** (b)
146. (b)	**147.** (a)	**148.** (b)	**149.** (c)	**150.** (b)					

Note *(*) None of the option is correct.*

Hints & Solutions

Physics

1. ***(c)*** Given, half life $(T_{1/2}) = 23.1$ days

If N_0 be the initial amount of a radioactive sample, then active nuclei at $t = 15$ days

$$N_1 = N_0 e^{-15\lambda} \quad \text{... (i)}$$

Active nuclei at $t = 16$ days

$$N_2 = N_0 e^{-16\lambda} \quad \text{...(ii)}$$

Disintegration constant,

$$\lambda = \frac{0.693}{T_{1/2}} = \frac{0.693}{23.1} = 0.03$$

$$\therefore \text{ \% decay in 15th to 16th days} = \frac{N_1 - N_2}{N_1} \times 100$$

$$= \frac{N_0 e^{-15\lambda} - N_0 e^{-16\lambda}}{N_0 e^{-15\lambda}} \times 100$$

[Using Eqs. (i) and (ii)]

$$= (1 - e^{-\lambda}) \times 100 = (1 - e^{-0.03}) \times 100$$

$$= \left(1 - \frac{1}{e^{0.03}}\right) \times 100 = \left(1 - \frac{1}{1.03}\right) \times 100 = 2.9\%$$

2. ***(a)*** Excess pressure inside first bubble of radius r_1,

$$p_1 = p_0 + \frac{4T}{r_1}$$

Excess pressure inside second bubble of radius r_2,

$$p_2 = p_0 + \frac{4T}{r_2}$$

Excess pressure inside double bubble,

$$p = p_2 - p_1$$

On putting values of p_1 and p_2, we get

$$p = 4T\left(\frac{r_1 - r_2}{r_1 r_2}\right) \quad \text{...(i)}$$

If R be the radius of double bubble, then

$$p = \frac{4T}{R}$$

or $$4T\left(\frac{r_1 - r_2}{r_1 r_2}\right) = \frac{4T}{R} \quad \text{[Using Eq. (i)]}$$

$$\Rightarrow \quad R = \frac{r_1 r_2}{r_1 - r_2} = \frac{3 \times 10^{-3} \times 2 \times 10^{-3}}{(3-2) \times 10^{-3}} = 6 \times 10^{-3} \text{ m}$$

$$\Rightarrow \quad R = 6 \times 10^{-3} \text{ m}$$

3. ***(b)*** Magnetic field at the centre of circular loop,

$$B = \frac{\mu_0 I}{2r} \quad \text{... (i)}$$

Period of revolution of electron is

$$T = \frac{2\pi}{\omega}$$

where, ω = angular velocity of electron.

Current produced due to movement of electron,

$$I = \frac{e}{T} = \frac{e}{\frac{2\pi}{\omega}} = \frac{e\omega}{2\pi}$$

On substituting the value of I in Eq. (i), we get

$$B = \frac{\mu_0}{2r} \cdot \frac{e\omega}{2\pi}$$

$$\Rightarrow \quad B = \frac{\mu_0 e\omega}{4\pi r}$$

4. ***(d)*** Given, elastic limit of steel wire, $Y = 2.2 \times 10^8$ m/s^2

Area of cross-section of steel wire,

$$A = 4 \text{ cm}^2 = 4 \times 10^{-4} \text{m}^2$$

Mass of elevator, $m = 1000$ kg

The maximum tension in the steel wire that can support elevator is given by

$$T = \frac{1}{4} \times \text{stress} \times \text{area of cross-section}$$

$$= \frac{1}{4} \times 2.2 \times 10^8 \times 4 \times 10^{-4}$$

$$= 2.2 \times 10^4 \text{ N/m}^2$$

If f be the maximum upward acceleration of the elevator, then tension is given by

$$\Rightarrow \quad T = m(g + f)$$

$$2.2 \times 10^4 = 1000\,(10 + f)$$

$$\Rightarrow \quad f = 12 \text{ m/s}^2$$

5. ***(c)*** For an ideal monoatomic gas, the heat capacity

$$\text{ratio, } \gamma = \frac{5}{3}$$

T_i, T_f = initial and final temperatures of gas.

In an adiabatic process,

$$TV^{\gamma - 1} = \text{constant}$$

i.e. $$T_f V_f^{\gamma - 1} = T_i V_i^{\gamma - 1}$$

$$\Rightarrow \quad T_f = T_i \left(\frac{V_i}{V_f}\right)^{\gamma - 1} = 300\left(\frac{1}{2}\right)^{\frac{5}{3} - 1}$$

$$= 300\left(\frac{1}{2}\right)^{2/3}$$

6. *(a)* Given, frequency of sound produced by sources A and B, $f_A = f_B = 400$ Hz

Speed of sound in air, $v = 340$ m/s

Apparent frequency heard by the observer when source A is moving towards him,

$$f_A' = \frac{v}{v - v_S} \times f_A = \frac{340}{340 - v_S} \times 400 \quad \ldots\text{(i)}$$

Apparent frequency heard by the observer when source B is moving away from him,

$$f_B' = \frac{v}{v + v_S} \times f_B = \frac{340}{340 + v_S} \times 400 \quad \ldots\text{(ii)}$$

Given, beats $= f_A' - f_B' = 4$

On substituting the values of Eq. (i) and Eq. (ii), we get

$$\frac{340}{340 - v_S} \times 400 - \frac{340}{340 + v_S} \times 400 = 4$$

$$\Rightarrow \quad \frac{34000 \times 2v_S}{340^2 - v_S^2} = 1$$

[Neglecting the value of v_S^2 $\because v_S^2 << 340^2$]

$$\frac{34000 \times 2v_S}{340^2} = 1$$

$$\Rightarrow \quad v_S = 1.7 \text{ m/s}$$

7. *(b)* Given, density of gas at point A, $\rho_A = \rho_0$

General equation for an ideal gas is

$$pV = nRT \Rightarrow pV = \frac{m}{M} . RT$$

where m = mass and M = molecular mass.

$$\Rightarrow \quad p = \frac{m}{V} . \frac{R}{M} . T$$

$$\Rightarrow \quad p = \rho \left(\frac{R}{M}\right) T \quad \ldots\text{(i)} \left[\because \rho = \frac{m}{V}\right]$$

Applying Eq. (i) at point A and B, respectively, we get

$$p = \rho_A \left(\frac{R}{M}\right) T_0 \quad \ldots\text{(ii)}$$

$$2p = \rho_B \left(\frac{R}{M}\right) 3T_0 \quad \ldots\text{(iii)}$$

From Eqs. (ii) and (iii), we get

$$\frac{1}{2} = \frac{\rho_A}{\rho_B} \cdot \frac{1}{3}$$

$$\Rightarrow \quad \rho_B = \frac{2}{3}\rho_A = \frac{2}{3}\rho_0 \quad [\because \rho_A = \rho_0]$$

8. *(c)* Force constant K of rubber is given by

$$K = \frac{YA}{l} = \frac{5 \times 10^8 \times 10^{-6}}{0.1} = 5 \times 10^3 \text{ N/m}$$

Now, from law of conservation of energy

elastic potential energy of cord = kinetic energy of particle

i.e. $\frac{1}{2}K(\Delta l)^2 = \frac{1}{2}mv^2$

$$\Rightarrow \quad v = \sqrt{\frac{K}{m}} . \Delta l = \sqrt{\frac{5 \times 10^3}{5 \times 10^{-3}}}(0.125 - 0.1)$$

$$= 10^3 \times 0.025 = 25 \text{ m/s}$$

9. *(a)* By work-energy theorem,

$$W = P \times t = \frac{1}{2}mv^2$$

$$\Rightarrow \quad v^2 = 2\frac{Pt}{m} \Rightarrow v = \left(\frac{2Pt}{m}\right)^{1/2}$$

As we know, $v = \frac{ds}{dt}$

or $$\frac{ds}{dt} = \left(\frac{2Pt}{m}\right)^{1/2}$$

$$\int ds = \int \left(\frac{2Pt}{m}\right)^{1/2} dt$$

$$s = \left(\frac{2P}{m}\right)^{1/2} . \frac{2}{3}t^{3/2}$$

i.e. $s \propto t^{3/2} \Rightarrow s^2 \propto t^3$

10. *(c)* Given, speed of electrons, $v = 5 \times 10^5$ m/s

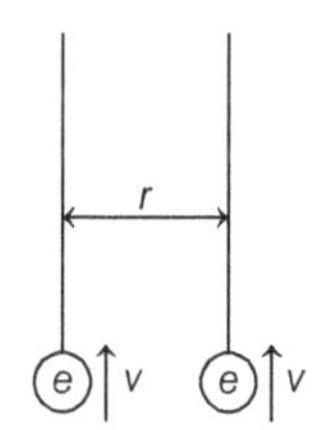

Electrostatic force between two electrons,

$$F_e = k.\frac{e^2}{r^2} \quad \ldots\text{(i)}$$

Magnetic field produced due to moving electron with velocity v,

$$B = \frac{\mu_0}{4\pi} . \frac{ev}{r^2} \quad \ldots\text{(ii)}$$

Magnetic force on electron due to magnetic field

$$F_m = Bev = \frac{\mu_0}{4\pi} . \frac{ev}{r^2} . ev \quad \text{(from Eq. (ii))}$$

$$F_m = \frac{\mu_0}{4\pi} . \frac{e^2v^2}{r^2} \quad \ldots\text{(iii)}$$

$\therefore$ From Eqs. (i) and (iii), we have

$$\frac{F_e}{F_m} = \frac{ke^2/r^2}{\frac{\mu_0 e^2 v^2}{4\pi . r^2}} = \frac{k . 4\pi}{\mu_0 v^2}$$

$$\Rightarrow \quad \frac{F_e}{F_m} = \frac{9 \times 10^9 \times 4 \times 3.14}{4\pi \times 10^{-7}(5 \times 10^5)^2} = 3.6 \times 10^5$$

11. ***(d)*** From the figure, It is clear that diodes D_1 and D_3 are reverse biased, therefore they will not conduct and resistance 11Ω and 5Ω will be not in use but diode D_2 is in forward biasing position. Therefore, current flows through diode D_2 is same as current through battery, according to the circuit shown after redrawing,

$\therefore$ Current, $I = \dfrac{20}{7+3} = 2\,\text{A}$

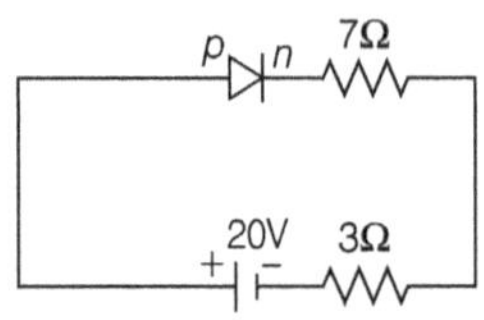

12. ***(b)*** The moment of inertia of disc of radius R about its axis is

$$I = \frac{M'R^2}{2}$$

where, M' = mass of complete disc $= 2M$

$$\therefore \quad I = \frac{2MR^2}{2}$$

$$I = MR^2 \qquad \ldots \text{(i)}$$

If I_1 be the moment of inertia of semicircular disc about its axis, then disc may be assumed as combination of two semi-circular parts,

Thus, $I = I_1 + I_1 \Rightarrow I_1 = \dfrac{I}{2} = \dfrac{MR^2}{2}$ [From Eq. (i)]

13. ***(b)*** Given, total energy given to electron in third orbit,

$$E = 12\,\text{eV}$$

Energy of third orbit of H-atom, $E' = \dfrac{-13.6}{n^2}$

$$= \frac{-13.6}{3^2} = -\,1.51\,\text{eV}$$

Ionisation energy of electron of third orbit, $E'' = -\,E'$

$$= -(-\,1.51\,\text{eV}) \Rightarrow E'' = 1.51\,\text{eV}$$

Hence, final energy of electron when it comes out of H-atom from third orbit

$$E_F = E - E''$$
$$= 12 - 1.51 = 10.49\,\text{eV}$$

14. ***(a)*** Given, $D = 2\,\text{m}$, $d = 0.25\,\text{mm} = 2.5 \times 10^{-4}\,\text{m}$

Velocity of screen, $\dfrac{dD}{dt} = 5\,\text{m/s}$

Speed of first maxima $= \dfrac{d\beta}{dt} = ?$

In Young's double slit experiment, fringe width is given by

$$\beta = \frac{D\lambda}{d}$$

On differentiating with respect to t, we get,

$$\frac{d\beta}{dt} = \frac{\lambda}{d} \cdot \frac{dD}{dt}$$
$$= \frac{800 \times 10^{-9}}{2.5 \times 10^{-4}} \cdot 5$$
$$= 16 \times 10^{-3}\,\text{m/s} = 16\,\text{mm/s}$$

15. ***(c)*** Given, current through the square loop is 2A.

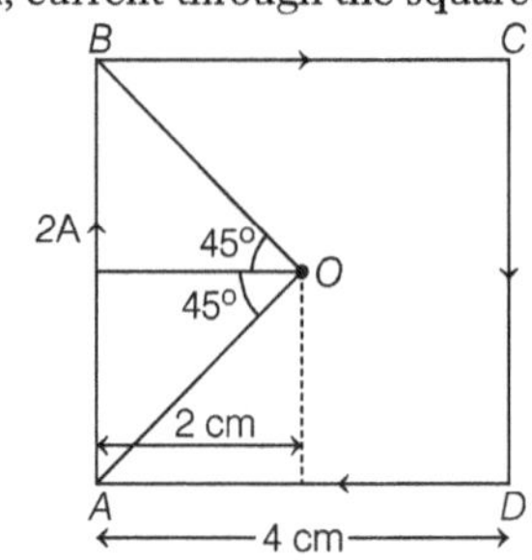

$I = 2\text{A}$

The given situation is shown in the figure. Magnetic field due to wire AB at the centre O is

$$B_{AB} = \frac{\mu_0}{4\pi} \cdot \frac{I}{r}[\sin 45° + \sin 45°]$$
$$= 10^{-7} \times \frac{2}{2 \times 10^{-2}}\left[\frac{1}{\sqrt{2}} + \frac{1}{\sqrt{2}}\right]$$
$$= \sqrt{2} \times 10^{-5}\,\text{T}$$

[Downward to the plane of square loop]

Since, magnetic field due to side AB, BC, CD and DA is same in magnitude and direction. Hence, net magnetic field at O

$$B = 4 \times B_{AB} = 4\sqrt{2} \times 10^{-5}\,\text{T}$$

16. ***(c)*** Wheatstone bridge in balanced condition maintains $\dfrac{P}{Q} = \dfrac{R}{S}$ i.e. no current flows through the galvanometer (i.e. $i_g = 0$)

The given circuit can be redrawn as

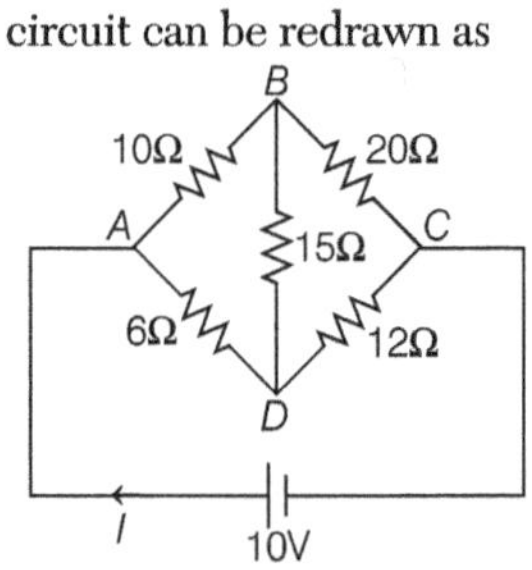

Since, $\dfrac{R_{AB}}{R_{BC}} = \dfrac{R_{AD}}{R_{DC}} = \dfrac{1}{2}$, hence it is a balanced Wheatstone bridge, therefore current flowing in arm $BD = 0$

$$\therefore \quad \frac{1}{R_{eq}} = \frac{1}{10+20} + \frac{1}{6+12} = \frac{1}{30} + \frac{1}{18}$$

$$\Rightarrow \quad R_{eq} = \frac{30 \times 18}{30 + 18} = 11.25\ \Omega$$

$$\therefore \quad I = \frac{V}{R_{eq}} = \frac{10}{11.25} = 0.89\text{ A} \simeq 0.9\text{ A}$$

17. ***(a)*** The flux linked with coil (made of wire of radius r) of area A and magnetic field B, is given by

$$\phi = BA\cos\theta = B\left(\frac{\pi r^2}{2}\right)\cos\omega t$$

$$\phi = \frac{\pi B r^2}{2}\cos\omega t \qquad \left[\because A = \frac{\pi r^2}{2} \text{ and } \theta = \omega t\right]$$

By Faraday's law of electromagnetic induction, induced emf, $e = -\frac{d\phi}{dt}$

Hence, after differentiating, we get,

$$e = -\frac{d}{dt}\left(\frac{\pi B r^2}{2}\cos\omega t\right) = \frac{\pi B r^2 \omega}{2}\sin\omega t$$

$$\therefore \quad \text{Power, } P = \frac{e^2}{R} = \frac{\pi^2 B^2 r^4 \omega^2 \sin^2\omega t}{4R}$$

(where, R = resistance of the circuit)

$$\therefore \text{ Mean power, } \overline{P} = \overline{\frac{\pi^2 B^2 r^4 \omega^2}{4R}} \cdot \overline{\sin^2\omega t}$$

$$= \frac{\pi^2 B^2 r^4 \omega^2}{4R} \cdot \frac{1}{2} \qquad \left[\because \overline{\sin^2\omega t} = \frac{1}{2}\right]$$

$$= \frac{(\pi B r^2 \omega)^2}{8R}$$

18. ***(b)*** For the given convex lens, focal length, $f = 25$ cm

Magnification of lens in term of u and f is given by

$$m = \frac{f}{f+u} = \frac{25}{25+u} \qquad \ldots \text{(i)}$$

When the image is real, then $m = -2$ and $u = -x_1$

$$\therefore \quad -2 = \frac{25}{25 - x_1} \Rightarrow x_1 = \frac{75}{2} \qquad \text{[from Eq. (i)]}$$

Again, when the image is virtual, then $m = 2$ and $u = -x_2$

$$\therefore \quad 2 = \frac{25}{25 - x_2} \Rightarrow x_2 = \frac{25}{2} \qquad \text{[from Eq. (i)]}$$

Hence, the ratio of x_2 and x_1 is

$$\therefore \quad \frac{x_2}{x_1} = \frac{25/2}{75/2} = \frac{1}{3}$$

19. ***(d)*** For perfectly elastic collision,

total momentum before collision

= total momentum after collision

$$\text{i.e.} \quad m \cdot u + \frac{3m}{4} \times 0 = mv_x + \frac{3m}{4} \cdot v_y$$

$$\Rightarrow \quad 4u = 4v_x + 3v_y \qquad \ldots \text{(i)}$$

Coefficient of restitution,

$$e = \frac{\text{Velocity of seperation } (v_y - v_x) \text{ after collision}}{\text{Velocity of approach } (u_x - u_y) \text{ before collision}}$$

$$\Rightarrow \quad 1 = \frac{v_y - v_x}{u - 0} \qquad \left[\because u_x = u \text{ and } u_y = 0\right]$$

$$u = v_y - v_x \qquad \ldots \text{(ii)}$$

Solving Eqs. (i) and (ii), we have

$$v_x = \frac{u}{7} \text{ and } v_y = \frac{8u}{7}$$

$$\therefore \quad \frac{\lambda_y}{\lambda_x} = \frac{h/mv_y}{h/\frac{3m}{4}.v_x} = \frac{3}{4}.\frac{v_x}{v_y} = \frac{3}{4}.\frac{u/7}{8u/7} = \frac{3}{32}$$

20. ***(c)*** We know that, velocity of light c is terms of ε_0 and μ_0 is given by

$$c = \frac{1}{\sqrt{\mu_0 \varepsilon_0}} \Rightarrow c^2 = \frac{1}{\mu_0 \varepsilon_0}\ c^2 = \frac{1}{\mu_0^2} \cdot \frac{\mu_0}{\varepsilon_0}$$

$$\Rightarrow c = \frac{1}{\mu_0}\sqrt{\frac{\mu_0}{\varepsilon_0}} \Rightarrow \sqrt{\frac{\mu_0}{\varepsilon_0}} = c\mu_0 \qquad \ldots \text{(i)}$$

$$\text{Dimension of magnetic field, } B = \frac{\tau}{NIA} = \frac{[ML^2T^{-2}]}{[A][L^2]} = [MA^{-1}T^{-2}]$$

Again, $B = \mu_0 n I$

$$\therefore \quad \mu_0 = \frac{B}{nI} = \frac{[MA^{-1}T^{-2}]}{[L^{-1}][A]} = [MLA^{-2}T^{-2}]$$

$$\mu_0 = [MLA^{-2}T^{-2}] \qquad \ldots \text{(ii)}$$

$$\text{and} \quad c = [LT^{-1}] \qquad \ldots \text{(iii)}$$

From Eqs. (i), (ii) and (iii), we get

$$\sqrt{\frac{\mu_0}{\varepsilon_0}} = [LT^{-1}][MLA^{-2}T^{-2}] = [ML^2T^{-3}A^{-2}]$$

21. ***(a)*** Given, $F = at$

when, $t = 0$, then linear momentum $= p$

when, $t = T$, then linear momentum $= 2p$

According to Newton's law of motion,

Applied force, $F = \frac{dp}{dt}$

$$\text{or,} \quad dp = F dt$$

$$dp = at\, dt$$

$$\text{or,} \quad \int_p^{2p} dp = a\int_0^T t\, dt$$

$$\Rightarrow \quad (2p - p) = a \cdot \frac{T^2}{2}$$

$$\Rightarrow \quad \frac{2p}{a} = T^2 \Rightarrow T = \sqrt{\frac{2p}{a}}$$

22. ***(d)*** Block does not move upto a maximum applied force of 3 N down the inclined plane.

Equation of motion is

$$3 + mg\sin 30° = f$$

$$3 + mg\sin 30° = \mu\, mg\cos 30° \qquad \ldots \text{(i)}$$

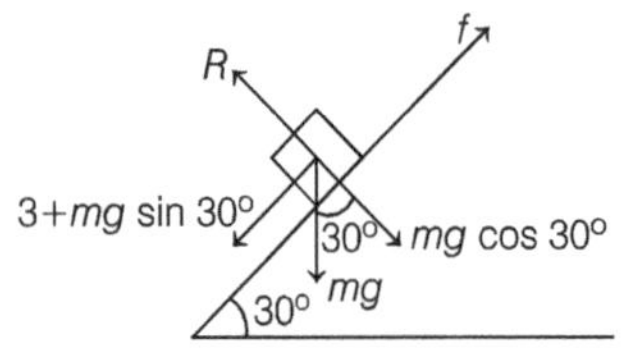

Similarly, block also does not move upto a maximum applied force of 12 N up the plane.

Now, equation of motion is

$$mg\sin 30° + f = 12$$

or, $\quad mg\sin 30° + \mu\, mg\cos 30° = 12 \qquad \ldots \text{(ii)}$

Solving Eqs. (i) and (ii), we have

$$mg\sin 30° = \frac{9}{2} \text{ and } \mu\, mg\cos 30° = \frac{15}{2}$$

Hence, dividing

$$\therefore \quad \frac{1}{\mu}\tan 30° = \frac{9/2}{15/2} = \frac{3}{2} \Rightarrow \mu = \frac{5}{3\sqrt{3}}$$

23. ***(a)*** If middle portion of lens is painted black, then less number of refracted rays will intersect, therefore intensity of the corresponding image of object will be less. Though, the rays coming from object refracted with small portion of lens, so complete image of object will be formed but of lesser intensity.

24. ***(d)*** Given, speed of sound, $v = 340$ m/s

Length of open organ pipe, $l = 2$ m

Frequency, $f = 1200$ Hz

For open organ pipe,

Fundamental frequency, $f_0 = \frac{v}{2l} = \frac{340}{2 \times 2} = 85$

$\therefore$ Number of tones present in the open organ pipe

$$= \frac{f}{f_0} = \frac{1200}{85} = 14.11 \simeq 14$$

25. ***(*)*** Given, charge on proton,

$$q = e = 1.6 \times 10^{-19} \text{ C}$$

Radius, $\quad r = 2\text{ mm} = 2 \times 10^{-3}$ cm

Frequency, $\quad f = 10$ Hz

Current associated by proton is given by

$$I = \frac{q}{T} \qquad [T \to \text{time period}]$$

$$= qf \qquad \left[\because T = \frac{1}{f}\right]$$

$$= 1.6 \times 10^{-19} \times 10 = 1.6 \times 10^{-18} \text{ A}$$

$\therefore$ Magnetic dipole moment, $M = IA$

$$= I \cdot \pi r^2 = 1.6 \times 10^{-18} \times 3.14 \times (2 \times 10^{-3})^2$$

$$= 20.09 \times 10^{-24}$$

$$\simeq 2 \times 10^{-23} \text{ A-m}^2$$

26. ***(b)*** When capacitor is fully charged, then it acts like a open switch, therefore 8 kΩ resistor has no use. So, rest of the resistances 4 kΩ, 7kΩ and 9 kΩ will be in series.

Hence, current drawn from cell,

$$I = \frac{10}{(4+7+9) \times 10^3} = \frac{10}{20 \times 10^3} = 0.5 \text{ mA}$$

27. ***(c)*** The charge on the capacitor at time t in a discharging circuit is given by

$$q = q_0 e^{-t/\tau}$$

where, $\tau = RC$ = time constant

$$\therefore \quad q = q_0\, e^{-t/RC}$$

when, $\quad q = \frac{q_0}{2}$

then $\quad \frac{q_0}{2} = q_0\, e^{-t/RC}$

$$\frac{1}{2} = e^{-t/RC}$$

Taking natural log on both side, we get

$$\log_e\left(\frac{1}{2}\right) = \log_e e^{-t/RC}$$

$$-\log_e 2 = -\frac{t}{RC}\log_e e \Rightarrow t = RC\log_e 2$$

$$[\because \log_e e = 1]$$

28. ***(d)*** If v be the velocity in the inter planetary space and U_i, U_f initial and final potential energy

Then, by the law of conservation of energy,

$$U_i + K_i = U_f + K_f$$

$$\frac{-GM_e m}{R_e} + \frac{1}{2}m(3v_e)^2 = 0 + \frac{1}{2}mv^2$$

$$\frac{-GM_e}{R_e} + \frac{9}{2}v_e^2 = \frac{1}{2}v^2$$

$$\Rightarrow \quad \frac{-2GM_e}{R_e} + 9v_e^2 = v^2$$

$$\Rightarrow \quad \frac{-2gR_e^2}{R_e} + 9v_e^2 = v^2 \qquad [\because GM_e = gR_e^2]$$

$$-2gR_e + 9 \times 2 \times gR_e = v^2, \qquad [\because v_e^2 = 2gR_e]$$

$$16\, gR_e = v^2$$

$$\Rightarrow \quad v = \sqrt{8}\,.\sqrt{2g\,R_e}$$

$$= 2\sqrt{2}\, v_e$$

29. *(d)* Given, density of bob, $\rho = \frac{4}{3} \times 10^3$ kg/m^3

Density of water, $\sigma = 10^3$ kg/m^3

If g' be gravitational acceleration in water, then

$$g' = g\left(1 - \frac{\sigma}{\rho}\right) = g\left(1 - \frac{10^3}{\frac{4}{3} \times 10^3}\right) = \frac{g}{4}$$

As, $T_{\text{air}} = 2\pi\sqrt{\frac{l}{g}}$

Similarly, $T_{\text{water}} = 2\pi\sqrt{\frac{l}{g'}}$ $\left[\because T_{\text{water}} = 2\text{s and } g' = \frac{g}{4}\right]$

$$\Rightarrow \quad 2 = 2\pi\sqrt{\frac{l}{g/4}} = 2\pi\sqrt{\frac{l}{g}}.2$$

$$2 = 2.T_{\text{air}} \Rightarrow T_{\text{air}} = 1 \text{ s}$$

30. *(c)* By Newton's law of motion,

$$F - f_s = ma_{CM}$$

$$\Rightarrow \quad F - f_s = ma \quad \ldots \text{(i)}$$

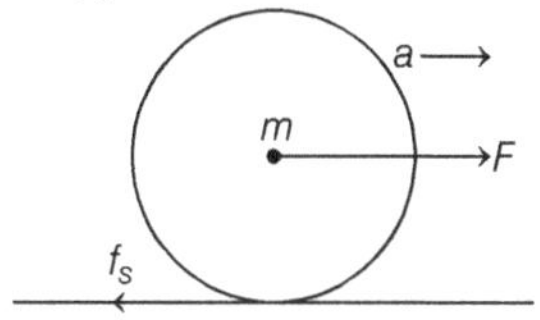

and torque, $\tau = I\alpha$

where, α = tangential component of acceleration

and I = moment of inertia.

$\Rightarrow \quad fR = I\alpha$

As the cylinder is rolling without slipping, the acceleration of point of contact will be zero.

$$a_{CM} - R\alpha = 0 \Rightarrow \alpha = \frac{a_{CM}}{R} = \frac{a}{R}$$

$$fR = \frac{MR^2}{2} \cdot \frac{a}{R} \quad \left[\therefore I = \frac{MR^2}{2}\right]$$

$$\Rightarrow \quad f = \frac{ma}{2}$$

Substituting in Eq. (i), we get

$$F = f + ma$$

$$\Rightarrow \quad F = \frac{ma}{2} + ma = \frac{3ma}{2}$$

31. *(a)* Velocity of projectile at time t is given by

$$\mathbf{v} = \mathbf{u} - \mathbf{g}t\hat{\mathbf{j}}$$

$$\mathbf{v} = \hat{\mathbf{i}} + \sqrt{3}\,\hat{\mathbf{j}} - 10t\,\hat{\mathbf{j}}$$

$$\therefore \quad \mathbf{r} = \int v\, dt$$

$$= \int(\hat{\mathbf{i}} + \sqrt{3}\,\hat{\mathbf{j}} - 10t\,\hat{\mathbf{j}})\, dt$$

$$= t\hat{\mathbf{i}} + \sqrt{3}\,t\,\hat{\mathbf{j}} - 10\frac{t^2}{2}\hat{\mathbf{j}}$$

$$= t\hat{\mathbf{i}} + \sqrt{3}\,t\hat{\mathbf{j}} - 5\,t^2\hat{\mathbf{j}}$$

$$\mathbf{r} = t\,\hat{\mathbf{i}} + (\sqrt{3}t - 5t^2)\,\hat{\mathbf{j}}$$

or $\quad x\hat{\mathbf{i}} + y\hat{\mathbf{j}} = t\hat{\mathbf{i}} + (\sqrt{3}t - 5t^2)\,\hat{\mathbf{j}}$

On comparing, we get

$$x = t \quad \ldots \text{(i)}$$

and $\quad y = \sqrt{3}\,t - 5t^2 \quad \ldots$ (ii)

From Eqs (i) and (ii), we get

$$y = \sqrt{3}\,x - 5x^2$$

32. *(d)* $\mathbf{F} = 20\hat{\mathbf{i}} + 10\hat{\mathbf{j}}$

$\therefore \quad F_x = 20$ N and $F_y = 10$ N

$\therefore$ Acceleration is expressed as

$$a_x = \frac{F_x}{m} = \frac{20}{1}$$

$$a_x = 20 \text{ m/s}^2$$

and $\quad a_y = \frac{F_y}{m} = \frac{10}{1} = 10 \text{ m/s}^2$

$\therefore$ Displacement of body in x-direction in 2s.

$$x = u_x t + \frac{1}{2}a_x t^2$$

$$= 0 + \frac{1}{2} \times 20 \times 2^2 = 40 \text{ m}$$

$\therefore$ Displacement of body in y-direction in 2s.

$$y = u_y t + \frac{1}{2}a_y t^2$$

$$= 0 + \frac{1}{2} \times 10 \times 2^2 = 20\text{m}$$

$\therefore$ Position of the body at $t = 2$ s

$$\mathbf{r} = x\hat{\mathbf{i}} + y\hat{\mathbf{j}}$$

$$\mathbf{r} = 40\hat{\mathbf{i}} + 20\hat{\mathbf{j}}$$

33. *(b)* Potential energy of cylinder at top position

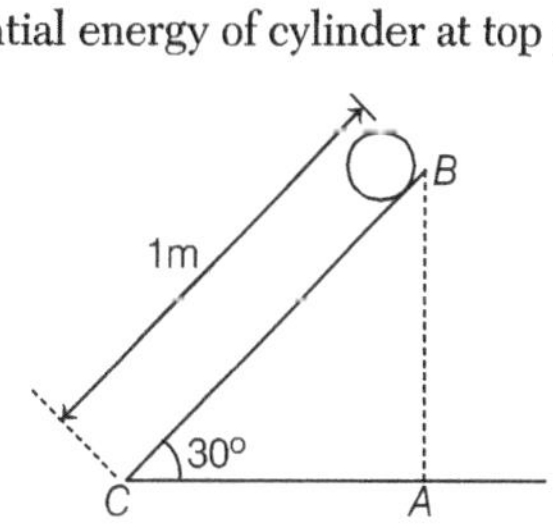

$$U = mg(AB)$$

$$= 2 \times 10\,[1.\sin 30°] \quad \left[\because \sin 30° = \frac{AB}{BC} \quad \therefore AB = 1 \sin 30°\right]$$

$$= 10\text{J}$$

If v is the linear speed of cylinder when it reaches at bottom, then at bottom, total potential energy of cylinder is converted into kinetic energy,

i.e. $\quad K = U$

As the cylinder undergoes both rotational and translational motion,

$$\therefore \quad \frac{1}{2}I\omega^2 + \frac{1}{2}mv^2 = 10$$

$$\frac{1}{2}\cdot\frac{mr^2}{2}\cdot\frac{v^2}{r^2} + \frac{1}{2}mv^2 = 10 \qquad \left[\because \omega = \frac{v}{r}\right]$$

$$\Rightarrow \quad \frac{3}{4}mv^2 = 10$$

$$v = \sqrt{\frac{40}{3m}} = \sqrt{\frac{40}{3\times 2}} = \sqrt{\frac{20}{3}} \text{ m/s}$$

34. ***(a)*** According to the question,

$$F_1 = -k_1x \text{ and } F_2 = -k_2x$$

$\therefore$ Frequency of oscillation,

$$\nu_1 = \frac{1}{2\pi}\sqrt{\frac{k_1}{m}}$$

$$5 = \frac{1}{2\pi}\sqrt{\frac{k_1}{m}} \Rightarrow k_1 = 100\,\pi^2 m$$

Similarly, $12 = \dfrac{1}{2\pi}\sqrt{\dfrac{k_2}{m}}$

$$\Rightarrow \quad k_2 = 576\,\pi^2 m$$

Now, $\quad F = F_1 + F_2 = -(k_1 + k_2)\,x$

$\therefore$ Frequency of oscillation,

$$\nu = \frac{1}{2\pi}\sqrt{\frac{k_1+k_2}{m}} = \frac{1}{2\pi}\sqrt{\frac{100\pi^2 m + 576\pi^2 m}{m}}$$

$$\Rightarrow \quad \nu = 13 \text{ Hz}$$

35. ***(c)*** Number of moles in 2×10^{-4} kg hydrogen

$$\therefore \quad n = \frac{\text{Mass of hydrogen in gram}}{\text{Molecules mass}}$$

$$= \frac{2\times 10^{-4}\times 10^3}{2} = 0.1 \text{ mole}$$

$$\therefore \quad pV = nRT$$

$$\frac{pV}{T} = nR = 0.1\times R = \frac{R}{10}$$

36. ***(a)*** Given, for an ammeter, $I_g = 1$ A, $R_g = 1.5\ \Omega$

Total current, $I = 4$ A

If R_s be the shunt resistance, then according to the question circuit diagram is given, ($R_s \parallel R_g$, hence have equal potential difference)

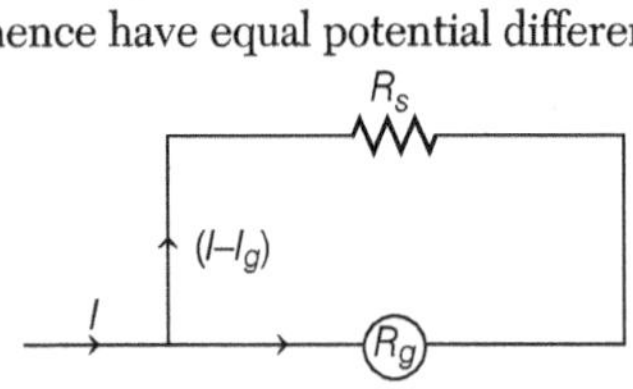

where, $I_gR_g = (I - I_g)\,R_s$

$$R_s = \frac{I_g}{I - I_g}\times R_g = \frac{1}{4-1}\times 1.5 = 0.5\ \Omega$$

37. ***(c)*** The electric field intensity at a point lying outside the sphere (non-conducting) is

$$E = \frac{1}{4\pi\varepsilon_0}\cdot\frac{q}{r^2}$$

i.e. $\quad E \propto \dfrac{1}{r^2} \quad \ldots$ (i)

The electric field intensity at the surface ($r = R$),

$$E = \frac{1}{4\pi\varepsilon_0}\frac{q}{R^2} \Rightarrow E \propto \frac{1}{R^2}$$

where, R being the radius of sphere.

The electric field intensity inside the sphere is

$$E = \frac{qr}{4\pi\varepsilon_0 R^3}$$

$$\Rightarrow \quad E \propto r$$

At the centre of sphere, $r = 0, \therefore E = 0$

Hence, electric field will increase linearly with distance till R as $E \propto r$ after that it decreases as it is inversely proportional to the square of distance.

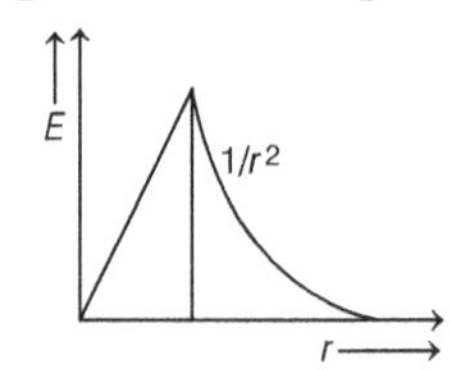

38. ***(a)*** Given, $R = (56 \pm 5.6)$ kΩ

$$= 56 \text{ k}\Omega \pm 10\% \text{ of } 56 \text{ k}\Omega$$

$$= 56\times 10^3 \pm 10\%\ \Omega$$

As per the colour code for carbon resistors, the colour assigned to numbers

5 → Green
6 → Blue
3 → Orange

For 10% accuracy or tolerance, the colour is silver.

Hence, the bands of colours of carbon resistor in sequence are green, blue, orange and silver.

39. ***(c)*** When two waves with same frequency and constant phase difference interfere to each other, then after interference, intensity of resultant wave is distributed such that it is maximum at some points [constructive interference or bright fringes] and minimum at another points. [destructive interference or dark fringes].

Hence, energy is distributed and remains constant [in the form of bright and dark fringes].

40. ***(d)*** According to the given graph $(I - t)$ induced current (I) is obtained from the slope.

Hence, $I = \dfrac{1}{R}\dfrac{d\phi}{dt}$

$\Rightarrow d\phi = I\,dt.\,R =$ area of triangle $POS \times R$

$$= \frac{8\times 0.2}{2}\times 50$$

$$= 40 \text{ Wb}$$

Chemistry

41. ***(a)*** Let atoms of Y in ccp structure = 100

Then, number of tetrahedral voids = 2×100

$\because$ Number of X atoms = $\frac{2}{3}$rd of tetrahedral voids

$\therefore$ Number of X atoms = $\frac{2}{3} \times 200 = \frac{400}{3}$

Thus, $\frac{X}{Y} = \frac{400}{3 \times 100} = \frac{400}{300} = \frac{4}{3}$

Hence, formula of compound can be = X_4Y_3.

42. ***(a)*** Given, density (d_1) at 30°C (T_1) = 1.35 kg/m^3

To find density (d_2) at STP, (i.e at 760 torr and 273 K (T_2) temperature), we use formula

$$\frac{d_2}{d_1} = \frac{p_2 T_1}{p_1 T_2}$$

where, p_1 = 768 torr (Pressure at 30°C, i.e. at 303 K)

p_2 = 760 torr (Pressure at 273 K)

$$\therefore\ d_2 = \frac{d_1 \times p_2 \times T_1}{p_1 \times T_2} = \frac{1.35 \times 760 \times 303}{768 \times 273} = 1.48\ \text{kg/m}^3$$

43. ***(c)*** $\Delta_o = (-0.4 \times nt_{2g}) + (0.6 \times ne_g)$

where, n = number of electrons in nt_{2g} and ne_g.

CFSE of given d-electron configurations are as follows

a. For high spin, d^6;

$\Delta_o\,(\text{CFSE}) = -0.4 \times 4 + 0.6 \times 2 = -0.4$

b. For low spin, d^4;

$\Delta_o(\text{CFSE}) = -0.4 \times 4 + 0.6 \times 0 = 1.6$

c. For low spin d^5

$\Delta_o\ (\text{CFSE}) = -0.4 \times 5 + 0.6 \times 0 = -2.0$

d. For high spin, d^7

$\Delta_o\,(\text{CFSE}) = -0.4 \times 5 + 0.6 \times 2 = -0.8$

Neglecting –ve sign (as it indicates that energy is released.

Hence, maximum Δ_o (CFSE) value is shown by low spin d^5 configuration.

44. ***(b)*** According to Rydberg's equation

$$\bar{\nu} = \frac{1}{\lambda} = RZ^2\left[\frac{1}{n_L^2} - \frac{1}{n_H^2}\right]$$

where,

$\bar{\nu}$ = wave number

λ = wavelength

R = Rydberg constant (= 109677 cm^{-1})

Z = atomic number

n_L = lower energy state

n_H = higher energy state.

i.e. $\frac{1}{\lambda} \propto Z^2$

Thus, $\frac{\lambda(\text{He}^+)}{\lambda(\text{H})} = \frac{Z^2(\text{H})}{Z^2(\text{He}^+)}$

To find $\lambda(\text{He}) = ?$

Given, $\lambda(\text{H}) = 91.2$ nm

$Z(\text{H}) = 1$

$Z(\text{He}) = 2$

$\therefore\ \lambda(\text{He}^+) = \frac{1 \times 91.2}{4} = 22.8$ nm

45. ***(b)*** $C_{18}H_8(s) + 12O_2(g) \longrightarrow 10CO_2(g) + 4H_2O(l)$

We know that, $\Delta H = \Delta E + \Delta n_g RT$

Δn_g = No. of moles of gaseous products – no. of moles of gaseous reactants.

$\Delta n_g = 10 - 12 = -2$

$\Delta H = -1228.2 \times 10^3 + (-2) \times 2 \times 298$

$= -1229393$ cal

$= -1229.39$ kcal

46. ***(a)*** Among the formation of B, C, D and E, the slowest step is the rate determining step. As formation of B, i.e. 0.002 mol/h per mole of A is the slowest step. Hence, step I is the rate determining step.

47. ***(d)*** In the given reaction as,

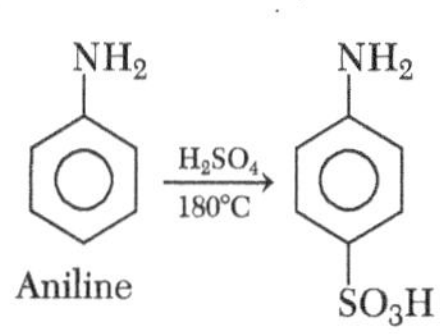

Aniline Sulphonic acid

Sulphanilic acid exist as dipolar ion (Zwitter ion). It act as inner salt and —SO_3H group diminishes the basic character of —NH_2. Thus, statement (d) is true.

48. ***(c)*** The structure of PO_4^{3-} is as follows

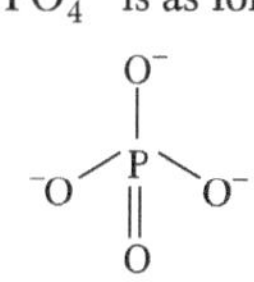

As phosphorus central atom forms four (4) sigma bonds and has no lone pair of electrons. It uses four hybrid orbitals and thus, hybridisation of P-atom is sp^3.

49. ***(c)*** Electronic configuration of the given species are

a. $Sm^{2+}(Z = 62) = [Xe]\,4f^6$

b. $Eu^{2+}\ (Z = 63) = [Xe]\,4f^7$

c. $Yb^{2+}(Z = 70) = [Xe]\,4f^{14}$

d. $Ce^{2+}(Z = 58) = [Xe]\,4f^1 5d^1$

On the basis of above electronic configuration, only $Yb^{2+}(Z = 70)$ has fully-filled orbitals i.e. no unpaired electrons. Thus, it is dimagnetic in nature.

50. ***(c)*** Alcohol in which —OH group is directly bonded to benzene are aromatic alcohols.

I. 

It is an aromatic alcohol, as —OH group is directly bonded with the benzene ring.

II. 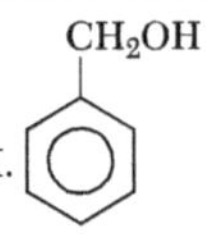

It is not an aromatic alcohol because —OH group is not bonded directly with the benzene group.

III. 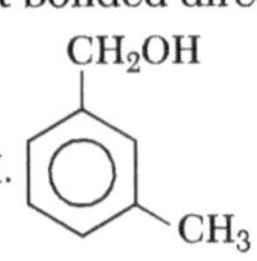

It is also not an aromatic alcohol because —OH group is not bonded directly with the benzene ring.

IV. 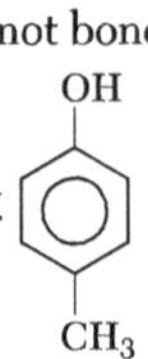

It is an aromatic alcohol because —OH group is directly bonded with the benzene ring.

51. ***(b)*** $RSiCl_3$ gives cross-linked silicon polymer on hydrolysis.

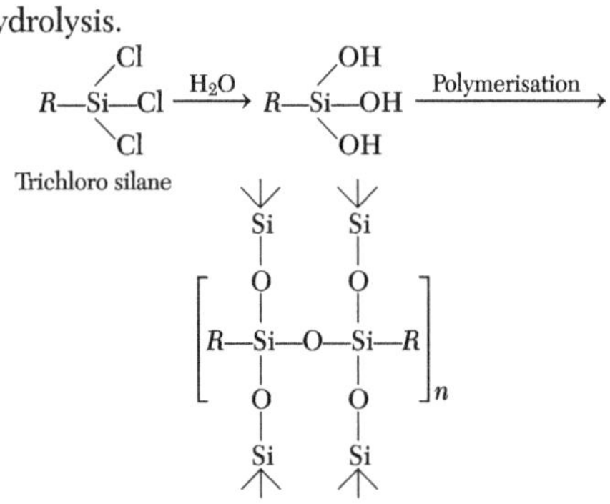

Cross-linked silicon polymer

52. ***(a)*** Let k be the rate constant of given reaction

$$2O_3 \xrightarrow{k} 3O_2$$

From the slowest step, $r = k'[O][O_3]$ ... (i)

To eliminate $[O]$, (From fast step)

$$k_{eq} = \frac{[O_2][O]}{[O_3]} \Rightarrow [O] = \frac{k_{eq}[O_3]}{[O_2]}$$

Now, substituting value of $[O]$ in Eq (i)

$$r = \frac{k' k_{eq}.[O_3][O_3]}{[O_2]}$$

Let $k' k_{eq} = k$

Thus, $r = k[O_3]^2[O_2]^{-1}$

53. ***(a)*** Depression in freezing point is given as $\Delta T_f = K_f m$

Molality of solution $(m) = \dfrac{\Delta T_f}{K_f}$

$$= \frac{0.6}{1.86} = 0.322\,\text{m}$$

Elevation in boiling point of solution is given as

$$\Delta T_b = K_b \times m$$
$$= 0.512 \times 0.322 = 0.165$$

$\therefore$ Boiling point of solution

$$= 373 + 0.16$$
$$= 373.16\text{ K}$$

54. ***(a)*** The strength of acidic character of boron trihalides depends upon $p\pi$ - $p\pi$ back bonding.

In boron trihalides, $p\pi$-$p\pi$ back bonding occurs due to empty orbital of boron and filled orbitals of halogen.

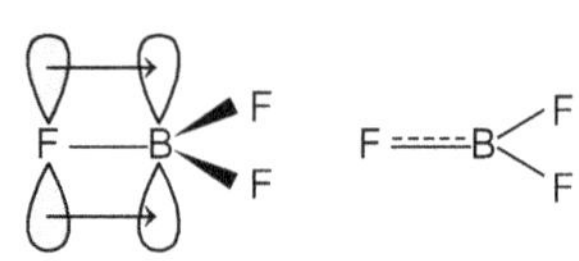

Phenomenon of back bonding in BF_3 molecule

The $p\pi$-$p\pi$ back bonding is shown maximum by BF_3, as the size of B and F are small and comparatively same. Due to this effect tendency of accepting lone-pair of electrons of boron decreases as size of halogen increases. The order of size of halogens are F < Cl < Br < I. Thus, acidic nature is in order

$$BF_3 < BCl_3 < BBr_3 < BI_3$$

55. ***(d)*** Mobility of alkali ions in aqueous solution $\propto$ size of ion

$$\propto \frac{1}{\text{size of hydrated ion}}.$$

The mobility of alkali ions is inversely proportional to the size of ion in hydrated state, because smaller the ion, more it will get hydrated and becomes larger in size, thus its ionic mobility decreases.

Hence, correct order of mobility is

$$Rb^+ > K^+ > Na^+ > Li^+$$

56. ***(c)*** Bakelite is a thermosetting plastic which is formed by the reaction of phenol with HCHO in the presence of H_2SO_4 (conc).

The reaction occurs as follows

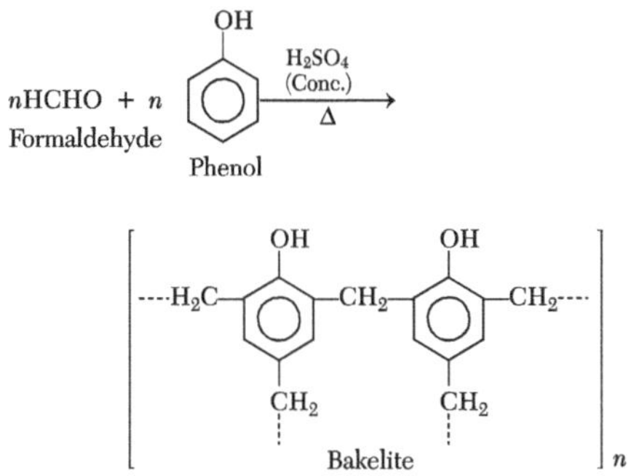

A cross-linked condensation polymer (at *o*-and *p*-position) bakelite is thus produced.

57. ***(c)*** Paracetamol is the drug, that can be used as an antipyretic as well as an analgesic, i.e. to reduce the fever and gives relieve from pain. Its structure is as follows

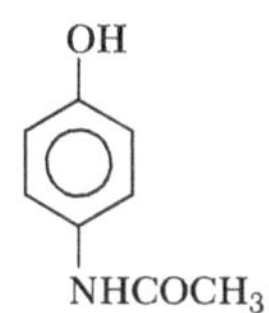

Paracetamol

58. ***(a)*** In the given structure, ring (I) has one O-atom and is a six membered ring, thus is called pyranose ring, whereas ring (II) has one O-atom and is a five membered ring, thus is called furanose.

The disaccharide is linked by a glycosidic bond between C-1 of glucose (in the α-position, α-linkage) and C-2 of fructose (in the β-position, β-linkage).

59. ***(c)*** Due to size and geometries of bases, the only possible pairing in DNA are between G (guanine) and C (cytosine) through three H-bonds and between A (adenine) and T (thymine) through two H-bonds.

Thus, the complementary strand of DNA will be

T A G C A T A C

(complementary strand)

60. ***(a)*** When Na_2SO_3 reacts with hot and dil. H_2SO_4 , it gives $SO_2(g)$ which decolourises bromine-water.

The reaction proceed as follows:

(i) $Na_2SO_3 + H_2SO_4 \longrightarrow Na_2SO_4 + SO_2 + H_2O$

(ii) $Br_2 + H_2O \longrightarrow 2HBr + [O]$

(iii) $SO_2 + [O] \longrightarrow SO_3$

(decolourisation of bromine water)

61. ***(b)*** The given reaction is for preparation of DDT (dichlorodiphenyl trichloroethane). Complete reaction is given as follows

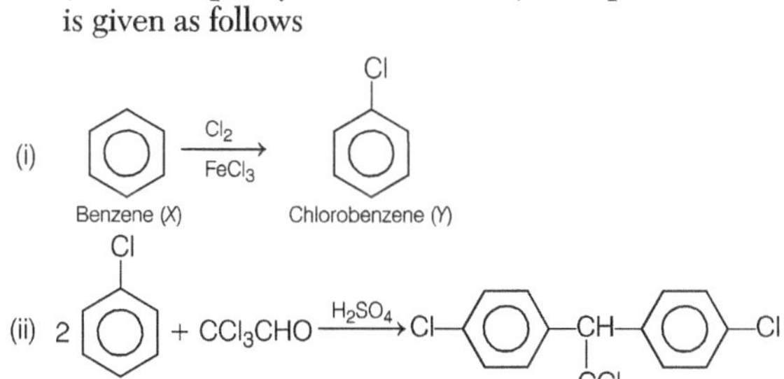

62. ***(d)*** The mixture of $CO(g)$ and $H_2(g)$ is known as water gas as it is used in the synthesis of other compounds,

$$\underbrace{CO(g) + 2H_2(g)}_{\text{Water gas}} \xrightarrow[\text{Catalyst}]{\text{Cobalt}} \underset{\text{Methanol}}{CH_3OH(l)}$$

63. ***(c)*** Chromium hydroxide, i.e. $Cr(OH)_3$ when reacts with NaOH and Na_2O_2, gives yellow colour of sodium chromate. The whole reaction is as follows

$$3Na_2O_2 + 2Cr(OH)_3 \longrightarrow \underset{\text{(Yellow)}}{2Na_2CrO_4} + 2H_2O + 2NaOH$$

64. ***(a)*** Cl is the good leaving group, i.e. the weakest nucleophile among the given options.

∴ The rate of reaction is faster, when 'Z' is 'Cl'.

65. ***(b)*** Use Nernst equation, i.e.

$$E(\text{emf}) = E^\circ_{\text{cell}} - \frac{0.0591}{2}\log\frac{[\text{Product}]}{[\text{Reactant}]}$$

For reaction (i)

$$E_1 = E^\circ_{\text{cell}} - \frac{0.0591}{2}\log\frac{0.1}{1}$$

$$E_1 = E^\circ_{\text{cell}} + \frac{0.0591}{2}$$

For reaction (ii)

$$E_2 = E^\circ_{\text{cell}} - \frac{0.0591}{2}\log\frac{1}{1}$$

$$E_2 = E^\circ_{\text{cell}} - 0$$

$$E_2 = E^\circ_{\text{cell}}$$

For reaction (iii)

$$E_3 = E^\circ_{\text{cell}} - \frac{0.0591}{2}\log\frac{1}{0.1}$$

$$E_3 = E^\circ_{\text{cell}} - \frac{0.0591}{2}$$

Hence, $E_1 > E_2 > E_3$.

66. ***(c)*** This test is used for the presence of sulphur in any compound. When sodium nitroprusside, react with sulphide ion, (e.g. Na_2S) purple colouration is produced, in which the following reaction takes place.

$$Na_2S + \underset{\text{(Sod. nitroprusside)}}{Na_2[Fe(CN)_5NO]} \longrightarrow \underset{\substack{\text{Purple colouration} \\ \text{(sod. thionitroprusside)}}}{Na_4[Fe(CN)_5NOS]}$$

67. ***(c)***

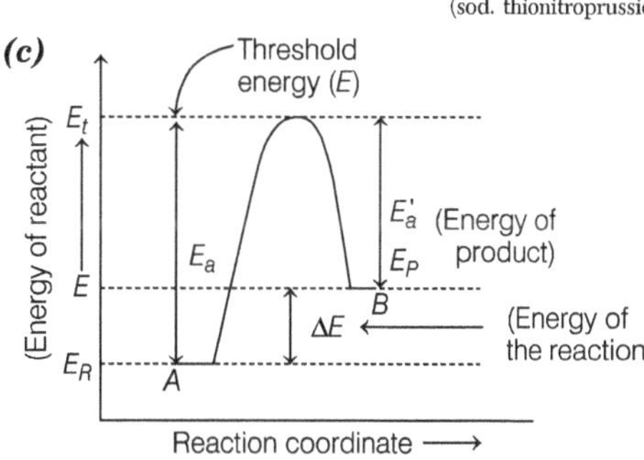

where, E_a = activation energy of forward reaction

E'_a = activation energy of backward reaction

The above energy profile diagram shows that $E_a > E'_a$.

The potential energy of the product is greater than that of the reactant, so the reaction is endothermic.

$$E_a = E_a' + \Delta E \Rightarrow E_t = E_a \text{ or } E_t > E_a'$$

∴ The threshold energy is equal or greater than activation energy.

68. ***(c)*** Ethene and benzene on reaction with anhy.$AlCl_3$ produce ethyl benzene as follows

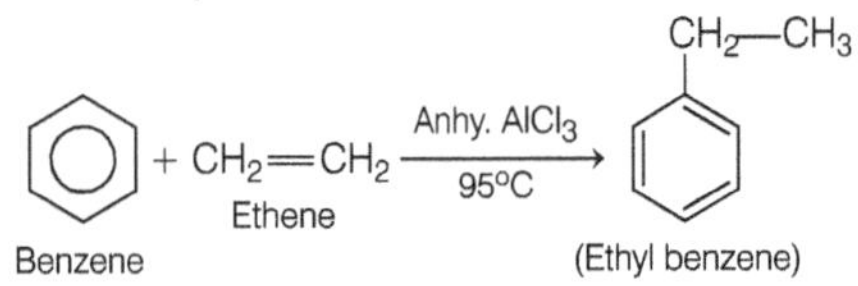

69. ***(b)*** Given, pH of solution $A = 3$

$$\text{pH} = -\log[\text{H}^+]$$

$$[\text{H}^+] = 10^{-\text{pH}}$$

$$[\text{H}^+] = 10^{-3}\text{M}$$

$[\text{H}^+]$ of $A = 10^{-3}$M and, pH of solution $B = 2$

$$[\text{H}^+] = 10^{-\text{pH}} = 10^{-2}\text{M}$$

$[\text{H}^+]$ of $B = 10^{-2}$ M

Thus, total $[\text{H}^+]$ are $= 10^{-3} + 10^{-2} = 1.1 \times 10^{-3}$ M

$$\therefore \quad \text{pH}_{(\text{final})} = -\log[\text{H}^+]$$

$$= -\log[1.1 \times 10^{-3}]$$

$$= 3 - \log 1.1$$

$$= 3 - 1.04 = 1.9$$

70. ***(a)*** Freundlich isotherm is related with the pressure of gas adsorbed as follows

$$\frac{x}{m} = k \cdot p^{1/n}$$

where, x/m = amount of gas (x) adsorbed over the surface of mass (m).

p = pressure of gas

k = Freundlich constant

Thus, value of 'n' is always greater than 1 and therefore value of $\frac{1}{n}$ lies between 0 and 1.

71. ***(c)*** Electronic configuration of Mn is $(Z = 25)$ $= [\text{Ar}]3d^5 4s^2$

∴ Electronic configuration of Mn^{4+} is $= [\text{Ar}]\, 3d^3$

Thus, Mn^{4+} has 3 unpaired electrons and its spin only magnetic moment (μ) is $= \sqrt{n(n+2)}$ BM

where, n = number of unpaired electrons.

$$\mu = \sqrt{3(3+2)} \text{ BM}$$

$$\Rightarrow \quad \mu = \sqrt{15} = 3.87 \text{ BM}$$

$$\approx 4 \text{ BM}$$

72. ***(a)*** ∵ Gram equivalents of an acid

$= $ gram equivalents of KOH i.e. $\frac{0.45}{E} = \frac{20 \times 0.5}{1000}$

where, E = equivalent mass $= \frac{0.45 \times 1000}{20 \times 0.5} = 45$

$$\text{Basicity} = \frac{\text{Molecular mass}}{\text{Equivalent mass}}$$

Thus, basicity $= \frac{90}{45} = 2$

73. ***(a)*** Given, $A + B \rightleftharpoons C + D$

	$A + B$		$C + D$	
Initial conc.	4	4	0	0
At equilibrium	(4 – 2)	(4 – 2)	2	2
i.e.	2	2	2	2

$$K_{\text{eq}} = \frac{[C][D]}{[A][B]}$$

$$\therefore \quad K_{\text{eq}} = \frac{2 \times 2}{2 \times 2} = 1$$

74. ***(b)*** Given,

125 mL of 1 M $AgNO_3$ = 13.5 g of Ag

∵ 1000 mL of $AgNO_3$ = 108 g of Ag

∴ 125 mL of $AgNO_3 = \frac{108 \times 125}{1000} = 13.5$ g

Also, 108 g of Ag is deposited by = 96500 C

∴ 13.5 g of Ag is deposited by $= \frac{96500 \times 13.5}{108}$

$= 12062.5$ C

$$\because \quad Q = it$$

∴ t (time) $= Q/i = \frac{12062.50}{241.25}$, $t = 50$ s

75. ***(b)*** Orbitals of nearly equal energy are called degenerated orbitals. As, energy of,

a. $\sigma 2s > \sigma 1s$

b. $\pi\, 2p_x \approx \pi 2p_y$

c. $\pi^*\, 2p_x < \sigma^*\, 2p_z$

d. $\sigma 2p_x < \sigma^*\, 2p_z$

Thus, only $\pi\, 2p_x$ and $\pi\, 2p_y$ are degenerate orbitals.

76. ***(c)*** Total energy (E) of any photon is given by the relation : $E = \frac{hc}{\lambda}$

where, h = Planck's constant $= 6.6 \times 10^{-34}$ J-s

c = velocity of light $= 3 \times 10^{-10}$ m/s

λ = wavelength = 150 pm

$= 1.5 \times 10^{-10}$ m.

Thus, $E = \frac{hc}{\lambda} = \frac{6.6 \times 10^{-34} \times 3 \times 10^8}{1.5 \times 10^{-10}}$;

$= 1.32 \times 10^{-15}$ J

and, energy of ejected electron

$$E' = \frac{1}{2}mv^2$$

where,

m = mass of electron = 9.1×10^{-31} kg

v = velocity of electron = 1.5×10^7 m/s

$$E' = \frac{1}{2}mv^2 = \frac{1}{2} \times 9.1 \times 10^{-31} \times (1.5 \times 10^7)^2$$

$$= 1.024 \times 10^{-16}$$

Thus, total energy of photon = binding energy of electron (B) + energy of ejected electron (E')

Thus, $1.32 \times 10^{-15} = B + E'$

$$\therefore \quad B = E - E'$$
$$= [1.32 \times 10^{-15}] - [1.024 \times 10^{-16}]$$
$$= 1.2176 \times 10^{-15} \text{ J}$$
$$= \frac{1.2176 \times 10^{-15}}{1.6 \times 10^{-19}} \text{ eV} = 7.6 \times 10^3 \text{ eV}$$

77. ***(b)*** The degree of unsaturation for compound C_7H_8 is three (3), as it reacts with Cl_2/Δ. Hence, C_7H_8 is toluene, i.e $C_6H_5CH_3$. The whole reaction occurs as follows:

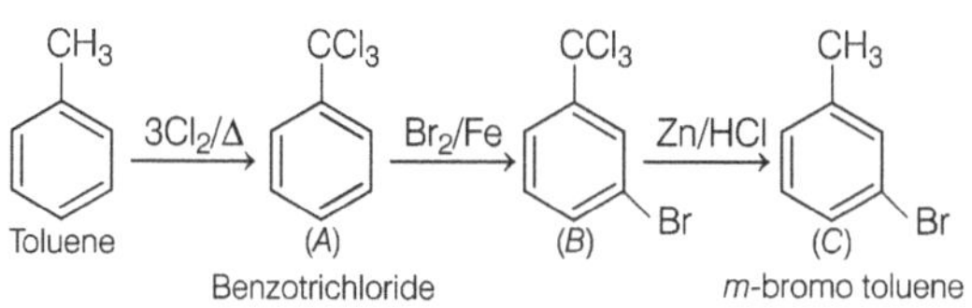

78. ***(c)***

Element	Ratio (by weight)	Atomic weight	Mol ratio	Simplest mol ratio
C	9	12	$\frac{9}{12} = 0.75$	$\frac{0.75}{0.25} = 3$
H	1	1	$\frac{1}{1} = 1.00$	$\frac{1.00}{0.25} = 4$
N	3.5	14	$\frac{3.5}{14} = 0.25$	$\frac{0.25}{0.25} = 1$

$\therefore$ Empirical formula = C_3H_4N

The empirical formula mass = $3 \times 12 + 4 \times 1 + 1 \times 14$
$= 54$ g

Given, molecular mass = 108 g

Thus, $n = \frac{108}{54} = 2$

Hence, molecular formula = (Empirical formula) $\times$ 2

$(C_3H_4N) \times 2 = C_6H_8N_2$

79. ***(a)*** π-bond do not participate in hybridisation

a. $H—\overset{sp}{C}\equiv\overset{sp}{C}—\overset{sp}{C}\equiv\overset{sp}{C}—H$ b. $\overset{sp^3}{CH_3}—\overset{sp}{C}\equiv\overset{sp}{C}—\overset{sp^3}{CH_3}$

c. $\underset{sp^2}{CH_2} = \underset{sp}{C} = \overset{sp^2}{CH_2}$ d. $\underset{sp^2}{CH_2} = \underset{sp^2}{CH}—\underset{sp}{C}\equiv\underset{sp}{CH}$

$\therefore$ Option (a) is a compound in which all carbon atoms are *sp* hybridised.

80. ***(a)*** In bassemer converter, copper sulphide is partially oxidised to cuprous oxide, which further reacts with the remaining copper sulphide to give copper and sulphur dioxide,

i.e. $Cu_2S + 2Cu_2O \longrightarrow 6Cu + SO_2$

a. English Proficiency

81. (***b***) 'Many a' takes singular noun, as well as singular verb. So, replace 'have' by 'has'.

82. (***a***) The correct syntax for interrogative sentences is 'wh' word + do/does+sub+verb+?

So, we should replace 'it takes' by 'does it take'.

83. (***c***) Replace 'his' by 'ones' to make the possessive correct.

84. (***a***) Assured is followed by preposition 'of'.

85. (***d***) Preposition 'to' should be used with 'submit' here.

86. ***(d)*** *Audacious* means bold or fearless. So, its correct antonym will be 'timid', which means 'lacking in courage or confidence.'

87. ***(c)*** Cogent, which means reasonable and convincing. The correct antonym will be 'dissuasive', which means divert from the measure or purpose.

88. ***(b)*** Taint means a mark of disgrace or blemish. The correct antonym will be 'clear', which means free of ambiguity or doubt.

89. ***(b)*** Vivacious means full of life and energy or animated.

So, its correct synonym will be 'Lively'.

90. ***(b)*** Sporadic means rare and scattered in occurrence. Here 'scattered' has the same sense. It is the correct synonym.

91. (***d***) Both official and corporate India is allergic to 'mention of clean technology.'

92. (***d***) According to the passage, 'Failure in crops' is a life and death question to many Indians.

93. ***(a)*** The most similar in meaning to the word 'profligacy' is 'wastefulness'.

94. ***(a)*** India cannot tolerate any further 'crop failure'.

95. ***(d)*** The reason could be 'US wants to use it as a handle against the developing countries in the forthcoming meet'.

b. Logical Reasoning

96. ***(d)*** In the 'College', education is given to 'Students', in the same way, treatment is given to the 'Patient' in 'Hospital'.

97. ***(b)*** Except 35, all others are multiples of 9, but 35 is the multiple of 5 and 7.

98. ***(c)*** Pattern of the series is as shown below

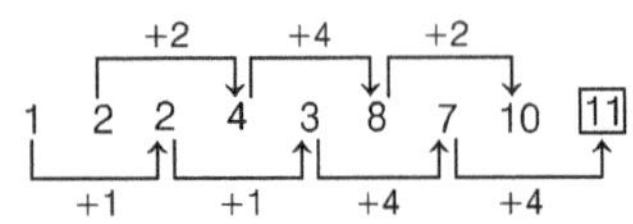

This series contains two separate series. In first series, 1 is added in first two steps and 4 is added in third and fourth steps.
In second series, 2 and 4 is added alternately.

99. ***(b)*** From given information,

Ramesh > Mohan > Shyam ...(i)

Gautam > Ramesh > Rajat ...(ii)

Shyam > Rajat ...(iii)

Combining Eqs. (i), (ii) and (iii), we get

Gautam > Ramesh > Mohan > Shyam > Rajat

Hence, Rajat is the shortest among all.

100. ***(b)*** If figure shown in option (b), is placed in the place of missing portion of the original figure, then it is completed as shown below

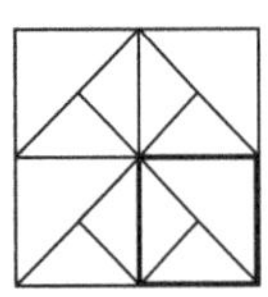

101. ***(b)*** Figure given in option (b) can be formed by joining the pieces given in question figure, as shown below.

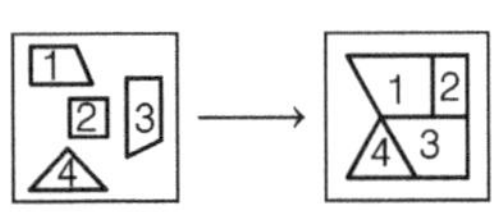

102. ***(d)*** In figure (X), the square sheet of paper is being folded along the vertical line of symmetry, so that right half of the sheet overlaps the left half.
In figure (Y), the sheet is folded further to a quarter.
In figure (Z), two squares are punched in the folded sheet. Clearly, the punched squares will be created in each quarter of the paper and after unfolding the first fold the figure will look like as,

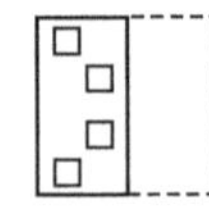

then after unfolding the last fold the transparent sheet will look like as given in option (d).

i.e.

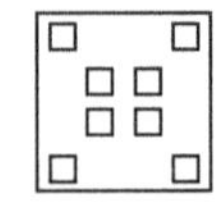

103. ***(a)*** In each row, the number of objects increases by 1 at each step from left to right. So, in third row, the missing segment contain 2 + 1 = 3 dots.

104. ***(b)*** One of the three dots occupies the region which is common to the circle, rectangle and triangle; another dot occupies the region which is common to the triangle, circle, rectangle and square and the third dot occupies the region which is common to the circle, rectangle and square. These three characteristics as shown by these three dots are found in figure (b). It possesses region which is common to the circle, rectangle and triangle, a region which is common to the triangle circle, rectangle, and square and a region which is common to the circle, rectangle and square

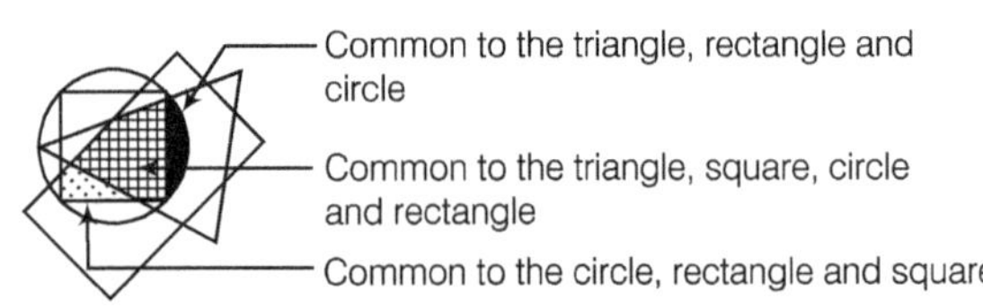

105. ***(a)*** Naming the figure,

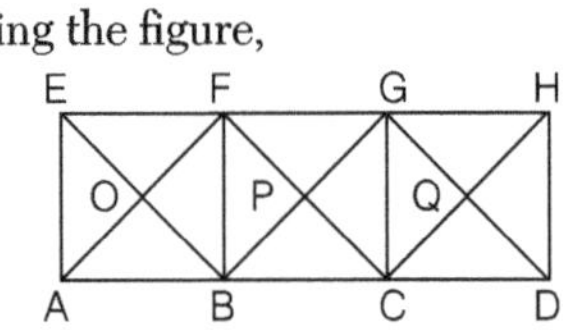

Clearly, there are 28 triangles in the given figure namely, ΔEOF, ΔAOE, ΔAOB, ΔBOF, ΔABF, ΔBEF, ΔABE, ΔAEF, ΔBPF, ΔFPG, ΔCPG, ΔBPC, ΔBFG, ΔBCG, ΔCFG, ΔBCF, ΔGQC, ΔCDQ, ΔDQH, ΔGQH, ΔGDC, ΔGDH, ΔGHC, ΔCDH, ΔAFC, ΔBGD, ΔEBG and ΔFCH.

Mathematics

106. *(d)* Given, $a = 3b$

$$e = \sqrt{1 - \frac{b^2}{a^2}} \Rightarrow e = \sqrt{1 - \frac{b^2}{9b^2}}$$

$$\Rightarrow \quad e = \sqrt{1 - \frac{1}{9}} = \frac{\sqrt{8}}{3} = \frac{2\sqrt{2}}{3}$$

107. *(a)* We have, $\sum_{m=1}^{n} \tan^{-1}\left(\frac{2m}{m^4 + m^2 + 2}\right)$

$$= \sum_{m=1}^{n} \tan^{-1}\left(\frac{(m^2 + m + 1) - (m^2 - m + 1)}{1 + (m^2 + m + 1)(m^2 - m + 1)}\right)$$

$$= \sum_{m=1}^{n} [\tan^{-1}(m^2 + m + 1) - \tan^{-1}(m^2 - m + 1)]$$

$$\left[\because \tan^{-1} x - \tan^{-1} y = \tan^{-1}\left(\frac{x - y}{1 + xy}\right)\right]$$

$$= (\tan^{-1} 3 - \tan^{-1} 1) + (\tan^{-1} 7 - \tan^{-1} 3) + (\tan^{-1}(13) - \tan^{-1} 7) + \ldots + (\tan^{-1}(n^2 + n + 1) - \tan^{-1}(n^2 - n + 1))$$

$$= \tan^{-1}(n^2 + n + 1) - \tan^{-1} 1$$

$$= \tan^{-1}\left(\frac{n^2 + n + 1 - 1}{1 + (n^2 + n + 1)}\right) = \tan^{-1}\left(\frac{n^2 + n}{n^2 + n + 2}\right)$$

108. *(b)* Given, $(2x^2 + 3x + 4)^{10} = \sum_{r=0}^{20} a_r x^r$ ...(i)

Replacing x by $\frac{2}{x}$, we get

$$\left(\frac{8}{x^2} + \frac{6}{x} + 4\right)^{10} = \sum_{r=0}^{20} a_r \left(\frac{2}{x}\right)^r$$

$$\Rightarrow 2^{10}(2x^2 + 3x + 4)^{10} = \sum_{r=0}^{20} a_r 2^r x^{20-r}$$

$$\Rightarrow \quad 2^{10} \sum_{r=0}^{20} a_r x^r = \sum_{r=0}^{20} a_r 2^r x^{20-r} \quad \text{[from Eq. (i)]}$$

Comparing coefficient of x^8 both sides, we get

$$2^{10} a_8 = a_{12} 2^{12} \Rightarrow \frac{a_8}{a_{12}} = \frac{2^{12}}{2^{10}} = 4$$

109. *(d)* In ΔADB,

$$\angle ABD = \angle ADB$$

[Angles opposite to equal sides are equal]

$$\therefore \quad \angle ADC = \pi - \angle ABD$$

Using 2nd formula in $m - n$ cot theorem

Here, $m = 1$ and $n = 1$

$$(1 + 1)\cot(\pi - B) = 1 \cot B - 1 \cot C$$

$$\Rightarrow \quad 3 \cot B = \cot C$$

$$\Rightarrow \quad \frac{\tan B}{\tan C} = 3$$

110. *(a)* We have,

$$L = \lim_{n \to \infty} \sqrt{n} \int_0^1 \frac{dx}{(1 + x^2)^n}$$

We know that, $(1 + x^2)^n > 1 + nx^2$

$$\Rightarrow \quad \frac{1}{(1 + x^2)^n} < \frac{1}{1 + nx^2}$$

$$\Rightarrow \quad \int_0^1 \frac{dx}{(1 + x^2)^n} < \int_0^1 \frac{1}{1 + nx^2} dx$$

$$\Rightarrow \quad \int_0^1 \frac{1}{(1 + x^2)^n} < \frac{1}{\sqrt{n}} [\tan^{-1} \sqrt{n}\, x]_0^1$$

$$= \frac{1}{\sqrt{n}} \tan^{-1} \sqrt{n}$$

$$\therefore L = \lim_{n \to \infty} \sqrt{n} \int_0^1 \frac{1}{(1 + x^2)^n} dx < \lim_{n \to \infty} \sqrt{n} \frac{1}{\sqrt{n}} \tan^{-1} \sqrt{n}$$

$$\Rightarrow \quad L < \frac{\pi}{2}$$

$$\therefore \quad \frac{1}{2} < L < 2$$

111. *(d)* We have,

$$\sqrt{x + 3 - 4\sqrt{x - 1}} + \sqrt{x + 8 - 6\sqrt{x - 1}} = 1$$

$$\sqrt{(\sqrt{x - 1})^2 - 2 \times 2\sqrt{x - 1} + 4} + \sqrt{(\sqrt{x - 1})^2 - 2 \times 3\sqrt{x - 1} + 9} = 1$$

$$\Rightarrow \quad \sqrt{(\sqrt{x - 1} - 2)^2} + \sqrt{(\sqrt{x - 1} - 3)^2} = 1$$

$$\Rightarrow \quad |\sqrt{x - 1} - 2| + |\sqrt{x - 1} - 3| = 1$$

$$\Rightarrow \quad |\sqrt{x - 1} - 2| + |\sqrt{x - 1} - 3|$$

$$= (\sqrt{x - 1} - 2) - (\sqrt{x - 1} - 3)$$

We know that,

If $\quad |x - a| + |x - b| = (x - a) - (x - b)$

Then, $\quad (x - a)(x - b) < 0$

$$\therefore \quad (\sqrt{x - 1} - 2)(\sqrt{x - 1} - 3) < 0$$

$$\Rightarrow \quad 2 < \sqrt{x - 1} < 3 \Rightarrow 5 < x < 10$$

$\therefore$ Equation have infinite many solutions.

112. *(d)* We have,

$$ax + y + z = a \quad \ldots \text{(i)}$$

$$x + by + z = b \quad \ldots \text{(ii)}$$

$$x + y + cz = c \quad \ldots \text{(iii)}$$

are inconsistent

$$\therefore \quad \begin{vmatrix} a & 1 & 1 \\ 1 & b & 1 \\ 1 & 1 & c \end{vmatrix} = 0 \text{ and } \begin{vmatrix} a & 1 & 1 \\ b & b & 1 \\ c & 1 & c \end{vmatrix} \neq 0$$

Now, $\begin{vmatrix} a & 1 & 1 \\ 1 & b & 1 \\ 1 & 1 & c \end{vmatrix} = 0$

$$a(bc-1)-1(c-1)+1(1-b)=0$$
$$\Rightarrow \quad abc-a-c+1+1-b=0$$
$$\Rightarrow \quad abc-a-b-c+2=0$$

and $\begin{vmatrix} a & 1 & 1 \\ b & b & 1 \\ c & 1 & c \end{vmatrix} \neq 0$

It is possible only when $a \neq 1$, $b \neq 1$ and $c \neq 1$.

113. *(a)* P = Probability of getting an even number $= \frac{1}{2}$

$$q = 1 - \frac{1}{2} = \frac{1}{2}$$

For atleast 4 times an even number, probability is

$${}^7C_4\left(\frac{1}{2}\right)^4\left(\frac{1}{2}\right)^3 + {}^4C_5\left(\frac{1}{2}\right)^5\left(\frac{1}{2}\right)^2 + {}^7C_6\left(\frac{1}{2}\right)^6\left(\frac{1}{2}\right)^1 + {}^7C_7\left(\frac{1}{2}\right)^7$$

$$= \frac{1}{2}$$

114. *(b)* Equation of plane containing line $x-y-1=0$ and $z-1=0$ is $(x-y-1)+\lambda(z-1)=0$

$$\Rightarrow \quad x-y+\lambda z-1-\lambda=0 \quad \ldots\text{(i)}$$

Since, this plane is parallel to line

$$\frac{x}{2}-\frac{z}{3}=1 \text{ and } y=3$$
$$\frac{x-1}{2}=\frac{y-3}{0}=\frac{z}{3} \quad \ldots\text{(ii)}$$
$$\therefore \quad 2(1)+(0)(-1)+3(\lambda)=0$$
$$\Rightarrow \quad \lambda=-2/3$$

Putting the value of λ in Eq. (i), we get

$$x-y-\frac{2}{3}z-1+\frac{2}{3}=0 \text{ and } 3x-3y-2z=1$$

115. *(d)* We have, $(1+x+x^2)^{20} = \sum_{r=0}^{40} a_r x^r$

x is replaced by $-\frac{1}{x}$, we get

$$\left(1-\frac{1}{x}+\frac{1}{x^2}\right)^{20} = \sum_{r=0}^{40}(-1)^r a_r\left(\frac{1}{x}\right)^r$$
$$\Rightarrow \quad (1-x+x^2)^{20} = \sum_{r=0}^{40}(-1)^r a_r x^{40-r}$$

$\sum_{r=0}^{39}(-1)^r a_r \cdot a_{r+1}$ is coefficient of x in product of

$$(1+x+x^2)^{20}\left(1-\frac{1}{x}+\frac{1}{x^2}\right)^{20} = 0$$

116. *(a)* Let

$$L = \lim_{x \to \frac{3\pi}{4}} \frac{4\sin^2 x\cos x - \cos x + \sin x}{\sin x + \cos x}$$
$$L = \lim_{x \to \frac{3\pi}{4}} \frac{2\sin x(\sin^2 x + \cos^2 x + 2\sin x\cos x - 1) - \cos x + \sin x}{\sin x + \cos x}$$
$$L = \lim_{x \to \frac{3\pi}{4}} \frac{2\sin x[(\sin x + \cos x)^2 - 1] - \cos x + \sin x}{\sin x + \cos x}$$
$$L = \lim_{x \to \frac{3\pi}{4}} \frac{2\sin x(\sin x + \cos x)^2 - 2\sin x - \cos x + \sin x}{\sin x + \cos x}$$
$$L = \lim_{x \to \frac{3\pi}{4}} \frac{(\sin x + \cos x)(2\sin x(\sin x + \cos(x-1)}{\sin x + \cos x}$$
$$L = \lim_{x \to \frac{3\pi}{4}} 2\sin x(\sin x + \cos x) - 1 = -1$$

117. *(b)* Given,

$$\mathbf{a} \times \mathbf{b} = \mathbf{c}$$
$$\mathbf{b} \times \mathbf{c} = \mathbf{a}$$
$$\mathbf{c} \times \mathbf{a} = \mathbf{b}$$
$$(\mathbf{b} \times \mathbf{c}) \cdot \mathbf{a} = \mathbf{a} \cdot \mathbf{a}$$
$$(\mathbf{c} \times \mathbf{a}) \cdot \mathbf{b} = \mathbf{b} \cdot \mathbf{b}$$

Now, $\mathbf{a}^2 - \mathbf{b}^2 = (\mathbf{b} \times \mathbf{c}) \cdot \mathbf{a} - (\mathbf{c} \times \mathbf{a}) \cdot \mathbf{b}$

$$= [\mathbf{b}\,\mathbf{c}\,\mathbf{a}] - [\mathbf{c}\,\mathbf{a}\,\mathbf{b}] = 0$$

$\therefore \quad |\mathbf{a}|^2 - |\mathbf{b}|^2 = 0$

$\Rightarrow \quad |\mathbf{a}| - |\mathbf{b}| = 0$

We know that,

$$|\mathbf{a} \times \mathbf{b}\;\; \mathbf{b} \times \mathbf{c}\;\; \mathbf{c} \times \mathbf{a}| = [\mathbf{a}\;\mathbf{b}\,\mathbf{c}]^2$$
$$[\mathbf{a}\;\mathbf{b}\,\mathbf{c}] = [\mathbf{a}\;\mathbf{b}\,\mathbf{c}]^2$$
$$|\mathbf{a}\,\mathbf{b}\mathbf{c}][-[\mathbf{a}\;\mathbf{b}\,\mathbf{c}] - 1] = 0$$
$$[\mathbf{a}\;\mathbf{b}\,\mathbf{c}] = 1$$
$$|\mathbf{a}||\mathbf{b}||\mathbf{c}| = 1$$
$$\therefore \quad |\mathbf{a}| = |\mathbf{b}| = |\mathbf{c}| \neq 2$$

118. *(a)* Let $z = x + iy$

Then $\arg z = \tan^{-1}\left(\frac{y}{x}\right)$

Given, $\arg(z) = \frac{\pi}{6}$

$$\therefore \quad \tan^{-1}\frac{y}{x} = \frac{\pi}{6}$$
$$\Rightarrow \quad \frac{y}{x} = \tan\frac{\pi}{6} = \frac{1}{\sqrt{3}}$$
$$\Rightarrow \quad x = \sqrt{3}\,y$$

BITSAT 2019

Also given, $|z - 2\sqrt{3}i| = \lambda$

$$|x + iy - 2\sqrt{3}\, i| = \lambda$$
$$x^2 + (y - 2\sqrt{3})^2 = \lambda^2$$

This is the equation of circle whose centre is $(0, 2\sqrt{3})$ and $r = \lambda$.

Now, $x - \sqrt{3}y = 0$ touches the circle.

$$\therefore \quad \lambda = \left|\frac{0 - \sqrt{3}\,(2\sqrt{3})}{\sqrt{1+3}}\right|$$
$$\Rightarrow \quad \lambda = \left|\frac{6}{2}\right| = 3$$
$$\therefore \quad \lambda = 3$$

119. ***(a)*** We have,

$$l + m + n = 0 \quad \text{...(i)}$$

and $$l^2 = m^2 + n^2 \quad \text{...(ii)}$$

Putting the value of l in Eq. (ii), we get

$$(m+n)^2 = m^2 + n^2$$
$$\Rightarrow \quad m^2 + n^2 + 2mn = m^2 + n^2$$
$$\Rightarrow \quad mn = 0$$
$$\Rightarrow \quad m = 0, n = 0$$

when $m = 0, l = -n$

$\therefore$ Direction cosines are $\left(-\frac{1}{\sqrt{2}}, 0, \frac{1}{\sqrt{2}}\right)$.

When $n = 0, l = -m$

$\therefore$ Direction cosines are $\left(-\frac{1}{\sqrt{2}}, \frac{1}{\sqrt{2}}, 0\right)$.

Angle between lines are

$$\cos\theta = \left(\frac{-1}{\sqrt{2}} \times -\frac{1}{\sqrt{2}}\right) + \left(0 \times \frac{1}{\sqrt{2}}\right) + \left(\frac{1}{\sqrt{2}} \times 0\right)$$
$$\cos\theta = \frac{1}{2} \Rightarrow \theta = \pi/3$$

120. ***(c)*** Let the foot of perpendicular be (h, k).

Equation of tangent with slope m passing (h, k) is $y = mx \pm \sqrt{6m^2 + 2}$, where $m = -h/k$

$$\Rightarrow \quad \sqrt{\frac{6h^2}{k^2} + 2} = \frac{h^2 + k^2}{k}$$
$$6h^2 + 2k^2 = (h^2 + k^2)^2$$

So, required locus is $6x^2 + 2y^2 = (x^2 + y^2)^2$.

121. ***(d)*** Let

$$h(x) = f(x) - 2g(x)$$

Now, $$h(0) = f(0) - 2g(0)$$
$$\Rightarrow \quad h(0) = 2 - (0)$$
$$[\because f(0) = 2, g(0) = 0]$$
$$\Rightarrow \quad h(0) = 2$$

Also, $$h(1) = f(1) - 2g(1)$$
$$\Rightarrow \quad h(1) = 6 - 2(2)$$
$$[\because f(1) = 6, g(1) = 2]$$
$$\Rightarrow \quad h(1) = 2$$
$$\therefore \quad h(0) = h(1)$$
$$h'(x) = f'(x) - 2g'(x)$$

By Rolle's theorem $h'(c) = 0$

$\therefore f'(c) - 2g'(c) = 0, f'(c) = 2g'(c)$

122. ***(b)*** We have,

$$I = \int\left(1 + x - \frac{1}{x}\right)e^{x + \frac{1}{x}}\,dx$$
$$I = \int e^{x+\frac{1}{x}}dx + \int\left(x - \frac{1}{x}\right)e^{x+\frac{1}{x}}\,dx$$
$$I = \int e^{x+\frac{1}{x}}dx + \int x\left(1 - \frac{1}{x^2}\right)e^{x+\frac{1}{x}}dx$$
$$I = \int e^{x+\frac{1}{x}}\,dx + xe^{x+\frac{1}{x}} - \int e^{x+\frac{1}{x}}\,dx$$
$$I = xe^{x+\frac{1}{x}} + C$$

123. ***(d)*** We have,

$$(1 + ax + bx^2)(1 - 2x)^{18}$$
$$\Rightarrow (1 + ax + bx^2)(1 - {}^{18}C_1 2x + {}^{18}C_2(2x^2) - {}^{18}C_3(2x)^3 + {}^{18}C_4(2x)^4 \ldots)$$

Coefficient of x^3 is $-{}^{18}C_3(2)^3 + a \cdot {}^{18}C_2(2)^2 - b{}^{18}C_1(2)$

and coefficient of x^4 is

$${}^{18}C_4(2)^4 - {}^{18}C_3(2)^3 a + {}^{18}C_2(2)^2 b$$

Since, coefficient of x^3 and x^4 are zero.

$$\therefore -{}^{18}C_3(2)^3 + {}^{18}C_2(2)^2(a) - {}^{18}C_1(2)\,b = 0$$
$$\Rightarrow \quad \frac{4 \times 17 \times 16}{3 \times 2} - 17a + b = 0 \quad \text{... (i)}$$

and $$80 - \frac{32}{3}a + b = 0 \quad \text{...(ii)}$$

Solving Eqs. (i) and (ii), we get

$$a = 16 \text{ and } b = \frac{272}{3}$$

124. ***(a)*** Equation of plane parallel to the plane $x - 2y + 2z - 5 = 0$ is $x - 2y + 2z + \lambda = 0$.

Since, this plane is unit distance from origin.

$$\therefore \quad 1 = \left|\frac{\lambda}{\sqrt{1+4+4}}\right|$$
$$|\lambda| = 3 \Rightarrow \lambda = \pm 3$$

$\therefore$ Equation of plane is

$$x - 2y + 2z = 3$$

125. *(b)* Let A be the event that maximum is 6.

B be the event that minimum is 3.

$$P(A) = \frac{{}^5C_2}{{}^8C_3} \text{ (number less than 6 is 5)}$$

$$P(B) = \frac{{}^5C_2}{{}^8C_3} \text{ (number greater than 3 is 5)}$$

$$P(A \cap B) = \frac{{}^2C_1}{{}^8C_3}$$

$\therefore$ Required probability

$$= P(B/A) = \frac{P(A \cap B)}{P(A)} = \frac{{}^2C_1}{{}^5C_2} = \frac{2}{10} = \frac{1}{5}$$

126. *(b)* We have, $\frac{\sin(\lambda\alpha)}{\sin\alpha} - \frac{\cos(\lambda\alpha)}{\cos\alpha} = \lambda - 1$

$$\Rightarrow \sin(\lambda\alpha)\cos\alpha - \cos(\lambda\alpha)\sin\alpha = (\lambda - 1)\sin\alpha\cos\alpha$$

$$\Rightarrow \sin(\lambda - 1)\alpha = (\lambda - 1)\sin\alpha\cos\alpha$$

$$\Rightarrow \sin(\lambda - 1)\alpha = \frac{\lambda - 1}{2}\sin 2\alpha$$

It is possible, when $\lambda - 1 = 0 \Rightarrow \lambda = 1$

and $\frac{\lambda - 1}{2} = 1 \Rightarrow \lambda = 3$

$\therefore \lambda$ has only two solution.

127. *(b)* Given, $y = ax^2 + bx + c$ has two x-intercepts one is positive and one is negative and vertex is $(2, -2)$. The graph of $y = ax^2 + bx + c$ is

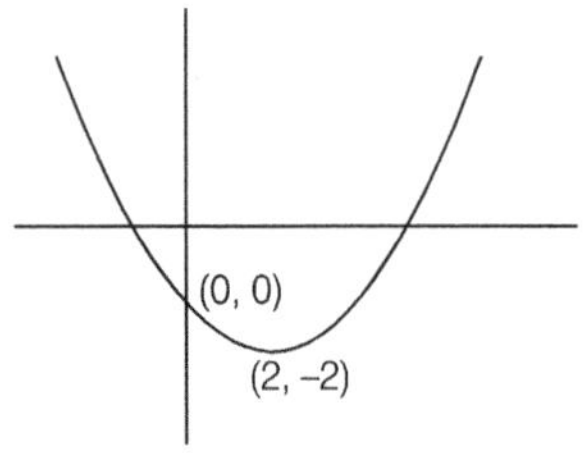

Clearly from graph

$$c > 0, a > 0$$

and $-\frac{b}{a} > 0 \Rightarrow -b > 0$

$$\Rightarrow b < 0$$

$$f(1) < 0 \Rightarrow a + b + c < 0$$

$$ab < 0, ac < 0, bc > 0$$

128. *(b)* Let $A = (1, 2)$ equation of line is

$$\frac{x-1}{\cos\theta} = \frac{y-2}{\sin\theta} = r = \frac{\sqrt{6}}{3}$$

$$\therefore \quad x = \frac{\sqrt{6}}{3}\cos\theta + 1, y = \frac{\sqrt{6}}{3}\sin\theta + 2$$

This point lies on line $x + y = 4$

$$\therefore \frac{\sqrt{6}}{3}\cos\theta + \frac{\sqrt{6}}{3}\sin\theta = 1$$

$$\Rightarrow \sin\theta + \cos\theta = \frac{\sqrt{3}}{\sqrt{2}}$$

$$\Rightarrow \sin\left(\frac{\pi}{4} + \theta\right) = \frac{\sqrt{3}}{2}$$

$$\Rightarrow \frac{\pi}{4} + \theta = 60^\circ \text{ or } 120^\circ$$

$$\Rightarrow \theta = 15^\circ \text{ or } 75^\circ$$

$$\therefore \text{Larger angle} = 75^\circ.$$

129. *(d)* $[P + 1] = [P] + 1$, Let $[P] = n$, then n is integer

$\therefore ([P + 1], [P]) = (n + 1, n)$ lie inside the region of circles

$S_1 = x^2 + y^2 - 2x - 15 = 0, C_1 = (1, 0), r_1 = 4$

and $S_2 = x^2 + y^2 - 2x - 7 = 0, C_2 = (1, 0), r_2 = 2\sqrt{2}$

Both circles are concentric.

$\therefore (n + 1)^2 + n^2 - 2(n + 1) - 7 > 0$

and $(n + 1)^2 + n^2 - 2(n + 1) - 15 < 0 \Rightarrow 4 < n^2 < 8$

Which is not possible for any integer.

130. *(a)* We have,

$$\frac{dy}{dx} = e^{x-y}(e^x - e^y)$$

$$\Rightarrow e^y\frac{dy}{dx} + e^y e^x = e^{2x}$$

Put $e^y = v$

$$\therefore e^y\frac{dy}{dx} = \frac{dv}{dx}$$

$$\therefore \frac{dv}{dx} + ve^x = e^{2x}$$

This is linear equation

$$\text{I.F.} = e^{\int e^x dx} = e^{e^x}$$

$\therefore$ Required solution is

$$v \cdot e^{e^x} = \int e^{e^x} \cdot e^{2x}dx + C$$

$$v \cdot e^{e^x} = e^{e^x}(e^x - 1) + C$$

$$\Rightarrow e^y \cdot e^{e^x} = e^{e^x}(e^x - 1) + C$$

$$\Rightarrow e^y = e^x - 1 + ce^{-e^x}$$

131. *(d)* Given, curve is $(y - x)^2 = x^3$

$$\Rightarrow y - x = \pm x\sqrt{x}$$

$$\Rightarrow y = x \pm x\sqrt{x}$$

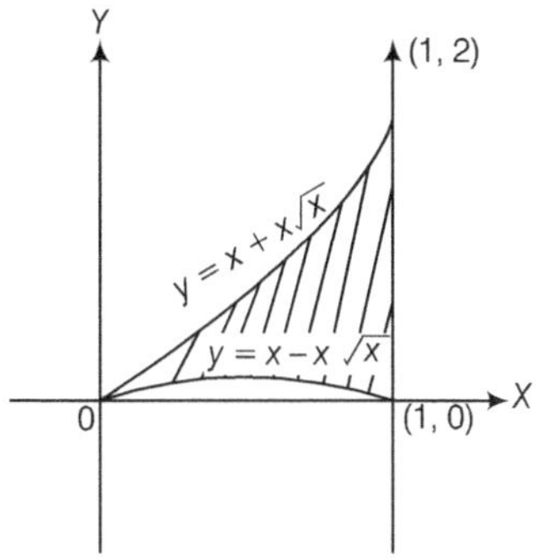

$\therefore$ Area bounded $= \left|\int_0^1 \{(x + x\sqrt{x}) - (x - x\sqrt{x})\} dx\right|$

$= \left|2\int_0^1 x^{3/2} dx\right| = \frac{4}{5}$ sq unit

132. ***(d)*** We have, $f(x) = \int_0^x (3\sin x + 4\cos x) dx$

$\Rightarrow \quad f'(x) = 3\sin x + 4\cos x = 5\sin\left(x + \tan^{-1}\frac{4}{3}\right)$

Since, $\frac{5\pi}{4} < x < \frac{4\pi}{3}$

$\Rightarrow \quad \pi < x + \tan^{-1} 4/3 < 2\pi$

$\because f'(x) < 0$, i.e. $f(x)$ is decreasing.

$\because$ Least value of $f(x) is \left(\frac{5\pi}{4}, \frac{4\pi}{3}\right) = f\left(\frac{4\pi}{3}\right)$

Least $f(x) = \int_0^{4\pi/3} (3\sin x + 4\cos x)\, dx$

$= [-3\cos x + 4\sin x]_0^{4\pi/3}$

$= -3\left(-\frac{1}{2} - 1\right) + 4\left(-\frac{\sqrt{3}}{2}\right)$

$= \frac{9 - 4\sqrt{3}}{2}$

133. ***(a)*** Let $z_1 = re^{i\theta}$

Then, $\quad \bar{z}_1 = re^{-i\theta}$

$\frac{z_1}{\bar{z}_1} = e^{2i\theta} = e^{2\pi i/n}$

$\Rightarrow \quad \theta = \frac{\pi}{n}$

Also, $\frac{I_m(z_1)}{R_e(z_1)} = \tan\theta = \tan\frac{\pi}{n} = \sqrt{2} - 1$

$\Rightarrow \quad \tan\frac{\pi}{8} = \sqrt{2} - 1$

$\therefore \quad n = 8$

134. ***(b)*** $(1 + x + x^3 + x^4)^{10} = [(1 + x)(1 + x^3)]^{10}$

$= (1 + x)^{10}(1 + x^3)^{10}$

$= (1 + {}^{10}C_1 + {}^{10}C_2\, x^2 + {}^{10}C_3\, x^3 + {}^{10}C_4\, x^4 \ldots)$

$\times (1 + {}^{10}C_1\, x^3 + {}^{10}C_2\, x^6 + \ldots)$

$\therefore$ Coefficient of $x^4 = ({}^{10}C_1)({}^{10}C_1) + {}^{10}C_4$

$= 100 + 210 = 310$

135. ***(c)*** Since only one letters can not be in wrong envelope so that at least two letters are in wrong envelope means all the letters are not in right envelopes.

$\therefore \; x = 6! - 1 = 720 - 1 = 719$

Y = number of ways, so that all the letters are in wrong envelopes

$= 6!\left(1 - \frac{1}{1!} + \frac{1}{2!} - \frac{1}{3!} + \frac{1}{4!} - \frac{1}{5!} + \frac{1}{6!}\right)$

$= 360 - 120 + 30 - 6 + 1 = 265$

$\therefore \quad x - y = 719 - 265 = 454$

136. ***(c)*** We have, $x = \log_5 3 + \log_7 5 + \log_9 7$

Since, AM $\geq$ GM

$\therefore \quad \frac{\log_5 3 + \log_7 5 + \log_9 7}{3}$

$\geq (\log_5 3 \cdot \log_7 5 \cdot \log_9 7)^{1/3}$

$\Rightarrow \quad x \geq 3\left(\frac{\log 3}{\log 5} \times \frac{\log 5}{\log 7} \times \frac{\log 7}{\log 9}\right)^{1/3}$

$\Rightarrow \quad x > 3\left(\frac{1}{2}\right)^{1/3} \quad \Rightarrow x > \frac{3}{\sqrt[3]{2}}$

137. ***(c)*** $\quad x^2 - 2x + A = 0$

$\Rightarrow \quad x = 1 \pm \sqrt{1 - A}$

$\therefore \quad p = 1 - \sqrt{1 - A},\; q = 1 + \sqrt{1 - A}$

Similarly, $r = 9 - \sqrt{81 - B},\; s = 9 + \sqrt{81 - B}$

Since, $\quad p, q, r, s$ are in AP.

$\therefore \quad q - p = s - r$

$\Rightarrow \quad 2\sqrt{1 - A} = 2\sqrt{81 - B}$

$\Rightarrow \quad B = 80 + A$

and $\quad q - p = r - q$

$\Rightarrow \quad 3\sqrt{1 - A} = 8 - \sqrt{81 - B}$

$\Rightarrow \quad A = -3, B = 77$

138. ***(c)*** We have,

$(1 - a)x + y + z = 0$

$x + (1 - b)\,y + z = 0$

$x + y + (1 - c)\,z = 0$

has many solutions

$\therefore \quad \begin{vmatrix} 1 - a & 1 & 1 \\ 1 & 1 - b & 1 \\ 1 & 1 & 1 - c \end{vmatrix} = 0$

$\Rightarrow \quad ab + ac + bc = abc$

$\therefore \quad \frac{ab+bc+ac}{3} \geq (a^2 b^2 c^2)^{1/3} = (abc)^{1/3} \geq 3$

$\therefore \quad$ Minimum of $abc = (3)^3 = 27$

139. ***(c)*** Given,

$$AA' = A'A$$

and $B = A^{-1}A'$

So,
$$BB' = (A^{-1}A')(A^{-1}A')'$$
$$= (A^{-1}A')[(A')'(A^{-1})']$$
$$= (A^{-1}A')[A(A')^{-1}]$$
$$[\because (A')' = A, (A^{-1})' = (A')^{-1}]$$
$$= A^{-1}(A'A)(A')^{-1}$$
$$= A^{-1}(AA'(A')^{-1}$$
$$= (A^{-1}A)(A^{-1}A)'$$
$$= I \cdot I = I$$

140. ***(a)*** We have,

$$\sin^{-1}x + \tan^{-1}x = \frac{\pi}{2} \Rightarrow 0 < x < 1$$

$\therefore \quad \tan^{-1}x = \frac{\pi}{2} - \sin^{-1}x = \cos^{-1}x$

$\Rightarrow \quad \tan^{-1}x = \tan^{-1}\left(\frac{\sqrt{1-x^2}}{x}\right)$

$\Rightarrow \quad x^2 = \sqrt{1-x^2}$

$\Rightarrow \quad x^4 + x^2 - 1 = 0 \Rightarrow x^2 = \frac{-1 \pm \sqrt{5}}{2}$

$\Rightarrow \quad x^2 > 0$

$\therefore \quad x^2 = \frac{\sqrt{5}-1}{2}$

$\Rightarrow \quad = 2x^2 + 1 = \sqrt{5}$

141. ***(a)*** Given, $|\cos x| = 2[x]$

$$0 \leq |\cos x| \leq 1$$

$\therefore \quad 0 \leq 2[x] \leq 1$

$$0 \leq [x] \leq \frac{1}{2}$$
$$x = 0$$

$\therefore \quad |\cos x| = 0$

$$x : n\pi + \frac{\pi}{2}$$

For any value of $n \in N, [n] \neq 0$

Hence, the equation has no solution.

142. ***(c)*** $0 \leq \alpha, \beta \leq \frac{\pi}{4} \Rightarrow \alpha + \beta \leq \frac{\pi}{2}$

$\cos(\alpha+\beta) = 4/5 \Rightarrow \tan(\alpha+\beta) = 3/4$

$\sin(\alpha-\beta) = 5/13 \Rightarrow \alpha - \beta \in \theta_1$

and $\tan(\alpha-\beta) = 5/12$

Now, $\tan 2\alpha = \tan(\alpha+\beta+\alpha-\beta)$

$$= \frac{\tan(\alpha+\beta)+\tan(\alpha-\beta)}{1-\tan(\alpha+\beta)\tan(\alpha-\beta)}$$
$$= \frac{3/4+5/12}{1-3/4 \times \frac{5}{12}} = \frac{56}{33}$$

143. ***(c)*** 18 draws are required for 2 aces means in the first 17 draws, there is one ace and 16 non-ace and 18th draw 2nd ace.

$\therefore$ Required probability $= \frac{^{48}C_{16} \times {}^4C_1}{^{52}C_{17}} \times \frac{3}{35} = \frac{561}{15925}$

144. ***(c)*** Since, $(\mathbf{a}+p\mathbf{b})$ is normal to $\mathbf{c}$.

$\therefore \quad (\mathbf{a}+p\mathbf{b}) \cdot \mathbf{c} = 0$

$$((1-p)\hat{\mathbf{i}} + (2+2p)\hat{\mathbf{j}} + (3+p)\hat{\mathbf{k}}) \cdot (3\hat{\mathbf{i}} + \hat{\mathbf{j}}) = 0$$
$$3 - 3p + 2 + 2p = 0 \Rightarrow p = 5$$

145. ***(b)*** We have,

$$f(x) = x[x]$$
$$f(x) = px \qquad [\because [x] = p]$$
$$f'(x) = p = [x]$$

$\therefore \quad f'(x) = [x]$

146. ***(b)*** We have,

$$f(x) = x^2 e^{-x^2}$$

$\Rightarrow \quad f'(x) = 2xe^{-x^2} - 2x^3 e^{-x^2}$

$\Rightarrow \quad f'(x) = 2xe^{-x^2}(1-x^2)$

$\Rightarrow \quad f'(x) = 0 \Rightarrow x = 0, -1, 1$

$f'(-h) < 0, f'(h) > 0$, So minimum at $x = 0$

$f'(1-h) > 0, f'(1+h) < 0$, So maximum at $x = 1$

$f'(-1-h) > 0, f'(-1+h) < 0$, So maximum at $x = -1$

$\therefore$ Minimum $f(x) = f(0) = 0$

Maximum $f(x) = e^{-1} = \frac{1}{e}$

Since, $f(x) \geq 0 \ \forall \ x$

So, difference between maximum and minimum values $= \frac{1}{e} - 0 = \frac{1}{e}$

147. ***(a)*** Given,

$$f'(x) = f(x) + \int_0^1 f(x)\,dx, f(0) = 1 \qquad \ldots(i)$$

$\Rightarrow \quad f''(x) = f'(x) + 0 \Rightarrow \frac{f''(x)}{f'(x)} = 1$

$\Rightarrow \quad \int \frac{f''(x)}{f'(x)}\,dx = \int dx$

$\Rightarrow \quad \log f'(x) = x + C$

$\Rightarrow \quad f'(x) = Ae^x$

$\Rightarrow \quad f(x) = Ae^x + K$

BITSAT 2019

$$f(0) = A + K = 1 \quad \ldots\text{(ii)}$$

$$\therefore \quad Ae^x = Ae^x + K + \int_0^1 (Ae^x + k)\, dx$$

$$\Rightarrow \quad k + [Ae^x + Kx]_0^1 = 0$$

$$\Rightarrow \quad k + Ae - A + K = 0$$

$$\Rightarrow \quad A(e-1) + 2k = 0 \quad \ldots\text{(iii)}$$

From Eqs. (ii) and (iii), we get

$$A = \frac{2}{3-e}, k = \frac{1-e}{3-e}$$

$$\therefore \quad f(x) = \frac{2e^x}{3-e} + \frac{1-e}{3-e}$$

148. *(b)* We have,

$$y = e^{2x} + x^2$$

$$\Rightarrow \quad \frac{dy}{dx} = 2e^{2x} + 2x$$

$$\left(\frac{dy}{dx}\right)_{x=0} = 2$$

$\therefore$ Equation of normal at (0, 1) is

$$y - 1 = -\frac{1}{2}(x)$$

$$\Rightarrow \quad x + 2y - 2 = 0$$

$\therefore$ Distance from (0, 0) from normal $= \frac{2}{\sqrt{5}}$

149. *(c)* Given, line $\frac{x-1}{-1} = \frac{y-2}{3} = \frac{z-4}{1}$ is passes through $P(1, 2, 4)$.

To find the reflection of line we need one more point on the line clearly $M(0, 5, 5)$ also lie on line.

Let $N(\alpha, \beta, \gamma)$ be the reflection M in the

$$x + y + z = 7$$

$$\therefore \quad \frac{\alpha}{2} + \frac{\beta+5}{2} + \frac{\gamma+5}{2} + 7$$

$$\Rightarrow \quad \alpha + \beta + \gamma = 4$$

Also, MN is perpendicular, i.e. parallel to the normal of the plane

$$\therefore \quad \frac{\alpha}{1} = \frac{\beta-5}{1} = \frac{\gamma-5}{1} = \lambda$$

$$\Rightarrow \quad \alpha = \lambda, \beta = \lambda + 5, \gamma = \lambda - 5$$

$$\therefore \lambda + \lambda + 5 + \lambda + 5 = 4$$

$$\Rightarrow \quad \lambda = -2$$

$$\therefore \quad N = (-2, 3, 5)$$

Equation PN is $\frac{x-1}{-3} = \frac{y-2}{1} = \frac{z-4}{-1}$

150. *(b)* We have,

$$x + y + z = 5$$

$$x^2 + y^2 + z^2 = 9$$

$$\Rightarrow \quad x^2 + y^2 + (5 - x - y)^2 = 9$$

$$\Rightarrow \quad y^2 + x^2 - 5x - 5y + xy + 8 = 0$$

$$\Rightarrow \quad y^2 + y(x-5) + (x^2 - 5x + 8) = 0$$

Since $y \in R$

$$\therefore \quad (x-5)^2 - 4(x^2 - 5x + 8) \geq 0$$

$$\Rightarrow \quad x^2 - 10x + 25 - 4x^2 + 20x - 32 \geq 0$$

$$\Rightarrow \quad 3x^2 - 10x + 7 \leq 0$$

$$\Rightarrow \quad 3x^2 - 7x - 3x + 7 \leq 0$$

$$\Rightarrow \quad (3x-7)(x-1) \leq 0$$

$$\Rightarrow \quad x \in \left[1, \frac{7}{3}\right]$$

Length of interval $= \left(\frac{7}{3} - 1\right) = \frac{4}{3}$

BITSAT

Birla Institute of Technology & Science Admission Test

Solved Paper 2018*

Instructions

- There are 150 questions in all. The number of questions in each part is as given below.

		No. of Questions
Part I	Physics	1-40
Part II	Chemistry	41-80
Part III	a. English Proficiency	81-95
	b. Logical Reasoning	96-105
Part IV	Mathematics	106-150

- All questions are Multiple Choice Questions having four options out of which **only one** is correct.
- Each correct answer fetches 3 marks while incorrect answer has a penalty of 1 mark.
- Time allotted to complete this paper is 3 hrs.

PART I

Physics

1. Two inductors L_1 and L_2 are connected in parallel and a time varying current flows as shown in figure. The ratio of current $\frac{i_1}{i_2}$ at any time t is

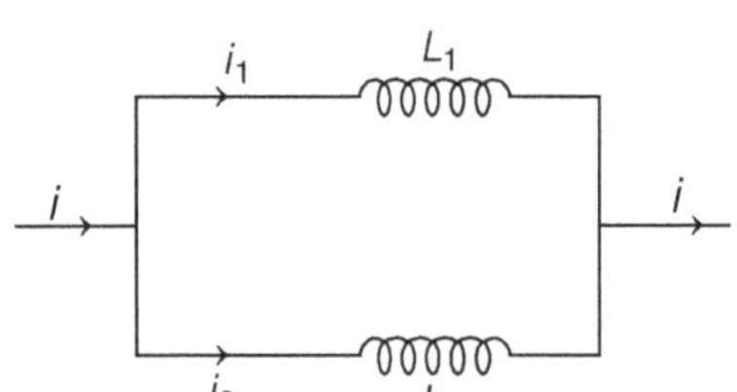

a. $\frac{L_2}{L_1}$ *b.* $\frac{L_1}{L_2}$

c. $\frac{L_2^2}{(L_1+L_2)^2}$ *d.* $\frac{L_1^2}{(L_1+L_2)^2}$

2. A parallel plate capacitor C with plates of unit area and separation d is filled with a liquid of dielectric constant $K = 2$, the level of liquid is $d/3$, initially.

Suppose, the liquid level decreases at a constant speed v, the time constant as a function of time is

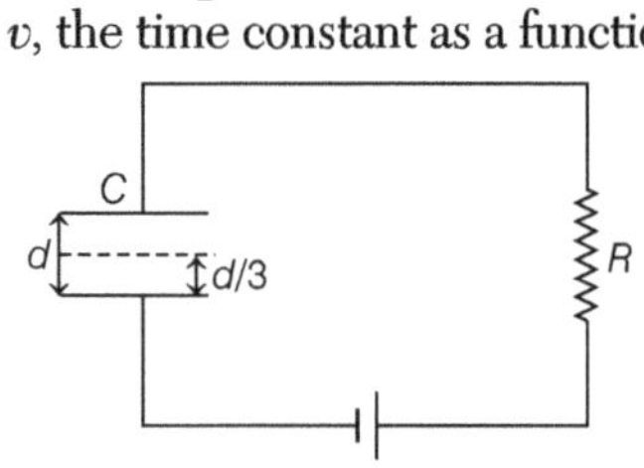

a. $\frac{6\varepsilon_0 R}{5d+3vt}$ *b.* $\frac{(15d+9vt)\varepsilon_0 R}{2d^3-3dvt-gv^2t^2}$

c. $\frac{6\varepsilon_0 R}{5d-3vt}$ *d.* $\frac{(15d-9vt)\varepsilon_0 R}{2d^3+3dvt-9v^2t^2}$

3. Steam at 100°C is passed into 1.1 kg of water contained in a calorimeter of water equivalent to 0.2 kg at 15°C till the temperature of the calorimeter and its contents rises to 80°C. The mass of steam condensed (in kg) is

(Given, latent heat of steam = 540 cal/g)

a. 0.130 *b.* 0.065 *c.* 0.260 *d.* 0.135

***Memory Based**

4. Dimension of which base quantity corresponds to that of $\sqrt{Gh/c^3}$ =?

a. Time *b.* Length *c.* Mass *d.* Temperature

5. A reservoir is at 827°C and Carnot's engine takes a thousand kilocalories of heat from it and exhausts it to a sink at 27°C. What is the amount of work and the efficiency of the engine?

a. 2.7×10^5 cal, 70.70% *b.* 2.72×10^5 cal, 72.72%

c. 2.70×10^5 cal, 80.70% *d.* 3.70×10^5 cal, 70.70%

6. A train moves towards stationary observer with speed 34 m/s. The train blows whistle and its frequency is registered by the observer as f_1. If the train's speed is reduced to 17 m/s, the frequency registered is f_2. If the speed of sound is 340 m/s, then the ratio f_1/f_2 is

a. 19/18 *b.* 18/19 *c.* 2 *d.* 1/2

7. An object of mass 5 kg is projected with a velocity 20 ms^{-1} at an angle 60°, to the horizontal. At the highest point of its path, the projectile explodes and breaks up into two fragments of masses 1 kg and 4 kg. The fragments separate horizontally after the explosion, which releases internal energy such that the KE of the system at the highest point is doubled. The separation between the two fragments when they reach the ground is

a. 52.25 m *b.* 44.25 m *c.* 65.32 m *d.* 78.76 m

8. An automobile moving with a speed of 36 km/h reaches an upward inclined road of angle 30°, its engine becomes switch off. If the coefficient of friction is 0.1, then how much distance will automobile move before coming to rest?

a. 12.53 m *b.* 21.42 m *c.* 15.43 m *d.* 8.53 m

9. In the circuit given below, the value of resistance X, when the potential difference between the points B and D is zero, will be

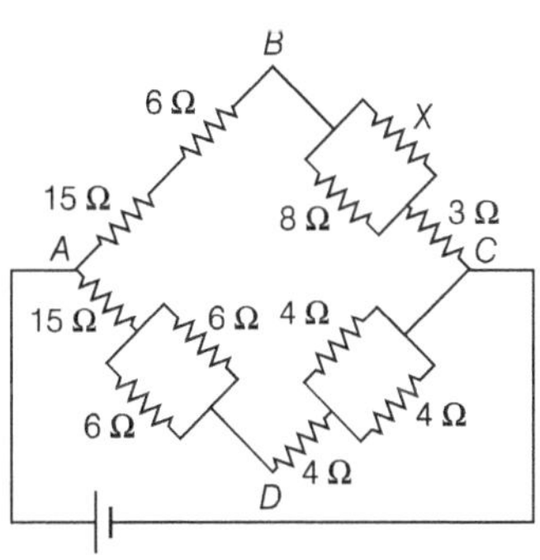

a. 9 Ω *b.* 8 Ω *c.* 4 Ω *d.* 6 Ω

10. A block of wood floats in water with (4/5) th of its volume submerged. If the same block just floats in a liquid, the density of liquid (in kg m^{-3}) is

a. 1250 *b.* 600

c. 400 *d.* 800

11. As current i flowing through the loop as shown in figure. The magnetic field at the centre O is

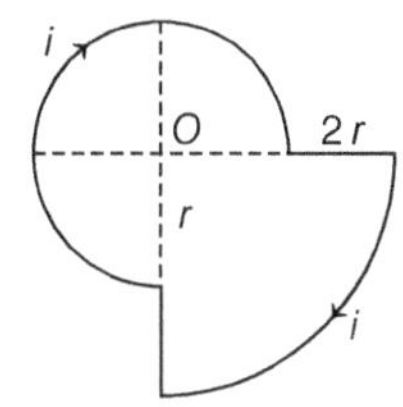

a. $\frac{7\mu_0 i}{12r}$ (acting downwards)

b. $\frac{5\mu_0 i}{12r}$ (acting upwards)

c. $\frac{7\mu_0 i}{12r}$ (acting upwards)

d. $\frac{5\mu_0 i}{12r}$ (acting downwards)

12. The ratio of angular momentum L to the atomic dipole moment μ_l for hydrogen like atoms and ions is

a. always constant and is equal to the ratio of mass to the charge of electron

b. always constant and is equal to twice the ratio of mass to the charge of electron

c. proportional to the principal quantum number n

d. proportional to $\frac{1}{n^2}$

13. Three rods of identical cross-sectional area and made from the same metal, form the sides of an isosceles triangle ABC right angled at B as shown in figure. The point A and B are maintained at temperature T and $\sqrt{2}\,T$ respectively, in the steady state. Now, assuming that only heat conduction takes place. The temperature of point C will be

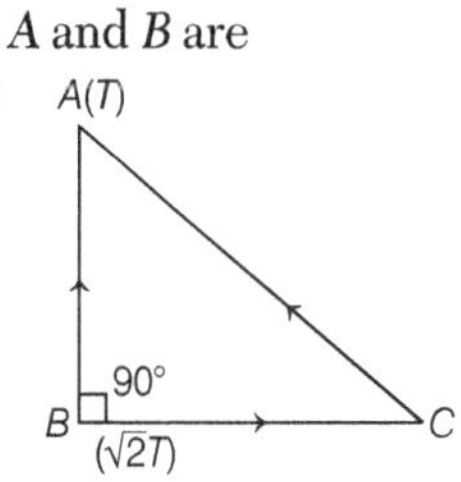

a. $\frac{T}{\sqrt{2}+1}$ *b.* $\frac{T}{\sqrt{2}-1}$

c. $\frac{3T}{\sqrt{2}+1}$ *d.* $\frac{\sqrt{3}\,T}{(\sqrt{2}+1)}$

14. A resistor R, an inductor L, a capacitor C and voltmeters V_1, V_2 and V_3 are connected to an oscillator in the circuit as shown in the adjoining diagram. When the frequency of the oscillation is increased, then at the resonant frequency, the voltmeter reading is zero in the case of

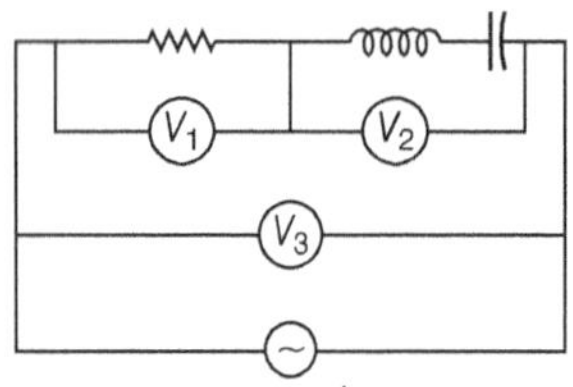

$e = e_0 \sin \omega t$

a. voltmeter V_1 only *b.* voltmeter V_2 only
c. voltmeter V_3 only *d.* All the three voltmeters

15. In Young's double slit experiment, intensity at a point is $\left(\frac{1}{4}\right)$ of the maximum intensity. Angular position of this point is

a. $\sin^{-1}\left(\frac{\lambda}{d}\right)$ *b.* $\sin^{-1}\left(\frac{\lambda}{2d}\right)$
c. $\sin^{-1}\left(\frac{\lambda}{3d}\right)$ *d.* $\sin^{-1}\left(\frac{\lambda}{4d}\right)$

16. The bob of simple pendulum is a spherical hollow ball filled with water. A plugged hole near the bottom of the oscillating bob get suddenly unplugged. During observation, till water is coming out, the time period of oscillation would

a. first increase and then decrease to the original value
b. first decrease and then increase to the original value
c. remain unchanged
d. increase towards a saturation value

17. At a certain temperature, the number density of charge carriers in a semiconductor is n. When an electric field is applied to it, the charge carriers drift with an average speed v. If the temperature of the semiconductor is raised, then

a. n will increase but v will decrease
b. n will decrease but v will increase
c. Both n and v will increase
d. Both n and v will decrease

18. Consider the acceleration, velocity and displacement of a tennis ball as its falls to the ground and bounces back. Directions of which of these change in the process?

a. Velocity only
b. Displacement and velocity
c. Acceleration, velocity and displacement
d. Displacement and acceleration

19. A convex lens shown in the figure is made up of two types of transparent materials. A point sources of light is placed on its principal axis. If reflections from the boundaries between layers are ignored, the lens will form

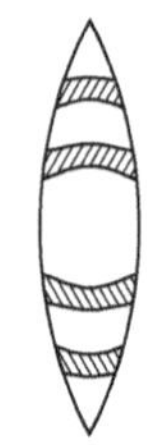

a. only one image *b.* two images
c. infinite images *d.* no image at all

20. If the time period is doubled, then the angular momentum of the body will (provided the moment of inertia of the body is constant)

a. remain constant *b.* quadruple
c. become half *d.* double

21. Breaking stress of a steel wire is p and the density of steel is ρ. The greatest length of steel wire that can hang vertically without breaking is

a. $\frac{p}{\rho g}$ *b.* $\frac{p}{2\rho g}$
c. $\frac{2p}{\rho g}$ *d.* None of these

22. A stone is projected with velocity $2\sqrt{gh}$, so that it just clears two walls of equal height h, at distance of $2h$ from each other. The time interval of passing between the two walls is

a. $\sqrt{h/g}$ *b.* $\sqrt{2h/g}$ *c.* $2\sqrt{h/g}$ *d.* $2h/g$

23. An object takes n times as much time to slide down a 45° rough inclined plane as it takes to slide down a perfectly smooth 45° inclined plane. The coefficient of kinetic friction between the rough plane and the object is

a. $n^2 - 1$ *b.* $1 - \frac{1}{n^2}$ *c.* $n^2 + 1$ *d.* $1 + \frac{1}{n^2}$

24. Some amount of a radioactive substance (half-life = 10 days) is spread inside a room and consequently the level of radiation becomes 50 times the permissible level for normal occupancy of the room. After how many days will the room be safe for occupation?

a. 20 days *b.* 34.8 days
c. 56.4 days *d.* 62.9 days

25. Helium gas goes through a cycle $ABCDA$ (consisting of two isochoric and two isobaric lines) as shown in figure. The efficiency of this cycle is approximately

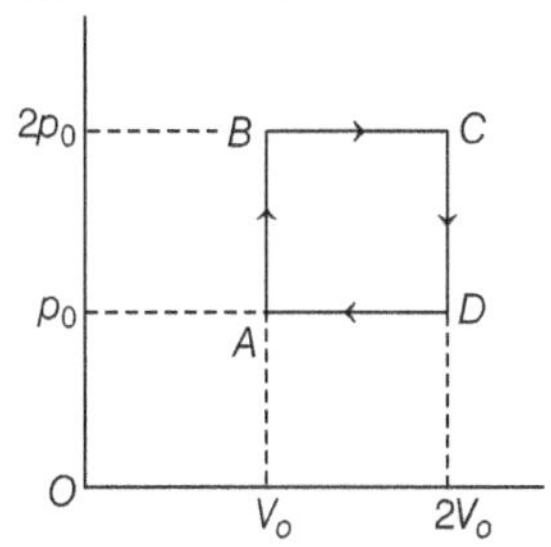

a. 15.4% *b.* 9.1% *c.* 10.5% *d.* 12.5%

26. The frequency and the intensity of incident beam of light falling on the surface of a photoelectric material is increased by a factor of two. This will

a. increase the maximum kinetic energy of the ejected photoelectrons by a factor of more than two and would increase the photoelectric current by a factor of two

b. increase the maximum kinetic energy of the photo electrons and would increase the photoelectric current both by a factor of two

c. increase the maximum kinetic energy of the photoelectrons by a factor of two and will have no effect on the magnitude of the photoelectric current produced

d. not produce any effect on the kinetic energy of the emitted photoelectrons but will increase the photoelectric current by a factor of two

27. A copper rod of mass m slides under gravity on two smooth parallel rails l distance apart and set an angle θ to the horizontal. At the bottom, the rails are joined by a resistance R in figure. There is a uniform magnetic field B prependicular to the plane of the rails. The terminal velocity of rod is

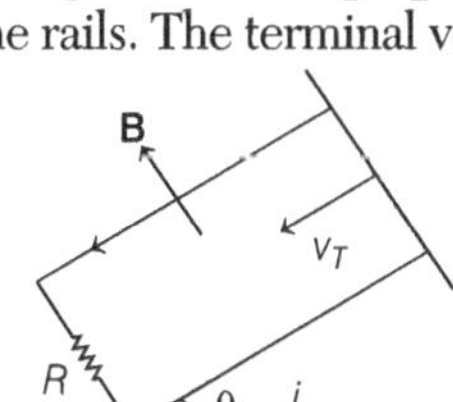

a. $\frac{mgR\tan\theta}{B^2l^2}$ *b.* $\frac{mgR\cot\theta}{B^2l^2}$

c. $\frac{mgR\sin\theta}{B^2l^2}$ *d.* $\frac{mgR\cos\theta}{B^2l^2}$

28. An asteroid of mass m is approaching earth, initially at a distance $10\,R_e$ with speed v_i. It hits earth with a speed v_f (R_e and M_e are radius and mass of earth), then

a. $v_f^2 = v_i^2 + \frac{2Gm}{R_e}\left(1+\frac{1}{10}\right)$

b. $v_f^2 = v_i^2 + \frac{2GM_e}{R_e}\left(1+\frac{1}{10}\right)$

c. $v_f^2 = v_i^2 + \frac{2GM_e}{R_e}\left(1-\frac{1}{10}\right)$

d. $v_f^2 = v_i^2 + \frac{2Gm}{R_e}\left(1-\frac{1}{10}\right)$

29. When photon of energy 4.0 eV strikes the surface of a metal A, the ejected photoelectrons have maximum kinetic energy T_A eV and de-Broglie wavelength λ_A. The maximum kinetic energy of photoelectrons liberated from another metal B by photon of energy 4.50 eV is $T_B = (T_A - 1.50)$ eV. If the de-Broglie wavelength of these photoelectrons $\lambda_B = 2\lambda_A$, then choose the correct statement(s).

a. The work function of A is 1.50 eV.

b. The work function of B is 4.0 eV.

c. $T_A = 3.2$ eV

d. All of the above

30. A pulley of radius 2 m is rotated about its axis by a force $= (20t - 5t^2)$ N (where t is measured in seconds) applied tangentially. If the moment of inertia of the pulley about its axis of rotation is 10 kg m^2, then the number of rotation made by the pulley before its direction of motion is reversed is

a. more than 3 but less than 6

b. more than 6 but less than 9

c. more than 9

d. less than 3

31. A planoconvex lens has thickness of 4 cm. When placed on a horizontal table, with the curved surface in contact with it, the apparent depth of the bottom most point of the lens is found to be 3 cm. If the lens is inverted such that the plane face is in contact with the table, the apparent depth of the centre of the plane face is found to be 25/8 cm. The focal length of the lens is (assume thickness of lens to be negligible)

a. 85 cm *b.* 59 cm *c.* 75 cm *d.* 7.5 cm

32. A spherically symmetric gravitational system of particles has mass density

$$\rho = \begin{cases} \rho_0 & \text{for } r \le R \\ 0 & \text{for } r > R \end{cases}$$

where, ρ_0 is a constant. A test mass can undergo circular motion under the influence of the gravitational field of particles. Its speed v as a function of distance $r(0 < r < \infty)$ from the centre of the system is represented by

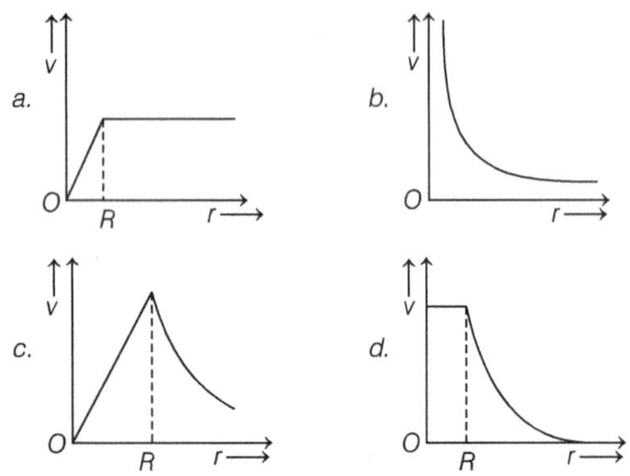

33. In the given circuit,

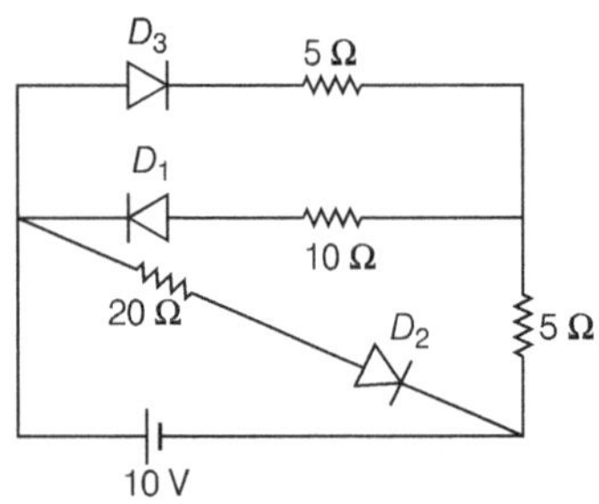

The current through battery is

a. 0.5 A *b.* 1 A *c.* 1.5 A *d.* 2.5 A

34. A trolley having mass of 200 kg moves with uniform speed of 36 kmh^{-1} on a frictionless track. A child of mass 20 kg runs on the trolley from one end to the other (10 m away) with a speed of 4 ms^{-1} relative to the trolley in a direction opposite to its motion and ultimately jumps out of the trolley. With how much velocity has the trolley moved from the time the child begins to run?

a. 10.36 ms^{-1} *b.* 11.36 ms^{-1}

c. 12.36 ms^{-1} *d.* 14.40 ms^{-1}

35. A gas has molar heat capacity $C = 37.55$ J $\text{mol}^{-1}\text{K}^{-1}$, in this process $pT = \text{constant}$. The number of degree of freedom of the molecule of gas is

a. 2 *b.* 3 *c.* 5 *d.* 7

36. If x, v and a denote the displacement, the velocity and the acceleration of a particle executing SHM of time period T. Then, which of the following does not change with time?

a. $\frac{aT}{x}$ *b.* $aT + 2\pi v$

c. $\frac{aT}{v}$ *d.* $a^2T^2 + 4\pi^2v^2$

37. To increase the current sensitivity of a moving coil galvanometer by 50%, its resistance is increased, so that new resistance becomes twice its initial resistance, by what factor does its voltage sensitivity change?

a. Increases by 15% *b.* Decreases by 15%

c. Increases by 25% *d.* Decreases by 25%

38. A ball is dropped vertically from a height d above the ground. It hits the ground and bounces up vertically to a height $d/2$. Neglecting subsequent motion and air resistance, its velocity v varies with height h above the ground as

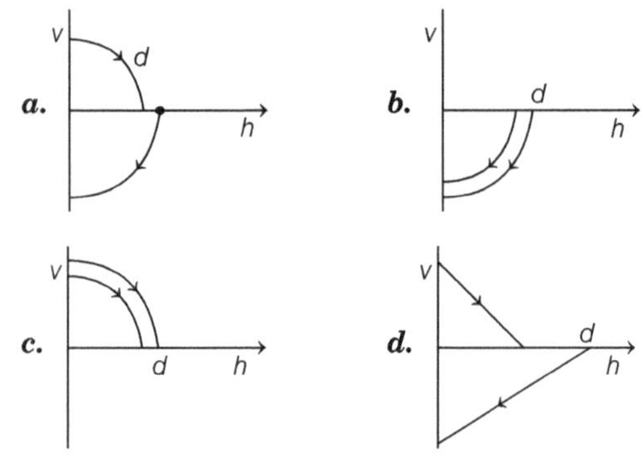

39. In the given circuit, it is observe that the current I is independent of the value of resistance R_5. Then, the resistance value must satisfy

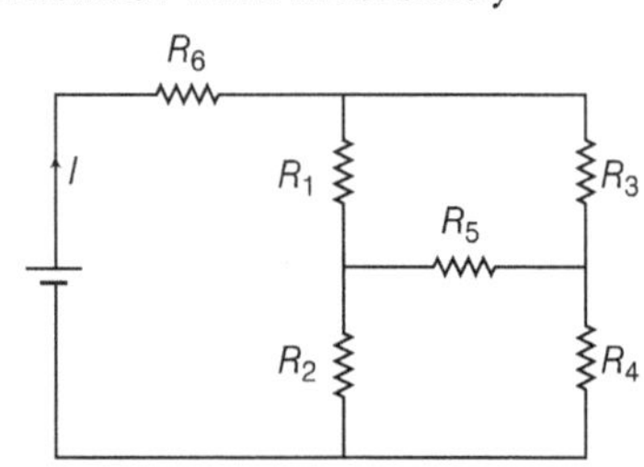

a. $\frac{1}{R_5} + \frac{1}{R_6} = \frac{1}{R_1 + R_2} \cdot \frac{1}{R_3 + R_4}$

b. $R_1R_4 = R_2R_3$

c. $R_1R_2R_5 = R_3R_4R_6$

d. $R_1R_3 = R_2R_4 = R_5R_6$

40. If a drop of liquid breaks into smaller droplets, it results in lowering of temperature of the droplets. Let a drop of radius R, break into N small droplets each of radius r, then decrease (drop) in temperature Q' (given, specific heat of liquid drop = S and surface tension = T)

a. $\frac{3T}{\rho S}\left[\frac{1}{r} - \frac{1}{R}\right]$ *b.* $-\frac{2T}{\rho S}\left[\frac{1}{r} - \frac{1}{R}\right]$

c. $\frac{2T}{\rho S}\left[\frac{1}{R} - \frac{1}{r}\right]$ *d.* $\frac{3T}{\rho S}\left[\frac{1}{R} - \frac{1}{r}\right]$

PART II

Chemistry

41. A metallic element has a cubic lattice. Each edge of the unit cell is 2Å and the density of metal is 2.5 g cm^{-3}. The unit cell in 200 g of metal are

a. 1×10^{24} *b.* 1×10^{22} *c.* 1×10^{20} *d.* 1×10^{25}

42. Four gases *P*, *Q*, *R* and *S* have almost same values of '*b*' but their '*a*' values (*a* and *b* are van der Waals' constant) are in the order $Q<R<S<P$. At a particular temperature, among the four gases, the most easily liquefiable one is

a. *P* *b.* *Q* *c.* *R* *d.* *S*

43. If $\left[\dfrac{0.51\times10^{-10}}{4}\right]$ metre is the radius of smallest electron orbit in hydrogen like atom, then the atom is

a. hydrogen atom *b.* He^+

c. Li^{2+} *d.* Be^{3+}

44. Which of the following is the correct order for the wavelength, of absorption in the visible region?

a. $[Ni(NO_2)_6]^{4-}<[Ni(NH_3)_6]^{2+}<[Ni(H_2O)_6]^{2+}$

b. $[Ni(NO_2)_6]^{4-}<[Ni(H_2O)_6]^{2+}<[Ni(NH_3)_6]^{2+}$

c. $[Ni(H_2O)_6]^{2+}<[Ni(NH_3)_6]^{2+}<[Ni(NO_2)_6]^{4-}$

d. $[Ni(NH_3)_6]^{2+}<[Ni(H_2O)_6]^{2+}<[Ni(NO_2)_6]^{4-}$

45. The empirical formula and molecular mass of a compound are CH_2O and 180g respectively. The molecular formula of the compound will be

a. $C_9H_{18}O_9$ *b.* CH_2O *c.* $C_6H_{12}O_6$ *d.* $C_2H_4O_2$

46. In the context of the Hall-Heroult process for the extraction of Al, which of this following statement is incorrect?

a. CO and CO_2 are produced in this process.

b. Al_2O_3 is mixed with CaF_2 which lowers the melting point of the mixture and brings conductivity.

c. Al^{3+} is reduced at the cathode to Al.

d. Na_3AlF_6 serves as the electrolyte.

47. Which test among the following is not used for the distinction among 1°, 2° and 3° aliphatic amine?

a. Hinsberg's reagent test

b. Carbylamine reaction

c. Azo dye test

d. Action with nitrous acid

48. The incorrect statement about carbonate (CO_3^{2-}) ion is,

a. It has planar structure

b. It has one coordinate bond

c. It has three resonating structure

d. Hydrolysis of CO_3^{2-} ion gives basic solution

49. Under the same reaction conditions, initial concentration of 1.386 mol dm^{-3} of a substance becomes half in 40 s and 20 s through first order and zero order kinetics, respectively. Ratio $\left(\dfrac{k_1}{k_0}\right)$ of the rate constants for first order (k_1) and zero order (k_0) of the reaction is

a. 0.5 $mol^{-1}dm^3$ *b.* 1.0 mol dm^{-3}

c. 1.5 mol dm^{-3} *d.* 2.0 $mol^{-1}dm^3$

50. Which substance has a dipole moment?

a. CCl_4 *b.* CH_2Cl_2

c. C_2Cl_2 *d.* C_2Cl_4

51. Enthalpy of combustion of methane and ethane are –210 kcal/mol and –368 kcal/mol respectively. The enthalpy of combustion of decane is

a. –1582 kcal *b.* –1632 kcal

c. –1700 kcal *d.* –1480 kcal

52. The correct sequence of reagents for the following conversion will be

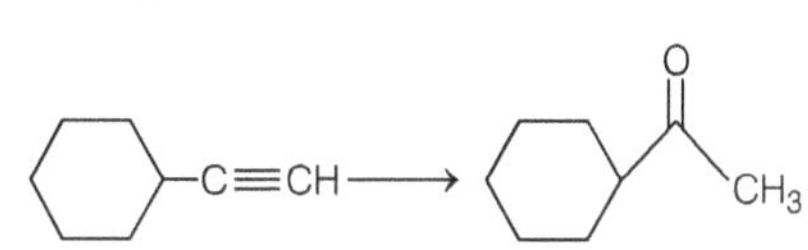

a. O_3 / Red P, $AlCl_3$, MeCOOH

b. $H_2SO_4+HgSO_4$, H_2O / Heat

c. O_3 / Zn - AcOH, $H_2SO_4+HgSO_4$ / H_2O / Heat

d. CH_3COOH, $H_2O_2+\bar{O}H$ / H_2O

53. In an atom, an electron is moving with a speed of 600 m/s with an accuracy of 0.005%, certainty with which the position of the electron can be located is ($h=6.6\times10^{-34}$ kg m^2s^{-1}, mass of electron, $e_m=9.1\times10^{-31}$ kg)

a. 1.52×10^{-4} m *b.* 5.10×10^{-3} m

c. 1.92×10^{-3} m *d.* 3.84×10^{-3} m

54. Which of the following diatomic molecules would be stabilised by the removal of an electron?

a. C_2 *b.* CN
c. N_2 *d.* O_2

55. 0.5F of electricity is passed through 500 mL of copper sulphate solution. The amount of copper which can be deposited will be

a. 63.5g
b. 31.75g
c. 15.80g
d. unpredictable

56. Consider the following sequence of reactions.

$$Z \xrightarrow{PCl_5} X \xrightarrow{\text{Alc. KOH}} Y \xrightarrow[2H_2O/\Delta]{\text{Conc. } H_2SO_4} Z$$

'Z' is

a. $CH_3CH_2CH_2OH$
b. $(CH)_2COH$
c. $CH_3—CH(OH)—CH_3$
d. None of the above

57. An equilibrium mixture at 300K contains N_2O_4 and NO_2 at 0.28 and 1.1 atm pressure, respectively. If the volume of the container is doubled, the new equilibrium pressure of these two gases are respectively.

a. 0.064 atm and 0.095 atm
b. 0.64 atm and 0.095 atm
c. 0.095 atm and 0.632 atm
d. 0.095 atm and 0.64 atm

58. Which among the following actinoids does not have stable electronic configuration?

a. Protactinium
b. Nobelium
c. Americium
d. Lawrencium

59. Which of the following statement is incorrect regarding the equation?

$$MeCHO + [Ag(NH_3)_2]^+ + \bar{O}H \longrightarrow MeCOO^- + Ag$$

a. The equivalent weight of MeCHO is 22.
b. Three moles of $\bar{O}$H are required in the reaction.
c. MeCHO acts as an oxidising agent.
d. $[Ag(NH_3)_2]^+$ gets reduced.

60. Which of the following graph is correctly represented according to Freundlich isotherm?

a.
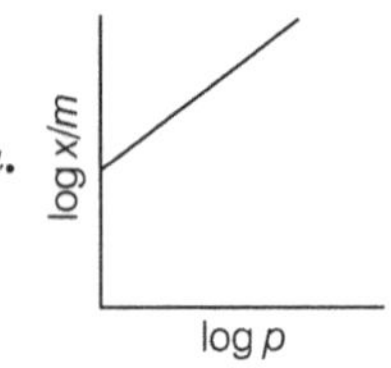

b.
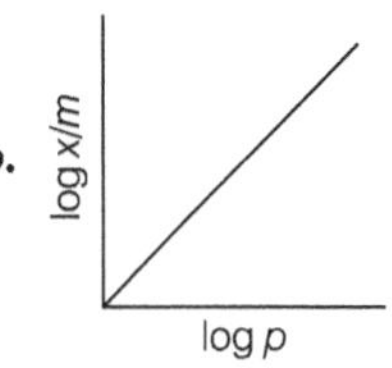

c.
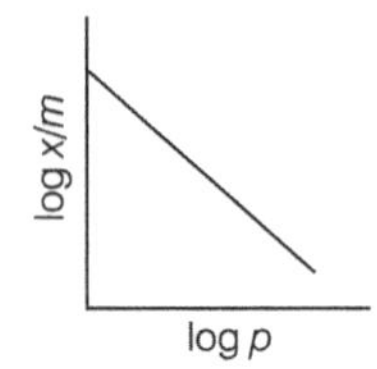

d.
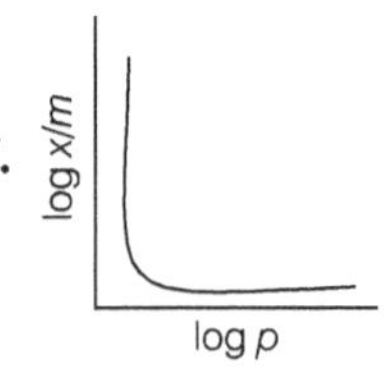

61. pH of a saturated solution of $Ba(OH)_2$ is 12. The value of its K_{sp} is

a. 3.3×10^{-7} M *b.* 5.0×10^{-7}M
c. 4.0×10^{-6} M *d.* 5.0×10^{-6}M

62. An unsaturated hydrocarbon 'X' gives white precipitate with Tollen's reagent. If X is gaseous in nature, the molecular formula of X is

a. C_3H_6 *b.* C_2H_4
c. C_2H_2 *d.* C_4H_8

63. The rate of a reaction triples when temperature changes from 20°C to 50°C. The energy of activation for the reaction is
$(R = 8.314\ JK^{-1}\ mol^{-1})$

a. 181.327 J mol^{-1} *b.* 428.141 J mol^{-1}
c. 32.4321 kJ mol^{-1} *d.* 28.8118 kJ mol^{-1}

64. Which of the following compound will give blood red colour while doing the Lassaigne's test for N?

a. $(NH_2)C{=}O$ *b.* $H_2N(C_6H_4)SO_3H$
c. $C_6H_5SO_3H$ *d.* $CHCl_3$

65. For a reaction, $A + B^{2+} \longrightarrow B + A^{2+}$, at 25°C $E° = 0.2955$ V. The value of K_{eq} is

a. 10 *b.* 10^{10}
c. −10 *d.* 10^{-10}

66. Which of the following is the correct order of stability of conformations for $NH_2—CH_2—CH_2—OH$?

a. Gauche > Eclipsed > Anti
b. Gauche > Anti > Eclipsed
c. Eclipsed > Gauche > Anti
d. Anti > Eclipsed > Gauche

67. When H_2S gas is passed into a mixture of Mn^{2+}, Ni^{2+}, Cu^{2+} and Hg^{2+} ion in an acidified aqueous solution, the precipitates formed are

a. CuS and HgS
b. MnS and CuS
c. MnS and NiS
d. NiS and HgS

68. A mixture of bromo trichloride and hydrogen is subjected to silent electric discharge to form *x* and HCl. *x* is mixed with NH_3 and heated to 200°C to form *y*. The formula of *y* is

a. B_2O_3 *b.* $B_3N_3H_6$
c. H_3BO_3 *d.* B_2H_6

69. Which of the following reactions increase the production of dihydrogen from synthesis gas?

a. $CH_4(g) + H_2O(g) \xrightarrow[\text{Ni}]{1270\text{K}} CO(g) + 3H_2(g)$

b. $C(s) + H_2O(g) \xrightarrow{1270\text{ K}} CO(g) + H_2(g)$

c. $CO(g) + H_2O(g) \xrightarrow[\text{Catalyst}]{673\text{ K}} CO_2(g) + H_2(g)$

d. $C_2H_6 + 2H_2O \xrightarrow[\text{Ni}]{1270\text{K}} 2CO + 5H_2$

70. When the heat of a reaction at constant pressure is -2.5×10^3 cal and entropy change for the reaction is 7.4 cal deg^{-1}, it is predicted that the reaction at 25°C is

a. reversible *b.* spontaneous
c. non-spontaneous *d.* irreversible

71. The addition of HBr to 1-butene gives a mixture of products *x*, *y* and *z*.

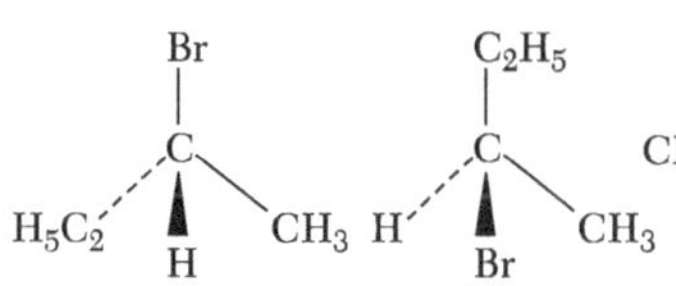

(*x*) (*y*)

$CH_3—CH_2—CH_2—CH_2Br$

(*z*)

The mixture consists of

a. *x* and *y* as major and *z* as minor products
b. *y* as major, *x* and *z* as minor products
c. *y* as minor, *x* and *z* as major products
d. *x* and *y* as minor and *z* as major products

72. The correct statement about silicone is

a. they are ketones with silyl group (SiH_3) similar to alkyl, $(SiH_3)_2CO$.
b. they are synthetic polymer containing repeated R_2SiO_2 units.
c. they are formed by hydrolysis of R_2SiCl_2.
d. All of the above

73. When dil. sulphuric acid reacts with aqueous solution of potassium chromate, the colour changes from yellow to orange. This shows that

a. chromate ions reduced
b. chromate ions are oxidised
c. monocentric complex is converted into dicentric complex
d. oxygen gets removed from chromate ions

74. Valence electrons in the element *A* are 3 and that in element *B* are 6. Most probable compound formed from A and *B* is

a. A_2B *b.* AB_2
c. A_6B_3 *d.* A_2B_3

75. In non-reducing disaccharide, the reducing group of monosaccharides i.e. aldehydic or ketonic group are bounded.

Which of the following disaccharide is a non-reducing sugar?

a.
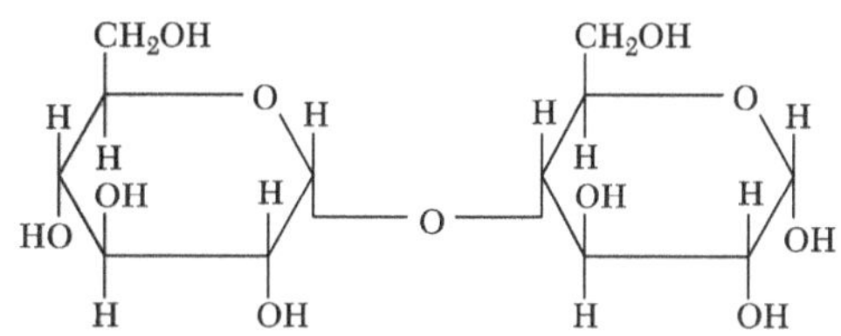

b.
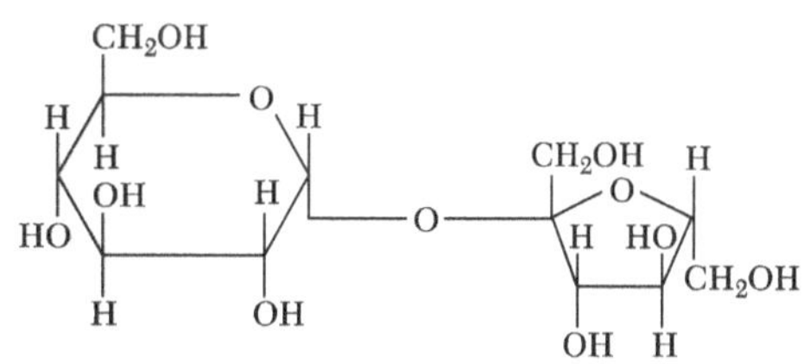

c.
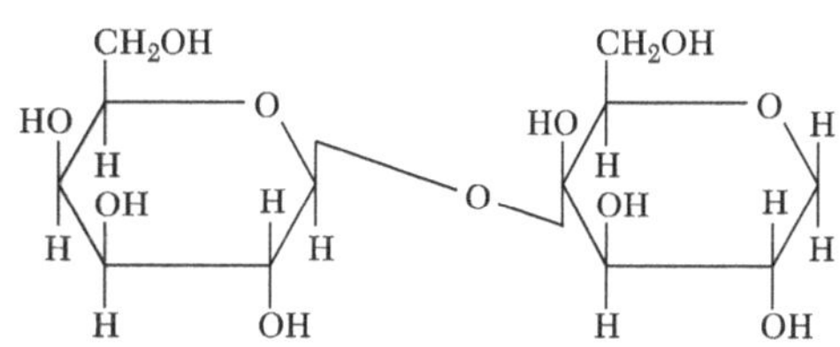

d.
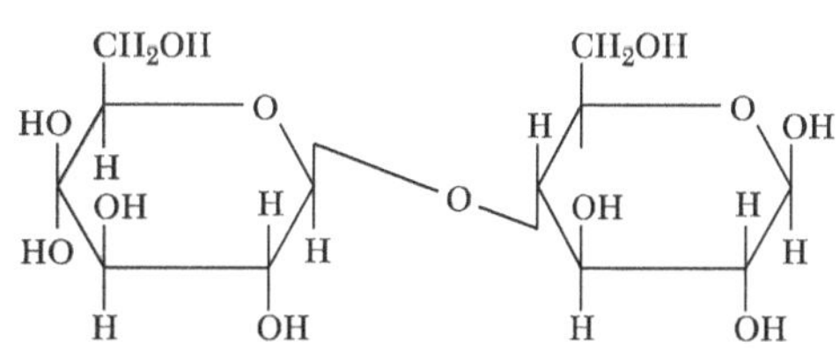

76. Which of the following 0.1 M aqueous solution will have lowest freezing point?

a. Potassium sulphate
b. Sodium chloride
c. Urea
d. Glucose

77. A penicillin is a member of a family of drugs that have a

a. four membered cyclic amide fused to a five membered thiazole ring.
b. three membered cyclic amide fused to a five-membered thiazole ring
c. four-membered cyclic amide fused to have a four membered thiazole ring.
d. five-membered cyclic amide fused to have a five membered thiazole ring

78. Which of the following property of alkaline earth metal increases with their atomic number?

a. electronegativity
b. solubility of their hydroxides in water
c. solubility of their sulphate in water
d. ionisation energy

79. Which of the following expression is correct for the rate of reaction given below ?

$$5Br^-(aq) + BrO_3^-(aq) + 6H^+(aq) \longrightarrow 3Br_2(aq) + 3H_2O(l)$$

a. $\frac{\Delta[Br^-]}{\Delta t} = \frac{5}{6}\frac{\Delta[H^+]}{\Delta t}$
b. $\frac{\Delta[Br^-]}{\Delta t} = \frac{6}{5}\frac{\Delta[H^+]}{\Delta t}$
c. $\frac{\Delta[Br^-]}{\Delta t} = \frac{5\Delta[H^+]}{\Delta t}$
d. $\frac{\Delta[Br^-]}{\Delta t} = \frac{6\Delta[H^+]}{\Delta t}$

80. Match the polymer given in column I with correct monomer of column II and choose the correct option.

	Column I		Column II
A.	Neoprene	I.	Isoprene
B.	Natural rubber	II.	Tetrafluoro ethane
C.	Teflon	III.	Chloroprene
D.	Acrilan	IV.	Acryl nitrite

Codes

	A	B	C	D
a.	IV	III	II	I
b.	I	II	III	IV
c.	III	I	II	IV
d.	II	IV	I	III

PART III

a. English Proficiency

Directions (Q. Nos. 81-83) *In the following questions, the sentences may or may not be grammatically correct. Find out which part of a sentence has an error and mark that part. If there is no error, mark part 'd' as your answer.*

81. The captain alongwith his team ***(a)***/are practising very hard ***(b)***/ for the forthcoming match. ***(c)***/ No error ***(d)***

82. I am going ***(a)***/to have this certificate ***(b)***/attest by the director. ***(c)***/No error ***(d)***

83. He is ***(a)***/ having many ***(b)***/friends here.***(c)***/No error ***(d)***

Directions (Q. Nos. 84-85) *Fill in the blanks with suitable preposition from the alternatives given under each sentence.*

84. Is not learning superior .... wealth?

a. than *b.* from
c. by *d.* to

85. He could not cope .... the heavy workload.

a. in with *b.* up with
c. up *d.* with

Directions (Q. Nos. 86-88) *Select the word or the phrase which is closest to the opposite in meaning of the italicized word or phrase.*

86. He was in a *dejected* mood.

a. jubilant *b.* rejected *c.* irritable *d.* romantic

87. The attack on the freedom of the press is a *retrograde* step.

a. progressive *b.* stubborn
c. punitive *d.* aggressive

88. We should not *belittle* the value of small things.

a. exalt *b.* praise
c. inflate *d.* expand

Directions (Q. Nos. 89-90) *Choose the word nearest in meaning to the underlined word.*

89. Before I could make out anything he had spoken again.

a. find out *b.* apprehend
c. explain *d.* reveal

90. He wrote a scathing review of the prize winning novel.

a. biased *b.* scornful
c. unbalanced *d.* subjective

Directions (Q. Nos. 91-95) *Read the passage given below and answer the questions that follow.*

The megalomaniac differs from the narcissist by the fact that he wishes to be powerful rather than charming and seeks to be feared rather than loved. To this type belong many lunatics and most of the great men in history. Love of power, like vanity, is a strong element in normal human nature and as such is to be accepted; it becomes deplorable only when it is excessive or associated with an insufficient sense of reality. Where this occurs, it makes a man unhappy or foolish, if not both. The lunatic who thinks he is crowned head may be, in a sense, happy, but his happiness is not of a kind that any sane person would envy. Alexander the Great was psychologically of the same type as the lunatic, though he possessed the talent to achieve the lunatic's dream.

He could not, however, achieve his own dream, which enlarged his scope as his achievement grew. When it became clear that he was the greatest conqueror known to fame, he decided that he was a God. Was he a happy man? His drunkenness, his furious rages, his indifference to women and his claim to divinity, suggest that he was not. There is no ultimate satisfaction in the cultivation of one element of human nature at the expense of all the others, nor in viewing all the world as raw material for the magnificence of one's own ego.

91. What is the difference between an ordinary megalomaniac and a megalomaniac like Alexander the Great?

a. The ordinary megalomaniac does not have excessive desire for power which Alexander the Great had.

b. The ordinary megalomaniac does not have the talent to realise his wish which Alexander the Great had.

c. The ordinary megalomaniac is a lunatic while Alexander the Great was not a lunatic.

d. The ordinary megalomaniac is not great while Alexander the Great was great.

92. How does a megalomaniac differ from a narcissist?

a. By wishing to be charming and feared

b. By wishing to be loved and not feared

c. By wishing to be powerful and not feared

d. By wishing to be powerful and feared

93. In "Where this occurs it makes a man..... if not both," 'this' refers to

a. vanity *b.* lunacy

c. love of power *d.* excessive lover of power

94. Which among the following is the reason for unhappiness?

a. Dealing with the raw material of the world

b. Realising one's dream as a megalomaniac

c. The nurturing of only one element in human nature

d. Being indifferent towards women

95. Why has love of power to be accepted?

a. Because it can become unreal

b. Because it is an excess in human nature

c. Because it is a part of human nature

d. Because it is vanity in human nature

b. Logical Reasoning

96. 'Umpire' is related to 'Match' in the same way as 'Judge' is related to

a. Bar council *b.* Lawyer

c. Judgement *d.* Lawsuit

97. Find the odd one from the following options.

a. 9, 49 *b.* 13, 121

c. 10, 61 *d.* 7, 25

98. Complete the series by replacing question mark '?'.

10, 9, 16, 45, 176, ?

a. 815 *b.* 222 *c.* 555 *d.* 875

99. In a cricket team, Dhoni is taller than Virat but not as tall as Raina, Rohit is shorter than Dhoni but taller than Shikhar. Who among them is the shortest?

a. Dhoni *b.* Virat

c. Shikhar *d.* Cannot be determined

100. Identify the missing part of the question figure and select it from given answer figures.

Question figure

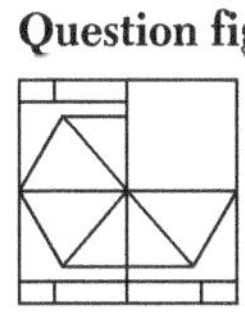

Answer figures

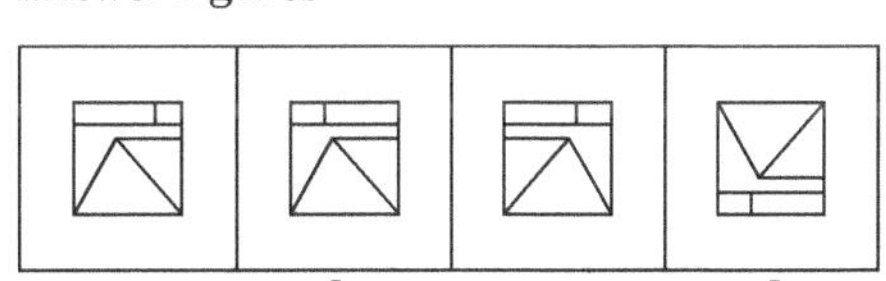

a. *b.* *c.* *d.*

101. Find out which of the figure (a), (b), (c) and (d) can be formed from the pieces given in question figures.

Question figure

Answer figures

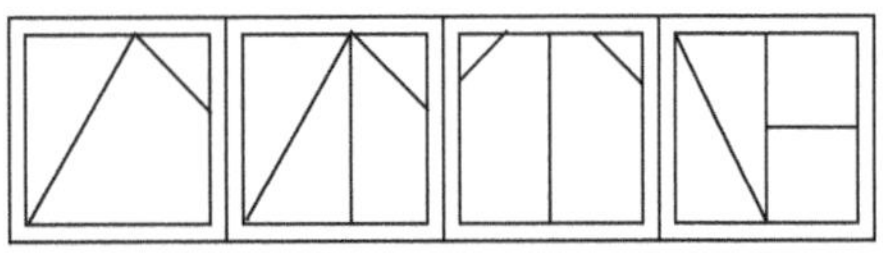

a. *b.* *c.* *d.*

102. In this question, a piece of paper is folded and then cut as shown below. The dotted lines shown are the portion which have been folded. The curve arrow shows the directions of folding. And the number of scissors beneath the figure show the number of portions cut. From the given responses, indicate how it will appear when opened. The opening is in the same order as folding.

Question figures

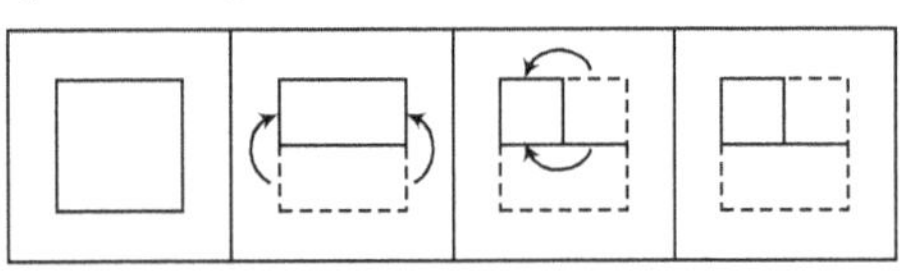

Answer figures

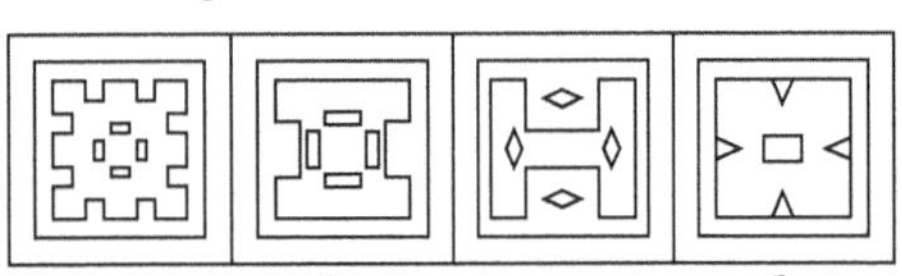

a. *b.* *c.* *d.*

103. Which of the answer figures (a), (b), (c) or (d) completes the figure matrix ?

Questions figures

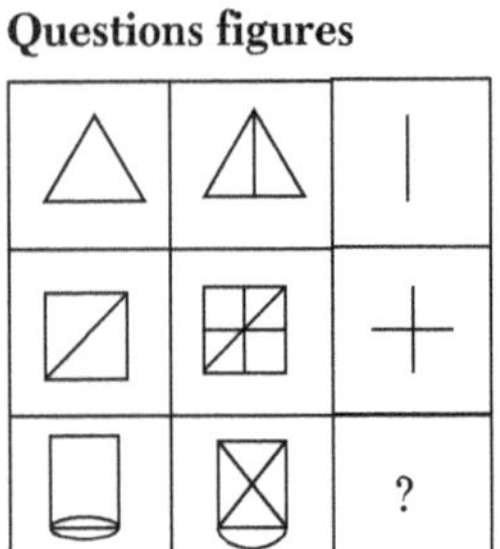

Answer figures

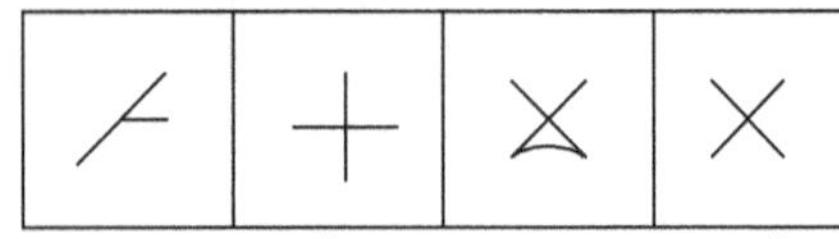

a. *b.* *c.* *d.*

104. In the following question, one or more dots are place in the question figure.

This figure is followed by four alternatives marked as (a), (b), (c) and (d). One out of these four options contains region(s) common to circle, square and triangles, similar to that marked by the dot in question figure. Find that figure.

Question figure

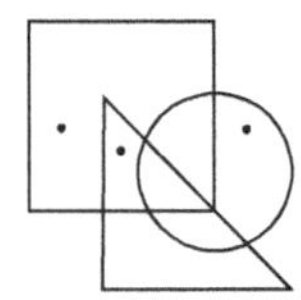

Answer figures

a.

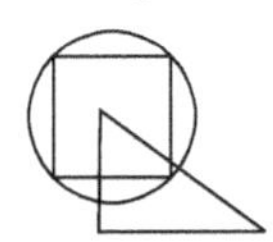

b.

c.

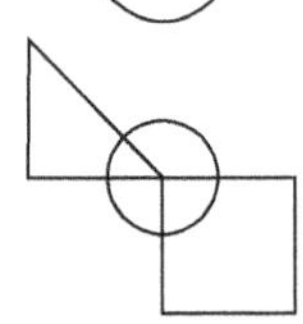

d.

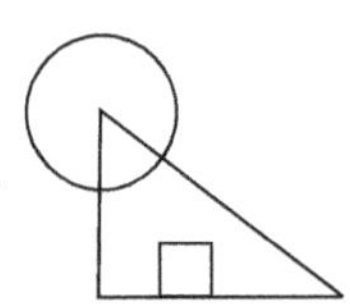

105. How many triangles are there in the following figure?

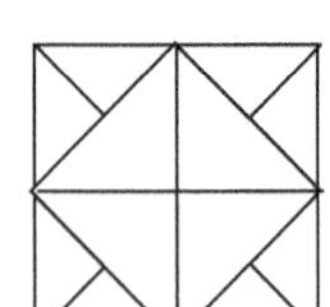

a. 16 *b.* 20 *c.* 12 *d.* 22

PART IV

Mathematics

106. The coefficient of x^{-n} in $(1+x)^n\left(1+\frac{1}{x}\right)^n$ is

a. 0 *b.* 1 *c.* 2^n *d.* $2n$

107. The greatest term in the expansion of $\sqrt{3}\left(1+\frac{1}{\sqrt{3}}\right)^{20}$ is

a. $\frac{26840}{9}$ *b.* $\frac{24840}{9}$
c. $\frac{25840}{9}$ *d.* None of these

108. The nth roots of unity are in

a. AP *b.* GP
c. HP *d.* None of these

109. If $P=\begin{bmatrix} a & b & c \\ b & c & a \\ c & a & b \end{bmatrix}$, $abc=P^TP=I$, then the value of $a^3+b^3+c^3$ is

a. 2 *b.* 1 *c.* 0 *d.* 5

110. If $x^2=\begin{vmatrix} \sin\theta & \cos\theta & 0 \\ -\cos\theta & \sin\theta & 1 \\ \sin\theta & \cos\theta & 2 \end{vmatrix}$, then the value of $4x^2+x\sin\frac{3\pi}{2}+5$ is

a. $13-\sqrt{2}$ *b.* $13+\sqrt{2}$
c. $\sqrt{2}-13$ *d.* Both (a) and (b)

111. If a, b, c are in GP and $\log a-\log 2b$, $\log 2b-\log 3c$ and $\log 3c-\log a$ are in AP, then a, b and c are the lengths of the sides of a triangle, which is

a. equilateral *b.* right angled
c. acute-angled *d.* obtuse angled

112. If $\sum_{r=1}^{n} t_r=\frac{n(n+1)(n+2)(n+3)}{8}$, where t_r denotes the rth term of a series, then $\lim_{n\to\infty}\sum_{r=1}^{n}\frac{1}{t_r}$ is

a. $\frac{1}{8}$ *b.* $\frac{1}{4}$ *c.* $\frac{1}{2}$ *d.* 1

113. Which of the following statement is a tautology?

a. $(p\vee q)\vee(\sim p)$ *b.* $(\sim q\wedge p)\vee(p\vee\sim p)$
c. Both (a) and (b) *d.* None of these

114. If a parallelogram is cut by two sets of m lines parallel to its sides, then the number of parallelogram thus formed, is

a. ${}^mC_2\times{}^mC_2$ *b.* $2({}^{m+2}C_2)$
c. $({}^{m+2}C_2)^2$ *d.* None of these

115. The inverse of the function $f(x)=\log_a(x+\sqrt{x^2+1})$ (where, $a<0$, $a\neq1$) is

a. $\frac{1}{2}(a^x-a^{-x})$ *b.* not defined for all x
c. defined for $x>0$ *d.* None of these

116. The value of $S=\sum_{n=1}^{\infty}\tan^{-1}\frac{2n}{n^4+n^2+2}$ is equal to

a. $\frac{\pi}{2}$ *b.* π
c. $\frac{\pi}{4}$ *d.* None of these

117. Equation $\sin x+\cos(t+x)+\cos(t-x)=2$ has real solution, then $\sin t$ can be

a. 1/2 *b.* 1/5 *c.* 3/4 *d.* –3/4

118. A line makes angles α, β, γ with the coordinate axes. If $\alpha+\beta=\frac{\pi}{2}$, then $(\cos\alpha+\cos\beta+\cos\gamma)^2$ is equal to

a. $1+\cos 2\alpha$ *b.* $1-\sin 2\alpha$
c. $1+\sin 2\alpha$ *d.* None of these

119. Straight lines $3x+4y=5$ and $4x-3y=15$ intersect at the point A. If point B and C are chosen on these two lines such that $AB=AC$, then the possible equation of the line BC passing through the point (1, 2) is

a. $x+7y+13=0$ or $7x+y+9=0$
b. $x+7y+13=0$ or $7x+2y+7=0$
c. $x-7y+13=0$ or $7x+y-9=0$
d. None of the above

120. Normals drawn to $y^2=4ax$ at the points where it is intersected by the line $y=mx+c$ intersected at P. Coordinates of foot of the another normal drawn to the parabola from the point 'P' is

a. $\left(\frac{a}{m^2}, -\frac{2a}{m}\right)$ *b.* $\frac{9}{m}, -\frac{6a}{m}$
c. $(am^2, -2am)$ *d.* $\left(\frac{4a}{m^2}, -\frac{4a}{m}\right)$

121. The area of the triangle formed by joining the origin to the point of intersection of the line $x\sqrt{5}+2y=3\sqrt{5}$ and circle $x^2+y^2=10$ is

a. 3 **b.** 4 **c.** 5 **d.** 6

122. Radius of the largest circle which passes through the focus of the parabola $y^2=4x$ and contained in it, is

a. 8 **b.** 4 **c.** 2 **d.** 5

123. A tangent drawn to hyperbola $\frac{x^2}{a^2}-\frac{y^2}{b^2}=1$ at $P\left(\frac{\pi}{6}\right)$ forms a triangle of area $3a^2$ sq. units, with coordinate axes. If the eccentricity of hyperbola is e, then the value of e^2-9 is

a. 9 **b.** 10 **c.** 11 **d.** 8

124. If the sum of squares of distances of a point from the planes $x+y+z=0$, $x-z=0$ and $x-2y+z=0$ is p^2, then locus of the point is

a. $x^2+z^2=p^2$

b. $x^2+2xy+y^2+z^2=p^2$

c. $x+y+z=p^2$

d. $x^2+y^2+z^2=p^2$

125. Line $\frac{(x+1)}{\lambda}=y-1=\frac{(z+2)}{-4}$ is perpendicular to $2x+2y-8z+5=0$, then λ is

a. 1 **b.** -4

c. -5 **d.** -3

126. $OPQR$ is a square and M, N are the middle points of the sides PQ and QR respectively, then the ratio of the areas of the square and ΔOMN is

a. 4 : 1 **b.** 2 : 1

c. 8 : 3 **d.** 4 : 3

127. The line passing through the extremity A of the major axis and extremity B of the minor axis of the ellipse $x^2+9y^2=9$ meets its auxiliary circle at the point M. Then, the area of the triangle with vertices at A, M and the origin O is

a. $\frac{31}{10}$ **b.** $\frac{29}{10}$

c. $\frac{21}{10}$ **d.** $\frac{27}{10}$

128. If e_1 and e_2 are the eccentricities of a hyperbola $3x^2-3y^2=25$ and its conjugate, then

a. $e_1^2+e_2^2=2$ **b.** $e_1^2+e_2^2=4$

c. $e_1+e_2=4$ **d.** $e_1+e_2=\sqrt{2}$

129. Let $f:R\to R$ be a function satisfying $f(x+y)=f(x)+2y^2+kxy$ for all $x, y\in R$. If $f(1)=2$ and $f(2)=8$, then $f(x)$ is equal to

a. $2x^2$ **b.** $6x-4$

c. x^2+3x-2 **d.** $-x^2+9x-6$

130. If the planes $\mathbf{r}(2\hat{\mathbf{i}}-\lambda\hat{\mathbf{j}}+3\hat{\mathbf{k}})=0$ and $\mathbf{r}\cdot(\lambda\hat{\mathbf{i}}+5\hat{\mathbf{j}}-\hat{\mathbf{k}})=5$ are perpendicular to each other, then the value of $\lambda^2+\lambda$ is

a. 0 **b.** 2

c. 1 **d.** 3

131. Solution of the differential equation $\frac{dy}{dx}=\sin(x+y)+\cos(x+y)$ is equal to

a. $\log\left(2+\sec\frac{x+y}{2}\right)=x+C$

b. $\log(1+\tan(x+y))=x+C$

c. $\log\left(1+\tan\frac{x+y}{2}\right)=y+C$

d. $\log\left(1+\tan\frac{x+y}{2}\right)=x+C$

132. The value of α, so that $\lim_{x\to 0}\frac{1}{x^2}(e^{\alpha x}-e^x-x)=\frac{3}{2}$, is

a. 1 **b.** 0

c. 4 **d.** 2

133. An inverted conical flask is being filled with water at the rate of 3 cm^3/s. The height of the flask is 10 cm and the radius of the base is 5 cm. How fast is the water level rising when the level is 4 cm?

a. $\frac{4}{3}\pi$ cm/s **b.** $\frac{3}{4\pi}$ cm/s

c. $\frac{3\pi}{4}$ cm/s **d.** $\frac{4}{3\pi}$ cm/s

134. The equation of the curve whose slope at any point is equal to $y+2x$ and which passes through the origin is

a. $y=2(x-1)$ **b.** $y=2(e^x-x-1)$

c. $y=2(e^x-1)$ **d.** $y=2(e^x x-1)$

135. Let $f(x)=\begin{cases} x^p\sin\frac{1}{x}, & x\neq 0 \\ 0, & x=0\end{cases}$, then $f(x)$ is continuous but not differentiable at $x=0$, if

a. $p<0$ **b.** $p=0$

c. $0<p\leq 1$ **d.** $p\geq 1$

136. The solution of the differential equation $\frac{d^2y}{dx^2} = \sin 3x + e^x + x^2$ when $y_1(0) = 1$ and $y(0) = 0$, is

a. $-\frac{\sin 3x}{9} + e^x + \frac{x^4}{12} + \frac{1}{3}x - 1$

b. $-\frac{\sin 3x}{9} + e^x + \frac{x^4}{12} + \frac{1}{3}x$

c. $-\frac{\cos 3x}{9} + e^x + \frac{x^4}{12} + \frac{1}{3}x + 1$

d. None of the above

137. For which interval the given function $f(x) = -2x^3 - 9x^2 - 12x + 1$ is decreasing?

a. $(-2, \infty)$,

b. $(-2, -1)$

c. $(-\infty, -1)$

d. $(-\infty, -2)$ or $(-1, \infty)$

138. If θ is the angle between the vectors $4(\hat{i} - \hat{k})$ and $\hat{i} + \hat{j} + \hat{k}$, then $(\sin\theta + \cos\theta)$ equals to

a. 0 b. $\frac{1}{2}$

c. 1 d. 2

139. In a ΔABC, D, E, F are the mid-points of the sides BC, CA and AB respectively, the vector AD is equal to

a. **BE + CF** b. **BE – CF**

c. **CF – BE** d. **– BE – CF**

140. The arithmetic mean of a set of observation is $\overline{X}$. If each observation is divided by α and increased by 10, then the mean of the new series is

a. $\frac{\overline{X}}{\alpha}$ b. $\frac{\overline{X} + 10}{\alpha}$

c. $\frac{\overline{X} + 10\alpha}{\alpha}$ d. $\alpha\overline{X} + 10$

141. If h is the altitude of a parallelopiped determined by the vectors **a**, **b**, **c** and the base is taken to be the parallelogram determined by **a** and **b** where $\mathbf{a} = \hat{i} + \hat{j} + \hat{k}$, $\mathbf{b} = 2\hat{i} + 4\hat{j} - \hat{k}$ and $\mathbf{c} = \hat{i} + \hat{j} + 3\hat{k}$, then the value of $19h^2$ is

a. 19 b. 16

c. 8 d. None of these

142. The mean and variance of a binomial distribution **BD** for 3 trials is 2.7, then the **BD** is given by

a. $(0.2 + 0.8)^5$ b. $(0.3 + 0.7)^5$

c. $(0.4 + 0.6)^5$ d. None of these

143. Let $P(x) = \int \frac{dx}{e^x + 8e^{-x} + 4e^{-3x}}$,

$Q(x) = \int \frac{dx}{e^{3x} + 8e^x + 4e^{-x}}$ and

$R(x) = P(x) - 2Q(x)$.

If $R(x) = \frac{1}{2}A\left(\frac{B + 2e^{-x}}{C}\right) + K$, then the value of (A, B, C) is

a. $(\tan^{-1}, 2, e^x)$ b. $(\tan^{-1}, e^x, 2)$

c. $\left(\tan^{-1}, \frac{1}{2}, \frac{1}{e^x}\right)$ d. $\left(\tan^{-1}, \frac{1}{e^x}, \frac{1}{2}\right)$

144. The value of $\int_0^1 \cot^{-1}(1 - x + x^2)dx$ is

a. $\log 2$ b. $\frac{\pi}{2} - \log 2$

c. $\frac{\pi}{2} + \log 2$ d. $-\log 2$

145. The area of the region included between the curves $x^2 + y^2 = a^2$ and $\sqrt{|x|} + \sqrt{|y|} = \sqrt{a}$ $(a > 0)$, is

a. $\left(\pi - \frac{2}{3}\right)a^2$ sq units

b. $\left(\frac{2}{3} - \pi\right)a^2$ sq units

c. $\frac{2}{3}\pi a^2$ sq units

d. $\left(\pi + \frac{2}{3}\right)a^2$ sq units

146. Let A and B are two independent events. If the probability that both A and B occur together is $\frac{1}{6}$ and the probability that neither of them occurs is $\frac{1}{3}$, then the probability of occurrence of A is

a. 0 or 1 b. $\frac{1}{2}$ or $\frac{1}{3}$

c. $\frac{1}{2}$ or $\frac{1}{4}$ d. $\frac{1}{3}$ or $\frac{1}{4}$

147. In a test an examiner either guesses or copies or knows the answer to a multiple choice question with 4 choices. The probability that he/she makes a guess is $\frac{1}{3}$. The probability that he/she copies the answer is $\frac{1}{6}$. If the probability that the answer is correct, given that he/she copied, it is $\frac{1}{8}$, then the probability that he/she knows the answer to a question given that he/she correctly answered it, is

a. $\frac{27}{29}$ b. $\frac{26}{29}$

c. $\frac{25}{29}$ d. $\frac{24}{29}$

148. If $p: 4$ is an even prime number, $q: 6$ is a divisor of 12 and $r:$ the HCF of 4 and 6 is 2, then which of the following is correct?

a. $(p \wedge q)$ b. $(p \vee q) \wedge \sim r$

c. $\sim (q \wedge r) \vee p$ d. $\sim p \vee (q \wedge r)$

149. The maximum value of $Z = 4x + 2y$ subject to constraints $2x + 3y \leq 18$, $x + y \geq 10$ and $x, y \geq 0$ is

a. 20 b. 36

c. 40 d. None of these

150. The coordinates of the point at which minimum value of $Z = 7x - 8y$ subject to constraints $x + y - 20 \leq 0$, $y \geq 5$, $x \geq 0$, $y \geq 0$ is attained, is

a. (20, 0) b. (15, 5)

c. (0, 5) d. (0, 20)

Answers

Physics

1. (a)	2. (a)	3. (a)	4. (b)	5. (b)	6. (a)	7. (b)	8. (d)	9. (b)	10. (d)
11. (d)	12. (b)	13. (c)	14. (b)	15. (c)	16. (a)	17. (a)	18. (b)	19. (b)	20. (c)
21. (a)	22. (c)	23. (b)	24. (c)	25. (a)	26. (a)	27. (c)	28. (c)	29. (b)	30. (a)
31. (c)	32. (c)	33. (c)	34. (a)	35. (c)	36. (a)	37. (d)	38. (a)	39. (b)	40. (d)

Chemistry

41. (d)	42. (a)	43. (d)	44. (a)	45. (c)	46. (d)	47. (c)	48. (b)	49. (a)	50. (b)
51. (b)	52. (b)	53. (c)	54. (d)	55. (c)	56. (c)	57. (d)	58. (a)	59. (c)	60. (a)
61. (b)	62. (c)	63. (d)	64. (b)	65. (b)	66. (b)	67. (a)	68. (b)	69. (c)	70. (b)
71. (a)	72. (d)	73. (c)	74. (d)	75. (b)	76. (a)	77. (a)	78. (b)	79. (a)	80. (c)

English Proficiency

81. (a)	82. (c)	83. (b)	84. (d)	85. (d)	86. (a)	87. (a)	88. (a)	89. (b)	90. (b)
91. (b)	92. (d)	93. (d)	94. (c)	95. (c)					

Logical Reasoning

96. (d)	97. (c)	98. (d)	99. (d)	100. (c)	101. (b)	102. (a)	103. (d)	104. (b)	105. (b)

Mathematics

106. (b)	107. (c)	108. (b)	109. (a)	110. (d)	111. (d)	112. (c)	113. (c)	114. (c)	115. (a)
116. (c)	117. (a)	118. (c)	119. (c)	120. (d)	121. (c)	122. (b)	123. (d)	124. (d)	125. (a)
126. (c)	127. (d)	128. (b)	129. (a)	130. (a)	131. (d)	132. (d)	133. (b)	134. (b)	135. (c)
136. (a)	137. (d)	138. (c)	139. (d)	140. (c)	141. (c)	142. (b)	143. (b)	144. (b)	145. (a)
146. (b)	147. (d)	148. (d)	149. (d)	150. (d)					

Hints & Solutions

Physics

1. ***(a)*** As the inductors are in parallel, induced emf across the two inductors is the same, i.e.

$$e_1 = e_2$$
$$L_1\left(\frac{di_1}{dt}\right) = L_2\left(\frac{di_2}{dt}\right)$$

On integrating both sides, we get

$$L_1\int\frac{di_1}{dt}dt = L_2\int\frac{di_2}{dt}dt$$
$$\Rightarrow \quad L_1\int di_1 = L_2\int di_2$$
$$L_1 i_1 = L_2 i_2$$
$$\Rightarrow \quad \frac{i_1}{i_2} = \frac{L_2}{L_1}$$

2. ***(a)*** We have, the time constant, $\tau = RC'$ ...(i)

$$\text{Now, } C' = \frac{C_1C_2}{C_1 + C_2}$$
$$= \frac{\left(\frac{A\varepsilon_0}{d-x}\right)\left(\frac{KA\,\varepsilon_0}{x}\right)}{\frac{A\varepsilon_0}{d-x} + \frac{KA\varepsilon_0}{x}} \quad \left[\because C_1 = \frac{A\varepsilon_0}{d-x} \text{ and } C_2 = \frac{kA\varepsilon_0}{x}\right]$$
$$\therefore \quad C' = \frac{KA\varepsilon_0}{x + K(d-x)}$$

Putting the value of C' in Eq. (i), we get

$$\tau = R\left(\frac{KA\varepsilon_0}{x + K(d-x)}\right) \quad \text{...(ii)}$$

At time t, the level of liquid is $\frac{d}{3} - vt$.

$\therefore$ Putting $x = \frac{d}{3} - vt$ in Eq. (ii), we get

$$\tau = \frac{R\,K\,A\varepsilon_0}{\frac{d}{3} - vt + k\left(d - \frac{d}{3} + vt\right)}$$

Given, $A = 1$ and $K = 2$

$$\therefore \quad \tau = \frac{3 \times 2\,\varepsilon_0 \times R}{d - 3vt + 6d - 2d + 6vt} = \frac{6\,R\varepsilon_0}{5d + 3vt}$$

3. ***(a)*** According to principle of calorimetry,

Heat gained = Heat lost

Heat is lost by steam in two stages

(i) Change of state from steam to water at 100°C.

(ii) To change water at 100°C to water at 80°C.

Now we have,

The mass of water and the calorimeter

$$= (1.1 + 0.02)\text{kg} = 1.12\,\text{kg}$$

Specific heat capacity of water $= 4.184 \times 10^3$ Jkg^{-1}K^{-1}

Now, by using, $Q = mc\Delta T$

Heat gained by the calorimeter and water is

$$Q = 1.12 \times 4.18 \times 10^3 \times (80 - 15)$$
$$= 1.12 \times 4.18 \times 10^3 \times 65 \text{ Jkg}^{-1}\text{K}^{-1} \quad \text{...(i)}$$

Let the mass of the steam be m kg.

So, latent heat of the vaporisation of water at 100°C is 540 cal/g $= 540 \times 4.184 \times 10^3$ J/kg

$\therefore$ Heat lost by the steam,

$$Q' = m(540 \times 4.184 \times 10^3)$$
$$+ m(4.184 \times 10^3 \times (100 - 80)) \text{ J/kgK} \quad \text{...(ii)}$$

From Eq. (i) and (ii), we get

$$m(540 + 20) = 1.12 \times 65$$
$$\Rightarrow \quad m = \frac{1.12 \times 65}{560} = 0.13 \text{ kg}$$

4. ***(b)*** As, gravitational constant,

$$[G] = [\text{M}^{-1}\text{L}^3\text{T}^{-2}]$$

Planck's constant, $[h] = [\text{ML}^2\text{T}^{-1}]$

$$\text{Hence,} \quad [G][h] = [\text{M}^{-1}\text{L}^3\text{T}^{-2}][\text{ML}^2\text{T}^{-1}]$$
$$= [\text{M}^0\text{L}^5\text{T}^{-3}]$$

Velocity of light, $[c] = [\text{LT}^{-1}]$

$$\text{Now,} \quad \left[\frac{Gh}{c^3}\right]^{1/2} = \frac{[\text{L}^5\text{T}^{-3}]^{1/2}}{[\text{L}^3\text{T}^{-3}]^{1/2}}$$
$$= [\text{L}^2]^{1/2} = [\text{L}]$$

Hence, $\quad [\text{L}] =$ length

5. ***(b)*** Given, $Q = 10^6$ cal

$$T_1 = 827°\text{C} = (827 + 273) = 1100 \text{ K}$$
$$T_2 = 27°\text{C} = (27 + 273) = 300 \text{ K}$$
$$\text{As,} \quad \frac{Q_1}{T_1} = \frac{Q_2}{T_2}$$
$$\therefore \quad Q_2 = \frac{T_2}{T_1}Q_1 = \frac{300}{1100} \times 10^6$$
$$= 2.72 \times 10^5 \text{ cal}$$

Efficiency of the engine,

$$\eta = \left(1 - \frac{T_2}{T_1}\right) \times 100$$
$$\eta = \left(1 - \frac{300}{1100}\right) \times 100$$
$$= 72.72\%$$

6. ***(a)*** According to Doppler's effect, the approximate frequency heard by the stationary observer,

$$\nu = \frac{v}{v - v_s}\nu_0$$

Case (i) $v_s = 34$ m/s

where, v = speed of sound in air,

v_s = speed of source

and v_0 = frequency of the source.

$$\therefore \quad \nu_1 = \frac{340}{340 - 34}\nu_0$$

$$= \frac{340}{306}\nu_0 \quad \ldots \text{(i)}$$

Case (ii) $v_s = 17$ m/s

$$\therefore \quad \nu_2 = \frac{340}{340 - 17}\nu_0 = \frac{340}{323}\nu_0 \quad \ldots \text{(ii)}$$

From Eqs. (i) and (ii), we get

$$\therefore \quad \frac{\nu_1}{\nu_2} = \frac{340/306}{340/323} \Rightarrow \frac{323}{306} \approx \frac{19}{18}$$

7. ***(b)*** Given, $m = 5$ kg, $v = 20\ \text{ms}^{-1}$, $\theta = 60°$

Vertical component of velocity, $v_y = v \sin 60°$

$$= 20 \times \frac{\sqrt{3}}{2} = 10\sqrt{3}\ \text{ms}^{-1}$$

Time taken to reach the highest point = Time taken to reach the ground from highest point.

$$t = \frac{v \sin\theta}{g} = \frac{v_y}{g} = \frac{10\sqrt{3}}{9.8} = 1.77\ \text{s}$$

If the highest point, m splits up into two parts of masses $m_1 = 1$ kg and $m_2 = 4$ kg.

If their velocities v_1 and v_2 respectively, then applying the principle of conservation of linear momentum, we get

$$m_1 v_1 + m_2 v_2 = mv \cos\theta$$

$$v_1 + 4v_2 = 5 \times 20 \times \frac{1}{2} \quad [\because \theta = 60°]$$

$$\Rightarrow \quad v_1 + 4v_2 = 5 \times 10 = 50 \quad \ldots \text{(i)}$$

$$\text{Initial KE} = \frac{1}{2} m (v\cos\theta)^2$$

$$= \frac{1}{2} \times 5 \times (10)^2 = 250\ \text{J}$$

Final KE = 2 (initial KE) = 2 × 250 = 500 J

$$\therefore \quad \frac{1}{2} m_1 v_1^2 + \frac{1}{2} m_2 v_2^2 = 500$$

$$\text{or} \quad \frac{1}{2} \times 1 \times v_1^2 + \frac{1}{2} \times 4 \times v_2^2 = 500$$

$$\text{or} \quad v_1^2 + 4v_2^2 = 1000 \quad \ldots \text{(ii)}$$

Solving Eqs. (i) and (ii), we get

$$v_1 = 30\ \text{m/s}, \quad v_2 = 5\ \text{m/s}$$

Hence, the separation between the two fragments

$$= (v_1 - v_2) \times t = (30 - 5) \times 1.77\ \text{m} = 44.25\ \text{m}$$

8. ***(d)*** Given, initial speed, $u = 36$ km/h

$$= \frac{36 \times 1000}{60 \times 60} = 10\ \text{ms}^{-1}$$

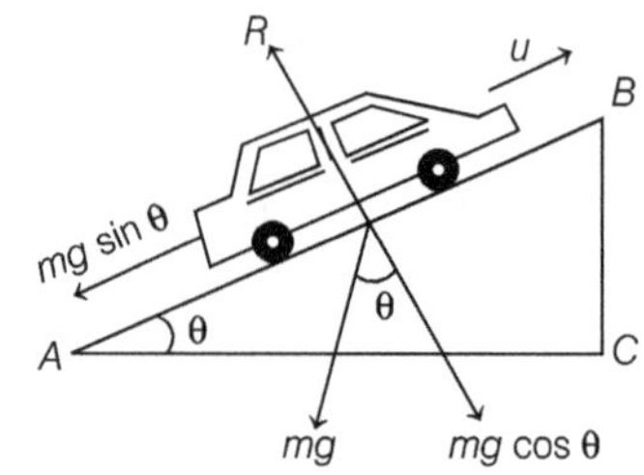

$$\theta = 30°, \mu = 0.1, s = ?$$

Here, work done in moving up the inclined road

= KE of the vehicle

$$(mg \sin\theta + F)s = \frac{1}{2} mu^2$$

$$(mg \sin\theta + \mu R) \times s = \frac{1}{2} mu^2$$

$$(mg \sin\theta + \mu\, mg \cos\theta) \times s = \frac{1}{2} mu^2$$

$$s = \frac{\frac{1}{2} mu^2}{mg(\sin\theta + \mu \cos\theta)} = \frac{u^2}{2g(\sin\theta + \mu\cos\theta)}$$

$$= \frac{10 \times 10}{2 \times 10 \times (\sin 30° + 0.1 \cos 30°)} = 8.53\ \text{m}$$

9. ***(b)*** $P = 15 + 6 = 21\ \Omega$,

$$Q = \frac{8X}{8 + X} + 3$$

$$R = 15 + \frac{6 \times 6}{6 + 6} = 18\ \Omega$$

$$S = 4 + \frac{4 \times 4}{4 + 4} = 6\ \Omega$$

As, potential difference between B and D is zero, only when the bridge is balanced i.e.,

$$\frac{P}{Q} = \frac{R}{S}, \text{ so } Q = \frac{PS}{R} = \frac{21 \times 6}{18} = 7\ \Omega$$

$$\Rightarrow \quad 3 + \frac{8X}{8 + X} = 7 \Rightarrow X = 8\Omega$$

10. ***(d)*** Let V be the volume of the block. When block floats in water, then $V\rho_{\text{block}}\, g = \left(\frac{4}{5}V\right)\rho_{\text{water}}\, g$

$$\text{or} \quad \rho_{\text{block}} = \frac{4}{5}\rho_{\text{water}} \quad \ldots \text{(i)}$$

When block floats in liquid.

$$V\rho_{\text{block}}\, g = V\rho_{\text{liquid}}\, g$$

$$\rho_{\text{block}} = \rho_{\text{liquid}}$$

$$\rho_{liquid} = \frac{4}{5}\rho_{water} \quad \text{[from Eq. (i)]}$$
$$= \frac{4}{5} \times 10^3 \text{ kg m}^{-3}$$
$$\Rightarrow \quad \rho_{liquid} = 800 \text{ kg m}^{-3}$$

11. ***(d)*** The angle, subtended by arc DE at O is $\frac{\pi}{2}$ and FC at $O = \frac{3\pi}{2}$ the effective magnetic field at O is

$$B = B_{DE} + B_{FC}$$

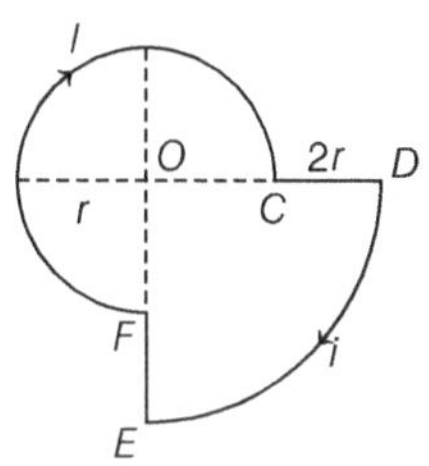

$$= \frac{\mu_0 i}{4\pi 3r} \times \frac{\pi}{2} + \frac{\mu_0 i}{4\pi r} \times \left(\frac{3\pi}{2}\right) \quad [\because OD = 3r \text{ and } OC = r]$$
$$= \frac{\mu_0 i}{8r}\left[\frac{1}{3} + 3\right]$$
$$= \frac{5\mu_0 i}{12r} \text{ (acting downwards)}$$

12. ***(b)*** The electron moving with a speed v in the circular Bohr orbit of radius r constitutes a current of magnitude.

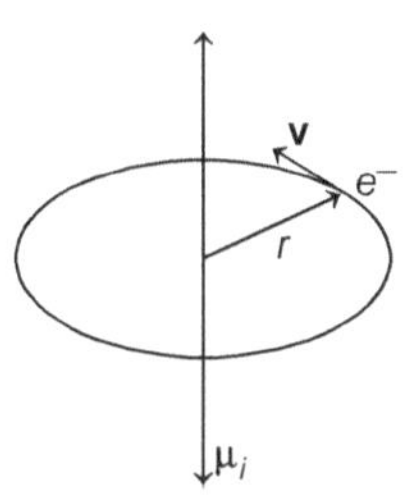

$i = \frac{e}{T}$, where T is the orbital period of the electron.

$$T = 2\pi r / v \Rightarrow i = ev / 2\pi r$$

By definition of atomic dipole moment μ_l, its magnitude is given by

$$\mu_l = iA = \frac{ev}{2\pi r} \cdot \pi r^2 = \frac{evr}{2}$$

$\therefore$ Angular momentum, $L = mvr$

$$\therefore \quad \frac{L}{\mu_l} = \frac{2mvr}{evr} = \frac{2m}{e} = \text{constant}$$

13. ***(c)*** Let T_0 be the temperature of point C and x be the length of rod AB or BC.

Then, $CA = \sqrt{x^2 + x^2} = \sqrt{2}\,x$

At steady state, the rate of heat flowing from B to C = rate of heat flowing from C to A.

$$\text{So,} \quad \frac{KA(\sqrt{2}T - T_0)}{x} = \frac{KA(T_0 - T)}{\sqrt{2}\,x}$$
$$\Rightarrow \quad \sqrt{2}(\sqrt{2}T - T_0) = T_0 - T$$

By solving, we get

$$T_0 = \frac{3T}{(\sqrt{2}+1)}$$

14. ***(b)*** At resonance, $X_L = X_C$ or $\omega L = \frac{1}{\omega C}$

Voltage across the series L-C combination,

$$V_2 = i(X_L - X_C) = 0$$

15. ***(c)*** $I = I_{max}\cos^2\left(\frac{\phi}{2}\right)$

$$\therefore \quad \frac{I_{max}}{4} = I_{max}\cos^2\frac{\phi}{2} \quad \left[\text{Given } I, = \frac{I_{max}}{4}\right]$$
$$\Rightarrow \quad \cos\frac{\phi}{2} = \frac{1}{2}$$
$$\Rightarrow \quad \frac{\phi}{2} = \frac{\pi}{3}$$
$$\Rightarrow \quad \phi = \frac{2\pi}{3} = \left(\frac{2\pi}{\lambda}\right) \times \Delta x \quad \ldots \text{(i)}$$

where, $\Delta x = d\sin\theta$

Putting the value of Δx in Eq. (i), we get

$$\sin\theta = \frac{\lambda}{3d}$$
$$\theta = \sin^{-1}\left(\frac{\lambda}{3d}\right)$$

16. ***(a)*** The bob filled completely with water, has its centre of mass at its centre. The time period of oscillation is

$$T = 2\pi\sqrt{\frac{l}{g}}$$

As the water starts coming out of the bob, its centre of mass shifts vertically downward as a result effective length of the pendulum increases and hence its period also increases. When the bob is empty, again its centre of mass appears at its centre and as a result, the period of oscillation again reaches to its original value.

17. ***(a)*** On raising the temperature of the semiconductor, covalent bonds start breaking up and thus more charge carriers release. This will increase the value of n and as a result the rate of collision of charge carriers will increase. This will decrease the drift speed v.

18. ***(b)*** When a tennis ball falls on the ground and bounces back, its velocity and displacement changes in reverse direction while acceleration remains unchanged.

19. ***(b)*** Since, the lens is made up of two kinds of transparent material, it has two refractive indices for the incident beam of light. Hence, there will be two focal lengths of the lens and therefore two images will be observed.

20. ***(c)*** As we know,

Angular momentum, $L = I\omega$

$$\Rightarrow \quad L = I \times \frac{2\pi}{T}$$

$$L \propto \frac{1}{T}$$

Given, $T_2 = 2T$

Hence, $$\frac{L_1}{L_2} = \frac{2T}{T}$$

$$\Rightarrow \quad L_2 = \frac{L_1}{2}$$

21. ***(a)*** Maximum stress $= \dfrac{\text{Maximum weight}}{\text{Cross - sectional area}}$

Now, maximum weight of steel wire

$= \text{Volume} \times \text{Density} \times g$

$= Al\rho g$

where, l is the maximum length of steel wire that can hang vertically without breaking, ρ is the density of steel and A is the cross-sectional area of steel wire.

$\therefore$ Maximum stress, $p = \dfrac{Al\rho g}{A} = l\rho g$

$$\Rightarrow \quad l = \frac{p}{\rho g}$$

22. ***(c)*** $2h = ut$

$$u_x = \frac{2h}{\Delta t} \qquad [\because u = u_x,\ t = \Delta t] \ \ldots \text{(i)}$$

By equation of motion,

$$h = u_y t - \frac{1}{2} g t^2$$

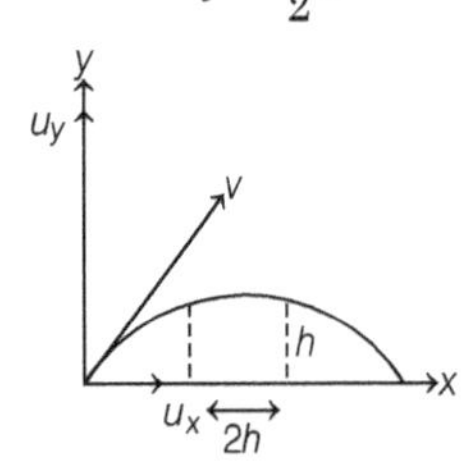

$$\Rightarrow \quad gt^2 - 2u_y t + 2h = 0$$

$$\Rightarrow \quad t_1 = \frac{2u_y + \sqrt{4u_y^2 - 8gh}}{2g}$$

$$\Rightarrow \quad t_2 = \frac{2u_y - \sqrt{4u_y^2 - 8gh}}{2g}$$

$$\Rightarrow \quad \Delta t = t_1 - t_2 = \frac{\sqrt{4u_y^2 - 8gh}}{g}$$

$$\Rightarrow \quad u_y^2 = \frac{g^2(\Delta t)^2}{4} + 2gh \qquad \ldots\text{(ii)}$$

As, $$u_x^2 + u_y^2 = u^2 = (2\sqrt{gh})^2$$

From Eqs. (i) and (ii), we get

$$\frac{4h^2}{(\Delta t)^2} + \frac{g^2(\Delta t)^2}{4} + 2gh = 4gh$$

$$\frac{g^2}{4}(\Delta t)^4 - 2gh(\Delta t)^2 + 4h^2 = 0$$

$$(\Delta t)^2 = \frac{2gh \pm \sqrt{4g^2h^2 - 4g^2h^2}}{g^2/2} = \frac{4h}{g}$$

$$\Rightarrow \quad \Delta t = 2\sqrt{\frac{h}{g}}$$

23. ***(b)*** $s = ut + \frac{1}{2}at^2$, $a = g\sin\theta$, $u = 0$

Then, $$s = 0 + \frac{1}{2}at^2$$

or $$t = \sqrt{\frac{2s}{a}} = \sqrt{\frac{2s}{g\sin\theta}} \quad \text{[for smooth plane]}$$

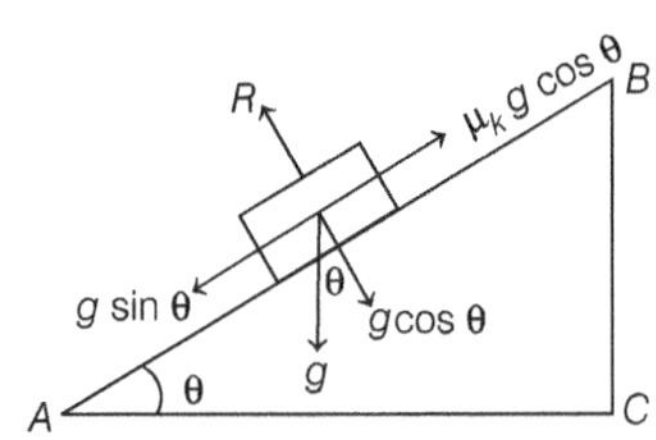

For the rough plane, the effective value of acceleration along the incline is

$$a' = g\sin\theta - \mu_k g\cos\theta$$

and $$t' = \sqrt{\frac{2s}{a}} = \sqrt{\frac{2s}{g\sin\theta - \mu_k g\cos\theta}}$$

Now, $$\frac{t'}{t} = n = \sqrt{\frac{g\sin\theta}{g\sin\theta - \mu_k g\cos\theta}}$$

$$\Rightarrow \quad n = \sqrt{\frac{\sin 45^\circ}{\sin 45^\circ - \mu_k \cos 45^\circ}}$$

$$\Rightarrow \quad n = \sqrt{\frac{1}{1 - \mu_k}}$$

$$\Rightarrow \quad \frac{1}{n^2} = 1 - \mu_k$$

$$\Rightarrow \quad \mu_k = 1 - \frac{1}{n^2}$$

24. ***(c)*** Since, the initial activity is 50 times the activity for safe occupancy, therefore $R_0 = 50R$, where $R = \lambda N$.

Since, $R \propto N$

$$\frac{R}{R_0} = \frac{N}{N_0} = \left(\frac{1}{2}\right)^n = \left(\frac{1}{2}\right)^{t/T_{1/2}}$$

or $$\left(\frac{1}{2}\right)^{t/10} = \frac{1}{50} \qquad [\because T_{1/2} = 10]$$

$$(2)^{t/10} = 50$$

Taking log both sides, we get

$$\Rightarrow \frac{t}{10}\log_{10} 2 = \log_{10} 50$$

$$\Rightarrow t = \frac{10\log_{10} 50}{\log_{10} 2} = \frac{10 \times 1.699}{0.301} = 56.4 \text{ days}$$

25. ***(a)*** Helium is monoatomic gas, for which

$$C_V = \frac{3}{2}R, C_p = \frac{5}{2}R$$

Work done by the gas in one complete cycle
W = area $ABCDA = p_0V_0$

From A to B,

Heat given to gas $= nC_V\Delta T$

$$= 1 \times \left(\frac{3}{2}R\right) \times \Delta T = \frac{3}{2}V_0(\Delta p) = \frac{3}{2}V_0p_0$$

From B to C, heat given to gas $= nC_p\Delta T$

$$= 1 \times \left(\frac{5}{2}R\right) \times \Delta T = \frac{5}{2}(2p_0)\Delta V = 5p_0V_0$$

$$\text{Efficiency of cycle} = \frac{\text{Work done by the gas / cycle}}{\text{Total heat given to gas / cycle}}$$

$$= \frac{p_0V_0}{\frac{3}{2}p_0V_0 + 5p_0V_0} = \frac{2}{13}$$

$$\text{Efficiency (\%)} = \frac{2}{13} \times 100 = 15.4\%$$

26. ***(a)*** Using Einstein's photoelectric equation

$$h\nu = \phi + eV_s$$

$$eV_s = h\nu - \phi \qquad \ldots \text{(i)}$$

Thus, when the frequency of incident light radiation (photons) ν increases to 2ν, the stopping potential V_s changes to V_s'.

$$eV_s' = h(2\nu) - \phi \qquad \ldots \text{(ii)}$$

On dividing Eq. (ii) by Eq. (i), we get

$$\frac{V_s'}{V_s} = \frac{2h\nu - \phi}{h\nu - \phi} = \frac{2h\nu - 2\phi + \phi}{h\nu - \phi}$$

$$\Rightarrow = \frac{2(h\nu - \phi) + \phi}{h\nu - \phi} = 2 + \frac{\phi}{h\nu - \phi}$$

$$\therefore \frac{V_s'}{V_s} > 2$$

$$\Rightarrow V_s' > 2V_s$$

27. ***(c)*** Terminal velocity of the rod is attained when magnetic force on the rod (Bil) balances the component of weight of the rod ($mg\sin\theta$), as in figure.

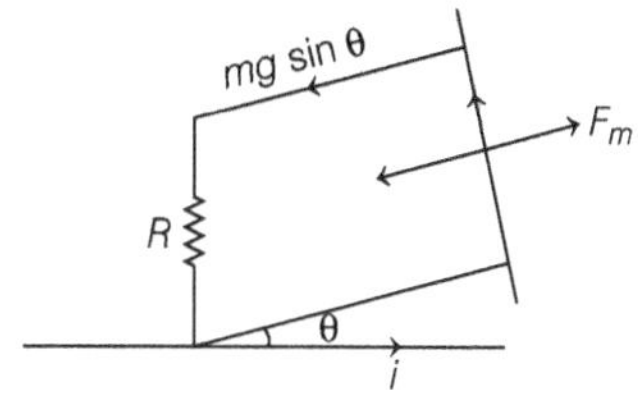

So, $Bil = mg\sin\theta$

$$B\left(\frac{e}{R}\right)l = mg\sin\theta \qquad \left[\because i = \frac{e}{R}\right]$$

$$\Rightarrow \frac{Ble}{R} = mg\sin\theta$$

$$\Rightarrow \frac{Bl(Blv_T)}{R} = mg\sin\theta \qquad [\because e = Blv]$$

$$\Rightarrow v_T = \frac{mg\,R\sin\theta}{B^2l^2}$$

28. ***(c)*** Initial energy of the asteroid is

$$E_i = K_i + U_i = \frac{1}{2}mv_i^2 - \frac{GM_em}{10R_e}$$

Final energy of the asteroid,

$$E_f = \frac{1}{2}mv_f^2 - \frac{GM_em}{R_e}$$

According to law of conservation of energy, $E_i = E_f$

$$\frac{1}{2}mv_i^2 - \frac{GM_em}{10R_e} = \frac{1}{2}mv_f^2 - \frac{GM_em}{R_e}$$

$$v_f^2 - \frac{2GM_e}{R_e} = v_i^2 - \frac{2GM_e}{10R_e}$$

$$\Rightarrow v_f^2 = v_i^2 + \frac{2GM_e}{R_e}\left(1 - \frac{1}{10}\right)$$

29. ***(b)*** From Einstein photoelectric equation,

$$E = \phi_0 + \text{KE}_{\text{max}}$$

For metal A, $\quad 4 = \phi_A + T_A \qquad \ldots$ (i)

For metal B, $\quad 4.5 = \phi_B + (T_A - 1.5) \qquad \ldots$(ii)

From Eqs. (i) and (ii), we get

$$\phi_B - \phi_A = 2$$

Now, according to de-Broglie hypothesis,

$$\lambda_A = \frac{h}{mv} = \frac{h}{\sqrt{2mT_A}}$$

Similarly, $\lambda_B = \dfrac{h}{\sqrt{2mT_B}}$

$$\therefore \frac{\lambda_A}{\lambda_B} = \sqrt{\frac{T_B}{T_A}} = \sqrt{\frac{T_A - 15}{T_A}} = \left(1 - \frac{1.5}{T_A}\right)^{1/2}$$

$$\left(\frac{1}{2}\right)^2 = 1 - \frac{1.5}{T_A}$$

On solving, we get $T_A = 2.0$ eV

So, $\phi_A = 4 - T_A = 4 - 2 = 2.0$ eV

$\phi_B = 6 - T_A = 6 - 2 = 4.0$ eV

30. ***(a)*** We have,

$$\text{Torque} = r \times F = I\alpha$$

So, $$2(20t - 5t^2) = 10\alpha$$

$$\alpha = 4t - t^2$$

But $\alpha = \dfrac{d\omega}{dt}$, so

$$\frac{d\omega}{dt} = 4t - t^2$$

$\Rightarrow$ $$d\omega = (4t - t^2)\,dt$$

Integrating, both sides, we get

$$\omega = 2t^2 - \frac{t^3}{3}$$

$\Rightarrow \omega$ will be zero at $t = 6$ s.

So, $$\omega = \frac{d\theta}{dt} = 2t^2 - \frac{t^3}{3}$$

or $$d\theta = \left(2t^2 - \frac{t^3}{3}\right) dt$$

Again, integrating both sides, we get

$$\theta = \frac{2t^3}{3} - \frac{t^4}{12}$$

Since, $t = 6$ s

so, $$\theta = \frac{2 \times 6^3}{3} - \frac{6^4}{12} = 36$$

Number of turns, $n = \dfrac{\theta}{2\pi} = \dfrac{36}{2\pi} = 5.73$

So, option (a) is the correct

31. ***(c)***

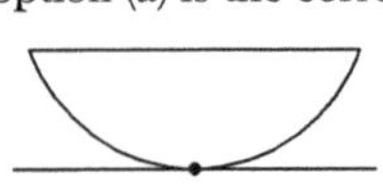

Fig. (a)

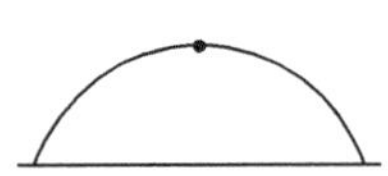

Fig. (b)

In first case, as from Fig. (a), refraction of the rays, takes place from a plane surface, so we can use

$$d_{\text{app}} = \frac{d_{\text{actual}}}{\mu}$$

$$3 = 4/\mu \quad \Rightarrow \quad \mu = 4/3$$

where, d_{app} and d_{actual} = apparent and real depth, respectively.

Now, in second case, as from Fig. (b), refraction takes place from a spherical surface, so

$$\frac{\mu_2}{v} - \frac{\mu_1}{u} = \frac{\mu_2 - \mu_1}{R}$$

or $$\frac{1}{(-25/8)} - \frac{4/3}{(-4)} = \frac{1 - 4/3}{-R}$$

or $$\frac{1}{3R} = \frac{1}{3} - \frac{8}{25} = \frac{1}{75}$$

$\Rightarrow$ $$R = 25 \text{ cm}$$

Now, using lens Maker's formula, to calculate focal length,

$$\frac{1}{f} = (\mu - 1)\left(\frac{1}{R_1} - \frac{1}{R_2}\right)$$

$\Rightarrow$ $$\frac{1}{f} = \left(\frac{4}{3} - 1\right)\left(\frac{1}{\infty} - \frac{1}{(-25)}\right) = \frac{1}{75}$$

$\Rightarrow$ $$f = 75 \text{ cm}$$

32. ***(c)*** When $r \le R$, then force on the test mass m at the surface of the sphere = mg

Force on the test mass at distance r from the centre of sphere is

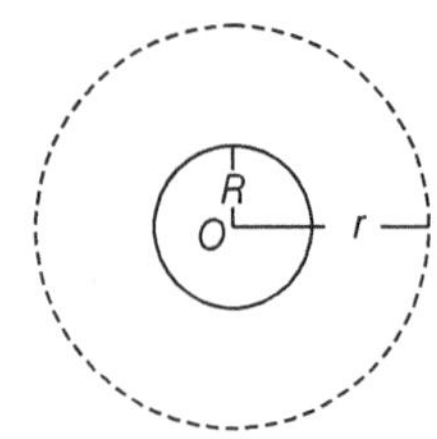

If $r < R$, then $$F = \frac{GMm}{R^2} \cdot \frac{r}{R} = \frac{GMm}{R^3} \cdot r$$

$\therefore$ $$\frac{mv^2}{r} = \frac{GMm}{R^3} \cdot r$$

$\therefore$ $$v \propto r$$

If $r > R$, then $$F = \frac{GMm}{r^2}$$

$\therefore$ $$\frac{Mv^2}{r} = \frac{GMm}{r^2}$$

$\therefore$ $$v \propto \frac{1}{\sqrt{r}}$$

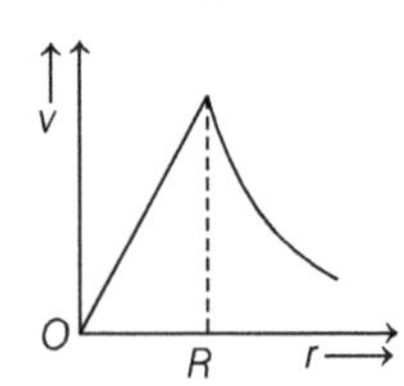

Hence, option (c) is correct.

33. ***(c)*** In the given circuit, diode D_1 is reverse biased, so it will not conduct but D_2, D_3 are forward biased, so they will conduct, hence corresponding equivalent circuit. For the given circuit is

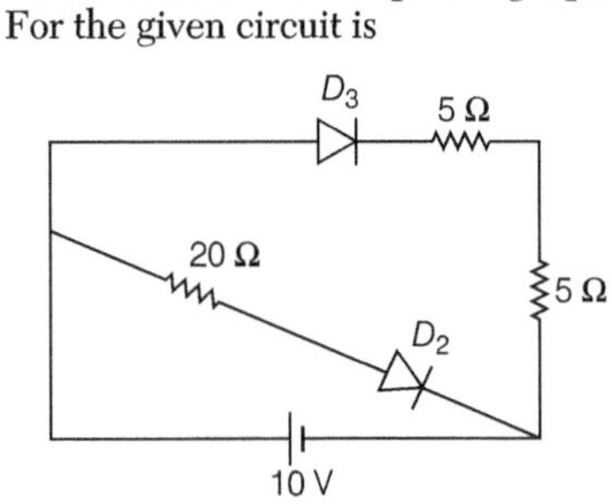

Now, the equivalent resistance of the circuit is

$$R_{eq} = \frac{(5+5)\times 20}{(5+5)+20} = \frac{20}{3}\,\Omega$$

Current through battery, $i = \dfrac{10}{\frac{20}{3}} = 1.5\text{ A}$

34. ***(a)*** Since, no external force is acting on the system, we can apply conservation of linear momentum.

Speed of 200 kg trolley $= \dfrac{36\times 1000}{60\times 60} = 10\text{ ms}^{-1}$

If u be the initial velocity of trolley, v_b be the absolute velocity of the boy after the beginning of journey of the boy, their relative velocity is 4.

So, $\quad v' - v_b = 4 \Rightarrow v_b = (v' - 4)$

Now, applying law of conservation of momentum,

Momentum before the boy begins to run

= Momentum after the beginning of boy's running

$$\Rightarrow \quad 220\times 10 = 200\,v' + 20(v' - 4)$$

$$2200 = 220v' - 80$$

$$\Rightarrow \quad 220v' = 2280$$

$$v' = \frac{2280}{220} = 10.36\text{ ms}^{-1}$$

35. ***(c)*** Given, $C = 37.55\text{ J mol}^{-1}\text{ K}^{-1}$

Also, $pT = \text{constant}\,(k)$ $\quad\quad$ ... (i)

According to ideal gas equation, $pV = RT$

$$\Rightarrow \quad p = \frac{RT}{V} \quad\quad \text{... (ii)}$$

Putting the value of p in Eq. (i), we get

$$\frac{RT}{V}\times T = k \Rightarrow V = \frac{RT^2}{k}$$

On differentiating above equation both sides, we get

$$\frac{dV}{dT} = \frac{2RT}{k} \quad\quad \text{... (iii)}$$

But $\quad \dfrac{T}{k} = \dfrac{1}{p}$ $\quad\quad$ [from Eq. (i)]

Hence, Eq. (iii) becomes

$$\frac{dV}{dT} = \frac{2R}{p}$$

So, $\quad C = C_V + \dfrac{pdV}{dT}$

or $\quad C = C_V + \dfrac{p\times 2R}{p} = C_V + 2R$

or $\quad C_V = C - 2R$ $\quad\quad$ ... (iv)

As $C_V = \dfrac{nR}{2}$, where n = number of degrees of freedom.

Putting the value of C_V in Eq. (iv), we get

$$\frac{nR}{2} = C - 2R$$

$$n = \frac{2(C-2R)}{R} = \frac{2\,(37.55 - 2\times 8.3)}{8.3}$$

$$= 5.048 = \simeq 5$$

36. ***(a)*** We know, instantaneous displacement, $x = r\sin\omega t$

$\because$ Instantaneous velocity, $v = \dfrac{dx}{dt} = r\omega\cos\omega t$

$\because$ Instantaneous acceleration, $a = \dfrac{dv}{dt}$

$$= -r\omega^2\sin\omega t = -\omega^2 x$$

So, $\quad \dfrac{aT}{x} = \dfrac{-\omega^2 x\times T}{x} = -\omega^2 T = \dfrac{-4\pi^2}{T^2}\times T$

$$= \frac{-4\pi^2}{T} = \text{constant}$$

$\Rightarrow \quad \dfrac{aT}{v} = \dfrac{-\omega^2 r\sin\omega t\times T}{\omega r\cos\omega t} = -\omega T\tan\omega T$

$$= -\frac{2\pi}{T}\times T\tan\omega t = \text{not constant}$$

Similarly, $aT + 2\pi v$ and $a^2T^2 + 4\pi^2 v^2$ is also not constant, i.e. both are function of t.

37. ***(d)*** New, current sensitivity, $I_s' = I_s + \dfrac{50}{100}I_s$

$$= \frac{150}{100}I_s = \frac{3}{2}I_s$$

New resistance, $R' = 2R$

Initial voltage sensitivity, $V_s = \dfrac{I_s}{R}$

Now, new voltage sensitivity,

$$V_s' = \frac{I_s'}{R'} = \frac{\frac{3}{2}I_s}{2R} = \frac{3}{4}V_s$$

% decrease in voltage sensitivity

$$= \frac{V_s - V_s'}{V_s}\times 100$$

$$\Rightarrow \quad = \left(1 - \frac{V_s'}{V_s}\right)\times 100$$

$$\therefore \quad = \left(1 - \frac{3}{4}\right)\times 100 = 25\%$$

38. ***(a)*** For the uniformly accelerated/decelerated motion,

$$v^2 = u^2 \pm 2gh$$

So, from this equation, we can say v-h graph is parabola.

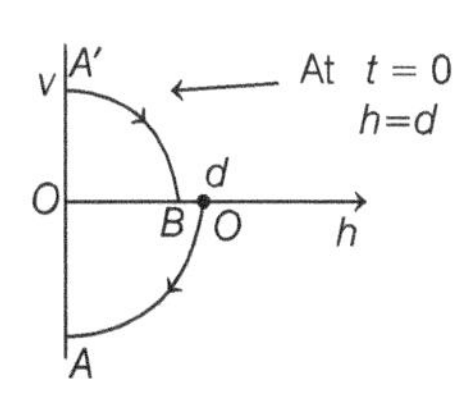

Initially, velocity is downwards (– ve), after collision, it reverses the direction with smaller magnitude and velocity is upwards (+ ve). So, graph (a) satisfies these conditions.

Also,

When $t = 0, h = d$

Velocity increases downwards $(0 \rightarrow A)$

When $t = 1$, velocity reverses its direction $(A' \rightarrow B)$.

39. ***(b)*** From the given circuit, we can say that I is independent of resistance R_5, so no current flows through R_5. This required that the junction of R_1 and R_2 is at the same potential as the junction of R_3 and R_4 ($\because$ Wheatstone bridge condition)

So, $$\frac{R_1}{R_2} = \frac{R_3}{R_4} \text{ or } R_1 R_4 = R_3 R_2$$

40. ***(d)*** Since, volume remains unchanged, during this phenomenon, so

$$\frac{4}{3}\pi R^3 = N \times \frac{4}{3}\pi r^3$$

$$N = \frac{R^3}{r^3}$$

Now, change in surface area $= 4\pi R^2 - N 4\pi r^2$

$$= 4\pi (R^2 - Nr^2)$$

Energy released $(\Delta U) = T \times$ change in surface area

$$= T \times 4\pi [R^2 - Nr^2]$$

Here, all this energy released is at the cost of lowering the temperature and mass of the big drop of liquid $= \frac{4}{3}\pi R^2 \rho$.

Now, change in temperature,

$$\Delta\theta = \frac{\Delta U}{MS} = \frac{T \times 4\pi(R^2 - Nr^2)}{\left(\frac{4}{3}\pi R^3 \rho\right) S}$$

$$\Rightarrow \quad = \frac{3T}{\rho S}\left(\frac{1}{R} - \frac{Nr^2}{R^3}\right)$$

$$\Rightarrow \quad = \frac{3T}{\rho S}\left(\frac{1}{R} - \frac{R^3 \times r^2}{r^3 \times R^3}\right)$$

$$\Rightarrow \quad \Delta\theta = \frac{3T}{\rho S}\left(\frac{1}{R} - \frac{1}{r}\right)$$

Chemistry

41. ***(d)*** Edge length of the unit cell = $2\text{Å} = 2\times10^{-8}$ cm

Volume of unit cell = $a^3 = (2\times10^{-8})^3 = 8\times10^{-24}\ \text{cm}^3$

Mass of one unit cell = volume × density

$$= 8\times10^{-24} \times 2.5$$

Number of unit cell in 200 g of metal

$$= \frac{\text{Mass of the metal}}{\text{Mass of one unit cell}} = \frac{200}{8\times10^{-24}\times2.5} = 1\times10^{25}$$

42. ***(a)*** Higher the value of 'a', more will be the tendency to get liquefy. Since, value of a is highest for gas P. Thus, it is the most liquefiable gas among the given gases.

43. ***(d)*** Radius of hydrogen like atom,

$$r_n = \frac{n^2}{Z} r_o$$

where, $r_o = 0.51\times10^{-10}$ m

and $r_n = \frac{0.51\times10^{-10}\ \text{m}}{4}$

In the ground state, $n=1$

$$\therefore \quad \frac{0.51\times10^{-10}}{4} = \frac{(1)^2}{Z}\times0.51\times10^{-10}$$

$$\therefore \quad Z = 4$$

So, the atom is triply ionised beryllium (Be^{3+}).

44. ***(a)*** When central metal ion is same (here Ni^{2+}), the absorption of colour depends on the ligand.

According to spectrochemical series, CFSE values of the ligands increases is as follows

$$I^- < Br^- < Cl^- < NO_3^- < F^- < H_2O$$
$$< NH_3 < NO_2^- < CN^- < CO$$

Thus, H_2O is the weakest ligand among these, therefore the absorbed energy will be lowest in $[Ni(H_2O)_6]^{2+}$, so it will absorb highest wavelength (red light).

As, $$E \propto \frac{1}{\lambda}$$

$\therefore$ The order of increasing wavelength is

$$[Ni(NO_2)_6]^{4-} < [Ni(NH_3)_6]^{2+} < [Ni(H_2O)_6]^{2+}.$$

45. ***(c)*** Empirical formula mass $= CH_2O = 12+2+1\times16$

$$= 30 \text{ g}$$

Molecular mass = 180 g

$$n = \frac{\text{Molecular mass}}{\text{Empirical formula mass}} = \frac{180}{30} = 6$$

$\because$ Molecular formula $= n\times$ empirical formula

$\therefore$ Molecular formula $= 6\times CH_2O = C_6H_{12}O_6$

46. ***(d)*** Statements (d) is incorrect. Whereas all other statements are correct.

Corrected statement

Al_2O_3 serves as the electrolyte, undergoing the redox process. Na_3AlF_6 although is an electrolyte but serves as a solvent, not electrolyte.

47. ***(c)*** Azo dye test is not used for the distinction of 1°, 2° and 3° aliphatic amines. The other tests can be easily used to distinguish between 1°, 2° and 3° amines.

	Test	Primary amine	Secondary amine	Tertiary amine
(a)	Hinsberg's reagent test	Forms a sulphonamide soluble in alkali	Forms a sulphonamide insoluble in alkali	No reaction
(b)	Carbylamine reaction	Forms a carbylamine with unpleasant smell	No reaction	No reaction
(d)	Reaction with nitrous acid	Forms primary alcohol and evolves nitrogen gas with effervescence	Forms nitrosamine which gives Libermann's nitroso amine reaction	Forms nitrite salt

48. ***(b)*** Resonating structures of carbonate CO_3^{2-} ion are

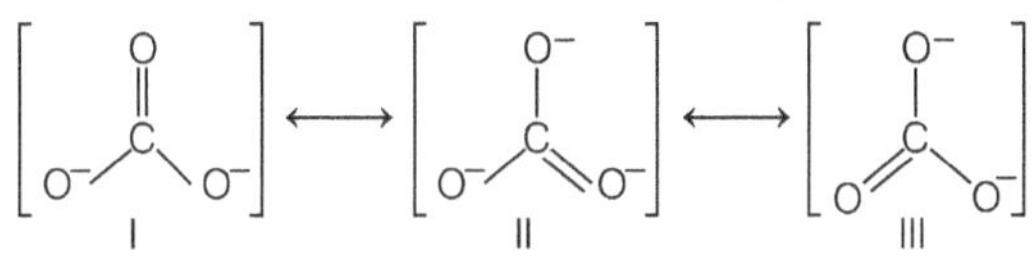

Thus, there is no coordinate bond present in CO_3^{2-}
Hence, the second statement is incorrect.

49. ***(a)*** For first order reaction,

$$t_{1/2} = \frac{\ln 2}{k_1} = 40\,\text{s} \quad \ldots\text{(i)}$$

For zero order reaction,

$$t_{1/2} = \frac{[A]_0}{2k_0} = 20\,\text{s} \quad \ldots\text{(ii)}$$

Divide Eq. (i) by Eq. (ii), we get

$$\frac{1}{2} = \frac{[A]_0}{2k_0} \times \frac{k_1}{\ln 2}$$

$$\frac{k_1}{k_0} = \frac{\ln 2}{[A]_0} = \frac{0.693}{1.386} = 0.5\ \text{mol}^{-1}\ \text{dm}^3$$

50. ***(b)*** CH_2Cl_2 is similar to CH_4 (tetrahedral) but it has three different bond angles.

H—C—H, H—C—Cl, Cl—C—Cl, which are close to but not equal to 109.5°. Since, electronegativity of Cl > C > H the bond dipole moments do not cancel and the molecule has a dipole moment.

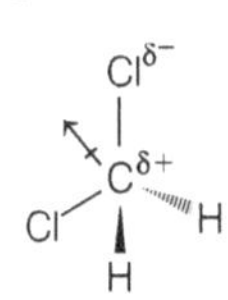

51. ***(b)*** Given, $CH_4 + 2O_2 \longrightarrow CO_2 + 2H_2O$;

$$\Delta H = -210\ \text{kcal / mol} \quad \ldots\text{(i)}$$

$$C_2H_6 + \frac{7}{2}O_2 \longrightarrow 2CO_2 + 3H_2O\ ;$$

$$\Delta H = -368\ \text{kcal / mol} \quad \ldots\text{(ii)}$$

On subtracting Eq. (i) from Eq (ii), we get

$$CH_2 + \frac{3}{2}O_2 \longrightarrow CO_2 + H_2O\ ;$$

$$\Delta H = -158\ \text{kcal / mol}$$

$\therefore$ Enthalpy of combustion of one CH_2 unit

$$= -158\ \text{kcal / mol}$$

tt$\Delta H_{comb}(C_{10}H_{22}) = \Delta H_{comb}(CH_4) + 9 \times \Delta H_{comb}(CH_2)$

$$= -210 + (9 \times -158)$$
$$= -1632\ \text{kcal}$$

52. ***(b)*** R—C≡CH is converted to ketone by catalytic hydration with (i) $H_2SO_4 + HgSO_4$, (ii) H_2O, heat

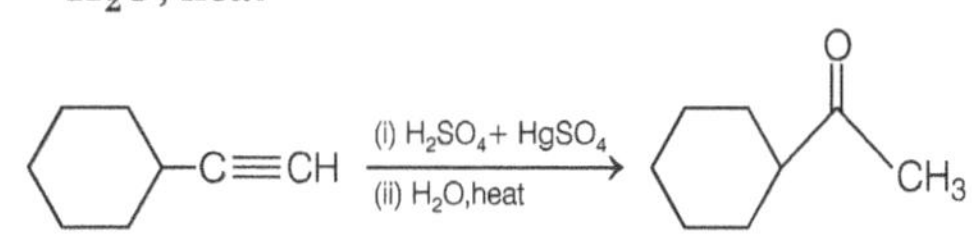

53. ***(c)*** By Heisenberg's uncertainty principle,

$$\Delta x \times \Delta P \geq h / 4\pi$$
$$\Delta x \times m\Delta v \geq h / 4\pi$$

So, $\Delta x \geq \dfrac{h}{4\pi m \Delta v}$ ...(i)

Given, Δv = 0.005% or 600 m/s

$$= \frac{600 \times 0.005}{100} = 0.03$$

On putting all values we get,

$$\therefore\ \Delta x \times 9.1 \times 10^{-31} \times 0.03 = \frac{6.6 \times 10^{-34}}{4 \times 3.14}$$

$$\Rightarrow \Delta x = \frac{6.6 \times 10^{-34}}{4 \times 3.14 \times 0.03 \times 9.1 \times 10^{-31}} = 1.92 \times 10^{-3}\ \text{m}$$

54. ***(d)*** In O_2 molecule, two electrons are present in antibonding orbitals.

$$O_2(8+8=16) = \sigma 1s^2, \sigma^* 1s^2, \sigma 2s^2 \sigma^* 2s^2,$$
$$\sigma(2p_z)^2, \pi 2p_x^2 = \pi 2p_y^2, \pi^* 2p_x^1 = \pi^* 2p_y^1$$

Removal of one electron from the O_2 molecule gives O_2^+ in which the number of antibonding electrons is one less and hence, BO (Bond order) increases. Thus, removal of one electron from O_2 stabilises the molecule.

55. ***(c)*** According to Faraday's first law, when an electric current is passed through an electrolyte, amount of substance deposited is directly proportional to the quantity of electric charge passed through the electrolyte.

For the reaction,

$$Cu^{2+} + 2e^- \longrightarrow Cu \quad \ldots\text{(i)}$$

1 mol of 63.5g of Cu is obtained by passing 2F of electricity.

Mass of Cu obtained by passing 1F of electricity $= \dfrac{63.5}{2}$ g

$\therefore$ Mass of Cu obtained by passing 0.5F of electricity

$$= \frac{63.5}{2} \times 0.5 = 15.80 \text{ g}$$

56. ***(c)*** The sequence of reactions are as follows :

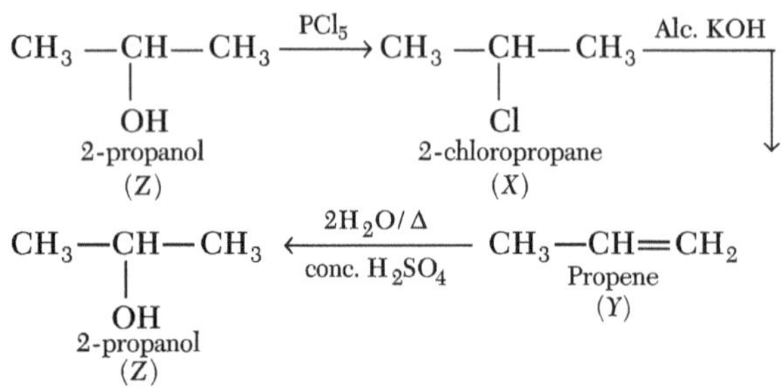

57. ***(d)*** $\underset{(0.28)}{N_2O_4(g)} \rightleftharpoons \underset{(1.1)}{2NO_2(g)}$

Pressure at equilibrium

$$K_p = \frac{p^2_{NO_2}}{p_{N_2O_4}} = \frac{(1.1)^2}{0.28} = 4.32 \text{ atm}$$

If volume of the container is doubled, the pressure will reduced to half

$$N_2O_4 \rightleftharpoons 2N_2O$$

$$p_{N_2O_4} = \frac{0.28}{2} = 0.14 - x \text{ atm}$$

$$p_{NO_2} = \frac{1.1}{2} = 0.55 + 2x \text{ atm}$$

$$K_p = \frac{[p_{NO_2}]}{[p_{N_2O_4}]}$$

$$K_p = \frac{(0.55 + 2x)^2}{(0.14 - x)} = 4.32$$

$$0.30 + 1.1x + 4x^2 = 0.605 - 4.32x$$

$$4x^2 + 5.42x - 0.302 = 0$$

This is quadratic equation with solution

$$x = -\frac{b \pm \sqrt{b^2 - 4ac}}{2a}$$

$$x = \frac{(5.42) \pm \sqrt{(5.42)^2 - 4(4)(-0.302)}}{2(4)}$$

$$x = -\frac{5.42 \pm 5.85}{8}$$

$$x = 0.045$$

or $\quad x = -1.41$

(Negative value is discarded as it will lead to negative value of pressure.)

So, $\quad x = 0.045$

$$p_{N_2O_4} = 0.14 - x = 0.14 - 0.045 = 0.095 \text{ atm}$$

$$p_{NO_2} = 0.55 + 2x = (0.55 + 2 \times 0.045) = 0.64 \text{ atm}$$

The new equilibrium pressure of the two gases are 0.095 and 0.64 atm respectively.

58. ***(a)*** Protactinium (Pa) is the element that does not have stable electronic configuration.

Protactinium (Pa) = $[Rn]5f^2 6d^1 7s^2$

(Atomic number = 91)

Nobelium (No) = $[Rn]5f^{14} 6d^0 7s^2$

(Atomic number = 102)

Lawrencium (Lr) = $[Rn]5f^{14} 6d^1 7s^2$

(Atomic number = 103)

Amercium (Am) = $[Rn]5f^7 6d^0 7s^2$

(Atomic number = 95)

59. ***(c)*** **Oxidation**

$$H_3C—CHO + 3\bar{O}H \longrightarrow CH_3COO^- + 2e^- + 2H_2O \quad \ldots(i)$$

Reduction $[Ag(NH_3)_2]^+ + e^- \longrightarrow Ag + 2NH_3] \times 2$

$$\therefore CH_3CHO + 2[Ag(NH_3)_2]^+ + 3\bar{O}H \longrightarrow CH_3CO\bar{O} + 2H_2O + 2Ag + 4NH_3$$

Molecular mass (M) of CH_3CHO = 44 g mol^{-1}

$$\text{Equivalent weight} = \frac{M}{n\text{-factor}}$$

$$\omega = 44/2 = 22 \text{ g mol}^{-1}$$

$\therefore$ Acetaldehyde is reducing agent.

$\therefore$ Statement (c) is incorrect.

60. ***(a)*** According to Freundlich adsorption isotherm,

$$x/m = kp^{1/n}$$

where, x = mass of gas adsorbed on mass 'm' of the absorbent at pressure p.

k and n = constants.

Taking logarithm on both sides, we get

$$\log x/m = \log k + \frac{1}{n}\log p$$

This is equation of a straight line with slope of $1/n$ and intercept of $\log k$.

61. ***(b)*** Given, pH of $Ba(OH)_2 = 12$

$\because \quad pH + pOH = 14$

$\therefore \quad pOH = 14 - pH = 14 - 12 = 2$

Now, $\quad pOH = -\log[OH^-]$

$\Rightarrow \quad 2 = -\log[OH^-]$

$$[OH^-] = 10^{-2}$$

As conc. of Ba^{2+} is half of OH^- i.e. $[Ba^{2+}] = 0.5 \times 10^{-2}$

$$\underset{S\text{ mol}^{-1}}{Ba(OH)_2(s)} \rightleftharpoons \underset{S}{Ba^{2+}} + \underset{2S}{2OH^-}$$

$$K_{sp} = [Ba^{2+}][OH^-]^2$$

$\Rightarrow \quad [0.5 \times 10^{-2}][10^{-2}]^2 = 0.5 \times 10^{-6}$

$$= 5 \times 10^{-7} M$$

62. ***(c)*** With Tollen's reagent (ammoniacal $AgNO_3$), a white precipitate of silver salt is obtained on reaction with C_2H_2 (alkyne).

$$\underset{\text{Alkyne}}{H-C \equiv C-H} + 2AgNO_3 + 2NH_4OH \longrightarrow \underset{\text{Silver acetylide (white ppt.)}}{Ag-C \equiv C-Ag} + 2NH_4NO_3 + H_2O$$

63. ***(d)*** Arrhenius equation is given by,

$$\log_{10}\frac{K_2}{K_1} = \frac{E_a}{2.303 \times R}\left[\frac{T_2 - T_1}{T_1 T_2}\right]$$

Given, $\frac{K_2}{K_1} = 3$; $R = 8.314\ JK^{-1}mol^{-1}$

$T_1 = 20 + 273 = 293$ K

and $T_2 = 50 + 273 = 323$ K

Substituting the given values in Arrhenius equation,

$$\log_{10} 3 = \frac{E_a}{8.314 \times 2.303}\left[\frac{323 - 293}{323 \times 293}\right]$$

$$E_a = \frac{2.303 \times 8.314 \times 323 \times 293 \times 0.477}{30}$$

$$= 28811.8\ J\ mol^{-1}$$

$$= 28.8118\ kJ\ mol^{-1}$$

64. ***(b)*** Compounds containing both N and S give blood red colour in Lassaigne's test due to the formation of $Fe(SCN)_2$. Thus, $H_2N\ (C_6H_4)SO_3H$ gives blood red colour in Lassaigne's test of nitrogen.

65. ***(b)*** Given that,

$$A + B^{2+} \longrightarrow B + A^{2+},\ E° = 0.2955V$$

$$n = 2,\ T = 298\,K,\ E° = 0.2955\,V$$

Using Nernst equation,

$$E° = \frac{2.303nRT}{nF}\log K_{eq}$$

$$E° = \frac{0.0591}{n}\log K_{eq}\ (\text{at } 25°C)$$

$$\Rightarrow \quad 0.2955 = \frac{0.0591}{2}\log K_{eq}$$

$$K_{eq} = 10^{10}$$

66. ***(b)***

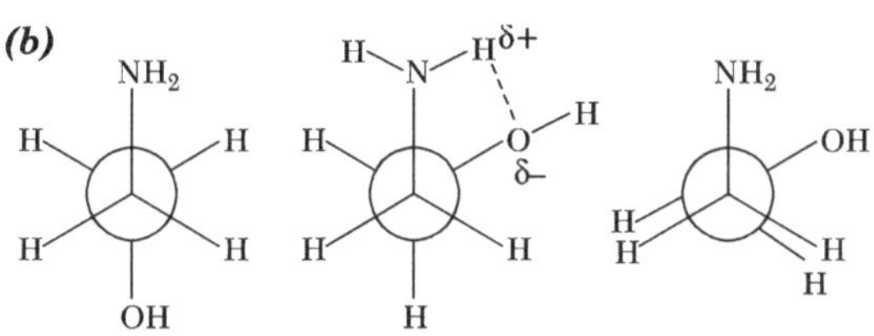

Anti *Gauche* *Eclipsed*

Since, gauche form is stabilised by intermolecular hydrogen bonding, hence it is more stable than anti form while eclipsed form is least stable due to high angular strain.

∴ The correct order is, *gauche* > *anti* > *eclipsed.*

67. ***(a)*** In acidic medium, H_2S gas is very feebly ionised giving very small concentration of sulphide ion for precipitation. Therefore, the most insoluble salts CuS and HgS are precipitated. Since the solubility product of Mn^{2+}, Ni^{2+} are higher thus they remain unaffected.

68. ***(b)*** When a mixture of BCl_3 and H_2 is subjected to silent electric discharge, diborane and HCl are formed.

$$\underset{\text{Boron trichloride}}{2BCl_3} + 6H_2 \xrightarrow[\text{discharge}]{\text{Electric}} \underset{(x)}{B_2H_6} + 6HCl$$

Diborane reacts with NH_3 at 200°C to give borazine (inorganic benzene).

$$\underset{(x)}{3B_2H_6} + \underset{\text{Ammonia}}{6NH_3} \xrightarrow{200°C} \underset{\text{Borazine }(y)}{2B_3N_3H_6} + 12H_2$$

69. ***(c)*** The process of producing *syn* gas or synthesis gas from coal is called 'coal gasification'.

$$\underset{\text{Coal}}{C(s)} + \underset{\text{Steam}}{H_2O(g)} \xrightarrow[\text{Ni}]{1270\text{ K}} \underbrace{CO(g) + H_2(g)}_{\text{Synthesis gas}}$$

The production of hydrogen can be increased by reaching carbon monoxide of the *syn* gas with steam in the presence of iron chromate as a catalyst at 673 K.

$$CO(g) + H_2O(g) \xrightarrow{FeCrO_4,\ 673\text{ K}} CO_2(g) + H_2(g)$$

CO_2 is removed by scrubbing with a solution of sodium arsenite.

70. ***(b)*** Heat at constant pressure means enthalpy. i.e.

$$\Delta H = -2.5 \times 10^3 \text{ cal}$$

$$\Delta S = 7.4 \text{ cal deg}^{-1}$$

$$T = 298\text{ K}$$

$$\Delta G = \Delta H - T\Delta S$$

$$= -2.5 \times 10^3 - 298 \times 7.4$$

$$= -4705 \text{ cal}$$

The value of change in Gibbs free energy (ΔG) is negative.

Hence, the process is spontaneous.

71. ***(a)*** The alkene is unsymmetrical, hence will follow Markownikoff's rule to give major product.

$$\underset{\text{But-1-ene}}{CH_3-CH_2-CH=CH_2} + H-Br \longrightarrow$$

$$\underset{\text{Major}}{\underset{(x)}{CH_3-CH_2-CHBr-CH_3}} + \underset{\text{Major}}{\underset{(y)}{C_2H_5-\overset{\overset{Br}{|}}{\underset{\underset{H}{|}}{C^*}}-CH_3}}$$

$$+ \underset{(z)\text{ Minor}}{CH_3-CH_2-CH_2-CH_2Br}$$

Since, *y* contains, a chiral carbon, it exists in two enantiomers (*x* and *y*) which are mirror images of each other.

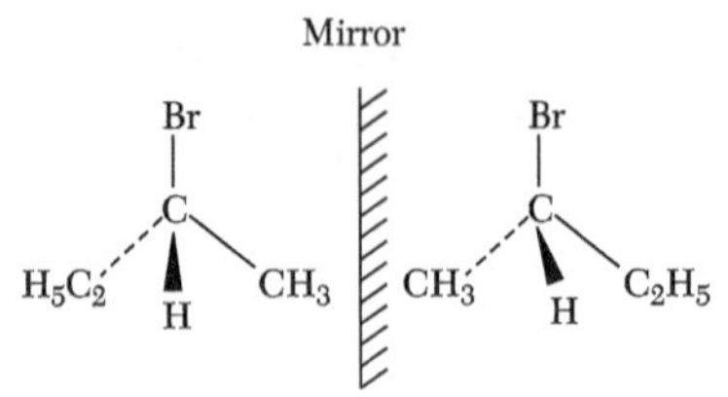

72. ***(d)*** Silicones are synthetic organo-silicon polymers containing repeated R_2SiO units. Since the empirical formula is same as that of a ketone (R_2CO), the name silicon has been given to these materials. They can be formed by hydrolysis of dichlorosilanes (R_2SiCl_2).

$$R_2SiCl_2 + 2H_2O \longrightarrow R_2Si(OH)_2 + 2HCl$$

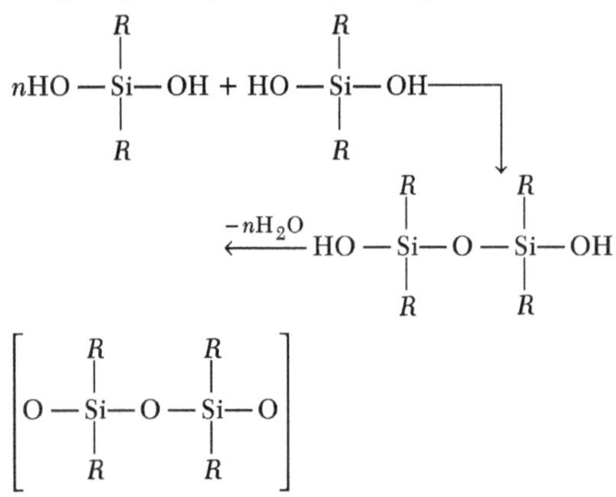

73. ***(c)*** On acidification of potassium chromate solution, yellow colour changes to orange colour due to formation of dichromate which suggests that monocentric complex is converted into dicentric complex.

$$\underset{\text{Potassium chromate}}{2K_2CrO_4} + \underset{\text{(Dil.)}}{H_2SO_4} \longrightarrow \underset{\text{Potassium dichromate}}{K_2Cr_2O_7} + K_2SO_4 + H_2O$$

74. ***(d)*** Valence electrons in A = 3.

Valence electron in B = 6.

Thus, A is electropositive and B is electronegative, A can loose three electrons and B can gain two electrons to attain stable configuration

Hence, A exist as A^{3+} and B as B^{2-}

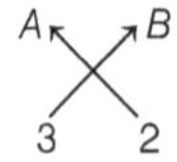

$\therefore$ Compound formed is A_2B_3.

75. ***(b)*** The non-reducing sugar is

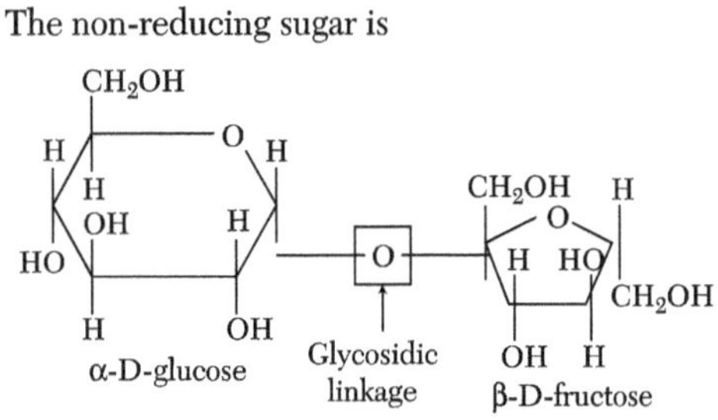

This structure represents sucrose in which α-D-glucose and β-D fructose is attached to each other by $C_1 - C_2$ glycosidic linkage. Since reducing groups of glucose and fructose are involved in glycosidic bond formation, this is considered as non-reducing sugar.

76. ***(a)*** Potassium sulphate, $i = 3$

Sodium chloride, $i = 2$

Urea, $i = 1$

Glucose, $i = 1$

Depression is freezing point is given by

$$\Delta T_f = imK_f$$

$\therefore$ Greater the value of i, greater lowering in freezing point and hence, lower will be the freezing temperature.

Therefore, potassium sulphate (K_2SO_4) solution has the lowest freezing point.

77. ***(a)*** A penicillin is a member of a family of drugs that have a four membered cyclic amide fused to a five membered thiazole ring. It is narrow spectrum antibiotic

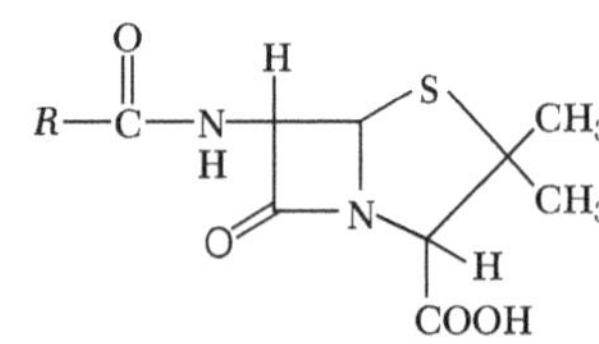

Penicillin

78. ***(b)*** The lattice energy of the hydroxides of alkaline earth metal decreases more rapidly than their hydration energy leading to more negative value of ΔH_{sol} down the group. More negative is ΔH_{sol}, more is the solubility of compounds. Hence, the solubility of hydroxides of alkaline earth metal increases with their atomic number.

79. ***(a)*** The given chemical reaction is

$$5Br^-(aq) + BrO_3^-(aq) + 6H^+(aq) \longrightarrow 3Br_2(aq) + 3H_2O\,(l)$$

Rate law expression for the given reaction can be written as

$$-\frac{1}{5}\frac{\Delta[Br]^-}{\Delta t} = -\frac{\Delta[BrO_3^-]}{\Delta t}$$

$$-\frac{1}{6}\frac{\Delta[H^+]}{\Delta t} = +\frac{1}{3}\frac{\Delta[Br_2]}{\Delta t}$$

$$\Rightarrow \quad -\frac{\Delta[Br^-]}{\Delta t} = -\frac{5\Delta[BrO_3^-]}{\Delta t}$$

$$= -\frac{5}{6}\frac{\Delta[H^+]}{\Delta t}$$

$$\Rightarrow \quad \frac{\Delta[Br^-]}{\Delta t} = \frac{5}{6}\frac{\Delta[H^+]}{\Delta t}$$

80. ***(c)*** The polymerisation reaction of the given monomers to form respective polymers are given below :
A-III, B-I, C-II, D-IV

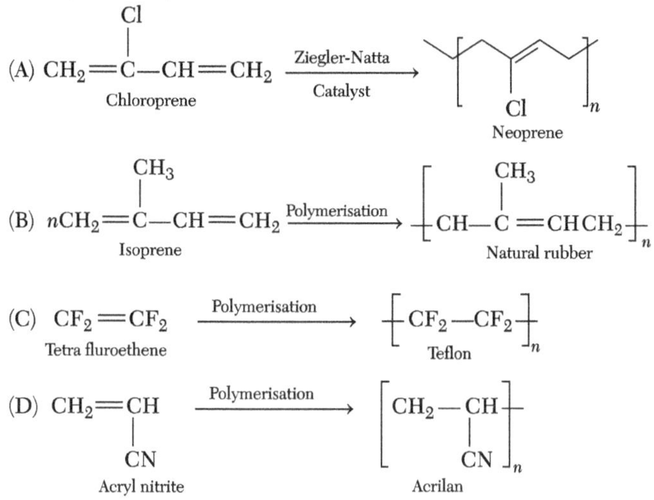

a. English Proficiency

81. ***(a)*** Use 'is' in place of 'are' as when two subjects are joined 'along with', then helping verb is used according to first subject.

82. ***(c)*** Use 'attested' in place of 'attest' as sentence is in passive voice.

83. ***(b)*** Use 'making' in place of 'having' as 'have' is not used in progressive sense.

84. ***(d)*** 'Superior' is followed by 'to' preposition.

85. ***(d)*** 'Cope' always agree with preposition 'with' and use of any other auxiliary term is improper.

86. ***(a)*** 'Dejected' means very unhappy. 'Jubilant' is its opposite meaning word which means extremely happy.

87. ***(a)*** 'Retrograde' means making a situation worse. 'Progressive' is its opposite meaning word which means happening or developing steadily.

88. ***(a)*** 'Belittle' means to decrease the importance of something of somebody. 'Exalt' is its opposite meaning word which means to make somebody rise to a higher rank or position.

89. ***(b)*** 'Apprehend' means to grasp with the understanding. So, it is the nearest, meaning word of 'make out anything'.

90. ***(b)*** 'Scathing' means expressing a very strong negative opinion about somebody or something. 'Scornful' is also nearest meaning word of 'scathing' as its means feeling or showing disgust and anger to somebody or something.

91. ***(b)*** The ordinary megalomaniac does not have the talent to realise his wish which Alexander the Great had.

92. ***(d)*** Megalomaniac differ from a narcissist by wishing to be powerful and feared.

93. ***(d)*** In the given line, 'this' refers to excessive lover of power.

94. ***(c)*** The nurturing of only one element in human nature, is the reason for unhappiness.

95. ***(c)*** Love of power to be accepted because it is a part of human nature.

b. Logical Reasoning

96. ***(d)*** 'Umpire' is required to give decision in 'Match'. Likewise 'Judge' is required to give decision in a 'Lawsuit'.

97. ***(c)*** As, $9-2=7 \Rightarrow 7^2 = 49$

$13-2=11 \Rightarrow 11^2 = 121$

$10-2=8 \Rightarrow 8^2 = 64 \neq 61$

$7-2=5 \Rightarrow 5^2 = 25$

Thus, option (c) is odd.

98. ***(d)*** The pattern is as follows

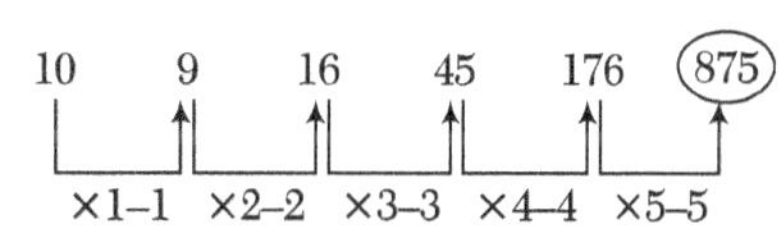

99. ***(d)*** The data is inadequate because it is not given that who is taller between Virat and Shikhar. Hence, answer cannot be determined.

100. ***(c)*** Option (c) will complete the given pattern as follow.

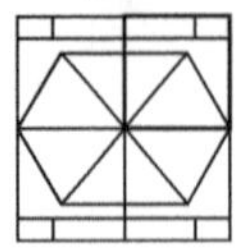

101. ***(b)*** Options (b) can be formed from the pieces given in problem figures.

102. ***(a)*** After folding and cutting the paper, answer figure (a) will appear.

103. ***(d)*** The third figure is each row comprises of parts which are not common in the first two figures.

104. ***(b)*** Correct answer figure is (b).

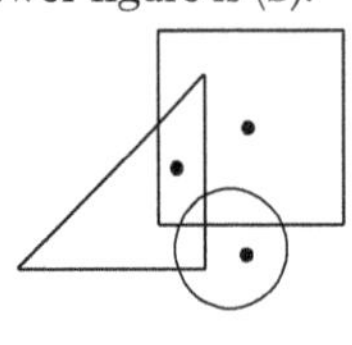

105. ***(b)*** The figure in the question may be labelled as shown below

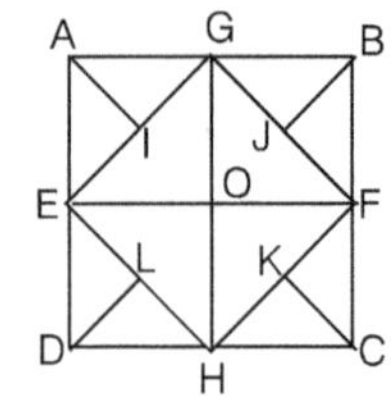

There are nine triangles in the upper half of the figure ABFE = ΔAEI, ΔAIG, ΔAEG, ΔGEO, ΔGBJ, ΔBFJ, ΔGBF, ΔGOF and ΔGEF.

Similarly, there are nine triangles in the lower half figure, i.e. EFCD.

There are two more triangles ΔEGH and ΔFGH.

Hence, there are a total of 20 triangles.

Mathematics

106. ***(b)*** $(1+x)^n\left(1+\frac{1}{x}\right)^n = \left[(2+x)+\frac{1}{x}\right]^n$

$$[\because a^n \cdot b^n = (a\cdot b)^n]$$

$$= {}^nC_0(2+x)^n \cdot \left(\frac{1}{x}\right)^0 + \ldots + {}^nC_n \cdot \frac{1}{x^n}$$

$\therefore$ Coefficient of $x^{-n} = {}^nC_n = 1$

107. ***(c)*** $T_{r+1} = \sqrt{3}\,.{}^{20}C_r\left(\frac{1}{\sqrt{3}}\right)^r$

$$\Rightarrow \quad T_r = \sqrt{3}\,.{}^{20}C_{r-1}\left(\frac{1}{\sqrt{3}}\right)^{r-1}$$

Now, $\dfrac{T_{r+1}}{T_r} = \dfrac{20-r+1}{r}\left(\dfrac{1}{\sqrt{3}}\right)$

Since, $T_{r+1} \geq T_r$

$\Rightarrow \quad 20 - r + 1 \geq \sqrt{3}r$

$\Rightarrow \quad r \leq \dfrac{21}{\sqrt{3}+1} = \dfrac{21}{2.73}$

$\Rightarrow \quad r \leq 7.692 \Rightarrow r = 7$

$\therefore$ The greatest term is $T_8 = \sqrt{3}\cdot{}^{20}C_7\left(\frac{1}{\sqrt{3}}\right)^7$

$$= \frac{25840}{9}$$

108. ***(b)*** Let $x = 1$

$$= \cos 0° + i\sin 0°$$

$$= \cos 2r\pi + i\sin 2r\pi = e^{i\,2r\pi}$$

$$\Rightarrow \quad x^{1/n} = e^{i(2r\pi)/n};\ r = 0, 1, 2, \ldots$$

Then, the roots are 1, $e^{\frac{2\pi i}{n}}, e^{\frac{4\pi i}{n}}, \ldots\ldots$

Which are clearly in GP with common ratio $e^{2\pi i/n}$.

109. ***(a)*** Given, $P = \begin{bmatrix} a & b & c \\ b & c & a \\ c & a & b \end{bmatrix}$

$$\therefore \quad P^T = \begin{bmatrix} a & b & c \\ b & c & a \\ c & a & b \end{bmatrix}$$

Also given, $P^TP = I$

$\therefore \quad |P^TP| = |I| = 1$

or $\quad \begin{vmatrix} a & b & c \\ b & c & a \\ c & a & b \end{vmatrix}^2 = 1 \Rightarrow \begin{vmatrix} a & b & c \\ b & c & a \\ c & a & b \end{vmatrix} = \pm 1$

or $\quad 3abc - a^3 - b^3 - c^3 = \pm 1$

or $\quad 3 \mp 1 = a^3 + b^3 + c^3$

$\therefore \quad a^3 + b^3 + c^3 = 2 \text{ or } 4$

110. ***(d)*** Given, $x^2 = \begin{vmatrix} \sin\theta & \cos\theta & 0 \\ -\cos\theta & \sin\theta & 1 \\ \sin\theta & \cos\theta & 2 \end{vmatrix}$

On expanding $\begin{vmatrix} \sin\theta & \cos\theta & 0 \\ -\cos\theta & \sin\theta & 1 \\ \sin\theta & \cos\theta & 2 \end{vmatrix}$ along C_3,

we get

$$x^2 = 0 - 1\begin{vmatrix} \sin\theta & \cos\theta \\ \sin\theta & \cos\theta \end{vmatrix} + 2\begin{vmatrix} \sin\theta & \cos\theta \\ -\cos\theta & \sin\theta \end{vmatrix}$$

$$= -1(\sin\theta\cos\theta - \sin\theta\cos\theta) + 2(\sin^2\theta + \cos^2\theta)$$

$$= -1 \times 0 + 2 \times 1$$

$\Rightarrow x^2 = 2 \Rightarrow x = \pm\sqrt{2}$

If $x = \sqrt{2}$, then

$$4x^2 + x\sin\frac{3\pi}{2} + 5 = 4 \times (\sqrt{2})^2 - \sqrt{2} + 5$$

$$= 8 - \sqrt{2} + 5 = (13 - \sqrt{2})$$

If $x = -\sqrt{2}$, then $4x^2 + x\sin\frac{3\pi}{2} + 5$

$$= 4 \times (-\sqrt{2})^2 + \sqrt{2} + 5$$

$$= 8 + \sqrt{2} + 5 = (13 + \sqrt{2})$$

111. ***(d)*** Given, a, b, c are in GP and $\log a - \log 2b$, $\log 2b - \log 3c$ and $\log 3c - \log a$ are in AP.

$\therefore \quad b^2 = ac$

and $2(\log 2b - \log 3c) = \log a - \log 2b + \log 3c - \log a$

$\Rightarrow \quad b^2 = ac$ and $2b = 3c$

$\Rightarrow \quad c = \frac{4a}{9}$ and $b = \frac{2a}{3}$

$\because \quad a + b = \frac{5a}{3} > c, b + c = \frac{10a}{9} > a$

and $c + a = \frac{13a}{9} > b.$

$\therefore$ a, b, c are the sides of a triangle.

Also, a is the greatest side

$$\therefore \quad \cos A = \frac{b^2 + c^2 - a^2}{2bc} = -\frac{29}{48} < 0$$

$\therefore$ ΔABC is an obtuse angled triangle.

112. ***(c)*** Given, $\sum_{r=1}^{n} t_r = \frac{n(n+1)(n+2)(n+3)}{8} = S_n$ (say)

$$\therefore \quad \sum_{r=1}^{n-1} t_r = \frac{(n-1)n(n+1)(n+2)}{8} = S_{n-1}$$

Now, $\quad t_n = S_n - S_{n-1} = \frac{n(n+1)(n+2)}{2}$

$$\therefore \lim_{n\to\infty}\sum_{r=1}^{n}\frac{1}{t_r} = \lim_{n\to\infty}\sum_{r=1}^{n}\frac{2}{n(n+1)(n+2)}$$

$$= \lim_{n\to\infty}\sum_{r=1}^{n}\left(\frac{1}{n(n+1)} - \frac{1}{(n+1)(n+2)}\right)$$

$$= -\lim_{n\to\infty}\sum_{r=1}^{n}\left(\frac{1}{(n+1)(n+2)} - \frac{1}{n(n+1)}\right)$$

$$= -\lim_{n\to\infty}\left(\frac{1}{(n+1)(n+2)} - \frac{1}{2}\right) = -\left(0 - \frac{1}{2}\right) = \frac{1}{2}$$

113. ***(c)*** Truth table

p	q	$\sim p$	$\sim q$	$p \vee q$	$\sim q \wedge p$	$p \vee \sim p$	$(p \vee q) \vee (\sim p)$	$(\sim q \wedge p) \vee (p \vee \sim p)$
T	T	F	F	T	F	T	T	T
T	F	F	T	T	T	T	T	T
F	T	T	F	T	F	T	T	T
F	F	T	T	F	F	T	T	T

114. ***(c)*** The two sets of m parallel lines along with two sets of two parallel lines of the given parallelogram will form two sets of $(m+2)$ parallel lines. Each parallelogram is formed by choosing two parallel lines from each of the above.

$\therefore$ Total number of parallelograms

$$= {}^{m+2}C_2 \times {}^{m+2}C_2 = ({}^{m+2}C_2)^2$$

115. ***(a)*** Let $f(x) = y$, then

$$a^y = x + \sqrt{x^2 + 1}$$

$$\Rightarrow \quad a^{-y} = \frac{1}{x + \sqrt{x^2 + 1}}$$

$$\Rightarrow \quad a^{-y} = \frac{x - \sqrt{x^2 + 1}}{-1} \quad \text{[rationalising]}$$

$$\therefore \quad a^y - a^{-y} = 2x$$

$$\Rightarrow \quad x = \frac{1}{2}(a^y - a^{-y})$$

$$\therefore \quad f^{-1}(y) = \frac{1}{2}(a^y - a^{-y})$$

$[\because f(x) = y \Rightarrow x = f^{-1}(y)]$

$$\Rightarrow \quad f^{-1}(x) = \frac{1}{2}(a^x - a^{-x}) \quad \text{[replacing } y \text{ by } x]$$

116. ***(c)*** Given,

$$S = \sum_{n=1}^{\infty} \tan^{-1}\frac{2n}{n^4 + n^2 + 2} \quad \text{...(i)}$$

Let $n^4 + n^2 + 1 = [(n^2)^2 + 1^2 + 2(n^2)(1)] - n^2$

$$= (n^2 + 1)^2 - n^2$$

$$= (n^2 + n + 1)(n^2 - n + 1) \quad \text{...(ii)}$$

Let $\quad u_n = \tan^{-1}\left(\frac{2n}{n^4 + n^2 + 2}\right)$

$$= \tan^{-1}\frac{2n}{1 + (n^4 + n^2 + 1)}$$

$$= \tan^{-1}\frac{(n^2 + n + 1) - (n^2 - n + 1)}{1 + (n^2 + n + 1)(n^2 - n + 1)}$$

$$u_n = \tan^{-1}(n^2 + n + 1) - \tan^{-1}(n^2 - n + 1) \quad \text{...(iii)}$$

On putting $n = 1, 2, 3, \ldots$ successively in Eq. (iii), we get

$$u_1 = \tan^{-1} 3 - \tan^{-1} 1$$
$$u_2 = \tan^{-1} 7 - \tan^{-1} 3$$
$$u_3 = \tan^{-1} 13 - \tan^{-1} 7$$
$$\ldots\ldots\ldots\ldots\ldots\ldots$$
$$u_n = \tan^{-1}(n^2 + n + 1) - \tan^{-1}(n^2 - n + 1)$$

On adding vertically, we get

$$\sum_{n=1}^{\infty} u_n = \tan^{-1}(n^2 + n + 1) - \tan^{-1} 1$$

$$S = \lim_{n \to \infty} \sum_{n=1}^{\infty} u_n \quad \text{[from Eq. (i)]}$$
$$= \lim_{n \to \infty} \tan^{-1}(n^2 + n + 1) - \tan^{-1} 1$$
$$= \frac{\pi}{2} - \frac{\pi}{4} = \frac{\pi}{4}$$

117. ***(a)*** Given, $\sin x + \cos(t + x) + \cos(t - x) = 2$

$$\Rightarrow \quad \sin x + 2\cos t \cdot \cos x = 2$$

For real solution

$$\sqrt{1 + 4\cos^2 t} \geq 2$$
$$\Rightarrow \quad \cos^2 t \geq \frac{3}{4} \Rightarrow -\frac{1}{2} \leq \sin t \leq \frac{1}{2}$$

118. ***(c)*** We have,

$$\cos^2\alpha + \cos^2\beta + \cos^2\gamma = 1$$
$$\Rightarrow \quad \cos^2\alpha + \cos^2\left(\frac{\pi}{2} - \alpha\right) + \cos^2\gamma = 1 \quad \left[\text{given, } \alpha + \beta = \frac{\pi}{2}\right]$$
$$\Rightarrow \quad \cos^2\alpha + \sin^2\alpha + \cos^2\gamma = 1$$
$$\Rightarrow \quad 1 + \cos^2\gamma = 1$$
$$\Rightarrow \quad \cos^2\gamma = 0 \Rightarrow \cos\gamma = 0$$
$$\therefore \quad (\cos\alpha + \cos\beta + \cos\gamma)^2 = (\cos\alpha + \sin\alpha)^2$$
$$= 1 + 2\sin\alpha \cdot \cos\alpha = 1 + \sin 2\alpha$$

119. ***(c)*** The given straight lines are $3x + 4y = 5$ and $4x - 3y = 15$. Clearly, these straight lines are perpendicular to each other $(m_1 m_2 = -1)$ and intersect at A. Now, B and C are points on these lines such that $AB = AC$ and BC passes through $(1, 2)$.

From figure it is clear that $\angle B = \angle C = 45°$

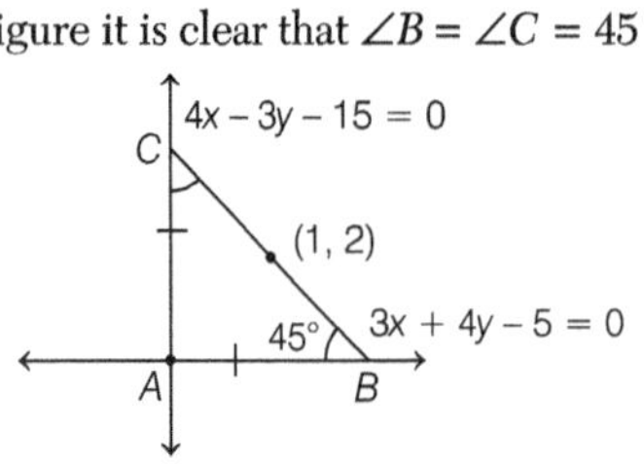

Let slope of BC be m. Then,

$$\tan 45° = \left|\frac{m + \frac{3}{4}}{1 - \frac{3}{4}m}\right|$$
$$\Rightarrow \quad \pm 1 = \frac{4m + 3}{4 - 3m}$$
$$\Rightarrow \quad 4m + 3 = \pm(4 - 3m)$$
$$\Rightarrow \quad 4m + 3 = 4 - 3m$$
$$\text{or} \quad 4m + 3 = -4 + 3m$$
$$\Rightarrow \quad m = \frac{1}{7} \text{ or } m = -7$$

Hence, equation of BC is

$$y - 2 = \frac{1}{7}(x - 1)$$
$$\text{or} \quad y - 2 = -7(x - 1) \Rightarrow 7y - 14 = x - 1$$
$$\text{or} \quad y - 2 = -7x + 7 \Rightarrow x - 7y + 13 = 0$$
$$\text{or} \quad 7x + y - 9 = 0$$

120. ***(d)*** Let $y = mx + c$, intersect $y^2 = 4ax$ at $A(at_1^2, 2at_1)$ and $B(at_2^2, 2at_2)$

Then, $$\frac{2}{t_1 + t_2} = m \Rightarrow t_1 + t_2 = \frac{2}{m}$$

Let the foot of another normal be $C(at_3^2, 2at_3)$.

Then, $$t_1 + t_2 + t_3 = 0$$
$$t_3 = (t_1 + t_2) = -\frac{2}{m}$$

Thus, other foot is $\left(\frac{4a}{m^2}, \frac{-4a}{m}\right)$.

121. ***(c)*** Length of perpendicular from origin to the line $x\sqrt{5} + 2y = 3\sqrt{5}$ is

$$OL = \frac{3\sqrt{5}}{\sqrt{(\sqrt{5})^2 + 2^2}} = \frac{3\sqrt{5}}{\sqrt{9}} = \sqrt{5}$$

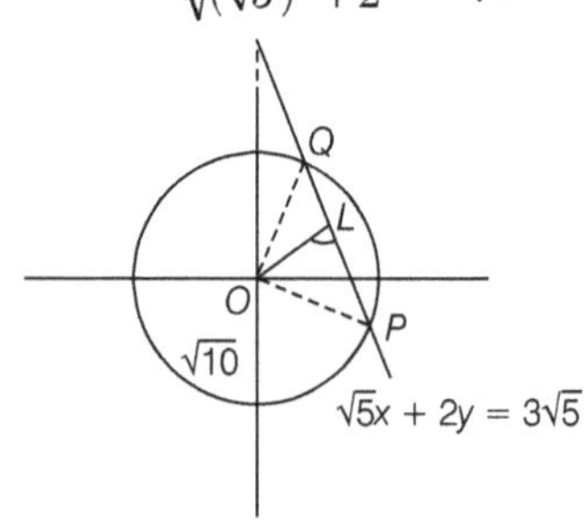

Radius of the given circle $= \sqrt{10} = OQ = OP$

$$PQ = 2QL$$
$$= 2\sqrt{OQ^2 - OL^2}$$
$$= 2\sqrt{10 - 5} = 2\sqrt{5}$$

Thus, area of $\Delta OPQ = \frac{1}{2} \times PQ \times OL$

$$= \frac{1}{2} \times 2\sqrt{5} \times \sqrt{5} = 5 \text{ sq units}$$

122. ***(b)*** Let r be the radius of the largest circle passing through the focus $(1, 0)$ of $y^2 = 4x$

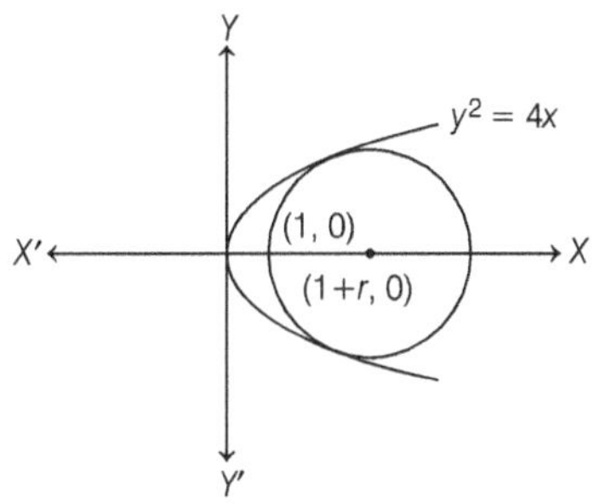

Clearly, centre of the circle will be on X-axis and its coordinates are $(1 + r, 0)$.

The equation of the circle, is $(x - 1 - r)^2 + y^2 = r^2$.

It touches $y^2 = 4x$. Therefore, the equation $(x - r - 1)^2 + 4x = r^2$ must have equal roots

$$\therefore \quad 4(1 - r)^2 - 4(2r + 1) = 0$$

$$\Rightarrow \quad r = 4$$

123. ***(d)*** The point $P\left(\frac{\pi}{6}\right)$ is $\left(a \sec\frac{\pi}{6}, b \tan\frac{\pi}{6}\right)$

or $P\left(\frac{2a}{\sqrt{3}}, \frac{b}{\sqrt{3}}\right)$

$\therefore$ Equation of tangent at P is $\dfrac{x}{\frac{\sqrt{3}a}{2}} - \dfrac{y}{\sqrt{3}\,b} = 1$

$\therefore$ Area of the triangle $= \frac{1}{2} \times \frac{\sqrt{3}a}{2} \times \sqrt{3}b = 3a^2$

$$\therefore \quad \frac{b}{a} = 4$$

$$\therefore \quad e^2 = 1 + \frac{b^2}{a^2} = 17$$

Now, $e^2 - 9 = 17 - 9 = 8$

124. ***(d)*** Let distances of a point $p(x, y, z)$ from the planes $x + y + z = 0$, $x - z = 0$ and $x - 2y + z = 0$ are $\frac{x + y + z}{\sqrt{3}}, \frac{x - z}{\sqrt{2}}$ and $\frac{x - 2y + z}{\sqrt{6}}$ respectively, then the sum of the squares of distances, is as

$$\Rightarrow \left(\frac{x + y + z}{\sqrt{3}}\right)^2 + \left(\frac{x - z}{\sqrt{2}}\right)^2 + \left(\frac{x - 2y + z}{\sqrt{6}}\right)^2 = p^2$$

$$\Rightarrow 2(x + y + z)^2 + 3(x - z)^2 + (x - 2y + z)^2 = 6p^2$$

$$\Rightarrow 2x^2 + 2y^2 + 2z^2 + 4xy + 4yz + 4zx + 3x^2 + 3z^2 - 6xz + x^2 + 4y^2 + z^2 - 4xy - 4yz + 2xz = 6p^2$$

$$\Rightarrow \quad 6x^2 + 6y + 6z^2 = 6p^2 \Rightarrow x^2 + y^2 + z^2 = p^2$$

125. ***(a)*** When line

$$\frac{x - x_1}{a_1} = \frac{y - y_1}{b_1} = \frac{z - z_1}{c_1}$$

is perpendicular to plane $ax + by + cz + d = 0$, then

$$\frac{a_1}{a} = \frac{-b_1}{b} = \frac{c_1}{c}$$

$$\therefore \quad \frac{\lambda}{2} = \frac{1}{2} = \frac{-4}{-8} \Rightarrow \lambda = 1$$

126. ***(c)*** Taking the coordinates of vertices O, P, Q, R as $(0, 0), (a, 0), (a, a), (0, a)$, respectively.

$\therefore$ The coordinates of M is $\left(a, \frac{a}{2}\right)$ and N is $\left(\frac{a}{2}, a\right)$.

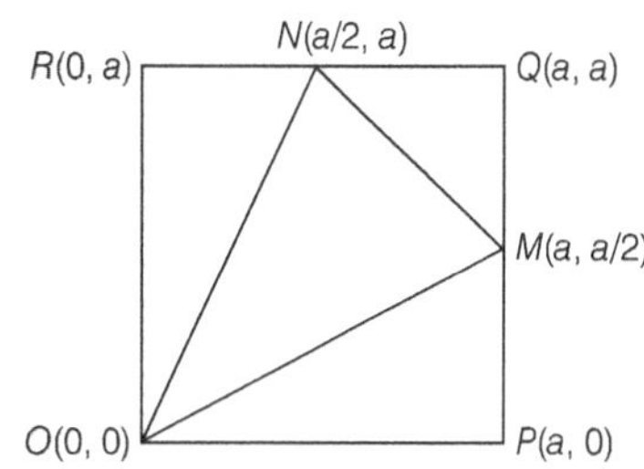

$$\therefore \quad \text{Area of } \Delta OMN = \frac{1}{2}\left\|\begin{matrix} 0 & 0 & 1 \\ a & a/2 & 1 \\ a/2 & a & 1 \end{matrix}\right\| = \frac{3a^2}{8}$$

and area of the square $= a^2$

$\therefore$ The required ratio is $8 : 3$.

127. ***(d)*** Equation of auxiliary circle is

$$x^2 + 9y^2 = 9 \qquad \ldots\text{(i)}$$

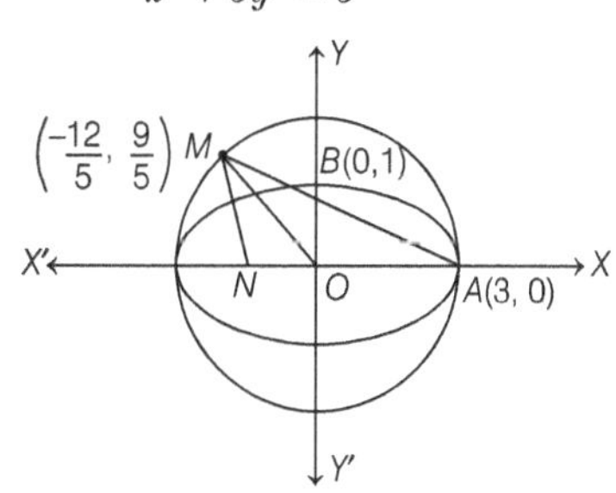

$\therefore$ Equation of AM is $\frac{x}{3} + \frac{y}{1} = 1 \qquad \ldots\text{(ii)}$

On solving Eqs. (i) and (ii), we get

$$M\left(-\frac{12}{5}, \frac{9}{5}\right)$$

Now, area of $\Delta\, AOM = \frac{1}{2} OA \times MN = \frac{27}{10}$ sq units

128. ***(b)*** Given equation can be written as

$$x^2 - y^2 = \frac{25}{3}$$

$$\therefore \quad e_1 = \sqrt{1 + \frac{b^2}{a^2}} = \sqrt{1+1} = \sqrt{2}$$

The equation of conjugate hyperbola is

$$-x^2 + y^2 = \frac{25}{3}$$

$$\therefore \quad e_2 = \sqrt{1 + \frac{b^2}{a^2}} = \sqrt{1+1} = \sqrt{2}$$

$$\therefore \quad e_1^2 + e_1^2 = (\sqrt{2})^2 + (\sqrt{2})^2 = 4$$

129. ***(a)*** We have,

$$f(x+y) = f(x) + 2y^2 + kxy \text{ for all } x, y \in R$$

$$\Rightarrow \quad \frac{f(x+y) - f(x)}{y} = 2y + kx \text{ for all } x \in R$$

$$\Rightarrow \lim_{y \to 0} \frac{f(x+y) - f(x)}{y} = \lim_{y \to 0} (2y + kx)$$

$$\Rightarrow f'(x) = kx \text{ for all } x \in R$$

$$\Rightarrow f(x) = \frac{kx^2}{2} + C \text{ for all } x \in R \quad \text{[by integration]}$$

But, $f(1) = 2$ and $f(2) = 8$.

$$\therefore \quad 2 = \frac{k}{2} + C \text{ and } 8 = 2k + C$$

$$\Rightarrow \quad k = 4 \text{ and } C = 0$$

Hence, $f(x) = 2x^2$ for all $x \in R$

130. ***(a)*** Since, given planes are perpendicular to each other, i.e. its normal are perpendicular.

$$\therefore \quad 2(\lambda) - \lambda(5) + 3(-1) = 0$$

$$\Rightarrow \quad -3\lambda - 3 = 0 \Rightarrow \lambda = -1$$

$$\therefore \quad \lambda^2 + \lambda = (-1)^2 - 1 = 0$$

131. ***(d)*** Put $x + y = v$ and $\frac{dv}{dx} = 1 + \frac{dy}{dx}$ in the given differential equation.

$$\therefore \frac{dv}{dx} = 1 + \sin v + \cos v$$

$$\Rightarrow \frac{dv}{2\cos^2\frac{v}{2} + 2\sin\frac{v}{2}\cos\frac{v}{2}} = dx \Rightarrow \frac{\frac{1}{2}\sec^2\frac{v}{2}}{1 + \tan\frac{v}{2}} dv = dx$$

$$\Rightarrow \quad \log\left(1 + \tan\frac{x+y}{2}\right) = x + C \text{ [by integration]}$$

132. ***(d)*** Since, the numerator tends to ∞ as $x \to 0$,

$$\text{so} \quad \lim_{x \to 0} \frac{1}{x^2}(e^{\alpha x} - e^x - x)$$

$$= \frac{1}{2} \lim_{x \to 0} \frac{(\alpha e^{\alpha x} - e^x - 1)}{x}$$

For the last limit to exist we must have,

$$\lim_{x \to 0} (\alpha e^{\alpha x} - e^x - 1) = 0$$

$$\therefore \quad \alpha - 1 - 1 = 0$$

$$\Rightarrow \quad \alpha = 2$$

For $\alpha = 2$ the last limit and equal to

$$= \frac{1}{2} \lim_{x \to 0} \frac{(2e^{2x} - e^x - 1)}{x}$$

$$= \frac{1}{2} \lim_{x \to 0} (4e^{2x} - e^x) = \frac{3}{2}$$

133. ***(b)*** Let depth of water at time t be h and the radius of the base of water level be r.

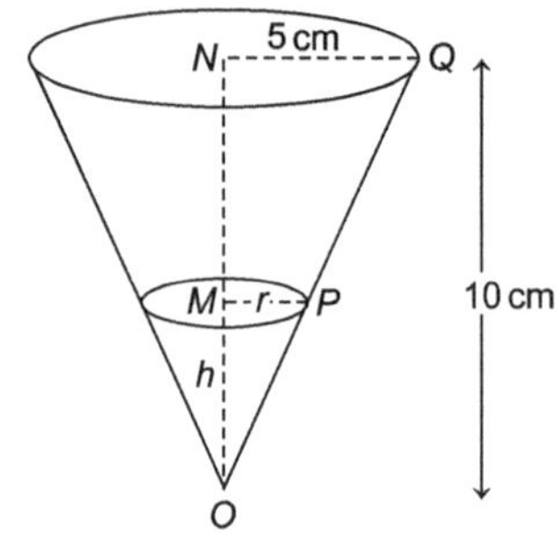

From similar ΔOMP and ΔONQ, we have

$$\frac{OM}{ON} = \frac{PM}{QN}$$

$$\Rightarrow \quad \frac{h}{10} = \frac{r}{5}$$

$$\Rightarrow \quad r = \frac{5h}{10} \Rightarrow r = \frac{h}{2}$$

We have, $\quad V = \frac{1}{3}\pi r^2 h = \frac{1}{3}\pi\left(\frac{h}{2}\right)^2 h$

$$\Rightarrow \quad V = \frac{\pi h^3}{12}$$

On differentiating both sides, we get

$$\frac{dV}{dt} = \frac{\pi}{12} \cdot 3h^2 \frac{dh}{dt} = \frac{\pi h^2}{4}\frac{dh}{dt}$$

Given, $\frac{dV}{dt} = 3 \text{ cm}^3/\text{s}$ when $h = 4$ cm, so we get

$$3 = \frac{\pi \times 4^2}{4}\frac{dh}{dt} \Rightarrow \frac{dh}{dt} = \frac{3}{4\pi} \text{ cm/s}$$

Hence, the water level is rising at $\frac{3}{4\pi}$ cm/s.

134. ***(b)*** We have,

$$\frac{dy}{dx} = y + 2x$$

$$\Rightarrow \quad \frac{dy}{dx} - y = 2x$$

Now, $\quad \text{IF} = e^{-\int 1\, dx} = e^{-x}$

$\therefore y \cdot e^{-x} = \int 2x\, e^{-x}\, dx + k$, k be the constant of integration

$$= 2[x \int e^{-x} dx - \int 1 \cdot (-e^{-x})\, dx] + k$$

[using integration by parts]

$$\Rightarrow \quad y \cdot e^{-x} = -2xe^{-x} - 2e^{-x} + k \quad \ldots \text{(i)}$$

As curve (i) passes through (0, 0)

$$\therefore \quad 0 = 0 - 2 + k$$

$$\Rightarrow \quad k = 2$$

Thus, the curve is

$$ye^{-x} = -2xe^{-x} - 2e^{-x} + 2$$

$$\therefore \quad y = 2(e^x - x - 1)$$

135. ***(c)*** $f(0) = 0$

For $f(x)$ to be continuous at $x = 0$

$$\lim_{x \to 0} f(x) = 0$$

$$\therefore \quad \lim_{x \to 0} x^p \sin \frac{1}{x} = 0$$

This possible only when $p > 0 \quad \ldots$(i)

$$f'(0) = \lim_{h \to 0} \frac{f(h) - f(0)}{h}$$

$$= \lim_{h \to 0} \frac{h^p \sin \frac{1}{h} - 0}{h}$$

$$= \lim_{h \to 0} h^{p-1} \sin \frac{1}{h}$$

$\Rightarrow \quad f'(0)$ will exist only when $p > 0$

$\therefore \quad f(x)$ will not be differentiable if

$$p \leq 1 \quad \ldots \text{(ii)}$$

From Eqs. (i) and (ii), for $f(x)$ to be not differentiable but continuous at $x = 0$, possible values of p are given by $0 < p \leq 1$.

136. ***(a)*** Integrating the given differential equation, we have

$$\frac{dy}{dx} - \frac{-\cos 3x}{3} + e^x + \frac{x^3}{3} + C_1$$

But $y_1(0) = 1$

So, $\quad 1 = \left(-\frac{1}{3}\right) + 1 + C_1$

$$\Rightarrow \quad C_1 = \frac{1}{3}$$

$$\therefore \quad \frac{dy}{dx} = \frac{-\cos 3x}{3} + e^x + \frac{x^3}{3} + \frac{1}{3}$$

Again integrating, we get

$$y = -\frac{\sin 3x}{9} + e^x + \frac{x^4}{12} + \frac{1}{3}x + C_2$$

But $y(0) = 0$, so $0 = 0 + 1 + C_2$

$$\Rightarrow \quad C_2 = -1$$

Thus, $\quad y = -\dfrac{\sin 3x}{9} + e^x + \dfrac{x^4}{12} + \dfrac{1}{3}x - 1$

137. ***(d)*** Given, $f(x) = -2x^3 - 9x^2 - 12x + 1$

$$\Rightarrow \quad f'(x) = -6x^2 - 18x - 12$$

To be decreasing $f'(x) < 0$

$$\Rightarrow \quad -6x^2 - 18x - 12 < 0$$

$$\Rightarrow \quad x^2 + 3x + 2 > 0$$

$$\Rightarrow \quad (x+2)(x+1) > 0$$

Therefore, either $x < -2$ or $x > -1$

$$\Rightarrow \quad x \in (-1, \infty) \text{ or } (-\infty, -2)$$

138. ***(c)*** Let $\mathbf{a} = 4\hat{\mathbf{i}} - 4\hat{\mathbf{k}}$ and $\mathbf{b} = \hat{\mathbf{i}} + \hat{\mathbf{j}} + \hat{\mathbf{k}}$

Since, θ is the angle between **a** and **b**

$$\therefore \quad \cos\theta = \frac{\mathbf{a} \cdot \mathbf{b}}{|\mathbf{a}||\mathbf{b}|}$$

$$= \frac{(4\hat{\mathbf{i}} - 4\hat{\mathbf{k}}) \cdot (\hat{\mathbf{i}} + \hat{\mathbf{j}} + \hat{\mathbf{k}})}{|\mathbf{a}||\mathbf{b}|}$$

$$= \frac{4 + 0 - 4}{|\mathbf{a}||\mathbf{b}|} = 0$$

$$\cos\theta = \cos 90° \Rightarrow \theta = 90°$$

$\therefore \quad \sin\theta + \cos\theta = \cos 90° + \sin 90° = 0 + 1 = 1$

139. ***(d)***

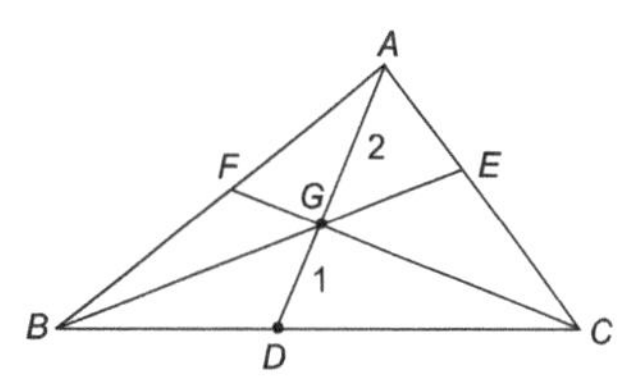

$$\therefore \quad \mathbf{AD} = 3\mathbf{GD}$$

$$= 3 \cdot \frac{1}{2}(\mathbf{GB} + \mathbf{GC})$$

$$= \frac{3}{2}\left(\frac{2}{3}\mathbf{EB} + \frac{2}{3}\mathbf{FC}\right)$$

$$= -\mathbf{BE} - \mathbf{CF}$$

140. ***(c)*** Let $x_1, x_2, \ldots, x_n$ be n observations.

Then, $\quad \bar{x} = \dfrac{1}{n}\Sigma x_i$

Let $\quad y_i = \dfrac{x_i}{\alpha} + 10$

Then, $\quad \dfrac{1}{n}\sum_{i=1}^{n} y_i = \dfrac{1}{\alpha}\left(\dfrac{1}{n}\Sigma x_i\right) + \dfrac{1}{n}(10n)$

$$\Rightarrow \quad \bar{x}_{\text{new}} = \frac{1}{\alpha}\bar{X} + 10 = \frac{\bar{X} + 10\alpha}{\alpha}$$

141. ***(c)*** Volume of the parallelopiped $= |[\mathbf{a}\ \mathbf{b}\ \mathbf{c}]|$

$\Rightarrow$ (area of the base parallelogram) $\times h = |[\mathbf{a}\ \mathbf{b}\ \mathbf{c}]|$

$$\Rightarrow \quad |\mathbf{a} \times \mathbf{b}|\, h = |[\mathbf{a}\ \mathbf{b}\ \mathbf{c}]|$$

$$\Rightarrow \quad |-5\hat{\mathbf{i}} + 3\hat{\mathbf{j}} + 2\hat{\mathbf{k}}|\, h = \left\| \begin{matrix} 1 & 1 & 1 \\ 2 & 4 & -1 \\ 1 & 1 & 3 \end{matrix} \right\|$$

$$\Rightarrow \quad \sqrt{38}\, h = 4 \Rightarrow h = \frac{4}{\sqrt{38}}$$

$$\Rightarrow \quad 38h^2 = 4^2 \Rightarrow 19h^2 = 8$$

142. ***(b)*** Mean (m) for BD $= np$

and variance (σ^2) for BD $= npq$

Given, $np + npq = 2.7$ and $n = 3$

$$\therefore \quad np(1+q) = 2.7 \Rightarrow p(1+q) = \frac{2.7}{3}$$

$$\Rightarrow \quad p(1+q) = 0.9$$

$$\Rightarrow \quad (1-q)(1+q) = 0.9$$

[$\because$ for BD probability standard deviation (SD), $p + q = 1$]

$$\Rightarrow \quad 1 - q^2 = 0.9$$

$$\Rightarrow \quad q^2 = 0.1 \Rightarrow q = \pm 0.3$$

$$\therefore \quad q = 0.3$$

[$\because q = -0.3$ can't possible as $0 \le q \le 1$]

$$\Rightarrow \quad p = 0.7 \text{ and } q = 0.3$$

$$\therefore \quad \text{BD} = (0.3 + 0.7)^5$$

143. ***(b)*** We have,

$$R(x) = \int \frac{dx}{e^x + 8e^{-x} + 4e^{-3x}} - 2\int \frac{dx}{e^{3x} + 8e^x + 4e^{-x}}$$

$$\Rightarrow \quad R(x) = \int \frac{e^x(e^{2x} - 2)}{e^{4x} + 8e^{2x} + 4}\, dx$$

On substituting

$$e^x = t \Rightarrow e^x\, dx = dt,$$

we get

$$R(t) = \int \frac{(t^2 - 2)\, dt}{t^4 + 8t^2 + 4} = \int \frac{(1 - 2t^{-2})\, dt}{(t + 2t^{-1})^2 + 4}$$

$$= \frac{1}{2}\tan^{-1}\left(\frac{t + 2t^{-1}}{2}\right) + K$$

$$\Rightarrow \quad R(x) = \frac{1}{2}\tan^{-1}\left(\frac{e^x + 2e^{-x}}{2}\right) + K$$

Hence, $(A, B, C) = (\tan^{-1}, e^x, 2)$

144. ***(b)*** $\cot^{-1}(1 - x + x^2) = \tan^{-1}\left(\dfrac{1}{1 - x + x^2}\right)$

$$= \tan^{-1}\left(\frac{1}{1 - x(1-x)}\right)$$

$$= \tan^{-1}\left(\frac{x + (1-x)}{1 - x(1-x)}\right)$$

$$\Rightarrow \cot^{-1}(1 - x + x^2) = \tan^{-1}x - \tan^{-1}(1-x)$$

$$\therefore \int_0^1 \cot^{-1}(1 - x + x^2)\, dx$$

$$= \int_0^1 \tan^{-1}x\, dx - \int_0^1 \tan^{-1}(1-x)\, dx$$

$$= \int_0^1 \tan^{-1}x\, dx + \int_0^1 \tan^{-1}x\, dx$$

$$\left[\because \int_0^a f(x)\, dx = -\int_0^a f(a-x)\, dx\right]$$

$$= 2\int_0^1 \tan^{-1}x\, dx$$

On evaluating by integration by parts, we have

$$= 2\left\{[\tan^{-1}x \cdot x]_0^1 - \int_0^1 \frac{x}{1+x^2}\, dx\right\}$$

$$= 2\left\{\frac{\pi}{4} - \left[\frac{1}{2}\ln(1+x^2)\right]_0^1\right\}$$

$$= 2\left[\frac{\pi}{4} - \frac{1}{2}\log 2\right] = \frac{\pi}{2} - \log 2$$

Hence, $\displaystyle\int_0^1 \cot^{-1}(1 - x + x^2)\, dx = \frac{\pi}{2} - \log 2$

145. ***(a)*** Given curves are

$$\sqrt{|x|} + \sqrt{|y|} = \sqrt{a} \quad \ldots\text{(i)}$$

and

$$x^2 + y^2 = a^2 \quad \ldots\text{(ii)}$$

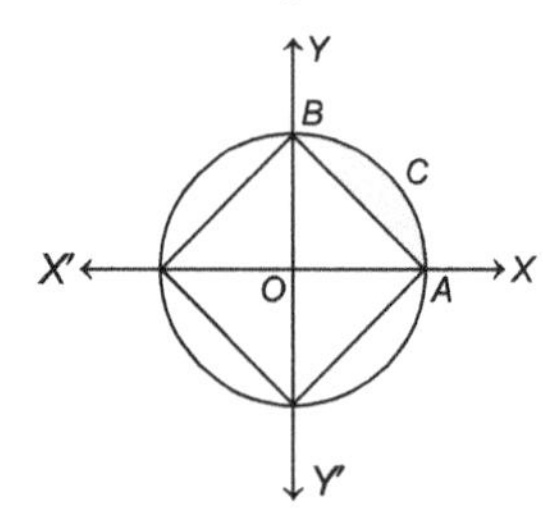

Now, required areas = 4 [shaded area in the first quadrant]

$$= 4\left[\frac{\pi a^2}{4} - \int_0^a (\sqrt{a} - \sqrt{x})^2\, dx\right]$$

[from Eqs. (i) and (ii)]

$$= 4\left[\frac{\pi a^2}{4} - \int_0^a (a + x - 2\sqrt{a}\sqrt{x})\, dx\right]$$

$$= 4\frac{\pi a^2}{4} - 4\left[ax + \frac{x^2}{2} - \frac{4}{3}\sqrt{a}\, x^{3/2}\right]_0^a$$

$$= \left(\pi - \frac{2}{3}\right)a^2 \text{ sq units}$$

146. ***(b)*** $P(A \cap B) = \frac{1}{6}$ and $P(A^C \cap B^C) = \frac{1}{3}$

Now, $P(A \cup B)^C = P(A^C \cap B^C) = \frac{1}{3}$

$\Rightarrow \quad 1 - P(A \cup B) = \frac{1}{3}$

$\Rightarrow \quad P(A \cup B) = \frac{2}{3}$

But $P(A \cup B) = P(A) + P(B) - P(A \cap B)$

$\Rightarrow \quad P(A) + P(B) = \frac{5}{6} \quad \dots(i)$

$\because$ A and B are independent events

$\therefore \quad P(A \cap B) = P(A)\,P(B)$

$\Rightarrow \quad P(A)\,P(B) = \frac{1}{6}$

$[P(A) - P(B)]^2 = [P(A) + P(B)]^2 - 4P(A)\,P(B)$

$= \frac{25}{36} - \frac{4}{6} = \frac{1}{36}$

$\Rightarrow \quad P(A) - P(B) = \pm\frac{1}{6} \quad \dots(ii)$

On solving Eqs. (i) and (ii), we get

$P(A) = \frac{1}{2}$ or $\frac{1}{3}$

147. ***(d)*** Let A be the event that the examinee gives the correct answer.

Let G, C, K stand for guessing, copying and knowing, respectively.

Given, $P(C) = \frac{1}{6}$, $P(G) = \frac{1}{3}$

and $P\left(\frac{A}{C}\right) = \frac{1}{8}$

$\Rightarrow \quad P(C) + P(G) + P(K) = 1$

$\Rightarrow \quad \frac{1}{6} + \frac{1}{3} + P(K) = 1$

$\therefore \quad P(K) = \frac{1}{2}$

Also, $P\left(\frac{A}{K}\right) = 1$, for if the examinee knows, he/she will correctly answer it and $P\left(\frac{A}{G}\right) = \frac{1}{4}$, since there are four choices.

Now, total probability

$$P(A) = P(G)\,P\left(\frac{A}{G}\right) + P(C)\,P\left(\frac{A}{C}\right) + P(K)\,P\left(\frac{A}{K}\right)$$

$$= \frac{1}{3}\cdot\frac{1}{4} + \frac{1}{6}\cdot\frac{1}{8} + \frac{1}{2}\cdot 1$$

$$= \frac{1}{12} + \frac{1}{48} + \frac{1}{2} = \frac{4+1+24}{48}$$

$$= \frac{29}{48}$$

Now, $P\left(\frac{K}{A}\right) = \dfrac{P(K)\,P\left(\frac{A}{K}\right)}{P(A)}$

$$= \frac{\frac{1}{2}\cdot 1}{\frac{29}{48}} = \frac{1}{2}\cdot\frac{48}{29}$$

$$= \frac{24}{29}$$

148. ***(d)*** Since, p : 4 is an even prime number, q : 6 is a divisor of 12 and r : the HCF of 4 and 6 is 2.

So, $\sim p \vee (q \wedge r)$ is correct.

149. ***(d)*** From the figure, it is clear that there is no common area. So, we cannot find maximum value of Z.

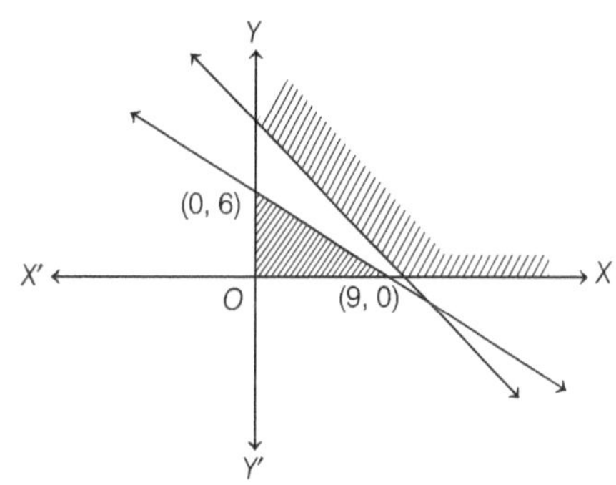

150. ***(d)*** Feasible region is $ABCA$ and $Z = 7x - 8y$

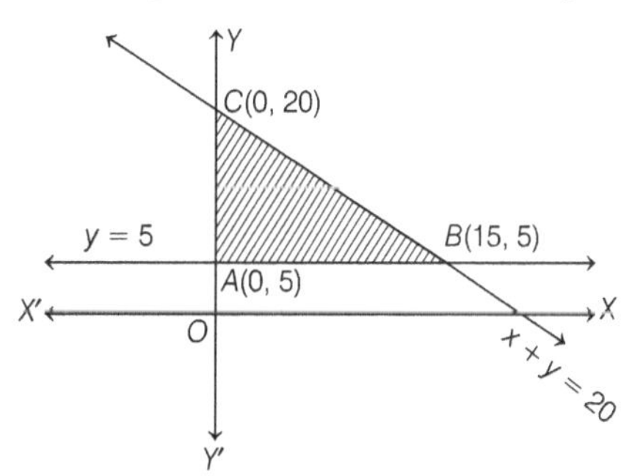

At $A(0, 5)$, $Z = 7 \times 0 - 8 \times 5 = -40$

At $B(15, 5)$, $Z = 7(15) - 8(5) = 65$

At $C(0, 20)$, $Z = 7(0) - 8(20) = -160$

Hence, the minimum value of Z is attained at point (0, 20).

BITSAT

Birla Institute of Technology & Science Admission Test

Solved Paper 2017*

Instructions

- There are 150 questions in all. The number of questions in each part is as given below.

		No. of Questions
Part I	Physics	1-40
Part II	Chemistry	41-80
Part III	a. English Proficiency	81-95
	b. Logical Reasoning	96-105
Part IV	Mathematics	106-150

- All questions are Multiple Choice Questions having four options out of which **only one** is correct.
- Each correct answer fetches 3 marks while incorrect answer has a penalty of 1 mark.
- Time allotted to complete this paper is 3 hrs.

PART I

Physics

1. If temperature of a black body increases from 300 K to 900 K, then the rate of energy radiation increases by

a. 81 *b.* 3
c. 9 *d.* 2

2. A whistle of frequency 500 Hz tied to the end of a string of length 1.2 m revolves at 400 rev/min. A listener standing some distance away in the plane of rotation of whistle hears frequencies in the range.

(Speed of sound = 340 m/s)

a. 436 Hz to 574 Hz *b.* 426 Hz to 586 Hz
c. 426 Hz to 574 Hz *d.* 436 Hz to 586 Hz

3. The focal length of a thin convex lens for red and blue rays are 100 cm and 96.8 cm, respectively. Then, the dispersive power of the material of the lens is

a. 0.968 *b.* 0.98
c. 0.0325 *d.* 0.325

4. Two metal plates having a potential difference of 800 V are 2 cm apart. It is found that a particle of mass 1.96×10^{-15} kg remains suspended in the region between the plates. The charge on the particle must be (e = elementary charge).

a. $2e$ *b.* $3e$ *c.* $6e$ *d.* $8e$

5. At what angle θ to the horizontal should an object is projected, so that the maximum height reached is equal to the horizontal range?

a. $\tan^{-1}(2)$ *b.* $\tan^{-1}(4)$
c. $\tan^{-1}\left(\frac{2}{3}\right)$ *d.* $\tan^{-1}(3)$

6. A body of mass 1 kg is executing simple harmonic motion. Its displacement y at t seconds is given by $y = 6\sin\left(100t + \frac{\pi}{4}\right)$ cm.

Its maximum kinetic energy is

a. 6 J *b.* 18 J *c.* 24 J *d.* 36 J

*Memory Based

7. A positive charge q is projected in magnetic field of width $\frac{mv}{\sqrt{2}\, qB}$ with velocity v. Then, the time taken by charged particle to emerge from the magnetic field is

a. $\frac{m}{\sqrt{2}\, qB}$ *b.* $\frac{\pi m}{4qB}$

c. $\frac{\pi m}{2qB}$ *d.* $\frac{\pi m}{\sqrt{2} \cdot qB}$

8. In Young's double slit experiment, the slits are 2 mm apart and are illuminated by photons of two wavelengths $\lambda_1 = 12000$ Å and $\lambda_2 = 10000$ Å. At what minimum distance from the common central bright fringe on the screen 2m from the slit will a bright fringe from one interference pattern coincide with a bright fringe from the other?

a. 8 mm *b.* 6 mm

c. 4 mm *d.* 3 mm

9. Two blocks A and B are placed one over the other on a smooth horizontal surface. The maximum horizontal force that can be applied on lower block B, so that A and B move without separation is 49 N. The coefficient of friction between A and B is

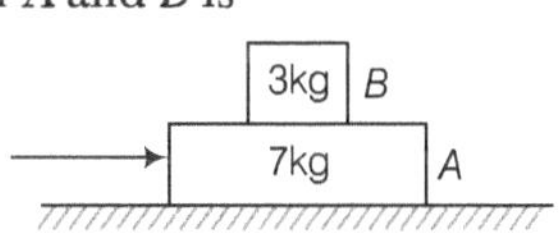

a. 0.2 *b.* 0.3

c. 0.5 *d.* 0.8

10. An aeroplane is flying in a horizontal direction with a velocity u and at a height of 2000 m. When it is vertically below a point A on the ground a food packet is released from it. The packet strikes the ground at point B. If $AB = 3$ km and $g = 10$ m/s^2, then the value of u is

a. 54 km/h *b.* 540 km/h

c. 150 km/h *d.* 300 km/h

11. A conducting circular loop is placed in a uniform magnetic field, $B = 0.025$ T with its plane perpendicular to the loop. The radius of the loop is made to shrink at a constant rate of 1 mm/s. The induced emf when the radius is 2 cm, is

a. $2\pi\,\mu$V *b.* $\pi\,\mu$V

c. $\frac{\pi}{2}\,\mu$V *d.* $2\,\mu$V

12. A mild steel wire of length $2L$ and cross-sectional area A is stretched, well with in the elastic limit, horizontally between two pillars as shown in figure. A mass m is suspended from the mid-point of the wire, strain in the wire is

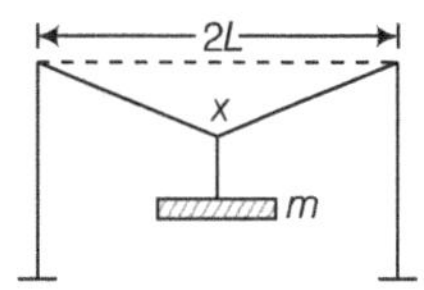

a. $\frac{x^2}{2L^2}$ *b.* $\frac{x}{L}$ *c.* $\frac{x^2}{L}$ *d.* $\frac{x^2}{2L}$

13. The resistance of a wire at 20 °C is 20Ω and 500 °C is 60 Ω. At which temperature, its resistance will be 25Ω?

a. 50 °C *b.* 60 °C

c. 70 °C *d.* 80 °C

14. The de-Broglie wavelength of a proton (charge $= 1.6 \times 10^{-19}$ C, $m = 1.6 \times 10^{-27}$ kg) accelerated through a potential difference of 1 kV is

a. 600 Å *b.* 0.9×10^{-12} m

c. 7 Å *d.* 0.9 nm

15. An ice-berg of density 900 kgm^{-3} is floating in water of density 1000 kgm^{-3}. The percentage of volume of ice-berg outside the water is

a. 20% *b.* 35%

c. 10% *d.* 11%

16. The total energy of an electron in the first excited state of hydrogen is about – 3.4 eV. Its kinetic energy in this state is

a. – 3.4 eV *b.* – 6.8 eV *c.* 6.8 eV *d.* 3.4 eV

17. A common emitter amplifier has a voltage gain of 50, an input impedance of 100Ω and an output impedance of 200Ω. The power gain of the amplifier is

a. 500 *b.* 1000

c. 1250 *d.* 50

18. The horizontal range and maximum height attained by a projectile are R and H, respectively. If a constant horizontal acceleration $a = \frac{g}{4}$ is imparted to the projectile due to wind, then its horizontal range and maximum height will be

a. $(R + H), \frac{H}{2}$ *b.* $\left(R + \frac{H}{2}\right), 2H$

c. $(R + 2H), H$ *d.* $(R + H), H$

19. A balloon is filled at 27° C and 1 atm pressure by 500 m^3 He. At $-3°$ C and 0.5 atm pressure, the volume of He will be

a. 700 m^3 *b.* 900 m^3
c. 1000 m^3 *d.* 500 m^3

20. The ratio of intensity at the centre of a bright fringe to the intensity at a point distance one-fourth of the distance between two successive bright fringes will be

a. 4 *b.* 3 *c.* 2 *d.* 1

21. A rectangular block of mass m and area of cross-section A floats in a liquid of density ρ. If it is given a vertical displacement from equilibrium, it undergoes oscillation with a time period T. Then,

a. $T \propto \sqrt{\rho}$ *b.* $T \propto \frac{1}{\sqrt{A}}$
c. $T \propto \sqrt{A}$ *d.* $T \propto \frac{1}{\sqrt{m}}$

22. Three charges are placed at the three vertices of an equilateral triangle of side a as shown in the figure. The force experienced by the charge placed at the vertex A in a direction normal to BC is

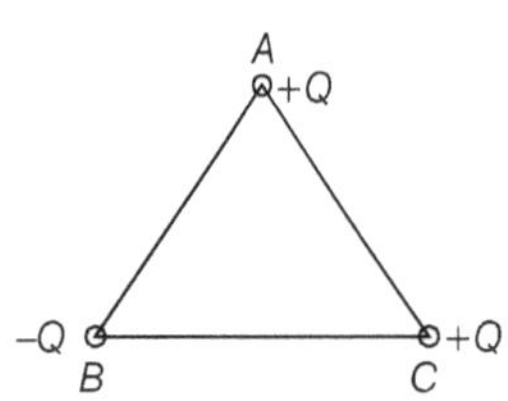

a. $\frac{Q^2}{4\pi\varepsilon_0 a^2}$ *b.* $-\frac{Q^2}{4\pi\varepsilon_0 a^2}$ *c.* zero *d.* $\frac{Q^2}{2\pi\varepsilon_0 a^2}$

23. A load of mass m falls from a height h on the scale pan hung from a spring as shown. If the spring constant is k and the mass of the scale pan is zero and the mass m does not bounce relative to the pan, then the amplitude of vibration is

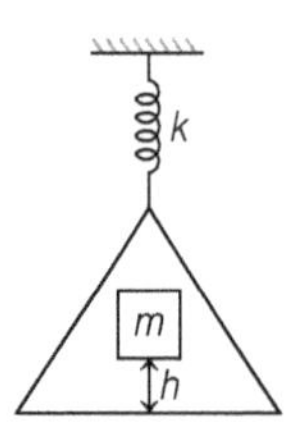

a. $\frac{mg}{k}$ *b.* $\frac{mg}{k}\sqrt{1+\frac{2hk}{mg}}$
c. $\frac{mg}{k}+\frac{mg}{k}\sqrt{\frac{1+2hk}{mg}}$ *d.* None of these

24. The activity of a radioactive sample is measured as N_0 counts per minute at $t = 0$ and $\frac{N_0}{e}$ counts per minute at $t = 5$ min. The time (in minutes) at which the activity reduces to half its value is

a. $\log_e \frac{2}{5}$ *b.* $\frac{5}{\log_e 2}$
c. $5 \log_{10} 2$ *d.* $5 \log_e 2$

25. A plano-convex lens fits exactly into a plano-concave lens. Their plane surfaces are parallel to each other. If lenses are made of different materials of refractive indices μ_1 and μ_2, R is the radius of curvature of the curved surface of the lenses, then the focal length of combination is

a. $\frac{R}{2(\mu_1+\mu_2)}$ *b.* $\frac{R}{2(\mu_1-\mu_2)}$
c. $\frac{R}{(\mu_1-\mu_2)}$ *d.* $\frac{2R}{\mu_1-\mu_2}$

26. In the given circuit diagram,

$E = 5$ V, $r = 1\,\Omega$, $R_2 = 4\Omega$, $R_1 = R_3 = 1\,\Omega$ and $C = 3\mu$F

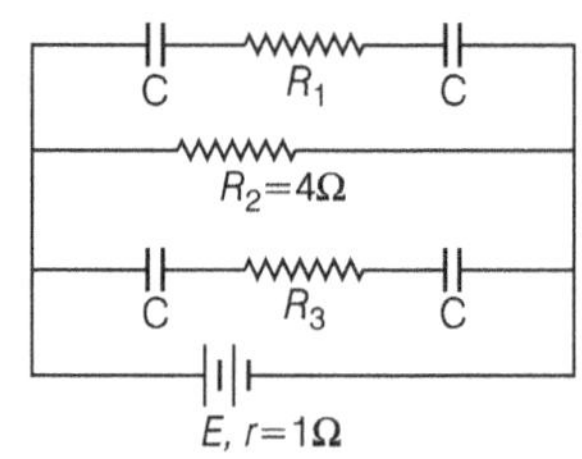

Then, what will be the numerical value of charge on each plates of the capacitor?

a. 24 μC *b.* 12 μC
c. 6 μC *d.* 3 μC

27. A block A of mass 100 kg rests on another block B of mass 200 kg and is tied to a wall as shown in the figure. The coefficient of friction between A and B is 0.2 and that between B and ground is 0.3. The minimum force required to move the block B is ($g = 10\,ms^{-2}$)

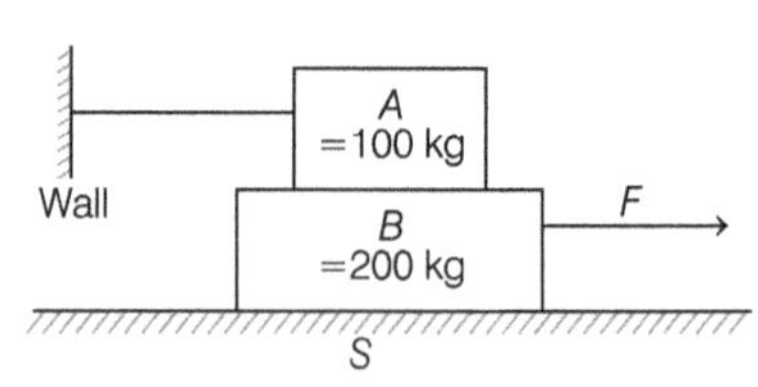

a. 900 N *b.* 200 N
c. 1100 N *d.* 700 N

28. A uniform rod of length l and mass m is free to rotate in a vertical plane about A. The rod initially in horizontal position is released. The initial angular acceleration of the rod is (Moment of inertia of rod about A is $\frac{ml^2}{3}$)

a. $\frac{3g}{2l}$ *b.* $\frac{2l}{3g}$ *c.* $\frac{3g}{2l^2}$ *d.* $mg\frac{l}{2}$

29. Monochromatic radiation of wavelength λ is incident on a hydrogen sample in ground state. Hydrogen atom absorbs a friction of light and subsequently emits radiations of six different wavelengths. The wavelength λ is

a. 97.2 nm *b.* 121.6 nm
c. 110.3 nm *d.* 45.2 nm

30. A coil in the shape of an equilateral triangle of side l is suspended between the pole pieces of a permanent magnet such that B is in plane of the coil. If due to a current i in the triangle, a torque τ rests on it, the side l of the triangle is

a. $2\left(\frac{\tau}{\sqrt{3}\,Bi}\right)^{\frac{1}{2}}$ *b.* $\frac{2}{\sqrt{3}}\left(\frac{\tau}{Bi}\right)$
c. $2\left(\frac{\tau}{Bi}\right)^{\frac{1}{2}}$ *d.* $\frac{1}{\sqrt{3}}\left(\frac{\tau}{Bi}\right)$

31. Work done in increasing the size of a soap bubble from radius of (3 to 5) cm is nearly (surface tension of soap solution = 0.03 Nm^{-1})

a. $0.2\,\pi$ mJ *b.* 2π mJ
c. 0.4 mJ *d.* 0.4π mJ

32. The velocity of a projectile at the initial point A is $(2\hat{\mathbf{i}} + 3\hat{\mathbf{j}})$ m/s. Its velocity (in m/s) at point B is

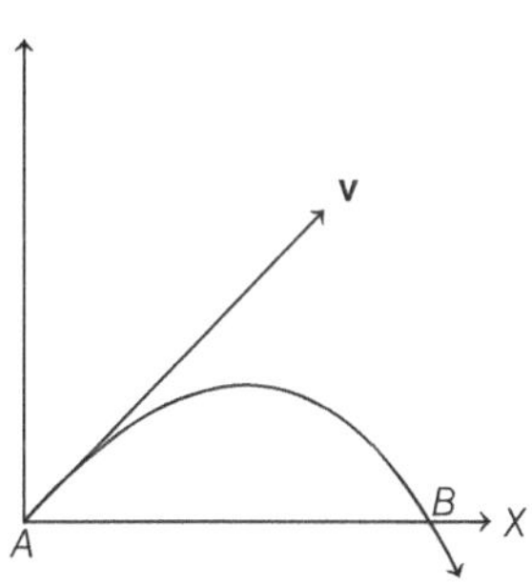

a. $-2\hat{\mathbf{i}} - 3\hat{\mathbf{j}}$ *b.* $-2\hat{\mathbf{i}} + 3\hat{\mathbf{j}}$
c. $2\hat{\mathbf{i}} - 3\hat{\mathbf{j}}$ *d.* $2\hat{\mathbf{i}} + 3\hat{\mathbf{j}}$

33. In the circuit shown, the heat produced in 5Ω resistor is 10 cal s^{-1}. The heat produced per sec in 4Ω resistor will be

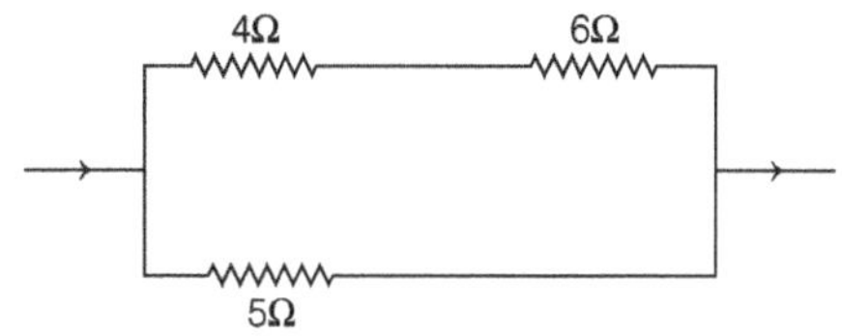

a. 1 cal *b.* 2 cal
c. 3 cal *d.* 4 cal

34. An α-particle after passing through potential difference of V volt collides with a nucleus. If the atomic number of the nucleus is Z, then distance of closest approach is

a. $14.4\cdot\frac{Z}{V}$ Å
b. $14.4\frac{Z}{V}$ m
c. $14.4\frac{V}{Z}$ m
d. $14.4\frac{V}{Z}$ Å

35. Two simple pendulums of lengths 5m and 20m respectively are given small displacement in one direction at the same time. They will again be in the same sense when the pendulum of shorter length has completed n oscillations. Then, n is

a. 5 *b.* 1 *c.* 2 *d.* 3

36. A parallel plate capacitor with air between the plates has a capacitance of 9 pF. The separation between the plates is d. The space between the plates is now filled with two dielectrics constant $K_1 = 3$ and thickness $\frac{d}{3}$ while the other one has dielectric constant $K_2 = 6$ and thickness $\frac{2d}{3}$. Capacitance of the capacitor is now

a. 1.8 pF *b.* 45 pF
c. 40.5 pF *d.* 20.25 pF

37. A particle moving along X-axis has acceleration f, at time t given by $f = f_0\left(1 - \frac{t}{T}\right)$, where f_0 and T are constants. The particle at $t = 0$ and the instant when $f = 0$, the particle's velocity v_X is

a. f_0T *b.* $\frac{1}{2}f_0T^2$
c. f_0T^2 *d.* $\frac{1}{2}f_0T$

38. A geostationary satellite orbits around the earth in a circular orbit of radius 36000 km. Then, the time period of a sky satellite orbiting a few 100 km above the earth's surface ($R = 6400$ km) will approximately be

a. $\frac{1}{2}$ h *b.* 1 h *c.* 2 h *d.* 4 h

39. A transverse wave propagating on a stretched string of linear density 3×10^{-4} kgm^{-1} is represented by the equation

$$y = 0.2 \sin(1.5x + 60t)$$

where, x in metres and t is in seconds. The tension in the string (in Newton) is

a. 0.24 *b.* 0.48
c. 1.20 *d.* 1.80

40. What is the magnetic field at the centre of arc in the figure below?

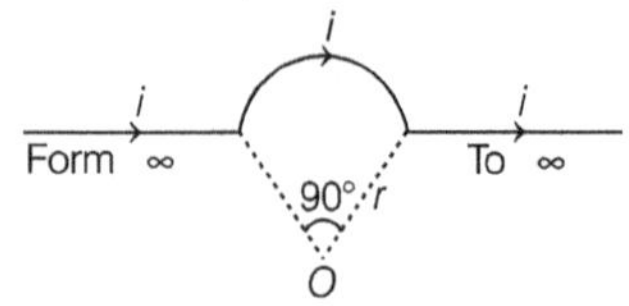

a. $\frac{\mu_0}{4\pi} \cdot \frac{2i}{r} [\sqrt{2} + \pi]$

b. $\frac{\mu_0}{4\pi} \cdot \frac{2i}{r} \left[\sqrt{2} + \frac{\pi}{4}\right]$

c. $\frac{\mu_0}{4\pi} \cdot \frac{i}{r} [\sqrt{2} + \pi]$

d. $\frac{\mu_0}{4\pi} \cdot \frac{i}{r} \left[\sqrt{2} + \frac{\pi}{4}\right]$

PART II

Chemistry

41. 4 g of copper was dissolved in conc. HNO_3. The copper nitrate thus obtained gave 5g of its oxide on strong heating. The equivalent weight of copper is

a. 23 *b.* 32
c. 12 *d.* 20

42. Choose the incorrect statement.

a. Sodium borohydride reacts very slowly with cold water.

b. Sodium borohydride reacts violently with cold water to give $H_2(g)$.

c. Solubility of sodium borohydride in water at 25°C is 10.05 g/mL.

d. Melting point of sodium borohydride is 500°C.

43. The orbital angular momentum of an electron in $2s$ orbital is

a. $\frac{h}{4\pi}$ *b.* zero

c. $\frac{h}{2\pi}$ *d.* $\sqrt{2} . \frac{h}{2\pi}$

44. Which of the following are isoelectronic species?

I. CH_3^+ II. NH_2^-

III. NH_4^+ IV. NH_3

a. I, II and III *b.* II, III and IV
c. I, II and IV *d.* II and I

45. Two elements A and B have electronegativities 1.2 and 3.0 respectively. The nature of bond between A and B would be

a. ionic *b.* covalent
c. coordinate *d.* metallic

46. In the compound,

$$\underset{(I)}{CH_2} = \underset{(II)}{CH} - C \equiv \underset{(III)}{CH}$$

The most acidic hydrogen atom is

a. Only I *b.* Only II
c. Only III *d.* All are equally acidic

47. Reductive ozonolysis of $(CH_3)_2C = C(CH_3)_2$ followed by hydrolysis gives

a. only one type of ketone
b. only one type of aldehyde
c. two types of ketone
d. two types of aldehyde

48. Which of the following reactions does not involved absorption energy?

I. $O(g) + e^- \longrightarrow O^-(g)$

II. $S(g) + e^- \longrightarrow S^-(g)$

III. $O^-(g) + e^- \longrightarrow O^{2-}(g)$

IV. $Cl(g) + e^- \longrightarrow Cl^-(g)$

a. Only II *b.* I and III
c. I, II and III *d.* I, II and IV

49. The most reactive amine towards reaction with dil. HCl is

a. $CH_3—NH_2$ *b.* 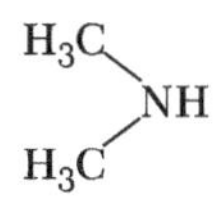 *c.* $(CH_3)_3 \cdot N$ *d.* 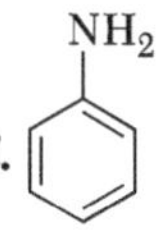

50. Blocks of magnesium are fixed to the bottom of a ship to

a. block hole in the ship
b. acidity of sea water
c. make the ship lighter
d. prevent the action of water and salt

51. On electrolysis of water, a total of 1 mole of gases is evolved. The amount of water decomposed is

a. 1 mol *b.* 2 mol *c.* $\frac{1}{3}$ mol *d.* $\frac{2}{3}$ mol

52. How many moles of Fe^{2+} ions are formed when excess of iron react with 500 mL 0.4 NHCl, under inert atmosphere?
(assume no change in volume)

a. 0.4 *b.* 0.1 *c.* 0.2 *d.* 0.8

53. Sodium sulphate is soluble in water but barium sulphate is insoluble because

a. hydration energy of Na_2SO_4 is more than of its lattice energy
b. lattice energy of $BaSO_4$ is more than its hydration energy
c. Both (a) and (b)
d. None of the above

54. In the reaction,

$$CH_3CH_2Cl \xrightarrow{KCN} (A) \xrightarrow{LiAlH_4} (P)$$; end product (P) is

a. $CH_3CH_2NO_2$ *b.* $CH_3CH_2CH_2NO_2$
c. $CH_3CH_2NH_2$ *d.* $CH_3CH_2CH_2NH_2$

55. Which of the following reactions is an example of calcination process?

a. $2Ag + 2HCl + [O] \longrightarrow 2AgCl + H_2O$
b. $2Zn + O_2 \longrightarrow 2ZnO$
c. $2ZnS + 3O_2 \longrightarrow 2ZnO + 2SO_2$
d. $MgCO_3 \xrightarrow{\Delta} MgO + CO_2$

56. For an endothermic reaction, where ΔH represent the enthalpy of reaction in kJ/mol, the minimum value for energy of activation (for forward reaction) will be

a. less than ΔH *b.* zero
c. more than ΔH *d.* equal to ΔH

57. Which of the following metal is leached by cyanide process?

a. Ag *b.* Na *c.* Al *d.* Cu

58. Which of the following is a diamagnetic complex?

a. $[Co(NH_3)_6]^{3+}$ *b.* $[NiCl_4]^{2-}$
c. $[CuCl_4]^{2-}$ *d.* $[Fe(H_2O)_6]^{3+}$

59. Neoprene is a

a. monomer of rubber *b.* synthetic rubber
c. a natural rubber *d.* vulcanised rubber

60. Among the following, which have highest melting point?

a. Ionic solids *b.* Pseudo solids
c. Molecular solids *d.* Amorphous solids

61. The night-blindness is developed due to deficiency of vitamin

a. B_6 *b.* C *c.* B_{12} *d.* A

62. The transfer RNA anticodon for the messenger RNA codon G – C – A is

a. C – G – U *b.* C – C – U
c. U – C – C *d.* G – U – C

63. 0.765 g of an acid gives 0.535 g of CO_2 and 0.138 g of H_2O. Then, the ratio of percentage of carbon and hydrogen is

a. 19 : 2 *b.* 18 : 11 *c.* 20 : 17 *d.* 1 : 7

64. Maximum pK_b value of

a. 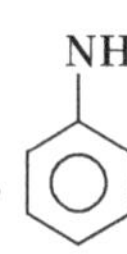 *b.* $(CH_3CH_2)_2NH$

c. $(CH_3)_2NH$ *d.* 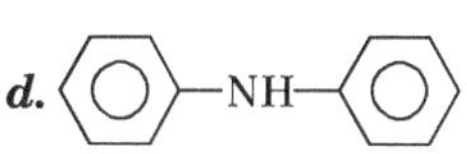

65. Which of the following is an incorrect set of quantum members?

a. $n = 2, l = 0, m = 0$ *b.* $n = 1, l = 0, m = 0$
c. $n = 3, l = 3, m = 0$ *d.* $n = 2, l = 1, m = 1$

66. The most acidic oxide for nitrogen is

a. NO_2 *b.* N_2O *c.* NO *d.* N_2O_5

67. Which of the following show maximum bond order?

a. O_2 *b.* O_2^- *c.* O_2^+ *d.* O_2^{2-}

68. Which of the following show an increase in entropy?

I. Boiling of water II. Melting of ice
III. Freezing of water

IV. Formation of hydrogen gas from water

a. (I) and (II) *b.* Only (III) *c.* (I), (II) and (IV) *d.* (III) and (IV)

69. BF_3 is an acid, according to

a. Lewis *b.* Arrhenius *c.* Bronsted and Lowery *d.* All of these

70. For the reaction,

$$N_2O_4(g) \longrightarrow 2NO_2(g)$$

a. $\Delta H > \Delta E$ *b.* $\Delta H < \Delta E$ *c.* $\Delta H = \Delta E$ *d.* $\Delta H = 0$

71. Which of the following elements mostly form covalent compounds?

a. Cs *b.* Rb *c.* K *d.* Li

72. When aqueous solution of borax is acidified with HCl, we get

a. B_2H_6 *b.* H_3BO_3 *c.* B_2O_3 *d.* All of these

73. Which of the following compound does not follow Huckel's rule?

a.  *b.*  *c.*  *d.* 

74. A graph is ploted between $\log k$ *versus* $\frac{1}{T}$ for calculation of activation energy (E_a). The correct plot is

a. 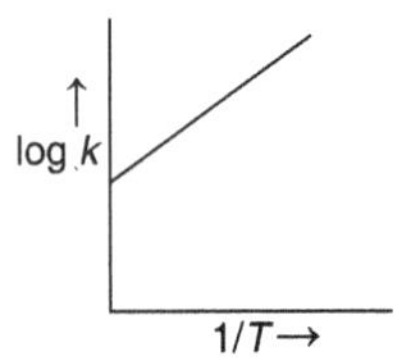

b. 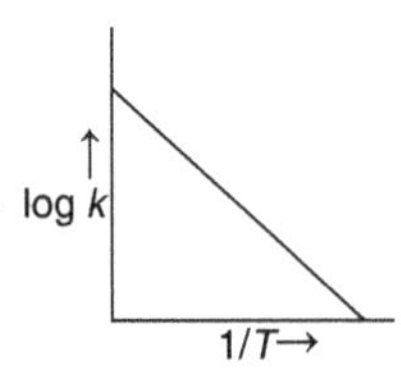

c. 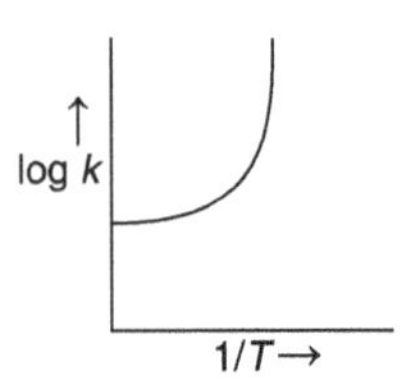

d. 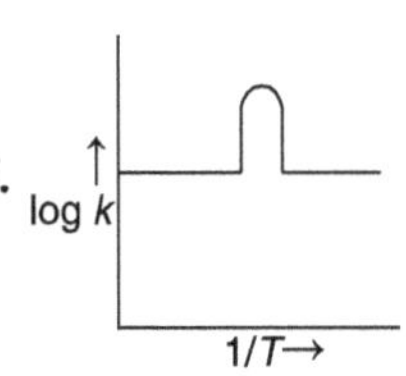

75. The hybridisation of Fe in $[K_3Fe(CN)_6]$ is

a. sp^3 *b.* dsp^3 *c.* sp^3d^2 *d.* d^2sp^3

76. Which of the following shows maximum magnetic moment?

a. Mg^{2+} *b.* Ti^{3+} *c.* V^{3+} *d.* Fe^{2+}

77. Consider the following radioactive decays.

I. ${}_{92}U \xrightarrow{-\alpha} {}_{90}Th$ and II. ${}_{90}Th \xrightarrow{-\alpha} {}_{88}Ra$

In which case group of parent and daughter elements remain unchanged?

a. In (I) *b.* In (II) *c.* Both in (I) and (II) *d.* None of these

78. Chlorobenzene $\xrightarrow{\text{Reaction }(X)}$ Phenol $\xrightarrow{\text{Reaction }(Y)}$ Salicylaldehyde.

The reaction(s) '*X*' and '*Y*' respectively are

a. Fries rearrangement and Kolbe
b. Cumene and Reimer-Tiemann
c. Dow and Reimer-Tiemann
d. Dow and Sandmeyer

79. Which of the following has largest number of moles?

a. 8g of oxygen atoms
b. 16 g of oxygen gas
c. 14 g of nitrogen gas (N_2)
d. All have same number of moles

80. One mole each of four ideal gases are kept as follows.

I. 5 L of gas (*A*) at 2 atm pressure
II. 2.5 L of gas (*B*) at 2 atm pressure
III. 1.25 L of gas (*C*) at 2 atm pressure
IV. 2.5 L of gas (*D*) at 2.5 atm pressure

Which of the above gases is kept at highest temperature?

a. Gas (*A*) *b.* Gas (*B*) *c.* Gas (*C*) *d.* Gas (*D*)

PART III

a. English Proficiency

Directions (Q. Nos. 81-83) *In the following questions, the sentences may or may not be grammatically correct. Find out which part of a sentence has an error and mark that part. If there is no error, mark part 'd' as your answer.*

81. Along the northern frontier of India (*a*) / is seen (*b*) / the Himalayas mighty in their splendour. (*c*) / No error (*d*)

82. The father with the son were (*a*) / mysteriously missing (*b*) / from the house. (*c*) / No error (*d*)

83. It is not advisable (*a*) / to take heavy luggages (*b*) / while on journey these days. (*c*) / No error (*d*)

Directions (Q. Nos. 84-85) *Fill in the blanks with suitable preposition from the alternatives given under each sentence.*

84. The problem of communal harmony cannot be glossed .... by the government.

a. at *b.* on
c. over *d.* for

85. She could not muster .... courage to stand against the maltreatment.

a. to *b.* up
c. about *d.* on

Directions (Q. Nos. 86-88) *The following sentences consist of a word or a phrase which is written in italicised letters. Each sentence is followed by four words or phrases. Select the word or the phrase which is closest to the opposite in meaning of the italicised word or phrase.*

86. Philosophers say that the world is *an illusion*.

a. a fact *b.* a reality
c. an actuality *d.* a truth

87. She used to *disparage* her neighbours every now and then.

a. please *b.* praise
c. belittle *d.* denigrate

88. The momentum of the movement *slackened* in course of time.

a. stopped
b. quickened
c. multiplied
d. recovered

Directions (Q. Nos. 89-90) *In the following sentences, a word or a phrase is written in italicised letters. For each italicised part, four words/phrases are listed below each sentence. Choose the word nearest in meaning to the italicised word/phrase.*

89. The opposition criticised the ruling party for the *deteriorating* law and order situation in the state.

a. disrupting
b. worsening
c. crumbling
d. eroding

90. The two opposing parties have reached *stalemate*.

a. dilemma
b. deadlock
c. exhaustion
d. settlement

Directions (Q. Nos. 91-95) *Read the passage given below and answer the questions that follow.*

A pioneering scheme has been started recently in Southampton of England's south coast to educate tourists who have been convicted of drunken driving.

The penalty for drunken driving might be the loss of the driving licence and a heavy fine. But under the new scheme, convicted drivers do not pay the fine. Instead they have to attend eight training sessions; one a week organised by the local authority probation service.

Designed to demonstrate the damage alcohol can do, the scheme was devised by senior probation officer John Cook. He said that about a quarter of the people who came to him had a drinking problem, and had not realised how much they were drinking. One way of getting the message across was to make the drivers pour out their usual ration of alcohol and then measure it. Almost everyone poured out not a single measure, but a double atleast, an example of how easy it is to have more than just one drink and to encourage other people to do the same. The instructors on the course are giving clinical evidence of the effects of alcohol on the body and brain. The sober truth is that drinking badly affects driving skills, although the drinker might like to believe otherwise.

91. The southampton scheme requires convicted drivers

a. to pay a heavy fine
b. to attend eight driving sessions–one a week
c. to undergo a probation service
d. to surrender their driving licence

92. John Cook devised the scheme

a. as a demonstration technique for driving
b. to deny the harmful effects of alcohol
c. to show that Southampton was concerned about drivers
d. to prove that alcohol does influence driving

93. The problem with a quarter of the people who went to John Cook was that they

a. did not want to stop drinking
b. were unaware of the fact that they could get drunk
c. would not admit that they had a drinking problem
d. did not know how much they were drinking

94. Most drivers start off with atleast

a. a double measure
b. a single measure
c. a little less than a single measure
d. two doubles

95. The truth is that alcohol

a. does not affect the body but only the brain
b. affects only the brain
c. affects the body and the brain
d. has no effect on the body or the brain

b. Logical Reasoning

96. 'Shoes' is related to 'Leather', in the same way as 'Rubber' is related to

a. Plastic *b.* Polythene
c. Latex *d.* Chappal

97. Find the odd one from the following options

a. 81 : 243 *b.* 25 : 75
c. 64 : 192 *d.* 16 : 64

98. Complete the series by replacing '?' mark.

4, 11, 30, 67, 128, ?

a. 219 *b.* 228
c. 237 *d.* 240

99. Lakshmi is elder than Meenu. Leela is elder than Meenu but younger than Lakshmi. Latha is younger than both Meenu and Hari but Hari is younger than Meenu. Who is the youngest?

a. Lakshmi *b.* Meenu
c. Leela *d.* Latha

100. In the following question a part of problem figure is missing. Find out from the given answer figures *a*, *b*, *c* and *d* that can replace the question mark (?) to complete the figure.

Question Figure

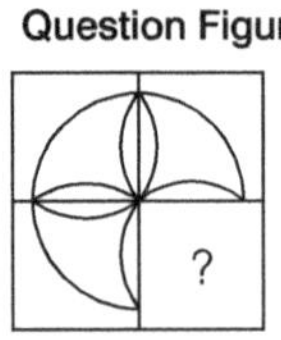

Answer Figures

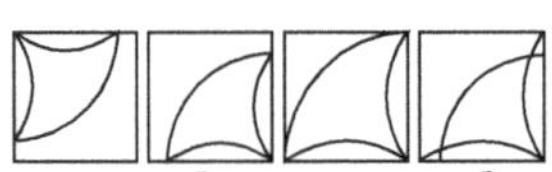

a. *b.* *c.* *d.*

101. In the following question, five figures are given. Out of them find the three figures that can be joined to form a square.

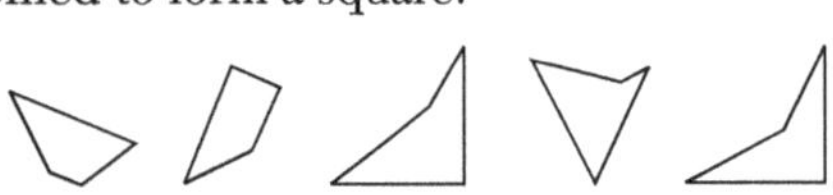

(A) (B) (C) (D) (E)

a. ACD *b.* BCD
c. BDE *d.* CDE

102. The three problem figures marked X, Y and Z show the manner in which a piece of paper is folded step by step and then cut. From the answer figures *a*, *b*, *c* and *d* select the one showing the unfolded pattern of the paper after the cut.

Question Figures

X Y Z

Answer Figures

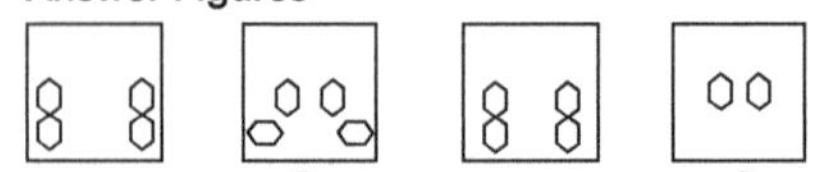

a. *b.* *c.* *d.*

103. Choose the answer figure which completes the problem figure matrix.

Question Figures

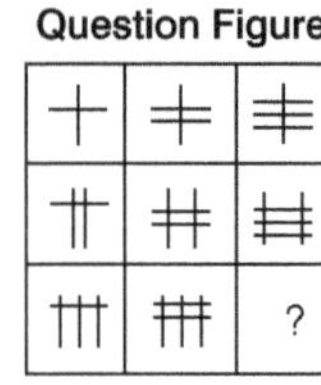

Answer Figures

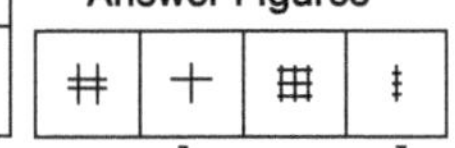

a. *b.* *c.* *d.*

104. In the following question, one or more dots are place in the figure marked as (A). This figure is followed by four alternatives marked as *a*, *b*, *c* and *d*. One out of these four options contains region(s) common to circle, square and triangles, similar to that marked by the dot in figures (A). Find that figure

Question Figure

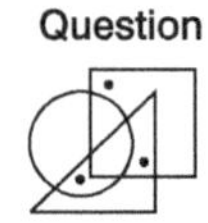

(A)

Answer Figures

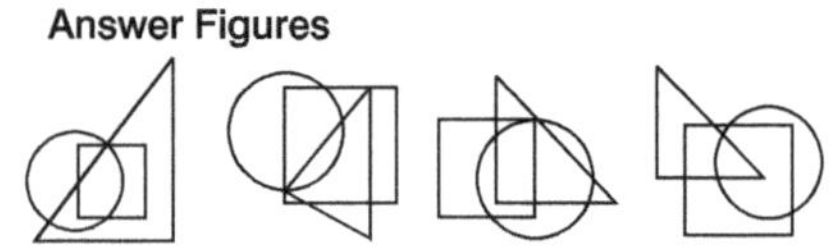

a. *b.* *c.* *d.*

105. How many different triangles are there in the figures shown below?

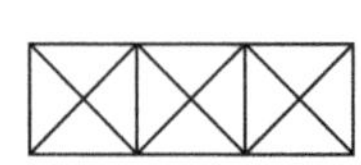

a. 28 *b.* 24
c. 20 *d.* 16

PART IV

Mathematics

106. The coefficient of x^5 in the expansion of $(1+x)^{21}+(1+x)^{22}+\ldots+(1+x)^{30}$ is

a. ${}^{51}C_5$ *b.* ${}^{9}C_5$
c. ${}^{31}C_6-{}^{21}C_6$ *d.* ${}^{30}C_5+{}^{20}C_5$

107. If $z=a+ib$ satisfies $\arg(z-1)=\arg(z+3i)$, then $(a-1):b=$

a. $2:1$ *b.* $1:3$
c. $-1:3$ *d.* None of these

108. If p and p' denote the lengths of the perpendicular from a focus and the centre of an ellipse with semi-major axis of length a, respectively, on a tangent to the ellipse and r denotes the focal distance of the point, then

a. $ap=rp'$ *b.* $rp=ap'$
c. $ap=rp'+1$ *d.* $ap'+rp=1$

109. The value of $\sum_{r=1}^{10} r\cdot\dfrac{{}^nC_r}{{}^nC_{r-1}}$ is equal to

a. $5(2n-9)$ *b.* $10n$
c. $9(n-4)$ *d.* None of these

110. The numbers $3^{2\sin 2\alpha-1}$, 14 and $3^{4-2\sin 2\alpha}$ form first three terms of an AP, its fifth term is

a. -25 *b.* -12 *c.* 40 *d.* 53

111. For the equation $3x^2+px+3=0$, $p>0$, if one of the roots is square of the other, then p is equal to

a. 1/2 *b.* 1 *c.* 3 *d.* 2/3

112. If $a=\log_2 3$, $b=\log_2 5$ and $c=\log_7 2$, then $\log_{140} 63$ in terms of a, b, c is

a. $\dfrac{2ac+1}{2c+abc+1}$ *b.* $\dfrac{2ac+1}{2a+c+a}$
c. $\dfrac{2ac+1}{2c+ab+a}$ *d.* None of these

113. If $\cos(x-y)$, $\cos x$ and $\cos(x+y)$ are in HP, then $\cos x\sec(y/2)$ is equal to

a. $\pm\sqrt{2}$ *b.* $\pm 1/\sqrt{2}$
c. ± 2 *d.* None of these

114. Let $A=\{1, 2, 3, 4, 5\}$ and R be a relation defined by $R=\{(x, y): x, y\in A, x+y=5\}$. Then, R is

a. reflexive and symmetric but not transitive
b. an equivalence relation
c. symmetric but neither reflexive nor transitive
d. neither reflexive nor symmetric but transitive

115. The number of times the digit 5 will be written when listing the integers from 1 to 1000, is

a. 271 *b.* 272
c. 300 *d.* None of these

116. Let A and B be two sets such that $A\cap X=B\cap X=\phi$ and $A\cup X=B\cup X$ for same set X. Then,

a. $A=B$ *b.* $A=X$
c. $B=X$ *d.* $A\cup B=X$

117. Let $A=[-1, 1]$ and $f: A\to A$ be defined as $f(x)=x|x|$ for all $x\in A$, then $f(x)$ is

a. many-one and into function
b. one-one and into function
c. many-one and into function
d. one-one and onto function

118. The general solution of $\sin x-3\sin 2x+\sin 3x=\cos x-3\cos 2x+\cos 3x$ is

a. $n\pi+\dfrac{\pi}{8}$ *b.* $\dfrac{n\pi}{2}+\dfrac{\pi}{8}$
c. $(-1)^n\dfrac{n\pi}{2}+\dfrac{\pi}{8}$ *d.* $2n\pi+\cos^{-1}\dfrac{3}{2}$

119. Two equal sides of an isosceles triangle are $7x-y+3=0$ and $x+y-3=0$ and its third side passes through the point $(1, -10)$. Find the equation of the third side

a. $x-3y=-31$ *b.* $x-3y=31$
c. $x+3y=31$ *d.* $x+3y=-31$

120. If two distinct chords drawn from the point (p, q) on the circle $x^2+y^2=px+qy$ (where $pq\neq 0$) are bisected by the X-axis, then

a. $p^2=q^2$ *b.* $p^2=8q^2$
c. $p^2<8q^2$ *d.* $p^2>8q^2$

121. The length of perpendicular drawn from the point (2, 3, 4) to line $\dfrac{4-x}{2}=\dfrac{y}{6}=\dfrac{1-z}{3}$, is

a. $\dfrac{3}{7}\sqrt{101}$ *b.* $\dfrac{2}{7}\sqrt{101}$
c. $\dfrac{2}{7}\sqrt{103}$ *d.* $\dfrac{3}{7}\sqrt{103}$

122. The image of the point $(1, 6, 3)$ in the line $\dfrac{x}{1}=\dfrac{y-1}{2}=\dfrac{z-2}{3}$ is

a. $(-1, 0, 7)$ *b.* $(-1, 0, -7)$
c. $(1, 0, 7)$ *d.* $(2, 0, 7)$

BITSAT 2017

123. The distances of the point $(1, -5, 9)$ from the plane $x - y + z = 5$ measured along a straight line $x = y = z$ is $2\sqrt{3}\,k$, then the value of k is

a. 5 *b.* 6
c. $\sqrt{3}$ *d.* 4

124. $\lim_{n \to \infty} \sin[\pi\sqrt{n^2+1}]$ is equal to

a. ∞ *b.* 0
c. does not exist *d.* None of these

125. If $f(x) = \begin{cases} e^x, & x \le 0 \\ |1-x|, & x > 0 \end{cases}$, then

a. $f(x)$ is differentiable at $x = 0$
b. $f(x)$ is continuous at $x = 0, 1$
c. $f(x)$ is differentiable at $x = 1$
d. None of these

126. If a function $f: R \to R$ satisfy the equation $f(x+y) = f(x) + f(y), \forall\, x, y$ and the function $f(x)$ is continuous at $x = 0$, then

a. $f(x)$ is continuous for all positive real values of x
b. $f(x)$ is continuous for all x
c. $f(x) = 0$ for all x
d. None of these

127. The value of $f(0)$, so that the function $f(x) = \dfrac{2x - \sin^{-1} x}{2x + \tan^{-1} x}$ is continuous at each point in its domain, is

a. $\frac{1}{3}$ *b.* $-\frac{1}{3}$ *c.* $\frac{2}{3}$ *d.* $-\frac{2}{3}$

128. Consider the greatest integer function, defined by $f(x) = [x], 0 \le x < 2$. Then,

a. f is derivable at $x = 1$
b. f is not derivable at $x = 1$
c. f is derivable at $x = 2$
d. None of these

129. Let $f(x) = -2x^3 + 21x^2 - 60x + 41$, then

a. $f(x)$ is decreasing in $(-\infty, 1)$
b. $f(x)$ is decreasing in $(-\infty, 2)$
c. $f(x)$ is increasing in $(-\infty, 1)$
d. $f(x)$ is increasing in $(-\infty, 2)$

130. Rolle's theorem is not applicable for the function $f(x) = |x|$ in the interval $[-1, 1]$ because

a. $f'(1)$ does not exist
b. $f'(-1)$ does not exist
c. $f(x)$ is discontinuous at $x = 0$
d. $f'(0)$ does not exist

131. If the curve $y = a^x$ and $y = b^x$ intersect at angle α, then $\tan\alpha$ is equal to

a. $\dfrac{a-b}{1+ab}$ *b.* $\dfrac{\log a - \log b}{1 + \log a \log b}$
c. $\dfrac{a+b}{1-ab}$ *d.* $\dfrac{\log a + \log b}{1 - \log a \log b}$

132. The expression $\dfrac{\int_0^n [x]\,dx}{\int_0^n \{x\}\,dx}$, where $[x]$ and $\{x\}$ are integral and fractional part of x and $n \in N$, is equal to

a. $\frac{1}{n-1}$ *b.* $\frac{1}{n}$ *c.* n *d.* $n-1$

133. The maximum value of $f(x) = x + \sin 2x$, $x \in [0, 2\pi]$ is

a. $\frac{\pi}{2}$ *b.* 2π *c.* $\frac{3\pi}{4}$ *d.* $\frac{3\pi}{2}$

134. The area under the curve $y = |\cos x - \sin x|$, $0 \le x \le \frac{\pi}{2}$ and above X-axis, is

a. $2\sqrt{2}$ *b.* $2\sqrt{2} - 2$ *c.* $2\sqrt{2} + 2$ *d.* 0

135. The solution of $\dfrac{dy}{dx} = \cos(x+y) + \sin(x+y)$, is

a. $\log\left[1 + \tan\left(\frac{x+y}{2}\right)\right] + C = 0$
b. $\log\left[1 + \tan\left(\frac{x+y}{2}\right)\right] = x + C$
c. $\log\left[1 - \tan\left(\frac{x+y}{2}\right)\right] = x + C$
d. None of these

136. The area enclosed by the curves $y = x^3$ and $y = \sqrt{x}$ is

a. $\frac{5}{3}$ sq units *b.* $\frac{5}{4}$ sq units
c. $\frac{5}{12}$ sq units *d.* $\frac{12}{5}$ sq units

137. If $\lim_{x \to \infty}\left(an - \dfrac{1+n^2}{1+n}\right) = b$, where a is finite number, then

a. $a = 2$ *b.* $a = 0$ *c.* $b = 1$ *d.* $b = -1$

138. If the papers of 4 students can be checked by anyone of the 7 teachers, then the probability that all the 4 papers are checked by exactly 2 teachers, is equal to

a. 12/49 *b.* 6/49 *c.* 9/49 *d.* 15/49

139. If equation $(10x-5)^2+(10y-4)^2 = \lambda^2(3x+4y-1)^2$ represents a hyperbola, then

a. $-2<\lambda<2$
b. $\lambda>2$
c. $\lambda<-2$ or $\lambda>2$
d. $0<\lambda<2$

140. Let $\hat{a}$ and $\hat{b}$ be two non-collinear unit vectors. If $u=\hat{a}-(\hat{a}\cdot\hat{b})\hat{b}$ and $v=\hat{a}\times\hat{b}$, then $|v|$ is equal to

a. $|u|$ *b.* $|u|+|v\cdot\hat{a}|$
c. $2|v|$ *d.* $|v|+u\cdot(\hat{a}+\hat{b})$

141. If the variance of the observations $x_1, x_2, \ldots, x_n$ is σ^2, then the variance of $\alpha x_1, \alpha x_2, \ldots, \alpha x_n, \alpha\neq 0$ is

a. σ^2 *b.* $\alpha\sigma^2$
c. $\alpha^2\sigma^2$ *d.* $\dfrac{\sigma^2}{\alpha^2}$

142. Coefficient of variation of two distributions are 50 and 60 and their arithmetic means are 30 and 25, respectively. Difference of their standard deviation is

a. 0 *b.* 1
c. 1.3 *d.* 2.5

143. The maximum value of $z=9x+13y$ subject to constraints $2x+3y\leq 18$, $2x+y\leq 10$, $x\geq 0$, $y\geq 0$ is

a. 130 *b.* 81
c. 79 *d.* 99

144. A coin is tossed 7 times. Each time a man calls head. The probability that he wins the toss atleast 4 occassions is

a. $\dfrac{1}{4}$ *b.* $\dfrac{5}{8}$
c. $\dfrac{1}{2}$ *d.* $\dfrac{1}{6}$

145. If $\sin^{-1}x+\sin^{-1}y+\sin^{-1}z=\dfrac{3\pi}{2}$ and $f(1)=2$, $f(p+q)=f(p)\cdot f(q)$, $\forall\, p, q\in R$, then $x^{f(1)}+y^{f(2)}+z^{f(3)}-\dfrac{(x+y+z)}{x^{f(1)}+y^{f(2)}+z^{f(3)}}$ is equal to

a. 0 *b.* 1
c. 2 *d.* 3

146. The value of $\dfrac{2}{1!}+\dfrac{2+4}{2!}+\dfrac{2+4+6}{3!}+\ldots$ is

a. e *b.* $2e$
c. $3e$ *d.* None of these

147. If z_1, z_2 and z_3 represent the vertices of an equilateral triangle such that $|z_1|=|z_2|=|z_3|$, then

a. $z_1+z_2=z_3$ *b.* $z_1+z_2+z_3=0$
c. $z_1z_2=\dfrac{1}{z_3}$ *d.* $z_1-z_2=z_3-z_2$

148. If $\int\dfrac{(\sqrt{x})^5\,dx}{(\sqrt{x})^7+x^6}=\lambda\log\left(\dfrac{x^a}{x^a+1}\right)+C$, then $a+\lambda$ equal to

a. 2 *b.* >2
c. <2 *d.* >3

149. Line joining the points $(0, 3)$ and $(5, -2)$ is a tangent to the curve $y=\dfrac{ax}{1+x}$, then

a. $a=1\pm\sqrt{3}$ *b.* $a\in\phi$
c. $a=-1\pm\sqrt{3}$ *d.* $a=-2\pm 2\sqrt{3}$

150. The shortest distance between the parabolas $y^2=4x$ and $y^2=2x-6$ is

a. 2 *b.* $\sqrt{5}$
c. 3 *d.* None of these

Answers

Physics

1. (a)	**2.** (d)	**3.** (c)	**4.** (b)	**5.** (b)	**6.** (b)	**7.** (b)	**8.** (b)	**9.** (c)	**10.** (b)
11. (b)	**12.** (a)	**13.** (d)	**14.** (b)	**15.** (c)	**16.** (d)	**17.** (c)	**18.** (d)	**19.** (b)	**20.** (c)
21. (b)	**22.** (c)	**23.** (b)	**24.** (d)	**25.** (c)	**26.** (c)	**27.** (c)	**28.** (a)	**29.** (a)	**30.** (a)
31. (d)	**32.** (c)	**33.** (b)	**34.** (a)	**35.** (c)	**36.** (c)	**37.** (d)	**38.** (c)	**39.** (b)	**40.** (b)

Chemistry

41. (b)	**42.** (b)	**43.** (b)	**44.** (b)	**45.** (a)	**46.** (c)	**47.** (a)	**48.** (d)	**49.** (b)	**50.** (d)
51. (d)	**52.** (b)	**53.** (c)	**54.** (d)	**55.** (d)	**56.** (c)	**57.** (a)	**58.** (a)	**59.** (b)	**60.** (a)
61. (d)	**62.** (a)	**63.** (a)	**64.** (d)	**65.** (c)	**66.** (d)	**67.** (c)	**68.** (c)	**69.** (a)	**70.** (a)
71. (d)	**72.** (b)	**73.** (d)	**74.** (b)	**75.** (d)	**76.** (d)	**77.** (a)	**78.** (c)	**79.** (d)	**80.** (a)

English Proficiency

81. (b)	**82.** (a)	**83.** (b)	**84.** (c)	**85.** (b)	**86.** (b)	**87.** (b)	**88.** (b)	**89.** (b)	**90.** (b)
91. (b)	**92.** (d)	**93.** (d)	**94.** (a)	**95.** (c)					

Logical Reasoning

96. (c)	**97.** (d)	**98.** (a)	**99.** (d)	**100.** (a)	**101.** (a)	**102.** (b)	**103.** (c)	**104.** (c)	**105.** (a)

Mathematics

106. (c)	**107.** (b)	**108.** (a)	**109.** (a)	**110.** (d)	**111.** (c)	**112.** (d)	**113.** (a)	**114.** (c)	**115.** (c)
116. (a)	**117.** (d)	**118.** (b)	**119.** (b)	**120.** (d)	**121.** (a)	**122.** (c)	**123.** (a)	**124.** (b)	**125.** (b)
126. (b)	**127.** (a)	**128.** (b)	**129.** (b)	**130.** (d)	**131.** (b)	**132.** (d)	**133.** (b)	**134.** (b)	**135.** (b)
136. (c)	**137.** (c)	**138.** (b)	**139.** (c)	**140.** (a)	**141.** (c)	**142.** (a)	**143.** (c)	**144.** (c)	**145.** (c)
146. (c)	**147.** (b)	**148.** (b)	**149.** (b)	**150.** (b)					

Hints & Solutions

Physics

1. ***(a)*** According to Stefan's law, $\frac{E}{t} = \sigma A e T^4$

So, we can write as

$$\frac{E_2}{E_1} = \left(\frac{T_2}{T_1}\right)^4$$

$$\frac{E_2}{E_1} = \left(\frac{900}{300}\right)^4 \Rightarrow \frac{E_2}{E_1} = (3)^4$$

$$E_2 = 81\, E_1$$

$$\Rightarrow \quad \frac{E_2}{E_1} = 81$$

2. *(d)* $v_s = r\omega = 1.2 \times 2\pi f \qquad [\because \omega = 2\pi f]$

$$= 1.2 \times 2 \times 3.14 \times \left(\frac{400}{60}\right) = 50.24$$

$$\simeq 50 \text{ m/s}$$

$$\nu_{\min} = \frac{v}{v + v_s} \cdot \nu$$

$$= \frac{340}{340 + 50} \times 500 = 436 \text{ Hz}$$

$$\nu_{\max} = \frac{v}{v - v_s} \cdot \nu = \frac{340}{340 - 50} \times 500$$

$$= 586 \text{ Hz}$$

3. *(c)* We have, $f_R - f_B = \omega f_Y$

$$\omega = \frac{f_R - f_B}{f_Y}$$

Here, $f_R = 100$ cm

$f_B = 96.8$ cm

$f_Y = \sqrt{f_B \times f_R}$

$= \sqrt{96.8 \times 100}$

$= 98.4$ cm

$\therefore$ Dispersive power

$$\omega = \frac{f_R - f_B}{f_Y}$$

$$\Rightarrow \quad = \frac{100 - 96.8}{98.4}$$

$$\Rightarrow \quad \omega = \frac{3.2}{98.4} = 0.0325$$

4. *(b)* $V = 800$ V

$$\therefore \quad E = \frac{V}{d} = \frac{800}{2 \times 10^{-2}} \text{ V/m}$$

$$= 4 \times 10^4 \text{ V/m}$$

Force due to E is balanced by the weight of particle, i.e.,

$$qE = mg$$

$$\Rightarrow \quad q = \frac{mg}{E}$$

$$= \frac{1.96 \times 10^{-15} \times 9.8}{4 \times 10^4}$$

$$= 4.8 \times 10^{-19}$$

$$= 3 \times 1.6 \times 10^{-19}$$

$$= 3e$$

5. *(b)* Given, $H = R$

$$\text{or} \quad \frac{u^2 \sin^2 \theta}{2g} = \frac{u^2 \sin 2\theta}{g}$$

$$= \frac{2u^2 \sin \theta \cos \theta}{g}$$

or $\tan \theta = 4$

$\Rightarrow \quad \theta = \tan^{-1}(4)$

6. *(b)* Given, $A = 6 \text{ cm} = 6 \times 10^{-2}$ m

$\omega = 100$ rad/s, $m = 1$ kg

$$K_{\max} = \frac{1}{2} m\omega^2 A^2$$

$$= \frac{1}{2} \times 1 \times (100)^2 \times (6 \times 10^{-2})^2$$

$$= 18 \text{ J}$$

7. *(b)* When a charge q is projected in a perpendicular magnetic field B with velocity v, then radius of path followed is given by

$$R = \frac{mv}{qB}$$

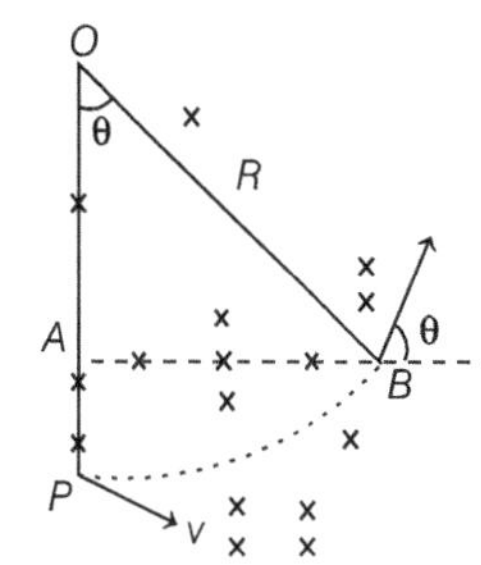

As width $\left(\frac{mv}{\sqrt{2}qB}\right)$ of field is less than radius, so path will be a part of circle.

In ΔOAB, $\sin\theta = \frac{AB}{OB}$

$$\sin\theta = \frac{\frac{mv}{\sqrt{2}qB}}{\frac{mv}{qB}} \qquad \left[\because OB = R = \frac{mv}{qB} \text{ and } AB = \frac{mv}{\sqrt{2}qB}\right]$$

$$\sin\theta = \frac{1}{\sqrt{2}} \Rightarrow \theta = \frac{\pi}{4}$$

$\therefore$ Time taken to cover 2π angle is $T = \frac{2\pi m}{qB}$

Hence, time taken to cover $\frac{\pi}{4}$ angle, is

$$t = \frac{\left(\frac{2\pi m}{qB}\right)}{2\pi} \times \frac{\pi}{4} = \frac{\pi m}{4qB}$$

8. *(b)* Given, $\lambda_1 = 12000$ Å

and $\lambda_2 = 10000$ Å

$D = 2$ m

$d = 2 \text{ mm} = 2 \times 10^{-3}$ m

We have,

$$\frac{\lambda_1}{\lambda_2} = \frac{n_2}{n_1} = \frac{12000}{10000} = \frac{6}{5}$$

5th and 6th fringes will coincide, respectively. The minimum distance is given as

$$X = \frac{n_1 \lambda_1 D}{d} \qquad (n_1 = 5)$$

$$= \frac{5 \times 12000 \times 10^{-10} \times 2}{2 \times 10^{-3}}$$

$$= 5 \times 1.2 \times 10^4 \times 10^{-10} \times 10^3 = 6 \text{ mm}$$

9. ***(c)*** Equation of motion of lower block is

$$F = \mu(M_A + M_B)g$$

$$\mu = \frac{F}{(M_A + M_B)g}$$

$$\Rightarrow \quad \mu = \frac{49}{(3+7)9.8} = \frac{1}{2} = 0.5$$

10. ***(b)*** Given, $R = 3$ km $= 3000$ m

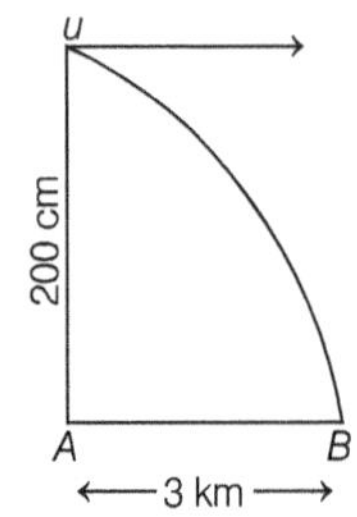

Range, $R = u\sqrt{\frac{2h}{g}}$

$$\Rightarrow \quad u = R\sqrt{\frac{g}{2h}} \quad \text{or} \quad u = 3000 \times \sqrt{\frac{10}{2 \times 2000}}$$

$$= 3000 \times \frac{1}{20}$$

$$= 150 \text{ m/s}$$

$$= 150 \times \frac{18}{5} \text{ km/h}$$

$$= 540 \text{ km/h}$$

11. ***(b)*** Here, magnetic field, $B = 0.025$ T

$\therefore$ Magnetic flux $\phi = B \cdot A = B \cdot \pi r^2$

Induced emf $|e| = \frac{d\phi}{dt}$

$$\Rightarrow \quad = B\pi\, 2r \cdot \frac{dr}{dt}$$

$$\Rightarrow \quad = 0.025 \times \pi \times 2 \times 2 \times 10^{-2} \times 1 \times 10^{-3}$$

$$\Rightarrow \quad e = \pi\mu \text{ V}$$

12. ***(a)*** Increase in length,

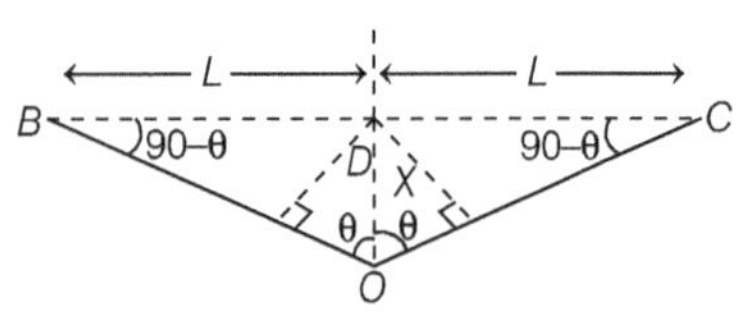

$$\Delta L = BO + OC - (BC)$$

$[\because BO = OC$ and $BC = 2BD = 2L]$

$$= 2BO - 2BD$$

$$\Rightarrow \quad \Delta L = 2BO - 2L$$

$$= 2[L^2 + x^2]^{\frac{1}{2}} - 2L$$

or $\quad \Delta L = 2L\left[1 + \frac{x^2}{L^2}\right]^{\frac{1}{2}} - 2L$

[$\because$ using Bionomial theorem]

$$\Rightarrow \quad \Delta L \approx 2L\left[1 + \frac{1}{2}\frac{x^2}{L^2} - 1\right]$$

$$= \frac{x^2}{L} \qquad [\because x << L]$$

$$\therefore \text{ Strain} = \frac{\Delta L}{2L} = \frac{\frac{x^2}{L}}{2L} = \frac{x^2}{2L^2}$$

13. ***(d)*** Given, $t_1 = 20°\text{C}$

$R_1 = 20\Omega$

$t_2 = 500°\text{C}$

$R_2 = 60\Omega$

We have, $R_t = R_0(1 + \alpha t)$

Here, $20 = R_0(1 + 20\alpha)$ ...(i)

$60 = R_0(1 + 500\alpha)$ ...(ii)

From Eqs. (i) and (ii), we get

$$\Rightarrow \quad \frac{60}{20} = \frac{R_0(1+500\alpha)}{R_0(1+20\alpha)}$$

$$\Rightarrow \quad 3 = \frac{1+500\alpha}{1+20\alpha}$$

$$\Rightarrow \quad 3(1 + 20\alpha) = 1 + 500\alpha$$

$$\Rightarrow \quad 3 + 60\alpha = 1 + 500\alpha$$

$$\Rightarrow \quad 440\alpha = 2$$

$$\Rightarrow \quad \alpha = \frac{2}{440} = \frac{1}{220}\,°\text{C}$$

Also given $R_t = 25\ \Omega$, $t = ?$

Now, $\quad \frac{20}{25} = \frac{R_0\left(1 + \frac{20}{220}\right)}{R_0\left(1 + \frac{t}{220}\right)}$ [Using Eq. (i)]

$$\Rightarrow \quad 4\left(1 + \frac{t}{220}\right) = 5\left(1 + \frac{20}{220}\right)$$

On solving, we get

$$t = 80°\text{ C}$$

14. ***(b)*** de-Broglie wavelength,

$$\lambda = \frac{h}{p} = \frac{h}{\sqrt{2mE}} = \frac{h}{\sqrt{2mqV}}$$

$$\Rightarrow \quad \lambda = \frac{6.6 \times 10^{-34}}{\sqrt{2 \times 1.6 \times 10^{-27} \times 1.6 \times 10^{-19} \times 1000}}$$

$$\Rightarrow \quad \lambda = \frac{6.6 \times 10^{-34}}{7.16 \times 10^{-22}} = 0.9 \times 10^{-12} \text{ m}$$

15. ***(c)*** Let the volume of ice-berg is V and its density is ρ. If this ice-berg floats in water with volume V_{in} inside it, then

$$V_{in}\sigma g = V\rho g \quad [\sigma = \text{density of water}]$$

$$\Rightarrow \quad V_{in} = \left(\frac{\rho}{\sigma}\right)\cdot V$$

$$\Rightarrow \quad V_{out} = V - V_{in}$$

$$\Rightarrow \quad = \left[\frac{\sigma - \rho}{\sigma}\right]V = \frac{1000 - 900}{1000}V$$

$$\Rightarrow \quad V_{out} = \frac{V}{10}$$

$$\Rightarrow \quad V_{out}/V = 0.1 = 10\%$$

16. ***(d)*** We have, kinetic energy of electron,

$$K = \frac{Ze^2}{8\pi\varepsilon_0 \cdot r}$$

also, potential energy of electron,

$$U = -\frac{1}{4\pi\varepsilon_0}\cdot\frac{Ze^2}{r}$$

$\therefore$ Total energy,

$$E = K + U = \frac{Ze^2}{8\pi\varepsilon_0 \cdot r} - \frac{Ze^2}{4\pi\varepsilon_0 \cdot r}$$

$$E = -\frac{Ze^2}{8\pi\varepsilon_0 r} \text{ or } K = -E = -(-3.4)$$

$$= 3.4 \text{ eV}$$

17. ***(c)*** Voltage gain $= \beta \times$ impedance gain

$$\Rightarrow \quad 50 = \beta \times \frac{200}{100}$$

$$\left[\because \text{Impedance gain} = \frac{\text{Output impedance}}{\text{Input impedance}}\right]$$

$$\Rightarrow \quad \beta = 25$$

Also power gain $= \beta^2 \times$ impedance gain

$$= (25)^2 \times \frac{200}{100} = 1250$$

18. ***(d)*** Time of flight, $T = \dfrac{2u_y}{g}$

Maximum height, $H = \dfrac{u_y^2}{2g}$

and horizontal range, $R = u_x \times T$

When a horizontal acceleration is given to the projectile, then u_y, T and H will remains unchanged while range will become

$$R' = u_x \times T + \frac{1}{2}aT^2$$

$$= R + \frac{1}{2}\frac{g}{4}\left(\frac{4u_y^2}{g^2}\right) = R + H \quad \left[\because a = \frac{g}{4}\right]$$

19. ***(b)*** We have, $\dfrac{p_1V_1}{T_1} = \dfrac{p_2V_2}{T_2}$ $\quad \left[\because \dfrac{pV}{T} = \text{constant}\right]$

$$\text{or } V_2 = \frac{p_1V_1T_2}{p_2T_1}$$

$$= \frac{1 \times 500 \times (273 - 3)}{0.5 \times (273 + 27)} = \frac{1 \times 500 \times 270}{0.5 \times 300}$$

$$V_2 = 900 \text{ m}^3$$

20. ***(c)*** Intensity at centre of bright fringe,

$$I_0 = I + I + 2\sqrt{II}\cdot\cos 0$$

$$I_0 = 2I + 2I \quad [\because \cos 0° = 1]$$

$$\Rightarrow \quad I_0 = 4I$$

Similarly, intensity at point distance one-fourth of the distance

(with phase difference $= \dfrac{2\pi}{4} = \dfrac{\pi}{2}$)

$$I' = I + I + 2\sqrt{II}\cos\frac{\pi}{2} = 2I + 2\sqrt{II} \times 0$$

$$I' = 2I$$

$$\therefore \quad \frac{I_0}{I'} = \frac{4I}{2I} = 2$$

21. ***(b)*** Let block is displaced through x m, then weight of displaced water or upthrust (upwards)

$$= -Ax\rho g$$

where, A is area of a cross-section of the block and ρ is density. This must be equal to force ($= ma$) applied where m is mass of the block and a is acceleration.

$$\therefore \quad ma = -Ax\rho g$$

$$\text{or} \quad a = -\frac{A\rho g}{m}x = -\omega^2 x$$

This is equation of SHM.

So, the time period of oscillation is given as

$$T = \frac{2\pi}{\omega} = 2\pi\sqrt{\frac{m}{A\rho g}} \Rightarrow T \propto \frac{1}{\sqrt{A}}$$

22. ***(c)*** Resultant force,

$$F' = \sqrt{F^2 + F^2 + 2FF\cos 120°} = F \quad \left[\because \cos 120° = \frac{-1}{2}\right]$$

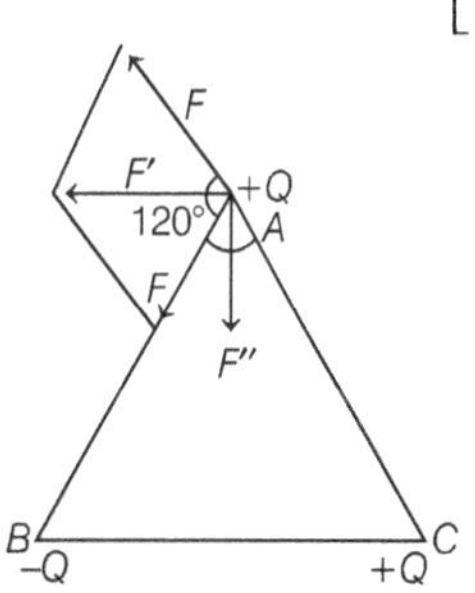

Now, from figure

$$F = \sqrt{F'^2 + F''^2 + 2F'F''\cos 90°}$$

Now, the force normal to BC at vertex A is

$$F'' = \sqrt{F^2 - F'^2} = 0 \qquad (\because F' = F)$$

23. *(b)* From the conservation principle,

$$mgh = \frac{1}{2}kX_0^2 - mgX_0$$

where, X_0 is maximum elongation in spring.

$$\Rightarrow \quad \frac{1}{2}kX_0^2 - mgX_0 - mgh = 0$$

$$\Rightarrow \quad X_0^2 - \frac{2mg}{k}X_0 - \frac{2mg}{k}h = 0$$

$$X_0 = \frac{2\frac{mg}{2} \pm \sqrt{\left(\frac{2mg}{k}\right)^2 + 4 \times \frac{2mg}{k}h}}{2}$$

Amplitude = elongation in spring for lowest extreme position – elongation in spring for equilibrium position

$$= X_0 - X_1 = \frac{mg}{k}\sqrt{1 + \frac{2hk}{mg}} \qquad \left[\because X_1 = \frac{mg}{k}\right]$$

24. *(d)* Fraction remains after n half-lives

$$\frac{N}{N_0} = \left(\frac{1}{2}\right)^n = \left(\frac{1}{2}\right)^{\frac{t}{T}}$$

Given, $N = \dfrac{N_0}{e} \Rightarrow \dfrac{N_0}{eN_0} = \left(\dfrac{1}{2}\right)^{\frac{5}{T}}$

$$\Rightarrow \quad \frac{1}{e} = \left(\frac{1}{2}\right)^{\frac{5}{T}}$$

Taking log on both sides, we get

$$\log 1 - \log e = \frac{5}{T}\log\frac{1}{2}$$

$$\Rightarrow \quad -1 = \frac{5}{T}(-\log 2)$$

$$\Rightarrow \quad T = 5\log_e 2$$

Now, let t' be the time after which activity reduces to half

$$\left(\frac{1}{2}\right) = \left(\frac{1}{2}\right)^{t'/5\log_e 2}$$

$$\Rightarrow \quad t' = 5\log_e 2$$

25. *(c)* Focal length of combination,

$$\frac{1}{f} = \frac{1}{f_1} + \frac{1}{f_2} \qquad \text{...(i)}$$

Using lens Maker's formula,

$$\frac{1}{f_2} = (\mu_2 - 1)\left(\frac{1}{-R} - \frac{1}{\infty}\right) = -\frac{(\mu_2 - 1)}{R}$$

$$\frac{1}{f_1} = (\mu_1 - 1)\left(\frac{1}{\infty} - \frac{1}{-R}\right) = \frac{\mu_1 - 1}{R}$$

or $$\frac{1}{f_1} = \frac{(\mu_1 - 1)}{R}$$

$$\frac{1}{f_2} = -\frac{(\mu_2 - 1)}{R}$$

Putting these values in Eq (i), we get

$$\Rightarrow \quad \frac{1}{f} = \frac{(\mu_1 - 1)}{R} - \frac{(\mu_2 - 1)}{R}$$

$$\Rightarrow \quad \frac{1}{f} = \frac{(\mu_1 - 1 - \mu_2 + 1)}{R}$$

$$\Rightarrow \quad \frac{1}{f} = \frac{\mu_1 - \mu_2}{R}$$

$$\Rightarrow \quad f = \frac{R}{(\mu_1 - \mu_2)}$$

26. *(c)* Current will flow only through the branch containing resistance R_2. (Because at stedy state, capacitor works as open circuit, hence branch containing resistances R_1 and R_2 is ineffective).

$$\therefore \quad i = \frac{E}{R_2 + r} = \frac{5}{4 + 1} = 1\text{ A}$$

Potential difference across R_2

$$= 1 \times 4 = 4\text{ V}$$

If q be the charge on each plate of capacitor, then

$$\frac{q}{C} + \frac{q}{C} = 4$$

or $$\frac{2q}{C} = 4$$

or $$\frac{2q}{3 \times 10^{-6}} = 4$$

$$q = 6\,\mu\text{C}$$

27. *(c)* Friction force between A and B and between block B and surface S will oppose F

$$\therefore \quad F = F_{AB} + F_{BS}$$

$$= \mu_{AB}m_Ag + \mu_{BS}(m_A + m_B)g$$

$$= 0.2 \times 100 \times 10 + 0.3(100 + 200) \times 10$$

$$= 200 + 900 = 1100\text{ N}$$

28. *(a)* The moment of inertia of the uniform rod about an axis through one end and perpendicular to its length is

$$= \frac{ml^2}{3}$$

where, m is the mass and l its length.

Torque $(\tau = I \cdot \alpha)$ acting on centre of gravity of rod is given by

$$\tau = mg\frac{l}{2}$$

also, $$\frac{ml^2}{3} \cdot \alpha = mg\frac{l}{2}$$

$$\Rightarrow \quad \alpha = \frac{3g}{2l}$$

BITSAT 2017

29. *(a)* As H-atom emits 6 spectral lines, so $\frac{n(n-1)}{2} = 6$

$\therefore \quad n = 4$

Now, $\frac{1}{\lambda} = R\left[1 - \frac{1}{4^2}\right]$

$\Rightarrow \quad \frac{1}{\lambda} = R\left[\frac{15}{16}\right] \Rightarrow \lambda = \frac{16}{109677 \times 15}$ cm

$\Rightarrow \quad \lambda = 97.2\,\text{nm}$

30. *(a)* Torque acting on equilateral triangle in a magnetic field B is

$\tau = i\,AB\sin\theta$

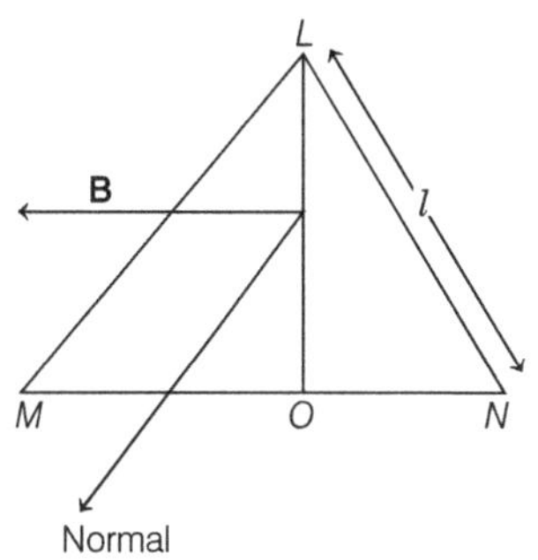

Area of triangle LMN,

$$A = \frac{\sqrt{3}}{2}l^2 \text{ and } \theta = 90°$$

Substituting the given values in expression for torque, we get

$$\tau = i \times \frac{\sqrt{3}}{4}l^2 B\sin 90°$$

$$\tau = \frac{\sqrt{3}}{4}il^2 B \qquad (\because \sin 90° = 1)$$

Here, $l = 2\left(\frac{\tau}{\sqrt{3}\,Bi}\right)^{1/2}$

31. *(d)* Work done = Change in surface energy

$r_2 = 5$ cm

$r_1 = 3$ cm

$\Rightarrow \quad W = 2T \times 4\pi(r_2^2 - r_1^2)$

$= 2 \times 0.03 \times 4\pi[(5)^2 - (3)^2] \times 10^{-4}$

$= 0.4\pi$ mJ

32. *(c)* From the figure, the X-component remains unchanged while the Y-component is reverse. Then, the velocity at point B is $(2\hat{\mathbf{i}} - 3\hat{\mathbf{j}})$ m/s.

33. *(b)* The equivalent circuits diagram is

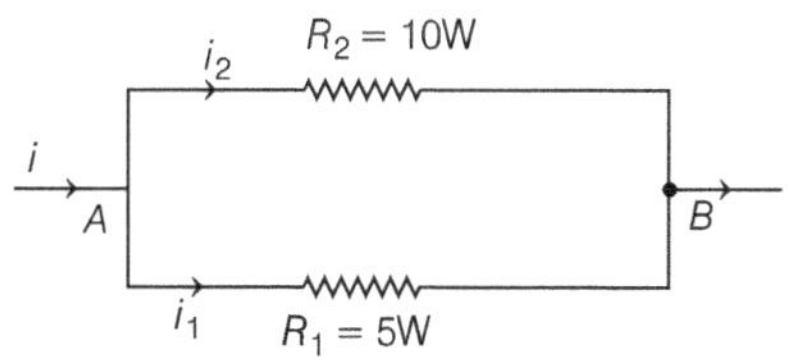

Potential drop is same for both braveles

$\therefore \quad i_1R_1 = i_2R_2$

$i_1 \times 5 = i_2 \times 10$

$\Rightarrow \quad \frac{i_1}{i_2} = 2$

Now, heat produced is given by

$$H = i^2RT$$

$\therefore \quad \frac{H_1}{H_2} = \left(\frac{i_1}{i_2}\right)^2 \times \left(\frac{R_1}{R_2}\right) \qquad [\because t = 1s]$

or $\quad \frac{10}{H_2} = (2)^2 \times \left(\frac{5}{4}\right) \qquad [\because R_2 = 4\Omega]$

$\Rightarrow \quad H_2 = 2$ cal

34. *(a)* Given KE of α-particle = 2 eV

$$r = \frac{(Ze)(e)}{4\pi\varepsilon_0 \cdot (\text{KE})}$$

$$= \frac{2Ze \times 9 \times 10^9}{2V}$$

$\Rightarrow \quad r = \frac{2 \times Z \times 1.6 \times 10^{-19} \times 9 \times 10^9}{2V}$

$\Rightarrow \quad r = 14.4 \cdot \frac{Z}{V}$ Å

35. *(c)* According to question,

Time period of oscillation of a simple series is

$$T = 2\pi\sqrt{\frac{l}{g}} \Rightarrow T \propto \sqrt{l}$$

$\therefore$ Ratio to time period, $\frac{T_1}{T_2} = \frac{1}{2}$

$\Rightarrow \quad T_2 = 2T_1$

Thus, the shorter pendulum would have completely exactly 2 oscillations when the longer pendulum complete first oscillation.

36. *(c)* The two capacitors formed are in series, hence capacitance of the combination

$$C = \frac{C_1C_2}{C_1 + C_2} \qquad \text{...(i)}$$

where, $\quad C_1 = \frac{K_1\varepsilon_0 A}{\frac{d}{3}} \qquad \left(\because C = \frac{\varepsilon_0 A}{d}\right)$ ...(ii)

$$C_2 = \frac{K_2\varepsilon_0 A}{\frac{2d}{3}} \qquad \text{...(iii)}$$

$$C_{eq} = \frac{\frac{3K_1\varepsilon_0 A}{d} \times \frac{K_2\varepsilon_0 A \cdot 3}{2d}}{\frac{3K_1\varepsilon_0 A}{d} + \frac{3K_2\varepsilon_0 A}{2d}}$$

[Using Eq. (i) and (ii)]

$$\Rightarrow \quad C_{eq} = \frac{9}{2}\frac{K_1K_2}{(6K_1+3K_2)}\cdot\left(\frac{\varepsilon_0 A}{d}\right)$$

It is given

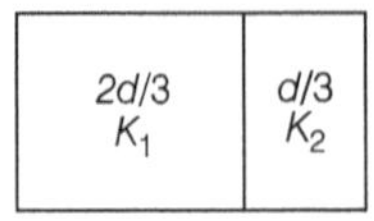

$$\frac{\varepsilon_0 A}{d} = 9\,\text{pF}$$

Using given values, $C_{eq} = 40.5$ pF

37. ***(d)*** Acceleration

$$f = \frac{dv}{dt} = f_0\left(1-\frac{t}{T}\right)$$

or $\quad dv = f_0\left(1-\frac{t}{T}\right)\cdot dt \quad$ ...(i)

Integrating Eq. (i) on both sides, we get

$$v = f_0 t - \frac{f_0}{T}\cdot\frac{t^2}{2} + C \quad \text{...(ii)}$$

After applying boundary conditions

$v = 0$ at $t = 0$

We get, $C = 0$

$$\Rightarrow \quad v = f_0 t - \frac{f_0}{T}\cdot\frac{t^2}{2} \quad \text{...(iii)}$$

As $f = f_0\left(1-\frac{t}{T}\right)$

When $f_0 = 0, t = T$

Substituting $t = T$ in Eq. (iii), then velocity

$$v_X = f_0 T - \frac{f_0}{T}\cdot\frac{T^2}{2} = \frac{1}{2}f_0 T$$

38. ***(c)*** By Kepler's law of planetary motion,

$T^2 \propto r^3$, hence $T_1^2 \propto r_1^3$

Simiarly $T_2^2 \propto r_2^3$

Given, $r_2 = 6400$ km

$r_1 = 36000$ km

For a geostationary satellite, $T = 24$ h

$\therefore \quad T_1^2 \propto (36000)^3$

$T_2^2 \propto (6400+h)^3$

Therefore,

$$(T_2)^2 = T_1^2\left(\frac{6400+h}{36000}\right)^3$$

$$T_2 = T_1\left(\frac{6400}{36000}\right)^{3/2} \text{ Neglecting } h\ [\because h <<< R]$$

$$\therefore \quad T_2 = (24)\left(\frac{64}{360}\right)^{3/2} \Rightarrow T_2 = 1.8\text{ h}$$

$T_2 \simeq 2$ hr

39. ***(b)*** Equation of wave,

$$y = 0.2\sin[1.5x + 60t]$$

Comparing with standard equation,

$$y = A\sin[kx + \omega t]$$

we get $\quad k = 1.5 = \frac{2\pi}{\lambda}$ and $\omega = 60 = \frac{2\pi}{T}$

$\therefore$ Velocity of wave, $v = \frac{\omega}{k} = \frac{60}{1.5} = 40$ m/s

Velocity of wave in stretched string, $v = \sqrt{T/m}$,

where, m is the linear density, T is tension in the string.

So, $T = v^2 m = (40)^2 \times 3\times 10^{-4}$

$= 0.48$ N

40. ***(b)*** Given, radius of circular loop is r

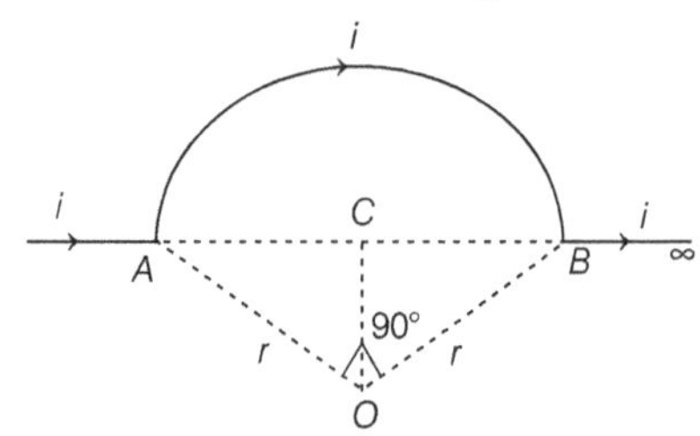

In ΔAOB

$$AB^2 = AO^2 + OB^2$$

$$AB^2 = r^2 + r^2$$

$$\Rightarrow \quad AB = \sqrt{2}r$$

In ΔAOC,

$$AO^2 = OC^2 + AC^2$$

$$\Rightarrow \quad OC^2 = r^2 - \left(\frac{r}{\sqrt{2}}\right)^2 \quad \left[\because AC = \frac{AB}{2} = \frac{r}{\sqrt{2}}\right]$$

$$\Rightarrow \quad OC = AC = \frac{r}{\sqrt{2}}$$

Now, magnetic field due to straight wire is

$$B_1 = \frac{\mu_0 i}{4\pi a}[\sin\phi_1 + \sin\phi_2] = \frac{\mu_0 i}{4\pi\times\frac{r}{\sqrt{2}}}$$

[sin 90°+sin 90°]

$$\Rightarrow \quad B_1 = \frac{\mu_0 i\times\sqrt{2}}{2\pi r} \quad \text{...(i)}$$

And magnetic field due to circular path is

$$B_2 = \frac{\mu_0 i}{2r}\times\frac{1}{4} \quad \text{...(iii)}$$

[$\because$ Circular path is quadrant of circle]

Net magnetic field at point O is $B = B_1 + B_2$

$$\Rightarrow \quad B = \frac{\sqrt{2}\mu_0 i}{2\pi r} + \frac{\mu_0 i}{8r} = \frac{\mu_0}{4\pi}\cdot\frac{2i}{r}\left[\sqrt{2}+\frac{\pi}{4}\right]$$

Chemistry

41. ***(b)*** $\because$ 4 g of copper gave 5g of its oxide means,

1 g of oxygen combine with 4 g of copper.

$\because$ Eq. wt of oxygen = 8,

Therefore, 8g of oxygen combine with 4×8 g of copper = 32 g

Hence, equivalent weight of copper = 32

42. ***(b)*** Option (b) is the incorrect statement and can be corrected as, reaction of $NaBH_4$ with cold water is very slow. It is not violent. All other are correct statements.

43. ***(b)*** $\because$ For s-subshell, $l = 0$

Hence, orbital angular momentum of an electron in 2s orbital is $= \sqrt{l(l+1)} \cdot h / 2\pi = \sqrt{0(0+1)^{\frac{h}{2\pi}}} = 0$

Thus, the angular orbital momentum of an electron in 2s orbital is 0.

44. ***(b)*** I. Number of electrons in $\overset{+}{C}H_3 = 6 + 3 - 1 = 8$

II. Number of electrons in $NH_2^- = 7 + 2 + 1 = 10$

III. Number of electrons in $NH_4^+ = 7 + 4 - 1 = 10$

IV. Number of electrons in $NH_3 = 7 + 3 = 10$

Since, species with same number of electrons are called isoelectronic species. Hence, II, III and IV are isoelectronic.

45. ***(a)*** $\because$ Difference of electronegativity between (A) and (B) $= 3.0 - 1.2 = 1.8$ and a bond having electronegativity difference greater than 1.65 is of ionic nature.

Hence, bond between (A) and (B) would be of ionic nature.

46. ***(c)*** The hydrogen, attached with sp-hybrid C-atom is most acidic because of highest electronegativity of sp hybrid carbon atom, H atom becomes relatively more acidic.

47. ***(a)*** Reductive ozonolysis of $(H_3C)_2C{=}C(CH_3)_2$ followed by hydrolysis gives only one type of ketones because of symmetry across the C = C and bonding of two same alkyl groups (CH_3) with both the carbons bonded through double bond.

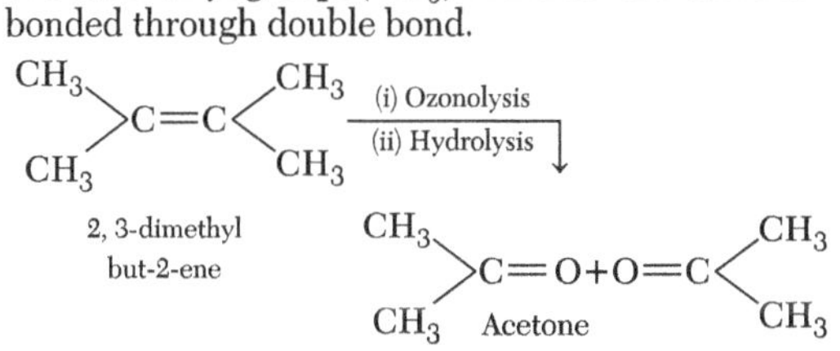

48. ***(d)*** Reaction (I), (II) and (IV) are of exothermic nature, thus do not involve absorption energy.

49. ***(b)*** More be the ability of N-atom of —NH_2 group to donate lone pair of electrons to HCl (dil.), more be the reactivity of amine.

In (b) two alkyl groups are attached to the —NH_2 (having electron releasing nature) which increases the electron donating ability of N-atom of —NH_2.

Hence (b) is most reactive.

Note (i) (b) i.e. $(CH_3)_3N$ has three alkyl groups attached with the N-atom, but due to steric-effect, N-atom cannot able to donate lone pair of electrons.

(ii) In (d), due to resonance, lone pair of electrons get delocalised.

50. ***(d)*** Blocks of magnesium prevents the action of water and salt of iron (of ship) by sacrificing itself.

This method of protection of ship (i.e. of iron) is known as sacrificial protection.

51. ***(d)*** $\because$ Only one mole of total gases are evolved and H_2O decomposes as

$$H_2O \xrightarrow{\text{Decomposition}} H_2 + \frac{1}{2}O_2$$

i.e. total moles $= 1 + \frac{1}{2} = \frac{3}{2}$ mole by one mole of H_2O.

Now,

$\because$ 3/2 moles of gases are evolved by = 1 mole of H_2O

$\therefore$ 1 mole of gases are evolved by $= \frac{1}{3/2} = \frac{2}{3}$ mole of H_2O.

52. ***(b)*** $\because$ Reaction between Fe and HCl is as follows

$$Fe + 2HCl \longrightarrow \underset{(+2)}{FeCl_2} + H_2$$

n (moles of Fe) = n (moles of HCl)

n(Fe) = Normality (HCl) × Volume of HCl

$$n(\text{Fe}) = \frac{0.4 \times 500}{1000} = 0.2 \text{ mol}$$

As each Fe consumer (2 electrons) to change the Fe^{2+} ion.

$\therefore$ Number of Fe^{2+} ions produced $= \frac{0.2}{2} = 0.1$ mol

53. ***(c)*** For dissolution hydration energy must be more higher (–ve) than of lattice energy. Thus, (c) is the correct answer.

54. ***(d)*** In the given reaction the end product is propyl amine. It is obtained as follows

$$CH_3CH_2Cl \xrightarrow{KCN} \underset{(A)}{CH_3CH_2CN} \xrightarrow[\text{(Reduction)}]{LiAlH_4} \underset{\text{propyl amine } (P)}{CH_3CH_2CH_2NH_2}$$

55. ***(d)*** Conversion of carbonate ore into its oxide (by heating in absence of air) is known as calcination.

$$MgCO_3 \xrightarrow{\Delta} MgO + CO_2(g)$$

56. ***(c)*** We know that, $\Delta H = E_f - E_b$

where, E_b and E_f = activation energy for backward and forward respectively.

Thus, $E_f = \Delta H + E_b$, means $E_f > \Delta H$.

57. ***(a)*** Ag is leached by cyanide process by making the complex using NaCN/KCN as follows

$$Ag_2S + 4NaCN \longrightarrow \underset{\text{(Soluble)}}{2Na[Ag(CN)_2]} + Na_2S$$

58. ***(a)*** Among the given option (a) is a diamagnetic complex.

(a) $[Co(NH_3)_6]^{3+} \Rightarrow Co\ (27) \Rightarrow Co^{3+} = 3d^6 4s^0$. Thus, has no unpaired electrons and is a diamagnetic complex.

(b) $[NiCl_4]^{2-} \Rightarrow Ni\ (28) \Rightarrow Ni^{2+} = 3d^9 4s^0$

Thus, has two unpaired electron and is paramagnetic.

(c) $[CuCl_4]^{2-} \Rightarrow Cu\ (29) \Rightarrow Cu^{2+} = 3d^9 4s^0$

Thus, has one unpaired electron and is paramagnetic.

(d) $[Fe(H_2O)_6]^{3+} \Rightarrow Fe\ (26) \Rightarrow Fe^{3+} = 3d^5 4s^0$

Thus, has five unpaired electrons and is paramagnetic.

59. ***(b)*** Neoprene is a synthetic rubber. It is a polymer of chloroprene (2-chlorobuta,1, 3-diene).

60. ***(a)*** Ionic solids have highest melting point due to strong electrostatic force of attraction.

61. ***(d)*** Night-blindness is developed due to deficiency of vitamin A.

62. ***(a)*** The transfer RNA anticodon for messanger RNA codon G – C – A is

C – G – U, because in anticodon of RNA,

G replaced by C, C replaced by G and A replaced by U

63. ***(a)*** For % of CO_2,

$\because$ 0.765 g of acid gives CO_2 = 0.535 g

$\therefore$ 100 g of acid gives $CO_2 = \dfrac{0.535 \times 100}{0.765}$

% of CO_2 = 70.00

For % of carbon,

$\because$ 44 g of CO_2 gives, C = 12 g

$\therefore$ 70 g of CO_2 gives, $C = \dfrac{12 \times 70}{44} = 19.09$ g

$\therefore$ % of carbon = 19.00

Similarly, for % of $H_2O = \dfrac{0.138 \times 100}{0.765}$

= 18.03 = 18.00

% of hydrogen $= \dfrac{2 \times 18}{18} = 2.00$

Hence, ratio of carbon and hydrogen is 19 : 2.

64. ***(d)*** Least basic has maximum value of pK_b.
Since, in option (d), lone pair of electrons over N-atom are delocalised to two benzene rings, thus basic.

65. ***(c)*** For principal quantum number (n), l (azimuthal quantum number) can be upto ($n-1$) and thus, $m = \pm l$, i.e. $+l, 0, -l$

(a) $n = 2, l = 0, m = 0$ (correct)

(b) $n = 1, l = 0, m = 0$ (correct)

(c) $n = 3, l = 3, m = 0$ (not correct)

(d) $n = 2, l = 1, m = 1$ (correct)

66. ***(d)*** More higher be the positive oxidation number of N-atom in its oxide more acidic be the nature of oxide. Thus, N_2O_5 is most acidic oxide.

Oxide	Oxidation state of N
NO_2	+4
N_2O	1
NO	2
N_2O_5	5

67. ***(c)*** $\because$ Bond order (BO) $= \dfrac{N_b - N_a}{2}$

Thus, for O_2, BO $= \dfrac{10-6}{2} = 2$

for O_2^-, BO $= \dfrac{10-7}{2} = 1.5$; for O_2^+, BO $= \dfrac{10-5}{2} = 2.5$

for O_2^{2-}, BO $= \dfrac{10-8}{2} = 1.0$

Hence, maximum value of BO is for O_2^+.

68. ***(c)*** Option (c) is the correct option as in (I), (II) and (IV) the state changes towards more random state, thus entropy increases.

69. ***(a)*** BF_3 is a Lewis acid, as it accepts lone pair of electrons. It is not an Arrhenius acid because, it does not furnish H^+ ion in its aqueous solution. It is also not a Bronsted Lowry acid as it does not give H^+ ions in any solvent.

70. ***(a)*** $\because \Delta H = \Delta E + \Delta n_g RT$

and Δn_g (for the given reaction) = 2 – 1 = +1

$\therefore \Delta H > \Delta E$

71. ***(d)*** First member of 1st group i.e. lithium, mostly form covalent compounds due to its small size and comparatively high I.E. (with respect to other members of its group).

72. ***(b)*** When aqueous solution of borax ($Na_2B_4O_7$) is acidified with HCl, we get H_3BO_3 and NaCl as follows

$$\underset{\text{Borox}}{Na_2B_4O_7} + 2HCl + 5H_2O \longrightarrow 2NaCl + \underset{\text{Boric acid}}{4H_3BO_3}$$

73. ***(d)*** An aromatic compound can follow Huckel's rule, if it has $(4n+2)\pi$ electrons.

(where, n = integer 0, 1, 2, 3, ... etc.)

Thus,

(a) $n = 0$; number of π-electrons $= 4 \times 0 + 2 = 2\pi$

$\therefore$ Follow Huckel's rule.

(b) $n = 1$; number of π-electrons = 6π

$\therefore$ Follows Huckel's rule.

(c) $n = 1$; number of π-electrons = 6π

$\therefore$ Follows Huckel's rule.

(d) $n = 1$, number of π-electrons = 8π

$(4n + 2 = 4 \times 1 + 2 = 6)$

But this compound has 8 electrons which distorts the plane and hence does not follows Huckel's rule.

74. ***(b)*** According to Arrhenius equation's

$$\log k = \log A - \frac{E_a}{2.303R} \cdot \frac{1}{T}$$

On comparing it with the equation of straight line with negative slope, $y = mx + c$

The value of $\log k$ should increase uniformily with T or decreases with $\frac{1}{T}$.

Thus, plot (b) is the correct answer.

75. ***(d)*** CN^- is a strong ligand and form six coordinate bonds with Fe element. Thus, Fe shows d^2sp^3 hybridisation in $[K_3Fe(CN)_6]$.

76. ***(d)*** More the number of unpaired electrons more is the magnetic moment.

Number of unpaired electrons in $Mg^{2+} = 0$ [Ne]

Number of unpaired electrons in $Ti^{3+} = 1$ $[Ar]3d^1$

Number of unpaired electrons in $V^{3+} = 2$ $[Ar]3d^2$

Number of unpaired electrons in $Fe^{2+} = 4$ $[Ar]3d^6$

Thus, Fe^{2+} show maximum value of magnetic moment (μ). Also, $\mu = \sqrt{n(n+2)}$ BM(where, n = number of unpaired electrons).

77. ***(a)*** Elements with atomic number 90 to 103 belongs to same group, i.e. III B group (i.e. actinoids).

78. ***(c)***

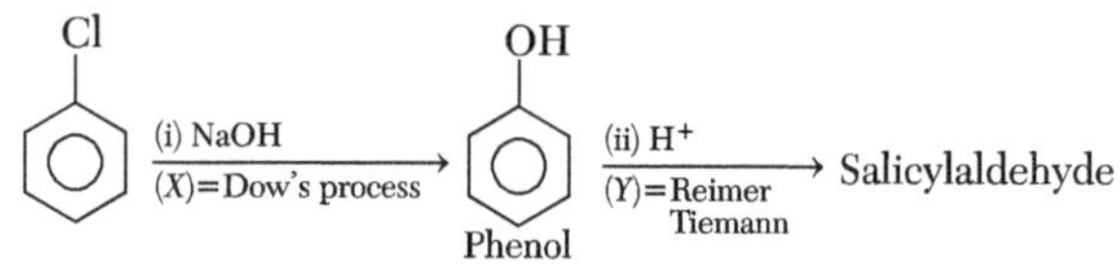

79. ***(d)*** $\because n = \frac{10}{m} = \frac{\text{actual mass}}{\text{molar mass}}$

For

(a) $n = \frac{w}{m} = \frac{8}{16} = \frac{1}{2}$ mol

(b) $n = \frac{16}{32} = \frac{1}{2}$ mol

(c) $n = \frac{14}{28} = \frac{1}{2}$ mol

(d) Thus, all have same number of moles.

80. ***(a)*** $\because$ For ideal gas,

$$pV = nRT \quad \text{or} \quad T = \frac{pV}{nR}$$

$\because$ $n = 1$ = constant (Given)

R = gas constant

$\therefore$ $T \propto pV$

Thus, the gas that shows highest product of $p \times V$, has highest temperature.

For gas $(A) = p \times V = 2 \times 5 = 10$

$(B) = p \times V = 2 \times 2.5 = 5$

$(C) = p \times V = 2 \times 1.25 = 2.5$

$(D) = p \times V = 2.5 \times 2.5 = 6.25$

Hence, gas (A) shows highest temperature.

BITSAT 2017

a. English Proficiency

81. ***(b)*** 'Are' should be used in place of 'is'.

82. ***(a)*** 'Was' should be used in place of 'were'.

83. ***(b)*** 'Luggage' should be used in place of 'luggages'.

84. ***(c)*** Gloss over something is used.

85. ***(b)*** Preposition *up* is used with muster.

86. ***(b)*** **Illusion** means a false idea or belief; a deceptive appearance or impression. Its antonym will be **reality.**

87. ***(b)*** **Disparage** means to criticise someone in a way that shows that you do not respect or value him. Its antonym will be **praise.**

88. ***(b)*** **Slackened** means reduce or decrease in speed or intensity. Its antonym will be **quickened**.

89. ***(b)*** **Deteriorating** means becoming progressively worse. So, 'worsening' is its nearest meaning word.

90. ***(b)*** **Stalemate** means a situation in which neither group involved in an argument or an act can win or get an advantage and no action can be taken.

91. ***(b)*** The southampton scheme requires convicted drivers to attend eight driving sessions-one a week.

92. ***(d)*** John Cook devised the scheme to prove that alcohol does influence driving.

93. ***(d)*** The problem with a quarter of the people who went to John Cook was that they did not know how much they were drinking.

94. ***(a)*** Most drivers start off with atleast a double measure.

95. ***(c)*** The truth is that alcohol affects the body and the brain.

b. Logical Reasoning

96. ***(c)*** 'Leather' is a raw material used to make 'Shoes'. Same as, 'Rubber' is made using 'Latex' which is a raw material.

97. ***(d)*** Here, $81 \times 3 = 243$

$25 \times 3 = 75$

$\Rightarrow \quad 64 \times 3 = 192$

but $\quad 16 \times 3 = 48 \neq 64$

We can see that in all options second number is three times the first number except 16 : 64.

Hence, 16 : 64 is odd one.

98. ***(a)*** The sequence to given series is as follows

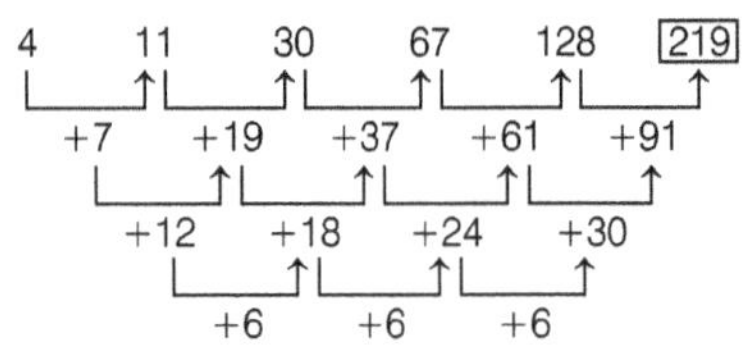

Hence, 219 will come in place of question mark'?'.

99. ***(d)*** From given information, we have

Lakshmi > Meenu ...(i)

Lakshmi > Leela > Meenu ...(ii)

Meenu > Hari > Latha ...(iii)

After arranging the equations,

Lakshmi > Leela > Meenu > Hari > Latha

Hence, Latha is the youngest.

100. ***(a)*** Here, all the three equal parts have same design. So, we can obtain the answer figure for the missing portion by rotating question figure by 90° clockwise. Hence, option(a) is the right answer.

101. ***(a)*** Out of five given figures, we can take figure (A), (C) and (D) to make a square.

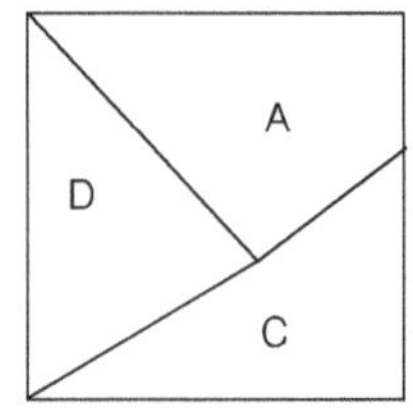

102. ***(b)*** After folding, paper step by step and then cut as shown in the question, it will show as an option (b).

103. ***(c)*** Moving row wise the number of horizontal lines increases by one in each step continuing the process, we will get figure of option (c) which will replace the question mark'?'.

104. ***(c)*** Required figure given in option (c).

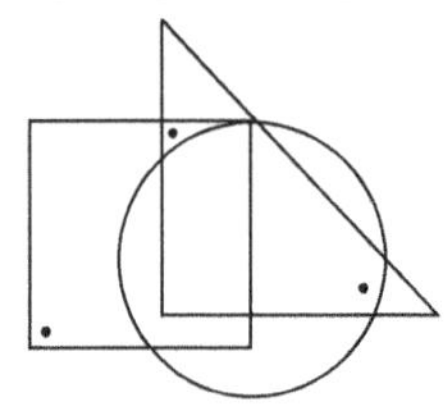

105. ***(a)*** Naming the figure,

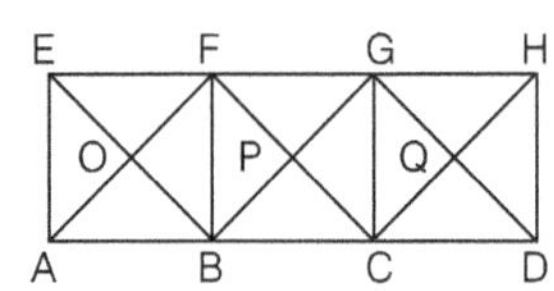

Clearly, there are 28 triangles in the given figure namely, ΔEOF, ΔAOE, ΔAOB, ΔBOF, ΔABF, ΔBEF, ΔABE, ΔAEF, ΔBPF, ΔFPG, ΔCPG, ΔBPC, ΔBFG, ΔBCG, ΔCFG, ΔBCF, ΔGQC, ΔCDQ, ΔDQH, ΔGQH, ΔGDC, ΔGDH, ΔGHC, ΔCDH, ΔAFC, ΔBGD, ΔEBG and ΔFCH.

Mathematics

106. ***(c)*** $(1+x)^{21} + (1+x)^{22} + \ldots\ldots + (1+x)^{30}$

$$= (1+x)^{21}\,[1 + (1+x)^1 + \ldots. + (1+x)^9]$$

It forms a GP series with $r = (1+x)$

$$= (1+x)^{21}\left[\frac{(1+x)^{10} - 1}{(1+x) - 1}\right] = \frac{1}{x}[(1+x)^{31} - (1+x)^{21}]$$

$\therefore$ Coefficient of x^5 in the given expression

$$= \text{Coefficient of } x^5 \text{ in } \frac{1}{x}[(1+x)^{31} - (1+x)^{21}]$$

$$= \text{Coefficient of } x^6 \text{ in} [(1+x)^{31} - (1+x)^{21}]$$

$$= {}^{31}C_6 - {}^{21}C_6$$

107. ***(b)*** We have,

$\arg(z-1) = \arg(z+3i) \qquad [\because z = a+ib]$

$\Rightarrow \arg((a-1)+ib) = \arg(a+(b+3)\,i)$

$$\Rightarrow \tan^{-1}\left(\frac{b}{a-1}\right) = \tan^{-1}\left(\frac{b+3}{a}\right)$$

$$\Rightarrow \quad \text{ttt } \frac{b}{a-1} = \frac{b+3}{a} \Rightarrow ab = (a-1)(b+3)$$

$\Rightarrow \quad 3(a-1) = b$

$\Rightarrow \quad (a-1) : b = 1 : 3$

108. ***(a)*** Let the ellipse be $\frac{x^2}{a^2} + \frac{y^2}{b^2} = 1$

and let $\quad \frac{x\cos\theta}{a} + \frac{y\sin\theta}{b} = 1 \qquad \ldots(i)$

be a tangent to it at point $(a\cos\theta, b\sin\theta)$.
Then, p = length of the perpendicular from $S(ae, 0)$ on Eq. (i)

$$\Rightarrow \quad p = \left| \frac{e\cos\theta - 1}{\sqrt{\frac{\cos^2\theta}{a^2} + \frac{\sin^2\theta}{b^2}}} \right| \Rightarrow ap = \frac{ae\cos\theta - a}{\sqrt{\frac{\cos^2\theta}{a^2} + \frac{\sin^2\theta}{b^2}}}$$

p' = length of the perpendicular from $O(0, 0)$ on Eq.(i)

$$\Rightarrow \quad p' = \left| \frac{1}{\sqrt{\frac{\cos^2\theta}{a^2} + \frac{\sin^2\theta}{b^2}}} \right| \text{ and } r = ae\cos\theta - a$$

$$\Rightarrow \quad rp' = \frac{ae\cos\theta - a}{\sqrt{\frac{\cos^2\theta}{a^2} + \frac{\sin^2\theta}{b^2}}}$$

Clearly, $rp' = ap$

109. ***(a)*** We have,

$$\sum_{r=1}^{10} r \cdot \frac{{}^nC_r}{{}^nC_{r-1}} = \sum_{r=1}^{10} (n - r + 1)$$

$$\Rightarrow \quad \sum_{r=1}^{10} r \cdot \frac{{}^nC_r}{{}^nC_{r-1}} = \sum_{r=1}^{10} \{(n+1) - r\}$$

$$\Rightarrow \quad \sum_{r=1}^{10} r \cdot \frac{{}^nC_r}{{}^nC_{r-1}} = 10(n+1) - \sum_{r=1}^{10} r$$

$$\Rightarrow \quad \sum_{r=1}^{10} r \cdot \frac{{}^nC_r}{{}^nC_{r-1}} = 10(n+1) - 55$$

$$= 10n - 45 = 5(2n - 9)$$

110. ***(d)*** Since, $3^{2\sin 2\alpha - 1}$, 14 and $3^{4 - 2\sin 2\alpha}$ are in AP.
Therefore,

$$2 \times 14 = 3^{2\sin 2\alpha - 1} + 3^{4 - 2\sin 2\alpha}$$

$$\Rightarrow \quad 28 = \frac{a}{3} + \frac{3^4}{a}, \text{ where } a = 3^{2\sin 2\alpha}$$

$$\Rightarrow \quad a^2 - 84a + 243 = 0$$

$$\Rightarrow \quad (a - 81)(a - 3) = 0 \Rightarrow a = 81, a = 3$$

$$\Rightarrow \quad 3^{2\sin 2\alpha} = 3^4 \text{ or } 3^{2\sin 2\alpha} = 3$$

$$\Rightarrow \quad 2\sin 2\alpha = 1 \qquad [\because 2\sin 2\alpha \neq 4]$$

$$\Rightarrow \quad \sin 2\alpha = 1/2 \Rightarrow 2\alpha = 30° \qquad [\because \sin 30° = 1/2]$$

Thus, the first three terms of the AP are 1, 14, 27.

Hence, its fifth term $a_5 = a_1 + (5 - 1)d$

$$= 1 + 4 \times 13 = 1 + 52 = 53$$

111. ***(c)*** Let α, α^2 be the roots of $3x^2 + px + 3$

$\therefore$ Product of the roots,

$$\alpha \cdot \alpha^2 = 3/3 \Rightarrow \alpha^3 = 1$$

$$\Rightarrow \quad \alpha = 1, \omega, \omega^2$$

Where, $\omega = -\frac{1 + \sqrt{3}i}{2}$ and $\omega^2 = \frac{-1 - \sqrt{3}i}{2}$

Again, $\quad \alpha + \alpha^2 = -\frac{p}{3}$

$$\Rightarrow \quad 1 + 1 = -\frac{p}{3} \qquad (\text{if } \alpha = 1)$$

$$\Rightarrow \quad p = -6$$

But $p > 0$

$\therefore \ \alpha = 1$ is not possible.

If $\alpha = \omega$, then $\alpha + \alpha^2 = \omega + \omega^2 = -1$

$$\{\because 1 + \omega + \omega^2 = 0\}$$

$$\therefore -1 = -\frac{p}{3} \Rightarrow p = 3$$

Again, if $\alpha = \omega^2$, then

$$\alpha + \alpha^2 = \omega^2 + \omega^4 = \omega^2 + \omega = -1$$

$$[\because \omega^3 = 1, \omega^4 = \omega.\omega^3 = \omega]$$

$$\therefore \quad -1 = -\frac{p}{3} \Rightarrow p = 3$$

112. ***(d)*** We have,

$$\log_{140} 63 = \log_{2^2 \times 5 \times 7}(3 \times 3 \times 7)$$

$$= \frac{\log_2(3 \times 3 \times 7)}{\log_2(2^2 \times 5 \times 7)} = \frac{\log_2 3 + \log_2 3 + \log_2 7}{2\log_2 2 + \log_2 5 + \log_2 7}$$

$$= \frac{2a + \frac{1}{c}}{2 + b + \frac{1}{c}} = \frac{2ac + 1}{2c + bc + 1} \quad \left[\because \log_2 7 = \frac{1}{\log_7 2}\right]$$

113. ***(a)*** Given that, $\cos(x - y)$, $\cos x$, $\cos(x + y)$ are in HP.

Then, $\cos x = \frac{2\cos(x - y)\cos(x + y)}{\cos(x - y) + \cos(x + y)}$

$$\left[\because a, b, c \text{ are in HP } b = \frac{2ac}{a + c}\right]$$

$$\Rightarrow \cos x = \frac{2(\cos^2 x - \sin^2 y)}{2\cos x \cos y}$$

$$\Rightarrow \quad \cos^2 x \cos y = \cos^2 x - \sin^2 y$$

$$\Rightarrow \cos^2 x(\cos y - 1) = -\sin^2 y$$

$$\Rightarrow \cos^2 x(1 - \cos y) = 1 - \cos^2 y$$

$$\Rightarrow \cos^2 x(1 - \cos y) = (1 - \cos y)(1 + \cos y)$$

$$\Rightarrow \quad \cos^2 x = 1 + \cos y$$

$$\Rightarrow \quad \cos^2 x = 2\cos^2(y/2)$$

$$\Rightarrow \cos^2 x \sec^2(y/2) = 2$$

$$\Rightarrow \cos x \sec(y/2) = \pm\sqrt{2}$$

114. ***(c)*** We first write the elements of the set R.

i.e. $R = \{(1, 4), (4, 1), (2, 3), (3, 2)\}$

Since, $(1, 1) \notin R \Rightarrow R$ is not reflexive.

Now as, $(1, 4) \in R \Rightarrow (4, 1) \in R$ and $(2, 3) \in R$

$$\Rightarrow \quad (3, 2) \in R$$

So, R is symmetric

Now, $(1, 4) \in R$ and $(4, 1) \in R$

$$\not\Rightarrow \quad (1, 1) \in R$$

Thus, R is not transitive.

Hence, R is symmetric but neither reflexive nor transitive.

BITSAT 2017

115. ***(c)*** Any number from 1 to 999 is of the form *abc* when $0 \le a, b, c \le 9$. Let us first count the number in which 5 occurs exactly once.

Since, 5 can occur at one place in $1 \times {}^3C_1 \times 9 \times 9 = 243$ ways, next 5 can occur in exactly two places in ${}^3C_2 \times 9 = 27$. Lastly, 5 can occur in all three digits in only one way.

Hence, the number of times 5 occurs

$$= 1 \times 243 + 27 \times 2 + 1 \times 3$$
$$= 243 + 54 + 3 = 300$$

116. ***(a)*** Given, that, $A \cup X = B \cup X$

$\Rightarrow A \cap (A \cup X) = A \cap (B \cup X)$

$\Rightarrow (A \cap A) \cup (A \cap X) = (A \cap B) \cup (A \cap X)$ [Using distributive law]

$\Rightarrow A \cup \phi = (A \cap B) \cup \phi [\because A \cap X = \phi]$

$\Rightarrow \quad A = A \cap B \quad \ldots$(i)

Again, consider $A \cup X = B \cup X$

$\Rightarrow B \cap (A \cup X) = B \cap (B \cup X)$

$\Rightarrow (B \cap A) \cup (B \cap X) = (B \cap B) \cup (B \cap X)$ [Using distributive law]

$\Rightarrow (B \cap A) \cup \phi = B \cup \phi \ [B \cap X = \phi]$

$\Rightarrow A \cap B = B \quad \ldots$(ii)

Thus from Eq. (i) and (ii), we get

$A = B$

117. ***(d)*** $f(x) = x|x| = \begin{cases} x^2, & x \ge 0 \\ -x^2, & x < 0 \end{cases}$

and $f[-1,1] \to [-1,1]$

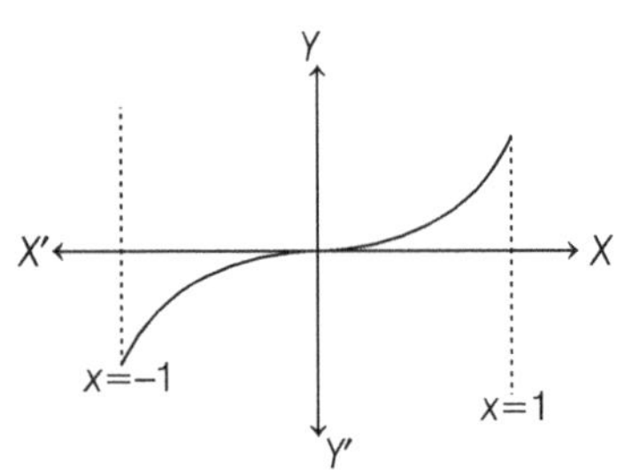

Since, $-1 \le x \le 1$, therefore $-1 \le f(x) \le 1$

$\Rightarrow$ codomain range.

$\therefore$ Function is one-one and onto.

118. ***(b)*** We have,

$$\sin x - 3\sin 2x + \sin 3x = \cos x - 3\cos 2x + \cos 3x$$

$\Rightarrow \sin x + \sin 3x - 3\sin 2x = \cos x + \cos 3x - 3\cos 2x$

$\Rightarrow 2\sin 2x \cos x - 3\sin 2x - 2\cos 2x \cos x + 3\cos 2x = 0$

$$\left[\because \sin x + \sin y = 2\sin\frac{x+y}{2}\cos\frac{x-y}{2} \text{ and } \cos x + \cos y = 2\cos\frac{x+y}{2}\cos\frac{x-y}{2}\right]$$

$\Rightarrow \sin 2x\,(2\cos x - 3) - \cos 2x\,(2\cos x - 3) = 0$

$\Rightarrow (\sin 2x - \cos 2x)(2\cos x - 3) = 0$

$\Rightarrow \quad \sin 2x = \cos 2x \quad [\because \cos x \ne 3/2]$

$\because \quad \tan 2x = 1$

$\Rightarrow \quad \tan 2x = \tan\dfrac{\pi}{4}$

$\Rightarrow \quad 2x = n\pi \pm \dfrac{\pi}{4}$

$\Rightarrow \quad x = \dfrac{n\pi}{2} + \dfrac{\pi}{8} \quad [\because$ neglect – ve sign]

119. ***(b)*** The third side is parallel to a bisector of the angle between equal sides.

The bisectors are $7x - y + 3 = \pm 5\,(x + y - 3)$

$\Rightarrow 2x - 6y + 18 = 0$ or $12x + 4y - 12 = 0$

$\Rightarrow x - 3y + 9 = 0$ or $3x + y - 3 = 0$

Let the third side be $x - 3y = k$ or $3x + y = L$

It passes through $(1, -10)$.

$k = 31, L = -7$

Hence, required lines are $x - 3y = 31$, $3x + y = -7$

120. ***(d)*** Let (t, m) be the other end of the chord drawn from the point (p, q) on the circle

$$x^2 + y^2 = px + qy$$

Their mid-point is $\left(\dfrac{t+p}{2}, \dfrac{m+q}{2}\right)$

Since, mid-point lies on X-axis i.e. $y = 0$

$$m + q = 0 \quad \ldots\text{(i)}$$

Also, (t, m) lies on the circle.

$\therefore \quad t^2 + m^2 - pt - qm = 0 \quad \ldots$(ii)

From Eqs. (i) and (ii), we get $t^2 - pt + 2q^2 = 0$

Which is quadratic in t such that, Discriminant > 0

$\Rightarrow \quad p^2 - 8q^2 > 0 \Rightarrow p^2 > 8q^2$

121. ***(a)*** Let P be the foot of the perpendicular drawn from $A(2, 3, 4)$ to the given line l.

$$\frac{x-4}{-2} = \frac{y}{6} = \frac{z-1}{-3}$$

Now, any point on the line l is given by

$$x = 4 - 2\lambda,\ y = 6\lambda,\ z = 1 - 3\lambda$$

The coordinates of P are $(4 - 2\lambda, 6\lambda, 1 - 3\lambda)$

The direction ratios of AP are

$(4 - 2\lambda - 2, 6\lambda - 3, 1 - 3\lambda - 4)$

i.e. $(2 - 2\lambda, 6\lambda - 3, -3 - 3\lambda)$

And the direction ratios of l are -2, 6 and -3.

Given, $AP \perp l$

$\therefore -2(2 - 2\lambda) + 6(6\lambda - 3) - 3(-3 - 3\lambda) = 0$

$\Rightarrow \quad \lambda = \dfrac{13}{49}$

$\therefore \quad AP^2 = (4 - 2\lambda - 2)^2 + (6\lambda - 3)^2 + (1 - 3\lambda - 4)^2$

$= 22 - 26\lambda + 49\lambda^2$

Put $\lambda = \frac{13}{49}$, we get

$$AP^2 = \frac{909}{49} \Rightarrow AP = \frac{3}{7}\sqrt{101}$$

122. ***(c)*** Let $P(1, 6, 3)$ be the given point and L be the foot of perpendicular from P to the given line. The coordinates of a general point on the given line are

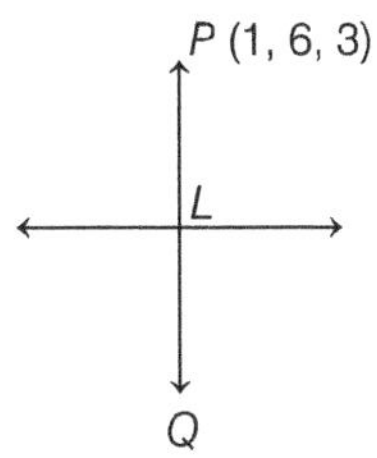

$$\frac{x-0}{1} = \frac{y-1}{2} = \frac{z-2}{3} = \lambda \text{ (say)}$$

i.e. $x = \lambda$

$$y = 2\lambda + 1, \quad z = 3\lambda + 2$$

If the coordinates of L are $(\lambda, 2\lambda + 1, 3\lambda + 2)$, then the direction ratios of PL are $(\lambda - 1, 2\lambda - 5, 3\lambda - 1)$.

Since, the direction ratios of given line which is perpendicular to PL, are 1, 2 and 3. Therefore, $(\lambda - 1)1 + (2\lambda - 5)2 + (3\lambda - 1)3 = 0$, which gives $\lambda = 1$.

Hence, coordinates of L are $(1, 3, 5)$.

Let $Q(x_1, y_1, z_1)$ be the image of $P(1, 6, 3)$ in the given line. Then, L is the mid-point of PQ.

Therefore, $\frac{x_1+1}{2} = 1, \frac{y_1+6}{2} = 3, \frac{z_1+3}{2} = 5$

$$\Rightarrow \quad x_1 = 1, y = 0, z = 7$$

Hence, the image of $(1, 6, 3)$ in the given line is $(1, 0, 7)$.

123. ***(a)*** Given equation of plane is $x - y + z = 5$.

The distance measured along the line $x = y = z$.

$$\Rightarrow \quad x/1 = y/1 = z/1$$

Direction ratios of the given line is $(1, 1, 1)$.

So, the equation of line PQ is [where, $P = (1,-5,9)$]

$$\frac{x-1}{1} = \frac{y+5}{1} = \frac{z-9}{1} = \lambda$$

$$\Rightarrow \quad x = \lambda + 1, y = \lambda - 5, z = \lambda + 9$$

Since, it lies on the plane $x + y + z = 5$

$\therefore \ (\lambda + 1) - (\lambda - 5) + (\lambda + 9) = 5 \Rightarrow \lambda = -10$

The coordinate of Q is $(-9, -15, -1)$ and the coordinate of P is $(1, -5, 9)$

$$PQ = \sqrt{(10)^2 + (10)^2 + (10)^2} = 10\sqrt{3}$$

$\therefore \ 2\sqrt{3}k = 10\sqrt{3} \Rightarrow k = 5$

124. ***(b)*** $\lim\limits_{n \to \infty} \sin\left\{n\pi\left(1 + \frac{1}{n^2}\right)^{1/2}\right\}$

$$= \lim_{n \to \infty} \sin\left\{n\pi\left(1 + \frac{1}{2n^2} - \frac{1}{8n^4} + \ldots\right)\right\}$$

$$= \lim_{n \to \infty} \sin\left\{n\pi + \frac{\pi}{2n} - \frac{\pi}{8n^3} + \ldots\right\}$$

$$= \lim_{n \to \infty} (-1)^n \sin\pi\left(\frac{1}{2n} - \frac{1}{8n^3} + \ldots\right) = 0$$

125. ***(b)*** $f(x) = \begin{cases} e^x, & x \le 0 \\ 1 - x, & 0 < x \le 1 \\ x - 1, & x > 1 \end{cases}$

At $x = 0$

$$\lim_{x \to 0} f(x) = \lim_{x \to 0^+}(1 - x) = 1$$

$$\lim_{x \to 0^-} f(x) = \lim_{x \to 0^-} e^x = k = f(0)$$

$\Rightarrow f(x)$ is continuous at $x = 0$

At $x = 1$

$$\lim_{x \to 1^+}(x - 1) = 1 - 1 = 0$$

$$\lim_{x \to 1^-}(1 - x) = 1 - 1 = 0$$

$$f(1) = 1 - 1 = 0$$

$f(x)$ is continuous at $x = 1$

$$Rf'(1) = \lim_{h \to 0} \frac{f(1+h) - f(1)}{h}$$

$$= \lim_{h \to 0} \frac{1 + h - 1 - 0}{h} = 1$$

$$Lf'(1) = \lim_{h \to 0} \frac{f(1-h) - f(1)}{-h} = \lim_{h \to 0} \frac{1 - (1-h) - 0}{-h}$$

$$= \lim_{h \to 0} \frac{h}{-h} = -1$$

$\therefore \quad Rf'(1) \ne Lf'(1)$

$\Rightarrow f(x)$ is not differentiable at $x = 1$

$$Rf'(0) = \lim_{h \to 0} \frac{f(0+h) - f(0)}{h}$$

$$= \lim_{h \to 0} -\left[\frac{1 - h - 1}{h}\right] = -1$$

$$Lf'(0) = \lim_{h \to 0} \frac{f(0-h) - f(0)}{-h}$$

$$= \lim_{h \to 0} \frac{e^{-h} - 1}{-h} = 1$$

So, it is not differentiable at $x = 0$.

Similarly, it is not differentiable at $x = 1$ but it is continuous at $x = 0$ and 1.

126. ***(b)*** Since, $f(x)$ is continuous at $x = 0$

$\therefore \lim\limits_{x \to 0} f(x) = f(0)$

Let a be any point

Now, at $x = a$, $\lim\limits_{x \to a} f(x) = \lim\limits_{h \to 0} f(a + h)$

$$= \lim_{h \to 0} f(a) + \lim_{h \to 0} f(h)$$

$$= f(a) + f(0) = f(a + 0) = f(a)$$

$\therefore \ f(x)$ is continuous at $x = a$, where a is any arbitrary point.

Hence, $f(x)$ is continuous for all x.

127. ***(a)*** Since, x, $\sin^{-1} x$, $\tan^{-1} x$ are continuous functions, so the function f is clearly continuous at each point of its domain except possibly at $x = 0$.

So, for f to be continuous at $x = 0$

$$f(0) = \lim_{x\to 0} f(x) = \lim_{x\to 0} \frac{2 - \frac{\sin^{-1} x}{x}}{2 + \frac{\tan^{-1} x}{x}} = \frac{1}{3}$$

$$\left[\because \lim_{x\to 0} \frac{\sin^{-1} x}{x} = 1, \lim_{x\to 0} \frac{\tan^{-1} x}{x} = 1\right]$$

128. ***(b)*** The given function is

$$f(x) = \begin{cases} 0, & 0 \le x < 1 \\ 1, & 1 \le x < 2 \end{cases}$$

Now, $\text{LHD} = \lim_{x\to 1^-} \frac{f(x) - f(1)}{x-1} = \lim_{x\to 1^{-1}} \frac{0-1}{x-1}$

Which does not exist

$\therefore$ f is not derivable at $x = 1$.

129. ***(b)*** Given, $f(x) = -2x^3 + 21x^2 - 60x + 41$ ...(i)

On differentiating Eq. (i) w.r.t. x, we get

$f'(x) = -6x^2 + 42x - 60$

$= -6(x^2 - 7x + 10) = -6(x-2)(x-5)$

If $x < 2$, $f'(x) < 0$, i.e., $f(x)$ is decreasing.

130. ***(d)*** Rolle's theorem is applicable only if function is continuous and differentiable. But $f(x) = |x|$ is not differentiable at $x = 0$.

i.e. $f'(0)$ does not exist.

131. ***(b)*** Clearly, the point of intersection of curves is $(0, 1)$.

Now, slope of tangent of first curve

$$m_1 = \frac{dy}{dx} = a^x \log a$$

$$\Rightarrow \left(\frac{dy}{dx}\right)_{(0,1)} = m_1 = \log a$$

Slope of tangent of second curve,

$$m_2 = \frac{dy}{dx} = b^x \log b$$

$$\Rightarrow m_2 = \left(\frac{dy}{dx}\right)_{(0,1)} = \log b$$

$$\therefore \tan\alpha = \frac{m_1 - m_2}{1 + m_1 m_2} = \frac{\log a - \log b}{1 + \log a \log b}$$

132. ***(d)*** We have, $\int_0^n [x]\, dx$

$$= \int_0^1 0\, dx + \int_1^2 1\, dx + \int_2^3 2\, dx + \dots + \int_{n-1}^n (n-1)\, dx$$

$$= 1(2-1) + 2(3-2) + 3(4-3) + \dots + (n-1)\{n-(n-1)\}$$

$$= 1 + 2 + 3 + \dots + (n-1) = \frac{n(n-1)}{2}$$

and $\int_0^n \{x\}\, dx = \int_0^n (x - [x])\, dx = \int_0^n x dx - \int_0^n [x] dx$

$$= \frac{n^2}{2} - \frac{n(n-1)}{2} = \frac{n}{2}$$

$$\therefore \frac{\int_0^n [x]\, dx}{\int_0^n \{x\}\, dx} = n - 1$$

133. ***(b)*** Given, $f(x) = x + \sin 2x$

$\Rightarrow f'(x) = 1 + 2\cos 2x$

For maximum or minimum value, $f'(x) = 0$

$\Rightarrow 1 + 2\cos 2x = 0 \Rightarrow \cos 2x = -\frac{1}{2} = \cos\frac{2\pi}{3}$

$\Rightarrow 2x = 2n\pi \pm \frac{2\pi}{3}$

$\Rightarrow x = n\pi \pm \frac{\pi}{3}$

$\Rightarrow x = \frac{4\pi}{3}, \frac{2\pi}{3}, \frac{5\pi}{3}$

Find $f(0), f\left(\frac{2\pi}{3}\right), f\left(\frac{4\pi}{3}\right), f\left(\frac{5\pi}{3}\right), f(2\pi)$

$\Rightarrow f(0) = 0, f\left(\frac{2\pi}{3}\right) = \frac{2\pi}{3} - 0.8$

$\Rightarrow f\left(\frac{4\pi}{3}\right) = \frac{4\pi}{3} + 0.8, f\left(\frac{5\pi}{3}\right) = \frac{5\pi}{3} - 0.8$

$\Rightarrow f(2\pi) = 2\pi + 0 = 2\pi$

$\therefore$ Maximum value of $f(x) = 2\pi$.

134. ***(b)*** Given, $y = |\cos x - \sin x|, 0 \le x \le \frac{\pi}{2}$

$$= \begin{cases} \cos x - \sin x, & 0 \le x \le \frac{\pi}{4} \\ \sin x - \cos x, & \frac{\pi}{4} \le x \le \frac{\pi}{2} \end{cases}$$

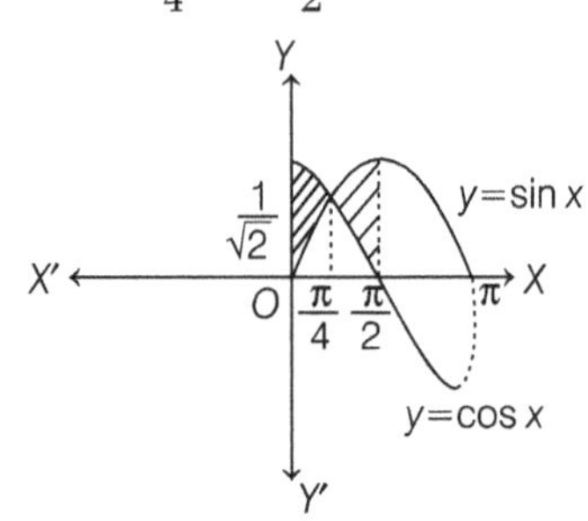

$$\text{Required area} = \int_0^{\pi/2} |\cos x - \sin x|\, dx$$

$$= \int_0^{\pi/4} (\cos x - \sin x)\, dx + \int_{\pi/4}^{\pi/2} (\sin x - \cos x)\, dx$$

$$= [\sin x + \cos x]_0^{\pi/4} + [-\cos x - \sin x]_{\pi/4}^{\pi/2}$$

$$= (\sqrt{2} - 1) - (1 - \sqrt{2}) = 2\sqrt{2} - 2$$

135. ***(b)*** Put $x + y = v$ and $1 + \frac{dy}{dx} = \frac{dv}{dx}$

Therefore, the differential equation reduces to

$$\frac{dv}{dx} = (1 + \cos v) + \sin v = 2\cos^2\frac{v}{2} + 2\sin\frac{v}{2}\cos\frac{v}{2}$$

$$= 2\cos^2\frac{v}{2}\left(1+\frac{\tan v}{2}\right)$$

$$\Rightarrow \quad \int \frac{\sec^2\frac{v}{2}}{2\left[1+\tan\left(\frac{v}{2}\right)\right]}\,dv = \int dx$$

$$\therefore \quad \log\left[1+\tan\left(\frac{x+y}{2}\right)\right] = x + C$$

136. ***(c)*** Since, the intersection of two curves $y = x^3$ and $y = \sqrt{x}$ are $x = 0$ and $x = 1$.

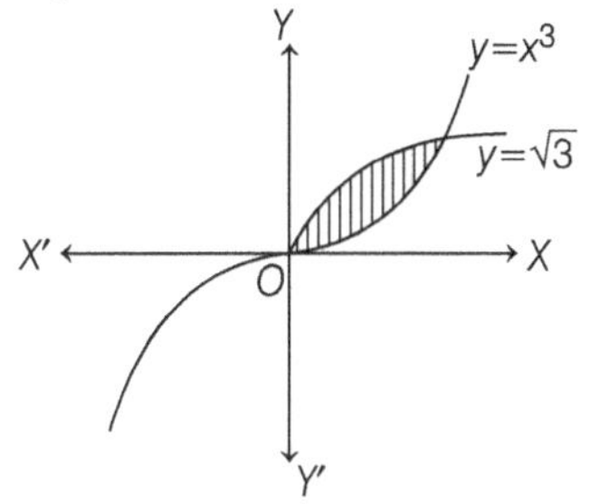

$$\therefore \quad A = \left|\int_0^1 (x^3 - \sqrt{x})\,dx\right| = \left|\left[\frac{x^4}{4} - \frac{2x^{3/2}}{3}\right]_0^1\right|$$

$$= \left|\left[\frac{1}{4} - \frac{2}{3}\right]\right| = \frac{5}{12} \text{ sq units}$$

137. ***(c)*** $\lim_{n\to\infty} \frac{an(1+n)-(1+n^2)}{1+n}$

$$= \lim_{n\to\infty} \frac{(a-1)n^2 + an - 1}{n+1} = \infty,$$

If $a - 1 \neq 0$ limit does not exist and if $a - 1 = 0$, then

$$\lim_{n\to\infty} \frac{an-1}{n+1} = a = b \Rightarrow a = b = 1$$

138. ***(b)*** Total ways in which papers can be checked is equal to 7^4. Now, two teachers who have to check all the papers can be selected in ${}^7C_2$ ways and papers can be checked by them is $(2^4 - 2)$ favourable ways.

Thus, required probability $= \frac{{}^7C_2\,(2^4-2)}{7^4} = \frac{6}{49}$

139. ***(c)*** Given, equation of hyperbola is $(10x-5)^2 + (10y-4)^2 = \lambda^2(3x+4y-1)^2$ can be rewritten as

$$\frac{\sqrt{\left(x-\frac{1}{2}\right)^2+\left(y-\frac{2}{5}\right)^2}}{\left|\frac{3x+4y-1}{5}\right|} = \left|\frac{\lambda}{2}\right|$$

This is of the form of $\frac{PS}{PM} = e$

Where, P is any point on the hyperbola and S is a focus and M is the point of directrix.

Here, $\left|\frac{\lambda}{2}\right| > 1 \Rightarrow |\lambda| > 2 \qquad (\because e > 1)$

$\Rightarrow \quad \lambda < -2$ or $\lambda > 2$

140. ***(a)*** $v = \hat{a} \times \hat{b} = |a||b|\sin\theta\,\hat{n} = \sin\theta\,\hat{n}$

[$\because \hat{a}$ and $\hat{b}$ are unit vectors]

Where, θ is the angle between a and b.

$[\because |a| = 1 = |b|]$

$\therefore \quad |u| = \sin\theta$

Now, $u = \hat{a} - (\hat{a}\,.\,\hat{b})\,\hat{b} = \hat{a} - (\cos\theta)\,\hat{b}$

$$\Rightarrow |u|^2 = (\hat{a} - (\cos\theta)\,\hat{b})\,.\,(\hat{a} - \cos\theta)\,\hat{b})$$

$$= \mathbf{a}^2 + \cos^2\theta\,\mathbf{b}^2 - 2\cos\theta\,(\hat{a}\cdot\hat{b})$$

$$[|\mathbf{a}|^2 = |\mathbf{b}|^2 = |\mathbf{a}||\mathbf{b}| = 1]$$

$$= 1 + \cos^2\theta - 2\cos^2\theta \quad [\because \hat{a}\cdot\hat{b} = |\hat{a}||\hat{b}|\cos\theta]$$

$$= 1 - \cos^2\theta = \sin^2\theta$$

$\therefore \quad |u| = \sin\theta$

Thus $|v| = |u|$

141. ***(c)*** Variance $= \frac{1}{n}\Sigma(x-\bar{x})^2 = \sigma^2$

New variance $= \frac{1}{n}\Sigma(\alpha x - \alpha\bar{x})^2$

$$= \alpha^2\frac{1}{n}\Sigma(x-\bar{x})^2 = \alpha^2\sigma^2$$

142. ***(a)*** Given, coefficient of variation $C_1 = 50$ and coefficient of variation $C_2 = 60$.

We have, $\bar{x}_1 = 30$ and $\bar{x}_2 = 25$

$\because \quad C = \frac{\sigma}{\bar{x}} \times 100$

$\Rightarrow \quad 50 = \frac{\sigma_1}{30} \times 100 \Rightarrow \sigma_1 = 15$

and $\quad 60 = \frac{\sigma_2}{25} \times 100 \Rightarrow \sigma_2 = 15$

$\therefore$ Required difference $\sigma_1 - \sigma_2 = 15 - 15 = 0$

143. ***(c)*** Given, constraints are $2x + 3y \leq 18$, $2x + y \leq 10$, $x \geq 0$, $y \geq 0$.

The feasible region is $OABCO$.

Also, $z = 9x + 13y$

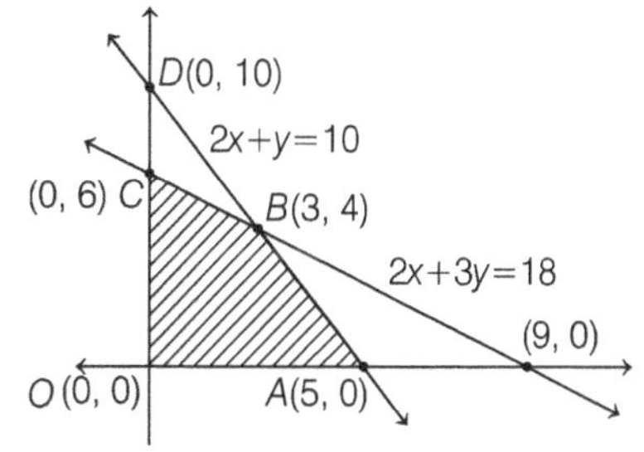

At $O(0, 0)$, $z = 0$

At $A(5, 0)$, $z = 45$

At $B(3, 4)$, $z = 27 + 52 = 79$

At $C(0, 6)$, $z = 78$

$\therefore$ Maximum value of z is 79.

Probability of getting head = 1/2

144. ***(c)*** The man has to win atleast 4 times.

$\because$ Probability of getting head = 1/2

$\therefore$ Required probability

$$= {}^7C_4\left(\frac{1}{2}\right)^4\left(\frac{1}{2}\right)^3 + {}^7C_5\left(\frac{1}{2}\right)^5\left(\frac{1}{2}\right)^2 + {}^7C_6\left(\frac{1}{2}\right)^6\left(\frac{1}{2}\right) + {}^7C_7\left(\frac{1}{2}\right)^7$$

$$= ({}^7C_4 + {}^7C_5 + {}^7C_6 + {}^7C_7) \times \frac{1}{2^7} = \frac{64}{2^7} = \frac{1}{2}$$

145. ***(c)*** $\because \frac{-\pi}{2} \le \sin^{-1} x \le \frac{\pi}{2}, \frac{-\pi}{2} \le \sin^{-1} y \le \frac{\pi}{2}$

and $\frac{-\pi}{2} \le \sin^{-1} z \le \frac{\pi}{2}$

Given that,

$$\sin^{-1}x + \sin^{-1}y + \sin^{-1}z = \frac{3\pi}{2}$$

Which is possible only when

$$\sin^{-1}x = \sin^{-1}y = \sin^{-1}z = \frac{\pi}{2}$$

$\Rightarrow \quad x = y = z = 1$

Also given that $f(p+q) = f(p)\cdot f(q) \forall\, p, q \in R$ $\quad [\because f(1) = 2]$

Put $p = q = 1$

Then, $f(2) = f(1)\; f(1) = 2 \times 2 = 4$

and put $p = 1, q = 2$

then, $f(3) = f(1)\, f(2) = 2 \cdot 2^2 = 8$

$$\therefore \quad x^{f(1)} + y^{f(2)} + z^{f(3)} - \frac{x + y + z}{x^{f(1)} + y^{f(2)} + z^{f(3)}}$$

$$= 1 + 1 + 1 - \frac{3}{1+1+1} = 3 - 1 = 2$$

146. ***(c)*** Here,

$$T_n = \frac{n(n+1)}{n!} = \frac{n-1+2}{(n-1)!} = \frac{1}{(n-2)!} + \frac{2}{(n-1)!}$$

$$\therefore \quad S = \sum_{n=1}^{\infty} T_n = \sum_{n=1}^{\infty}\frac{1}{(n-2)!} + 2\sum_{n=1}^{\infty}\frac{1}{(n-1)!}$$

$$= e + 2e = 3e$$

147. ***(b)*** Since, $|z_1| = |z_2| = |z_3|$

$\therefore$ 0 is the circumcentre of an equilateral ΔABC.

$$\therefore \frac{x_1 + x_2 + x_3}{3} = 0 = \frac{y_1 + y_2 + y_3}{3}$$

where, $z_1 = x_1 + iy_1$

$$\Rightarrow \frac{x_1 + x_2 + x_3}{3} + i\left(\frac{y_1 + y_2 + y_3}{3}\right) = 0$$

$$\Rightarrow (x_1 + iy_1) + (x_2 + iy_2) + (x_3 + iy_3) = 0$$

$$\Rightarrow \quad z_1 + z_2 + z_3 = 0$$

148. ***(b)*** Let $I = \int \frac{dx}{(\sqrt{x})^2 + (\sqrt{x})^7} = \int \frac{dx}{(\sqrt{x})^7\left(\frac{1}{(\sqrt{x})^5} + 1\right)}$

Put $1 + \frac{1}{(\sqrt{x})^5} = t$

$$\Rightarrow -\frac{5}{2}\cdot\frac{1}{(\sqrt{x})^7}\,dx = dt$$

$$\therefore I = -\frac{2}{5}\int\frac{dt}{t} = \frac{-2}{5}\log t + C$$

$$= \frac{-2}{5}\log\left[1 + \frac{1}{(\sqrt{x})^5}\right] + C = -\frac{2}{5}\log\left(\frac{x^{5/2}+1}{x^{5/2}}\right) + C$$

$$\Rightarrow \lambda = \frac{-2}{5} \text{ and } a = \frac{5}{2}$$

$\therefore \quad a + \lambda = 2.1 > 2$

149. ***(b)*** Equation of line joining the points $(0, 3)$ and $(5, -2)$ is

$$y - 3 = \frac{-2-3}{5-0}(x - 0).$$

$$\Rightarrow \quad y - 3 = -x \Rightarrow y = 3 - x$$

If this line is tangent to $y = \frac{ax}{(x+1)}$, then

$(3 - x)(x - 11) = ax$ should have equal roots.

$\Rightarrow \quad x^2 + (a-2)x - 3 = 0$

$D = 0$ for equal roots

Thus, $\quad (a-2)^2 + 12 = 0$

$\Rightarrow$ No value of $a \Rightarrow a \in \phi$

150. ***(b)*** Shortest distance between two curves occured along the common normal.

$\therefore$ Normal to $y^2 = 4x$ at $(m^2, 2m)$ is

$\Rightarrow \quad y - 2m = -m(x - m^2)$

$\quad y + mx - 2m - m^3 = 0$

Normal to $y^2 = 2(x-3)$ at $\left(\frac{m^2}{2} + 3, m\right)$ is

$$y - m = -m\left(x - \frac{1}{2}m^2 - 3\right)$$

$$y + m(x - 3) - m - \frac{m^3}{2} = 0$$

Both normals are same, if

$$-2m - m^3 = -4m - \frac{1}{2}m^3 \Rightarrow m = 0, 2, -2$$

So, points will be $(4, 4)$ and $(5, 2)$ or $(4, -4)$ and $(5, -2)$.

Hence, shortest distance will be

$$\sqrt{(1+4)} = \sqrt{5}$$

BITSAT

Birla Institute of Technology & Science Admission Test

Solved Paper 2016*

Instructions

- There are 150 questions in all. The number of questions in each part is as given below.

		No. of Questions
Part I	Physics	1–40
Part II	Chemistry	41–80
Part III	a. English Proficiency	81–95
	b. Logical Reasoning	96–105
Part IV	Mathematics	106–150

- All questions are Multiple Choice Questions having four options out of which **only one** is correct.
- Each correct answer fetches 3 marks while incorrect answer has a penalty of 1 mark.
- Time allotted to complete this paper is 3 hrs.

PART I

Physics

1. You are given two spheres of same material and radii 10 cm and 20 cm. They are heated to the same temperature and placed in the same environment. The ratio of their rates of cooling will be

a. 1 : 2 *b.* 2 : 1
c. 1 : 4 *d.* 4 : 1

2. A man is standing unsymmetrically between two mountains. He fires a gun and hears the first echo after 1.5 s and the second echo after 2.5 s. If the speed of sound is 340 m/s, what is the distance between the mountains?

a. 340 m *b.* 410 m
c. 640 m *d.* 680 m

3. A thin plano-convex lens of focal length f is split into two halves. One of the halves a shifted along the optical axes. The separation between object and image plane is 1.8 m. The magnification of the image formed by one of the half lens is 2. Find the focal length of lens.

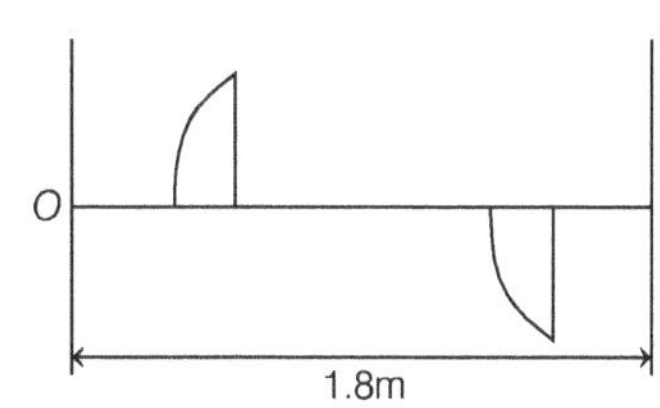

a. 0.1 m *b.* 0.4 m *c.* 0.9 m *d.* 1 m

4. A test charge q_0 is placed at the centre of a spherical conductor of radius R. A charge Q is placed on the spherical conductor. What will be the electrostatic force on the charge Q due to q_0?

a. Zero *b.* $\frac{1}{4\pi\varepsilon_0} \cdot \frac{Qq_0}{R^2}$
c. $\frac{1}{4\pi\varepsilon_0} \cdot \frac{Qq_0}{2R^2}$ *d.* $\frac{1}{4\pi\varepsilon_0} \cdot \frac{Qq_0}{4R^2}$

5. A particle is projected from A at the angle of 90° with the plane AB as shown in figure. If again touches the plane at B after time T, then what is the length of AB?

*Memory Based

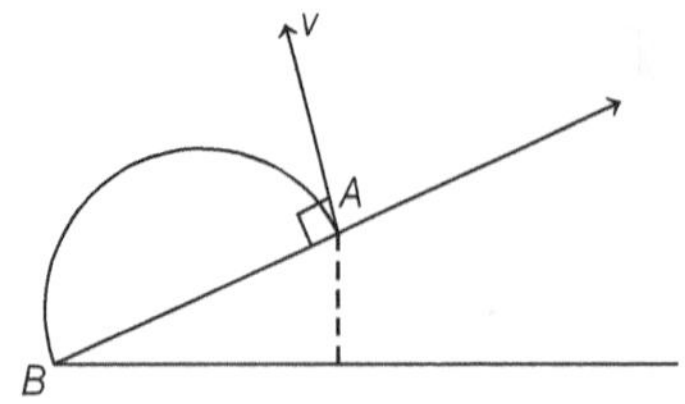

a. $Tu \sin\theta$ *b.* $Tu \cos\theta$
c. $Tu \tan\theta$ *d.* $Tu \cot\theta$

6. A pendulum is executing SHM. The velocity of the bob at the mean position is 3 cm/s. If without changing the length, the amplitude is doubled, then what will be the velocity of the bob in the mean position?

a. 1.5 cm/s *b.* 3 cm/s
c. 4.5 cm/s *d.* 6 cm/s

7. An α-particle describes a circular path of radius r_0 in a magnetic field B. What will be the radius of the circular path described by the proton of same energy in the same magnetic field?

a. $\frac{r_0}{2}$ *b.* r_0 *c.* $\sqrt{2}$ *d.* $2r_0$

8. In an interference pattern, the position of zeroth order maxima is 4.8 mm from a certain point P on the screen. The fringe width is 0.2 mm. The position of second order minima from point P is

a. 5.1 mm *b.* 5 mm
c. 40 mm *d.* 5.2 mm

9. A 40 kg slab rests on a frictionless floor. A 10 kg block rests on top of the slab. The static coefficient of friction between the block and the slab is 0.6, while the kinetic coefficient of friction is 0.4. The 10 kg blocked is acted upon by a horizontal force of 100 N. If $g = 9.8 \text{ ms}^{-2}$, the resulting acceleration of the slab will be

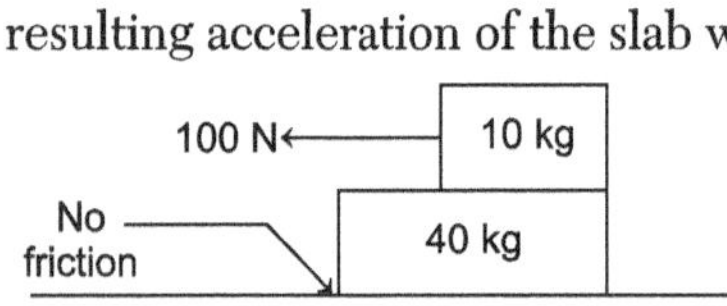

a. 1.47 ms^{-2} *b.* 1 ms^{-2}
c. 9.8 ms^{-2} *d.* 0.98 ms^{-2}

10. A bomber plane moves horizontally with a speed of 500 ms^{-1} and a bomb released from it, strikes the ground in 10 s. Angle at which it strikes the ground will be ($g = 10 \text{ ms}^{-2}$)

a. $\tan^{-1}\left(\frac{1}{5}\right)$ *b.* $\tan^{-1}\left(\frac{1}{4}\right)$
c. $\tan^{-1}(1)$ *d.* $\tan^{-1}(5)$

11. A square loop of wire of side 5 cm is lying on a horizontal table. An electromagnet wave, above and to one side of the loop is turned on causing a uniform magnetic field downward at an angle of 60° to the vertical as shown in the figure. The magnetic induction is 0.5 T. The average induced emf in the loop, if the field increases from zero to its final value in 0.2 s is

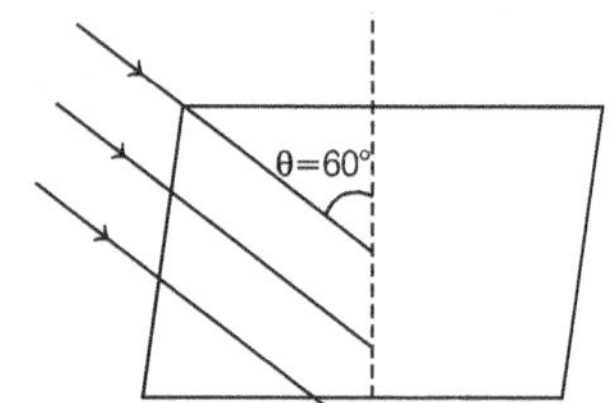

a. 5.4×10^{-3} V *b.* 3.12×10^{-3} V
c. zero *d.* 0.25×10^{-3} V

12. One end of a uniform rod of mass m_1 and uniform area of cross-section A is suspended from the roof and mass m_2 is suspended from the other end. What is the stress at the mid-point of the rod?

a. $(m_1 + m_2) g/A$ *b.* $(m_1 - m_2) g/A$
c. $\left[\frac{(m_1/2) + m_2}{A}\right] g$ *d.* $\left[\frac{m_1 + (m_2/2)}{A}\right] g$

13. The temperature coefficient of resistance for a wire is $0.00125°\text{C}^{-1}$. At 300 K its resistance is 1 Ω. The temperature at which the resistance becomes 1.5Ω is

a. 450 K *b.* 713.5 K *c.* 454 K *d.* 900 K

14. An electron of mass m and charge e initially at rest gets accelerated by a constant electric field E. The rate of change of de-Broglie wavelength of this electron at time t (ignoring relativistic effect) is

a. $\frac{h}{eEt^2}$ *b.* $\frac{eEt}{h}$ *c.* $\frac{mh}{eEt^2}$ *d.* $-\frac{h}{eE}$

15. Figure shows the vertical cross-section of a vessel filled with a liquid of density ρ. The normal thrust per unit area on the wall of the vessel at point P, as shown will be

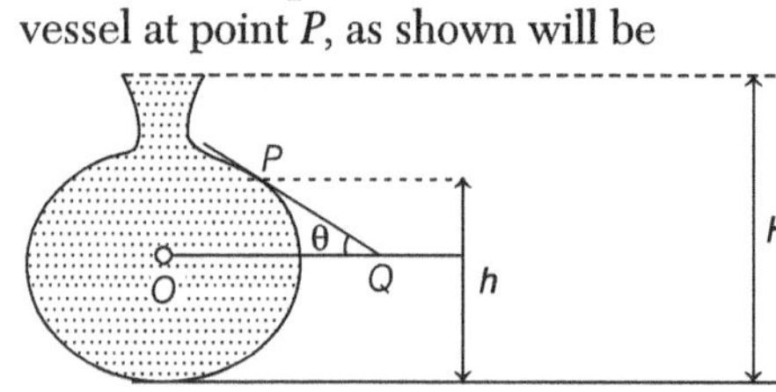

a. $h\rho g$ *b.* $H\rho g$
c. $(H - h)\rho g$ *d.* $(H - h)\rho g \cos\theta$

16. In the Bohr model of the hydrogen atom, let R, v and E represent the radius of the orbit, the speed of electron and the total energy of the electron, respectively. Which of the following quantities is proportional to quantum number n?

a. $\frac{R}{E}$ *b.* $\frac{E}{v}$ *c.* RE *d.* vR

17. In the common emitter configuration of *n-p-n* transistor 10^{10} electrons enter the emitter in 1 μs and 2% of electrons are lost to the base. What is the current gain of the amplifier?

a. 2 *b.* 98 *c.* 1 *d.* 49

18. A projectile is given an initial velocity of $(\hat{\mathbf{i}} + 2\hat{\mathbf{j}})$ m/s, where $\hat{\mathbf{i}}$ is along the ground and $\hat{\mathbf{j}}$ is along the vertical. If $g = 10 \text{ m/s}^2$, then the equation of its trajectory is

a. $y = x - 5x^2$ *b.* $y = 2x - 5x^2$

c. $4y = 2x - 5x^2$ *d.* $4y = 2x - 25x^2$

19. 1 kg of diatomic gas is at a pressure of $8 \times 10^4 \text{ Nm}^{-2}$. The density of the gas is 4 kg m^{-3}. What is the energy of the gas due to its thermal motion?

a. 5×10^4 J *b.* 6×10^4 J

c. 7×10^4 J *d.* 3×10^4 J

20. In the Young's double slit experiment, the interference pattern is found to have an intensity ratio between bright and dark fringes as 9. This implies that

a. the intensities at the screen due to two slits are 5 units and 4 units respectively

b. the intensities at the screen due to two slits are 4 units and 1 unit respectively

c. the amplitude ratio is 3

d. the amplitude ratio is 2

21. The potential energy of a particle in motion along X-axis is given by $U = U_0 - U_0 \cos ax$. The period of small oscillations is

a. $2\pi\sqrt{\frac{ma}{U_0}}$ *b.* $2\pi\sqrt{\frac{U_0}{ma}}$ *c.* $\frac{2\pi}{a}\sqrt{\frac{m}{U_0}}$ *d.* $2\pi\sqrt{\frac{m}{aU_0}}$

22. A charged particle q of mass m is released from rest in a uniform electron field E. Neglecting the effect of gravity, the kinetic energy of the charged particle after time t second is

a. $\frac{Eqm}{t}$ *b.* $\frac{E^2q^2t^2}{2m}$

c. $\frac{2E^2t^2}{mq}$ *d.* $\frac{Eq^2m}{2t^2}$

23. A body of mass m hangs from three springs each of spring constant k shown in the figure. If the mass is slightly displaced and let go, the time period of oscillations of system is

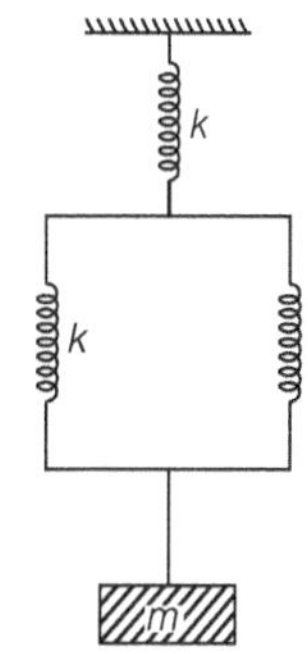

a. $2\pi\sqrt{\frac{m}{3k}}$ *b.* $2\pi\sqrt{\frac{3m}{2k}}$

c. $2\pi\sqrt{\frac{2m}{3k}}$ *d.* $2\pi\sqrt{\frac{3k}{m}}$

24. A radioactive material decays by simultaneous emission of two particles with respective half-lives T_1 and T_2. The effective half-life of material is

a. $\sqrt{T_1 T_2}$ *b.* T_1 / T_2

c. $\frac{T_1 T_2}{T_1 + T_2}$ *d.* $\frac{T_1 + T_2}{2}$

25. The focal length of a concave mirror is f and the distance of the object to the principal focus is p. Then, the ratio of the size of the image to the size of the object is

a. f / p *b.* p / f

c. fp *d.* $\sqrt{p / f}$

26. In the given circuit, with steady state, the potential drop across the capacitor must be

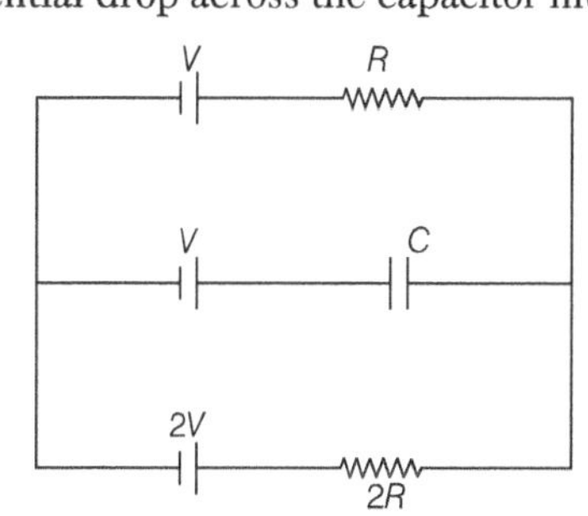

a. V *b.* $\frac{V}{2}$

c. $\frac{V}{3}$ *d.* $\frac{2V}{3}$

27. A block of mass m is placed on another block of mass M. The arrangement is placed on a horizontal table. All surfaces are smooth. If $m = 2$ kg, $M = 4$ kg and constant force $F = 3$ N, then initial acceleration of masses M and m will be (in m/s^2)

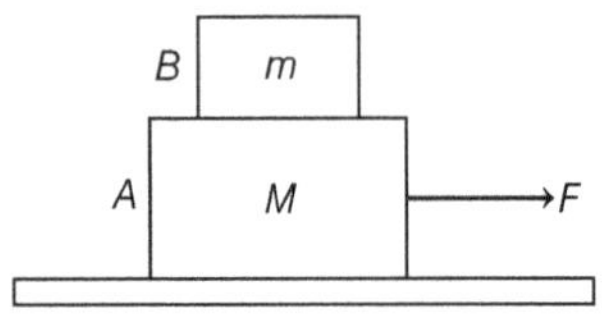

a. 0.5, 0.25 **b.** 0.5, 0
c. 0.5, −0.5 **d.** 0.75, 0

28. A thin rod has moment of inertia I_1 about an axis passing through its centre and perpendicular to length. It is bent to form a ring. The ring so formed has moment of inertia I_2 about an axis passing through centre and perpendicular to the plane of the ring. Then, the ratio I_1 / I_2 is

a. 1 **b.** $\frac{\pi}{4}$ **c.** $\frac{\pi^2}{3}$ **d.** $\frac{3}{5}$

29. In the hydrogen spectrum, λ_3 and λ_2 are the wavelengths of radiations emitted due to transition from second and first excited states to the ground state. Then, the ratio $\frac{\lambda_3}{\lambda_2}$ is

a. $\frac{27}{32}$ **b.** $\frac{32}{27}$ **c.** $\frac{4}{9}$ **d.** $\frac{9}{4}$

30. A conductive circular loop of radius r carries a constant current i. It is placed in a uniform magnetic field B such that B is perpendicular to the plane of loop. The magnetic force acting on the loop is

a. Bir **b.** $2\pi irB$
c. zero **d.** πirB

31. A soap bubble in vacuum has a radius 3 cm and another soap bubble in vacuum has radius 4 cm. If two bubbles coalesce under isothermal condition. Then, the radius of new bubble is

a. 2.3 cm **b.** 4.5 cm
c. 5 cm **d.** 7 cm

32. A projectile is thrown with an initial velocity of $(a\hat{i} + b\hat{j})$ m/s. If the range of the projectile is twice the maximum height reached by it, then

a. $b = \frac{a}{2}$ **b.** $b = a$
c. $b = 2a$ **d.** $b = 4a$

33. Find the power consumed by given circuit.

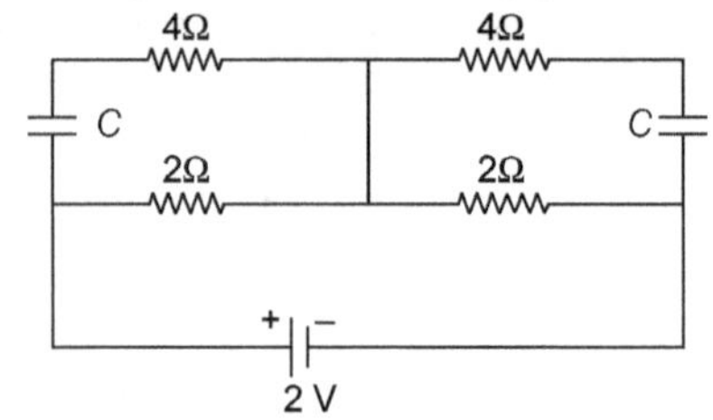

a. 1.5 W **b.** 2 W
c. 1 W **d.** insufficient data

34. An α-particle of mass m suffers one dimensional elastic collision with a nucleus of unknown mass. After the collision α-particle is scattered directly backwards losing 75% of its initial kinetic energy. Then, the mass of the nucleus is

a. m **b.** $2m$ **c.** $3m$ **d.** $5m$

35. A meter stick of mass m, initially in vertical position, is displaced through an angle 45°, keeping upper end pivoted. What is the change in potential energy?

a. $\frac{mg}{2\sqrt{2}}$ **b.** $\frac{mg}{\sqrt{2}}$
c. $\frac{mg}{2}\left(1 - \frac{1}{\sqrt{2}}\right)$ **d.** $mg\left(1 - \frac{1}{\sqrt{2}}\right)$

36. A parallel plate capacitor of value 1.77 μF is to be designed using a dielectric material (dielectric constant = 200) breakdown strength of 3×10^6 V m^{-1}. In order to make such a capacitor, which can withstand a potential difference of 20 V across the plates, the separation between the plates d and the area A of the plates can be
$(\varepsilon_0 = 8.85 \times 10^{-12}\ C^2N^{-1}m^{-2})$

a. $d = 10^{-6}$ m, $A = 10^{-3}$ m^2
b. $d = 10^{-5}$ m, $A = 10^{-2}$ m^2
c. $d = 10^{-4}$ m, $A = 10^{-4}$ m^2
d. $d = 10^{-4}$ m, $A = 10^{-5}$ m^2

37. A particle starting from rest undergoes a rectilinear motion with acceleration a. The variation of a with time is shown in figure. The maximum velocity attained by the particle is

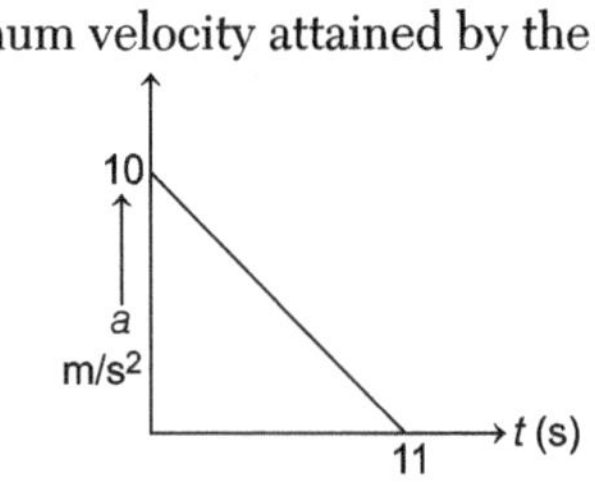

a. 55 m/s **b.** 550 m/s
c. 110 m/s **d.** 650 m/s

38. A satellite of mass m is circulating around the Earth with constant angular velocity. If the radius of the orbit is R_0 and mass of the Earth is M, the angular momentum of satellite about the centre of Earth is

a. $m\sqrt{\frac{GM}{R_0}}$

b. $m\sqrt{GMR_0}$

c. $m\sqrt{\frac{Gm}{R_0}}$

d. $M\sqrt{\frac{GM}{R_0}}$

39. In a sonometer wire the tension is maintained by suspending a 50.7 kg mass from the free end of a wire. The suspended mass has a volume $0.0075\ m^3$. The fundamental frequency of vibration of wire is 260 Hz. If the suspended mass is completely submerged in water, the fundamental frequency becomes

a. 480 Hz

b. 260 Hz

c. 240 Hz

d. 280 Hz

40. A current of 5 A is flowing in a wire of length 1.5 m. A force of 7.5 N acts on it. When it is placed in a uniform magnetic field of induction 2 T. The angle between the magnetic field and the direction of current is

a. 30° b. 45°
c. 60° d. 90°

PART II

Chemistry

41. A 400 mg iron capsule contains 100 mg of ferrous fumarate, $(CHCOO)_2Fe$. The percentage of iron present in it, is approximately

a. 33% b. 25% c. 14% d. 8%

42. The ratio of the difference between 1st and 2nd Bohr's orbit energy to that between 2nd and 3rd Bohr's orbit energy is

a. $\frac{1}{2}$ b. $\frac{1}{3}$ c. $\frac{27}{5}$ d. $\frac{5}{27}$

43. Hydrogen resembles halogen in many respects for which several factors are responsible. Of the following factors, which one is the most important in this respect?

a. Its tendency to lose an electron to form a cation

b. Its tendency to gain a single electron in its valence shell to attain stable electronic configuration

c. Its low negative electron gain enthalpy value

d. Its small size

44. From the following sets, which one does not contain isoelectronic species?

a. $BO_3^{3-}, CO_3^{2-}, NO_3^-$ b. $SO_3^{2-}, CO_3^{2-}, NO_3^-$

c. CN^-, N_2, C_2^{2-} d. $PO_4^{3-}, SO_4^{2-}, ClO_4^-$

45. The bond length of HCl bond is 2.29×10^{-10} m. The percentage ionic character of HCl, if measured dipole moment is 6.226×10^{-30} C-m, is

a. 8% b. 20%

c. 17% d. 50%

46. How many carbon atoms in the molecule HOOC—$(CHOH)_2$—COOH are asymmetric?

a. 1 b. 2

c. 3 d. None of these

47. Which of the following on reductive ozonolysis gives only glyoxal?

a. Ethene b. Acetylene

c. Benzene d. Both (b) and (c)

48. Which of the following transitions involve maximum amount of energy?

a. $M^-(g) \longrightarrow M(g)$ b. $M(g) \longrightarrow M^+(g)$

c. $M^+(g) \longrightarrow M^{2+}(g)$ d. $M^{2+}(g) \longrightarrow M^{3+}(g)$

49. The boiling point of alcohols and phenols increases with increase in the number of

a. oxygen atom b. carbon atom

c. Both (a) and (b) d. hydrogen atom

50. A mixture of dilute NaOH and aluminium pieces is used to open drain because

a. the gas obtained creates higher pressure

b. they react explosively

c. they create fire

d. the given statement is not true

51. The conductivity of strong electrolyte

a. increases on dilution slightly

b. decreases on dilution

c. does not change with dilution

d. depends upon density of electrolyte itself

52. What is the freezing point of a 10% (by weight) solution of CH_3OH in water? [K_f of $CH_3OH = 1.86°$ C/m]

a. 90°C *b.* 10°C
c. 6.45°C *d.* −6.45° C

53. Why are potassium and cesium, rather than lithium used in photoelectric cells?

a. Because of their lower IE
b. Because of their lower EA
c. Both (a) and (b)
d. Because of their higher IE

54. Alkyl iodides can be prepared by

a. $RCH_2COOAg + I_2 \xrightarrow[\Delta]{CCl_4} RCH_2I$

b. $RCH_2Cl + NaI \xrightarrow[\Delta]{Acetone} RCH_2I + NaCl$

c. $R—OH + I_2 \longrightarrow RI + H_2O$

d. $CH_4 + I_2 \longrightarrow CH_3I$

55. Copper matte contains

a. Cu_2S and FeS *b.* $FeSiO_3$ and Cu_2O
c. Fe_2O_3 and Cu *d.* Cu_2O and Fe

56. For which type of reaction, order and molecularity have the same value?

a. First order
b. Bimolecular reaction
c. Trimolecular reaction
d. Elementary reaction

57. Metal which can be extracted from all the three dolomite, magnesite and carnallite is

a. Na *b.* K
c. Mg *d.* Ca

58. In the complex, $K_2Fe[Fe(CN)_6]$

a. the complex is high spin complex
b. both Fe atoms are in the same oxidation state
c. the coordination number of iron is 4
d. both Fe atoms are in different oxidation state

59. Novolac on heating with formaldehyde undergoes cross-linking to form

a. bakelite *b.* melamine
c. dacron *d.* resin

60. Which of the following is a network solid?

a. SO_2 (solid) *b.* I_2
c. Diamond *d.* H_2O (ice)

61. The compound 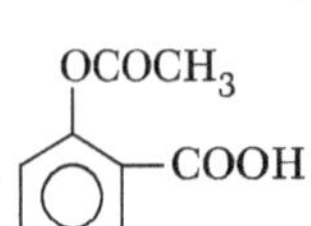 is used as a/an

a. antiseptic *b.* antibiotic
c. analgesic *d.* pesticide

62. DNA and RNA contain four bases each. Which of the following bases is not present in RNA?

a. Adenine *b.* Thymine
c. Uracil *d.* Cytosine

63. In Duma's method, 0.3 g of an organic compound gave 45 mL of nitrogen at STP. The percentage of nitrogen is

a. 16.9 *b.* 18.7
c. 29.6 *d.* 23.2

64. Which one of the following will be most readily dehydrated in acidic condition?

a. 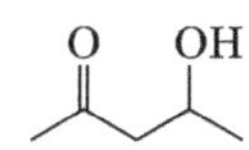 *b.* 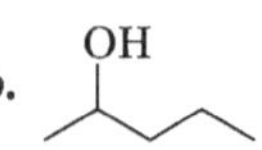

c. 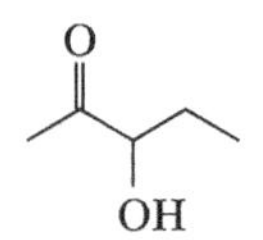 *d.* 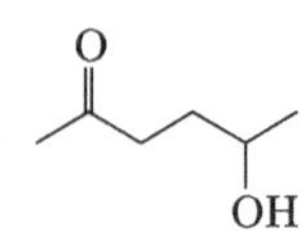

65. The number of spherical nodes in 3*p*-orbital is

a. 0 *b.* 1
c. 2 *d.* 3

66. An element has $[Ar]\ 3d^1 4s^2$ configuration in its oxidation state. It's position in the periodic table is

a. period-3, group-3 *b.* period-3, group-7
c. period-4, group-3 *d.* period-3, group-9

67. The HOMO in CO is

a. π-bonding *b.* π-antibonding
c. σ-antibonding *d.* σ-bonding

68. If a gas at constant temperature and pressure expands then its

a. internal energy increases and then decreases
b. internal energy increases
c. internal energy remains the same
d. internal energy decreases

69. What is the conjugate base of OH^-?

a. O^{2-} *b.* O^- *c.* H_2O *d.* O_2

70. Adsorption is accompanied by

a. ΔH is negative
b. decrease in entropy of the system
c. $T\Delta S$ for the process is negative
d. All of the above

71. The difference of water molecules in gypsum and plaster of Paris is

a. $\frac{5}{2}$ *b.* 2
c. $\frac{1}{2}$ *d.* $1\frac{1}{2}$

72. Boron when treated with carbon forms

a. B_4C *b.* BC_4
c. B_4C_3 *d.* B_2C_3

73. Which of the following compounds is anti-aromatic?

a. Benzene *b.* Cyclobutadiene
c. Cyclodecapentene *d.* Cyclooctatetraene

74. A schematic plot of $\ln K_{eq}$ *versus* inverse of temperature for a reaction is shown below :

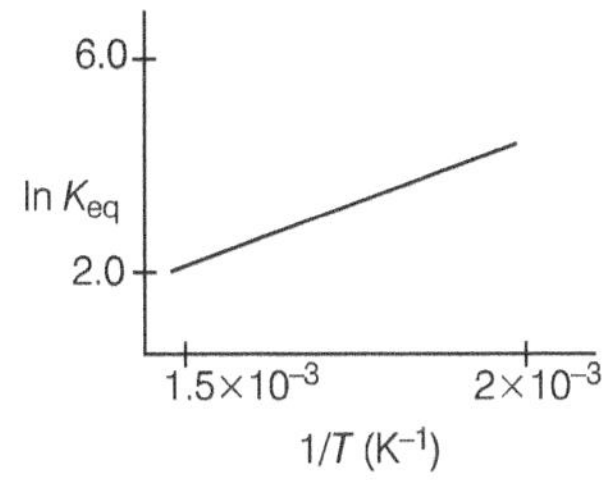

The reaction must be

a. highly spontaneous at ordinary temperature
b. one with negligible enthalpy change
c. endothermic
d. exothermic

75. Clathrates are

a. non-stoichiometric compounds
b. complex compounds
c. interstitial compounds
d. ionic compounds

76. Transition elements are coloured

a. due to unpaired *d*-electrons
b. due to small size
c. due to metallic structure
d. All of the above

77. The colour of $_{62}Sm^{3+}$ is yellow. The expected colour of $_{66}Dy^{3+}$ is

a. yellow *b.* red
c. blue *d.* green

78. Which of the following compounds is the weakest Bronsted base?

a. 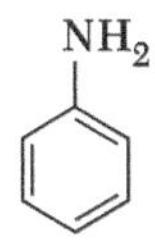

b. 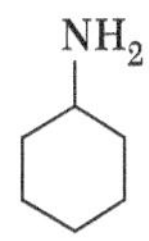

c. 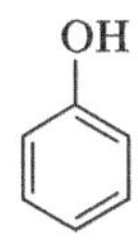

d. 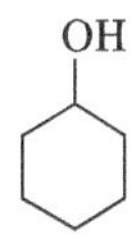

79. Which one of the following will have the largest number of atoms?

a. 1 g Au(*s*) *b.* 1 g Na(*s*)
c. 1 g Li(*s*) *d.* 2 g $Cl_2(g)$

80. The density of neon will be maximum at

a. STP *b.* 0°C, 2 atm
c. 273°C, 1 atm *d.* 273°C, 2 atm

PART III

a. English Proficiency

Directions (Q. Nos. 81 to 83) *Choose the word which best expresses the meaning of the underlined word in the sentences.*

81. Have you traced the addresses that I have told you earlier?

a. Searched *b.* Found
c. Prognosised *d.* Fathomed

82. Terrorism is a gargantuan problem in our country.

a. Gigantic *b.* Callous
c. Spurious *d.* Lee

83. Discipline will always propitiate other's wishes to you.

a. Enrages *b.* Appease
c. Futilitate *d.* Reverberate

Directions (Q. Nos. 84 to Q. No. 86) *Fill in the blanks with suitable words from the options given below.*

84. A dog was run .... by the car.

a. down
b. up
c. with
d. by

85. The old man .... to the hospital after the car hit him near the bridge.

a. is taken *b.* are taken
c. was taken *d.* has taken

86. The rise in .... was not limited to the prosperous agricultural zones in the rural area.

a. salary *b.* pay
c. increment *d.* wages

Directions (Q. Nos. 87 to 89) *Choose the word which is closest to the opposite in the meaning of the sentence.*

87. Why did he called a man of bad reputation to his house in the night ?

a. Gallant *b.* Notorious
c. Famous *d.* Pervert

88. The whole world, with its flora and fauna, is considered to be perishable.

a. Sustainable *b.* Immortal
c. Adorable *d.* Clinical

89. Despite his penury, he tried to teach many children on his own resources.

a. Scarcity *b.* Dearth
c. Opulence *d.* Affluence

Directions (Q. Nos. 90 to 92) *In each of the following questions, out of the four options, choose the one which can be substituted for the given words or sentences.*

90. One who does not consume alcohol.

a. Rationalist *b.* Atheist
c. Incalcucable *d.* Teetotaller

91. An animal having both male and female sex, inside single body.

a. Hermaphrodite *b.* Aquatic
c. Acerbic *d.* Camouflage

92. A text without the name of author.

a. Unnamed *b.* Pseudonym
c. Unanimous *d.* Concealed

Directions (Q. Nos. 93 to 95) *Choose the order of the sentences marked A, B, C, D, E and F to form a logical paragraph and then choose the most suitable option.*

A. The researchers in these companies claim that they could do better by allowing their employees to doze off at work place.

B. The dreams, while at work, are thus helpful to solve crucial problems.

C. Would you believe that some UK based companies are arranging for bed at the work place ?

D. The reason, they claim, could be that dreams produce creative solutions.

E. We only hope that these crucial problems in UK are different from those of ours.

F. But it is true and is considered as a step to improve quality of their products.

93. Which of the following should be the fourth sentence after rearrangement?

a. A *b.* B
c. C *d.* D

94. Which of the following should be the fifth sentence after rearrangement ?

a. A *b.* B
c. C *d.* D

95. Which of the following should be the sixth sentence after rearrangement ?

a. A *b.* B
c. C *d.* E

b. Logical Reasoning

96. 'Ribs' are related to 'Lungs' in the same way 'Shell' is related to

a. Breathe *b.* Almond
c. Nut *d.* Respiration

97. Find the odd one.

a. 20, 40, 46 *b.* 25, 50, 56
c. 12, 24, 30 *d.* 18, 36, 40

98. Complete the series by replacing '?' mark.

30, 68, 130, 222, 350, ?

a. 420
b. 460
c. 520
d. 580

99. Choose the answer figure which completes the problem figure matrix.

Problem Figure | Answer Figures

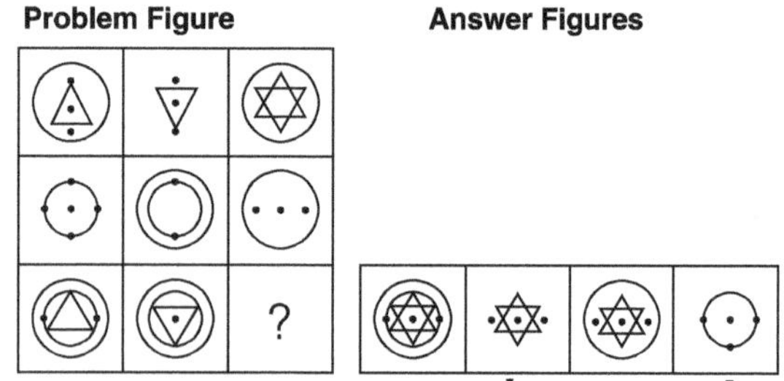

a. *b.* *c.* *d.*

100. The three problem figures marked X, Y and Z show the manner in which a piece of paper P is folded step by step and then cut. Form the answer figures (a) (b) (c) and (d) select the one showing the unfolded pattern of the paper after the cut.

Problem Figure

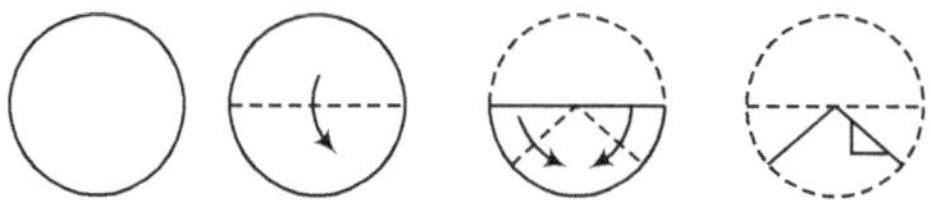

P X Y Z

Answer Figures

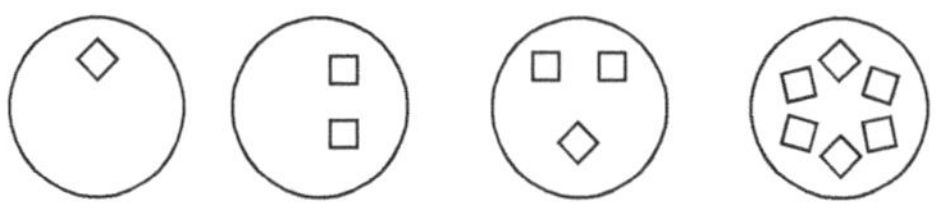

a. *b.* *c.* *d.*

101. In the following question, five figures are given. Out of them, find the three figures that can be joined to form an equilateral triangle.

A. B. C. D. E.

a. ABE *b.* BCE
c. ADE *d.* BDE

102. In the following question, a part of problem figure is missing. Find out from the given answer figures *a, b, c* and *d*, that can replace the '?' to complete the figure.

Problem Figure

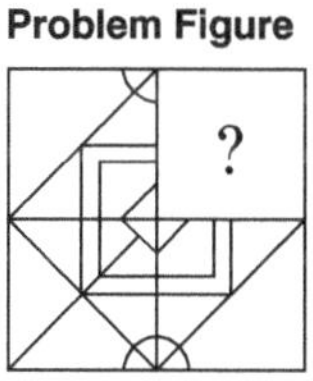

Answer Figures

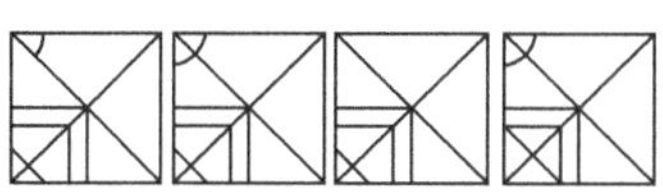

a. *b.* *c.* *d.*

103. In the following question, one or more dots are placed in the figure marked as (A). This figure is followed by four alternatives marked as *a, b, c* and *d*. One out of these four options contains region(s) common to circle, square, triangle, similar to that marked by the dot in figure (A). Find that figure.

Problem Figure

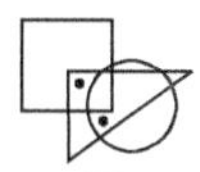

(A)

Answer Figures

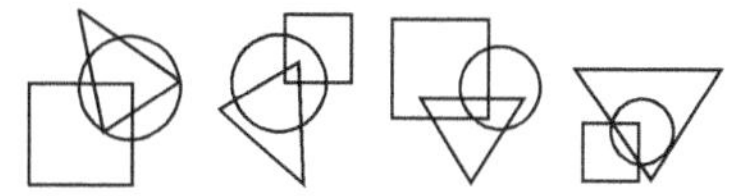

a. *b.* *c.* *d.*

104. On a shelf , there are placed six volumes side-by-side labelled A, B, C, D, E and F. B, C, E and F have green covers while others have yellow covers. A, D, B are new volumes while the rest are old volumes. A, C, B are law reports while the rest are medical extracts. Which two volumes are old medical extracts and have green covers?

a. B, C *b.* C, D
c. C, E *d.* E, F

105. Choose the set of figures which follows the given rule.

Rule Closed figures gradually become open and open figures gradually become closed.

1.
2.
3.
4.

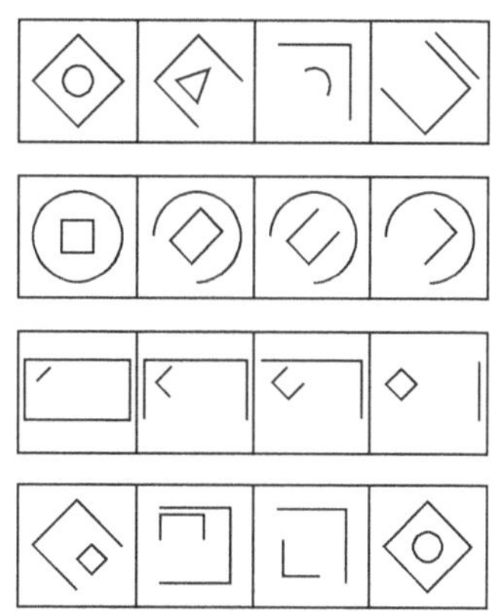

a. 1 *b.* 2 *c.* 3 *d.* 4

PART IV

Mathematics

106. The value of $\lim_{x \to a} \left(2 - \frac{a}{x}\right)^{\tan \frac{\pi x}{2a}}$ is

a. $e^{-1/\pi}$ *b.* $e^{2/\pi}$ *c.* $e^{-2/\pi}$ *d.* $e^{1/\pi}$

107. If $|z_1| = 2, |z_2| = 3, |z_3| = 4$ and $|2z_1 + 3z_2 + 4z_3| = 4$, then $|8z_2z_3 + 27z_3z_1 + 64z_1z_2|$ is equal to

a. 24 *b.* 48 *c.* 72 *d.* 96

108. Let $\omega = \frac{-1}{2} + i\frac{\sqrt{3}}{2}$. Then, the value of the determinant $\begin{vmatrix} 1 & 1 & 1 \\ 1 & -1-\omega^2 & \omega^2 \\ 1 & \omega^2 & \omega^4 \end{vmatrix}$ is

a. 3ω *b.* $3\omega(\omega - 1)$
c. $3\omega^2$ *d.* $3\omega(1 - \omega)$

BITSAT 2016

109. The sum of n terms of the series $\frac{1}{1\cdot3}+\frac{1}{3\cdot5}+\frac{1}{5\cdot7}+\ldots$ is

a. $\frac{1}{2n+1}$ b. $\frac{2n}{2n+1}$ c. $\frac{n}{2n+1}$ d. $\frac{2n}{n+1}$

110. The period of the function $f(x)=|\sin x|-|\cos x|$, is

a. $\pi/2$ b. π
c. 2π d. None of these

111. Let $f(x)=x^3-x^2+x+1$ and $g(x)$ be a function defined by

$$g(x)=\begin{cases}\text{Max }\{f(t):0\le t\le x\}, & 0\le x\le 1\\ 3-x & ,1<x\le 2\end{cases}, \text{ then}$$

$g(x)$ is

a. continuous and differentiable on $[0, 2]$
b. continuous but not differentiable on $[0, 2]$
c. Neither continuous nor differentiable on $[0, 2]$
d. None of the above

112. The area enclosed between the curve $y=\log_e(x+e)$ and the coordinate axes, is

a. 4 b. 3 c. 2 d. 1

113. Four numbers are multiplied together. Then, the probability that the product will be divisible by 5 or 10 is

a. $\frac{369}{625}$ b. $\frac{399}{625}$
c. $\frac{123}{625}$ d. $\frac{133}{625}$

114. Let X denotes the number of times head occur in n tosses of a fair coin. If $P(X=4)$, $P(X=5)$ and $P(X=6)$ are in AP, then the value of n is

a. 7, 14 b. 10, 14
c. 12, 7 d. 14, 12

115. On the set N of natural numbers, defined the relation R by $a\,R\,b$ if the GCD of a and b is 2, then R is

a. Reflexive but not symmetric
b. Symmetric only
c. Equivalence
d. Neither reflexive nor symmetric and transitive

116. The coefficient of the middle term in the binomial expansion, in powers of x, of $(1+\alpha x)^4$ and of $(1-\alpha x)^6$ is same, if α equals

a. $\frac{3}{5}$ b. $\frac{10}{3}$
c. $\frac{-3}{10}$ d. $\frac{-5}{3}$

117. Greatest integral value of 'n' for which the equation $2\cos x(\sin x-\cos x)=n$ has atleast one solution is

a. -2 b. 0
c. 1 d. 2

118. Let S and S′ be two foci of the ellipse $\frac{x^2}{a^2}+\frac{y^2}{b^2}=1$. If a circle described on SS' as diameter intersects the ellipse in real and distinct points, then the eccentricity e of the ellipse satisfies

a. $e=1/\sqrt{2}$ b. $e\in(1/\sqrt{2}, 1)$
c. $e\in(0, 1/\sqrt{2})$ d. None of these

119. It has been found that if A and B play a game 12 times, A wins 6 times, B wins 4 times and they draw twice. A and B take part in a series of 3 games. The probability that they will win alternately is

a. $\frac{5}{72}$ b. $\frac{5}{36}$
c. $\frac{19}{27}$ d. None of these

120. Let a, b, c be in AP and $|a|<1, |b|<1, |c|<1$. If

$x=1+a+a^2+\ldots$ to ∞,

$y=1+b+b^2+\ldots$ to ∞ and

$z=1+c+c^2+\ldots$ to ∞, then, x, y, z are in

a. AP b. GP
c. HP d. None of these

121. If the sides of a triangle are in GP and its larger angle is twice the smallest, then the common ratio (r) satisfies the inequality

a. $0<r>\sqrt{2}$ b. $1<r<\sqrt{2}$
c. $1<r<2$ d. None of these

122. The maximum value of $\mu=3x+4y$, subjected to the conditions $x+y\le 40$, $x+2y\le 60$, $x, y\ge 0$, is

a. 130 b. 140 c. 40 d. 120

123. Let A and B be two sets containing 2 elements and 4 elements respectively. The number of subsets of $A\times B$ having 3 as more elements is

a. 256 b. 220 c. 219 d. 211

124. If $a_1, a_2, a_3, \ldots, a_n$ are in HP, then the expression $a_1 a_2+a_2 a_3+\ldots a_{n-1} a_n$ is equal to

a. $n(a_1-a_n)$ b. $(n-1)(a_1-a_n)$
c. $na_1 a_n$ d. $(n-1)a_1 a_n$

125. The equation of the plane through the intersection of the planes $x + y + z = 1$ and $2x + 3y - z + 4 = 0$ parallel to X-axis, is

a. $y - 3z + 6 = 0$ *b.* $3y - z + 6 = 0$
c. $y + 3z + 6 = 0$ *d.* $3y - 2z + 6 = 0$

126. Let the straight line $x = b$ divide the area enclosed by $y = (1-x)^2$, $y = 0$ and $x = 0$ into two parts R_1 $(0 \le x \le b)$ and R_2 $(b \le x \le 1)$ such that $R_1 - R_2 = \frac{1}{4}$. Then, b equals

a. $\frac{3}{4}$ *b.* $\frac{1}{2}$ *c.* $\frac{1}{3}$ *d.* $\frac{1}{4}$

127. If three vectors **a**, **b**, **c** are such that $\mathbf{a} \ne 0$ and $\mathbf{a} \times \mathbf{b} = 2(\mathbf{a} \times \mathbf{c})$, $|\mathbf{a}| = |\mathbf{c}| = 1$, $|\mathbf{b}| = 4$ and the angle between **b** and **c** is $\cos^{-1}\left(\frac{1}{4}\right)$, then $\mathbf{b} - 2\mathbf{c} = \lambda \mathbf{a}$, where λ is equal to

a. ± 4 *b.* -2 *c.* ± 3 *d.* 2

128. z_1, z_2 and z_3 are the affixes of the vertices of a triangle having its circumcentre at the origin. If z is the affix of its orthocentre, then

a. $z_1 + z_2 + z_3 + z = 0$
b. $z_1 + z_2 + z_3 - z = 0$
c. $z_1 - z_2 + z_3 + z = 0$
d. $z_1 + z_2 - z_3 + z = 0$

129. The total number of 5-digit numbers of different digits in which digit in the middle is the highest is

a. $30 \times 3!$
b. $33 \times 3!$
c. $\sum_{n=4}^{9} {}^nC_4 \times 4!$
d. None of the above

130. If $\alpha \le \sin^{-1} x + \cos^{-1} x + \tan^{-1} x \le \beta$, then

a. $\alpha = \frac{\pi}{4}, \beta = \frac{3\pi}{4}$ *b.* $\alpha = -\pi, \beta = 2\pi$
c. $\alpha = 0, \beta = \pi$ *d.* None of these

131. The solution of the differential equation

$$\frac{dy}{dx} = \frac{y}{x} + \frac{\phi\left(\frac{y}{x}\right)}{\phi'\left(\frac{y}{x}\right)}$$ is

a. $\phi\left(\frac{y}{x}\right) = kx$ *b.* $x\,\phi\left(\frac{y}{x}\right) = k$
c. $\phi\left(\frac{y}{x}\right) = ky$ *d.* $y\,\phi\left(\frac{y}{x}\right) = k$

132. If a variable takes value 0, 1, 2, ..., n with frequencies q^n, ${}^nC_1 q^{n-1}p$, ${}^nC_2 q^{n-2}p^2$, ... ${}^nC_n p^n$, where $p + q = 1$, then the mean is

a. np *b.* nq
c. $n(p+q)$ *d.* None of these

133. If the mean and standard deviation of 10 observations $x_1, x_2, ..., x_{10}$ are 2 and 3 respectively, then the mean of $(x_1+1)^2$, $(x_2+1)^2, ..., (x_{10}+1)^2$ is equal to

a. 18 *b.* 13.5
c. 14.4 *d.* 16

134. The tangent to the curve $y = e^x$ drawn at the point (c, e^c) intersects the line joining the point $(c-1, e^{c-1})$ and $(c+1, e^{c+1})$

a. On the left of $x = c$ *b.* On the right of $x = c$
c. At no point *d.* At all points

135. The lines $\frac{x-2}{1} = \frac{y-3}{1} = \frac{z-4}{-k}$ and $\frac{x-1}{k} = \frac{y-4}{2} = \frac{z-5}{1}$ are coplanar, if

a. $k = 3$ or -3 *b.* $k = 0$ or -1
c. $k = 1$ or -1 *d.* $k = 0$ or -3

136. If the direction ratios of two lines are given by $a + b + c = 0$ and $2ab + 2ac - bc = 0$, then the angle between the lines is

a. $\frac{\pi}{3}$ *b.* $\frac{\pi}{2}$
c. $\frac{2\pi}{3}$ *d.* π

137. If the function $f(x) = x^3 - 6x^2 + ax + b$ satisfies Rolle's theorem in the interval $[1, 3]$ and $f'\left(\frac{2\sqrt{3}+1}{\sqrt{3}}\right) = 0$, then

a. $a = 11$ *b.* $a = -11$
c. $a = 6$ *d.* $a = -6$

138. Let f be a bijection satisfying $f'(x) = f(x)$. Then, $(f^{-1})''(x)$ is equal to

a. $f^{-1}(x)$ *b.* $f(x)$
c. $-\frac{1}{x^3}$ *d.* $-\frac{1}{x^2}$

139. The interval in which $2x^3 + 5$ increases less rapidly than $9x^2 - 12x$, is

a. $(-\infty, 1)$ *b.* $(2, \infty)$
c. $(1, 2)$ *d.* None of these

140. The greatest value of the function $f(x) = \sin^{-1} x^2$ in the interval $\left[-\frac{1}{\sqrt{2}}, \frac{1}{\sqrt{2}}\right]$, is

a. $\frac{\pi}{6}$ b. $\frac{\pi}{3}$ c. $\frac{\pi}{2}$ d. $-\frac{\pi}{3}$

141. Let $f : R \to R$ be given by $f(x) = [x]^2 + [x+1] - 3$, where $[x]$ denotes the greatest integer less than or equal to x. Then, $f(x)$ is

a. many-one and into b. one-one and onto
c. many-one and onto d. one-one and into

142. Point (2, 4) is translated through a distance of $3\sqrt{2}$ units measured parallel to the line $y - x = 1$, in the direction of decreasing ordinates, to reach at Q. If R is the image of Q with respect to the line $y - x = 1$, then coordinates of R are

a. (0, 0) b. (− 1, 1) c. (6, 6) d. (5, 7)

143. If a chord of the circle $x^2 + y^2 = 32$ make equal intercepts of length l on the coordinate axes, then

a. $l \in (0, 8)$ b. $l \in (-8, 0)$
c. $l \in (-8, 8)$ d. $l \in (-4\sqrt{2}, 4\sqrt{2})$

144. The value of a so that the sum of the squares of the roots of the equation $x^2 - (a-2)x - a + 1 = 0$ assume the least value, is

a. 1 b. 2 c. 3 d. 0

145. The slope of the tangent to the curve $y = \int_x^{x^2} \cos^{-1} t^2 \, dt$, at $x = 2^{-\frac{1}{4}}$ is

a. $\left(\frac{\sqrt[5]{8}}{4} - \frac{1}{3}\right)\pi$ b. $\left(\frac{\sqrt[4]{8}}{2} - \frac{3}{4}\right)\pi$
c. $\left(\frac{\sqrt[4]{8}}{3} - \frac{1}{4}\right)\pi$ d. None of these

146. If $\lim_{x \to \frac{\pi}{4}} = \frac{4\sqrt{2} - (\cos x + \sin x)^5}{1 - \sin 2x}$, then x is equal to

a. $\sqrt{2}$ b. $3\sqrt{2}$
c. $5\sqrt{2}$ d. None of these

147. Let f be a differentiable function satisfying $f(x) + f(y) + f(z) + f(x) f(y) f(z) = 14$ for all $x, y, z \in R$. Then,

a. $f'(x) < 0$ for all $x \in R$
b. $f'(x) > 0$ for all $x \in R$
c. $f'(x) = 0$ for all $x \in R$
d. None of the above

148. Let $f : [1, \infty) \to [2, \infty)$ be a differentiable function such that $f(1) = 2$. If $6\int_1^x f(t)\, dt = 3x f(x) - x^3$ for all $x \ge 1$, then the value of $f(2)$ is

a. 3 b. 4
c. 5 d. 6

149. If g is inverse of f and $f'(x) = \frac{1}{1+x^n}$, then $g'(x)$ equals

a. $1 + x^n$ b. $1 + [g(x)]^n$
c. $1 + [f(x)]^n$ d. None of these

150. The value of k for which

$$f(x) = \begin{cases} \left(1 + xe^{-1/x^2} \sin \frac{1}{x^4}\right)^{e^{1/x^2}}, & x \ne 0 \\ k, & x = 0 \end{cases}$$

is continuous at $x = 0$, is

a. 4 b. 3
c. 2 d. 1

Answers

Physics

1. (b)	**2.** (d)	**3.** (b)	**4.** (a)	**5.** (c)	**6.** (d)	**7.** (b)	**8.** (a)	**9.** (d)	**10.** (a)
11. (b)	**12.** (c)	**13.** (b)	**14.** (a)	**15.** (c)	**16.** (d)	**17.** (d)	**18.** (b)	**19.** (a)	**20.** (b)
21. (c)	**22.** (b)	**23.** (b)	**24.** (c)	**25.** (a)	**26.** (c)	**27.** (d)	**28.** (c)	**29.** (a)	**30.** (c)
31. (c)	**32.** (c)	**33.** (c)	**34.** (c)	**35.** (c)	**36.** (a)	**37.** (a)	**38.** (b)	**39.** (c)	**40.** (a)

Chemistry

41. (d)	**42.** (c)	**43.** (b)	**44.** (b)	**45.** (c)	**46.** (b)	**47.** (d)	**48.** (d)	**49.** (b)	**50.** (a)
51. (b)	**52.** (d)	**53.** (a)	**54.** (b)	**55.** (a)	**56.** (d)	**57.** (c)	**58.** (b)	**59.** (a)	**60.** (c)
61. (c)	**62.** (b)	**63.** (b)	**64.** (a)	**65.** (b)	**66.** (c)	**67.** (d)	**68.** (c)	**69.** (a)	**70.** (d)
71. (d)	**72.** (a)	**73.** (b)	**74.** (d)	**75.** (a)	**76.** (a)	**77.** (a)	**78.** (c)	**79.** (c)	**80.** (b)

English Proficiency

81. (a)	**82.** (a)	**83.** (b)	**84.** (a)	**85.** (c)	**86.** (d)	**87.** (c)	**88.** (b)	**89.** (c)	**90.** (d)
91. (a)	**92.** (c)	**93.** (d)	**94.** (b)	**95.** (d)					

Logical Reasoning

96. (c)	**97.** (d)	**98.** (c)	**99.** (c)	**100.** (b)	**101.** (c)	**102.** (b)	**103.** (c)	**104.** (d)	**105.** (c)

Mathematics

106. (c)	**107.** (d)	**108.** (b)	**109.** (c)	**110.** (b)	**111.** (b)	**112.** (d)	**113.** (a)	**114.** (a)	**115.** (b)
116. (c)	**117.** (b)	**118.** (b)	**119.** (b)	**120.** (c)	**121.** (b)	**122.** (b)	**123.** (c)	**124.** (d)	**125.** (a)
126. (b)	**127.** (a)	**128.** (b)	**129.** (d)	**130.** (a)	**131.** (a)	**132.** (a)	**133.** (a)	**134.** (a)	**135.** (d)
136. (c)	**137.** (a)	**138.** (d)	**139.** (c)	**140.** (a)	**141.** (a)	**142.** (a)	**143.** (c)	**144.** (a)	**145.** (c)
146. (c)	**147.** (c)	**148.** (d)	**149.** (b)	**150.** (d)					

Hints & Solutions

Physics

1. ***(b)*** Rate of cooling, $R = \frac{dT}{dt} = \frac{eA\sigma}{mc}(T^4 - T_0^4)$

where, symbols have their standard meanings.

As both spheres made from same material,

$$\Rightarrow \quad \frac{R_1}{R_2} = \frac{\frac{dT_1}{dt}}{\frac{dT_2}{dt}} = \frac{\frac{A_1}{m_1}}{\frac{A_2}{m_2}} = \frac{\frac{A_1}{V_1\rho}}{\frac{A_2}{V_2\rho}} = \frac{A_1}{V_1} \cdot \frac{V_2}{A_2}$$

$$= \frac{4\pi r_1^2}{\frac{4}{3}\pi r_1^3} \cdot \frac{\frac{4}{3}\pi r_2^3}{4\pi r_2^2} = \frac{r_2}{r_1} = \frac{20}{10} \Rightarrow \frac{R_1}{R_2} = 2:1$$

2. ***(d)*** First echo is heared at time, $t_1 = 2d_1/v$

where, d_1 = distance of the man from first mountain and v = speed of sound.

Second echo is heared at time, $t_2 = \frac{2d_2}{v}$

$$\text{Hence,} \quad t_1 + t_2 = \frac{2}{v}(d_1 + d_2)$$

$$\Rightarrow \quad d_1 + d_2 = \frac{v}{2}(t_1 + t_2)$$

$$= \frac{340}{2} \times 4 \quad [\because t_1 + t_2 = 1.5 + 2.5 = 4 \text{ s}]$$

$$\Rightarrow \quad d_1 + d_2 = 680 \text{ m}$$

3. ***(b)*** Given, arrangement is equivalent to focal length determination by displacement method.

Let the separation between object and image plane be a and separation between two halves be d.

So, $a = 1.8\,\text{m}$

$\therefore$ Magnification $m = \dfrac{a+d}{a-d} = \dfrac{2}{1}$

$$\Rightarrow \frac{1.8+d}{1.8-d} = \frac{2}{1} \Rightarrow 1.8 + d = 3.6 - 2d$$

$$\Rightarrow 3d = 3.6 - 1.8 \Rightarrow d = \frac{1.8}{3} = 0.6\text{ m}$$

We know that, focal length of the lens, $f = \dfrac{a^2 - d^2}{4a}$

$$= \frac{(1.8)^2 - (0.6)^2}{4 \times 1.8} = \frac{3.24 - 0.36}{7.2} = \frac{2.88}{7.2} = 0.4\text{ m}$$

4. ***(a)*** The charge on the conductor Q will behave as if it is concentrated at its centre. Both the charges are located at the same point. So, the electrostatic force between them is zero.

5. ***(c)*** Taking component of g, along and perpendicular to the plane, we get

Acceleration along the inclined plane (towards B) $= g\sin\theta$

Acceleration perpendicular to the inclined plane AB (in downward direction) $= g\cos\theta$

Time taken by the particle to attain maximum height is

$$t = \frac{u}{g\cos\theta}$$

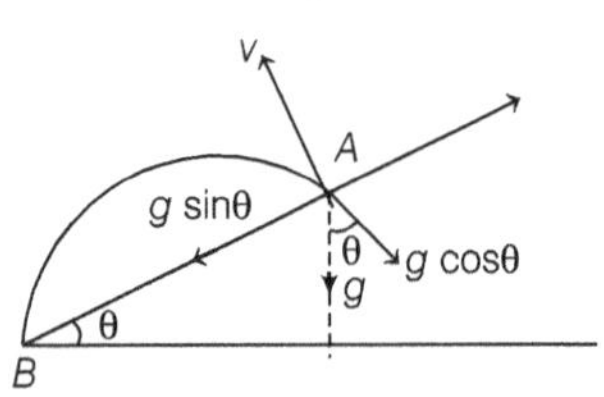

$\therefore$ Time of flight, $T = 2t$

$$= \frac{2u}{g\cos\theta} \quad \ldots\text{(i)}$$

$\therefore$ Distance moved by the particle along AB,

$$R = \frac{1}{2} g\sin\theta \cdot t_1^2$$

$$= \frac{1}{2} g\sin\theta \cdot \frac{4u^2}{g^2\cos^2\theta} \quad [\text{using Eq. (i), where } t_1 = T]$$

$$= \frac{\sin\theta \cdot 2u^2}{g\cos^2\theta}$$

$$= \frac{2u}{g\cos\theta} \cdot \frac{u}{\cos\theta} \cdot \sin\theta$$

$$\Rightarrow R = T \cdot \tan\theta$$

6. ***(d)*** Since, the length of the pendulum is not changed therefore its time period T as well as angular frequency ω remains unchanged.

Now, $v_{\max} = A\omega \Rightarrow v_{\max} \propto A$

Therefore, when amplitude is doubled. The maximum velocity at the mean position also gets doubled 6 cm/s.

7. ***(b)*** Radius of the circular path in magnetic field B is given by

$$r = \frac{mv}{qB}$$

$$\left[\because \text{KE} = \frac{1}{2}mv^2 = \frac{1}{2}\frac{m^2v^2}{m} = E \Rightarrow m^2v^2 = 2mE \Rightarrow mv = \sqrt{2mE}\right]$$

$$\therefore \quad r = \frac{\sqrt{2mE}}{qB}$$

$$\therefore \quad \frac{r_\alpha}{r_p} = \frac{\sqrt{2m_\alpha E} / q_\alpha B}{\sqrt{2m_p E} / q_p B}$$

$$= \sqrt{\frac{m_\alpha}{m_p}} \cdot \frac{q_p}{q_\alpha} = \sqrt{\frac{4 \cdot m_p}{m_p}} \cdot \frac{q_p}{2q_p} \quad \left\{\because m_\alpha \simeq 4m_p \text{ and } q_\alpha = 2q_p\right\}$$

$$\Rightarrow \frac{r_\alpha}{r_p} = \frac{2}{2} = 1$$

$$\therefore \quad r_\alpha = r_p = r_0$$

8. ***(a)*** The distance between zeroth order maxima and second order minima is

$$y_1 = \frac{\beta}{2} + \beta = \frac{3\beta}{2}$$

$$= \frac{3}{2} \times 0.2 = 0.3\text{ mm}$$

$\therefore$ The distance of second order minima from point P is

$$y = (4.8 + 0.3) = 5.1\text{ mm}$$

9. ***(d)*** Maximum static frictional force between the block and the slab is

From, $f = \mu_s mg$

$$f_{\text{ms}} = 0.6 \times 10 \times 9.8\text{ N} = 58.8\text{ N}$$

Since, the applied force is greater than f_{ms}. Therefore, the block will be in motion. So, we should consider kinetic friction f_k.

$$\therefore \quad f_k = \mu_k mg = 0.4 \times 10 \times 9.8$$

$$\Rightarrow \quad f_k = 4 \times 9.8$$

This would cause acceleration in 40 kg slab.

$\therefore$ Acceleration of the slab,

$$a = \frac{f_k}{40} = \frac{4 \times 9.8}{40} = 0.98\text{ ms}^{-2}$$

10. ***(a)*** Horizontal component of velocity of the bomb is $v_x = 500\text{ ms}^{-1}$ and vertical component of velocity of the bomb while striking the ground is given as

$$v_y = u_y + a_y t \quad \text{[By equation of motion]}$$
$$= 0 + 10 \times 10 = 100 \text{ ms}^{-1}$$

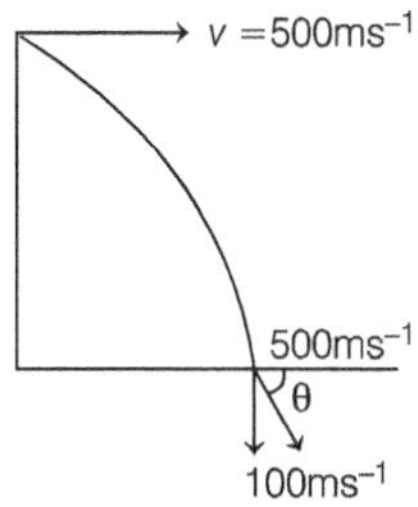

∴ Angle with which the bomb strikes the ground is

$$\theta = \tan^{-1}\left(\frac{v_y}{v_x}\right) = \tan^{-1}\left(\frac{100}{500}\right) \quad \text{or} \quad \theta = \tan^{-1}\left(\frac{1}{5}\right)$$

11. ***(b)*** Average value of induced emf, $\bar{e} = \frac{\Delta\phi}{\Delta t}$

$$= \frac{(NBA\cos\theta - 0)}{\Delta t}$$
$$= \frac{(1 \times 0.5 \times 25 \times 10^{-4} \cos 60° - 0)}{0.2}$$
$$\Rightarrow \quad \bar{e} = 3.12 \times 10^{-3} \text{ V}$$

12. ***(c)*** Stress = (weight due to mass m_2 + half of the weight of rod)/area

$$= (m_2 g + m_1 g / 2) / A$$
$$= [(m_1 / 2) + m_2]g / A$$

13. ***(b)*** $R_{t_1} = R_0(1 + \alpha\, t_1)$ and $R_{t_2} = R_0(1 + \alpha\, t_2)$

$$\therefore \quad \frac{R_{t_2}}{R_{t_1}} = \frac{R(1 + \alpha\, t_2)}{R(1 + \alpha\, t_1)}$$

or

$$\frac{1.5}{1} = \frac{1 + 0.00125 \times t_2}{1 + 0.00125 \times 27}$$

[∵ Given, $t_1 = 300 \text{ K} = 300 - 273°\text{C} = 27°\text{C}$]

On solving, we get

$$t_2 = 440.5°\text{C}$$
$$= 440.5 + 273 = 713.5 \text{ K}$$

14. ***(a)*** Here, $u = 0$, $a = \frac{F}{m} = \frac{-eE}{m}$

$$\therefore \quad v = u + at \quad \text{(From equation of motion)}$$
$$= 0 + \frac{-eE}{m}t$$
$$\Rightarrow \quad v = \frac{-eEt}{m}$$

de-Broglie wavelength,

$$\lambda = \frac{h}{mv} = \frac{h}{-m(eEt/m)} = \frac{h}{-eEt}$$

Rate of change of de-Broglie wavelength,

$$\frac{d\lambda}{dt} = \frac{d}{dt}\left(\frac{h}{-eEt}\right) = \frac{-h}{eE}\left(-\frac{1}{t^2}\right) = \frac{h}{eEt^2}$$

15. ***(c)*** Depth of point P below the free surface of water in the vessel = $(H - h)$, since the liquid exerts equal pressure in all direction at one level, hence the pressure at $P = (H - h)\rho g$.

16. ***(d)*** As, $R \propto n^2$

$$v \propto \frac{1}{n} \text{ and } E \propto \frac{1}{n^2}$$
$$\therefore \quad vR \propto \left(\frac{1}{n} \times n^2\right)$$

i.e. $vR \propto n$

17. ***(d)*** Current gain of CE amplifier, $\beta = \frac{I_C}{I_B}$

$$\therefore \quad \beta = \frac{98\% \text{ of } qt}{2\% \text{ of } qt} = \frac{98\% \text{ of net}}{2\% \text{ of net}}$$
$$= \frac{(98\%) \times 10^{10} \times 1.6 \times 10^{-19} \times 10^{-6}}{(2\%) \times 10^{10} \times 1.6 \times 10^{-19} \times 10^{-6}} = 49$$

18. ***(b)*** Initial velocity, $u = (\hat{\mathbf{i}} + 2\hat{\mathbf{j}})$ m/s

Magnitude of velocity, $u = \sqrt{(1)^2 + (2)^2} = \sqrt{5}$ m/s

Equation of trajectory of projectile,

$$y = x\tan\theta - \frac{gx^2}{2u^2}(1 + \tan^2\theta) \quad \left(\because \tan\theta = \frac{u_y}{u_x} = \frac{2}{1} = 2\right)$$
$$= x \times 2 - \frac{10\,x^2}{2(\sqrt{5})^2}[1 + (2)^2]$$
$$= 2x - \frac{10\,x^2}{2 \times 5}(1 + 4) = 2x - 5x^2$$

19. ***(a)*** For diatomic gas, number of degrees of freedom per molecules is $n = 5$.

Average kinetic energy of diatomic gas due to thermal motion is

$$E = \frac{5}{2}(nR)\,T$$
$$= \frac{5}{2}pV \quad (\because pV = nRT)$$
$$= \frac{5}{2} \times p \times \frac{m}{\rho} \quad \left(\because V = \frac{m}{\rho}\right)$$
$$= \frac{5}{2} \times \frac{(8 \times 10^4) \times 1}{4}$$
$$\Rightarrow \quad E = 5 \times 10^4 \text{ J}$$

20. ***(b)*** As, $\frac{I_{\max}}{I_{\min}} = \frac{(a+b)^2}{(a-b)^2} = 9$

$$\Rightarrow \quad \frac{a+b}{a-b} = 3$$
$$\Rightarrow \quad 3a - 3b = a + b \Rightarrow a = 2b$$
$$\therefore \quad \frac{I_1}{I_2} = \frac{a^2}{b^2} = \frac{4b^2}{b^2} = 4:1$$

21. ***(c)*** Force acting on the particle is given by

$$F = -\frac{dU}{dx} = -\frac{d}{dx}(U_0 - U_0 \cos ax)$$
$$= -aU_0 \sin ax$$

For small oscillations,

$$\sin ax \approx ax$$

$$\therefore \quad F \simeq -a\,U_0(a\cdot x) = -a^2U_0x$$

$$\therefore \text{Acceleration, } \alpha = -\frac{a^2U_0}{m}\cdot x \quad \ldots\text{(i)}$$

as $\qquad \alpha \propto -x$

Motion is simple harmonic motion.

Standard equation of SHM is $\alpha = -\omega^2 x$ ...(ii)

∴ Comparing Eqs. (i) and (ii) we get

$$\omega^2 = \frac{a^2U_0}{m}$$

$$\therefore \text{Time period, } T = \frac{2\pi}{\omega} = 2\pi\sqrt{\frac{m}{a^2U_0}}$$

or $$T = \frac{2\pi}{a}\sqrt{\frac{m}{U_0}}$$

22. ***(b)*** From work-energy theorem,

$$\text{Kinetic energy} = Fs = (qE)\cdot\left(\frac{1}{2}at^2\right) \quad \ldots\text{(i)}$$

where, $$a = \frac{F}{m} = \frac{qE}{m}$$

Put in Eq. (i), we get

$$\therefore \quad \text{KE} = (qE)\cdot\frac{1}{2}\left(\frac{qE}{m}\right)t^2$$

or $$\text{KE} = \frac{1}{2}\cdot\frac{q^2E^2t^2}{m}$$

23. ***(b)*** Effective spring constant of the three springs is

$$k_{\text{eff}} = \frac{k\times 2k}{k+2k} = \frac{2}{3}k$$

$$\therefore \text{Time period, } T = 2\pi\sqrt{\frac{m}{(2k/3)}}$$

$$\Rightarrow \quad T = 2\pi\sqrt{\frac{3m}{2k}}$$

24. ***(c)*** If a sample decays simultaneously through emission of two particles, then the net decay constant

$$\lambda = \lambda_1 + \lambda_2$$
$$\frac{\log_e 2}{T} = \frac{\log_e 2}{T_1} + \frac{\log_e 2}{T_2}$$

or $$\frac{1}{T} = \frac{1}{T_1} + \frac{1}{T_2}$$

or $$T = \frac{T_1 T_2}{T_1 + T_2}$$

25. ***(a)*** From Newton's formula,

$$x_1 x_2 = f^2$$

Given, $\quad x_1 = p$

$$\therefore \quad x_2 = \frac{f^2}{p}$$

Also, $\quad u = f + x_1$

and $\quad v = f + x_2$

$$\therefore \text{Magnification, } m = \frac{\text{Height of image}}{\text{Height of object}} = \frac{v}{u} = \frac{f + x_2}{f + x_1}$$
$$= \frac{f + f^2/p}{f + p} = f/p$$

26. ***(c)*** In steady state, there is no current in capacitor branch.

Net voltage of remaining circuit = $2V - V = V$

Net resistance = $R + 2R = 3R$

Current in the circuit, $i = V/3R$

Potential across lowest row, $V' = 2V - i\cdot 2R$

$$\Rightarrow \quad V' = 2V - \frac{V}{3R}\cdot 2R = \frac{4}{3}V$$

The right end terminal being at higher potential. The potential difference across each row is same.

∴ Potential difference across capacitor,

$$V_C = V' - V$$

$$\Rightarrow \quad V_C = \frac{4}{3}V - V = \frac{V}{3}$$

27. ***(d)*** Initial acceleration of block A is

$$a_1 = \frac{F}{M} = \frac{3}{4} = 0.75 \text{ m/s}^2$$

No horizontal force acts on block B.

∴ Acceleration of block B is zero.

28. ***(c)*** Moment of inertia of the rod, $I_1 = \dfrac{ML^2}{12}$

Moment of inertia of the ring, $I_2 = MR^2$

$$\therefore \quad \frac{I_1}{I_2} = \frac{ML^2/12}{MR^2} = \frac{L^2}{12R^2}$$

Given, length of rod = circumference of the ring

$$L = 2\pi R$$

$$\therefore \quad \frac{I_1}{I_2} = \frac{(2\pi R)^2}{12R^2} = \frac{\pi^2}{3}$$

29. ***(a)*** From hydrogen spectra, $\dfrac{1}{\lambda} = R\left(\dfrac{1}{n_1^2} - \dfrac{1}{n_2^2}\right)$

For second excited state,

$$n_2 = 3, n_1 = 1, \lambda \to \lambda_3$$

$$\frac{1}{\lambda_3} = R\left(\frac{1}{1^2} - \frac{1}{3^2}\right) = \frac{8}{9}R$$

Similarly, for first excited state,

$$n_2 = 2, n_1 = 1, \lambda \to \lambda_2$$

$$\frac{1}{\lambda_2} = R\left(\frac{1}{1^2} - \frac{1}{2^2}\right) = \frac{3}{4}R$$

$$\therefore \quad \frac{\lambda_3}{\lambda_2} = \frac{3/4}{8/9} = \frac{27}{32}$$

30. *(c)* No magnetic force acts on a closed current carrying circular coil in a uniform magnetic field.

$$\therefore \quad F = 0$$

31. *(c)* Let $r_1 = 3$ cm, $r_2 = 4$ cm be the radii of two bubbles.

As bubbles coalesce in vacuum, then the total pressure of air inside the bubble is same as excess pressure $p = 4T/r$.

As process is isothermal, so we have from Boyle's law

$$p_1V_1 + p_2V_2 = pV \quad \ldots(i)$$

If r is the radius of coalesced system, then equation gives

$$\left(\frac{4T}{r_1}\right)\frac{4}{3}\pi r_1^3 + \left(\frac{4T}{r_2}\right)\frac{4}{3}\pi r_2^3 = \frac{4T}{R}\left(\frac{4}{3}\pi r^3\right)$$

$$\Rightarrow \quad r_1^2 + r_2^2 = r^2$$

$$\text{or} \quad r = \sqrt{r_1^2 + r_2^2}$$

$$\text{or} \quad r = \sqrt{3^2 + 4^2} = \sqrt{25}$$

$$\Rightarrow \quad r = 5 \text{ cm}$$

32. *(c)* Given, $u_x = a$

and $u_y = b$

$$\therefore \text{ Range,} \quad R = \frac{u^2 \sin 2\theta}{g}$$

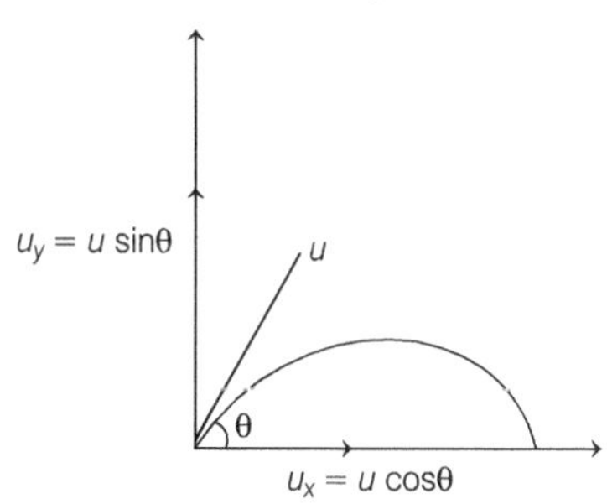

$$= 2\frac{(u\cos\theta)(u\sin\theta)}{g} = \frac{2u_x \cdot u_y}{g}$$

$$\Rightarrow \quad R = \frac{2ab}{g}$$

Maximum height, $H = \dfrac{u^2 \sin^2\theta}{2g} \Rightarrow H = \dfrac{b^2}{2g}$

Given, $R = 2H$

$$\Rightarrow \quad \frac{2ab}{g} = 2 \times \frac{b^2}{2g} \Rightarrow b = 2a$$

33. *(c)* In steady state, capacitor does not allow DC current to flow, therefore equivalent circuit is as shown in figure.

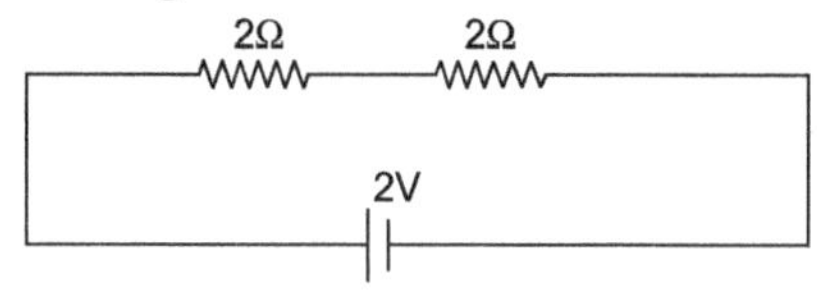

$$\therefore \text{ Power,} \quad P = \frac{V^2}{R} = \frac{(2)^2}{4} \quad [R_{eq} = 2 + 2 = 4\Omega]$$

$$\Rightarrow \quad P = 1 \text{ W}$$

34. *(c)* From law of conservation of linear momentum,

$$mu = -mv_1 + Mv_2 \quad \ldots(i)$$

Remaining energy of α-particle,

$$\frac{1}{2}mv_1^2 = 0.25 \times \frac{1}{2}mu^2$$

$$v_1 = 0.5\, u \quad \ldots(ii)$$

Kinetic energy delivered to nucleus,

$$\frac{1}{2}Mv_2^2 = 0.75 \times \frac{1}{2}mu^2$$

$$v_2 = \sqrt{0.75\frac{m}{M}} \cdot u$$

Substituting v_1 and v_2 in Eq. (i), we get

$$mu = -m \times 0.5\,u + \sqrt{0.75 mM} \cdot u$$

On solving, we get $(1.5\, m)^2 = 0.75 mM$

$$\Rightarrow \quad M = \frac{(1.5)^2}{0.75}m = 3m$$

35. *(c)* The point marked on the ruler at a distance $l/2$ is the centre of mass of ruler. Now, this point is raised by a distance $\left(\frac{l}{2} - h\right)$ change in PE of the whole body $= \Delta U$

$$\text{As,} \quad h = \frac{l\cos 45°}{2}$$

$$\Rightarrow \quad \Delta U = mg\left(\frac{l}{2} - h\right) \quad \ldots(i)$$

$$\therefore \quad \Delta U = mg\left(\frac{l}{2} - \frac{l}{2}\cos 45°\right)$$

$$\Delta U = \frac{mg}{2}\left(1 - \frac{1}{\sqrt{2}}\right) \quad [\because l = 1\text{m}]$$

From the given figure, C is the point of centre of mass,

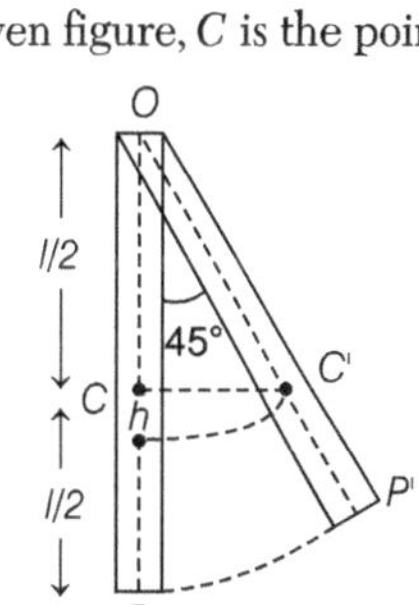

36. ***(a)*** Electric field between the plates, $E = \frac{V}{d}$

$$\Rightarrow \quad d_{min} = \frac{V}{E_{max}} = \frac{V}{\text{Dielectric strength}}$$
$$= \frac{20}{3 \times 10^6} = 6.6 \times 10^{-6} \text{ m}$$

This shows that the value of d in alternative (a) is allowed.

Now, $$C = \frac{K\varepsilon_0 A}{d} \Rightarrow \frac{A}{d} = \frac{C}{K\varepsilon_0}$$

$$\Rightarrow \quad \frac{A}{d} = \frac{1.77 \times 10^{-6}}{200 \times 8.85 \times 10^{-12}} = 10^3$$

or $$A = 10^{-3} \text{ m}$$

37. ***(a)*** In given graph, $a > 0$, so velocity increases.

Change in velocity = v-u = area enclosed by graph
$$= \frac{1}{2} \times 10 \times 11 = 55 \text{ m/s}$$

As $u = 0$

So, maximum velocity, $v = 55$ m/s

38. ***(b)*** The angular momentum,

$$L = mvR_0 = m\sqrt{\frac{GM}{R_0}} \cdot R_0 = m\sqrt{\frac{GM}{R_o} \cdot R_o^2}$$

$$\therefore \quad L = m\sqrt{GMR_0}$$

39. ***(c)*** $\because$ Frequency,

$$n = \frac{1}{2l}\sqrt{\frac{T}{\mu}}$$

where, T = tension in wire

and μ = linear mass density.

Density of mass, $\rho = \frac{m}{v} = \frac{50.7}{0.0075} = 6760 \text{ kg/m}^3$

$$\therefore \quad \frac{n_1}{n_2} = \frac{\frac{1}{2l}\sqrt{\frac{T_1}{\mu}}}{\frac{1}{2\lambda}\sqrt{\frac{T_2}{\mu}}} = \sqrt{\frac{T_1}{T_2}} = \sqrt{\frac{mg}{mg - B}}$$

where, B = force of buoyancy,

$$= \sqrt{\frac{\rho V g}{(\rho V - \rho_1 V)g}} = \sqrt{\frac{\rho}{\rho - \rho_1}}$$

where, ρ = density of mass

ρ_1 = density of water

$= 10^3 \text{ kg/m}^3$

$$\Rightarrow \quad \frac{n_1}{n_2} = \sqrt{\frac{6760}{6760 - 1000}}$$
$$= \sqrt{\frac{6760}{5760}} = \sqrt{1.1736} = 1.0833$$

$$\therefore \quad n_2 = \frac{n_1}{1.0833} = \frac{260}{1.0833} = 240 \text{ Hz}$$

40. ***(a)*** The force acting on current carrying wire,

$$F = iBl\sin\theta$$

$$\therefore \quad 7.5 = 5 \times 2 \times 1.5 \sin\theta$$

$$\sin\theta = \frac{7.5}{2 \times 5 \times 1.5} = \frac{1}{2}$$

$$\Rightarrow \quad \theta = 30°$$

Chemistry

41. ***(d)*** Molecular mass of $(CHCOO)_2Fe = 170$

$\therefore$ In 100 mg $(CHCOO)_2Fe$, iron present $= \frac{56}{170} \times 100$
$= 32.9$ mg

Since, this quantity of Fe is present in 400 mg of capsule.

$\therefore$ % of Fe in capsule $= \frac{32.9}{400} \times 100 = 8.2 \approx 8\%$

42. ***(c)*** $\Delta E_I = E_2 - E_1$

$$= \frac{E_H}{(n_2)^2} - \frac{E_H}{(n_1)^2} = E_H\left(\frac{1}{2^2} - \frac{1}{1^2}\right) = -\frac{E_H \times 3}{4}$$

$$\Delta E_{II} = E_3 - E_2 = E_H\left[\frac{1}{3^2} - \frac{1}{2^2}\right] = -E_H \times \frac{5}{36}$$

$$\frac{\Delta E_I}{\Delta E_{II}} = E_H \times \frac{3}{4} \times \frac{36}{E_H \times 5} = \frac{108}{20} = \frac{27}{5}$$

43. ***(b)*** Hydrogen like halogen accepts an electron readily to achieve nearest inert gas configuration.

44. ***(b)***

Species	Electrons in central element	Electrons in other element	Charge gained	Total electrons
BO_3^{3-}	5	$3 \times 8 = 24$	+ 3	32
CO_3^{2-}	6	$3 \times 8 = 24$	+ 2	32
NO_3^-	7	$3 \times 8 = 24$	+ 1	32
SO_3^{2-}	16	$3 \times 8 = 24$	+ 2	42
CN^-	6	7	+1	14
N_2	7	7	0	14
C_2^{2-}	6	6	+ 2	14
PO_4^{3-}	15	$4 \times 8 = 32$	+ 3	50
SO_4^{2-}	16	$4 \times 8 = 32$	+ 2	50
ClO_4^-	17	$4 \times 8 = 32$	+ 1	50

Thus, SO_3^{2-}, CO_3^{2-} and NO_3^- are not isoelectronic species.

45. ***(c)*** $\mu_{cal} = e \times d = 1.6 \times 10^{-19} \times 2.29 \times 10^{-10}$

$$= 3.664 \times 10^{-29} \text{ C-m}$$

$$\% \text{ ionic character} = \frac{\mu_{exp}}{\mu_{cal}} \times 100 = \frac{6.226 \times 10^{-30}}{3.664 \times 10^{-29}} \times 100 = 16.9\% \simeq 17\%$$

46. ***(b)*** In the given compound, there are two chiral carbon atoms. i.e. two carbon atoms are asymmetric.

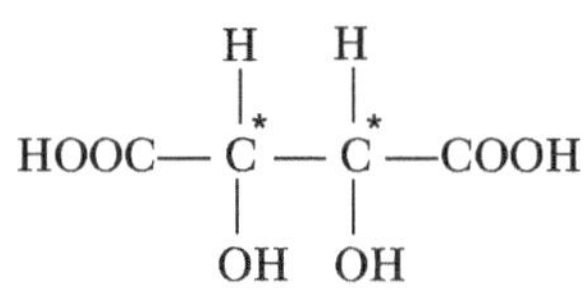

47. ***(d)***

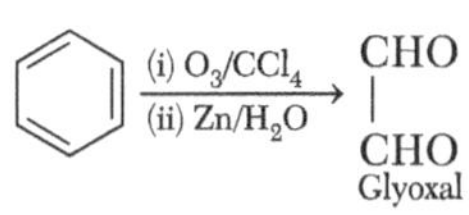

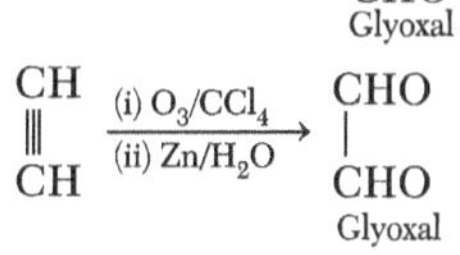

Ethene on reductive ozonolysis gives two molecules of formaldehyde.

$$H_2C{=}CH_2 \xrightarrow[\text{(ii) Zn/H}_2\text{O}]{\text{(i) O}_3\text{/CCl}_4} 2HCHO$$

48. ***(d)*** $M^{2+} \longrightarrow M^{3+}$, after the removal of second electron, the effective nuclear charge per electron increases due to which high energy is required to remove third electron.

49. ***(b)*** Boiling point of alcohols and phenols increases with increase in the number of carbon atoms (increase in van der Waals' forces). In alcohols, the boiling points decrease with increase of branching in carbon chain (because of decrease in van der Waals' forces with decrease in surface area).

50. ***(a)*** NaOH reacts with Al to evolve dihydrogen gas. The higher pressure of the hydrogen gas can be used to open drains.

$$2Al(s) + 2NaOH(aq) + 2H_2O(l) \longrightarrow 2NaAlO_2(aq) + 3H_2(g)$$

51. ***(b)*** Because the number of current carrying particles, i.e. ions present per centimetre cube of the solution become lesser on dilution. Hence, the conductivity of strong electrolytes decreases on dilution.

52. ***(d)*** 1.00 kg of solution contains 0.100 kg of CH_3OH and 0.900 kg H_2O.

Number of moles in 100 g of

$$CH_3OH = \left(\frac{100}{32.0}\right) = 3.12 \text{ mol}$$

$$\therefore \text{Molality of } CH_3OH \text{ in 0.90 kg } H_2O = \frac{3.12 \text{ mol}}{0.900 \text{ kg}} = 3.47 \text{ m}$$

$$\Delta T_f = K_f \times m = (1.86° \text{ C/m})(3.47 \text{ m}) = 6.45°C$$

Freezing point of the solution $(T_f) = T_f° - \Delta T_f$

$$= 0 - 6.45 = -6.45°C$$

53. ***(a)*** Metals having very high tendency to lose electrons are used in photoelectric cells. Lower the ionisation energy, higher is the tendency to lose electrons.

Potassium and cesium have much lower ionisation enthalpy (IE) than that of lithium. Therefore, these metals on exposure to light emit electrons easily but lithium does not. That's why, K and Cs rather than Li are used in photoelectric cells.

54. ***(b)*** $RCH_2Cl + NaI \xrightarrow[\Delta]{\text{Acetone}} RCH_2I + NaCl$

This reaction is known as Finkelstein reaction or halogen exchange reaction.

All other reactions are not feasible.

55. ***(a)*** Copper matte contains Cu_2S and FeS.

56. ***(d)*** If the reaction is an elementary reaction, then order and molecularity have same value.

57. ***(c)*** Dolomite = $MgCO_3 \cdot CaCO_3$

Magnesite = $MgCO_3$

Carnallite = $KCl \cdot MgCl_2 \cdot 6H_2O$

Therefore, Mg metal is extracted from the above minerals.

58. ***(b)*** In the complex, $K_2Fe[Fe(CN)_6]$, both the iron (Fe) atoms are present in the same oxidation state, i.e. +2.

59. ***(a)*** Novolac on heating with formaldehyde undergoes cross-linking to form an infusible solid mass called bakelite.

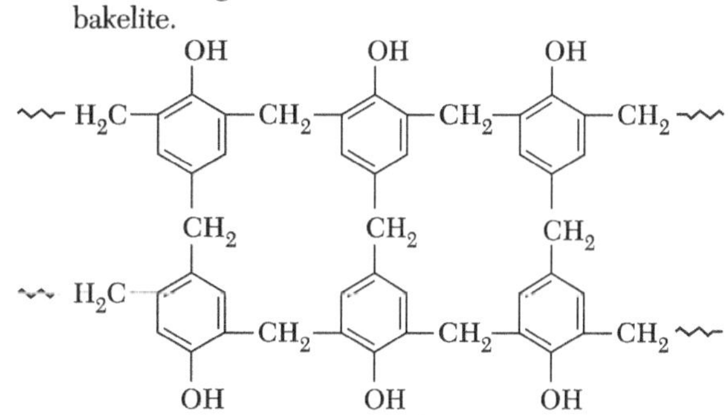

Bakelite

60. ***(c)*** In diamond, the constituent atoms are held together by strong covalent bonds. It is a giant molecule. Thus, classified as network solid.

Note *SO_2, H_2O and I_2 are examples of molecular solids.*

61. ***(c)*** 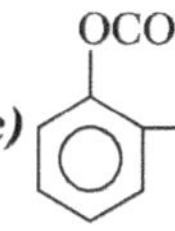 —COOH is aspirin. It is an acetylated product of salicylic acid and used as an analgesic.

62. ***(b)*** Thymine is not present in RNA.

As DNA contains four bases *viz.* adenine ***(A)***, guanine (G), cytosine (C) and thymine (T). RNA also contains four bases, the first three bases are same as in DNA but the fourth one is uracil (U).

63. ***(b)*** In Duma's method, % of nitrogen

$$= \frac{28 \times V \times 100}{22400 \times W}$$

Given, $W = 0.3$ g, $V = 45$ mL

$$\% \text{ Nitrogen } = \frac{28 \times 45 \times 100}{22400 \times 0.3} = 18.75$$

64. ***(a)*** Greater the conjugation, greater the stability and hence, easier the dehydration.

65. ***(b)*** Number of spherical nodes in $3p$-orbital

$$= n - l - 1 = 3 - 1 - 1 = 1$$

66. ***(c)*** $Sc = [Ar]\, 3d^1\, 4s^2$

$n = 4$

Atomic number of element = 21

Group = total number of valence shell electrons = 3

So, the element belongs to group-3, period-4.

67. ***(d)*** CO (14 electrons)

CO

$$= \sigma 1s^2, \overset{*}{\sigma} 1s^2, \sigma 2s^2, \overset{*}{\sigma} 2s^2, \pi\, 2p_x^2 \approx \pi 2p_y^2, \sigma\, 2p_z^2$$

σ-bonding molecular orbital is HOMO.

68. ***(c)*** We know that, internal energy of a gas depends upon its pressure and temperature. Thus, if a gas expands at constant temperature and pressure then its internal energy remains same.

69. ***(a)*** Conjugate base is formed by loss of H^+.

$$OH^- \longrightarrow \underset{\text{(Conjugate base)}}{O^{2-}} + H^+$$

O^{2-} is the conjugate base of OH^-.

70. ***(d)*** In adsorption due to forces of attraction, ΔH is negative (heat is released) and as the particles come closer, entropy of the system decreases, i.e. ΔS is negative, hence $T\Delta S$ is also negative.

71. ***(d)*** Formula of gypsum = $CaSO_4 \cdot 2H_2O$

Formula of plaster of Paris = $CaSO_4 \cdot \frac{1}{2} H_2O$

Difference of water molecules = $2H_2O - \frac{1}{2}\, H_2O$

$$= 1\frac{1}{2}\, H_2O$$

72. ***(a)*** Boron when treated with carbon forms boron carbide.

$$4B + C \xrightarrow{\Delta} B_4C$$

73. ***(b)*** Cyclodecapentene and cyclooctatetraene both are non-aromatic. Both follow the Huckel rule of aromaticity but their rings are not flat. Cyclobutadiene is anti-aromatic. While benzene having 6π-electrons is aromatic.

74. ***(d)*** $$\underbrace{\log K_{eq}}_{(A)} = \underbrace{- \frac{\Delta H^\circ}{2.303\, RT} + \frac{\Delta S^\circ}{R}}_{(B)}$$

Slope of the given line indicating that term A is positive, thus ΔH° is negative. Thus, reaction is exothermic.

75. ***(a)*** Clathrates are non-stoichiometric compounds where the ratio of guest and host molecules does not correspond to ideal chemical formula.

76. ***(a)*** Due to availability of unpaired d-electrons, d-d transition occurred and hence, transition elements are coloured.

77. ***(a)*** The electronic configuration of ${}_{62}Sm^{3+}$ is $[Xe]4f^5 5d^0 6s^0$ and that of ${}_{66}Dy^{3+}$ is $[Xe]4f^9 5d^0 6s^0$. The colour of f^n and f^{14-n} are often identical, i.e. yellow.

78. ***(c)*** Phenol because of the resonance stabilisation of phenoxide ion, have higher tendency to give a proton instead of accepting. It is the weakest Bronsted base or strongest Bronsted acid.

79. ***(c)***

(a) $1 \text{g Au} = \frac{1}{197}$ mol atoms of Au

$= \frac{1}{197} \times 6.022 \times 10^{23}$ atoms of Au

(b) $1 \text{g Na} = \frac{1}{23}$ mol atoms of Na

$= \frac{1}{23} \times 6.022 \times 10^{23}$ atoms of Na

(c) $1 \text{g Li} = \frac{1}{7}$ mol atoms of Li

$= \frac{1}{7} \times 6.022 \times 10^{23}$ atoms of Li

(d) $1 \text{g } Cl_2 = \frac{1}{71}$ mol molecules of Cl_2

$= \frac{1}{71} \times 6.022 \times 10^{23}$ molecules of Cl_2

$= \frac{2}{71} \times 6.022 \times 10^{23}$ atoms of Cl

($\because$ 1 molecule of chlorine contains 2 atoms)

Therefore, 1 g of Li has largest number of atoms.

Note If a substance exists in atomic form, it contains mol atoms and if a substance exists in molecular form, it contains mol molecules.

80. ***(b)*** We know that, density $(d) = pM/RT$

$$d \propto \frac{1}{T} \text{ and } d \propto p$$

Thus, density of neon is maximum at 0°C and 2 atm.

a. English Proficiency

81. ***(a)*** 'Traced' means 'searched'. Hence, 'searched' is the correct choice.

82. ***(a)*** 'Gargantuan' means 'gigantic' or 'colossal' which has a sense of being too big.

83. ***(b)*** 'Propitiate' means 'to appease' or 'to bring favourable remarks'.

84. ***(a)*** 'Run down' is a phrasal verb which means 'to be crushed' or 'to be trampled' which makes the sense here.

85. ***(c)*** The event is of past tense. So, the verb must be used in past tense. Hence, 'was taken' is correct choice. Also, the mode is passive.

86. ***(d)*** 'Wages' is the correct choice here. Salary, pay are not suitable usage because in agricultural context, it is improper to be used. These are basically used as corporate terms. But, 'wages' is suitable here.

87. ***(c)*** 'A man of bad reputation' is called to be 'notorious'. Hence, the opposite will be 'famous'.

88. ***(b)*** 'Perishable' means 'which has a destined end'. Hence, its opposite will be 'immortal' which means 'that will never end'.

89. ***(c)*** 'Penury' means 'the state of being poor'. Hence, the opposite will be 'opulence' which means 'having no scarcity' or 'being in plenty'.

90. ***(d)*** One who does not consume alcohol is called 'teetotaller'.

91. ***(a)*** 'Hermaphrodite' is an animal that has both sex inside its body.

92. ***(c)*** A text that doesn't bear the name of an author is called 'unanimous'.

Q. Nos. 93 to 95

On rearranging the given sentences, the correct and logical order of the sentences will be CAFDBF. Taking the reference, we can answer the questions

93. ***(d)*** D **94.** ***(b)*** B

95. ***(d)*** E

b. Logical Reasoning

96. ***(c)*** 'Ribs' protect the 'Lungs'. Similarly, 'Shell' protects the 'Nut'.

97. ***(d)*** All except 18, 36, 40 have second number double the first number and third number 6 more than the second number.

98. ***(c)*** The pattern is as follows,

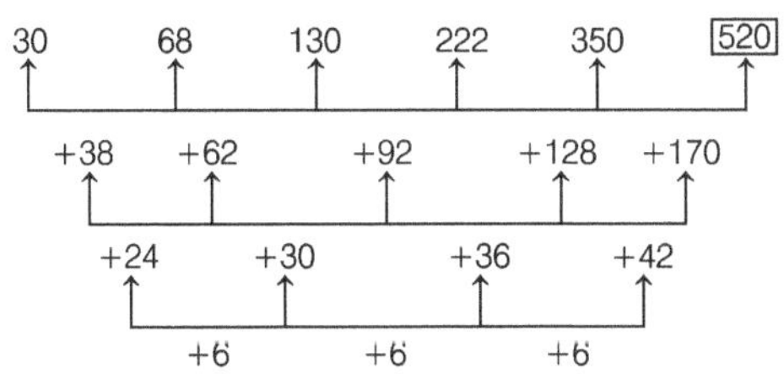

∴ ? = 520

99. ***(c)*** The third figure in each row comprises of parts which are not common to the first two figures.

100. ***(b)*** Figure (X) is the first step in which a circular piece of paper is folded from upper to the lower half along the diameter. In figure (Y), both the extreme ends of the figure (X) (after folding) have been folded to form a triangle and then as given in figure (Z), a cut has been marked on the right side.

It is clear that this cut will result into two marks, one in the lower half and one in the upper half of the paper, when it is unfolded. Answer figure (b), represents the unfolded shape of the paper after the cut.

101. ***(c)*** As done in above example, we have to closely approximate the different formations. So, first we combine the figures (E) and (A) which will look like as

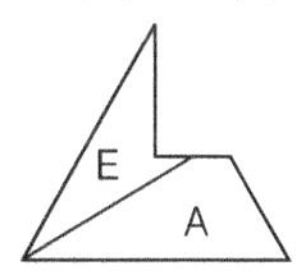

Now, the missing figure which will complete the above figure is figure (D). The complete figure will be as

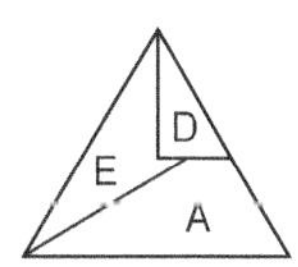

102. ***(b)*** All the four parts are similar except for a diagonal line passing through the upper right and lower left parts. Also, each part comprises of a quarter circle in such a way that there is a semi-circle on a pair of opposite sides of the question figure. Clearly, answer figure (b) completes the pattern.

103. ***(c)***

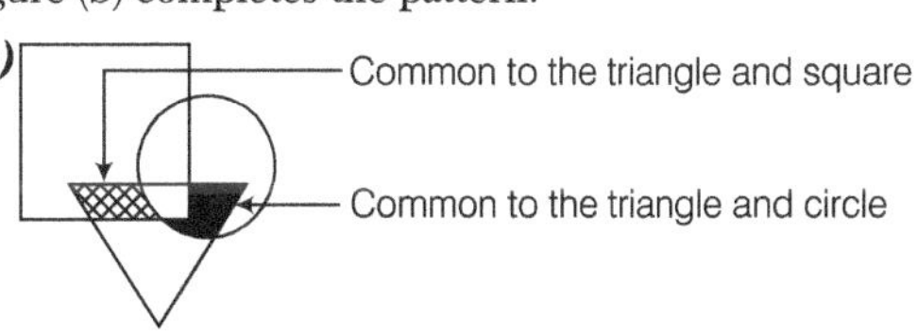

Hence, figure shown in option (c) is correct figure.

104. *(d)*

Volumes	Green cover	Yellow cover	New volume	Old volume	Law reports	Medical reports
A	—	✓	✓	—	✓	—
B	✓	—	✓	—	✓	—
C	✓	—	—	✓	✓	—
D	—	✓	✓	—	—	✓
E	✓	—	—	✓	—	✓
F	✓	—	—	✓	—	✓

Clearly, E and F are old volumes which have green covers and are medical extracts.

105. *(c)* According to question in figure 3, closed figures gradually become open and open figure gradually become closed.

Mathematics

106. *(c)* We have,

$$\lim_{x \to a}\left(2-\frac{a}{x}\right)^{\tan\frac{\pi x}{2a}} \qquad [\because \text{form } 1^\infty]$$

$$= \lim_{x \to a}\left\{1+\left(1-\frac{a}{x}\right)\right\}^{\tan\frac{\pi x}{2a}} = e^{\lim_{x\to a}\left(\frac{x-a}{x}\right)\tan\frac{\pi x}{2a}}$$

$$= e^{\lim_{h\to 0}\left(\frac{h}{a+h}\tan\frac{\pi}{2a}(a+h)\right)} = e^{\lim_{h\to 0}\left(\frac{h}{a+h}\tan\left(\frac{\pi}{2}+\frac{\pi h}{2a}\right)\right)}$$

$$= e^{\lim_{h\to 0}\left(-\frac{h}{a+h}\cot\frac{\pi h}{2a}\right)} = e^{\lim_{h\to 0}\left(\frac{-h}{a\tan \pi h/2a}\right)}$$

$$= e^{\lim_{h\to 0}\left(-\frac{2}{\pi}\left\{\frac{\frac{\pi h}{2a}}{\tan \pi h/2a}\right\}\right)} = e^{-2/\pi}$$

107. *(d)* We have,

$$|8z_2z_3+27z_3z_1+64z_1z_2|$$

$$= \left|z_1z_2z_3\left(\frac{8}{z_1}+\frac{27}{z_2}+\frac{64}{z_3}\right)\right|$$

$$= |z_1z_2z_3|\left|\frac{8}{z_1}+\frac{27}{z_2}+\frac{64}{z_3}\right|$$

$$= |z_1||z_2||z_3|\left|\frac{8\bar{z}_1}{|z_1|^2}+\frac{27\bar{z}_2}{|z_2|^2}+\frac{64\bar{z}_3}{|z_3|^2}\right|$$

$$= 24\,|2\bar{z}_1+3\bar{z}_2+4\bar{z}_3|$$

$$= 24\,|\overline{2z_1+3z_2+4z_3}|$$

$$= 24 \times 4 = 96$$

108. *(b)* We have,

$$\Delta = \begin{vmatrix} 1 & 1 & 1 \\ 1 & -1-\omega^2 & \omega^2 \\ 1 & \omega^2 & \omega^4 \end{vmatrix}$$

$$\Rightarrow \quad \Delta = \begin{vmatrix} 1 & 0 & 0 \\ 1 & -2-\omega^2 & \omega^2-1 \\ 1 & \omega^2-1 & \omega^4-1 \end{vmatrix}$$

[Applying $C_2 \to C_2 - C_1, C_3 \to C_3 - C_1$]

$$\Rightarrow \Delta = (-2-\omega^2)(\omega-1)-(\omega^2-1)^2 \quad [\because \omega^3 = 1]$$

$$\Rightarrow \Delta = -(2\omega+\omega^3-2-\omega^2)-(\omega^4-2\omega^2+1)$$

$$\Rightarrow \Delta = -(2\omega-1-\omega^2)-(\omega-2\omega^2+1)$$

$$\Rightarrow \Delta = -(2\omega+\omega)-(-3\omega^2) = -3\omega+3\omega^2 = 3\omega(\omega-1)$$

109. *(c)* Let T_r be the rth term of the given series. Then,

$$T_r = \frac{1}{(2r-1)(2r+1)}, \; r = 1, 2, 3 \ldots\ldots n$$

$$\Rightarrow \quad T_r = \frac{1}{2}\left(\frac{1}{2r-1}-\frac{1}{2r+1}\right)$$

Required sum S is given by

$$S = \sum_{r=1}^{n} T_r$$

$$\Rightarrow \quad S = \frac{1}{2}\left[\left(\frac{1}{1}-\frac{1}{3}\right)+\left(\frac{1}{3}-\frac{1}{5}\right)+\left(\frac{1}{5}-\frac{1}{7}\right)+ \ldots + \left(\frac{1}{2n-1}-\frac{1}{2n+1}\right)\right]$$

$$\Rightarrow \quad S = \frac{1}{2}\left[1-\frac{1}{2n+1}\right] = \frac{n}{2n+1}$$

110. *(b)* We have, $f(x) = |\sin x| - |\cos x|$

$\therefore \quad f(x+\pi) = |\sin(x+\pi)| - |\cos(x+\pi)|$

$\Rightarrow \quad f(x+\pi) = |-\sin x| - |-\cos x|$

$\Rightarrow \quad f(x+\pi) = |\sin x| - |\cos x| = f(x)$ for all $x \in R$

$\Rightarrow \quad f(x+\pi) = f(x)$ for all $x \in R$

So, $f(x)$ is periodic with period π.

BITSAT 2016

111. ***(b)*** We have, $f(x) = x^3 - x^2 + x + 1$

$$f'(x) = 3x^2 - 2x + 1 > 0 \text{ for all } x$$

$[\because$ Discriminant < 0 and coefficient of $x^2 > 0]$

$\Rightarrow$ $f(x)$ is increasing for all $x \in R$.

$$\therefore \quad g(x) = \begin{cases} x^3 - x^2 + x + 1, & 0 \le x \le 1 \\ 3 - x, & 1 < x \le 2 \end{cases}$$

Clearly, $g(x)$ is everywhere continuous and differentiable except possible at $x = 1$.

Clearly, $\lim_{x \to 1^-} g(x) = \lim_{x \to 1^+} g(x) = g(1)$

So, it is continuous at $x = 1$.

We observe that

(LHD at $x = 1$)

$$= \frac{d}{dx}(x^3 - x^2 + x + 1)\}_{x=1}$$

$$= (3x^2 - 2x + 1)_{\text{at } x=1} = 2$$

and (RHD at $x = 1$) $= \frac{d}{dx}(3 - x)_{x=1} = -1$

Clearly, $g(x)$ is not differentiable at $x = 1$.

112. ***(d)*** The area enclosed by $y = \log_e(x + e)$ and the coordinate axes is shaded in figure.

Let A be the required area. Then,

$$A = \int_{1-e}^{0} y\,dx = \int_{1-e}^{0} \log_e(x + e)\,dx$$

$$\Rightarrow \quad A = [x \log_e(x + e)]_{1-e}^{0} - \int_{1-e}^{0} \frac{x}{x + e}\,dx$$

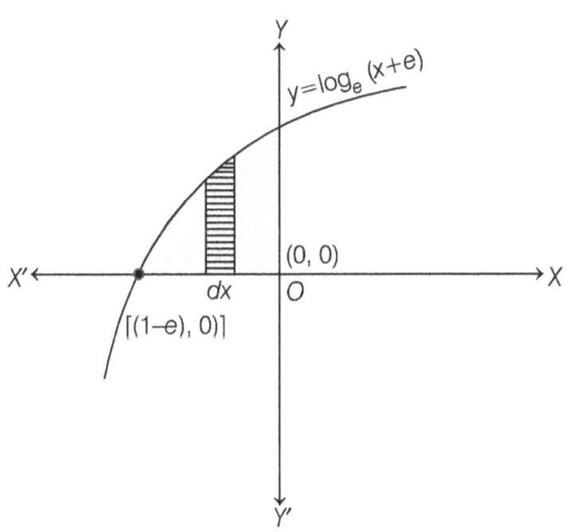

We have to find the area of shaded region.

$$A = [x \log_e(x + e)]_{1-e}^{0} - [x - e \log_e(x + e)]_{1-e}^{0}$$

$$\Rightarrow \quad A = 0 - [(0 - e) - (1 - e) + 0] = 1 \text{ sq. unit}$$

113. ***(a)*** The divisibility of the product of four numbers depends upon the value of the last digit of each number. The last digit of a number can be any one of the ten digits 0, 1, ... 9. So, the total number of ways of selecting last digit of four numbers is $10 \times 10 \times 10 \times 10 = 10^4$

If the product of the 4 numbers is not divisible by 5 or 10, then the number of choices for the last digit of each number is 8 (excluding 0 or 5).

So, favourable number of ways $= 8^4$

$\therefore$ The probability that the product is not divisible by 5 or 10 is $\left(\frac{8}{10}\right)^4$.

Hence, required probability $= 1 - \left(\frac{8}{10}\right)^4 = \frac{369}{625}$

114. ***(a)*** Clearly, X is a binomial variate with parameters n and $p = 1/2$, such that

$$P(X = r) = {}^nC_r\, p^r q^{n-r} = {}^nC_r \left(\frac{1}{2}\right)^r \left(\frac{1}{2}\right)^{n-r} = {}^nC_r \left(\frac{1}{2}\right)^n$$

Now,

$P(X = 4)$, $P(X = 5)$ and $P(X = 6)$ are in AP.

$$\Rightarrow \quad 2P(X = 5) = P(X = 4) + P(X = 6)$$

$$\Rightarrow \quad 2\,{}^nC_5 \left(\frac{1}{2}\right)^n = {}^nC_4 \left(\frac{1}{2}\right)^n + {}^nC_6 \left(\frac{1}{2}\right)^n$$

$$\Rightarrow \quad 2.\,{}^nC_5 = {}^nC_4 + {}^nC_6$$

$$\Rightarrow \quad 2\frac{n!}{(n-5)!\,5!} = \frac{n!}{(n-4)!\,4!} + \frac{n!}{(n-6)!\,6!}$$

$$\Rightarrow \quad \frac{2}{5(n-5)} = \frac{1}{(n-4)(n-5)} + \frac{1}{6 \times 5}$$

$$\Rightarrow \quad n^2 - 21n + 98 = 0$$

$$\Rightarrow (n - 7)(n - 14) = 0 \Rightarrow n = 7, 14$$

115. ***(b)*** For any $a \in N$, we have

(GCD of a and a) $= a$

So, R is not reflexive.

Let $(a, b) \in R$. Then,

$a\,R\,b$

$\Rightarrow$ GCD of a and b is 2.

$\Rightarrow$ GCD of b and a is 2.

$\Rightarrow$ $b\,R\,a$

So, R is symmetric.

We observe that,

GCD of 6 and 4 is 2 and GCD of 4 and 18 is also 2.

But GCD of 6 and 18 is 6.

i.e. $6R4$ and $4\,R\,18$ but $6 \not{R}\, 18$.

So, R is not transitive.

116. ***(c)*** We have,

In the expansion of $(1 + \alpha x)^4$, There are 5 terms.

middle term $T_3 = {}^4C_2\,(\alpha x)^2 = 6\alpha^2 x^2$

In the expansion of $(1 - \alpha x)^6$, There are 7 terms.

middle term $T_4 = {}^6C_3(-\alpha x)^3 = -20\alpha^3x^3$

Since, coefficient of middle term in $(1+\alpha x)^4$

$= \text{Coefficient of middle term in } (1-\alpha x)^6$

$\Rightarrow \quad 6\alpha^2 = -20\alpha^3$

$\Rightarrow \alpha \neq 0, \alpha = \dfrac{-3}{10}$

117. ***(b)*** Given equation is $2\cos x \sin x - 2\cos^2 x = n$

$\Rightarrow \quad \sin 2x - 1 - \cos 2x = n \quad [\because \sin 2x = 2\sin x \cos x]$

$\Rightarrow \quad \sin 2x - \cos 2x = n+1, \quad -\sqrt{2} \leq n+1 \leq \sqrt{2}$

$\therefore$ Greatest integral value of $n = 0$

118. ***(b)*** The equation of the circle described on SS' as a diameter is

$$(x-ae)(x+ae)+(y-0)(y-0)=0$$

$$\Rightarrow \quad x^2 - a^2e^2 + y^2 = 0 \Rightarrow x^2+y^2 = a^2e^2$$

The abscissae of the points of intersection of the ellipse and this circle are the roots of the equation.

$$\frac{x^2}{a^2} + \frac{a^2e^2 - x^2}{b^2} = 1$$

$$\Rightarrow \quad x^2\left(\frac{1}{a^2} - \frac{1}{b^2}\right) = 1 - \frac{a^2e^2}{b^2}$$

$$\Rightarrow \quad x^2\left(\frac{b^2-a^2}{a^2b^2}\right) = \frac{b^2-a^2e^2}{b^2} \quad [\because b^2 = a^2(1-e^2)]$$

$$\Rightarrow \quad -\frac{x^2a^2e^2}{a^2} = a^2 - 2a^2e^2$$

$$\Rightarrow \quad x^2 = a^2\left(\frac{2e^2-1}{e^2}\right)$$

$$\Rightarrow \quad x = \pm\frac{a}{e}\sqrt{2e^2-1}$$

This will give distinct value of x, if $2e^2 - 1 > 0$ i.e. $e > \dfrac{1}{\sqrt{2}}$.

Hence, $\quad e \in (1/\sqrt{2}, 1)$

119. ***(b)*** Let E be the event that A wins the game and F be the event that B wins the game. Then,

$$P(E) = \frac{6}{12} = \frac{1}{2} \text{ and } P(F) = \frac{4}{12} = \frac{1}{3}$$

$\therefore$ Required probability

$$= P\{(E \cap F \cap E) \cup (F \cap E \cap F)\}$$
$$= P\{(E \cap F \cap E) + P(F \cap E \cap F)\}$$
$$= P(E)\,P(F)\,P(E) + P(F)\,P(E)\,P(F)$$
$$= \frac{1}{2}\times\frac{1}{3}\times\frac{1}{2}+\frac{1}{3}\times\frac{1}{2}\times\frac{1}{3} = \frac{5}{36}$$

120. ***(c)*** We have,

$$x = \frac{1}{1-a}, y = \frac{1}{1-b}, z = \frac{1}{1-c}$$

Since, a, b, c are in AP.

$\Rightarrow \quad 1-a, 1-b, 1-c$ are in AP.

$\Rightarrow \quad \dfrac{1}{1-a}, \dfrac{1}{1-b}, \dfrac{1}{1-c}$ are in HP.

$\Rightarrow \quad x, y, z$ are in HP.

121. ***(b)*** Let the sides of the triangle be a/r, a and ar, with $a > 0$ and $r > 1$. Let α be the smallest angle, so that the largest angle is 2α. Then, α is opposite to the side a/r, and 2α is opposite to the side ar.

Applying sine rule, we get

$$\frac{a/r}{\sin\alpha} = \frac{ar}{\sin 2\alpha}$$

$$\Rightarrow \quad \frac{\sin 2\alpha}{\sin\alpha} = r^2 \Rightarrow 2\cos\alpha = r^2$$

$\Rightarrow \quad r^2 < 2$

$\Rightarrow \quad r < \sqrt{2} \quad [\because \alpha \neq 0, \therefore 2\cos\alpha < 2]$

Hence, $\quad 1 < r < \sqrt{2}$

122. ***(b)*** Feasible region is $OACBO$ and $\mu = 3x + 4y$.

At point O (0, 0)

$$\mu = 0 + 0 = 0$$

At point A (40, 0)

$$\mu = 3(40) + 4(0) = 120$$

At point B (0, 30)

$$\mu = 0 + 4(30) = 120$$

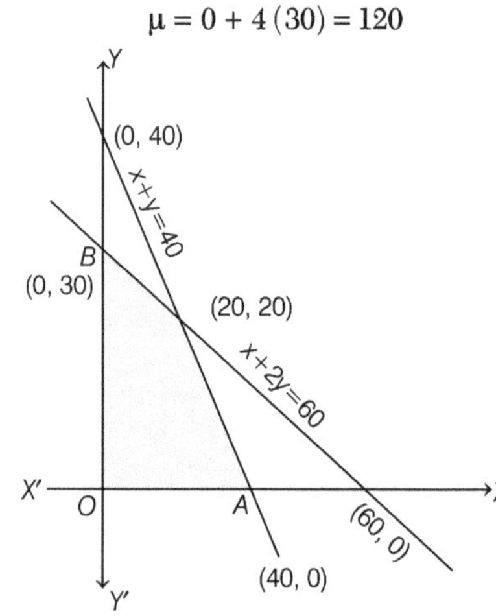

At point C (20, 20)

$$\mu = 3(20) + 4(20) = 140$$

$\therefore$ The maximum value at the point C (20, 20) is 140.

123. ***(c)*** We have, $n(A) = 2$ and $n(B) = 4$

$\therefore \quad n(A \times B) = n(A) \times n(B) = 2 \times 4 = 8$

The number of subsets at $A \times B$ having 3 as more elements

$$= {}^8C_3 + {}^8C_4 + {}^8C_5 + {}^8C_6 + {}^8C_7 + {}^8C_8$$
$$= ({}^8C_0 + {}^8C_1 + \ldots + {}^8C_8) - ({}^8C_0 + {}^8C_1 + {}^8C_2)$$
$$= 2^8 - (1 + 8 + 28) = 219$$

124. ***(d)*** Since, $a_1, a_2, a_3 \ldots, a_n$ are in HP.

$\therefore \frac{1}{a_1}, \frac{1}{a_2}, \frac{1}{a_3}, \ldots, \frac{1}{a_n}$ are in AP.

Let d be the common difference of the AP then,

$$\frac{1}{a_2} - \frac{1}{a_1} = d, \frac{1}{a_3} - \frac{1}{a_2} = d, \ldots \frac{1}{a_n} - \frac{1}{a_{n-1}} = d$$

$$\Rightarrow \quad a_1 - a_2 = d(a_1 a_2), a_2 - a_3 = d(a_2 a_3),$$
$$a_3 - a_4 = d(a_3 a_4), \ldots, a_{n-1} - a_n = d(a_{n-1} a_n)$$

$$\Rightarrow \quad (a_1 - a_2) + (a_2 - a_3) + \ldots + (a_{n-1} - a_n)$$
$$= d(a_1 a_2 + a_2 a_3 + \ldots + a_{n-1} a_n)$$

$$\Rightarrow \quad a_1 - a_n = d(a_1 a_2 + a_2 a_3 + \ldots + a_{n-1} a_n) \quad \ldots \text{(i)}$$

Now, $\frac{1}{a_1}, \frac{1}{a_2}, \frac{1}{a_3}, \ldots, \frac{1}{a_n}$ are in AP with common difference d.

$$\Rightarrow n\text{th term } \frac{1}{a_n} = \frac{1}{a_1} + (n-1)d$$

$$\Rightarrow \frac{1}{a_n} - \frac{1}{a_1} = (n-1)d \Rightarrow \frac{a_1 - a_n}{a_1 a_n} = (n-1)d$$

$$\Rightarrow \quad a_1 - a_n = (n-1)d(a_1 a_n) \quad \ldots \text{(ii)}$$

From Eqs. (i) and (ii), we get

$$(n-1)d(a_1 a_n) = d(a_1 a_2 + a_2 a_3 + \ldots + a_{n-1} a_n)$$
$$\Rightarrow \quad (n-1) a_1 a_n = a_1 a_2 + a_2 a_3 + \ldots + a_{n-1} a_n$$

125. ***(a)*** The equation of the plane through the intersection of the planes $x + y + z = 1$ and $2x + 3y - z + 4 = 0$ is

$$(x + y + z - 1) + \lambda(2x + 3y - z + 4) = 0$$

As, $(2\lambda + 1)x + (3\lambda + 1)y + (1 - \lambda)z + 4\lambda - 1 = 0 \quad \ldots \text{(i)}$

It is parallel to X-axis i.e. $\frac{x}{1} = \frac{y}{0} = \frac{z}{0} \Rightarrow$ DR's $= \langle 1, 0, 0 \rangle$

$$\therefore \quad 1(2\lambda + 1) + 0 \times (3\lambda + 1) + 0(1 - \lambda) = 0$$

$$\Rightarrow \quad \lambda = -\frac{1}{2}$$

Substituting $\lambda = -\frac{1}{2}$ in Eq. (i), we get

$y - 3z + 6 = 0$ as the equation of the required plane.

126. ***(b)*** Clearly,

$$R_1 = \int_0^b (1-x)^2 \, dx \text{ and } R_2 = \int_b^1 (1-x)^2 \, dx$$

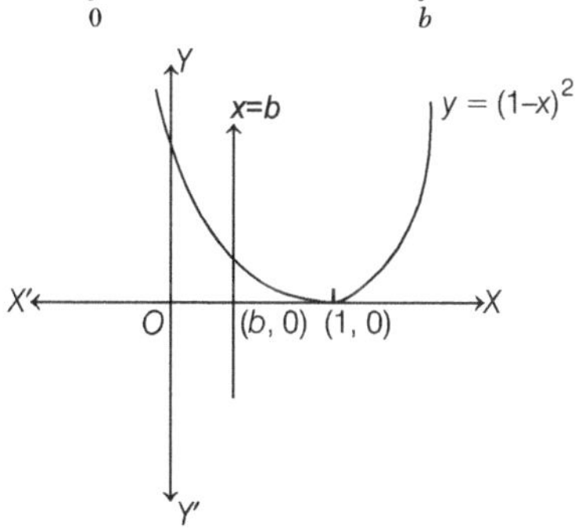

$$\Rightarrow \quad R_1 = \left[\frac{(x-1)^3}{3}\right]_0^b \text{ and } R_2 = \left[\frac{(x-1)^3}{3}\right]_b^1$$

$$\Rightarrow \quad R_1 = \frac{(b-1)^3}{3} + \frac{1}{3} \text{ and } R_2 = -\frac{(b-1)^3}{3}$$

$$\therefore \quad R_1 - R_2 = \frac{1}{4}$$

$$\Rightarrow \quad \frac{2}{3}(b-1)^3 + \frac{1}{3} = \frac{1}{4} \Rightarrow \frac{2}{3}(b-1)^3 = -\frac{1}{12}$$

$$\Rightarrow \quad (b-1)^3 = -\frac{1}{8} \Rightarrow b - 1 = -\frac{1}{2} \Rightarrow b = \frac{1}{2}$$

127. ***(a)*** We have,

$$\mathbf{a} \times \mathbf{b} = 2(\mathbf{a} \times \mathbf{c}) \Rightarrow \mathbf{a} \times (\mathbf{b} - 2\mathbf{c}) = 0$$

$\Rightarrow$ **a** is parallel to $\mathbf{b} - 2\mathbf{c}$.

Now, $\quad (\mathbf{b} - 2\mathbf{c}) = \lambda \mathbf{a}$

$$\Rightarrow \quad (\mathbf{b} - 2\mathbf{c})^2 = \lambda^2 |\mathbf{a}|^2$$

$$|\mathbf{b}|^2 + 4|\mathbf{c}|^2 - 4(\mathbf{b} \cdot \mathbf{c}) = \lambda^2 |\mathbf{a}|^2$$

$$\Rightarrow \quad 16 + 4 - 4 \times |\mathbf{b}||\mathbf{c}| \times \frac{1}{4} = \lambda^2 \Rightarrow 20 - 4 = \lambda^2$$

$$\Rightarrow \quad \lambda^2 = 16 \Rightarrow \lambda = \pm 4$$

128. ***(b)*** We know that the centroid G, circumcentre O and orthocentre O' of a triangle are collinear such that G divides OO' in the ratio $1:2$. Since, affix at G is $\frac{z_1 + z_2 + z_3}{3}$ and O is the origin.

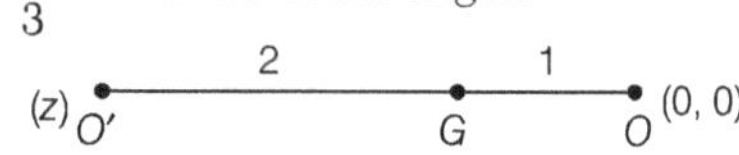

$$\frac{z_1 + z_2 + z_3}{3} = \frac{1 \cdot z + 2 \times 0}{1 + 2}$$

$$\Rightarrow \quad z = z_1 + z_2 + z_3$$

$$\Rightarrow \quad z_1 + z_2 + z_3 - z = 0$$

129. ***(d)*** The largest digit in the middle of the five digit number can be 4 (when number are formed with digits 0, 1, 2, 3, 4) 5 (when 0, 1, 2, 3, 4, 5 are used) …, 9 (when 0, 1, 2, …, 9 are used).

$\therefore$ Number of five digit numbers with largest digit 4 in the middle

$$= ({}^4C_4 \times 4! - {}^3C_3 \times 3!)$$

Number of five digit numbers with largest digit 5 in the middle

$$= ({}^5C_4 \times 4! - {}^4C_3 \times 3!) \text{ and so on.}$$

Hence, required number of numbers =

$$({}^4C_4 \times 4! - {}^3C_3 \times 3!) + ({}^5C_4 \times 4! - {}^4C_3 \times 3!)$$
$$+ ({}^6C_4 \times 4! - {}^5C_3 \times 3!) + \ldots + ({}^9C_4 \times 4! - {}^8C_3 \times 3!)$$

$$= 4! \sum_{n=4}^{9} {}^nC_4 - 3! \sum_{n=3}^{8} {}^nC_3 \quad = 4!\,{}^{10}C_5 - 3!\,{}^9C_4$$

$$= 3![4 \times {}^{10}C_5 - {}^9C_4] = 3! \times 882$$

130. *(a)* We have,

$$\sin^{-1}x + \cos^{-1}x = \frac{\pi}{2} \text{ for all } x \in [-1,1]$$

Also, $$-\frac{\pi}{4} \le \tan^{-1}x \le \frac{\pi}{4} \text{ for all } x \in [-1,1]$$

$\therefore$ $$-\frac{\pi}{4} + \frac{\pi}{2} \le \sin^{-1}x + \cos^{-1}x + \tan^{-1}x \le \frac{\pi}{4} + \frac{\pi}{2}$$

$\Rightarrow$ $$\frac{\pi}{4} \le \sin^{-1}x + \cos^{-1}x + \tan^{-1}x \le \frac{3\pi}{4}$$

$\Rightarrow$ $\alpha = \frac{\pi}{4}$ and $\beta = \frac{3\pi}{4}$

131. *(a)* Given, $$\frac{dy}{dx} = \frac{y}{x} + \frac{\phi(y/x)}{\phi'(y/x)} \quad \text{...(i)}$$

Substituting $y = vx$ and $\frac{dy}{dx} = v + x\frac{dv}{dx}$ in Eq. (i), we get

$$v + x\frac{dv}{dx} = v + \frac{\phi(v)}{\phi'(v)}$$

$\Rightarrow$ $$\frac{\phi'(v)}{\phi(v)}\,dv = \frac{1}{x}\,dx$$

On integrating both sides, we get

$$\log \phi(v) = \log x + \log k$$

$\Rightarrow$ $$\phi(v) = kx \Rightarrow \phi(y/x) = kx$$

132. *(a)* The required mean $\bar{x}$ is given by

$$\bar{x} = \frac{0 \times {}^nC_0\, q^n p^0 + 1 \times {}^nC_1 q^{n-1}p + \dots + n \times {}^nC_n\, q^0 p^n}{{}^nC_0 q^n p^0 + {}^nC_1 q^{n-1}p^1 + \dots + {}^nC_n q^{n-n} p^n}$$

$\Rightarrow$ $$\bar{x} = \frac{\sum_{r=0}^{n} r \times {}^nC_r q^{n-r} p^r}{\sum_{r=0}^{n} {}^nC_r q^{n-r} p^r}$$

$\Rightarrow$ $$\bar{x} = \frac{\sum_{r=1}^{n} r \times \frac{n}{r}\, {}^{n-1}C_{r-1}\, q^{n-r} \times p \times p^{r-1}}{\sum_{r=0}^{n} {}^nC_r q^{n-r} p^r}$$

$\Rightarrow$ $$\bar{x} = \frac{np\left\{\sum_{r=1}^{n} {}^{n-1}C_{r-1}\, p^{r-1} q^{(n-1)-(r-1)}\right\}}{\sum_{r=0}^{n} {}^nC_r q^{n-r} p^r}$$

$\Rightarrow$ $$\bar{x} = \frac{np(q+p)^{n-1}}{(q+p)^n}$$

$\Rightarrow$ $\bar{x} = np$ $\quad [\because q + p = 1]$

133. *(a)* We have, mean = 2 and SD = 3

$\Rightarrow$ $$\frac{1}{10}\sum_{i=1}^{10} x_i = 2 \text{ and } \frac{1}{10}\sum_{i=1}^{10}(x_i^2 - 2^2) = 3^2$$

$\Rightarrow$ $$\sum_{i=1}^{10} x_i = 20 \text{ and } \sum_{i=1}^{10} x_i^2 = 130$$

Let $\bar{x}$ be the mean of $(x_1+1)^2, (x_2+1)^2, \dots (x_{10}+1)^2$. Then,

$$\bar{x} = \frac{1}{10}\sum_{i=1}^{10}(x_i+1)^2$$

$\Rightarrow$ $$\bar{x} = \frac{1}{10}\left(\sum_{i=1}^{10} x_i^2\right) + \frac{2}{10}\left(\sum_{i=1}^{10} x_i\right) + \frac{1}{10}\sum_{i=1}^{10} 1$$

$\Rightarrow$ $$\bar{x} = \frac{1}{10} \times 130 + \frac{2}{10} \times 20 + \frac{10}{10} = 18$$

134. *(a)* The equation of the tangent to the curve $y = e^x$ at (c, e^c).

Slope of tangent $= dy/dx = e^x$

$\Rightarrow$ $$y - e^c = e^c(x - c) \quad \text{...(i)}$$

The equation of the line joining $(c-1, e^{c-1})$ and $(c+1, e^{c+1})$ is

$$y - e^{c-1} = \frac{e^c(e - e^{-1})}{2}(x - c + 1) \quad \text{...(ii)}$$

On subtracting Eq. (i) from Eq. (ii), we get

$$e^c - e^{c-1} = \frac{e^c(e - e^{-1})}{2}(x - c + 1) - e^c(x - c)$$

$$\Rightarrow e^c - e^{c-1} = (x - c)\, e^c\left(\frac{e - e^{-1}}{2} - 1\right) + e^c\left(\frac{e - e^{-1}}{2}\right)$$

$\Rightarrow$ $$x - c = \frac{(e-1)^2}{2 - (e-1)^2} < 0 \Rightarrow x < c$$

Hence, Eqs. (i) and (ii) intersect at a point on the left of $x = c$.

135. *(d)* We know that the lines,

$$\frac{x - x_1}{l_1} = \frac{y - y_1}{m_1} = \frac{z - z_1}{n_1}$$

and $$\frac{x - x_2}{l_2} = \frac{y - y_2}{m_2} = \frac{z - z_2}{n_2}$$

are coplanar, iff

$$\begin{vmatrix} x_2 - x_1 & y_2 - y_1 & z_2 - z_1 \\ l_1 & m_1 & n_1 \\ l_2 & m_2 & n_2 \end{vmatrix} = 0$$

So, the given lines will be coplanar, if

$$\begin{vmatrix} 1-2 & 4-3 & 5-4 \\ 1 & 1 & -k \\ k & 2 & 1 \end{vmatrix} = 0$$

$\Rightarrow$ $$k^2 + 3k = 0 \Rightarrow k = 0, -3$$

136. *(c)* We have,

$$a + b + c = 0 \text{ and } 2ab + 2ac - bc = 0$$

$\Rightarrow$ $$a = -(b + c) \text{ and } 2a(b + c) - bc = 0$$

$\Rightarrow$ $$-2(b + c)^2 - bc = 0$$

$\Rightarrow 2b^2 + 5bc + 2c^2 = 0 \Rightarrow (2b + c)(b + 2c) = 0$

$\Rightarrow \quad 2b + c = 0$ or $b + 2c = 0$

If $2b + c = 0$, then $a = -(b + c) \Rightarrow a = b$

$\therefore \quad a = b$ and $c = -2b \Rightarrow \frac{a}{1} = \frac{b}{1} = \frac{c}{-2}$

If $b + 2c = 0$, then $a = -(b + c) \Rightarrow a = c$

$\therefore \quad a = c$ and $b = -2c \Rightarrow \frac{a}{1} = \frac{b}{-2} = \frac{c}{1}$

Thus, the direction ratio of two lines are proportional to 1, 1, − 2 and 1, − 2, 1 respectively. So, the angle θ between them is given by

$$\cos\theta = \frac{1-2-2}{\sqrt{1+1+4}\sqrt{1+4+1}} = -\frac{1}{2} = \cos\frac{2\pi}{3}$$

$$\Rightarrow \quad \theta = \frac{2\pi}{3}$$

137. ***(a)*** Since $f(x)$ satisfies the conditions of Rolle's theorem.

$\therefore \quad f(1) = f(3)$,

where $f(x) = x^3 - 6x^2 + ax + b$

$\Rightarrow \quad 1 - 6 + a + b = 27 - 54 + 3a + b$

$\Rightarrow \quad a = 11$

Now, $f'(x) = 3x^2 - 12x + a = 3x^2 - 12x + 11$

Clearly, $f'\left(\frac{2\sqrt{3}+1}{\sqrt{3}}\right) = 0.$

Hence, $a = 11$

138. ***(d)*** We have, $f^{-1}(f(x)) = x$ for all x

On differentiating w.r.t. 'x'

$(f^{-1})'(f(x))\, f'(x) = 1$ for all x

$(f^{-1})'(f(x)) = \frac{1}{f(x)}$ for all x $[\because f'(x) = f(x)]$

Differentiating both sides w.r.t. x

$$(f^{-1})''(f(x))\, f'(x) = -\frac{1}{\{f(x)\}^2} f'(x)$$

$$\Rightarrow \quad (f^{-1})''(f(x)) = -\frac{1}{\{f(x)\}^2}$$

$$\Rightarrow \quad (f^{-1})''(x) = -\frac{1}{x^2}$$

139. ***(c)*** Let $f(x) = 2x^3 + 5$ and $g(x) = 9x^2 - 12x$.

Then, $f(x)$ increases less rapidly than $g(x)$ means that

$$\frac{d}{dx}(f(x)) < \frac{d}{dx}(g(x))$$

$$\Rightarrow \quad \frac{d}{dx}(f(x) - g(x)) < 0$$

$$\Rightarrow \quad f'(x) - g'(x) < 0 \Rightarrow 6x^2 - (18x - 12) < 0$$

$$\Rightarrow \quad x^2 - 3x + 2 < 0$$

$$\Rightarrow \quad x \in (1, 2)$$

140. ***(a)*** We have,

$$f(x) = \sin^{-1} x^2 \Rightarrow f'(x) = \frac{2x}{\sqrt{1 - x^4}}$$

$\therefore \quad f'(x) = 0 \Rightarrow x = 0$

Now, $f\left(-\frac{1}{\sqrt{2}}\right) = \sin^{-1}\left(\frac{1}{2}\right) = \frac{\pi}{6}$,

$f\left(\frac{1}{\sqrt{2}}\right) = \sin^{-1}\left(\frac{1}{2}\right) = \frac{\pi}{6}$

and $f(0) = \sin^{-1} 0 = 0$

Hence, the greatest value is $\pi/6$.

141. ***(a)*** We have,

$$f(x) = [x]^2 + [x + 1] - 3$$

$$\Rightarrow \quad f(x) = [x]^2 + [x] + 1 - 3$$

$[\because [x + n] = [x] + n$, where $n \in Z]$

$$\Rightarrow \quad f(x) = [x]^2 + [x] - 2$$

$$\Rightarrow \quad f(x) = ([x] + 2)([x] - 1)$$

Clearly, $f(x) = 0$ for all $x \in [1, 2) \cup [-2, -1)$

So, f is a many-one function.

Also, $f(x)$ assumes only integral values

$\therefore \quad$ Range of $f \neq R$.

Hence, $f(x)$ is a many-one into function.

142. ***(a)*** The equation of a line through $P(2, 4)$ and parallel to $y - x = 1$ is

$$\frac{x-2}{\cos\frac{\pi}{4}} = \frac{y-4}{\sin\frac{\pi}{4}}$$

The coordinates of Q are given by

$$\frac{x-2}{\cos\frac{\pi}{4}} = \frac{y-4}{\sin\frac{\pi}{4}} = -3\sqrt{2} \Rightarrow x = -1, y = 1$$

Thus, the coordinates of Q are $(-1, 1)$.

R is the image of Q with respect to the line $y - x = 1$

Therefore, coordinates of R are given by

$$\frac{x+1}{-1} = \frac{y-1}{1} = \frac{-2(1+1-1)}{(-1)^2 + (1)^2}$$

$$\Rightarrow \quad \frac{x+1}{-1} = \frac{y-1}{1} = -1$$

$$\Rightarrow \quad x = 0 \text{ and } y = 0$$

Hence, the coordinates of R are $(0, 0)$.

143. ***(c)*** The equation of lines making equal intercepts of length l on the coordinate axes is

$$x \pm y = \pm l$$

$$\Rightarrow \quad r = \sqrt{32} = 4\sqrt{2}$$

This will be a chord of the circle $x^2 + y^2 = 32$, if length of the perpendicular from the centre is less than radius.

i.e. $$\left|\frac{\pm l}{\sqrt{1+l}}\right| < 4\sqrt{2}$$

$$\Rightarrow \quad |l| < 8 \Rightarrow l \in (-8, 8)$$

144. *(a)* $x^2 - (a-2)x - a + 1 = 0$

Let α and β be the roots of the given equation, so that

$$\alpha + \beta = a - 2 \text{ and } \alpha\beta = -(a-1)$$

Let $S = (\alpha^2 + \beta^2)$. Then,

$$S = (\alpha + \beta)^2 - 2\alpha\beta$$

$$= (a-2)^2 + 2(a-1) = a^2 - 2a + 2$$

$\therefore \quad dS/da = 2a - 2$

Now, $\quad dS/da = 0 \Rightarrow a = 1$

Also, $\quad d^2S/da^2 = 2 > 0$ for all a

Hence, S is minimum when $a = 1$.

145. *(c)* We have,

$$y = \int_x^{x^2} \cos^{-1} t^2 \, dt$$

$$\Rightarrow \quad dy/dx = 2x \cos^{-1}(x^4) - \cos^{-1}(x^2)$$

[by Leibnitz's rule]

$$\Rightarrow \left(\frac{dy}{dx}\right)_{x = 2^{-\frac{1}{4}}} = 2 \times 2^{-\frac{1}{4}} \cos^{-1}\left(\frac{1}{2}\right) - \cos^{-1}\left(\frac{1}{\sqrt{2}}\right)$$

$$\Rightarrow \left(\frac{dy}{dx}\right)_{x = 2^{-\frac{1}{4}}} = 2^{\frac{3}{4}} \times \frac{\pi}{3} - \frac{\pi}{4} = \left(\frac{\sqrt[4]{8}}{3} - \frac{1}{4}\right)\pi$$

146. *(c)* $\displaystyle \lim_{x \to \frac{\pi}{4}} \frac{4\sqrt{2} - (\cos x + \sin x)^5}{1 - \sin 2x}$

$$= \lim_{x \to \frac{\pi}{4}} \frac{\{(\cos x + \sin x)^2\}^{5/2} - (2)^{5/2}}{(1 + \sin 2x) - 2}$$

$$= \lim_{x \to \frac{\pi}{4}} \frac{(1 + \sin 2x)^{5/2} - 2^{5/2}}{(1 + \sin 2x) - 2}$$

$$= \lim_{y \to 2} \frac{y^{5/2} - 2^{5/2}}{y - 2}, \text{ where } y = 1 + \sin 2x$$

$$= \frac{5}{2} \times 2^{5/2 - 1} = 5\sqrt{2}$$

147. *(c)* We have,

$f(x) + f(y) + f(z) + f(x)\,f(y)\,f(z) = 14$

for all $x, y, z \in R$ ...(i)

Putting $x = y = z = 0$ in Eq. (i), we get

$$3f(0) + \{f(0)\}^3 = 14$$

$$\Rightarrow \quad \{f(0)\}^3 + 3f(0) - 14 = 0$$

$$\Rightarrow \quad f(0) = 2$$

Now, putting $y = z = x$ in Eq. (i), we get

$3f(x) + \{f(x)\}^3 = 14$ for all $x \in R$

On differentiating w.r.t. x, we get

$$3f'(x) + 3\{f(x)\}^2 f'(x) = 0 \text{ for all } x \in R$$

$$\Rightarrow \quad [\{f(x)\}^2 + 1]\, f'(x) = 0 \text{ for all } x \in R$$

$$\Rightarrow \quad f'(x) = 0 \text{ for all } x \in R$$

148. *(d)* We have, $\displaystyle 6\int_1^x f(t)\,dt = 3x\,f(x) - x^3$

On differentiating both sides w.r.t. x, we get

$$6f(x) = 3f(x) + 3xf'(x) - 3x^2$$

$$\Rightarrow \quad 3f(x) = 3x\,f'(x) - 3x^2$$

$$\Rightarrow \quad f'(x) - \frac{1}{x} f(x) = x$$

$$\Rightarrow \quad \frac{d}{dx} f(x) + \left(-\frac{1}{x}\right) f(x) = x$$

This is a linear differential equation with integrating factor

$$e^{\int -\frac{1}{x} dx} = e^{-\log x} = 1/x$$

Solution is given by,

$$\frac{1}{x} f(x) = \int x \times \frac{1}{x}\, dx + C$$

$$\Rightarrow \quad f(x) = x^2 + Cx$$

$$\therefore \quad f(1) = 1 + C \Rightarrow 2 = 1 + C$$

$$\Rightarrow \quad C = 1$$

Thus, $\quad f(x) = x^2 + x$

$$\therefore \quad f(2) = 4 + 2 = 6$$

149. *(b)* Since g is the inverse of f.

$$\therefore \quad fog\,(x) = x \text{ for all } x$$

$$\Rightarrow \quad \frac{d}{dx}\{fog\,(x)\} = 1 \text{ for all } x$$

$$\Rightarrow \quad f'(g(x)) \cdot g'(x) = 1 \Rightarrow f'\{g(x)\} = \frac{1}{g'(x)}$$

$$\Rightarrow \quad \frac{1}{1 + [g(x)]^n} = \frac{1}{g'(x)} \quad t\left[\because f'(x) = \frac{1}{1 + x^n}\right]$$

$$\Rightarrow \quad g'(x) = 1 + [g(x)]^n$$

150. *(d)* For $f(x)$ to be continuous at $x = 0$, we must have $\displaystyle \lim_{x \to 0} f(x) = f(0)$

$$\Rightarrow \quad \lim_{x \to 0}\left(1 + xe^{-1/x^2} \sin\frac{1}{x^4}\right)^{e^{1/x^2}} = k$$

$$\Rightarrow \quad e^{\lim_{x \to 0}\left(xe^{1/x^2} e^{-1/x^2} \sin\frac{1}{x^4}\right)} = k$$

$$\Rightarrow \quad e^{\lim_{x \to 0} x \sin\frac{1}{x^4}} = k \Rightarrow e^{\lim_{x \to 0} x \sin\frac{1}{x^4}} = k$$

$$\Rightarrow \quad e^0 = k \Rightarrow k = 1$$

BITSAT

Birla Institute of Technology & Science Admission Test

Solved Paper 2015*

Instructions

- There are 150 questions in all. The number of questions in each part is as given below.

	No. of Questions
Part I Physics	1-40
Part II Chemistry	41-80
Part III a. English Proficiency	81-95
b. Logical Reasoning	96-105
Part IV Mathematics	106-150

- All questions are Multiple Choice Questions having four options out of which **only one** is correct.
- Each correct answer fetches 3 marks while incorrect answer has a penalty of 1 mark.
- Time allotted to complete this paper is 3 hrs.

PART I

Physics

1. A body cools from 60°C to 50°C in 10 min. What will be its temperature at the end of the next 10 min, if the room temperature is 25°C? (Assume Newton's law of cooling holds.)

a. 49°C *b.* 42.8°C
c. 44°C *d.* 45.8°C

2. Two tuning forks with natural frequencies 340 Hz each move relative to a stationary observer. One fork moves away from the observer, while the other moves towards the observer at the same speed. The observer hears beats of frequency of 3 Hz. Find the speed of the tuning forks.

a. 1.5 m/s
b. 2 m/s
c. 1 m/s
d. 2.5 m/s

3. Two spherical mirrors, one convex and other concave are placed coaxially at a distance $2R$ from each other. Both the mirrors have same radius of curvature R. What is the radius of 3rd image from first three images of the circle, if the small circle of radius a is drawn on the convex mirror in the figure?

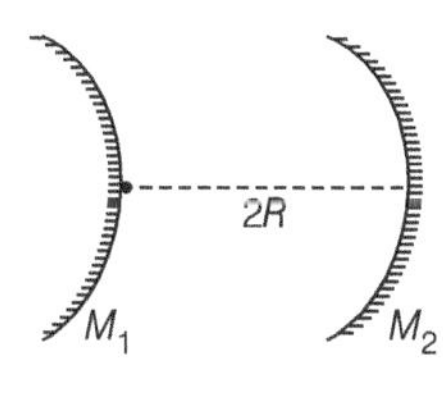

a. $a/41$ *b.* $a/43$ *c.* $a/56$ *d.* $a/53$

4. In the given figure, two point charges q_1 and q_2 are placed at distance a and b from the centre of a metallic sphere having charge Q. Find the electric field due to the metallic sphere at the point P.

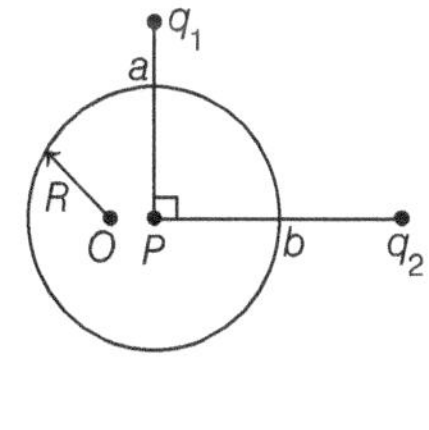

*Memory Based

a. $\frac{1}{4\pi\varepsilon_0}\sqrt{\left(\frac{q_1}{a^2}\right)^2+\left(\frac{q_2}{b^2}\right)^2}$

b. $\frac{1}{4\pi\varepsilon_0}\cdot\frac{Q}{R^2}$

c. $\frac{1}{4\pi\varepsilon_0}\sqrt{\left(\frac{Q}{R^2}\right)^2+\left(\frac{q_1}{a^2}+\frac{q_2}{b^2}\right)^2}$

d. None of the above

5. A body is thrown vertically upwards from A, the top of the tower, it reaches the ground in time t_1. If it is thrown vertically downwards from A with the same speed, it reaches the ground in time t_2. If it is allowed to fall freely from A, then the time it takes to reach the ground is given by

a. $t=\frac{t_1+t_2}{2}$ b. $t=\frac{t_1-t_2}{2}$

c. $t=\sqrt{t_1t_2}$ d. $t=\sqrt{\frac{t_1}{t_2}}$

6. A simple pendulum is vibrating with an angular amplitude of 90° as shown in the figure. For what value of α, the acceleration is directed horizontally?

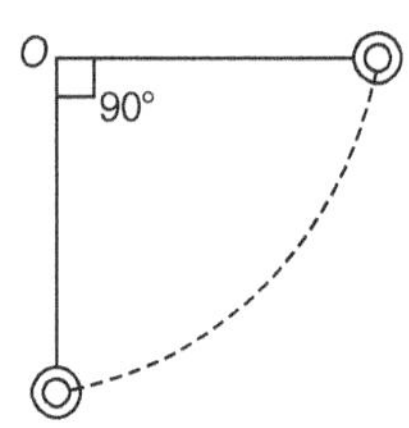

a. Zero b. 90°

c. $\cos^{-1}(1/\sqrt{3})$ d. $\sin^{-1}(1/\sqrt{3})$

7. A charged particle moving in a uniform magnetic field and losses 4% of its kinetic energy. The radius of curvature of its path changes by

a. 2% b. 4%

c. 10% d. None of these

8. Calculate the wavelength of light used in an interference experiment from the following data : fringe width = 0.03 cm. Distance between the slits and eye-piece through which the interference pattern in observed is 1m. Distance betweeen the images of the virtual source when a convex lens of focal length 16 cm is used at a distance of 80 cm from the eye-piece is 0.8 cm.

a. 0.0006 Å b. 0.0006 m

c. 600 cm d. 6000 Å

9. The masses of blocks A and B are m and M, respectively. Between A and B, there is a constant frictional force F and B can slide on a smooth horizontal surface. A is set in motion with velocity, while B is at rest. What is the distance moved by A relative to B before they move with the same velocity?

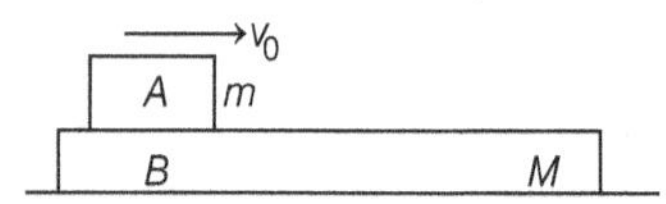

a. $\frac{mMv_0^2}{F(m-M)}$ b. $\frac{mMv_0^2}{2F(m-M)}$

c. $\frac{mMv_0^2}{F(m+M)}$ d. $\frac{mMv_0^2}{2F(M+m)}$

10. A particle of mass m is projected from the surface of the earth with a speed (v_0 > escape velocity). Find the speed of particle at height $h=R$ (radius of the earth).

a. $\sqrt{gR}$ b. $\sqrt{v_0^2-2gR}$

c. $\sqrt{v_0^2-gR}$ d. None of these

11. An elasticised conducting band is around a spherical balloon. Its plane passes through the centre of balloon. A uniform magnetic field of magnitude 0.04 T is directed perpendicular to the plane of band. Air is let out of the balloon at 100 cm^3/s at an instant, when the radius of the balloon is 10 cm. The induced emf in the band is

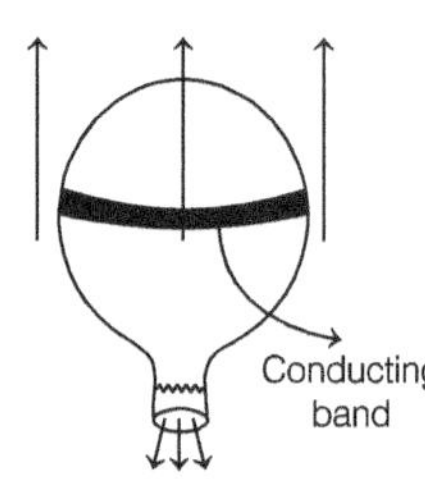

a. 20 μV b. 25 μV

c. 10μV d. 15 μV

12. An elastic string of unstretched length L and force constant k is stretched by a small length x. It is further stretched by another small length y. The work done in the second stretching is

a. $1/2\,ky^2$ b. $1/2\,ky\,(2x+y)$

c. $1/2\,k(x^2+y^2)$ d. $1/2\,k\,(x+y)^2$

13. The resistance of a 50 cm long wire is 10 Ω. The wire is stretched to uniform wire of length 100 cm. The new resistance will be

a. 15 Ω b. 30 Ω c. 20 Ω d. 40 Ω

14. A beam of light has three wavelengths 4144 Å, 4972 Å and 6216 Å with a total intensity of 3.6×10^{-3} Wm^{-2} equally distributed amongst the three wavelengths. The beam falls normally on the area 1 cm^2 of a clean metallic surface of work function 2.3 eV. Assume that there is no loss of light by reflection and that each energetically capable photon ejects one electron. Calculate the number of photoelectrons liberated in 2s.

a. 2×10^9 *b.* 1.075×10^{12}
c. 9×10^8 *d.* 3.75×10^6

15. A square gate of size $1\text{ m}\times 1\text{ m}$ is hinged at its mid-point. A fluid of density ρ fills the space to the left of the gate. The force *F* required to hold the gate stationary is

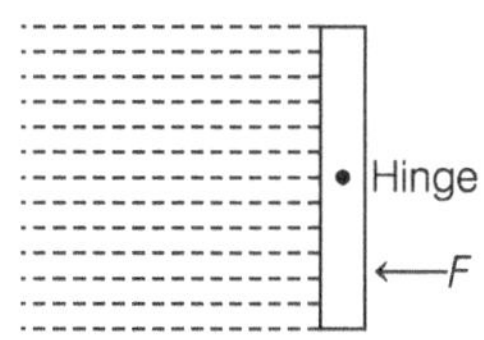

a. $\frac{\rho g}{3}$ *b.* $\frac{\rho g}{2}$
c. $\frac{\rho g}{6}$ *d.* $\frac{\rho g}{8}$

16. When 0.50 Å X-rays strike a material, the photoelectrons from the *k*-shell are observed to move in a circle of radius 23 mm in a magnetic field of 2×10^{-2} T acting perpendicularly to the direction of emission of photoelectrons. What is the binding energy of *k*-shell electrons?

a. 3.5 keV *b.* 6.2 keV
c. 2.9 keV *d.* 5.5 keV

17. In *CE*-transistor amplifier, the audio signal voltage across the collector resistance of 2 kΩ is 2 V. If the base resistance is 1 kΩ and the current amplification of the transistor is 100, the input signal voltage is

a. 2 mV *b.* 3 mV
c. 10 mV *d.* 0.1 mV

18. The equation of motion of projectile is

$$y = 12x - \frac{3}{4}x^2$$

The horizontal component of velocity is 3 ms^{-1}. What is the range of the projectile?

a. 18 m *b.* 21.6 m
c. 12 m *d.* 16 m

19. A vessel of volume 20 L contains a mixture of hydrogen and helium at temperature of 27°C and pressure 2 atm. The mass of mixture is 5 g. Assuming the gases to be ideal, the ratio of mass of hydrogen to that of helium in the given mixture will be

a. 1 : 2 *b.* 2 : 3 *c.* 2 : 1 *d.* 2 : 5

20. In a YDSE, light of wavelength $\lambda = 5000$ Å is used, which emerges in phase from two slits a distance $d = 3\times 10^{-7}$ m apart. A transparent sheet of thickness $t = 1.5\times 10^{-7}$ m, refractive index $\mu = 1.17$ is placed over one of the slits. What is the new angular position of the central maxima of the interference pattern, from the centre of the screen? find the value of *y*.

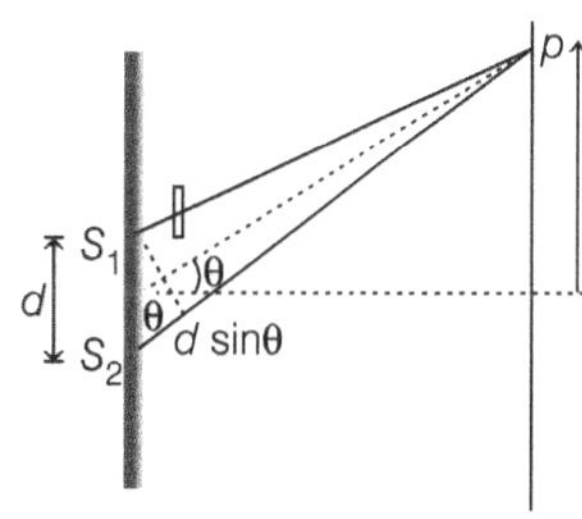

a. 4.9° and $\frac{D(\mu-1)t}{2d}$ *b.* 4.9° and $\frac{D(\mu-1)t}{d}$
c. 3.9° and $\frac{D(\mu+1)t}{d}$ *d.* 2.9° and $\frac{2D(\mu-1)t}{d}$

21. A particle moves with a simple harmonic motion in a straight line. In the first second starting from rest, it travels a distance *a* and in the next second, it travels a distance *b* in the same direction. The amplitude of motion is

a. $\frac{2a^2}{3a-b}$ *b.* $\frac{3a^2}{3b-a}$ *c.* $\frac{2a^3}{3b-a}$ *d.* $\frac{3a^3}{3a-b}$

22. At the corners of an equilateral triangle of side a (1 m), three point charges are placed (each of 0.1 C). If this system is supplied energy at the rate of 1 kW, then calculate the time required to move one of the charges to the mid-point of the line joining the other two.

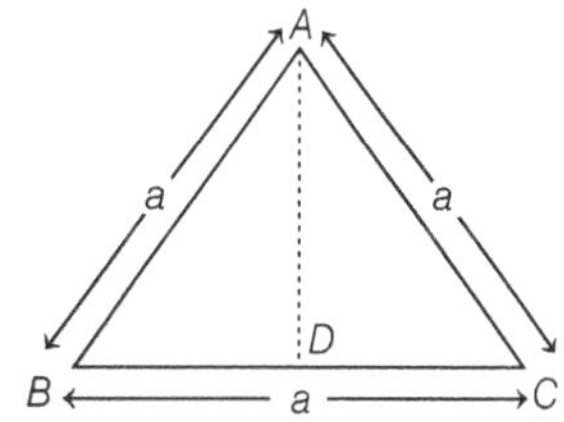

a. 50 h *b.* 60 h
c. 48 h *d.* 54 h

BITSAT 2015

23. A load of mass m falls from a height h onto the scale pan hung from the spring as shown in the Figure. If the spring constant is k and mass of the scale pan is zero and the mass m does not bounce relative to the pan, then the amplitude of vibration is

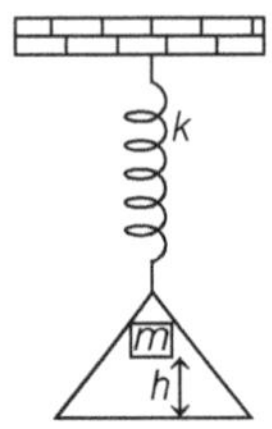

a. mg/k **b.** $\frac{mg}{k}\sqrt{\left(1+\frac{2hk}{mg}\right)}$

c. $\frac{mg}{k}+\frac{mg}{k}\sqrt{\left(\frac{1+2hk}{mg}\right)}$ **d.** $\frac{mg}{k}\sqrt{\left(\frac{1+2hk}{mg}-\frac{mg}{k}\right)}$

24. In an ore containing uranium, the ratio of U^{238} to Pb^{206} is 3. Calculate the age of the ore, assuming that all the lead present in the ore is the final stable product of U^{238}. Take, the half-life of U^{238} to be 4.5×10^9 yrs.

a. 1.6×19^3 yrs **b.** 1.5×10^4 yrs

c. 1.867×10^9 yrs **d.** 2×10^5 yrs

25. A concave mirror is palced on a horizontal table with its axis directed vertically upward. Let O be the pole of the mirror and C its centre of curvature. A point object is placed at C. It has real image, also located at C. If the mirror is now filled with water, the image will be

a. real and will remain at C

b. real and located at a point between C and ∞

c. real and located at a point between C and O

d. None of the above

26. A direct current of 5A is superposed on an alternating current $I = 10\sin\omega t$ flowing through the wire. The effective value of the resulting current will be

a. (15/2)A **b.** $5\sqrt{3}$ A **c.** $5\sqrt{5}$ A **d.** 15 A

27. Two particles A and B, each of mass m, are kept stationary by applying a horizontal force $F = mg$ on the particle B as shown in the figure. Then,

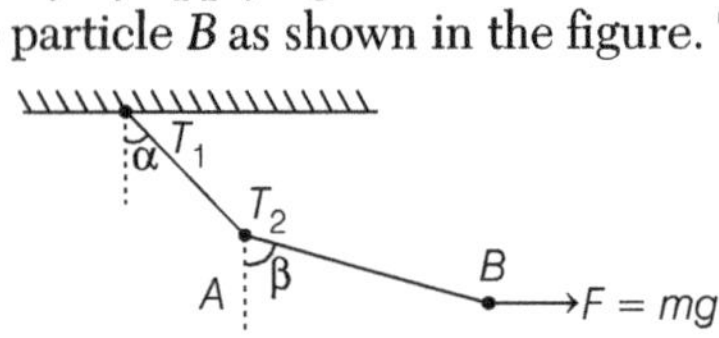

a. $2\tan\beta = \tan\alpha$ **b.** $2T_1 = 5T_2$

c. $T_1\sqrt{2} = T_2\sqrt{5}$ **d.** None of these

28. A thin rod of length $4l$ and mass $4M$ is bent at the points as shown in figure. What is the moment of inertia of the rod about the axis passes through point O and perpendicular to the plane of paper?

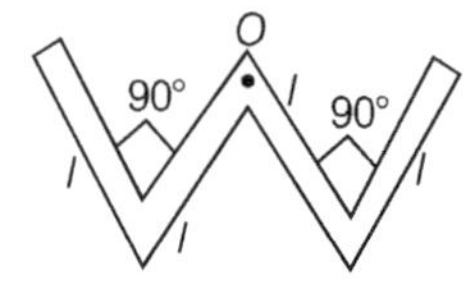

a. $\frac{Ml^2}{3}$ **b.** $\frac{10Ml^2}{3}$

c. $\frac{Ml^2}{12}$ **d.** $\frac{Ml^2}{24}$

29. The wavelength of the second line of Balmer series in the hydrogen spectrum is 4861 Å. The wavelength of the first line is

a. $\frac{27}{20}\times 4861$ Å

b. $\frac{20}{27}\times 4861$ Å

c. 20×4861 Å

d. 4861 Å

30. Two particles X and Y having equal charges, after being accelerated through the same potential difference, enter a region of uniform magnetic field and describe circular paths of radii R_1 and R_2, respectively. The ratio of masses of X and Y is

a. $(R_1/R_2)^{1/2}$ **b.** (R_2/R_1)

c. $(R_1/R_2)^2$ **d.** (R_1/R_2)

31. A glass capillary tube of internal radius $r = 0.25$ mm is immersed in water. The top end of the tube projects by 2cm above the surface of the water. At what angle does the liquid meet the tube?

Surface tension of water = 0.07 N/m.

a. $\theta = 90°$ **b.** $\theta = 70°$

c. $\theta = 45°$ **d.** $\theta = 35°$

32. A particle of mass 2 m is projected at an angle of 45° with the horizontal with a velocity of $20\sqrt{2}$ m/s. After 1 s, explosion takes place and the particle is broken into two equal pieces. As a result of explosion, one part comes to rest. The maximum height from the ground attained by the other part is

a. 50 m **b.** 25 m

c. 40 m **d.** 35 m

33. Maximum power developed across variable resistance R in the circuit as shown in the figure.

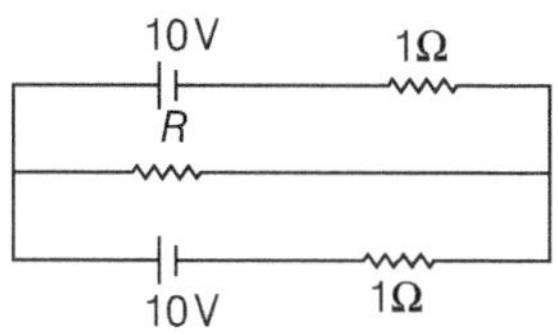

a. 50 W *b.* 75 W *c.* 25 W *d.* 60 W

34. A neutron moving with speed v makes a head on collision with a hydrogen atom in ground state kept at rest. The minimum kinetic energy of the neutron for which inelastic collision takes place is

a. 10.2 eV *b.* 20.4 eV
c. 12.1 eV *d.* 16.8 eV

35. Vertical displacement of a plank with a body of mass m on it is varying according to law $y = \sin \omega t + \sqrt{3} \cos \omega t$. The minimum value of ω for which the mass just breaks off the plank and the moment it occurs first after $t = 0$, are given by

a. $\sqrt{g/2}, \frac{\sqrt{2}}{6} \frac{\pi}{\sqrt{g}}$ *b.* $\frac{g}{\sqrt{2}}, \frac{2}{3}\sqrt{\pi/g}$
c. $\sqrt{g/2}, \frac{\pi}{3}\sqrt{2/g}$ *d.* $\sqrt{2g}, \sqrt{2\pi/3g}$

36. A parallel plate capacitor of capacitance C is connected to a battery and is charged to a potential difference V. Another capacitor of capacitance $2C$ is similarly charged to a potential difference $2V$. The charging battery is now disconnected and the capacitors are connected in parallel to each other in such a way that the positive terminal of one is connected to the negative terminal of the other. The final energy of the configuration is

a. Zero *b.* $\frac{3}{2}CV^2$
c. $\frac{25}{6}CV^2$ *d.* $\frac{9}{2}CV^2$

37. A man drives a car y towards x at speed 60 km/h. A car leaves station x for station y every 10 min. The distance between x and y is 60 km. The car travels at the speed of 60 km/h. A man drives a car from y towards x at speed of 60 km/h. If he starts at the moment when first car leaves the station y, then how many cars would be meet on the route?

a. 20 *b.* 7
c. 10 *d.* 2

38. The ratio of the energy required to raise a satellite upto a height h above the earth to the kinetic energy of the satellite in the orbit is

a. $h : R$ *b.* $R : 2h$
c. $2h : R$ *d.* $R : h$

39. The frequency of a sonometer wire is 100 Hz. When the weights producing the tensions are completely immersed in water, the frequency becomes 80 Hz and on immersing the weights in a certain liquid, the frequency becomes 60 Hz. The specific gravity of the liquid is

a. 1.42 *b.* 1.77
c. 1.82 *d.* 1.21

40. A wire of length l is bent in the form of a circular coil of some turns. A current i flows through the coil. The coil is placed in a uniform magnetic field B. The maximum torque on the coil can be

a. $\frac{iBl^2}{4\pi}$ *b.* $\frac{iBl^2}{\pi}$
c. $\frac{iBl^2}{2\pi}$ *d.* $\frac{2iBl^2}{\pi}$

BITSAT 2015

PART II

Chemistry

41. Which of the following pollutants is main product of automobiles exhaust?

a. CO *b.* CO_2
c. NO *d.* Hydrocarbons

42. The harmful chemical present in tobacco is

a. morphine *b.* zinc
c. arsenic *d.* phosphate

43. The disease causes by the high concentration of hydrocarbon pollutants in atmosphere is

a. silicosis *b.* TB *c.* cancer *d.* asthma

44. In disulphide bond formation of protein, the functional group that participates is

a. amino group *b.* thio ether
c. thiol *d.* thio ester

45. The correct structure of tyrosine is

a. 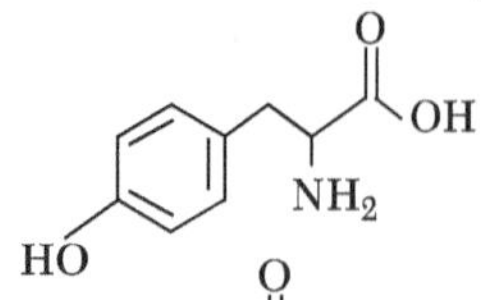 *b.* 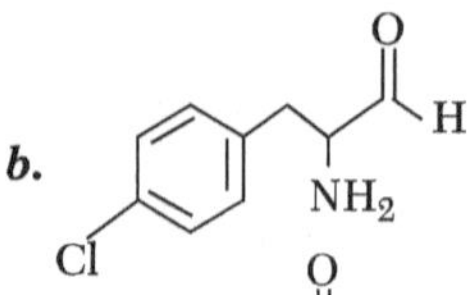

c. 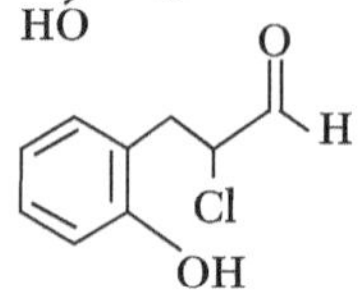 *d.* 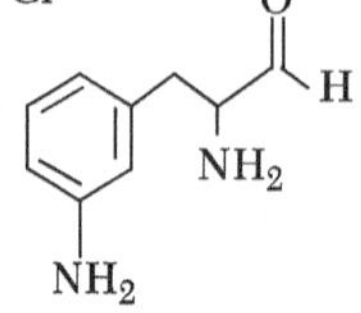

46. Which law of the thermodynamics helps in calculating the absolute entropies of various substances at different temperatures?

a. First law *b.* Second law
c. Third law *d.* Zeroth law

47. The boiling point of alkyl halides are higher than those of corresponding alkanes because of

a. dipole-dipole interaction
b. dipole-induced dipole interaction
c. H-bonding
d. None of the above

48. Lucas test is given by

a. *n*-alcohol *b.* *sec*-alcohol
c. *tert*-alcohol *d.* Both (b) and (c)

49. The carbylamine reaction is

a. 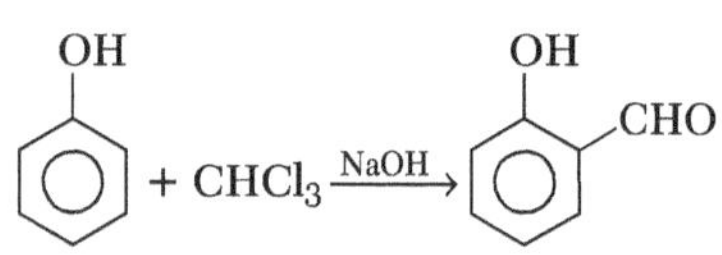

b. 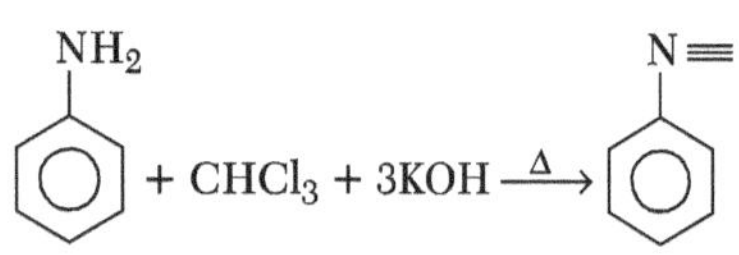

c. 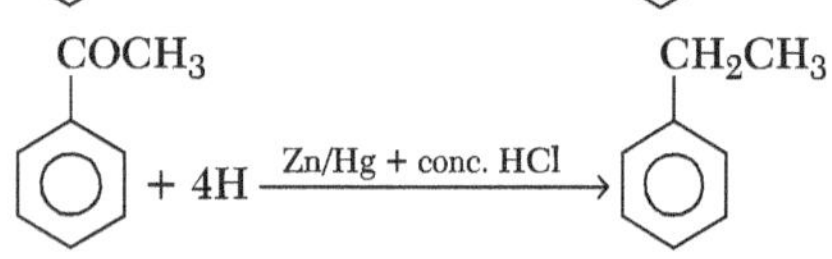

d. 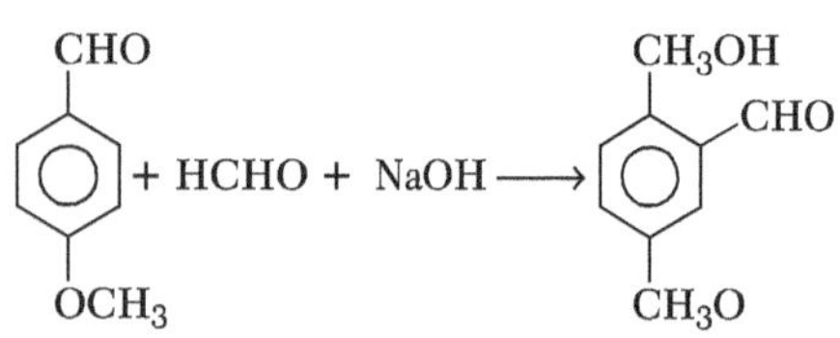

50. The colour of $CoCl_3 \cdot 5NH_3 \cdot H_2O$ is

a. red *b.* orange
c. yellow *d.* pink

51. Some salts containing two different metallic elements give test for only one of them in solution, such salts are

a. double salts *b.* normal salts
c. complex salts *d.* None of these

52. The metal present in the chlorophyll is

a. magnesium *b.* zinc
c. cobalt *d.* copper

53. The common name of $K[PtCl_3(\eta^2\text{-}C_2H_4)]$ is

a. potassium salt *b.* Zeise's salt
c. complex salt *d.* None of these

54. The metal present in vitamin B_{12} is

a. magnesium *b.* cobalt *c.* copper *d.* zinc

55. The anthracene is purified by

a. crystallisation *b.* filtration
c. distillation *d.* sublimation

56. Which is an intensive property?

a. Mass *b.* Volume
c. Energy *d.* Temperature

57. Laughing gas is

a. nitrogen pentoxide *b.* nitrous oxide
c. nitrogen trioxide *d.* nitric oxide

58. Brown ring test is used to detect

a. nitrate *b.* bromide *c.* iron *d.* iodide

59. Semiconductor materials, like Si and Ge are usually purified by

a. distillation *b.* zone refining
c. liquation *d.* electrolytic refining

60. Which of the following is not a method of concentration of metal?

a. Smelting *b.* Froth floatation
c. Distillation *d.* Electromagnetic method

61. Which of the following according to Le-Chatelier's principle is correct?

a. Increase in temperature favours the endothermic reaction.
b. Increase in temperature favours the exothermic reaction.
c. Increase in pressure shifts the equilibrium in that side in which number of gaseous moles increases
d. All of the above

62. The van't Hoff equation is

a. $\log \frac{K_1}{K_2} = \frac{\Delta H}{2.303R}\left[\frac{T_2 - T_1}{T_1T_2}\right]$

b. $\log \frac{K_2}{K_1} = \frac{\Delta H}{2.303R}\left[\frac{T_2 - T_1}{T_1T_2}\right]$

c. $\log \frac{K_2}{K_1} = \frac{\Delta H}{2.303R}\left[\frac{T_1 - T_2}{T_1T_2}\right]$

d. None of the above

63. Which of the following is a strong base?
a. PH_3 *b.* AsH_3
c. NH_3 *d.* SbH_3

64. What is the correct increasing order of Bronsted bases ?
a. $ClO_4^- < ClO_3^- < ClO_2^- < ClO^-$
b. $ClO_4^- > ClO_3^- > ClO_2^- > ClO^-$
c. $ClO_3^- < ClO_4^- < ClO_2^- < ClO^-$
d. $ClO^- > ClO_3^- > ClO_2^- > ClO_4^-$

65. The mass of the substance deposited when one Faraday of charge is passed through its solution is equal to
a. relative equivalent weight
b. gram equivalent weight
c. specific equivalent weight
d. None of the above

66. The efficiency of fuel cell is given by the expression, η is
a. $\eta = -\frac{nFE_{cell}}{\Delta H} \times 100$
b. $\eta = -\frac{nFE_{cell}}{\Delta S} \times 100$
c. $\eta = -\frac{nFE_{cell}}{\Delta A} \times 100$
d. None of the above

67. For a gaseous reaction, the units of rate of reaction is/ are
a. atm $time^{-1}$ *b.* bar $time^{-1}$
c. atm^{-1} time *d.* Both (a) and (b)

68. The unit of rate constant for reactions of second order is
a. $L\ mol^{-1}\ s^{-1}$ *b.* $L^{-1}\ mol\ s^{-1}$
c. $L\ mol\ s^{-1}$ *d.* s^{-1}

69. In a first order reaction with time the concentration of the reactant decreases
a. linearly *b.* exponentially
c. no change *d.* None of these

70. *Aqua-regia* is a mixture of
a. conc. HNO_3(3 parts) + conc. HCl (1 part)
b. conc. HNO_3 (1 part) + conc. HCl (3 parts)
c. conc. HNO_3 (1 part) + conc. H_2SO_4 (3 parts)
d. conc. HNO_3 (3 parts) + conc. H_2SO_4 (1 part)

71. The P—P—P angle in P_4 molecule and S—S—S angle in S_8 molecule is (in degree) respectively
a. 60°, 107°
b. 107°, 60°
c. 40°, 60°
d. 60°, 40°

72. What is the formula of borax?
a. $NaAlSi_3O_8$
b. $Na_2B_4O_4 \cdot 10H_2O$
c. $Na_2SO_4 \cdot 10H_2O$
d. $NaNO_3$

73. The byproduct of Solvay-ammonia process is
a. CO_2 *b.* NH_3
c. $CaCl_2$ *d.* $CaCO_3$

74. The number of elements present in the *d*-block of the periodic table is
a. 40 *b.* 41
c. 45 *d.* 46

75. Chromite ore has the formula
a. $Cr_2O_3 \cdot FeO$ *b.* $FeOCr_2O_3$
c. Cr_2O_3 *d.* None of these

76. Which of the following represents hexadentate ligand?
a. EDTA *b.* DMG
c. Ethylenediamine *d.* None of these

77. Prussian blue is
a. $K_4\ [Fe(CN)_6]$
b. $K\ Fe[Fe(CN)_6]$
c. $K_3[Fe(CN)_6]$
d. $Fe_4[Fe(CN)_6]_3$

78. Which one of given elements shows maximum number of different oxidation states in its compounds?
a. Am *b.* Fu
c. La *d.* Gd

79. $K_4[Fe(CN)_6]$ is used in detecting
a. Fe^{3+} ion *b.* Cu^+ ion
c. Cu^{3+} ion *d.* Fe^{2+} ion

80. For a process to be spontaneous
a. $\Delta G = +$ ve *b.* $\Delta H = -$ ve
c. $\Delta S = -$ ve *d.* $\Delta G = -$ ve

PART III

a. English Proficiency

Directions (Q. Nos. 81-83) *Choose the word which best expresses the meaning of the underlined word in the sentence.*

81. He is a good-looking but insipid young man.
a. Arrogant
b. Unscrupulous
c. Sick
d. Lacking in spirit

82. It was an astute move to sell the property at that stage.
a. Shrewd *b.* Unwise
c. Dishonest *d.* Inexplicable

83. Even the most careful researcher cannot predict the possible future ramifications of his findings.
a. Uses *b.* Developments
c. Consequences *d.* Conclusions

Directions (Q. Nos. 84-86) *Fill in the blanks.*

84. Do you prefer .... or traditional art forms?
a. archaic *b.* contemporary
c. foreign *d.* simultaneous

85. She remained .... all her life.
a. spinster *b.* bachelor *c.* unmarried *d.* single

86. He was accused of bringing money into the country in .... of Foreign Exchange Rules.
a. anticipation *b.* compensation
c. perpetration *d.* violation

Directions (Q. Nos. 87-89) *Choose the word which is closest to the opposite in meaning of the following sentence.*

87. Do not stay in the grasslands after dark, as some animals become .... when they see humans.
a. provoked *b.* alerted
c. aggressive *d.* threatened

88. In all places and at all times, there is a profusion of talents.
a. Plenty *b.* Scarcity
c. Aversion *d.* Generosity

89. Unlike the other candidates, his manner was entirely languid.
a. energetic *b.* lazy
c. liquid *d.* slow

Directions (Q. Nos. 90-92) *In each of the following questions, out of the four alternatives, choose the one which can be substituted for the given words/sentences.*

90. A person who loves everybody.
a. Egoist *b.* Fatalist
c. Humanist *d.* Altruist

91. A small village or a group of houses
a. Community *b.* Settlement
c. Hamlet *d.* Colony

92. Wild Imagination
a. Whim *b.* Fantasy
c. Fancy *d.* Memory

Directions (Q. Nos. 93-95) *Choose the order of the sentences marked A, B, C, D and E to form a logical paragraph.*

93. Technology is
A : investors, workers, companies and policymakers
B : are clueless about how to safeguard jobs
C : difficult to comprehend that
D : changing work at a pace so
a. DCBA *b.* CDAB
c. CDBA *d.* DCAB

94. Since Independence,
A : on the necessity of
B : application of Uniform Civil Code
C : We have been debating
D : for all the citizens
a. CADB *b.* ABCD
c. CABD *d.* ACBD

95. The ruling party,
A : for power than by the sincere desire
B : in a democracy is obviously
C : to serve the people in a real sense
D : guided more by the lust
a. DBCA *b.* BDCA
c. DBAC *d.* BDAC

b. Logical Reasoning

96. Three positions of a dice are given below. Identify the number on the face opposite to 1.

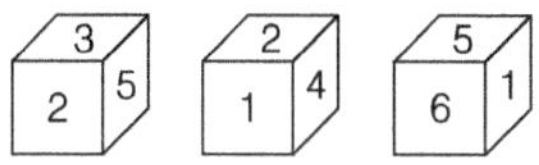

a. 3 *b.* 4 *c.* 5 *d.* 2

97. In the following question, choose the missing terms out of the given alternatives.

2	9	11	7
8	5	13	−3
7	?	10	− 4
6	4	10	?

a. 3 and −2 *b.* −3 and −2
c. 3 and 2 *d.* −3 and 2

98. On evening before sunset, 2 friends Raman and Arjun were talking to each other face to face. If Raman's shadow was exactly to his left side, which direction was Arjun facing?

a. West *b.* East *c.* North *d.* South

99. Sunil is the son of Keshav. Simran, Keshav's sister has a son Maruti and daughter Sita. Prem is the maternal uncle of Maruti. How is Sunil related to Maruti?

a. Cousin *b.* Uncle *c.* Brother *d.* Nephew

100. If the following equation has to be balance, then the signs of which of the following options will be used?

24_6_12_16=0

a. ÷, + and − *b.* −, ÷ and +
c. −, − and − *d.* ÷, + and ÷

101. A piece of paper is folded and punched as shown below in the question figures. From the given answer figures, indicate how it will appear when opened.

Question figures

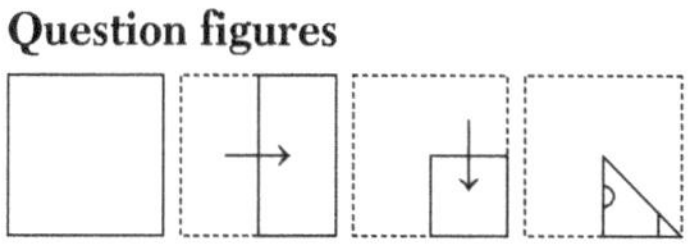

Answer figures

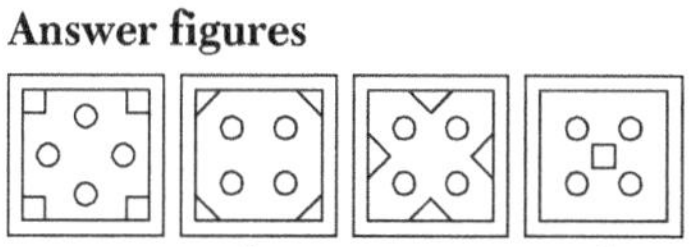

a. *b.* *c.* *d.*

102. How many squares are there in the following figure?

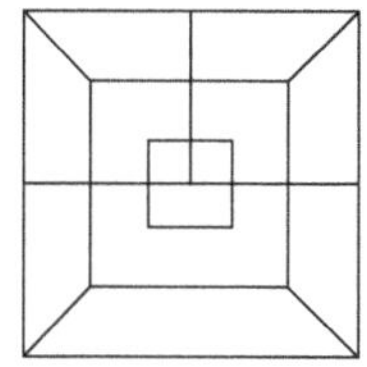

a. 5 *b.* 9
c. 7 *d.* 8

103. Arrange the following words in a logical and meaningful order.

1. Country
2. Furniture
3. Forest
4. Wood
5. Trees

a. 1, 3, 5, 4, 2
b. 1, 4, 3, 2, 5
c. 2, 4, 3, 1, 5
d. 5, 2, 3, 1, 4

104. From the given answer figures, select the one in which the question figure is hidden/embedded.

Question figure

Answer figures

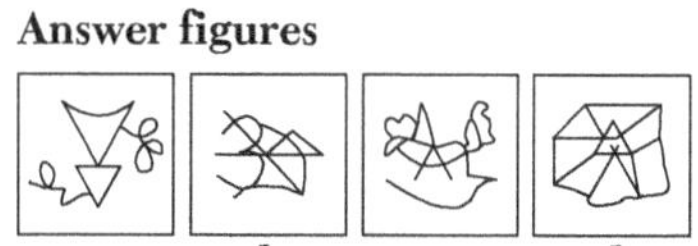

a. *b.* *c.* *d.*

105. In the question given below, find out which of the figures can be formed from the pieces gives in the problem figure.

Problem figure

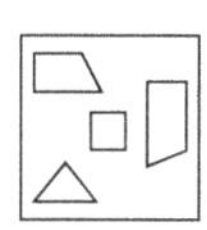

Answer figures

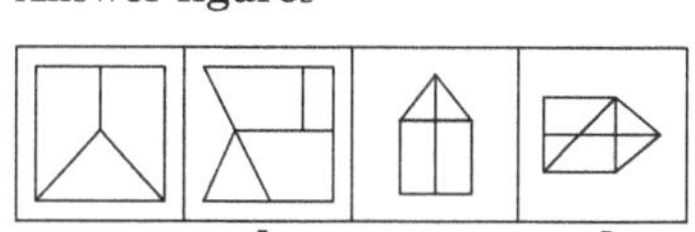

a. *b.* *c.* *d.*

PART IV

Mathematics

106. $\lim_{x \to \infty} \dfrac{\int_0^{2x} xe^{x^2}\,dx}{e^{4x^2}}$ equals

a. 0 *b.* ∞ *c.* 2 *d.* $\frac{1}{2}$

107. If ω is the complex cube root of unity, then the value of $\omega + \omega^{\left(\frac{1}{2}+\frac{3}{8}+\frac{9}{32}+\frac{27}{128}+\dots\right)}$ is

a. -1 *b.* 1 *c.* $-i$ *d.* i

108. Let $a = \cos\frac{2\pi}{7} + i\sin\frac{2\pi}{7}$, $\alpha = a + a^2 + a^4$ and $\beta = a^3 + a^5 + a^6$. Then, the equation whose roots are α and β, is

a. $x^2 - x + 2 = 0$ *b.* $x^2 + x - 2 = 0$

c. $x^2 - x - 2 = 0$ *d.* $x^2 + x + 2 = 0$

109. The value of $\frac{3}{4} + \frac{15}{16} + \frac{63}{64} + \dots$ upto n terms is

a. $n - \frac{4^n}{3} - \frac{1}{3}$ *b.* $n + \frac{4^{-n}}{3} - \frac{1}{3}$

c. $n + \frac{4^n}{3} - \frac{1}{3}$ *d.* $n - \frac{4^{-n}}{3} + \frac{1}{3}$

110. The period of the function $f(x) = \cos x + \{x\}$ is

a. 2π *b.* 1

c. π *d.* non-existent

111. If a function $f(x)$ is given by

$$f(x) = \frac{x}{1+x} + \frac{x}{(x+1)(2x+1)} + \frac{x}{(2x+1)(3x+1)} + \dots \infty,$$ then at $x = 0$, $f(x)$

a. has no limit

b. is not continuous

c. is continuous but not differentiable

d. is differentiable

112. If g is the inverse of function f and $f'(x) = \sin x$, then $g'(x)$ is equal to

a. $\operatorname{cosec}\{g(x)\}$ *b.* $\sin\{g(x)\}$

c. $\frac{1}{\sin\{g(x)\}}$ *d.* None of these

113. A bag contains $(2n + 1)$ coins. It is known that n of these coins have a head on both sides, whereas the remaining $(n + 1)$ coins are fair. A coin is picked up at random from the bag and tossed. If the probability that the toss results in a head is 31/42, then n is equal to

a. 10 *b.* 11 *c.* 12 *d.* 13

114. If $\phi(x)$ is a differentiable function, then the solution of the differential equation $dy + \{y\,\phi'(x) - \phi(x)\,\phi'(x)\}\,dx = 0$, is

a. $y = \{\phi(x) - 1\} + Ce^{-\phi(x)}$

b. $y\phi(x) = \{\phi(x)\}^2 + C$

c. $ye^{\phi(x)} = \phi(x)\,e^{\phi(x)} + C$

d. $y - \phi(x) = \phi(x)\,e^{-\phi(x)}$

115. The area of the region $R = \{(x, y): |x| \le |y|$ and $x^2 + y^2 \le 1\}$ is

a. $\frac{3\pi}{8}$ sq units *b.* $\frac{5\pi}{8}$ sq units

c. $\frac{\pi}{2}$ sq units *d.* $\frac{\pi}{8}$ sq unit

116. A and B are any two non-empty sets and A is proper subset of B. If $n(A) = 5$, then find the minimum possible value of $n(A \Delta B)$.

a. 1 *b.* 5

c. Cannot be determined *d.* None of these

117. If $\alpha = 2\tan^{-1}(\sqrt{2} - 1)$,

$\beta = 3\sin^{-1}\frac{1}{\sqrt{2}} + \sin^{-1}\left(-\frac{1}{2}\right)$ and

$\gamma = \cos^{-1}\frac{1}{3}$, then

a. $\alpha < \beta < \gamma$ *b.* $\alpha < \gamma < \beta$

c. $\beta < \gamma < \alpha$ *d.* $\gamma < \beta < \alpha$

118. If $\frac{e^x + e^{5x}}{e^{3x}} = a_0 + a_1x + a_2x^2 + a_3x^3 + \dots$, then the value of $2a_1 + 2^3a_3 + 2^5a_5 + \dots$ is

a. $e^2 + e^{-2}$ *b.* $e^4 - e^{-4}$

c. $e^4 + e^{-4}$ *d.* 0

119. Let **a**, **b** and **c** be three vectors satisfying $\mathbf{a} \times \mathbf{b} = 2(\mathbf{a} \times \mathbf{c})$, $|\mathbf{a}| = |\mathbf{c}| = 1$, $|\mathbf{b}| = 4$ and $|\mathbf{b} \times \mathbf{c}| = \sqrt{15}$. If $\mathbf{b} - 2\mathbf{c} = \lambda\mathbf{a}$, then λ equals

a. 1 *b.* -1 *c.* 2 *d.* -4

120. The total number of 4-digit numbers in which the digits are in descending order, is

a. ${}^{10}C_4 \times 4!$ *b.* ${}^{10}C_4$

c. $\frac{10!}{4!}$ *d.* None of these

BITSAT 2015

121. The line which is parallel to X-axis and crosses the curve $y = \sqrt{x}$ at an angle of 45°, is

a. $x = \frac{1}{4}$ b. $y = \frac{1}{4}$

c. $y = \frac{1}{2}$ d. $y = 1$

122. In a ΔABC, the lengths of the two larger sides are 10 and 9 units, respectively. If the angles are in AP, then the length of the third side can be

a. $5 \pm \sqrt{6}$ b. $3\sqrt{3}$

c. 5 d. None of these

123. The arithmetic mean of the data $0, 1, 2, \ldots, n$ with frequencies $1, {}^nC_1, {}^nC_2, \ldots, {}^nC_n$ is

a. n b. $\frac{2^n}{n}$

c. $n+1$ d. $\frac{n}{2}$

124. The mean square deviation of a set of n observations $x_1, x_2, \ldots, x_n$ about a point c is defined as $\frac{1}{n}\sum_{i=1}^{n}(x_i - c)^2$.

The mean square deviations about -2 and 2 are 18 and 10 respectively, the standard deviation of this set of observations is

a. 3 b. 2

c. 1 d. None of these

125. Let S be the focus of the parabola $y^2 = 8x$ and PQ be the common chord of the circle $x^2 + y^2 - 2x - 4y = 0$ and the given parabola. The area of ΔPQS is

a. 4 sq units b. 3 sq units

c. 2 sq units d. 8 sq units

126. The number of real roots of the equation $e^{x-1} + x - 2 = 0$ is

a. 1 b. 2

c. 3 d. 4

127. Minimise $Z = \sum_{j=1}^{n}\sum_{i=1}^{m} c_{ij}\, x_{ij}$

Subject to $\sum_{i=1}^{m} x_{ij} = b_j,\ j = 1, 2, \ldots, n$

$\sum_{j=1}^{n} x_{ij} = b_j,\ i = 1, 2, \ldots, m$ is a LPP with number of constraints

a. $m - n$ b. mn

c. $m + n$ d. m/n

128. India plays two matches each with West Indies and Australia. In any match, the probabilities of India getting points 0, 1 and 2 are 0.45, 0.05 and 0.50, respectively. Assuming that the outcomes are independent, the probability of India getting atleast 7 points is

a. 0.8750 b. 0.0875 c. 0.0625 d. 0.0250

129. Let M be a 3×3 non-singular matrix with $\det(M) = \alpha$. If M^{-1} adj (adj A) $= KI$, then the value of K is

a. α^3 b. α

c. α^2 d. None of these

130. Tangents are drawn from the origin to the curve $y = \cos x$. Their points of contact lie on

a. $x^2y^2 = y^2 - x^2$ b. $x^2y^2 = x^2 + y^2$

c. $x^2y^2 = x^2 - y^2$ d. None of these

131. The slope of the tangent to the curve $y = e^x \cos x$ is minimum at $x = a$, $0 \le a \le 2\pi$, then the value of a is

a. 0 b. π

c. 2π d. $3\pi/2$

132. Two lines $L_1 : x = 5, \frac{y}{3-\alpha} = \frac{z}{-2}$,

$L_2 : x = \alpha, \frac{y}{-1} = \frac{z}{2-\alpha}$

are coplanar. Then, α can take value (s)

a. 1, 4, 5 b. 1, 2, 5 c. 3, 4, 5 d. 2, 4, 5

133. Let P be a point on the ellipse $\frac{x^2}{a^2} + \frac{y^2}{b^2} = 1$ of eccentricity e. If A, A' are the vertices and S, S' are the foci of the ellipse, then area of $\Delta PSS'$: area of $\Delta APA'$ is equal to

a. $e^3 : 1$ b. $e^2 : 1$

c. $e : 1$ d. $1/e : 1$

134. The largest interval for which

$$x^{12} - x^9 + x^4 - x + 1 > 0 \text{ is}$$

a. $(-4, 0]$ b. $(-100, 100)$

c. $(0, 1)$ d. $(-\infty, \infty)$

135. If $f(\theta) = \sin\left[\tan^{-1}\left(\frac{\sin\theta}{\sqrt{\cos 2\theta}}\right)\right]$, where $-\frac{\pi}{4} < \theta < \frac{\pi}{4}$. Then, the value of $\frac{d}{d(\tan\theta)} f(\theta)$ is

a. 1 b. 2

c. 3 d. 4

BITSAT 2015

136. If $(1+3+5+\ldots+p)+(1+3+5+\ldots+q)$ $=(1+3+5+\ldots+r)$, then the least possible value of $p+q+r$, given $p>6$, is

a. 12 *b.* 21
c. 45 *d.* 54

137. For $x>0$, $\lim_{x\to 0}\left\{(\sin x)^{1/x}+\left(\frac{1}{x}\right)^{\sin x}\right\}$ is equal to

a. 0 *b.* -1
c. 1 *d.* 2

138. If $0<x<\frac{\pi}{2}$, then

a. $\tan x<x<\sin x$ *b.* $x<\sin x<\tan x$
c. $\sin x<\tan x<x$ *d.* None of these

139. The degree of the differential equation satisfying the relation $\sqrt{1+x^2}+\sqrt{1+y^2}$ $=\lambda(x\sqrt{1+y^2}-y\sqrt{1+x^2})$ is

a. 1 *b.* 2
c. 3 *d.* 4

140. Let $f(x)$ be a polynomial of degree three satisfying $f(0)=-1$ and $f(1)=0$. Also, 0 is a stationary point of $f(x)$. If $f(x)$ does not have an extremum at $x=0$, then the value of $\int \frac{f(x)}{x^3-1}dx$ is

a. $\frac{x^2}{2}+C$ *b.* $x+C$
c. $\frac{x^3}{6}+C$ *d.* None of these

141. If $e^x+e^{f(x)}=e$, then the domain of $f(x)$ is

a. $(-\infty, 1)$ *b.* $(-\infty, 0)$
c. $(1, \infty)$ *d.* None

142. The equation of the line with gradient $-3/2$, which is concurrent with the lines $4x+3y-7=0$ and $8x+5y-1=0$, is

a. $3x+2y-2=0$
b. $3x+2y-63=0$
c. $2y-3x-2=0$
d. $2y-3x-63=0$

143. Area of the circle in which a chord of length $\sqrt{2}$ makes an angle $\pi/2$ at the centre, is

a. $\pi/2$ sq units *b.* 2π sq units
c. π sq units *d.* $\pi/4$ sq units

144. The ratio of the greatest value of $2-\cos x+\sin^2 x$ is to its least value, is

a. $7/4$ *b.* $11/4$
c. $13/4$ *d.* None of these

145. If z_1, z_2, z_3 and z_4 represent the vertices of a rhombus taken in the anti-clockwise order, then

a. $z_1+z_2+z_3+z_4=0$
b. $z_1+z_2=z_3+z_4$
c. $\arg\left(\frac{z_2-z_4}{z_1-z_3}\right)=\frac{\pi}{2}$
d. $\arg\left(\frac{z_1-z_2}{z_3-z_4}\right)=\frac{\pi}{4}$

146. If $\rho=\{(x, y)\,|\,x^2+y^2=1;\, x, y\in R\}$. Then, ρ is

a. reflexive
b. symmetric
c. transitive
d. anti-symmetric

147. A line makes the same angle θ with each of the X and Z-axes. If the angle β, which it makes with Y-axis, is such that $\sin^2\beta=3\sin^2\theta$, then $\cos^2\theta$ equals

a. 2/5 *b.* 1/5
c. 3/5 *d.* 2/3

148. If in a binomial distribution $n=4$, $P(X=0)=\frac{16}{81}$, then $P(X=4)$ equals

a. $\frac{1}{16}$ *b.* $\frac{1}{81}$
c. $\frac{1}{27}$ *d.* $\frac{1}{8}$

149. Let $f: R\to R$ be a function such that
$$f(x+y)=f(x)+f(y), \forall\, x, y\in R$$
If $f(x)$ is differentiable at $x=0$, then which one of the following is incorrect?

a. $f(x)$ is continuous, $\forall\, x\in R$
b. $f'(x)$ is constant, $\forall\, x\in R$
c. $f(x)$ is differentiable, $\forall\, x\in R$
d. $f(x)$ is differentiable only in a finite interval containing zero

150. If binomial coefficients of three consecutive terms of $(1+x)^n$ are in HP, then the maximum value of n is

a. 1 *b.* 2
c. 0 *d.* None of these

Answers

Physics

1. (b)	**2.** (a)	**3.** (a)	**4.** (a)	**5.** (c)	**6.** (c)	**7.** (a)	**8.** (d)	**9.** (d)	**10.** (c)
11. (a)	**12.** (b)	**13.** (d)	**14.** (b)	**15.** (c)	**16.** (b)	**17.** (c)	**18.** (d)	**19.** (d)	**20.** (b)
21. (a)	**22.** (a)	**23.** (b)	**24.** (c)	**25.** (c)	**26.** (b)	**27.** (c)	**28.** (b)	**29.** (a)	**30.** (c)
31. (b)	**32.** (d)	**33.** (a)	**34.** (b)	**35.** (a)	**36.** (b)	**37.** (b)	**38.** (c)	**39.** (b)	**40.** (a)

Chemistry

41. (c)	**42.** (c)	**43.** (c)	**44.** (c)	**45.** (a)	**46.** (c)	**47.** (a)	**48.** (d)	**49.** (b)	**50.** (d)
51. (c)	**52.** (a)	**53.** (b)	**54.** (b)	**55.** (d)	**56.** (d)	**57.** (b)	**58.** (a)	**59.** (b)	**60.** (a)
61. (a)	**62.** (b)	**63.** (c)	**64.** (a)	**65.** (b)	**66.** (a)	**67.** (d)	**68.** (a)	**69.** (b)	**70.** (b)
71. (a)	**72.** (b)	**73.** (c)	**74.** (a)	**75.** (b)	**76.** (a)	**77.** (d)	**78.** (a)	**79.** (a)	**80.** (d)

English Proficiency

81. (d)	**82.** (a)	**83.** (c)	**84.** (b)	**85.** (a)	**86.** (d)	**87.** (c)	**88.** (b)	**89.** (a)	**90.** (d)
91. (c)	**92.** (b)	**93.** (d)	**94.** (c)	**95.** (d)					

Logical Reasoning

96. (a)	**97.** (a)	**98.** (c)	**99.** (a)	**100.** (a)	**101.** (a)	**102.** (b)	**103.** (a)	**104.** (d)	**105.** (b)

Mathematics

106. (d)	**107.** (a)	**108.** (d)	**109.** (b)	**110.** (d)	**111.** (b)	**112.** (c)	**113.** (a)	**114.** (a)	**115.** (c)
116. (a)	**117.** (b)	**118.** (d)	**119.** (d)	**120.** (b)	**121.** (c)	**122.** (a)	**123.** (d)	**124.** (a)	**125.** (a)
126. (a)	**127.** (c)	**128.** (b)	**129.** (b)	**130.** (c)	**131.** (b)	**132.** (a)	**133.** (c)	**134.** (d)	**135.** (a)
136. (b)	**137.** (c)	**138.** (d)	**139.** (a)	**140.** (b)	**141.** (a)	**142.** (a)	**143.** (c)	**144.** (c)	**145.** (c)
146. (b)	**147.** (c)	**148.** (b)	**149.** (d)	**150.** (d)					

Hints & Solutions

Physics

1. ***(b)*** According to Newton's law of cooling

$$\frac{d\theta}{dt} = -k(\theta - \theta_0)$$

$$\Rightarrow \quad \log(\theta - \theta_0) = -kt + C \quad \dots(i)$$

Given, at $t = 0$, $\theta = 60°C$

$$\log(60 - 25) = C$$

$$\Rightarrow \quad C = \log 35 \quad \dots(ii)$$

From Eqs. (i) and (ii), we get

$$\Rightarrow \quad kt = \log 35 - \log(\theta - 25)$$

$$\log(\theta - 25) = -kt + \log 35$$

$$kt = \log 35 - \log(\theta - 25) \quad \dots(iii)$$

At, $t = 10$ min, $\theta = 50°C$

$$\therefore \quad k \times 10 = \log 35 - \log 25$$

$$\Rightarrow \quad k = \frac{1}{10}\log\left(\frac{7}{5}\right)$$

$$\left(\frac{1}{10}\log\frac{7}{5}\right)t = \log 35 - \log(\theta - 25)$$

When $t = 20$ min, $\theta = ?$

$$\therefore \quad \left(\frac{1}{10}\log\frac{7}{5}\right)20 = \log\left(\frac{35}{\theta - 25}\right)$$

$$\Rightarrow \quad 2\log\left(\frac{7}{5}\right) = \log\left(\frac{35}{\theta - 25}\right)$$

$$\Rightarrow \quad \log\left(\frac{7}{5}\right)^2 = \log\left(\frac{35}{\theta - 25}\right)$$

$$\Rightarrow \quad \left(\frac{7}{5}\right)^2 = \frac{35}{\theta - 25} \Rightarrow \theta = 42.8\ °C$$

2. ***(a)*** Let v be the speed of sound and v_s be the speed of forks. The apparent frequency of fork which moves towards the observer is

$$n_1 = \left(\frac{v}{v - v_s}\right) n$$

The apparent frequency of the fork which moves away from observer is

$$n_2 = \left(\frac{v}{v + v_s}\right) n$$

If x is the number of beats heard per second,

then $x = n_1 - n_2$

$$\Rightarrow \quad x = \left(\frac{v}{v - v_s}\right) n - \left(\frac{v}{v + v_s}\right) n$$

$$\Rightarrow \quad x = \frac{v(v + v_s) - v(v - v_s)}{v^2 - v_s^2}(n)$$

$$\Rightarrow \quad \frac{2vv_s n}{v^2 - v_s^2} = x \Rightarrow 2\left(\frac{v_s}{v}\right) n \approx x \qquad \{\because \text{ if } v_s << v\}$$

$$\Rightarrow \quad v_s = \frac{xv}{2n}$$

Given, $v = 340$ m / s, $x = 3$, $n = 340$ Hz.

$$\therefore \quad v_s = \frac{340 \times 3}{2 \times 340} = 1.5 \text{ m/s}$$

3. ***(a)*** First image is formed by reflection from the concave mirror (M_2).

$$\frac{1}{v_1} + \frac{1}{u_1} = \frac{1}{f}$$

$$\Rightarrow \quad \frac{1}{v_1} + \frac{1}{-2R} = \frac{-2}{R} \quad \left[\because f_1 = \frac{-R}{2}, u_1 = -2R\right]$$

$$\Rightarrow \quad v_1 = -\frac{2R}{3}$$

$$m = -\frac{v_1}{u_1}$$

$$\Rightarrow \quad \frac{-I}{O} = -\left(\frac{-2R}{3}\right) \cdot \frac{1}{(-2R)}$$

$$\Rightarrow \quad \frac{-I}{a} = \frac{-1}{3}$$

$$\Rightarrow \quad I = \frac{a}{3}$$

Radius of image circle = $a/3$ and image is formed by the convex mirror (M_1).

Object distance = $2R - \frac{2R}{3} = \frac{4R}{3}$

$$f = \frac{-R}{2}, \quad v = ?$$

$$\frac{3}{4R} + \frac{1}{v_2} = -\frac{2}{R} \Rightarrow v_2 = -\frac{4R}{11}$$

$$\therefore \quad m_2 = -\frac{v_2}{u_2} = -\left(\frac{-4R/11}{4R/3}\right)$$

$$\Rightarrow \quad m_2 = \frac{3}{11}$$

New, object distance = $2R - \left(-\frac{4R}{11}\right) = \frac{26R}{11}$

$$\frac{1}{v_3} - \frac{1}{26R} = -\frac{2}{R} \Rightarrow v_3 = -\frac{26R}{41}$$

$$m_3 = -\frac{v_3}{u_3} = -\frac{\left(-\frac{26R}{41}\right)}{\left(\frac{26R}{41}\right)} = \frac{11}{41}$$

$\because$ Radius of 3rd image $= \frac{a}{11} \times \frac{11}{41} = \frac{a}{41}$

4. ***(a)*** The net electric field at P is zero.

$$\therefore \quad E_P = E_1 + E_2 + E_{\text{sphere}} = 0$$

$$0 = E_1 + E_2 + E_{\text{sphere}}$$

$$\therefore \quad E_{\text{sphere}} = -E_1 - E_2$$

$$\therefore \quad |E_{\text{sphere}}| = \sqrt{E_1^2 + E_2^2}$$

$$= \sqrt{\left(\frac{q_1}{4\pi\varepsilon_0 a^2}\right)^2 + \left(\frac{q_2}{4\pi\varepsilon_0 b^2}\right)^2}$$

$$= \frac{1}{4\pi\varepsilon_0}\sqrt{\left(\frac{q_1}{a^2}\right)^2 + \left(\frac{q_2}{b^2}\right)^2}$$

$$\therefore \quad E_{\text{sphere}} = \frac{1}{4\pi\varepsilon_0}\sqrt{\left(\frac{q_1}{a^2}\right)^2 + \left(\frac{q_2}{b^2}\right)^2}$$

5. ***(c)*** Condsider that the body is projected vertically upwards from A with a speed u_0.

Using equation, $s = ut + \frac{1}{2}at^2$, we get

For first case, $-h = u_0 t_1 - \frac{1}{2}gt_1^2$ ...(i)

For second case, $-h = -u_0 t_2 - \frac{1}{2}gt_2^2$ ...(ii)

On subtracting Eq. (ii) from Eq. (i), we get

$$0 = u_0(t_2 + t_1) + \frac{1}{2}g(t_2^2 - t_1^2)$$

$$\Rightarrow \quad u_0 = \frac{1}{2}g(t_1 - t_2) \qquad \text{...(iii)}$$

Putting the value of u_0 in Eq. (ii), we get

$$-h = -\left(\frac{1}{2}\right)g(t_1 - t_2)t_2 - \left(\frac{1}{2}\right)gt_2^2$$

$$\Rightarrow \quad h = \frac{1}{2}g(t_1 t_2) \qquad \text{...(iv)}$$

For third case, $u = 0$, $t = ?$

$$-h = 0 \times t - \left(\frac{1}{2}\right)gt^2$$

$$\Rightarrow \quad h = \left(\frac{1}{2}\right)gt^2 \quad \ldots(v)$$

On comparing Eqs. (iv) and (v), we get

$$\frac{1}{2}gt^2 = \frac{1}{2}gt_1t_2 \Rightarrow t = \sqrt{t_1 t_2}$$

6. ***(c)*** For horizontal acceleration, net force in vertical direction should be zero.

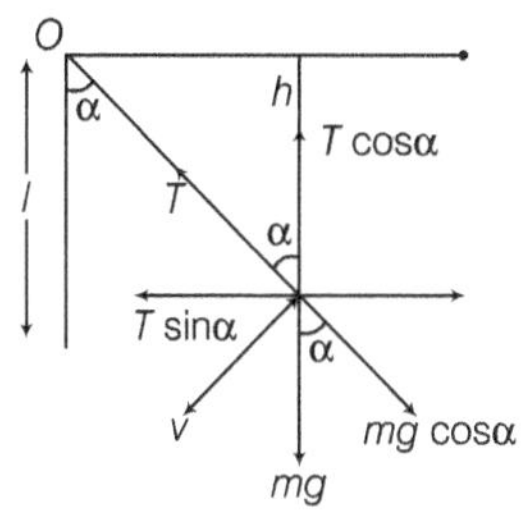

$$T \cos \alpha = mg$$

Loss in potential energy = Gain in kinetic energy.

$$\Rightarrow \quad mgh = \frac{1}{2}mv^2$$

$$\Rightarrow \quad mgl \cos \alpha = \frac{1}{2} mv^2$$

$$\therefore \quad v = \sqrt{2gl \cos \alpha}$$

The net force towards O, is

$$T - mg \cos \alpha = mv^2/l$$

$$T - mg \cos\alpha = \frac{m\, 2gl \cos \alpha}{l} = 2mg \cos \alpha$$

$$\Rightarrow \quad T = 3\, mg \cos \alpha$$

$$\therefore \quad T \cos \alpha = mg$$

$$3\, mg \cos^2 \alpha = mg \Rightarrow \alpha = \cos^{-1}(1/\sqrt{3})$$

7. ***(a)*** Since, we know that, $r = \dfrac{mv}{Bq}$ ...(i)

$$\Rightarrow \quad \text{KE} = K = \frac{1}{2} mv^2$$

$$\therefore \quad mv - \sqrt{2\, Km}$$

$\therefore$ From Eq. (i), $r = \dfrac{mv}{qB} = \dfrac{\sqrt{2\, Km}}{qB}$

$\Rightarrow r \propto \sqrt{K}$ or $r = cK^{1/2}$ [where, c is a constant.]

$$\frac{dr}{dr} = c\frac{dK^{1/2}}{dr}$$

or $$\frac{c\Delta k}{\Delta r} = 2\sqrt{K}$$

or $$\frac{\Delta r}{r} = \frac{c\Delta K}{2\sqrt{K}\, c\sqrt{K}} = \frac{\Delta K}{2K}$$

or $$\frac{\Delta r}{r} \times 100 = \frac{\Delta K}{2K} \times 100 = \frac{1}{2} \times 4\% = 2\%$$

8. ***(d)*** $\beta = 0.03$ cm, $D = 1$ m $= 100$ cm

Distance between images of the source = 0.8 cm

Distance of image from lens, $v = 80$ cm

Convex lens of focal length, $f = 16$ cm

Distance of slit from lens = u

By lens formula,

$$\frac{1}{v} + \frac{1}{u} = \frac{1}{f} \Rightarrow \frac{1}{60} + \frac{1}{u} = \frac{1}{16}$$

$$\Rightarrow \quad u = 20 \text{ cm}$$

$$\because \quad \text{Magnification, } m = \frac{v}{u} = \frac{80}{20} = 4$$

$$\text{Magnification} = \frac{\text{Distances between images of slits}}{\text{Distance between slits}}$$

$$= \frac{0.8}{d} = 4$$

$$\Rightarrow \quad d = 0.2 \text{ cm}$$

$$\because \quad \beta = \frac{D\lambda}{d}$$

$$\Rightarrow \quad \beta = \frac{100\lambda}{0.2} \times 0.03 \times 10^{-2}$$

or $$\lambda = 6000 \text{ Å}$$

9. ***(d)*** Free body diagram for the blocks A and B are as shown below

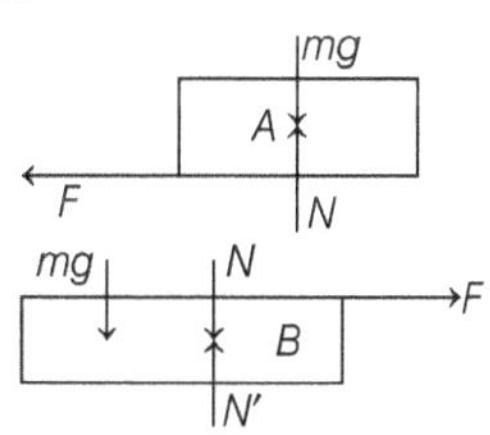

Equations of motion, $a_B = \dfrac{F}{M}$ (in $+x$- direction)

$$a_A = \frac{F}{m} (\text{in} - x\text{- direction}).$$

Relative velocity of A w.r.t. B,

$$a_{A,\,B} = a_A - a_B = -\frac{F}{m} - \frac{F}{M}$$

$$= -F\left(\frac{M + m}{Mm}\right) (\text{along} - x\text{-direction})$$

Initial relative velocity of A w.r.t. B,

$$u_{AB} = v_0$$

Final relative velocity of A w.r.t. B,

$$v^2 = u^2 + 2as$$

$$0 = v_0^2 - \frac{2F(m + M)s}{Mm}$$

$$\Rightarrow \quad s = \frac{Mmv_0^2}{2F(m + M)} \quad [\because GM = gR^2]$$

10. ***(c)*** During motion of the particle, total mechanical energy remains constant.

At the surface of the earth, total mechanical energy is

$$E_i = -\frac{GmM}{R} + \frac{1}{2}mv_0^2 = -\frac{GMmR}{R^2} + \frac{1}{2}mv_0^2$$

$$= -gmR + \frac{1}{2}mv_0^2 \qquad [\because GM = gR^2]$$

Total mechanical energy at height $h = R$ is

$$E_f = -\frac{GmM}{2R} + \frac{1}{2}mv^2 = -\frac{gmR}{2} + \frac{1}{2}mv^2.$$

According to the principle of conservation of energy,

$$E_i = F_F$$

$$\therefore \quad -gmR + \frac{1}{2}mv_0^2 = -\frac{gmR}{2} + \frac{1}{2}mv^2$$

$$\Rightarrow \quad -2gR + v_0^2 = -gR + v^2$$

$$\Rightarrow \quad v = \sqrt{v_0^2 - gR}$$

11. ***(a)*** Volume of the balloon at any instant, when radius is r

$$V = \frac{4}{3}\pi r^3$$

Time rate of change of volume, $\frac{dV}{dt} = 4\pi r^2 . \frac{dr}{dt}$

Time rate of change of radius of balloon,

$$\frac{dr}{dt} = \frac{1}{4\pi r^2} . \frac{dV}{dt}$$

Flux through rubber band at given instant,

$$\phi = B(\pi r^2)$$

Induced emf,

$$E_{\text{induced}} = -\frac{d\phi}{dt} = -\frac{d}{dt}(B\pi r^2) = -2\pi r B\left(\frac{dr}{dt}\right)$$

$$= -2\pi r B\left(\frac{1}{4\pi r^2}.\frac{dV}{dt}\right) = -\frac{B}{2r}.\frac{dV}{dt}$$

As, volume of the balloon is decreasing, $\frac{dV}{dt}$ is negative.

$$\Rightarrow \quad E_{\text{induced}} = -\frac{(0.04)}{2\times 10\times 10^{-2}}\times(-100\times 10^{-6})$$

$$= 20\,\mu V$$

12. ***(b)*** Elastic force in the string is conservative in nature.

$$\therefore \quad W = -\Delta U$$

where, W = work done by elastic force of string.

$$\therefore \quad W = -(U_F - U_i) = U_i - U_F$$

$$= \frac{1}{2}kx^2 - \frac{k}{2}(x+y)^2$$

$$= \frac{1}{2}kx^2 - \frac{1}{2}k(x^2 + y^2 + 2xy)$$

$$= \frac{1}{2}kx^2 - \frac{1}{2}ky^2 - \frac{1}{2}kx^2 - \frac{1}{2}k(2xy)$$

$$= -kxy - \frac{1}{2}ky^2$$

$\therefore$ The work done against elastic force is

$$W_{\text{external}} = -W = \frac{ky}{2}(2x+y)$$

13. ***(d)*** When the wire is stretched, its area and length changes in such a way that volume always remains same. Wire A has resistance R, length = l, radius = r.

When wire A is stretched, Resistance = R', length = l', radius = r_1

$\therefore$ Volume of original wire = Volume of stretched wire

$$V = V'$$

$$\Rightarrow \quad Al = A'l'$$

$$\text{or} \quad \pi r^2 l = \pi r_1^2 l' \quad \text{or} \quad \frac{l}{l'} = \frac{r_1^2}{r^2} = \frac{50}{100} = \frac{1}{2}$$

$$\therefore \quad R = \rho\frac{l}{A},\ R' = \rho\frac{l'}{A'}$$

$$\text{or} \quad \frac{R}{R'} = \frac{l/A}{l'/A'} = \frac{l}{A}\times\frac{A'}{l'}$$

$$= \frac{l}{l'}\times\frac{r_1^2}{r^2} = \frac{1}{2}\times\frac{1}{2} = \frac{1}{4}$$

$$R' = 4R = 4\times 10$$

$$= 40\,\Omega$$

14. ***(b)*** We know that, threshold wavelength, $\lambda_0 = \frac{hc}{\phi}$

$$\Rightarrow \quad \lambda_0 = \frac{(6.63\times 10^{-34})\times 3\times 10^8}{2.3\times(1.6\times 10^{-19})}$$

$$= 5.404\times 10^{-7}\,\text{m} = 5404\,\text{Å}$$

Thus, wavelength 4144 Å and 4972 Å will emit electron from the metal surface.

Energy incident on the surface per unit time for each wavelength

= Intensity of each × Area of the surface wavelength

$$= \frac{3.6\times 10^{-3}}{3}\times 1\times 10^{-4} = 1.2\times 10^{-7}$$

Energy incident on the surface for each wavelength in 2 s

$$E = (1.2\times 10^{-7})\times 2 = 2.4\times 10^{-7}\,\text{J}$$

Number of photons n_1 due to the wavelength 4144 Å

$$n_1 = \frac{E}{hc/\lambda_1} = \frac{E\lambda_1}{hc} = \frac{(2.4\times 10^{-7})(4144\times 10^{-10})}{(6.63\times 10^{-34})(3\times 10^8)}$$

$$= 0.5\times 10^{12}$$

Number of photons n_2 due to wavelength 4972 Å

$$n_2 = \frac{E}{hc/\lambda_2} = \frac{E\lambda_2}{hc} = \frac{(2.4\times 10^{-7})(4972\times 10^{-10})}{(6.63\times 10^{-34})(3\times 10^8)}$$

$$N = n_1 + n_2 = 0.5\times 10^{12} + 0.575\times 10^{12}$$

$$= 1.075\times 10^{12}$$

15. ***(c)*** At a depth y from the surface of the fluid, the net force acting on the gate element of width dy is

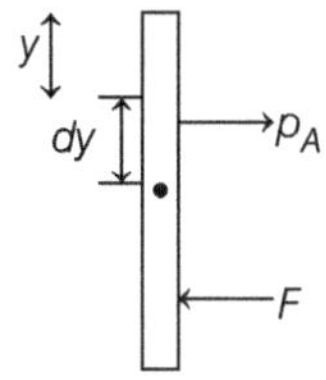

$dF = (p_0 + \rho g y - p_o) \times 1\, dy = \rho g y\, dy$

Torque of this force about the hinge is

$$d\tau = dF \times \left(\frac{1}{2} - y\right) = \rho g y\, dy \times \left(\frac{1}{2} - y\right)$$

Net torque experienced by the gate is

$$\tau_{net} = \int d\tau + F \times \frac{1}{2}$$

$$= \int_0^1 \rho g\, y\, dy \left(\frac{1}{2} - y\right) + F \times \frac{1}{2} = 0$$

$$\Rightarrow \quad F = \frac{\rho g}{6}$$

16. ***(b)*** The velocity of the photoelectrons is found by the relation,

$$evB = m\frac{v^2}{R} \Rightarrow v = \frac{e}{m}BR$$

The kinetic enregy of the photoelectrons is

$$k = \frac{1}{2}mv^2 = \frac{1}{2}\frac{e^2B^2R^2}{m}$$

$$= \frac{1}{2}\frac{(1.6\times10^{-19})^2(2\times10^{-2})^2(23\times10^{-3})^2}{(9.1\times10^{-31})}$$

$$= 2.97\times10^{-15}\text{ J}$$

or $$k = \frac{2.97\times10^{-15}}{1.6\times10^{-19}} = 18.36\text{ keV}$$

The energy of the incident photon is $E_v = \frac{hc}{\lambda}$.

$$\Rightarrow \quad E_v = \frac{12.4}{0.50} = 24.8\text{ keV}$$

Since, the binding energy is the difference between these two values.

$$\text{BE} = E_v - k = 24.8 - 18.6 = 6.2\text{ keV}$$

17. ***(c)***

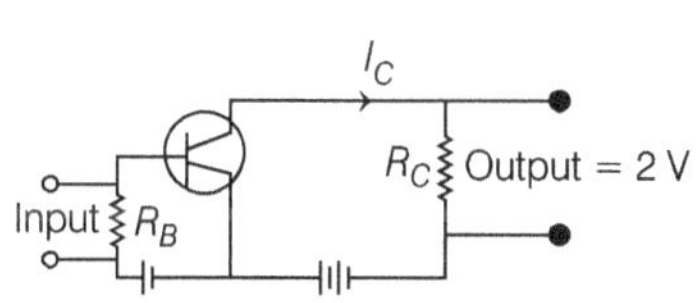

$$V_0 = I_C R_C = 2$$

$$\Rightarrow \quad I_C = \frac{2}{2\times10^3} = 10^{-3}\text{ A}$$

Current gain $= \frac{I_C}{I_B} = 100$

$$\Rightarrow \quad I_B = \frac{I_C}{100} = \frac{10^{-3}}{100} = 10^{-5}\text{ A}$$

$$V_i = R_B I_B$$

$$= 1\times10^3\times10^{-5} = 10^{-2}\text{ V}$$

$$\Rightarrow \quad V_i = 10\text{ mV}$$

18. ***(d)*** Given, $y = 12x - \frac{3}{4}x^2$, $u_x = 3\text{ ms}^{-1}$.

$$v_y = \frac{dy}{dt} = 12\frac{dx}{dt} - \frac{3}{4}\times(2x)\left(\frac{dx}{dt}\right)$$

$$= 12\frac{dx}{dt} - \frac{3}{2}x\frac{dx}{dt}$$

At $x = 0$, $v_y = u_y = 12\frac{dx}{dt}$

$$= 12u_x = 12\times3 = 36\text{ ms}^{-1}$$

and $a_y = \frac{d}{dt}v_y = \frac{d}{dt}\left(\frac{dy}{dt}\right)$

$$= 12\frac{d^2x}{dt^2} - \frac{3}{2}\left(\left(\frac{dx}{dt}\right)^2 + x\frac{d^2x}{dt^2}\right)$$

But, $\frac{d^2x}{dt^2} = a_x = 0$, hence

$$a_y = -\frac{3}{2}\left(\frac{dx}{dt}\right)^2 = -\frac{3}{2}u_x^2$$

$$\Rightarrow \quad a_y = -\frac{3}{2}\times(3)^2 = -\frac{27}{2}\text{ms}^{-1}$$

$$\Rightarrow \quad |a_y| = \frac{27}{2}\text{ms}^{-1}$$

Range, $$R = \frac{2u_x u_y}{|a_y|}$$

$$= \frac{2\times3\times36}{27/2} = 16\text{ m} = 16\text{ m}$$

19. ***(d)*** Let us consider that there are n_1 moles of hydrogen and n_2 moles of helium in the given mixture. Then, the pressure of the mixture will be

$$p = \frac{n_1RT}{V} + \frac{n_2RT}{V} = (n_1 + n_2)\frac{RT}{V}$$

$$\Rightarrow 2\times101.3\times10^3 = (n_1 + n_2)\times\frac{(8.3\times300)}{20\times10^{-3}}$$

or $$(n_1 + n_2) = \frac{2\times101.3\times10^3\times20\times10^{-3}}{(8.3)(300)}$$

or $$n_1 + n_2 = 1.62 \quad \ldots\text{(i)}$$

The mass of the mixture is (in grams),

$$n_1\times m_1 + n_2\times m_2 = 5$$

$$\Rightarrow \quad n_1\times2 + n_2\times4 = 5$$

$$\Rightarrow \quad (n_1 + 2n_2) = 2.5 \quad \ldots\text{(ii)}$$

Solving the Eqs. (i) and (ii), we get

$$n_1 = 0.74, n_2 = 0.88$$

Hence, $\dfrac{m_H}{m_{He}} = \dfrac{0.74 \times 2}{0.88 \times 4} = \dfrac{1.48}{3.52} = \dfrac{2}{5}$

20. ***(b)*** The path difference introduced due to the introduction of transparent sheet is given by $\Delta x = (\mu - 1)t$

If the central maxima occupies position of nth fringe, then $(\mu - 1)\, t = n\lambda = d \sin\theta$

$$\Rightarrow \sin\theta = \frac{(\mu - 1)\, t}{d} = \frac{(1.17 - 1) \times 1.5 \times 10^{-7}}{3 \times 10^{-7}} = 0.085$$

Hence, the angular position of central maxima is

$$\theta = \sin^{-1}(0.085) = 4.88° \approx 4.9$$

For small angles, $\sin\theta \approx \theta \approx \tan\theta$

$$\Rightarrow \quad \tan\theta = \frac{y}{D}$$

$$\therefore \quad \frac{y}{D} = \frac{(\mu - 1)\, t}{d}$$

Shift of central maxima is $y = \dfrac{D(\mu - 1)\, t}{d}$

21. ***(a)*** Let the acceleration be a, hence $a = -\omega^2 x$.

Therefore, distance of the particle from the centre at any time t is given by

$x = r\cos(\omega t)$, where r is the amplitude

when $t = 1$ s, $x = r - a$

$\therefore (r - a) = r\cos\omega$

$$\cos\omega = \frac{r - a}{r} \quad \ldots\text{(i)}$$

when $t = 2$s, $x = r - a - b$

Therefore, $r - a - b = \cos 2\omega$

$$\therefore \quad r - a - b = r(2\cos^2\omega - 1) \quad \ldots\text{(ii)}$$

On substituting the value of $\cos\omega$ from Eq. (i) to Eq. (ii), we get

$$r - a - b = r\left[\frac{2(r - a)^2}{r^2} - 1\right]$$

$$\Rightarrow \quad r - a - b = \frac{2(r - a)^2}{r} - r$$

$$\therefore \quad r(3a - b) = 2a^2 \Rightarrow r = \frac{2a^2}{3a - b}$$

22. ***(a)*** U_i = intial potential energy of the system

$$= \frac{1}{4\pi\varepsilon_0}\left[\frac{q^2}{a} + \frac{q^2}{a} + \frac{q^2}{a}\right] = \frac{1}{4\pi\varepsilon_0}\left(\frac{3q^2}{a}\right)$$

$$= 9 \times 10^9 \left(3 \times \frac{(0.1)^2}{1}\right) = 27 \times 10^7 \text{ J}$$

Let charge at A is moved to mid-point D.

Then, $U_f = \dfrac{1}{4\pi\varepsilon_0}\left[\dfrac{2q^2}{(a/2)} + \dfrac{q^2}{a}\right]$

$$= 5 \times \frac{1}{4\pi\varepsilon_0}\left(\frac{q^2}{a^2}\right) = 5 \times 9 \times 10^9 \times \frac{(0.1)^2}{1^2}$$

$$= 45 \times 10^7 \text{ J}$$

Work done

$= U_f - U_i = 95 \times 10^7 - 27 \times 10^7 = 18 \times 10^7$ J

Also, energy supplied per second = 1000 J

Time taken $= \dfrac{18 \times 10^7}{1000} = 18 \times 10^4$ s

$$= \frac{18 \times 10^4}{60 \times 60}\text{ h} = 50 \text{ h}$$

23. ***(b)*** According to the conservation principle,

$$mgh = \frac{1}{2}kx^2 - mgx$$

where, x is maximum elongation in the spring (when the particle is in its lowest extreme position).

$$\Rightarrow \quad \frac{1}{2}kx^2 - mgx - mgh = 0$$

or $\quad x^2 - \dfrac{2mg}{k}x - \dfrac{2mg}{k}\cdot h = 0$

$$\therefore \quad x = \frac{\dfrac{2mg}{k} \pm \sqrt{\left[\left(\dfrac{2mg}{k}\right)^2 + 4 \times \dfrac{2mg}{k}h\right]}}{2}$$

$\because$ Amplitude = elongation in the spring for lowest extreme position – elongation in spring for equilibrium position.

$$= x - x_0 = \frac{mg}{k}\sqrt{\left(1 + \frac{2hk}{mg}\right)} \quad \left[\because x_0 = \frac{mg}{k}\right]$$

24. ***(c)*** Given, $\dfrac{M_U}{M_{Pb}} = 3$

Let the initial mass of uranium be M_0.

Final mass of uranium after time t, $M = \dfrac{3}{4}M_0$.

According to the law of disintegration,

$$\frac{M}{M_0} = \left(\frac{1}{2}\right)^{t/T} \Rightarrow \frac{M_0}{M} = (2)^{t/T}$$

$$\therefore \log_{10}\left(\frac{M_0}{M}\right) = \frac{t}{T}\log_{10}(2)$$

$$t = T\,\frac{\log_{10}\left(\dfrac{M_0}{M}\right)}{\log_{10}(2)} = \frac{T\log_{10}\left(\dfrac{4}{3}\right)}{\log_{10}(2)}$$

$$= \frac{T\log_{10}(1.333)}{\log_{10}(2)} = 4.5 \times 10^9\left(\frac{0.1249}{0.3010}\right)$$

$$\Rightarrow \quad t = 1.867 \times 10^9 \text{ yrs}$$

25. ***(c)*** When there is no water in the mirror, the rays of light are incident normally on the mirror and retrace their path. So, we get an image coincident with the object as shown in the figure.

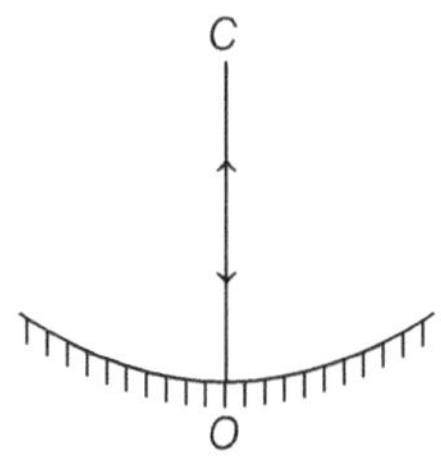

When the mirror is filled with the water, then equivalent focal length f is given by

$$\frac{1}{f}=\frac{1}{f_{\text{water lens}}}+\frac{1}{f_{\text{concave mirror}}}+\frac{1}{f_{\text{water lens}}}$$

$$\Rightarrow \quad \frac{1}{f}=2\times\left(\frac{1}{f_{\text{water lens}}}\right)+\left(\frac{1}{f_{\text{concave mirror}}}\right)$$

$$\Rightarrow \quad \frac{1}{f}=2(\mu-1)\left(\frac{1}{R}\right)+\frac{2}{R}$$

$$\Rightarrow \quad \frac{1}{f}=\frac{2(\mu-1)}{R}+\frac{2}{R}$$

$$\Rightarrow \quad \frac{1}{f}=\frac{2\mu}{R}\Rightarrow f=\frac{R}{2\mu}$$

Clearly, focal length of the new optical system is less than the original. So, the object is effectively at a distance greater than twice the focal length. So, the real image will be formed between F and $2F$ or we can say between C and O.

26. ***(b)*** Total current when direct current of 5A is superimposed on alternating current,

$$I=(5+10\sin\omega t)$$

$$\Rightarrow I_{\text{eff}}=\left[\frac{\int_0^T I^2dt}{\int_o^T dt}\right]^{1/2}=\left[\frac{1}{T}\int_0^T(5+10\sin\omega t)^2\,dt\right]^{1/2}$$

$$=\left[\frac{1}{T}\int_0^T(25+100\sin\omega t+100\sin^2\omega t)\right]^{1/2}$$

But, $\frac{1}{T}\int_0^T \sin\omega t\cdot dt=0$ and $\frac{1}{T}\int_0^T\sin^2\omega t\cdot dt=\frac{1}{2}$

So, $$I_{\text{eff}}=\left[25+\frac{1}{2}\times 100\right]^{1/2}$$

$$=5\sqrt{3}\text{ A}$$

27. ***(c)*** Consider FBD of A and B as shown below.

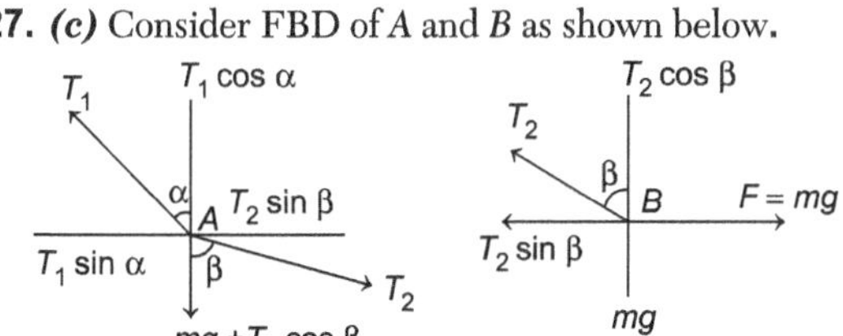

Balancing forces, we have

From FBD of A,

$$T_1\sin\alpha=T_2\sin\beta \quad \ldots\text{(i)}$$

$$T_1\cos\alpha=T_2\cos\beta+mg \quad \ldots\text{(ii)}$$

From FBD of B,

$$T_2\cos\beta=mg \quad \ldots\text{(iii)}$$

$$T_2\sin\beta=mg \quad \ldots\text{(iv)}$$

$$\Rightarrow \quad \tan\beta=1 \quad \text{[From Eqs. (iii) and (iv)]}$$

$$\therefore \quad \sin\beta=\frac{1}{\sqrt{2}}$$

Hence, $T_2=\frac{mg}{\sin\beta}=\sqrt{2}$

Using Eqs. (iv) in (i) and (iii) in (ii), we get

$$T_1\sin\alpha=mg \text{ and } T_1\cos\alpha=mg+mg=2\,mg$$

$$\tan\alpha=\frac{1}{2}\Rightarrow\cos\alpha=\frac{2}{\sqrt{5}}$$

$$\therefore \quad T_1=\frac{2mg}{\cos\alpha}=\frac{2mg}{\frac{2}{\sqrt{5}}}$$

$$\Rightarrow \quad T_1=mg\sqrt{5}$$

$$\Rightarrow \quad \frac{T_1}{T_2}=\frac{\sqrt{5}}{\sqrt{2}}\Rightarrow\sqrt{2}T_1=\sqrt{5}\,T_2$$

28. ***(b)*** Total moment of inertia

$$=I_1+I_2+I_3+I_4=2I_1+2I_2$$

$$=2(I_1+I_2)[I_3=I_2,\ I_1=I_4]$$

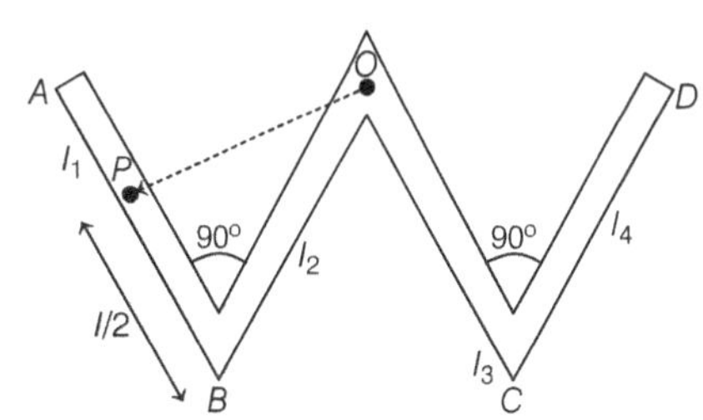

Now, $$I_2=I_3=\frac{Ml^2}{3}$$

Using parallel axis theorem, we have

$$I=I_{\text{CM}}+Mx^2$$

and $$x=\sqrt{l^2+\frac{l^2}{4}}$$

$$I_1=I_4=\frac{Ml^2}{12}+M\left[\sqrt{l^2+\left(\frac{l}{2}\right)^2}\right]^2$$

Putting all the values, we get

$$I=10\left(\frac{Ml^2}{3}\right)$$

29. ***(a)*** Wavelength of second line of Balmer series is

$$\frac{1}{\lambda_2} = R\left[\frac{1}{2^2} - \frac{1}{4^2}\right] \Rightarrow \frac{1}{\lambda_2} = R\left[\frac{1}{4} - \frac{1}{16}\right]$$

$$\frac{1}{\lambda_2} = R\left[\frac{3}{16}\right] \Rightarrow \lambda_2 = \frac{16}{3R}$$

For first line of Balmer series,

$$\frac{1}{\lambda_1} = R\left[\frac{1}{2^2} - \frac{1}{3^2}\right]$$

$$\Rightarrow \quad \frac{1}{\lambda_1} = R\left[\frac{1}{4} - \frac{1}{9}\right]$$

$$\Rightarrow \quad \frac{1}{\lambda_1} = \frac{5R}{36} \Rightarrow \lambda_1 = \frac{36}{5R}$$

$$\therefore \quad \frac{\lambda_1}{\lambda_2} = \frac{36}{5R} \times \frac{3R}{16} = \frac{27}{20}$$

$$\Rightarrow \quad \lambda_1 = \frac{27}{20} \times 4861 \text{ Å}$$

30. ***(c)*** If v be the velocity of charge particle of charge q accelerated through potential difference V, then

$$\frac{1}{2}mv^2 = qV$$

$$\Rightarrow \quad v = \sqrt{\frac{2qV}{m}}$$

As we know that, when a charge particle is allowed to move in a uniform magnetic field, then it describes spiral (or circular) path.

Centripetal force, $\frac{mv^2}{R} = qvB$

$$\therefore \quad v = \left(\frac{qB}{m}\right)R$$

Hence, $$\sqrt{\frac{2qV}{m}} = \left(\frac{qB}{m}\right)R$$

$$\Rightarrow \quad R = \left(\frac{2mV}{q}\right)^{1/2} \times \frac{1}{B}$$

Here, V, q and B are constant.

Hence, $m \propto R^2$

So, $$\frac{m_1}{m_2} = \left(\frac{R_1}{R_2}\right)^2$$

31. ***(b)*** Water wets glass and so the angle of contact is zero. Neglecting the small mass in the meniscus, for full rise,

$$2\pi rT = \pi r^2 h\rho g$$

$$\Rightarrow \quad h = \frac{2T}{r\rho g}$$

$$= \frac{2 \times 0.07}{0.25 \times 10^{-3} \times 1000 \times 9.8}$$

$$= 0.057 \text{ m} = 5.7 \text{ cm}$$

But here the tube is only 2 cm above the water and so, water will rise by 2 cm and meet the tube at an angle θ such that

$$2\pi rT \cos\theta = \pi r^2 h' \rho g$$

$$\Rightarrow \quad 2T\cos\theta = h' r\rho g$$

$$\Rightarrow \quad \cos\theta = \frac{h' r\rho g}{2T}$$

$$\Rightarrow \quad \cos\theta = \frac{2 \times 10^{-2} \times 0.25 \times 10^{-3} \times 1000 \times 9.8}{2 \times 0.07}$$

$$= 0.35$$

$$\Rightarrow \quad \theta = \cos^{-1}(0.35)$$

$$= 69.5 \simeq 70$$

32. ***(d)*** Horizontal and vertical components of initial velocity are

$$u_x = 20\sqrt{2}\cos 45° = 20 \text{ m/s}$$

and $$u_y = 20\sqrt{2}\sin 45° = 20 \text{ m/s}$$

After 1s, horizontal component remains unchanged, while the vertical component becomes

$$v_y = u_y - gt$$

Due to explosion, one part comes to rest. Hence, from the conservation of linear momentum, vertical component of second part will become $v_y' = 20$ m/s. Therefore, maximum height attained by the second part will be

$$H = h_1 + h_2$$

Here, h_1 = height attained in 1 s.

Using, $h = ut + \frac{1}{2}at^2$

$$\Rightarrow \quad h_1 = (20 \times 1) - \frac{1}{2} \times 10 \times (1)^2$$

$$h_1 = 15 \text{ m}$$

Now, $a = g = 10 \text{ m/s}^2$

$$h_2 = \frac{v_y'^2}{2g}$$

$$= \frac{(20)^2}{2 \times 10} = 20 \text{ m}$$

$$\Rightarrow \quad H = 20 + 15 = 35 \text{ m}$$

33. ***(a)***

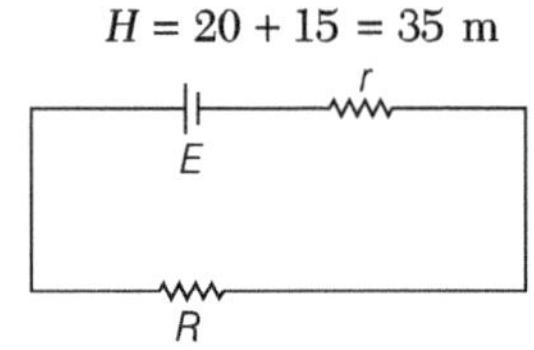

Here, $r = \frac{r_1 r_2}{r_1 + r_2} = \frac{(1)(1)}{1+1} = \frac{1}{2}\,\Omega$

and $$E = \frac{\Sigma E/r}{\Sigma 1/r} = \frac{\left(\frac{10}{1}\right) + \left(\frac{10}{1}\right)}{1+1} = 10 \text{ V}$$

Maximum power will be developed across R, when

$$R = r = \frac{1}{2}\,\Omega$$

$$i_{max} = \frac{E}{R+r} = \frac{10}{1/2+1/2} = 10\text{ A}$$

$$P_{max} = i_{max}^2 R = (10)^2\left(\frac{1}{2}\right) = 50\text{ W}$$

34. ***(b)*** Let v = speed of neutron before collision.

v_1 = speed of neutron after collision.

v_2 = speed of proton or hydrogen atom after collision.

and ΔE = energy of excitation.

From the conservation of linear momentum,

$$mv = mv_1 + mv_2$$

$$\Rightarrow \quad v = v_1 + v_2 \quad \ldots\text{(i)}$$

From conservation of energy,

$$\frac{1}{2}mv^2 = \frac{1}{2}mv_1^2 + \frac{1}{2}mv_2^2$$

$$v^2 = v_1^2 + v_2^2 \quad \ldots\text{(ii)}$$

From squaring Eq. (i), we get

$$v^2 = v_1^2 + v_2^2 + 2v_1v_2$$

From Eq. (ii), we get

$$v^2 = v_1^2 + v_2^2 + \frac{2\Delta E}{m}$$

$$\therefore \quad 2v_1v_2 = \frac{2\Delta E}{m}$$

$$\therefore \quad (v_1 - v_2)^2 = (v_1 + v_2)^2 - 4v_1v_2 = v^2 - \frac{4\Delta E}{m}$$

As, $v_1 - v_2$ must be real, $v^2 - \frac{4\Delta E}{m} \geq 0$

$$\Rightarrow \quad \frac{1}{2}mv^2 \geq 2\Delta E$$

The minimum energy that can be absorbed by the hydrogen atom in the ground state to go into the excited state is 2eV. Therefore, the maximum kinetic energy needed is

$$\frac{1}{2}mv_{min}^2 = 2 \times 10.2 = 20.4\text{ eV}$$

35. ***(a)*** $A_R = \sqrt{(\sqrt{3})^2 + (1)^2} = 2$

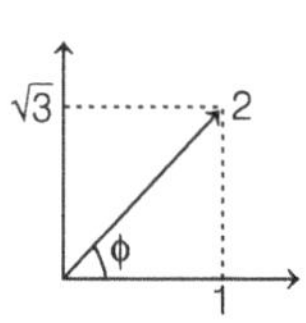

$$\theta = \tan^{-1}\left(\frac{\sqrt{3}}{1}\right) = \frac{\pi}{3}$$

$$\therefore \quad y = 2\sin\left(\omega t + \frac{\pi}{3}\right)$$

$$\frac{d^2y}{dt^2} = a = -2\omega^2\sin\left(\omega t + \frac{\pi}{3}\right)$$

$$a_{max} = -2\omega^2 = g$$

The minimum value of ω for which mass just breaks off the plank.

$$\omega = \sqrt{\frac{g}{2}}$$

This will be happen for the first time when

$$\omega t + \frac{\pi}{3} = \frac{\pi}{2} \text{ or } \omega t = \frac{\pi}{6}$$

$$\therefore \quad t = \frac{\pi}{6\omega} = \frac{\pi}{6}\sqrt{\frac{2}{g}}$$

36. ***(b)*** The diagrammatic representation of the given problem is shown in the figure.

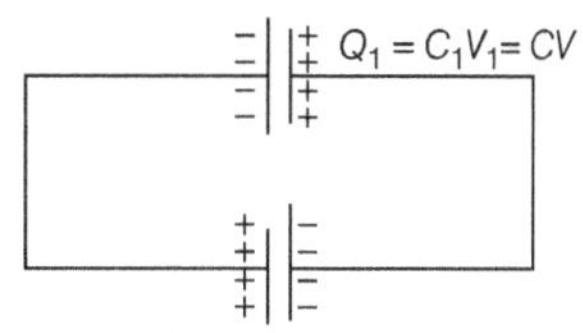

The net charge shared between the two capacitors

$$Q' = Q_2 - Q_1 = 4\,CV - CV = 3\,CV$$

The two capacitors will have same potential, say V'.

The net capacitance of the parallel combination of the two capacitors will be

$$C' = C_1 + C_2 = C + 2C = 3C$$

The potential of the capacitors will be

$$V' = \frac{Q'}{C'} = \frac{3CV}{3C} = V$$

The electrostatic energy of the capacitors will be

$$E' = \frac{1}{2}C'V'^2 = \frac{1}{2}(3C)V^2 = \frac{3}{2}CV^2$$

37. ***(b)*** Distance between two cars leaving from the station x is

$$d = \frac{1}{6} \times 60 = 10\text{ km}$$

Man meets the first car after time,

$$t_1 = \frac{60}{60+60} = \frac{1}{2}\text{ h}$$

He will meet the next car after time,

$$t_2 = \frac{10}{60+60} = \frac{1}{12}\text{ h}$$

In the remaining half an hour, the number of cars he will meet again is, $n = \frac{1/2}{1/12} = 6$

$\therefore$ Total number of cars would be meet on route will be 7.

38. ***(c)*** Energy required to raise a satellite upto height h,

$$E_1 = \Delta U = \frac{mgh}{1 + \frac{h}{R}} \quad \ldots\text{(i)}$$

E_2 = energy required to put in orbit $= \frac{1}{2} m v_0^2$

$$= \frac{1}{2} m \left(\frac{GM}{r}\right) \left[\text{as, } v_0 \text{ orbital speed} = \sqrt{\frac{GM}{r}}\right]$$

$$= \frac{1}{2} m \left(\frac{GM}{R+h}\right) = \frac{1}{2} m \left(\frac{GM}{R^2}\right) \times \frac{R}{1+\frac{h}{R}}$$

$$= \frac{1}{2} m . \frac{gR^2}{R^2} \times \frac{R}{1+h/R} \qquad [\because GM = gR^2]$$

$$E_2 = \frac{mgR}{2\left(1+\frac{h}{R}\right)} \qquad \dots\text{(ii)}$$

From Eqs. (i) and (ii), we get

$$\frac{E_1}{E_2} = \frac{2h}{R}$$

39. ***(b)*** As, frequency,

$$f \propto \sqrt{mg} \text{ or } f \propto \sqrt{g}$$

In water, $f_w = 0.8 f_{\text{air}}$

$$\frac{g'}{g} = (0.8)^2 = 0.64 \Rightarrow 1 - \frac{\rho_w}{\rho_m} = 0.64$$

$$\Rightarrow \quad \frac{\rho_w}{\rho_m} = 0.36 \qquad \dots\text{(i)}$$

In liquid, $\frac{g'}{g} = (0.6)^2 = 0.36$

$$1 - \frac{\rho_l}{\rho_m} = 0.36 \Rightarrow \frac{\rho_l}{\rho_m} = 0.64 \qquad \dots\text{(ii)}$$

From Eqs. (i) and (ii), we get

$$\frac{\rho_l}{\rho_w} = \frac{0.64}{0.36} \Rightarrow \rho_l = 1.77$$

Here, ρ_w = relative density of water = 1

ρ_m = relative density of mass

ρ_l = relative density of liquid.

40. ***(a)*** Let N be the number of turns and R be the radius of the coil.

Then, $\quad l = 2\pi RN$

$$\Rightarrow \quad R = \frac{l}{2\pi N} \qquad \dots\text{(i)}$$

Now, magnetic moment of the coil is

$$M = NiA = Ni(\pi R^2)$$

$$\Rightarrow \quad M = (Ni\pi)\left(\frac{l^2}{4\pi^2 N^2}\right) = \frac{il^2}{4\pi N}$$

Maximum value of M can be

$$M_{\max} = \frac{il^2}{4\pi} \quad \text{at} \quad N = 1$$

$$\therefore \quad \tau_{\max} = M_{\max} B \sin 90° = \frac{iBl^2}{4\pi}$$

$$\Rightarrow \quad \tau_{\max} = \frac{iBl^2}{4\pi}$$

Chemistry

41. ***(c)*** NO pollutant is the main product of automobiles exhaust.

42. ***(c)*** The harmful chemical present in tobacco is arsenic.

43. ***(c)*** The disease causes by the high concentration of hydrocarbon pollutants in atmosphere is cancer. Hydrocarbon causes irritation and hence, abnormal growth of tissues.

44. ***(c)*** Thiol, i.e. $R—SH$ participates in disulphide bond formation of protein.

45. ***(a)*** Structure of tyrosine is

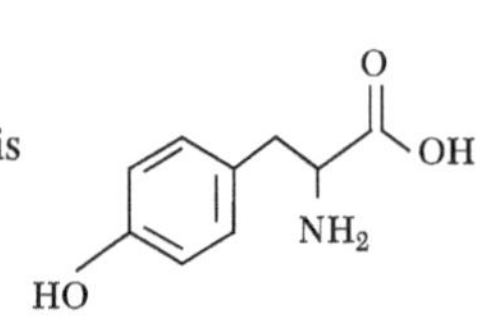

46. ***(c)*** The third law of thermodynamics helps to calculate absolute entropies of pure substance at different temperature.

$$\Delta S = S_T - S_0 = \int_0^T \frac{C_p \cdot dt}{T} \Rightarrow \Delta S = C_p \ln T$$

$$\Delta S = 2.303\, C_p \log T$$

47. ***(a)*** Dipole-dipole interaction occur in the alkyl halides. Due to this interaction, the boiling point of alkyl halide increases as compared to other alkanes.

48. ***(d)*** Lucas test is given by both (b) and (c) but not by *n*-alcohol.

e.g.

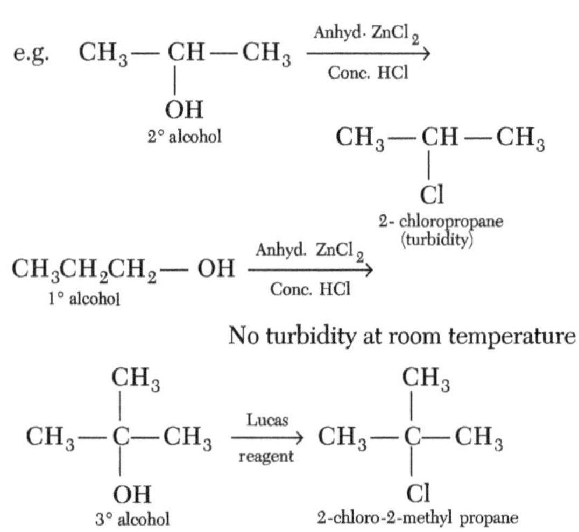

The precipitate formed is based on the difference in reactivity of the three classes of alcohol. The difference in reactivity reflects the different case of formation of the corresponding carbocations. Tertiary carbocation is far more stable than 2° and 1°, while primary carbocation is least stable.

49. ***(b)*** When a primary amine (aromatic or aliphatic) is warmed with chloroform and alc. KOH, it forms an isocyanide or carbylamine having offensive smell. This reaction is called carbylamine reaction.

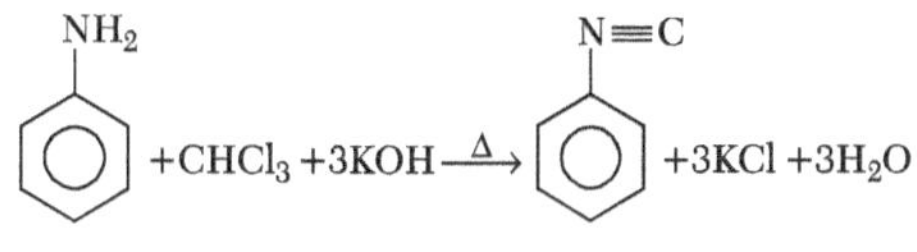

50. ***(d)*** The colour of $CoCl_3 \cdot 5NH_3 \cdot H_2O$ is pink.

51. ***(c)*** Complex salts has two different metallic elements but give test for only one element.

e.g. $K_4[Fe(CN)_6]$ gives test for K^+ only not for Fe^{2+} ions. $K_4[Fe(CN)_6] \rightleftharpoons 4K^+ + [Fe(CN)_6]^{4-}$

52. ***(a)*** Mg metal is present in chlorophyll.

53. ***(b)*** Zeise's salt is the common name of $K[PtCl_3(\eta^2 - C_2H_4)]$.

54. ***(b)*** Cobalt metal is present in vitamin B_{12}.

55. ***(d)*** Anthracene is purified by sublimation because in this method, a substance directly changes from (or transient from) the solid to the gas phase without passing through the intermediate liquid phase.

56. ***(d)*** Temperature is an intensive property. Intensive property is that which does not depend upon the quantity of matter present in the system. e.g. temperature, density, refractive index, boiling point, etc.

57. ***(b)*** Nitrous oxide is the laughing gas (i.e. N_2O).

58. ***(a)*** Brown ring test is used to detect nitrate.

$$NO_3^- + 3Fe^{2+} + 4H^+ \longrightarrow 3Fe^{3+} + NO + 2H_2O$$

$$[Fe(H_2O)_6]^{2+} + NO \longrightarrow \underset{\text{(Brown ring)}}{[Fe(H_2O)_5.(NO)]^{2+}} + H_2O$$

59. ***(b)*** Semiconductor materials, like Si and Ge are usually purified by zone refining. Zone refining method of purifying crystal or to prepare high pure materials, mainly semiconductor, in which a nano region of a crystal is melted and this molten zone is moved along the crystal.

The molten region will melt impure solid at its forward edge and leaves a pure materials behind. The impurity will be removed at one end.

60. ***(a)*** Smelting method is not used for concentration of metal. In smelting, oxides of less electropositive metals are reduced by strong heating with coal/coke. Flux is added to reduce melting point of impurities and to change them into slag.

61. ***(a)*** The Le-Chatelier's principle states that, if a system at equilibrium is subjected to a change in temperature, pressure or concentration, the equilibrium shifts in the direction that tends to under the effect of change.

According to this, only statement (a) is correct, i.e. increase in temperature favours the endothermic reaction while decrease in temperature favours the exothermic reaction.

Increase of pressure shifts, the equilibrium in that side in which number of gaseous moles decreases.

62. ***(b)*** van't Hoff equation shows variation of chemical equilibrium with temperature.

$$\log \frac{K_2}{K_1} = \frac{\Delta H}{2.303R}\left[\frac{T_2 - T_1}{T_1 \cdot T_2}\right]$$

where, ΔH = enthalpy change, K_1 and K_2 = equilibrium constants at temperature T_1 and T_2 respectively and R = universal gas constant.

63. ***(c)*** NH_3 is strong base as it has one lone pair. Basic character decreases from N to Bi due to increase in atomic size.

64. ***(a)*** $ClO_4^- < ClO_3^- < ClO_2^- < ClO^-$ is the correct order of Bronsted base because $HClO_4 > HClO_3 > HClO_2 > HClO$ is the correct order of conjugated acid.

(Since, $HClO_4$ can easily give H^+ ion.) In ClO_4^-, the negative charge is not localised because of the resonance.

But in ClO^-, there is no resonance. Thus, the basicity will increase.

65. ***(b)*** The mass of the substance deposited when one Faraday of charge is passed through its solution is equal to gram equivalent weight.

66. ***(a)*** $\eta = -\dfrac{nFE_{cell}}{\Delta H} \times 100$

67. ***(d)*** For a gaseous reaction, the units of rate of reaction are atm $time^{-1}$ or bar $time^{-1}$.

68. ***(a)*** Unit of rate constant for reactions of second order is L mol^{-1} s^{-1}.

General formula = $(concentration)^{1-n}$ $time^{-1}$

$= [mol\ L^{-1}]^{1-n}$ $time^{-1}$

where, n = order of reaction

69. ***(b)*** The formula of first order reaction is $A = [A_0]e^{-kt}$.

$\therefore$ The concentration of reactants will exponentially decreases with time.

$$A \propto e^{-t}$$

70. ***(b)*** *Aqua-regia* is a mixture of conc. HNO_3 (1 part) + conc. HCl (3 parts). Its IUPAC name is nitric acid hydrochloride.

71. ***(a)*** P_4 **molecule** The four sp^3-hybridised phosphorus atoms lie at the corners of a regular tetrahedron with $\angle PPP = 60°$. Each phosphorus atom is linked to three other P-atoms by covalent bonds.

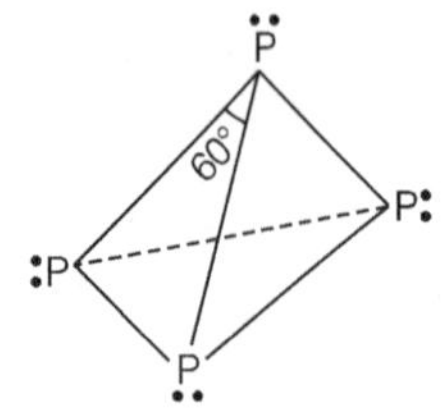

So, that each P-atom complete its octet.

S_8 molecule Sulphur exists as octa-atomic solid because sulphur does not form $p\pi$-$p\pi$ multiple bonds due to its larger size and hence, do not exist as diatomic molecule instead S form complex structure by forming single bond resulting a puckered 8-membered crown shaped rings. Thus $\angle SSS = 107°$.

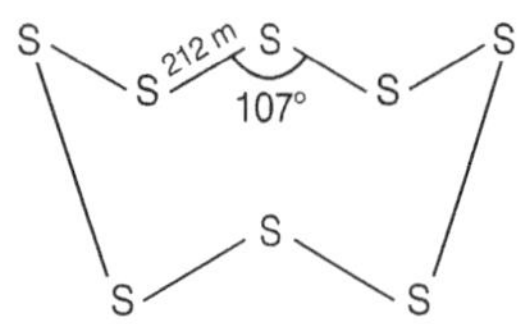

72. ***(b)*** The formula of borax is $Na_2B_4O_4 \cdot 10H_2O$.

$$NaAlSi_3O_8 \longrightarrow \text{Albite}$$
$$Na_2SO_4 \cdot 10H_2O \longrightarrow \text{Glauber's salt}$$
$$NaNO_3 \longrightarrow \text{Chile saltpetre}$$

73. ***(c)*** The Solvay-ammonia process is used for the preparation of sodium carbonate. $CaCl_2$ is produced as a byproduct which has no use.

The main steps of Solvay-ammonia process are

(i) $NaCl + CO_2 + NH_3 + H_2O \longrightarrow NaHCO_3 + NH_4Cl$

(ii) $CaCO_3 \longrightarrow CO_2 + CaO$

(iii) $2NH_4Cl + CaO \longrightarrow 2NH_3 + \underset{\text{(By product)}}{CaCl_2} + H_2O$

74. ***(a)*** 40 elements are present in *d*-block.

75. ***(b)*** $FeOCr_2O_3 \rightarrow$ Chromite

76. ***(a)*** Ethylenediammine tetraacetate (EDTA) represent the hexadentar ligand.

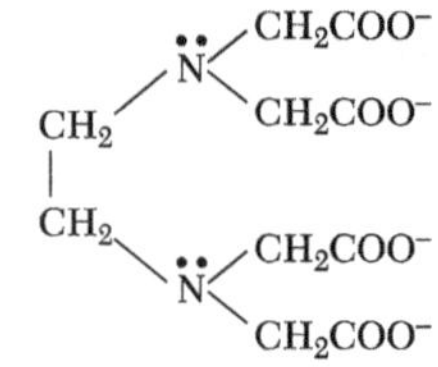

77. ***(d)*** Prussian blue is $Fe_4[Fe(CN)_6]_3$.

78. ***(a)*** Am has maximum number of oxidation states, i.e. + 3, + 4, + 5, + 6.

79. ***(a)*** $K_4[Fe(CN)_6]$ is used in detecting Fe^{3+} ion only not other.

$$4Fe^{3+} + 3K_4[Fe(CN)_6] \longrightarrow Fe_4[Fe(CN)_6]_3 + 12K^+$$

80. ***(d)*** For a process to be spontaneous, $\Delta G = -$ ve.

ΔH	ΔS	$\Delta G = T \cdot \Delta S$	Nature of reaction
(–)	(+)	(–)	Spontaneous at all temperatures
(–)	(–)	(–) at low temperature (+) at high temperature	Spontaneous at low temperatures non-spontaneous at high temperature
(+)	(+)	(+) at low temperature (–) at high temperature	Non-spontaneous at low temperatures, spontaneous at high temperatures
(+)	(–)	(+)	Non-spontaneous at all temperatures

a. English Proficiency

81. ***(d)*** 'Insipid' means lacking in spirit.

82. ***(a)*** 'Astute' means having or showing shrewdness and an ability to notice and understand things clearly. So, 'shrewd' would be its correct answer.

83. ***(c)*** 'Ramification' means a complex or unwelcome consequence of an action or event. So, 'consequence' would be its correct synonym.

84. ***(b)*** The word 'contemporary' means belonging to the same time. So, option (b) is the suitable word for the given blank.

85. ***(a)*** For women, 'spinster' is used. 'Spinster' means a woman who is not married.

86. ***(d)*** The word 'violate' means to go against or refuse to obey a law. Hence, 'violation' should be used in the given blank.

87. ***(c)*** The word 'aggressive' means angry and behaving in a threatening way. So, 'aggressive' should be used in the given blank.

88. ***(b)*** 'Profusion' means in plenty or in ample. Hence, 'scarcity' is its correct antonym that means less in quantity.

89. ***(a)*** 'Languid' means lazy and thus 'energetic' would be its correct antonym.

90. ***(d)*** A person who loves everybody is called 'Altruist'.

91. ***(c)*** A small village or a group of houses is called 'Hamlet'.

92. ***(b)*** The one which can be substituted for the given words/sentence is given in option (b) i.e. fantasy.

93. ***(d)*** The correct sequence is DCAB.

94. ***(c)*** The correct sequence is CABD.

95. ***(d)*** The correct sequence is BDAC.

b. Logical Reasoning

96. ***(a)*** 1 is adjacent to 2, 4, 5 and 6. So, 1 is opposite to 3.

97. ***(a)*** The pattern is as follows,

$$2+9=11$$
$$9-2=7$$

Also, $$8+5=13$$
$$5-8=-3$$

Now, $$7+?=10$$
$$\boxed{?=3}$$
$$3-7=-4$$

Also, $$6+4=10$$
$$\boxed{4-6=-2}$$

So, the correct options are (3, – 2).

98. ***(c)*** According to the question,

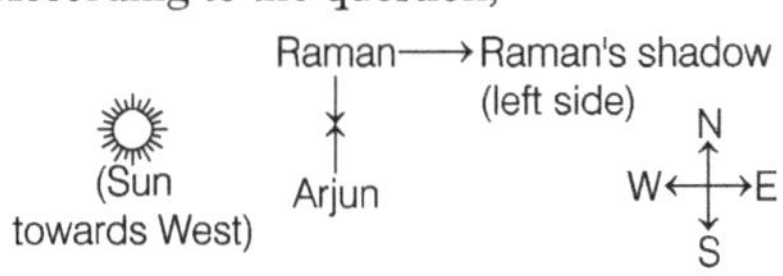

∵ Raman's shadow towards East, so Arjun was facing North.

99. ***(a)***

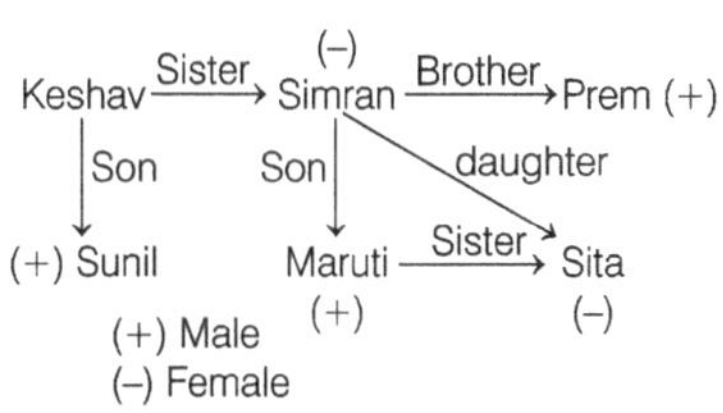

From the above family diagram, it is clear that Sunil and Maruti are cousins.

100. ***(a)*** From option (a),

$$24\div 6+12-16=0$$

$\Rightarrow$ $$4+12-16=0 \Rightarrow 16-16=0$$

$\therefore$ $$\text{LHS} = \text{RHS}$$

101. ***(a)*** The unfolded figure of question figure is depicted in the answer figure (a).

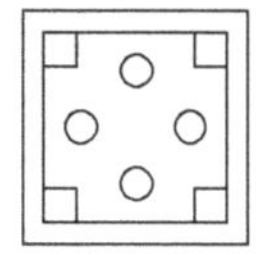

102. ***(b)***

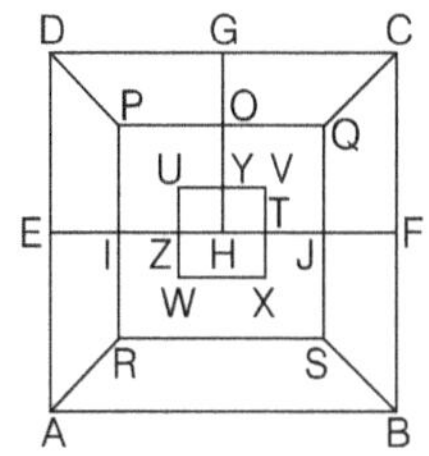

Number of squares

DGHE, GCFH, PQSR, UVXW, UYHZ, VTHY, POHI, QOHJ, ABCD = 9

103. ***(a)*** From the given words, it is deduced that a country contains forests, a forest has trees, trees give wood that is used to make furniture.

So, the correct logical order is 1, 3, 5, 4, 2.

104. ***(d)*** Clearly, question figure is embedded in the answer figure (d).

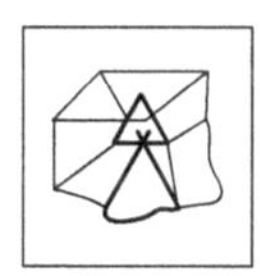

105. ***(b)*** Figure given in option (b) can be formed by joining pieces given in the question figure as shown below.

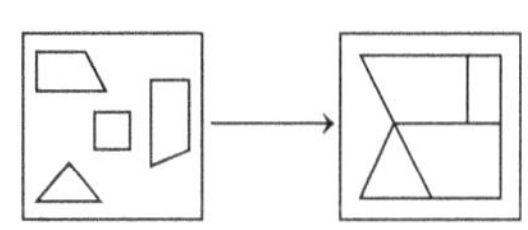

Mathematics

106. ***(d)*** $\lim\limits_{x\to\infty} \dfrac{\int_0^{2x} xe^{x^2}\,dx}{e^{4x^2}} = \lim\limits_{x\to\infty} \dfrac{\int_0^{2x} e^{x^2}\,d(x^2)}{2e^{4x^2}}$

$$= \lim_{x\to\infty} \frac{[e^{x^2}]_0^{2x}}{2e^{4x^2}} = \lim_{x\to\infty} \frac{e^{4x^2}-1}{2e^{4x^2}}$$

$$= \lim_{x\to\infty}\left(\frac{1}{2}-\frac{1}{2e^{4x^2}}\right) = \frac{1}{2}$$

Alternate Method

Using $\dfrac{d}{dx}\left(\int_{a(x)}^{b(x)} f(x)dx\right) = f(b)b'(x) - f(a)a'(x)$,

we get, $\lim\limits_{x\to\infty} \dfrac{\int_0^{2x} xe^{x^2}dx}{e^{4x^2}} = \lim\limits_{x\to\infty} \dfrac{2xe^{(2x)^2}(2)-0}{e^{4x^2}(8x)}$

$$= \lim_{x\to\infty} \frac{1}{2}\frac{e^{4x^2}}{e^{4x^2}} = \lim_{x\to\infty}\frac{1}{2} = \frac{1}{2}$$

107. ***(a)*** We have, $\dfrac{1}{2}+\dfrac{3}{8}+\dfrac{9}{32}+\dfrac{27}{128}+\ldots$

$$= \frac{3^0}{2^1}+\frac{3^1}{2^3}+\frac{3^2}{2^5}+\frac{3^3}{2^7}+\ldots$$

$$= \frac{1}{2}\left[1+\frac{3}{2^2}+\frac{3^2}{2^4}+\frac{3^3}{2^6}+\ldots\right]$$

$$= \frac{1}{2}\left[\frac{1}{1-\frac{3}{2^2}}\right] = \frac{1}{2}\left[\frac{1}{1-\frac{3}{4}}\right] = 2$$

Now, $\omega + \omega^{\left(\frac{1}{2}+\frac{3}{8}+\frac{9}{32}+\frac{27}{128}+\ldots\right)} = \omega + \omega^2$

$= -1$ $\quad [\because 1 + \omega + \omega^2 = 0]$

108. ***(d)*** We have, $a = \cos\frac{2\pi}{7} + i\sin\frac{2\pi}{7}$

$$\Rightarrow \quad a^7 = \left(\cos\frac{2\pi}{7} + i\sin\frac{2\pi}{7}\right)^7$$

$= \cos 2\pi + i\sin 2\pi$ [by De-Moivre's theorem]

$= 1 + i\cdot 0 = 1$

Now, $\alpha + \beta = a + a^2 + a^4 + a^3 + a^5 + a^6$

$$= a + a^2 + a^3 + a^4 + a^5 + a^6$$

$$= a\left(\frac{1-a^6}{1-a}\right) = \frac{a-a^7}{1-a} = \frac{a-1}{1-a} \quad [\because a^7 = 1]$$

$$= \frac{a-1}{-(a-1)} = -1$$

and $\alpha\cdot\beta = (a + a^2 + a^4)\cdot(a^3 + a^5 + a^6)$

$$= a^4(1 + a + a^3)(1 + a^2 + a^3)$$

$$= a^4(1 + a^2 + a^3 + a + a^3 + a^4 + a^3 + a^5 + a^6)$$

$$= a^4(1 + a + a^2 + 3a^3 + a^4 + a^5 + a^6)$$

$$= a^4 + a^5 + a^6 + 3a^7 + a^8 + a^9 + a^{10}$$

$$= 3 + a + a^2 + a^3 + a^4 + a^5 + a^6$$

$[\because a^7 = 1, \text{ so } a^8 = a^7\cdot a = a, a^9 = a^7\cdot a^2 = a^2, a^{10} = a^7\cdot a^3 = a^3]$

$$= 3 + a\left(\frac{1-a^6}{1-a}\right) = 3 + \frac{a-a^7}{1-a}$$

$$= 3 + \frac{a-1}{1-a} = 3 - 1 = 2$$

So, the required equation is

$$x^2 - (\alpha+\beta)x + \alpha\beta = 0 \Rightarrow x^2 + x + 2 = 0$$

109. ***(b)*** We have, $\frac{3}{4} + \frac{15}{16} + \frac{63}{64} + \ldots$ upto n terms

$$= \frac{2^2-1}{2^2} + \frac{2^4-1}{2^4} + \frac{2^6-1}{2^6} + \ldots \text{ upto } n \text{ terms}$$

$$= \left(1-\frac{1}{2^2}\right) + \left(1-\frac{1}{2^4}\right) + \left(1-\frac{1}{2^6}\right) + \ldots \text{ upto } n \text{ terms}$$

$= (1 + 1 + 1 + \ldots$ upto n terms$)$

$$-\left(\frac{1}{2^2} + \frac{1}{2^4} + \frac{1}{2^6} - + \ldots n \text{ terms}\right)$$

$$= n - \frac{1}{2^2}\left[\frac{1-\left(\frac{1}{2^2}\right)^n}{1-\frac{1}{2^2}}\right] = n - \frac{1}{3}(1 - 4^{-n})$$

$$= n + \frac{4^{-n}}{3} - \frac{1}{3}$$

110. ***(d)*** We have, $f(x) = \cos x + \{x\}$

$$\Rightarrow \quad f(x) = \cos x + x - [x]$$

Here, $\cos x$ is a periodic function with period 2π and $x - [x]$ is periodic function with period 1. But LCM of 2π and 1 does not exist.

Hence, $f(x)$ is not a periodic function and its period does not exist.

111. ***(b)*** We have,

$$f(x) = \frac{x}{1+x} + \frac{x}{(x+1)(2x+1)} + \frac{x}{(2x+1)(3x+1)} + \ldots\infty$$

$$= \lim_{n\to\infty}\sum_{r=1}^{n}\frac{x}{[(r-1)x+1](rx+1)}$$

$$= \lim_{n\to\infty}\sum_{r=1}^{n}\left[\frac{1}{[(r-1)x+1]} - \frac{1}{rx+1}\right]$$

$$= \lim_{n\to\infty}\left[1 - \frac{1}{nx+1}\right] = 1$$

For $x = 0$, we have $f(x) = 0$

Thus, we have $f(x) = \begin{cases} 1, & x \neq 0 \\ 0, & x = 0 \end{cases}$

Clearly, $\lim_{x\to 0^-} f(x) = \lim_{x\to 0^+} f(x) \neq f(0)$

So, $f(x)$ is not continuous at $x = 0$.

112. ***(c)*** Since, g is the inverse of function f. Therefore, $fog(x) = x$, for all x

$$\Rightarrow \quad \frac{d}{dx}\{fog(x)\} = \frac{d}{dx}(x), \text{ for all } x$$

$$\Rightarrow \quad \frac{d}{dx} f'\{g(x)\}\cdot g'(x) = 1, \text{ for all } x$$

$$\Rightarrow \sin\{g(x)\}\, g'(x) = 1, \text{ for all } x$$

$$\Rightarrow \quad g'(x) = \frac{1}{\sin\{g(x)\}}$$

113. ***(a)*** Consider the following events:

E_1 = Getting a coin having head on both sides from the bag

E_2 = Getting a fair coin from the bag

A = Toss results in a head

Then, $P(A) = P(E_1)\,P(A/E_1) + P(E_2)\,P(A/E_2)$

$$\Rightarrow \quad \frac{31}{42} = \frac{n}{2n+1}\times 1 + \frac{n+1}{2n+1}\times\frac{1}{2}$$

$$\Rightarrow \quad \frac{31}{21} = \frac{2n}{2n+1} + \frac{n+1}{2n+1}$$

$$\Rightarrow \quad \frac{31}{21} = \frac{3n+1}{2n+1}$$

$$\therefore \quad n = 10$$

114. ***(a)*** We have, $dy + \{y\phi'(x) - \phi(x)\,\phi'(x)\}\,dx = 0$

$$\Rightarrow \quad \frac{dy}{dx} + \phi'(x)\,y = \phi(x)\,\phi'(x)$$

which is a linear differential equation with

$$\text{If} = e^{\int \phi'(x)\,dx} = e^{\phi(x)}$$

$\therefore$ Solution is $y \cdot e^{\phi(x)} = \int \phi(x) \cdot \phi'(x)\, e^{\phi(x)} dx + C$

$$\Rightarrow \quad y \cdot e^{\phi(x)} = \int \underbrace{\phi(x) \cdot e^{\phi(x)}}_{\text{I}}\ \underbrace{\phi'(x)}_{\text{II}}\, dx + C$$

$$\Rightarrow \quad y \cdot e^{\phi(x)} = \phi(x)\, e^{\phi(x)} - \int \phi'(x)\, e^{\phi(x)}\, dx + C$$

$$\Rightarrow \quad y \cdot e^{\phi(x)} = \phi(x)\, e^{\phi(x)} - e^{\phi(x)} + C$$

$$\Rightarrow \quad y = \{\phi(x) - 1\} + Ce^{-\phi(x)}$$

115. ***(c)***

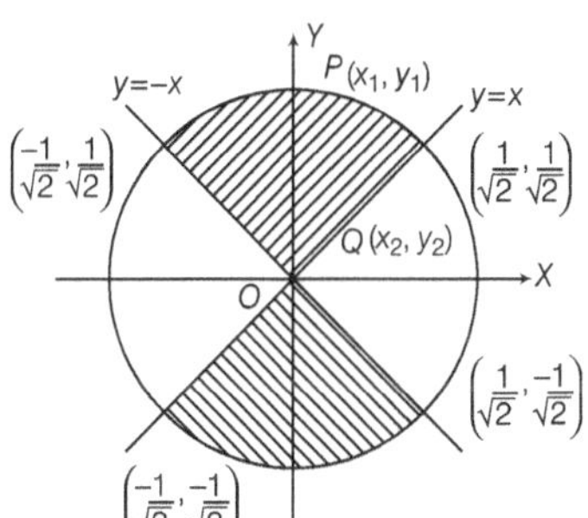

Required area = 4 (Area of the shaded region in first quadrant)

$$= 4\int_0^{1/\sqrt{2}} (y_1 - y_2)\, dx = 4\int_0^{1/\sqrt{2}} (\sqrt{1-x^2} - x)\, dx$$

$$= 4\left[\frac{x}{2}\sqrt{1-x^2} + \frac{1}{2}\sin^{-1} x - \frac{x^2}{2}\right]_0^{1/\sqrt{2}}$$

$$= 4\left[\frac{1}{2\sqrt{2}} \times \frac{1}{\sqrt{2}} + \frac{1}{2} \times \frac{\pi}{4} - \frac{1}{4}\right]$$

$$= \frac{\pi}{2} \text{ sq units}$$

116. ***(a)*** It is given that A is a proper subset of B.

$$\therefore \quad A - B = \phi$$

$$\Rightarrow \quad n(A - B) = 0$$

We have, $n(A) = 5$, so the minimum number of elements in B is 6.

Minimum possible value of

$$n\,(A \Delta B) = n(B) - n(A)$$
$$= 6 - 5 = 1$$

117. ***(b)*** Given, $\alpha = 2\tan^{-1}(\sqrt{2} - 1)$

$$= 2\tan^{-1}\left(\tan\frac{\pi}{8}\right) = \frac{\pi}{4} = \cos^{-1}\frac{1}{\sqrt{2}}$$

$$\beta = 3\sin^{-1}\left(\frac{1}{\sqrt{2}}\right) + \sin^{-1}\left(-\frac{1}{2}\right)$$

$$= 3 \times \frac{\pi}{4} - \frac{\pi}{6} = \frac{7\pi}{12}$$

$$\gamma = \cos^{-1}\left(\frac{1}{3}\right)$$

Clearly, $\alpha < \beta$

Also, $\cos^{-1} x$ is a decreasing function on $[-1, 1]$ and $\frac{1}{3} < \frac{1}{\sqrt{2}}$.

$$\therefore \quad \cos^{-1}\frac{1}{3} > \cos^{-1}\frac{1}{\sqrt{2}} \Rightarrow \gamma > \alpha$$

We observe that, $0 < \gamma < \frac{\pi}{2}$ and $\frac{\pi}{2} < \beta < \pi$

$$\Rightarrow \quad \gamma < \beta$$

Thus, we have $\alpha < \gamma < \beta$.

118. ***(d)*** We have, $\dfrac{e^x + e^{5x}}{e^{3x}} = a_0 + a_1x + a_2x^2 + a_3x^3 + \ldots$

$$\Rightarrow \quad e^{-2x} + e^{2x} = a_0 + a_1x + a_2x^2 + a_3x^3 + \ldots$$

$$\Rightarrow \quad 2\left[1 + \frac{(2x)^2}{2!} + \frac{(2x)^4}{4!} + \ldots\right]$$

$$= a_0 + a_1x + a_2x^2 + a_3x^3 + \ldots$$

$$\Rightarrow \quad a_1 = a_3 = a_5 = \ldots = 0$$

Hence, $2a_1 + 2^3a_3 + 2^5a_5 + \ldots = 0$

119. ***(d)*** Let θ be the angle between **b** and **c**.

Then, $|\mathbf{b} \times \mathbf{c}| = \sqrt{15}$

$$\Rightarrow \quad |\mathbf{b}||\mathbf{c}| \sin\theta = \sqrt{15}$$

$$\Rightarrow \quad \sin\theta = \frac{\sqrt{15}}{4 \times 1} = \frac{\sqrt{15}}{4}$$

$$\therefore \quad \cos\theta = \frac{1}{4}$$

Now, $\mathbf{b} - 2\mathbf{c} = \lambda\,\mathbf{a} \Rightarrow |\mathbf{b} - 2\mathbf{c}|^2 = |\lambda\,\mathbf{a}|^2$

$$\Rightarrow \quad |\mathbf{b}|^2 + 4|\mathbf{c}|^2 - 4(\mathbf{b} \cdot \mathbf{c}) = \lambda^2\ |\mathbf{a}|^2$$

$$\Rightarrow \quad 16 + 4 - 4|\mathbf{b}||\mathbf{c}| \cos\theta = \lambda^2$$

$$\Rightarrow \quad 20 - 16\cos\theta = \lambda^2$$

$$\Rightarrow \quad 20 - 4 = \lambda^2$$

$$\Rightarrow \quad \lambda^2 = 16$$

$$\Rightarrow \quad \lambda = \pm 4$$

120. ***(b)*** Total number of arrangements of 10 digits 0, 1, 2, ..., 9 by taking 4 at a time $= {}^{10}C_4 \times 4!$

We observe that in every arrangement of 4 selected digits there is just one arrangement in which the digits are in descending order.

$\therefore$ Required number of 4-digit numbers

$$= \frac{{}^{10}C_4 \times 4!}{4!} = {}^{10}C_4$$

121. ***(c)*** Given equation of a line parallel to X-axis is $y = k$. On solving this equation with $y = \sqrt{x}$, we get $x = k^2$. Thus, the line $y = k$ intersects with $y = \sqrt{x}$ at (k^2, k). It is given that the line $y = k$ intersects the curve $y = \sqrt{x}$ at an angle of $\pi/4$. This means that the slope of the tangent to $y = \sqrt{x}$ at (k^2, k) is $\tan\left(+\frac{\pi}{4}\right) = +1$

$$\Rightarrow \left(\frac{dy}{dx}\right)_{(k^2,\, k)} = +1 \Rightarrow \left(\frac{1}{2\sqrt{x}}\right)_{(k^2,\, k)} = +1$$

$$\Rightarrow k = +\frac{1}{2}$$

Hence, the required line is $y = +\frac{1}{2}$.

122. ***(a)*** Let $a = 10$ and $b = 9$

It is given that the angles are in AP.

$\therefore \quad 2B = A + C$

$\Rightarrow \quad 3B = A + B + C$

$\Rightarrow \quad 3B = 180°$

$\Rightarrow \quad B = 60°$

Now, $\cos B = \dfrac{a^2 + c^2 - b^2}{2ac}$

$$\Rightarrow \cos 60° = \frac{10^2 + c^2 - 9^2}{2 \times 10 \times c}$$

$$\Rightarrow \frac{1}{2} = \frac{100 + c^2 - 81}{20\,c}$$

$$\Rightarrow c^2 - 10c + 19 = 0 \Rightarrow c = 5 \pm \sqrt{6}$$

123. ***(d)*** The required mean is given by

$$\overline{X} = \frac{0 \cdot 1 + 1 \cdot {}^nC_1 + 2 \cdot {}^nC_2 + \ldots + n \cdot {}^nC_n}{1 + {}^nC_1 + {}^nC_2 + \ldots + {}^nC_n}$$

$$= \frac{\sum_{r=0}^{n} r \cdot {}^nC_r}{\sum_{r=0}^{n} {}^nC_r} = \frac{\sum_{r=1}^{n} r \cdot \frac{n}{r}\, {}^{n-1}C_{r-1}}{\sum_{r=0}^{n} {}^nC_r}$$

$$= \frac{n \sum_{r=1}^{n} {}^{n-1}C_{r-1}}{\sum_{r=0}^{n} {}^nC_r} = \frac{n \cdot 2^{n-1}}{2^n} = \frac{n}{2}$$

124. ***(a)*** We have, $\dfrac{1}{n}\sum_{i=1}^{n}(x_i + 2)^2 = 18$

and $\quad \dfrac{1}{n}\sum_{i=1}^{n}(x_i - 2)^2 = 10$

$$\Rightarrow \sum_{i=1}^{n}(x_i + 2)^2 = 18n$$

and $\quad \sum_{i=1}^{n}(x_i - 2)^2 = 10n$

$$\Rightarrow \sum_{i=1}^{n}(x_i + 2)^2 + \sum_{i=1}^{n}(x_i - 2)^2 = 28n$$

and $\quad \sum_{i=1}^{n}(x_i + 2)^2 - \sum_{i=1}^{n}(x_i - 2)^2 = 8n$

$$\Rightarrow 2\sum_{i=1}^{n}(x_i^2 + 4) = 28n \text{ and } 2\sum_{i=1}^{n} 4x_i = 8n$$

$$\Rightarrow \sum_{i=1}^{n} x_i^2 + 4n = 14n \text{ and } \sum_{i=1}^{n} x_i = n$$

$$\Rightarrow \sum_{i=1}^{n} x_i^2 = 10n \text{ and } \sum_{i=1}^{n} x_i = n$$

$$\therefore \sigma = \sqrt{\frac{1}{n}\sum_{i=1}^{n} x_i^2 - \left(\frac{1}{n}\sum_{i=1}^{n} x_i\right)^2}$$

$$= \sqrt{\frac{10n}{n} - \left(\frac{n}{n}\right)^2} = 3$$

125. ***(a)*** The parametric equations of the parabola $y^2 = 8x$ are $x = 2t^2$ and $y = 4t$.

On putting $x = 2t^2$ and $y = 4t$ in

$x^2 + y^2 - 2x - 4y = 0$, we get

$$4t^4 + 16t^2 - 4t^2 - 16t = 0$$

$$4t^4 + 12t^2 - 16t = 0$$

$$\Rightarrow 4t(t^3 + 3t - 4) = 0$$

$$\Rightarrow t\,(t - 1)(t^2 + t + 4) = 0$$

$$\Rightarrow t = 0, t = 1 \quad [\because t^2 + t + 4 \neq 0]$$

Thus, the coordinates of points of intersection of the circle and the parabola are $O(0, 0)$ and $P(2, 4)$.

Clearly, these are diametrically opposite points on the circle.

The coordinates of the focus S of the parabola are $(2, 0)$ which lies on the circle. So, OSP is a right triangle.

$\therefore$ Area of $\Delta OSP = \dfrac{1}{2} \times OS \times SP$

$$= \frac{1}{2} \times 2 \times 4 = 4 \text{ sq units}$$

126. *(a)* Let $f(x) = e^{x-1} + x - 2$

Then, $f(1) = e^0 + 1 - 2 = 0$

So, $x = 1$ is a real roots of the equation $f(x) = 0$.
Let $x = \alpha$ be the other root of $f(x) = 0$ such that $\alpha > 1$ or $\alpha < 1$.

Consider, the interval $[1, \alpha]$ or $[\alpha, 1]$. Clearly, $f(1) = f(\alpha) = 0$

By Rolle's theorem, $f'(x) = 0$ has a root in $(1, \alpha)$ or in $(\alpha, 1)$.

But $f'(x) = e^{x-1} + 1 > 0$, for all x. Thus, $f'(x) \neq 0$, for any $x \in (1, \alpha)$ or $x \in (\alpha, 1)$, which is a contradiction.
Hence, $f(x) = 0$ has no real root other than 1.

127. *(c)* Constraints will be

$$\begin{aligned} x_{11} + x_{21} + \ldots + x_{m1} &= b_1 \\ x_{12} + x_{22} + \ldots + x_{m2} &= b_2 \\ \vdots \quad \vdots \quad & \quad \vdots \\ x_{1n} + x_{2n} + \ldots + x_{mn} &= b_n \\ x_{11} + x_{12} + \ldots + x_{1n} &= b_1 \\ x_{21} + x_{22} + \ldots + x_{2n} &= b_2 \\ \vdots \quad \vdots \quad & \quad \vdots \\ x_{m1} + x_{m2} + \ldots + x_{mn} &= b_n \end{aligned}$$

So, total number of constraints $= m + n$

128. *(b)* We have, probability of getting atleast seven points

= Probability of getting 7 points or 8 points

Seven points in four matches can be obtained in the following four different ways

2, 2, 2, 1 ; 2, 2, 1, 2; 2, 1, 2, 2; 1, 2, 2, 2

The probability of each of these ways

$$= (0.50)^3 \times (0.05) = 0.00625$$

$\therefore$ Probability of getting 7 points

$$= 4 \times 0.00625 = 0.0250$$

Eight points in four matches can be obtained only in one way i.e. 2, 2, 2, 2.

$\therefore$ Probability of getting 8 points

$$= (0.50)^4 = 0.0625$$

Thus, the required probability

$$= 0.0250 + 0.0625 = 0.0875$$

129. *(b)* We know that, $M\,(\text{adj}\, M) = |M|\, I$

Replacing M by adj M, we get

$$\text{adj}\, M\, [\text{adj}(\text{adj}\, M)] = \det(\text{adj}\, M)\, I$$

$$\Rightarrow \det(M)\, M^{-1}\, [\text{adj}(\text{adj}\, M)] = \alpha^2 I$$

$$\left[\because M^{-1} = \frac{1}{|M|}\text{adj}(M)\right]$$

$$\Rightarrow \quad \alpha M^{-1} \cdot [\text{adj}(\text{adj}\, M)] = \alpha^2 I$$

$$\Rightarrow \quad M^{-1}\, [\text{adj}(\text{adj}\, M)] = \alpha I$$

But $\quad M^{-1}\, [\text{adj}(\text{adj}\, M)] = KI$

Hence, $\quad K = \alpha$

130. *(c)* Let (x_1, y_1) be one of the points of contact. Then, the equation of the tangent at (x_1, y_1) is

$$y - y_1 = \left(\frac{dy}{dx}\right)_{(x_1, y_1)} (x - x_1)$$

$$\Rightarrow \quad y - y_1 = -\sin x_1\, (x - x_1)$$

which passes through the origin.

$$\therefore \quad 0 - y_1 = -\sin x_1\, (0 - x_1)$$

$$\Rightarrow \quad y_1 = -x_1 \sin x_1 \quad \ldots\text{(i)}$$

Also, point (x_1, y_1) lies on $y = \cos x$.

$$\therefore \quad y_1 = \cos x_1 \quad \ldots\text{(ii)}$$

From Eqs. (i) and (ii), we get

$$\sin^2 x_1 + \cos^2 x_1 = \frac{y_1^2}{x_1^2} + y_1^2 = 1$$

$$\Rightarrow \quad x_1^2 = y_1^2 + y_1^2 x_1^2$$

Hence, the locus of (x_1, y_1) is

$$x^2 = y^2 + y^2x^2 \Rightarrow x^2y^2 = x^2 - y^2$$

131. *(b)* Let m be the slope of the tangent to the curve $y = e^x \cos x$.

Then, $\quad m = dy/dx = e^x (\cos x - \sin x)$

$$\Rightarrow \quad \frac{dm}{dx} = e^x (\cos x - \sin x) + e^x(-\sin x - \cos x)$$

$$= -2e^x \sin x$$

and $\quad \dfrac{d^2m}{dx^2} = -2e^x(\sin x + \cos x)$

$$\therefore \quad \frac{dm}{dx} = 0 \Rightarrow \sin x = 0$$

$$\Rightarrow \quad x = 0, \pi, 2\pi$$

Clearly, $\dfrac{d^2m}{dx^2} > 0$ for $x = \pi$

Thus, slope of the tangent m is minimum at $x = \pi$.

Hence, $a = \pi$.

132. *(a)* The equations of given lines are

$$L_1 : \frac{x-5}{0} = \frac{y}{3-\alpha} = \frac{z}{-2},$$

$$L_2 : \frac{x-\alpha}{0} = \frac{y}{-1} = \frac{z}{2-\alpha}$$

Since, these lines are coplanar.

Therefore, $\begin{vmatrix} 5-\alpha & 0-0 & 0-0 \\ 0 & 3-\alpha & -2 \\ 0 & -1 & 2-\alpha \end{vmatrix} = 0$

$$\Rightarrow \quad (5-\alpha)[(3-\alpha)(2-\alpha) - 2] = 0$$

$$\Rightarrow \quad (5-\alpha)(6 - 3\alpha - 2\alpha + \alpha^2 - 2) = 0$$

$$\Rightarrow \quad (5-\alpha)(\alpha^2 - 5\alpha + 4) = 0$$

$$\Rightarrow \quad (5-\alpha)(\alpha - 1)(\alpha - 4) = 0$$

$$\Rightarrow \quad \alpha = 1, 4, 5$$

133. ***(c)*** Let $P(a\cos\theta, b\sin\theta)$ be a point on the ellipse $\frac{x^2}{a^2}+\frac{y^2}{b^2}=1$ of eccentricity e. Then, the coordinates of A, A′, S and S′ are $(a, 0), (-a, 0), (ae, 0)$ and $(-ae, 0)$, respectively.

Now, area of $\Delta PSS'$

$$=\frac{1}{2}\left\|\begin{matrix} a\cos\theta & b\sin\theta & 1\\ ae & 0 & 1\\ -ae & 0 & 1\end{matrix}\right\|$$

$$= abe\sin\theta$$

$$\text{and area of } \Delta APA' = \frac{1}{2}\left\|\begin{matrix} a\cos\theta & b\sin\theta & 1\\ a & 0 & 1\\ -a & 0 & 1\end{matrix}\right\|$$

$$= ab\sin\theta$$

$\therefore$ Area of $\Delta PSS'$: Area of $\Delta APA' = e:1$

134. ***(d)*** $x^{12}-x^9+x^4-x+1$

$$= x^9(x^3-1)+x(x^3-1)+1=(x^3-1)(x^9+x)+1$$

$$= x(x^3-1)(x^8+1)+1\ge 1>0, \text{ for all } x\ge 1$$

Also, $x^{12}-x^9+x^4-x+1$

$$= x^{12}+x^4(1-x^5)+(1-x)>0, \text{ for all } 0\le x<1$$

Clearly, $x^{12}-x^9+x^4-x+1>0, \forall\, x<0$

$$\therefore \quad x^{12}-x^9+x^4-x+1>0, \forall x\in(-\infty,\infty)$$

135. ***(a)*** We have, $f(\theta)=\sin\left[\tan^{-1}\left(\frac{\sin\theta}{\sqrt{1-2\sin^2\theta}}\right)\right]$

$$= \sin\left[\sin^{-1}\left(\frac{\sin\theta}{\sqrt{\sin^2\theta+1-2\sin^2\theta}}\right)\right]$$

$$= \sin[\sin^{-1}(\tan\theta)] = \tan\theta$$

$$\because \quad \tan^{-1}x=\sin^{-1}\frac{x}{\sqrt{1+x^2}}$$

$$\therefore \quad \frac{d}{d(\tan\theta)}f(\theta)=1$$

136. ***(b)*** Since, p, q and r are odd natural numbers.

Let $p=2x-1$, $q=2y-1$ and $r=2z-1$.

Then, $(1+3+5+\ldots+p)+(1+3+5+\ldots+q)$

$$=(1+3+5+\ldots+r)\Rightarrow x^2+y^2=z^2$$

$\Rightarrow x, y$ and z are pythagorean triplets.

It is given that, $p>6\Rightarrow x\ge 4$.

First pythagorean triplet containing a number greater than or equal to 4 is (4, 3, 5).

$$\therefore \quad x=4, y=3 \text{ and } z=5$$

$$\Rightarrow \quad p=7, q=5 \text{ and } r=9$$

$$\Rightarrow \quad p+q+r=21$$

137. ***(c)*** Let $a=\lim_{x\to 0}\left\{(\sin x)^{\frac{1}{x}}+\left(\frac{1}{x}\right)^{\sin x}\right\}$

$$\Rightarrow a=\lim_{x\to 0}(\sin x)^{1/x}+\lim_{x\to 0}\left(\frac{1}{x}\right)^{\sin x} \Rightarrow a=l+m$$

Now, $\quad l=\lim_{x\to 0}(\sin x)^{1/x}$ and $m=\lim_{x\to 0}\left(\frac{1}{x}\right)^{\sin x}$

$$l=0$$

$$\Rightarrow \log m=\lim_{x\to 0}(-\sin x\log x)=\lim_{x\to 0}\frac{-\log x}{\operatorname{cosec} x}$$

$$=\lim_{x\to 0}\frac{-1/x}{-\operatorname{cosec} x\cot x}=\lim_{x\to 0}\frac{\sin^2 x}{x\cos x}$$

$$=\lim_{x\to 0}\frac{\tan x}{x}\times\sin x=1\times 0=0$$

$$\Rightarrow \quad m=e^0=1 \Rightarrow a=0+1=1$$

138. ***(d)*** Consider the functions $f(x)$ and $g(x)$ given by $f(x)=\tan x-x$

and $\quad g(x)=x-\sin x$, for $0<x<\frac{\pi}{2}$

We have, $f'(x)=\sec^2 x-1$ and

$$g'(x)=1-\cos x$$

$$\Rightarrow \quad f'(x)>0 \text{ and } g'(x)>0, \forall\, x\in\left(0,\frac{\pi}{2}\right)$$

$$\Rightarrow \quad f(x)>f(0) \text{ and } g(x)>g(0), \forall\, x\in\left(0,\frac{\pi}{2}\right)$$

$$\Rightarrow \quad \tan x-x>0 \text{ and } x-\sin x>0, \forall\, x\in\left(0,\frac{\pi}{2}\right)$$

$$\Rightarrow \quad \tan x>x \text{ and } x>\sin x, \forall\, x\in\left(0,\frac{\pi}{2}\right)$$

$$\Rightarrow \quad \sin x<x<\tan x, \forall\, x\in\left(0,\frac{\pi}{2}\right)$$

139. ***(a)*** On putting $x=\tan A$ and $y=\tan B$ in the given relation, we get $\sqrt{1+\tan^2 A}+\sqrt{1+\tan^2 B}$

$$=\lambda[\tan A\sqrt{1+\tan^2 B}-\tan B\sqrt{1+\tan^2 A}]$$

$$\Rightarrow \sec A+\sec B=\lambda[\tan A\sec B-\tan B\sec A]$$

$$[\because \sec^2\theta-\tan^2\theta=1]$$

$$\Rightarrow \cos A+\cos B=\lambda[\sin A-\sin B]$$

$$\Rightarrow 2\cos\frac{A+B}{2}\cos\frac{A-B}{2}$$

$$=\lambda\left(2\cos\frac{A+B}{2}\sin\frac{A-B}{2}\right)$$

$$\left[\begin{aligned}&\because \cos x+\cos y=2\cos\frac{x+y}{2}\cos\frac{x-y}{2}\\ &\text{and } \sin x-\sin y=2\cos\frac{x+y}{2}\sin\frac{x-y}{2}\end{aligned}\right]$$

BITSAT 2015

$\Rightarrow \quad \tan \frac{A-B}{2} = \frac{1}{\lambda} \Rightarrow A - B = 2\tan^{-1}(1/\lambda)$

$\Rightarrow \quad \tan^{-1} x - \tan^{-1} y = 2\tan^{-1}(1/\lambda)$

On differentiating w.r.t. x, we get

$$\frac{1}{1+x^2} - \frac{1}{1+y^2}\frac{dy}{dx} = 0 \Rightarrow \frac{dy}{dx} = \frac{1+y^2}{1+x^2}$$

Hence, the degree is 1.

140. ***(b)*** Let $f(x) = ax^3 + bx^2 + cx + d$

Then, $f(0) = -1$ and $f(1) = 0$

$\Rightarrow \quad d = -1$ and $a + b + c + d = 0$

So, $\quad a + b + c = 1 \quad \ldots(i)$

It is given that $x = 0$ is a stationary point of $f(x)$ but it is not a point of extremum.

Therefore, $f'(0) = 0 = f''(0)$ and $f'''(0) \neq 0$

Now, $\quad f(x) = ax^3 + bx^2 + cx + d$

$\Rightarrow \quad f'(x) = 3ax^2 + 2bx + c,$

$\Rightarrow \quad f''(x) = 6ax + 2b$ and $f'''(x) = 6a$

$\because \quad f'(0) = 0, f''(0) = 0$ and $f'''(0) \neq 0$

$\Rightarrow \quad c = 0, b = 0$ and $a \neq 0 \quad \ldots(ii)$

From Eqs. (i) and (ii), we get

$a = 1, b = c = 0$ and $d = -1$

$\therefore \quad f(x) = x^3 - 1$

Hence, $\int \frac{f(x)}{x^3 - 1}dx = \int \frac{x^3 - 1}{x^3 - 1}dx = \int 1\, dx = x + C$

141. ***(a)*** We have, $e^x + e^{f(x)} = e$

$\Rightarrow \quad e^{f(x)} = e - e^x$

$\Rightarrow \quad f(x) = \log_e(e - e^x) = \log[e(1 - e^{x-1})]$

$= \log_e e + \log_e(1 - e^{x-1})$

$= 1 + \log_e(1 - e^{x-1})$

Clearly, for $f(x)$ to be real, we must have

$1 - e^{x-1} > 0$

$\Rightarrow \quad e^x < e \Rightarrow x < 1 \Rightarrow x \in (-\infty, 1)$

142. ***(a)*** The equation of a line concurrent with the lines $4x + 3y - 7 = 0$ and $8x + 5y - 1 = 0$ is

$(4x + 3y - 7) + \lambda(8x + 5y - 1) = 0$

$\Rightarrow (4 + 8\lambda)x + (3 + 5\lambda)y - 7 - \lambda = 0$

The gradient of this line is $-\frac{3}{2}$, therefore

$$-\frac{8\lambda + 4}{5\lambda + 3} = -\frac{3}{2} \Rightarrow \lambda = 1$$

So, the required line is

$12x + 8y - 8 = 0 \Rightarrow 3x + 2y - 2 = 0$

which has gradient $\frac{-3}{2}$.

143. ***(c)*** Let AB be the chord of length $\sqrt{2}$, O be the centre of the circle and let OC be the perpendicular from O on AB.

Then, $AC = BC = \frac{\sqrt{2}}{2} = \frac{1}{\sqrt{2}}$

$\because \quad \angle AOB = \frac{\pi}{2} = \angle BOC = \frac{1}{2}\left(\frac{\pi}{2}\right) = \frac{\pi}{4}$

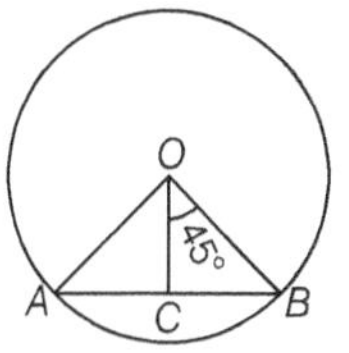

In ΔOBC, we have

$$OB = BC \operatorname{cosec} 45° = \frac{1}{\sqrt{2}} \times \sqrt{2} = 1$$

$\therefore$ Area of the circle $= \pi(OB)^2 = \pi$ sq units

144. ***(c)*** Let $y = 2 - \cos x + \sin^2 x$

Then, $\quad y = 2 - \cos x + 1 - \cos^2 x$

$= 3 - (\cos^2 x + \cos x)$

$= \frac{13}{4} - \left(\cos x + \frac{1}{2}\right)^2$

Now, $-1 \leq \cos x \leq 1$, for all x.

$\Rightarrow \quad -\frac{1}{2} \leq \cos x + \frac{1}{2} \leq \frac{3}{2}$

$\Rightarrow \quad 0 \leq \left(\cos x + \frac{1}{2}\right)^2 \leq \frac{9}{4}$

$\Rightarrow \quad -\frac{9}{4} \leq -\left(\cos x + \frac{1}{2}\right)^2 \leq 0$

$\Rightarrow \quad \frac{13}{4} - \frac{9}{4} \leq \frac{13}{4} - \left(\cos x + \frac{1}{2}\right)^2 \leq \frac{13}{4}$

$\Rightarrow \quad 1 \leq y \leq \frac{13}{4}$

$\therefore \quad y_{\max} = \frac{13}{4}$ and $y_{\min} = 1$

Hence, the required ratio is 13/4.

145. ***(c)*** Since, the diagonals of a rhombus bisect each other at right angle.

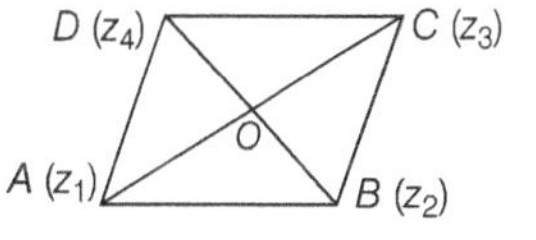

$\therefore \quad \frac{z_1 + z_3}{2} = \frac{z_2 + z_4}{2} \Rightarrow z_1 + z_3 = z_2 + z_4$

Also, $\quad \angle AOB = \frac{\pi}{2} \Rightarrow \arg\left(\frac{z_2 - z_4}{z_1 - z_3}\right) = \frac{\pi}{2}$

146. ***(b)*** Let $x = 2,\ y = 2$ but $2^2 + 2^2 \neq 1 \Rightarrow p$ is not reflexive

Let $(0, 1)$ and $(1,0) \in p$

$\Rightarrow$ but $(0, 0) \notin p$

$\Rightarrow p$ is not transitive the relation is not reflexive and transitive, but it is symmetric, because

$$x^2 + y^2 = 1$$

$$\Rightarrow \quad y^2 + x^2 = 1$$

147. ***(c)*** Let l, m and n be the direction cosines.

Then, $\quad l = \cos\theta,\ m = \cos\beta,\ n = \cos\theta$

$$\therefore \quad l^2 + m^2 + n^2 = 1$$

$$\Rightarrow \quad \cos^2\theta + \cos^2\beta + \cos^2\theta = 1$$

$$\Rightarrow \quad 2\cos^2\theta + 1 - \sin^2\beta = 1$$

$$\Rightarrow \quad 2\cos^2\theta - \sin^2\beta = 0 \qquad [\because \cos^2\theta + \sin^2\theta = 1]$$

$$\Rightarrow \quad 2\cos^2\theta - 3\sin^2\theta = 0 \qquad [\because \sin^2\beta = 3\sin^2\theta \text{ (given)}]$$

$$\Rightarrow \quad \tan^2\theta = 2/3$$

$$\therefore \cos^2\theta = \frac{1}{1+\tan^2\theta} = \frac{1}{1+2/3} = \frac{3}{5}$$

148. ***(b)*** We have, $n = 4$ and $P(X = 0) = \dfrac{16}{81}$

Let p be the probability of success and q that of failure in a trial.

Then, $\quad P(X = 0) = \dfrac{16}{81}$

$$\Rightarrow \quad {}^4C_0\, p^0 q^4 = \frac{16}{81}$$

$$\Rightarrow \quad q^4 = \left(\frac{2}{3}\right)^4$$

$$\Rightarrow \quad q = \frac{2}{3} \Rightarrow p = \frac{1}{3} \quad \{\because p + q = 1\}$$

$$\therefore \quad P(X = 4) = {}^4C_4 p^4 q^0$$

$$= p^4 = \left(\frac{1}{3}\right)^4 = \frac{1}{81}$$

149. ***(d)*** We have,

$$f(x + y) = f(x) + f(y),\ \forall\ x, y \in R$$

$$\Rightarrow \quad f(0) = f(0) + f(0)$$

[replacing x and y both by zero]

$$\Rightarrow \quad f(0) = 0$$

Now, $\quad f'(0) = \lim\limits_{h \to 0} \dfrac{f(0+h) - f(0)}{h}$

$$f'(0) = \lim_{h \to 0} \frac{f(h)}{h}$$

Now, $\quad f'(x) = \lim\limits_{h \to 0} \dfrac{f(x+h) - f(x)}{h}$

$$= \lim_{h \to 0} \frac{f(x) + f(h) - f(x)}{h}$$

$$\Rightarrow \quad f'(x) = \lim_{h \to 0} \frac{f(h)}{h} = f'(0)$$

$$\Rightarrow \quad f(x) = x f'(0) + C$$

But $\quad f(0) = 0$

$$\therefore \quad C = 0$$

Hence, $f(x) = x f'(0),\ \forall\ x \in R$.

Clearly, $f(x)$ is everywhere continuous and differentiable and $f'(x)$ is constant, $\forall\ x \in R$.

150. ***(d)*** Let the coefficients of rth, $(r + 1)$th and $(r + 2)$th terms be in HP.

$\Rightarrow$ Coefficients of $\dfrac{1}{r\text{th}}, \dfrac{1}{(r+1)\text{th}}, \dfrac{1}{(r+2)\text{th}}$ term will be in AP.

Then, $\quad \dfrac{2}{{}^nC_r} = \dfrac{1}{{}^nC_{r-1}} + \dfrac{1}{{}^nC_{r+1}}$

$$\Rightarrow \quad 2 = \frac{{}^nC_r}{{}^nC_{r-1}} + \frac{{}^nC_r}{{}^nC_{r+1}}$$

$$\Rightarrow \quad 2 = \frac{n - r + 1}{r} + \frac{r + 1}{n - r}$$

$$\Rightarrow \quad n^2 - 4nr + 4r^2 + n = 0$$

$$\Rightarrow \quad (n - 2r)^2 + n = 0$$

which is not possible for any value for n.

BITSAT 2015

BITSAT

Birla Institute of Technology & Science Admission Test

Solved Paper 2014*

Instructions

- There are 150 questions in all. The number of questions in each part is as given below.

		No. of Questions
Part I	Physics	1-40
Part II	Chemistry	41-80
Part III	a. English Proficiency	81-95
	b. Logical Reasoning	96-105
Part IV	Mathematics	106-150

- All questions are Multiple Choice Questions having four options out of which **only one** is correct.
- Each correct answer fetches 3 marks while incorrect answer has a penalty of 1 mark.
- Time allotted to complete this paper is 3 hrs.

PART I

Physics

1. In process of amplitude modulation of signal to be transmitted. Signal to be modulated is given by $m(t) = A_m \sin \omega_m t$, carrier wave is given by $C(t) = A_c \sin \omega_c t$, modulated signal $C_m(t)$ is given by

a. $C_m(t) = A_c \sin\omega_c t + A_m \sin\omega_m t$
b. $C_m(t) = (A_c + A_m) \sin\omega_c t$
c. $C_m(t) = [A_c + m(t)] \sin\omega_c t$
d. None of the above

2. Graph of stopping potential for most energetic emitted photoelectron (V_S) with frequency of incident radiation on metal is given below

Value of $\frac{AB}{BC}$, in graph is

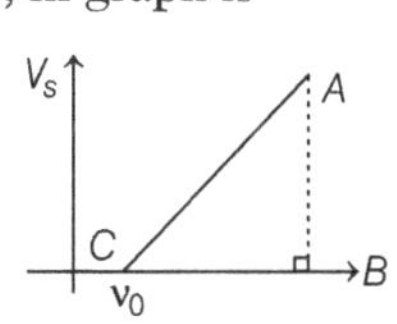

a. h *b.* e *c.* $\frac{h}{e}$ *d.* $\frac{e}{h}$

(h = Planck's constant, e = electronic charge)

3. A block of mass 0.18 kg is attached to a spring of force constant 2 N/m. The coefficient of friction between the block and the floor is 0.1. Initially the block is at rest and the spring is unstretched. An impulse is given to the block. The block slides a distance of 0.06 m and comes to rest for the first time the initial velocity of the block in m/s is $v = N/10$. Then, N is

a. 2 *b.* 3 *c.* 4 *d.* 6

4. Ohm's law says

a. $V = IR$
b. V/I = constant
c. Both (a) and (b) are correct
d. Both (a) and (b) are incorrect

5. A train accelerating uniformly from rest attains a maximum speed of 40 ms^{-1} in 20 s. It travels at this speed for 20 s and is brought to rest with uniform retardation in further 40 s. What is the average velocity during this period?

a. 80 m/s *b.* 25 m/s *c.* 40 m/s *d.* 30 m/s

*Memory Based

6. The strain-stress curves of three wires of different materials are shown in the figure. P, Q, R are the elastic limits of the wires, the figure shows that

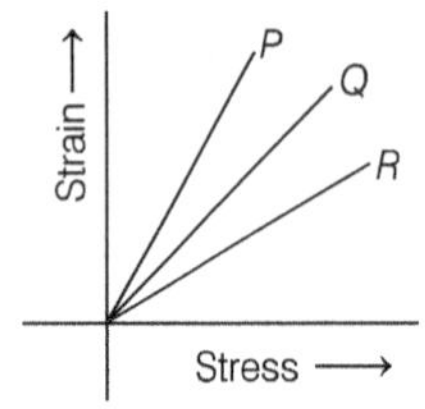

a. elasticity of wire P is maximum
b. elasticity of wire Q is maximum
c. tensile strength of R is maximum
d. None of the above

7. For the equation $F \propto A^a v^b d^c$, where F is the force, A is the area, v is the velocity and d is the density, the values of a, b and c are respectively

a. 1, 2, 1 *b.* 2, 1, 1 *c.* 1, 1, 2 *d.* 0, 1,1

8. In hydrogen atom, an electron jumps from bigger orbit to smaller orbit so that radius of smaller orbit is one-fourth of radius of bigger orbit. If speed of electron in bigger orbit was v then speed in smaller orbit is

a. $v/4$ *b.* $v/2$ *c.* v *d.* $2v$

9. A steel wire of length 4.7 m and cross-section 3.0×10^{-5} m^2 stretches by the same amount as a copper wire of length 3.5 m and cross-section 4.0×10^{-5} m^2 under a given load. What is the ratio of the Young's modulus of steel, so that of copper ?

a. 1.5 : 2 *b.* 1.8 : 2 *c.* 1.5 : 1 *d.* 1.8 : 1

10. A ring shaped conductor with radius a carries a net positive charge q uniformly distributed on it as shown in figure. A point P is situated at a distance x from its centre. Which of the following graph shows the correct variation of electric field (E) with distance (x)?

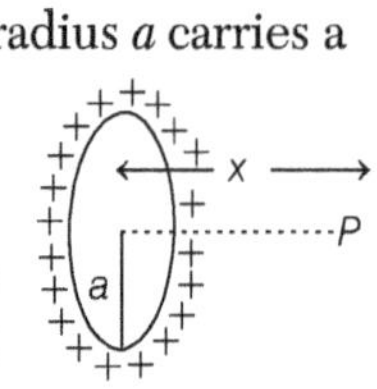

a.

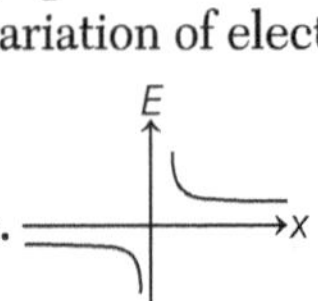

b.

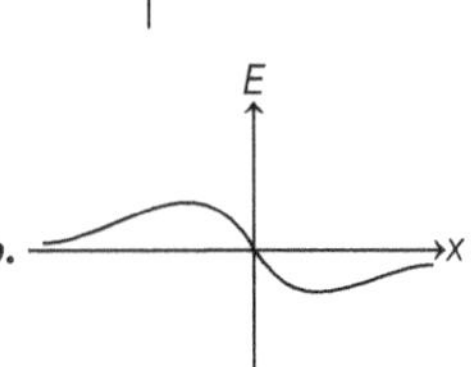

c.

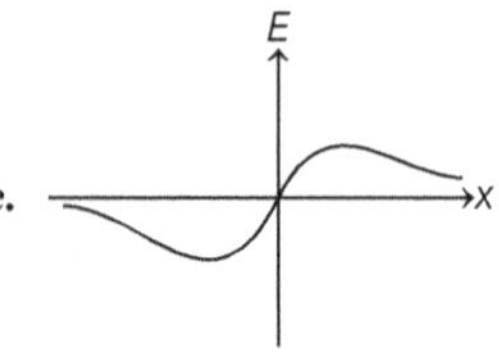

d. None of the above

11. .....A..... is the essential condition for coherent sources. Here, A refers to

a. constant phase difference
b. equal amplitude
c. Both (a) and (b) are correct
d. Both (a) and (b) are incorrect

12. The angular size of the central maxima due to a single slit diffraction is (a = slit width)

a. $\frac{\lambda}{a}$ *b.* $\frac{2\lambda}{a}$
c. $\frac{3\lambda}{2a}$ *d.* $\frac{\lambda}{2a}$

13.

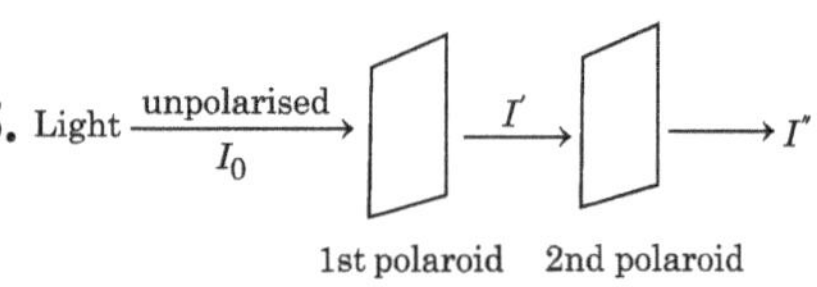

Find the final intensity of light I'', if the angle between the axes of two polaroids is 60°.

a. $\frac{3I_0}{2}$ *b.* $\frac{I_0}{2}$
c. $\frac{I_0}{4}$ *d.* $\frac{I_0}{8}$

14. Resolving power of a telescope will be more, if the diameter a of the objective is

a. larger
b. smaller
c. resolving power does not depend on a
d. None of the above

15. Let binding energy per nucleon of nucleus is denoted as E_{bn} and radius of nucleus is denoted as r. If mass number of nuclei A, B are 64 and 125 respectively, then

a. $r_A < r_B, E_{bnA} < E_{bnB}$
b. $r_A > r_B, E_{bnA} > E_{bnB}$
c. $r_A = \frac{4}{5} r_B, E_{bnA} < E_{bnB}$
d. $r_A < r_B, E_{bnA} > E_{bnB}$

16. The heat energy

a. is a state variable
b. does not depend on the state of the system
c. is equal to internal energy of the system
d. None of the above

17. A current 4.0 A exists in a wire of cross-sectional area $2.0\ mm^2$. If each cubic metre of the wire contains 12.0×10^{28} free electrons, then the drift speed is

a. 2×10^{-8} m/s

b. 0.5×10^{-3} m/s

c. 1.04×10^{-4} m/s

d. None of the above

18. In an experiment on the specific heat of a metal a 0.2 kg block of the metal at 150°C is dropped in a copper calorimeter (of water equivalent 0.025 kg) containing 150 cc of water at 27°C. The final temperature is 40°C. Calculate the specific heat of the metal. If heat losses to the surroundings are not negligible, is our answer greater or smaller than the actual value of specific heat of the metal?

a. 0.02 b. 0.2

c. 0.01 d. 0.1

19. Electric field in a region is given by $\mathbf{E} = \left(\frac{M}{x^3}\right)\hat{\mathbf{i}}$, then the correct expression for the potential in the region is
(assume potential at infinity is zero).

a. $\frac{M}{2x^2}$ b. Mx^2

c. $\frac{M}{3x^4}$ d. None of these

20. A ball is projected upwards from the top of tower with a velocity 50 m/s making an angle 30° with the horizontal. The height of tower is 70 m. After how many seconds from the instant of throwing will the ball reach the ground?

a. 2 s b. 5 s

c. 7 s d. 9 s

21. Three capacitors $X = 1\,\mu F$, $Y = 2\,\mu F$ and $Z = 3\,\mu F$ are connected as shown in figure, then the equivalent capacitance between points A and B is

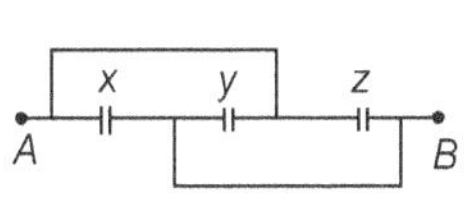

a. $6\,\mu F$ b. $12\,\mu F$

c. $3\,\mu F$ d. None of these

22. The work done in blowing a soap bubble of surface tension $0.06\ Nm^{-1}$ from 2 cm radius to 5 cm radius is

a. 0.004168 J b. 0.003168 J

c. 0.003158 J d. 0.004568 J

23. The sum of the magnitudes of two forces acting at a point is 16 N. The resultant of these forces is perpendicular to the smaller force which has a magnitude of 8 N. If the smaller force is magnitude x, then the value of x is

a. 2 N b. 4 N c. 6 N d. 7 N

24. What will be the value of current i in the circuit shown?

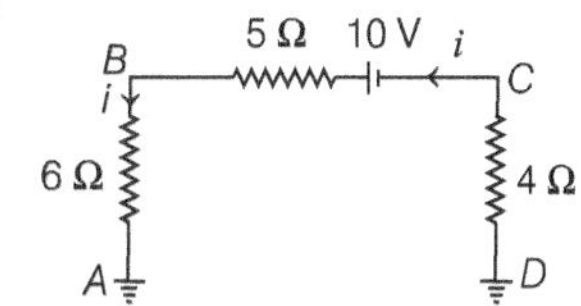

a. 0.67 A b. 1 A

c. 0.32 A d. None of these

25. The average depth of Indian Ocean is about 3000 m. Bulk modulus of water is $2.2 \times 10^4\ N/m^2$, $g = 10\ m/s^2$, then fractional compression $\frac{\Delta V}{V}$ of water at the bottom of the Indian Ocean will be

a. 1.36% b. 20.6% c. 13.9% d. 0.52%

26. If $\mathbf{A}$ and $\mathbf{B}$ denote the sides of a parallelogram and its area is $\frac{1}{2}AB$ (A and B are magnitude of $\mathbf{A}$ and $\mathbf{B}$ respectively), the angle between $\mathbf{A}$ and $\mathbf{B}$ is

a. 30° b. 45° c. 60° d. 90°

27. If edge lengths of a cuboid are measured to be 1.2 cm, 1.5 cm and 1.8 cm, then volume of the cuboid is

a. $3.240\ cm^3$ b. $3.24\ cm^3$

c. $3.2\ cm^3$ d. $3.0\ cm^3$

28. A ball thrown upwards from the top of a tower with speed v reaches the ground in t_1 second. If this ball is thrown downwards from the top of the same tower with speed v, it reaches the ground in t_2 sec. In what time the ball shall reach the ground, if it is allowed to fall freely under gravity from the top of the tower?

a. $\frac{t_1 + t_2}{2}$ b. $\frac{t_1 - t_2}{2}$ c. $\sqrt{t_1 t_2}$ d. $t_1 + t_2$

29. One end of steel wire is fixed to ceiling of an elevator moving up with an acceleration $2\ m/s^2$ and a load of 10 kg hangs from other end. Area of cross-section of the wire is $2\ cm^2$. The longitudinal strain in the wire is ($g = 10\ m/s^2$ and $Y = 2 \times 10^{11}\ Nm^{-2}$)

a. 4×10^{11} b. 3×10^{-6}

c. 8×10^{-6} d. 2×10^{-6}

30. The electrostatic force of repulsion between two positively charged ions carrying equal charge is 3.7×10^{-9} N, when they are separated by a distance of 5 Å. How much electrons are missing from each ion?

a. 10 *b.* 8 *c.* 2 *d.* 1

31. A narrow beam of protons and deuterons, each having the same momentum, enters a region of uniform magnetic field directed perpendicular to their direction of momentum. The ratio of the radii of the circular paths described by them is

a. $1:2$ *b.* $1:1$ *c.* $2:1$ *d.* $1:3$

32. A solenoid of length 1.0 m has a radius of 1 cm and has a total of 1000 turn wound on it. It carries a current of 5 A. If an electron were to move with a speed of 10^4 ms^{-1} along the axis of this current carrying solenoid, the force experienced by this electron is

a. 2 N *b.* 1.2 N
c. zero *d.* 2.5 N

33. Magnetic field

a. can increase the speed of charge particle
b. can accelerate a charge particle
c. Both (a) and (b) are correct
d. Both (a) and (b) are incorrect

34. A square coil of side 10 cm has 20 turns and carries a current of 12 A. The coil is suspended vertically and the normal to the plane of the coil, makes an angle θ with the direction of a uniform horizontal magnetic field of 0.80 T. If the torque, experienced by the coil, equals 0.96 N-m, the value of θ is

a. 0° *b.* $\frac{\pi}{2}$ rad *c.* $\frac{\pi}{3}$ rad *d.* $\frac{\pi}{6}$ rad

35. In the circuit (figure) the current is to be measured. The ammeter shown is a galvanometer with a resistance $R_G = 60.00\,\Omega$ converted to an ammeter by a shunt resistance $r_s = 0.02\,\Omega$. The value of the current is

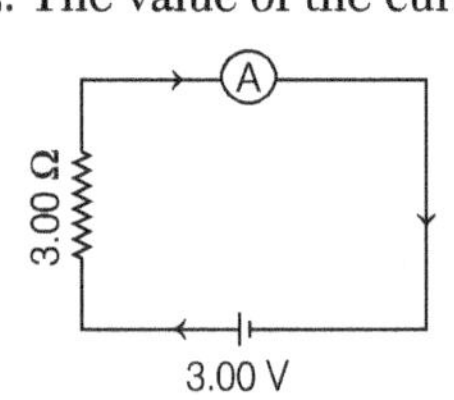

a. 0.79 A *b.* 0.29 A
c. 0.99 A *d.* 0.8 A

36. The susceptibility of a magnetism at 300 K is 1.2×10^{-5}. The temperature at which the susceptibility increases to 1.8×10^{-5} is

a. 150 K *b.* 200 K
c. 250 K *d.* 20 K

37. Figure shows some of the equipotential surface of the magnetic scalar potential. The magnetic field **B** at a point in the region is

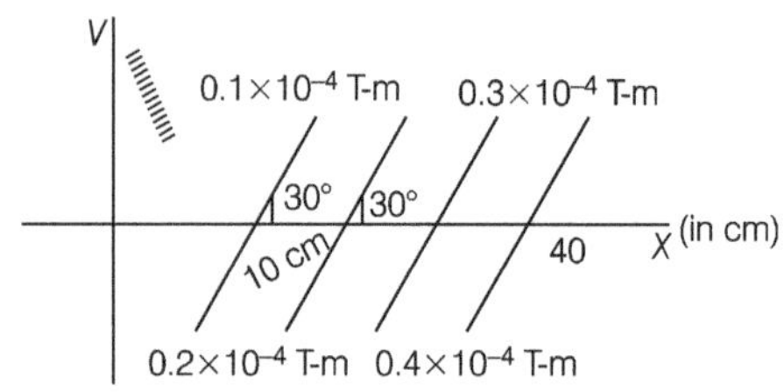

a. 1×10^{-4} T
b. 2×10^{-4} T
c. 3×10^{-4} T
d. 4×10^{-4} T

38. In a thermodynamic process, the pressure of a fixed mass of a gas is changed in such a manner that the gas release 20 J of heat and 8 J of work is done on the gas. If the initial internal energy of the gas was 30 J, then the final internal energy will be

a. 2 J *b.* 42 J
c. 18 J *d.* 58 J

39. Two balloons are filled one with pure He gas and other with air respectively. If the pressure and temperature of these balloons are same, then the number of molecules per unit volume is

a. more in He filled balloon
b. same in both balloons
c. more in air filled balloon
d. in the ratio 1 : 4

40. The equation of a simple harmonic motion is given by $y = 3\sin\frac{\pi}{2}(50t - x)$, where x and y are in metres and t is in seconds, the ratio of maximum particle velocity to the wave velocity is

a. 2π *b.* $\frac{3}{2}\pi$
c. 3π *d.* $\frac{2}{3}\pi$

PART II

Chemistry

41. What will be the product of the reaction?

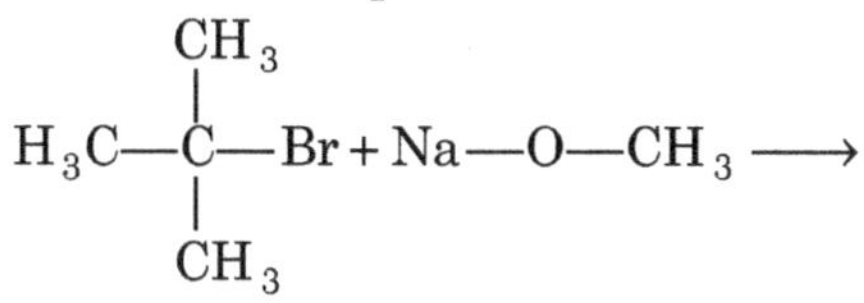

a. 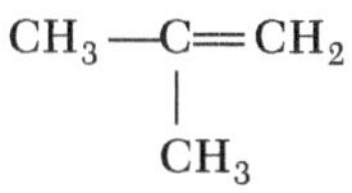

b. 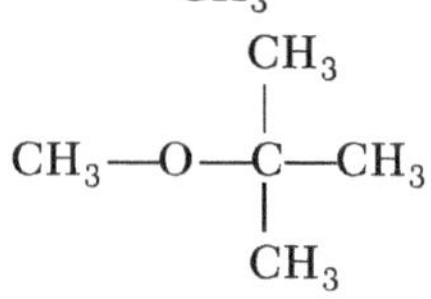

c. $CH_3-CH_2-CH_2-CH_3$

d. 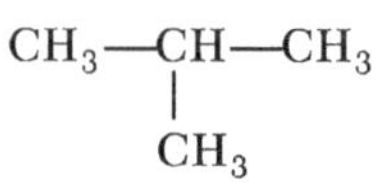

42. Which of the following relation is incorrect regarding Bohr's theory?

a. Velocity of electron $\propto \frac{1}{n}$

b. Frequency of revolution $\propto \frac{1}{n^2}$

c. Radius of orbit $\propto n^2Z$

d. Force on electron $\propto \frac{1}{n^4}$

43. Which of the following pair have identical shape?

a. CH_4, SF_4 *b.* BCl_3, ClF_3

c. XeF_2, $ZnCl_2$ *d.* SO_2, CO_2

44. 10 g of sample of mixture of $CaCl_2$ and NaCl is treated to precipitate all the calcium as $CaCO_3$. This $CaCO_3$ is heated to convert all the Ca to CaO and the final mass of CaO is 1.62 g. The percent by mass of $CaCl_2$ in the original mixture is

a. 32.1% *b.* 16.2% *c.* 21.8% *d.* 12.0%

45. How many chiral centre are possible for the product of following reaction?

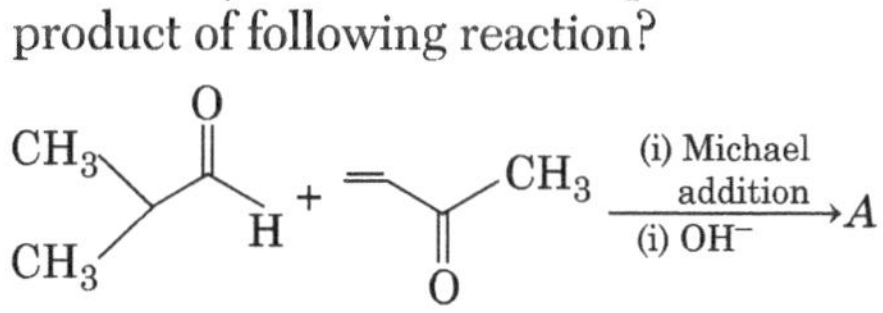

a. 1 *b.* 0 *c.* 3 *d.* 2

46. Elements/ions having same number of electrons are known as isoelectronic species. Arrange the following elements in correct order of atomic/ionic radii and choose the correct choice from the four choices given below

$$O^{2-}, Na^+, Mg^{2+}, F^-, Al^{3+}$$

a. $Al^{3+} < Mg^{2+} < Na^+ < F^- < O^{2-}$

b. $Al^{3+} < Na^+ < Mg^{2+} < F^- < O^{2-}$

c. $Al^{3+} > Mg^{2+} > Na^+ > F^- > O^{2-}$

d. None of the above

47. The ratio of oxidation states of Cl in potassium chloride to that in potassium chlorate is

a. $+\frac{1}{5}$ *b.* $-\frac{1}{5}$ *c.* $-\frac{2}{5}$ *d.* $+\frac{3}{5}$

48. A reaction, $Cu^{2+} + 2e^- \longrightarrow Cu$ is given. For this reaction, graph between E_{red} *versus* $\ln[Cu^{2+}]$ is a straight line of intercept 0.34 V, then the electrode oxidation potential of the half cell Cu/Cu^{2+} (0.1 M) will be

a. 0.34 *b.* $0.34 + \frac{0.0591}{2}$

c. $-0.34 - \frac{0.0591}{2}$ *d.* $-0.34 + \frac{0.0591}{2}$

49. Which one of the following silanes on hydrolysis produces cross linked polymers ?

a. $RSiCl_3$ *b.* R_2SiCl_2 *c.* R_3SiCl *d.* R_4Si

50. Identify the correct product formed during the following reaction.

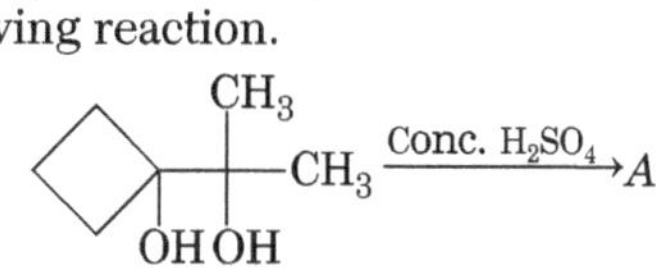

a. 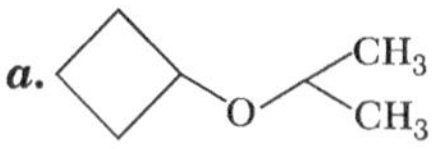

b. 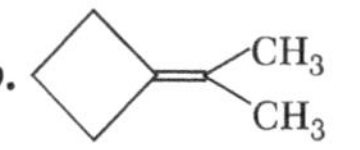

c. 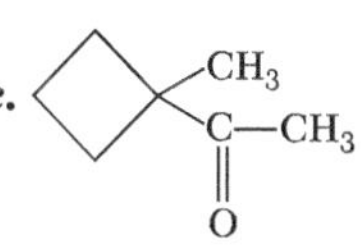

d. 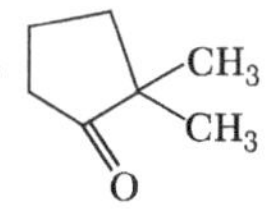

51. Usually, $CaCl_2$ is preferred over NaCl for cleaning snow on roads particulary in very cold countries. This is because

a. NaCl makes the road slippery but $CaCl_2$ does not
b. $CaCl_2$ is hygroscopic but NaCl is not
c. $CaCl_2$ is less soluble in H_2O than NaCl
d. eutectic mixture of $CaCl_2/H_2O$ freezes at −55 °C while that of NaCl/H_2O freezes at −18°C.

52. The gold numbers of a few protective colloids are given

x	0.005
y	3.5
z	40

The protective nature of these colloidal solutions follow the order

a. $z > x > y$ *b.* $x < y > z$
c. $z > y > x$ *d.* $x > y > z$

53. Consider the following Ellingham diagram for carbon.

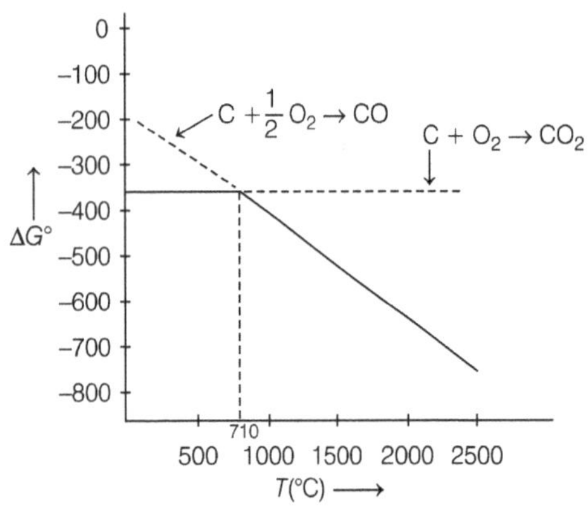

Which of the statement is incorrect for the above Ellingham diagram?

a. Upto 710°C, the reaction of formation of CO_2 is energetically more favourable but above 710°C, the formation of CO is preferred.
b. Carbon can be used to reduce any metal oxide at a sufficiently high temperature.
c. Carbon reduces many oxides at elevated temperature because $\Delta G°$ *vs* temperature line has a negative slope.
d. $\Delta S°[C(s) + \frac{1}{2}O_2(g) \longrightarrow CO(g)] < \Delta S°[C(s) + O_2(g) \longrightarrow CO_2(g)]$

54. Arrange the following in correct order of basicity.

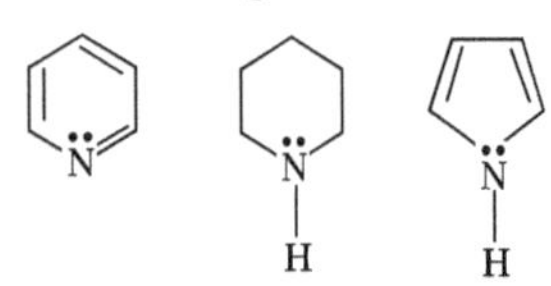

I II III

a. I > II > III
b. III > II > I
c. II > I > III
d. I > III > II

55. What is the density of Na_2O having antifluorite type crystal structure, if the edge length of the cube is 100 pm and what is the effect on density by 0.05% Frenkel defect ?

a. 823.5 g cm^{-3}, density increases
b. 414.16 g cm^{-3}, density decreases
c. 823.5 g cm^{-3}, density remains same
d. 414.16 g cm^{-3}, density remains same

56. A swimmer coming out from a pool is covered with a film of water weiging about 18 g. Calculate the internal energy of vaporisation at 100°C.
[$\Delta_{vap}H^\ominus$ for water at 373 K = 40.66 kJ mol^{-1}]
The correct option is

a. 35.67 kJ mol^{-1}
b. 37.56 kJ mol^{-1}
c. 36.57 kJ mol^{-1}
d. 38.75 kJ mol^{-1}

57. Which of the following is correct order of stability of carbocation?

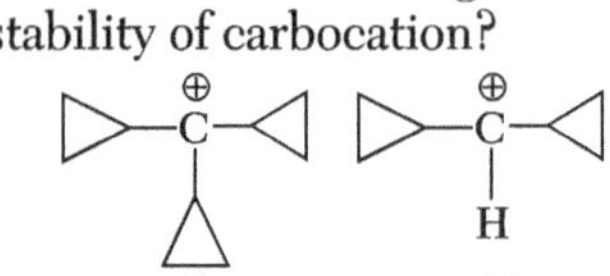

I II

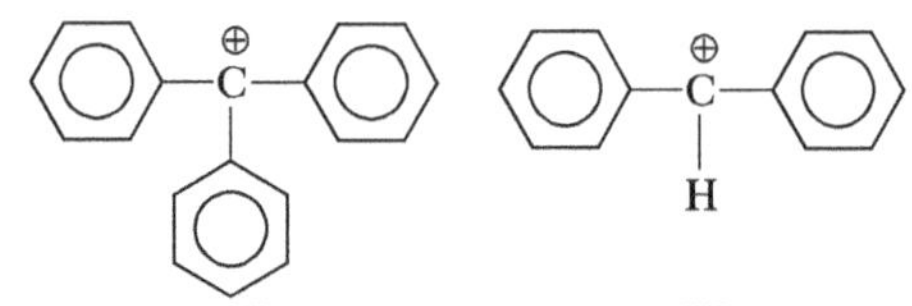

III IV

a. IV > III > II > I *b.* I > II > III > IV
c. III > II > I > IV *d.* I > III > II > IV

58. Mercury is a liquid metal because

a. it has a completely filled *d*-orbital that causes *d-d* overlapping
b. it has completely filled *d*-orbital that prevents *d-d* overlapping
c. it has a completely filled *s*-orbital
d. it has a small atomic size

59. The volume of 10 N and 4 N HCl required to make 1 L of 7 N HCl are

a. 0.75 L of 10 N HCl and 0.25 L of 4 N HCl
b. 0.50 L of 10 N HCl and 0.50 L of 4 N HCl
c. 0.65 L of 10 N HCl and 0.35 L of 4 N HCl
d. 0.85 L of 10 N HCl and 0.15 L of 4 N HCl

BITSAT 2014

60. Following is the graph between $\log T_{50}$ and $\log a$ (a = initial concentration) for a given reaction at 27° C. Hence, order is

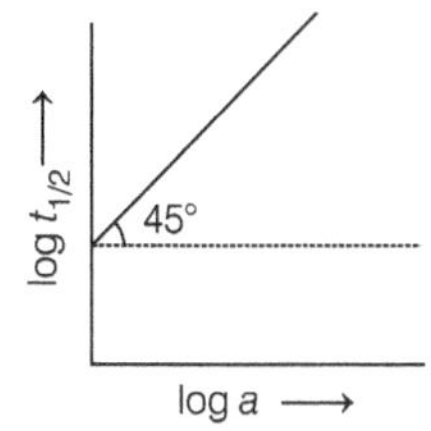

a. 1 *b.* 2 *c.* 3 *d.* 0

61. The catalyst used for olefin polymerisation is

a. Ziegler-Natta catalyst
b. Raney nickel catalyst
c. Wilkinson catalyst
d. Merrified resin

62. Which one of the following is a covalent hydride?

a. CaH_2 *b.* NaH
c. BH_3 *d.* BeH_2

63. Which one of the following is used for the separation of noble gas mixture from air ?

a. Charcoal
b. 90% CaC_2 + 10% $CaCl_2$
c. Soda lime + potash solution
d. 90% $CaCO_3$ + 10% urea

64. Consider the following statements.

I. NCl_5 does not exist while PCl_5 does.
II. Both O_2^+ and NO are paramagnetic.
III. The three C—O bonds are not equal in carbonate ion.
IV. Lead prefers to form tetravalent compound.

Which of the above statements are incorrect?

a. I and III *b.* I, III and IV
c. II and III *d.* III and IV

65. The liquefied metal that expand on solidification is

a. Al *b.* Zn
c. Ga *d.* Cu

66. Point out the correct statement for the set of characteristics of ZnS crystal.

a. Coordination number (4 : 4) : ccp ; Zn^{2+} ion in the alternate tetrahedral voids.
b. Coordination number (6 : 6); hcp; Zn^{2+} ion in all tetrahedral voids.
c. Coordination number (6 : 4); hcp; Zn^{2+} ion in all octahedral voids.
d. Coordination number (4 : 4); ccp; Zn^{2+} ion in all tetrahedral voids.

67. Arrange the following compounds in the increasing order of nucleophilic addition reaction.

I. HCHO
II. CH_3COCH_3
III. $C_6H_5COCH_3$
IV. $C_2H_5COC_6H_5$

a. I< II< III< IV *b.* IV< III< II< I
c. IV< II< III< I *d.* III< IV< II< I

68. The heat of combustion of sucrose, $C_{12}H_{22}O_{11}(s)$ at constant volume is 1348.9 kcal mol^{-1} at 25°C, then the heat of reaction at constant pressure when steam is produced

a. −1348.9 kcal *b.* −1342.34 kcal
c. + 1250 kcal *d.* None of these

69. Arrange the following compounds in increasing order of their boiling points.

I. $(CH_3)_2$CH—CH_2Br
II. CH_3CH_2—CH_2—CH_2—Br
III. CH_3—C(CH_3)(Br)—CH_3

a. II < I < III *b.* I < II < III
c. III < I < II *d.* III < II < I

70. Which of the following compounds will give positive iodoform test with I_2 and NaOH?

a. $C_6H_5COC_6H_5$
b. CH_3CH_2CHO
c. $C_6H_5COCH_2CH_3$
d. C_6H_5—CH(OH)—CH_3

71. What will be the product when most acidic species among following will react with 3-chloroprop-1-ene?

a. 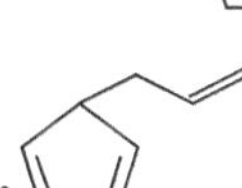 *b.* 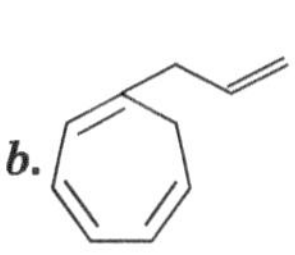

c.  *d.* 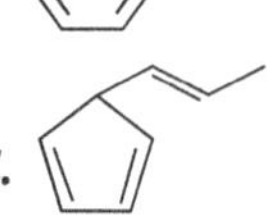

72. $CaCO_3(s) \xrightarrow{\text{Heat}} X(s) + Z(g)$

$X(s) \xrightarrow{\text{Carbon, heat}} C(s) + D(g)$

$C(s) + H_2O \longrightarrow E(g)$.

Then, compound $E(g)$ and $C(s)$ will be

a. C_2H_2, CaO *b.* C_2H_2, CaC_2
c. CH_4, CaC_2 *d.* CH_4, CaO

73. Which of the following will not form optical isomers ?

a. $[Co(en)_3]^{3+}$ *b.* $[Co(NH_3)_3(NO_2)_3]$
c. $[Pt(en)_2Cl_2]^{2+}$ *d.* $[CrCl_2(ox)_2]^{3-}$

74. $C_6H_5NH_2 \xrightarrow[180°C]{H_2SO_4} \underset{\textit{Para}\text{-form}}{NH_2C_6H_4(SO_3H)}$

The true statement about the product is

a. It does not exist as Zwitter ion
b. It does not act as inner salt
c. $—SO_3$ diminishes the basic character of $—NH_2$
d. $—NH_2$ displays a powerful basic character

75. A copolymer of ethene and vinyl chloride contains alternate monomers of each type. What is the mass percentage of vinyl chloride in this copolymer ?

a. 38% *b.* 69% *c.* 72% *d.* 82%

76. The number of disulphide linkages present in insulin are

a. 1 *b.* 2 *c.* 3 *d.* 4

77. Which of the following statement is not true about the drug barbital?

a. It is used in sleeping pills.
b. It is a non-hypnotic drug.
c. It is transquillizer.
d. It causes addiction.

78. Calculate the pH at the equivalence point during the titration of 0.1 M, 25 mL CH_3COOH with 0.05 M NaOH solution.
$[K_a(CH_3COOH) = 1.8 \times 10^{-5}]$

a. 9.63 *b.* 8.63 *c.* 10.63 *d.* 11.63

79. The temperature 30.98°C is called critical temperature (T_c) of carbon dioxide. The critical temperature is the

a. lowest temperature at which liquid carbon dioxide is observed
b. highest temperature at which gas carbon dioxide is observed
c. highest temperature at which solid carbon dioxide is observed
d. highest temperature at which liquid carbon dioxide is observed

80. The types of reactions for these are

I. II. III. IV.

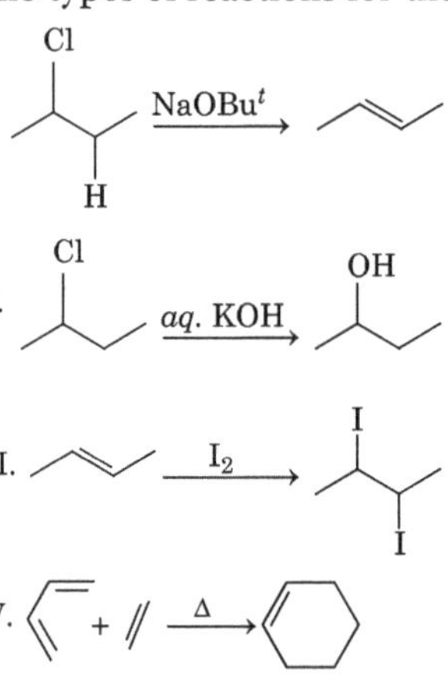

a. elimination, substitution, addition, addition
b. addition, elimination, addition, substitution
c. elimination, addition, substitution, addition
d. substitution, elimination, addition, addition

PART III

a. English Proficiency

Directions (Q. Nos. 81-83) *Out of the four alternatives, choose the one which expresses the right meaning of the word.*

81. Augment

a. Increase *b.* Decrease *c.* Save *d.* Mention

82. Consolation

a. Comfort *b.* Problem *c.* Sadness *d.* Solution

83. Auxiliary

a. Chief *b.* Supplemental
c. Negligible *d.* Separate

Directions (Q. Nos. 84-86) *Choose the word apposite is meaning to the given word.*

84. Auspicious

a. Prosperous *b.* Unfavourable
c. Improper *d.* New

85. Recompense

a. Emolument *b.* Reward
c. Payment *d.* Penalty

86. Impede

a. Block *b.* Delay *c.* Push *d.* Freeze

Directions (Q. Nos. 87-89) *A part of sentence is underlined. Balance are given alternatives to the underlined part a, b, c and d which may improve the sentence. Choose the correct alternative.*

87. They <u>requested</u> me to follow them.
a. ordered
b. urged
c. asked
d. No improvement

88. She did not <u>believed</u> me.
a. believing *b.* believe to
c. believe *d.* No improvement

89. I am fine, what about <u>you</u>?
a. your
b. your's
c. yours
d. No improvement

Directions (Q. Nos. 90-92) *Sentence Completion*

90. They were afraid .... the lion, so they dropped the idea of hunting in jungle.
a. in *b.* to
c. from *d.* of

91. Our company signed a profitable ..... last month.
a. issue *b.* agenda
c. deal *d.* paper

92. What is your ...... for tonight?
a. principle *b.* motto
c. plan *d.* objective

Directions (Q. Nos. 93-95) *Arrange the following sentences in correct pattern and mark at the correct combination.*

93. 1. Today we live in modern technology era.
P. We have a lot of problems now.
Q. We want to get everything in one day.
R. Ancient time was quite pleasant.
S. We had no problems then.
C. Perhaps greed is the main cause for this.
a. PQRS
b. PRSQ
c. SRQP
d. RPQS

94. 1. He is a common man.
P. Yesterday our city saw a brutal crime.
Q. Police is trying to arrest innocent persons.
R. The criminals are well known.
S. Police as well as whole system in corrupt.
C. Police will arrest him as he is an easy target because of being a common man.
a. PRSQ *b.* PQSR *c.* PQRS *d.* PSQR

95. 1. I want to change the room.
P. Last month I got a job.
Q. I had been living there for six months.
R. The office is far from the room.
S. I want to cut expenses of travelling.
C. Hopefully I will do this next week.
a. PQRS *b.* PRSQ
c. QPRS *d.* PQSR

b. Logical Reasoning

96. In a certain code language, 'SAFER' is written as '5@3#2' and 'RIDE' is written as '2©%#', how would 'FEDS' be written in that code?
a. 3#©5 *b.* 3@%5 *c.* 3#%5 *d.* 3#%2

97. Find the missing number from the given response.

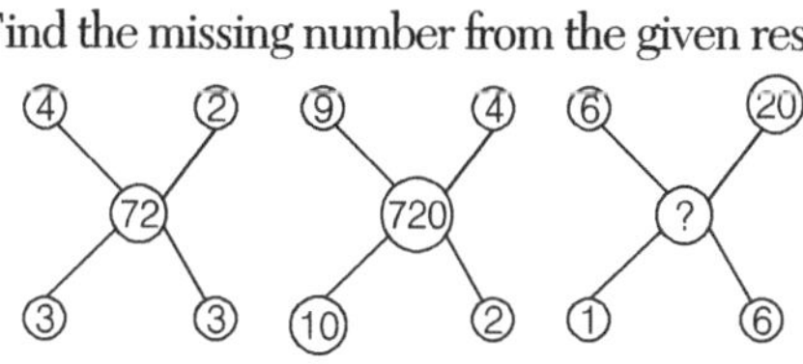

a. 72 *b.* 720
c. 7200 *d.* 38

98. If the first and second letters in the word DEPRESSION were interchanged, also the third and fourth letters, the fifth and the sixth letters and so on, then which of the following would be seventh letter from the right?
a. O *b.* P *c.* R *d.* S

99. Today is Thrusday. The day after 59 days will be
a. Sunday *b.* Monday
c. Tuesday *d.* Wednesday

100. Which of the following represents coal mines, factories and fields?

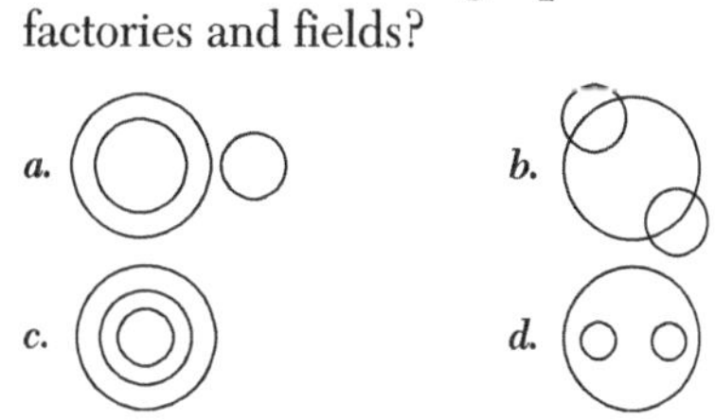

101. Find out the missing term in the series
1, 8, 27, ?, 125, 216
a. 52 *b.* 58 *c.* 64 *d.* 65

102. If '+' means '×', '–' means '+', '×' means '÷' and '÷' means '–', then $6 - 9 + 8 \times 3 \div 20 = ?$
a. –2 *b.* 6 *c.* 10 *d.* 12

103. What is the water image of

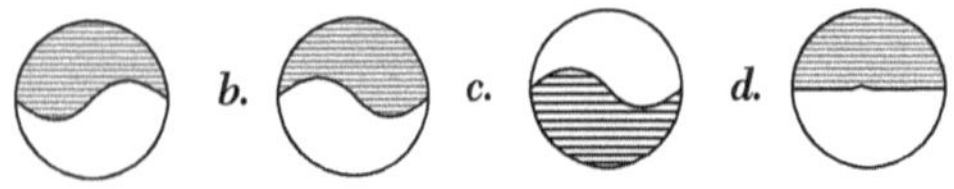

a. b. c. d.

104. A piece of paper is folded and penched as shown in the figure below

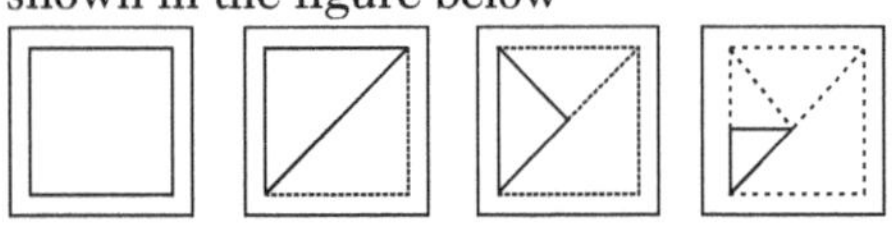

a. b. c. d.

How will it appear when unfolded?

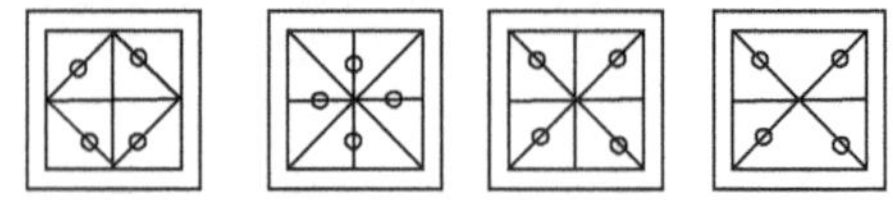

(a) (b) (c) (d)

105. Here are some words translated from an artificial language.
mallon piml means blue light
mallon tifl means blue berry
arpan tifl means rasp berry
Which word could means 'light house'?

a. tifl mallon
b. piml arpan
c. mallon arpan
d. piml doken

PART IV

Mathematics

106. If p, q, r and s are positive real numbers such that $p+q+r+s=2$, then $M=(p+q)(r+s)$ satisfies the relation

a. $0<M\leq 1$ *b.* $1\leq M\leq 2$
c. $2\leq M\leq 3$ *d.* $3\leq M\leq 4$

107. The complex number $z=x+iy$ which satisfies the equation $\left|\dfrac{z-3i}{z+3i}\right|=1$, lie on

a. the X-axis
b. the straight line $y=3$
c. a circle passing through origin
d. None of the above

108. If $f(x)$ is an odd periodic function with period 2, then $f(4)$ equals to

a. -4 *b.* 4
c. 2 *d.* 0

109. The solution of the differential equation

$$\frac{x+\dfrac{x^3}{3!}+\dfrac{x^5}{5!}+\ldots}{1+\dfrac{x^2}{2!}+\dfrac{x^4}{4!}+\ldots}=\frac{dx-dy}{dx+dy}\text{ is}$$

a. $2ye^{2x}=Ce^{2x}+1$ *b.* $2ye^{2x}=Ce^{2x}-1$
c. $ye^{2x}=Ce^{2x}+2$ *d.* None of these

110. The value of k such that the lines $2x-3y+k=0$, $3x-4y-13=0$ and $8x-11y-33=0$ are concurrent, is

a. 20 *b.* -7 *c.* 7 *d.* -20

111. Two lines, whose equations are $\dfrac{x-3}{2}=\dfrac{y-2}{3}=\dfrac{z-1}{\lambda}$ and $\dfrac{x-2}{3}=\dfrac{y-3}{2}=\dfrac{z-2}{3}$ lie in the same plane. Then, the value of $\sin^{-1}\sin\lambda$ is equal to

a. 3 *b.* $\pi-3$
c. 4 *d.* $\pi-4$

112. If $\dfrac{e^x}{1-x}=B_0+B_1x+B_2x^2+\ldots+B_nx^n+\ldots$, then the value of B_n-B_{n-1} is

a. 1 *b.* $\dfrac{1}{n}$
c. $\dfrac{1}{n!}$ *d.* None of these

113. $2^{3n}-7n-1$ is divisible by

a. 64 *b.* 36
c. 49 *d.* 25

114. If $\int_0^{25} e^{x-[x]}\,dx=k(e-1)$, then the value of k is equal to

a. 12 *b.* 25
c. 23 *d.* 24

115. A variable chord PQ of the parabola $y^2=4ax$ subtends a right angle at the vertex, then the locus of the points of intersection of the normal at P and Q is

a. a parabola *b.* a hyperbola
c. a circle *d.* None of these

116. If $\omega \neq 1$ is a cube root of unity, then

$$A = \begin{bmatrix} 1+2\omega^{100}+\omega^{200} & \omega^2 & 1 \\ 1 & 1+2\omega^{100}+\omega^{200} & \omega \\ \omega & \omega^2 & 2+\omega^{100}+2\omega^{200} \end{bmatrix}$$

a. A is singular
b. $|A| \neq 0$
c. A is symmetric
d. None of the above

117. $\lim_{x \to \tan^{-1} 3} \frac{\tan^2 x - 2\tan x - 3}{\tan^2 x - 4\tan x + 3}$ equals to

a. 1 b. 2
c. 0 d. 3

118. The locus of the points of intersection of the tangents at the extremities of the chords of the ellipse $x^2 + 2y^2 = 6$ which touches the ellipse $x^2 + 4y^2 = 4$, is

a. $x^2 + y^2 = 4$ b. $x^2 + y^2 = 6$
c. $x^2 + y^2 = 9$ d. None of these

119. Number of roots of the equation $|\sin x \cdot \cos x| + \sqrt{2 + \tan^2 x + \cot^2 x} = \sqrt{3}$, where $x \in [0, 4\pi]$, are

a. 1 b. 2
c. 3 d. None of these

120. If $f(x) = \begin{cases} \left[\tan\left(\frac{\pi}{4} + x\right)\right]^{1/x}, & x \neq 0 \\ k, & x = 0 \end{cases}$

For what value of k, $f(x)$ is continuous at $x = 0$?

a. 1 b. 0
c. e d. e^2

121. The period of $\sin^2 \theta$ is

a. π^2 b. π
c. 2π d. $\frac{\pi}{2}$

122. Five persons A, B, C, D and E are in queue of a shop. The probability that A and E are always together, is

a. $\frac{1}{4}$ b. $\frac{2}{3}$
c. $\frac{2}{5}$ d. $\frac{3}{5}$

123. $\lim_{x \to \infty} \frac{x^4 \cdot \sin\left(\frac{1}{x}\right) + x^2}{1 + |x|^3}$ equals to

a. 0 b. -1 c. 2 d. 1

124. The sum of the series $\log_4 2 - \log_8 2 + \log_{16} 2 \ldots$ is

a. e^2 b. $\log_e 2 + 1$
c. $\log_e 3 - 2$ d. $1 - \log_e 2$

125. The mean of n terms is $\bar{x}$. If the first term is increased by 1, second by 2 and so on, then the new mean is

a. $\bar{x} + n$ b. $\bar{x} + \frac{n}{2}$
c. $\bar{x} + \frac{n+1}{2}$ d. None of these

126. The greatest and least values of $(\sin^{-1} x)^2 + (\cos^{-1} x)^2$ are respectively

a. $\frac{\pi^2}{4}$ and 0 b. $\frac{\pi}{2}$ and $\frac{-\pi}{2}$
c. $\frac{5\pi^2}{4}$ and $\frac{\pi^2}{8}$ d. $\frac{\pi^2}{4}$ and $\frac{-\pi^2}{4}$

127. The probability of simultaneous occurrence of atleast one of two events A and B is p. If the probability that exactly one of A and B occurs is q, then $P(A') + P(B')$ is equal to

a. $2 - 2p + q$ b. $2 + 2p - q$
c. $3 - 3p + q$ d. $2 - 4p + q$

128. The solution of the differential equation
$$x = 1 + xy\frac{dy}{dx} + \frac{x^2y^2}{2!}\left(\frac{dy}{dx}\right)^2 + \frac{x^3y^3}{3!}\left(\frac{dy}{dx}\right)^3 + \ldots$$
is

a. $y = \log x + C$ b. $y^2 = (\log x)^2 + C$
c. $y = \log x + xy$ d. $xy = x^y + C$

129. If n is a positive integer, then $n^3 + 2n$ is divisible by

a. 2 b. 6
c. 15 d. 3

130. The sum of the coefficients in the expansion of $(5x - 4y)^n$, where n is a positive integer, is

a. 0 b. n
c. 1 d. -1

131. The distance of the point $(1, -5, 9)$ from the plane $x + y + z = 5$ measured along a straight line $x = y = z$ is $2\sqrt{3}k$, then the value of k is

a. 5 b. 6
c. $\sqrt{3}$ d. 4

BITSAT 2014

132. $\lim_{x \to 1} \frac{x^m - 1}{x^n - 1}$ is equal to

a. $\frac{n}{m}$ b. $\frac{m}{n}$

c. $\frac{2m}{n}$ d. $\frac{2n}{m}$

133. The value of $x > 1$ satisfying the equation $\int_1^x t \log t \, dt = \frac{1}{4}$, is

a. $\sqrt{e}$ b. $e^{3/2}$

c. e^2 d. $2e - 1$

134. $a = \sum_{n=0}^{\infty} \frac{x^{3n}}{3n!}, b = \sum_{n=1}^{\infty} \frac{x^{3n-2}}{(3n-2)!}$ and

$c = \sum_{n=1}^{\infty} \frac{x^{3n-1}}{(3n-1)!}$, then the value of

$a^3 + b^3 + c^3 - 3abc$ is

a. 1 b. 0

c. -1 d. -2

135. The unit vector perpendicular to the vectors $\hat{i} - \hat{j}$ and $\hat{i} + \hat{j}$ forming a right handed system is

a. $\hat{k}$ b. $-\hat{k}$

c. $\frac{\hat{i} - \hat{j}}{\sqrt{2}}$ d. $\frac{\hat{i} + \hat{j}}{\sqrt{2}}$

136. The number of solutions of the equation $x^3 + 2x^2 + 5x + 2 \cos x = 0$ in $[0, 2\pi]$ are

a. 0 b. 1

c. 2 d. 3

137. $\lim_{x \to 0} \frac{(1+x)^8 - 1}{(1+x)^2 - 1}$ is equal to

a. 8 b. 6

c. 4 d. 2

138. The area bounded by the curves $y = -\sqrt{-x}$ and $x = -\sqrt{-y}$, where $x, y \le 0$, is

a. 1/3 b. 1/4

c. 1/5 d. 1/2

139. If α and β are the roots of the equation $x^2 - px + q = 0$, then the value of

$(\alpha + \beta)x - \left(\frac{\alpha^2 + \beta^2}{2}\right)x^2 + \left(\frac{\alpha^3 + \beta^3}{3}\right)x^3 + \ldots$, is

a. $\log(1 - px + qx^2)$

b. $\log(1 + px - qx^2)$

c. $\log(1 + px + qx^2)$

d. None of the above

140. If $\cos^{-1} x > \sin^{-1} x$, then

a. $\frac{1}{\sqrt{2}} < x \le 1$ b. $0 \le x < \frac{1}{\sqrt{2}}$

c. $-1 \le x < \frac{1}{\sqrt{2}}$ d. $x > 0$

141. A student is allowed to select atmost n books from a collection of $(2n + 1)$ books. If the number of ways in which he can do this, is 64, then the value of n is

a. 6 b. n

c. 3 d. None of these

142. Let $R = \{(3, 3), (6, 6), (9, 9), (12, 12), (6, 12), (3, 9), (3, 12), (3, 6)\}$ be a relation on the set $A = \{3, 6, 9, 12\}$. The relation is

a. an equivalence relation

b. reflexive and symmetric

c. reflexive and transitive

d. only reflexive

143. The value of a, so that the sum of squares of the roots of the equation $x^2 - (a - 2)x - a + 1 = 0$ assume the least value, is

a. 2 b. 0 c. 3 d. 1

144. The form of the differential equation of the central conics $ax^2 + by^2 = 1$ is

a. $x = y \frac{dy}{dx}$

b. $x\left(\frac{dy}{dx}\right)^2 + xy \frac{d^2y}{dx^2} - y \frac{dy}{dx} = 0$

c. $x + y \frac{d^2y}{dx^2} = 0$

d. None of the above

145. A furniture dealer deals in only two items namely tables and chairs. He has ₹ 5000 to invest and space to store at the most 60 pieces. A table cost him ₹ 250 and a chair ₹ 60. He can sell a table at a profit of ₹ 15. Assume that he can sell all the items that he produced. The number of constraints in the problem are

a. 2 b. 3 c. 4 d. 5

146. $10^n + 3(4^{n+2}) + 5$ is divisible by $(n \in N)$

a. 7 b. 5 c. 9 d. 17

147. The total number of subsets of a finite set A has 56 more elements than the total number of subsets of another finite set B. What is the number of elements in the set A?

a. 5 b. 6

c. 7 d. 8

BITSAT 2014

148. A sample of 35 observations has the mean 80 and standard deviation as 4. A second sample of 65 observations from the same population has mean 70 and standard deviation 3. Then, the standard deviation of the combined sample is

a. 5.85 *b.* 5.58
c. 34.2 *d.* None of these

149. Three straight lines $2x+11y-5=0$, $24x+7y-20=0$ and $4x-3y-2=0$

a. from a triangle
b. are only concurrent
c. are concurrent with one line bisecting the angle between the other two
d. None of the above

150. If the plane $x+y+z=1$ is rotated through an angle 90° about its line of intersection with the plane $x-2y+3z=0$, then the new position of the plane is

a. $x-5y+4z=1$
b. $x-5y+4z=-1$
c. $x-8y+7z=2$
d. $x-8y+7z=-2$

Answers

Physics

1. (c)	**2.** (c)	**3.** (c)	**4.** (b)	**5.** (b)	**6.** (c)	**7.** (a)	**8.** (d)	**9.** (d)	**10.** (c)
11. (a)	**12.** (b)	**13.** (d)	**14.** (a)	**15.** (d)	**16.** (b)	**17.** (c)	**18.** (*)	**19.** (a)	**20.** (c)
21. (a)	**22.** (b)	**23.** (c)	**24.** (a)	**25.** (a)	**26.** (a)	**27.** (c)	**28.** (c)	**29.** (b)	**30.** (c)
31. (b)	**32.** (c)	**33.** (b)	**34.** (d)	**35.** (c)	**36.** (b)	**37.** (b)	**38.** (c)	**39.** (b)	**40.** (b)

Chemistry

41. (a)	**42.** (c)	**43.** (c)	**44.** (a)	**45.** (a)	**46.** (a)	**47.** (b)	**48.** (d)	**49.** (a)	**50.** (d)
51. (d)	**52.** (d)	**53.** (d)	**54.** (c)	**55.** (d)	**56.** (b)	**57.** (d)	**58.** (b)	**59.** (b)	**60.** (d)
61. (a)	**62.** (c)	**63.** (b)	**64.** (d)	**65.** (c)	**66.** (a)	**67.** (c)	**68.** (b)	**69.** (c)	**70.** (d)
71. (a)	**72.** (b)	**73.** (b)	**74.** (c)	**75.** (b)	**76.** (b)	**77.** (b)	**78.** (b)	**79.** (d)	**80.** (a)

English Proficiency

81. (a)	**82.** (a)	**83.** (b)	**84.** (b)	**85.** (d)	**86.** (c)	**87.** (a)	**88.** (c)	**89.** (d)	**90.** (d)
91. (c)	**92.** (c)	**93.** (b)	**94.** (a)	**95.** (c)					

Logical Reasoning

96. (c)	**97.** (b)	**98.** (b)	**99.** (a)	**100.** (d)	**101.** (c)	**102.** (c)	**103.** (b)	**104.** (b)	**105.** (d)

Mathematics

106. (a)	**107.** (a)	**108.** (d)	**109.** (b)	**110.** (b)	**111.** (d)	**112.** (c)	**113.** (c)	**114.** (b)	**115.** (a)
116. (a)	**117.** (b)	**118.** (c)	**119.** (d)	**120.** (d)	**121.** (b)	**122.** (c)	**123.** (b)	**124.** (d)	**125.** (c)
126. (c)	**127.** (a)	**128.** (b)	**129.** (d)	**130.** (c)	**131.** (a)	**132.** (b)	**133.** (a)	**134.** (a)	**135.** (a)
136. (a)	**137.** (c)	**138.** (a)	**139.** (a)	**140.** (c)	**141.** (c)	**142.** (c)	**143.** (d)	**144.** (b)	**145.** (c)
146. (c)	**147.** (b)	**148.** (a)	**149.** (c)	**150.** (d)					

Note *(*) None of the option is correct.*

Hints & Solutions

Physics

1. ***(c)*** In amplitude modulation, amplitude of carrier wave varies with signal to be modulated, so modulated signal $C_m(t)$ is given by

$$C_m(t) = [A_c + m(t)]\sin\omega_c t$$

$$\Rightarrow \quad C_m(t) = (A_c + A_m \sin\omega_m t)\sin\omega_c t$$

$$[\because \text{given } m(t) = A_m \sin\omega_m t]$$

2. ***(c)*** By Einstein's photoelectric effect equation,

$$KE_{max} = eV_S = h\nu - h\nu_0$$

$$\Rightarrow \quad V_S = \left(\frac{h}{e}\right)\nu - \frac{h\nu_0}{e}$$

Graph of V_s with ν_0 is straight line whose slope $= \frac{h}{e}$

From given graph slope $= \frac{AB}{BC} \Rightarrow \frac{AB}{BC} = \frac{h}{e}$

3. ***(c)*** Here, $m = 0.18$ kg, $K = 2$ N/m, $\mu = 0.1$, $x = 0.06$ m.

According to conservation of mechanical energy principle, we know

Decrease in mechanical energy = Work done against friction

$$\frac{1}{2}mv^2 - \frac{1}{2}kx^2 = \mu mgx$$

$$\Rightarrow \quad v = \sqrt{\frac{2\mu mgx + Kx^2}{m}}$$

Substituting the values of m, μ, g, x and K, we get

$$v = \sqrt{\frac{2\times 0.1\times 0.18\times 9.8\times 0.06 + 2\times 0.06}{0.18}}$$

$$v = (4/10)\text{ m/s} = \frac{N}{10} \text{ (given)}$$

So, $N = 4$

4. ***(b)*** Ohm's law says

$$\frac{V}{I} = R = \text{constant}$$

5. ***(b)*** From equation of motion,

(i) $v = u + at_1$

$$40 = 0 + a \times 20$$

$$\therefore \quad a = 2\text{ m/s}^2$$

and $\quad v^2 - u^2 = 2as$

$$40^2 - 0 = 2 \times 2s_1$$

$$\therefore \quad s_1 = 400\text{ m}$$

(ii) $s_2 = v \times t_2 = 40 \times 20 = 800$ m

(iii) $v = u + at$

$$0 = 40 + a \times 40$$

$$\therefore \quad a = -1\text{ m/s}^2$$

$$0^2 - 40^2 = 2(-1)\,s_3$$

$$\therefore \quad s_3 = 800\text{ m}$$

Total distance travelled

$$= s_1 + s_2 + s_3$$

$$= 400 + 800 + 800 = 2000\text{ m}$$

Total time taken = 20 + 20 + 40 = 80 s

$\therefore$ Average velocity $= \frac{2000}{80} = 25$ m/s

6. ***(c)*** As stress is shown on X-axis and strain on Y-axis.

So, we can say that Young's modulus of elasticity,

$$Y = \frac{\text{stress}}{\text{strain}} = \cot\theta = \frac{1}{\tan\theta} = \frac{1}{\text{slope}}$$

So, elasticity of wire P is minimum and R is maximum. Hence, we can say tensile strength of R is maximum.

7. ***(a)*** Given, $F \propto A^a v^b d^c$

$$F = kA^a v^b d^c$$

$$\Rightarrow \quad [F] = [A]^a[v]^b[d]^c$$

$$[MLT^{-2}] = [L^{2a}] \times [L^bT^{-b}]\,[M^cL^{-3c}]$$

$$= [M^cL^{2a+b-3c}T^{-b}]$$

Comparing powers of M, L and T, we get

$$c = 1,\ 2a + b - 3c = 1$$

$$-b = -2 \text{ or } b = 2$$

Also, $\quad 2a + 2 - 3(1) = 1$

$$2a = 2 \text{ or } a = 1$$

$\therefore$ This is 1, 2, 1

8. ***(d)*** Radius of nth orbit, $r_n \propto n^2$

$$\frac{r_{n\text{ big}}}{r_{n\text{ small}}} = \frac{n^2_{\text{big}}}{n^2_{\text{small}}} = \frac{4}{1} \text{ (given)}$$

$$\Rightarrow \quad \frac{n_{\text{big}}}{n_{\text{small}}} = 2$$

$$\Rightarrow \quad \frac{n_{\text{small}}}{n_{\text{big}}} = \frac{1}{2}$$

Velocity of electron in nth orbit,

$$v_n \propto \frac{1}{n}$$

$$\frac{v_{n\text{ big}}}{v_{n\text{ small}}} = \frac{n_{\text{small}}}{n_{\text{big}}} = \frac{1}{2}$$

$$v_{n\text{ small}} = 2(v_{n\text{ big}}) = 2v$$

9. ***(d)*** As given for steel wire,

$$A_1 = 3\times 10^{-5}\ \text{m}^2,\ l_1 = 4.7\ \text{m},\ \Delta l_1 = \Delta l;\ F_1 = F$$

For copper wire,

$$A_2 = 4\times 10^{-5}\ \text{m}^2,\ l_2 = 3.5\ \text{m},\ \Delta l_2 = \Delta l,\ F_2 = F$$

Let Y_1 and Y_2 be the Young's modulus of steel wire and copper wire, respectively.

So, $$Y_1 = \frac{F_1}{A_1}\times\frac{l_1}{\Delta l_1} = \frac{F}{3\times 10^{-5}}\times\frac{4.7}{\Delta l}$$

and $$Y_2 = \frac{F_2\times l_2}{A_2\times \Delta l_2} = \frac{F\times 3.5}{4\times 10^{-5}\times \Delta l}$$

$\therefore$ $$\frac{Y_1}{Y_2} = \frac{4.7\times 4\times 10^{-5}}{3.5\times 3.0\times 10^{-5}} = 1.8$$

So, $Y_1 : Y_2 = 1.8 : 1$

10. ***(c)*** The net electric field at point P is given by

$$\mathbf{E} = \frac{qx}{4\pi\varepsilon_0(x^2+a^2)^{3/2}}\hat{\mathbf{i}}$$

$\therefore$ At centre of ring, $x = 0 \Rightarrow \mathbf{E} = 0$

It $x >> a$, $$\mathbf{E} = \frac{1}{4\pi\varepsilon_0}\frac{q}{x^2}$$

$\Rightarrow$ $$E \propto \frac{L}{x^2}$$

At, $x << a$, $E = \dfrac{qx}{4\pi\varepsilon_0 a^3}$

$\Rightarrow$ $$E \propto x$$

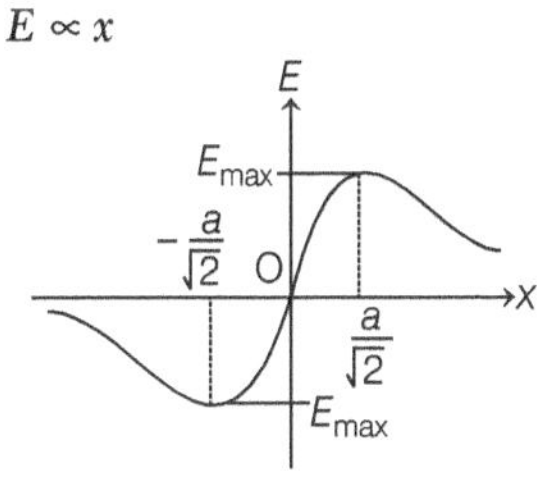

and E will be maximum, where

$$\frac{dE}{dx} = 0$$

or $$x = \pm\frac{a}{\sqrt{2}}$$

and $$E_{max} = \frac{1}{4\pi\varepsilon_0}\frac{2q}{3\sqrt{3}a^2}$$

11. ***(a)*** Constant phase difference is the essential condition for coherent sources.

12. ***(b)***

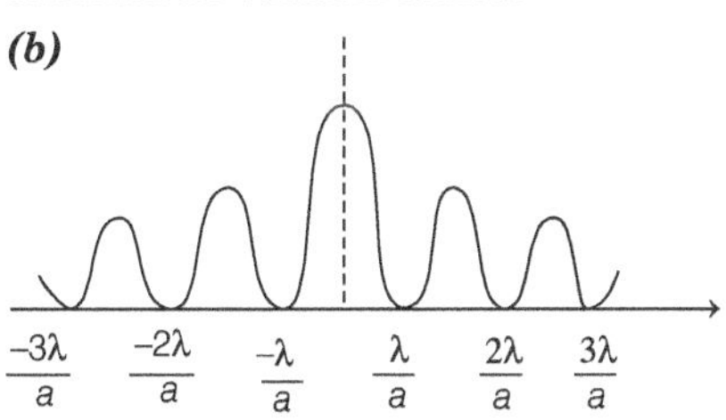

So, angular size of central maxima is

$$= \frac{\lambda}{a} - \left(-\frac{\lambda}{a}\right) = 2\left(\frac{\lambda}{a}\right) = \frac{2\lambda}{a}$$

13. ***(d)*** From first polaroid, the unpolarised light will become polarised with half the intensity.

So, $$I' = \frac{I_0}{2}$$

From second polaroid,

$$I'' = I'\cos^2\theta = \frac{I_0}{2}\cos^2(60) = \frac{I_0}{2}\frac{1}{4} = \frac{I_0}{8}$$

14. ***(a)*** Resolving power, $\text{RP} = \dfrac{a}{1.22\,\lambda}$

$$\text{RP} \propto a$$

15. ***(d)*** $r = r_0\,(A)^{1/3}$

r increases with increasing A (= mass number).

So, $r_A < r_B$ as mass number of A is smaller.

E_{bn} decreases with increasing A for $A > 56$.

So, ^{56}Fe has highest E_{bn} value.

So, E_{bn} for $A = 64$ is larger as compared to E_{bn} for nucleus with $A = 125$

$$E_{bnA} > E_{bnB}$$

16. ***(b)*** Heat energy is not a state variable, i.e., unlike temperature, pressure and volume, it is not a intrinsic property of a system. It has meaning only as it describes the transfer of energy into or out of a system.

17. ***(c)*** The current density in the wire is

$$J = \frac{I}{A} = \frac{4}{2\times 10^{-6}} = 2\times 10^6\ \text{Am}^{-2}$$

We know that, $I = neA\,v_d$

$\Rightarrow$ $$v_d = \frac{I}{neA} = \frac{J}{ne}$$

$$= \frac{2\times 10^6}{12\times 10^{28}\times 1.6\times 10^{-19}}$$

$\Rightarrow$ $$v_d = \frac{10^6}{6\times 1.6\times 10^9} = 1.04\times 10^{-4}\ \text{m/s}$$

18. (*) Mass of metal, $m = 0.2$ kg $= 200$ g

Fall in temperature of metal

$$\Delta T = 150 - 40 = 110°\text{C}$$

If s is specific heat of metal, the heat lost by the metal

$$\Delta Q = ms\,\Delta T = 200\times 110\times s \qquad \ldots\text{(i)}$$

Volume of water = 150 cc

Mass of water $m' = 150$ g

Water equivalent of calorimeter

$$w = 0.025\ \text{kg} = 25\ \text{g}$$

Rise in temperature of water in calorimeter

$$\Delta T' = 40 - 27 = 13°\text{C}$$

Heat gained by water and calorimeter

$$\Delta Q' = (m' + w)\,\Delta T' \times s_w$$
$$= (150 + 25) \times 13 \times 4.18$$
$$\Delta Q' = 175 \times 13 \times 4.18 \quad \text{...(ii)}$$

From Eqs. (i) and (ii), we get

$$\Delta Q' = \Delta Q$$

So, $200 \times s \times 100 = 175 \times 13 \times 4.18$

$$\Rightarrow \quad s = \frac{175 \times 13 \times 4.18}{200 \times 100} \approx 0.42$$

Hence, no option is correct.

19. ***(a)*** $V(x, y, z) = -\int_{\infty}^{(x, y, z)} \mathbf{E} \cdot d\mathbf{r}$

$$= -\int_{\infty}^{(x, y, z)} \frac{Mdx}{x^3} = \frac{M}{2x^2}$$

20. ***(c)*** Taking vertical downward motion of projectile from point of projection to ground, we have

$$u = -50 \sin 30° = -25 \text{ m/s}$$
$$a = g = +10 \text{ m/s}^2, s = 70 \text{ m}, t = ?$$

From equation of motion,

$$s = ut + \frac{1}{2}at^2$$

So, $\quad 70 = -25 \times t + \frac{1}{2} \times 10 \times t^2$

or $\quad 5t^2 - 25t - 70 = 0$

or $\quad t^2 - 5t - 14 = 0$

On solving, we get

$$t = 7\text{s}$$

21. ***(a)*** The equivalent circuit of the given figure is as follow.

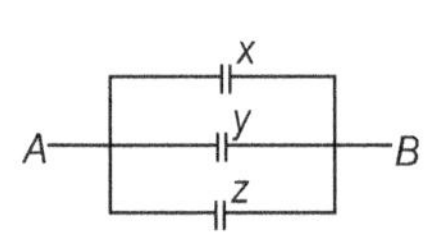

$\therefore \quad C_{eq} = X + Y + Z = 1 + 2 + 3 = 6\,\mu\text{F}$

22. ***(b)*** As given surface tension, $S = 0.06 \text{ Nm}^{-1}$,

$r_1 = 2 \text{ cm} = 0.02 \text{ m}, r_2 = 5 \text{ cm} = 0.05 \text{ m}$

Since, bubble has two surface

Initial surface area of the bubble $= 2 \times 4\pi r_1^2$

$$= 2 \times 4\pi \times (0.02)^2 = 32\,\pi \times 10^{-4} \text{ m}^2$$

Final surface area of the bubble $= 2 \times 4\pi r_2^2$

$$= 2 \times 4 \times \pi \times (0.05)^2$$
$$= 200 \times \pi \times 10^{-4} \text{ m}^2$$

So, work done $= S \times$ increase in surface

$$= 0.06 \times (200\,\pi \times 10^{-4} - 32\,\pi \times 10^{-4})$$
$$= 0.06 \times 168\,\pi \times 10^{-4}$$
$$= 0.003168 \text{ J}$$

23. ***(c)*** The given situation is shown below

$$x + y = 16$$

Also, $\quad y^2 = 8^2 + x^2$

or $\quad y^2 = 64 + (16 - y)^2 \quad (\because x = 16 - y)$

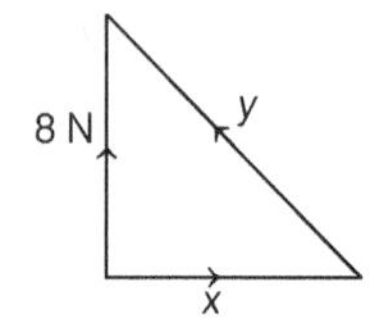

or $\quad y^2 = 64 + 256 + y^2 - 32y$

or $\quad 32y = 320$

$$Y = 10 \text{ N}$$

$\therefore \quad x + 10 = 16 \Rightarrow x = 6 \text{ N}$

24. ***(a)*** By Kirchhoff's laws,

$$V_A - V_D = -6i - 5i + 10 - 4i$$

Here, $\quad V_A = V_D$

Since, points A and D are centred.

$$-6i - 5i + 10 - 4i = 0 \Rightarrow 15i = 10$$

$$\Rightarrow \quad i = \frac{10}{15} = 0.67 \text{ A}$$

25. ***(a)*** The pressure exerted by a 3000 m column of water on the bottom layer.

$$p = h\rho g = 3000 \times 1000 \times 10$$
$$= 3 \times 10^7 = 3 \times 10^7 \text{ N/m}^2$$

Fractional compression $\left(\frac{\Delta V}{V}\right)$

$$= \frac{\text{Stress}}{B} = \frac{3 \times 10^7}{2.2 \times 10^9}$$
$$= 1.36 \times 10^{-2}$$
$$\frac{\Delta V}{V} \times 100 = 1.36\%$$

26. ***(a)*** Area of parallelogram $= |\mathbf{A} \times \mathbf{B}|$

$$\Rightarrow \quad AB \sin\theta = \frac{1}{2} AB$$
$$\Rightarrow \quad \sin\theta = \frac{1}{2}, \theta = 30°$$

27. ***(c)*** Volume of cuboid $= l \times b \times h$

$$= 1.8 \times 1.5 \times 1.2 \text{ cm}^3$$
$$= 2.70 \times 1.2 = 3.240 \text{ cm}^3$$

Using concept of significant figures, product is reported in number of significant figures present in measurement which has least number of significant figures, here all measurement have 2 significant figures.

So, volume $= 3.2 \text{ cm}^3$

(keeping 2 significant figures only.)

28. ***(c)*** According to first condition,

$$h = -vt_1 + \frac{1}{2}gt_1^2 \text{ or } \frac{h}{t_1} = -v + \frac{1}{2}gt_1 \quad \ldots\text{(i)}$$

Similarly, according to second condition,

$$h = vt_2 + \frac{1}{2}gt_2^2 \text{ or } \frac{-h}{t_2} = +v + \frac{1}{2}gt_2 \quad \ldots\text{(ii)}$$

$$\therefore \quad \frac{h}{t_1} + \frac{h}{t_2} = \frac{1}{2}g(t_1 + t_2) \quad \text{[From Eqs. (i) and (ii)]}$$

or $$h = \frac{1}{2}gt_1t_2$$

For falls under gravity from the top of the tower,

$$h = \frac{1}{2}gt^2$$

$$\therefore \quad \frac{1}{2}gt_1t_2 = \frac{1}{2}gt^2$$

$$\Rightarrow \quad t = \sqrt{t_1t_2}$$

29. ***(b)*** When elevator is moving up with acceleration a_0, then tension in the wire.

$$T = m(g + a_0) = 10(10 + 2) = 120 \text{ N}$$

$$\text{Stress} = \frac{T}{A} = \frac{120}{2 \times 10^{-4}} = 60 \times 10^4 \text{ Nm}^{-2}$$

and $$Y = \frac{\text{Stress}}{\text{Strain}} \Rightarrow \text{Strain} = \frac{\text{Stress}}{Y}$$

$$= \frac{60 \times 10^4}{2 \times 10^{11}}$$

$$= 30 \times 10^{-7} = 3 \times 10^{-6}$$

30. ***(c)*** Here, $F = 3.7 \times 10^{-9}$ N

Let, $q_1 = q_2 = q$

$r = 5 \text{ Å} = 5 \times 10^{-10}$ m

$$\because \quad F = \frac{1}{4\pi\varepsilon_0}\frac{q_1q_2}{r^2}$$

$$\Rightarrow \quad 3.7 \times 10^{-9} = 9 \times 10^9 \frac{q \times q}{(5 \times 10^{-10})^2}$$

$$q^2 = \frac{3.7 \times 10^{-9} \times 25 \times 10^{-20}}{9 \times 10^9}$$

$$= 10.28 \times 10^{-38}$$

or $$q = 3.2 \times 10^{-19} \text{ C}$$

Now, $$q = ne$$

$$\therefore \quad n = \frac{q}{e} = \frac{3.2 \times 10^{-19}}{1.6 \times 10^{-19}} = 2$$

31. ***(b)*** Since, the radius of circular path of charge particle in magnetic field, $r = \frac{mv}{qB} = \frac{p}{qB}$

Now, the radius of circular path of charge particle of given momentum p and magnetic field B is given by

$$r \propto \frac{1}{q}$$

But charge on both charge particles protons and deuterons, is same. Therefore,

$$\frac{r_p}{r_D} = \frac{q_D}{q_p} = \frac{1}{1}$$

32. ***(c)*** Here, $L = 1$ m, $N = 1000$

The number of turn per unit length

$$n = N/L = 1000 \text{ turn/m}$$

Magnetic field inside the solenoid

$$B = \mu_0 nI = \mu_0 \times 1000 \times 5$$

$$= 2\pi \times 10^{-3} \text{ T}$$

The direction of magnetic field is along the axis of solenoid.

For electron $q = -e$, $v = 10^4$ ms^{-1}

Magnetic Lorentz force, $F = -evB \sin 0° = 0$ as the angle between B and v is 0°.

33. ***(b)*** Magnetic field can accelerate a charge particle by changing the direction of its velocity but it cannot change the speed of charged particle as magnetic force always acts perpendicular to the velocity of charged particle.

34. ***(d)*** Area of coil, $A = \text{side}^2 = (0.1)^2 = 0.01 \text{ m}^2$

Number of turns, $N = 20$

Current, $I = 12$ A

Normal to the coil make an angle θ with the direction of B, magnetic field $B = 0.80$ T

Torque, experienced by the coil, $\tau = 0.96$ N-m

Since, total torque on the coil $\tau = (NIA)B \sin\theta$

Substituting the values in above formula,

$$0.96 \text{ N-m} = 20 \times 12 \text{ A} \times 0.01 \text{ m}^2 \times 0.80 \text{ T} \times \sin\theta$$

$$\Rightarrow \quad \sin\theta = \frac{0.96}{1.92} = \frac{1}{2}$$

$$\Rightarrow \quad \theta = \frac{\pi}{6} \text{ rad}$$

35. ***(c)*** $R_G = 60.00\ \Omega$; shunt resistance, $r_s = 0.02\ \Omega$

Total resistance in the circuit is $R_G + 3 = 63\ \Omega$

Hence, $I = \frac{3}{63} = 0.048$ A

Resistance of the galvanometer converted to an ammeter is

$$\frac{R_G r_s}{R_G + r_s} = \frac{60\ \Omega \times 0.02\ \Omega}{(60 + 0.02)\ \Omega} = 0.02\ \Omega$$

Total resistance in the circuit is

$$= 0.02 + 3 = 3.02\ \Omega$$

Hence, current, $I = \frac{3}{3.02} = 0.99$ A

36. *(b)* $\chi = \frac{C}{T} \Rightarrow \frac{\chi_1}{\chi_2} = \frac{T_2}{T_1}$

$$\Rightarrow \quad \frac{1.2 \times 10^{-5}}{1.8 \times 10^{-5}} = \frac{T_2}{300}$$

$$\Rightarrow \quad T_2 = \frac{12}{18} \times 300 = 200 \text{ K}$$

37. *(b)* Since perpendicular distance between two equipotentials is

$$dx = 10 \sin 30° \text{ cm} = 5 \text{ cm}$$

Since the potential gradient gives magnetic field (**B**) as

$$\frac{dV}{dx} = B$$

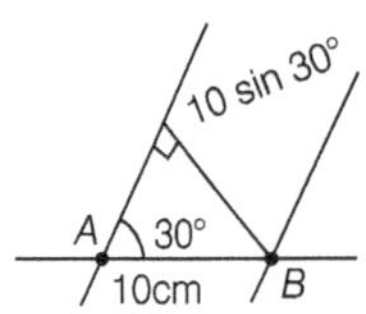

Substituting the values,

$$\frac{dV}{dx} = B = \frac{0.1 \times 10^{-4} \text{ T-m}}{5 \times 10^{-2} \text{ m}}$$

$$B = 2 \times 10^{-4} \text{ T}$$

Since B is perpendicular to equipotential surface. Here, it is at angle 120° with (+ ve)X -axis.

38. *(c)* Given, $Q = -20$ J, $W = -8$ J

Using first law of thermodynamics,

$$Q = \Delta U + W$$

$$\Rightarrow \quad -20 = \Delta U + (-8)$$

$$\Rightarrow \quad \Delta U = -12 \text{ J}$$

$$\Rightarrow \quad U_f - U_i = -12 \text{ J}$$

$$\Rightarrow \quad U_f - 30 = -12 \text{ J}$$

$$\Rightarrow \quad U_f = 18 \text{ J}$$

39. *(b)* Assuming the balloons have the same volume, as $pV = nRT$. If p, V and T are the same, n the number of moles present will be the same, whether it is He or air. Hence, number of molecules per unit volume will be same in both the balloons.

40. *(b)* The given wave equation is

$$y = 3 \sin \frac{\pi}{2}(50t - x)$$

$$\Rightarrow \quad y = 3 \sin\left(25\pi t - \frac{\pi}{2}x\right) \quad \ldots\text{(i)}$$

Comparing with standard equation,

$$y = a \sin(\omega t - kx) \quad \ldots\text{(ii)}$$

We get, $\omega = 25\pi,\ k = \pi/2$

Wave velocity, $v = \frac{\omega}{K} = \frac{25\pi}{\pi/2} = 50$ m/s

Particle velocity,

$$v_P = \frac{dy}{dt} = \frac{d}{dt}\left[3 \sin\left(25\pi t - \frac{\pi}{2}x\right)\right]$$

$$v_P = 75\pi \cos\left(25\pi t - \frac{\pi}{2}x\right)$$

Maximum particle velocity, $(v_P)_{max} = 75\pi$ m/s

$$\Rightarrow \quad \frac{(v_P)_{max}}{v} = \frac{75\pi}{50}$$

$$\Rightarrow \quad \frac{(v_p)_{max}}{v} = \frac{3}{2}\pi$$

Chemistry

41. *(a)* When tertiary alkyl halide is treated with sodium alkoxide than elimination reaction competes over substitution reaction because alkoxides are not only nucleophiles but strong base as well. Therefore, alkenes are formed instead of ethers as follows.

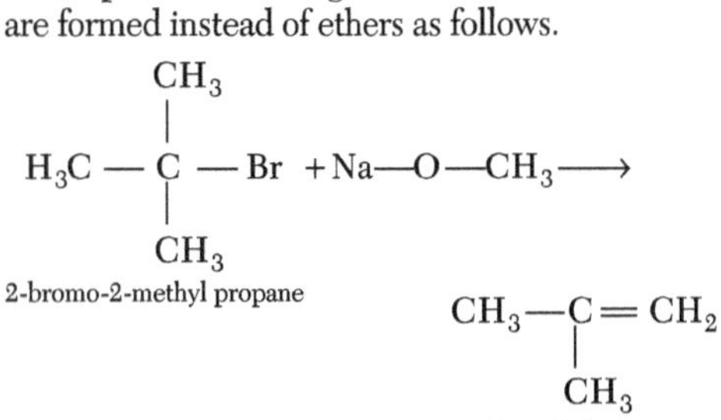

2-methyl propene

42. *(c)* Radius of orbit is directly proportional to ratio of square of principal quantum number (n) and atomic number (z).

i.e., Radius of orbit $\propto \frac{n^2}{Z}$

$\therefore$ Option (c) is incorrect regarding Bohr's theory.

43. *(c)* Molecules/compounds and their shape can be arranged as

Compound	Shape
XeF_2	Linear
$ZnCl_2$	Linear
BCl_3	Triangular planar
ClF_3	T-shaped
CH_4	Tetrahedral
SF_4	*See-saw*
SO_2	Bent
CO_2	Linear

$\therefore$ XeF_2 and $ZnCl_2$ both have same shape (linear).

44. *(a)* $CaCl_2 + NaCl = 10$ g

Let weight of $CaCl_2 = x$ g

$$\underset{\substack{1 \text{ mol} \\ \frac{x}{111} \text{ mol}}}{CaCl_2} \longrightarrow \underset{\substack{1 \text{ mol} \\ \frac{x}{111} \text{ mol}}}{CaCO_3} \longrightarrow \underset{\substack{1 \text{ mol} \\ \frac{x}{111} \text{ mol}}}{CaO}$$

Mole of CaO $= \frac{1.62}{56}$

$$\therefore \quad \frac{x}{111} = \frac{1.62}{56}$$

$$x = 3.21 \text{ g}$$

$$\% \text{ of } CaCl_2 = \frac{3.21}{10} \times 100 = 32.1\%$$

45. ***(a)*** This problems includes conceptual mixing of Michael addition and number of chiral centre.

- Complete the reaction using the concept of Michael addition.
- Then, complete intramolecular aldol condensation.
- Now, count number of chiral centre.

Michael addition Addition of nucleophile to enone system is done in such a way that the addition looks like addition at 1st and 4th position of enone is known as Michael addition.

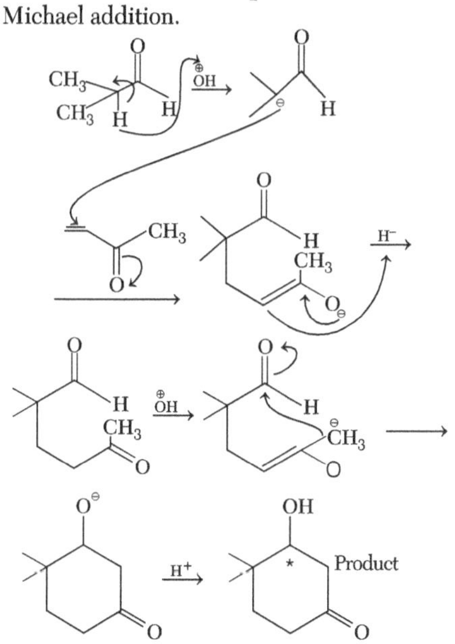

The number of chiral centre in product is 1 represented by asterisk (*).

46. ***(a)*** Among isoelectronic species, cations having highest charge are smallest while anion having highest charge are largest.

Cation < Neutral atom < Anion

Hence, correct choice is

$$Al^{3+} < Mg^{2+} < Na^{+} < F^{-} < O^{2-}$$

47. ***(b)*** Oxidation state of Cl in KCl = – 1

Oxidation state of Cl in $KClO_3$ = + 5

$\therefore$ Ratio of oxidation states of Cl $= \frac{-1}{5}$

48. ***(d)*** $Cu^{2+} + 2e^{-} \longrightarrow Cu$

$$E_{Cu^{2+}/Cu} = E_{Cu^{2+}/Cu} - \frac{0.059}{2} \log \frac{1}{[Cu^{2+}]}$$

$$= E^{\circ}_{Cu^{2+}/Cu} - \frac{RT}{2F} \ln [Cu^{2+}]$$

Intercept = 0.34 $\Rightarrow E^{\circ}_{Cu^{2+}/Cu} = 0.34$

$$E_{Cu^{2+}/Cu} = 0.34 + \frac{0.059}{2} \log 0.1 = 0.31 \text{ V}$$

$$E_{Cu/Cu^{2+}} = -E^{\circ}_{Cu^{2+}/Cu} = -0.34 + \frac{0.059}{2} \text{V}$$

49. ***(a)*** $RSiCl_3$ on hydrolysis produces cross linked polymer.

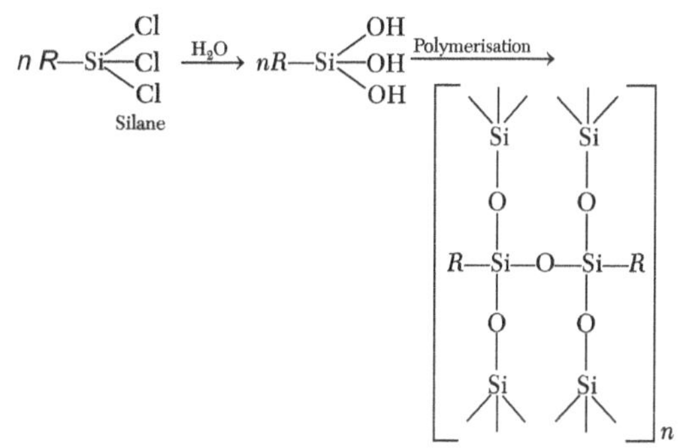

50. ***(d)*** **Pinacol-pinacolone rearrangement** The diol is converted into α-hydroxy ketone when reacted in presence of acid is believed to proceeds through rearrangement of carbocation as shown.

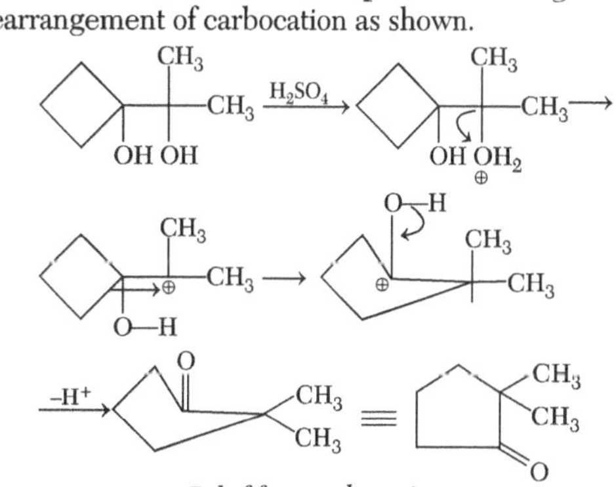

Relief from angle strain

Basic of carbocationic rearrangement is due to relief from angle strain.

51. ***(d)*** A mixture of chemical compounds having a single chemical composition, solidifies at a lower temperature than any other composition made up of the same ingredients. This mixture is called **eutectic mixture.**

Freezing point of eutectic mixture of NaCl/H_2O is only – 18°C but the ambient temperature of very cold countries is much lower than – 18°C. In such situations, NaCl will be ineffective. Thus, for such situations eutectic mixture $CaCl_2 / H_2O$ is used because it has freezing point of $-55\,°C$ which is much lower than NaCl.

This mixture lowers the freezing point of ice that allows street snow or ice to melt at lower temperature.

52. ***(d)*** **Caution Point** Gold number is used for calculating the protective powers of lyophillic colloids.

Smaller the gold number, greater is its protective power. Hence, the order would be

$$x > y > z$$

53. ***(d)*** Since, $\Delta G° = \Delta H° - T\Delta S°$

where, $\Delta G°$ = standard Gibbs' free energy of the reaction

$\Delta S°$ = standard entropy of the reaction

$\Delta H°$ = standard enthalpy of the reaction

T = temperature

From, the above equation, it is clear that, $\Delta G°$ will be more negative when $\Delta S°$ is less negative (or $\Delta S°$ is high].

In the diagram, $\Delta G°$ value for $C + \frac{1}{2}O_2 \longrightarrow CO$ is less negative, (lower) than that for

$$C + O_2 \longrightarrow CO_2$$

Therefore, $\Delta S°$ would be higher for

$$C + \frac{1}{2}O_2 \longrightarrow CO$$

than that for $C + O_2 \longrightarrow CO_2$.

Hence, $\Delta S°\ [C(s) + \frac{1}{2}O_2(g) \longrightarrow CO(g)]$

$$> \Delta S°\ [C(s) + O_2(g) \longrightarrow CO_2(g)]$$

54. ***(c)*** This problem includes conceptual mixing of basic strength, hybridisation of nitrogen atom and extent of conjugation. While solving such problem, students are advised to draw the structure and mark the type of hybridisation on N-atom, then answer the question by using combined concept of hybridisation and conjugation.

Hybridisation of N-atom in below compounds are sp^2, sp^3 and sp^2 respectively.

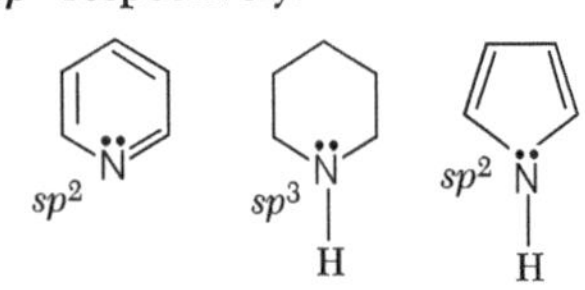

I II III

Greater the s-character more will be electronegativity of N-atom and lesser will be its basicity. On this basis, I and III are less basic than II.

Conjugation If lone pairs of electron of N is involved in conjugation causes decrease in basicity of compound due to lesser availability of lone pair for donation to show basic nature.

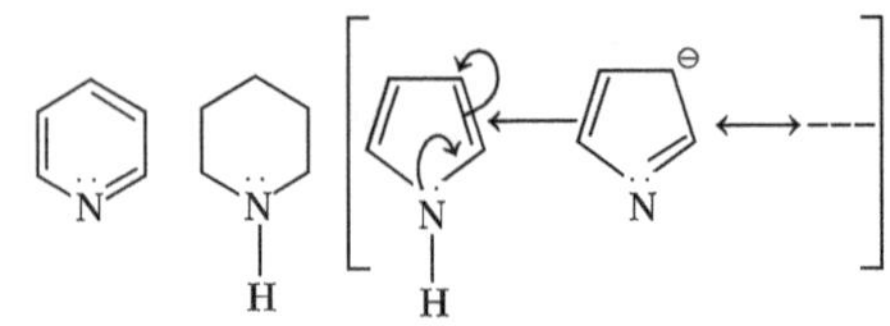

Lone pair involved in formation of aromatic sextet of 6π-electron (least basic).

$\therefore$ Correct order of basicity is II > I > III.

55. ***(d)*** As, density $(\rho) = \dfrac{Z_{eff} \times \text{molecular weight}}{N_A \times a^3}$

(For antifluorite, $Z_{eff} = \dfrac{4}{\text{unit cell}}$

$$\rho = \frac{4 \times (23 \times 2 + 16)}{6 \times 10^{23} \times (10^{-10}\,\text{cm})^3}$$

$$= 414.16\ \text{g cm}^{-3}$$

$[1 \text{ picometre (pm)} = 10^{-12}\ \text{m} = 10^{-10}\ \text{cm}]$

Frenkel defect is the type of stoichiometric defect in which density of the crystal does not change.

56. ***(b)*** We can represent the process of evaporation as

$$18\ \text{g}\ H_2O(l) \xrightarrow{\text{vaporisation}} 18\ \text{g}\ H_2O(g)$$

Number of moles in 18 g $H_2O(l)$ is

$$= \frac{18\ \text{g}}{18\ \text{g mol}^{-1}} = 1\ \text{mol}$$

$$\Delta_{vap}U = \Delta_{vap}H^{\ominus} - p\Delta V$$

$$= \Delta_{vap}H^{\ominus} - \Delta n_g RT$$

Assume steam behave as an ideal gas.

$$\Delta_{vap}U = (40.66) - (1)(8.314 \times 10^{-3})(373)$$

$$= 40.66 - 3.10$$

$$= 37.56\ \text{kJ mol}^{-1}$$

57. ***(d)*** In case of cyclopropyl carbocation, stability of carbocation depends upon conjugation between bent orbitals of cyclopropyl ring and vacant *p*-orbital of cationic carbon. This type of bonding is known as banana bonding.

In case of carbocation III and IV, stability is because of the resonance effect. Three *R*-effect dominates over two banana bondings so, III is more stable than II.

$\therefore$ Correct order of stability of carbocation is

$$I > III > II > IV$$

58. ***(b)*** The electronic configuration of mercury is $[Xe]4f^{10}, 5d^{10}, 6s^2$. Its *d*-subshell is completely filled thus, it prevents the overlapping of *d*-orbitals (*d-d* overlapping). Hence, it is liquid metal at room temperature.

59. ***(b)*** Let V litre of 10 N HCl be mixed with $(1-V)$ litre of 4 N HCl to give $(V+1-V)=1$ L of 7 N HCl

As we know that,

$$N_1V_1 + N_2V_2 = NV$$
$$10V + 4(1-V) = 7 \times 1$$
$$10V + 4 - 4V = 7$$
$$6V = 7 - 4$$
$$V = \frac{3}{6} = 0.50 \text{ L}$$

Volume of 10 N HCl = 0.50 L

Volume of 4 N HCl = 1 − 0.50 = 0.50 L

60. ***(d)*** $t_{1/2} \propto \left(\frac{1}{a}\right)^{n-1}$ or $t_{1/2} = k(a)^{1-n}$

$$\log t_{1/2} = \log k + (1-n)\log a$$

(It represents straight line equation; $y = c + mx$)

slope, $m = (1-n) = \tan 45° = 1$

$\therefore \quad (1-n) = 1$

$\Rightarrow \quad n = 0$

61. ***(a)*** Ziegler-Natta catalyst $[TiCl_4 + Al(C_2H_5)_3]$ is used as a catalyst in the polymerisation of olefins.

62. ***(c)*** Hydrides are binary compounds of hydrogen. These can be classified into four groups

(i) Ionic hydrides : NaH, CaH_2, LiH

(ii) Covalent hydrides : B_2H_6 , NH_3, SbH_3

(iii) Polynuclear hydrides : $LiAlH_4$, $NaBH_4$

(iv) Interstitial hydrides : those in which hydrogen is trapped in the interstitial spaces of transition metals.

Here, B_2H_6 is a dimeric form of BH_3. BH_3 covalently combined with another BH_3 molecule to form B_2H_6. B_2H_6 contain 3 centre 2 electrons bond.

63. ***(b)*** The method used to separate noble gas mixture from air is called **Fischer-Ringe's** method. When air free from moisture and CO_2 is passed over a heated mixture (800°C) of 90% CaC_2 + 10% $CaCl_2$ in an iron sealed tube, the following reactions take place

$$CaC_2 + N_2 \xrightarrow{800°C} CaCN_2 + C$$
$$2C + O \longrightarrow 2CO$$
$$C + O_2 \longrightarrow CO_2$$
$$2CaC_2 + 3CO_2 \longrightarrow 2CaCO_3 + 5C$$
$$CuO + CO \longrightarrow Cu + CO_2$$

CO_2 gas is absorbed by KOH solution. Thus, a mixture of inert gases are obtained.

64. ***(d)*** I. In nitrogen, d-orbitals are absent, hence, it does not form NCl_5. Thus, NCl_5 does not exist but PCl_5 does.

II. O_2^+ and NO are isoelectronic and contains one unpaired electron each, thus both are paramagnetic.

III. In carbonate ion, CO_3^{2-} all three C—O bonds are identical due to resonance.

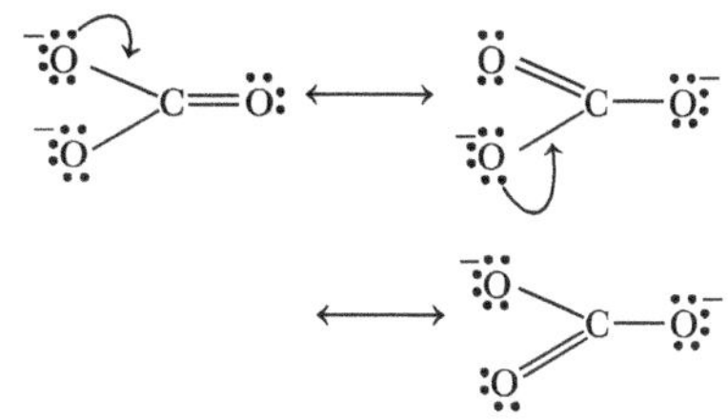

IV. Pb^{2+} is more stable than Pb^{4+} due to inert pair effect, hence prefers to form divalent compounds.

Thus, the incorrect statements are III and IV.

65. ***(c)*** Gallium, Ga is a soft silvery white metal and is liquid at room temperature. When it solidifies, expands by 3.1%. Thus, it should not be stored in glass or metal containers.

66. ***(a)*** ZnS has zinc blende type structure (*i.e.*, ccp structure). The S^{2-} ions are present at the corners of the cube and at the centre of each face. Zinc ions occupy half of the tetrahedral sites. Each zinc ion is surrounded by four sulphide ions which are disposed towards the corner of regular tetrahedron. Similarly, S^{2-} ion is surrounded by four Zn^{2+} ions.

∴ Option (a) is correct statement.

67. ***(c)*** Reactivity of nucleophilic addition reaction depends upon the electron deficiency of carbonyl group and steric hinderance. Steric hinderance decreases the rate of reaction. This steric hinderance is minimum in methanal and maximum in benzophenone.

∴ Increasing order of nucleophilic addition reaction of the given compound is

$$\text{IV} < \text{II} < \text{III} < \text{I}$$

68. ***(b)*** The combustion equation of sucrose is

$$C_{12}H_{22}O_{11}(s) + 12O_2(g) \longrightarrow 12CO_2(g) + 11H_2O(g)$$

Here, $\Delta n = 12 + 11 - 12 = 11$

As we know,

$$\Delta H = \Delta E + \Delta nRT$$
$$\Delta H = (-1348.9 \times 10^3) + 11 \times 2 \times 298$$
$$= -1348900 + 6556$$
$$= -1342344 \text{ cal}$$
$$= -1342.344 \text{ kcal}$$

69. ***(c)*** Boiling point decreases with increase in branching. Compound (III) has two branches, compound (I) has one branch and compound (II) is a normal alkyl halide with no branch. So, the boiling point is minimum for compound (III) and maximum for compound (II).

70. ***(d)*** Aldehydes and ketones possess $\left(-\overset{\overset{\displaystyle O}{\|}}{C}-CH_3\right)$ group will give positive iodoform test. Apart from methylated carbonyl compounds, alcohols with $\left(CH_3-\underset{\underset{\displaystyle OH}{|}}{C}H-\right)$ group also give positive iodoform test.

$$C_6H_5-\underset{\underset{\displaystyle OH}{|}}{C}H-CH_3 \xrightarrow[\text{NaOH}]{I_2 \text{ and}} \underset{\text{Iodoform}}{CHI_3} + C_6H_5COONa$$

∴ Option (d) will give iodoform test.

71. ***(a)*** This problem includes conceptual mixing of acidic character, aromaticity and nucleophilic substitution reaction.
Students are advised to identify the most stable intermediate obtained among all (after the removal of H^+) keeping in mind the concept of conjugation and aromaticity. Then complete the reaction further using concept of nucleophilic substitution reaction.

Acidic character The species which easily donate its hydrogen and produces stable conjugate base is acid. The species which produces more stable conjugate base is more stronger acid.

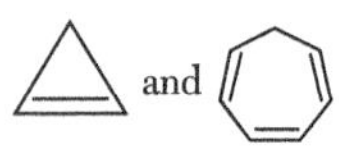

do not looses H^+ hence are not acidic.

looses the H^+ easily and produces more stable aromatic cyclopentadienyl anion.

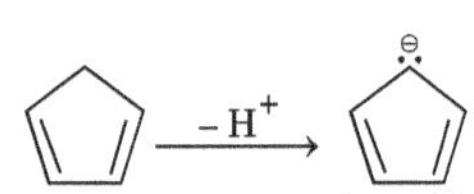

Now, cyclopentadienyl anion on reaction with 3-chloro prop-1-ene produces the product *via* nucleophilic substitution reaction.

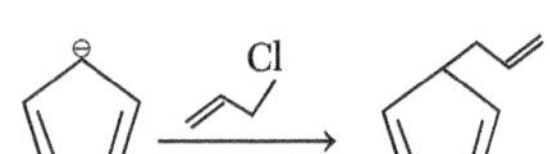

72. ***(b)*** The given reaction take place as follows :

$$CaCO_3 \xrightarrow{\text{Heat}} \underset{(X)}{CaO} + \underset{(Z)}{CO_2}$$

$$CaO + 3C \xrightarrow{\text{Heat}} \underset{(C)}{CaC_2} + \underset{(D)}{CO}$$

$$\underset{(C)}{CaC_2} + 2H_2O \longrightarrow Ca(OH)_2 + \underset{(E)}{C_2H_2}$$

73. ***(b)*** Optical isomerism is shown by only those complexes which lack symmetry. Complex $[Co(NH_3)_3(NO_2)_3]$ shows facial and meridional isomerism. Both isomers of this complex contain plane of symmetry. So, it will not form optical isomers.

74. ***(c)***

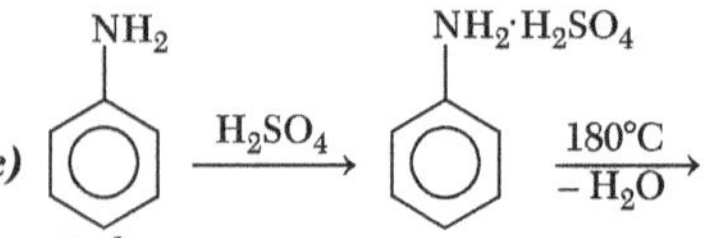

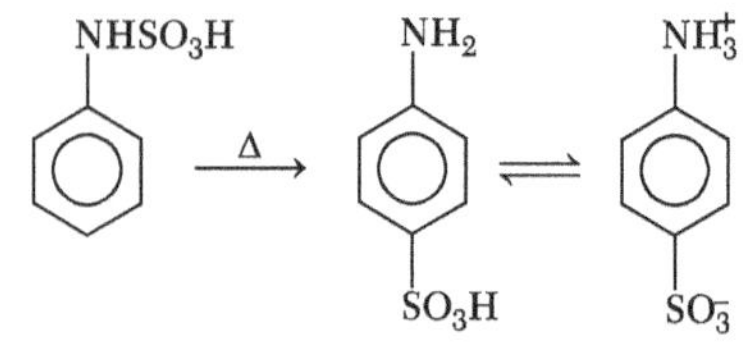

Sulphanilic acid exists as a dipolar ion which has acidic as well as basic groups in the same molecule. Such ions are called Zwitter ions or inner salts.

Thus, $-SO_3$ diminishes the basic character of $-NH_2$.

75. ***(b)*** The structure of copolymer of ethene and vinyl chloride is shown below

$$\underset{\text{Ethene}}{n(CH_2=CH_2)} + \underset{\text{Vinyl chloride}}{n(CH_2=CHCl)}$$

$$\longrightarrow \underset{\text{Copolymer}}{\left(CH_2-CH_2-CH_2-\underset{\underset{}{}}{\overset{\overset{\displaystyle Cl}{|}}{C}H}-\right)_n}$$

Molecular weight of ethene $(CH_2CH_2) = 28$

Molecular weight of vinyl chloride (CH_2CHCl)
$= 62.5$

Empirical formula weight of copolymer
$= 28 + 62.5 = 90.5$

Mass % of vinyl chloride in the copolymer

$$= \frac{62.5 \times 100}{90.5} = 69.06 \approx 69\%$$

76. ***(b)*** Insulin is composed of two peptide chains referred to chain *A* and *B*. Chain *A* of 21 residues and chain *B* of 30 residues are cross linked by two disulphide bridges.

77. ***(b)*** Barbital is a sleep-producing drug, *i.e.*, hypnotic tranquillizer. It causes addiction.

78. ***(b)*** Since, at equivalence point

(for acid) $N_1V_1 = N_2V_2$ (for base)

∴ Volume of NaOH required to reach

$$\text{equivalence point} = \frac{0.1 \times 25}{0.05} = 50 \text{ mL}$$

$$\therefore \text{ Concentration of salt formed} = \frac{\text{millimoles of acid}}{\text{total volume in mL}}$$

$$= \frac{25 \times 0.1}{75} = \frac{0.1}{3}$$

$$\text{Since, } [H^+] = \sqrt{\frac{K_w \times K_a}{C}} = \sqrt{\frac{10^{-14} \times 1.8 \times 10^{-5} \times 3}{0.1}}$$

$$\therefore \quad = 2.32 \times 10^{-9} \text{ M}$$

$$pH = \log [H^+]$$

$$= -\log (2.32 \times 10^{-9})$$

$$= 9 - \log 2.32$$

$$= 9 - 0.37$$

$$pH = 8.63$$

79. ***(d)*** Critical temperature of a gas is highest temperature at which liquification of the gas first occurs. The temperature 30.98°C is called critical temperature of carbon dioxide because this is the highest temperature at which liquid carbon dioxide is observed. Above this temperature it is gas.

80. ***(a)*** I. **Elimination reaction**

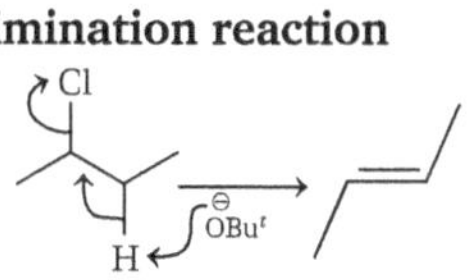

The reaction in which smaller neutral molecule are removed during the reaction is known as elimination reaction.

II. **Substitution reaction** The reaction in which one nucleophilic group is replaced by another nucleophile is known as nucleophilic substitution reaction.

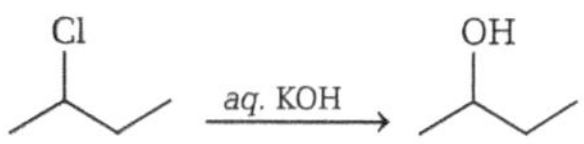

Here, OH replaces Cl.

III. **Addition reaction** The reaction in which reactant undergo addition with reagent to give a single product.

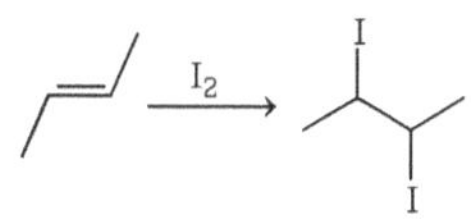

IV. **Addition reaction**

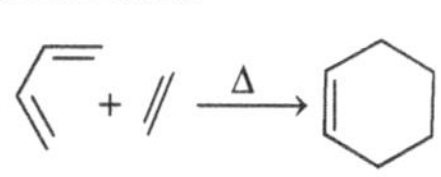

Hence, correct option is (a).

a. English Proficiency

81. ***(a)*** Augment means make bigger, so 'increase' is its correct synonoym.

82. ***(a)*** Consolation means 'comfort received by a person after a loss', so 'comfort' is the correct option.

83. ***(b)*** Auxiliary means 'providing additional help', so 'supplemental' is the correct option.

84. ***(b)*** Auspicious means 'favourable', so 'unfavourable' is best opposite word for it.

85. ***(d)*** Recompense means 'payment', so 'penalty' is the correct opposite word for it.

86. ***(c)*** Impede means delay or prevent (someone or something) by obstructing them, so 'push' is correct opposite word for it. Push means a vigorous effort to do or obtain something.

87. ***(a)*** Here a sense of command is depicted in sentence, so we should use 'ordered' for proper meaning of sentence.

88. ***(c)*** Sentence is in past tense and first form is used in those negative sentence which contain 'did', so option (c) is correct.

89. ***(d)*** No improvement is needed in the given sentence, as it is right.

90. ***(d)*** Afraid agrees with preposition 'of', so option (d) is correct.

91. ***(c)*** Normally, company signs a contract or deal, so use of 'deal' is proper here.

92. ***(c)*** The question gives a sense of query about normal routine of some special/specific day, so use of 'plan' is more proper here.

93. ***(b)*** According to the events of sentence, PRSQ is best arrangement.

94. ***(a)*** According to events of sentence, PRSQ is best arrangement.

95. ***(c)*** According to sequence of events in the sentence, QPRS is best arrangement.

b. Logical Reasoning

96. ***(c)*** Given,

SAFER = 5@3#2 and RIDE = 2©%#

S ⟶ 5	R ⟶ 2
A ⟶ @	I ⟶ ©
F ⟶ 3	D ⟶ %
E ⟶ #	E ⟶ #
R ⟶ 2	

∴ Code for FEDS = 3#%5

97. ***(b)*** From the given responses,

$$4 \times 2 \times 3 \times 3 = 72$$

$$9 \times 4 \times 2 \times 10 = 720$$

Similarly, $6 \times 20 \times 1 \times 6 = \boxed{720}$

98. ***(b)*** Since, consecutive two letters are interchanged. Therefore,

DE PR ES SI ON

↓ ↓ ↓ ↓ ↓

ED R[P]SE IS NO ← On counting

Now, on counting from right hand side P is the 7th letter from right.

99. ***(a)*** Every day of week repeats after seven days.

Hence, $59 = 7 \times 8 + 3$ = 56 days + 3 odd days

∴ It will be Thrusday after 56 days.

∴ 57th day = Thrusday

⇒ 58th day = Friday

59th day = Saturday

⇒ 60th day = Sunday

∴ It will be Sunday after 59 days.

100. ***(d)*** Both coal mines and factories are located in the fields.

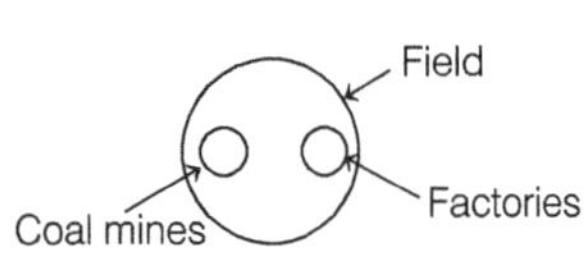

101. ***(c)*** From the given series,

$$1^3 \longrightarrow 1$$

$$2^3 \longrightarrow 8$$

$$3^3 \longrightarrow 27$$

$$\boxed{4^3 \longrightarrow 64}$$

$$5^3 \longrightarrow 125$$

$$6^3 \longrightarrow 216$$

Therefore, 64 will come in place of questions mark.

102. ***(c)*** Interchanging the symbols as given in the above question, the above equation becomes

$$6 + 9 \times 8 \div 3 - 20 = 6 + 9 \times \frac{8}{3} - 20$$

$$= 6 + 24 - 20 = 10$$

103. ***(b)*** Water image is the reflection of image in water.

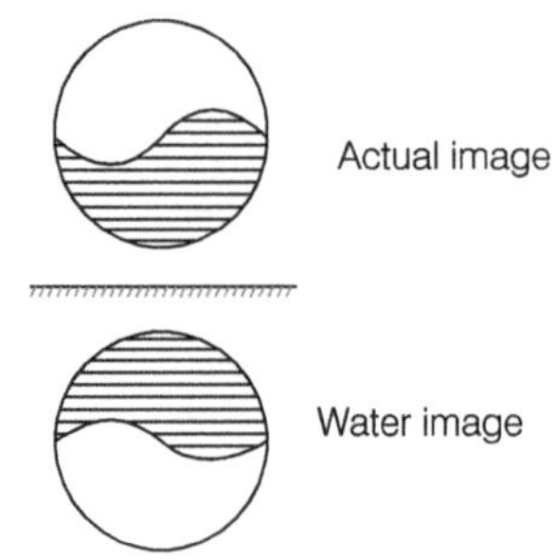

104. ***(b)*** On unfolding layer 1,

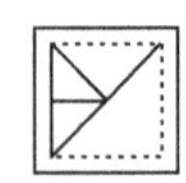

On unfolding layer 2,

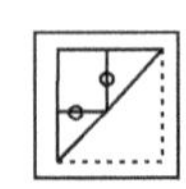

On unfolding layer 3,

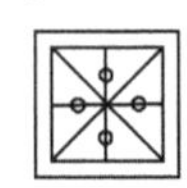

105. ***(d)***

[mallon] piml = [blue] light

[mallon] (tifl) = [blue] (berry)

arpan (tifl) = rasp (berry)

∴ From the above analysis, we see codes as

blue = Mallon

Light = piml

berry = tifl

arpan = rasp

∴ Light *i.e.*, piml is given in two options (b) and (d) but in option (b) other code given is of rasp, hence it cannot be the code of house. So, option (d) is correct.

Mathematics

106. ***(a)*** Since, AM $\geq$ GM, then

$$\frac{(p+q)+(r+s)}{2} \geq \sqrt{(p+q)(r+s)}$$

$\Rightarrow \quad 2/2 \geq \sqrt{M} \Rightarrow \sqrt{M} \leq 1 \Rightarrow M \leq 1$

Also, $(p+q)(r+s) > 0 \qquad (\because p, q, r, s > 0)$

$\therefore \qquad M > 0$

Hence, $\qquad 0 < M \leq 1$

107. ***(a)*** Given, $\left|\frac{z-3i}{z+3i}\right| = 1 \Rightarrow |z-3i| = |z+3i|$

(if $|z - z_1| = |z - z_2|$, then it is a perpendicular bisector of z_1 and z_2)

Hence, perpendicular bisector of $(0, 3)$ and $(0, -3)$ is X-axis.

108. ***(d)*** Since, $f(x)$ is an odd periodic function with period 2.

$\therefore \quad f(-x) = -f(x)$ and $f(x+2) = f(x)$

$\therefore \quad f(2) = f(0+2) = f(0)$

and $\quad f(-2) = f(-2+2) = f(0)$

Now, $\quad f(0) = f(-2) = -f(2) = -f(0)$

$\Rightarrow \quad 2f(0) = 0$, *i.e.*, $f(0) = 0$

$\therefore \quad f(4) = f(2+2) = f(2) = f(0) = 0$

Thus, $\quad f(4) = 0$

109. ***(b)*** We have, $\dfrac{x + \frac{x^3}{3!} + \frac{x^5}{5!} + \dots}{1 + \frac{x^2}{2!} + \frac{x^4}{4!} + \dots} = \dfrac{dx - dy}{dx + dy}$

On applying componendo and dividendo, we get

$$\frac{\left(x + \frac{x^3}{3!} + \frac{x^5}{5!} + \dots\right) + \left(1 + \frac{x^2}{2!} + \frac{x^4}{4!} + \dots\right)}{\left(x + \frac{x^3}{3!} + \frac{x^5}{5!} + \dots\right) - \left(1 + \frac{x^2}{2!} + \frac{x^4}{4!} + \dots\right)}$$

$$= \frac{(dx - dy) + (dx + dy)}{(dx - dy) - (dx + dy)}$$

$$\Rightarrow \frac{\left(1 + x + \frac{x^2}{2!} + \frac{x^3}{3!} + \frac{x^4}{4!} + \dots\right)}{-\left(1 - x + \frac{x^2}{2!} - \frac{x^3}{3!} + \dots\right)} = \frac{2dx}{-2dy}$$

$$\Rightarrow \frac{e^x}{-e^{-x}} = -\frac{dx}{dy}$$

$$\Rightarrow \frac{dy}{dx} = \frac{e^{-x}}{e^x} \Rightarrow \frac{dy}{dx} = e^{-2x}$$

$$\Rightarrow dy = e^{-2x} dx$$

On integrating both sides, we get

$$y = \frac{e^{-2x}}{(-2)} + C_1 \Rightarrow 2y = -e^{-2x} + 2C_1$$

$\Rightarrow \quad 2y = -e^{-2x} + C$ (where, $C = 2C_1$)

$\Rightarrow \quad 2ye^{2x} = -1 + e^{2x}C \Rightarrow 2ye^{2x} = Ce^{2x} - 1$

110. ***(b)*** Given lines are concurrent,

So, $\begin{vmatrix} 2 & -3 & k \\ 3 & -4 & -13 \\ 8 & -11 & -33 \end{vmatrix} = 0$

$\Rightarrow 2(132 - 143) + 3(-99 + 104) + k(-33 + 32) = 0$

$\Rightarrow -22 + 15 - k = 0 \Rightarrow k = -7$

111. ***(d)*** Given lines are

$$\frac{x-3}{2} = \frac{y-2}{3} = \frac{z-1}{\lambda} \qquad \dots\text{(i)}$$

and $$\frac{x-2}{3} = \frac{y-3}{2} = \frac{z-2}{3} \qquad \dots\text{(ii)}$$

These lines lie in the same plane. So, both are coplanar.

$\therefore \quad \begin{vmatrix} 2 & 3 & \lambda \\ 3 & 2 & 3 \\ 1 & -1 & -1 \end{vmatrix} = 0$

$\Rightarrow 2(-2+3) - 3(-3-3) + \lambda(-3-2) = 0$

$\Rightarrow \quad 2 + 18 - 5\lambda = 0 \Rightarrow 5\lambda = 20 \Rightarrow \lambda = 4$

$\therefore \sin^{-1}\sin\lambda = \sin^{-1}\sin 4$

$= \sin^{-1}\sin(\pi - 4)$

$= \pi - 4$

112. ***(c)*** We have,

$$e^x = (1-x)(B_0 + B_1x + B_2x^2 + \dots + B_{n-1}x^{n-1} + B_nx^n + \dots)$$

By the expansion of e^x, we get

$$1 + \frac{x}{1!} + \frac{x^2}{2!} + \dots + \frac{x^n}{n!} + ..$$

$$= (1-x)(B_0 + B_1x + B_2x^2 + \dots + B_{n-1}x^{n-1} + B_nx^n + \dots)$$

Equating the coefficient of x^n on both sides, we get

$$B_n - B_{n-1} = \frac{1}{n!}$$

113. ***(c)*** Let $P(n) = 2^{3n} - 7n - 1 \Rightarrow P(1) = 0, P(2) = 49$

$P(1)$ and $P(2)$ are divisible by 49.

Let $P(k) = 2^{3k} - 7k - 1 = 49I$

$P(k+1) = 2^{3k+3} - 7k - 8$

$= 8(49I + 7k + 1) - 7k - 8$

$= 49(8I) + 49k = 49\lambda$

(where, $\lambda = 8I + k$, which is an integer.)

BITSAT 2014

114. ***(b)*** We know that, $x-[x]$ is periodic function with period one.

$\therefore\ e^{x-[x]}$ has period one

[since, $f(x)$ is periodic with period T, then

$$\int_0^{nT} f(x)dx = n\int_0^T f(x)dx]$$

$$\therefore \int_0^{25\times 1} e^{x-[x]}dx = 25\int_0^1 e^{x-[x]}dx$$

$$= 25\int_0^1 e^{x-0}dx$$

$$= 25[e^x]_0^1 = 25[e-1]$$

$$\therefore \quad 25(e-1) = k(e-1)$$

Hence, $k = 25$

115. ***(a)*** Let P be $(at_1^2, 2at_1)$ and Q be $(at_2^2, 2at_2)$. Since, PQ subtends a right angle at the vertex $(0, 0)$.

Hence, $$t_1t_2 = -4 \quad \ldots(i)$$

If (h, k) is the point of intersection of normals at P and Q, then $$h = 2a + a(t_1^2+t_2^2+t_1t_2) \quad \ldots(ii)$$

and $$k = -at_1t_2(t_1+t_2) \quad \ldots(iii)$$

In order to find the locus of (h, k), we have to eliminate t_1 and t_2 between Eqs. (i), (ii) and (iii),

$$k = 4a(t_1+t_2) \quad \ldots(iv)$$

[from Eqs. (i) and (iii)]

and $$h - 2a = a[(t_1+t_2)^2 - t_1t_2]$$

$$\Rightarrow h - 2a = a\left[\frac{k^2}{16a^2}+4\right] \quad \text{[from Eq. (iv)]}$$

$$\Rightarrow h - 6a = \frac{k^2}{16a}$$

Hence, the required locus is $y^2 = 16a(x-6a)$.

Which represents a parabola.

116. ***(a)*** Given matrix can be rewritten as

$$A = \begin{bmatrix} 1+2\omega+\omega^2 & \omega^2 & 1 \\ 1 & 1+\omega^2+2\omega & \omega \\ \omega & \omega^2 & 2+\omega+2\omega^2 \end{bmatrix}$$

$$\therefore |A| = \begin{vmatrix} \omega & \omega^2 & 1 \\ 1 & \omega & \omega \\ \omega & \omega^2 & 1+\omega^2 \end{vmatrix} \quad (\because 1+\omega+\omega^2=0)$$

$$= \omega\begin{vmatrix} \omega & \omega & 1 \\ 1 & 1 & \omega \\ \omega & \omega & -\omega \end{vmatrix} \quad (\because 1+\omega+\omega^2=0)$$

$$\Rightarrow \omega(0) = 0$$

117. ***(b)*** $$\lim_{x\to\tan^{-1}3} \frac{\tan^2 x - 2\tan x - 3}{\tan^2 x - 4\tan x + 3}$$

$$= \lim_{\tan x\to 3} \frac{(\tan x-3)(\tan x+1)}{(\tan x-3)(\tan x-1)}$$

$$= \lim_{\tan x\to 3} \frac{\tan x+1}{\tan x-1}$$

$$= \frac{3+1}{3-1} = \frac{4}{2} = 2$$

118. ***(c)*** The given equation of second ellipse can be rewritten as

$$\frac{x^2}{4}+\frac{y^2}{1}=1$$

Equation of tangent to this ellipse is

$$\frac{x}{2}\cos\theta + y\sin\theta = 1 \quad \ldots(i)$$

Equation of the first ellipse can be rewritten as

$$\frac{x^2}{6}+\frac{y^2}{3}=1 \quad \ldots(ii)$$

Let ellipse (i) meets the first ellipse at P and Q and the tangents at P and Q to the second ellipse intersected at (h, k), then Eq. (i) is the chord of contact of (h, k) with respect to the ellipse (ii) and thus, its equation is

$$\frac{hx}{6}+\frac{ky}{3}=1 \quad \ldots(iii)$$

Since, Eqs. (i) and (iii) represent the same line

$$\frac{h/6}{\frac{1}{2}\cos\theta} = \frac{k/3}{\sin\theta} = 1$$

$$h = 3\cos\theta$$

and $$k = 3\sin\theta$$

Hence, locus is $x^2+y^2=9$.

119. ***(d)*** We have, $|\sin x\cos x| + |\tan x+\cot x| = \sqrt{3}$

$$\Rightarrow |\sin x\cos x| + \frac{1}{|\sin x\cdot\cos x|} = \sqrt{3}$$

But $$|\sin x\cdot\cos x| + \frac{1}{|\sin x\cdot\cos x|} \geq 2$$

Hence, there is no solution.

120. ***(d)*** $$\lim_{x\to 0}\left[\tan\left(\frac{\pi}{4}+x\right)\right]^{1/x} = \lim_{x\to 0}\left[\frac{1+\tan x}{1-\tan x}\right]^{1/x}$$

$$= \lim_{x\to 0}\left[(1+\tan x)^{\frac{1}{\tan x}}\right]^{\frac{\tan x}{x}} \times \lim_{x\to 0}\left[(1-\tan x)^{-\frac{1}{\tan x}}\right]^{\frac{\tan x}{x}}$$

$$= e\cdot e = e^2$$

121. ***(b)*** Since, $\sin^2\theta = \dfrac{1-\cos 2\theta}{2} = \dfrac{1}{2}-\dfrac{1}{2}\cos 2\theta$

$\therefore$ Period of $\sin^2\theta = \dfrac{2\pi}{2} = \pi$

122. ***(c)*** $\because$ Total number of ways $= 5!$

and favourable number of ways $= 2\cdot 4!$

$\therefore$ Required probability $= \dfrac{2\cdot 4!}{5!} = \dfrac{2}{5}$

BITSAT 2014

123. ***(b)*** $\lim_{x \to \infty} \dfrac{x^4 \cdot \sin\left(\frac{1}{x}\right) + x^2}{1 + |x|^3} = \lim_{x \to \infty} \left[\dfrac{x \sin\left(\frac{1}{x}\right) + \frac{1}{x}}{\frac{1}{x^3} + \frac{|x|^3}{x^3}} \right]$

[dividing numerator and denominator by x^3]

$$= \frac{\lim_{x \to \infty} \frac{\sin\left(\frac{1}{x}\right)}{\frac{1}{x}} + \lim_{x \to \infty} \frac{1}{x}}{\lim_{x \to \infty} \frac{1}{x^3} + \lim_{x \to \infty} \frac{|x|^3}{x^3}} = \frac{1-0}{0-1} = -1$$

124. ***(d)*** Given that, $\log_4 2 - \log_8 2 + \log_{16} 2 - \dots$

$$= \frac{1}{\log_2 4} - \frac{1}{\log_2 8} + \frac{1}{\log_2 16} - \dots$$

$$= \frac{1}{2} - \frac{1}{3} + \frac{1}{4} - \dots$$

$$= 1 - \left[1 - \frac{1}{2} + \frac{1}{3} - \frac{1}{4} + \frac{1}{5} - \dots\right]$$

$$= 1 - \log_e 2$$

125. ***(c)*** Let the observation be $x_1, x_2, x_3, \dots, x_n$.

Now, mean $(\bar{x}) = \dfrac{x_1 + x_2 + \dots + x_n}{n}$

When first term increased by 1, second by 2 and so on, then observations will be

$(x_1 + 1), (x_1 + 2), (x_1 + 3), \dots, (x_n + n)$

Then, new mean

$$(\bar{x}_1) = \frac{(x_1+1) + (x_2+2) + \dots + (x_n+n)}{n}$$

$$= \frac{(x_1 + x_2 + \dots + x_n) + (1 + 2 + 3 + \dots + n)}{n}$$

$$\Rightarrow \bar{x}_1 = \frac{x_1 + x_2 + \dots + x_n}{n} + \frac{1 + 2 + 3 + \dots + n}{n}$$

$$\Rightarrow \bar{x}_1 = \bar{x} + \frac{n(n+1)}{2n} \quad \left[\because 1 + 2 + 3 + \dots + n = \frac{n(n+1)}{2}\right]$$

$$\Rightarrow \bar{x}_1 = \bar{x} + \frac{n+1}{2}$$

126. ***(c)*** We have, $(\sin^{-1} x)^2 + (\cos^{-1} x)^2$

$$= (\sin^{-1} x + \cos^{-1} x)^2 - 2\sin^{-1} x \cdot \cos^{-1} x$$

$$= \frac{\pi^2}{4} - 2\sin^{-1} x\left(\frac{\pi}{2} - \sin^{-1} x\right)$$

$$= \frac{\pi^2}{4} - \pi \sin^{-1} x + 2(\sin^{-1} x)^2$$

$$= 2\left[(\sin^{-1} x)^2 - \frac{\pi}{2}\sin^{-1} x + \frac{\pi^2}{8}\right]$$

$$= 2\left[\left(\sin^{-1} x - \frac{\pi}{4}\right)^2 + \frac{\pi^2}{16}\right]$$

Thus, the least value is $2\left(\dfrac{\pi^2}{16}\right)$ *i.e.*, $\dfrac{\pi^2}{8}$ and the greatest value is $2\left[\left(\dfrac{-\pi}{2} - \dfrac{\pi}{4}\right)^2 + \dfrac{\pi^2}{16}\right]$ *i.e.*, $\dfrac{5\pi^2}{4}$.

127. ***(a)*** Since, P (exactly one of A, B occurs) $= q$ (given), we get

$$P(A \cup B) - P(A \cap B) = q$$

$$\Rightarrow p - P(A \cap B) = q$$

$$\Rightarrow P(A \cap B) = p - q$$

$$\Rightarrow 1 - P(A' \cup B') = p - q$$

$$\Rightarrow P(A' \cup B') = 1 - p + q$$

$$\Rightarrow P(A') + P(B') - P(A' \cap B') = 1 + q - p$$

$$\Rightarrow P(A') + P(B') = (1 - p + q) + [1 - P(A \cup B)]$$

$$= (1 - p + q) + (1 - p)$$

$$= 2 - 2p + q$$

128. ***(b)*** The given equation is reduced to $x = e^{xy\left(\frac{dy}{dx}\right)}$

$$\Rightarrow \log x = xy\frac{dy}{dx} \Rightarrow y\,dy = \frac{\log x}{x}dx$$

$$\Rightarrow \int y\,dy = \int \frac{1}{x}\log x\,dx \Rightarrow \frac{y^2}{2} = \frac{(\log x)^2}{2} + C'$$

$$\Rightarrow y^2 = (\log x)^2 + 2C'$$

$$\Rightarrow y^2 = (\log x)^2 + C \quad \text{(where, } C = 2C'\text{)}$$

129. ***(d)*** Let $P(n) = n^3 + 2n$

$$P(1) = 1 + 2 = 3$$

$$P(2) = 8 + 4 = 12$$

$$P(3) = 27 + 6 = 33$$

Clearly, we see that all these numbers are divisible by 3.

130. ***(c)*** $\because$ Using Binomial theorem,

$$(5x - 4y)^n = {}^nC_0(5x)^n + {}^nC_1(5x)^{n-1}(-4y) + {}^nC_2(5x)^{n-2}(-4y)^2 + \dots + {}^nC_n(-4y)^n$$

Sum of coefficients

$$= {}^nC_0 5^n + {}^nC_1 5^{n-1}(-4) + {}^nC_2 5^{n-2} \cdot (-4)^2 + \dots + {}^nC_n(-4)^n$$

$$= (5-4)^n = 1^n = 1$$

131. ***(a)*** Given equation of plane is $x + y + z = 5$.

The distance measured along the line $x = y = z$.

Direction ratio's of the given line is (1, 1, 1).

So, the equation of line PQ is

$$\frac{x-1}{1} = \frac{y+5}{1} = \frac{z-9}{1}$$

Now, let $\dfrac{x-1}{1} = \dfrac{y+5}{1} = \dfrac{z-9}{1} = \lambda$

$$x = \lambda + 1,\ y = \lambda - 5,\ z = \lambda + 9$$

BITSAT 2014

lies on the plane $x + y + z = 5$

$$\lambda + 1 - \lambda + 5 + \lambda + 9 = 5 \Rightarrow \lambda = -10$$

The coordinate of Q is $(-9, -15, -1)$ and the coordinate of P is $(1, -5, 9)$.

$$PQ = \sqrt{(10)^2 + (10)^2 + (10)^2} = 10\sqrt{3}$$

$$\therefore \quad 2\sqrt{3}k = 10\sqrt{3} \quad \Rightarrow k = 5$$

132. *(b)* $\lim_{x \to 1} \dfrac{x^m - 1}{x^n - 1} = \lim_{x \to 1} \dfrac{mx^{m-1}}{nx^{n-1}}$ [by L' Hospital rule]

$$= m/n$$

133. *(a)* Consider that, $I = \int_1^x t \log t \, dt$

$$= \left[\log t \cdot \frac{t^2}{2}\right]_1^x - \int_1^x \frac{1}{t} \cdot \frac{t^2}{2} dt = \frac{x^2}{2} \log x - \frac{1}{2}\left[\frac{t^2}{2}\right]_1^x$$

$$= \frac{x^2}{2} \log x - \frac{1}{2}\left[\frac{x^2}{2} - \frac{1}{2}\right]$$

$$\Rightarrow \quad \frac{1}{4} = \frac{x^2}{2} \log x - \frac{1}{4}(x^2 - 1)$$

$$\Rightarrow \quad \frac{1}{2}x^2 \log x - \frac{1}{4}x^2 = 0 \qquad (\text{as } x > 1)$$

$$\Rightarrow \quad x^2(2\log x - 1) = 0 \Rightarrow 2\log x - 1 = 0$$

$$\Rightarrow \quad \log x = \frac{1}{2} \Rightarrow x = e^{1/2} \Rightarrow x = \sqrt{e}$$

134. *(a)* We have,

$$a = \sum_{n=0}^{\infty} \frac{x^{3n}}{(3n)!}, \; b = \sum_{n=1}^{\infty} \frac{x^{3n-2}}{(3n-2)!} \text{ and } c = \sum_{n=1}^{\infty} \frac{x^{3n-1}}{(3n-1)!}$$

$$\text{Now, } a + b + c = \sum_{n=0}^{\infty} \frac{x^{3n}}{3n!} + \sum_{n=1}^{\infty} \frac{x^{3n-2}}{(3n-2)!} + \sum_{n=1}^{\infty} \frac{x^{3n-1}}{(3n-1)!}$$

$$= 1 + x + \frac{x^2}{2!} + \frac{x^3}{3!} + \ldots = e^x$$

$$a + b\omega + c\omega^2 = 1 + \omega x + \frac{\omega^2 x^2}{2!} + \frac{\omega^3 x^3}{3!} + \ldots = e^{\omega x}$$

and $a + b\omega^2 + c\omega = e^{\omega^2 x}$, ω is imaginary cube root of unity.

Now, $a^3 + b^3 + c^3 - 3abc$

$$= (a + b + c)(a + b\omega + c\omega^2)(a + b\omega^2 + c\omega)$$

$$= e^x \cdot e^{\omega x} \cdot e^{\omega^2 x} = e^{x(1 + \omega + \omega^2)} = e^{0 \cdot x} = 1$$

135. *(a)* Required unit vector is $\dfrac{(\hat{\mathbf{i}} - \hat{\mathbf{j}}) \times (\hat{\mathbf{i}} + \hat{\mathbf{j}})}{|(\hat{\mathbf{i}} - \hat{\mathbf{j}}) \times (\hat{\mathbf{i}} + \hat{\mathbf{j}})|}$

$$= \frac{\hat{\mathbf{k}} + \hat{\mathbf{k}}}{2} = \frac{2\hat{\mathbf{k}}}{2} = \hat{\mathbf{k}}$$

136. *(a)* $f(x) = x^3 + 2x^2 + 5x + 2\cos x$

$$f'(x) = 3x^2 + 4x + 5 - 2 \cdot \sin x$$

$$= 3\left(x + \frac{2}{3}\right)^2 + \frac{11}{3} - 2 \cdot \sin x$$

$$\Rightarrow \quad f'(x) > 0, \forall x$$

$f(x)$ is increasing for all $x \in R$.

Also, $f(0) = 2 \Rightarrow f(x) = 0$

So, $f(x)$ has no solution.

137. *(c)* $\lim_{x \to 0} \dfrac{(1 + x)^8 - 1}{(1 + x)^2 - 1}$

$$= \lim_{x \to 0} \frac{[(1 + x)^4 + 1][(1 + x)^2 + 1][(1 + x)^2 - 1]}{(1 + x)^2 - 1}$$

$$= 2 \times 2 = 4$$

138. *(a)* Given that, $y = -\sqrt{-x}$

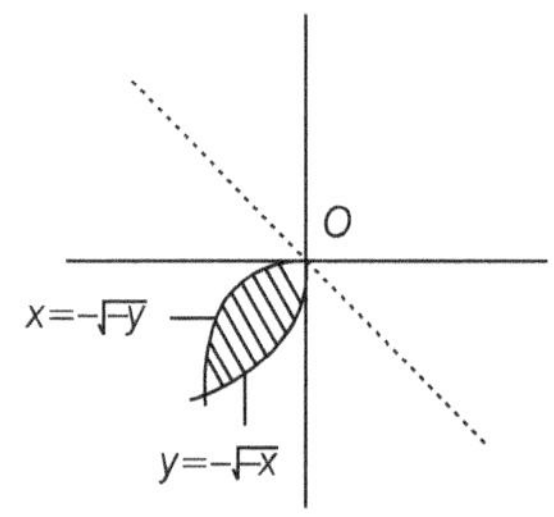

$\Rightarrow \quad y^2 = -x$, where x and y both negative.

Now, $x = -\sqrt{-y}$

$\Rightarrow \quad x^2 = -y$, where x and y both negative.

$$\therefore \text{ Area} = \left| \int_{-1}^{0} -\sqrt{-x} \, dx - \int_{-1}^{0} -x^2 \, dx \right| = \frac{1}{3}$$

139. *(a)* Given series

$$\left[\alpha x - \frac{1}{2}(\alpha x)^2 + \frac{1}{3}(\alpha x)^3 - \ldots\right] + \left[\beta x - \frac{1}{2}(\beta x)^2 + \frac{1}{3}(\beta x)^3 - \ldots\right]$$

$$= \log(1 - \alpha x) + \log(1 - \beta x)$$

$$= \log[1 - (\alpha + \beta)x + \alpha\beta x^2]$$

Now, $\alpha + \beta = p$ and $\alpha\beta = q$

Given series $= \log(1 - px + qx^2)$

140. *(c)* We know that,

$$\sin^{-1} x + \cos^{-1} x = \pi/2, \forall x \in [-1, 1]$$

$$\therefore \quad \cos^{-1} x > \sin^{-1} x$$

$$\Rightarrow \quad \frac{\pi}{2} - \sin^{-1} x > \sin^{-1} x \Rightarrow \frac{\pi}{2} > 2\sin^{-1} x$$

$$\Rightarrow \quad \sin^{-1} x < \frac{\pi}{4} \Rightarrow x < \frac{1}{\sqrt{2}}$$

$$\therefore \quad -1 \le x < \frac{1}{\sqrt{2}}$$

141. ***(c)*** Number of ways

$$= {}^{2n+1}C_0 + {}^{2n+1}C_1 + {}^{2n+1}C_2 + \dots + {}^{2n+1}C_n$$
$$= \frac{1}{2}(2^{2n+1}) = 2^{2n}$$

Thus, $2^{2n} = 64$ i.e., $2^{2n} = 2^6$

On comparing, $2n = 6 \Rightarrow n = 3$

142. ***(c)*** R is reflexive as $(3, 3), (6, 6), (9, 9), (12, 12) \in R$.

R is not symmetric as $(6, 12) \in R$ but $(12, 6) \notin R$.

R is transitive as the only pair which needs verification is $(3, 6)$ and $(6, 12) \in R \Rightarrow (3, 12) \in R$.

143. ***(d)*** Let α and β be the roots of the equation

$$x^2 - (a-2)x - a + 1 = 0$$

$\therefore \quad \alpha + \beta = a - 2$ and $\alpha\beta = -(a-1)$

$$s = \alpha^2 + \beta^2 = (\alpha+\beta)^2 - 2\alpha\beta = (a-2)^2 + 2(a-1)$$
$$= a^2 - 4a + 4 + 2a - 2 = a^2 - 2a + 2$$

Now, $ds/da = 2a - 2$

For maximum and minimum, $ds/da = 0$

$\Rightarrow \quad 2a - 2 = 0 \Rightarrow a = 1$

Also, $\quad d^2s/da^2 = 2 > 0$

Hence, at $a = 1$, s will have minimum value.

144. ***(b)*** We have, $ax^2 + by^2 = 1$

On differentiating both sides w.r.t. x, we get

$$2ax + 2by\frac{dy}{dx} = 0 \Rightarrow ax + by\frac{dy}{dx} = 0 \quad \dots(i)$$

$$\Rightarrow \quad \frac{-a}{b} = \frac{y}{x}\frac{dy}{dx} \quad \dots(ii)$$

Again, differentiating Eq. (i) w.r.t. x, we get

$$a + b\left(\frac{dy}{dx}\right)^2 + by\frac{d^2y}{dx^2} = 0$$

$$\Rightarrow \quad -\frac{a}{b} = \left(\frac{dy}{dx}\right)^2 + y\frac{d^2y}{dx^2} \quad \dots(iii)$$

From Eqs. (ii) and (iii), we get

$$\frac{y}{x}\frac{dy}{dx} = \left(\frac{dy}{dx}\right)^2 + y\frac{d^2y}{dx^2}$$

$$\Rightarrow \quad y\left(\frac{dy}{dx}\right) = x\left(\frac{dy}{dx}\right)^2 + xy\frac{d^2y}{dx^2}$$

or $\quad x\left(\frac{dy}{dx}\right)^2 + xy\left(\frac{d^2y}{dx^2}\right) - y\frac{dy}{dx} = 0$

145. ***(c)*** If x tables and y chairs are purchased for maximum profit.

Then, $\quad x + y \le 60$

$$5x + \frac{6y}{5} \le 100 \Rightarrow x \ge 0, y \ge 0$$

So, number of constraints are four.

146. ***(c)*** For $n = 1$, $10^n + 3\cdot 4^{n+2} + 5 = 10 + 3\cdot 4^3 + 5$

$= 207$, which is divisible by 9

So, by induction, the result is divisible by 9.

147. ***(b)*** Let sets A and B have m and n elements, respectively.

$$2^m - 2^n = 56$$
$$2^n(2^{m-n} - 1) = 56 = 8 \times 7 = 2^3 \times 7$$

Comparing both sides, we get

$$2^n = 2^3 \text{ and } 2^{m-n} - 1 = 7$$

$\Rightarrow \quad n = 3$ and $2^{m-n} = 8$

$\Rightarrow \quad 2^{m-n} = 2^3 \Rightarrow m - n = 3$

$\Rightarrow \quad m - 3 = 3 \Rightarrow m = 6$

Number of the elements in A is 6.

148. ***(a)*** Mean of the combined sample

$$\overline{X} = \frac{n_1\bar{x}_1 + n_2\bar{x}_2}{n_1 + n_2} = \frac{35 \times 80 + 65 \times 70}{35 + 65} = 73.5$$

Standard deviation of the combined sample is given by

$$\sigma^2 = \frac{n_1(\sigma_1^2 + d_1^2) + n_2(\sigma_2^2 + d_2^2)}{n_1 + n_2}$$

where, $d_1^2 = (\bar{x}_1 - \bar{x})^2 = (80 - 73.5)^2 = 42.25$

$d_2^2 = (\bar{x}_2 - \bar{x})^2 = (70 - 73.5)^2 = 12.25$

$$\therefore \quad \sigma^2 = \frac{35(4^2 + 42.25) + 65(3^2 + 12.25)}{35 + 65}$$

$\Rightarrow \quad \sigma^2 = 34.2 \Rightarrow \sigma = 5.85$

149. ***(c)*** For the two lines $24x + 7y - 20 = 0$ and $4x - 3y - 2 = 0$, the angle bisectors are given by

$$\frac{24x + 7y - 20}{25} = \pm\frac{4x - 3y - 2}{5}$$

Taking positive sign, we get $2x + 11y - 5 = 0$

So, the given three lines are concurrent with one line bisecting the angle between the other two.

150. ***(d)*** The new position of plane is

$$(x - 2y + 33) + \lambda(x + y + z - 1) = 0$$

$\Rightarrow \quad (1+\lambda)x + (\lambda - 2)y + (3+\lambda)z - \lambda = 0$

Given that this is perpendicular to

$$x + y + z = 1$$

$\therefore \quad (1+\lambda)\cdot 1 + (\lambda - 2)\cdot 1 + (3+\lambda)\cdot 1 = 0$

$\Rightarrow \quad 1 + \lambda + \lambda - 2 + 3 + \lambda = 0$

$\Rightarrow \quad 3\lambda + 2 = 0$

$\Rightarrow \quad \lambda = \frac{-2}{3}$

Hence, the new position of the plane is

$$(x - 2y + 3z) - \frac{2}{3}(x + y + z - 1) = 0$$

$\Rightarrow \quad 3x - 2x - 6y - 2y + 9z - 2z + 2 = 0$

$\therefore \quad x - 8y + 7z = -2$

BITSAT 2014

BITSAT

Birla Institute of Technology & Science Admission Test

Solved Paper 2013*

Instructions

- There are 150 questions in all. The number of questions in each part is as given below.

	No. of Questions
Part I Physics	1-40
Part II Chemistry	41-80
Part III a. English Proficiency	81-95
b. Logical Reasoning	96-105
Part IV Mathematics	106-150

- All questions are Multiple Choice Questions having four options out of which **only one** is correct.
- Each correct answer fetches 3 marks while incorrect answer has a penalty of 1 mark.
- Time allotted to complete this paper is 3 hrs.

PART I

Physics

1. A square shape current loop of side length l and carrying current I lies in a uniform magnetic field B acting perpendicular to the plane of square loop and directed inward. The net magnetic force acting on current loop is

a. lBL *b.* $4IBL$ *c.* zero *d.* $2IBL$

2. Two parallel conductors carry current in opposite directions, as shown in figure. One conductor carries a current of 10.0 A. Point C is a distance $d/2$ to the right of the 10.0 A current. If $d = 18$ cm and l is adjusted so that the magnetic field at C is zero, the value of the current I is

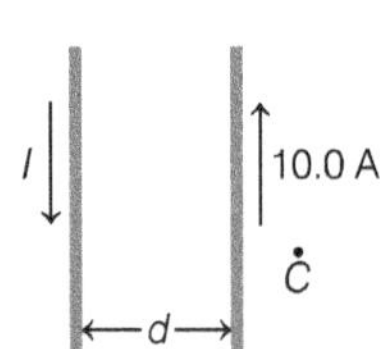

a. 10.0 A *b.* 30.0 A
c. 8.0 A *d.* 18.0 A

3. Two long, parallel conductors carry currents in the same direction, as shown in figure. Conductor A carries a current of 100 A and is held firmly in position. Conductor B carries a current I_B and is allowed to slide freely up and down (parallel to A) between a set of non-conducting guides. The mass per unit length of conductor B is 0.1 g/cm and the distance between the two conductors is 5 cm. If system of conductors is in equilibrium, then the value of current I_B is

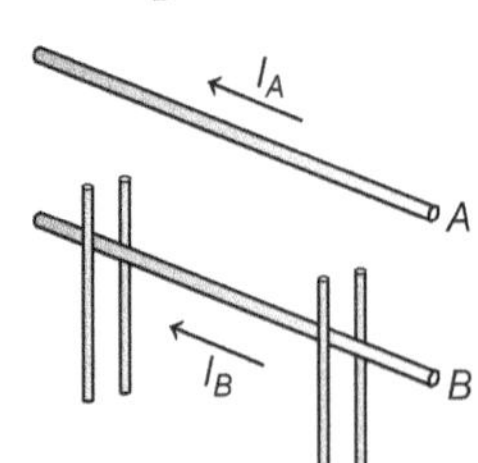

a. 250 A *b.* 240 A *c.* 220 A *d.* 230 A

*Memory Based

4. The number of photo electrons in a photoelectric effect experiment depends on the

a. frequency of light *b.* intensity of light
c. Both (a) and (b) *d.* Neither (a) nor (b)

5. In hydrogen atom, if $\lambda_1, \lambda_2, \lambda_3$ are shortest wavelengths in Lyman, Balmer and Paschen series respectively, then $\lambda_1 : \lambda_2 : \lambda_3$ equals

a. $1:4:9$ *b.* $9:4:1$
c. $1:2:3$ *d.* $3:2:1$

6. Half-lives of elements *A* and *B* are 1 h and 2 h, respectively. Which one of the following is correct?

a. Element *A* decays slower
b. Decay constant of *A* is smaller
c. If initial number of nuclei are same, then activity of *A* is more
d. Mean life of *A* is more

7. A glass piece is dipped in a liquid of refractive index 4/3, it gets dissappeared in the liquid. The refractive index of the glass piece is

a. $\frac{3}{4}$ *b.* $\frac{5}{3}$ *c.* $\frac{4}{5}$ *d.* $\frac{4}{3}$

8.

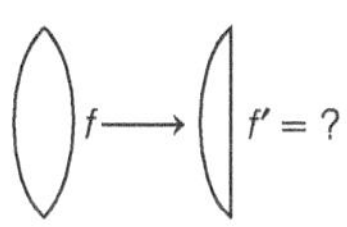

If the bio-convex lens is cut as shown in the figure, then the new focal length f' is

a. $2f$ *b.* f *c.* $f/2$ *d.* infinite

9. Refractive index of a medium depends

a. on the medium only
b. on the incident light only
c. Both (a) and (b)
d. None of the above

10. A point object is placed at the focus of a convex mirror. The image will be formed at

a. infinity *b.* centre of curvature
c. at focus itself *d.* None of these

11.

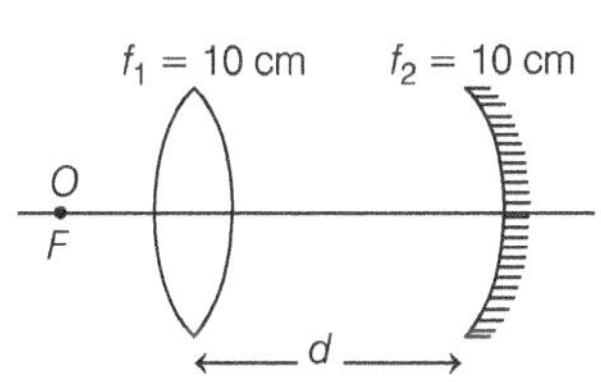

A point object is placed at the focus of the bi-convex lens. What should be the value of *d*, so the final image forms at infinity?

a. 10 cm *b.* 20 cm
c. 15 cm *d.* None of these

12. The image formed by a concave spherical mirror

a. is always virtual
b. is always real
c. is always inverted
d. may be erect

13. The total energy of a revolving satellite around the earth is $-k$ J. The minimum energy required to throw it out of earth's gravitational fields, is

a. k J *b.* $\frac{k}{2}$ J
c. $2k$ J *d.* None of these

14.

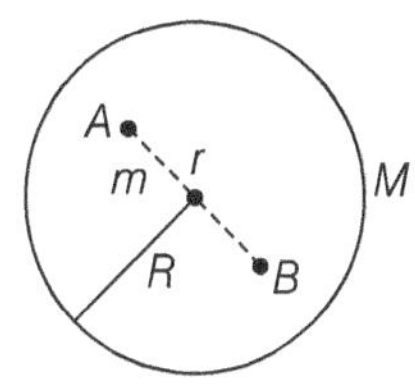

There is a shell of mass *M* and density of shell is uniform. The work done to take a point mass from point *A* to *B* is ($AB = r$)

a. $\frac{GmM}{r}$ *b.* $\frac{GmM}{R}$
c. $-\frac{GmM}{r}$ *d.* zero

15. A body of mass $m = 20$ g is attached to an elastic spring of length $L = 50$ cm and spring constant $k = 2\ \text{Nm}^{-1}$. The system is revolved in a horizontal plane with a frequency $v = 30$ rev/min. Find the radius of the circular motion and the tension in the spring.

a. 0.25 m, 0.1 N
b. 0.5 m, 0.52 N
c. 0.55 m, 0.1 N
d. 0.9 m, 0.2 N

16. A gramophone record of mass *M* and radius *R* is rotating at an angular velocity ω. A coin of mass *m* is gently placed on the record at a distance $r = R/2$ from its centre. The new angular velocity of the system is

a. $\frac{2\omega M}{(2M + m)}$ *b.* $\frac{2\omega M}{(M + 2m)}$
c. ω *d.* $\frac{\omega M}{m}$

17. A block of mass $m = 1$ kg is placed over a plank *Q* of mass $M = 6$ kg, placed over a smooth horizontal surface as shown in figure. Block *P* is given a velocity $v = 2\ \text{m/s}^2$ to the right.

If the coefficient of friction between P and Q is $\mu = 0.3$. Find the acceleration of Q relative to P.

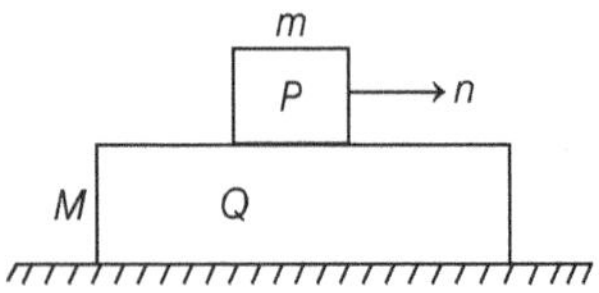

a. $4\ m/s^2$ *b.* $3.5\ m/s^2$
c. $2\ m/s^2$ *d.* $10.0\ m/s^2$

18. A man runs at a speed of 4 m/s to overtake a standing bus. When he is 6 m behind the door at $t = 0$, the bus moves forward and continuous with a constant acceleration of $1.2\ m/s^2$. The man reaches the door in time t. Then

a. $4t = 6 + 0.6\ t^2$
b. $1.2\ t^2 = 4t$
c. $4t^2 = 1.2\ t$
d. $6 + 4t = 0.2\ t^2$

19. In completly inelastic collision

a. the complete KE of the medium must lost
b. the linear momentum of the system must remain conserved during collision
c. Both (a) and (b)
d. Neither (a) nor (b)

20. The number of particles per unit volume is given by $n = -\dfrac{D(n_2 - n_1)}{x_2 - x_1}$ are crossing a unit area perpendicular to X-axis in unit time, when n_1 and n_2 are the number of particles per unit volume for the values x_1 and x_2 of x, respectively. Then, the dimensional formula of diffusion constant D is

a. $[LT^0]$ *b.* $[L^2T^{-4}]$
c. $[LT^{-3}]$ *d.* $[L^2T^{-1}]$

21. In the given circuit (as shown in figure). Each capacitor has a capacity of 3 μF. What will be the net charge on each capacitor?

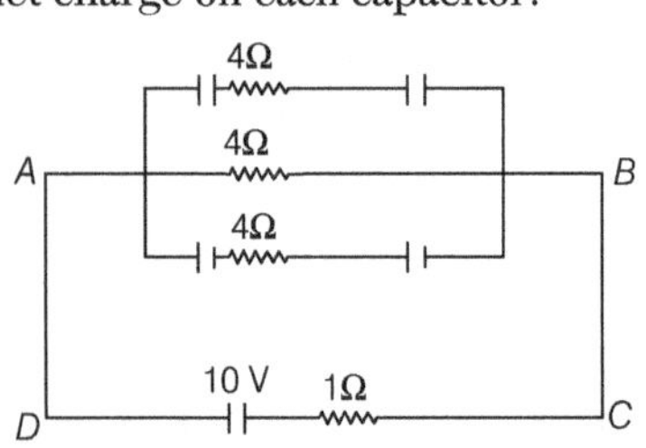

a. 48 μC *b.* 24 μC
c. 12 μC *d.* None of these

22. A solid conductor is placed in an uniform electric field as shown in figure. Which path will the lines of force follow?

a.
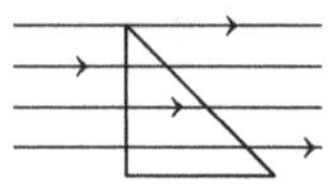
b.
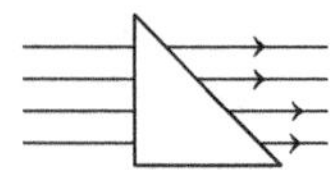
c.
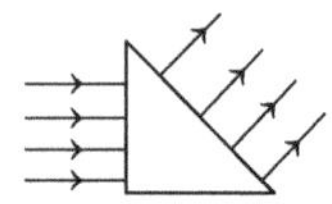
d.
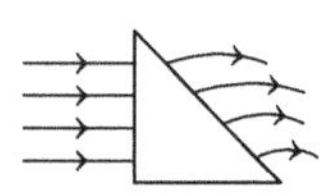

23. A bomb at rest explodes into three parts of the same mass. The linear momentum of two parts are $-2P\,\hat{i}$ and $P\,\hat{j}$. The magnitude of momentum of third part is $P\sqrt{x}$. Find x.

a. 1 *b.* 5 *c.* 2 *d.* 10

24. Block is placed on an inclined plane. The block is moving towards right horizontally with an acceleration $a_0 = g$. The length of the inclined plane (AC) is equal to 1 m. Whole the situation are shown in the figure. Assume that, all the surfaces are frictionless. The time taken by the block to reach from C to A is (Take, $g = 10\ m/s^2$)

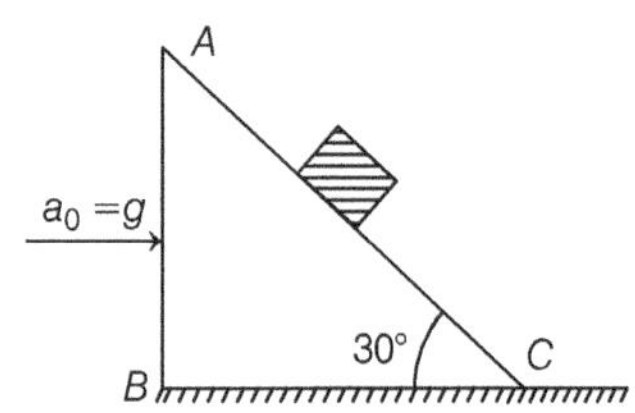

a. 0.74 s *b.* 0.9 s *c.* 0.52 s *d.* 1.24 s

25. Pseudo force is

a. electromagnetic in nature *b.* a nuclear force
c. a gravitational force *d.* None of these

26. A light in extensible string that goes over a smooth fixed pulley as shown in the figure connect two blocks of masses 0.36 kg and 0.72 kg. Taking, $g = 10\ m/s^2$. Find the work done by string on the block of mass 0.36 kg during the first second after the system is released from rest.

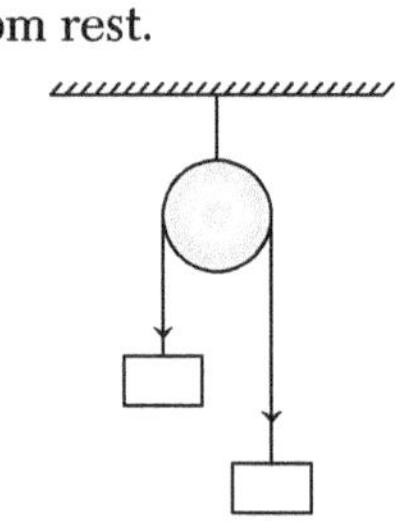

a. 4 J *b.* 2 J *c.* 8 J *d.* 10 J

27. In the equation $A = 3BC^2$, A and C have dimensions of capacitance and magnetic induction respectively. In MKS system, the dimensional formula of B is

a. $[M^{-3}L^{-2}T^{-2}Q^{-4}]$ *b.* $[ML^{-2}]$
c. $[M^{-3}L^{-2}Q^{4}T^{8}]$ *d.* $[M^{-3}L^{-2}Q^{4}T^{4}]$

28. Infinite number of masses each 1 kg are placed along the x-axis at $x = \pm 1$ m, ± 2 m, ± 4m, ± 8m, ± 16m ...... The magnitude of the resultant gravitational potential in terms of gravitational constant G at the origin ($x = 0$) is

a. $G/2$ *b.* G *c.* $2G$ *d.* $4G$

29. Three bulbs X, Y and Z are connected as shown in figure. The bulbs Y and Z are identical. If bulb Z gets fused then,

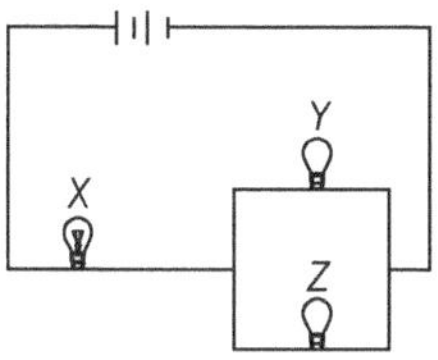

a. Both X and Y will glow more brightly
b. Both X and Y will glow less brightly
c. X will glow less brightly and Y will glow more brightly
d. X will glow more brightly and Y will glow less brightly

30. Active state of *n-p-n* transistor, in circuit is achieved by

a. low input voltage *b.* high input voltage
c. Both (a) and (b) *d.* Neither (a) nor (b)

31. A turntable of radius $R = 10$ m is rotation making 98 rev in 10 s with a boy of mass $m = 60$ kg standing at its centre. He starts running along a radius. Find the frequency of the turntable when the boy is 4 m from the centre. The moment of inertia of the turntable about its axis 1000 kg-m^2.

a. 10 Hz *b.* 2.5 Hz *c.* 5 Hz *d.* 4 Hz

32. To transmit a signal, if height of transmitting signal above surface of the earth is H, this signal can be received on surface of the earth upto distance d from transmitter. Then

a. $d \propto H$ *b.* $d \propto H^2$ *c.* $d \propto H^{1/2}$ *d.* $d \propto H^{3/2}$

33. The circuit is equivalent to

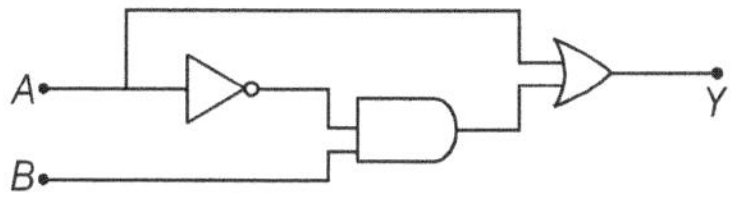

a. AND gate *b.* OR gate
c. Not gate *d.* None of these

34. Length of 20 cm (exact) long pipe is measured by two instruments and reported as 19.65 cm and 20.1 cm

a. 19.65 cm is more accurate
b. both measurements are equally precise
c. both measurements are equally accurate
d. 20.1 cm is less precise

35. An electric pump on the ground floor of a building takes 10 min to fill a tank of volume 2000 L with water. If the tank is 40 m above the ground and the efficiency of the pump is 40%, how much electric power is consumed by the pump in filling the tank?
[Take, $g = 10$ m/s^2]

a. 2 kW *b.* 3.33 kW
c. 4 kW *d.* 6 kW

36. A vessel containing 1 mole of O_2 gas (molar mass 32) at temperature T. The pressure of the gas is p. An identical vessel containing one mole of He gas (molar mass 4) at temperature $2T$ has a pressure of

a. $p/8$ *b.* p
c. $2p$ *d.* $8p$

37. The temperature of an ideal gas is increased from 27°C to 127°C, then percentage increase in v_{rms} is

a. 37 % *b.* 11%
c. 33% *d.* 15.5%

38. A particle of mass $m = 5$ g is executing simple harmonic motion with an amplitude 0.3 m and time period $\pi/5$ s. The maximum value of force acting on the particle is

a. 5 N *b.* 4 N
c. 0.5 N *d.* 0.15 N

39. A partition wall has two layers of different materials A and B in contact with each other. They have the same thickness but the thermal conductivity of layer A is twice that of B. At steady state if the temperature difference across the layer B is 50 K, then the corresponding difference across the layer A is

a. 50 K *b.* 12.5 K
c. 25 K *d.* 60 K

40. Pulse rate of a normal person is 75 per min. The time period of heart is

a. 0.8 s *b.* 0.75 s *c.* 1.25 s *d.* 1.75 s

PART II

Chemistry

41. For the properties mentioned, the correct trend for the different species is in

a. strength of Lewis acid—$BCl_3 > AlCl_3 > GaCl_3$
b. inert pair effect— Al> Ga> In
c. oxidising property— $Al^{3+} > In^{3+} > Tl^{3+}$
d. first ionisation enthalpy— B> Al> Tl

42. Bohr's theory is applicable to

a. He *b.* Li^{2+}
c. He^{2+} *d.* None of these

43. Using MOT, which of the following pair denotes paramagnetic species

a. B_2 and C_2 *b.* B_2 and O_2
c. N_2 and C_2 *d.* O_2 and O_2^{2-}

44. 0.1 g of metal combines with 46.6 mL of oxygen at STP. The equivalent weight of metal is

a. 12 *b.* 24
c. 18 *d.* 36

45. Which of the following choice represent correct order of first ionisation enthalpy ?

a. $B < C < N < O < F$ *b.* $B > C > N > O > F$
c. $B < C < N > O < F$ *d.* $B < C < N > O > F$

46. Which of the following reactant produces most stable alkene on treatment with base?

a. 2-chlorobutane
b. 2, 3-dichlorobutane
c. 2, 2-dichlorobutane
d. 2, 3-dichloro-2, 3-dimethylbutane

47. Which of the following is more acidic among the given halogen compounds?

a. CHF_3 *b.* CHI_3 *c.* $CHCl_3$ *d.* $CHBr_3$

48. What will be the final product of the reaction ?

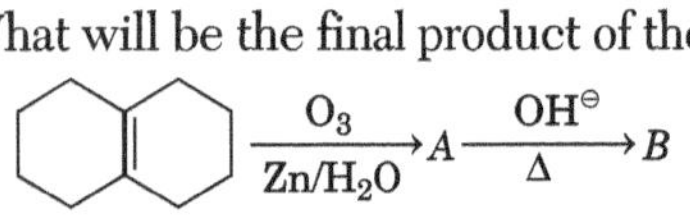

a. 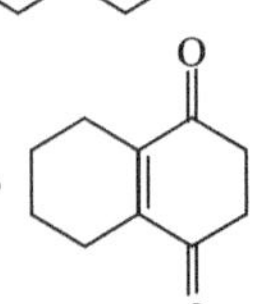 *b.* 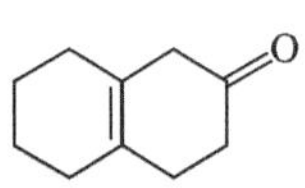

c. 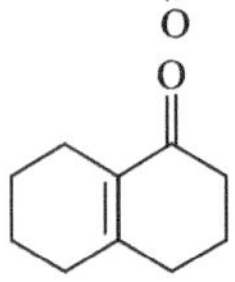 *d.* 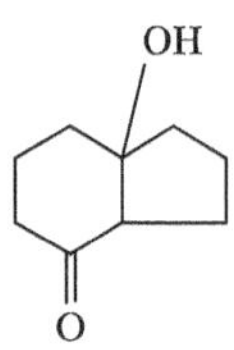

49. The vapour pressure of a solvent decreased by 10 mm of Hg when a non-volatile solute was added to the solvent. The mole fraction of solute in solution is 0.2, what would be the mole fraction of solvent, if the decrease in vapour pressure is 20 mm of Hg ?

a. 0.8 *b.* 0.6 *c.* 0.4 *d.* 0.3

50. Choose the law that corresponds to data shown for the following reaction, $A + B \rightarrow$ products

Exp.	[A]	[B]	Initial rate
1	0.012	0.035	0.1
2	0.024	0.070	0.8
3	0.024	0.035	0.1
4	0.012	0.070	0.8

a. Rate $= k[B]^3$ *b.* Rate $= k[B]^4$
c. Rate $= k[A][B]^3$ *d.* Rate $= k[A]^3[B]$

51. The magnitude of Δ_o will be highest in which of the following complex?

a. $[Cr(CN)_6]^{3-}$ *b.* $[Cr(H_2O)_6]^{3+}$
c. $[Cr(NH_3)_6]^{3+}$ *d.* $[Cr(C_2O_4)_3]^{3-}$

52. Arrange these in correct order of decreasing reactivity.

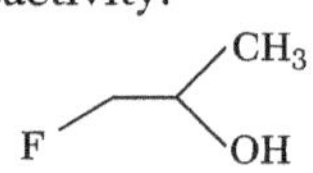

I

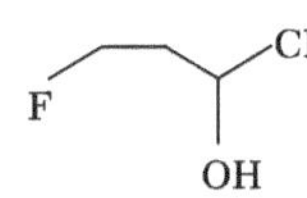

II

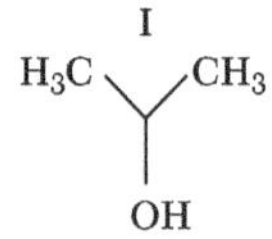

III

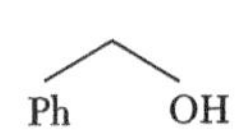

IV

a. I > II > III > IV *b.* I > III > II > IV
c. IV > III > II > I *d.* IV > III > I > II

53. When 2-methyl propan-1-ol is treated with a mixture of conc. HCl and $ZnCl_2$, turbidity appears immediately due to the formation of

a. 2-methyl propane
b. 2-methyl propene
c. 2-chloro-2-methyl propane
d. 2-chlorobutane

54. Gastric juice in human stomach has pH value about 1.8 and pH of small intestine is about 7.8. The pK_a value of aspirin is 3.5. Aspirin will be
a. ionised in the small intestine and stomach
b. ionised in the stomach and almost unionised in the small intestine
c. unionised in small intestine and stomach
d. completely ionised in small intestine and stomach

55. When a solution of potassium chromate is treated with an excess of dilute nitric acid
a. Cr^{3+} and $Cr_2O_7^{2-}$ are formed
b. $Cr_2O_7^{2-}$ and H_2O are formed
c. CrO_4^{2-} reduced to Cr^{3+}
d. CrO_4^{2-} oxidised to $Cr_2O_7^{2-}$ only

56. Calcium carbide reacts with heavy water to form
a. C_2D_2 *b.* CaD_2
c. CaD_2O *d.* CD_2

57. Fluorine acts as strongest oxidising agent because of its high
a. electron affinity *b.* ionisation enthalpy
c. hydration enthalpy *d.* bond enthalpy

58. The reaction of P_4 with X leads selectively to P_4O_6. The X is
a. dry O_2
b. moist O_2
c. mixture of O_2 and N_2
d. O_2 in presence of aqueous NaOH

59. The acidic strength for the hydrides of group 15 follows the order
a. $NH_3 > PH_3 > AsH_3 > SbH_3$
b. $NH_3 < PH_3 < AsH_3 < SbH_3$
c. $NH_3 > PH_3 > SbH_3 > AsH_3$
d. $NH_3 < PH_3 < SbH_3 < AsH_3$

60. Which of the following statements are incorrect in context of borax ?
a. It is made up of two triangular BO_3 units and two tetrahedral BO_4 units
b. One mole of borax can be used as a buffer
c. It is a useful primary standard for titrating against acids
d. Aqueous solution of borax can be used as buffer

61. Salt $(A) + S \longrightarrow B \xrightarrow{BaCl_2}$ white precipitate
A is paramagnetic in nature and contains about 55% K. Thus, A is
a. K_2O *b.* K_2O_2
c. KO_2 *d.* K_2SO_4

62. When equal volume each of two sols of AgI, one obtained by adding $AgNO_3$ to slight excess of KI and another obtained by adding KI to slight excess of $AgNO_3$ are mixed together. It is observed that
a. the sol particles acquired more electric charge
b. the sols coagulated each other mutually
c. a true solution is obtained
d. the two sols stabilised each other

63. In the extraction of Ag, Zn is removed from (Zn-Ag) alloy through
a. cupellation *b.* fractional crystallisation
c. distillation *d.* electrolytic refining

64. A reaction takes place in three steps. The rate constants are k_1, k_2 and k_3. The overall rate constant $k = \frac{k_1 k_3}{k_2}$. If E_1, E_2 and E_3 (energy of activation) are 60, 30 and 10 kJ, respectively, the overall energy of activation is
a. 40 *b.* 30 *c.* 400 *d.* 300

65. If $E^\circ_{Fe^{3+}/Fe}$ and $E^\circ_{Fe^{2+}/Fe}$ are -0.36 V and -0.439 V respectively, then the value of $E^\circ_{Fe^{3+}/Fe^{2+}}$ is
a. $(-0.036 - 0.439)$ V
b. $[3(-0.36) + 2(-0.439)]$ V
c. $(-0.36 - 0.439)$ V
d. $[3(-0.36) - 2(-0.439)]$ V

66. KCl crystallises in the same type of lattice as does NaCl. Given that, $r_{Na^+}/r_{Cl^-} = 0.55$ and $r_{K^+}/r_{Cl^-} = 0.74$. Determine the ratio of the side of the unit cell for KCl to that of NaCl.
a. 0.124 *b.* 1.123 *c.* 0.891 *d.* 1.414

67. The compound formed as a result of oxidation of propyl benzene by $KMnO_4$ is
a. benzaldehye *b.* benzyl alcohol
c. benzoic acid *d.* acetophenone

68. Which of the following is an outer *d*-orbital or high spin complex?
a. $[Co(NH_3)_6]^{3+}$ *b.* $[Ni(CN)_4]^{2-}$
c. $[NiCl_4]^{2-}$ *d.* $[CoF_6]^{3-}$

69. The monosaccharide having anomeric carbon atoms are
a. geometrical isomers
b. α and β-optical isomers
c. having symmetrical carbon atoms
d. None of the above

70. Amine is not formed in the reaction

I. Hydrolysis of RCN

II. Reduction of $RCH{=}NOH$

III. Hydrolysis of RNC

IV. Hydrolysis of $RCONH_2$

The correct answer is

a. I, II and IV *b.* I and IV

c. II and III *d.* I, II and III

71. In vulcanisation of rubber

a. sulphur reacts to form a new compound

b. sulphur cross-links are introduced

c. sulphur forms a very thin protective layer on rubber

d. All of the above

72. What will be the correct structural formula of product for the following reaction ?

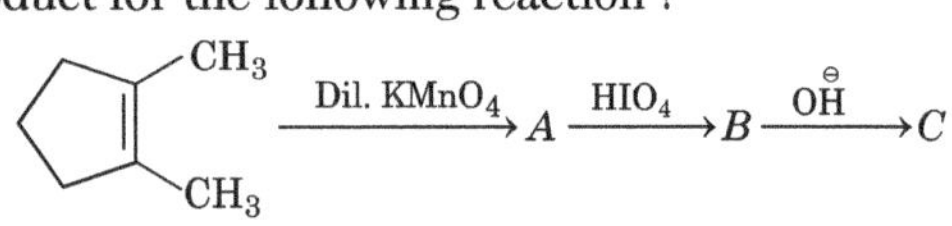

a.

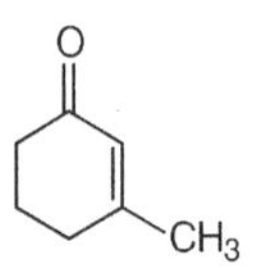

b.

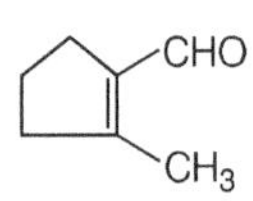

c.

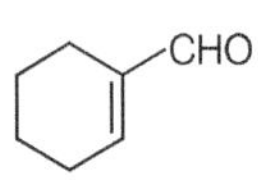

d.

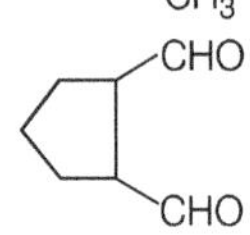

73. What will be the correct relation between product when 2-methyl cyclohexene is treated with (i) B_2H_6 in presence of H_2O_2/OH and (ii) H_2O/H_2SO_4 (also consider stereochemistry of product)?

a. They are metamers

b. They are tautomers

c. They are functional isomers

d. They are positional isomers

74. The equilibrium constant K_p for the reaction, $N_2(g) + 3H_2(g) \rightleftharpoons 2NH_3(g)$ is $1.6 \times 10^{-4}\,(\text{atm})^{-2}$ at 400°C. What will be the equilibrium constant at 500°C, if heat of the reaction in this temperature range is -25.14 kcal ?

a. $1.231 \times 10^{-4}\,(\text{atm})^{-2}$

b. $1.876 \times 10^{-7}\,(\text{atm})^{-2}$

c. $1.462 \times 10^{-5}\,(\text{atm})^{-2}$

d. $3.462 \times 10^{-5}\,(\text{atm})^{-2}$

75. At 27°C, K_p value for reaction

$CaCO_3(s) \rightleftharpoons CaO(s) + CO_2(g)$ is 0.1 atm

The K_C value for this reaction is

a. 4×10^{-3} *b.* 6×10^{-3}

c. 2×10^{-3} *d.* 9×10^{-3}

76. At constant temperature and pressure which one of the following statements is correct for the reaction ?

$$CO(g) + \frac{1}{2}O_2(g) \longrightarrow CO_2(g)$$

a. $\Delta H = \Delta E$

b. $\Delta H < \Delta E$

c. $\Delta H > \Delta E$

d. ΔH is independent of physical state of reactant

77. IUPAC name and degree of unsaturation of compound X is

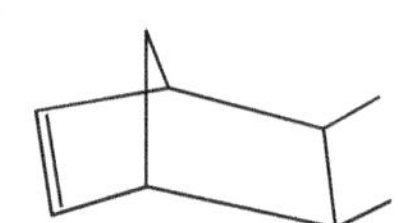

X

a. 2, 3-dimethyl bicyclo [2,2,1] hept-5 ene, 2

b. 1, 2-dimethyl bicyclo [2, 2, 1] hept-4 ene, 3

c. 5, 6-dimethyl bicyclo [2, 2, 1] hept-2 ene, 3

d. 4, 5-dimethyl bicyclo [2, 2, 1] hept-1 ene, 2

78. The oxidation state of sulphur in $Na_2S_4O_6$ is

a. $+6$ *b.* $+\frac{3}{2}$

c. $+\frac{5}{2}$ *d.* -2

79. Which of the following antibiotic contains nitro group attached to aromatic nucleus in its structure

a. tetracyclin *b.* penicillin

c. streptomycin *d.* chloramphenicol

80. The behaviour of the gas becomes more ideal at

I. very low pressure

II. value of Z is unity

III. very high pressure

IV. value of Z is greater than one

Choose the correct option.

a. I and II are correct

b. I and IV are correct

c. I and III are correct

d. III and IV are correct

PART III

a. English Proficiency

Directions (Q.Nos. 81-83) *Out of the four alternatives, choose the one which expresses the right meaning of the word.*

81. Sagacious

a. Shameless *b.* Wise
c. Powerless *d.* Foolish

82. Remedial

a. Corrective *b.* Proficient
c. General *d.* Optional

83. Reticent

a. Confident *b.* Sad
c. Truthful *d.* Secretive

Directions (Q. Nos. 84-86) *Choose the word apposite in meaning to the given word.*

84. Fidelity

a. Faith
b. Devotedness
c. Allegiance
d. Treachery

85. Infrangible

a. Complecated *b.* Breakable
c. Weird *d.* Software

86. Progeny

a. Kid *b.* Parent
c. Friend *d.* Enemy

Directions (Q. Nos. 87-89) *A part of sentence is underlined. Balance are given alternatives to the underlined part a, b, c and d which may improve the sentence. Choose the correct alternative.*

87. It was not possible to drag any conclusion so he left the case.

a. Fetch
b. Find
c. Draw
d. No improvement

88. I am looking after my pen which is missing.

a. Looking for
b. Looking in
c. Looking back
d. No improvement

89. "Mind your language !" he shouted.

a. change
b. inspect
c. hold
d. No improvement

Directions (Q. Nos. 90-92) *Sentence Completion*

90. I ...... to go there when I was student.

a. liked *b.* used
c. prefer *d.* denied

91. She was angry ...... me.

a. at *b.* about
c. with *d.* in

92. You should not laugh ...... the poor.

a. on *b.* at
c. with *d.* over

Directions (Q. Nos. 93-95) *Sentence rearrangement*

93. 1. He is a famous doctor.
P. Once I had to consult with him.
Q. I never believed him.
R. He suggested me a proper remedy.
S. I become completely fine.
6. Now I also admit this fact.

a. P Q R S *b.* Q P S R
c. Q P R S *d.* R Q S P

94. 1. We don't know the plan of Ram.
P. He cares for his friends.
Q. He is a complete person.
R. We want some help and advice.
S. As we are in a trouble.
6. We hope he will do his best for us.

a. P R S Q *b.* Q P R S
c. P Q R S *d.* P S R Q

95. 1. It is not my problem.
P. All residents of this society are careless.
Q. I am unable to convince anyone.
R. They don't want to do some good.
S. Every one seems to be unwise here.
6. We all have to suffer one day.

a. P R S Q *b.* P R Q S
c. P Q R S *d.* P S R Q

b. Logical Reasoning

96. In a certain code language 'DOME' is written as '8943' and 'MEAL' is written as '4321'. What group of letters can be formed for the code '38249'?

a. EOADM *b.* MEDOA
c. EMDAO *d.* EDAMO

97. Find the missing number from the given response.

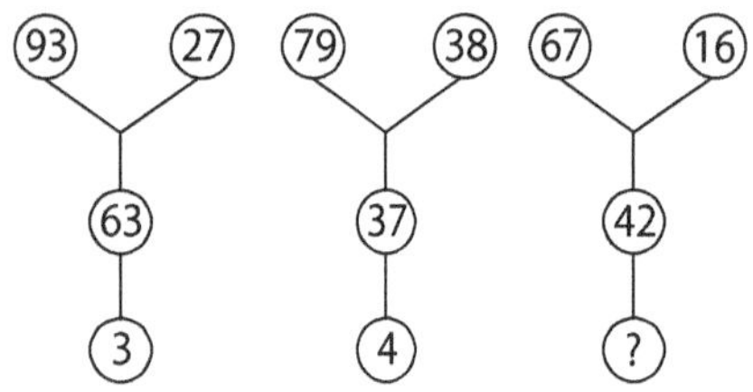

fig. 1 fig. 2 fig. 3

a. 5 *b.* 6
c. 8 *d.* 9

98. Which of the following correctly represents the relationship among illiterates, poor people and unemployed?

a. 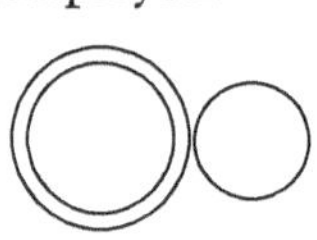 *b.* 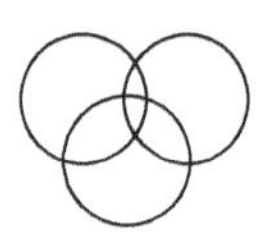

c. 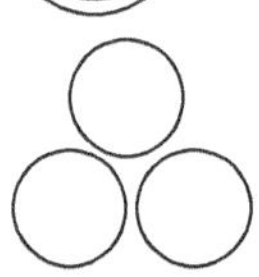 *d.* 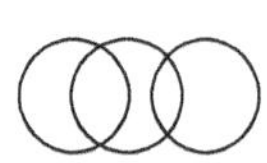

99. 'A' starts crossing the field diagonally from North-West. After walking half the distance, he turns right, walks some distance and turns left. Which direction is 'A' facing now?

a. North-East *b.* North-West
c. South-East *d.* South-West

100. In a classroom, there are 5 rows and 5 children A, B, C, D and E are seated one behind the other in 5 separate rows as follows

- A is sitting behind C but in front of B.
- C is sitting behind E and A is sitting in front of B.
- C is sitting behind E and D is sitting in front of E.
- The order in which they are sitting from the first row to the last is

a. D E C A B *b.* B A C E D
c. A C B D E *d.* A B E D C

101. Which of the following will fill the series?

$$2, 9, 28, \underline{?}, 126$$

a. 64 *b.* 65
c. 72 *d.* 56

102. Which one of the following figures completes the original figure?

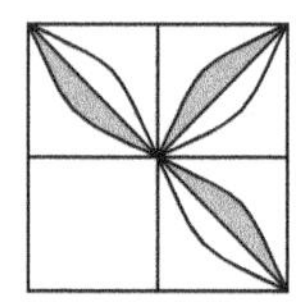

a. 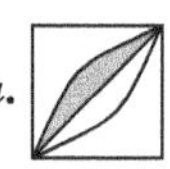 *b.* 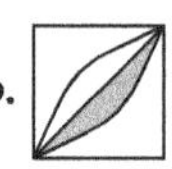 *c.* 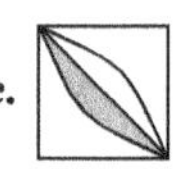 *d.* 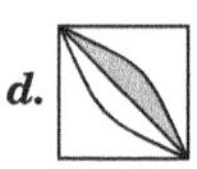

103. How many squares are there in the following figure?

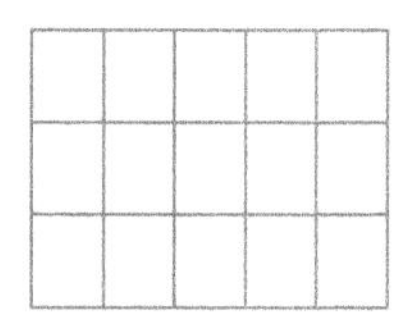

a. 24 *b.* 25
c. 26 *d.* 27

104. Two signs in the equations have been interchanged, find out the two signs to make equation correct.

$$3 \div 5 \times 8 + 2 - 10 = 13$$

a. + and − *b.* × and ÷
c. ÷ and − *d.* ÷ and +

105. **Assertion** [A] = India is a democratic country.
Reason [R] = India has a constitution of its own.

Choose the correct alternative from the given options.

a. Both (A) and (R) are true and (R) is correct explanation of (A).
b. Both (A) and (R) are true but (R) is not the correct explanation of (A).
c. (A) is true (R) is false.
d. (A) is false (R) is true.

PART IV

Mathematics

106. The value of $\sum_{k=1}^{6}\left(\frac{\sin 2\pi k}{7}-\frac{i \cos 2\pi k}{7}\right)$ is

a. -1 *b.* 0
c. $-i$ *d.* i

107. The mean life of a sample of 60 bulbs was 650 and the standard deviation was 8 h. A second sample of 80 bulbs has a mean life of 660 h and standard deviation 7 h. Find the over all standard deviation.

a. 8.97
b. 8.98
c. 8.94
d. None of the above

108. Let R be the relation on the set R of all real numbers, defined by aRb iff $|a-b|\le 1$. Then, R is

a. reflexive and symmetric only
b. reflexive and transitive only
c. equivalence
d. None of the above

109. The value of $\int_0^{10\pi}([\sec^{-1}x]+[\cot^{-1}x])\,dx$, where $[.]$ denotes the greatest integer function, is

a. $10\pi-\tan^{-1}x$ *b.* $8\pi-\sec 1$
c. $10\pi-\sec 1$ *d.* $10\pi+\sec 1$

110. The value of the expression

$\sin[\cot^{-1}(\cos(\tan^{-1}1))]$ is

a. 0 *b.* 1
c. $\frac{1}{\sqrt{3}}$ *d.* $\sqrt{\frac{2}{3}}$

111. The sum of the series

$1+2\cdot 2+3\cdot 2^2+4\cdot 2^3+\dots+100\cdot 2^{99}$ is

a. $100\cdot 2^{100}+1$ *b.* $99\cdot 2^{100}+1$
c. $99\cdot 2^{99}-1$ *d.* $100\cdot 2^{100}-1$

112. The shaded region given below represents the constraints (other than $x\ge 0, y\ge 0$)

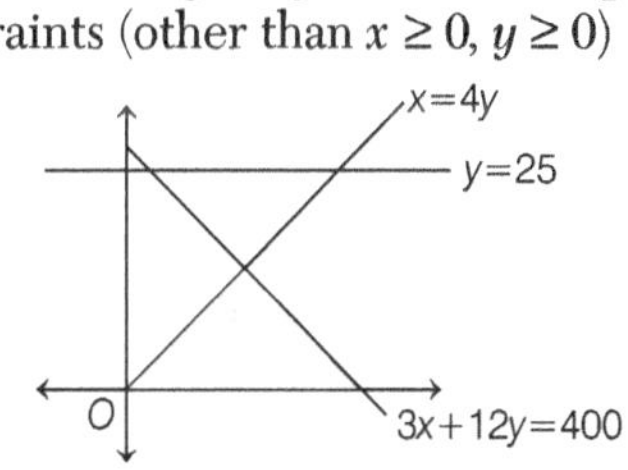

a. $3x+4y\le 400,\ y\le 25,\ x\le 4y$
b. $3x+12y\ge 400,\ y\le 25,\ x\ge 4y$
c. $3x+12y\le 400,\ y\le 25,\ x\ge 4y$
d. None of the above

113. The coefficient of x^n in the expansion of $\log_e\left(\frac{1}{1+x+x^2+x^3}\right)$, when n is odd, is

a. $-\frac{2}{n}$ *b.* $-\frac{1}{n}$
c. $\frac{1}{n}$ *d.* None of these

114. The maximum value of $f(x)=\frac{\log x}{x}$ is

a. 1 *b.* $\frac{2}{e}$
c. e *d.* $\frac{1}{e}$

115. Let **a**, **b** and **c** be non-zero vectors such that no two are collinear and $(\mathbf{a}\times\mathbf{b})\times\mathbf{c}=\frac{1}{3}|\mathbf{b}||\mathbf{c}|\mathbf{a}$. If θ is the acute angle between the vectors **b** and **c**, then $\sin\theta$ is equal to

a. $\frac{2\sqrt{2}}{3}$ *b.* $\frac{\sqrt{2}}{3}$
c. $\frac{2}{3}$ *d.* $\frac{1}{3}$

116. The value of $\lim_{x\to 0}\left(\frac{1+5x^2}{1+3x^2}\right)^{1/x^2}$ is

a. e^2 *b.* e
c. $1/e$ *d.* $1/e^2$

117. An object is observed from the points A, B and C lying in a horizontal straight line which passes directly underneeth the object. The angular elevation at B is twice that at A and at C three times that at A. If $AB=a$, $BC=b$, then the height of the object is

a. $\frac{3a}{2b}\sqrt{(a+b)(3b-a)}$
b. $\frac{3b}{2a}\sqrt{(a+b)(3a-b)}$
c. $\frac{a}{2b}\sqrt{(a+b)(3b-a)}$
d. None of the above

118. Function $f:(-\infty, -1] \to (0, e^5]$ defined by $f(x) = e^{x^3 - 3x + 2}$ is

a. many-one and onto *b.* many-one and into
c. one-one and onto *d.* one-one and into

119. The foci of the conic section $25x^2 + 16y^2 - 150x = 175$ are

a. $(0, \pm 3)$ *b.* $(0, \pm 2)$
c. $(3, \pm 3)$ *d.* $(0, \pm 1)$

120. The system of equations

$$x - y + 3z = 4$$
$$x + z = 2$$
$$x + y - z = 0$$ has

a. a unique solution
b. finitely many solution
c. infinitely many solutions
d. None of the above

121. The sum of the sequence 5, 55, 555, ... upto n infinite terms is

a. $\frac{5}{9}\left[\frac{10(10^n - 1) + n}{9}\right]$

b. $\frac{5}{9}\left[\frac{10(10^n - 1)}{9} - n\right]$

c. $\frac{5}{9}\left[\frac{10(10^{n+1} - 1)}{9} - n\right]$

d. $\frac{5}{9}\left[\frac{10(10^{n-1} - 1)}{9} - n\right]$

122. A plane passes through the point $(1, -2, 3)$ and is parallel to the plane $2x - 2y + z = 0$. The distance of the point $(-1, 2, 0)$ from the plane is

a. 2 *b.* 3 *c.* 4 *d.* 5

123. The distance between the pair of lines represented by the equation $x^2 - 6xy + 9y^2 + 3y - 9y - 4 = 0$ is

a. $\frac{15}{\sqrt{10}}$ *b.* $\frac{1}{2}$

c. $\sqrt{\frac{5}{2}}$ *d.* $\frac{1}{\sqrt{10}}$

124. If $A = \{x \in \mathbb{C} : x^4 - 1 = 0\}$

$B = \{x \in \mathbb{C} : x^2 - 1 = 0\}$

$C = \{x \in \mathbb{C} : x^2 + 1 = 0\}$

where $\mathbb{C}$ is complex plane.

a. $A = B \cup C$ *b.* $C = A \cap B$
c. $B = A \cap C$ *d.* $A = B \cap C$

125. The general solution of the differential equation $\frac{dy}{dx} + \sin\left(\frac{x+y}{2}\right) = \sin\left(\frac{x-y}{2}\right)$ is

a. $\log \tan\left(\frac{y}{2}\right) = C - 2\sin x$

b. $\log \tan\left(\frac{y}{4}\right) = C - 2\sin\left(\frac{x}{2}\right)$

c. $\log \tan\left(\frac{y}{2} + \frac{\pi}{4}\right) = C - 2\sin x$

d. None of the above

126. The set of all real x satisfying the inequality $\frac{3 - |x|}{4 - |x|} \ge 0$

a. $[-3, 3] \cup (-\infty, -4) \cup (4, \infty)$
b. $(-\infty, -4) \cup (4, \infty)$
c. $(-\infty, -3) \cup (4, \infty)$
d. $(-\infty, -3) \cup (3, \infty)$

127. If N is the any four digit number say x_1, x_2, x_3, x_4, then the maximum value of $\frac{N}{x_1 + x_2 + x_3 + x_4}$ is equal to

a. 1000 *b.* $\frac{1111}{4}$
c. 800 *d.* None of these

128. If A and B are two events such that

$P(A) = 0.6$, $P(B) = 0.2$ and $P\left(\frac{A}{B}\right) = 0.5$, then

$P\left(\frac{A'}{B'}\right)$ equals to

a. $\frac{1}{10}$ *b.* $\frac{3}{10}$ *c.* $\frac{3}{8}$ *d.* $\frac{6}{7}$

129. The quartile deviation for the data

x:	2	3	4	5	6
y:	3	4	8	4	1

is

a. 0 *b.* $\frac{1}{4}$ *c.* $\frac{1}{2}$ *d.* 1

130. If $\int f(x) \cos x \, dx = \frac{1}{2} f^2(x) + C$, then $f(x)$ can be

a. x *b.* 1
c. $\cos x$ *d.* $\sin x$

131. There are 10 points in a plane, out of these 6 are collinear. If 'n' is the number of triangles formed by joining these points, then

a. $n \le 100$ *b.* $100 < n < 140$
c. $140 < n \le 190$ *d.* $n > 190$

132. $\frac{{}^8C_0}{6} - {}^8C_1 + {}^8C_2 \cdot 6 - {}^8C_3 \cdot 6^2 + {}^8C_4 \cdot 6^3 + \ldots + {}^8C_8 \cdot 6^7$ equals to

a. 0 b. 6^7

c. 6^8 d. $\frac{5^8}{6}$

133. A committee of 4 students is selected at random from a group consisting 8 boys and 4 girls. Given that there is atleast one girl on the committee, then the probability that there are exactly 2 girls on the committee is

a. $\frac{68}{125}$ b. $\frac{56}{165}$

c. $\frac{63}{625}$ d. $\frac{168}{425}$

134. What are the values of c for which Rolle's theorem for the function $f(x) = x^3 - 3x^2 + 2x$ in the interval $[0, 2]$ is verified ?

a. $c = \pm 1$ b. $c = 1 \pm \frac{1}{\sqrt{3}}$

c. $c = \pm 2$ d. None of these

135. If $\int \frac{4}{\sin^4 x + \cos^4 x} dx = a \tan^{-1}\left(\frac{\tan x - \frac{1}{\tan x}}{b}\right) + c$, then find the value of a and b, respectively.

a. $2\sqrt{2}, \sqrt{2}$ b. $\sqrt{2}, 2$

c. $\sqrt{3}, \sqrt{2}$ d. $\sqrt{2}, 4$

136. If $A = \begin{bmatrix} -5 & -8 & 0 \\ 3 & 5 & 0 \\ 1 & 2 & -1 \end{bmatrix}$, then A is

a. idempotent b. nilpotent

c. involutory d. periodic

137. The radius of the circle passing through the foci of the ellipse $\frac{x^2}{16} + \frac{y^2}{19} = 1$ and having its centre (0, 3) is

a. 4 b. $\frac{3}{7}$ c. $\sqrt{12}$ d. $\frac{7}{2}$

138. Let L be the line of intersection of the planes $2x + 3y + z = 1$ and $x + 3y + 2z = 2$. If L makes an angle α with the positive X-axis, then $\cos\alpha$ is equal to

a. $1/2$ b. 1 c. $1/\sqrt{2}$ d. $1/\sqrt{3}$

139. If OAB is an equilateral triangle inscribed in the parabola $y^2 = 4ax$ with O as the vertex, then the length of the side of the ΔOAB is

a. $8a\sqrt{3}$ b. $4a\sqrt{3}$

c. $2a\sqrt{3}$ d. $a\sqrt{3}$

140. If $f(x+y) = f(x) \cdot f(y)$ for all real x, y and $f(0) \neq 0$, then function $g(x) = \frac{f(x)}{1 + \{f(x)\}^2}$ is

a. even function b. odd function

c. odd, if $f(x) > 0$ d. neither even nor odd

141. If $f(x) = (\tan^{-1} x)^2 + \frac{2}{\sqrt{x^2+1}}$, then $f(x)$ is increasing in

a. $(0, \infty)$ b. $(-\infty, 0)$

c. $(-\infty, -5)$ d. None of these

142. The number of solutions of $\cos x = |1 + \sin x|$, $0 \le x \le 3\pi$ is

a. 1 b. 2

c. 3 d. 4

143. If a, b, c are in GP and $a^{1/x} = b^{1/y} = c^{1/z}$, then x, y, z are in

a. AP b. GP

c. HP d. None of these

144. The acute angle between the lines, whose direction cosines are given by $2l - m + 2n = 0$, $lm + mn + nl = 0$, is

a. $\frac{\pi}{6}$ b. $\frac{\pi}{4}$

c. $\frac{\pi}{3}$ d. $\frac{\pi}{2}$

145. The equation of the lines through (1, 1) and making angles of 45° with the line $x + y = 0$ are

a. $x - 1 = 0, x - y = 0$

b. $x - y = 0, y - 1 = 0$

c. $x + y - 2 = 0, y - 1 = 0$

d. $x - 1 = 0, y - 1 = 0$

146. The area of the figure bounded by two branches of the curve $(y - x)^2 = x^3$ and straight line $x = 1$ is

a. $\frac{4}{5}$ sq unit b. $\frac{4}{7}$ sq unit

c. $\frac{4}{9}$ sq unit d. $\frac{4}{11}$ sq unit

147. If $(x + iy)^{1/3} = 2 + 3i$, then $3x + 2y$ is equal to

a. -20 b. -60 c. -120 d. 60

148. In a town of 10000 families it was found that 40% family buy newspaper A, 20% buy newspaper B and 10% families buy newspaper C, 5% families buy A and B, 3% buy B and C and 4% buy A and C. If 2% families buy all the three newspaper, then the number of families which buy A only is

a. 3100
b. 3300
c. 2900
d. 1400

149. If $|\mathbf{a}|=2, |\mathbf{b}|=5$ and $|\mathbf{a}\times\mathbf{b}|=8$, then $|\mathbf{a}\cdot\mathbf{b}|$ is equal to

a. 3 *b.* 4 *c.* 5 *d.* 6

150. The equation of circle which passes through the origin and cuts off intercepts 5 and 6 from the positive parts of the X-axis and Y-axis respectively is $\left(x-\frac{5}{2}\right)^2+(y-3)^2=\lambda$, where λ is

a. 61/4 *b.* 4/6 *c.* 1/4 *d.* 0

Answers

Physics

1. (c)	2. (b)	3. (a)	4. (b)	5. (a)	6. (c)	7. (d)	8. (a)	9. (c)	10. (d)
11. (b)	12. (d)	13. (a)	14. (d)	15. (c)	16. (a)	17. (b)	18. (a)	19. (b)	20. (d)
21. (c)	22. (c)	23. (b)	24. (a)	25. (d)	26. (c)	27. (d)	28. (d)	29. (c)	30. (d)
31. (c)	32. (c)	33. (b)	34. (d)	35. (b)	36. (c)	37. (d)	38. (d)	39. (c)	40. (a)

Chemistry

41. (d)	42. (b)	43. (b)	44. (a)	45. (c)	46. (d)	47. (c)	48. (d)	49. (b)	50. (a)
51. (a)	52. (c)	53. (c)	54. (a)	55. (b)	56. (a)	57. (c)	58. (c)	59. (b)	60. (b)
61. (c)	62. (b)	63. (c)	64. (a)	65. (d)	66. (b)	67. (c)	68. (d)	69. (b)	70. (b)
71. (b)	72. (a)	73. (d)	74. (c)	75. (a)	76. (b)	77. (c)	78. (c)	79. (d)	80. (a)

English Proficieny

81. (b)	82. (a)	83. (d)	84. (d)	85. (b)	86. (b)	87. (c)	88. (a)	89. (d)	90. (b)
91. (c)	92. (b)	93. (c)	94. (b)	95. (a)					

Logical Reasoning

96. (d)	87. (d)	98. (b)	99. (c)	100. (a)	101. (b)	102. (b)	103. (c)	104. (d)	105. (b)

Mathematics

106. (d)	107. (c)	108. (a)	109. (c)	110. (d)	111. (b)	112. (c)	113. (b)	114. (d)	115. (a)
116. (a)	117. (c)	118. (d)	119. (c)	120. (c)	121. (b)	122. (d)	123. (c)	124. (a)	125. (b)
126. (a)	127. (a)	128. (c)	129. (d)	130. (d)	131. (a)	132. (d)	133. (d)	134. (b)	135. (a)
136. (c)	137. (a)	138. (d)	139. (a)	140. (a)	141. (a)	142. (c)	143. (a)	144. (d)	145. (d)
146. (a)	147. (c)	148. (b)	149. (d)	150. (a)					

Hints & Solutions

Physics

1. ***(c)*** Let the current is flowing in anti-clockwise direction as shown in figure.

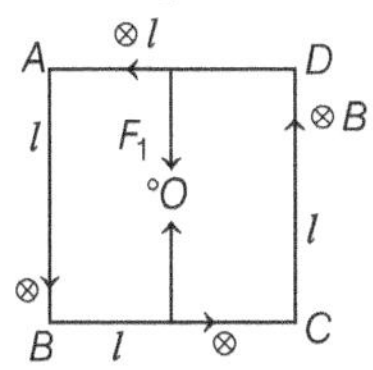

Now, magnetic force on $AD = F_1 = ilB$ towards centre O (By Fleming's left hand rule)

Similarly, magnetic force on $BC = F_2 = ilB$ towards centre O.

Since, two forces are equal in magnitude and opposite in direction and lie in same line of action, therefore they cancel out each other.

Also, magnetic force on $CD = F_3 = ilB$ towards centre O (By Fleming's left hand rule)

Similarly, magnetic force on $AB = F_4 = ilB$ towards centre O.

Again, two forces are equal in magnitude and opposite in direction and lie in same line of action, therefore they cancel out each other.

So, the net force on the current loop is zero.

2. ***(b)*** The magnetic field at C due to first conductor is

$B_1 = \frac{\mu_0}{2\pi} \frac{i}{3d/2}$ (Since, point C is separated by $d + \frac{d}{2} = \frac{3d}{2}$ from 1st conductor)

The direction of field is perpendicular to the plane of paper and directed outward.

The magnetic field at C due to second conductor is

$B_2 = \frac{\mu_0}{2\pi} \frac{10}{d/2}$ (Since, point C is separated by $\frac{d}{2}$ from 2nd conductor)

The direction of field is perpendicular to the plane of paper and directed inward.

Since, direction of B_1 and B_2 at point C is in opposite direction and the magnetic field at C is zero, therefore,

$$B_1 = B_2$$

$$\frac{\mu_0}{2\pi} \frac{i}{3d/2} = \frac{\mu_0}{2\pi} \frac{10}{d/2}$$

On solving, we get

$$i = 30.0 \text{ A}$$

3. ***(a)*** When system of conductors is in equilibrium,

The magnetic force of attraction per unit length between conductors = weight of conductor B per unit length.

$$\frac{\mu_0}{2\pi} \frac{I_A \times I_B}{d} = \frac{mg}{L} = \left(\frac{m}{L}\right)g$$

$$\frac{\mu_0}{2\pi} \frac{I_A \times I_B}{d} = \left(\frac{m}{L}\right)g$$

$$2 \times 10^{-7} \times \frac{100 \times I_B}{0.05} = (0.01 \text{ kg/m}) \times 10$$

On solving, we get

$$I_B = 250 \text{ A}$$

4. ***(b)*** Number of photo electrons depends on the intensity of incident light only.

5. ***(a)*** For hydrogen atom,

$$\frac{1}{\lambda} = R\left(\frac{1}{n_1^2} - \frac{1}{n_2^2}\right), n_2 > n_1$$

For the shortest wavelength of Lyman series,

$n_1 = 1, n_2 = \infty \Rightarrow \frac{1}{\lambda_1} = R$

For the shortest wavelength of Balmer series,

$n_1 = 2, n_2 = \infty \Rightarrow \frac{1}{\lambda_2} = \frac{R}{4}$

For the shortest wavelength of Paschen series,

$n_1 = 3, n_2 = \infty \Rightarrow \frac{1}{\lambda_3} = \frac{R}{9}$

So, $\lambda_1 = \frac{1}{R}, \lambda_2 = \frac{4}{R}, \lambda_3 = \frac{9}{R}$

$\Rightarrow \lambda_1 : \lambda_2 : \lambda_3 = 1 : 4 : 9$

6. ***(c)*** Let, initial number of nuclei of each element = N_o

Decay constants,

$$\lambda_A - \frac{0.693}{1} \text{hr}^{-1}, \lambda_B - \frac{0.693}{2} \text{hr}^{-1}$$

$$\lambda_A > \lambda_B$$

Activities, $R_A = \lambda_A N_0$

$R_B = \lambda_B N_0$

$\Rightarrow R_A > R_B$ as $\lambda_A > \lambda_B$

Less half-life of element A implies faster decay

Mean life, $\tau = \frac{1}{\lambda}$

$$\tau_A = \frac{1}{\lambda_A}, \tau_B = \frac{1}{\lambda_B}$$

$$\tau_A < \tau_B \text{ as } \lambda_A > \lambda_B$$

7. ***(d)*** The glass piece will dissappear only, if the refractive index of the glass and liquid is same.

So, refractive index of glass piece must be 4/3.

8. ***(a)*** Focal length of bi-convex lens,

$$\frac{1}{f} = (\mu - 1)\left(\frac{1}{R_1} - \frac{1}{R_2}\right)$$

$$\Rightarrow \quad \frac{1}{f} = (\mu - 1)\left(\frac{1}{R} + \frac{1}{R}\right)$$

[Since, $R_1 = R$ and $R_2 = -R$]

$$f = \frac{R}{2(\mu - 1)} \quad \ldots(i)$$

and new focal length of lens,

$$\frac{1}{f'} = (\mu - 1)\left(\frac{1}{R} - \frac{1}{\infty}\right)$$

$$\Rightarrow \quad f' = \frac{R}{\mu - 1} \quad \ldots(ii)$$

From Eqs. (i) and (ii), we get

$$f' = 2f$$

9. ***(c)*** Refractive index of a medium depends on the medium as well as on the wavelength of the incident light.

10. ***(d)*** Image will not form, because object is placed on the side from where reflection is not possible.

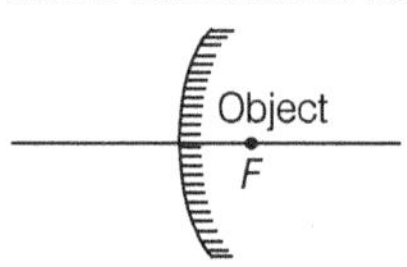

11. ***(b)*** The ray diagram for the given situation is shown below.

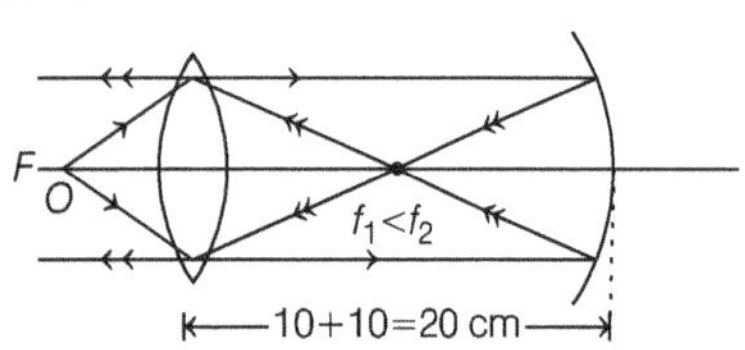

The final image will be at infinity only, if the focii of lens and mirror coincides. The situation could be understood on the basis of given diagram. So, the value of $d = 10 + 10 = 20$ cm.

12. ***(d)*** The image formed by a concave spherical mirror could be real, virtual, erect and inverted.

13. ***(a)*** Minimum energy required to through the satellite out of earth's gravitational field is equal to its binding energy.

$\therefore$ Binding energy, BE = –TE

$= -(-k)$ [Given, TE = $-k$J]

$= k$J

14. ***(d)***

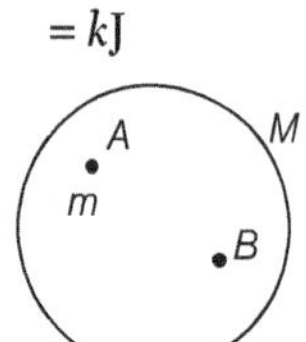

The gravitational field at A and B are zero because inside the spherical shell, gravitational field is zero. So, no work required to make change between the points A and B.

15. ***(c)*** Angular velocity, $\omega = 2\pi f = 2\pi \times \frac{30}{60} = \pi$ rad/s

For an elastic spring force, $F = k \times x$, where x is the extension.

Radius of circular motion, $r = L + x$

Centripetal force $= mr\omega^2 = F$

$$\Rightarrow m(L + x)\,\omega^2 = kx$$

$$\Rightarrow \quad x = \frac{mL\omega^2}{k - m\omega^2} = \frac{0.02 \times 0.5 \times (3.14)^2}{2 - 0.02 \times (3.14)^2}$$

$$\approx 0.05 \text{ m}$$

Radius of the circular motion(r)

$$= L + x = 0.5 + 0.05$$

$$= 0.55 \text{ m}$$

Tension in the spring,

$$T = kx = 2 \times 0.05 \approx 0.1 \text{ N}$$

16. ***(a)*** The initial angular momentum of the rotating record is

$$L = I\omega$$

where, $I = \frac{1}{2} MR^2$

Let ω' be the angular velocity of the record when the coin of mass m is placed on it at a distance r from its centre.

The new angular momentum of the system becomes

$$L' = (I + mr^2)\,\omega'$$

Since, no external torque acts on the system, the angular momentum is conserved *i.e.*,

$$L' = L \text{ or } (I + mr^2)\,\omega' = I\omega$$

or $$\omega' = \frac{I\omega}{I + mr^2} = \frac{\frac{1}{2}MR^2\omega}{\frac{1}{2}MR^2 + mr^2}$$

or $$\omega' = \frac{\omega}{1 + \frac{2mr^2}{MR^2}} \quad \ldots(i)$$

Putting $r = \frac{R}{2}$ in Eq. (i), we get

$$\omega' = \left[\frac{\omega}{1 + \frac{2m \times \left(\frac{R}{2}\right)^2}{MR^2}}\right]$$

$$\Rightarrow \quad \omega' = \frac{2\omega M}{2M + m}$$

17. *(b)* Frictional force between P and Q is $f = \mu mg$ which will retard P and accelerate Q.

Retardation of P, $a_P = -\frac{f}{m} = \frac{-\mu mg}{m} = -\mu g$

Acceleration of Q, $a_Q = \frac{+f}{M} = \frac{\mu mg}{M}$

Acceleration of Q relative to P is

$$a_{QP} = a_Q - a_P = \frac{\mu mg}{M} - (-\mu g)$$

$$= \mu g\left[1 + \frac{m}{M}\right] = 0.3 \times 10\left[1 + \frac{1}{6}\right]$$

$$\Rightarrow \quad a_{QP} = 3.5 \text{ m/s}^2$$

18. *(a)* Let us draw the figure for given situation,

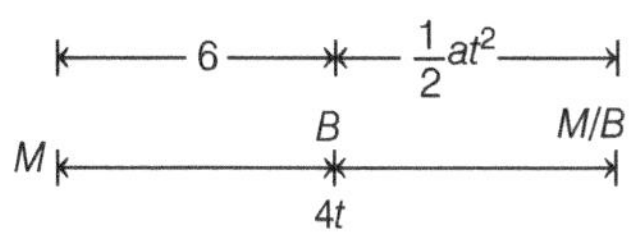

The man needs to cover the distance by bus is time t and 6 m to catch the bus.

$$\Rightarrow \quad 4t = 6 + \frac{1}{2} \times 1.2 \times t^2$$

$$\Rightarrow \quad 4t = 6 + 0.6\, t^2$$

19. *(b)* In any type of collision, the linear momentum of the system remain conserved even during collision.

20. *(d)* Given that, $n = -\frac{D(n_2 - n_1)}{x_2 - x_1}$

$$\Rightarrow \quad D = -\frac{n(x_2 - x_1)}{n_2 - n_1}$$

Here, $[n] = \left[\frac{1}{\text{area} \times \text{time}}\right] = \frac{1}{[\text{L}^2\text{T}]} = [\text{L}^{-2}\text{T}^{-1}]$

$x_2 - x_1 = [\text{L}]$ and $n_2 - n_1 = \left[\frac{1}{\text{volume}}\right] = \left[\frac{1}{\text{L}^3}\right]$

$= [\text{L}^{-3}]$

So, $[D] = \frac{[\text{L}^{-2}\text{T}^{-1}\text{L}]}{[\text{L}^{-3}]} = [\text{L}^2\text{T}^{-1}]$

21. *(c)* Since, capacitors work as open circuit for DC supply, hence it can be drawn as

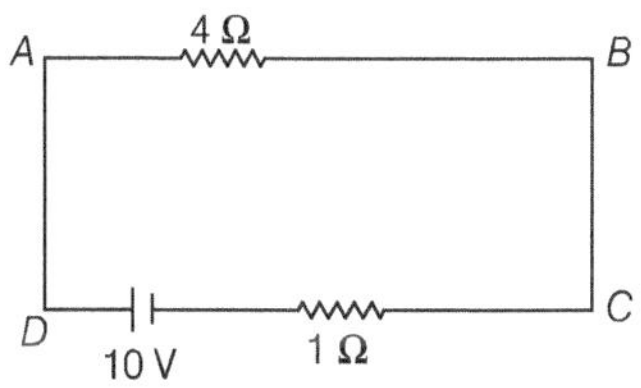

$\therefore$ Equivalent resistance $= 4 + 1 = 5\,\Omega$

$\therefore$ Current, $I = \frac{V}{R} = \frac{10}{5} = 2\text{A}$

Potential difference between point A and B

$= I \times 4 = 2 \times 4 = 8\text{V}$

$\because$ Two capacitors of $3\,\mu\text{F}$ each are in series

$\therefore$ Potential difference across each capacitor

$= \frac{8}{2} = 4\text{ V}$

Charge on each capacitor, $q = CV = 3 \times 4$

$= 12\,\mu\text{C}$

22. *(c)* The electric field inside a conductor is zero and is always perpendicular to the surface of a conductor.

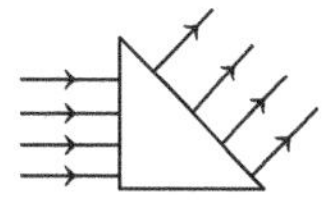

23. *(b)* Given, $\mathbf{p}_1 = -2p\hat{\mathbf{i}} = 2p$ along negative X-axis.

$\mathbf{p}_2 = p\,\hat{\mathbf{j}} = p$ along Y-axis.

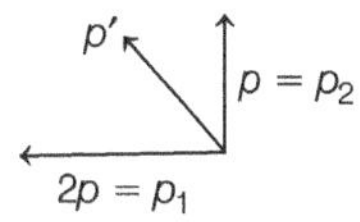

The magnitude of resultant momentum of two parts

$$p' = \sqrt{p_1^2 + p_2^2} = \sqrt{(2p)^2 + p^2} = p\sqrt{5}$$

As the bomb was initially at rest, hence final momentum of all the three parts must be zero

$$\mathbf{p}_3 + \mathbf{p}' = 0$$

$$\mathbf{p}_3 = -\mathbf{p}' = -\mathbf{p}\sqrt{5}$$

$\therefore$ Magnitude of $\mathbf{p}_3$,

$= \sqrt{5}\,\mathbf{p}$

$= \sqrt{x} \cdot \mathbf{p}$ (Given)

$\therefore \quad x = 5$

24. *(a)* The forces on smaller block is given as

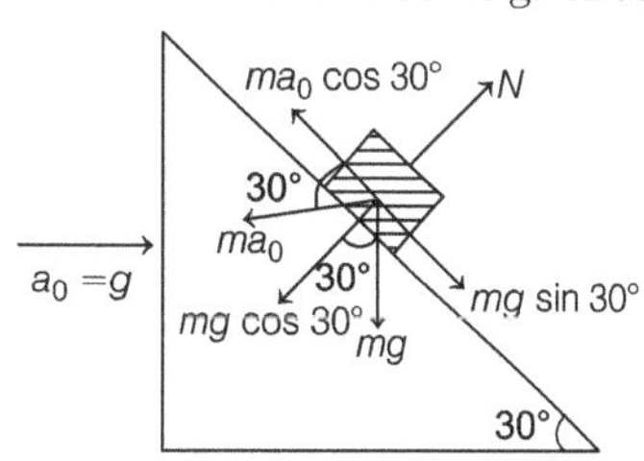

For the motion of the block along the incline plane in upward direction.

Net force on the block = mass × acceleration of the block

$$ma_0 \cos 30° - mg \sin 30° = ma$$

$$\Rightarrow \quad mg \cos 30° - mg \sin 30° = ma \qquad (\because a_0 = g)$$

$$\Rightarrow \quad a = \left(\frac{\sqrt{3} - 1}{2}\right)g = 3.66 \text{ m/s}^2$$

Now, from equation of motion, $s = \frac{1}{2}at^2$

$$\Rightarrow \quad t = \sqrt{\frac{2s}{a}} = \sqrt{\frac{2 \times 1}{3.66}} = 0.74 \text{ s}$$

25. ***(d)*** Pseudo force is not a real force.

26. ***(c)*** The given situation is shown below. Here, a is common acceleration of the blocks and T is tension in string.

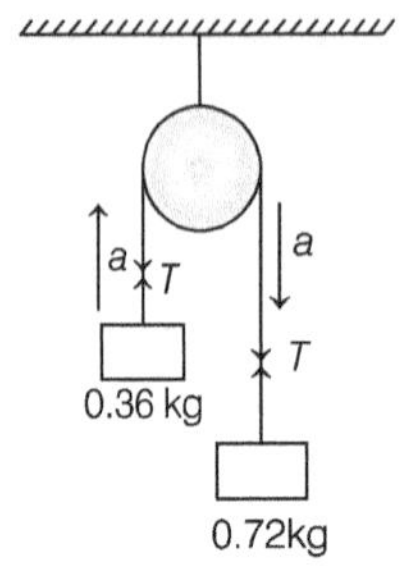

So, acceleration, $a = \dfrac{\text{Net pulling force}}{\text{Total mass}}$

$$= \frac{0.72\,g - 0.36\,g}{0.72 + 0.36} = \frac{g}{3}$$

Distance, travelled in 1 $s = ut + \dfrac{1}{2}at^2$

$$= 0 + \frac{1}{2} \times \frac{g}{3} \times (1)^2 = \frac{g}{6}$$

So, $\quad T - mg = ma$

$$T - 0.36\,g = 0.36\,a$$

$$T = 0.48\,g$$

Now, work done by string on the block

$$W_T = Ts\cos 0° \qquad \text{(on 0.36 kg of mass)}$$

$$= (0.48\,g)\left(\frac{g}{6}\right)(1) = 0.08\,g^2$$

$$= 0.08 \times (10)^2 = 8\text{ J}$$

27. ***(d)*** We know that, capacitance, $C = \dfrac{Q}{V}$

$$= \frac{Q}{(W/Q)} \qquad \left[\because V = \frac{W}{Q}\right]$$

$$= \frac{Q_2}{W}$$

$$\therefore \qquad [C] = \frac{[Q^2]}{[W]}$$

$$\Rightarrow \qquad [A] = \frac{[Q^2]}{[ML^2T^{-2}]} \qquad [\text{Given, } C = A]$$

$$\Rightarrow \qquad [A] = [M^{-1}L^{-2}T^2Q^2]$$

Again we know that, $F = BQv$

$$\Rightarrow \qquad B = \frac{F}{Qv} \Rightarrow [B] = \frac{[F]}{[Q][v]}$$

$$\Rightarrow \qquad [C] = \frac{[MLT^{-2}]}{[Q][LT^{-1}]} \qquad [\text{Given, } B = C]$$

$$\Rightarrow \qquad [C] = [MT^{-1}Q^{-1}]$$

Given, $A = 3BC^2 \Rightarrow B = \dfrac{A}{3C^2} \Rightarrow [B] = \dfrac{[A]}{[C]^2}$

$$= \frac{[M^{-1}L^{-2}T^2Q^2]}{[MT^{-1}Q^{-1}]^2} = [M^{-3}L^{-2}T^4Q^4]$$

28. ***(d)*** The given situation is shown below.

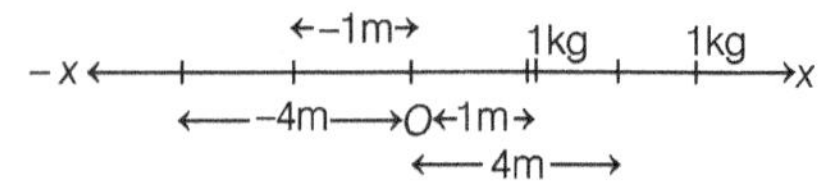

Hence, resultant gravitational potential at origin O is given as

$$V = 2GM\left(\frac{1}{r_1} + \frac{1}{r_2} + \frac{1}{r_3} \ldots\right)$$

$$= 2G \times 1\left(\frac{1}{1} + \frac{1}{2} + \frac{1}{4} + \frac{1}{8} + \frac{1}{16}\right)$$

It forms a GP, so

Sum of GP $= \dfrac{a}{1-r}$

Hence, $V = 2G\left(\dfrac{1}{1 - \dfrac{1}{2}}\right) = 4G$

29. ***(c)*** If bulb Z is fused, the current stops flowing through Z. The effective resistance of the circuit due to bulbs X and Y in series becomes more as compared to before. Due to which, the current in the circuit decrease.

$\therefore$ Brightness $\propto$ (Current)^2

So, the brightness of bulb X decreases, because total current decreases.

Now, bulb Y gets more current than before fusing the bulb, Z. Hence, more current will flow through it in comparision to before.

$\therefore$ Brightness of bulb Y will increase.

30. ***(d)*** Transfer characteristic of transistor is

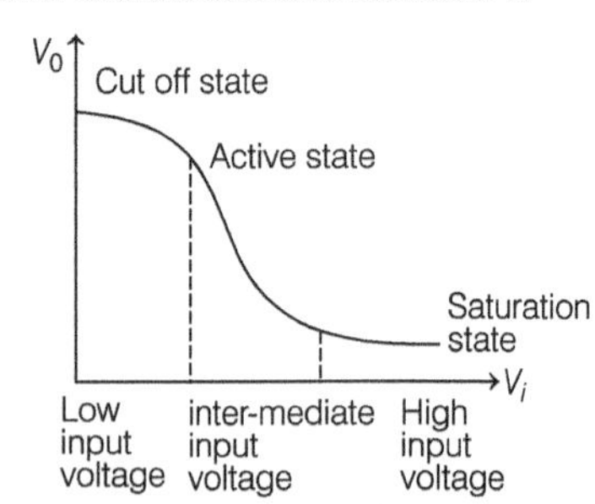

Active state is achieved at inter-mediate input voltage.

31. ***(c)*** Initial moment of inertia of the system is

M_1 = moment of inertia of turntable + Moment of inertia of boy at the centre

$$= 1000 + 0 = 1000\text{ kgm}^2$$

Initial frequency, $\nu_1 = \frac{98}{10} = 9.8 \text{ rev/s}$

Final moment of the system

M_2 = MI of turntable + MI of boy at a distance 4 m from the centre of turntable.

$$= 1000 + 60 \times (4)^2 = 1960 \text{ kgm}^2$$

Since no external torque acts, the angular momentum of the system is conserved *i.e.*,

$$I_2\omega_2 = I_1\omega_1$$

$$\Rightarrow \quad I_2 \cdot 2\pi\nu_2 = I_1 \cdot 2\pi\nu_1$$

$$\Rightarrow \quad I_2\nu_2 = I_1\nu_1$$

$$\Rightarrow \quad \nu_2 = \frac{I_1\nu_1}{I_2} = \frac{1000 \times 9.8}{1960}$$

$$\Rightarrow \quad \nu_2 = 5 \text{ rev/s} = 5 \text{ Hz}$$

32. ***(c)*** The given situation is shown below.

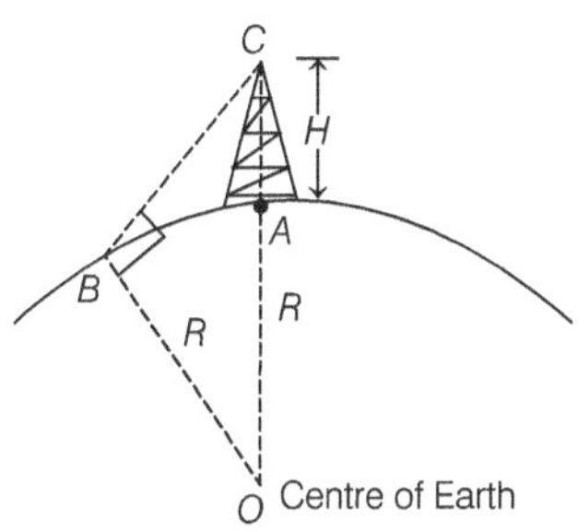

In ΔBOC,

$$OB^2 + BC^2 = OC^2$$

$$R^2 + BC^2 = (R+H)^2$$

$$\Rightarrow \quad BC = \sqrt{2RH + H^2} = \sqrt{RH\left(2 + \frac{H}{R}\right)}$$

Here, $\quad \frac{H}{R} << 2$

So, $\quad BC = \sqrt{2RH}$

$$BC \approx AB = d = \text{distance of reach}$$

$$\Rightarrow \quad d = \sqrt{2RH} \propto H^{1/2}$$

where, $\quad R$ = constant = 6400 km = radius of the earth

33. ***(b)*** Output of given logic circuit is

$$Y = A + (\overline{A} \cdot B) = (A + \overline{A}) \cdot (A + B)$$

Since, $X + YZ = (X+Z)(X+Z)$ [By Boolean alzebra]

$$= 1 \cdot (A+B) = A + B \qquad [\because (A + \overline{A}) = 1]$$

This is the output of OR gate.

34. ***(d)*** Accuracy is closeness to true/exact value.

Precision is based on instrument, more decimal places in measurement indicate more precision.

So, 20.1 cm is less precise as compared to 19.65 cm, because 20.1 cm has lesser number of decimal places.

35. ***(b)*** Volume of tank, $V = 2000$ L

$$= 2000 \times 10^{-3} \text{ m}^3 = 2\text{m}^3$$

Mass of water, $m = \rho_w \times V = 1000 \times 2 = 2 \times 10^3$ kg

Work done to lift this mass to a height *i.e.*, $h = 40$ m is

$W = mgh = 2 \times 10^3 \times 10 \times 40 = 8 \times 10^5$ J

$$\text{Power needed} = \frac{\text{Work done}}{\text{Time taken}} = \frac{8 \times 10^5}{10 \times 60}$$

$$P' = \frac{4}{3} \times 10^3 \text{ W}$$

If P is the total power consumed, the useful power available = 40% of P

i.e. $\quad P' = 0.4\,P$

$$0.4\,P = \frac{4}{3} \times 10^3$$

$$\Rightarrow \quad P = 3.33 \times 10^3 \text{ W} = 3.33 \text{ kW}$$

36. ***(c)*** Applying gas equation, $pV = nRT$

We can write; $p_1V = n_1RT_1$

and $\quad p_2V = n_2RT_2$

$$\Rightarrow \quad \frac{p_2}{p_1} = \frac{n_2}{n_1} \times \frac{T_2}{T_1} = \frac{1}{1} \times \frac{2T}{T} = 2$$

$$\Rightarrow \quad p_2 = 2p$$

37. ***(d)*** Given, $T_1 = 27 + 273 = 300$ K

$$T_2 = 127 + 273 = 400 \text{ K}$$

We know, $v_{rms} = \sqrt{\frac{3RT}{M}}$

$$\Rightarrow \% \text{ increase in } v_{rms} = \frac{\sqrt{\frac{3RT_2}{M}} - \sqrt{\frac{3RT_1}{M}}}{\sqrt{\frac{3RT_1}{M}}} \times 100$$

$$= \frac{\sqrt{T_2} - \sqrt{T_1}}{\sqrt{T_1}} \times 100 = \frac{\sqrt{400} - \sqrt{300}}{\sqrt{300}} \times 100$$

$$= \frac{20 - 17.32}{17.32} \times 100$$

$\Rightarrow$ % increase in v_{rms} = 15.5%

38. ***(d)*** We know,

Maximum acceleration in SHM, $a_{max} = \omega^2 A = \frac{4\pi^2}{T^2} A$

$$= \frac{4\pi^2}{\left(\frac{\pi}{5}\right)^2} \times 0.3 = 30 \text{ m/s}^2$$

Maximum force, $F_{max} = ma_{max} = \frac{5}{1000} \times 30$

$$= 0.15 \text{ N}$$

39. ***(c)*** Let T be the junction temperature

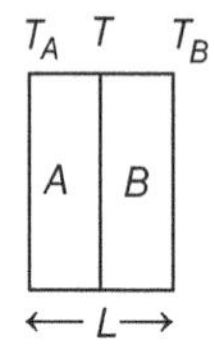

Here, $K_A = 2K_B$, $T - T_B = 50\text{K}$

At the steady state, $H_A = H_B$

$$\Rightarrow \quad \frac{K_A A(T_A - T)}{L} = \frac{K_B A(T - T_B)}{L}$$

$$\Rightarrow \quad 2K_B(T_A - T) = K_B(T - T_B)$$

$$\Rightarrow \quad T_A - T = \frac{T - T_B}{2} = \frac{50}{2} = 25\text{ K}$$

40. ***(a)*** The beat frequency of heart is

$$\nu = \frac{75}{(1\text{ min})} = \frac{75}{60\text{ s}}$$

$$= 1.25\text{ s}^{-1}$$

$$= 1.25\text{ Hz}$$

The time period of heart is

$$T = \frac{1}{\nu} = \frac{1}{1.25\text{ s}^{-1}} = 0.8\text{ s}$$

Chemistry

41. ***(d)*** As we know, on moving down the group first ionisation enthalpy decreases top to bottom, therefore order of first ionisation enthalpy for group 13 element is

$$\text{B} > \text{Al} > \text{Ga} > \text{In} > \text{Tl}$$

42. ***(b)*** Bohr's theory is applicable to H-like species containing one electron only, *e.g.*, Li^{2+}.

43. ***(b)*** Among given four pairs, B_2 and O_2 are paramagnetic due to presence of unpaired electron.

MO(EC) of $B_2 = \sigma 1s^2 \sigma^* 1s^2\ \sigma 2s^2 \sigma^* 2s^2\ \pi 2p_x^1 = \pi 2p_y^1$

MO(EC) of $O_2 = \sigma 1s^2\ \sigma^* 1s^2\ \sigma 2s^2 \sigma^* 2s^2\ \sigma 2p_z^2$

$$\pi 2p_x^2 = \pi 2p_y^2\ \pi^* 2p_x^1 = \pi^* 2p_y^1$$

44. ***(a)*** 1 mole of O_2 = 4 eq. of oxygen

22400 mL of O_2 = 4 eq. of oxygen

46.6mL of $O_2 = \frac{4}{22400} \times 46.6 = 0.00832$ eq.

Equivalent of metal = Equivalent of oxygen

$$\frac{\text{Weight}}{\text{Equivalent}} = 0.00832$$

$$\frac{0.1}{\text{Equivalent}} = 0.00832$$

$$\therefore \quad \text{Equivalent} = \frac{0.1}{0.00832} = 12.0$$

45. ***(c)*** Ionisation enthalpy is the minimum amount of enthalpy required to remove the outermost electron from an isolated gaseous atom. Quantitatively, it depends on the attraction between electron present on outermost shell and nucleus. Greater the interaction between outermost electron and nucleus, higher will be its ionisation enthalpy. So correct order of first ionisation enthalpy must be

$$\text{B} < \text{C} < \text{N} < \text{O} < \text{F}$$

But due to extra stable half-filled electronic configuration of p-orbital of N has more value of first ionisation enthalpy than oxygen, hence correct order is

$$\text{B} < \text{C} < \text{N} > \text{O} > \text{F}$$

46. ***(d)*** Molecular structure of given names of organic compounds are written as

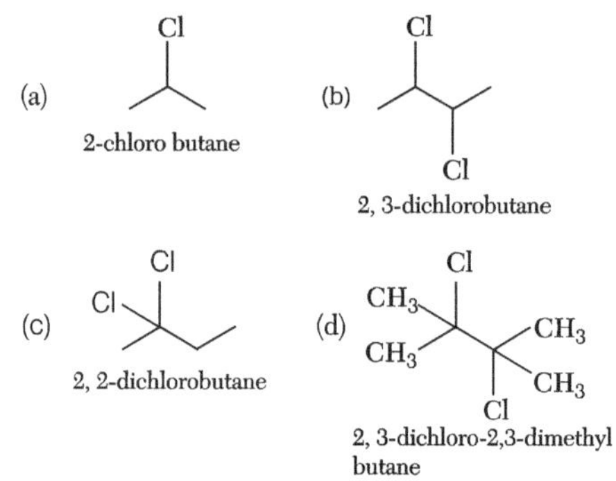

According to Saytzeff's rule, more substituted (alkylated) alkene are more stable. When the alkyl halide is treated with base, it undergo elimination reaction and produces alkene as follows :

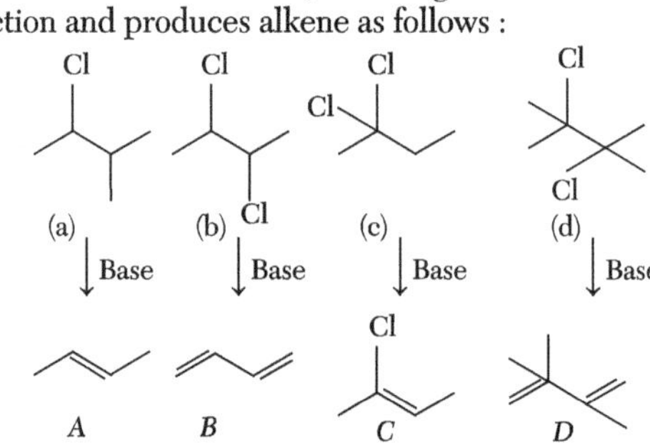

Conjugation Greater the conjugation greater will be the stability of product.

Hence, (d) has maximum stability, the correct choice is (d), which is stabilised by conjugation as well as Saytzeff's rule.

47. ***(c)*** Due to stronger $-I$ effect of F than Cl, CHF_3 should be more acidic than $CHCl_3$. But actually reverse is true. This is due to $:CCl_3^-$ left after the removal of a proton from $CHCl_3$ is stabilised by resonance due to presence of d-orbitals in Cl while $:CF_3^-$ left after the removal of a proton from CHF_3 is not stabilised by resonance due to the absence of d-orbitals on F.

$\therefore$ $CHCl_3$ is more acidic among the given halogen compound.

48. ***(d)*** **Ozonolysis** On ozonolysis the given alkene undergo ozonide formation followed by reduction to produce bicarbonyl compound as

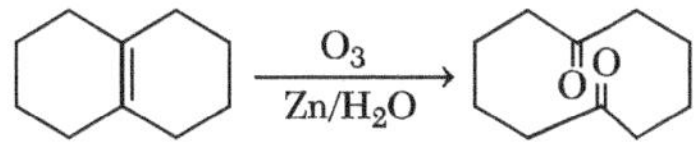

Bicarbonyl (*A*)

Now, this bicarbonyl compound undergoes interamolecular aldol condensation as follows

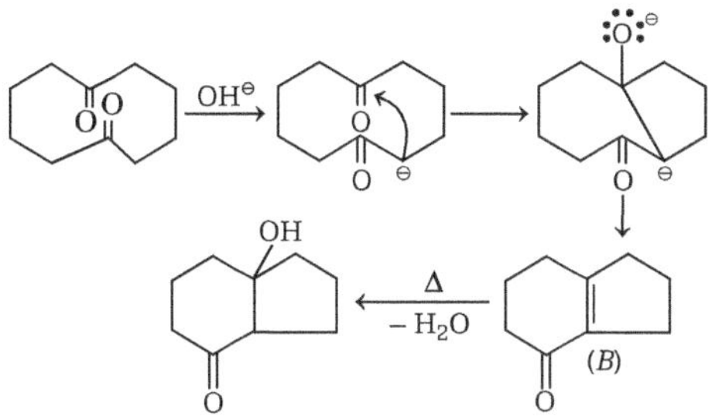

49. ***(b)*** This question is based on Raoult's law. It represent that, the partial pressure of each component in the solution is directly proportional to its mole fraction for a solution *i.e.*, $p_A \propto \chi_A$ and $p_B \propto \chi_B$.

From Raoult's law,

$$p^\circ - p_s = p^\circ \times \text{mole fraction of solute}$$

$$10 = p^\circ \times 0.2 \quad \ldots(i)$$

$$20 = p^\circ \times \chi_2 \quad \ldots(ii)$$

$$\therefore \quad \chi_2 = 0.4 \quad \text{(on comparing)}$$

and $\quad \chi_1 = 1 - 0.4 = 0.6$

where, χ_1, χ_2 = mole fractions of solvent and solute respectively.

50. ***(a)*** It is seen that, in experiments (3) and (2), $[A]$ is constant and $[B]$ is doubled and rates becomes 8 times, so order w.r.t, $[B] = 3$. In experiments (1) and (3), $[B]$ is constant and $[A]$ is doubled, but rate does not change, so order w.r.t. $[A] = 0$

Thus, rate $= k\,[B]^3$.

51. ***(a)*** The crystal field splitting, Δ_o depends upon the field produced by the ligand and charge on the metal ion. In all these complexes of chromium, charge acquired by metal ion is + 3. Therefore, Δ_o depends upon the field produced by the ligand.

In accordance with the spectrochemical series, the increasing order of field strength is

$$C_2O_4^{2-} < H_2O < NH_3 < CN^-$$

Thus, CN^- is the strong field ligand and will produce highest magnitude of Δ_o.

52. ***(c)*** This problem includes conceptual mixing of carbocation stability and reactivity of alcohol.

- Remove the —OH^- group by dehydration and then arrange the carbocation in increasing order of stability
- Order of carbocation stability is same as S_N1 reactivity of alkyl halides.

The carbocation is formed during reaction of alcohol by removal of —OH group. More stable the carbocation more will be its reactivity. During formation of carbocation, reactions are as follows :

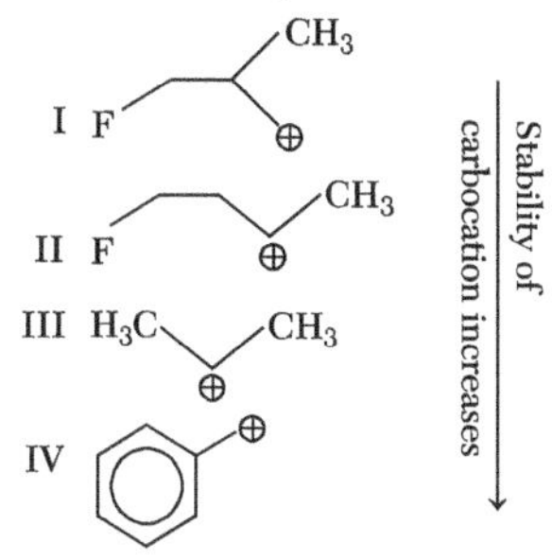

Benzyl carbocation is more stable due to conjugation with phenyl ring.

Hence, the correct order of decreasing reactivity is IV> III> II> I.

53. ***(c)*** When 2-methyl propan-1-ol is treated with a mixture of conc. HCl and $ZnCl_2$ (Lucas reagent) then *tert*-alkyl halide is formed and produce turbidity due to its less solubility.

$$\underset{\text{2-methyl propan-1-ol}}{CH_3{-}\overset{\overset{\displaystyle CH_3}{|}}{CH}{-}CH_2OH} \xrightarrow[ZnCl_2]{\text{Conc. HCl}} \underset{\text{2-chloro-2-methyl propane (turbidity)}}{CH_3{-}\overset{\overset{\displaystyle CH_3}{|}}{\underset{\underset{\displaystyle Cl}{|}}{C}}{-}CH_3}$$

54. ***(a)*** Aspirin is a moderate acid ($pK_a = 3.5$). Therefore, it is almost unionised in stomach due to its strong acidic medium. It happens due to common ion effect. On the other hand, in small intestine, the medium is alkaline, hence, aspirin will be sufficiently ionised in it.

55. ***(b)*** The reaction of K_2CrO_4 with dilute nitric acid is represented as

$$2K_2CrO_4 + 2HNO_3 \longrightarrow K_2Cr_2O_7 + KNO_3 + H_2O$$

It's ionic equation is

$$2CrO_4^{2-} + 2H^+ \longrightarrow Cr_2O_7^{2-} + H_2O$$

Hence, $Cr_2O_7^{2-}$ and H_2O are formed.

56. ***(a)*** When CaC_2 reacts with water molecule to form acetylene. Similarly, it reacts with D_2O to form C_2D_2.

$$CaC_2 + 2D_2O \longrightarrow C_2D_2 + Ca(OD)_2$$

57. ***(c)*** Fluorine acts as strongest oxidising agent due to

(a) low enthalpy of dissociation of F—F bond

(b) high hydration enthalpy of F^-.

58. ***(c)*** The reaction of P_4 with X leads selectively to P_4O_6. The reaction is as follows

$$P_4 + 3O_2 \xrightarrow{O_2 + N_2} P_4O_6$$

N_2 prevents the further reaction of P_4O_6 into P_4O_{10} .

59. ***(b)*** The acidic strength of hydrides is inversely proportional to their stability. Since, the stability of hydrides decreases from N to Sb. Therefore, the acidic strength increases from N to Sb. Hence, the correct order would be

$$NH_3 < PH_3 < AsH_3 < SbH_3$$

Caution point As the stability decreases from NH_3 to BiH_3, the reducing character of hydrides increases.

60. ***(b)*** Borax is $Na_2B_4O_7 \cdot 10H_2O$ in which 2 molecules of water among 10 molecules form a part of structure and exists as $Na_2[B_4O_5(OH)_4].8H_2O$,

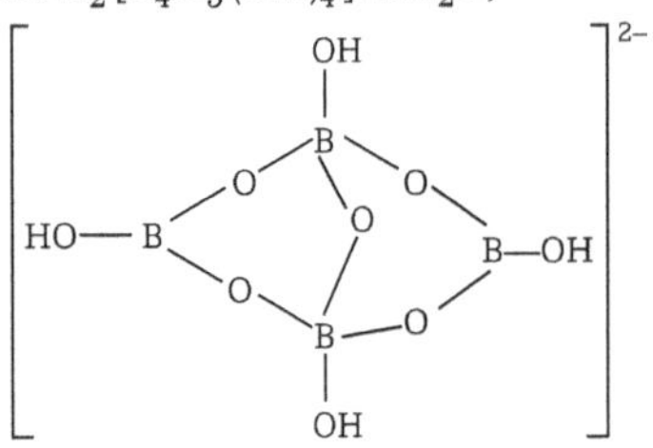

$$Na_2[B_4O_5(OH)_4] \cdot 8H_2O + 2HCl \longrightarrow 2NaCl + 4H_3BO_3 + 5H_2O$$

Methyl orange with pH value of 3.7 is used to detect end point. Aqueous solution of borax acts as buffer, because borax is salt of strong base NaOH and weak acid H_3BO_3.

61. ***(c)*** Among the given oxides, only KO_2 *i.e.*, potassium superoxide is paramagnetic in nature. This is because peroxide ion, O_2^- has three electron bond which makes it paramagnetic and coloured.

O^{2-}	$^-O—O^-$	$[O \equiv O]^{2-}$
Oxide ion	Peroxide ion	Superoxide ion

Hence, A is KO_2.

62. ***(b)*** The two sols prepared contains not only AgI but also KI and $AgNO_3$ as these are taken in excess amounts. When these sols are mixed, the sols being oppositely charged coagulates each other.

63. ***(c)*** The extraction of Ag using (Zn-Ag) alloy is called Parke's process.

As zinc is volatile at 920°C while Ag is not. Thus, on heating (Zn + Ag) alloy, zinc vapourises while Ag remains at the bottom of the vessel. Hence, Zn is removed from (Zn / Ag) alloy through distillation.

64. ***(a)*** $k_1 = Ae^{E_{a_1}/RT}$; $k_2 = Ae^{-E_{a_2}/RT}$

Given that, $k_3 = Ae^{-E_{a_3}/RT}$

Overall rate, $k = \dfrac{k_1 k_3}{k_2}$

Therefore, overall, $E_a = E_{a_1} + E_{a_3} - E_{a_2}$

$= 60 + 10 - 30 = 40$ kJ

65. ***(d)*** Given that, $E^\circ_{Fe^{3+}/Fe} = -0.36$ V;

$E^\circ_{Fe^{2+}/Fe} = -0.439$ V

$$Fe^{3+} + 3e^- \longrightarrow Fe; \quad \Delta G_1 = -3F(E^\circ_{Fe^{3+}/Fe})$$

$$Fe^{2+} + 2e^- \longrightarrow Fe; \quad \Delta G_2 = -2F(E^\circ_{Fe^{2+}/Fe})$$

$$Fe^{3+} + e^- \longrightarrow Fe^{2+}; \Delta G_3 = -F(E^\circ_{Fe^{3+}/Fe^{2+}})$$

$$E^\circ_{Fe^{3+}/Fe^{2+}} = 3E^\circ_{Fe^{3+}/Fe} - 2E^\circ_{Fe^{2+}/Fe}$$

$$= [3(-0.36) - 2(-0.439)] \text{ V}$$

66. ***(b)*** Given that, $r_{Na^+}/r_{Cl^-} = 0.55$

$r_{K^+}/r_{Cl^-} = 0.74$

$\dfrac{r_{KCl}}{r_{NaCl}} = ?$

$\because \quad \dfrac{r_{Na^+}}{r_{Cl^-}} = 0.55$

$\therefore \quad \dfrac{r_{Na^+}}{r_{Cl^-}} + 1 = 0.55 + 1$

$\dfrac{r_{Na^+} + r_{Cl^-}}{r_{Cl^-}} = 1.55$ ...(i)

$\therefore \quad \dfrac{r_{K^+}}{r_{Cl^-}} = 0.74$

$\therefore \quad \dfrac{r_{K^+}}{r_{Cl^-}} + 1 = 0.74 + 1$

$\dfrac{r_{K^+} + r_{Cl^-}}{r_{Cl^-}} = 1.74$ ...(ii)

Dividing Eq. (ii) by Eq. (i), we get

$$\frac{r_{K^+} + r_{Cl^-}}{r_{Na^+} + r_{Cl^-}} = \frac{1.74}{1.55} = 1.1226 \approx 1.23$$

67. ***(c)*** When alkyl benzene is treated with acidic or alkaline $KMnO_4$, the entire side chain is oxidised to the carboxylic acid irrespective of length of the side chain.

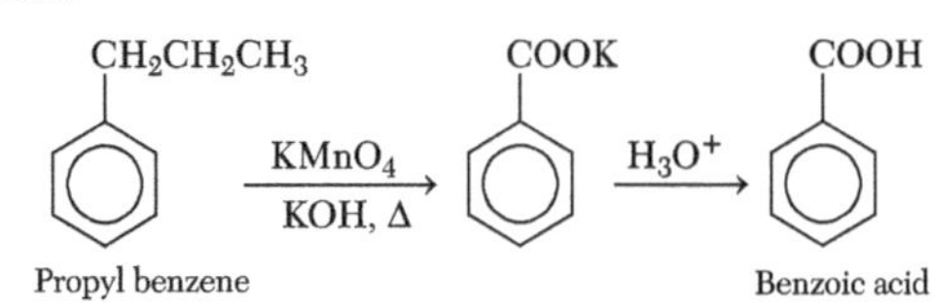

68. ***(d)*** $[Co(NH_3)_6]^{3+}$, $[Ni(CN)_4]^{2-}$ and $[NiCl_4]^{2-}$ are inner d-orbital or low spin complex while $[CoF_6]^{3-}$ is outer d-orbital or high spin complex.

$[CoF_6]^{3-}$, Co = [Ar] $4s^2\, 3d^7,\, 4p^0$

Co^{3+} = [Ar] $3d^6,\, 4s^0,\, 4p^0$

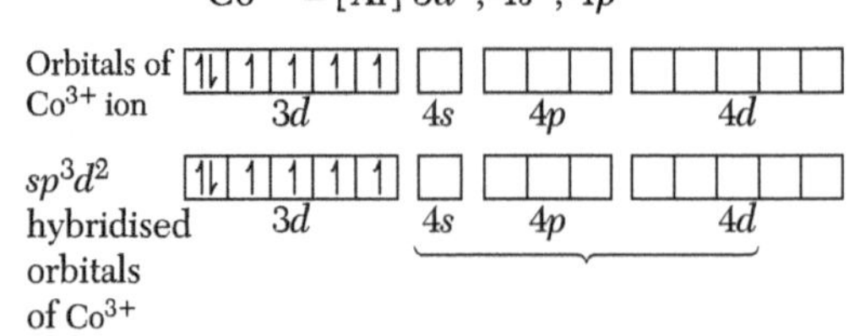

Here, F^- is a weak legand so, no pairing of electron takes place.

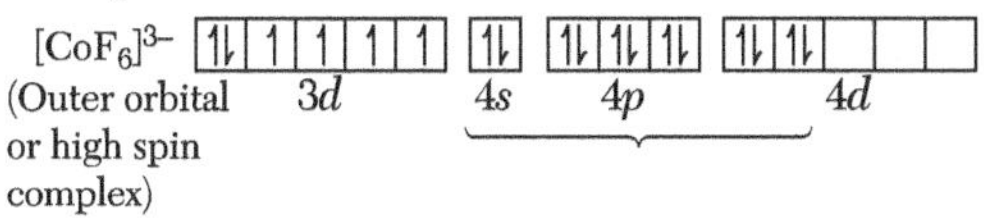

69. ***(b)*** C_1 carbon of monosaccharides is called anomeric carbon. When —OH group attached with C_1 carbon is towards right, it is called α-form and when —OH group is towards left, it is called β-form. Such pair or optical isomers which differ in the configuration only around anomeric carbon are called anomers.

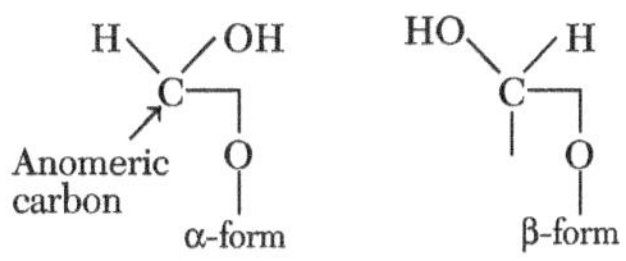

70. ***(b)*** In I and IV, amine is not formed.

I. Hydrolysis of *R*CN;

$$RCN \longrightarrow RCOOH + NH_3$$

II. Reduction of *R*CH═NOH;

$$RCH{=}NOH \longrightarrow RCH_2NH_2 + H_2O$$

III. Hydrolysis of *R*NC ;

$$RNC + 2H_2O \xrightarrow{\Delta} RNH_2 + HCOOH$$

IV. Hydrolysis of $RCONH_2$;

$$RCONH_2 \longrightarrow RCOOH + NH_3$$

71. ***(b)*** In vulcanisation of rubber, sulphur cross-links are introduced at the reactive sites of double bonds.

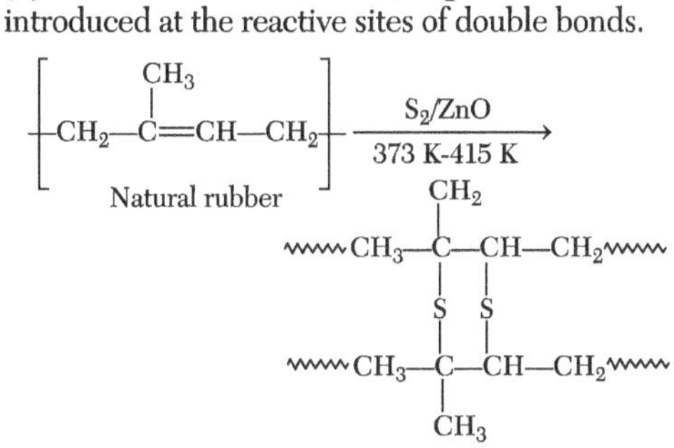

Vulcanised rubber

72. ***(a)*** **Hydroxylation reaction** The alkene on treatment with dil. $KMnO_4$ produces *vicinal cis* diol.

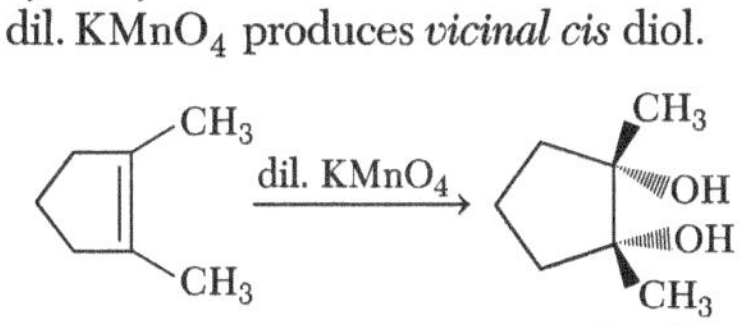

vicinal cis diol

Malaprade oxidation *cis* diol undergo malaprade oxidation in presence of HIO_4 and believe to proceed as

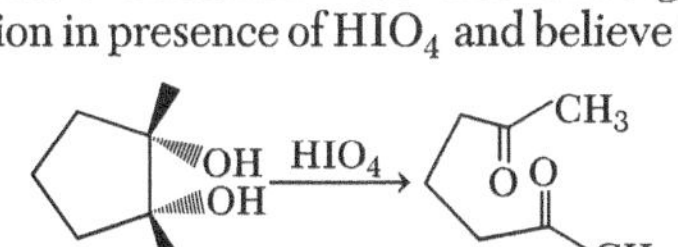

(*A*) (*B*)

Intramolecular aldol reaction This diketone undergo intramolecular aldol condensation to produce the cyclic α, β-unsaturated ketone as follows

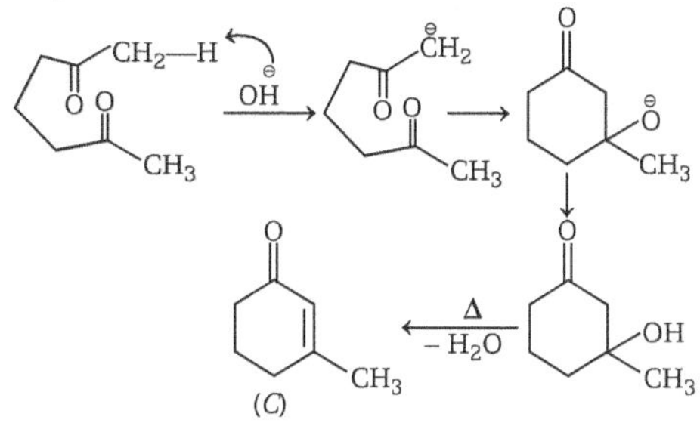

(*C*)

73. ***(d)*** This problem includes conceptual mixing of hydroboration. This problem can be solved by using the skill of electrophilic addition reaction in hydroboration oxidation reaction including isomerism. The steps to solve this problems are complete reaction with acid and then identify the isomerism in the out come products.

(i) **Hydroboration-oxidation** When alkene or substituted alkene is treated with B_2H_6, it gives alkyl borane which on treatment with H_2O_2/OH^- causes oxidation of alkyl borane to give alcohol. The above reaction is believed to proceed as

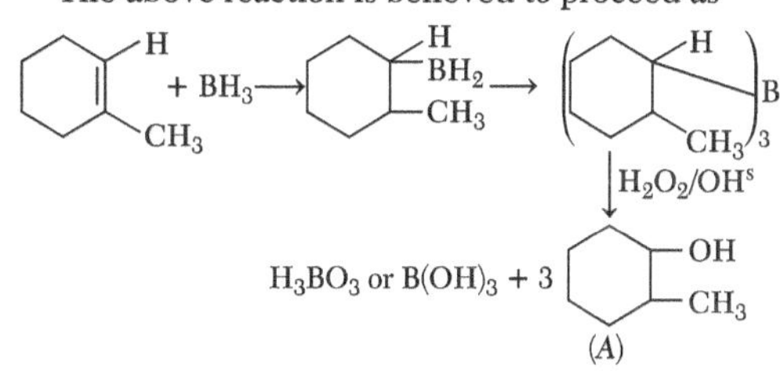

(*A*)

(ii) When 2-methyl cyclohexene is treated with H_2O/H_2SO_4.

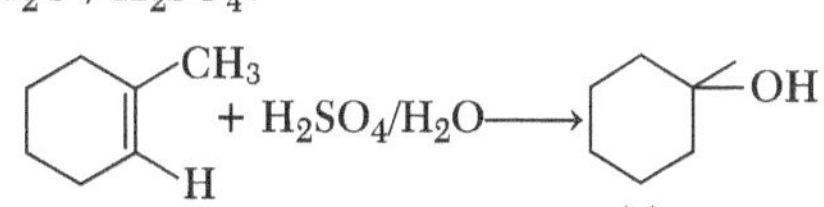

(*B*)

A and *B* have difference in position of OH only, so *A* and *B* are position isomers.

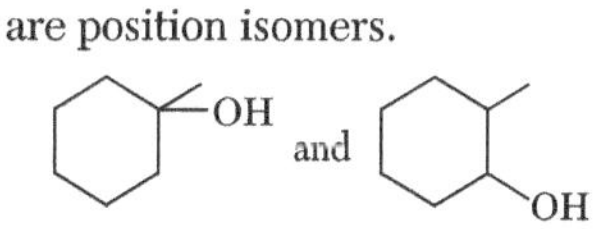

74. ***(c)*** Equilibrium constants at different temperature and heat of the reaction are related by the equation.

$$\ln \frac{K_{p_2}}{K_{p_1}} = \frac{\Delta H^\circ}{R}\left[\frac{1}{T_1} - \frac{1}{T_2}\right]$$

$$2.303 \log \frac{K_{p_2}}{K_{p_1}} = \frac{\Delta H^\circ}{R}\left[\frac{T_2 - T_1}{T_1 T_2}\right]$$

$$\log K_{p_2} = \frac{-25140}{2.303 \times 2}\left[\frac{773 - 673}{773 \times 673}\right] + \log(1.6 \times 10^{-4})$$

$$\log K_{p_2} = -4.835$$

$$\therefore\ K_{p_2} = 1.462 \times 10^{-5}\ (\text{atm})^{-2}$$

75. *(a)* $K_p = K_C (RT)^{\Delta n}$ $[\because \Delta n = 1]$

$$\Rightarrow \quad K_C = \frac{K_p}{RT} = \frac{0.1}{0.082 \times 300} = 4 \times 10^{-3}$$

76. *(b)* As we know, $\Delta H = \Delta E + \Delta n\, RT$

where, Δn = gaseous product moles – gaseous reactant moles

For the reaction,

$$CO(g) + \frac{1}{2} O_2(g) \longrightarrow CO_2(g)$$

$$\Delta n = 1 - \left(1 + \frac{1}{2}\right) = -\frac{1}{2}$$

$$\therefore \quad \Delta H = \Delta E - \frac{1}{2} RT$$

Hence, $\Delta H < \Delta E$

77. *(c)* This problem contains conceptual mixing of nomenclature of cyclic hydrocarbon and degree of unsaturation. This problem can be solved by identifying the parent chain, functional group, position of functional group, substituent and their position one by one and then write the name of compound according to IUPAC names then calculate degree of unsaturation.

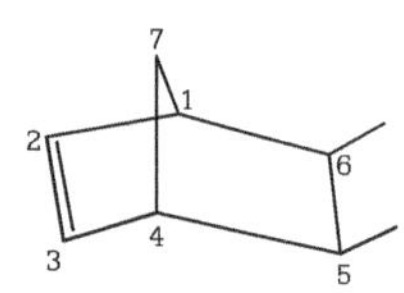

Total carbon atom forming the bicyclic ring = 7 (hept.)

functional group $\Rightarrow$ double bond (ene)

Position of double bond $\Rightarrow$ 2, 2-ene

Substituents $\Rightarrow$ 2-methyl group $\Rightarrow$ dimethyl

Position of substituents = 5, 6 $\longrightarrow$ 5, 6-dimethyl

Number of cyclic chain = 2 $\longrightarrow$ Bicyclo

3-bridges are of 2 carbons, 2 carbons and one carbon hence,

IUPAC name = 5, 6-dimethyl bicyclo [2, 2, 1] hept-2-ene

Molecular formula of compound is C_9H_{14}.

Degree of unsaturation can be calculated as

$$u = (C + 1) - \frac{H}{2} + \frac{N}{2}$$

where, u = degree of unsaturation

C = number of carbons

H = number of hydrogens

N = number of nitrogens

Hence, for a compound having molecular formula C_9H_{14}, the degree of unsaturation may be calculated as

$$u = (9 + 1) - \frac{14}{2} = 10 - 7 = 3$$

78. *(c)* Oxidation number of Na = + 1

Oxidation number of O = – 2

Let oxidation number of S = x

$\therefore$ 2 (O.N. of Na) + 4(O.N. of S) + 6(O.N. of O) = 0

$2(+1) + 4x + 6(-2) = 0$

$+2 + 4x - 12 = 0 \Rightarrow 4x = +12 - 2$

$$\Rightarrow \quad x = +\frac{10}{4}$$

$$\Rightarrow \quad x = +\frac{5}{2}$$

79. *(d)* Among the given antibiotics, only chloramphenicol contains a nitro group attached to aromatic ring. Its structure is as follows

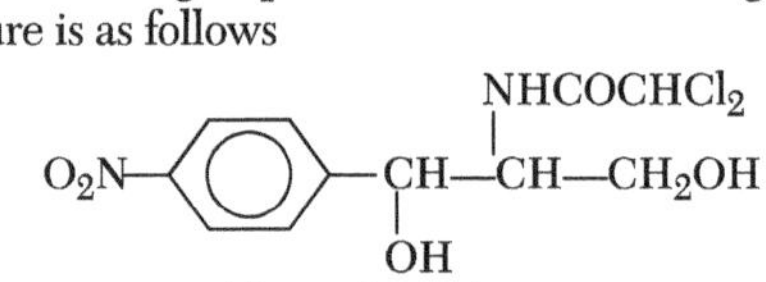

Chloramphenicol

80. *(a)* The deviation from ideal behaviour can be measured in terms of compressibility factor Z.

$$Z = \frac{pV}{nRT}$$

For ideal gas $Z = 1$,

i.e., $pV = nRT$

At very low pressures all gases shown have $Z \approx 1$ and behave as ideal gas. At high pressure all the gases have $Z > 1$ which show the deviation from ideality.

At intermediate pressures, most gases have $Z < 1$ which also show the deviation from ideality.

a. English Proficiency

81. *(b)* Sagacious having or showing keen mental discernment and good judgement; wise, shrewd, so 'wise' is correct answer.

82. *(a)* Remedial means done to correct or improve something, so 'corrective' is correct answer.

83. *(d)* Reticent means not revealing one's thoughts or feelings readily, so 'secretive' is correct answer.

84. *(d)* Fidelity means 'faithfulness in relations', so 'treachery' is its correct antonym.

85. *(b)* Infrangible means 'strong', so 'breakable' is its correct antonym.

86. *(b)* Progeny means 'child', so 'parent' is its correct antonym.

87. *(c)* Use of 'draw' is more suitable for using before word 'conclusion', so option (c) is correct.

88. ***(a)*** Use of 'looking for' is proper because 'look for' means 'to search for something' which suits here.

89. ***(d)*** 'Mind your language' is proper to use here because it gives proper sense of sentence.

90. ***(b)*** 'Use to' is used when any habit is to be shown, so use of option (b) is proper.

91. ***(c)*** 'Angry' agrees with preposition 'with', so use of option (c) is correct here.

92. ***(b)*** Laugh agrees with preposition 'at', so use of option (b) is correct here.

93. ***(c)*** According to sequence of events, so option (c) is best answer.

94. ***(b)*** According to sequence of events, option (b) best suits here.

95. ***(a)*** According to sequence of events, option (a) is proper here.

b. Logical Reasoning

96. ***(d)*** As

D ⟶ 8 and M ⟶ 4
O ⟶ 9 E ⟶ 3
M ⟶ 4 A ⟶ 2
E ⟶ 3 L ⟶ 1

In the same way 38249 will be coded as

3 ⟶ E
8 ⟶ D
2 ⟶ A
4 ⟶ M
9 ⟶ O

Hence, option (d) is correct.

97. ***(d)*** From fig. 1, 93 – (27 + 63) = 93 – 90 = 3
From fig. 2, 79 – (38 + 37) = 79 – 75 = 4
From fig. 3, 67 – (16 + 42) = 67 – 58 = 9
Hence, option (d) is correct.

98. ***(b)*** Some poor people can be unemployed, some unemployed people can be illiterates and some illiterates can be poor. Hence, correct diagram is

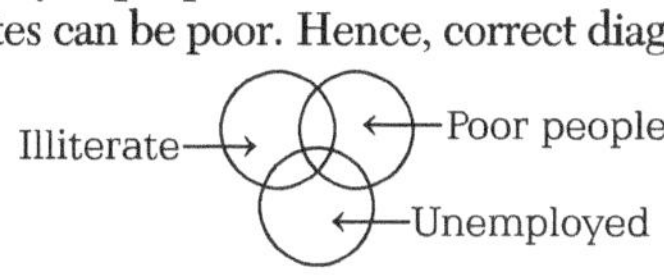

Hence, option (b) is correct.

99. ***(c)*** Starting point

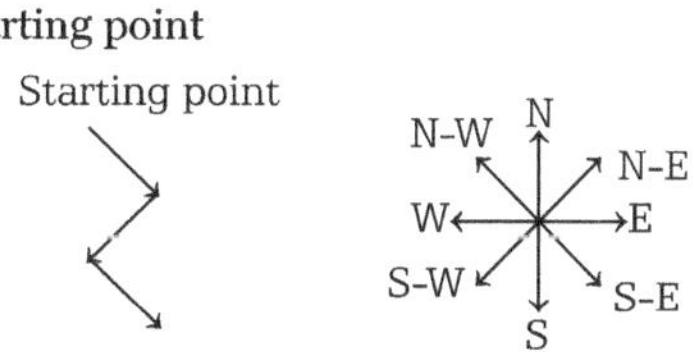

Hence, 'A' is moving in South-East direction.

Hence option (c) is correct.

100. ***(a)*** From the information given in the question the arrangement of students is

1st ⟶ D
2nd ⟶ E
3rd ⟶ C
4th ⟶ A
5th ⟶ B

Hence, option (a) is correct.

101. ***(b)*** The given series follows the pattern

$$1^3 + 1 = 1 + 1 = 2$$
$$2^3 + 1 = 8 + 1 = 9$$
$$3^3 + 1 = 27 + 1 = 28$$
$$4^3 + 1 = 64 + 1 = \boxed{65}$$
$$5^3 + 1 = 125 + 1 = 126$$

Hence, option (b) is correct.

102. ***(b)*** Clearly, option figure (b) completes the original figure which looks like the figure given below

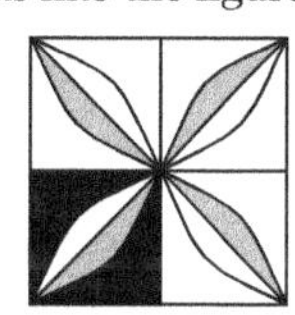

Hence, option (b) is correct.

103. ***(c)*** On labelling the figure, we get

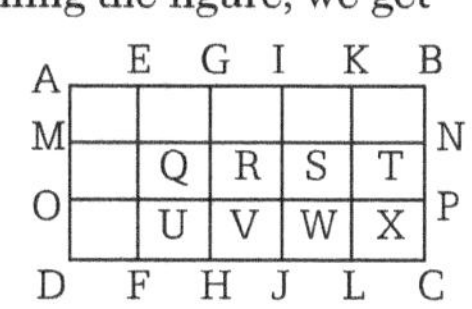

Each row contains five squares.

∴ Total number of single squares = 5 × 3 = 15

Now, combination of 4 small squares will be

= 8 (*i.e.*, AOVG, EUWI, GVXK, IWPB, MDHR, QFJS, RHLT and SJCN)

Now, combination of 9 small squares will be

= 3 (*i.e.*, ADJI, EFLK and GHCB)

∴ Total number of squares

= 15 + 8 + 3 = 26 squares

Hence, option (c) is correct.

Alternate method : Given

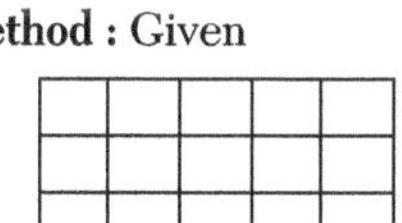

m = number of rows = 3 and n = number of columns = 5

BITSAT 2013

So, total number of squares

$= m \times n + (m-1) \times (n-1) + \dots$

$= 3 \times 5 + 2 \times 4 + 1 \times 3 = 15 + 8 + 3 = 26$

104. ***(d)*** Interchanging symbols + and – as given in option (a) the above equation becomes

$3 \div 5 \times 8 - 2 + 10$

$= \frac{3}{5} \times 8 - 2 + 10 = \frac{24}{5} + 8 \neq 13$

Interchanging symbols × and ÷ as given in option (b), we get

$3 \times 5 \div 8 + 2 - 10$

$= 3 \times \frac{5}{8} + 2 - 10 = \frac{15}{8} - 8 \neq 13$

Interchanging symbols ÷ and – as given in option (c), we get

$3 - 5 \times 8 + 2 \div 10$

$= 3 - 5 \times 8 + \frac{2}{10}$

$= 3 - 40 + \frac{2}{10} \neq 13$

Interchanging symbols ÷ and + as given in option (d), we get

$3 + 5 \times 8 \div 2 - 10$

$= 3 + 5 \times \frac{8}{2} - 10$

$= 3 + 20 - 10 = 13$

Hence, option (d) is correct.

105. ***(b)*** Both Assertion and Reason are correct but India is a democratic country because the government is elected by its citizens and not because India has its own constitution.

Hence, option (b) is correct.

Mathematics

106. ***(d)*** Given, $\sum_{k=1}^{6}\left(\frac{\sin 2\pi k}{7} - \frac{i\cos 2\pi k}{7}\right)$

$= \sum_{k=1}^{6} - i\left(\frac{\cos 2\pi k}{7} + \frac{i\sin 2\pi k}{7}\right)$

$= -i\sum_{k=1}^{6} e^{i2\pi k/7} = -i[e^{i2\pi/7} + e^{i4\pi/7} + \dots e^{i12\pi/7}]$

$= -i\left[e^{i2\pi/7}\frac{(1-e^{i12\pi/7})}{1-e^{i2\pi/7}}\right] = -i\left[\frac{e^{i2\pi/7} - e^{i14\pi/7}}{1-e^{i2\pi/7}}\right]$

$= -i\left[\frac{e^{i2\pi/7} - e^{2\pi i}}{1-e^{i2\pi/7}}\right] = -i\left[\frac{e^{i2\pi/7} - 1}{1-e^{i2\pi/7}}\right] = i$

107. ***(c)*** Given, $n_1 = 60, \bar{x}_1 = 650, \sigma_1 = 8,$ $n_2 = 80, \bar{x}_2 = 660, \sigma_2 = 7$

∴ Combined SD

$= \sqrt{\frac{n_1\sigma_1^2 + n_2\sigma_2^2}{n_1+n_2} + \frac{n_1n_2(\bar{x}_1 - \bar{x}_2)^2}{(n_1+n_2)^2}}$

$= \sqrt{\frac{60 \times 64 + 80 \times 49}{60+80} + \frac{60 \times 80(650-660)^2}{(60+80)^2}}$

$= \sqrt{\frac{3840 + 3920}{140} + \frac{(4800 \times 100)}{(140)^2}}$

$= \sqrt{\frac{7760}{140} + \frac{480000}{19600}} = \sqrt{\frac{776}{14} + \frac{4800}{196}}$

$= \sqrt{55.42 + 24.49} = \sqrt{79.91} = 8.94$

108. ***(a)*** Since, $|a - a| = 0 < 1$, so $aRa, \forall\ a \in R$

∴ R is reflexive.

Now, $aRb \Rightarrow |a - b| \le 1 \Rightarrow |b - a| \le 1 \Rightarrow bRa$

∴ R is symmetric.

But R is not transitive

As, $1R2$, $2R3$ but $1 \not{R} 3$ $\quad [\because |1 - 3| = 2 > 1]$

109. ***(c)*** Given that,

$I = \int_0^{10\pi}([\sec^{-1}x] + [\cot^{-1}x])dx$

$= \int_0^{\sec 1}([\sec^{-1}x] + [\cot^{-1}x])dx + \int_{\sec 1}^{10\pi}([\sec^{-1}x] + [\cot^{-1}x])dx$

$= \int_0^{\sec 1}(0 + 0)dx + \int_{\sec 1}^{10\pi}(1 + 0)dx$

$= 0 + [x]_{\sec 1}^{10\pi} = 10\pi - \sec 1$

110. ***(d)*** $\sin\left[\cot^{-1}\left(\cos\frac{\pi}{4}\right)\right]$

$= \sin\left[\cot^{-1}\frac{1}{\sqrt{2}}\right] = \sin\left[\sin^{-1}\sqrt{\frac{2}{3}}\right] = \sqrt{\frac{2}{3}}$

111. ***(b)*** Let

$S = 1 + 2 \cdot 2 + 3 \cdot 2^2 + 4 \cdot 2^3 + \dots + 100 \cdot 2^{99}$ ...(i)

It is an arithmetico-geometric series. On multiplying Eq. (i) by 2 and then subtracting, we get

$S = 1 + 2 \cdot 2 + 3 \cdot 2^2 + 4 \cdot 2^3 + \dots + 100 \cdot 2^{99}$

$2S = 1 \cdot 2 + 2 \cdot 2^2 + 3 \cdot 2^3 + \dots + 99 \cdot 2^{99} + 100 \cdot 2^{100}$

$-S = 1 + 2 + 2^2 + 2^3 + \dots + 2^{99} - 100 \cdot 2^{100}$

$\Rightarrow -S = \frac{1(2^{100} - 1)}{2 - 1} - 100 \cdot 2^{100}$

$\Rightarrow -S = 2^{100} - 1 - 100 \cdot 2^{100}$

$\Rightarrow -S = -1 - 99 \cdot 2^{100}$

$\Rightarrow \ S = 99 \cdot 2^{100} + 1$

112. ***(c)*** Consider a point (2, 0) on the X-axis.

Substituting $x = 2$, $y = 0$ in

$$3x + 12y = 6 < 400.$$

Hence, one constraints is $3x + 12y \le 400$

Again, substituting $x = 2$, $y = 0$ in

$$x - 4y = 2 - 0 > 0$$

$\therefore$ $x - 4y \ge 0$ is other constraints or $x \ge 4y$

and also the third constraint from the figure is $y \le 25$.

Hence, the correct alternative is (c).

113. ***(b)*** $\log\left(\dfrac{1}{1+x+x^2+x^3}\right) = \log\left(\dfrac{1-x}{1-x^4}\right)$

$$= \log_e(1-x) - \log(1-x^4) = -\sum_{r=1}^{\infty}\frac{x^r}{r} + \sum_{r=1}^{\infty}\frac{x^{4r}}{r}$$

When n is odd, there is no term in the second series containing x^n, therefore the coefficient x^n is zero in the second series and in the first series the coefficient of x^n is $-\dfrac{1}{n}$. Hence, when n is odd, then the coefficient of x^n in the whole expansion is $-\dfrac{1}{n} + 0 = -\dfrac{1}{n}$.

114. ***(d)*** For maximum value, find $f'(x)$

$$f'(x) = \frac{1}{x^2}(1 - \log x)$$

$f'(x) > 0$ for $x < e$ and $f'(x) < 0$ for $x > e$

$\Rightarrow f(x)$ is increasing for $x < e$ and decreasing for $x > e$

$\Rightarrow x = e$ is the point of local maxima.

$\therefore$ Maximum value of $f(x) = 1/e$

Hence, the answer is (d).

115. ***(a)*** We have, $(\mathbf{a} \times \mathbf{b}) \times \mathbf{c} = \dfrac{1}{3}|\mathbf{b}||\mathbf{c}|\mathbf{a}$

$$\Rightarrow \quad (\mathbf{a}\cdot\mathbf{c})\mathbf{b} - (\mathbf{b}\cdot\mathbf{c})\mathbf{a} = \frac{1}{3}|\mathbf{b}||\mathbf{c}|\mathbf{a}$$

$$\Rightarrow (\mathbf{a}\cdot\mathbf{c})\mathbf{b} - \{(\mathbf{b}\cdot\mathbf{c}) + \frac{1}{3}|\mathbf{b}||\mathbf{c}|\}\mathbf{a} = 0$$

$$\Rightarrow \quad (\mathbf{a}\cdot\mathbf{c}) - 0 \text{ and } \mathbf{b}\cdot\mathbf{c} + \frac{1}{3}|\mathbf{b}||\mathbf{c}| - 0$$

($\because \theta$ is the angle between $\mathbf{b}$ and $\mathbf{c}$)

$$\Rightarrow |\mathbf{b}||\mathbf{c}|\cos\theta + \frac{1}{3}|\mathbf{b}||\mathbf{c}| = 0 \Rightarrow \cos\theta = -\frac{1}{3}$$

$$\therefore \quad \sin\theta = \sqrt{\frac{8}{9}} = \frac{2\sqrt{2}}{3}$$

116. ***(a)*** $\lim\limits_{x\to 0}\left(\dfrac{1+5x^2}{1+3x^2}\right)^{1/x^2} = \lim\limits_{x\to 0}\left(1 + \dfrac{2x^2}{1+3x^2}\right)^{1/x^2}$

$$= e^{\lim\limits_{x\to 0}\left(\frac{2x^2}{1+3x^2}\right)} = e^2$$

117. ***(c)*** Let $ED = h$, $\angle EAB = \alpha$

$\therefore$ $\angle EBD = 2\alpha$, $\angle ECD = 3\alpha$

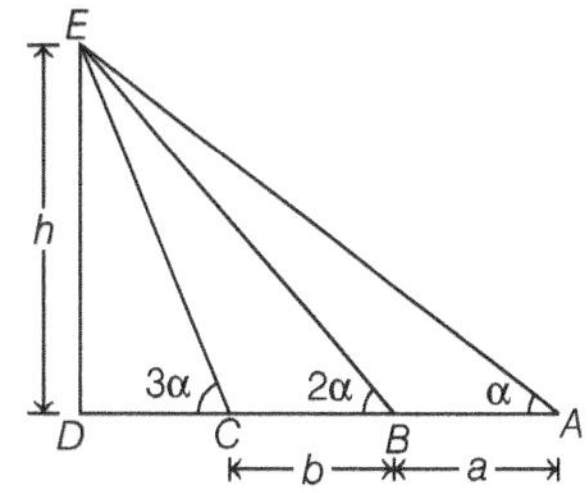

Now, $\quad \angle DBE = \angle EAB + \angle BEA$

$\Rightarrow \quad 2\alpha = \alpha + \angle BEA$

$\Rightarrow \quad \angle BEA = \alpha = \angle EAB$

$\Rightarrow \quad AB = EB = a$

Similarly, $\quad \angle BEC = \alpha$

From ΔEBC, $\dfrac{BC}{\sin\alpha} = \dfrac{EB}{\sin(180^\circ - 3\alpha)}$

$$\Rightarrow \quad \frac{b}{\sin\alpha} = \frac{a}{\sin 3\alpha}$$

$$\Rightarrow \quad \frac{a}{b} = \frac{\sin 3\alpha}{\sin\alpha}$$

$$\Rightarrow \quad \frac{a}{b} = \frac{3\sin\alpha - 4\sin^3\alpha}{\sin\alpha} = 3 - 4\sin^2\alpha$$

$$\Rightarrow \quad 4\sin^2\alpha = 3 - \frac{a}{b} = \frac{3b-a}{b}$$

$$\Rightarrow \quad \sin\alpha = \sqrt{\frac{3b-a}{4b}}$$

From ΔEBD, $\sin 2\alpha = \dfrac{ED}{EB}$

$$\Rightarrow \quad ED = a \cdot 2\sin\alpha \cdot \cos\alpha$$

$$\Rightarrow \quad h = 2a\sqrt{\frac{3b-a}{4b}} \cdot \sqrt{1 - \frac{3b-a}{4b}}$$

$$= 2a\sqrt{\frac{3b-a}{4b}}\sqrt{\frac{b+a}{4b}}$$

$$= \frac{a}{2b}\sqrt{(a+b)(3b-a)}$$

118. ***(d)*** We have, $\quad f(x) = e^{x^3 - 3x + 2}$

Let $\quad h(x) = x^3 - 3x + 2$

$\therefore \quad h'(x) = 3x^2 - 3 = 3(x^2 - 1)$

$\Rightarrow \quad h'(x) \ge 0$ for $x \in (-\infty, -1]$

$\therefore$ $f(x)$ is increasing function.

$\therefore$ $f(x)$ is one-one.

Now, range of $f(x) = (0, e^4]$

But codomain of $f(x) = (0, e^5]$

$\therefore$ $f(x)$ is an into function.

119. *(c)* Given equation can be rewritten as

$$\frac{(x-3)^2}{16}+\frac{y^2}{25}=1$$

$\therefore \quad a^2=16$ and $b^2=25$

Now, $$e=\sqrt{1-\frac{16}{25}}=\sqrt{\frac{25-16}{25}}=\sqrt{\frac{9}{25}}=\frac{3}{5}$$

Hence, the foci of conic section are $(3, \pm be)$ *i.e.*, $(3, \pm 3)$.

So, option (c) is correct.

120. *(c)* Let $D=\begin{vmatrix}1 & -1 & 3\\ 1 & 0 & 1\\ 1 & 1 & -1\end{vmatrix}$

$$=1(0-1)-1(1+1)+3(1)=-1-2+3=0$$

$$D_1=\begin{vmatrix}4 & -1 & 3\\ 2 & 0 & 1\\ 0 & 1 & -1\end{vmatrix}=4(0-1)-1(0+2)+3(2)$$

$$=-4-2+6=0$$

$$D_2=\begin{vmatrix}1 & 4 & 3\\ 1 & 2 & 1\\ 1 & 0 & -1\end{vmatrix}$$

$$=1(-2)+4(1+1)+3(0-2)=-2+8-6=0$$

$$D_3=\begin{vmatrix}1 & -1 & 4\\ 1 & 0 & 2\\ 1 & 1 & 0\end{vmatrix}$$

$$=1(0-2)-1(2-0)+4(1-0)$$

$$=-2-2+4=0$$

Hence, the given system of equations has infinitely many solutions.

121. *(b)* $S_n=5+55+555+\dots$ upto n terms

$=5[1+11+111+\dots$ upto n terms$]$

$=\frac{5}{9}[9+99+999+\dots$ upto n terms$]$

$=\frac{5}{9}[(10-1)+(10^2-1)+(10^3-1)+\dots$ upto n terms$]$

$=\frac{5}{9}[(10+10^2+10^3+\dots$ upto n terms$)$

$-(1+1+1+1+\dots$ upto n terms$)]$

$$=\frac{5}{9}\left[\frac{10(10^n-1)}{10-1}-n\right]=\frac{5}{9}\left[\frac{10(10^n-1)}{9}-n\right]$$

122. *(d)* Let the parallel plane to $2x-2y+z=0$ is $2x-2y+z+\lambda=0$

It passes through $(1,-2,3)$.

$\therefore \quad 2+4+3+\lambda=0$

$\Rightarrow \quad \lambda=-9$

The distance of $(-1,2,0)$ from the plane

$2x-2y+z-9=0$ is $\left|\frac{-2-4-9}{\sqrt{4+4+1}}\right|=\left|\frac{-15}{3}\right|=5$

123. *(c)* Clearly, we have

$a=1, h=-3, b=9, g=\frac{3}{2}, f=\frac{-9}{2}$ and $c=-4$

$$\text{Required distance}=\left|2\sqrt{\frac{f^2-bc}{b(a+b)}}\right|$$

$$=\left|2\sqrt{\frac{\left(\frac{-9}{2}\right)^2+9\times 4}{9(9+1)}}\right|$$

$$=\left|2\sqrt{\frac{225}{4\times 90}}\right|=\left|\frac{2\sqrt{5}}{2\sqrt{2}}\right|=\sqrt{\frac{5}{2}}$$

124. *(a)* $A=\{1,-1,i,-i\}$

$B=\{1,-1\},\ C=\{i,-i\}$

Now, $B\cup C=\{1,-1,i,-i\}=A$

125. *(b)* We have, $\frac{dy}{dx}+\sin\left(\frac{x+y}{2}\right)=\sin\left(\frac{x-y}{2}\right)$

$$\Rightarrow \frac{dy}{dx}=\sin\left(\frac{x-y}{2}\right)-\sin\left(\frac{x+y}{2}\right)$$

$$\Rightarrow \frac{dy}{dx}=-2\cos\left(\frac{x}{2}\right)\sin\left(\frac{y}{2}\right)$$

$$\Rightarrow \frac{dy}{\sin\left(\frac{y}{2}\right)}=-2\cos\left(\frac{x}{2}\right)dx$$

On integrating both sides, we get

$$\int\frac{dy}{\sin\left(\frac{y}{2}\right)}=-2\int\cos\left(\frac{x}{2}\right)dx$$

$$\Rightarrow \frac{1}{2}\int\operatorname{cosec}\left(\frac{y}{2}\right)dy=-\int\cos\left(\frac{x}{2}\right)dx$$

$$\Rightarrow \frac{1}{2}\cdot\left[\frac{\log\left\{\operatorname{cosec}\left(\frac{y}{2}\right)-\cot\left(\frac{y}{2}\right)\right\}}{\frac{1}{2}}\right]=-\frac{\sin\left(\frac{x}{2}\right)}{\frac{1}{2}}+C$$

$$\Rightarrow \log\left[\frac{1}{\sin\left(\frac{y}{2}\right)}-\frac{\cos\left(\frac{y}{2}\right)}{\sin\left(\frac{y}{2}\right)}\right]=-2\sin\left(\frac{x}{2}\right)+C$$

$$\Rightarrow \log\left[\frac{2\sin^2\left(\frac{y}{4}\right)}{2\sin\left(\frac{y}{4}\right)\cos\left(\frac{y}{4}\right)}\right]=-2\sin\left(\frac{x}{2}\right)+C$$

$(\because 1-\cos x=2\sin^2\frac{x}{2}$ and $\sin x=2\sin x/2\ \cos x/2)$

$$\Rightarrow \log\tan\left(\frac{y}{4}\right)=C-2\sin\left(\frac{x}{2}\right)$$

126. ***(a)*** Given, $\frac{3-|x|}{4-|x|} \geq 0$

$$\Rightarrow \quad 3-|x| \leq 0 \text{ and } 4-|x| < 0$$
$$\text{or} \quad 3-|x| \geq 0 \text{ and } 4-|x| > 0$$
$$\Rightarrow \quad |x| \geq 3 \text{ and } |x| > 4$$
$$\text{or} \quad |x| \leq 3 \text{ and } |x| < 4$$
$$\Rightarrow \quad |x| > 4 \text{ or } |x| \leq 3$$
$$\Rightarrow x \in (-\infty, -4) \cup [-3, 3] \cup (4, \infty)$$

127. ***(a)*** $\frac{N}{x_1 + x_2 + x_3 + x_4}$

$$= \frac{1000x_1 + 100x_2 + 10x_3 + x_4}{x_1 + x_2 + x_3 + x_4}$$
$$= 1000 - \left(\frac{900x_2 + 990x_3 + 999x_4}{x_1 + x_2 + x_3 + x_4}\right)$$

$\Rightarrow$ Maximum value is 1000.

Hence, answer is (a).

128. ***(c)*** $\because P(A \cap B) = P(A/B) \cdot P(B)$

$$= 0.5 \times 0.2 = 0.1$$
$$\therefore \quad P(A'/B') = \frac{P(A' \cap B')}{P(B')} = \frac{P(A \cup B)'}{P(B')}$$
$$= \frac{1 - P(A \cup B)}{1 - P(B)}$$
$$= \frac{1 - P(A) - P(B) + P(A \cap B)}{1 - 0.2}$$
$$= \frac{1 - 0.6 - 0.2 + 0.1}{0.8} = \frac{3}{8}$$

129. ***(d)*** Here, $N = \Sigma f = 20$

$$Q_1 = \frac{(N+1)\text{th}}{4} \text{ observation}$$
$$= \left(\frac{21}{4}\right)\text{th observation}$$
$$= 3$$
$$\text{Similarly, } Q_3 = \frac{3(N+1)\text{th}}{4} \text{ observation}$$
$$= \left(\frac{63}{4}\right)\text{th observation}$$
$$= 5$$
$$\text{Now, quartile deviation } = \frac{1}{2}(Q_3 - Q_1)$$
$$= \frac{1}{2}(5 - 3) = 1$$

130. ***(d)*** Given that, $\int f(x) \cos x \, dx = \frac{1}{2} f^2(x) + C$

On differentiating w.r.t. x, we get

$$f(x) \cos x = \frac{1}{2} \cdot 2f(x) \cdot f'(x)$$
$$\Rightarrow \quad \cos x = f'(x) \Rightarrow \cos x = \frac{d}{dx}(f(x))$$
$$\int \cos x \, dx = f(x)$$
$$f(x) = \sin x + C$$

131. ***(a)*** **Case 1** Taking 2 points from collinear points and one from non-collinear.

i.e., number of triangles so formed = ${}^6C_2 \times {}^4C_1$

$$= \frac{6 \cdot 5}{1 \cdot 2} \times \frac{4}{1} = 60$$

Case 2 Taking 1 point from collinear and two from non-collinear points

i.e., number of triangles so formed = ${}^6C_1 \times {}^4C_2$

$$= \frac{6}{1} \times \frac{4 \cdot 3}{1 \cdot 2} = 36$$

Case 3 All the three points from non-collinear points.

i.e., number of triangles so formed

$$= {}^4C_3 = {}^4C_1 = \frac{4}{1} = 4$$

Total number of triangles = 60 + 36 + 4 = 100

Alternatively Number of triangles

$$= {}^{10}C_3 - {}^6C_3 = \frac{10 \cdot 9 \cdot 8}{1 \cdot 2 \cdot 3} - \frac{6 \cdot 5 \cdot 4}{1 \cdot 2 \cdot 3}$$
$$= 120 - 20 = 100$$

Hence, option (a) is correct answer.

132. ***(d)*** $\frac{{}^8C_0}{6} - {}^8C_1 + {}^8C_2 \cdot 6 - {}^8C_3 \cdot 6^2 + {}^8C_4 \cdot 6^3 + \ldots + {}^8C_8 \cdot 6^7$

$$= \frac{1}{6}[{}^8C_0 - {}^8C_1 \cdot 6^1 + {}^8C_2 \cdot 6^2 - {}^8C_3 \cdot 6^3 + \ldots + {}^8C_8 \cdot 6^8]$$
$$= \frac{1}{6}[1 - 6]^8 = \frac{1}{6} \times (-5)^8 = \frac{5^8}{6}$$

Hence, the answer is (d).

133. ***(d)*** Let A denote the event that atleast one girl will be chosen and B be the event that exactly 2 girls will be chosen. We require $P(B/A)$.

Since, A denotes the event that atleast one girl will be chosen, A' denotes that no girl is chosen, *i.e.*, 4 boys are chosen.

$$\text{Then, } P(A') = \frac{{}^8C_4}{{}^{12}C_4} = \frac{70}{495} = \frac{14}{99}$$
$$\Rightarrow \quad P(A) = 1 - \frac{14}{99} = \frac{85}{99}$$

Now, $P(A \cap B) = P$ (2 boys and 2 girls)

$$= \frac{{}^8C_2 \cdot {}^4C_2}{{}^{12}C_4} = \frac{28 \times 6}{495} = \frac{56}{165}$$
$$\therefore \quad P\left(\frac{B}{A}\right) = \frac{P(A \cap B)}{P(A)} = \frac{56 \times 99}{165 \times 85} = \frac{168}{425}$$

134. ***(b)*** Here, we observe that

(a) $f(x)$ is a polynomial, so it is continuous in the interval $[0, 2]$.

(b) $f'(x) = 3x^2 - 6x + 2$ exists for all $x \in (0, 2)$.

So, $f(x)$ is differentiable for all $x \in (0, 2)$

(c) $f(0) = 0$, $f(2) = 2^3 - 3(2)^2 + 2(2) = 0$

$\therefore \quad f(0) = f(2)$

Thus, all the three conditions of Rolle's theorem are satisfied.

So, there must exist $c \in (0, 2)$ such that

$$f'(c) = 0$$

$$\Rightarrow f'(c) = 3c^2 - 6c + 2 = 0$$

$$\Rightarrow \quad c = 1 \pm \frac{1}{\sqrt{3}} \Rightarrow c \in (0, 2)$$

135. ***(a)*** Consider that, $I = \int \frac{4}{\sin^4 x + \cos^4 x} dx$

$$I = \int \frac{4}{\cos^4 x (\tan^4 x + 1)} dx$$

$$= \int \frac{4 \sec^4 x}{1 + \tan^4 x} dx$$

$$= 4 \int \frac{\sec^2 x (1 + \tan^2 x)}{1 + \tan^4 x} dx$$

Put $\tan x = t \Rightarrow \sec^2 x \, dx = dt$

$$I = 4 \int \frac{1 + t^2}{1 + t^4} dt = 4 \int \frac{1 + 1/t^2}{t^2 + 1/t^2} dt$$

$$\Rightarrow \quad I = 4 \int \frac{1 + 1/t^2}{\left(t - \frac{1}{t}\right)^2 + 2} dt$$

Now, put $t - \frac{1}{t} = z \Rightarrow \left(1 + \frac{1}{t^2}\right) dt = dz$

$$\therefore \quad I = 4 \int \frac{dz}{z^2 + (\sqrt{2})^2} = \frac{4}{\sqrt{2}} \tan^{-1}\left(\frac{z}{\sqrt{2}}\right) + C$$

$$\Rightarrow I = 2\sqrt{2} \tan^{-1} \left(\frac{\tan x - \frac{1}{\tan x}}{\sqrt{2}} \right) + C$$

$\therefore \quad a = 2\sqrt{2}$ and $b = \sqrt{2}$

136. ***(c)*** Given, $A = \begin{bmatrix} -5 & -8 & 0 \\ 3 & 5 & 0 \\ 1 & 2 & -1 \end{bmatrix}$

$$A^2 = \begin{bmatrix} -5 & -8 & 0 \\ 3 & 5 & 0 \\ 1 & 2 & -1 \end{bmatrix} \begin{bmatrix} -5 & -8 & 0 \\ 3 & 5 & 0 \\ 1 & 2 & -1 \end{bmatrix}$$

$$= \begin{bmatrix} 25 - 24 + 0 & 40 - 40 + 0 & 0 + 0 + 0 \\ -15 + 15 + 0 & -24 + 25 + 0 & 0 + 0 + 0 \\ -5 + 6 - 1 & -8 + 10 - 2 & 0 + 0 + 1 \end{bmatrix}$$

$$= \begin{bmatrix} 1 & 0 & 0 \\ 0 & 1 & 0 \\ 0 & 0 & 1 \end{bmatrix} = I$$

As $\quad A^2 = I$

$\therefore$ A is involutory.

137. ***(a)*** Given, $\frac{x^2}{16} + \frac{y^2}{9} = 1$

$$\therefore \quad e = \sqrt{1 - \frac{9}{16}} = \sqrt{\frac{16 - 9}{16}} = \sqrt{\frac{7}{16}} = \frac{\sqrt{7}}{4}$$

$\therefore$ Coordinates of foci are $(\pm \sqrt{7}, 0)$.

Since, centre of circle is $(0, 3)$ and passing through foci $(\pm \sqrt{7}, 0)$.

$$\therefore \quad \text{Radius of the circle} = \sqrt{(0 \pm \sqrt{7})^2 + (3 - 0)^2}$$

$$= \sqrt{7 + 9} = 4$$

Hence, option (a) is correct.

138. ***(d)*** The two normal vectors are $\mathbf{m} = 2\hat{\mathbf{i}} + 3\hat{\mathbf{j}} + \hat{\mathbf{k}}$

and $\mathbf{n} = \hat{\mathbf{i}} + 3\hat{\mathbf{j}} + 2\hat{\mathbf{k}}$

The line L is along, $\mathbf{m} \times \mathbf{n} = \begin{vmatrix} \hat{\mathbf{i}} & \hat{\mathbf{j}} & \hat{\mathbf{k}} \\ 2 & 3 & 1 \\ 1 & 3 & 2 \end{vmatrix}$

$$= \hat{\mathbf{i}}(6 - 3) - \hat{\mathbf{j}}(4 - 1) + \hat{\mathbf{k}}(6 - 3)$$

$$= 3\hat{\mathbf{i}} - 3\hat{\mathbf{j}} + 3\hat{\mathbf{k}} = 3(\hat{\mathbf{i}} - \hat{\mathbf{j}} + \hat{\mathbf{k}})$$

Now, the direction cosines of X-axis are $(1, 0, 0)$.

$$\therefore \cos \alpha = \frac{3(\hat{\mathbf{i}} - \hat{\mathbf{j}} + \hat{\mathbf{k}}) \cdot \hat{\mathbf{i}}}{\sqrt{3^2(1^2 + 1^2 + 1^2)}\sqrt{1}} = \frac{3}{3\sqrt{3}} = \frac{1}{\sqrt{3}}$$

$$\Rightarrow \quad \cos \alpha = \frac{1}{\sqrt{3}}$$

139. ***(a)*** In ΔOCA, $\tan 30° = \frac{AC}{OC}$

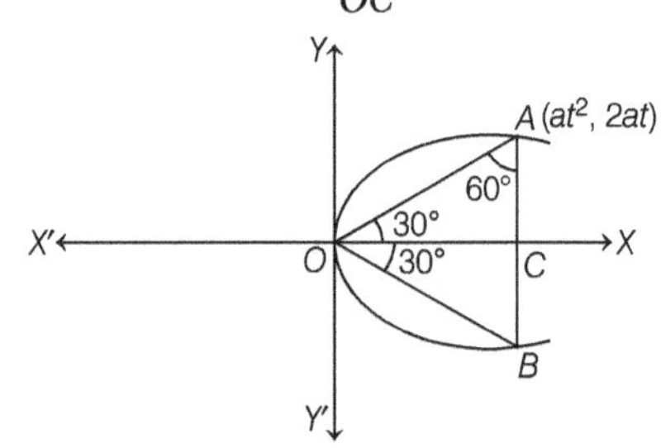

$$\Rightarrow \quad \frac{1}{\sqrt{3}} = \frac{2at}{at^2} \Rightarrow t = 2\sqrt{3}$$

Again, in ΔOCA,

$$OA = \sqrt{OC^2 + AC^2} = \sqrt{(at^2)^2 + (2at)^2}$$

$$= \sqrt{[(2\sqrt{3})^2]^2 a^2 + 4a^2 (2\sqrt{3})^2}$$

$$= \sqrt{192a^2} = 8a\sqrt{3}$$

Hence, option (a) is correct.

140. ***(a)*** Given, $f(x+y)=f(x)\cdot f(y)$

Put $x=y=0$, then $f(0)=1$

and put $y=-x$, then $f(0)=f(x)\,f(-x)$

$$\Rightarrow \quad f(-x)=\frac{1}{f(x)}$$

Now consider, $g(x)=\dfrac{f(x)}{1+\{f(x)\}^2}$

$$\Rightarrow \quad g(-x)=\frac{f(-x)}{1+\{f(-x)\}^2}=\frac{\frac{1}{f(x)}}{1+\left\{\frac{1}{f(x)}\right\}^2}$$

$$=\frac{f(x)}{1+\{f(x)\}^2}=g(x)$$

141. ***(a)*** Given $f(x)=(\tan^{-1}x)^2+\dfrac{2}{\sqrt{x^2+1}}$

$$f'(x)=\frac{2}{1+x^2}\left[\tan^{-1}x-\frac{x}{\sqrt{1+x^2}}\right]$$

Let $\quad g(x)=\tan^{-1}x-\dfrac{x}{\sqrt{x^2+1}}$

$$\Rightarrow \quad g'(x)=\frac{1}{1+x^2}\left[1-\frac{1}{\sqrt{x^2+1}}\right]>0 \text{ for all } x\in R$$

$\Rightarrow g(x)$ is increasing for all $x\in R$.

But $\quad g(0)=0 \Rightarrow g(x)>0$ for $x>0$

So, $\quad f'(x)>0$ for $x>0$

Hence, $f(x)$ is increasing in $(0,\infty)$.

142. ***(c)*** Clearly, $1+\sin x\geq 0$

$\therefore$ The given equation becomes

$$\cos x-\sin x=1$$

$$\Rightarrow \quad \cos x\cdot\frac{1}{\sqrt{2}}-\sin x\cdot\frac{1}{\sqrt{2}}=\frac{1}{\sqrt{2}}$$

$$\Rightarrow \quad \cos\left(x+\frac{\pi}{4}\right)=\frac{1}{\sqrt{2}}$$

$$\Rightarrow \quad x+\frac{\pi}{4}=\frac{\pi}{4},\frac{7\pi}{4},\frac{9\pi}{4},\frac{15\pi}{4}$$

$$\Rightarrow \quad x=0,\frac{3\pi}{2},2\pi,\frac{7\pi}{2}$$

$$\because \quad 0\leq x\leq 3\pi$$

$$\Rightarrow \quad x=0,\frac{3\pi}{2},2\pi$$

143. ***(a)*** Given, $\quad a^{\frac{1}{x}}=b^{\frac{1}{y}}=c^{\frac{1}{z}}$

Let $\quad a^{\frac{1}{x}}=b^{\frac{1}{y}}=c^{\frac{1}{z}}=k$

$$\therefore \quad a=k^x,\ b=k^y,\ c=k^z$$

$\because a, b, c$ are in GP.

Therefore, $\quad b^2=ac \Rightarrow (k^y)^2=k^x\cdot k^z$

$$\Rightarrow \quad k^{2y}=k^{x+z}\Rightarrow 2y=x+z$$

$\therefore$ x, y and z are in AP.

144. ***(d)*** Given that, $\quad 2l-m+2n=0 \quad$ ...(i)

and $\quad lm+mn+nl=0 \quad$ ...(ii)

From Eq. (i), $m=2(l+n)$ put in Eq. (ii),

$$2l(l+n)+2n(l+n)+nl=0$$

$$\Rightarrow \quad 2l^2+2nl+2nl+2n^2+nl=0$$

$$\Rightarrow \quad 2l^2+5nl+2n^2=0$$

$$\Rightarrow \quad 2l^2+4nl+nl+2n^2=0$$

$$\Rightarrow \quad 2l(l+2n)+n(l+2n)=0$$

$$\Rightarrow \quad (l+2n)(n+2l)=0$$

$$\Rightarrow \quad l=-2n \text{ and } n=-2l$$

If $\quad l=-2n$, then $m=2(-2n+n)=-2n$

and if $n=-2l$, then $m=2(l-2l)=-2l$

The DR's are $1,-2,-2$ and $-2,-2,1$.

Now, $1(-2)-(2)(-2)-2(1)=-2+4-2=0$

Hence, lines are perpendicular, so angle between them is $\pi/2$.

145. ***(d)*** Let m be the slope of required line.

$$\therefore \quad \left|\frac{m-(-1)}{1+m(-1)}\right|=1$$

$$\Rightarrow \quad \frac{m+1}{1-m}=\pm 1$$

$$\Rightarrow \quad m+1=1-m$$

and $\quad m+1=-1+m$

$$\Rightarrow \quad m=0 \text{ and } m=\infty$$

$\therefore$ Equation of line through $(1,1)$ is

$$y-1=0,\ x-1=0$$

Hence, option (d) is correct.

146. ***(a)*** Given curve is

$$(y-x)^2=x^3$$

$$\Rightarrow \quad y-x=\pm x\sqrt{x}$$

$$\Rightarrow \quad y=x+x\sqrt{x} \quad \text{...(i)}$$

$$\Rightarrow \quad y=x-x\sqrt{x} \quad \text{...(ii)}$$

and $\quad x=1$

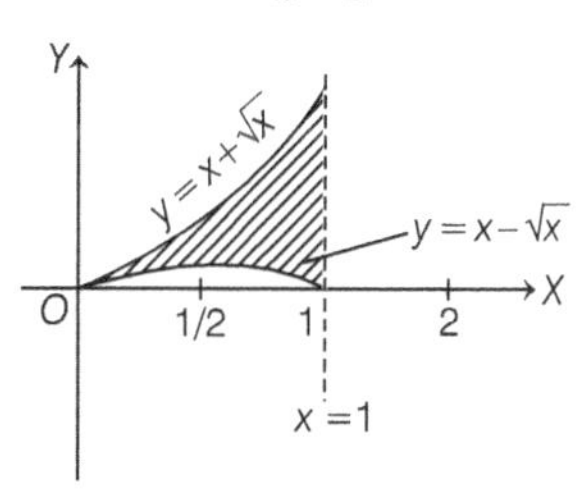

From the figure, required area

$$= \int_0^1 \{(x + x\sqrt{x}) - (x - x\sqrt{x})\}\, dx$$

$$= \int_0^1 2x\sqrt{x}\, dx = 2\int_0^1 x^{3/2}\, dx$$

$$= 2\left[\frac{x^{5/2}}{5/2}\right]_0^1 = \frac{4}{5}[1-0]$$

$$= \frac{4}{5} \text{ sq unit}$$

147. *(c)* $(x + iy)^{1/3} = 2 + 3i$

On cubing both sides, we get

$$x + iy = (2 + 3i)^3$$

$\Rightarrow \quad x + iy = (2)^3 + (3i)^3 + 3 \times 2 \times 3i(2 + 3i)$

$\Rightarrow \quad x + iy = 8 - 27i + 18i(2 + 3i)$

$\Rightarrow \quad x + iy = 8 - 27i + 36i - 54$

$\Rightarrow \quad x + iy = -46 + 9i$

On comparing real and imaginary both sides, we get

$$x = -46,\ y = 9$$

Then, $3x + 2y = 3(-46) + 2(9)$

$$= -138 + 18 = -120$$

148. *(b)* $n(A) = 40\%$ of $10000 = 4000$

$n(B) = 20\%$ of $10000 = 2000$

$n(C) = 10\%$ of $10000 = 1000$

$n(A \cap B) = 5\%$ of $10000 = 500$

$n(B \cap C) = 3\%$ of $10000 = 300$

$n(A \cap C) = 4\%$ of $10000 = 400$

$n(A \cap B \cap C) = 2\%$ of $10000 = 200$

To find $n[A \cap B^C \cap C^C]$

$$= n[A \cap (B \cup C)^C]$$

$$= n(A) - n[A \cap (B \cup C)]$$

$$= n(A) - [n(A \cap B) \cup n(A \cap C)]$$

$$= n(A) - [n(A \cap B) + n(A \cap C) - n(A \cap B \cap C)]$$

$$= 4000 - [500 + 400 - 200] = 3300$$

149. *(d)* Since, $\sin\theta = \dfrac{|\mathbf{a} \times \mathbf{b}|}{|\mathbf{a}||\mathbf{b}|} = \dfrac{8}{10} = \dfrac{4}{5}$

$\therefore \quad \cos\theta = \pm\dfrac{3}{5}$

$\therefore \quad \mathbf{a} \cdot \mathbf{b} = |\mathbf{a}||\mathbf{b}|\cos\theta = 2 \times 5 \times \left(\pm\dfrac{3}{5}\right) = \pm 6$

$\Rightarrow |\mathbf{a} \cdot \mathbf{b}| = 6$

150. *(a)* From figure, we have

$$OP = 5,\ OQ = 6 \text{ and } OM = \frac{5}{2},\ CM = 3$$

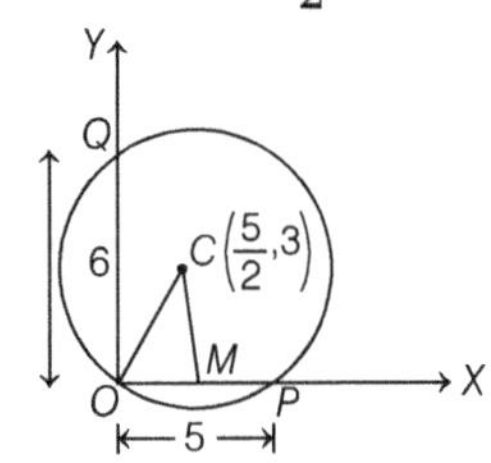

$\therefore$ In ΔOMC, $OC^2 = OM^2 + MC^2$

$\Rightarrow \quad OC^2 = \left(\dfrac{5}{2}\right)^2 + (3)^2 \Rightarrow OC = \dfrac{\sqrt{61}}{2}$

Thus, the required circle has its centre $\left(\dfrac{5}{2}, 3\right)$ and radius $\dfrac{\sqrt{61}}{2}$.

Hence, its equation is

$$\left(x - \frac{5}{2}\right)^2 + (y - 3)^2 = \left(\frac{61}{4}\right).$$

Hence, $\lambda = \dfrac{61}{4}$

BITSAT

Birla Institute of Technology & Science Admission Test

Solved Paper 2012*

Instructions

- There are 150 questions in all. The number of questions in each part is as given below.

		No. of Questions
Part I	Physics	1–40
Part II	Chemistry	41–80
Part III	a. English Proficiency	81–95
	b. Logical Reasoning	96–105
Part IV	Mathematics	106–150

- All questions are Multiple Choice Questions having four options out of which **only one** is correct.
- Each correct answer fetches 3 marks while incorrect answer has a penalty of 1 mark.
- Time allotted to complete this paper is 3 hrs.

PART I

Physics

1. A straight wire of mass 200 g and length 1.5 m carries a current of 2 A. It is suspended in mid-air by a uniform horizontal magnetic field B. The magnitude of B (in tesla) is (Assume, $g = 9.8 \text{ ms}^{-2}$)

a. 2 *b.* 1.5 *c.* 0.55 *d.* 0.65

2. In the circuit shown the value of I in ampere is

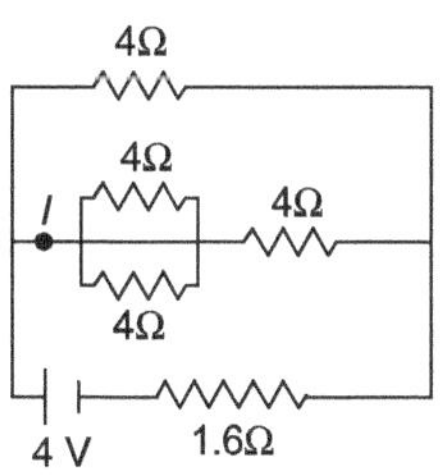

a. 1 *b.* 0.60 *c.* 0.4 *d.* 1.5

3. When light of wavelength 300 nm falls on a photoelectric emitter, photoelectrons are liberated. For another emitter, light of wavelength 600 nm is sufficient for liberating photoelectrons. The ratio of the work-function of the two emitters' is

a. 1 : 2 *b.* 2 : 1
c. 4 : 1 *d.* 1 : 4

4. A mono atomic gas is suddenly compressed to (1/8)th of its initial volume adiabatically. The ratio of its final pressure to the initial pressure is (Given, the ratio of the specific heats of the given gas to be 5/3)

a. 32 *b.* $\frac{40}{3}$
c. $\frac{24}{5}$ *d.* 8

5. The intensity of the magnetic induction field at the centre of a single turn circular coil of radius 5 cm carrying current of 0.9 A is

a. $36\pi \times 10^{-7}$ T
b. $9\pi \times 10^{-7}$ T
c. $36\pi \times 10^{-6}$ T
d. $9\pi \times 10^{-6}$ T

*Memory Based

6. A capacitor of capacity $0.1\,\mu F$ connected in series to a resistor of $10\ M\ \Omega$ is charged to a certain potential and then made to discharge through resistor. The time in which the potential will take to fall to half its original value is (Given, $\log_{10} 2 = 0.3010$)

a. 2 s *b.* 0.693 s *c.* 0.5 s *d.* 1.0 s

7. If the force is given by $F = at + bt^2$ with t as time. The dimensions of a and b are

a. $[MLT^{-4}], [MLT^{-2}]$ *b.* $[MLT^{-3}], [MLT^{-4}]$
c. $[ML^2T^{-3}], [ML^2T^{-2}]$ *d.* $[ML^2T^{-3}], [ML^3T^{-4}]$

8. A ray of light is incident on the interface between water and glass at an angle i and refracted parallel to the water surface, then value of μ_g will be

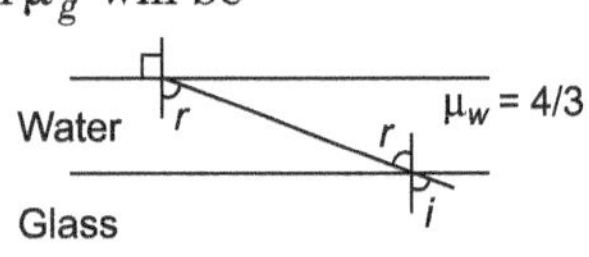

a. $(4/3) \sin i$ *b.* $\frac{1}{\sin i}$ *c.* $\frac{4}{3}$ *d.* 1

9. A body is moved in straight line by constant power of machine. What will be the relation between the travelling distance and time?

a. $s^2 \propto t^3$ *b.* $s \propto t^2$ *c.* $s^3 \propto t^2$ *d.* $s \propto t^3$

10. Magnetic moment of bar magnet is M. The work done to turn the magnet by 90° of magnet in direction of magnetic field B will be

a. zero *b.* $\frac{1}{2} MB$ *c.* $2\,MB$ *d.* MB

11. Voltage V and current i in AC circuit are given by

$V = 50 \sin (50\,t)$ V

$i = 50 \sin \left(50\,t + \frac{\pi}{3}\right)$ mA

The power dissipated in circuit is

a. 5.0 W *b.* 2.5 W *c.* 1.25 W *d.* zero

12. A simple wave motion represents by $y = 5 (\sin 4\pi t + \sqrt{3} \cos 4\pi t)$. Its amplitude is

a. 5 *b.* $5\sqrt{3}$ *c.* $10\sqrt{3}$ *d.* 10

13. A large open tank has two holes in the wall. One is a square hole of side L at a depth y from the top and the other is a circular hole of radius R at a depth $4y$ from the top. When the tank is completely filled with water, the quantities or water flowing out per second from the two holes are the same. Then, the value of R is

a. $L/\sqrt{2\pi}$ *b.* $2\pi L$
c. $L\sqrt{2/\pi}$ *d.* $L/2\pi$

14. In the circuit shown below, the ammeter reading is zero. Then, the value of the resistance R is

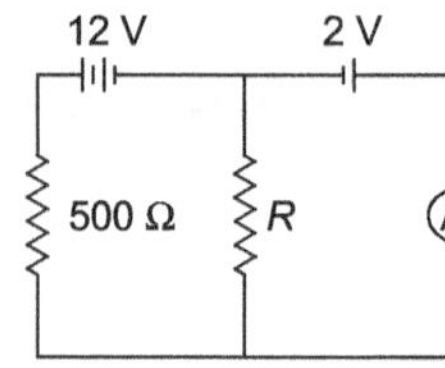

a. 50 Ω *b.* 100 Ω
c. 200 Ω *d.* 400 Ω

15. The dimensional formula for inductance is

a. $[ML^2T^{-2}A^{-2}]$ *b.* $[ML^2TA^{-2}]$
c. $[ML^2T^{-1}A^{-2}]$ *d.* $[ML^2T^{-2}A^{-1}]$

16. The maximum current that can be measured by a galvanometer of resistance 40 Ω is 10 mA. It is converted into a voltmeter that can read upto 50 V. The resistance to be connected in series with the galvanometer (in ohms) is

a. 2010 *b.* 4050
c. 5040 *d.* 4960

17. For a given velocity, a projectile has the same range R for two angles of projection, if t_1 and t_2 are the time of flight in the two cases, then

a. $t_1 t_2 \propto R$ *b.* $t_1 t_2 \propto R^2$
c. $t_1 t_2 \propto \frac{1}{R^2}$ *d.* $t_1 t_2 \propto \frac{1}{R}$

18. A sample of ideal mono atomic gas is taken round the cycle $ABCA$ as shown in the figure. The work done during the cycle is

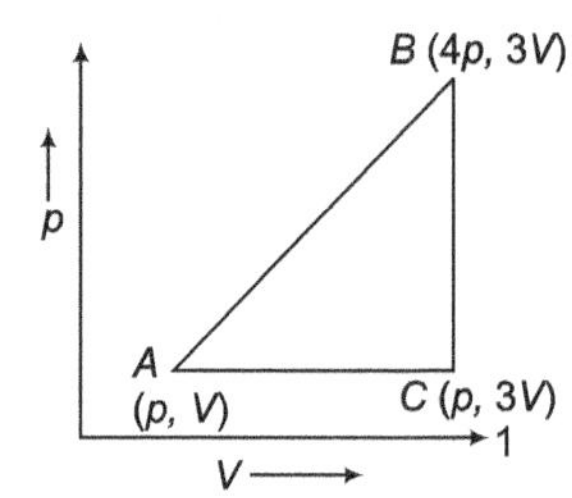

a. $3\,pV$ *b.* zero
c. $9\,pV$ *d.* $6\,pV$

19. A sound source is moving towards stationary listener with $\frac{1}{10}$th of the speed of sound. The ratio of apparent to real frequency is

a. $\left(\frac{9}{10}\right)^2$ *b.* $\frac{10}{9}$
c. $\frac{11}{10}$ *d.* $\left(\frac{11}{10}\right)^2$

20. A satellite is in a circular orbit round the Earth at an altitude R above the Earth's surface, where R is the radius of the earth. If g is the acceleration due to gravity on the surface of the Earth, then the speed of the satellite is

a. $\sqrt{2Rg}$ *b.* $\sqrt{Rg}$

c. $\sqrt{\frac{Rg}{2}}$ *d.* $\frac{\sqrt{Rg}}{4}$

21. A 10 kg stone is suspended with a rope of breaking strength 30 kg-wt. The minimum time in which the stone can be raised through a height 10 m starting from rest is (Taking, $g = 10\,\text{ms}^{-2}$)

a. 0.5 s *b.* 1.0 s

c. $\sqrt{\frac{2}{3}}$ s *d.* 2.0 s

22. How much work must be done by a force on 50 kg body in order to accelerate it from rest to 2 m/s^2 in 10 s?

a. 10^3 J *b.* 10^4 J

c. 2×10^3 J *d.* 4×10^4 J

23. A and B are two metals with threshold frequencies 1.8×10^{14} Hz and 2.2×10^{14} Hz. Two identical photons of energy 0.825 eV each are incident on them. Then, photoelectrons are emitted by (Taking, $h = 6.6 \times 10^{-34}$ J-s)

a. B only *b.* A only

c. Neither A nor B *d.* Both A and B

24. The square of resultant of two equal forces is three times their product. Angle between the forces is

a. π *b.* $\frac{\pi}{2}$ *c.* $\frac{\pi}{4}$ *d.* $\frac{\pi}{3}$

25. An object placed on a ground is in stable equilibrium. If the objects is given a slight push, then initially the position of centre of gravity

a. moves nearer to ground

b. rises higher above the ground

c. remains as such

d. may remain at same level

26. The maximum height attained by a projectile when thrown at an angle θ with the horizontal is found to be half the horizontal range. Then, θ is equal to

a. $\tan^{-1}(2)$ *b.* $\frac{\pi}{6}$

c. $\frac{\pi}{4}$ *d.* $\tan^{-1}\left(\frac{1}{2}\right)$

27. A shell of mass 20 kg at rest explodes into two fragments whose masses are in the ratio 2 : 3. The smaller fragment moves with a velocity of 6 ms^{-1}. The kinetic energy of the larger fragment is

a. 96 J *b.* 216 J

c. 144 J *d.* 360 J

28. If the displacement of simple pendulum at any time is 0.02 m and acceleration is 2 m/s^2, then in this time angular velocity will be

a. 100 rad/s *b.* 10 rad/s

c. 1 rad/s *d.* 0.1 rad/s

29. Which is constant, the Earth revolving around the Sun?

a. Angular momentum

b. Linear momentum

c. Rotational kinetic energy

d. Kinetic energy

30. In non-elastic collision,

a. momentum is conserved

b. energy is conserved

c. momentum and energy are conserved

d. momentum and energy are non-conserved

31. A mica slit of thickness t and refractive index μ is introduced in the ray from the first source S_1. By how much distance of fringes pattern will be displaced?

a. $\frac{d}{D}(\mu - 1)t$ *b.* $\frac{D}{d}(\mu - 1)t$

c. $\frac{d}{(\mu - 1)D}$ *d.* $\frac{D}{d}(\mu - 1)$

32. The refractive index of water is 4/3 and that of glass is 5/3. What will be the critical angle for the ray of light entering water from the glass?

a. $\sin^{-1}\left(\frac{4}{5}\right)$ *b.* $\sin^{-1}\left(\frac{5}{4}\right)$

c. $\sin^{-1}\left(\frac{1}{2}\right)$ *d.* $\sin^{-1}\left(\frac{2}{1}\right)$

33. The produced rays in sonography are

a. microwaves

b. infrared waves

c. sound waves

d. ultra sound

34. The ratio of secondary to primary turns of step up transformer is 4 : 1. If a current of 4 A is applied to the primary, then the induced current in secondary will be

a. 8 A *b.* 2 A

c. 1 A *d.* 0.5 A

35. The minimum force required to move a body up an inclined plane is three times the minimum force required to prevent it from sliding down the plane. If the coefficient of friction between the body and the inclined plane is $\frac{1}{2\sqrt{3}}$, then the angle of the inclined plane is

a. 60° *b.* 45° *c.* 30° *d.* 15°

36. If k_s and k_p respectively are effective spring constant in series and parallel combination of springs as shown in figure, find $\frac{k_s}{k_p}$

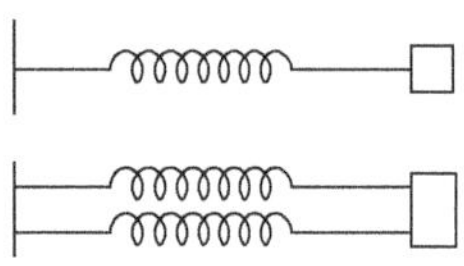

a. $\frac{9}{2}$ *b.* $\frac{3}{7}$ *c.* $\frac{2}{9}$ *d.* $\frac{7}{3}$

37. The power dissipated across resistance R which is connected across a battery of potential V is P. If resistance is doubled, then the power becomes

a. 1/2 *b.* 2
c. 1/4 *d.* 2

38. A body moves with uniform acceleration, then which of the following graph is correct?

a.
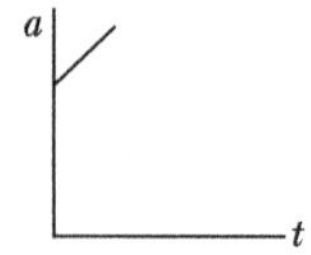

b.
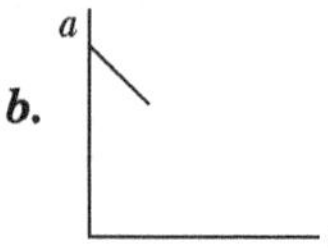

c.
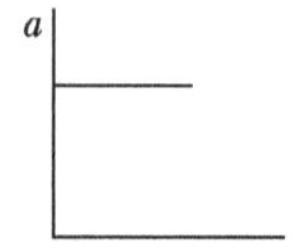

d.
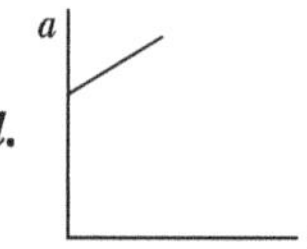

39. The rate at which a black body emits radiation at a temperature T is proportional to

a. $\frac{1}{T}$ *b.* T
c. T^3 *d.* T^4

40. Two equal charges q are kept fixed at a and $+a$ along the X-axis. A particle of mass m and charge $\frac{q}{2}$ is brought to the origin and given a small displacement along the X-axis, then

a. the particle executes oscillatory motion
b. the particle remains stationary
c. the particle executes, SHM along X-axis
d. the particle executes SHM along Y-axis

PART II

Chemistry

41. The ionic conductance of Ba^{2+} and Cl^- are respectively 127 and 76 Ω^{-1} cm^2 at infinite dilution. The equivalent conductance (in Ω^{-1} cm^2) of $BaCl_2$ at infinite dilution will be

a. 330 *b.* 203 *c.* 139.5 *d.* 51

42. If the elevation in boiling point of a solution of 10 g of solute (mol. wt. = 100) in 100 g of water is ΔT_b, the ebullioscopic constant of water is

a. $\frac{\Delta T_b}{10}$ *b.* ΔT_b *c.* $10\,\Delta T_b$ *d.* $100\,\Delta T_b$

43. Given that;

$$H_2O(l) \longrightarrow H^+(aq) + OH^-(aq); \quad \Delta H = 57.32\,kJ$$

$$H_2(g) + \frac{1}{2}O_2(g) \longrightarrow H_2O(l); \quad \Delta H = -286.02\,kJ$$

Then calculate the enthalpy of formation of OH^- at 25°C.

a. – 228.8 kJ *b.* – 343.52 kJ
c. + 228.8 kJ *d.* + 343.52 kJ

44. Calculate the amount of heat evolved when 500 cm^3 of 0.1 M HCl is mixed with 200 cm^3 of 0.2 M NaOH.

a. 57.3 kJ *b.* 2.865 kJ
c. 2.292 kJ *d.* 0.573 kJ

45. Which of the following will be the most effective in the coagulation of $Fe(OH)_3$ sol?

a. $Mg_3(PO_4)_2$ *b.* $BaCl_2$
c. NaCl *d.* KCN

46. Identify 'C' in the following reaction.

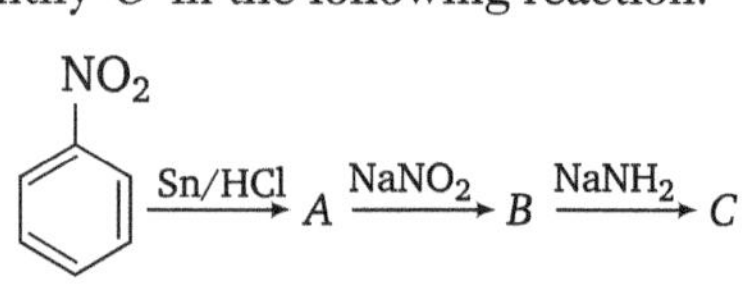

a. Benzamide
b. Benzoic acid
c. Chlorobenzene
d. Aniline

47. The following reaction is known as

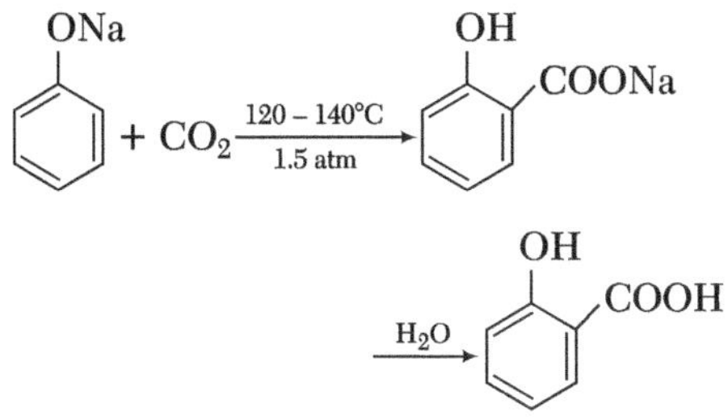

a. Friedel-Craft reaction
b. Kolbe reaction
c. Reimer-Tiemann reaction
d. Wittig reaction

48. Which of the following is isoelectronic with carbon?

a. Na^+ *b.* Al^{3+}
c. O^{2-} *d.* N^+

49. In which of the following species only one type of hybridisation is present?

a. $CH_3—CH_2—CH{=}CH_2$
b. $CH_3—CH{=}CH—\overset{+}{C}H_2$
c. $CH_2{=}CH—CH{=}CH_2$
d. $CH_3—CH{=}CH—CH_2^-$

50. $2MnO_4^- + 5H_2O_2 + 6H^+ \longrightarrow 2Z + 5O_2 + 8H_2O$

Identify Z in the above reaction

a. Mn^{2+} *b.* Mn^{4+}
c. Mn *d.* MnO_2

51. In the titration of NaOH and HCl, which of the following indicator will be used?

a. Methyl orange *b.* Methyl red
c. Both (a) and (b) *d.* None of these

52. Which of the following is correct IUPAC name for $K_2[Cr(CN)_2O_2(O)_2NH_3]$?

a. Potassiumamminecyanoperoxodioxochromatic (IV)
b. Potassiumamminecyanoperoxodioxochromium (V)
c. Potassiumamminecyanoperoxodioxochromium (VI)
d. Potassiumamminedicyanodioxoperoxochromate (VI)

53. Which of the following process is used for the preperation of acetone?

a. Waber process
b. Wacker process
c. Wolf-Kishner reduction
d. Gattermann-Koch synthesis

54. Lindane can be obtained by the reaction of benzene with

a. CH_3Cl/anhydrous $AlCl_3$
b. C_2H_5I/anhydrous $AlCl_3$
c. CH_3COCl/anhydrous $AlCl_3$
d. Cl_2 in sunlight

55. The structure of *cis-bis* (propenyl) ethene is

a. 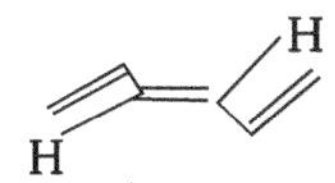 *b.* 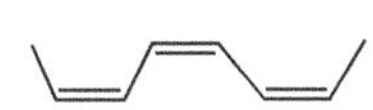

c. 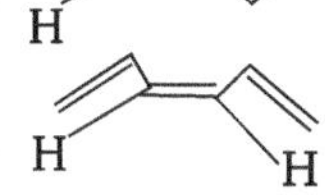 *d.* 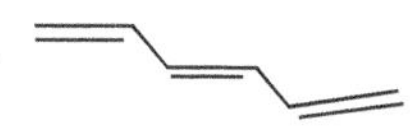

56. 5 moles of $Ba(OH)_2$ are treated with excess of CO_2. How much $BaCO_3$ will be formed?

a. 39.4 g *b.* 197 g *c.* 591 g *d.* 985 g

57. A diatomic molecule has a dipole moment of 1.2 D. If its bond distance is 1.0 Å, what fraction of an electronic charge, *e* exist on each atom?

a. 25% of *e* *b.* 50% of *e*
c. 60% of *e* *d.* 75% of *e*

58. A gas is heated through 1°C in a closed vessel and so, the pressure increases by 0.4%. The initial temperature of the gas was

a. – 23°C *b.* + 23°C
c. 250°C *d.* 523°C

59. For $2NOBr(g) \rightleftharpoons 2NO(g) + Br_2(g)$ at equilibrium, $p_{Br_2} = \frac{p}{q}$ and p is the total pressure, the ratio $\frac{K_p}{p}$ will be

a. $\frac{1}{3}$ *b.* $\frac{1}{9}$ *c.* $\frac{1}{27}$ *d.* $\frac{1}{81}$

60. The decomposition temperature is maximum for

a. $MgCO_3$ *b.* $CaCO_3$
c. $BaCO_3$ *d.* $SrCO_3$

61. When some amount of zinc is treated separately with excess of sulphuric acid and excess of sodium hydroxide solution, the ratio of volumes of hydrogen evolved is

a. 1 : 1 *b.* 1 : 2 *c.* 2 : 1 *d.* 2 : 3

62. A compound (*A*) when treated with PCl_5 and then ammonia gave (*B*). (*B*) when treated with bromine and caustic potash produced (*C*). (*C*) on treatment with $NaNO_2$/HCl at 0°C and then boiling produce *ortho*-cresol. Compound (*A*) is

a. *o*-chlorotoluene *b.* *o*- toluic acid
c. *m*-toluic acid *d.* *o*-bromotoluene

63. Alizarin is an example of
a. triaryl dye *b.* azo dye
c. vat dye *d.* anthraquinone dye

64. What will be the main product when acetylene reacts with hypochlorous acid?
a. Trichloroacetaldehyde
b. Acetaldehye
c. Dichloroacetaldehyde
d. Chloroacetaldehyde

65. Barium titanate has the perovskite structure, i.e. *i.e.*, a cubic lattice with Ba^{2+} ions at the corners of the unit cell, oxide ions at the face centres and titanium ions at the body centre. The molecular formula of barium titanate is
a. $BaTiO_3$ *b.* $BaTiO_4$
c. $BaTiO_2$ *d.* BaTiO

66. Which of the following hormone, is responsible for the growth of animals?
a. Auxin *b.* Insulin
c. Adrenaline *d.* Somatotropin

67. Which of the following have the largest ionic size?
a. F^- *b.* O^{2-}
c. Na^+ *d.* Mg^{2+}

68. If the radius of H is 0.53 Å then what will be the radius of ${}_3Li^{2+}$?
a. 0.17 Å *b.* 0.36 Å
c. 0.53 Å *d.* 0.59 Å

69. Which of the following will have highest value of pK_a?
a. FCH_2CH_2COOH *b.* $CH_3CH\cdot F\cdot COOH$
c. $CH_3CH\cdot Br\cdot COOH$ *d.* CH_3CH_2COOH

70. Gas (A) + NaOH $\longrightarrow B \xrightarrow{\Delta} C \xrightarrow{H^+} D$
C and D decolourises acidified $KMnO_4$. Identify C and D.
a. Na_2CO_3, NaOH *b.* $(COOH)_2$, $(COONa)_2$
c. $(COONa)_2$, $(COOH)_2$ *d.* None of these

71. The polymer polyurethanes are formed by treating di-isocyanate with
a. butadiene *b.* isoprene
c. glycol *d.* acrylonitrile

72. What will be the volume of O_2 at NTP liberated by 5 A current flowing for 193 s through acidulated water ?
a. 56 mL *b.* 112 mL
c. 158 mL *d.* 965 mL

73. CO_2 goes to air, causes green house effect and gets dissolved in water. What will be the effect on soil fertility and pH of the water?
a. Increase *b.* Decrease
c. Remain same *d.* None of these

74. $2N_2O_5 \rightleftharpoons 4NO_2 + O_2$
If rate and rate constant for above reaction are 2.40×10^{-5} mol L^{-1} s^{-1} and 3×10^{-5} s^{-1} respectively, then calculate the concentration of N_2O_5.
a. 1.4 *b.* 1.2 *c.* 0.04 *d.* 0.8

75. The molecules BF_3 and NF_3 both are covalent compounds, but BF_3 is non-polar and NF_3 is polar. The reason is that
a. boron is a metal and nitrogen is a gas in uncombined state.
b. BF_3 bonds have no dipole moment whereas NF_3 bond have dipole moment.
c. atomic size of boron is smaller than that of nitrogen.
d. BF_3 is symmetrical molecule whereas NF_3 is unsymmetrical.

76. 1.2% NaCl solution is isotonic with 7.2% glucose solution. What will be the van't Hoff factor, *i* ?
a. 0.5 *b.* 1 *c.* 2 *d.* 6

77. Green vitriol is
a. ferrous sulphate *b.* tin oxide
c. zinc oxide *d.* ferrous carbonate

78. 2-bromopentane with alcoholic KOH yields a mixture of three alkenes. Which of the following alkene is predominant?
a. 1-pentene *b.* *cis*-2-pentene
c. *trans*-2-pentene *d.* *cis*-1-pentene

79. In which of the following compounds, the bond length between hybridised carbon atom and other carbon atom is minimum?
a. Butane *b.* Propyne
c. Propene *d.* Butene

80. Which of the following is IUPAC name of compound?

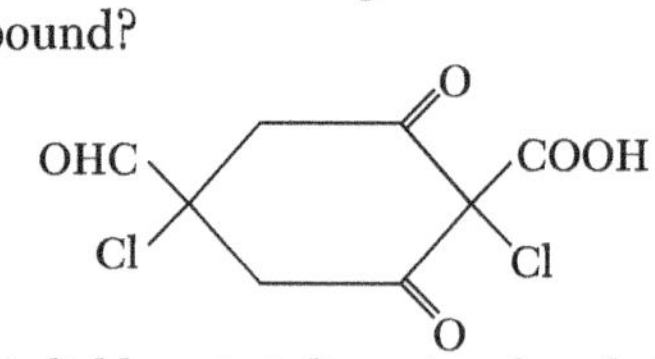

a. 1, 4-dichloro-2, 6-dioxo-4-carbonyl- 1-oic acid.
b. 2,4-dioxo-1, 4-dichlorohexane-1- carboxylic acid
c. 1,4-dichloro-2, 4, 6-dioxocyclohexane -1- carboxylic acid.
d. 1, 4-dichloro-4-formyl-2, 6-dioxy-cyclohexane -1-carboxylic acid.

PART III

a. English Proficiency

Directions (Q.Nos. 81 to 83) *Out of the four alternatives, choose the one which expresses the right meaning of the given word.*

81. Dubious
a. Doubtful *b.* Disputable
c. Duplicate *d.* Dangerous

82. Flabbergasted
a. Scared *b.* Embarrassed
c. Dumbfounded *d.* Humiliated

83. Eternal
a. Innumerable *b.* Unmeasurable
c. Prolonged *d.* Perpetual

Directions (Q.Nos. 84 to 86) *Choose the word opposite in meaning to the given word.*

84. Despair
a. Belief *b.* Trust *c.* Hope *d.* Faith

85. In toto
a. Bluntly *b.* Partially
c. Entirely *d.* Strongly

86. Protean
a. Amateur *b.* Catholic
c. Unchanging *d.* Rapid

Directions (Q.Nos. 87 to 89) *A part of the sentence is underlined. Below are given alternatives to the underlined part as a, b and c which may improve the sentence. Choose the correct alternative. In case no improvement is needed, your answer is 'd'.*

87. He <u>declined</u> all the allegations against him.
a. spurned *b.* refused
c. refuted *d.* No improvement

88. It is time we <u>leave.</u>
a. left *b.* have to leave
c. would leave *d.* No improvement

89. We spent an hour discussing <u>about his character</u>.
a. on his character *b.* of his character
c. his character *d.* No improvement

Directions (Q.Nos. 90 to 92) *Sentences are given with blanks to be filled in with an appropriate and suitable word. Four alternatives are suggested for each question. Choose the correct alternative out of the four.*

90. Are your really desirous ...... visiting Japan?
a. of *b.* in *c.* to *d.* about

91. When Indians from the South move North, they find certain aspects of life quite ...... from their own.
a. strange *b.* separate
c. different *d.* divergent

92. The sky is overcast, we .... the storm will soon burst.
a. expect
b. hope
c. trust
d. suspect

Directions (Q. Nos. 93 to 95) *The first and the last parts of the sentence are numbered 1 to 6. The rest of the sentence is spelt into four parts and named P, Q, R and S. These four parts are not given in their proper order. Read the parts and find out which of the four combinations is correct. Then find the correct answer.*

93. 1. Early to bed, early to rise, makes a man healthy, wealthy and wise.
P. But for the morning tea, I had to wait for someone to get up before me.
Q. This saying inspired me to rise early.
R. That day I was the first to get up.
S. One day I got up early in the morning.
6. Then I realised that it was a waste of time to get up early and wait for the morning tea.
a. QSRP *b.* QPRS
c. PQRS *d.* SPQR

94. 1. A wood-cutter was cutting a tree on a river bank.
P. He knelt down and prayed.
Q. His axe slipped and fell into the water.
R. God mercury appeared before him and asked about the matter.
S. He could not get it back as the river was very deep.
6. He dived into the water and came up with an axe of good.
a. RPQS *b.* RPSQ
c. QSRP *d.* QSPR

95. 1. A dog stole a piece of meat from a butcher's shop.
P. He barked in anger.
Q. He ran to the jungle with the piece of meat.
R. He saw his reflection.
S. He crossed a river on the way.
6. He lost his piece of meat.
a. QPSR *b.* QSRP
c. QPRS *d.* SRPQ

b. Logical Reasoning

96. In a certain code MONKEY is XDJMNL. How is 'TIGER' written as?

a. QDFHS *b.* SDFHS
c. SHFDQ *d.* UJHFS

97. Find the missing number from the given responses.

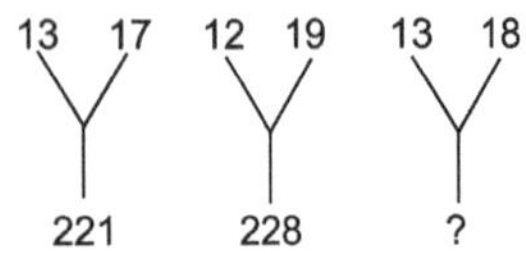

a. 31 *b.* 229
c. 234 *d.* 312

98. If the day before yesterday was Thursday, when will Sunday be?

a. tomorrow
b. day after tomorrow
c. today
d. two days after today

99. In a row of children Ravi is fourth from right and Shyam is second from left. When they interchange positions Ravi is ninth from right. What will be Shyam position from left?

a. Fifth *b.* Sixth
c. Seventh *d.* Eighth

100. A piece of paper is folded and a cut is made as shown below. From the given responses indicate, how it will appear when opened?

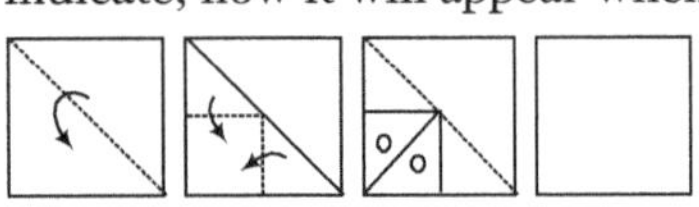

Answer Figures

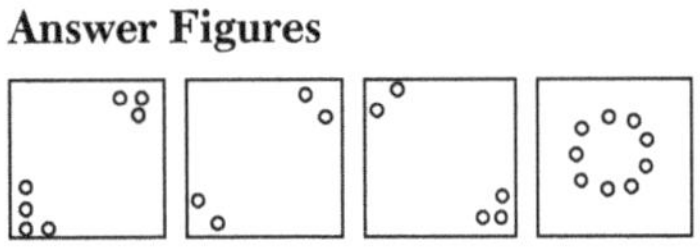

a. *b.* *c.* *d.*

101. Which represents carrot, food, vegetable?

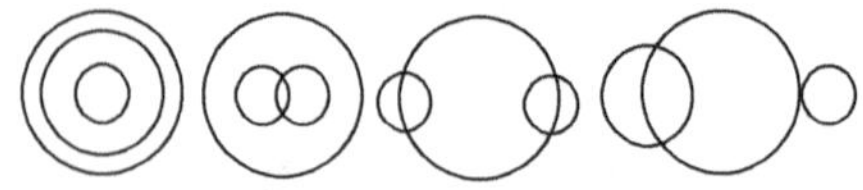

a. *b.* *c.* *d.*

102. "All the members of the Tennis club are members of the badminton club too". No woman plays badminton?

a. Some women play Tennis
b. No member of the Tennis club plays badminton
c. Some women are members of the Tennis club
d. No woman is a member of Tennis club

103. Which number is wrong in the given series?
1, 9, 25, 50, 81

a. 1 *b.* 25 *c.* 50 *d.* 81

104. In the following question and Δ stands for any of the mathematical signs at different places, which are given as choices under each question. Select the choice with the correct sequence of signs which when substituted makes the question as correct equation?

$$24 \Delta 4 \Delta 5 \Delta 4$$

a. $\times + =$ *b.* $= \times +$ *c.* $+ \times =$ *d.* $= + \times$

105. Which answer figure is the exact mirror image of the given figure when the mirror is held from the right at PQ ?

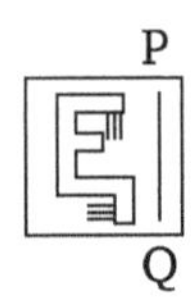

Answer Figures

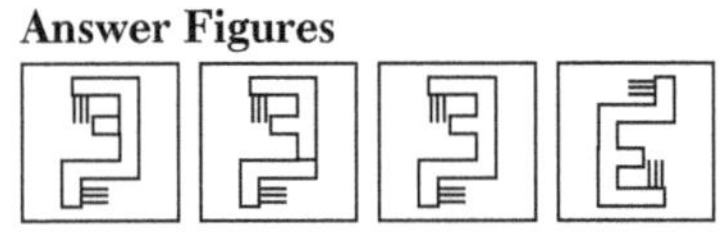

a. *b.* *c.* *d.*

PART IV

Mathematics

106. The equation of the base BC of an equilateral ΔABC is $x + y = 2$ and A is $(2, -1)$. The length of the side of the triangle is

a. $\sqrt{2}$ *b.* $\left(\frac{3}{2}\right)^{1/2}$
c. $\left(\frac{1}{2}\right)^{1/2}$ *d.* $\left(\frac{2}{3}\right)^{1/2}$

107. The equation of the circle circumscribing the triangle formed by the lines $x + y = 6$, $2x + y = 4$ and $x + 2y = 5$ is

a. $x^2 + y^2 + 17x + 19y - 50 = 0$
b. $x^2 + y^2 - 17x - 19y - 50 = 0$
c. $x^2 + y^2 + 17x - 19y - 50 = 0$
d. $x^2 + y^2 - 17x - 19y + 50 = 0$

108. The length of the tangent from (5, 1) to the circle $x^2+y^2+6x-4y-3=0$ is

a. 7 *b.* 49 *c.* 63 *d.* 21

109. If the length of the major axis of the ellipse $\left(\frac{x^2}{a^2}\right)+\left(\frac{y^2}{b^2}\right)=1$ is three times the length of minor axis, its eccentricity is

a. $\frac{1}{3}$ *b.* $\frac{1}{\sqrt{3}}$ *c.* $\sqrt{\frac{2}{3}}$ *d.* $\frac{2\sqrt{2}}{3}$

110. S and T are the foci of the ellipse $\left(\frac{x^2}{a^2}\right)+\left(\frac{y^2}{b^2}\right)=1$ and B is an end of the minor axis. If STB is an equilateral triangle, then eccentricity of the ellipse is

a. $\frac{1}{4}$ *b.* $\frac{1}{3}$ *c.* $\frac{1}{2}$ *d.* $\sqrt{\frac{3}{2}}$

111. The difference of the focal distance at any point on the hyperbola is equal to its

a. latusrectum
b. eccentricity
c. length of the transverse axis
d. half the length of the transverse axis

112. If $A+B+C=180°$, then $\frac{\cot A+\cot B+\cot C}{\cot A\cot B\cot C}$ is equal to

a. 1 *b.* $\cot A\cos B\cot C$
c. −1 *d.* 0

113. The angles of a triangle are in AP and the least angle is 30°. The greatest angle in radians is

a. $\frac{7\pi}{12}$ *b.* $\frac{2\pi}{3}$
c. $\frac{5\pi}{6}$ *d.* $\frac{\pi}{2}$

114. If $\tan 20°=p$, then $\frac{\tan 160°-\tan 110°}{1+\tan 160°\tan 110°}$ is equal to

a. $\left(\frac{1-p^2}{2p}\right)$ *b.* $\left(\frac{2p}{1+p^2}\right)$ *c.* $\left(\frac{1+p}{2p}\right)$ *d.* $\left(\frac{1-p}{2p}\right)$

115. If $4\sin^{-1}x+\cos^{-1}x=\pi$, then x is equal to

a. 1/2 *b.* 2
c. 1 *d.* 1/3

116. In a ΔABC, $a=2$, $b=3$ and $\sin A=\frac{2}{3}\cdot$ Then, $\cos C$ is equal to

a. $\frac{1}{2}$ *b.* $\frac{2}{3}$ *c.* $\frac{2}{\sqrt{13}}$ *d.* $\frac{1}{\sqrt{13}}$

117. The vector equation

$$\mathbf{r}=\hat{\mathbf{i}}-2\hat{\mathbf{j}}-\hat{\mathbf{k}}+t(6\hat{\mathbf{j}}-\hat{\mathbf{k}})$$

represents a straight line passing through the points

a. $(0, 6, -1)$ and $(1, -2, -1)$
b. $(0, 6, -1)$ and $(-1, -4, -2)$
c. $(1, -2, -1)$ and $(1, 4, -2)$
d. $(1, -2, -1)$ and $(0, -6, 1)$

118. The work done by the force $4\hat{\mathbf{i}}-3\hat{\mathbf{j}}+2\hat{\mathbf{k}}$ in moving a particle along a straight line from the point $(3, 2, -1)$ to $(2, -1, 4)$ is

a. 0 units *b.* 4 units *c.* 15 units *d.* 19 units

119. $\lim_{x\to 0}\left(\frac{(2+x)\sin(2+x)-2\sin 2}{x}\right)$ is equal to

a. sin 2 *b.* cos 2
c. 1 *d.* 2cos 2 + sin 2

120. If $f(x)=\frac{3x+\tan^2 x}{x}$ is continuous at $x=0$, then $f(0)$ is equal to

a. 3 *b.* 2 *c.* 4 *d.* 0

121. If x is measured in degree, then $\frac{d}{dx}(\cos x)$ is equal to

a. $-\sin x$ *b.* $-\frac{180}{\pi}\sin x$ *c.* $-\frac{\pi}{180}\sin x$ *d.* $\sin x$

122. $\left(\frac{d}{dx}\right)[\log(\sec x-\tan x)]$ is equal to

a. $-\sec x$ *b.* $\sec x+\tan x$
c. $\sec x$ *d.* $\sec x-\tan x$

123. If $x=\cos^3\theta$ and $y=\sin^3\theta$, then $1+\left(\frac{dy}{dx}\right)^2$ is equal to

a. $\tan^2\theta$ *b.* $\cot^2\theta$ *c.* $\sec^2\theta$ *d.* $\text{cosec}^2\theta$

124. If $x=at^2$, $y=2at$, then $\frac{d^2y}{dx^2}$ is equal to

a. $-\frac{1}{t^2}$ *b.* $-\frac{1}{2at^3}$ *c.* $\frac{1}{t^2}$ *d.* $-\frac{a}{2t^3}$

125. If the rate of change in the circumference of a circle of 0.3 cm/s, then the rate of change in the area of the circle when the radius is 5 cm, is

a. 1.5 sq cm/s *b.* 0.5 sq cm/s
c. 5 sq cm/s *d.* 3 sq cm/s

126. If $y=x^3-ax^2+48x+7$ is an increasing function for all real values of x, then a lies in

a. $(-14, 14)$ *b.* $(-12, 12)$
c. $(-16, 16)$ *d.* $(-21, 21)$

127. Rolle's theorem is not applicable for the function $f(x)=|x|$ in the interval $[-1, 1]$ because

a. $f'(1)$ does not exist
b. $f'(-1)$ does not exist
c. $f(x)$ is discontinuous at $x=0$
d. $f'(0)$ does not exist

128. $\int \frac{2dx}{(e^x+e^{-x})^2}$ is equal to

a. $-\frac{e^{-x}}{(e^x+e^{-x})}+C$ b. $-\frac{1}{(e^x+e^{-x})}+C$
c. $\frac{1}{(e^x+1)^2}+C$ d. $\frac{1}{(e^x-e^{-x})^2}+C$

129. $\int_0^{\pi/2} \frac{\sin^n\theta}{\sin^n\theta+\cos^n\theta}d\theta$ is equal to

a. 1 b. 0 c. $\frac{\pi}{2}$ d. $\frac{\pi}{4}$

130. $\int_0^{\pi} \cos^{101}x\,dx$ is equal to

a. $\frac{\pi}{4}$ b. $\frac{1}{102}$ c. $\left(\frac{\pi}{3}\right)^{101}$ d. 0

131. $\lim_{n\to\infty}\left[\frac{1}{n+1}+\frac{1}{n+2}+\ldots+\frac{1}{5n}\right]$

a. $\log 2$ b. $\log(1+\sqrt{5})$
c. $\log 6$ d. 0

132. By eliminating the arbitrary constant A and B from $y=Ax^2+Bx$, we get the differential equation

a. $\frac{d^3y}{dx^3}=0$ b. $x^2\frac{d^2y}{dx^2}-2x\frac{dy}{dx}+2y=0$
c. $\frac{d^2y}{dx^2}=0$ d. $x^2\frac{d^2y}{dx^2}+y=0$

133. If $f(x)=\frac{\log(1+ax)-\log(1-bx)}{x}$ for $x\neq 0$ and $f(0)=k$ and $f(x)$ is continuous at $x=0$, then k is equal to

a. $a+b$ b. $a-b$ c. a d. b

134. If $4-5i$ is a root of the quadratic equation $x^2+ax+b=0$, then (a, b) is equal to

a. $(8, 41)$ b. $(-8, 41)$ c. $(41, 8)$ d. $(-41, 8)$

135. If α and β are the roots of the quadratic equation $4x^2+3x+7=0$, then the value of $\frac{1}{\alpha}+\frac{1}{\beta}$ is

a. $-\frac{3}{4}$ b. $-\frac{3}{7}$ c. $\frac{3}{7}$ d. $\frac{4}{7}$

136. If α, β are the roots of $ax^2+bx+c=0$ and $\alpha+k, \beta+k$ are the roots of $px^2+qx+r=0$, then $\frac{b^2-4ac}{q^2-4pr}$ is equal to

a. $\frac{a}{p}$ b. 1 c. $\left(\frac{a}{p}\right)^2$ d. 0

137. Area of the triangle in the argand diagram formed by the complex numbers z, iz, $z+iz$, where $z=x+iy$ is

a. $|z|$ b. $|z|^2$ c. $2|z|^2$ d. $\frac{1}{2}|z|^2$

138. The sum of n terms of the series $\frac{1}{2}+\frac{3}{4}+\frac{7}{8}+\frac{15}{16}+\ldots$ is

a. $n-1+2^{-n}$ b. 1
c. $n-1$ d. $1+2^{-n}$

139. $0.2+0.22+0.222+\ldots$ to n terms is equal to

a. $\left(\frac{2}{9}\right)-\left(\frac{2}{81}\right)(1-10^{-n})$
b. $n-\left(\frac{1}{9}\right)(1-10^{-n})$
c. $\left(\frac{2}{9}\right)\left[n-\left(\frac{1}{9}\right)(1-10^{-n})\right]$
d. $\left(\frac{2}{9}\right)$

140. The number of ways in which a team of 11 players can be selected from 22 players including 2 of them and excluding 4 of them is

a. ${}^{16}C_{11}$ b. ${}^{16}C_5$
c. ${}^{16}C_9$ d. ${}^{20}C_8$

141. The number of ways four boys can be seated around a round table in four chairs of different colours is

a. 24 b. 12 c. 23 d. 64

142. If the coefficient of second, third and fourth terms in the expansion of $(1+x)^n$ are in AP, then n is equal to

a. 7 b. 4 c. 5 d. 6

143. If $\Delta=\begin{vmatrix}1 & a & a^2\\ 1 & b & b^2\\ 1 & c & c^2\end{vmatrix}=k(a-b)(b-c)(c-a)$, then k is equal to

a. -2 b. 1
c. 2 d. abc

BITSAT 2012

144. $\begin{vmatrix} a+b & a & b \\ a & a+c & c \\ b & c & b+c \end{vmatrix}$ is equal to

a. $4abc$ *b.* abc *c.* $a^2b^2c^2$ *d.* $4a^2bc$

145. If $\Delta_1 = \begin{vmatrix} x & a & b \\ b & x & a \\ a & b & x \end{vmatrix}$ and $\Delta_2 = \begin{vmatrix} x & b \\ a & x \end{vmatrix}$ are the given determinants, then

a. $\Delta_1 = 3(\Delta_2)^2$ *b.* $\left(\frac{d}{dx}\right)(\Delta_1) = 3\Delta_2$

c. $\left(\frac{d}{dx}\right)(\Delta_1) = 3(\Delta_2)^2$ *d.* $\Delta_1 = 3(\Delta_2)^{3/2}$

146. The system
$x + 4y - 2z = 3,\ 3x + y + 5z = 7$
and $2x + 3y + z = 5$ has

a. infinite number of solutions
b. unique solution
c. trivial solution
d. no solution

147. If the three points $(k, 2k), (2k, 3k), (3, 1)$ are collinear, then k is equal to

a. -2 *b.* 1
c. $\frac{1}{2}$ *d.* $-\frac{1}{2}$

148. The foot of the perpendicular from the point $(3, 4)$ on the line $3x - 4y + 5 = 0$ is

a. $\left(\frac{81}{25}, \frac{92}{25}\right)$ *b.* $\left(\frac{92}{25}, \frac{81}{25}\right)$
c. $\left(\frac{46}{26}, \frac{54}{24}\right)$ *d.* $\left(-\frac{81}{25}, -\frac{92}{25}\right)$

149. A kite is flying at an inclination of 60° with the horizontal. If the length of the thread is 120 m, then the height of the kite is

a. $60\sqrt{3}$ m *b.* 60 m *c.* $\frac{60}{\sqrt{3}}$ m *d.* 120 m

150. If the focus of parabola is at $(0, -3)$ and its directrix is $y = 3$, then its equation is

a. $x^2 = -12y$ *b.* $x^2 = 12y$
c. $y^2 = -12y$ *d.* $y^2 = 12x$

Answers

Physics

1. (d)	2. (c)	3. (b)	4. (a)	5. (a)	6. (b)	7. (b)	8. (b)	9. (a)	10. (d)
11. (c)	12. (d)	13. (a)	14. (b)	15. (a)	16. (d)	17. (a)	18. (a)	19. (b)	20. (c)
21. (b)	22. (b)	23. (b)	24. (d)	25. (b)	26. (a)	27. (a)	28. (b)	29. (a)	30. (a)
31. (b)	32. (a)	33. (d)	34. (c)	35. (c)	36. (c)	37. (a)	38. (c)	39. (d)	40. (c)

Chemistry

41. (c)	42. (b)	43. (a)	44. (c)	45. (a)	46. (d)	47. (b)	48. (d)	49. (c)	50. (a)
51. (c)	52. (d)	53. (b)	54. (d)	55. (b)	56. (d)	57. (a)	58. (a)	59. (d)	60. (c)
61. (a)	62. (b)	63. (d)	64. (c)	65. (a)	66. (d)	67. (b)	68. (a)	69. (d)	70. (c)
71. (c)	72. (a)	73. (b)	74. (d)	75. (d)	76. (c)	77. (a)	78. (c)	79. (b)	80. (d)

Logical Reasoning

81. (a)	82. (c)	83. (d)	84. (c)	85. (b)	86. (c)	87. (d)	88. (a)	89. (c)	90. (a)
91. (c)	92. (d)	93. (a)	94. (d)	95. (b)					

English Proficiency

96. (a)	97. (c)	98. (a)	99. (c)	100. (d)	101. (a)	102. (d)	103. (c)	104. (d)	105. (c)

Mathematics

106. (d)	107. (d)	108. (a)	109. (d)	110. (c)	111. (c)	112. (a)	113. (d)	114. (a)	115. (a)
116. (b)	117. (c)	118. (c)	119. (d)	120. (a)	121. (c)	122. (a)	123. (c)	124. (b)	125. (a)
126. (b)	127. (d)	128. (a)	129. (d)	130. (d)	131. (c)	132. (b)	133. (a)	134. (b)	135. (b)
136. (c)	137. (d)	138. (a)	139. (c)	140. (c)	141. (a)	142. (a)	143. (b)	144. (a)	145. (b)
146. (d)	147. (a)	148. (a)	149. (a)	150. (a)					

Hints & Solutions

Physics

1. ***(d)*** Magnetic force on straight wire

$$F = Bil\sin\theta = Bil\sin 90° = Bil$$

For equilibrium of wire in mid-air,

$$F = mg$$

$$\Rightarrow \quad Bil = mg$$

$$\therefore \quad B = \frac{mg}{il} = \frac{200 \times 10^{-3} \times 9.8}{2 \times 1.5} = 0.65\,\text{T}$$

2. ***(c)*** We can simplify the network as shown

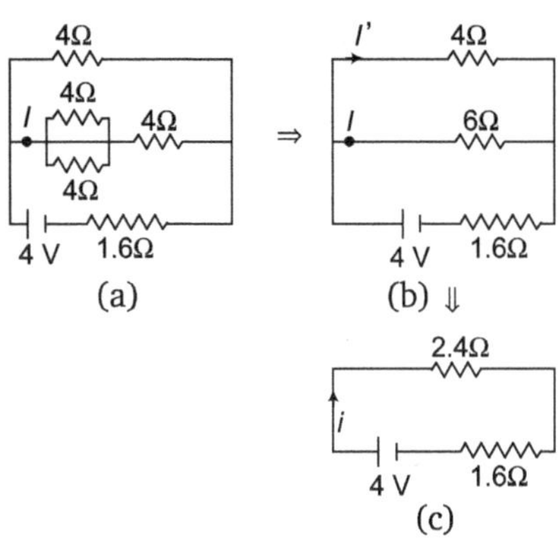

So, net resistance,

$$R = 2.4 + 1.6 = 4.0\,\Omega$$

Therefore, current from the battery,

$$i = \frac{V}{R} = \frac{4}{4} = 1\,\text{A}$$

Now, from the circuit (b),

$$4I' = 6I \Rightarrow I' = \frac{3}{2}I$$

But $i = I + I' = I + \frac{3}{2}I = \frac{5}{2}I$

$$\therefore \quad 1 = \frac{5}{2}I$$

$$\Rightarrow \quad I = \frac{2}{5} = 0.4\,\text{A}$$

3. ***(b)*** Work-function is given by

$$\phi = \frac{hc}{\lambda} \text{ or } \phi \propto \frac{1}{\lambda}$$

$$\because \quad \frac{\phi_1}{\phi_2} = \frac{\lambda_2}{\lambda_1} = \frac{600}{300} = \frac{2}{1}$$

4. ***(a)*** In an adiabatic process,

$$pV^{\gamma} = \text{constant}$$

$$\Rightarrow \quad \frac{p_1}{p_2} = \left(\frac{V_2}{V_1}\right)^{\gamma}$$

$$\Rightarrow \quad \frac{p_1}{p_2} = \left(\frac{1}{8}\right)^{5/3}$$

$$\Rightarrow \quad \frac{p_1}{p_2} = \left(\frac{1}{2^3}\right)^{5/3} = \frac{1}{32}$$

$$\therefore \quad \frac{p_2}{p_1} = 32$$

5. ***(a)*** The intensity of magnetic induction field,

$$B = \frac{\mu_0 I}{2r}$$

$$= \frac{4\pi \times 10^{-7} \times 0.9}{2 \times 5 \times 10^{-2}}$$

$$= 36\pi \times 10^{-7}\,\text{T}$$

6. ***(b)*** By equation of charging,

$$q = q_0(1 - e^{-t/CR})$$

According to question,

$$\frac{q}{q_0} = \frac{1}{2} = 0.50$$

$$\therefore \quad 0.50 = 1 - e^{-t/CR}$$

$$e^{-t/CR} = 1 - 0.50 = 0.5$$

$$\Rightarrow \quad e^{t/CR} = 2$$

or $$\frac{t}{CR} = \log_e 2$$

or $$\frac{t}{CR} = 2.3026\log_{10} 2$$

or $$t = CR \times 2.3026\log_{10} 2$$

or $$t = 0.1 \times 10^{-6} \times 10 \times 10^{6} \times 2.3026\log_{10} 2$$

or $$t = 2.3026 \times 0.3010$$

or $$t = 0.693\,\text{s}$$

7. ***(b)*** Dimension of at = Dimension of F

$$[at] = [F]$$

$$\Rightarrow \quad [a] = \left[\frac{F}{t}\right]$$

$$\Rightarrow \quad [a] = \left[\frac{\text{MLT}^{-2}}{\text{T}}\right] \Rightarrow [a] = [\text{MLT}^{-3}]$$

Dimension of bt^2 = Dimension of F

$$\Rightarrow \quad [bt^2] = [F]$$

$$\Rightarrow \quad [b] = \left[\frac{F}{t^2}\right]$$

$$\Rightarrow \quad [b] = \left[\frac{\text{MLT}^{-2}}{\text{T}^2}\right]$$

$$\Rightarrow \quad [b] = [\text{MLT}^{-4}]$$

8. ***(b)*** According to gives situation,

$${}_g\mu_w = \frac{\sin i}{\sin r} \quad \ldots\text{(i)}$$

and $${}_g\mu_a = \frac{\sin r}{\sin 90°} \quad \ldots\text{(ii)}$$

From Eqs. (i) and (ii), we get

$${}_g\mu_w \times {}_w\mu_a = \frac{\sin i}{\sin r} \times \frac{\sin r}{\sin 90°} = \sin i$$

or $$\frac{\mu_w}{\mu_g} \times \frac{\mu_a}{\mu_w} = \sin i \Rightarrow \mu_g = \frac{1}{\sin i}$$

9. ***(a)*** Power $= [\text{ML}^2\text{T}^{-3}] =$ constant

$$\therefore \quad \left[\frac{\text{ML}^2}{\text{T}^3}\right] = \text{constant}$$

$$\therefore \quad [\text{L}^2] \propto [\text{T}^3] \quad \text{or} \quad s^2 \propto t^3$$

10. ***(d)*** Work done, $W = MB(1 - \cos\theta)$

$$\theta = 90°$$

$$\therefore \quad W = MB$$

11. ***(c)*** Given, $V = 50\sin(50\,t)$ V

Maximum voltage, $V_0 = 50$ V

$$i = i_0 \sin\left(50\,t + \frac{\pi}{3}\right)\text{mA}$$

Maximum current, $i_0 = 50 \text{ mA} = 50 \times 10^{-3}$ A

Power dissipated, $$P = \frac{i_0}{\sqrt{2}} \times \frac{V_0}{\sqrt{2}}$$

$$= \frac{50 \times 50 \times 10^{-3}}{2}$$

$$= \frac{2500 \times 10^{-3}}{2} = 1.25 \text{ W}$$

12. ***(d)*** $y = 5(\sin 4\pi t + \sqrt{3}\cos 4\pi t)$

$$y = 5\sin 4\pi t + 5\sqrt{3}\cos 4\pi t$$

Amplitude, $$A = \sqrt{A_1^2 + A_2^2}$$

$$= \sqrt{(5)^2 + (5\sqrt{3})^2}$$

$$= \sqrt{25 + 75} = \sqrt{100} = 10$$

13. ***(a)*** By the principle of continuity,

$$A_1 v_1 = A_2 v_2$$

According to question, $A_1 = L^2$ [for square hole]

$$v_1 = \sqrt{2gy}$$

and $A_2 = \pi R^2$ [for circular hole]

$$v_2 = \sqrt{2g4y}$$

$$\therefore \quad L^2\sqrt{2gy} = \pi R^2\sqrt{2g4y}$$

or $$L^2 = 2\pi R^2 \text{ or } R = L/\sqrt{2\pi}$$

14. ***(b)*** Given circuit

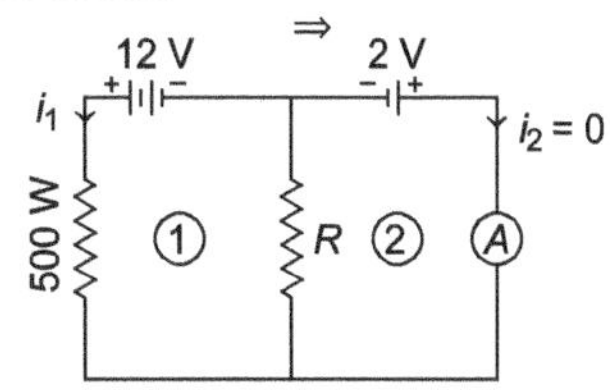

In loop (1), applying Kirchhoff's law,

$$12 - 500i_1 - Ri_1 = 0$$

$$\Rightarrow \quad 12 = i_1(500 + R) \quad \ldots\text{(i)}$$

In loop (2),

$$12 - 500i_1 - 2 = 0$$

$$\Rightarrow \quad 10 = 500\,i_1$$

or $$i_1 = \frac{1}{50}\text{ A} \quad \ldots\text{(ii)}$$

From Eqs. (i) and (ii), we get

$$12 \times \frac{1}{i_1} = (500 + R)$$

$$\Rightarrow \quad 12 \times 50 = 500 + R$$

$$\Rightarrow \quad R = 100\,\Omega$$

15. ***(a)*** EMF induced in an electrical circuit

$$\varepsilon = L\frac{dI}{dt} \quad \text{(Numerically)}$$

or $$L = \varepsilon\frac{dt}{dI} = \frac{W}{Q}\cdot\frac{dt}{dI} \quad \left(\because \varepsilon = V = \frac{W}{Q}\right)$$

$$= \frac{Wdt}{it \cdot dI}$$

$$\Rightarrow \quad [L] = \frac{[W][dt]}{[i][t][dI]} = \frac{[\text{ML}^2\text{T}^{-2}][\text{T}]}{[\text{A}][\text{T}][\text{A}]} \quad [\because Q = it]$$

$$= [\text{ML}^2\text{T}^{-2}\text{A}^{-2}]$$

16. ***(d)*** To convert a galvanometer into voltmeter, the necessary value of resistance to be connected in series with the galvanometer is

$$R = \frac{V}{I_g} - G = \frac{50}{10 \times 10^{-3}} - 40$$

$$= 5000 - 40 = 4960\,\Omega$$

17. ***(a)*** $t_1 = \dfrac{2u\sin\alpha}{g}$

$$t_2 = \frac{2u\sin(90° - \alpha)}{g} = \frac{2u\cos\alpha}{g}$$

So, $$t_1 \cdot t_2 = \frac{2u\sin\alpha}{g}\cdot\frac{2u\cos\alpha}{g}$$

$$= \frac{2u^2 \cdot 2\sin\alpha\cos\alpha}{g} = 2\frac{u^2}{g^2}\sin 2\alpha$$

or $\quad t_1 \cdot t_2 = \frac{2R}{g} \qquad \left(\because R = \frac{u^2 \sin 2\alpha}{g}\right)$

$$t_1 \cdot t_2 \propto R$$

18. ***(a)*** The work done = area of p-V graph

$$= \text{area of triangle } ABC$$
$$= \frac{1}{2} \times BC \times AC$$
$$= \frac{1}{2}(4p - p) \times (3V - V)$$
$$= \frac{1}{2} \times 3p \times 2V = 3pV$$

19. ***(b)*** Given, $v_s = \frac{v}{10}$

Apparent frequency, $n' = n\left(\frac{v}{v - v_s}\right)$

where, n = real frequency of source

v = velocity of sound

and $\quad v_s$ = velocity of source.

So, $\quad \frac{n'}{n} = \frac{v}{v - \frac{v}{10}} = \frac{10}{9}$

20. ***(c)*** Orbital velocity (v_o) at a height h above the earth's surface is given by

$$v_o = R_e\sqrt{\frac{g}{R_e + h}}$$

Given, $\quad h = R_e = R$

$\therefore \quad v_o = R\sqrt{\frac{g}{2R}} = \sqrt{\frac{Rg}{2}}$

21. ***(b)*** Tension in the string, $T = mg$

$$= 30 \times 10 = 300\,\text{N}$$

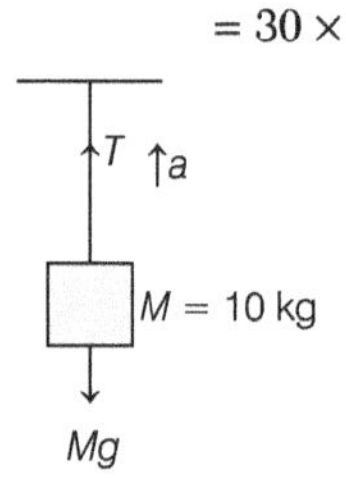

From the figure

$$T - Mg = Ma$$

$\Rightarrow \quad 300 - 10 \times 10 = 10a$

$\therefore \quad a = 20\ \text{ms}^{-2}$

Thus, the maximum acceleration with which the stone can be raised is 20 ms^{-2}.

Given, $\quad s = 10$ m

and $\quad u = 0$

$\therefore \quad s = \frac{1}{2}at^2 \Rightarrow 10 = \frac{1}{2}(20)t^2$

$\Rightarrow \quad t = 1$ s

22. ***(b)*** Now, by equation of motion,

$$s = ut + \frac{1}{2}at^2$$
$$= 0 + \frac{1}{2} \times 2 \times 10 \times 10 = 100\text{ m}$$

Hence, work done,

$$W = F \times s = ma \times s$$

$\therefore \quad = 50 \times 2 \times 100 = 10000 = 10^4$ J

23. ***(b)*** Threshold energy of A is

$$E_A = h\nu_A$$
$$= 6.6 \times 10^{-34} \times 1.8 \times 10^{14}$$
$$= 11.88 \times 10^{-20}\text{ J} = \frac{11.88 \times 10^{-20}}{1.6 \times 10^{-19}}\text{ eV}$$

$\Rightarrow \quad E_A = 0.74$ eV

Similarly, $E_B = 0.91$ eV

Since, the incident photons have energy greater than E_A but less than E_B.

So, photoelectrons will be emitted from metal A only.

24. ***(d)*** Let θ be the angle between vectors **P** and **Q** whose resultant is **R**.

Hence, $P = Q$ and $R^2 = 3PQ = 3P^2$

As $\quad R^2 = P^2 + Q^2 + 2PQ\cos\theta$

$\therefore \quad 3P^2 = P^2 + P^2 + 2P^2\cos\theta$

or $\quad 3P^2 - 2P^2 = 2P^2\cos\theta$

or $\quad P^2 = 2P^2\cos\theta$

or $\quad 1 = 2\cos\theta$

$\therefore \quad \cos\theta = \frac{1}{2}$, thus $\cos\theta = \cos 60°$

or $\quad \theta = 60° = \pi/3$

25. ***(b)*** In stable equilibrium, the centre of gravity of object, lies at minimum height from ground. As the object is given a slight push, its centre of gravity rises because it comes in unstable equilibrium.

26. ***(a)*** Maximum height, $H_0 = \frac{u^2\sin^2\theta}{2g}$

Range, $\quad R = \frac{u^2 \sin 2\theta}{g}$

Given, $\quad H_0 = R/2$

$\therefore \quad \frac{u^2\sin^2\theta}{2g} = \frac{u^2 2\sin\theta\cos\theta}{2g}$

$\Rightarrow \quad \sin\theta = 2\cos\theta \Rightarrow \tan\theta = 2$

$\therefore \quad \theta = \tan^{-1}(2)$

27. *(a)* Total mass of the shell = 20 kg
Ratio of the masses of the fragments = 2 : 3
∴ Masses of the fragments are

$$m_1 = 20 \times \frac{2}{2+3} = 8 \text{ kg} \Rightarrow m_2 = 20 \times \frac{3}{2+3} = 12 \text{ kg}$$

According to the conservation of momentum

$$m_1 v_1 = m_2 v_2$$

$$\therefore \quad 8 \times 6 = 12 \times v \Rightarrow V = 4 \text{ m/s}$$

v (velocity of the larger fragment) = 4 m/s

$$\text{Kinetic energy} = \frac{1}{2} mv^2 = \frac{1}{2} \times 12 \times (4)^2 = 96 \text{ J}$$

28. *(b)* Angular acceleration $|\alpha| = \omega^2 x$

$$\text{or angular velocity} \quad \omega = \sqrt{\frac{\alpha}{x}} = \sqrt{\frac{2}{0.02}} = 10 \text{ rad/s}$$

29. *(a)* According to Kepler's second law,

$$\frac{dA}{dt} = \frac{L}{2m} = \text{constant}$$

∴ Angular momentum L = constant

30. *(a)* Momentum is conserved is non-elastic collision but kinetic energy is not conserved.

31. *(b)* Fringe displacement, $x_0 = \dfrac{D(\mu - 1)t}{d}$

32. *(a)* $\omega_{H_g} = \dfrac{a_{H_g}}{a_{H_\omega}} = \dfrac{5/3}{4/3} = \dfrac{5}{4}$

$$\therefore \text{Critical angle, } \sin C = \frac{1}{\omega_{H_g}} = \frac{4}{5} \Rightarrow C = \sin^{-1}\left(\frac{4}{5}\right)$$

33. *(d)* The produced rays in sonography are ultrasound.

34. *(c)* $\dfrac{I_s}{I_p} = \dfrac{N_p}{N_s} = \dfrac{1}{4} \Rightarrow I_s = \dfrac{1}{4} \times 4 = 1 \text{ A}$

35. *(c)* Minimum force required to move a body up a rough inclined plane, $F_1 = mg(\sin\theta + \mu\cos\theta)$
Minimum force required to prevent the body form sliding down the rough inclined plane,

$$F_2 = \mu\, mg \cos\theta$$

According to question,

$$F_1 = 3F_2$$

$$\therefore \quad mg(\sin\theta + \mu\cos\theta) = 3(\mu\, mg\cos\theta)$$

$$\Rightarrow \quad \sin\theta + \mu\cos\theta = 3\mu\cos\theta$$

$$\sin\theta = 2\mu\cos\theta$$

$$\Rightarrow \quad \tan\theta = 2\mu = 2 \times \frac{1}{2\sqrt{3}} = \frac{1}{\sqrt{3}}$$

$$\Rightarrow \quad \tan\theta = \tan 30° \Rightarrow \theta = 30°$$

36. *(c)* The effective spring constant k_s of this arrangement is

$$\frac{1}{k_s} = \frac{1}{k} + \frac{1}{2k}$$

$$\frac{1}{k_s} = \frac{2+1}{2k} = \frac{3}{2k} \Rightarrow k_s = \frac{2k}{3}$$

The effective spring constant k_p of this arrangement is

$$k_p = k_1 + k_2 = k + 2k = 3k$$

$$\therefore \quad \frac{k_s}{k_p} = \frac{2k/3}{3k} = \frac{2}{9}$$

37. *(a)* Electric power, $P = \dfrac{V^2}{R}$ or $P \propto \dfrac{1}{R}$

$$\Rightarrow \quad \frac{P_2}{P_1} = \frac{R_1}{R_2} = \frac{R}{2R} \Rightarrow P_2 = \frac{P}{2}$$

38. *(c)* An object is said to be moving with a uniform acceleration, if its velocity change by equal amount in equal intervals of time. The velocity-time graph of uniformly accelerated motion is a straight line inclined to time axis.
Acceleration of an object in a uniformly accelerated motion in one dimension is equal to the slope of the velocity-time graph with time axis.
When a body moves with uniform acceleration, that means its acceleration remains constant with time, so graph between uniform acceleration and time is

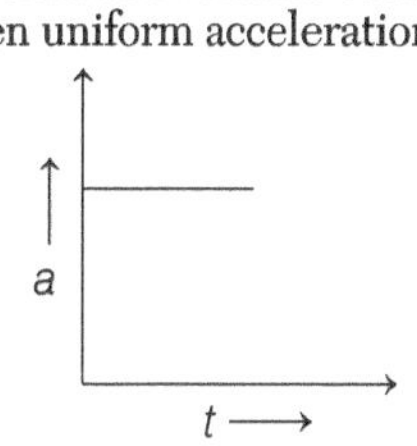

39. *(d)* From Stefan's law, the rate of emission of energy per unit surface area of a black body is directly proportional to the fourth power of absolute temperature (T) of the body.

i.e. $\quad E \propto T^4 \Rightarrow E = \sigma T^4$

40. *(c)* From Coulomb's law, $F = \dfrac{1}{4\pi\varepsilon_0}\dfrac{q_1 q_2}{r^2}$

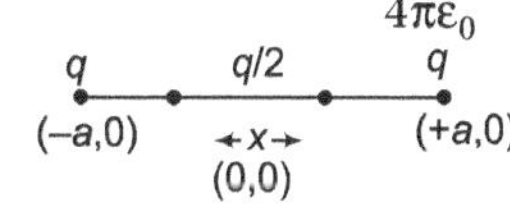

$$F = \frac{1}{4\pi\varepsilon_0}\frac{q \times \frac{q}{2}}{(a+x)^2} - \frac{1}{4\pi\varepsilon_0}\cdot\frac{q \times \frac{q}{2}}{(a-x)^2}$$

$$= \frac{1}{4\pi\varepsilon_0}\frac{q^2}{2}\left[\frac{1}{(a+x)^2} - \frac{1}{(a-x)^2}\right]$$

$$= \frac{1}{4\pi\varepsilon_0}\cdot\frac{q^2}{2}\left[-\frac{4ax}{(a^2 - x^2)^2}\right]$$

When $x << a$, then

$$F = -\frac{2q^2}{4\pi\varepsilon_0 a^3}x$$

$$\Rightarrow \quad F \propto -x$$

Hence, SHM along X-axis.

Chemistry

41. ***(c)*** λ_m^∞ for $BaCl_2 = \lambda_m^\infty\ Ba^{2+} + 2\lambda_m^\infty\ Cl^-$

$\therefore\ \lambda_{eq}^\infty$ for $BaCl_2 = \frac{1}{2}\lambda_m^\infty\ Ba^{2+} + \lambda_m^\infty\ Cl^-$

$= \frac{1}{2} \times 127 + 76$

$= 139.5\ \Omega^{-1}\ cm^2$

42. ***(b)*** $m = \frac{1000 \times K_b \times w}{W \times \Delta T_b}$

or $K_b = \frac{m \times W \times \Delta T_b}{1000 \times w}$

$= \frac{100 \times 100 \times \Delta T_b}{1000 \times 10} = \Delta T_b$

43. ***(a)*** Consider the formation of H_2O,

$H_2(g) + \frac{1}{2}O_2(g) \longrightarrow H_2O(l);\ \Delta H = -\ 286.20$ kJ

$\Delta H_r = \Delta H_f[H_2O(l)] - \Delta H_f[H_2(g)] - \frac{1}{2}\Delta H_f[O_2(g)]$

$-\ 286.20 = \Delta H_f[H_2O(l)] - 0 - 0$

$\therefore\ \Delta H_f[H_2O(l)] = -\ 286.20$

Now consider the ionisation of H_2O,

$H_2O(l) \longrightarrow H^+(aq) + OH^-(aq);\ \Delta H = 57.32$ kJ

$\Delta H_r = \Delta H_f[H^+(aq)] + \Delta H_f[OH^-(aq)] - \Delta H_f[H_2O(l)]$

$57.32 = 0 + \Delta H_f[OH^-(aq)] - (-\ 286.20)$

$\Rightarrow \Delta H_f[OH^-(aq)] = 57.32 - 286.20$

$= -\ 228.88$ kJ

44. ***(c)*** Reaction of HCl and NaOH is as follows

$HCl + NaOH \longrightarrow NaCl + H_2O$

At $t = 0$,

Number of moles of HCl $= \frac{500 \times 0.1}{1000}$

$= 0.05$

Number of moles of NaOH $= \frac{200 \times 0.2}{1000} = 0.04$

$\because$ During neutralisation of 1 mole of NaOH by 1 mole of HCl, heat evolved = 57.3 kJ

$\therefore$ To neutralise 0.04 moles of NaOH by 0.04 mole of HCl, heat evolved

$= 57.3 \times 0.04 = 2.292$ kJ

45. ***(a)*** According to Hardy-Schulze rule, coagulation power of ions is directly proportional to charge on ion.

$\because Fe(OH)_3$ is positively charged colloid.

$\therefore$ It will be coagulated by anion.

(a) $Mg_3(PO_4)_2 \rightleftharpoons 3Mg^{2+} + 2PO_4^{3-}$

(b) $BaCl_2 \rightleftharpoons Ba^{2+} + 2Cl^-$

(c) $NaCl \rightleftharpoons Na^+ + Cl^-$

(d) $KCN \rightleftharpoons K^+ + CN^-$

$\because PO_4^{3-}$ has highest charge among the given anions.

$\therefore Mg_3(PO_4)_2$ is the most effective in coagulation of $Fe(OH)_3$ sol.

46. ***(d)*** The given reaction is as follows

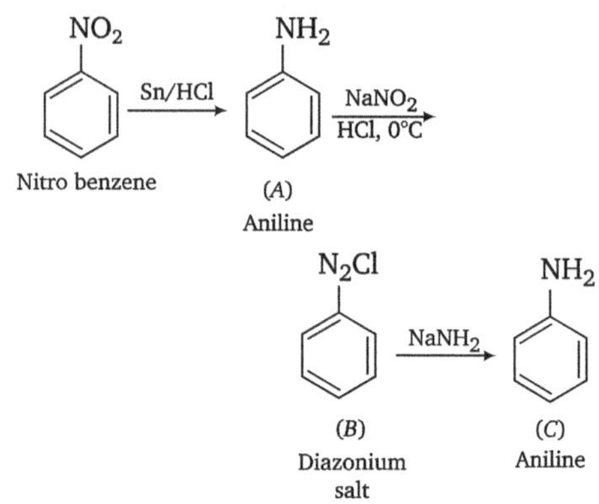

47. ***(b)*** At 120-140°C temperature and 1.5 atm pressure, sodium phenoxide reacts with CO_2 to yield sodium salicylate which on further hydrolysis gives salicylic acid. This reaction is known as Kolbe's reaction.

48. ***(d)*** Number of electrons in C = 6

(a) Number of electrons in $Na^+ = 11 - 1 = 10$

(b) Number of electrons in $Al^{3+} = 13 - 3 = 10$

(c) Number of electrons in $O^{2-} = 8 + 2 = 10$

(d) Number of electrons in $N^+ = 7 - 1 = 6$

$\therefore N^+$ is isoelectronic with C.

49. ***(c)*** (a) $\underset{sp^3}{CH_3}-\underset{sp^3}{CH_2}-\underset{sp^2}{CH}=\underset{sp^2}{CH_2}$

(b) $\underset{sp^3}{CH_3}-\underset{sp^2}{CH}=\underset{sp^2}{CH}-\underset{sp^2}{\overset{+}{CH_2}}$

(c) $\underset{sp^2}{CH_2}=\underset{sp^2}{CH}-\underset{sp^2}{CH}=\underset{sp^2}{CH_2}$

(d) $\underset{sp^3}{CH_3}-\underset{sp^2}{CH}=\underset{sp^2}{CH}-\underset{sp^3}{CH_2^-}$

Thus, in option (c), all carbon have same type of hybridisation i.e. sp^2.

50. ***(a)*** Permanganate ion is reduced to Mn^{2+} ion (from O.N. = 7 to O.N.1 in acidic medium).

$2MnO_4^- + 5H_2O_2 + 6H^+ \longrightarrow 2Mn^{2+} + 5O_2 + 8H_2O$

51. ***(c)*** In the titration of strong base with strong acid, we can use methyl orange, methyl red, phenolphthalein as indicator.

52. ***(d)*** The IUPAC name of $K_2[Cr(CN)_2O_2(O)_2NH_3]$ is potassiumamminedicyanodioxoperoxochromate (VI).

53. ***(b)*** In Wacker process, when mixture of propene and air is passed through mixture of Pd and $CuCl_2$ at high pressure, acetone is formed.

$$Pd + CuCl_2 \longrightarrow PdCl_2 + 2CuCl$$
$$4CuCl + 4HCl + O_2 \longrightarrow 4CuCl_2 + 2H_2O$$
$$\underset{\text{Propene}}{CH_3CH{=}CH_2} + PdCl_2 + H_2O \longrightarrow CH_3COCH_3 + Pd + 2HCl$$

54. ***(d)*** Lindane is γ-benzene hexachloride. It can be prepared by treating benzene with Cl_2 in sunlight.

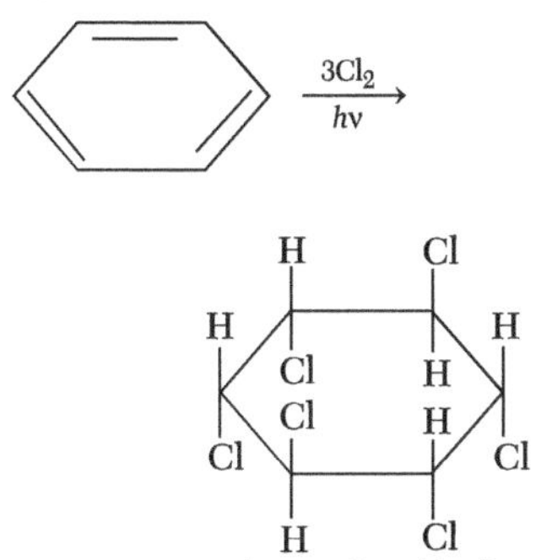

+ α- and β-isomers

γ-benzene hexachloride or Lindane

55. ***(b)*** The two propenyl groups are attached to 1, 2 position of carbon in *cis*-form.

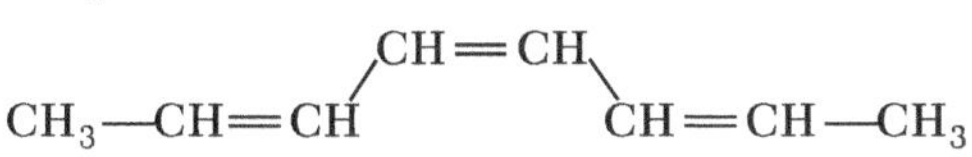

i.e. 

56. ***(d)*** $Ba(OH)_2 + CO_2 \longrightarrow BaCO_3 + H_2O$

$\because$ 5 moles of $Ba(OH)_3 \equiv$ 5 moles of $BaCO_3$

$\therefore$ Mass of $BaCO_3$ = moles of $BaCO_3 \times$ molecular mass of $BaCO_3$

$= 5 \times 197 = 985$ g

57. ***(a)*** $\delta = \dfrac{\text{Dipole moment}}{d} = \dfrac{1.2\text{ D}}{1.0 \times 10^{-8}\text{ cm}}$

$= \dfrac{1.2 \times 10^{-8}\text{ esu cm}}{1.0 \times 10^{-8}\text{ cm}}$

$= 1.2 \times 10^{-8}$ esu

The fraction of electronic charge, *e* is

$= \dfrac{1.2 \times 10^{-10}\text{ esu}}{4.8 \times 10^{-10}\text{ esu / e}}$

$= 0.25\ e = 25\%$ of e

58. ***(a)*** Let $T_1 = T$ and $T_2 = (T+1)$

$p_1 = p$ and $p_2 = p + \dfrac{0.4\,p}{100} = \dfrac{100.4}{100}\,p$

From $\dfrac{p_1V_1}{T_1} = \dfrac{p_2V_2}{T_2}$,

$$\frac{pV}{T} = \frac{100.4p}{100} \times \frac{V}{(T+1)}$$

$100T + 100 = 100.4\,T$

$0.4\,T = 100$

$T = \dfrac{100}{0.4} = 250$ K

$= (250 - 273)°C = -23°C$

59. ***(d)*** $2NOBr(g) \rightleftharpoons 2NO(g) + Br_2(g)$

$p - \left(\dfrac{2p}{9} + \dfrac{p}{9}\right) = \dfrac{6p}{9}$ $\qquad \dfrac{2p}{9} \qquad \dfrac{p}{9}$

and $K_p = \dfrac{(p_{NO})^2 \times (p_{Br_2})}{(p_{NOBr})^2}$

$= \dfrac{\left(\dfrac{2p}{9}\right)^2 \times \left(\dfrac{p}{9}\right)}{\left(\dfrac{6p}{9}\right)^2} = \dfrac{p}{81}$

$\Rightarrow K_p = \dfrac{p}{81}$

$\therefore \dfrac{K_p}{p} = \dfrac{1}{81}$

60. ***(c)*** Barium salts are quite stable because of great electropositive nature of Ba. Hence, Ba compounds possess high decomposition temperature.

61. ***(a)*** $Zn + H_2SO_4 \longrightarrow ZnSO_4 + H_2\uparrow$

$Zn + 2NaOH \longrightarrow Na_2ZnO_2 + H_2\uparrow$

Hence, ratio of volumes of hydrogen evolved is 1 : 1.

62. ***(b)*** The given reaction take place as follows

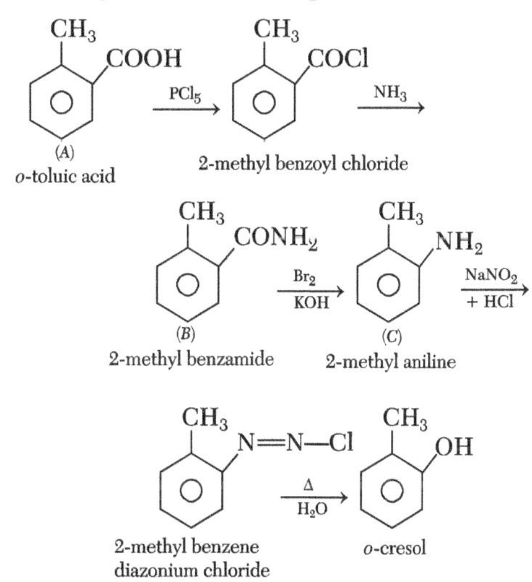

63. ***(d)*** Alizarin is an anthraquinone dye. It gives a bright red colour with aluminium and a blue colour with barium.

64. *(c)* Acetylene reacts with hypochlorous acid to give dichloroacetaldehyde.

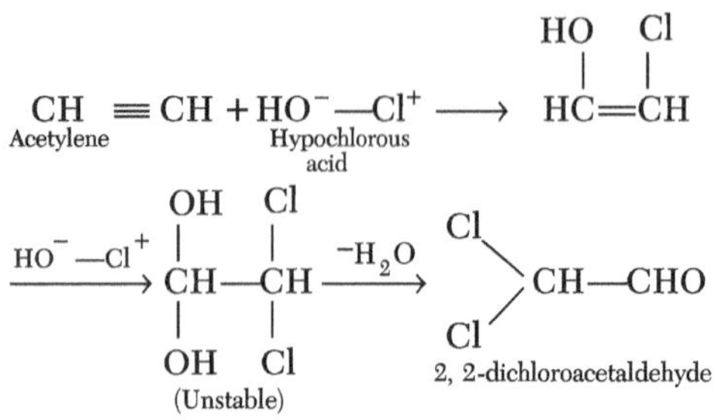

65. *(a)* Number of Ba^{2+} ions at the corner of unit cell

$$= 8 \times \frac{1}{8} = 1$$

Number of O^{2-} ions at the face centres $= 6 \times \frac{1}{2} = 3$

Number of Ti^{4+} ions at the body centre = 1

$\therefore$ Molecular formula of barium titanate is $BaTiO_3$.

66. *(d)* Somatotropin is the harmone, secreted by anterior lobe of pituitary gland. It is also called growth harmone as it stimulates protein synthesis, glycogensis and some other biological activities. Its deficiency causes midgets or dwarfism.

67. *(b)* Number of electrons in $F^- = 9 + 1 = 10$

Number of electrons in $O^{2-} = 8 + 2 = 10$

Number of electrons in $Na^+ = 11 - 1 = 10$

Number of electrons in $Mg^{2+} = 12 - 2 = 10$

Since, F^-, O^{2-}, Na^+, Mg^{2+} are isoelectric and the size of isoelectric species decreases with increase in nuclear charge (i.e., number of protons). Hence, correct order of size is $O^{2-} > F^- > Na^+ > Mg^{2+}$.

68. *(a)* The radius of hydrogen atom is 0.53 Å. ${}_3Li^{2+}$ ion also has only one electron but it has 3 protons in nucleus, hence its electron feels three times more attraction from nucleus in comparison of hydrogen atom. Thus, the radius of ${}_3Li^{2+}$ will be

$$= \frac{0.53}{3} = 0.17 \text{ Å}$$

69. *(d)* Stronger the acid, higher the k_a values and lower the pk_a value.

$$pk_a \propto \frac{1}{k_a}$$

$\therefore$ The order of acidity of given acids is as follows

$$CH_3CH\cdot F\cdot COOH > CH_3CH\cdot Br\cdot COOH > FCH_2CH_2COOH > CH_3CH_2COOH$$

Since, CH_3CH_2COOH is the weakest acid among the given compounds, therefore its pk_a value will be highest.

70. *(c)* The given reaction is as follows

$$\underset{(A)}{CO} + NaOH \longrightarrow \underset{(B)}{HCOONa} \xrightarrow[-H_2]{\Delta} \underset{C}{(COONa)_2} \xrightarrow{H^+} \underset{D}{(COOH)_2}$$

$\therefore$ *C* and *D* are $(COONa)_2$ and $(COOH)_2$.

71. *(c)* The polymer polyurethanes are formed by treating di-isocyanate with glycol as follows

$$\underset{\text{Di-isocyanate}}{O{=}C{=}N{-}R{-}N{=}C{=}O} + \underset{\text{Glycol}}{HO{-}R'{-}OH} \longrightarrow$$

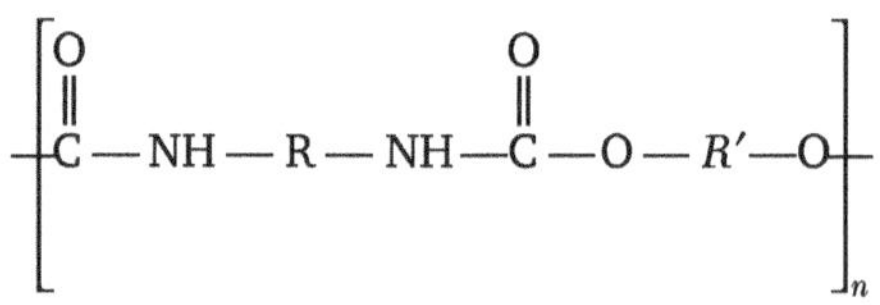

Polyurethanes

72. *(a)* Given, $C = 5\,A$

$$t = 193 \text{ s}$$

$$W = \frac{C \times t \times E}{F} = \frac{5 \times 193 \times 8}{96500} = 0.08 \text{g} O_2$$

$\because$ At NTP, volume of

$$32 \text{ g } O_2 = 22400 \text{ mL}$$

$$\therefore \text{Volume of } 0.08 \text{ g } O_2 = \frac{22400 \times 0.08}{32} = 56 \text{ mL}$$

73. *(b)* $CO_2 + H_2O \rightleftharpoons H_2CO_3 \longrightarrow H^+ + HCO_3^-$

Here, $[H^+]$ increases, hence pH decreases due to which soil fertility will also decrease.

74. *(d)* The reaction is of first order and for a first order reaction,

$$\text{Rate}(R) = k\,[N_2O_5]$$

$$2.4 \times 10^{-5} = 3 \times 10^{-5} \times [N_2O_5]$$

$$\Rightarrow [N_2O_5] = \frac{2.4 \times 10^{-5}}{3 \times 10^{-5}} = 0.8 \text{ mol L}^{-1}$$

75. *(d)* BF_3 is symmetrically planar, although it has polar bonds but resultant dipole moment is zero. In NF_3, lone pair cause distortion, hence, polarity arises.

76. *(c)* For NaCl, $pV = nRT \times i$

$$p_1 \times \frac{100}{1000} = \frac{1.2}{585} \times 0.0821 \times T \times i \quad \text{...(i)}$$

For glucose, $pV = nRT$

$$p_2 \times \frac{100}{1000} = \frac{7.2}{180} \times 0.0821 \times T \quad \text{...(ii)}$$

$\because$ NaCl and glucose solutions are isotonic.

$$\therefore \quad p_1 = p_2$$

On dividing Eq. (i) by (ii), we have

$$i = \frac{7.2}{180} \times \frac{58.5}{1.2} = 1.95 \approx 2$$

77. ***(a)*** $FeSO_4 \cdot 7H_2O$ (ferrous sulphate) is known as green vitriol.

78. ***(c)*** $\underset{\text{2-bromopentane}}{CH_3CH_2CH_2\underset{|}{\overset{}{C}}HCH_3}$ (Br on C-2) $\xrightarrow{C_2H_5OK}$

$\underset{\text{Pentene}}{CH_3CH_2CH_2CH{=\!=}CH_2} + \underset{\text{2-pentene (mix of cis and } trans \text{ form)}}{CH_3CH_2CH{=\!=}CHCH_3}$

By Saytzeff's rule, substituted alkenes are more stable. Out of *cis* and *trans* forms, *trans* product is more stable.

∴ *trans* 2-pentene is major product formed is given reaction.

79. ***(b)*** We know C—C bond length = 1.54 Å

that C=C bond length = 1.34 Å

C≡C bond length = 1.20 Å

Since, propyne has triple bond, therefore it has minimum bond length.

80. ***(d)*** IUPAC name of given compound is 1, 4-dichloro-4-formyl-2, 6-dioxy-cyclohexane-1-carboxylic acid.

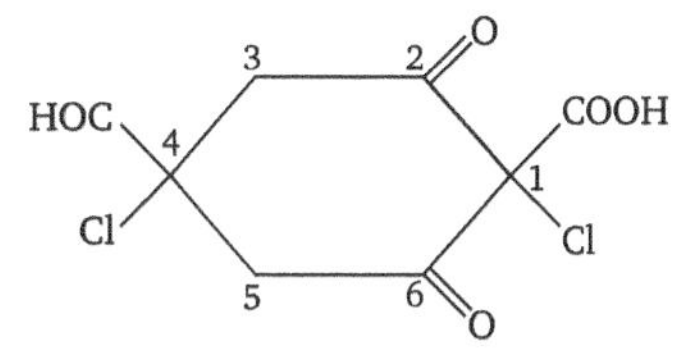

a. English Proficiency

81. ***(a)*** 'Dubious' means doubtful, uncertain, unsure. So, 'Doubtful' is its correct similar meaning word.

82. ***(c)*** The one which expresses the right meaning of the given word 'Flabbergasted' is Dumbfounded.

Flabbergasted means greatly surprised or astonished.

Dumbfounded means greatly astonished or amazed.

83. ***(d)*** 'Eternal' and 'Perpetual' both have same meaning i.e. lasting or existing forever; without end.

84. ***(c)*** 'Despair' means the complete loss or absence of hope. So, 'hope' would be its correct antonym.

85. ***(b)*** 'In toto' means as a whole, in all, or overall. So, 'Partially' would be its correct antonym.

86. ***(c)*** Protean means tending or able to change frequently or easily. So, 'Unchanging' would be its correct opposite meaning word.

87. ***(d)*** The given sentence needs no improvement.

88. ***(a)*** 'Left' should be used in place of 'leave' to make the sentence grammatically correct.

89. ***(c)*** 'About his character' is not correct in the given sentence. Only his character should be used, to make the sentence grammatically correct.

90. ***(a)*** Preposition 'of' is suitable for the given blank.

91. ***(c)*** 'Different' is appropriate to fill the given blank.

92. ***(d)*** 'Suspect' should be used in the given blank. Suspect means have an idea or impression of the existence, presence or truth of.

93. ***(a)*** The correct combination is QSRP, to make the meaningful sentence.

94. ***(d)*** The correct sequence is QSPR of the given jumbled sentences.

95. ***(b)*** The correct sequence is QSRP to make the meaningful sentence.

b. Logical Reasoning

96. ***(a)*** As

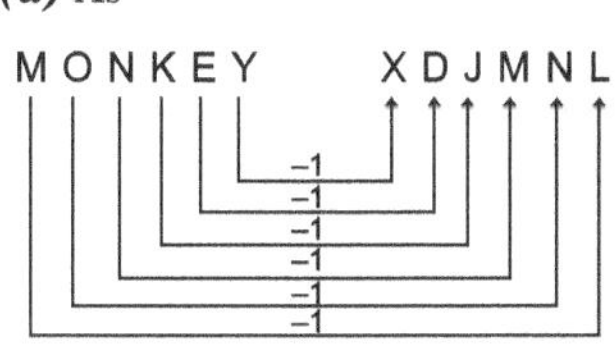

Similarly,

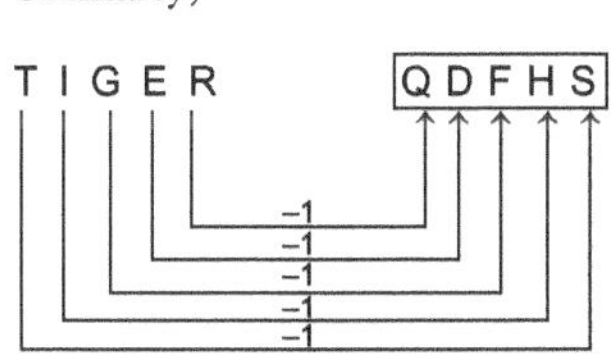

97. ***(c)*** As $13 \times 17 = 221$

and $12 \times 19 = 228$

Similarly, $13 \times 18 = \boxed{234}$

98. ***(a)*** If the day before yesterday was Thursday. Then, today will be Saturday and the Sunday will be tomorrow.

99. ***(c)*** When Ravi and Shyam interchange their position than Ravi's new position (ninth from right) is the same as Shyam's initial position (second from left)

∴ Total number of students in the row = 8 + 1 + 1 = 10

So, Shyam's new position is same as Ravi's initial position (fourth from right) third from left. So, Shyam's new position from left = 10 − 3 = 7th

100. ***(d)*** Answer figure (d) will appear when a piece of paper folded and cut as shown in the question figure.

101. ***(a)***

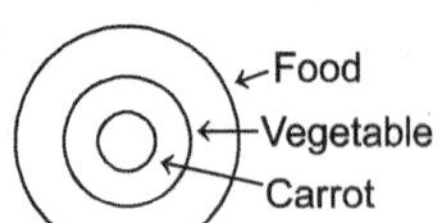

102. ***(d)***

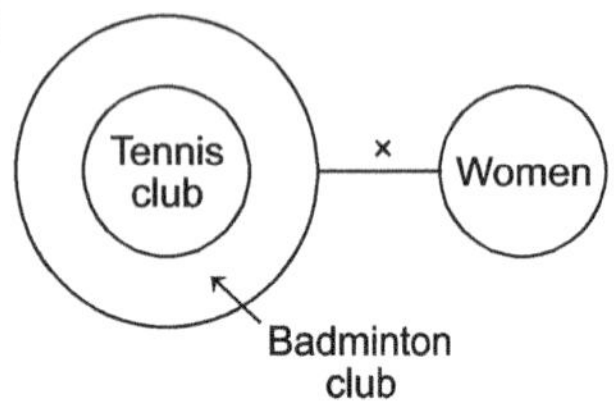

So, No woman is a member of Tennis club.

103. ***(c)*** The pattern of series is

1	9	25	49 [50]	81
↓	↓	↓	↓	↓
1^2	3^2	5^2	7^2	9^2

Hence, 50 is the wrong number.

104. ***(d)*** Given equation,

$$24 \Delta 4 \Delta 5 \Delta 4$$

From option (d),

$$24 = 4 + 5 \times 4$$
$$24 = 4 + 20$$
$$24 = 24$$

105. ***(c)*** Answer figure (c) is correct mirror image of question figure.

Mathematics

106. ***(d)*** Length of perpendicular from $A(2, -1)$ to the line $x + y - 2 = 0$ is

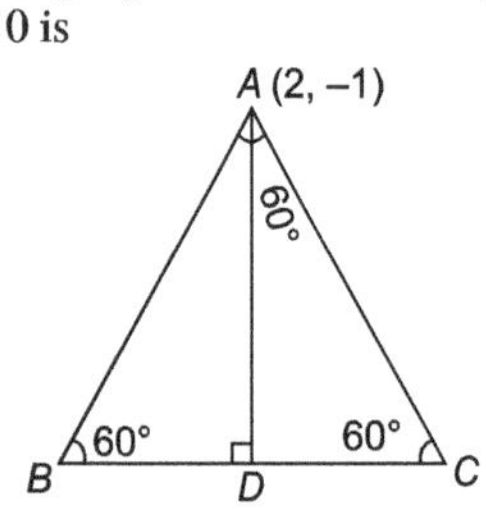

$$AD = \left|\frac{2-1-2}{\sqrt{1+1}}\right| = \frac{1}{\sqrt{2}}$$

In ΔABD, $\frac{AD}{AB} = \sin 60°$

$$\Rightarrow \quad \frac{1}{\sqrt{2}AB} = \frac{\sqrt{3}}{2}$$

$$\therefore \quad AB = \sqrt{\frac{2}{3}} = \left(\frac{2}{3}\right)^{\frac{1}{2}}$$

107. ***(d)*** Lines, $x + y = 6$, $2x + y = 4$ and $x + 2y = 5$ intersect at points $(-2, 8)$, $(7, -1)$ and $(1, 2)$.

Now, all these points lie on

$$x^2 + y^2 - 17x - 19y + 50 = 0$$

108. ***(a)*** Required length of tangent is $\sqrt{S_1}$, where

$$S_1 = 25 + 1 + 30 - 4 - 3 = 49$$

$$\therefore \quad \sqrt{S_1} = 7$$

109. ***(d)*** Length of minor axis $= 2b$ and according to the given condition length of major axis

$$= 3(2b) = 6b$$

$$\therefore \quad e = \sqrt{1 - \frac{b^2}{(3b)^2}} = \sqrt{\frac{8}{9}} = \frac{2\sqrt{2}}{3}$$

110. ***(c)*** In ΔBOT,

$$\frac{b}{ae} = \tan 60° \Rightarrow \quad b = ae\sqrt{3}$$

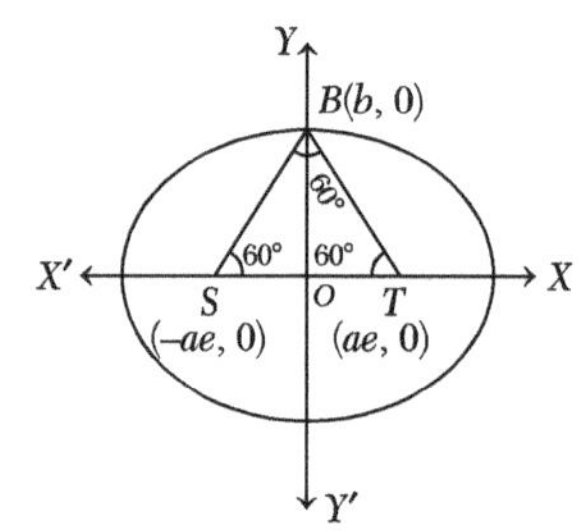

$$\therefore \quad e = \sqrt{1 - \frac{b^2}{a^2}} = \sqrt{1 - \frac{a^2e^2 3}{a^2}}$$

$$\Rightarrow \quad e^2 = 1 - 3e^2$$

$$\Rightarrow \quad 4e^2 = 1 \Rightarrow e = \pm\frac{1}{2}$$

$$\Rightarrow \quad e = \frac{1}{2} \qquad (\because e \text{ cannot be negative})$$

111. ***(c)*** The difference of the focal distance at any point on the hyperbola is same as length of the transverse axis. i.e., $2a$.

112. ***(a)*** Since, $A + B + C = 180°$

$$\Rightarrow \quad \cot(A + B + C)$$
$$= \frac{\Sigma \cot A \cot B - 1}{\cot A \cot B \cot C - \Sigma \cot A} = \frac{1}{0}$$

$$\Rightarrow \quad \cot A \cot B \cot C - \Sigma \cot A = 0$$

$$\Rightarrow \quad \frac{\cot A + \cot B + \cot C}{\cot A \cot B \cot C} = 1$$

113. ***(d)*** Now, $30° = 30° \times \frac{\pi}{180}\text{ rad} = \frac{\pi}{6}$

Let angle be a, $a + d$, $a + 2d$ are in AP.

Now, $3a + 3d = \pi \quad (\because A + B + C = \pi)$

$\Rightarrow \quad 3 \times \frac{\pi}{6} + 3d = \pi \Rightarrow 3d = \pi - \frac{\pi}{2}$

$\Rightarrow \quad d = \frac{1}{3}\left(\pi - \frac{\pi}{2}\right) = \frac{\pi}{6}$

$\therefore$ Greatest angle $= a + 2d$

$$= \frac{\pi}{6} + 2 \cdot \frac{\pi}{6} = \frac{\pi}{2}$$

114. ***(a)*** Given that, $\tan 20^\circ = p$

$\therefore \quad \frac{\tan 160^\circ - \tan 110^\circ}{1 + \tan 160^\circ \tan 110^\circ}$

$= \frac{\tan(180^\circ - 20^\circ) - \tan(90^\circ + 20^\circ)}{1 + \tan(180^\circ - 20^\circ)\tan(90^\circ + 20^\circ)}$

$= \frac{-\tan 20^\circ + \cot 20^\circ}{1 + \tan 20^\circ \cot 20^\circ}$

$= \frac{-p + \frac{1}{p}}{1+1} = \frac{1-p^2}{2p} \quad \left[\because \cot\theta = \frac{1}{\tan\theta} \text{ and } \cot\theta\tan\theta = 1\right]$

115. ***(a)*** Given, $4\sin^{-1} x + \cos^{-1} x = \pi$

$\Rightarrow \quad 4\sin^{-1} x + \frac{\pi}{2} - \sin^{-1} x = \pi$

$\Rightarrow \quad 3\sin^{-1} x = \frac{\pi}{2}$

$\Rightarrow \quad x = \sin\left(\frac{\pi}{6}\right) = \frac{1}{2}$

116. ***(b)*** $a = 2, b = 3, \sin A = \frac{2}{3} \Rightarrow \sin^2 A = \frac{4}{9}$

$\therefore \quad \cos^2 A = 1 - \sin^2 A = \frac{5}{9}$

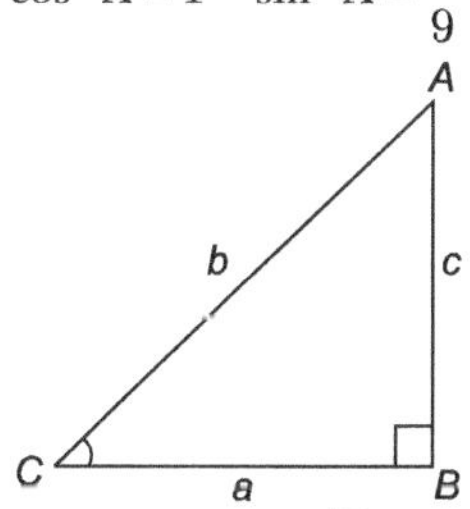

$\Rightarrow \quad \cos A = \frac{\sqrt{5}}{3}$

By sin rule, $\frac{\sin A}{a} = \frac{\sin B}{b}$

$\Rightarrow \quad \frac{2}{3} \times \frac{1}{2} = \frac{\sin B}{3} \Rightarrow \sin B = 1 \Rightarrow B = 90^\circ$

In $\Delta ABC, \cos C = \frac{a}{b} = \frac{2}{3}$

117. ***(c)*** Equation of the line passing through **a** and **b** is $\mathbf{a} + t(\mathbf{b} - \mathbf{a})$.

Here, $\mathbf{b} - \mathbf{a} = 6\hat{\mathbf{j}} - \hat{\mathbf{k}}$

$\Rightarrow \quad \mathbf{b} = 6\hat{\mathbf{j}} - \hat{\mathbf{k}} + \hat{\mathbf{i}} - 2\hat{\mathbf{j}} - \hat{\mathbf{k}} = \hat{\mathbf{i}} + 4\hat{\mathbf{j}} - 2\hat{\mathbf{k}}$

$\therefore$ Given, line passes through $(1, -2, -1)$ and $(1, 4, -2)$.

118. ***(c)*** Work done = (Force $\cdot$ Displacement)

$= (4\hat{\mathbf{i}} - 3\hat{\mathbf{j}} + 2\hat{\mathbf{k}}) \cdot \{(2\hat{\mathbf{i}} - \hat{\mathbf{j}} + 4\hat{\mathbf{k}}) - (3\hat{\mathbf{i}} + 2\hat{\mathbf{j}} - \hat{\mathbf{k}})\}$

$= (4\hat{\mathbf{i}} - 3\hat{\mathbf{j}} + 2\hat{\mathbf{k}}) \cdot (-\hat{\mathbf{i}} - 3\hat{\mathbf{j}} + 5\hat{\mathbf{k}})$

$= (-4 + 9 + 10) = 15$ units

119. ***(d)*** $\lim_{x \to 0} \frac{(2+x)\sin(2+x) - 2\sin 2}{x} \quad \left(\frac{0}{0} \text{ form}\right)$

$= \lim_{x \to 0} \frac{\sin(2+x) + (2+x)\cos(2+x)}{1} = \sin 2 + 2\cos 2$

120. ***(a)*** Now, $\lim_{x \to 0} f(x) = \lim_{x \to 0} \frac{3x + \tan^2 x}{x} \quad \left(\frac{0}{0} \text{ form}\right)$

$= \lim_{x \to 0} \frac{3 + 2\tan x \sec^2 x}{1} = 3$

Since, $f(x)$ is continuous at $x = 0$

$\therefore \quad f(0) = 3$

121. ***(c)*** $\frac{d}{dx}(\cos x) = -\frac{\pi}{180}\sin x$

122. ***(a)*** $\frac{d}{dx}[\log(\sec x - \tan x)]$

$= \frac{1}{\sec x - \tan x}[\sec x \tan x - \sec^2 x]$

$= \frac{\sec x[\tan x - \sec x]}{\sec x - \tan x} = -\sec x$

123. ***(c)*** Given, $x = \cos^3\theta, y = \sin^3\theta$

On differentiating w.r.t. θ, we get

$$\frac{dx}{d\theta} = -3\cos^2\theta\sin\theta$$

and $\quad \frac{dy}{d\theta} = 3\sin^2\theta\cos\theta$

Now, $\frac{dy}{dx} = -\frac{3\sin^2\theta\cos\theta}{3\cos^2\theta\sin\theta} = -\tan\theta$

$\therefore \quad 1 + \left(\frac{dy}{dx}\right)^2 = 1 + \tan^2\theta = \sec^2\theta$

124. ***(b)*** Given, $x = at^2, y = 2at$

On differentiating w.r.t. t, we get

$$\frac{dx}{dt} = 2at, \frac{dy}{dt} = 2a$$

$\therefore \quad \frac{dy}{dx} = \frac{2a}{2at} = \frac{1}{t}$

$\Rightarrow \quad \frac{d^2y}{dx^2} = -\frac{1}{t^2}\frac{dt}{dx}$

$$= -\frac{1}{t^2} \cdot \frac{1}{2at} = -\frac{1}{2at^3}$$

125. **(a)** Circumference of circle, $C = 2\pi r$

$\Rightarrow \quad \frac{dC}{dt} = 2\pi \frac{dr}{dt}$

$\Rightarrow \quad \frac{0.3}{2\pi} = \frac{dr}{dt} \left(\because \frac{dC}{dt} = 0.3 \text{ cm / s given}\right)$

Now, $\quad A = \pi r^2$

$\frac{dA}{dt} = 2\pi r \frac{dr}{dt} \Rightarrow \frac{dA}{dt} = r \times 0.3$

$\Rightarrow \quad \left[\frac{dA}{dt}\right]_{r=5} = 5 \times 0.3 = 1.5 \text{ sq cm/s}$

126. **(b)** $\frac{dy}{dx} = 3x^2 - 2ax + 48 > 0$

($\because$ y is an increasing function)

$\therefore$ Discriminant, $D < 0$

$\Rightarrow \quad 4a^2 - 4 \times 3 \times 48 < 0$

$\Rightarrow \quad a^2 - 144 < 0$

$\Rightarrow \quad a \in (-12, 12)$

127. **(d)** Rolle's theorem is not applicable for the function $f(x) = |x|$ in $[-1, 1]$

$\because f'(0)$ does not exist.

128. **(a)** Let $I = \int \frac{2dx}{(e^x + e^{-x})^2}$

$= \int \frac{2dx}{e^{2x} + e^{-2x} + 2} = \int \frac{2e^{2x} dx}{e^{4x} + 2e^{2x} + 1}$

Put $e^{2x} = t \Rightarrow 2e^{2x} dx = dt$

$\therefore I = \int \frac{dt}{t^2 + 2t + 1} = \int \frac{dt}{(t+1)^2} = -\frac{1}{t+1} + C$

$= -\frac{1}{e^{2x} + 1} = -\frac{e^{-x}}{e^x + e^{-x}} + C$

129. **(d)** Let $I = \int_0^{\pi/2} \frac{\sin^n \theta}{\sin^n \theta + \cos^n \theta} d\theta \quad \ldots \text{(i)}$

$= \int_0^{\pi/2} \frac{\sin^n (\pi/2 - \theta)}{\sin^n (\pi/2 - \theta) + \cos^n (\pi/2 - \theta)} d\theta$

$= \int_0^{\pi/2} \frac{\cos^n \theta}{\cos^n \theta + \sin^n \theta} d\theta \quad \ldots \text{(ii)}$

On adding Eqs. (i) and (ii), we get

$2I = \int_0^{\pi/2} \frac{\sin^n \theta + \cos^n \theta}{\cos^n \theta + \sin^n \theta} d\theta$

$= \int_0^{\pi/2} d\theta = \frac{\pi}{2} \Rightarrow I = \frac{\pi}{4}$

130. **(d)** Let $I = \int_0^{\pi} \cos^{101} x\, dx$

$\Rightarrow \quad I = \int_0^{\pi} [\cos(\pi - x)]^{101} dx$

$\Rightarrow \quad I = \int_0^{\pi} -\cos^{101} x\, dx$

$\therefore \quad 2I = \int_0^{\pi} (\cos^{101} x - \cos^{101} x) dx \Rightarrow I = 0$

131. **(c)** $\lim_{n \to \infty} \left[\frac{1}{n+1} + \frac{1}{n+2} + \ldots + \frac{1}{n+5n}\right]$

$= \lim_{n \to \infty} \sum_{r=0}^{5n} \frac{1}{n+r} = \lim_{n \to \infty} \left[\frac{1}{n} \sum_{r=0}^{5n} \frac{n}{n+r}\right]$

$= \lim_{n \to \infty} \left[\frac{1}{n} \sum_{r=0}^{5n} \frac{1}{1 + (r/n)}\right] = \int_0^5 \frac{1}{1+x} dx = [\log(1+x)]_0^5$

$= \log 6 - \log 1 = \log 6$

132. **(b)** Given, $y = Ax^2 + Bx \quad \ldots \text{(i)}$

On differentiating, w.r.t. x we get

$\frac{dy}{dx} = 2Ax + B$

$\Rightarrow \quad \frac{d^2y}{dx^2} = 2A$

$\therefore \quad \frac{dy}{dx} - \frac{d^2y}{dx^2} x = B$

From Eq. (i), we get

$y = \frac{1}{2} \frac{d^2y}{dx^2} x^2 + x\left[\frac{dy}{dx} - \frac{d^2y}{dx^2} x\right]$

$\Rightarrow \quad y = -\frac{1}{2} x^2 \frac{d^2y}{dx^2} + x \frac{dy}{dx}$

$\therefore \; x^2 \frac{d^2y}{dx^2} - 2x \frac{dy}{dx} + 2y = 0$

133. **(a)** Given, $f(x) = \frac{\log(1+ax) - \log(1-bx)}{x}$

$f(x)$ is continuous at $x = k$ and $f(0) = k$.

$\therefore \lim_{x \to 0} f(x) = \lim_{x \to 0} \frac{\log(1+ax) - \log(1-bx)}{x} \left(\frac{0}{0} \text{ form}\right)$

$= \lim_{x \to 0} \left(\frac{1}{1+ax} \cdot a + \frac{b}{1-bx}\right) = a + b$

$\therefore \; a + b = f(0) = k$

134. **(b)** If $4 - 5i$ is root of $x^2 + ax + b = 0$, then $4 + 5i$ is also the root.

$\therefore$ Sum of roots $= -a = 8 \Rightarrow a = -8$

and product of roots $= 16 + 25$

$\Rightarrow \quad b = 41$

$\therefore \quad (a, b) = (-8, 41)$

135. **(b)** Given equation can be rewritten as $x^2 + \frac{3}{4} x + \frac{7}{4} = 0$

$\Rightarrow \quad \alpha + \beta = -\frac{3}{4}, \alpha\beta = \frac{7}{4}$

Now, $\quad \frac{1}{\alpha} + \frac{1}{\beta} = \frac{\alpha + \beta}{\alpha\beta} = -\frac{3}{7}$

136. ***(c)*** Since, α, β are the roots of the equation $ax^2 + bx + c = 0$.

Then, $\alpha = -\dfrac{b}{2a} + \dfrac{\sqrt{b^2 - 4ac}}{2a}$

and $\beta = -\dfrac{b}{2a} - \dfrac{\sqrt{b^2 - 4ac}}{2a}$ ...(i)

and $\alpha + k, \beta + k$ are the roots of the equation $px^2 + qx + r = 0$.

Then, $\alpha + k = -\dfrac{q}{2p} + \dfrac{\sqrt{q^2 - 4pr}}{2p}$

and $\beta + k = -\dfrac{q}{2p} - \dfrac{\sqrt{q^2 - 4pr}}{2p}$

$$\Rightarrow \quad k = -\frac{q}{2p} + \frac{\sqrt{q^2 - 4pr}}{2p} + \frac{b}{2a} - \frac{\sqrt{b^2 - 4ac}}{2a}$$ [from Eq. (i)]

and $$k = -\frac{q}{2p} - \frac{\sqrt{q^2 - 4pr}}{2p} + \frac{b}{2a} + \frac{\sqrt{b^2 - 4ac}}{2a}$$ [from Eq. (i)]

$$\Rightarrow \quad \frac{\sqrt{q^2 - 4pr}}{2p} - \frac{\sqrt{b^2 - 4ac}}{2a} = \frac{-\sqrt{q^2 - 4pr}}{2p} + \frac{\sqrt{b^2 - 4ac}}{2a}$$

$$\Rightarrow \quad \frac{\sqrt{q^2 - 4pr}}{p} = \frac{\sqrt{b^2 - 4ac}}{a}$$

$$\Rightarrow \quad \frac{q^2 - 4pr}{p^2} = \frac{b^2 - 4ac}{a^2}$$

$$\therefore \quad \frac{b^2 - 4ac}{q^2 - 4qr} = \left(\frac{a}{p}\right)^2$$

137. ***(d)*** Since, $iz = ze^{i\pi/2}$

This implies that iz is the vector obtained by rotating vector z in anticlockwise direction through $90°$.

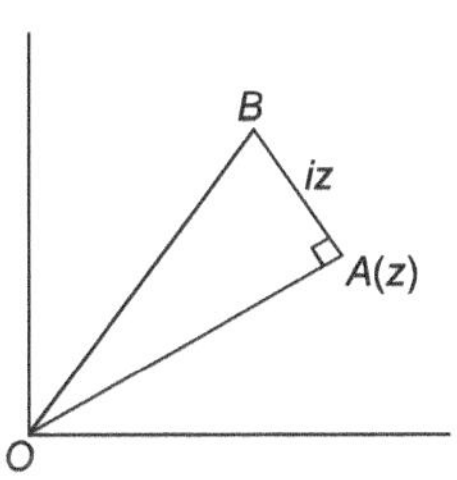

$\therefore OA \perp AB$

So, area of $\Delta OAB = \dfrac{1}{2} OA \times OB$

$$= \frac{1}{2}|z||iz| = \frac{1}{2}|z|^2$$

138. ***(a)*** Let S_n be the sum of first n terms of the series.

$$\therefore \quad S_n = \frac{1}{2} + \frac{3}{4} + \frac{7}{8} + \frac{15}{16} + \dots$$

$$\Rightarrow \quad S_n = \left(1 - \frac{1}{2}\right) + \left(1 - \frac{1}{4}\right) + \left(1 - \frac{1}{8}\right) + \dots$$

$$S_n = n - \frac{1(1 - 1/2)}{2(1 - 1/2)} = n - 1 + \frac{1}{2^n}$$

$$= n - 1 + 2^{-n}$$

139. ***(c)*** $0.2 + 0.22 + 0.222 + \dots n$ terms

$$= 2(0.1 + 0.11 + 0.111 + \dots n \text{ terms})$$

$$= 2\left(\frac{1}{10} + \frac{11}{100} + \frac{111}{1000} + \dots n \text{ terms}\right)$$

$$= \frac{2}{9}\left(\frac{9}{10} + \frac{99}{100} + \frac{999}{1000} + \dots n \text{ terms}\right)$$

$$= \frac{2}{9}\left(1 - \frac{1}{10} + 1 - \frac{1}{100} + 1 - \frac{1}{1000} + \dots n \text{ terms}\right)$$

$$= \frac{2}{9}\left[n - \left(\frac{1}{10} + \frac{1}{100} + \frac{1}{1000} + \dots n\right)\right]$$

$$= \frac{2}{9}\left[n - \frac{1}{10}\frac{\left\{1 - \left(\frac{1}{10}\right)^n\right\}}{\left(1 - \frac{1}{10}\right)}\right]$$

$$= \frac{2}{9}\left[n - \frac{1}{10} \times \frac{10}{9} \cdot \left(\frac{10^n - 1}{10^n}\right)\right] = \frac{2}{9}\left[n - \frac{1}{9}(1 - 10^{-n})\right]$$

140. ***(c)*** Number of required ways

$$= {}^{22-4-2}C_{11-2} = {}^{16}C_9$$

141. ***(a)*** $\therefore$ Required number of ways $= 4! = 24$

142. ***(a)*** Since, $T_2 = {}^nC_1$

and $T_3 = {}^nC_2 x, T_4 = {}^nC_3 x$

Also, coefficient of T_2, T_3, T_4 are in AP.

$$\Rightarrow \quad \frac{{}^nC_1 + {}^nC_3}{2} = {}^nC_2 \Rightarrow {}^nC_1 + {}^nC_3 = 2 \cdot {}^nC_2$$

$$\Rightarrow \quad \frac{n!}{(n-1)!(1)!} + \frac{n!}{(n-3)!3!} = \frac{2n!}{2!(n-2)!}$$

$$\Rightarrow \quad \frac{1}{(n-1)(n-2)} + \frac{1}{3!} = \frac{1}{(n-2)}$$

$$\Rightarrow \quad 1 + \frac{(n-1)(n-2)}{6} = (n-1)$$

$$\Rightarrow \quad 6 + n^2 - 3n + 2 = 6n - 6$$

$$\Rightarrow \quad n^2 - 3n - 6n + 8 + 6 = 0 \Rightarrow n^2 - 9n + 14 = 0$$

$$\Rightarrow \quad (n-7)(n-2) = 0 \Rightarrow n = 7 \text{ or } 2$$

$$\therefore \quad n = 7 \qquad (\because n \neq 2)$$

143. **(b)** $\Delta = \begin{vmatrix} 1 & a & a^2 \\ 1 & b & b^2 \\ 1 & c & c^2 \end{vmatrix} = \begin{vmatrix} 1 & a & a^2 \\ 0 & b-a & b^2-a^2 \\ 0 & c-a & c^2-a^2 \end{vmatrix}$

(using $R_2 \to R_2 - R_1$ and $R_3 \to R_3 - R_1$)

$$= (b-a)(c-a)\begin{vmatrix} 1 & a & a^2 \\ 0 & 1 & b+a \\ 0 & 1 & c+a \end{vmatrix}$$

$$= (b-a)(c-a)(c+a-b-a)$$

$\therefore \quad (b-a)(c-a)(c-b)$
$= k(a-b)(b-c)(c-a)$ (given)

$\Rightarrow \quad k = 1$

144. **(a)** $\Delta = \begin{vmatrix} a+b & a & b \\ a & a+c & c \\ b & c & b+c \end{vmatrix} = \begin{vmatrix} b & -c & b-c \\ a & a+c & c \\ b & c & b+c \end{vmatrix}$ (by $R_1 \to R_1 - R_2$)

$= \begin{vmatrix} 2b & 0 & 2b \\ a & a+c & c \\ b & c & b+c \end{vmatrix}$ (by $R_1 \to R_1 + R_3$)

$= \begin{vmatrix} 2b & 0 & 0 \\ a & a+c & c-a \\ b & c & c \end{vmatrix}$ (by $C_3 \to C_3 - C_1$)

$= 2b(ac + c^2 - c^2 + ac) = 4abc$

145. **(b)** Given, $\Delta_1 = \begin{vmatrix} x & a & b \\ b & x & a \\ a & b & x \end{vmatrix}, \Delta_2 = \begin{vmatrix} x & b \\ a & x \end{vmatrix}$

$\therefore \quad \frac{d}{dx}(\Delta_1) = \begin{vmatrix} 1 & 0 & 0 \\ b & x & a \\ a & b & x \end{vmatrix} + \begin{vmatrix} x & a & b \\ 0 & 1 & 0 \\ a & b & x \end{vmatrix} + \begin{vmatrix} x & a & b \\ b & x & a \\ 0 & 0 & 1 \end{vmatrix}$

$= \begin{vmatrix} x & a \\ b & x \end{vmatrix} + \begin{vmatrix} x & b \\ a & x \end{vmatrix} + \begin{vmatrix} x & a \\ b & x \end{vmatrix} = 3\Delta_2$

$\therefore \quad \frac{d}{dx}(\Delta_1) = 3\Delta_2$

146. **(d)** Given system of equations are

$x + 4y - 2z = 3$
$3x + y + 5z = 7$ and $2x + 3y + z = 5$

$\therefore \quad \Delta = \begin{vmatrix} 1 & 4 & -2 \\ 3 & 1 & 5 \\ 2 & 3 & 1 \end{vmatrix}$

$= 1(1-15) - 4(3-10) - 2(9-2)$
$= -14 + 28 - 14 = 0$

and $\quad \Delta_2 = \begin{vmatrix} 1 & 3 & -2 \\ 3 & 7 & 5 \\ 2 & 5 & 1 \end{vmatrix} = 1 \neq 0$

$\therefore$ No solution will exist.

147. **(a)** Area of triangle $= \frac{1}{2}\begin{vmatrix} k & 2k & 1 \\ 2k & 3k & 1 \\ 3 & 1 & 1 \end{vmatrix} = 0$

$\Rightarrow \quad \begin{vmatrix} k & 2k & 1 \\ k & k & 0 \\ 3-k & 1-2k & 0 \end{vmatrix} = 0$

(using $R_2 \to R_2 - R_1$ and $R_3 \to R_3 - R_1$)

$\Rightarrow \quad 1[k(1-2k) - k(3-k)] = 0$

$\Rightarrow \quad k - 2k^2 - 3k + k^2 = 0$

$\Rightarrow \quad -k^2 - 2k = 0$

$\Rightarrow \quad -k(k+2) = 0 \Rightarrow k = 0, -2$

148. **(a)** Let M be the foot of perpendicular from $P(3, 4)$ on the line $3x - 4y + 5 = 0$. Then, M is the point of intersection of $3x - 4y + 5 = 0$ and line passing through $P(3, 4)$ and perpendicular to

$$3x - 4y + 5 = 0 \quad \text{...(i)}$$

Equation of the line perpendicular to

$3x - 4y + 5 = 0$ is $4x + 3y + \lambda = 0$

This passes through $(3, 4)$

$\Rightarrow \quad 12 + 12 + \lambda = 0$

$\Rightarrow \quad \lambda = -24$

$\therefore$ Equation is $4x + 3y - 24 = 0$ ...(ii)

On solving Eqs. (i) and (ii), we get

$$y = 92/25 \Rightarrow x = \frac{81}{25}$$

$\therefore$ Required point is $\left(\frac{81}{25}, \frac{92}{25}\right)$.

149. **(a)** In $\Delta ABC, \sin 60° = h/120$

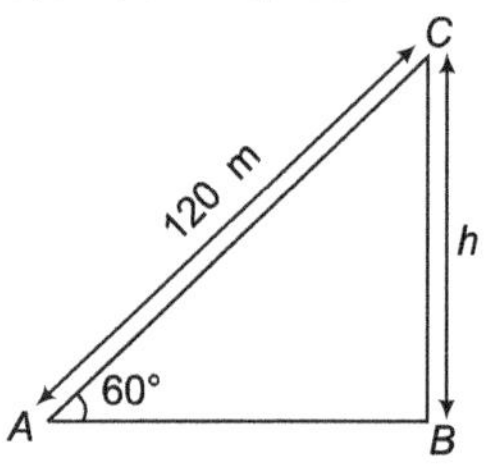

$\Rightarrow \quad h = 120 \times \frac{\sqrt{3}}{2}$

$\therefore \quad h = 60\sqrt{3}$ m

150. **(a)** Let $P(x, y)$ be any point on the parabola. Then, by definition

$$\sqrt{(x-0)^2 + (y+3)^2} = (y-3)$$

$\Rightarrow \quad x^2 + (y+3)^2 = (y-3)^2$

$\Rightarrow \quad x^2 + y^2 + 6y + 9 = y^2 - 6y + 9$

$\Rightarrow \quad x^2 = -12y$

BITSAT
MOCK TESTS
1-5

BITSAT

Birla Institute of Technology & Science Admission Test

Mock Test 1

Instructions

- There are 150 questions in all. The number of questions in each part is as given below.

	No. of Questions
Part I Physics	1-40
Part II Chemistry	41-80
Part III a. English Proficiency	81-95
b. Logical Reasoning	96-105
Part IV Mathematics	106-150

- All questions are Multiple Choice Questions having four options out of which **only one** is correct.
- Each correct answer fetches 3 marks while incorrect answer has a penalty of 1 mark.
- Time allotted to complete this paper is 3 hrs.

PART I

Physics

1. The energy E, angular momentum L and universal gravitational constant G are chosen as fundamental quantities. The dimensions of universal gravitational constant in the dimensional formula of Planck's constant h is

a. Zero *b.* -1 *c.* 5/3 *d.* 1

2. The component of vector $\mathbf{A} = a_x \hat{\mathbf{i}} + a_y \hat{\mathbf{j}} + a_z \hat{\mathbf{k}}$ along the direction of $\hat{\mathbf{i}} - \hat{\mathbf{j}}$ is

a. $a_x - a_y + a_z$ *b.* $a_x - a_y$

c. $(a_x - a_y)/\sqrt{2}$ *d.* $(a_x + a_y + a_z)$

3. A body thrown vertically up to reach its maximum height in t second. The total time from the time of projection to reach a point at half of its maximum height, while returning (in sec) is

a. $\sqrt{2}t$ *b.* $\left(t + \frac{t}{\sqrt{2}}\right)$ *c.* $\frac{3t}{2}$ *d.* $\frac{t}{\sqrt{2}}$

4. If a body is projected with an angle θ to the horizontal, then

a. its velocity is always perpendicular to its acceleration

b. its velocity becomes zero at its maximum height

c. its velocity makes zero angle with the horizontal at its maximum height

d. the body just before hitting the ground, the direction of velocity coincides with the acceleration

5. A river of salty water is flowing with a velocity 2 m/s. If the density of the water is 1.2 g/cc, then the kinetic energy of each cubic metre of water is

a. 2.4 J *b.* 24 J *c.* 2.4 kJ *d.* 4.8 kJ

6. A ball is dropped from a height h on a floor of coefficient of restitution e. The total distance covered by the ball just before second hit is

a. $h(1-2e^2)$ *b.* $h(1+2e^2)$ *c.* $h(1+e^2)$ *d.* he^2

7. Two particles A and B initially at rest, move towards each other, under mutual force of attraction. At an instance when the speed of A is v and speed of B is $2v$, then the speed of centre of mass is

a. Zero *b.* v *c.* $2.5v$ *d.* $4v$

8. Starting from rest, the time taken by a body sliding down on a rough inclined plane at 45° with the horizontal is, twice the time taken to travel on a smooth plane of same inclination and same distance. Then, the coefficient of kinetic friction is

a. 0.25 *b.* 0.33 *c.* 0.50 *d.* 0.75

9. A steel wire can withstand a load up to 2940 N. A load of 150 kg is suspended from a rigid support. The maximum angle through which the wire can be displaced from the mean position, so that the wire does not break when the load passes through the position of equilibrium, is

a. 30° *b.* 60° *c.* 80° *d.* 85°

10. The moment of inertia of a thin circular disc about an axis passing through its centre of gravity and perpendicular to its plane is I. Then, the moment of inertia of the disc about an axis parallel to its diameter and touching the edge of the rim is

a. I *b.* $2I$ *c.* $\frac{3}{2}I$ *d.* $\frac{5}{2}I$

11. The orbit of geo-stationary satellite is circular, the time period of satellite depends on

(i) mass of the satellite
(ii) mass of the earth
(iii) radius of the orbit
(iv) height of the satellite from the surface of earth

Which of the following is correct ?

a. Only (i) *b.* Both (i) and (ii)
c. (i), (ii) and (iii) *d.* (ii), (iii) and (iv)

12. A particle is executing simple harmonic motion with an amplitude A and time period T. The displacement of the particles after $2T$ period from its initial position is

a. A *b.* $4A$ *c.* $8A$ *d.* Zero

13. A load of 1 kg weight is a attached to one end of a steel wire of area of cross-section 3 mm^2 and Young's modulus 10^{11} N/m^2. The other end is suspended vertically from a hook on a wall, then the load is pulled horizontally and released. When the load passes through its lowest position the fractional change in length is $(g = 10 \, m/s^2)$

a. 0.3×10^{-4} *b.* 0.3×10^{-3}
c. 0.3×10^{3} *d.* 0.3×10^{4}

14. The surface tension of soap solution is 0.03 N/m. The work done in blowing to form a soap bubble of surface area 40cm^2 (in J) is

a. 1.2×10^{-4} *b.* 2.4×10^{-4}
c. 12×10^{-4} *d.* 24×10^{-4}

15. Two rain drops reach the earth with different terminal velocities having ratio 9 : 4. Then, the ratio of their volumes is

a. 3 : 2 *b.* 4 : 9 *c.* 9 : 4 *d.* 27 : 8

16. One litre of oxygen at a pressure of 1 atm and two litres of nitrogen at a pressure of 0.5 atm, are introduced into a vessel of volume 1 L. If there is no change in temperature, the final pressure of the mixture of gas (in atm) is

a. 1.5 *b.* 2.5 *c.* 2 *d.* 4

17. There is some change in length when a 33000 N tensile force is applied on a steel rod of area of cross-section 10^{-3} m^2. The change of temperature required to produce the same elongation, if the steel rod is heated, is (The modulus of elasticity is $3 \times 10^{11} N/m^2$ and the coefficient of linear expansion of steel is $1.1 \times 10^{-5} / °C$).

a. 20°C *b.* 15°C *c.* 10°C *d.* 0°C

18. In the adiabatic compression, the decrease in volume is associated with

a. increase in temperature and decrease in pressure
b. decrease in temperature and increase in pressure
c. decrease in temperature and decrease in pressure
d. increase in temperature and increase in pressure

19. Which of the following is true in the case of an adiabatic process, where $\gamma = C_p / C_V$?

a. $p^{1-\gamma}T^{\gamma}$ = constant *b.* $p^{\gamma}T^{1-\gamma}$ = constant
c. pT^{γ} = constant *d.* $p^{\gamma}T$ = constant

20. Two slabs A and B of equal surface area are placed one over the other such that their surfaces are completely in contact. The thickness of slab A is twice that of B. The coefficient of thermal conductivity of slab A is twice that of B. The first surface of slab A is maintained at 100°C, while the second surface of slab B is maintained at 25°C. The temperature at the contact of their surfaces is

a. 62.5°C *b.* 45°C *c.* 55°C *d.* 85°C

21. When a sound wave of wavelength λ is propagating in a medium, then the maximum velocity of the particle is equal to the wave velocity. The amplitude of wave is

a. λ *b.* $\frac{\lambda}{2}$ *c.* $\frac{\lambda}{2\pi}$ *d.* $\frac{\lambda}{4\pi}$

22. A car is moving with a speed of 72 km/h towards a hill. Car blows horn at a distance of 1800 m from the hill. If echo is heard after 10 s, the speed of sound (in m/s) is

a. 300 *b.* 320 *c.* 340 *d.* 360

23. The refractive index of a material of a plano-concave lens is 5/3, the radius of curvature is 0.3 m. The focal length of the lens in air is

a. – 0.45 m *b.* – 0.6 m
c. – 0.75 m *d.* – 1.0 m

24. **Statement (S)** Using Huygen's eye-piece measurements can be taken but are not correct.
Reason (R) The cross wires, scale and final image are not magnified proportionately because the image of the object is magnified by two lenses, whereas the cross wire scale is magnified by one lens only.
Identify the correct one of the following.
a. Both (S) and (R) are true, (R) explains (S).
b. Both (S) and (R) are true, but (R) cannot explain (S).
c. Only (S) is correct, but (R) is incorrect
d. Both (S) and (R) are incorrect

25. An achromatic combination of lenses produces
a. images in black and white
b. coloured images
c. images unaffected by variation of refractive index with wavelength
d. highly enlarged images are formed

26. In Fraunhofer diffraction experiment, L is the distance between screen and the obstacle, b is the size of obstacle and λ is wavelength of incident light. The general condition for the applicability of Fraunhofer diffraction is
a. $\frac{b^2}{L\lambda} >> 1$ *b.* $\frac{b^2}{L\lambda} = 1$ *c.* $\frac{b^2}{L\lambda} << 1$ *d.* $\frac{b^2}{L\lambda} \neq 1$

27. With a standard rectangular bar magnet the time period of a vibration magnetometer is 4 s. The bar magnet is cut parallel to its length into four equal pieces. The time period of vibration magnetometer when one piece is used (in second) (bar magnet breadth is small) is
a. 16 *b.* 8 *c.* 4 *d.* 2

28. The magnetised wire of moment M and length l is bent in the form of semicircle of radius r. Then, its magnetic moment is
a. $\frac{2M}{\pi}$ *b.* $2M$ *c.* $\frac{M}{\pi}$ *d.* zero

29. A charge of 1 μC is divided into two parts such that their charges are in the ratio of 2 : 3. These two charges are kept at a distance 1 m apart in vacuum. Then, the electric force between them (in N) is
a. 0.216 *b.* 0.00216 *c.* 0.0216 *d.* 2.16

30. Two charges $+q$ and $-q$ are kept apart. Then, at any point on the right bisector of line joining the two charges
a. the electric field strength is zero
b. the electric potential is zero
c. both electric potential and electric field strength are zero
d. both electric potential and electric field strength are non-zero

31. A current of 2 A flows in an electric circuit as shown in figure. The potential difference $(V_R - V_S)$, in volts (V_R and V_S are potentials at R and S respectively) is

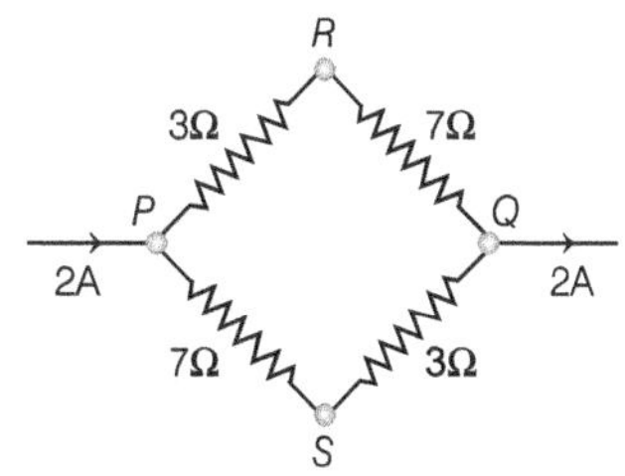

a. –4 *b.* +2 *c.* +4 *d.* –2

32. When a battery connected across a resistor of 16 Ω, the voltage across the resistor is 12 V. When the same battery is connected across a resistor of 10 Ω, voltage across it is 11 V. The internal resistance of the battery (in ohm) is
a. $\frac{10}{7}$ *b.* $\frac{20}{7}$ *c.* $\frac{25}{7}$ *d.* $\frac{30}{7}$

33. One junction of a certain thermoelectric couple is at a fixed temperature T_r and the other junction is at temperature T. The thermo-electromotive force for this is expressed by $E = k(T - T_r)\left[T_0 - \frac{1}{2}(T + T_r)\right]$. At temperature $T = \frac{1}{2}T_0$, the thermoelectric power is
a. $\frac{1}{2}kT_0$ *b.* kT_0
c. $\frac{1}{2}kT_0^2$ *d.* $\frac{1}{2}k(T_0 - T_r)^2$

34. In a galvanometer 5% of the total current in the circuit passes through it. If the resistance of the galvanometer is G, then the shunt resistance S connected to the galvanometer is
a. $19G$ *b.* $\frac{G}{19}$ *c.* $20\,G$ *d.* $\frac{G}{20}$

35. Two concentric coils of 10 turns each are placed in the same plane. Their radii are 20 cm and 40 cm and carry 0.2 A and 0.3 A current respectively in opposite directions. The magnetic induction (in T) at the centre is
a. $\frac{3}{4}\mu_0$ *b.* $\frac{5}{4}\mu_0$ *c.* $\frac{7}{4}\mu_0$ *d.* $\frac{9}{4}\mu_0$

36. The number of turns in primary and secondary coils of a transformer are 50 and 200, respectively. If the current in the primary coil is 4 A, then the current in the secondary coil is
a. 1 A *b.* 2 A *c.* 4 A *d.* 5 A

37. X-rays of wavelength 0.140 nm are scattered from a block of carbon. What will be the wavelengths of X-rays scattered at 90° ?

a. 0.140 nm
b. 0.142 nm
c. 0.144 nm
d. 0.146 nm

38. An X-ray tube produces a continuous spectrum of radiation with its shortest wavelength of 45×10^{-2} Å. The maximum energy of a photon in the radiation in eV is

$(h = 6.62 \times 10^{-34}$ J-s, $c = 3 \times 10^8$ m/s)

a. 27500 *b.* 22500 *c.* 17500 *d.* 12500

39. F_{pp}, F_{nn} and F_{np} are the nuclear forces between proton-proton, neutron-neutron and neutron-proton, respectively. Then, relation between them is

a. $F_{pp} = F_{nn} \neq F_{np}$ *b.* $F_{pp} \neq F_{nn} = F_{np}$
c. $F_{pp} = F_{nn} = F_{np}$ *d.* $F_{pp} \neq F_{nn} \neq F_{np}$

40. Which of the following statements is not correct when a junction diode is in forward bias ?

a. The width of depletion region decreases.
b. Free electrons on *n*-side will move towards the junction.
c. Holes on *p*-side move towards the junction.
d. Electron on *n*-side and holes on *p*-side will move away from junction.

PART II

Chemistry

41. An electronic transition in hydrogen atom results in the formation of H_α line of hydrogen in Lyman series, the energies associated with the electron in each of the orbits involved in the transition (in kcal mol^{-1}) are

a. −313.6, − 34.84 *b.* −313.6, − 78.4
c. −78.4, − 34.84 *d.* −78.4, − 19.6

42. The velocities of two particles *A* and *B* are 0.05 and 0.02 ms^{-1} respectively. The mass of *B* is five times the mass of A. The ratio of their de-Broglie's wavelength is

a. 2 : 1 *b.* 1 : 4
c. 1 : 1 *d.* 4 : 1

43. If the mass defect of $_5B^{11}$ is 0.081 u, its average binding energy (in MeV) is

a. 8.60 *b.* 6.85
c. 5.60 *d.* 5.86

44. The atomic numbers of elements *A*, *B*, *C* and *D* are Z − 1, Z, Z + 1 and Z + 2, respectively. If '*B*' is a noble gas, choose the correct answers from the following statements.

(1) *A* has higher electron affinity
(2) *C* exists in +2 oxidation state
(3) *D* is an alkaline earth metal

a. (1) and (2) *b.* (2) and (3)
c. (1) and (3) *d.* (1), (2) and (3)

45. The bond length of HCl molecule is 1.275 Å and its dipole moment is 1.03 D. The ionic character of the molecule (in percent) (charge of the electron = 4.8×10^{-10} esu) is

a. 100 *b.* 67.3 *c.* 33.66 *d.* 16.83

46. Which one of the following is a correct set ?

a. H_2O, sp^3, angular
b. BCl_3, sp^3, angular
c. NH_4^+, dsp^2, square planar
d. CH_4, dsp^2, tetrahedral

47. Match the following.

	List-I		List-II (At STP)
(A)	$10\text{ g } CaCO_3 \xrightarrow[\text{Decomposition}]{\Delta}$	(i)	0.224 L CO_2
(B)	$1.06\text{ g } Na_2CO_3 \xrightarrow{\text{Excess HCl}}$	(ii)	4.48 L CO_2
(C)	$2.4\text{ g C} \xrightarrow[\text{Combustion}]{\text{Excess } O_2}$	(iii)	0.448 L CO_2
(D)	$0.56\text{ g CO} \xrightarrow[\text{Combustion}]{\text{Excess } O_2}$	(iv)	2.24 L CO_2
		(v)	22.4 L CO_2

The correct match is

	A	B	C	D		A	B	C	D
a.	(iv)	(i)	(ii)	(iii)	*b.*	(v)	(i)	(ii)	(iii)
c.	(iv)	(i)	(iii)	(ii)	*d.*	(i)	(iv)	(ii)	(iii)

48. What is the temperature at which the kinetic energy of 0.3 moles of helium is equal to the kinetic energy of 0.4 moles of argon at 400 K ?

a. 400 K *b.* 873 K *c.* 533 K *d.* 300 K

49. When 25 g of a non-volatile solute is dissolved in 100 g of water, the vapour pressure is lowered by 2.25×10^{-1} mm. If the vapour pressure of water at 20°C is 17.5 mm, what is the molecular weight of the solute ?

a. 206 *b.* 302 *c.* 350 *d.* 276

50. 50 mL of H_2O is added to 50 mL of 1×10^{-3} M barium hydroxide solution. What is the pH of the resulting solution ?

a. 3.0 *b.* 3.3 *c.* 11.0 *d.* 11.7

51. **Assertion (A)** The aqueous solution of CH_3COONa is alkaline in nature.

Reason (R) Acetate ion undergoes anionic hydrolysis.

The correct answer is

a. both (A) and (R) are true and (R) is the correct explanation of (A).
b. both (A) and (R) are true but (R) is not the correct explanation of (A).
c. (A) is true but (R) is not true.
d. (A) is not true but (R) is true.

52. When same quantity of electricity is passed through aqueous $AgNO_3$ and H_2SO_4 solutions connected in series, 5.04×10^{-2} g of H_2 is liberated. What is the mass of silver (in grams) deposited ? (Eq. wts. of hydrogen = 1.008, silver = 108)

a. 54 *b.* 0.54 *c.* 5.4 *d.* 10.8

53. When electric current is passed through acidified water for 1930 s, 1120 mL of H_2 gas is collected (at STP) at the cathode. What is the current passed in amperes ?

a. 0.05 *b.* 0.50
c. 5.0 *d.* 50

54. For a crystal, the angle of diffraction (2θ) is 90° and the second order line has a d value of 2.28 Å. The wavelength (in Å) of X-rays used for Bragg's diffraction is

a. 1.612 *b.* 2.00
c. 2.28 *d.* 4.00

55. In a 500 mL flask, the degree of dissociation of PCl_5 at equilibrium is 40% and the initial amount is 5 moles. The value of equilibrium constant in mol L^{-1} for the decomposition of PCl_5 is

a. 2.33 *b.* 2.66
c. 5.32 *d.* 4.66

56. For a reversible reaction $A \rightleftharpoons B$, which one of the following statements is wrong from the given energy profile diagram ?

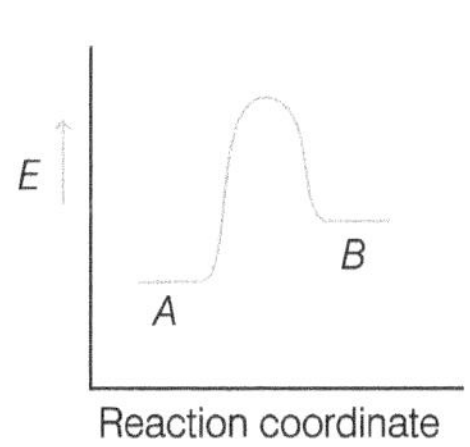

a. Activation energy of forward reaction is greater than backward reaction
b. The forward reaction is endothermic
c. The threshold energy is less than that of activation energy
d. The energy of activation of forward reaction is equal to the sum of heat of reaction and the energy of activation of backward reaction

57. Calculate ΔH in kJ for the following reaction

$$C(g) + O_2(g) \longrightarrow CO_2(g)$$

Given that,

$$H_2O(g) + C(g) \longrightarrow CO(g) + H_2(g); \quad \Delta H = +131 \text{ kJ}$$

$$CO(g) + \frac{1}{2}O_2(g) \longrightarrow CO_2(g); \Delta H = -282 \text{ kJ}$$

$$H_2(g) + \frac{1}{2}O_2(g) \longrightarrow H_2O(g); \Delta H = -242 \text{ kJ}$$

a. −393 *b.* +393 *c.* +655 *d.* −655

58. Which one of the following graphs represents Freundlich adsorption isotherm ?

a.
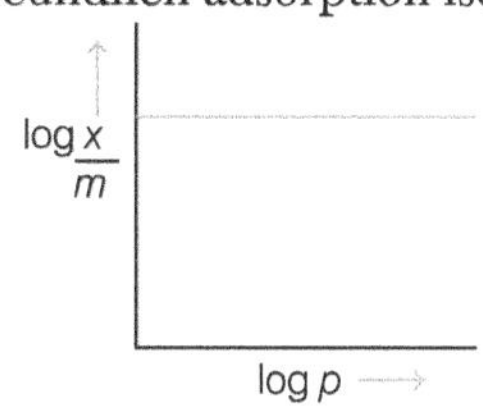

b.
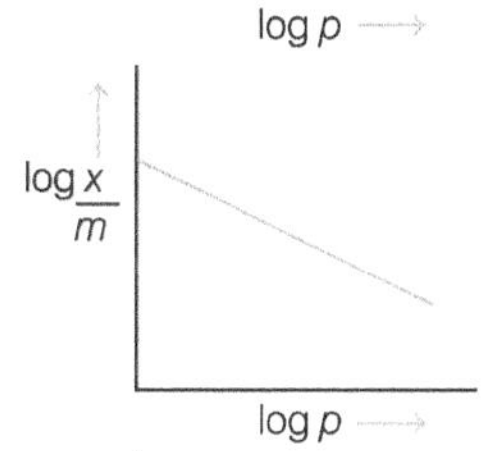

c.
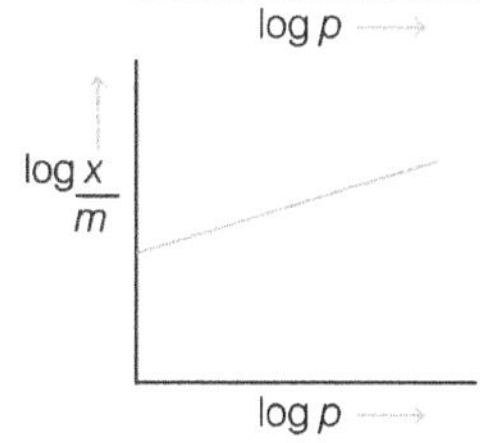

d.
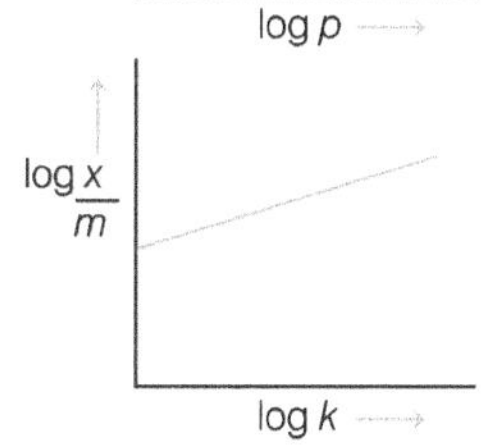

59. Which one of the following reactions represents the oxidising property of H_2O_2 ?

a. $2KMnO_4 + 3H_2SO_4 + 5H_2O_2 \longrightarrow K_2SO_4 + 2MnSO_4 + 8H_2O + 5O_2$
b. $2K_3[Fe(CN)_6] + 2KOH + H_2O_2 \longrightarrow 2K_4[Fe(CN)_6] + 2H_2O + O_2$
c. $PbO_2 + H_2O_2 \longrightarrow PbO + H_2O + O_2$
d. $2KI + H_2SO_4 + H_2O_2 \longrightarrow K_2SO_4 + I_2 + 2H_2O$

60. Which of the following statements are correct for alkali metal compounds ?

(i) Superoxides are paramagnetic in nature.
(ii) The basic strengths of hydroxides increases down the group.
(iii) The conductivity of chlorides in their aqueous solutions decreases down the group.
(iv) The basic nature of carbonates in aqueous solutions is due to cationic hydrolysis.

a. (i), (ii) and (iii) only *b.* (i) and (ii) only
c. (ii), (iii) and (iv) only *d.* (iii) and (iv) only

61. Boron halides behave as Lewis acids because of their .... nature.

a. proton donor *b.* covalent
c. electron deficient *d.* ionising

62. Identify B in the following reaction.

$$H_4SiO_4 \xrightarrow[-H_2O]{1000°C} A \xrightarrow[\Delta]{\text{Carbon}} B + CO$$

a. corundum *b.* quartz
c. silica *d.* carborundum

63. The correct order of reducing abilities of hydrides of V group elements is

a. $NH_3 < PH_3 < AsH_3 < SbH_3 < BiH_3$
b. $NH_3 > PH_3 > AsH_3 > SbH_3 > BiH_3$
c. $NH_3 < PH_3 > AsH_3 > SbH_3 > BiH_3$
d. $SbH_3 > BiH_3 > AsH_3 > NH_3 > PH_3$

64. The number of σ and π bonds in peroxodisulphuric acid are, respectively

a. 9 and 4 *b.* 11 and 4 *c.* 4 and 8 *d.* 4 and 9

65. Which one of the following reactions does not occur ?

a. $F_2 + 2Cl^- \longrightarrow 2F^- + Cl_2$
b. $Cl_2 + 2F^- \longrightarrow 2Cl^- + F_2$
c. $Br_2 + 2I^- \longrightarrow 2Br^- + I_2$
d. $Cl_2 + 2Br^- \longrightarrow 2Cl^- + Br_2$

66. The compound in which the number of $d\pi$-$p\pi$ bonds are equal to those present in ClO_4^-

a. XeF_4 *b.* XeO_3 *c.* XeO_4 *d.* XeF_6

67. $[Co(NH_3)_5SO_4]Br$ and $[Co(NH_3)_5Br]SO_4$ are a pair of .... isomers.

a. ionisation *b.* ligand
c. coordination *d.* hydrate

68. Among the following compounds, which one is not responsible for depletion of ozone layer ?

a. CH_4 *b.* $CFCl_3$
c. NO *d.* Cl_2

69. Which of the following compound(s) has 'Z' configuration ?

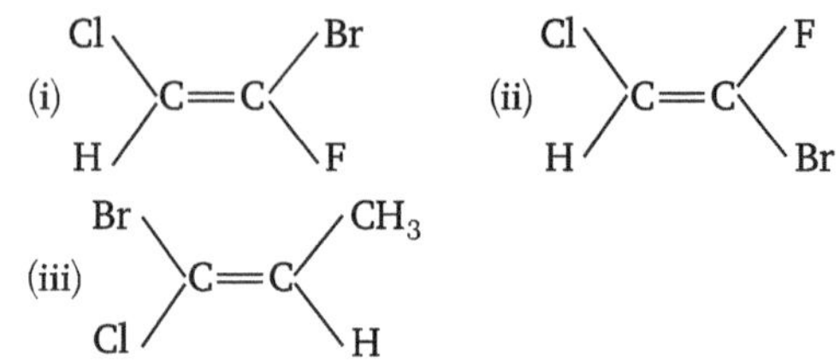

a. Only (i) *b.* Only (ii)
c. Only (iii) *d.* Both (i) and (iii)

70. According to Cahn-Ingold-Prelog sequence rules, the correct order of priority for the given groups is

a. $—COOH > —CH_2OH > —OH > —CHO$
b. $—COOH > —CHO > —CH_2OH > —OH$
c. $—OH > —CH_2OH > —CHO > —COOH$
d. $—OH > —COOH > —CHO > —CH_2OH$

71. What are X and Y respectively in the following reaction ?

$$Z\text{-product} \xleftarrow{Y} \text{2-butyne} \xrightarrow{X} E\text{-product}$$

a. Na/NH_3(liq.) and $Pd/BaSO_4 + H_2$
b. Ni/140°C and $Pd/BaSO_4 + H_2$
c. Ni/140°C and Na/NH_3(liq.)
d. $Pd/BaSO_4 + H_2$ and Na/NH_3(liq.)

72. In which of the following reactions, chlorine acts as an oxidising agent ?

(i) $CH_3CH_2OH + Cl_2 \longrightarrow CH_3CHO + HCl$
(ii) $CH_3CHO + Cl_2 \longrightarrow CCl_3 \cdot CHO + HCl$
(iii) $CH_4 + Cl_2 \xrightarrow{hv} CH_3Cl + HCl$

The correct answer is

a. Only (i)
b. Only (ii)
c. (i) and (iii)
d. (i), (ii) and (iii)

73. The correct order of reactivity of hydrogen halides with ethyl alcohol is

a. HF > HCl > HBr > HI
b. HCl > HBr > HF > HI
c. HBr > HCl > HI > HF
d. HI > HBr > HCl > HF

74. The IUPAC name of $C_2H_5—O—CH\langle^{CH_3}_{CH_3}$ is

a. ethoxy propane
b. 1,1-dimethyl ether
c. 2-ethoxy isopropane
d. 2-ethoxy propane

75. Acetone on addition to methyl magnesium bromide forms a complex, which on decomposition with acid gives *X* and Mg(OH)Br. Which one of the following is *X* ?

a. CH_3OH *b.* $(CH_3)_3COH$
c. $(CH_3)_2CHOH$ *d.* CH_3CH_2OH

76. Identify *A* and *B* in the following reaction

$$CH_3—CH_3 \xleftarrow{B} CH_3COOH \xrightarrow{A} CH_3CH_2OH$$

	A	*B*
a.	HI + red P	$LiAlH_4$
b.	Ni/Δ	$LiAlH_4$
c.	$LiAlH_4$	HI + red P
d.	Pd-$BaSO_4$	Zn + HCl

77. The structure of the compound formed, when nitrobenzene is reduced by lithium aluminium hydride ($LiAlH_4$) is

a. 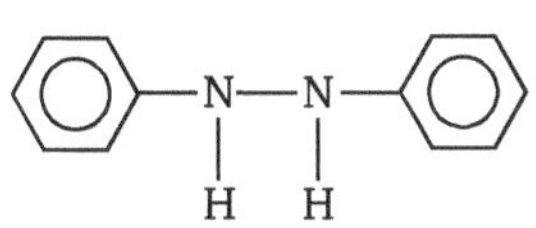
b. 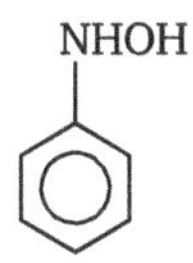
c. 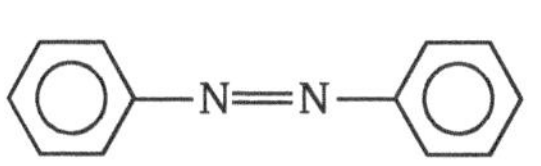
d. 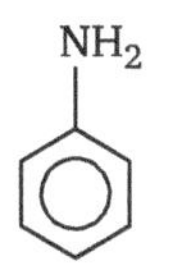

78. Match the following :

List-I		List-II	
(A)	Oxyhaemoglobin	(i)	Analgesic
(B)	Aspirin	(ii)	Oxygen carrier
(C)	Haemoglobin	(iii)	Photosynthesis
(D)	Chlorophyll	(iv)	Oil of winter green
		(v)	Fe^{2+} paramagnetic

The correct match is

	A	B	C	D
a.	(v)	(i)	(ii)	(iii)
b.	(iv)	(ii)	(i)	(iii)
c.	(iii)	(i)	(ii)	(iv)
d.	(v)	(ii)	(iii)	(i)

79. If $\overline{M}_w$ is the weight average molecular weight and $\overline{M}_n$ is the number average molecular weight of a polymer, the poly dispersity index (PDI) of the polymer is given by

a. $\frac{\overline{M}_n}{\overline{M}_w}$ *b.* $\frac{\overline{M}_w}{\overline{M}_n}$ *c.* $\overline{M}_w \times \overline{M}_n$ *d.* $\frac{1}{\overline{M}_w \times \overline{M}_n}$

80. Hydrolysis of sucrose with dilute aqueous sulphuric acid yields

a. 1 : 1 D-(+)-glucose; D-(–)-fructose
b. 1 : 2 D-(+)-glucose; D-(–)-fructose
c. 1 : 1 D-(–)-glucose; D-(+)-fructose
d. 1 : 2 D-(–)-glucose; D-(+)-fructose

Mock Test 1

PART III

a. English Proficiency

Directions (Q. Nos. 81-83) *In each of the following questions, choose the most appropriate alternative to fill in the blanks.*

81. The robbery......late at night.

a. happened *b.* occurred
c. look place *d.* appeared

82. I hope you must have......by now that failures are the stepping stones to success.

a. know *b.* felt *c.* decided *d.* realised

83. In a little published deal, Pepsi Cola has ......the entire soft drink market in Afghanistan.

a. conquered *b.* swallowed
c. captured *d.* occupied

Directions (Q. Nos. 84-86) *In each of the following questions, put the parts P, Q, R and S in their proper order to produce the correct sensentence.*

84. The Bible,

(P) has in many respects
(Q) the sacred book of all Christians
(R) among all the books of the world
(S) a unique character and position

a. QPSR *b.* QRPS
c. RPQS *d.* RQPS

85. The ultimate hope

(P) will force the nations
(Q) that the destructive nature of weapons
(R) to give up war
(S) has not been fulfilled

a. PQRS *b.* PRQS
c. QPRS *d.* RSQP

86. It was

(P) in keeping with my mood
(Q) a soft summer evening
(R as I walked sedately
(S) in the direction of the new house

a. QPRS *b.* QRPS
c. SQPR *d.* SRPQ

Directions (Q. Nos. 87-89) *In each of the following questions, choose the alternative which is most nearly the same in meaning to the word given in capital letters.*

87. EPHEMERAL
a. Uneral *b.* Mythical
c. Short-living *d.* Artificial

88. STUBBORN
a. Easy *b.* Obstinate
c. Willing *d.* Pliable

89. PROGNOSIS
a. Indentification *b.* Preface
c. Forecast *d.* Scheme

Directions (Q. Nos. 90-92) *In each of the following questions, choose the alternative which is opposite in meaning to the word given in capital letters.*

90. INFALLIBLE
a. Erring
b. Untrustworthy
c. Dubious
d. Unreliable

91. GATHER
a. Separate *b.* Suspend
c. Scatter *d.* Spend

92. EXALT
a. Depreciate *b.* Ennoble
c. Glorify *d.* Simplify

Directions (Q. Nos. 93-95) *In each of the following questions, choose the alternative which can be substituted for the given words/sentences.*

93. Elderly woman in charge of a girl on social occasions
a. Spinster *b.* Matron
c. Chaperon *d.* Chandler

94. Land so surrounded by water as to be almost an island
a. Archipelago *b.* Isthmus
c. Peninusula *d.* Lagoon

95. A place adjoining kitchen for washing dishes etc.
a. Cellar *b.* Wardrobe
c. Scullery *d.* Pantry

b. Logical Reasoning

96. Direction : In each of these questions, two figure/words are given to the left of the sign :: and one figures word to the right of the sign :: with four alternatives under it out of which one of the alternatives has the same relationship with the figures/words to the right of the sign :: as between the two figures/words to the left of the sign (::). Find the correct alternative.

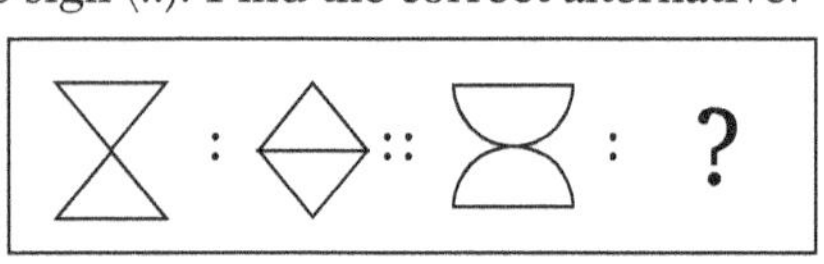

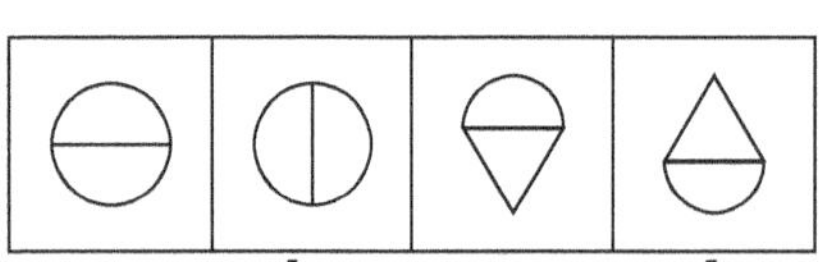

a. *b.* *c.* *d.*

97. Direction : In the question, three words are given. They are followed by four words one of which stands for the class to which these three words belong. Identify that word.
Newspaper, Hoarding. Television
a. Press *b.* Media *c.* Broadcast *d.* Rumour

98. Direction : Find out the number which will come next in the series.
2, 5, 14, 41, 122, 365, ?
a. 1029 *b.* 1094 *c.* 1059 *d.* 1000

99. Direction : In the given question, some satements are followed by one or more inferences. The inference or inferences may be wrongly or correctly drawn. Select one of the alternatives which contains the correctly drawn inference or inferences.

Which of the conclusions drawn from the given statements are correct?

Given statements :

Foreigners in Jordon without a valid work permit will be deported. A few Indian employees in the building industry in jordon do not possess valid work permits.

Inferences

(1) All Indians engaged in building industry in Jordon will be deported to India.

(2) A few Indians in building industry in Jordon will be deported.

(3) A bulk of Indians in Jordon will be deported to India.

(4) Indian employees in building industry without work permit will be deported from Jordon.

The inferences correctly drawn are
a. 1 and 3 *b.* 3 and 4
c. 2 and 4 *d.* 1 and 2

100. Select the series which obeys the given rule : Any figure can be traced by a single unbroken line without retracing

a. 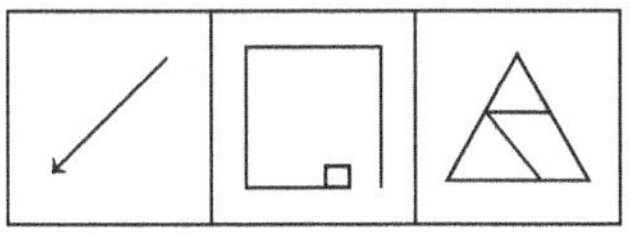

b. 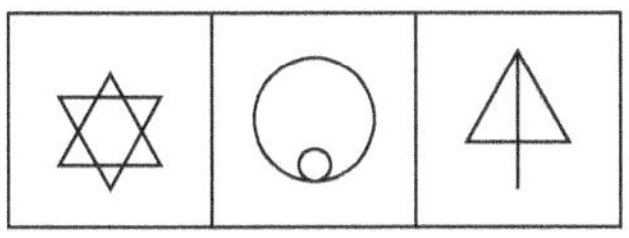

c. 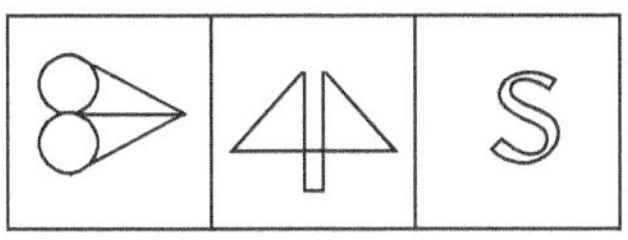

d. 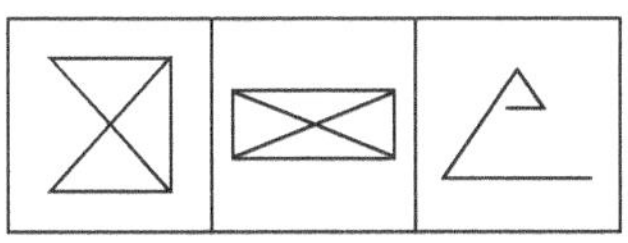

101. Select from amongst the four alternative figures, the one which complete the pattern in the problem figure.

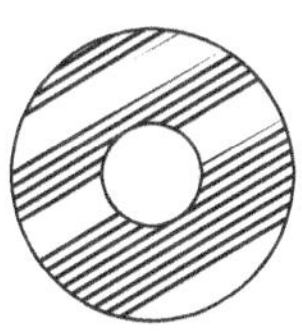

a. 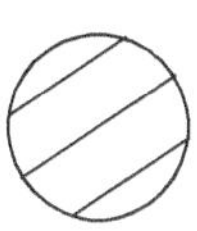 *b.* 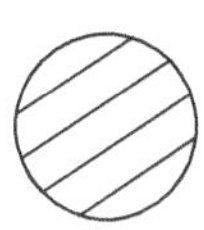

c. 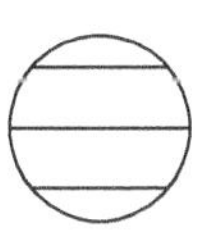 *d.* 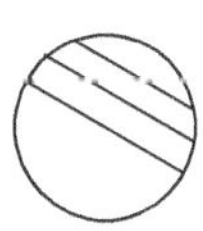

102. Direction In the following question a piece of paper is folded, cut and unfolded. One of the four figures given below is exactly like this unfolded paper. Find this out.

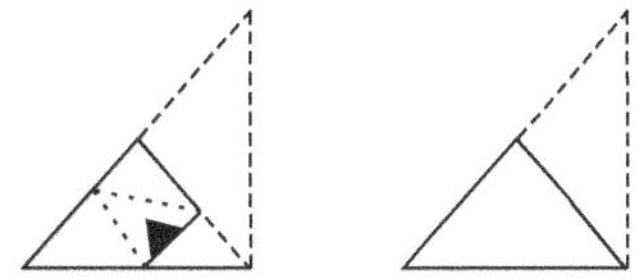

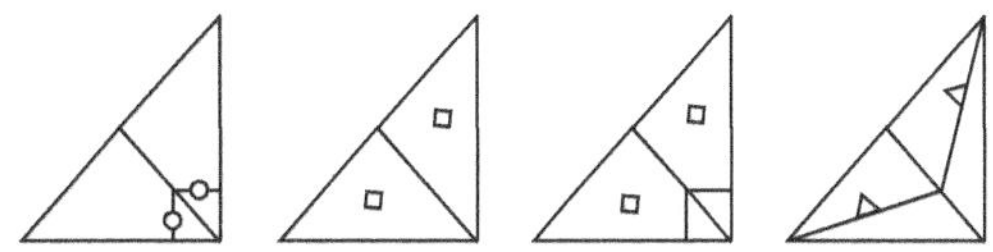

a. *b.* *c.* *d.*

103. Direction In the question, a figure is given, its components are given in one of the four alternative figures. Find this one

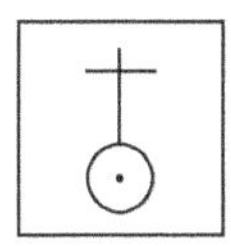

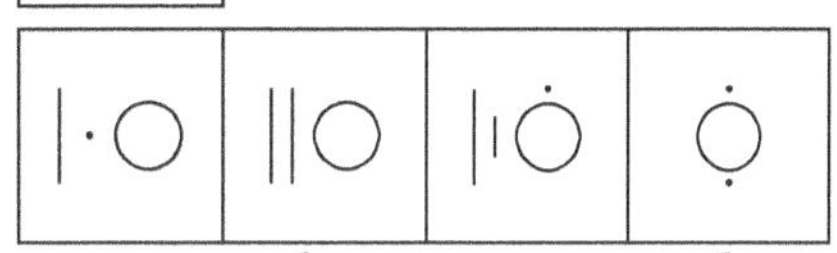

a. *b.* *c.* *d.*

104. Direction In the given question, four numbers/number-pairs are given. Select the one which is different from the other three.

a. 1234 *b.* 2345

c. 4567 *d.* 7896

105. Direction Find the group of letters from the four alternative which is obtained by applying the same rule to this given word to the right of the sign ::

FILM : ADGH :: MILK : ?

a. ADGF *b.* HDGE

c. HDGF *d.* HEGF

PART IV

Mathematics

106. If $f : R \to R$ and $g : R \to R$ are defined by $f(x) = |x|$ and $g(x) = [x - 3]$ for $x \in R$, then $\left\{g(f(x)) : -\frac{8}{5} < x < \frac{8}{5}\right\}$ is equal to

a. $\{0, 1\}$ *b.* $\{1, 2\}$

c. $\{-3, -2\}$ *d.* $\{2, 3\}$

107. For any integer $n \geq 1$, the sum $\sum_{k=1}^{n} k(k+2)$ is equal to

a. $\frac{n(n+1)(n+2)}{6}$ *b.* $\frac{n(n+1)(2n+1)}{6}$

c. $\frac{n(n+1)(2n+7)}{6}$ *d.* $\frac{n(n+1)(2n+9)}{6}$

108. 9 balls are to be placed in 9 boxes and 5 of the balls cannot fit into 3 small boxes. The number of ways of arranging one ball in each of the boxes is

a. 18720 *b.* 18270 *c.* 17280 *d.* 12780

109. If ${}^nP_r = 30240$ and ${}^nC_r = 252$, then the ordered pair (n, r) is equal to

a. (12, 6) *b.* (10, 5) *c.* (9, 4) *d.* (16, 7)

110. If $(1+x+x^2+x^3)^5 = \sum_{k=0}^{15} a_k x^k$, then $\sum_{k=0}^{7} a_{2k}$ is equal to

a. 128 *b.* 256 *c.* 512 *d.* 1024

111. If $\alpha+\beta=-2$ and $\alpha^3+\beta^3=-56$, then the quadratic equation whose roots are α and β is

a. $x^2+2x-16=0$ *b.* $x^2+2x+15=0$

c. $x^2+2x-12=0$ *d.* $x^2+2x-8=0$

112. The cubic equation whose roots are thrice to each of the roots of $x^3+2x^2-4x+1=0$ is

a. $x^3-6x^2+36x+27=0$

b. $x^3+6x^2+36x+27=0$

c. $x^3-6x^2-36x+27=0$

d. $x^3+6x^2-36x+27=0$

113. If $A=\begin{bmatrix}1 & -2\\ 4 & 5\end{bmatrix}$ and $f(t)=t^2-3t+7$, then $f(A)+\begin{bmatrix}3 & 6\\ -12 & -9\end{bmatrix}$ is equal to

a. $\begin{bmatrix}1 & 0\\ 0 & 1\end{bmatrix}$ *b.* $\begin{bmatrix}0 & 0\\ 0 & 0\end{bmatrix}$ *c.* $\begin{bmatrix}0 & 1\\ 1 & 0\end{bmatrix}$ *d.* $\begin{bmatrix}1 & 1\\ 0 & 0\end{bmatrix}$

114. $\begin{vmatrix}a-b-c & 2a & 2a\\ 2b & b-c-a & 2b\\ 2c & 2c & c-a-b\end{vmatrix}$ is equal to

a. 0 *b.* $a+b+c$

c. $(a+b+c)^2$ *d.* $(a+b+c)^3$

115. If ω is a complex cube root of unity, then $\sin\left\{(\omega^{10}+\omega^{23})\pi-\frac{\pi}{4}\right\}$ is equal to

a. $1/\sqrt{2}$ *b.* 1/2 *c.* 1 *d.* $\sqrt{3}/2$

116. $\sqrt{3}\operatorname{cosec} 20^\circ - \sec 20^\circ$ is equal to

a. 2 *b.* $2\sin 20^\circ \cdot \operatorname{cosec} 40^\circ$

c. 4 *d.* $4\sin 20^\circ \cdot \operatorname{cosec} 40^\circ$

117. If $\tan\theta+\tan\left(\theta+\frac{\pi}{3}\right)+\tan\left(\theta+\frac{2\pi}{3}\right)=3$, then which of the following is equal to 1 ?

a. $\tan 2\theta$ *b.* $\tan 3\theta$ *c.* $\tan^2\theta$ *d.* $\tan^3\theta$

118. $\{x\in R : \cos 2x + 2\cos^2 x = 2\}$ is equal to

a. $\left\{2n\pi+\frac{\pi}{3} : n\in Z\right\}$ *b.* $\left\{n\pi\pm\frac{\pi}{6} : n\in Z\right\}$

c. $\left\{n\pi+\frac{\pi}{3} : n\in Z\right\}$ *d.* $\left\{2n\pi-\frac{\pi}{3} : n\in Z\right\}$

119. If $\sin^{-1}\left(\frac{3}{x}\right)+\sin^{-1}\left(\frac{4}{x}\right)=\frac{\pi}{2}$, then x is equal to

a. 3 *b.* 5 *c.* 7 *d.* 11

120. In ΔABC, if $\frac{1}{b+c}+\frac{1}{c+a}=\frac{3}{a+b+c}$, then C is equal to

a. 90° *b.* 60° *c.* 45° *d.* 30°

121. In a triangle, if $r_1=2r_2=3r_3$, then $\frac{a}{b}+\frac{b}{c}+\frac{c}{a}$ is equal to

a. $\frac{75}{60}$ *b.* $\frac{155}{60}$ *c.* $\frac{176}{60}$ *d.* $\frac{191}{60}$

122. From the top of a hill h m high the angles of depressions of the top and the bottom of a pillar are α and β respectively. The height (in metres) of the pillar is

a. $\frac{h(\tan\beta-\tan\alpha)}{\tan\beta}$ *b.* $\frac{h(\tan\alpha-\tan\beta)}{\tan\alpha}$

c. $\frac{h(\tan\beta+\tan\alpha)}{\tan\beta}$ *d.* $\frac{h(\tan\beta+\tan\alpha)}{\tan\alpha}$

123. The position vectors of P and Q are respectively **a** and **b**. If R is a point on **PQ** such that **PR** = 5 **PQ**, then the position vector of R is

a. 5**b** − 4**a** *b.* 5**b** + 4**a**

c. 4**b** − 5**a** *d.* 4**b** + 5**a**

124. If the position vectors of A, B and C are respectively $2\hat{i}-\hat{j}+\hat{k}$, $\hat{i}-3\hat{j}-5\hat{k}$ and $3\hat{i}-4\hat{j}-4\hat{k}$, then $\cos^2 A$ is equal to

a. 0 *b.* 6/41 *c.* 35/41 *d.* 1

125. Let **a** be a unit vector, $\mathbf{b}=2\hat{i}+\hat{j}-\hat{k}$ and $\mathbf{c}=\hat{i}+3\hat{k}$. Then, maximum value of $[\mathbf{a}\ \mathbf{b}\ \mathbf{c}]$ is

a. −1 *b.* $\sqrt{10}+\sqrt{6}$

c. $\sqrt{10}-\sqrt{6}$ *d.* $\sqrt{59}$

126. If A and B are independent events of a random experiment such that $P(A\cap B)=\frac{1}{6}$ and $P(\overline{A}\cap\overline{B})=\frac{1}{3}$, then $P(A)$ is equal to

(Here, **E** is the complement of the event E)

a. 1/4 *b.* 1/3 *c.* 5/7 *d.* 2/3

Mock Test 1

127. For $k = 1, 2, 3$ the box B_k contains k red balls and $(k+1)$ white balls. Let $P(B_1) = \frac{1}{2}$, $P(B_2) = \frac{1}{3}$ and $P(B_3) = \frac{1}{6}$. A box is selected at random and a ball is drawn from it. If a red ball is drawn, then the probability that it has come from box B_2, is

a. $\frac{35}{78}$ *b.* $\frac{14}{39}$ *c.* $\frac{10}{13}$ *d.* $\frac{12}{13}$

128. If the sum of the distances of a point P from two perpendicular lines in a plane is 1, then the locus of P is a

a. rhombus *b.* circle
c. straight line *d.* pair of straight lines

129. The transformed equation of $3x^2 + 3y^2 + 2xy = 2$, when the coordinate axes are rotated through an angle of 45°, is

a. $x^2 + 2y^2 = 1$ *b.* $2x^2 + y^2 = 1$
c. $x^2 + y^2 = 1$ *d.* $x^2 + 3y^2 = 1$

130. If l, m and n are in arithmetic progression, then the straight line $lx + my + n = 0$ will pass through the point

a. $(-1, 2)$ *b.* $(1, -2)$ *c.* $(1, 2)$ *d.* $(2, 1)$

131. A pair of perpendicular straight lines passes through the origin and also through the point of intersection of the curve $x^2 + y^2 = 4$ with $x + y = a$. The set containing the value of a is

a. $\{-2, 2\}$ *b.* $\{-3, 3\}$ *c.* $\{-4, 4\}$ *d.* $\{-5, 5\}$

132. In ΔABC the mid points of the sides AB, BC and CA are respectively $(l, 0, 0)$, $(0, m, 0)$ and $(0, 0, n)$. Then, $\frac{AB^2 + BC^2 + CA^2}{l^2 + m^2 + n^2}$ is equal to

a. 2 *b.* 4 *c.* 8 *d.* 16

133. If the lines $2x - 3y = 5$ and $3x - 4y = 7$ are two diameters of a circle of radius 7, then the equation of the circle is

a. $x^2 + y^2 + 2x - 4y - 47 = 0$
b. $x^2 + y^2 = 49$
c. $x^2 + y^2 - 2x + 2y - 47 = 0$
d. $x^2 + y^2 = 17$

134. The inverse of the point $(1, 2)$ with respect to the circle $x^2 + y^2 - 4x - 6y + 9 = 0$, is

a. $\left(1, \frac{1}{2}\right)$ *b.* $(2, 1)$
c. $(0, 1)$ *d.* $(1, 0)$

135. If $2x + 3y + 12 = 0$ and $x - y + 4\lambda = 0$ are conjugate with respect to the parabola $y^2 = 8x$, then λ is equal to

a. 2 *b.* –2 *c.* 3 *d.* –3

136. The distance between the foci of the hyperbola $x^2 - 3y^2 - 4x - 6y - 11 = 0$ is

a. 4 *b.* 6 *c.* 8 *d.* 10

137. The radius of the circle with the polar equation $r^2 - 8r(\sqrt{3}\cos\theta + \sin\theta) + 15 = 0$ is

a. 8 *b.* 7 *c.* 6 *d.* 5

138. If $f: R \to R$ is defined by $f(x) = [x - 3] + |x - 4|$ for $x \in R$, then $\lim_{x \to 3^-} f(x)$ is equal to

a. –2 *b.* –1 *c.* 0 *d.* 1

139. If $f: R \to R$ is defined by

$$f(x) = \begin{cases} \dfrac{\cos 3x - \cos x}{x^2}, & \text{for } x \neq 0 \\ \lambda, & \text{for } x = 0 \end{cases}$$

and if f is continuous at $x = 0$, then λ is equal to

a. –2 *b.* –4 *c.* –6 *d.* –8

140. If $f(2) = 4$ and $f'(2) = 1$, then

$$\lim_{x \to 2} \frac{xf(2) - 2f(x)}{x - 2}$$

is equal to

a. –2 *b.* 1 *c.* 2 *d.* 3

141. If $x = a\left\{\cos\theta + \log\tan\left(\frac{\theta}{2}\right)\right\}$ and $y = a\sin\theta$, then $\frac{dy}{dx}$ is equal to

a. $\cot\theta$ *b.* $\tan\theta$ *c.* $\sin\theta$ *d.* $\cos\theta$

142. The equation of the normal to the curve $y^4 = ax^3$ at (a, a) is

a. $x + 2y = 3a$ *b.* $3x - 4y + a = 0$
c. $4x + 3y = 7a$ *d.* $4x - 3y = 0$

143. The length of the subtangent at $(2, 2)$ to the curve $x^5 = 2y^4$ is

a. $\frac{5}{2}$ *b.* $\frac{8}{5}$ *c.* $\frac{2}{5}$ *d.* $\frac{5}{8}$

144. If $\int e^x \left(\frac{1 - \sin x}{1 - \cos x}\right) dx = f(x) + c$, then $f(x)$ is equal to

a. $e^x \cot\left(\frac{x}{2}\right) + c$ *b.* $e^{-x} \cot\left(\frac{x}{2}\right) + c$
c. $-e^x \cot\left(\frac{x}{2}\right) + c$ *d.* $-e^{-x} \cot\left(\frac{x}{2}\right) + c$

145. If $\int e^x(1+x)\cdot\sec^2(xe^x)\,dx$

$= f(x)+c$, then $f(x)$ is equal to

a. $\cos(xe^x)$ b. $\sin(xe^x)$

c. $2\tan^{-1}(x)$ d. $\tan(x\,e^x)$

146. $\int_{-\pi/2}^{\pi/2} \sin|x|\,dx$ is equal to

a. 0 b. 1 c. 2 d. π

147. The area (in sq unit) of the region bounded by the curves $2x = y^2 - 1$ and $x = 0$ is

a. $\frac{1}{3}$ b. $\frac{2}{3}$ c. 1 d. 2

148. The solution of the differential equation

$$\frac{dy}{dx} = \frac{xy+y}{xy+x} \text{ is}$$

a. $x+y-\log\left(\frac{cy}{x}\right)$ b. $x+y=\log(cxy)$

c. $x-y-\log\left(\frac{cx}{y}\right)$ d. $y-x=\log\left(\frac{cx}{y}\right)$

149. The solution of the differential equation

$$\frac{dy}{dx} - y\tan x = e^x \sec x \text{ is}$$

a. $y = e^x\cos x + c$ b. $y\cos x = e^x + c$

c. $y = e^x\sin x + c$ d. $y\sin x = e^x + c$

150. The solution of the differential equation

$$xy^2\,dy - (x^3+y^3)\,dx = 0 \text{ is}$$

a. $y^3 = 3x^3 + c$ b. $y^3 = 3x^3\log(cx)$

c. $y^3 = 3x^3 + \log(cx)$ d. $y^3 + 3x^3 = \log(cx)$

Answers

Physics

1. (a)	2. (c)	3. (b)	4. (c)	5. (c)	6. (b)	7. (a)	8. (d)	9. (b)	10. (d)
11. (d)	12. (d)	13. (a)	14. (b)	15. (d)	16. (c)	17. (c)	18. (d)	19. (a)	20. (a)
21. (c)	22. (c)	23. (a)	24. (a)	25. (c)	26. (c)	27. (c)	28. (a)	29. (b)	30. (b)
31. (c)	32. (b)	33. (a)	34. (b)	35. (b)	36. (a)	37. (b)	38. (a)	39. (c)	40. (d)

Chemistry

41. (b)	42. (a)	43. (b)	44. (c)	45. (d)	46. (a)	47. (a)	48. (c)	49. (c)	50. (c)
51. (a)	52. (c)	53. (c)	54. (a)	55. (b)	56. (c)	57. (a)	58. (c)	59. (d)	60. (b)
61. (c)	62. (d)	63. (a)	64. (b)	65. (b)	66. (b)	67. (a)	68. (a)	69. (d)	70. (d)
71. (a)	72. (d)	73. (d)	74. (d)	75. (b)	76. (c)	77. (c)	78. (a)	79. (b)	80. (a)

English Proficiency

81. (b)	82. (d)	83. (c)	84. (a)	85. (c)	86. (a)	87. (c)	88. (b)	89. (c)	90. (a)
91. (c)	92. (a)	93. (c)	94. (c)	95. (c)					

Logical Reasoning

96. (a)	97. (b)	98. (b)	99. (c)	100. (b)	101. (b)	102. (b)	103. (c)	104. (d)	105. (c)

Mathematics

106. (c)	107. (c)	108. (c)	109. (b)	110. (c)	111. (d)	112. (d)	113. (b)	114. (d)	115. (a)
116. (c)	117. (b)	118. (b)	119. (b)	120. (b)	121. (d)	122. (a)	123. (a)	124. (c)	125. (d)
126. (b)	127. (b)	128. (a)	129. (b)	130. (b)	131. (a)	132. (c)	133. (c)	134. (c)	135. (d)
136. (c)	137. (b)	138. (c)	139. (b)	140. (c)	141. (b)	142. (c)	143. (b)	144. (c)	145. (d)
146. (c)	147. (b)	148. (d)	149. (b)	150. (b)					

Hints & Solutions

Physics

1. *(a)* $h \propto G^x L^y E^z$

$[M^1L^2T^{-1}] \propto [M^{-1}L^3T^{-2}]^x[M^1L^2T^{-1}]^y [M^1L^2T^{-2}]^z$

$[M^1L^2T^{-1}] = k[M^{-1}L^3T^{-2}]^x[M^1L^2T^{-1}]^y[M^1L^2T^{-2}]^z$

Comparing the powers on both sides, we get

$$1 = -x + y + z \quad \text{...(i)}$$

$$2 = 3x + 2y + 2z \quad \text{...(ii)}$$

$$-1 = -2x - y - 2z \quad \text{...(iii)}$$

On solving Eqs. (i), (ii) and (iii), we get

$$x = 0$$

2. *(c)* Let $\mathbf{B} = \hat{\mathbf{i}} - \hat{\mathbf{j}}$

Then component of vector **A** along **B**

$$= \frac{\mathbf{A} \cdot \mathbf{B}}{|\mathbf{B}|} = \frac{(a_x \hat{\mathbf{i}} + a_y \hat{\mathbf{j}} + a_z \hat{\mathbf{k}}) \cdot (\hat{\mathbf{i}} - \hat{\mathbf{j}})}{|\hat{\mathbf{i}} - \hat{\mathbf{j}}|}$$

$$= \frac{a_x - a_y}{\sqrt{2}}$$

3. *(b)* The ball is thrown vertically upwards, then according to equation of motion,

$$(0)^2 - u^2 = -2gh \quad \text{...(i)}$$

and
$$0 = u - gt \quad \text{...(ii)}$$

From Eqs. (i) and (ii), we get

$$h = \frac{gt^2}{2}$$

When the ball is falling downwards after reaching the maximum height

$$s = ut' + \frac{1}{2}g(t')^2$$

$$\frac{h}{2} = (0)\, t' + \frac{1}{2} g(t')^2$$

$$\Rightarrow \quad t' = \sqrt{\frac{h}{g}} = \frac{t}{\sqrt{2}}$$

Hence, the total time from the time of projection to reach a point at half of its maximum height, while returning $= t + t'$

$$= t + \frac{t}{\sqrt{2}}$$

4. *(c)* Direction of velocity is always tangent to the path, so at the top of trajectory, it is in horizontal direction.

5. *(c)* Given, velocity of river, $v = 2$ m/s

Density of water, $\rho = 1.2$ g/cc

Mass of each cubic metre

$$m = \frac{1.2 \times 10^{-3}}{(10^{-2})^3} = 1.2 \times 10^3 \text{ kg}$$

$\therefore$ Kinetic energy $= \frac{1}{2} mv^2$

$$= \frac{1}{2} \times 1.2 \times 10^3 \times (2)^2$$

$$= 2.4 \times 10^3 \text{ J} = 2.4 \text{ kJ}$$

6. *(b)*

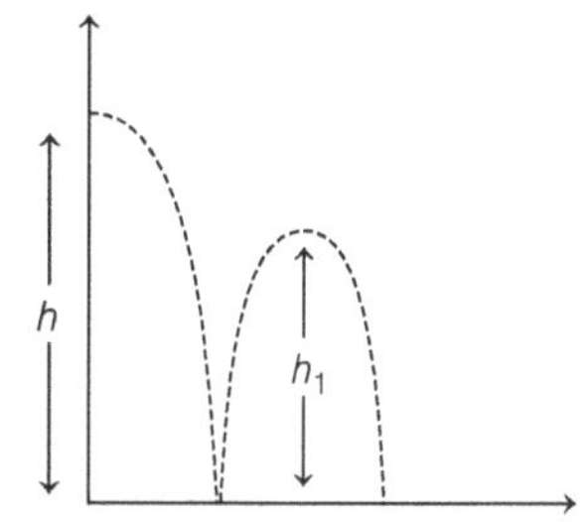

Total distance travelled by the ball before its second hit is

$$H = h + 2h_1$$

$$= h[1 + 2e^2] \quad [\because h_1 = he^2]$$

7. *(a)* As initially both the particles were at rest therefore velocity of centre of mass was zero and there is no external force on the system, so speed of centre of mass remains constant *i.e.* it should be equal to zero.

8. *(d)* The coefficient of friction is

$$\mu = \tan\theta \left[1 - \frac{1}{n^2}\right]$$

Here, $\theta = 45°$ and $n = 2$

$$\therefore \quad \mu = \tan 45° \left[1 - \frac{1}{2^2}\right]$$

$$= 1 - \frac{1}{4} = \frac{3}{4} = 0.75$$

9. *(b)* Now, at B

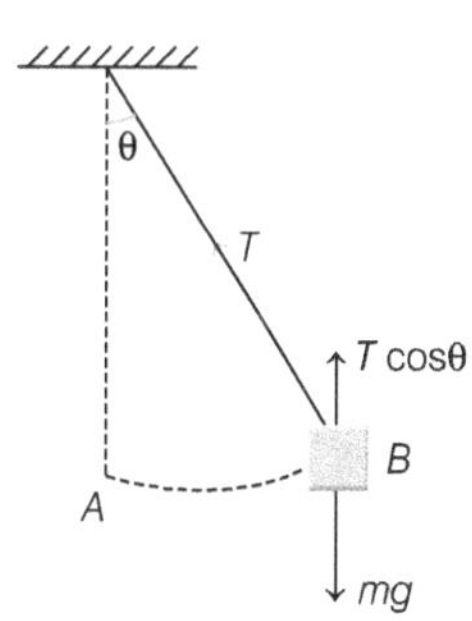

In equilibrium,

$$T\cos\theta = mg$$

$$\Rightarrow \quad \cos\theta = \frac{150\times 9.8}{2940}$$

$$\Rightarrow \quad \cos\theta = 0.5 = \frac{1}{2}$$

$$\Rightarrow \quad \theta = 60^\circ$$

10. ***(d)*** Moment of inertia of a circular disc about an axis passing through centre of gravity and perpendicular to its plane,

$$I = \frac{1}{2}MR^2 \quad \ldots\text{(i)}$$

From Eq. (i)

$$MR^2 = 2I$$

Then, moment of inertia of disc about tangent in a plane

$$= \frac{5}{4}MR^2$$

$$= \frac{5}{4}(2I) = \frac{5}{2}I$$

11. ***(d)*** Time period of satellite,

$T \propto \frac{1}{M^{1/2}}$, where M is mass of earth.

$\propto (R+h)^{3/2}$, where R is radius of the orbit, h is the height of satellite from the earth's surface.

12. ***(d)*** Time period T is the least interval of time after which the periodic motion of a body repeats itself. Therefore, displacement will be zero.

13. ***(a)***

$$Y = \frac{mgl}{A\,\Delta l}$$

$$\Rightarrow \quad \frac{\Delta l}{l} = \frac{mg}{AY}$$

$$\therefore \quad \frac{\Delta l}{l} = \frac{1\times 10}{3\times 10^{-6}\times 10^{11}} = 0.3\times 10^{-4}$$

14. ***(b)*** In case of soap bubble

Workdone, $W = T \times 2 \times \Delta A$

$= 0.03 \times 2 \times 40 \times 10^{-4}$

$= 2.4 \times 10^{-4}$ J

15. ***(d)*** Terminal velocity, $v_T \propto r^2$

$$\text{or} \quad \frac{v_{T_1}}{v_{T_2}} = \frac{r_1^2}{r_2^2}$$

$$\therefore \quad \sqrt{\frac{9}{4}} = \frac{r_1}{r_2}$$

$$\text{or} \quad \frac{r_1}{r_2} = \frac{3}{2}$$

$$\therefore \quad V = \frac{4}{3}\pi r^3$$

$$\text{or} \quad \frac{V_1}{V_2} = \frac{r_1^3}{r_2^3} = \frac{27}{8}$$

16. ***(c)*** Ideal gas equation is given by

$$pV = nRT \quad \ldots\text{(i)}$$

For oxygen, $p = 1$ atm, $V = 1$ L, $n = n_{O_2}$

Therefore Eq. (i) becomes

$$\therefore \quad 1\times 1 = n_{O_2}RT$$

$$\Rightarrow \quad n_{O_2} = \frac{1}{RT}$$

For nitrogen $p = 0.5$ atm, $V = 2$ L, $n = n_{N_2}$

$$\therefore \quad 0.5\times 2 = n_{N_2}RT$$

$$\Rightarrow \quad n_{N_2} = \frac{1}{RT}$$

For mixture of gas,

$$p_{mix}V_{mix} = n_{mix}RT$$

Here, $n_{mix} = n_{O_2} + n_{N_2}$

$$\therefore \quad \frac{p_{mix}V_{mix}}{RT} = \frac{1}{RT} + \frac{1}{RT}$$

$$\Rightarrow \quad p_{mix}V_{mix} = 2 \quad (\because V_{mix} = 1)$$

17. ***(c)*** Modulus of elasticity $= \frac{\text{Force}}{\text{Area}} \times \frac{l}{\Delta l}$

$$3\times 10^{11} = \frac{33000}{10^{-3}} \times \frac{l}{\Delta l}$$

$$\frac{\Delta l}{l} = \frac{33000}{10^{-3}} \times \frac{1}{3\times 10^{11}}$$

$$= 11\times 10^{-5}$$

Change in length, $\frac{\Delta l}{l} = \alpha\,\Delta T$

$$11\times 10^{-5} = 1.1\times 10^{-5} \times \Delta T$$

$$\Rightarrow \quad \Delta T = 10°\text{C}$$

18. ***(d)*** In adiabatic compression temperature and hence, internal energy of the gas increases. In compression pressure will also increase.

19. ***(a)*** For adiabatic change equation of state is

$$pV^\gamma = \text{constant}$$

It can also be re-written as

$$TV^{\gamma-1} = \text{constant} \quad \left[\text{as } p = \frac{nRT}{V}\right]$$

and

$$p^{1-\gamma}T^\gamma = \text{constant} \quad \left[\text{as } V = \frac{nRT}{p}\right]$$

20. ***(a)*** The temperature at the contact of the surface

$$= \frac{K_1d_2\theta_1 + K_2d_1\theta_2}{K_1d_2 + K_2d_1}$$

$$= \frac{2K_2d_2\times 100 + 2d_2\times K_2\times 25}{2K_2d_2 + K_2 2d_2}$$

$$= \frac{200+50}{4} = 62.5\,°\text{C}$$

21. ***(c)*** $v_{max} = v$

$\Rightarrow \quad A\omega = v \Rightarrow A \times 2\pi\nu = \nu\lambda$

or $\quad A = \dfrac{\lambda}{2\pi}$

22. ***(c)*** The speed of the car is 72 km/h

$$= 72 \times \frac{5}{18} = 20 \text{ m/s}$$

The distance travelled by car in 10 s

$$= 10 \times 20 = 200 \text{ m}$$

Hence, the distance travelled by sound in reaching the hill and coming back to the moving driver

$$= 1800 + (1800 - 200)$$
$$= 3400 \text{ m}$$

So, the speed of sound $= \dfrac{3400}{10} = 340$ m/s

23. ***(a)*** Lens Maker's formula,

$$\frac{1}{f} = (\mu - 1)\left[\frac{1}{R_1} - \frac{1}{R_2}\right]$$

where, $R_2 = \infty$, $R_1 = 0.3$ m

$\therefore \quad \dfrac{1}{f} = \left(\dfrac{5}{3} - 1\right)\left(\dfrac{1}{0.3} - \dfrac{1}{\infty}\right) \Rightarrow \dfrac{1}{f} = \dfrac{2}{3} \times \dfrac{1}{0.3}$

or $\quad f = 0.45$ m

Focal length is negative for concave lens.

$\therefore \quad f = -0.45$ m

24. ***(a)***Using Huygen's eye-piece, measurements can be taken but not accurately. The cross wires are viewed through the eye lens only, while the distant objects are viewed by rays refracted by both lenses. Due to this reason relative length of cross wire and image are disproportionate.

25. ***(c)*** The image of an object in white light formed by a lens is usually coloured and blurred. This defect of image is called chromatic aberration and arises due to the fact that focal length of a lens is different for different colours. In case of two thin lenses in contact, the combination will be free from chromatic aberration. The lens combination which satisfies this condition are called achromatic lenses.

Hence, in achromatic combination of lenses, images remains unaffected by variation of refrative index and wavelength.

26. ***(c)*** The general condition for Fraunhofer diffraction is $\dfrac{b^2}{L\lambda} << 1$.

27. ***(c)*** Time period of magnet, $T = 2\pi\sqrt{\dfrac{I}{MB}}$

When magnet is cut parallel to its length into four equal pieces. Then, new magnetic moment, $M' = \dfrac{M}{4}$

New moment of inertia, $I' = \dfrac{I}{4}$

$\therefore \quad$ New time period, $T' = 2\pi\sqrt{\dfrac{I'}{M'B}}$

$\Rightarrow \quad T = T' = 4$ s

28. ***(a)*** On bending a wire its pole strength remains unchanged, whereas its magnetic moment changes.

New magnetic moment,

$$M' = m(2r) = m\left(\frac{2l}{\pi}\right) = \frac{2M}{\pi}$$

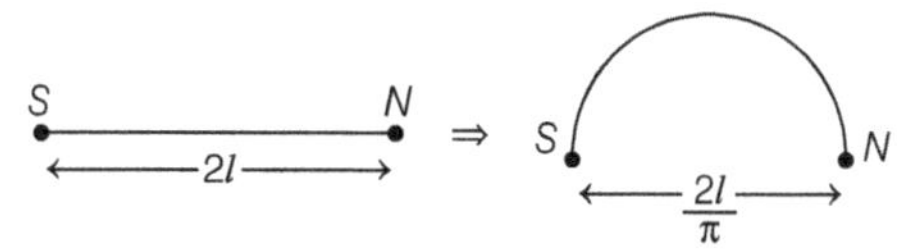

29. ***(b)*** Ratio of charges = 2 : 3

$\therefore \quad q_1 = \dfrac{2}{5} \times 1\,\mu\text{C} \quad \text{and} \quad q_2 = \dfrac{3}{5} \times 1\,\mu\text{C}$

Electrostatic force between the two charges,

$$F = \frac{1}{4\pi\varepsilon_0}\frac{q_1q_2}{r^2}$$
$$= \frac{9 \times 10^9 \times 2 \times 10^{-6} \times 3 \times 10^{-6}}{5 \times 5 \times (1)^2}$$
$$= 2.16 \times 10^{-3} \text{ N} = 0.00216 \text{ N}$$

30. ***(b)*** At equitorial point,

$$E_e = \frac{1}{4\pi\varepsilon_0}\frac{p}{r^3}$$

(directed from $+q$ to $-q$) and $V_e = 0$.

31. ***(c)*** Current through each arm

PRQ and $PSQ = 1$ A

($\because$ Resistance in each arm are same.)

$$V_P - V_R = 3 \text{ V} \quad \text{...(i)}$$
$$V_P - V_S = 7 \text{ V} \quad \text{...(ii)}$$

From Eqs. (i) and (ii), we get

$$V_R - V_S = +4 \text{ V}$$

32. ***(b)*** Here, $V < E$

$\therefore \quad E = V + Ir$

For first case,

$$E = 12 + \frac{12}{16}r \quad \text{...(i)}$$

For second case

$$E = 11 + \frac{11}{10}r \quad \text{...(ii)}$$

From Eqs. (i) and (ii), we get

$$12 + \frac{12}{16}r = 11 + \frac{11}{10}r$$

$\Rightarrow \quad r = \dfrac{20}{7}\,\Omega$

33. ***(a)*** We know that, thermoelectric power

$$S = \frac{dE}{dT}$$

Given, $E = k(T - T_r)\left[T_0 - \frac{1}{2}(T + T_r)\right]$

By differentiating the above equation w.r.t. T and putting $T = \frac{1}{2}T_0$, we get $S = \frac{1}{2}kT_0$

34. ***(b)*** Shunt of an ammeter,

$$S = \frac{I_g \times G}{I - I_g} = \frac{5 \times G}{100 - 5} = \frac{G}{19}$$

35. ***(b)*** Two coils carry currents in opposite directions, hence net magnetic field at centre will be difference of the two fields.

i.e. $B_{\text{net}} = \frac{\mu_0}{4\pi} \cdot 2\pi N\left[\frac{i_1}{r_1} - \frac{i_2}{r_2}\right]$

$$= \frac{10\,\mu_0}{2}\left[\frac{0.2}{0.2} - \frac{0.3}{0.4}\right] = \frac{5}{4}\mu_0$$

36. ***(a)*** In a transformer,

$\therefore \quad \frac{N_P}{N_S} = \frac{I_S}{I_P} \Rightarrow \frac{50}{200} = \frac{I_S}{4}$

$\Rightarrow \quad I_S = 1\text{ A}$

37. ***(b)*** For $\phi = 90°$, $\cos\phi = 0$

So, $\lambda' = \lambda + \frac{h}{m_e c}$

$$= 0.140 \times 10^{-9} + \frac{6.63 \times 10^{-34}}{(9.1 \times 10^{-31})(3 \times 10^8)}$$

$$= (0.140 \times 10^{-9} + 2.4 \times 10^{-12})\text{ m} = 0.142\text{ nm}$$

38. ***(a)*** $E = \frac{hc}{\lambda}$

$$= \frac{6.62 \times 10^{-34} \times 3 \times 10^8}{45 \times 10^{-12}}\text{ J}$$

$$= \frac{0.44 \times 10^{-14}}{1.6 \times 10^{-19}} \quad [\because 1\text{eV} = 1.6 \times 10^{-19}\text{J}]$$

$$= 0.275 \times 10^5\text{ eV} = 27500\text{ eV}$$

39. ***(c)*** Nuclear force between two particles is independent of charges of particle.

$\Rightarrow \quad F_{pp} = F_{nn} = F_{np}$

40. ***(d)*** In forward biasing both electrons and protons move towards the junction and hence the width of depletion region decreases.

Electron on n-side are repelled by opposite electric field and holes on p-side are repelled by the electric field, hence they move towards the junction.

Chemistry

41. ***(b)*** Energy of an electron in nth orbit,

$$E_n = -\frac{2\pi^2 k^2 m Z^2 e^4}{n^2 h^2}$$

On substituting the values of k, m, e and h, we get

$$E_n = -\frac{2.172 \times 10^{-18} Z^2}{m^2}\text{ J atom}^{-1}$$

or $= -\frac{1311.8 Z^2}{n^2}\text{ kJ mol}^{-1} = -\frac{313.52 Z^2}{n^2}\text{ kcal mol}^{-1}$

$[\because 1\text{ kcal} = 4.184\text{ kJ}]$

For H-atom, $Z = 1$

For Lyman series, $n_1 = 1$, $n_2 = 2$

Energy of electron in n_1 orbit

$$= -\frac{313.52 \times (1)^2}{(1)^2}\text{ kcal mol}^{-1}$$

$$= -313.52\text{ kcal mol}^{-1}$$

$$\approx -313.6\text{ kcal mol}^{-1}$$

Energy of electron in n_2 orbit

$$= -\frac{313.52 \times (1)^2}{(2)^2}\text{ kcal mol}^{-1}$$

$$= -\frac{313.52}{4}\text{ kcal mol}^{-1} = -78.38\text{ kcal mol}^{-1}$$

$$= -78.4\text{ kcal mol}^{-1}$$

42. ***(a)*** Given, velocity of particle $A = 0.05\text{ ms}^{-1}$

velocity of particle $B = 0.02\text{ ms}^{-1}$

Let then mass of particle $A = x$

then mass of particle $B = 5x$

de-Broglie's equation is

$$\lambda = \frac{h}{mv}$$

For particle A,

$$\lambda_A = \frac{h}{x \times 0.05} \quad \ldots(\text{i})$$

For particle B,

$$\lambda_B = \frac{h}{5x \times 0.02} \quad \ldots(\text{ii})$$

Dividing Eqs. (i) and (ii), we get

$$\frac{\lambda_A}{\lambda_B} = \frac{5x \times 0.02}{x \times 0.05}$$

$$\frac{\lambda_A}{\lambda_B} = \frac{2}{1}$$

or $\lambda_A : \lambda_B = 2:1$.

43. ***(b)*** Given, Δm for ${}_5\text{B}^{11} = 0.081$ u

Number of nucleons = 11

Binding energy $= 931 \times \Delta m$

$= 931 \times 0.081$

$= 75.411$ MeV

Average binding energy $= \dfrac{\text{binding energy}}{\text{number of nucleons}}$

$= \dfrac{75.411}{11}$

$= 6.85$ MeV

44. ***(c)*** Given,

Atomic number of element $B = Z$

($\because$ Noble gases i.e. belong to zero group)

Atomic number of element $A = Z - 1$ (i.e. halogens)

Atomic number of element $C = Z + 1$ (i.e. group IA)

Atomic number of element $D = Z + 2$ (i.e. group IIA)

Element A must be a halogen i.e. have highest electron affinity.

Element C must be an alkali metal and exist in $+1$ oxidation state but element D must be an alkaline earth metal with $+2$ oxidation state.

$\therefore$ Option (c) is correct.

45. ***(d)*** Given,

Observed dipole moment = 1.03 D

Bond length of HCl molecule, $d = 1.275$ Å

$= 1.275 \times 10^{-8}$ cm

Charge of electron, $e^- = 4.8 \times 10^{-10}$ esu

Percentage ionic character = ?

Theoretical value of dipole moment $= e \times d$

$= 4.8 \times 10^{-10} \times 1.275 \times 10^{-8}$ esu cm

$= 6.12 \times 10^{-18}$ esu cm

$= 6.12$ D

Percentage ionic character

$= \dfrac{\text{Observed dipole moment}}{\text{Theoretical value of dipole moment}} \times 100$

$= \dfrac{1.03}{6.12} \times 100 = 16.83\ \%$

46. ***(a)***

Molecule	*bp* + *lp*	Hybridisation	Shape
H_2O	2 + 2	sp^3	Angular
BCl_3	3 + 0	sp^2	Trigonal planar
NH_4^+	4 + 0	sp^3	Tetrahedral
CH_4	4 + 0	sp^3	Tetrahedral

47. ***(a)*** (A) $\underset{100\text{ g}}{CaCO_3} \xrightarrow[\text{Decomposition}]{\Delta} CaO + \underset{22.4\text{ L}}{CO_2}$

$\because$ 100 g $CaCO_3$ on decomposition gives

$= 22.4$ L CO_2

$\therefore$ 10 g $CaCO_3$ on decomposition will give

$= \dfrac{22.4 \times 10}{100}$ L $CO_2 = 2.24$ L CO_2

(B) $\underset{106\text{ g}}{Na_2CO_3} \xrightarrow{\text{Excess HCl}} 2NaCl + H_2O + \underset{22.4\text{ L}}{CO_2}$

106 g Na_2CO_3 gives $= 22.4$ L CO_2

1.06 g Na_2CO_3 will give $= \dfrac{22.4 \times 1.06}{106}$ L CO_2

$= 0.224$ L CO_2

(C) $\underset{12\text{ g}}{C} \xrightarrow[\text{Combustion}]{\text{Excess } O_2} \underset{22.4\text{ L}}{CO_2}$

12 g carbon on combustion gives

$= 22.4$ L CO_2

2.4 g carbon on combustion will give

$= \dfrac{22.4 \times 2.4}{12}$ L CO_2

$= 2 \times 2.24$ L $CO_2 = 4.48$ L CO_2

(D) $\underset{\substack{2\,[12+16] \\ 56\text{ g}}}{2CO} \xrightarrow[\text{Combustion}]{\text{Excess } O_2} \underset{2 \times 22.4\text{ L}}{2CO_2}$

56 g carbon monoxide on combustion gives $= 2 \times 22.4$ L CO_2

0.56 g carbon monoxide on combustion will give

$= \dfrac{2 \times 22.4 \times 0.56}{56}$ L $CO_2 = 0.448$ L CO_2

Hence, A— (iv), B— (i), C— (ii), D— (iii).

48. ***(c)*** Number of moles of helium = 0.3

Number of moles of argon = 0.4

We know that, KE = nRT

KE of helium $= 0.3 \times R \times T$ ...(i)

KE of argon $= 0.4 \times R \times 400$ ...(ii)

According to question,

KE of helium = KE of argon

$0.3 \times R \times T = 0.4 \times R \times 400$

$T = 533$ K

49. ***(c)*** Given,

weight of non-volatile solute,

$w = 25$ g

Weight of solvent, $W = 100$ g

Mock Test 1

Lowering of vapour pressure,

$$p° - p_s = 0.225 \text{ mm}$$

Vapour pressure of pure solvent,

$$p° = 17.5 \text{ mm}$$

Molecular weight of solvent (H_2O), $M = 18$ g

Molecular weight of solute, $m = ?$

According to Raoult's law,

$$\frac{p° - p_s}{p°} = \frac{w \times M}{m \times W}$$

$$\frac{0.225}{17.5} = \frac{25 \times 18}{m \times 100}$$

$$m = \frac{25 \times 18 \times 17.5}{22.5} = 350 \text{ g}$$

50. ***(c)*** Barium hydroxide is hydrolysed in water as follows

$$Ba(OH)_2 \rightleftharpoons Ba^{2+} + 2OH^-$$

$$\text{conc. of } Ba^{2+} = 1 \times 10^{-3} \text{ M}$$

$$\text{conc. of } [OH^-] = 2 \times 1 \times 10^{-3} \text{ M}$$

$$= 2 \times 10^{-3} \text{ M}$$

$$pOH = -\log[OH^-]$$

$$= -\log(2 \times 10^{-3}) = 2.69$$

$$pH + pOH = 14$$

$$pH = 14 - pOH$$

$$= 14 - 2.69 = 11.3 \approx 11.0$$

51. ***(a)*** $CH_3COONa + H_2O \longrightarrow CH_3COOH + NaOH$

The above process takes place in following steps

$$CH_3COONa \xrightleftharpoons[\text{in aqueous solution}]{\text{Ionisation}} \underset{\text{Strong base}}{CH_3COO^-} + Na^+$$

$$CH_3COO^- + H_2O \longrightarrow CH_3COOH + OH^-$$

Acetate ion undergoes anionic hydrolysis and the resulting solution is slightly basic due to excess of OH^- ions. Hence, both (A) and (R) are true and (R) is the correct explanation of (A).

52. ***(c)*** Given, weight of hydrogen liberated

$$= 5.04 \times 10^{-2} \text{ g}$$

Eq. wt. of hydrogen = 1.008

Eq. wt. of silver = 108

Weight of silver deposited, $w = ?$

According to Faraday's second law of electrolysis,

$$\frac{\text{Weight of silver deposited}}{\text{Weight of hydrogen liberated}} = \frac{\text{eq. wt. of silver}}{\text{eq. wt. of hydrogen}}$$

$$\frac{w}{5.04 \times 10^{-2}} = \frac{108}{1.008}$$

$$w = \frac{108 \times 5.04 \times 10^{-2}}{1.008} = 5.4 \text{ g}$$

53. ***(c)*** Electrolysis of water takes place as follows

$$H_2O \rightleftharpoons \underset{\text{Cathode}}{H^+} + \underset{\text{Anode}}{OH^-}$$

At anode

$$OH^- \xrightarrow{\text{Oxidation}} OH + e^-$$

$$4OH \longrightarrow 2H_2O + O_2$$

At cathode

$$2H^+ + 2e^- \xrightarrow{\text{Reduction}} H_2$$

Given, time, $t = 1930$ s

Number of moles of hydrogen collected

$$= \frac{1120 \times 10^{-3}}{22.4} \text{ moles}$$

$$= 0.05 \text{ moles}$$

$\because$ 1 mole of hydrogen is deposited by = 2 moles of electrons

$\therefore$ 0.05 moles of hydrogen will be deposited by

$$= 2 \times 0.05 = 0.10 \text{ mole of electrons}$$

Charge, $Q = nF = 0.1 \times 96500$

Charge, $Q = it$

$$0.1 \times 96500 = i \times 1930$$

$$i = \frac{0.1 \times 96500}{1930} = 5.0 \text{ A}$$

54. ***(a)*** Given, angle of diffraction $(2\theta) = 90°$

$$\theta = 45°$$

Distance between two planes, $d = 2.28$ Å

$n = 2$ [$\because$ second order diffraction]

Bragg's equation is

$$n\lambda = 2d \sin\theta$$

$$2 \times \lambda = 2 \times 2.28 \times \sin 45°$$

$$\lambda = 1.612 \text{ Å}$$

55. ***(b)***

PCl_5	$\rightleftharpoons$ PCl_3	+ Cl_2	
5	0	0	Initial moles
$5(1-\alpha)$	5α	5α	Moles at equilibrium
$\frac{5(1-\alpha)}{0.5}$	$\frac{5\alpha}{0.5}$	$\frac{5\alpha}{0.5}$	Conc. at equilibrium

$$\alpha = 40\% = 0.4, 1 - \alpha = 0.6$$

$$K_c = \frac{[PCl_3][Cl_2]}{[PCl_5]}$$

$$= \frac{\left(\frac{5 \times 0.4}{0.5}\right)\left(\frac{5 \times 0.4}{0.5}\right)}{\left(\frac{5 \times 0.6}{0.5}\right)} = \frac{16}{6}$$

$$= 2.66 \text{ mol/L}$$

Mock Test 1

56. *(c)*

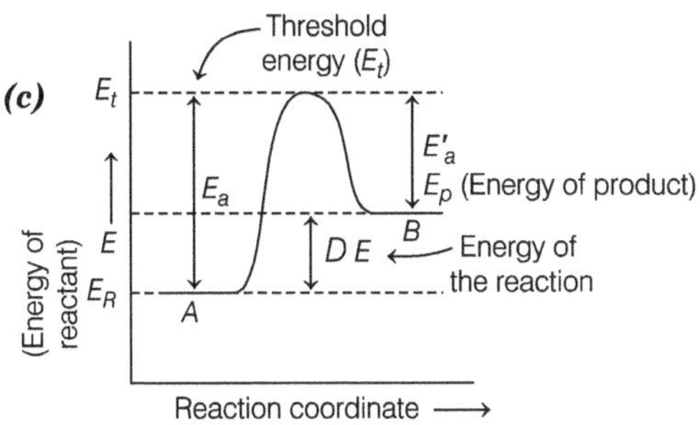

where, E_a = activation energy of forward reaction

E_a' = activation energy of backward reaction

The above energy profile diagram shows that

$$E_a > E_a'$$

The potential energy of the product is greater than that of the reactant, so the reaction is endothermic.

$$E_a = E_a' + \Delta E \Rightarrow E_t = E_a \text{ or } E_t > E_a'$$

$\therefore$ Option (c) is incorrect statement for the given energy profile diagram.

57. *(a)* Given,

$$H_2O\,(g) + C(g) \longrightarrow CO(g) + H_2(g); \quad \Delta H = 131 \text{ kJ} \quad \ldots\text{(i)}$$

$$CO(g) + \frac{1}{2}O_2(g) \longrightarrow CO_2(g);\ \Delta H = -282 \text{ kJ} \ \ldots\text{(ii)}$$

$$H_2(g) + \frac{1}{2}O_2(g) \longrightarrow H_2O(g);\ \Delta H = -242 \text{ kJ} \ \ldots\text{(iii)}$$

$$C(g) + O_2(g) \longrightarrow CO_2(g);\ \Delta H = ? \quad \ldots\text{(iv)}$$

On adding Eqs. (i), (ii) and (iii), we get Eq. (iv)

$$H_2O(g) + C(g) \longrightarrow CO(g) + H_2(g);\ \Delta H = +131 \text{ kJ}$$

$$CO(g) + \frac{1}{2}O_2(g) \longrightarrow CO_2(g);\ \Delta H = -282 \text{ kJ} \quad \ldots\text{(ii)}$$

$$H_2(g) + \frac{1}{2}O_2(g) \longrightarrow H_2O(g);\ \Delta H = -242 \text{ kJ} \quad \ldots\text{(iii)}$$

$$C(g) + O_2(g) \longrightarrow CO_2(g)$$

$$\Delta H = (131 - 282 - 242) \text{ kJ} = -393 \text{ kJ} \ \ldots\text{(iv)}$$

58. *(c)* When we plot a graph between log (x/m) and log p, a straight line with positive slope will be obtained. This graph represents the Freundlich adsorption isotherm.

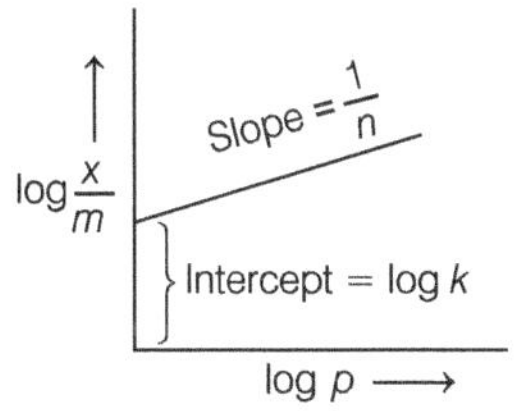

Graph of Freundlich

59. *(d)* The reaction in which H_2O_2 is reduced while the other reactant is oxidised, represents the oxidising property of H_2O_2.

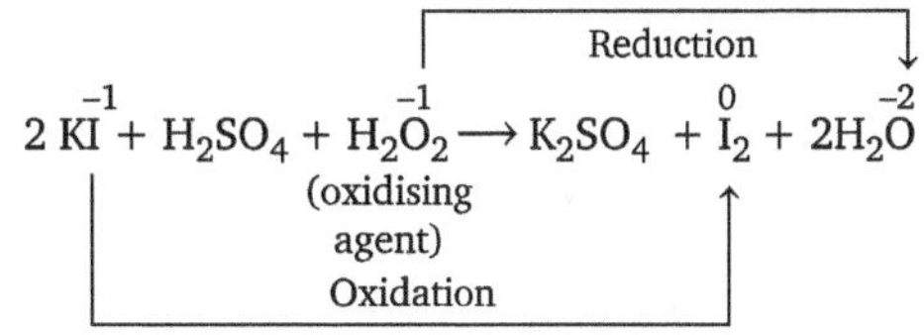

60. *(b)* (i) The alkali metal superoxides contain O_2^- ion, which has an unpaired electron, hence they are paramagnetic in nature.

(ii) The basic character of alkali metal hydroxides increases on moving down the group.

(iii) The conductivity of alkali metal chlorides in their aqueous solution increases on moving down the group because in aqueous solution alkali metal chlorides ionise to give alkali metal ions. On moving down the group the size of alkali metal ion increases, thus degree of hydration decreases, due to this reason their conductivity in aqueous solution increases on moving down the group.

(iv) $$M_2CO_3 \underset{\text{aq. solution}}{\overset{\text{Ionisation in}}{\rightleftharpoons}} 2M^+ + CO_3^{2-}$$

$$CO_3^{2-} + 2H_2O \longrightarrow H_2CO_3 + 2OH^-$$

$\therefore$ Statements (i) and (ii) are correct.

Thus, basic nature of carbonates in aqueous solution is due to anionic hydrolysis.

61. *(c)* According to Lewis, the compound which can accept a lone pair of electron, are called acids.

Boron halides, being electron deficient compounds, can accept a lone pair of electrons, so termed as Lewis acid.

62. *(d)* Orthosilicic acid (H_4SiO_4), on heating at high temperature, loses two water molecules and gives silica (SiO_2) which on reduction with carbon gives carborundum (SiC) and CO.

$$H_4SiO_4 \xrightarrow[-2H_2O]{1000°C} SiO_2 \xrightarrow[\Delta]{C} \underset{\text{Carborundum}}{SiC} + CO$$

63. *(a)* The reducing character of the hydrides of group V elements depends upon the stability of hydrides. With progressive decrease in stability, the reducing character of hydrides increases as we move down the group. Thus, ammonia, being stable, has least reducing ability. The order of reducing abilities of V group hydrides is

$$NH_3 < PH_3 < AsH_3 < SbH_3 < BiH_3$$

64. *(b)* The structure of peroxodisulphuric acid ($H_2S_2O_8$) is

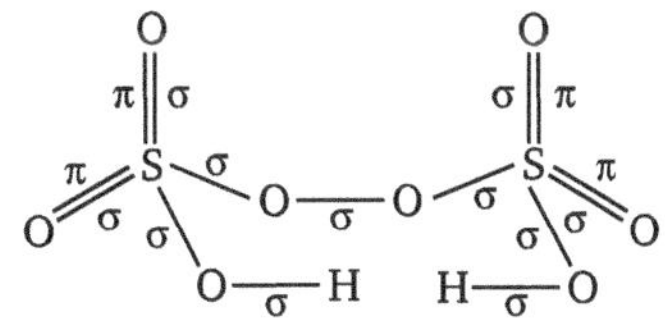

Hence, it contains 11σ and 4π bonds.

65. ***(b)*** With progressive increase in atomic number, the reduction potential of halogens decreases, thus oxidising power also decreases. Hence, a halogen with lower atomic number will oxidise the halide ion of higher atomic number and therefore, will liberate them from their salt solution.

Hence, the reaction $Cl_2 + 2F^- \longrightarrow 2Cl^- + F_2$

is not possible.

66. ***(b)*** The structure of ClO_4^- is

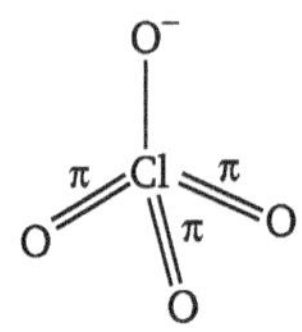

Thus, it contains 3 $d\pi$-$p\pi$ bonds.

XeO_3 also contains 3 $d\pi$-$p\pi$ bonds as,

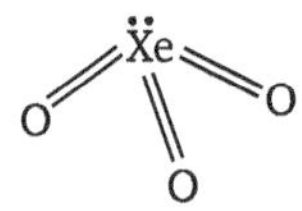

67. ***(a)*** $[Co(NH_3)_5SO_4]Br \rightleftharpoons [Co(NH_3)_5SO_4]^+ + Br^-$

$[Co(NH_3)_5Br]SO_4 \rightleftharpoons [Co(NH_3)_5Br]^{2+} + SO_4^{2-}$

The molecular formula of both of the above compounds is same but on ionisation, they give different ions in solution, so they are pair of ionisation isomers.

68. ***(a)*** In stratosphere, the following reactions takes place which are responsible for depletion of ozone layer.

$$NO + O_3 \longrightarrow NO_2 + O_2$$

$$CF_2Cl_2 \xrightarrow{h\nu} \dot{C}F_2Cl + \dot{C}l$$

$$CFCl_3 \xrightarrow{h\nu} \dot{C}FCl_2 + \dot{C}l$$

$$\dot{C}l + O_3 \longrightarrow Cl\dot{O} + O_2$$

$$Cl\dot{O} + O \longrightarrow \dot{C}l + O_2$$

Hence, methane (CH_4) is not responsible for ozone layer depletion.

69. ***(d)*** When the groups with higher priority (i.e. with high atomic number) are present on same side of double bond, then the configuration is Z but when present on opposite side of double bond, the configuration is *E*.

(i) 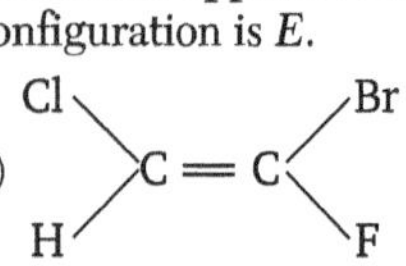

(Z)

(Priority : Cl > H and Br > F)

(ii) 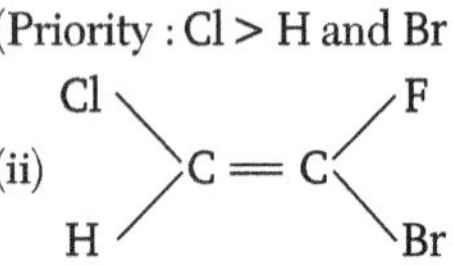

(*E*)

(Priority : Cl > H and Br > F)

(iii) 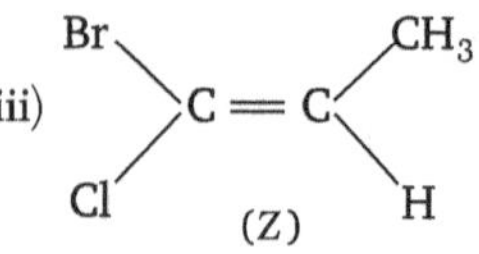

(Z)

(Priority : Br > Cl and CH_3 > H)

Hence, compound (i) and (iii) have Z configuration.

70. ***(d)*** According to Cahn-Ingold-Prelog sequence rules, the priority of groups is decided by the atomic number of their atoms. When the atom (which is directly attached to the asymmetric carbon atom) of a group has higher atomic number, then the group gets higher priority. Groups with atoms of comparable atomic number having double or triple bond, have high priority than those have single bond.

Hence, the order of priority of groups is

$$—OH > —COOH > —CHO > —CH_2OH$$

71. ***(a)*** The given reactions take place as follows

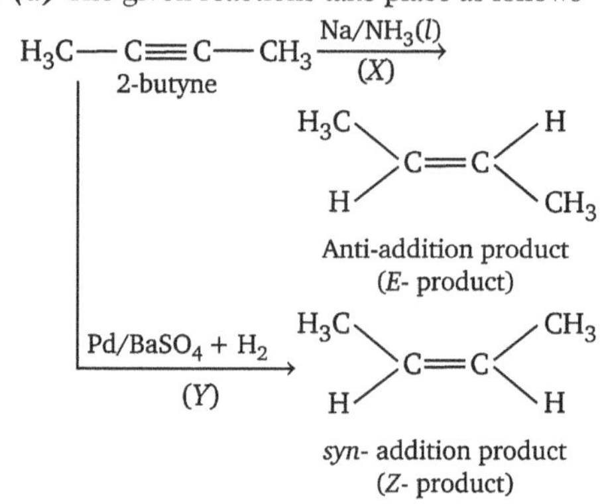

Hence, reagents *X* and *Y* are Na/NH_3 and Pd/$BaSO_4$ + H_2 respectively.

72. ***(d)*** In a reaction, the reagent, which is reduced or remove hydrogen from the other reactant (reagent), is termed as oxidising agent.

(i) $CH_3CH_2OH + \overset{0}{Cl_2} \longrightarrow CH_3CHO + H\overset{-1}{Cl}$

Reduction; Oxidising agent; Removal of hydrogen, i.e. oxidation

(ii) $CH_3CHO + \overset{0}{Cl_2} \longrightarrow CCl_3{\cdot}CHO + H\overset{-1}{Cl}$

Reduction; Oxidising agent; Oxidation (Removal of H)

(iii) $CH_4 + \overset{0}{Cl_2} \xrightarrow{h\nu} CH_3Cl + H\overset{-1}{Cl}$

Reduction; Oxidising agent; Oxidation

Hence, in all of the above reactions, chlorine acts as an oxidising agent.

73. ***(d)*** Among hydrogen halides, as the size of halide ion increases, its reactivity towards ethyl alcohol also increases. Thus, the order of reactivity of hydrogen halides is

$$HI > HBr > HCl > HF$$

74. ***(d)*** 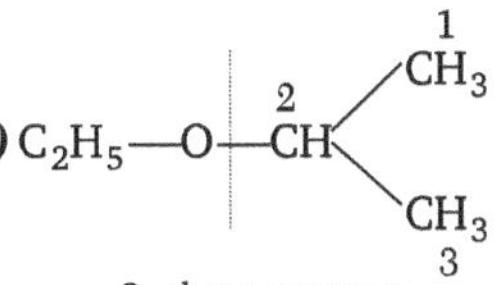

2-ethoxy propane

The above compound is an ether and its name is written as alkoxy alkane. Oxy is attached with the lower group. Hence, the IUPAC name of above compound is 2-ethoxy propane.

75. ***(b)*** The given reactions take place as follows

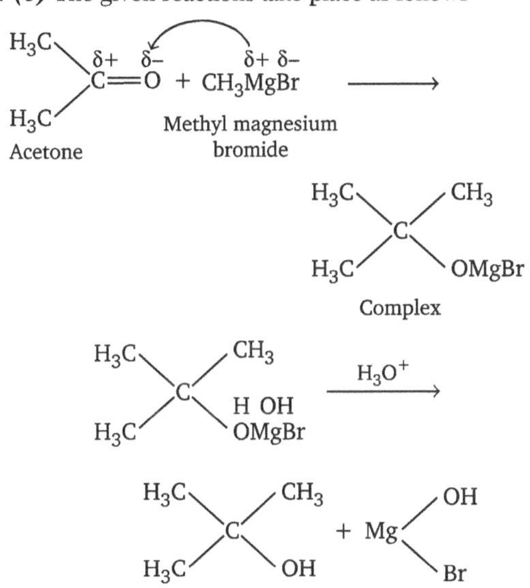

'X' or $(CH_3)_3COH$
2- methyl-propan-2-ol

$\therefore$ 'X' is 2-methyl propan-2-ol.

76. ***(c)*** Acetic acid on reduction with lithium aluminium hydride ($LiAlH_4$) gives ethyl alcohol while on reduction with HI and red P gives ethane.

$$\underset{\text{Acetic acid}}{CH_3COOH} \xrightarrow{LiAlH_4} \underset{\text{Ethyl alcohol}}{CH_3CH_2OH}$$

$$\underset{\text{Acetic acid}}{CH_3COOH} \xrightarrow{\text{Red P + HI}} \underset{\text{Ethane}}{CH_3—CH_3}$$

Hence, reagent *A* and *B* are respectively $LiAlH_4$ and HI/red P.

77. ***(c)*** Nitrobenzene on reduction with lithium aluminium hydride ($LiAlH_4$) gives azobenzene.

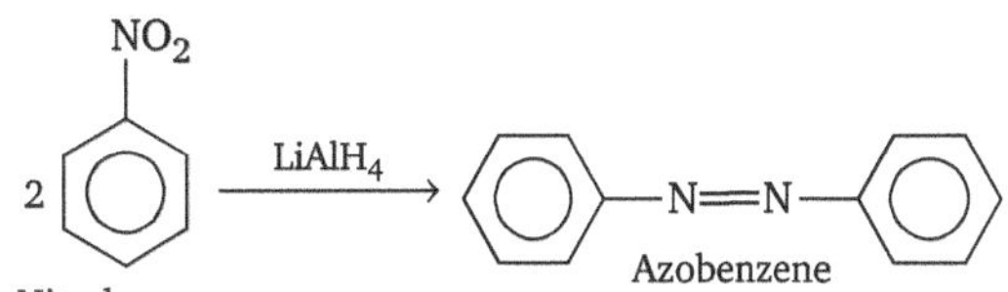

Nitrobenzene

78. ***(a)*** Aspirin is used as analgesics as well as antipyretics, i.e. it serve a dual purpose.

Chlorophyll is used in photosynthesis.

Oxyhaemoglobin contains Fe^{2+} ion, so it is paramagnetic and haemoglobin works as oxygen carrier.

Hence, A—(v), B—(i), C—(ii), D—(iii).

79. ***(b)*** The ratio of weight average molecular weight and the number average molecular weight is called poly dispersity index (PDI).

$$PDI = \frac{\overline{M}_w}{\overline{M}_n}$$

where, $\overline{M}_w$ = weight average molecular weight

$\overline{M}_n$ = number average molecular weight

PDI is unity for natural monodispersed polymer but for synthetic polymers it is always greater than unity.

80. ***(a)*** On hydrolysis with dilute aqueous sulphuric acid, sucrose gives a equimolar mixture of D-(+)- glucose and D-(–)-fructose.

$$\underset{\text{Sucrose}}{C_{12}H_{22}O_{11}} + H_2O \xrightarrow{H_2SO_4} \underset{\text{D-(+) glucose}}{C_6H_{12}O_6} + \underset{\text{D-(–) fructose}}{C_6H_{12}O_6}$$

$$1 \quad : \quad 1$$

Sucrose is dextrorotatory but after hydrolysis gives dextrorotatory glucose and laevorotatory fructose, laevorotatory fructose is more, so the mixture is laevorotatory.

a. English Proficiency

81. ***(b)*** The most appropriate word to fill the given blank, is 'took place' to make the sentence grammatically correct.

82. ***(d)*** 'Realised' is used in the given blank.

83. ***(c)*** 'Captured' is the most appropriate alternative to fill the blank.

84. ***(a)*** The correct sequence is QPSR.

85. ***(c)*** The correct sequence is QPRS.

86. ***(a)*** The correct sequence is QPRS.

87. ***(c)*** Ephemeral means lasting for a very short time. So 'short living' is most nearly the same in meaning to the word given.

88. ***(b)*** Stubborn means unreasonably obstinate or determined to do what he or she wants and refuses to do anything else. Hence, option (b) is is its correct answer.

89. ***(c)*** Prognosis means a forecast of the likely outcome of a situation.

So, option (c) is its correct answer.

90. ***(a)*** Infallible means incapable of making mistakes or being wrong. 'Erring' would be its apposite meaning word as erring means to make a mistake or to do something wrong.

91. ***(c)*** 'Gather' means come together or assemble. So, 'scatter' would be its correct opposite meaning word as its means separate and move off quickly in different directions.

92. ***(a)*** Exalt means to raise someone to a higher rank or more powerful position. So, 'depreciate' would be its correct opposite meaning word. Depreciate means reduce.

93. ***(c)*** Chaperone is an elder person, especially a woman, who stays with and take care of a younger one who is not married when she is in public. Hence, option (c) is its correct answer.

94. ***(c)*** A piece of land almost surrounded by water or projecting out into a body of water is called peninsula.

95. ***(c)*** A place adjoining kitchen for washing dishes and other dirty household work is called scullery.

b. Logical Reasoning

96. ***(a)*** From figure I to II and III to answer figure, both the symbols join in opposite way.

97. ***(b)*** Newspaper, Hoarding and Television are related to Media.

98. ***(b)*** Correct pattern of the series as follows

$2\times 3-1=5,\ 5\times 3-1=14,\ 14\times 3-1=41,$

$41\times 3-1=122,\ 122\times 3-1=365,$

$$365\times 3-1=\boxed{1094}$$

99. ***(c)*** The inferences correctly drawn are 2 and 4.

100. ***(b)*** The series shown in option (b) obeys the given rule.

101. ***(b)*** Figure shown is option (b) complete the pattern in the problem figure.

102. ***(b)*** Figure shown is option (b) exactly like this unfolded paper.

103. ***(c)*** Components of the given question figure are in the figure shown is option (c).

104. ***(d)*** Except 7896, all other options have consecutive digits.

105. ***(c)***

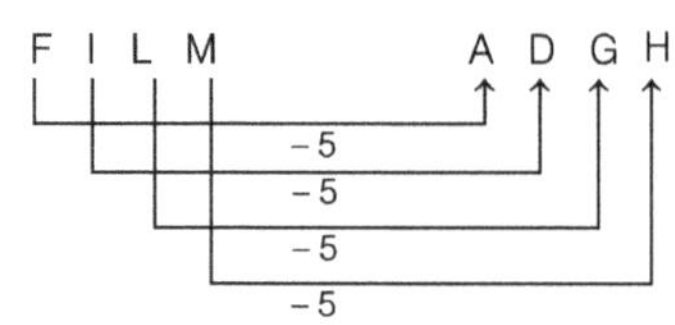

Similarly,

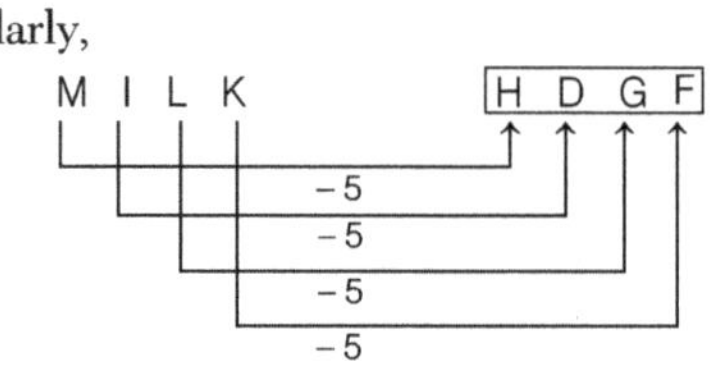

Mathematics

106. ***(c)*** Given that, $f(x)=|x|$

and $g(x)=[x-3]=[x]-3$

For $\quad -\dfrac{8}{5}<x<\dfrac{8}{5},\ 0\le f(x)<\dfrac{8}{5}$

Now, for $\quad 0\le f(x)<1,$

$$g(f(x))=[f(x)-3]$$

$$=-3\ [\because -3\le f(x)-3<-2]$$

Again, for $1\le f(x)<1.6$

$$g(f(x))=-2 \qquad [\because -2\le f(x)-3<-1.4]$$

Hence, required set is $\{-3,-2\}$.

107. ***(c)*** Now, $\sum_{k=1}^{n} k(k+2)$

$$=\sum_{k=1}^{n}(k^2+2k)=\sum_{k=1}^{n}k^2+2\sum_{k=1}^{n}k$$

$$=\frac{n(n+1)(2n+1)}{6}+\frac{2\cdot n(n+1)}{2}$$

$$=n(n+1)\left(\frac{2n+1}{6}+1\right)=\frac{n(n+1)(2n+7)}{6}$$

108. ***(c)*** Required number of arrangements

$$={}^6P_5\times 4!$$

$$=720\times 24$$

$$=17280$$

109. ***(b)*** Given that, ${}^nP_r=30240$ and ${}^nC_r=252$

$$\frac{n!}{(n-r)!}=30240 \text{ and } \frac{n!}{(n-r)!\,r!}=252$$

$$\Rightarrow \quad r!=\frac{30240}{252}=120$$

$$\Rightarrow \quad r=5$$

$$\therefore \quad \frac{n!}{(n-5)!}=30240$$

$$\Rightarrow n(n-1)(n-2)(n-3)(n-4)=30240$$

$$\Rightarrow n(n-1)(n-2)(n-3)(n-4)$$

$$=10(10-1)(10-2)(10-3)(10-4)$$

$$\Rightarrow n=10$$

Hence, required ordered pair is (10, 5).

110. *(c)* Given, $(1+x+x^2+x^3)^5 = \sum_{k=0}^{15} a_k x^k$

$$\Rightarrow \quad [(1+x)+x(1+x)]^5 = \sum_{k=0}^{15} a_k x^k$$

$$\Rightarrow (1+x)^{10} = a_0 x^0 + a_1 x + a_2 x^2 + \ldots + a_{15} x^{15}$$

$$\Rightarrow \quad {}^{10}C_0 + {}^{10}C_1 x + {}^{10}C_2 x^2 + \ldots + {}^{10}C_{10} x^{10}$$

$$= a_0 + a_1 x + a_2 x^2 + a_3 x^3 + \ldots + a_{15} x^{15}$$

On equating the coefficient of constant and even powers of x, we get

$$a_0 = {}^{10}C_0,\ a_2 = {}^{10}C_2,$$

$$a_4 = {}^{10}C_4, \ldots, a_{10} = {}^{10}C_{10},$$

$$a_{12} = a_{14} = 0$$

$$\therefore \quad \sum_{k=0}^{7} a_2 k = {}^{10}C_0 + {}^{10}C_2 + {}^{10}C_4 + {}^{10}C_6 + {}^{10}C_8 + {}^{10}C_{10} + 0 + 0$$

$$= 2^{10-1} = 2^9 = 512$$

111. *(d)* Given that, $\alpha + \beta = -2$ and $\alpha^3 + \beta^3 = -56$

$$\Rightarrow \quad (\alpha+\beta)(\alpha^2+\beta^2-\alpha\beta) = -56$$

$$\Rightarrow \quad \alpha^2+\beta^2-\alpha\beta = 28$$

Now, $\quad (\alpha+\beta)^2 = (-2)^2$

$$\Rightarrow \quad \alpha^2+\beta^2+2\alpha\beta = 4$$

$$\Rightarrow \quad 28 + 3\alpha\beta = 4$$

$$\Rightarrow \quad \alpha\beta = -8$$

$\therefore$ Required equation is

$$x^2 - (-2)x + (-8) = 0$$

$$\Rightarrow \quad x^2 + 2x - 8 = 0$$

112. *(d)* Given equation is

$$x^3 + 2x^2 - 4x + 1 = 0$$

Let α, β and γ be the roots of the given equation

$\therefore \quad \alpha+\beta+\gamma = -2,\ \alpha\beta+\beta\gamma+\gamma\alpha = -4$

and $\quad \alpha\beta\gamma = -1$

Let the required cubic equation has the roots 3α, 3β and 3γ.

$$\Rightarrow \quad 3\alpha+3\beta+3\gamma = -6,$$

$$3\alpha\cdot 3\beta + 3\beta\cdot 3\gamma + 3\gamma\cdot 3\alpha = -36$$

and $\quad 3\alpha\cdot 3\beta\cdot 3\gamma = -27$

$\therefore$ Required equation is

$$x^3 - (-6)x^2 + (-36)x - (-27) = 0$$

$$\Rightarrow \quad x^3 + 6x^2 - 36x + 27 = 0$$

113. *(b)* Given that,

$$A = \begin{bmatrix} 1 & -2 \\ 4 & 5 \end{bmatrix} \text{ and } f(t) = t^2 - 3t + 7$$

Now, $\quad A^2 = \begin{bmatrix} 1 & -2 \\ 4 & 5 \end{bmatrix}\begin{bmatrix} 1 & -2 \\ 4 & 5 \end{bmatrix} = \begin{bmatrix} -7 & -12 \\ 24 & 17 \end{bmatrix}$

Now, $\quad f(A) = A^2 - 3A + 7$

$$= \begin{bmatrix} -7 & -12 \\ 24 & 17 \end{bmatrix} - 3\begin{bmatrix} 1 & -2 \\ 4 & 5 \end{bmatrix} + 7\begin{bmatrix} 1 & 0 \\ 0 & 1 \end{bmatrix}$$

$$= \begin{bmatrix} -3 & -6 \\ 12 & 9 \end{bmatrix}$$

$$\therefore \quad f(A) + \begin{bmatrix} 3 & 6 \\ -12 & -9 \end{bmatrix} = \begin{bmatrix} -3 & -6 \\ 12 & 9 \end{bmatrix} + \begin{bmatrix} 3 & 6 \\ -12 & -9 \end{bmatrix}$$

$$= \begin{bmatrix} 0 & 0 \\ 0 & 0 \end{bmatrix}$$

114. *(d)* Let $\Delta = \begin{vmatrix} a-b-c & 2a & 2a \\ 2b & b-c-a & 2b \\ 2c & 2c & c-a-b \end{vmatrix}$

Applying $R_1 \to R_1 + R_2 + R_3$ and taking common $(a+b+c)$ from R_1

$$= (a+b+c)\begin{vmatrix} 1 & 1 & 1 \\ 2b & b-c-a & 2b \\ 2c & 2c & c-a-b \end{vmatrix}$$

Applying $C_2 \to C_2 - C_1$ and $C_3 \to C_3 - C_1$,

$$= (a+b+c)\begin{vmatrix} 1 & 0 & 0 \\ 2b & -b-c-a & 0 \\ 2c & 0 & -a-b-c \end{vmatrix}$$

$$= (a+b+c)[(-b-c-a)(-a-b-c)]$$

$$= (a+b+c)^3$$

115. *(a)* Since, ω is a cube root of unity.

$$\therefore \quad \sin\left\{(\omega^{10}+\omega^{23})\,\pi - \frac{\pi}{4}\right\}$$

$$= \sin\left\{(\omega+\omega^2)\,\pi - \frac{\pi}{4}\right\} \qquad [\because \omega^3 = 1]$$

$$= \sin\left(-\pi - \frac{\pi}{4}\right) \qquad (\because 1+\omega+\omega^2 = 0)$$

$$= -\sin\left(\pi + \frac{\pi}{4}\right) = \sin\frac{\pi}{4} = \frac{1}{\sqrt{2}}$$

116. *(c)* $\sqrt{3}$ cosec $20° -$ sec $20°$

$$= \frac{\tan 60°}{\sin 20°} - \frac{1}{\cos 20°}$$

$$= \frac{\sin 60° \cos 20° - \sin 20° \cos 60°}{\cos 60° \sin 20° \cos 20°}$$

$$= \frac{\sin 40°}{\cos 60° \sin 20° \cos 20°}$$

$$[\because \sin(x-y) = \sin x \cos y - \sin y \cos x]$$

$$= \frac{2\sin 20° \cos 20°}{\frac{1}{2}(\sin 20° \cos 20°)} = 4$$

$$[\because \sin 2\theta = 2\sin\theta\cos\theta]$$

117. ***(b)*** Given,

$$\tan\theta + \tan\left(\theta + \frac{\pi}{3}\right) + \tan\left(\theta + \frac{2\pi}{3}\right) = 3$$

$$\Rightarrow \tan\theta + \frac{\tan\theta + \sqrt{3}}{1 - \sqrt{3}\tan\theta} + \frac{\tan\theta - \sqrt{3}}{1 + \sqrt{3}\tan\theta} = 3$$

$$\left[\because \tan(x+y) = \frac{\tan x + \tan y}{1 - \tan x \tan y}\right]$$

$$\Rightarrow \tan\theta + \frac{8\tan\theta}{1 - 3\tan^2\theta} = 3$$

$$\Rightarrow \frac{9\tan\theta - 3\tan^3\theta}{1 - 3\tan^2\theta} = 3$$

$$\Rightarrow 3\tan 3\theta = 3 \Rightarrow \tan 3\theta = 1$$

Hence, option (b) is correct.

118. ***(b)*** Given equation is

$$\cos 2x + 2\cos^2 x = 2$$

$$\Rightarrow 2\cos^2 x - 1 + 2\cos^2 x = 2$$

$$\Rightarrow 4\cos^2 x = 3$$

$$\Rightarrow \cos^2 x = \frac{3}{4}$$

$$\Rightarrow \cos x = \pm\frac{\sqrt{3}}{2}$$

$$\therefore x = n\pi \pm \frac{\pi}{6} : n \in Z$$

119. ***(b)*** Given that,

$$\sin^{-1}\left(\frac{3}{x}\right) + \sin^{-1}\left(\frac{4}{x}\right) = \frac{\pi}{2}$$

$$\therefore \sin^{-1}\left(\frac{3}{x}\right) = \frac{\pi}{2} - \sin^{-1}\left(\frac{4}{x}\right)$$

$$\Rightarrow \sin^{-1}\left(\frac{3}{x}\right) = \cos^{-1}\left(\frac{4}{x}\right) \quad \left[\because \frac{\pi}{2} - \sin^{-1}\theta = \cos^{-1}\theta\right]$$

$$\Rightarrow \sin^{-1}\left(\frac{3}{x}\right) = \sin^{-1}\left(\frac{\sqrt{x^2 - 16}}{x}\right) \Rightarrow \frac{3}{x} = \frac{\sqrt{x^2 - 16}}{x}$$

$$\Rightarrow 9 = x^2 - 16 \Rightarrow x^2 = 25$$

$$\Rightarrow x = \pm 5$$

$$\Rightarrow x = 5$$

($\because -5$ is not satisfied the given equation)

120. ***(b)*** Given that,

$$\frac{1}{b+c} + \frac{1}{c+a} = \frac{3}{a+b+c}$$

$$\Rightarrow 1 + \frac{b}{a+c} + 1 + \frac{a}{b+c} = 3$$

$$\Rightarrow b(b+c) + a(a+c) = (a+c)(b+c)$$

$$\Rightarrow b^2 + bc + a^2 + ac = ab + ac + bc + c^2$$

$$\Rightarrow a^2 + b^2 - c^2 = ab$$

We know that, $\cos C = \dfrac{a^2 + b^2 - c^2}{2ab} = \dfrac{ab}{2ab} = \dfrac{1}{2}$

$$\Rightarrow C = 60°$$

121. ***(d)*** Given that, $r_1 = 2r_2 = 3r_3$

$$\therefore \frac{\Delta}{s-a} = \frac{2\Delta}{s-b} = \frac{3\Delta}{s-c} = \frac{\Delta}{k} \quad \text{(say)}$$

Then, $s - a = k, s - b = 2k, s - c = 3k$

$\Rightarrow 3s - (a+b+c) = 6k \Rightarrow s = 6k$

$$\therefore \frac{a}{5} = \frac{b}{4} = \frac{c}{3} = k$$

Now, $\dfrac{a}{b} + \dfrac{b}{c} + \dfrac{c}{a} = \dfrac{5}{4} + \dfrac{4}{3} + \dfrac{3}{5}$

$$= \frac{75 + 80 + 36}{60} = \frac{191}{60}$$

122. ***(a)*** Let AB be a hill whose height is h metres and CD be a pillar of height h' metres.

In $\Delta\, EDB$,

$$\tan\alpha = \frac{h - h'}{ED} \quad \ldots\text{(i)}$$

and in $\Delta\, ACB$,

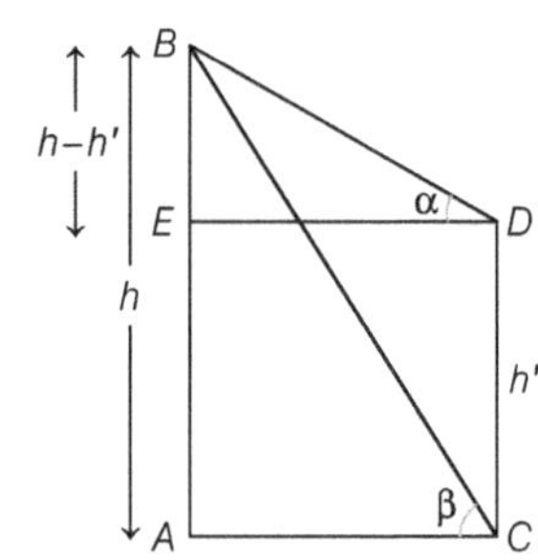

$$\tan\beta = \frac{h}{AC} = \frac{h}{ED} \quad \ldots\text{(ii)}$$

Eliminate ED from Eqs. (i) and (ii), we get

$$\tan\alpha = \frac{h - h'}{\dfrac{h}{\tan\beta}}$$

$$\Rightarrow h\frac{\tan\alpha}{\tan\beta} = h - h'$$

$$\Rightarrow h' = \frac{h(\tan\beta - \tan\alpha)}{\tan\beta}$$

123. ***(a)*** Given that,

PR = 5PQ

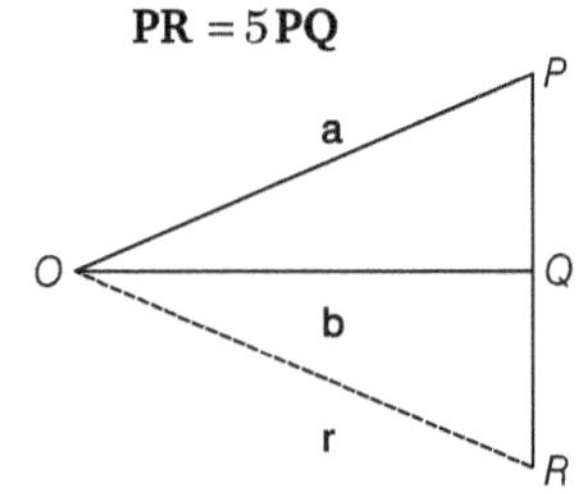

It mean R divides PQ externally in the ratio 5 : 4.

$\therefore$ Position vector of $R = \dfrac{5\mathbf{b} - 4\mathbf{a}}{5-4}$

$= 5\mathbf{b} - 4\mathbf{a}$

124. *(c)* Let $\mathbf{OA} = 2\hat{\mathbf{i}} - \hat{\mathbf{j}} + \hat{\mathbf{k}}$, $\mathbf{OB} = \hat{\mathbf{i}} - 3\hat{\mathbf{j}} - 5\hat{\mathbf{k}}$ and $\mathbf{OC} = 3\hat{\mathbf{i}} - 4\hat{\mathbf{j}} - 4\hat{\mathbf{k}}$

$\therefore \quad a = |\mathbf{OA}| = \sqrt{6}, b = |\mathbf{OB}| = \sqrt{35}$

and $\quad c = |\mathbf{OC}| = \sqrt{41}$

$$\therefore \quad \cos A = \frac{b^2 + c^2 - a^2}{2bc} = \frac{(\sqrt{35})^2 + (\sqrt{41})^2 - (\sqrt{6})^2}{2\sqrt{35}\sqrt{41}}$$

$$\Rightarrow \quad \cos A = \frac{70}{2\sqrt{35}\sqrt{41}} = \sqrt{\frac{35}{41}}$$

$$\Rightarrow \cos^2 A = \frac{35}{41}$$

125. *(d)* Given that,

$$\mathbf{b} = 2\hat{\mathbf{i}} + \hat{\mathbf{j}} - \hat{\mathbf{k}} \quad \text{and} \quad \mathbf{c} = \hat{\mathbf{i}} + 3\hat{\mathbf{k}}$$

$$\text{Now, } \mathbf{b} \times \mathbf{c} = \begin{vmatrix} \hat{\mathbf{i}} & \hat{\mathbf{j}} & \hat{\mathbf{k}} \\ 2 & 1 & -1 \\ 1 & 0 & 3 \end{vmatrix}$$

$$= \hat{\mathbf{i}}(3-0) - \hat{\mathbf{j}}(6+1) + \hat{\mathbf{k}}(0-1)$$

$$= 3\hat{\mathbf{i}} - 7\hat{\mathbf{j}} - \hat{\mathbf{k}}$$

Now, $[\mathbf{a}\ \mathbf{b}\ \mathbf{c}] = \mathbf{a} \cdot (\mathbf{b} \times \mathbf{c})$

$= |\mathbf{a}||\mathbf{b} \times \mathbf{c}| \cos\theta$

$= 1(\sqrt{3^2 + 7^2 + 1^2}) \cos\theta$

$= \sqrt{59} \cos\theta$

$\Rightarrow \quad [\mathbf{a}\ \mathbf{b}\ \mathbf{c}]_{\max} = \sqrt{59} \cdot 1$

($\because$ maximum value of $\cos\theta$ is 1)

Hence, maximum value is $\sqrt{59}$.

126. *(b)* Given that, $P(A \cap B) = \dfrac{1}{6}$ and $P(\overline{A} \cap \overline{B}) = \dfrac{1}{3}$

Since, A and B are independent.

$$\therefore \quad P(A)P(B) = \frac{1}{6} \quad \text{and} \quad P(\overline{A})\,P(\overline{B}) = \frac{1}{3}$$

$$\Rightarrow \quad [1 - P(A)][1 - P(B)] = \frac{1}{3}$$

$$\Rightarrow \quad 1 - [P(A) + P(B)] + P(A)P(B) = \frac{1}{3}$$

$$\Rightarrow \quad 1 + \frac{1}{6} - \frac{1}{3} = P(A) + P(B)$$

$$\Rightarrow \quad P(A) + P(B) = \frac{5}{6} \Rightarrow \ P(A) = \frac{1}{2}, P(B) = \frac{1}{3}$$

or $\quad P(A) = \dfrac{1}{3}, P(B) = \dfrac{1}{2}$

127. *(b)* In a box,

$$B_1 = 1R, 2W$$

$$B_2 = 2R, 3W \text{ and } B_3 = 3R, 4W$$

Also, given that,

$$P(B_1) = 1/2, P(B_2) = 1/3 \text{ and } P(B_3) = 1/6$$

$$\therefore P\left(\frac{B_2}{R}\right)$$

$$= \frac{P(B_2)\,P\left(\frac{R}{B_2}\right)}{P(B_1)\,P\left(\frac{R}{B_1}\right) + P(B_2)\,P\left(\frac{R}{B_2}\right) + P(B_3)\,P\left(\frac{R}{B_3}\right)}$$

$$= \frac{\frac{1}{3} \times \frac{2}{5}}{\frac{1}{2} \times \frac{1}{3} + \frac{1}{3} \times \frac{2}{5} + \frac{1}{6} \times \frac{3}{7}} = \frac{\frac{2}{15}}{\frac{1}{6} + \frac{2}{15} + \frac{1}{14}}$$

$$= \frac{2/15}{\frac{35 + 28 + 15}{210}} = \frac{2}{15} \times \frac{210}{78} = \frac{14}{39}$$

128. *(a)* Let $p(x_1, y_1)$ be *a* point such that the sum of the distance of P from two perpendicular lines $x + y = 0$ and $x - y = 0$ is 1.

$$\text{Then, } \left|\frac{x_1 + y_1}{\sqrt{2}}\right| + \left|\frac{x_1 - y_1}{\sqrt{2}}\right| = 1$$

$$\Rightarrow \quad \pm(x_1 + y_1) \pm (x_1 - y_1) = \sqrt{2}$$

Taking square from both sides, we get

$$(x_1 + y_1)^2 + (x_1 - y_1)^2 \pm 2(x_1 + y_1)(x_1 - y_1) = 2$$

$$\Rightarrow 2(x_1^2 + y_1^2) \pm (x_1^2 + y_1^2) \pm (x_1^2 - y_1^2) = 1$$

$$\Rightarrow 2x_1^2 = 1 \text{ or } 2y_1^2 = 1$$

$\therefore$ The locus of P is $(2x^2 - 1)(2y^2 - 1) = 0$

Which represents a rhombus.

129. *(b)* Since, the axes are rotated through an angle 45°, then we replace (x, y) by

$$(x \cos 45° - y \sin 45°, x \sin 45° + y \cos 45°)$$

$$ie, \quad \left(\frac{x}{\sqrt{2}} - \frac{y}{\sqrt{2}}, \frac{x}{\sqrt{2}} + \frac{y}{\sqrt{2}}\right)$$

in the given equation $3x^2 + 3y^2 + 2xy = 2$

$$\therefore \quad 3\left(\frac{x}{\sqrt{2}} - \frac{y}{\sqrt{2}}\right)^2 + 3\left(\frac{x+y}{\sqrt{2}}\right)^2 + 2\left(\frac{x-y}{\sqrt{2}}\right)\left(\frac{x+y}{\sqrt{2}}\right) = 2$$

$$\Rightarrow \quad \frac{3}{2}(x^2 + y^2 + 2xy) + \frac{3}{2}(x^2 + y^2 - 2xy) + \frac{2}{2}(x^2 - y^2) = 2$$

$$\Rightarrow \quad 4x^2 + 2y^2 = 2 \Rightarrow 2x^2 + y^2 = 1$$

130. ***(b)*** Since, l, m and n are in AP.

$\therefore \quad 2m = l + n$

Given equation of line is

$$lx + my + n = 0$$

Now, assume that the point (1, –2) satisfy the given equation.

$\therefore \quad l - 2m + n = 0$

$\Rightarrow \quad 2m = l + n$

$\Rightarrow \quad l, m, n$ are in AP.

Hence, option (b) is correct.

131. ***(a)*** To make the given curves $x^2 + y^2 = 4$ and $x + y = a$ homogeneous.

$$\therefore \quad x^2 + y^2 - 4\left(\frac{x+y}{a}\right)^2 = 0$$

$$\Rightarrow \quad a^2(x^2 + y^2) - 4(x^2 + y^2 + 2xy) = 0$$

$$\Rightarrow \quad x^2(a^2 - 4) + y^2(a^2 - 4) - 8xy = 0$$

Since, this is a perpendicular pair of straight lines.

$\therefore \quad a^2 - 4 + a^2 - 4 = 0$

$\Rightarrow \quad a^2 = 4 \Rightarrow a = \pm 2$

Hence, required set of a is $\{-2, 2\}$.

132. ***(c)*** From the figure,

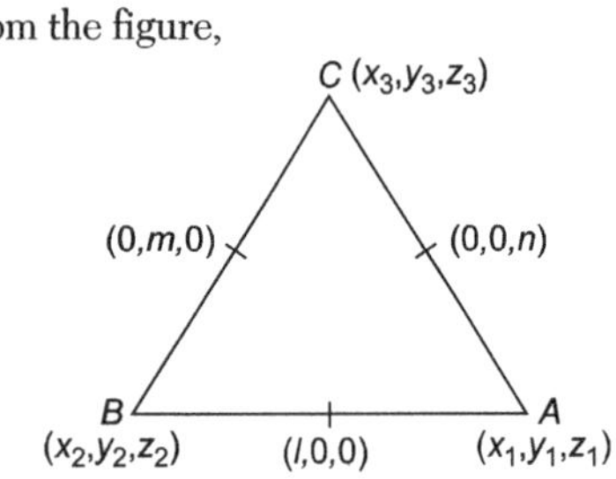

$$x_1 + x_2 = 2l,\ y_1 + y_2 = 0,\ z_1 + z_2 = 0,$$

$$x_2 + x_3 = 0,\ y_2 + y_3 = 2m,\ z_2 + z_3 = 0$$

and $\quad x_1 + x_3 = 0,\ y_1 + y_3 = 0,\ z_1 + z_3 = 2n$

On solving, we get

$$x_1 = l, x_2 = l, x_3 = -l,$$

$$y_1 = -m, y_2 = m, y_3 = m$$

and $\quad z_1 = n, z_2 = -n, z_3 = n$

$\therefore$ Coordinates are $A(l, -m, n)$, $B(l, m, -n)$ and $C(-l, m, n)$

$$\therefore \quad \frac{AB^2 + BC^2 + CA^2}{l^2 + m^2 + n^2}$$

$$= \frac{(4m^2 + 4n^2) + (4l^2 + 4n^2) + (4l^2 + 4m^2)}{l^2 + m^2 + n^2} = 8$$

133. ***(c)*** Since, the lines $2x - 3y = 5$ and $3x - 4y = 7$ are the diameters of a circle. Therefore, the point of intersection is the centre of the circle. On solving the given equations, we get $x = 1$ and $y = -1$ *ie*, the centre of the circle.

$\therefore$ Required equation of circle is

$$(x - 1)^2 + (y + 1)^2 = 7^2$$

$$\Rightarrow \quad x^2 + y^2 - 2x + 2y + 2 = 49$$

$$\Rightarrow \quad x^2 + y^2 - 2x + 2y - 47 = 0$$

134. ***(c)*** The equation of pole w.r.t. the point (1, 2) to the circle $x^2 + y^2 - 4x - 6y + 9 = 0$ is

$$x + 2y - 2(x + 1) - 3(y + 2) + 9 = 0$$

$$\Rightarrow \quad x + y - 1 = 0$$

Since, the inverse of the point (1, 2) is the foot (α, β) of the perpendicular from the point (1, 2) to the line $x + y - 1$.

$$\therefore \quad \frac{\alpha - 1}{1} = \frac{\beta - 2}{1} = -\frac{(1 \cdot 1 + 1 \cdot 2 - 1)}{1^2 + 1^2}$$

$$\Rightarrow \quad \alpha - 1 = \beta - 2 = -1$$

$$\Rightarrow \quad \alpha = 0, \beta = 1$$

Hence, required point is (0, 1).

135. ***(d)*** Using the condition that if two lines $l_1x + m_1y + n_1 = 0$ and $l_2x + m_2y + n_2 = 0$ are conjugate w.r.t. parabola $y^2 = 4ax$, then

$$l_1n_2 + l_2n_1 = 2\,am_1m_2 \quad \text{...(i)}$$

Given conjugate lines are $2x + 3y + 12 = 0$ and $x - y + 4\lambda = 0$ and equation of parabola is $y^2 = 8x$.

Here, $l_1 = 2, m_1 = 3, n_1 = 12$ and $l_2 = 1, m_2 = -1,$

$$n_2 = 4\lambda \quad \text{and} \quad a = 2$$

$\therefore$ From Eq. (i),

$$2 \times 4\lambda + 1 \times 12 = 2 \times 2 \times 3 \times (-1)$$

$$8\lambda = -12 - 12 \Rightarrow \lambda = -3$$

136. ***(c)*** Given, equation of hyperbola is

$$x^2 - 3y^2 - 4x - 6y - 11 = 0$$

$$\Rightarrow \quad (x^2 - 4x + 4) - 3(y^2 + 2y + 1) - 11 = 4 - 3$$

$$\Rightarrow \quad (x - 2)^2 - 3(y + 1)^2 = 12$$

$$\Rightarrow \quad \frac{(x-2)^2}{12} - \frac{(y+1)^2}{4} = 1$$

Now, $\quad e = \sqrt{1 + \frac{4}{12}} = \frac{2}{\sqrt{3}}$

$\therefore$ Distance between foci

$$= 2ae = 2 \times \sqrt{12} \times \frac{2}{\sqrt{3}} = 8$$

137. ***(b)*** Given polar equation of circle is

$$r^2 - 8r(\sqrt{3}\cos\theta + \sin\theta) + 15 = 0$$

or $\quad r^2 - 8(\sqrt{3}\, r\cos\theta + r\sin\theta) + 15 = 0$

where $r\cos\theta = x$ and $y = r\sin\theta$.

It can be rewritten in cartesian form as

$$x^2 + y^2 - 8(\sqrt{3}x + y) + 15 = 0$$

$$\Rightarrow \quad x^2 + y^2 - 8\sqrt{3}x - 8y + 15 = 0$$

Now, radius $= \sqrt{(4\sqrt{3})^2 + (4)^2 - 15}$

$$= \sqrt{48 + 16 - 15} = 7$$

138. *(c)* Given that,

$$f(x) = [x-3] + |x-4|$$

$$\therefore \lim_{x\to 3^-} f(x) = \lim_{x\to 3^-} ([x-3] + |x-4|)$$

$$= \lim_{h\to 0} ([3-h-3] + |3-h-4|)$$

$$= \lim_{h\to 0} ([-h] + 1 + h)$$

$$= -1 + 1 + 0 = 0$$

139. *(b)* Given that,

$$f(x) = \begin{cases} \dfrac{\cos 3x - \cos x}{x^2}, & \text{for } x \neq 0 \\ \lambda, & \text{for } x = 0 \end{cases}$$

Now, $\text{LHL} = \lim_{x\to 0^-} f(x)$

$$= \lim_{x\to 0^-} \frac{\cos 3x - \cos x}{x^2}$$

$$= \lim_{h\to 0} \frac{\cos 3(0-h) - \cos(0-h)}{(0-h)^2}$$

$$= \lim_{h\to 0} \frac{\cos 3h - \cos h}{h^2}$$

$$= \lim_{h\to 0} \frac{-3\sin 3h + \sin h}{2h}$$

(using L' Hospital's rule)

$$= \lim_{h\to 0} \frac{-9\cos 3h + \cos h}{2}$$

(using L' Hospital's rule)

$$= \frac{-9+1}{2} = -4$$

Since, $f(x)$ is continuous at $x = 0$

$$\therefore \lim_{x\to 0^-} f(x) = f(0)$$

$$\Rightarrow -4 = \lambda$$

$$\Rightarrow \lambda = -4$$

140. *(c)* Given that, $f(2) = 4$ and $f'(2) = 1$

$$\therefore \lim_{x\to 2} \frac{xf(2) - 2f(x)}{x-2}$$

$$= \lim_{x\to 2} \frac{xf(2) - 2f(2) + 2f(2) - 2f(x)}{x-2}$$

$$= \lim_{x\to 2} f(2) - 2\lim_{x\to 2} \frac{f(x) - f(2)}{x-2}$$

$$= f(2) - 2f'(2)$$

$$= 4 - 2(1) = 2$$

141. *(b)* Given that,

$x = a\left(\cos\theta + \log\tan\left(\dfrac{\theta}{2}\right)\right)$ and $y = a\sin\theta$

On differentiating w.r.t. θ, we get

$$\frac{dx}{d\theta} = a\left(-\sin\theta + \frac{1}{\tan\left(\frac{\theta}{2}\right)}\cdot\sec^2\frac{\theta}{2}\cdot\frac{1}{2}\right)$$

$$= a\left(-\sin\theta + \frac{1}{\sin\theta}\right) = \frac{a\cos^2\theta}{\sin\theta}$$

and $\dfrac{dy}{d\theta} = a\cos\theta$

$$\therefore \frac{dy}{dx} = \frac{dy/d\theta}{dx/d\theta} = \frac{a\cos\theta}{a\cos^2\theta/\sin\theta}$$

$$= \tan\theta$$

142. *(c)* Given curve is $y^4 = ax^3$

On differentiating w.r.t. x, we get

$$4y^3\frac{dy}{dx} = 3ax^2$$

$$\Rightarrow \left(\frac{dy}{dx}\right)_{(a,\,a)} = \frac{3a^3}{4a^3} = \frac{3}{4}$$

$\therefore$ Equation of normal at point (a, a) is

$$y - a = -\frac{4}{3}(x-a)$$

$$\Rightarrow 4x + 3y = 7a$$

143. *(b)* Given that,

$$2y^4 = x^5$$

On differentiating w.r.t. x, we get

$$8y^3\frac{dy}{dx} = 5x^4$$

$$\Rightarrow \left(\frac{dy}{dx}\right)_{(2,\,2)} = \frac{5(2)^4}{8(2)^3} = \frac{5}{4}$$

$\therefore$ Length of subtangent $= \dfrac{y}{dy/dx} = \dfrac{2}{5/4} = \dfrac{8}{5}$

144. *(c)* $\displaystyle\int e^x\left(\frac{1-\sin x}{1-\cos x}\right)dx$

$$= \int e^x\left(\frac{1 - 2\sin\frac{x}{2}\cos\frac{x}{2}}{2\sin^2\frac{x}{2}}\right)dx$$

$[\because \sin 2x = 2\sin x\cos x$ and $\cos 2x = 1 - 2\sin^2 x]$

$$= \frac{1}{2}\int \underset{\text{I}}{e^x}\left(\underset{\text{II}}{\text{cosec}^2\frac{x}{2}}\right)dx - \int e^x\cot\frac{x}{2}\,dx$$

$$= \frac{1}{2}\left[-e^x\cot\frac{x}{2}\cdot 2 + \int e^x\cot\frac{x}{2}\,2dx\right] - \int e^x\cot\left(\frac{x}{2}\right)dx + c$$

$$= -e^x\cot\left(\frac{x}{2}\right) + c$$

145. ***(d)*** Given that,

$$\int e^x (1+x) \cdot \sec^2 (xe^x)\, dx = f(x) + c$$

Put $xe^x = t$ in LHS, we get

$$\Rightarrow (e^x + xe^x)\, dx = dt$$

$$\therefore \quad \text{LHS} = \int \sec^2 t\, dt$$

$$= \tan t + c$$

$$\Rightarrow \tan (xe^x) + c = f(x) + c$$

$$\Rightarrow f(x) = \tan (xe^x)$$

146. ***(c)*** Let $I = \int_{-\pi/2}^{\pi/2} \sin |x|\, dx$

$$= 2 \int_0^{\pi/2} \sin x\, dx$$

$$= 2[-\cos x]_0^{\pi/2}$$

$$= 2$$

147. ***(b)*** Given, curve can be rewritten as

$$y^2 = 2\left(x + \frac{1}{2}\right)$$

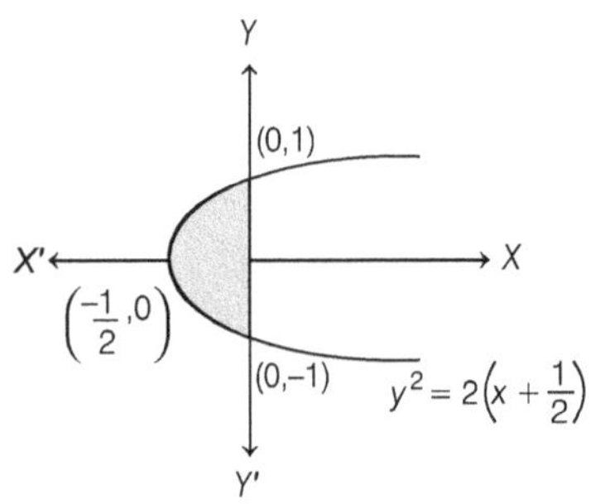

$$\therefore \text{Required area} = \int_{-1}^{1} x\, dy$$

$$= 2\int_0^1 \frac{y^2 - 1}{2}\, dy$$

$$= \left| \left[\frac{y^3}{3} - y \right]_0^1 \right|$$

$$= \left| \left[\frac{1}{3} - 1 \right] \right| = \frac{2}{3} \text{ sq unit}$$

148. ***(d)*** Given differential equation is

$$\frac{dy}{dx} = \frac{xy + y}{xy + x}$$

$$\Rightarrow \frac{dy}{dx} = \frac{y(1+x)}{x(1+y)}$$

$$\Rightarrow \frac{(1+y)}{y}\, dy = \frac{(1+x)}{x}\, dx$$

$$\Rightarrow \int \left(\frac{1}{y} + 1\right) dy = \int \left(\frac{1}{x} + 1\right) dx$$

$$\Rightarrow \log y + y = \log x + x + \log c$$

$$\Rightarrow y - x = \log\left(\frac{cx}{y}\right)$$

149. ***(b)*** Given linear differential equation is

$$\frac{dy}{dx} - y \tan x = e^x \sec x$$

$$\therefore \quad \text{IF} = e^{\int -\tan x\, dx} = e^{-\log \sec x}$$

$$= \frac{1}{\sec x}$$

$\therefore$ Solution is

$$y \cdot \frac{1}{\sec x} = \int e^x \sec x \cdot \frac{1}{\sec x}\, dx$$

$$\Rightarrow \frac{y}{\sec x} = e^x + c$$

$$\Rightarrow y \cos x = e^x + c$$

150. ***(b)*** Given differential equation can be rewritten as

$$\frac{dy}{dx} = \frac{x^3 + y^3}{xy^2}$$

It is a homogeneous differential equation.

Put $y = vx \Rightarrow \dfrac{dy}{dx} = v + x\dfrac{dv}{dx}$

$$\therefore \quad x\frac{dv}{dx} + v = \frac{x^3 + v^3x^3}{x^3v^2}$$

$$\Rightarrow x\frac{dv}{dx} + v = \frac{1 + v^3}{v^2}$$

$$\Rightarrow x\frac{dv}{dx} = \frac{1}{v^2} \Rightarrow v^2\, dv = \frac{dx}{x}$$

On integrating both sides, we get

$$\frac{v^3}{3} = \log x + \log c$$

$$\Rightarrow \frac{1}{3}\left(\frac{y}{x}\right)^3 = \log x + \log c$$

$$\Rightarrow y^3 = 3x^3 \log cx$$

BITSAT

Birla Institute of Technology & Science Admission Test

Mock Test 2

Instructions

- There are 150 questions in all. The number of questions in each part is as given below.

	No. of Questions
Part I Physics	1-40
Part II Chemistry	41-80
Part III a. English Proficiency	81-95
b. Logical Reasoning	96-105
Part IV Mathematics	106-150

- All questions are Multiple Choice Questions having four options out of which **only one** is correct.
- Each correct answer fetches 3 marks while incorrect answer has a penalty of 1 mark.
- Time allotted to complete this paper is 3 hrs.

PART I

Physics

1. The dimensions of the quantity $\hbar c$ $\left(\text{where } \hbar = \dfrac{h}{2\pi}\right)$ is

a. $[ML^2 T^{-1}]$ *b.* $[MLT^{-1}]$

c. $[ML^3T^{-2}]$ *d.* $[ML^2 T^{-1}]$

2. One second after the projection, a stone moves at an angle of 45° with the horizon. Two second from the start, it is travelling horizontally, its angle of projection with the horizon is (given, $g = 10\ ms^{-2}$)

a. 60° *b.* $\tan^{-1}(4)$

c. $\tan^{-1}(3)$ *d.* $\tan^{-1}(2)$

3. A sphere of mass m moving with a constant velocity u hits another stationary sphere of the same mass. If e is the coefficient of restitution, then the ratio of the velocity of two spheres after collision will be

a. $\dfrac{1-e}{1+e}$ *b.* $\dfrac{1+e}{1-e}$ *c.* $\dfrac{e+1}{e-1}$ *d.* $\dfrac{e-1}{e+1}t^2$

4. When the mass is rotated in a plane about a fixed point its angular momentum is directed along the

a. radius

b. tangent to orbit

c. axis of rotation

d. line at an angle of 45° to plane of rotation

5. If the radius of the earth were to shrink by one percent, its mass remaining the same, the value of g on the earth's surface would

a. increase by 0.5%

b. increase by 2%

c. decrease by 0.5%

d. decrease by 2%

6. Density of sea water is $1.03\ gcc^{-1}$. A ship passes from fresh water into sea water. It will

a. rise

b. sink

c. remain at the same depth

d. rise or sink depending on its shape and size

7. Water is flowing through a tube of non-uniform cross-section. If the radii of the tube at the entrance and the exit are in the ratio 3 : 2, then the ratio of the velocities of flow of water at the entrance and the exit is

a. 9 : 4 ***b.*** 4 : 9 ***c.*** 8 : 27 ***d.*** 27 : 8

8. The equation $\frac{d^2y}{dt^2}+\frac{bdy}{dt}+\omega^2 y=0$ represents the equation of motion for a

a. free vibration ***b.*** damped vibration
c. forced vibration ***d.*** resonant vibration

9. A star emits light of wavelength λ and it is receding from the earth with a speed v_s. The shift in the wavelength of the spectral line observed on the earth is

a. $\lambda\frac{v_s^2}{c^2}$ ***b.*** $-\lambda\frac{v_s^2}{c^2}$ ***c.*** $-\lambda\frac{v_s}{c}$ ***d.*** $\lambda\frac{v_s}{c}$

10. The end *A* of rod *AB* of length 1 m is maintained at 80°C and the end *B* at 0°C. The temperature at a distance of 60 cm from the end *A* is

a. 16°C ***b.*** 32°C ***c.*** 48°C ***d.*** 64°C

11. Two bulbs which consume powers P_1 and P_2 are connected in series. The power consumed by the combination is

a. P_1+P_2 ***b.*** $\sqrt{P_1P_2}$
c. $\frac{P_1P_2}{P_1+P_2}$ ***d.*** $\frac{2P_1P_2}{P_1+P_2}$

12. A current carrying circular coil is bent, so as to convert it into a double loop, both the loops being concentric and are carrying current in the same direction. If *B* is the initial magnetic field at the centre, then the final magnetic field at the centre will be

a. Zero ***b.*** *B* ***c.*** 2*B* ***d.*** 4*B*

13. Ions of different momenta (*p*), having the same charge, enter normally a uniform magnetic field. The radius of the orbit of an ion is proportional to

a. p ***b.*** $\frac{1}{p}$ ***c.*** p^2 ***d.*** $\frac{1}{p^2}$

14. The impedance of a circuit consists of 3 Ω resistance and 4 Ω reactance. The power factor of the circuit is

a. 0.4 ***b.*** 0.6 ***c.*** 0.8 ***d.*** 1.0

15. Of the following, the one which has the largest de-Broglie wavelength for the same speed is

a. Electron ***b.*** Proton
c. α-particle ***d.*** Oxygen atom

16. A block of mass 5 kg is placed on a rough inclined plane. The inclination of the plane is gradually increased till the block just begins to slide down. The inclination of the plane is than 3 in 5. The coefficient of friction between the block and the plane is ($g=10\text{ ms}^{-2}$)

a. $\frac{3}{5}$ ***b.*** $\frac{3}{4}$ ***c.*** $\frac{4}{5}$ ***d.*** $\frac{2}{3}$

17. A body covers a total distance of 3*S*. The first *S* is covered with a velocity *u*, the second *S* with *v* and the last *S* with ω. Then, the average velocity during the whole journey is

a. $\frac{u+v+\omega}{3}$ ***b.*** $\frac{3uv\omega}{u+v+\omega}$
c. $\frac{3uv\omega}{uv+v\omega+u\omega}$ ***d.*** Zero

18. A block weighing *w* is held against a vertical wall by pressing horizontally with a force *F*. Then, *F* needed to hold the block is

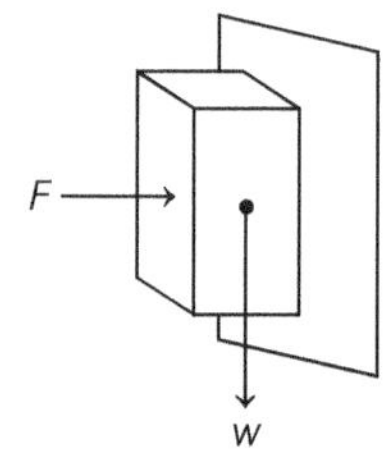

a. $=w$ ***b.*** $>\frac{w}{\mu}$ ***c.*** $<\frac{w}{\mu}$ ***d.*** $=\frac{w}{\mu}$

19. A solid cylinder of mass 2 kg rolls down (pure rolling) an inclined plane from a height of 4 m. Its rotational kinetic energy when it reaches the foot of the plane is ($g=10\text{ ms}^{-2}$)

a. 20 J ***b.*** 40 J ***c.*** $\frac{80}{3}$ J ***d.*** 80 J

20. The mean radius of the earth's orbit around the sun is 1.5×10^{11} m and that of the orbit of mercury is 6×10^{10} m. The mercury will revolve around the sun in nearly

a. $\sqrt{\frac{2}{5}}$ yr ***b.*** $\frac{2}{5}$ yr
c. $\left(\frac{2}{5}\right)^2$ yr ***d.*** $\left(\frac{2}{5}\right)^{3/2}$ yr

21. A liquid *X* of density 3.36 g/cm^3 is poured in a *U*-tube, which contains Hg. Another liquid *Y* is poured in left arm with height 8 cm, upper levels of *X* and *Y* are same. What is density of *Y*?

a. 0.8 g/cc ***b.*** 1.2 g/cc
c. 1.4 g/cc ***d.*** 1.6 g/cc

22. Three conductors draw respectively currents of 1 A, 2 A and 4 A when connected in turn across a battery. If they are connected in series across the same battery the current drawn will be

a. $\frac{2}{7}$ A *b.* $\frac{3}{7}$ A *c.* $\frac{4}{7}$ A *d.* $\frac{5}{7}$ A

23. One junction of a thermocouple is at a particular temperature T_r and another is at T. Its thermo emf is expressed as $E = k(T - T_r)\left\{T_0 - \frac{1}{2}(T + T_r)\right\}$. At temperature $T = \frac{T_0}{2}$, the value of thermo electric power will be

a. $\frac{1}{2}kT_0$ *b.* kT_0

c. $\frac{1}{2}kT_0^2$ *d.* $\frac{1}{2}k(T_0 - T_r)^2$

24. A convex lens makes a real image 4 cm long on a screen. When the lens is shifted to new position without disturbing the object, we again get a real image on the screen which is 16 cm tall. The length of the object must be

a. $\frac{1}{4}$ cm *b.* 8 cm *c.* 12 cm *d.* 20 cm

25. In germanium the energy gap is about 0.75 eV. The wavelength of light which germanium starts absorbing is

a. 5000 Å *b.* 1650 Å

c. 16500 Å *d.* 165000 Å

26. The transmitting antenna of a radio station is mounted vertically. At a point 10 km due north of the transmitter the peak electric field is 10^{-3} Vm^{-1}. The magnitude of the radiated magnetic field is

a. 3.33×10^{-10} T *b.* 3.33×10^{-12} T

c. 10^{-3} T *d.* 3×10^{5} T

27. In order to increase the current sensitivity of a moving coil galvanometer the

a. number of turns of the coil should be increased

b. strength of the magnetic field should be increased

c. area of the coil should be increased

d. All of the above

28. In a 10 m deep lake, the bottom is at a constant temperature of 4°C. The air temperature is constant at – 4°C. The thermal conductivity of ice is 3 times that of water. Neglecting the expansion of water on freezing, the maximum thickness of ice will be

a. 7.5 m *b.* 6 m *c.* 5 m *d.* 2.5 m

29. The ratio of the adiabatic bulk modulus to the isothermal bulk modulus of a perfect gas with f degrees of freedom is

a. $\frac{2}{f}$ *b.* $1 + \frac{1}{f}$

c. $1 + \frac{2}{f}$ *d.* $1 + \frac{(f-1)}{4}$

30. A pole is floating in a liquid with 80 cm of its length immersed. It is pushed down a certain distance and then released. Time period of vertical oscillation is

a. $\frac{4\pi}{7}$ s *b.* $\frac{3\pi}{7}$ s *c.* $\frac{2\pi}{7}$ s *d.* $\frac{\pi}{7}$ s

31. The radius of a wheel is R and its radius of gyration about its axis passing through its centre and perpendicular to its plane is K. If the wheel is rolling without slipping, the ratio of its rotational kinetic energy to its translational kinetic energy is

a. $\frac{K^2}{R^2}$ *b.* $\frac{R^2}{K^2}$

c. $\frac{R^2}{R^2 + K^2}$ *d.* $\frac{K^2}{R^2 + K^2}$

32. An inclined track ends in a circular loop of radius r. From what height on the track a particle should be released so that it completes the loop, assuming there is no friction?

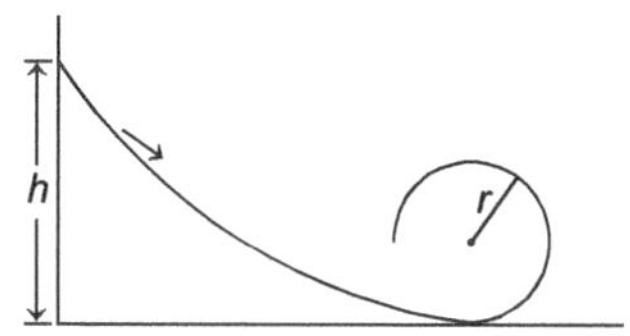

a. $\frac{r}{2}$ *b.* $\frac{3r}{2}$ *c.* $2r$ *d.* $\frac{5r}{2}$

33. Two loudspeakers L_1 and L_2, driven by a common oscillator and amplifier, are arranged as shown. The frequency of the oscillator is gradually increased from zero and the detector at D records a series of maxima and minima. If the speed of sound is 330 ms^{-1}, then the frequency at which the first maximum is observed is

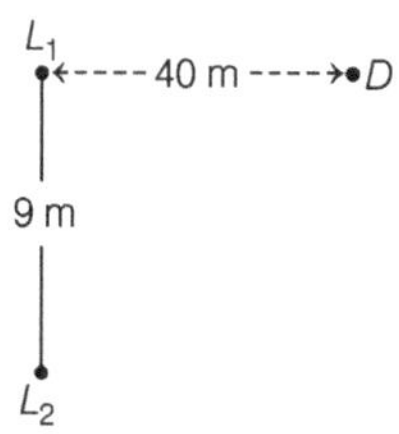

a. 165 Hz *b.* 330 Hz

c. 495 Hz *d.* 660 Hz

34. A stretched string of length 2 m vibrates in 4 segments. The distance between consecutive nodes is

a. 0.5 m *b.* 0.25 m
c. 1.0 m *d.* 0.75 m

35. A body of mass 2 kg is placed on a horizontal surface having coefficient of kinetic friction 0.4 and limiting coefficient of static friction 0.5. If a horizontal force 2.5 N is applied on the body, the frictional force acting on the body will be $(g = 10\ \text{ms}^{-2})$

a. 8 N *b.* 10 N
c. 20 N *d.* 2.5 N

36. The time taken by the current to rise to 0.63 of its maximum value in a DC-circuit containing inductance L and resistance R depends on

a. Only L *b.* Only R
c. $\frac{L}{R}$ *d.* LR

37. 24 identical cells, each of internal resistance 0.5 Ω, are arranged in a parallel combination of n rows, each row containing m cells in series. The combination is connected across a resistor of 3 Ω. In order to send maximum current through the resistor, we should have

a. $m = 12, n = 2$ *b.* $m = 8, n = 3$
c. $m = 2, n = 12$ *d.* $m = 3, n = 8$

38. If v is the molecular speed and l is the mean free path of molecule of a gas, then the collision frequency is

a. vl *b.* v/l *c.* l/v *d.* $\sqrt{vl}$

39. At $t = 0$, a transverse wave pulse in a wire is described by the function $y = \frac{6}{x^2 - 3}$, where x and y are in metre. The function $y(x, t)$ that describes this wave equation, if it is travelling in the positive x direction with a speed of $4.5\ \text{ms}^{-1}$ is

a. $y = \frac{6}{(x + 4.5\,t)^3 - 3}$ *b.* $y = \frac{6}{(x - 4.5\,t^2) + 3}$
c. $y = \frac{6}{(x + 4.5\,t)^2 - 3}$ *d.* $y = \frac{6}{(x - 4.5\,t)^2 - 3}$

40. A playground merry-go-round is at rest, pivoted about a frictionless axis. A child of mass m runs along the path tangential to the rim with speed v and jumps on to merry-go-round. If R be the radius of merry-go-round and I is the moment of inertia, then the angular velocity of the merry-go-round and the child is

a. $\frac{mvR}{mR^2 + I}$ *b.* $\frac{mvR}{I}$
c. $\frac{mR^2 + I}{mvR}$ *d.* $\frac{I}{mvR}$

PART II

Chemistry

41. The most strained cycloalkane is

a. cyclopropane *b.* cyclobutane
c. cyclopentane *d.* cyclohexane

42. Which of the following is an example of homogeneous catalysis ?

a. Haber process for the synthesis of NH_3.
b. Catalytic conversion of SO_2 to SO_3 in contact process.
c. Catalytic hydrogenation of oils.
d. Acid hydrolysis of methyl acetate.

43. A current of 0.5 A, when passed through $AgNO_3$ solution for 193s deposited 0.108 g Ag. The equivalent weight of silver is

a. 108 *b.* 54
c. 10.8 *d.* 5.4

44. van-Arkel method is based on

a. Cupellation method
b. Furnace refining method
c. Poling method
d. None of the above

45. In CaF_2, coordination number of Ca^{2+} ions is 8. The coordination number of F^- ion is

a. 2 *b.* 3 *c.* 4 *d.* 8

46. How many C—C bonds are present in $(CH_2)_6N_4$?

a. 0 *b.* 4
c. 6 *d.* 12

47. For the dissociation reaction,

$$H_2(g) \rightleftharpoons 2H(g); \quad \Delta H = 162\ \text{kcal};$$

heat of atomisation of H is

a. 81 kcal *b.* 162 kcal
c. 208 kcal *d.* 218 kcal

48. The highest bond strength is shown by

a. O—O bond *b.* S—S bond
c. Se—Se bond *d.* Te—Te bond

49. NaOH and CO react to form

a. HCHO *b.* HCOONa
c. CH_3CHO *d.* Na and CO_2

50. Oil paintings turn blackish after sometimes. The salt formed is

a. SnS *b.* CuS
c. PbS *d.* CdS

51. The compound which reacts with both HCl and NaOH is

a. CaO *b.* CO_2
c. ZnO *d.* N_2O_5

52. Lysergic acid diethylamide (LSD) and mescalin are

a. anti-malarials *b.* psychedlic drugs
c. hypnotics *d.* anaesthetics

53. Phenol can be tested by

a. Libermann nitroso test *b.* $FeCl_3$ solution
c. bromine water *d.* All of these

54. The bond dissociation enthalpies of C—H bond in $\equiv\underset{sp}{CH}$, $=\underset{sp^2}{CH}$ and $-\underset{sp^3}{CH}$ follows the order ?

a. $sp > sp^2 > sp^3$
b. $sp^3 > sp^2 > sp$
c. $sp^2 > sp > sp^3$
d. $sp > sp^3 > sp^2$

55. CH_3—CH—CH_2—CH_3 is

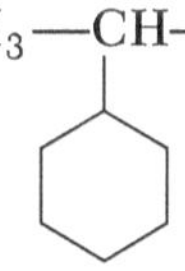

a. 2-phenylbutane *b.* 3-phenylbutane
c. 3-cyclohexylbutane *d.* 2-cyclohexylbutane

56. HCl does not behave as acid in

a. NH_3 *b.* H_2O
c. C_6H_6 *d.* C_2H_5OH

57. What is the energy of activation of a reaction if its rate doubles when the temperature is raised from 290 K to 300K ?

a. 12 kcal *b.* 15 kcal
c. 10 kcal *d.* 20 kcal

58. The active mass of solid is generally taken as

a. >1 *b.* $=1$
c. <1 *d.* 0

59. Aluminium is present in

a. gypsum *b.* carnallite
c. asbestos *d.* diaspore

60. The wave number of a spectral line is 5×10^5 m^{-1}. The energy corresponding to this line is

a. 3.49×10^{-23} kJ *b.* 4.45×10^{-24} J
c. 5.50×10^{-22} J *d.* 9.93×10^{-23} kJ

61. 1g of hydrogen is found to combine with 80 g of bromine. 1 g of calcium combines with 4 g of bromine. Eq. wt. of calcium is

a. 16 *b.* 20 *c.* 40 *d.* 80

62. Acetic acid on warming with hydrazoic acid in presence of conc. H_2SO_4 gives

a. CH_3NH_2 *b.* $CH_3CH_2NH_2$
c. CH_3CONH_2 *d.* CH_3COONH_4

63. The conversion of sucrose into CO_2 is

a. oxidation *b.* reduction
c. Both (a) and (b) *d.* None of these

64. Anhydrone, a drying agent is

a. $Mg(ClO_4)_2$ *b.* $Sr(ClO_4)_2$
c. $Ca(ClO_4)_2$ *d.* $Ba(ClO_4)_2$

65. $Mg + NO_2$ on burning give

a. N_2 *b.* NO *c.* N_2O *d.* N_2O_5

66. Which one of the following is a pseudohalogen ?

a. I_3^- *b.* IF_5 *c.* ICl *d.* CN^-

67. Compound used in petro industry is

a. $(C_2H_5)_4Pb$ *b.* $PbCl_2\cdot PbO$
c. Both (a) and (b) *d.* Neither (a) nor (b)

68. If dissociation constant of ammonia is 10^{-5}, its pK_b and pK_a values are respectively

a. 5,9 *b.* 9, 5 *c.* 7, 7 *d.* 4, 10

69. The molal freezing point for water is 1.86°C mol^{-1}. If 342 g of cane sugar is dissolved in 1000 mL of water, the solution will freeze at

a. 1.86°C *b.* -1.86°C *c.* 2.42°C *d.* -2.42°C

70. If the mean free path is l at 1 atm pressure then its value at 5 atm pressure is

a. $\frac{1}{25}l$ *b.* $\frac{1}{5}l$
c. $5\,l$ *d.* Cannot be calculated

71. A sample of radioactive substance shows an intensity of 2.3 milli Ci at a time t and an intensity of 1.62 milli Ci after 600 s.
The half-life period of the radioactive metal is

a. 1000 s *b.* 1187 s *c.* 1200 s *d.* 1500 s

72. Chain isomers of C_6H_{14} will be

a. 4 *b.* 5 *c.* 6 *d.* 7

73. Most stable carbonium ion is

(a) 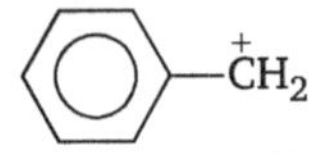

(b) 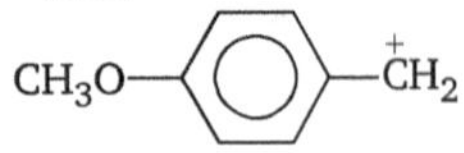

(c) 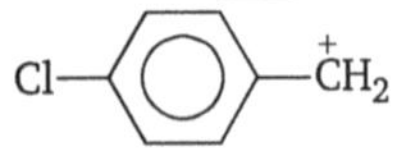

(d) 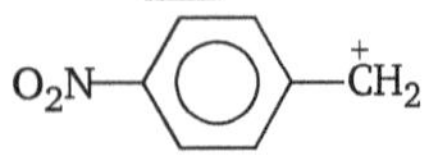

74. End product of reaction

$$H—C\equiv CH \xrightarrow[\text{excess}]{Na/NH_2} \xrightarrow[\text{excess}]{CH_3I} \xrightarrow[\Delta]{NaNH_2\text{-paraffin}}; \text{ is}$$

a. CH_3CHO *b.* $CH_3—CH=CH_2$
c. $CH_3—C\equiv C—CH_3$ *d.* $CH_3—CH_2—C\equiv CNa$

75. Correct statement(s) is are

a. Keratin is fibrous protein
b. Androsterone is male sex hormone
c. Vitamin B_1 is antineutritic factor
d. All of the above

76. $[H^+]$ of 0.2 N CH_3COOH which is 4% dissociated, is

a. 0.08 M *b.* 0.12 M
c. 0.008 M *d.* 0.8 M

77. Choose the correct match.

a. $H_2 + Cl_2 \xrightarrow{MnO_2} 2HCl$

b. $N_2 + 3H_2O \xrightarrow{V_2O_5} 2NH_3$

c. $4NH_3 + 5O_2 \xrightarrow{Pt} 4NO + 6H_2O$

d. $2SO_2 + O_2 \xrightarrow{Fe_2CO_3} 2SO_3$

78. How much water should be added to 200 mL of semi normal solution of NaOH to make it exactly decinormal?

a. 200 mL *b.* 800 mL
c. 1000 mL *d.* 11200 mL

79. Which cannot be stored in glass vessel?

a. XeF_2 *b.* XeF_4
c. XeF_6 *d.* All of these

80. 3-methyl-2-butanol on treatment with HCl gives (major product)

a. 2-chloro-2-methyl butane
b. 2-chloro-3-methyl butane
c. 2, 2-dimethyl pentane
d. None of the above

PART III

a. English Proficiency

Directions (Q. Nos. 81-83) *In the following questions, choose the alternative which can most appropriately replace the group of words italicised in the sentences.*

81. Despite being in the career of singing for the last 10 years, he has not been able to earn fame on account of his *practice of borrowing ideas and words from others and using them as his own.*

a. Adaptation *b.* Pantomime
c. Imitation *d.* Plagiarism

82. Every person is not allowed to enter *the place where public, government or historical records are kept*

a. Scullery *b.* Pantry *c.* Archives *d.* Coffer

83. The advertisement assured the public that the medicine would *give back* to the users, *their youthful vigour and appearance.*

a. Rejuvenate *b.* Restore
c. Replenish *d.* Render

Directions (Q. Nos. 84-87) *Rearrange the sentences labelled* A *to* E *to form a coherent paragraph and then answer the questions.*

(A) The large crowd, these myriad activities and the bright lights combine to transform the beach into a photographers' delight.
(B) Sand and spray are all pervasive on a beach.
(C) They love to construct fragile sand castles and splash about in the shallow waters.
(D) Even the elderly seem to forget their ageing worries here.
(E) These make beaches a favourite haunt of children.

84. Which sentence should come first in the paragraph?

a. A *b.* B
c. C *d.* D

85. Which sentence should come second in the paragraph?

a. D *b.* E
c. A *d.* B

86. Which sentence should come last in the paragraph?

a. D *b.* A *c.* E *d.* C

87. Which sentence should come third in the paragraph?

a. C *b.* D
c. E *d.* A

Directions (Q. Nos. 88-90) *Choose the alternative which is most similar in meaning to the word given in capital letters.*

88. PARAMOUR

a. Lover *b.* Companion
c. Friend *d.* Rival

89. REFECTORY

a. Dining Room *b.* Parlour
c. Living Room *d.* Restaurant

90. ASSENT

a. Compromise *b.* Judgement
c. Agreement *d.* Expression

Directions (Q. Nos. 91-93) *Choose the most appropriate alternative to fill in the blank.*

91. We must ...... our students on subjects like health and sanitation besides the usual subjects

a. learn *b.* teach
c. insist *d.* educate

92. The judge used his ...... power and let him off with a reprimand

a. residuary
b. official
c. legal
d. discretionary

93. He applied for and was ...... legal aid by the Labour Ministry

a. offered
b. granted
c. allowed
d. awarded

Directions (Q. Nos. 94-95) *Choose the alternative which best expresses the meaning of the given idiom/phrase.*

94. To show one's teeth

a. To ridicule
b. To face difficulties
c. To adopt a threatening attitude
d. To be humble

95. To pour oil in troubled water

a. To foment trouble
b. To add to the trouble
c. To instigate
d. To calm a quarrel with soothing words

b. Logical Reasoning

96. In the question given below choose the word which is least like the other words in the group.

a. Yen *b.* Lira *c.* Dollar *d.* Ounce

97. In the question given below choose the word which is least like the other words in the group.

a. Litchi *b.* Papaya
c. Jackfruit *d.* Guava

Directions (Q. Nos. 98-99) *In each of the following questions a part of the figure is missing. Find out from the given options* (a), (b), (c) *and* (d), *the right figure to fit in the missing figure.*

98.

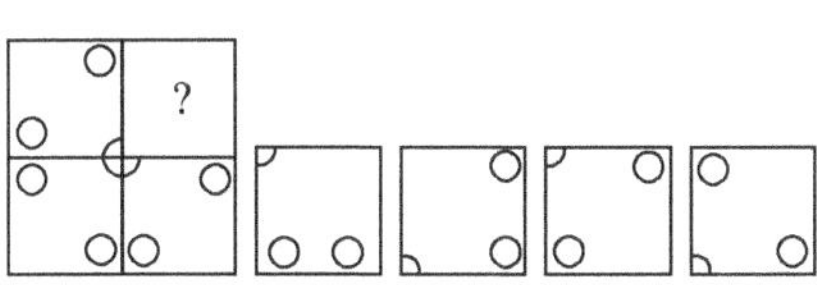

a. *b.* *c.* *d.*

99.

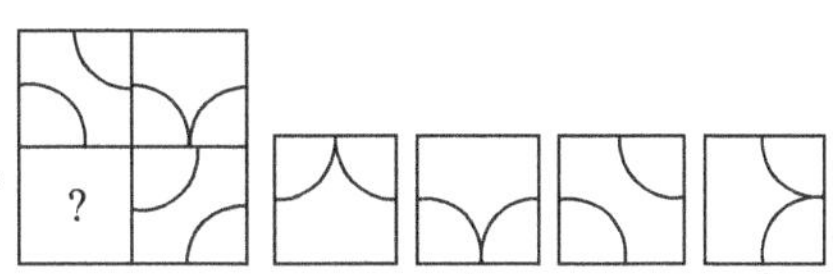

a. *b.* *c.* *d.*

Directions (Q. Nos. 100-101) *In each of the following questions a series is given with one number missing. Choose the correct alternative from the given ones that will complete the series.*

100. 4, 7, 13, ?, 34, 49

a. 20 *b.* 22
c. 24 *d.* 21

101. 5, 11, 24, ?, 106, 217

a. 52 *b.* 50
c. 49 *d.* 51

Directions (Q. Nos. 102-103) *Find the missing number in these questions from the given alternatives.*

102.

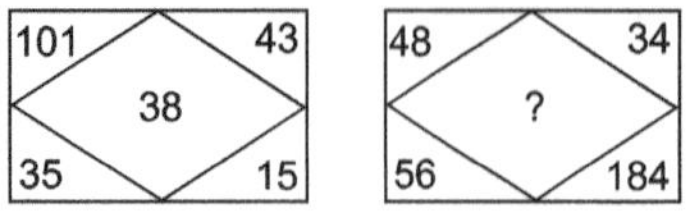

a. 127 *b.* 142 *c.* 158 *d.* 198

103.

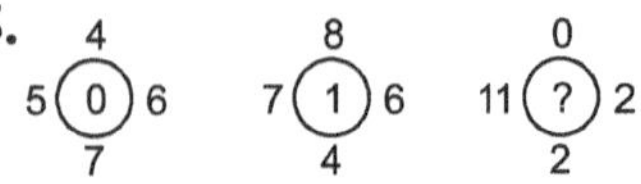

a. 0 *b.* 2 *c.* 11 *d.* 12

Directions (Q. Nos. 104-105) *In these questions a set of three figures* (i), (ii) *and* (iii) *are given that show a sequence of folding a piece of paper. Figure* (iii) *shows the manner in which the folded paper has been cut. Choose the correct alternative which most closely resemble the unfolded form of Figure* (iii).

104.

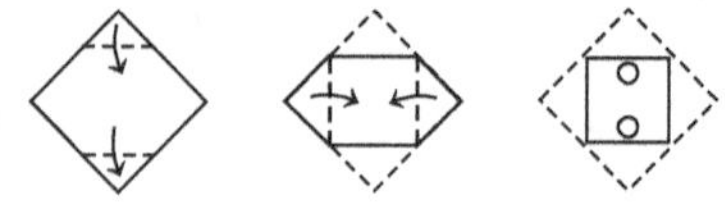

(i) (ii) (iii)

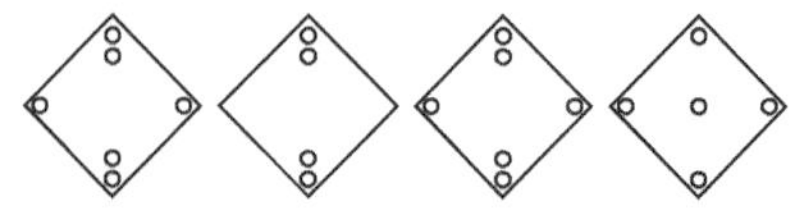

(a) (b) (c) (d)

105.

(i) (ii) (iii)

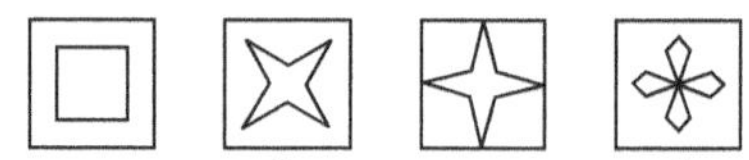

a. *b.* *c.* *d.*

PART IV

Mathematics

106. Let $f : R \to R, g : R \to R$ be two functions given by $f(x) = 2x - 3, g(x) = x^3 + 5$. Then, $(fog)^{-1}$ is equal to

a. $\left(\frac{x+7}{2}\right)^{1/3}$ *b.* $\left(x - \frac{7}{2}\right)^{1/3}$

c. $\left(\frac{x-2}{7}\right)^{1/3}$ *d.* $\left(\frac{x-7}{2}\right)^{1/3}$

107. Inverse of function $f(x) = \frac{10^x - 10^{-x}}{10^x + 10^{-x}}$ is

a. $\log_{10}(2 - x)$ *b.* $\frac{1}{2}\log_{10}\left(\frac{1+x}{1-x}\right)$

c. $\frac{1}{2}\log_{10}(2x - 1)$ *d.* $\frac{1}{4}\log_{10}\left(\frac{2x}{2-x}\right)$

108. Let $a_1, a_2, \ldots, a_{10}$ be in AP and $h_1, h_2, \ldots, h_{10}$ be in HP. If $a_1 = h_1 = 2$ and $a_{10} = h_{10} = 3$. Then, $a_4 h_7$ is

a. 2 *b.* 3 *c.* 5 *d.* 6

109. If $x^2 - 2x\cos\theta + 1 = 0$, then
$x^{2n} - 2x^n \cos n\theta + 1$ is equal to

a. $\cos 2n\theta$ *b.* $\sin 2n\theta$

c. 0 *d.* $R - \{0\}$

110. If the roots of the equation $ax^2 + bx + c = 0$ are real and distinct, then

a. both roots are greater than $-\frac{b}{2a}$

b. both roots are less than $-\frac{b}{2a}$

c. one of the roots exceeds $-\frac{b}{2a}$

d. None of the above

111. The number of the solutions of the equation $z^2 + \bar{z} = 0$ is

a. 1 *b.* 2

c. 3 *d.* 4

112. Find the least value of n for which $\left(\frac{1+i}{1-i}\right)^n = 1$

a. 4 *b.* 3

c. -4 *d.* 1

113. Rank of the matrix $A = \begin{bmatrix} 1 & 2 & 2 \\ 2 & 1 & 2 \\ 2 & 2 & 1 \end{bmatrix}$ is

a. 0 *b.* 1

c. 2 *d.* 3

114. If ω is a cube root of unity, then $\begin{vmatrix} 1 & \omega & \omega^2 \\ \omega & \omega^2 & 1 \\ \omega^2 & 1 & \omega \end{vmatrix}$ is equal to

a. 1 *b.* ω *c.* ω^2 *d.* 0

115. For the equations $x + 2y + 3z = 1$, $2x + y + 3z = 2$ and $5x + 5y + 9z = 4$

a. there is only one solution
b. there exists infinitely many solution
c. there is no solution
d. None of the above

116. The coefficient of x^5 in the expansion of $(2 - x + 3x^2)^6$ is

a. -4692 *b.* 4692 *c.* 2346 *d.* -5052

117. If $C_1, C_2, C_3, \ldots, C_n$ denote the coefficients in the binomial expansion of $(1 + x)^n$, then the value of $C_1 + 2C_2 + 3C_3 + \ldots + n^n c_n$ is

a. $(n-1)\,2^n$ *b.* $n \cdot 2^{n-1}$
c. $n \cdot 2^{n+1}$ *d.* None of these

118. The value of
$\frac{1}{3} + \frac{1}{3 \cdot 3^3} + \frac{1}{5 \cdot 3^5} + \frac{1}{7 \cdot 3^7} + \ldots + \infty$ is

a. $\frac{1}{2}\log_e 2$ *b.* $\log_e 2$
c. $\log_e 3$ *d.* None of these

119. Larger of $99^{50} + 100^{50}$ and 101^{50} is

a. 101^{50} *b.* $99^{50} + 100^{50}$
c. both are equal *d.* None of these

120. The number of positive integer satisfying the inequality ${}^{n+1}C_{n-2} - {}^{n+1}C_{n-1} \le 100$ is

a. 9 *b.* 8
c. 5 *d.* None of these

121. If $\frac{1+3p}{3}, \frac{1-p}{4}$ and $\frac{1-2p}{2}$ are mutually exclusive events. Then, range of p is

a. $\frac{1}{3} \le p \le \frac{1}{2}$ *b.* $\frac{1}{4} \le p \le \frac{1}{2}$
c. $\frac{1}{3} \le p \le \frac{2}{3}$ *d.* $\frac{1}{3} \le p \le \frac{2}{5}$

122. If A and B are two events such that $P(A) = \frac{3}{4}$ and $P(B) = \frac{5}{8}$, then

a. $P(A \cup B) \ge 3/4$ *b.* $P(A' \cap B) \le 1/4$
c. $\frac{3}{8} \le P(A \cap B) \le \frac{5}{8}$ *d.* All of these

123. The minimum value of $9\tan^2\theta + 4\cot^2\theta$ is

a. 13 *b.* 9 *c.* 6 *d.* 12

124. $\left(1 + \cos\frac{\pi}{8}\right)\left(1 + \cos\frac{3\pi}{8}\right)\left(1 + \cos\frac{5\pi}{8}\right)\left(1 + \cos\frac{7\pi}{8}\right)$ is equal to

a. $\frac{1}{2}$ *b.* $\frac{1}{8}$
c. $\cos\frac{\pi}{8}$ *d.* $\frac{1+\sqrt{2}}{2\sqrt{2}}$

125. The solution of the equation $\sec\theta - \operatorname{cosec}\theta = \frac{4}{3}$ is

a. $\frac{1}{2}[n\pi + (-1)^n \sin^{-1} 3/4]$
b. $\frac{n\pi}{2} + (-1)^n \sin^{-1} 3/4$
c. $n\pi + (-1)^n \sin^{-1} 3/4$
d. None of the above

126. The sum of inradius and circumradius of incircle and circumcircle of a regular polygon of side n is

a. $\frac{a}{4}\cot\frac{\pi}{2n}$ *b.* $a\cot\frac{\pi}{n}$
c. $\frac{a}{2}\cot\frac{\pi}{2n}$ *d.* $a\cot\frac{\pi}{2n}$

127. If $a \le \sin^{-1} x + \cos^{-1} x + \tan^{-1} x \le b$, then

a. $a = 0, b = \pi$ *b.* $a = 0, b = \frac{\pi}{2}$
c. $a = \frac{\pi}{2}, b = \pi$ *d.* None of these

128. If $\sin^{-1} x + \sin^{-1} y = \frac{2\pi}{3}$ and $\cos^{-1} x - \cos^{-1} y = \frac{\pi}{3}$. Then, (x, y) is equal to

a. $(0, 1)$ *b.* $(1/2, 1)$
c. $(1, 1/2)$ *d.* $(\sqrt{3}/2, 1)$

129. If $3f(x) - f\left(\frac{1}{x}\right) = \log_e x^4$ for $x > 0$, then $f(e^x)$ is

a. $2x$ *b.* x^2
c. x *d.* None of these

130. $\lim_{x \to 0} (\operatorname{cosec} x)^{1/\log x}$ is equal to

a. 0
b. 1
c. $\frac{1}{e}$
d. None of the above

131. The value of $\lim_{x\to\infty}\left(\frac{x+6}{x+1}\right)^{x+4}$ is

a. e b. e^2 c. e^4 d. e^5

132. The set of points, where the function $f(x)=x|x|$ is, differentiable is

a. $(-\infty, \infty)$ b. $(-\infty, 0)\cup(0, \infty)$
c. $(0, \infty)$ d. $[0, \infty]$

133. On which of the following intervals is the function $f(x)=2x^2-\log|x|, x\neq 0$ increasing in

a. $\left(\frac{1}{2}, \infty\right)$
b. $\left(-\infty, -\frac{1}{2}\right)\cup\left(\frac{1}{2}, \infty\right)$
c. $\left(-\infty, -\frac{1}{2}\right)\cup\left(0, \frac{1}{2}\right)$
d. $\left(-\frac{1}{2}, 0\right)\cup\left(\frac{1}{2}, \infty\right)$

134. The greatest value of $f(x)=(x+1)^{1/3}-(x-1)^{1/3}$ on [0, 1] is

a. 1 b. 2 c. 3 d. $\frac{1}{3}$

135. The intersection angle of the curve $xy=a^2$ and $x^2-y^2=a^2$ is

a. $\frac{\pi}{3}$ b. $\frac{\pi}{6}$ c. $\frac{\pi}{2}$ d. $\frac{5\pi}{6}$

136. If $f'(x)=(x-a)^{2n}(x-b)^{2m+1}$, where $m, n\in N$, then

a. $x=b$ is a point of minimum
b. $x=b$ is a point of maximum
c. $x=b$ is a point of inflexion
d. None of the above

137. If $\int\frac{1}{1+\sin x}dx=\tan\left(\frac{x}{2}+a\right)+b$, then

a. $a=-\frac{\pi}{4}, b\in R$
b. $a=\frac{\pi}{4}, b\in R$
c. $a=\frac{5\pi}{4}, b\in R$
d. None of the above

138. $\int_0^1|5x-3|\,dx$ is equal to

a. $\frac{10}{13}$ b. $\frac{31}{10}$
c. $\frac{13}{10}$ d. None of these

139. The area of the region bounded by $y=|x-1|$ and $y=1$ is

a. 1 b. 2
c. 1/2 d. None of these

140. The differential equation of all non-vertical lines in a plane is

a. $\frac{d^2y}{dx^2}=0$ b. $\frac{d^2x}{dy^2}=0$
c. $\frac{dy}{dx}=0$ d. $\frac{dx}{dy}=0$

141. The angle between the tangent drawn from origin to the circle $(x-7)^2+(y+1)^2=25$ is

a. $\frac{\pi}{3}$ b. $\frac{\pi}{6}$ c. $\frac{\pi}{2}$ d. $\frac{\pi}{8}$

142. If $y^2=P(x)$ be a cubic polynomial, then $2\frac{d}{dx}\left(y^3\frac{d^2y}{dx^2}\right)$ is equal to

a. $P'''(x)P'(x)$ b. $P''(x)P'''(x)$
c. $P(x)P'''(x)$ d. constant

143. The curve with parametric equations $x=1+4\cos\theta, y=2+3\sin\theta$ is an/a

a. ellipse b. parabola
c. hyperbola d. circle

144. The condition that the straight line joining the origin to the points of intersection of the line $4x+3y=24$ with the circle $(x-3)^2+(y-4)^2=25$

a. are coincident
b. are perpendicular
c. make equal angle with X-axis
d. None of the above

145. The line $\frac{x-2}{3}=\frac{y+1}{2}=\frac{z-1}{-1}$ intersects the curve $xy=c^2, z=0$, if c is equal to

a. ± 1 b. $\pm 1/3$
c. $\pm\sqrt{5}$ d. None of these

146. If $\mathbf{a}\cdot\hat{\mathbf{i}}=4$, then $(\mathbf{a}\times\hat{\mathbf{j}})\cdot(2\hat{\mathbf{j}}-3\hat{\mathbf{k}})$ is equal to

a. 12 b. 2 c. 0 d. -12

147. The vector $\mathbf{a}$ is equal to

a. $(\mathbf{a}\cdot\hat{\mathbf{i}})\hat{\mathbf{i}}+(\mathbf{a}\cdot\hat{\mathbf{j}})\hat{\mathbf{j}}+(\mathbf{a}\cdot\hat{\mathbf{k}})\hat{\mathbf{k}}$
b. $(\mathbf{a}\cdot\hat{\mathbf{j}})\hat{\mathbf{i}}+(\mathbf{a}\cdot\hat{\mathbf{k}})\hat{\mathbf{j}}+(\mathbf{a}\cdot\hat{\mathbf{i}})\hat{\mathbf{k}}$
c. $(\mathbf{a}\cdot\hat{\mathbf{k}})\hat{\mathbf{i}}+(\mathbf{a}\cdot\hat{\mathbf{i}})\hat{\mathbf{j}}+(\mathbf{a}\cdot\hat{\mathbf{j}})\hat{\mathbf{k}}$
d. $(\mathbf{a}\cdot\mathbf{a})(\hat{\mathbf{i}}+\hat{\mathbf{j}}+\hat{\mathbf{k}})$

148. The equation of the chord of $y^2 = 8x$ which is bisected at (2, – 3) is

a. $4x + 3y = 1$ *b.* $4x - 3y = 1$
c. $4x + 3y = 9$ *d.* None of these

149. If x_1 and x_2 are the means of two distribution such that $x_1 < \bar{x}_2$ and $\bar{x}$ is the mean of the combined distribution, then

a. $\bar{x} < \bar{x}_1$ *b.* $\bar{x} > \bar{x}_2$
c. $\bar{x} = \dfrac{\bar{x}_1 + \bar{x}_2}{2}$ *d.* $\bar{x}_1 < \bar{x} < \bar{x}_2$

150. Two opposite vertices of a rectangle are (1, 3) and (5, 1). If the equation of a diagonal of this rectangle is $y = 2x + c$. Then, the value of C is

a. – 1 *b.* – 3
c. – 4 *d.* – 9

Answers

Physics

1. (c)	2. (d)	3. (a)	4. (c)	5. (b)	6. (a)	7. (b)	8. (b)	9. (d)	10. (b)
11. (a)	12. (d)	13. (a)	14. (b)	15. (a)	16. (b)	17. (c)	18. (d)	19. (c)	20. (d)
21. (a)	22. (c)	23. (a)	24. (b)	25. (c)	26. (b)	27. (d)	28. (a)	29. (c)	30. (a)
31. (a)	32. (d)	33. (b)	34. (a)	35. (d)	36. (c)	37. (a)	38. (b)	39. (d)	40. (a)

Chemistry

41. (a)	42. (d)	43. (a)	44. (d)	45. (c)	46. (a)	47. (a)	48. (b)	49. (b)	50. (c)
51. (c)	52. (b)	53. (d)	54. (a)	55. (d)	56. (c)	57. (a)	58. (b)	59. (d)	60. (d)
61. (b)	62. (a)	63. (a)	64. (a)	65. (b)	66. (d)	67. (a)	68. (a)	69. (b)	70. (a)
71. (b)	72. (b)	73. (b)	74. (d)	75. (d)	76. (c)	77. (c)	78. (b)	79. (c)	80. (b)

English Proficiency

81. (d)	82. (c)	83. (a)	84. (a)	85. (d)	86. (a)	87. (a)	88. (a)	89. (a)	90. (c)
91. (d)	92. (d)	93. (b)	94. (c)	95. (d)					

Logical Reasoning

96. (d)	97. (c)	98. (d)	99. (d)	100. (b)	101. (d)	102. (b)	103. (c)	104. (b)	105. (b)

Mathematics

106. (d)	107. (b)	108. (d)	109. (c)	110. (c)	111. (d)	112. (a)	113. (d)	114. (d)	115. (a)
116. (d)	117. (b)	118. (a)	119. (a)	120. (b)	121. (a)	122. (d)	123. (d)	124. (b)	125. (a)
126. (c)	127. (a)	128. (b)	129. (c)	130. (c)	131. (d)	132. (a)	133. (d)	134. (b)	135. (c)
136. (a)	137. (a)	138. (c)	139. (a)	140. (a)	141. (c)	142. (c)	143. (a)	144. (b)	145. (c)
146. (d)	147. (a)	148. (d)	149. (d)	150. (c)					

Hints & Solutions

Physics

1. *(c)* $\hbar c = E\lambda$

$$= [ML^2 T^{-2}] \times [L] = [ML^3 T^{-2}]$$

2. *(d)* When stone travels horizontally, then it must be at the maximum height.

Time period, $T = \frac{u\sin\theta}{g}$ [for max. height]

So, $2 = \frac{u \sin \theta}{g}$

$\Rightarrow \quad u \sin \theta = 20 \quad \ldots\text{(i)}[\because g = 10 \text{ m/s}^2]$

Also, $\tan 45° = \frac{u \sin \theta - g(1)}{u \cos \theta}$

$\Rightarrow \quad u \cos \theta = 20 - 10$

$\Rightarrow \quad u \cos \theta = 10 \quad \ldots\text{(ii)}$

Squaring Eqs. (i) and (ii) and adding, we get

$$u = 10\sqrt{5} \text{ ms}^{-1} \text{ and } \tan \theta = 2$$

$$\theta = \tan^{-1}(2)$$

3. *(a)* Given, $m_1 = m_2 = m$

$$u_1 = u \text{ and } u_2 = 0$$

Let v_1 and v_2 be their velocities after the collision.

According to the momentum conservation,

$$mu = m(v_1 + v_2)$$

or $\quad u = v_1 + v_2 \quad \ldots\text{(i)}$

By definition, $e = \frac{v_2 - v_1}{u - 0}$

or $\quad v_2 - v_1 = eu \quad \ldots\text{(ii)}$

On solving Eqs. (i) and (ii), we get

$$v_1 = \frac{(1-e)u}{2}$$

and $\quad v_2 = \left(\frac{(1+e)u}{2}\right)$

$\Rightarrow \quad \frac{v_1}{v_2} = \frac{1-e}{1+e}$

4. *(c)* The angular momentum will be directed along the axis of rotation.

5. *(b)* $g = \frac{GM}{R^2}$

$$\frac{\Delta g}{g} \times 100 = 2\left(\frac{\Delta R}{R} \times 100\right)$$

$$= 2 \times 1\% = 2\%$$

Therefore value of g will increase by 2%.

6. (a) Fresh water has lesser density as compared to sea water. So, upthrust in sea water will be more as compared to fresh water.

Hence, when a ship passes from fresh water into sea water, it will rise.

7. *(b)* From equation of continuity,

$$A_1 v_1 = A_2 v_2$$

$\Rightarrow \quad \pi r_1^2 v_1 = \pi r_2^2 v_2$

$\Rightarrow \quad \frac{v_1}{v_2} = \frac{r_2^2}{r_1^2} = \frac{4}{9}$

8. *(b)* This equation represents damped vibration.

9. *(d)* $\frac{\Delta\lambda}{\lambda} = +\frac{v_s}{c} \Rightarrow \Delta\lambda = \frac{\lambda v_s}{c}$

When a star recedes, then λ increases (as f decreases).

10. *(b)* $\frac{\Delta T}{l} = \frac{80 - 0}{1} = 80 \text{ °C m}^{-1}$

$\therefore \quad \Delta T_1 = (0.60)(80) = 48°\text{C}$

Decrease in temperature after a distance of 60 cm is 48°C and hence actual temperature is $(80 - 48)°\text{C} = 32°\text{C}$.

11. *(a)* For a series circuit,

$$R_s = R_1 + R_2$$

$\Rightarrow \quad i^2 R_s = i^2 R_1 + i^2 R_2$

$\Rightarrow \quad P_s = P_1 + P_2$

12. *(d)* $B_{\text{initial}} = B = \frac{\mu_0 i}{2r}$

$$B_{\text{final}} = \frac{\mu_0 i}{2\left(\frac{r}{2}\right)} + \frac{\mu_0 i}{2\left(\frac{r}{2}\right)}$$

$\Rightarrow \quad B_{\text{final}} = \frac{2\mu_0 i}{r} = 4B_{\text{initial}} = 4B$

13. *(a)* Radius of ion in magnetic field is

$$r = \frac{mv}{qB} = \frac{p}{qB}$$

$\Rightarrow \quad r \propto p$

14. *(b)* $\tan \phi = \frac{X}{R} = \frac{4}{3}$

$\therefore$ Power factor $= \cos \phi = \frac{3}{5} = 0.6$

15. *(a)* $\lambda = \frac{h}{mv} \Rightarrow \lambda \propto \frac{1}{m}$

$\because$ Electron is the lightest particle of these four. So, its wavelength is maximum.

16. ***(b)*** $\sin\theta = \frac{3}{5}$ [given]

$\because \quad \mu = \tan\theta$

$\therefore \quad \mu = \frac{3}{4}$

17. ***(c)*** $v_{av} = \frac{\text{Total distance}}{\text{Total time}}$

$\Rightarrow \quad v_{av} = \frac{s+s+s}{\frac{s}{u}+\frac{s}{v}+\frac{s}{\omega}}$

$\Rightarrow \quad v_{av} = \frac{3\,uv\omega}{uv+v\omega+u\omega}$

18. ***(d)*** $F - N = 0 \Rightarrow F = N$

Further, $f = \mu N = w \Rightarrow \mu F = w$

$\Rightarrow \quad F = \frac{w}{\mu}$

$\Rightarrow \quad F \geq \frac{w}{\mu}$

$\therefore$ Minimum force required, $F = \frac{w}{\mu}$ applying energy conservation.

19. ***(c)*** At foot of the plane,

$$mgh = \frac{1}{2}I\omega^2 + \frac{1}{2}mu^2$$

$$2\times 10\times 4 = \frac{1}{2}\times\left(\frac{1}{2}mR^2\right)\times\left(\frac{u^2}{R^2}\right)+\frac{1}{2}mu^2 \quad \left[\because I = \frac{1}{2}mR^2,\ \omega = \frac{u}{R}\right]$$

$$80 = \frac{3}{4}mu^2 \quad \ldots(i)$$

$\therefore$ Rotational KE $= \frac{1}{2}I\omega^2 = \frac{1}{4}\times m(\omega R)^2$

$$= \frac{1}{4}mu^2 = \frac{1}{4}\times\frac{80\times 4}{3}$$

$$= \frac{80}{3}\text{ J}$$

20. ***(d)*** According to Kepler's law,

$$\frac{T_1^2}{T_2^2} = \frac{r_1^3}{r_2^3}$$

$$\frac{1}{T_2^2} = \left(\frac{1.5\times 10^{11}}{6\times 10^{10}}\right)^3$$

$$= \left(\frac{5}{2}\right)^3$$

$$T_2 = \left(\frac{2}{5}\right)^{3/2}\text{ yr}$$

21. ***(a)*** As shown in figure, in the two arms of a tube pressure remains same on level pp'. Hence,

$$8\times p_y\times g + 2p_{Hg}\times g = 10\times p_x\times g$$

$$8p_y + 2\times 13.6 = 10\times 3.36$$

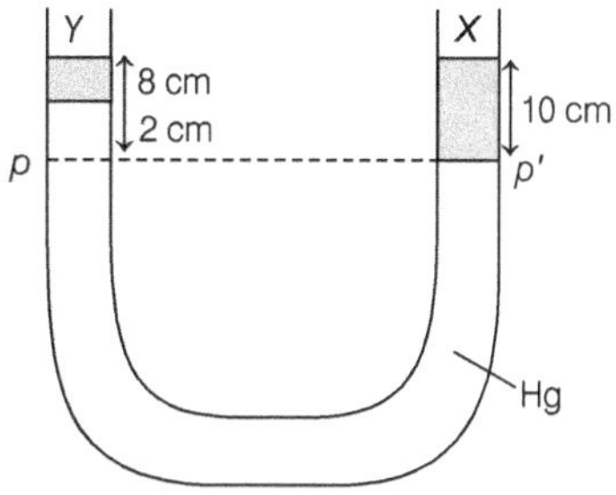

or $\quad p_y = \frac{33.6-27.2}{8}$

$= 0.8\text{ g/cc}$

22. ***(c)*** Let the potential of the battery be V, then

$$R_1 = \frac{V}{1},\ R_2 = \frac{V}{2},$$

$$R_3 = \frac{V}{4}$$

On connecting in series,

$$R_s = R_1 + R_2 + R_3$$

$\Rightarrow \quad R_s = V\left(1+\frac{1}{2}+\frac{1}{4}\right) \Rightarrow R_s = \frac{7}{4}V$

By definition,

$$I = \frac{V}{R_s} \Rightarrow I = \frac{4}{7}\text{ A}$$

23. ***(a)*** Thermoelectric power,

$$S = \frac{dE}{dT} = \frac{d}{dT}\left[k(T-T_r)\cdot\left\{T_0 - \frac{1}{2}(T+T_r)\right\}\right]$$

$$S = k(T-T_r)\times\left(-\frac{1}{2}\right)+\left[T_0-\frac{1}{2}(T+T_r)\right]\times k$$

When $T = \frac{T_0}{2}$,

$$S = -\frac{k}{2}\left(\frac{T_0}{2}-T_r\right)+\left\{T_0-\frac{1}{2}\left(\frac{T_0}{2}+T_r\right)\right\}\times k$$

$$= \frac{-kT_0}{4}+\frac{kT_r}{2}+kT_0-\frac{kT_0}{4}-\frac{kT_r}{2} = \frac{1}{2}kT_0$$

24. ***(b)*** Let the height of the object be h_1.

$\therefore$ At the first position,

$$m_1 = \frac{4}{h_1} = -\frac{v}{u} \quad \ldots(i)$$

and at the second position,

$$m_2 = \frac{16}{h_1} = -\frac{u}{v} \quad \ldots(ii)$$

Now, $m = m_1\times m_2$

$$\Rightarrow \left(-\frac{v}{u}\right)\times\left(-\frac{u}{v}\right)=\frac{4}{h_1}\times\frac{16}{h_1}$$

$$\Rightarrow h_1^2 = 64 \Rightarrow h_1 = 8\,\text{cm}$$

25. ***(c)*** $E = \dfrac{hc}{\lambda}$

$$\Rightarrow \lambda = \frac{hc}{E} = \frac{6.6\times 10^{-34}\times 3\times 10^8}{0.75\times 1.6\times 10^{-19}}$$

$$= 1.65\times 10^{-6}\ \text{m} = 16500\ \text{Å}$$

26. ***(b)*** Magnitude of magnetic field,

$$B_0 = \frac{E_0}{c} = \frac{10^{-3}}{3\times 10^8} = 3.33\times 10^{-12}\ \text{T}$$

27. ***(d)*** Current sensitivity of the galvanometer can be increased by all three given factors.

28. ***(a)*** Let us assume an intermediate temperature 0°C will lie at distance x from top.

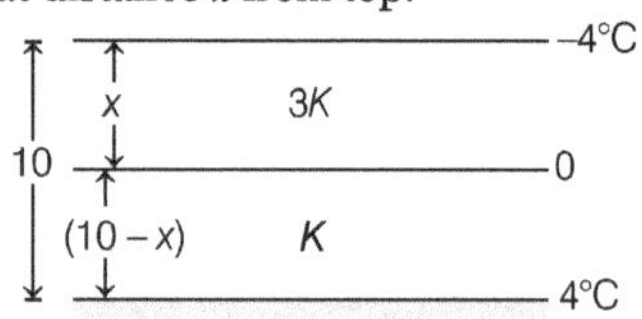

$\therefore$ Heat produced, for water bottom,

$$Q_1 = \frac{KA\Delta T}{L} = \frac{KA(40-0)}{10-x} \quad \text{...(i)}$$

For (ice) top, $Q_2 = \dfrac{3KA(0-(-4))}{x}$ ...(ii)

$[\because K_{ice} = 3\times K_{water}]$

So, equating (i) and (ii), we get

$$KA\left(\frac{4-0}{10-x}\right) = 3KA\left(\frac{0-(-4)}{x}\right)$$

$$\Rightarrow x = 7.5\,\text{m}$$

29. ***(c)*** $\dfrac{E_S}{E_T} = \dfrac{C_p}{C_V} = \gamma = 1 + \dfrac{2}{f}$

30. ***(a)*** $T = 2\pi\sqrt{\dfrac{l_{immersed}}{g}}$

$$\Rightarrow T = 2\pi\sqrt{\frac{0.80}{9.8}} = \frac{4\pi}{7}\text{s}$$

31. ***(a)*** $\dfrac{\text{Rotational KE}}{\text{Translational KE}} = \dfrac{\frac{1}{2}I\omega^2}{\frac{1}{2}Mv^2} = \dfrac{K^2}{R^2}$

$(\because I = MK^2, v = R\omega)$

32. ***(d)*** For completing the loop, $v = \sqrt{5gr}$

By law of conservation of energy,

Loss in potential energy = Gain in kinetic energy

$$\Rightarrow mgh = \frac{1}{2}mv^2 \Rightarrow h = \frac{5r}{2}$$

33. ***(b)*** Let Δx = path difference

$$\Rightarrow \Delta x = L_2D - L_1D$$

$$= \sqrt{40^2 + 9^2} - 40$$

$$= 41 - 40$$

$$= 1\ \text{m}$$

For first maximum,

$$\Delta x = (2n)\frac{\lambda}{2} \quad [\text{where, } n = 1]$$

$$\Rightarrow 1 = 2(1)\frac{\lambda}{2}$$

$$\Rightarrow \lambda = 1\ \text{m}$$

$$\therefore f = \frac{v}{\lambda} = 330\ \text{Hz}$$

34. ***(a)*** Since, it vibrates in 4 segments, so we must get 5 nodes and 4 anti-nodes.

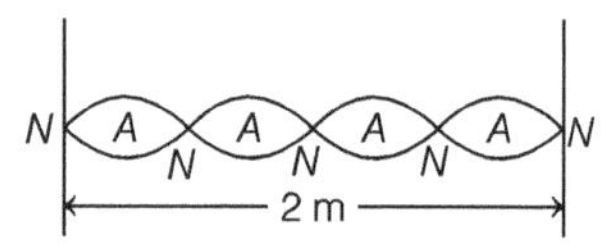

$$\Rightarrow 2 = 2\lambda \Rightarrow \lambda = 1\ \text{m}$$

Distance between two consecutive nodes is

$$\frac{\lambda}{2} = \frac{1}{2} = 0.5\ \text{m}$$

35. ***(d)*** $f_{static} = (\mu_{static})\, mg$ ...(i)

$$\Rightarrow f_{static} = (0.5)\,(2)\,(10)$$

$$\Rightarrow f_{static} = 10\ \text{N}$$

$\because$ Applied force is just 2.5 N < 10 N. So, the block will not move. As, we know that in static region, friction is a self adjusting force and has a value equal to that of applied force.

So, $f = F_{applied} = 2.5$ N

36. ***(c)*** It depends upon its time constant, $\tau = \dfrac{L}{R}$.

37. ***(a)*** $N = 24 = mn$

For current to be maximum,

$$R_{internal} = R_{external}$$

$$\Rightarrow \frac{mr}{n} = 3$$

$$\Rightarrow \frac{m}{n}(0.5) = 3$$

$$\Rightarrow \frac{m}{n} = 6$$

$$\Rightarrow m = 6n$$

Substituting this values in Eq. (i), we get

$$24 = 6n^2$$

$$\Rightarrow n = 2 \Rightarrow m = 12$$

38. ***(b)*** Collision frequency $= \dfrac{v}{l}$

Mock Test 2

39. ***(d)*** $y(x, t) = \dfrac{a}{(x \pm vt)^2 + b}$ is another form of progressive wave equation propagating with a speed v. Negative sign to be taken for propagation along $+X$-axis and positive sign to be taken for propagation along $-X$-axis.

Given, equation of pulse, $y = \dfrac{6}{x^2 - 3}$

Velocity of pulse = 4.5 m/s

∴ Equation of pulse with time,

$$y = \frac{6}{(x - 4.5t)^2 - 3}$$

40. ***(a)*** By law of conservation of angular momentum,

$$mvR = (I_{\text{system}})\omega$$

$$\Rightarrow \quad mvR = (I + mR^2)\omega$$

$$\Rightarrow \quad \omega = \frac{mvR}{I + mR^2}$$

Chemistry

41. ***(a)*** The greatest strain is involved in cycloalkane with bond angle 60°.

i.e.

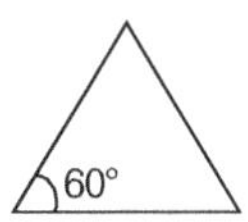

Cyclopropane

42. ***(d)*** During acid hydrolysis of methyl acetate, reactants and catalyst (H_2SO_4) are in the same phase. Thus, it is an example of homogeneous catalysis.

43. ***(a)*** $w = \dfrac{i \times t \times E}{F}$

$$\Rightarrow \quad E = \frac{w \times F}{i \times t} = \frac{0.108 \times 96500}{0.5 \times 193} = 108$$

44. ***(d)*** van-Arkel method is based on thermal decomposition of stable and volatile compounds.

45. ***(c)*** In CaF_2, the number of Ca^{2+} ion is double the number of F^- ion. Therefore, coordination number of F^- is, half of that of Ca^{2+}, i.e. 4.

46. ***(a)*** $(CH_2)_6N_4$ is urotropine and has following structure.

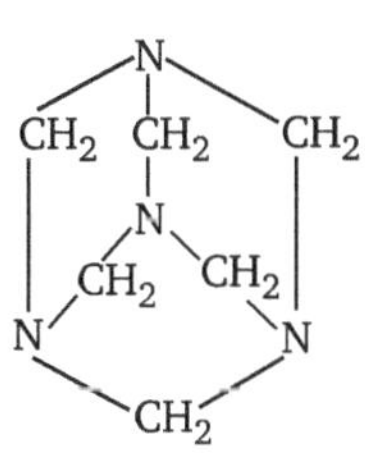

∴ It is clear from the structure that, it has no C—C bond.

47. ***(a)*** $\Delta H = \Delta H_{\text{(product)}} - \Delta H_{\text{(reactant)}}$

$$162 = 2 \times \Delta H_{\text{H}} - \Delta H_{\text{H}_2}$$

$$\Delta H_{\text{H}} = \frac{162}{2} = 81 \text{ kcal} \qquad (\because \Delta H_{\text{H}_2} = 0)$$

48. ***(b)*** As the size increases, bond strength decreases. But S—S bond is stronger than O—O bond because of inter-electronic repulsion due to smaller size of O atoms. Sulphur exhibits maximum catenation property.

49. ***(b)*** Sodium hydroxide reacts with carbon monoxide to form HCOONa as follows.

$$\text{NaOH} + \text{CO} \longrightarrow \text{HCOONa}$$

50. ***(c)*** Oil paintings contain white lead which reacts with atmospheric H_2S and converts into blackish PbS.

51. ***(c)*** ZnO (zinc oxide) reacts with both HCl and NaOH to give $ZnCl_2$ and Na_2ZnO_2 respectively, i.e.

$$ZnCl_2 \xleftarrow{\text{HCl}} ZnO \xrightarrow{\text{NaOH}} Na_2ZnO_2$$

52. ***(b)*** LSD and mescalin are psychedlic drugs. These produce visual and auditory hallucinations.

53. ***(d)*** Phenol may be tested by all the given reagents.

54. ***(a)*** Greater the s-character in the hybrid orbital, closer are the bonding electrons to the atom. Hence, shorter and stronger is the σ-bond.

55. ***(d)***

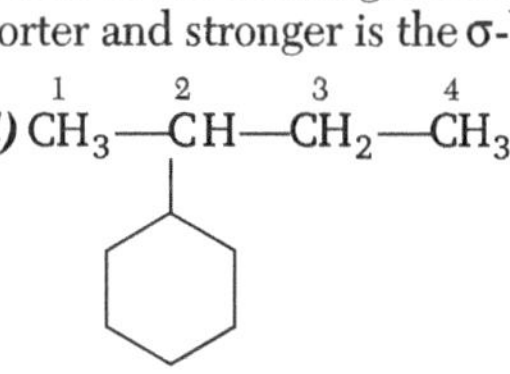

2-cyclohexyl butane

56. ***(c)*** HCl solution in benzene is not acidic because benzene cannot accept proton.

57. ***(a)*** $\log \dfrac{k_2}{k_1} = \dfrac{E_a}{2.303\,R}\left[\dfrac{T_2 - T_1}{T_1 T_2}\right]$

$$\log 2 = \frac{E_a}{2.303 \times 2}\left[\frac{300 - 290}{290 \times 300}\right]$$

$$E_a = 12062 \text{ cal} = 12.0 \text{ kcal}$$

58. ***(b)*** The active mass of solids is usually taken as unity.

59. ***(d)*** Aluminium is present in diaspore.

Gypsum—$CaSO_4 \cdot 2H_2O$
Carnallite — $KCl \cdot MgCl_2 \cdot 6H_2O$
Asbestos — $CaSiO_3 \cdot 3MgSiO_3$
Diaspore — $Al_2O_3 \cdot H_2O$

60. ***(d)*** Energy, $E = \dfrac{hc}{\lambda} = hc\bar{\nu} \qquad \left(\because \bar{\nu} = \dfrac{1}{\lambda}\right)$

$$= 6.62 \times 10^{-34} \times 3 \times 10^8 \times 5 \times 10^5$$

$$= 9.93 \times 10^{-23} \text{ kJ}$$

61. ***(b)*** 1g of H_2 combines with Br_2 = 80g of Br_2

Eq. wt. of $Br_2 = 80$

4g of Br_2 combined with Ca = 1 g

80 g of Br_2 combine with Ca $= \frac{1 \times 80}{4} = 20$ g

62. ***(a)*** $CH_3COOH \xrightarrow[H_2SO_4]{N_3H} CH_3NH_2 + CO_2 + N_2 \uparrow$

63. ***(a)*** Conversion of sucrose in CO_2.

$$\overset{0}{C_{12}H_{22}O_{11}} \xrightarrow{\text{Oxidation}} \overset{+4}{CO_2}$$

Here, oxidation number of carbon increases from 0 to + 4. Hence, oxidation occurs.

64. ***(a)*** Magnesium perchlorate, $Mg(ClO_4)_2$ is also called anhydrone.

65. ***(b)*** On burning Mg and NO_2 will give NO and MgO.

$Mg + NO_2 \xrightarrow{\Delta} MgO + NO$

66. ***(d)*** Pseudohalides are uni-negative groups which show certain characteristics of halide ions. e.g., CN^-.

67. ***(a)*** Tetra ethyl lead (TEL) is used as antiknocking agent in petro industry.

68. ***(a)*** K_b of $NH_3 = 10^{-5}$. (given)

$\therefore \quad pK_b = 5$ and $pK_a = 14 - 5 = 9$

69. ***(b)*** $m = \frac{1000\, K_f\, w}{\Delta T_f\, W}$

$$\Delta T_f = \frac{1000\, K_f w}{mW} = \frac{1000 \times 1.86 \times 342}{342 \times 1000} = 1.86°C$$

$\therefore$ The solution will freezes at $= 0 - 1.86°C$

$= - 1.86°C$

70. ***(a)*** $\lambda_1/\lambda_2 = p_2^2/p_1^2$

where, λ_1 and λ_2 are mean for path at pressure p_1 and p_2 respectively.

$$\Rightarrow \quad \frac{l}{\lambda_2} = \frac{5^2}{1} \Rightarrow \lambda_2 = \frac{l}{(5)^2} = \frac{l}{25}$$

71. ***(b)*** $\frac{N_0}{N} = \frac{2.30}{1.62}$

N_0 and N are intensity initially and after time t respectively

Now, rate constant for radioactive decay

$$\lambda = \frac{2.303}{600} \log \frac{2.30}{1.62} = 0.000584$$

$$\therefore \quad t_{1/2} = \frac{0.693}{\lambda} = \frac{0.693}{0.000584} = 1187 \text{ s}$$

72. ***(b)*** The chain isomers of C_6H_{14} are

$CH_3—CH_2—CH_2—CH_2—CH_2—CH_3$

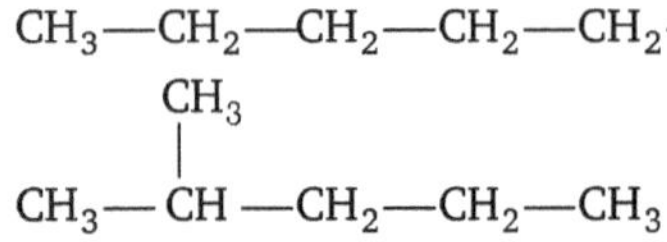

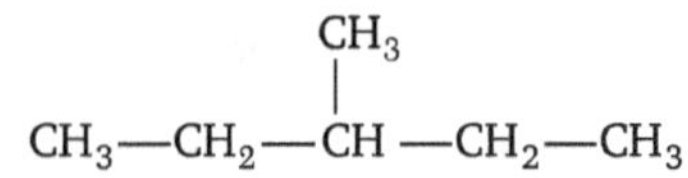

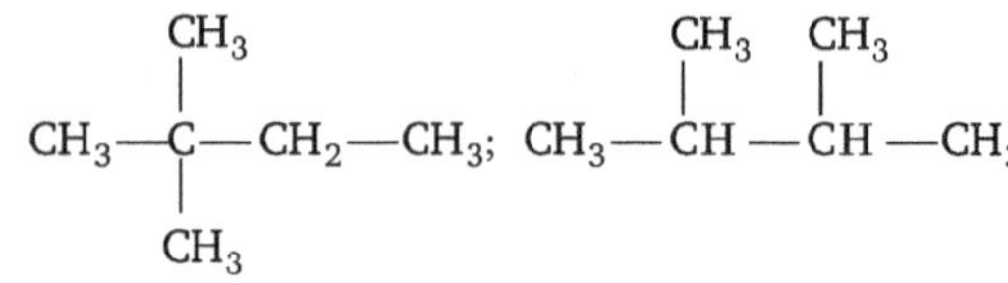

$\therefore$ Total number of chain isomers of C_6H_{14} are 5.

73. ***(b)*** The $—CH_3O$ group at *para* position stabilises the carbonium ion by + *M*-effect as it is electron donating group.

74. ***(d)*** The complete reactions take place as follows

$$\underset{\text{Ethyene}}{H—C{\equiv}C—H} \xrightarrow[\text{Excess}]{Na/NH_3} Na—C{\equiv}C—Na \xrightarrow[\text{excess}]{CH_3I}$$

$$\underset{\text{But-2-yne}}{CH_3—C{\equiv}C—CH_3} \xrightarrow[\text{Rearrangement}]{NaNH_2\text{-Paraffin}} \underset{\text{Butyne}}{[CH_3—CH_2—C{\equiv}CH]}$$

$$\xrightarrow{NaNH_2} CH_3—CH_2—C{\equiv}CNa$$

75. ***(d)*** All statements are correct.

76. ***(c)*** Degree of dissociation, $\alpha = 4\% = 4/100 = 0.04$

Normality = 0.2

$\therefore \quad [H^+] = \alpha \times N = 0.04 \times 0.2 = 0.008$ M

77. ***(c)*** $4NH_3 + 5O_2 \xrightarrow{Pt} 4NO + 6H_2O$

is the correct reaction.

78. ***(b)*** $N_1V_1 = N_2V_2$

$$\frac{1}{2} \times 200 = \frac{1}{10} \times V_2$$

$$V_2 = \frac{1}{2} \times 200 \times 10 = 1000 \text{ mL}$$

$\therefore$ Volume of water added = 1000 – 200 = 800 mL

79. ***(c)*** XeF_6 reacts with silica of glass and produces $XeOF_4$ as

$$2XeF_6 + \underset{\text{(of glass)}}{SiO_2} \longrightarrow 2XeOF_4 + SiF_4$$

80. ***(b)*** 3-methyl-2-butanol on treatment with HCl gives 2-chloro-3-methyl butane as major product.

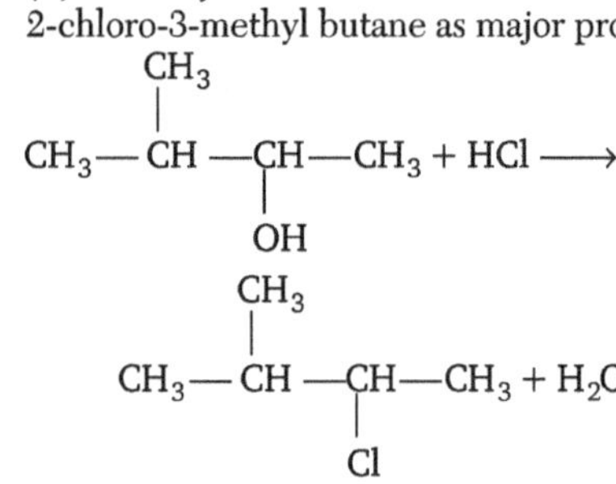

2-chloro-3-methyl butane

a. English Proficiency

81. ***(d)*** Plagiarism is the correct word for the given group of words. Plagiarism is the practice of taking someone else's work or ideas and passing them off as one's own.

82. ***(c)*** Archive is a place where public, government or historical records are kept.

83. ***(a)*** Rejuvenate is the most appropriate word that replaces the group of words italicised in the sentence.

84. ***(a)*** Sentence (A) should come first in the paragraph.

85. ***(d)*** Sentence (B) should come second in the paragraph.

86. ***(a)*** Sentence (D) should come last in the paragraph.

87. ***(a)*** Sentence (C) should come third in the paragraph.

88. ***(a)*** Paramour means a lover, especially the illicit partner of a married person. So, option (a) is its correct synonym.

89. ***(a)*** Refectory means - a dining hall (as in a monastery or college).

90. ***(c)*** Assent means the expression of approval or agreement. So, option (c) is its correct answer.

91. ***(d)*** 'Educate' is the correct word for the given blank.

92. ***(d)*** 'discretionary' should be used in the given sentence.

Discretionary means available to someone by choice, without having to get permission or authority.

93. ***(b)*** According to the given context, 'granted' is the correct word for the given blank.

94. ***(c)*** The idiom 'To show one's teeth means to display an angry, violent, and or threatening reaction to or against something or someone. Hence, option (c) is its correct answer.

95. ***(d)*** The idiom 'To pour oil in troubled water' means to soothe or calm down something or someone. Hence, option (d) is its correct answer.

b. Logical Reasoning

96. ***(d)*** Except 'Ounce' others are currencies of different countries.

97. ***(c)*** Jackfruit is a vegetable, while others are fruits.

98. ***(d)*** The missing figure is figure (d).

99. ***(d)*** The missing figure is figure (d).

100. ***(b)*** Pattern of the series as follows

4   7   13   [22]   34   49

Differences: +3, +6, +9, +12, +15

Second differences: +3, +3, +3, +3

101. ***(d)*** Pattern of the series as follows

5   11   24   [51]   106   217

×2+1, ×2+2, ×2+3, ×2+4, ×2+5

102. ***(b)*** $(101+15)-(43+35)=116-78=38$

$\therefore \quad (48+184)-(34+56)=232-90=\boxed{142}$

103. ***(c)*** In first figure, $(5+6)-(7+4)=11-11=0$

In second figure, $(7+6)-(8+4)=13-12=1$

In third figure, $(11+2)-(0+2)=13-2=\boxed{11}$

104. ***(b)*** When the figure (iii) is unfolded, it will look like figure (b).

105. ***(b)*** When the figure (iii) is unfolded, it will look like figure (b).

Mathematics

106. ***(d)*** Since, $f: R \rightarrow R$ and $g: R \rightarrow R$, given by $f(x)=2x-3$ and $g(x)=x^3+5$ respectively, are bijections, therefore f^{-1} and g^{-1} exists.

We have,

$$f(x)=2x-3$$

$$\therefore \quad f(x)=y$$

$$\Rightarrow \quad 2x-3=y \Rightarrow x=\frac{y+3}{2}$$

$$\Rightarrow \quad f^{-1}(y)=\frac{y+3}{2}$$

Thus, f^{-1} is given by

$$f^{-1}(x)=\frac{x+3}{3}, \forall\, n \in R$$

Similarly, $g^{-1}(x)=(x-5)^{1/3}, \forall\, x \in R$

Now, $(fog)^{-1}(x)=(g^{-1} o f^{-1})(x)=g^{-1}(f^{-1}(x))$

$$=g^{-1}\left(\frac{x+3}{2}\right)$$

$$=\left(\frac{x+3}{2}-5\right)^{1/3}=\left(\frac{x-7}{2}\right)^{1/3}$$

107. ***(b)*** Let $f(x)=y$, then $\dfrac{10^x-10^{-x}}{10^x+10^{-x}}=y$

$$\Rightarrow \quad \frac{10^{2x}-1}{10^{2x}+1}=y \Rightarrow 10^{2x}=\frac{1+y}{1-y}$$

$$\Rightarrow \quad x=\frac{1}{2}\log_{10}\left(\frac{1+y}{1-y}\right) \Rightarrow f^{-1}(y)=\frac{1}{2}\log_{10}\left(\frac{1+y}{1-y}\right)$$

$$\therefore \quad f^{-1}(x)=\frac{1}{2}\log_{10}\left(\frac{1+x}{1-x}\right)$$

108. ***(d)*** Let d be the common difference of AP.

Then, $a_{10} = 3$

$\Rightarrow a_1 + 9d = 3$

$\Rightarrow 2 + 9d = 3$

$\Rightarrow d = 1/9$

$\therefore a_4 = a_1 + 3d = 2 + \frac{1}{3} = \frac{7}{3}$

Let D be the common difference of $\frac{1}{h_1}, \frac{1}{h_2}, \ldots, \frac{1}{h_{10}}$, then

$$h_{10} = 3 \Rightarrow \frac{1}{h_{10}} = \frac{1}{3}$$

$$\Rightarrow \frac{1}{h_1} + 9D = \frac{1}{3} \Rightarrow \frac{1}{2} + 9D = \frac{1}{3}$$

$$\Rightarrow 9D = -\frac{1}{6} \Rightarrow D = -\frac{1}{54}$$

$$\therefore \frac{1}{h_7} = \frac{1}{h_1} + 6D = \frac{1}{2} - \frac{1}{9}$$

$$\Rightarrow h_7 = \frac{18}{7}$$

$$\therefore a_4 h_7 = \frac{7}{3} \times \frac{18}{7} = 6$$

109. ***(c)*** Given relation is

$$x^2 - 2x\cos\theta + 1 = 0$$

$$\Rightarrow x^2 - 2x\cos\theta + \cos^2\theta - \cos^2\theta + 1 = 0$$

$$\Rightarrow (x - \cos\theta)^2 + \sin^2\theta = 0$$

$$\Rightarrow x = \cos\theta \pm i\sin\theta$$

Take $x = \cos\theta + i\sin\theta$

$$x^n = \cos n\theta + i\sin n\theta$$

and $x^{-n} = \frac{1}{x^n} = \cos n\theta - i\sin n\theta$

$$\therefore x^n + (x^{-n}) = 2\cos n\theta$$

$$\therefore x^{2n} - 2x^n \cos n\theta + 1 = 0$$

110. ***(c)*** The roots of the given equation are

$$\alpha = \frac{-b - \sqrt{b^2 - 4ac}}{2a}$$

and $$\beta = \frac{-b + \sqrt{b^2 - 4ac}}{2a}$$

Since, α, β are real and distinct, therefore $b^2 - 4ac > 0$.

Now, if $a > 0$, then $\beta > -\frac{b}{2a}$ and if $a < 0$, then $a > -\frac{b}{2a}$.

Thus, one of the roots exceeds $-\frac{b}{2a}$.

111. ***(d)*** Let $z = x + iy$, then

$$z^2 + \bar{z} = 0$$

$$\Rightarrow (x^2 - y^2 + 2ixy) + (x - iy) = 0$$

$$\Rightarrow x^2 - y^2 + x = 0 \text{ and } 2xy - y = 0$$

Now, $2xy - y = 0 \Rightarrow y(2x - 1) = 0$

$$\Rightarrow y = 0 \text{ or } x = 1/2$$

If $y = 0$, then

$$x^2 - y^2 + x = 0 \Rightarrow x^2 + x = 0$$

$$\Rightarrow x = 0 \text{ or } x = -1$$

If $x = 1/2$, then $x^2 - y^2 + x = 0$

$$\Rightarrow y = \pm\frac{\sqrt{3}}{2}$$

Thus, the given equations has four solutions.

112. ***(a)*** $\because \frac{1+i}{1-i} = \frac{1+i}{1-i} \times \frac{1+i}{1+i}$

$$= \frac{(1+i)^2}{1-i^2} = \frac{1+2i+i^2}{1-i^2} = \frac{1+2i-1}{1+1} = i$$

$$\therefore \left(\frac{1+i}{1-i}\right)^n = 1$$

$$\Rightarrow i^n = 1$$

Thus, i^n will be equal to if $n = 4$.

113. ***(d)*** We have, $|A| = \begin{vmatrix} 1 & 2 & 2 \\ 2 & 1 & 2 \\ 2 & 2 & 1 \end{vmatrix}$

$$= 1(1-4) - 2(2-4) + 2(4-2)$$

$$= -3 + 4 + 4 = 5 \neq 0$$

Thus, A is non-singular matrix of order 3×3. Therefore, $r(A) = 3$

114. ***(d)*** We have, $\begin{vmatrix} 1 & \omega & \omega^2 \\ \omega & \omega^2 & 1 \\ \omega^2 & 1 & \omega \end{vmatrix}$

Applying $C_1 \to C_1 + C_2 + C_3$, we get

$$= \begin{vmatrix} 1+\omega+\omega^2 & \omega & \omega^2 \\ \omega+\omega^2+1 & \omega^2 & 1 \\ \omega^2+\omega+1 & 1 & \omega \end{vmatrix}$$

$$= \begin{vmatrix} 0 & \omega & \omega^2 \\ 0 & \omega^2 & 1 \\ 0 & 1 & \omega \end{vmatrix} = 0$$

115. ***(a)*** The determinant of the coefficient matrix of given system of equation is

$$\begin{vmatrix} 1 & 2 & 3 \\ 2 & 1 & 3 \\ 5 & 5 & 9 \end{vmatrix} = 1(9-15) - 2(18-5) + 3(10-5)$$

$$= 3 \neq 0$$

Hence, the system of given equation has unique solution.

Mock Test 2

116. ***(d)*** The general term in the expansion of $(2-x+3x^2)^6$ is $\frac{6!}{r!s!t!}2^r(-x)^s(3x^2)^t$, where $r+s+t=6$

$$=\frac{6!}{r!s!t!}2^r\times(-1)^s\times3^t\times x^{s+2t},$$

where $r+s+t=6$

For the coefficient of x^5, we must have $s+2t=5$.

But, $r+s+t=6$

$\therefore \quad s=5-2t$ and $r=1+t$,

where $0\le r,s,t\le 6$

Now, $t=0 \Rightarrow r=1, s=5$

$t=1 \Rightarrow r=2, s=3$

$t=2 \Rightarrow r=3, s=1$

Thus, there are three terms containing x^5 and coefficient of x^5

$$=\frac{6!}{1!5!0!}\times2^1\times(-1)^5\times3^0+\frac{6!}{2!3!1!}\times2^2\times(-1)^3\times3^1+\frac{6!}{3!1!2!}\times2^3\times(-1)^1\times3^2$$

$$=-12-720-4320=-5052$$

117. ***(b)*** Since, $(1+x)^n={}^nC_0+{}^nC_1x+{}^nC_2x^2+{}^nC_3x^3+\ldots+{}^nC_nx^n \quad \ldots(i)$

Differentiating Eq. (i) w.r.t. x, we get

$$n(1+x)^{n-1}={}^nC_1+2\,{}^nC_2\,x+3\,{}^nC_3\,x^2+\ldots+n\,{}^nC_n\,x^{n-1} \quad \ldots(ii)$$

Putting, $x=1$ is Eq.(ii), we get

$$n(1+1)^{n-1}={}^nC_1+2\,{}^nC_2+3\,{}^nC_3+\ldots+n\,{}^nC_n$$

$$\Rightarrow {}^nC_1+2\,{}^nC_2+3\,{}^nC_3+\ldots+n\,{}^nC_n=n\,2^{n-1}$$

118. ***(a)*** Given, series is

$$\frac{1}{3}+\frac{1}{3\cdot3^3}+\frac{1}{5\cdot3^5}+\frac{1}{7\cdot3^7}+\ldots\infty \quad \ldots(i)$$

Putting, $\frac{1}{3}=x$ in Eq. (i), we get

$$\frac{1}{3}+\frac{1}{3\cdot3^3}+\frac{1}{5\cdot3^5}+\frac{1}{7\cdot3^7}+\ldots\infty$$

$$=\left(x+\frac{x^3}{3}+\frac{x^5}{5}+\frac{x^7}{7}+\ldots\infty\right)$$

$$=\frac{1}{2}\log_e\left(\frac{1+x}{1-x}\right)$$

$$=\frac{1}{2}\log_e\left(\frac{1+\frac{1}{3}}{1-\frac{1}{3}}\right) \quad \left(\because x=\frac{1}{3}\right)$$

$$=\frac{1}{2}\log_e 2$$

119. ***(a)*** We have,

$$101^{50}-99^{50}$$

$$=(100+1)^{50}-(100-1)^{50}$$

$$=100^{50}\left(1+\frac{1}{100}\right)^{50}-100^{50}\left(1-\frac{1}{100}\right)^{50}$$

$$=100^{50}\left[2\left\{{}^{50}C_1\cdot\frac{1}{100}+{}^{50}C_3\left(\frac{1}{100}\right)^3+\ldots\right\}\right]$$

$$=\frac{101^{50}-99^{50}}{100^{50}}=1+2\left\{{}^{50}C_3\left(\frac{1}{100}\right)^3+\ldots\right\}$$

$$\Rightarrow \frac{101^{50}-99^{50}}{100^{50}}=1+\text{(positive quantity)}>1$$

$$\Rightarrow 101^{50}-99^{50}>100^{50}$$

$$\Rightarrow 101^{50}>100^{50}+99^{50}$$

120. ***(b)*** ${}^{n+1}C_{n-2}-{}^{n+1}C_{n-1}\le100$

$$\Rightarrow {}^{n+1}C_3-{}^{n+1}C_2\le100 \quad [\because {}^nC_r={}^nC_{n-r}]$$

$$\Rightarrow \frac{(n+1)n(n-1)}{6}-\frac{(n+1)n}{2}\le100$$

$$\Rightarrow (n+1)n(n-1)-3n(n+1)\le600$$

$$\Rightarrow (n+1)n(n-4)\le600$$

The values of n satisfying this inequality are 2, 3, 4, 5, 6, 7, 8, 9.

121. ***(a)*** Since, probability always lies between 0 and 1.

$$0\le\frac{1+3p}{3}\le1, 0\le\frac{1-p}{4}\le1, 0\le\frac{1-2p}{2}\le1$$

$$\Rightarrow 0\le1+3p\le3, 0\le1-p\le4, 0\le1-2p\le2$$

$$\Rightarrow -\frac{1}{3}\le p\le\frac{2}{3}, -3\le p\le1, -\frac{1}{2}\le p\le\frac{1}{2} \quad \ldots(i)$$

Again, the events are mutually exclusive

$$0\le\frac{1+3p}{3}+\frac{1-p}{4}+\frac{1-2p}{2}\le1$$

$$\Rightarrow 0\le13-3p\le12$$

$$\Rightarrow \frac{1}{3}\le p\le\frac{13}{5} \quad \ldots(ii)$$

From Eqs. (i) and (ii), we get

$$\max\left\{-\frac{1}{3}-3, \frac{-1}{2}, \frac{1}{3}\right\}\le p\le\min\left\{\frac{2}{3}, 1, \frac{1}{2}, \frac{13}{3}\right\}$$

$$\Rightarrow \frac{1}{3}\le p\le\frac{1}{2}$$

122. ***(d)*** $A\subseteq A\cup B\Rightarrow P(A)\le P(A\cup B)$

$$\Rightarrow P(A\cup B)\ge\frac{3}{4}$$

Also, $P(A\cap B)=P(A)+P(B)-P(A\cup B)$

$$\ge P(A)+P(B)-1$$

$$=\frac{3}{4}+\frac{5}{8}-1=\frac{3}{8}$$

Mock Test 2

Now, $A \cap B \subseteq B \Rightarrow P(A \cap B) \le P(B) = \frac{5}{8}$

$\therefore \quad \frac{3}{8} \le P(A \cap B) \le \frac{5}{8} \quad \ldots(i)$

Next, $P(A \cap B') = P(A) - P(A \cap B)$

$\Rightarrow \quad \frac{3}{4} - \frac{5}{8} \le P(A \cap B') \le \frac{3}{4} - \frac{3}{8}$

$\Rightarrow \quad \frac{1}{8} \le P(A \cap B') \le \frac{3}{8}$

$\because \quad P(A' \cap B) = P(B) - P(A \cap B)$

$\therefore \quad P(A \cap B) = P(B) - P(A' \cap B)$ [using Eq. (i)]

$\Rightarrow \quad \frac{3}{8} \le P(B) - P(A' \cap B) \le \frac{5}{8}$

$\Rightarrow \quad 0 \le P(A' \cap B) \le \frac{1}{4}$

123. ***(d)*** $\because$ AM > GM

$\Rightarrow \quad \frac{9\tan^2\theta + 4\cot^2\theta}{2} > \sqrt{4\cot^2\theta \cdot 9\tan^2\theta}$

$\Rightarrow \quad 9\tan^2\theta + 4\cot^2\theta > 12$

$\therefore$ Minimum value is 12.

124. ***(b)*** LHS $= \left(1 + \cos\frac{\pi}{8}\right)\left(1 + \cos\frac{3\pi}{8}\right)\left(1 + \cos\frac{5\pi}{8}\right)\left(1 + \cos\frac{7\pi}{8}\right)$

$= \left(1 - \cos^2\frac{\pi}{8}\right)\left(1 - \cos^2\frac{3\pi}{8}\right)$

$= \sin^2\frac{\pi}{8}\sin^2\frac{3\pi}{8}$

$= \frac{1}{4}\left[\cos\frac{\pi}{4} - \cos\frac{\pi}{2}\right]^2 = \frac{1}{8}$

125. ***(a)*** $\sec\theta - \operatorname{cosec}\theta = \frac{4}{3}$

$\Rightarrow \quad 3(\sin\theta - \cos\theta) = 4\sin\theta\cos\theta$

On squarring

$9(\sin\theta - \cos\theta)^2 = (4\sin\theta\cos\theta)^2$

$\Rightarrow \quad 9(1 - \sin 2\theta) = 4\sin^2 2\theta$

$[\because \sin^2 A + \cos^2 A = 1; \sin 2A = 2\sin A\cos A]$

$\Rightarrow \quad 4\sin^2 2\theta + 9\sin 2\theta - 9 = 0$

$\Rightarrow \quad \sin 2\theta = \frac{-9 \pm \sqrt{81 + 144}}{8} = \frac{3}{4}, -3$

But, $\sin 2\theta \ne -3$

$\therefore \quad \sin 2\theta = \frac{3}{4}$

$\Rightarrow \quad \theta = \frac{1}{2}\left[n\pi + (-1)^n \sin^{-1}\frac{3}{4}\right]$

126. ***(c)*** Let r and R be the inradius and circumradius of incircle and circumcircle of the given regular polygon of side n. Then we know that

$r + R = \frac{a}{2}\cot\frac{\pi}{n} + \frac{a}{2}\operatorname{cosec}\frac{\pi}{n}$

$= \frac{a}{2}\left(\frac{1 + \cos\pi/n}{\sin\pi/n}\right) = \frac{a}{2}\frac{2\cos^2\frac{\pi}{2n}}{2\sin\frac{\pi}{2n}\cdot\cos\frac{\pi}{2n}}$

$= \frac{a}{2}\cot\frac{\pi}{2n}$

127. ***(a)*** $\because \frac{\pi}{2} = \sin^{-1}x + \cos^{-1}x = \frac{\pi}{2} \quad \ldots(i)$

and $\quad -\frac{\pi}{2} < \tan^{-1}x < \frac{\pi}{2} \quad \ldots(ii)$

Adding Eqs. (i) and (ii), we get

$\frac{\pi}{2} - \frac{\pi}{2} < \sin^{-1}x + \cos^{-1}x + \tan^{-1}x < \frac{\pi}{2} + \frac{\pi}{2}$

$\Rightarrow \quad 0 < \sin^{-1}x + \cos^{-1}x + \tan^{-1}x < \pi$

Comparing it with given condition, we get

$a = 0, b = \pi$

128. ***(b)*** We have,

$\sin^{-1}x + \sin^{-1}y = \frac{2\pi}{3} \quad \ldots(i)$

and $\quad \cos^{-1}x - \cos^{-1}y = \frac{\pi}{3}$

$\Rightarrow \left(\frac{\pi}{2} - \sin^{-1}x\right) - \left(\frac{\pi}{2} - \sin^{-1}y\right) = \frac{\pi}{3}$

$\Rightarrow \quad -\sin^{-1}x + \sin^{-1}y = \frac{\pi}{3} \quad \ldots(ii)$

On adding Eqs. (i) and (ii), we get

$\sin^{-1}y = \frac{\pi}{2} \Rightarrow y = 1$

On subtracting Eq. (i) from Eq. (ii), we get

$\sin^{-1}x = \frac{\pi}{6} \Rightarrow x = \frac{1}{2}$

$\therefore \quad (x, y) = \left(\frac{1}{2}, 1\right)$

129. ***(c)*** We have, $3f(x) - f\left(\frac{1}{x}\right) = \log_e x^4$

$\Rightarrow \quad 3f(x) - f\left(\frac{1}{x}\right) = 4\log_e x \quad \ldots(i)$

$[\because \log m^n = n\log m]$

$\Rightarrow \quad 3f\left(\frac{1}{x}\right) - f(x) = 4\log_e\left(\frac{1}{x}\right)$

$\left(\text{Replacing } x \text{ by } \frac{1}{x}\right)$

$$\Rightarrow \quad 3f\left(\frac{1}{x}\right) - f(x) = -4\log_e x \quad \ldots\text{(ii)}$$

Solving Eqs. (i) and (ii), we get

$$8f(x) = 8\log_e x \Rightarrow f(x) = \log_e x$$

$$\Rightarrow \quad f(e^x) = x$$

130. ***(c)*** Let $p = \lim_{x\to 0} (\text{cosec } x)^{1/\log x}$ $(\infty^0 \text{ form})$

$$\Rightarrow \quad \log p = \lim_{x\to 0} \frac{1}{\log x}(\log \text{cosec } x)$$

$$= \lim_{x\to 0} \frac{\log(\text{cosec } x)}{\log x} \quad \left(\frac{\infty}{\infty}\text{ form}\right)$$

$$\therefore \quad \log p = \lim_{x\to 0}\left(\frac{-x \text{ cosec } x \cot x}{\text{cosec } x}\right)$$

[by L' Hospital's rule]

$$= \lim_{x\to 0}\left(-\frac{x}{\tan x}\right) = -1$$

$$\therefore \quad p = e^{-1} = \frac{1}{e}$$

131. ***(d)*** $\lim_{x\to\infty}\left(\frac{x+6}{x+1}\right)^{x+4} = \lim_{x\to\infty}\left(1+\frac{5}{x+1}\right)^{x+4}$

$$= \lim_{x\to\infty}\left\{\left(1+\frac{5}{x+1}\right)^{x+1}\right\}^{\frac{x+4}{x+1}}$$

$$= e^{\lim_{x\to\infty}\frac{5(x+4)}{(x+1)}} = e^5$$

132. ***(a)*** We have, $f(x) = \begin{cases} x^2, & x \ge 0 \\ -x^2, & x < 0 \end{cases}$

Clearly, $f(x)$ is differentiable for all $x > 0$ and for all $x < 0$.

So, we check the differentiability at $x = 0$

Now, (RHD at $x = a$)

$$-\left(\frac{d}{dx}(x^2)\right)_{x=0} - (2x)_{x=0} = 0$$

$\therefore$ (LHD at $x = 0$)

$$= \left(\frac{d}{dx}(-x^2)\right)_{x=0}$$

$$= (-2x)_{x=0} = 0$$

(LHD at $x = 0$) = (RHD at $x = 0$)

So, $f(x)$ is differentiable for all x i.e, the set of all points, where $f(x)$ is differentiable is $(-\infty, \infty)$ i.e, R.

133. ***(d)*** $f(x) = 2x^2 - \log|x|$

$$= \begin{cases} 2x^2 - \log x & x > 0 \\ 2x^2 - \log(-x) & x < 0 \end{cases}$$

$$\Rightarrow \quad f'(x) = 4x - \frac{1}{x}, \forall\, x \ne 0$$

For $f(x)$ to be increasing

$$f'(x) > 0 \Rightarrow 4x - \frac{1}{x} > 0$$

$$\Rightarrow \quad \frac{4x^2 - 1}{x} > 0$$

$$\Rightarrow \quad \frac{(2x-1)(2x+1)}{x} > 0$$

$$\Rightarrow \quad x \in \left(-\frac{1}{2}, 0\right) \cup \left(\frac{1}{2}, \infty\right)$$

134. ***(b)*** We have, $f(x) = (x+1)^{1/3} - (x-1)^{1/3}$

$$\therefore \quad f'(x) = \frac{1}{3}\left[\frac{1}{(x+1)^{2/3}} - \frac{1}{(x-1)^{2/3}}\right]$$

$$= \frac{(x-1)^{2/3} - (x+1)^{2/3}}{3(x^2-1)^{2/3}}$$

Clearly, $f'(x)$ does exist at $x = \pm 1$

$$\therefore \quad f'(x) = 0$$

$$\Rightarrow \quad (x-1)^{2/3} = (x+1)^{2/3}$$

$$\Rightarrow \quad x = 0$$

Clearly, $f'(x) \ne 0$ for any other value of $x \in [0, 1]$. The value of $f(x)$ at $x = 0$ is 2.

Hence, the greatest value of $f(x)$ is 2.

135. ***(c)*** $xy = a^2$

$$\therefore \quad x\frac{dy}{dx} + y = 0$$

$$\therefore \quad \frac{dy}{dx} = -\frac{y}{x}$$

$$\therefore \quad \left(\frac{dy}{dx}\right)_1 = -\frac{y}{x}$$

$$\Rightarrow \quad x^2 - y^2 = a^2$$

$$\Rightarrow \quad 2x - 2y\frac{dy}{dx} = 0$$

$$\Rightarrow \quad \frac{dy}{dx} = \frac{x}{y}$$

$$\therefore \quad \left(\frac{dy}{dx}\right)_2 = \frac{x}{y}$$

$$\Rightarrow \quad \left(\frac{dy}{dx}\right)_1 \times \left(\frac{dy}{dx}\right)_2 = -\frac{y}{x} \times \frac{x}{y} = -1$$

$\therefore$ Intersection angle $= \frac{\pi}{2}$

136. ***(a)*** We have, $f'(x) = (x-a)^{2n}(x-b)^{2m+1}$

$$\therefore \quad f'(x) = 0$$

$$\Rightarrow \quad x = a, b$$

Now, for $x = b - h$,

$$f'(x) = (b - h - a)^{2n}(-h)^{2m+1} < 0$$

and for $n = b + h$,

$$f'(x) = (b + h - a)^{2n}\, h^{2m+1} > 0$$

Thus, as x passes through b, $f'(x)$ changes sign from negative to positive. Hence, $x = b$ is a point of minimum.

137. ***(a)*** $\displaystyle\int \frac{1}{1+\sin x}\,dx = \int \frac{1}{1+\cos\left(\frac{\pi}{2}-x\right)}\,dx$

$$= \frac{1}{2}\int \sec^2\left(\frac{\pi}{4}-\frac{x}{2}\right)dx$$

$$= -\tan\left(\frac{\pi}{4}-\frac{x}{2}\right) + b$$

$$= \tan\left(-\frac{\pi}{4}+\frac{x}{2}\right) + b,$$

[where, b = constant]

$\therefore \quad a = -\dfrac{\pi}{4}, b \in R$

138. ***(c)*** $I = \int_0^1 |5x-3|\,dx$

Now, $|5x-3| = \begin{cases} 5x-3, & x \geq \dfrac{3}{5} \\ -(5x-3), & x < \dfrac{3}{5} \end{cases}$

$\therefore \quad I = \int_0^1 |5x-3|\,dx$

$$= \int_0^{3/5} -(5x-3)\,dx + \int_{3/5}^1 (5x-3)\,dx$$

$$= -\left[\frac{5x^2}{2} - 3x\right]_0^{3/5} + \left[\frac{5x^2}{2} - 3x\right]_{3/5}^1$$

$$= -\left[\left(\frac{5}{2}\times\frac{9}{25}-\frac{9}{5}\right)-(0-0)\right] + \left[\left(\frac{5}{2}-3\right)-\left(\frac{5}{2}\times\frac{9}{25}-\frac{9}{5}\right)\right]$$

$$= -\frac{45}{50}+\frac{9}{5}-\frac{1}{2}-\frac{45}{50}+\frac{9}{5} = \frac{13}{10}$$

139. ***(a)***

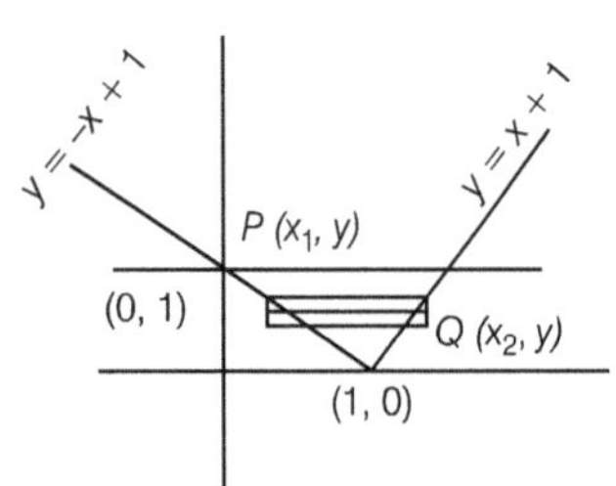

Required area $= \int_0^1 (x_1 - x_2)\,dx$

$$= \int_0^1 \{(-x+1)-(x-1)\}\,dx$$

$$= \int_0^1 (-2x+2)\,dx$$

$$= (-x^2+2x)_0^1$$

$$= -1+2-0+0 = 1$$

140. ***(a)*** The general equation of all non-vertical lines in a plane is

$$ax + by = 1, \qquad (\text{where } b \neq 0)$$

$$\Rightarrow \quad a + b\frac{dy}{dx} = 0 \qquad (\text{Differentiating w.r.t. } x)$$

$$\Rightarrow \quad b\frac{d^2y}{dx^2} = 0 \qquad (\text{Differentiating w.r.t. } x)$$

$$\Rightarrow \quad \frac{d^2y}{dx^2} = 0 \qquad (\because b \neq 0)$$

Hence, the required differential equation is

$$\frac{d^2y}{dx^2} = 0$$

141. ***(c)*** Let the equation of tangent drawn from (0, 0) to the circle be $y = mx$. Then,

$$p = a$$

$$\Rightarrow \quad \frac{7m+1}{\sqrt{m^2+1}} = 5$$

$$\Rightarrow \quad 24m^2 + 14m - 24 = 0$$

$$\Rightarrow \quad 12m^2 + 7m - 12 = 0$$

$$\Rightarrow \quad m_1 m_2 = \frac{-12}{12} = -1$$

$\therefore \quad$ Required angle $= \dfrac{\pi}{2}$

142. ***(c)*** $P'(x) = 2yy'$, $P''(x) = 2yy'' + 2y'^2$

$$\Rightarrow \quad P'''(x) = 2yy''' + 6y'y''$$

$$\therefore \quad 2\frac{d}{dx}(y^3y'') = 2[y^3y''' + 3y^2y'y'']$$

$$= y^2[2yy''' + 6y'y'']$$

$$= P(x)\,P'''(x)$$

143. ***(a)*** We have, $x = 1 + 4\cos\theta$, $y = 2 + 3\sin\theta$

$$\therefore \quad \left(\frac{x-1}{4}\right)^2 + \left(\frac{y-2}{3}\right)^2 = \cos^2\theta + \sin^2\theta$$

$$\Rightarrow \quad \frac{(x-1)^2}{16} + \frac{(y-2)^2}{9} = 1,$$

which is an ellipse.

144. ***(b)*** Since centre of the circle (3, 4) satisfies the line $4x \rightarrow 3y = 24$ therefore the gives line through the centre of the circle and is the diameter of the circle and the origin lies on the circle. So, required angle is the right angle in the semi-circle.

145. ***(c)*** At the point on the line, where it intersects the curve, we have $z = 0$, so that

$$\frac{x-2}{3}=\frac{y+1}{2}=\frac{0-1}{-1}$$

$\Rightarrow \quad \frac{x-2}{3}=1$

and $\quad \frac{y+1}{2}=1$

$\Rightarrow \quad x=5$

and $\quad y=1$

Putting these values of x and y in $xy=c^2$, we get

$c^2=5$

$\Rightarrow \quad c=\pm\sqrt{5}$

146. ***(d)*** $(\mathbf{a}\times\hat{\mathbf{j}})\cdot(2\hat{\mathbf{j}}-3\hat{\mathbf{k}})=\mathbf{a}\cdot\{\hat{\mathbf{j}}\times(2\hat{\mathbf{j}}-3\hat{\mathbf{k}})\}$

$=\mathbf{a}\cdot(-3\,i)$

$=-3\,(\mathbf{a}\cdot\hat{\mathbf{j}})=-12 \quad (\because \mathbf{a}\cdot\hat{\mathbf{i}}=4)$

147. ***(a)*** Let $\mathbf{a}=a_1\hat{\mathbf{i}}+a_2\hat{\mathbf{j}}+a_3\hat{\mathbf{k}}$, then

$\mathbf{a}\cdot\hat{\mathbf{i}}=a_1,\ \mathbf{a}\cdot\hat{\mathbf{j}}=a_2,\ \mathbf{a}\cdot\hat{\mathbf{k}}=a_3$

$\therefore \quad \mathbf{a}=(\mathbf{a}\cdot\hat{\mathbf{i}})\,\hat{\mathbf{i}}+(\mathbf{a}\cdot\hat{\mathbf{j}})\hat{\mathbf{j}}+(\mathbf{a}\cdot\hat{\mathbf{k}})\,\hat{\mathbf{k}}$

148. ***(d)*** The equations of the required chord is given by

$S'=T$

$\Rightarrow \quad 9-8\times 2=-3y-4(x+2)$

$\Rightarrow \quad 9-16=-4x-3y-8$

$\Rightarrow \quad 4x+3y+1=0$

149. ***(d)*** Let n_1 and n_2 be the number of observations in two groups having means $\bar{x}_1$ and $\bar{x}_2$ respectively.

Then, $\quad \bar{x}=\frac{n_1\bar{x}_1+n_2\bar{x}_2}{n_1+n_2}$

Now, $\quad \bar{x}-\bar{x}_1=\frac{n_1\bar{x}_1+n_2\bar{x}_2}{n_1+n_2}-\bar{x}_1$

$=\frac{n_2(\bar{x}_2-\bar{x}_1)}{n_1+n_2}>0$

$(\because \bar{x}_2>\bar{x}_1)$

$\Rightarrow \quad \bar{x}>\bar{x}_1 \quad \ldots$(i)

And $\quad \bar{x}-\bar{x}_2=\frac{n(\bar{x}_1-\bar{x}_2)}{n_1+n_2}<0 \quad (\because \bar{x}_2>\bar{x}_1)$

$\Rightarrow \quad \bar{x}<\bar{x}_2 \quad \ldots$(ii)

From Eqs. (i) and (ii), we get

$\bar{x}_1<\bar{x}<\bar{x}_2$

150. ***(c)*** The given diagonal does not passes through (1, 3) and (5, 1) as slope of the diagonal is 2 and that of the line joining (1, 3) and (5, 1) $=\frac{1-3}{5-1}=-\frac{1}{2}$.

The centre of the rectangle is mid-point of the join of (1, 3) and (5, 1) must lies on $y=2x+c$.

$\Rightarrow \quad \left(\frac{1+5}{2},\frac{3+1}{2}\right)=(3,2)$ lies on $y=2x+c$

$\Rightarrow \quad 2=2\times 3+c$

$\Rightarrow \quad c=-4$

BITSAT

Birla Institute of Technology & Science Admission Test

Mock Test 3

Instructions

- There are 150 questions in all. The number of questions in each part is as given below.

	No. of Questions
Part I Physics	1–40
Part II Chemistry	41–80
Part III a. English Proficiency	81–95
b. Logical Reasoning	96–105
Part IV Mathematics	106–150

- All questions are Multiple Choice Questions having four options out of which **only one** is correct.
- Each correct answer fetches 3 marks while incorrect answer has a penalty of 1 mark.
- Time allotted to complete this paper is 3 hrs.

PART I

Physics

1. When a wave traverses a medium, the displacement of a particle located at x at a time t is given by $y = a \sin(bt - cx)$, where a, b and c are constants of the wave, which of the following is a quantity with dimensions?

a. $\frac{y}{a}$ *b.* bt *c.* cx *d.* $\frac{b}{c}$

2. A body is projected vertically upwards at time $t = 0$ and it is seen at a height H at time t_1 and t_2 second during its flight. The maximum height attained is (g is acceleration due to gravity)

a. $\frac{g(t_2 - t_1)^2}{8}$ *b.* $\frac{g(t_1 + t_2)^2}{4}$

c. $\frac{g(t_1 + t_2)^2}{8}$ *d.* $\frac{g(t_2 - t_1)^2}{4}$

3. A particle is projected up from a point at an angle θ with the horizontal direction. At any time t, if p is the linear momentum, y is the vertical displacement, x is horizontal displacement, the graph among the following which does not represent the variation of kinetic energy KE of the particle is

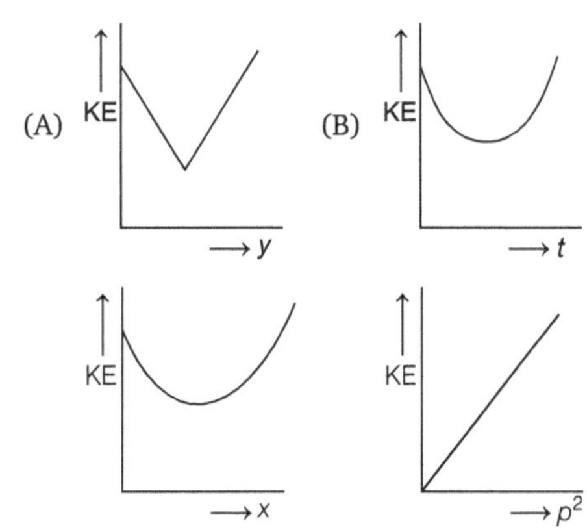

a. graph (A) *b.* graph (B)

c. graph (C) *d.* graph (D)

4. A motor of power P_0 is used to deliver water at a certain rate through a given horizontal pipe. To increase the rate of flow of water through the same pipe n times, the power of the motor is increased to P_1. The ratio of P_1 to P_0 is

a. $n:1$ *b.* $n^2:1$

c. $n^3:1$ *d.* $n^4:1$

5. A body of mass 5 kg makes an elastic collision with another body at rest and continues to move in the original direction after collision with a velocity equal to 1/10th of its original velocity. Then, the mass of the second body is

a. 4.09 kg *b.* 0.5 kg *c.* 5 kg *d.* 5.09 kg

6. A particle of mass 4 m explodes into three pieces of masses m, m and $2m$. The equal masses move along X-axis and Y-axis with velocities 4 ms^{-1} and 6 ms^{-1}, respectively. The magnitude of the velocity of the heavier mass is

a. $\sqrt{17} \text{ ms}^{-1}$ *b.* $2\sqrt{13} \text{ ms}^{-1}$

c. $\sqrt{13} \text{ ms}^{-1}$ *d.* $\frac{\sqrt{13}}{2} \text{ ms}^{-1}$

7. A body is projected vertically upwards from the surface of the Earth with a velocity equal to half the escape velocity. If R is the radius of the Earth, maximum height attained by the body from the surface of the Earth is

a. $R/6$ *b.* $R/3$ *c.* $2R/3$ *d.* R

8. The displacement of a particle executing SHM is given by

$$y = 5 \sin\left(4t + \frac{\pi}{3}\right)$$

If T is the time period and the mass of the particle is 2 g, then the kinetic energy of the particle when $t = \frac{T}{4}$ is given by

a. 0.4 J *b.* 0.5 J *c.* 3 J *d.* 0.3 J

9. If the ratio of lengths, radii and Young's modulus of steel and brass wires shown in the figure are a, b and c respectively, the ratio between the increase in lengths of brass and steel wires would be

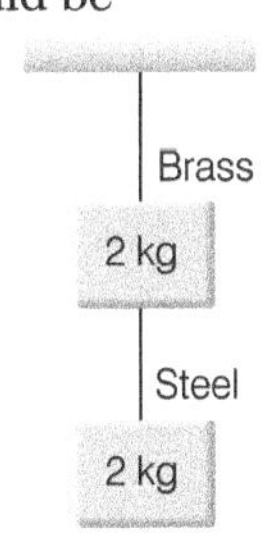

a. $\frac{b^2a}{2c}$ *b.* $\frac{bc}{2a^2}$ *c.* $\frac{ba^2}{2c}$ *d.* $\frac{a}{2b^2c}$

10. A soap bubble of radius r is blown up to form a bubble of radius $2r$ under isothermal conditions. If T is the surface tension of soap solution, then the energy spent in the blowing is

a. $3\pi Tr^2$ *b.* $6\pi Tr^2$

c. $12\pi Tr^2$ *d.* $24\pi Tr^2$

11. Eight spherical rain drops of the same mass and radius are falling down with a terminal speed of 6 cms^{-1}. If they coalesce to form one big drop, what will be the terminal speed of bigger drop? (Neglect the buoyancy of the air)

a. 1.5 cm s^{-1} *b.* 6 cm s^{-1}

c. 24 cm s^{-1} *d.* 32 cm s^{-1}

12. A clock pendulum made of invar has a period of 0.5 s, at 20°C. If the clock is used in a climate where the temperature averages to 30°C, how much time does the clock lose in each oscillation? (For invar, $\alpha = 9 \times 10^{-7}/°\text{C}$, g = constant)

a. 2.25×10^{-6} s *b.* 2.5×10^{-7} s

c. 5×10^{-7} s *d.* 1.125×10^{-6} s

13. A piece of metal weighs 45 g in air and 25 g in a liquid of density $1.5 \times 10^3 \text{ kg-m}^{-3}$ kept at 30°C. When the temperature of the liquid is raised to 40°C, the metal piece weighs 27 g. The density of liquid at 40°C is $1.25 \times 10^3 \text{ kg-m}^{-3}$. The coefficient of linear expansion of metal is

a. $1.3 \times 10^{-3}/°\text{C}$ *b.* $5.2 \times 10^{-3}/°\text{C}$

c. $2.6 \times 10^{-3}/°\text{C}$ *d.* $0.26 \times 10^{-3}/°\text{C}$

14. An ideal gas is subjected to a cyclic process $ABCD$ as depicted in the p-V diagram given below

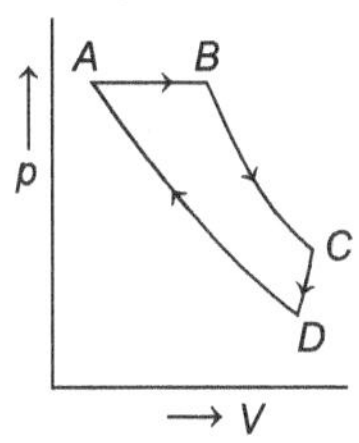

Which of the following curves represents the equivalent cyclic process?

a.
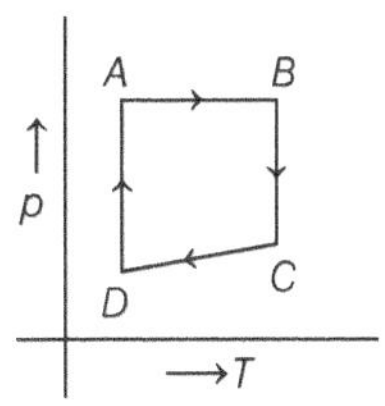

b.
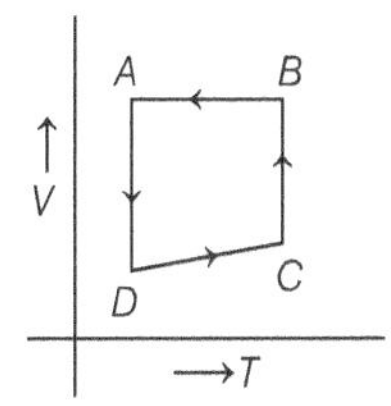

c.
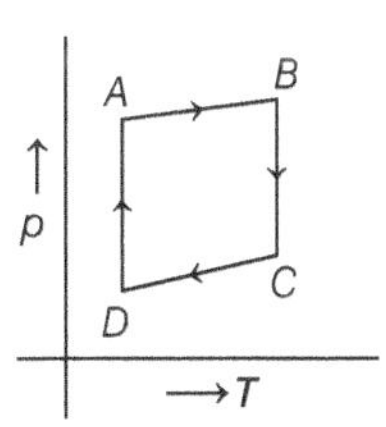

d.
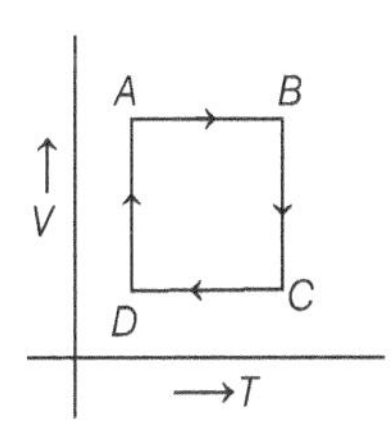

15. An ideal gas is subjected to cyclic process involving four thermodynamic states, the amounts of heat (Q) and work (W) involved in each of these states are

$Q_1 = 6000$ J, $Q_2 = -5500$ J; $Q_3 = -3000$ J; $Q_4 = 3500$ J

$W_1 = 2500$ J; $W_2 = -1000$ J; $W_3 = -1200$ J; $W_4 = x$ J.

The ratio of the net work done by the gas to the total heat absorbed by the gas is η. The values of x and η respectively are

a. 500; 7.5% **b.** 700; 10.5%
c. 1000; 21% **d.** 1500; 15%

16. Two cylinders A and B fitted with pistons contain equal number of moles of an ideal monoatomic gas at 400 K. The piston of A is free to move, while that of B is held fixed. Same amount of heat energy is given to the gas in each cylinder. If the rise in temperature of the gas in A is 42 K, the rise in temperature of the gas in B is

a. 21 K **b.** 35 K **c.** 42 K **d.** 70 K

17. Three rods of same dimensional have thermal conductivity 3 K, 2 K and K. They are arranged as shown in the figure below

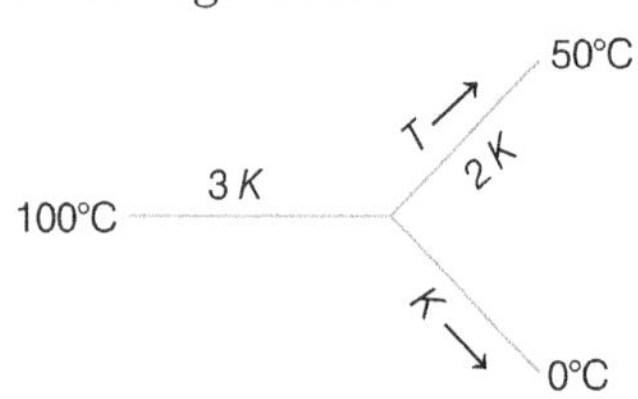

Then, the temperature of the junction in steady state is

a. $\frac{200}{3}$ °C **b.** $\frac{100}{3}$ °C **c.** 75° C **d.** $\frac{50}{3}$ °C

18. Two sources A and B are sending notes of frequency 680 Hz. A listener moves from A to B with a constant velocity u. If the speed of sound in air is 340 ms^{-1}, what must be the value of u, so that he hears 10 beats per second?

a. 2.0 ms^{-1} **b.** 2.5 ms^{-1}
c. 3.0 ms^{-1} **d.** 3.5 ms^{-1}

19. Two identical piano wires have a fundamental frequency of 600 cycle per second when kept under the same tension. What fractional increase in the tension of one wires will lead to the occurrence of 6 beats per second when both wires vibrate simultaneously?

a. 0.01 **b.** 0.02
c. 0.03 **d.** 0.04

20. In the Young's double slit experiment, the intensities at two points P_1 and P_2 on the screen are respectively I_1 and I_2. If P_1 is located at the centre of a bright fringe and P_2 is located at a distance equal to a quarter of fringe width from P_1, then $\frac{I_1}{I_2}$ is

a. 2 **b.** $\frac{1}{2}$
c. 4 **d.** 16

21. In Young's double slit experiment, the 10th maximum of wavelength λ_1 is at a distance of y_1 from the central maximum. When the wavelength of the source is changed to λ_2, 5th maximum is at a distance of y_2 from its central maximum. The ratio $\left(\frac{y_1}{y_2}\right)$ is

a. $\frac{2\lambda_1}{\lambda_2}$ **b.** $\frac{2\lambda_2}{\lambda_1}$
c. $\frac{\lambda_1}{2\lambda_2}$ **d.** $\frac{\lambda_2}{2\lambda_1}$

22. Four light sources produce the following four waves :

(i) $y_1 = a \sin(\omega t + \phi_1)$
(ii) $y_2 = a \sin 2\omega t$
(iii) $y_3 = a' \sin(\omega t + \phi_2)$
(iv) $y_4 = a' \sin(3\omega t + \phi)$

Superposition of which two waves give rise to interference?

a. (i) and (ii) **b.** (ii) and (iii)
c. (i) and (iii) **d.** (iii) and (iv)

23. The two lenses of an achromatic doublet should have

a. equal powers
b. equal dispersive powers
c. equal ratio of their power and dispersive power
d. sum of the product of their powers and dispersive power equal to zero

24. Two bar magnets A and B are placed one over the other and are allowed to vibrate in a vibration magnetometer. They make 20 oscillations per minute when the similar poles of A and B are on the same side, while they make 15 oscillations per minute when their opposite poles lie on the same side. If M_A and M_B are the magnetic moments of A and B and if $M_A > M_B$, the ratio of M_A and M_B is

a. 4 : 3 **b.** 25 : 7
c. 7 : 5 **d.** 25 : 16

25. A bar magnet is 10 cm long is kept with its north (N)-pole pointing north. A neutral point is formed at a distance of 15 cm from each pole. Given the horizontal component of earth's field is 0.4 G, the pole strength of the magnet is

a. 9 A-m *b.* 6.75 A-m
c. 27 A-m *d.* 1.35 A-m

26. An infinitely long thin straight wire has uniform linear charge density of $\frac{1}{3}\text{ Cm}^{-1}$. Then, the magnitude of the electric intensity at a point 18 cm away is (given, $\varepsilon_0 = 8.8\times 10^{-12}\text{ C}^2\text{ Nm}^{-2}$)

a. $0.33\times 10^{11}\text{ NC}^{-1}$ *b.* $3\times 10^{11}\text{ NC}^{-1}$
c. $0.66\times 10^{11}\text{ NC}^{-1}$ *d.* $1.32\times 10^{11}\text{ NC}^{-1}$

27. Two point charges $-q$ and $+q$ are located at points $(0, 0, -a)$ and $(0, 0, a)$ respectively. The electric potential at a point $(0, 0, z)$, where $z > a$ is

a. $\frac{qa}{4\pi\varepsilon_0 z^2}$ *b.* $\frac{q}{4\pi\varepsilon_0 a}$
c. $\frac{2qa}{4\pi\varepsilon_0(z^2 - a^2)}$ *d.* $\frac{2qa}{4\pi\varepsilon_0(z^2 + a^2)}$

28. In the adjacent shown circuit, a voltmeter of internal resistance R, when connected across B and C reads $\frac{100}{3}$ V. Neglecting the internal resistance of the battery, the value of R is

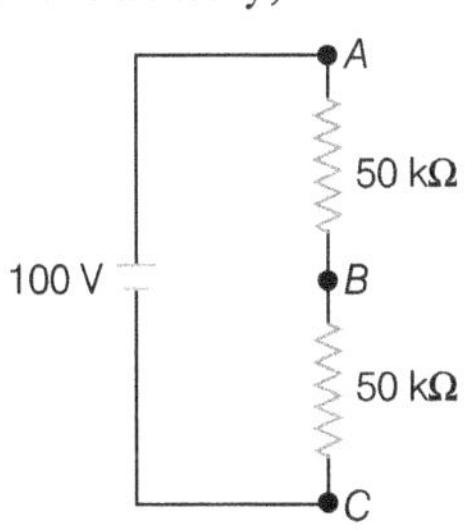

a. 100 kΩ *b.* 75 kΩ
c. 50 kΩ *d.* 25 kΩ

29. A cell in secondary circuit gives null deflection for 2.5 m length of potentiometer having 10 m length of wire. If the length of the potentiometer wire is increased by 1 m without changing the cell in the primary, the position of the null point now is

a. 3.5 m *b.* 3 m
c. 2.75 m *d.* 2.0 m

30. The following series L-C-R circuit, when driven by an emf source of angular frequency 70 k-rad/s per second, the circuit effectively behaves like

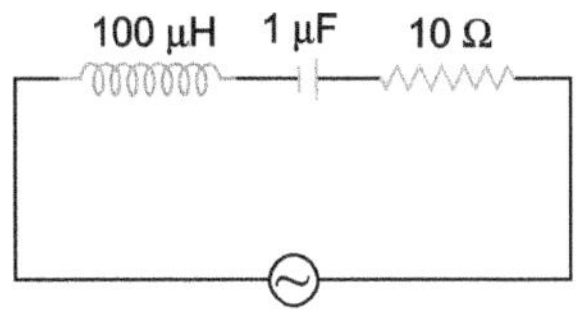

a. purely resistive circuit
b. series R-L circuit
c. series R-C circuit
d. series L-C circuit with $R = 0$

31. A wire of length l is bent into a circular loop of radius R and carries a current I. The magnetic field at the centre of the loop is B. The same wire is now bent into a double loop of equal radii. If both loops carry the same current I and it is in the same direction, the magnetic field at the centre of the double loop will be

a. Zero *b.* 2 B *c.* 4 B *d.* 8 B

32. An infinitely long straight conductor is bent into the shape as shown below. It carries a current of I ampere and the radius of the circular loop is R metre. Then, the magnitude of magnetic induction at the centre of the circular loop is

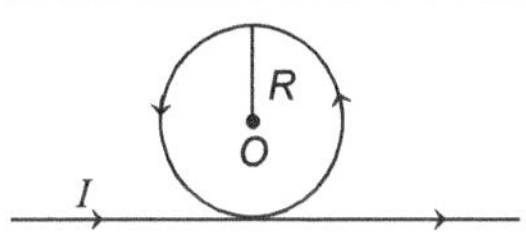

a. $\frac{\mu_0 I}{2\pi R}$ *b.* $\frac{\mu_0 nI}{2R}$
c. $\frac{\mu_0 I}{2\pi R}(\pi + 1)$ *d.* $\frac{\mu_0 I}{2\pi R}(\pi - 1)$

33. The work function of a certain metal is 3.31×10^{-19} J. Then, the maximum kinetic energy of photoelectrons emitted by incident radiation of wavelength 5000Å is

(Given, $h = 6.62\times 10^{-34}$ J-s, $c = 3\times 10^8\text{ ms}^{-1}$, $e = 1.6\times 10^{-19}$ C)

a. 2.48 eV *b.* 0.41 eV
c. 2.07 eV *d.* 0.82 eV

34. A photon of energy E ejects a photoelectron from a metal surface whose work-function is W_0. If this electron enters into a uniform magnetic field of induction B in a direction perpendicular to the field and describes a circular path of radius r, then the radius r is given by (in the usual notation)

a. $\frac{\sqrt{2m(W_0 - E)}}{eB}$ *b.* $\sqrt{2m(E - W_0)eB}$
c. $\frac{\sqrt{2e(E - W_0)}}{mB}$ *d.* $\frac{\sqrt{2m(E - W_0)}}{eB}$

35. Two radioactive materials X_1 and X_2 have decay constants 10λ and λ respectively. If initially they have the same number of nuclei, then the ratio of the number of nuclei of X_1 to that of X_2 will be $1/e$ after a time

a. $(1/10\lambda)$ *b.* $1/(11\lambda)$
c. $11/(10\lambda)$ *d.* $1/(9\lambda)$

36. Currents flowing in each of the following circuits A and B respectively are

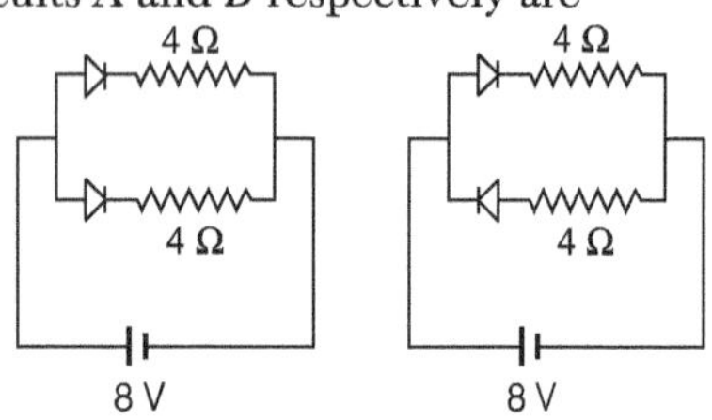

(Circuit *A*) (Circuit *B*)

a. 1 A, 2 A *b.* 2 A, 1 A
c. 4 A, 2 A *d.* 2 A, 4 A

37. A bullet of mass 0.02 kg travelling horizontally with velocity 250 ms^{-1} strikes a block of wood of mass 0.23 kg which rests on a rough horizontal surface. After the impact, the block and bullet move together and come to rest after travelling a distance of 40 m. The coefficient of sliding friction of the rough surface is $(g = 9.8 \text{ ms}^{-2})$

a. 0.75 *b.* 0.61 *c.* 0.51 *d.* 0.30

38. Two persons *A* and *B* are located in *X*-*Y* plane at the points (0, 0) and (0, 10) respectively. (The distances are measured in MKS unit). At a time $t = 0$, they start moving simultaneously with velocities $\mathbf{v}_A = 2\mathbf{j} \text{ ms}^{-1}$ and $\mathbf{v}_B = 2\hat{\mathbf{i}} \text{ ms}^{-1}$, respectively. The time after which *A* and *B* are at their closest distance is

a. 2.5 s *b.* 4 s
c. 1 s *d.* $\frac{10}{\sqrt{2}}$ s

39. A rod of length l is held vertically stationary with its lower end located at a point *P*, on the horizontal plane. When the rod is released to topple about *P*, the velocity of the upper end of the rod with which it hits the ground is

a. $\sqrt{\frac{g}{l}}$ *b.* $\sqrt{3gl}$
c. $3\sqrt{\frac{g}{l}}$ *d.* $\sqrt{\frac{3g}{l}}$

40. A wheel of radius 0.4 m can rotate freely about its axis as shown in the figure. A string is wrapped over its rim and a mass of 4 kg is hung. An angular acceleration of 8 rad s^{-2} is produced in it due to the torque. Then, moment of inertia of the wheel is $(g = 10 \text{ ms}^{-2})$

a. 2 kg-m^2 *b.* 1 kg-m^2
c. 4 kg-m^2 *d.* 8 kg-m^2

PART II

Chemistry

41. Given that, $\Delta H_f(\text{H}) = 218$ kJ/mol, express the H—H bond energy in kcal/mol.

a. 52.15 *b.* 911
c. 104 *d.* 52153

42. Identify the alkyne in the following sequence of reactions,

$$\text{Alkyne} \xrightarrow[\text{Lindlar's catalyst}]{H_2} A \xrightarrow{\text{Ozonolysis}} \underset{\text{only}}{B}$$

$$(B) \xleftarrow[\text{Process}]{\text{Wacker}} CH_2{=}CH_2$$

a. $H_3C—C{\equiv}C—CH_3$
b. $H_3C—CH_2—C{\equiv}CH$
c. $H_2C{=}CH—C{\equiv}CH$
d. $HC{\equiv}C—CH_2—C{\equiv}CH$

43. Fluorine reacts with dilute NaOH and forms a gaseous product *A*. The bond angle in the molecule of *A* is

a. 104°40′ *b.* 103°
c. 107° *d.* 109°28′

44. One mole of alkene *X* on ozonolysis gave one mole of acetaldehyde and one mole of acetone. The IUPAC name of *X* is

a. 2-methyl-2-butene *b.* 2-methyl-1-butene
c. 2-butene *d.* 1-butene

45. The number of $p\pi$-$d\pi$ bonds present in XeO_3 and XeO_4 molecules, respectively are

a. 3, 4 *b.* 4, 2
c. 2, 3 *d.* 3, 2

46. The wavelengths of electron waves in two orbits is 3 : 5. The ratio of kinetic energy of electrons will be

a. 25 : 9 *b.* 5 : 3
c. 9 : 25 *d.* 3 : 5

47. Which one of the following sets correctly represents the increase in the paramagnetic property of the ions?

a. $Cu^{2+} > V^{2+} > Cr^{2+} > Mn^{2+}$
b. $Cu^{2+} < Cr^{2+} < V^{2+} < Mn^{2+}$
c. $Cu^{2+} < V^{2+} < Cr^{2+} < Mn^{2+}$
d. $V^{2+} < Cu^{2+} < Cr^{2+} < Mn^{2+}$

48. Electrons with a kinetic energy of 6.023×10^4 J / mol are evolved from the surface of a metal, when it is exposed to radiation of wavelength of 600 nm. The minimum amount of energy required to remove an electron from the metal atom is

a. 2.3125×10^{-19} J *b.* 3×10^{-19} J
c. 6.02×10^{-19} J *d.* 6.62×10^{-34} J

49. The chemical entities present in thermosphere of the atmosphere are

a. O_2^+, O^+, NO^+ *b.* O_3
c. N_2, O_2, CO_2, H_2O *d.* O_3, O_2^+, O_2

50. The type of bonds present in sulphuric anhydride are

a. 3σ and three $p\pi$-$d\pi$
b. 3σ, one $p\pi$-$p\pi$ and two $p\pi$-$d\pi$
c. 2σ and three $p\pi$-$d\pi$
d. 2σ and two $p\pi$-$d\pi$

51. In Gattermann reaction, a diazonium group is replaced by X using Y. X and Y are

	X	Y
a.	$Cl^{\ominus}$	Cu/HCl
b.	$Cl^{\oplus}$	$CuCl_2$/HCl
c.	$Cl^{\ominus}$	$CuCl_2$/HCl
d.	Cl_2	Cu_2O/HCl

52. Which pair of oxyacids of phosphorus contains P—H bonds?

a. H_3PO_4, H_3PO_3 *b.* $H_3PO_5, H_4P_2O_7$
c. H_3PO_3, H_3PO_2 *d.* H_3PO_2, HPO_3

53. Dipole moment of HCl = 1.03 D, HI = 0.38 D. Bond length of HCl = 1.3 Å and HI = 1.6 Å. The ratio of fraction of electric charge, δ, existing on each atom in HCl and HI is

a. 12 : 1 *b.* 2.7 : 1
c. 3.3 : 1 *d.* 1 : 3.3

54. $SiCl_4$ on hydrolysis forms X and HCl. Compound X loses water at 1000°C and gives Y. Compounds X and Y respectively are

a. H_2SiCl_6, SiO_2
b. H_4SiO_4, Si
c. SiO_2, Si
d. H_4SiO_4, SiO_2

55. 1.5 g of $CdCl_2$ was found to contain 0.9 g of Cd. Calculate the atomic weight of Cd.

a. 118 *b.* 112
c. 106.5 *d.* 53.25

56. Aluminium reacts with NaOH and forms compound 'X'. If the coordination number of aluminium in 'X' is 6, the correct formula of X is

a. $[Al(H_2O)_4(OH)_2]^+$ *b.* $[Al(H_2O)_3(OH)_3]$
c. $[Al(H_2O)_2(OH)_4]^-$ *d.* $[Al(H_2O)_6](OH)_3$

57. The average kinetic energy of one molecule of an ideal gas at 27°C and 1 atm pressure is

a. 900 cal K^{-1} mol^{-1}
b. 6.21×10^{-21} JK^{-1} $molecule^{-1}$
c. 336.7 JK^{-1} $molecule^{-1}$
d. 3741.3 JK^{-1} mol^{-1}

58. **Assertion** (A) K, Rb and Cs form superoxides.

Reason (R) The stability of the superoxides increases from K to Cs due to decrease in lattice energy.

The correct answer is

a. Both (A) and (R) are true and (R) is the correct explanation of (A).
b. Both (A) and (R) are true but (R) is not the correct explanation of (A).
c. (A) is true but (R) is not true.
d. (A) is not true but (R) is true.

59. How many mL of perhydrol is required to produce sufficient oxygen which can be used to completely convert 2 L of SO_2 gas to SO_3 gas?

a. 10 mL *b.* 5 mL
c. 20 mL *d.* 30 mL

60. pH of a buffer solution decreases by 0.02 units when 0.12 g of acetic acid is added to 250 mL of a buffer solution of acetic acid and potassium acetate at 27°C. The buffer capacity of the solution is

a. 0.1 *b.* 10
c. 1 *d.* 0.4

61. Match the following.

List I	List II
(A) Feldspar	(I) $[Ag_3SbS_3]$
(B) Asbestos	(II) $Al_2O_3 \cdot H_2O$
(C) Pyrargyrite	(III) $MgSO_4 \cdot H_2O$
(D) Diaspore	(IV) $KAlSi_3O_8$
	(V) $CaMg_3(SiO_3)_4$

The correct answer is

	A	B	C	D
a.	IV	V	II	I
b.	IV	V	I	II
c.	IV	I	III	II
d.	II	V	IV	I

62. Which one of the following order is correct for the first ionisation energies of the elements?

a. B < Be < N < O b. Be < B < N < O
c. B < Be < O < N d. B < O < Be < N

63. What are X and Y in the following reaction sequence?

$$C_2H_5OH \xrightarrow{Cl_2} X \xrightarrow{Cl_2} Y$$

a. C_2H_5Cl, CH_3CHO
b. CH_3CHO, CH_3CO_2H
c. CH_3CHO, CCl_3CHO
d. C_2H_5Cl, CCl_3CHO

64. What are A, B, C in the following reactions?

(I) $(CH_3CO_2)_2Ca \xrightarrow{\Delta} A$

(II) $CH_3CO_2H \xrightarrow[\text{Red P}]{HI} B$

(III) $2CH_3CO_2H \xrightarrow{P_4O_{10}} C$

	A	B	C
a.	C_2H_6	CH_3COCH_3	$(CH_3CO)_2O$
b.	$(CH_3CO)_2O$	C_2H_6	CH_3COCH_3
c.	CH_3COCH_3	C_2H_6	$(CH_3CO)_2O$
d.	CH_3COCH_3	$(CH_3CO)_2O$	C_2H_6

65. One per cent composition of an organic compound A is, carbon 85.71% and hydrogen 14.29%. Its vapour density is 14. Consider the following reaction sequence

$$A \xrightarrow{Cl_2/H_2O} B \xrightarrow[\text{(ii) } H_3O^+]{\text{(i) KCN/EtOH}} C$$

and identify C,

a. $CH_3—CH(OH)—CO_2H$
b. $HO—CH_2—CH_2—CO_2H$
c. $HO—CH_2—CO_2H$
d. $CH_3—CH_2—CO_2H$

66. How many tripeptides can be prepared by linking the amino acids glycine, alanine and phenyl alanine?

a. One
b. Three
c. Six
d. Twelve

67. A codon has a sequence of A, and specifies a particular B that is to be incorporated into a C. What are A, B and C?

	A	B	C
a.	3 bases	amino acid	carbohydrate
b.	3 acids	carbohydrate	protein
c.	3 bases	protein	amino acid
d.	3 bases	amino acid	protein

68. Parkinson's disease is linked to abnormalities in the levels of dopamine in the body. The structure of dopamine is

a. 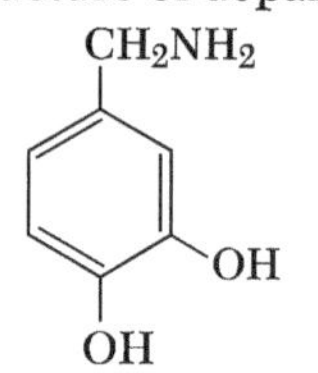

b. 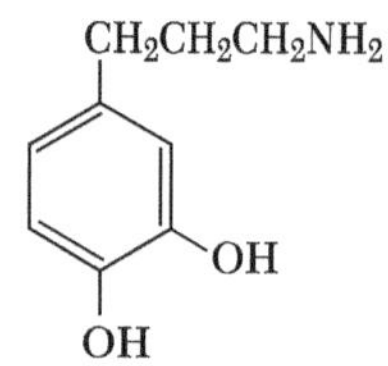

c. 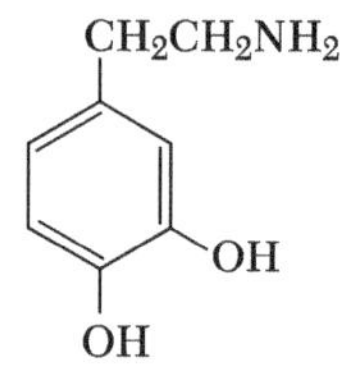

d. 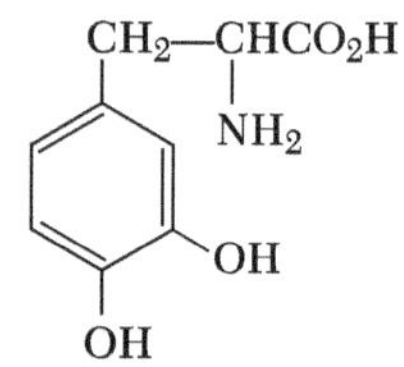

69. During the depression in freezing point experiment, an equilibrium is established between the molecules of

a. liquid solvent and solid solvent
b. liquid solute and solid solvent
c. liquid solute and solid solute
d. liquid solvent and solid solute

70. Consider the following reaction,

$$C_2H_5Cl + AgCN \xrightarrow{EtOH/H_2O} X \text{ (major)}$$

Which one of the following statements is true for X?

(I) It gives propionic acid on hydrolysis.
(II) It has an ester functional group.
(III) It has a nitrogen linked to ethyl carbon.
(IV) It has a cyanide group.

a. IV b. III c. II d. I

71. For the following cell reaction,

$$Ag \mid Ag^+ \mid AgCl \mid Cl^{\ominus} \mid Cl_2, Pt$$

$$\Delta G_f^\circ(AgCl) = -109 \text{ kJ/mol}$$

$$\Delta G_f^\circ(Cl^{\ominus}) = -129 \text{ kJ/mol}$$

$$\Delta G_f^\circ(Ag^+) = 78 \text{ kJ/mol}$$

E° of the cell is

a. −0.60 V b. 0.60 V
c. 6.0 V d. None of these

72. The synthesis of crotonaldehyde from acetaldehyde is an example of ...... reaction.

a. nucleophilic addition
b. elimination
c. electrophilic addition
d. nucleophilic addition elimination

73. At 25°C, the molar conductances at infinite dilution for the strong electrolytes NaOH, NaCl and $BaCl_2$ are 248×10^{-4}, 126×10^{-4} and 280×10^{-4} $Sm^2\ mol^{-1}$ respectively. λ°_m $Ba(OH)_2$ in $Sm^2\ mol^{-1}$ is

a. 52.4×10^{-4} *b.* 524×10^{-4}
c. 402×10^{-4} *d.* 262×10^{-4}

74. The cubic unit cell of a metal (molar mass $= 63.55\ g\ mol^{-1}$) has an edge length of 362 pm. Its density is $8.92\ g\ cm^{-3}$. The type of unit cell is

a. primitive
b. face centered
c. body centered
d. end centered

75. The equilibrium constant for the given reaction is 100.

$$N_2(g) + 2O_2(g) \rightleftharpoons 2NO_2(g)$$

What is the equilibrium constant for the reaction given below?

$$NO_2(g) \rightleftharpoons \frac{1}{2}N_2(g) + O_2(g)$$

a. 10 *b.* 1
c. 0.1 *d.* 0.01

76. For a first order reaction at 27°C, the ratio of time required for 75% completion to 25% completion of reaction is

a. 3.0 *b.* 2.303 *c.* 4.8 *d.* 0.477

77. The concentration of an organic compound in chloroform is 6.15 g per 100 mL of solution. A portion of this solution in a 5 cm polarimeter tube causes an observed rotation of –1.2°. What is the specific rotation of the compound?

a. +12° *b.* –3.9° *c.* –39° *d.* +61.5°

78. 20 mL of 0.1 M acetic acid is mixed with 50 mL of potassium acetate. pK_a of acetic acid $= 1.8 \times 10^{-5}$ at 27°C. Calculate concentration of potassium acetate if pH of the mixture is 4.8.

a. 0.1 M *b.* 0.04 M
c. 0.4 M *d.* 0.02 M

79. Calculate ΔH° for the reaction,

$$Na_2O(s) + SO_3(g) \longrightarrow Na_2SO_4(g)$$

given the following :

(A) $Na(s) + H_2O(l) \longrightarrow NaOH(s) + \frac{1}{2}H_2(g)$; $\Delta H^\circ = -146$ kJ

(B) $Na_2SO_4(s) + H_2O(l) \longrightarrow 2NaOH(s) + SO_3(g)$; $\Delta H^\circ = +418$ kJ

(C) $2Na_2O(s) + 2H_2(g) \longrightarrow 4Na(s) + 2H_2O(l)$; $\Delta H^\circ = +259$ kJ

a. +823 kJ *b.* –581 kJ
c. – 435 kJ *d.* +531 kJ

80. Which one of the following is most effective in causing the coagulation of an As_2S_3 sol?

a. KCl *b.* $AlCl_3$
c. $MgSO_4$ *d.* $K_3[Fe(CN)_6]$

Mock Test 3

PART III

a. English Proficiency

Directions (Q. Nos. 81-83) : *In each of the following questions, choose the most appropriate alternative to fill in the blanks.*

81. It is difficult to believe what he tells us because his account of any event is always full of .... of all sorts.

a. discrepancies *b.* differences
c. discretions *d.* distinctions

82. The bank clerk tried to ...... money from his friend's account.

a. empower *b.* embellish
c. embroil *d.* embezzle

83. Eight scientists have ...... the national awards for outstanding contribution and dedication to the profession.

a. bestowed *b.* picked *c.* bagged *d.* conferred

Directions (Q. Nos. 84-86) : *In the following questions, some parts have been jumbled up. You are required to rearrange these parts, which are labelled P, Q, R and S to produce the correct sentence.*

84. Freedom, is the restricted kind in the sense/(P), the rich and poor woman/(Q), that a wide gulf separates/(R), which a modern woman enjoys(S)

a. P S R Q *b.* S R Q P
c. R Q P S *d.* S P R Q

85. In life, some rules are/(P), as in business/(Q), they seem almost instinctive/(R), learnt so early that/(S)

a. R S P Q   *b.* Q P S R
c. R P S Q   *d.* Q S P R

86. Kapil, left in an aeroplane/(P), after reading a sailing magazine/(Q), had decided/(R), to build his own boat nine years earlier/(S)

a. P R Q S   *b.* R S Q P
c. R Q P S   *d.* P S R Q

Directions (Q. Nos. 87-89) *In each of the following questions,s choose the alternative which is most nearly the same in meaning to the word given in capital letters.*

87. DENOUEMENT

a. Outcome   *b.* Eschew
c. Action   *d.* Character

88. GAUCHE

a. Vain   *b.* Rich   *c.* Polished   *d.* Tactless

89. ACCOLADE

a. Honour   *b.* Appreciation
c. Greeting   *d.* Gift

Directions (Q. Nos. 90-92) : *In each of the following questions, choose the alternative which is opposite in meaning to the word given in capital letters.*

90. REPRIMAND

a. Reward   *b.* Appreciate
c. Encourage   *d.* Praise

91. IMPERTINENT

a. Polite   *b.* Indifferent
c. Unpleasant   *d.* Stubborn

92. EQUIVOCAL

a. Mistaken *b.* Quaint
c. Clear   *d.* Universal

Directions (Q. Nos. 93-95) *In each of the following questions, choose the alternative which can be substituted for the given words/sentences.*

93. Design made by putting together coloured pieces of glass or stones

a. Oleograph
b. Mosaic
c. Tracery
d. Relief

94. The doctrine that human soul passes from one body to another at the time of death

a. Metamorphosis
b. Transition
c. Transmigration
d. Extrapolation

95. A style in which a writer makes a display of his knowledge

a. Pedantic   *b.* Ornate
c. Verbose   *d.* Pompous

b. Logical Reasoning

Direction (Q. No. 96) *In each of these questions, two figure/words are given to the left of the sign :: and one figures word to the right of the sign :: with four alternatives under it out of which one of the alternatives has the same relationship with the figures/words to the right of the sign :: as between the two figures/words to the left of the sign (::). Find the correct alternative.*

96.

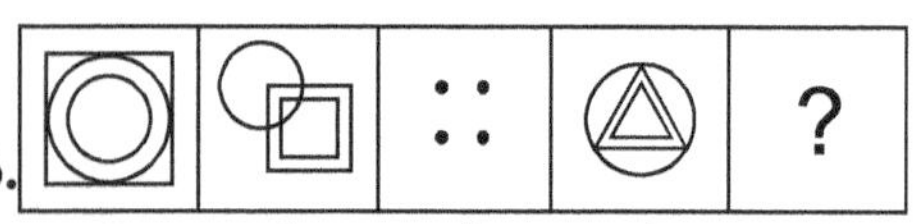

1   2   3   4   5

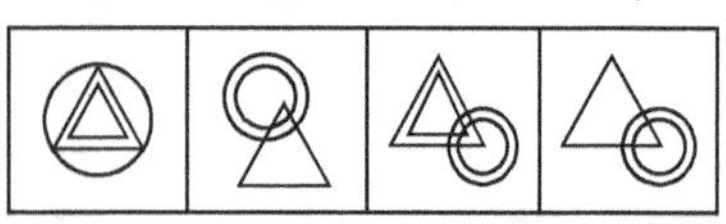

a.   *b.*   *c.*   *d.*

Directions (Q. Nos. 97-98) *In the following question, choose the missing word or sign (?) on the basis of the relationship between the words given on the left/right hand side of the signs.*

97. Doctor : Nurse :: ? : Follower

a. Worker   *b.* Employer
c. Union   *d.* Leader

98. One of the numbers does not fit into the series. Find the wrong number

1788, 892, 444, 220, 112, 52, 24

a. 52   *b.* 112   *c.* 220   *d.* 444

Directions (Q. 99-105) *In the question below is given a statement followed by three assumptions numbered I, II and III. An assumption is something supposed or taken for granted. You have to consider the statement and the following assumptions and decide which of the assumption(s) is/are implicit in the statement.*

99. Statement : Large number of people affected by the flood in the area gathered at the relief camp for food, water and shelter organised by the state government.

Assumptions :

I. The relief camp has enough supplies to provide food and water to the affected people in the area.
II. All those whose houses are submerged can be accommodated in the temporary shelters.
III. Many more affected people are yet to reach the relief camp.

a. Only I is implicit
b. Only I and II are implicit
c. Only II is implicit
d. Only II and III are implicit

100. Identify the missing part of the figure and select it from the given alternatives.

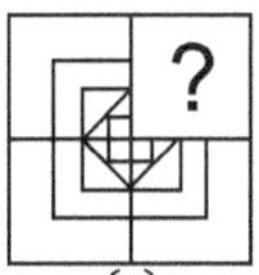

(x)

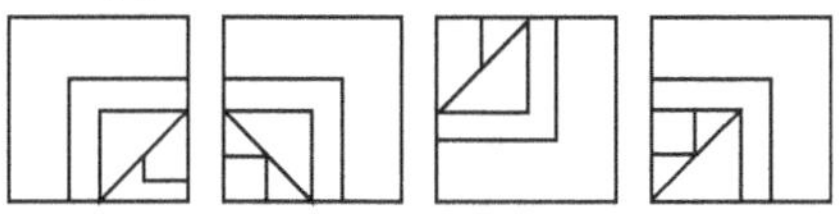

a. *b.* *c.* *d.*

101. Figure (x) is embedded in any one of the four alterntive figures. Choose the alternative which contains figure (x).

(x)

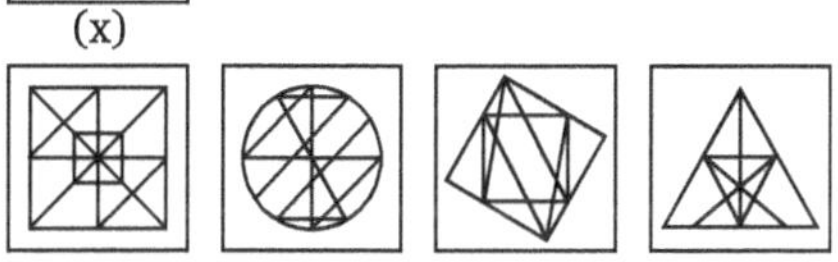

a. *b.* *c.* *d.*

102. Which symbol will appear on the opposite surface to the symbol × ?

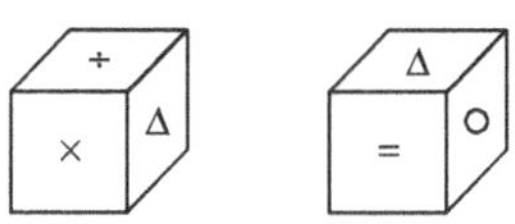

(i) (ii)

a. ÷ *b.* =
c. Δ *d.* 0

103. The three figures marked X, Y, Z show the manner in which a paper is folded step by step and then cut. From the answer figures (a), (b), (c), (d) select the one, showing the unfolded position of the paper after the cut.

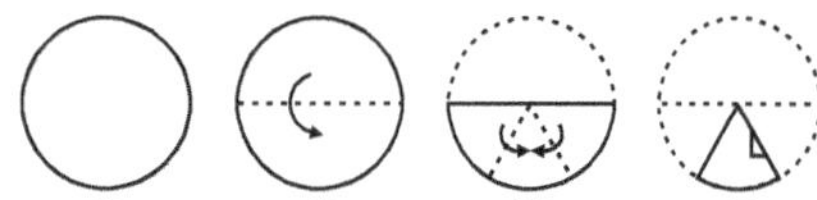

P X Y Z

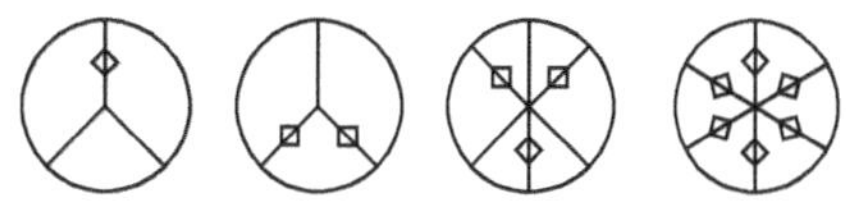

a. *b.* *c.* *d.*

104. Find the group of latters from the four alternatives which is obtained by applying the same rule to this given word to the right of the sign.

SERVANT : QGPXYPR :: KING ?

a. MKPI
b. IKLI
c. IGLE
d. IGPI

105. If P denotes '÷'
Q denotes '×'
R denotes '+'
S denotes '–',

then what is the value of 18 Q 12 P 4 R 5 S 6 =?

a. 64 *b.* 81
c. 53 *d.* 24

Mock Test 3

PART IV

Mathematics

106. $\left\{x \in R : \dfrac{2x-1}{x^3+4x^2+3x} \in R\right\}$ equals

a. $R-\{0\}$ *b.* $R-\{0, 1, 3\}$
c. $R-\{0, -1, -3\}$ *d.* $R-\left\{0, -1, -3, +\dfrac{1}{2}\right\}$

107. The number of subsets of $\{1, 2, 3, \ldots, 9\}$ containing at least one odd number is

a. 324 *b.* 396
c. 496 *d.* 512

108. The coefficient of x^{24} in the expansion of $(1+x^2)^{12}(1+x^{12})(1+x^{24})$ is

a. ${}^{12}C_6$ *b.* ${}^{12}C_6+2$
c. ${}^{12}C_6+4$ *d.* ${}^{12}C_6+6$

109. For $|x|<1$, the constant term in the expansion of $\dfrac{1}{(x-1)^2(x-2)}$ is

a. 2 *b.* 1
c. 0 *d.* $-\dfrac{1}{2}$

110. The roots of
$(x-a)(x-a-1)+(x-a-1)(x-a-2)$
$+(x-a)(x-a-2)=0, a\in R$ are always

a. equal b. imaginary
c. real and distinct d. rational and equal

111. Let $f(x)=x^2+ax+b$, where $a, b\in R$. If $f(x)=0$ has all its roots imaginary, then the roots of $f(x)+f'(x)+f''(x)=0$ are

a. real and distinct b. imaginary
c. equal d. rational and equal

112. If one of the roots of $\begin{vmatrix} 3 & 5 & x \\ 7 & x & 7 \\ x & 5 & 3 \end{vmatrix}=0$ is –10, then the other roots are

a. 3, 7 b. 4, 7
c. 3, 9 d. 3, 4

113. If x, y, z are all positive and are the pth, qth and rth terms of a geometric progression respectively, then the value of the determinant

$$\begin{vmatrix} \log x & p & 1 \\ \log y & q & 1 \\ \log z & r & 1 \end{vmatrix} \text{ equals}$$

a. $\log xyz$
b. $(p-1)(q-1)(r-1)$
c. pqr
d. 0

114. If $\begin{bmatrix} 1 & -1 & x \\ 1 & x & 1 \\ x & -1 & 1 \end{bmatrix}$ has no inverse, then the real value of x is

a. 2 b. 3 c. 0 d. 1

115. The locus of z satisfying the inequality $\left|\dfrac{z+2i}{2z+i}\right|<1$, where $z=x+iy$, is

a. $x^2+y^2<1$ b. $x^2-y^2<1$
c. $x^2+y^2>1$ d. $2x^2+3y^2<1$

116. The period of $\sin^4 x+\cos^4 x$ is

a. $\dfrac{\pi^4}{2}$ b. $\dfrac{\pi^2}{2}$ c. $\dfrac{\pi}{4}$ d. $\dfrac{\pi}{2}$

117. $\dfrac{\cos x}{\cos(x-2y)}=\lambda \Rightarrow \tan(x-y)\tan y$ is equal to

a. $\dfrac{1+\lambda}{1-\lambda}$ b. $\dfrac{1-\lambda}{1+\lambda}$
c. $\dfrac{\lambda}{1+\lambda}$ d. $\dfrac{\lambda}{1-\lambda}$

118. $\cos A \cos 2A \cos 4A \ldots \cos 2^{n-1} A$ equals

a. $\dfrac{\sin 2^n A}{2^n \sin A}$ b. $\dfrac{2^n \sin 2^n A}{\sin A}$
c. $\dfrac{2^n \sin A}{\sin 2^n A}$ d. $\dfrac{\sin A}{2^n \sin 2^n A}$

119. If $3\cos x \neq 2\sin x$, then the general solution of $\sin^2 x-\cos 2x=2-\sin 2x$ is

a. $n\pi+(-1)^n \dfrac{\pi}{2}, n\in Z$ b. $\dfrac{n\pi}{2}, n\in Z$
c. $(4n\pm 1)\dfrac{\pi}{2}, n\in Z$ d. $(2n-1)\pi, n\in Z$

120. In a ΔABC

$$\frac{(a+b+c)(b+c-a)(c+a-b)(a+b-c)}{4b^2c^2} \text{ equals}$$

a. $\cos^2 A$ b. $\cos^2 B$
c. $\sin^2 A$ d. $\sin^2 B$

121. P is a point on the segment joining the feet of two vertical poles of heights a and b. The angles of elevation of the tops of the poles from P are 45° each. Then, the square of the distance between the tops of the poles is

a. $\dfrac{a^2+b^2}{2}$ b. a^2+b^2
c. $2(a^2+b^2)$ d. $4(a^2+b^2)$

122. In a quadrilateral $ABCD$, the point P divides DC in the ratio $1:2$ and Q is the mid point of AC. If $\mathbf{AB}+2\mathbf{AD}+\mathbf{BC}-2\mathbf{DC}=k\mathbf{PQ}$, then k is equal to

a. –6 b. –4
c. 6 d. 4

123. If m_1, m_2, m_3 and m_4 are respectively the magnitudes of the vectors

$\mathbf{a}_1=2\hat{i}-\hat{j}+\hat{k}$, $\mathbf{a}_2=3\hat{i}-4\hat{j}-4\hat{k}$,
$\mathbf{a}_3=\hat{i}+\hat{j}-\hat{k}$ and $\mathbf{a}_4=-\hat{i}+3\hat{j}+\hat{k}$,

then the correct order of m_1, m_2, m_3 and m_4 is

a. $m_3<m_1<m_4<m_2$
b. $m_3<m_1<m_2<m_4$
c. $m_3<m_4<m_1<m_2$
d. $m_3<m_4<m_2<m_1$

124. The volume of the tetrahedron having the edges $\hat{i}+2\hat{j}-\hat{k}, \hat{i}+\hat{j}+\hat{k}, \hat{i}-\hat{j}+\lambda\hat{k}$ as coterminous, is $\dfrac{2}{3}$ cu unit. Then, λ equals

a. 1 b. 2 c. 3 d. 4

125. If A and B are events of a random experiment such that $P(A \cup B) = \frac{4}{5}$, $P(\bar{A} \cup \bar{B}) = \frac{7}{10}$ and $P(B) = \frac{2}{5}$, then $P(A)$ equals

a. 9/10 *b.* 8/10 *c.* 7/10 *d.* 3/5

126. If X is a binomial variate with the range $\{0, 1, 2, 3, 4, 5, 6\}$ and $P(X = 2) = 4P(X = 4)$, then the parameter p of X is

a. 1/3 *b.* 1/2 *c.* 2/3 *d.* 3/4

127. The area (in square unit) of the circle which touches the lines $4x + 3y = 15$ and $4x + 3y = 5$ is

a. 4π *b.* 3π *c.* 2π *d.* π

128. The point on the line $3x + 4y = 5$ which is equidistant from $(1, 2)$ and $(3, 4)$ is

a. $(7, -4)$ *b.* $(15, -10)$
c. $(1/7, 8/7)$ *d.* $(0, 5/4)$

129. The equation of the straight line perpendicular to the straight line $3x + 2y = 0$ and passing through the point of intersection of the lines $x + 3y - 1 = 0$ and $x - 2y + 4 = 0$ is

a. $2x - 3y + 1 = 0$ *b.* $2x - 3y + 3 = 0$
c. $2x - 3y + 5 = 0$ *d.* $2x - 3y + 7 = 0$

130. The value of λ with $|\lambda| < 16$ such that $2x^2 - 10xy + 12y^2 + 5x + \lambda y - 3 = 0$ represents a pair of straight lines, is

a. –10 *b.* –9 *c.* 10 *d.* 9

131. The area (in square unit) of the triangle formed by $x + y + 1 = 0$ and the pair of straight lines $x^2 - 3xy + 2y^2 = 0$ is

a. 7/12 *b.* 5/12 *c.* 1/12 *d.* 1/6

132. The pairs of straight lines $x^2 - 3xy + 2y^2 = 0$ and $x^2 - 3xy + 2y^2 + x - 2 = 0$ form a

a. square but not rhombus
b. rhombus
c. parallelogram
d. rectangle but not a square

133. The equations of the circle which pass through the origin and makes intercepts of lengths 4 and 8 on the X and Y-axes respectively are

a. $x^2 + y^2 \pm 4x \pm 8y = 0$ *b.* $x^2 + y^2 \pm 2x \pm 4y = 0$
c. $x^2 + y^2 \pm 8x \pm 16y = 0$ *d.* $x^2 + y^2 \pm x \pm y = 0$

134. The locus of centre of a circle which passes through the origin and cuts off a length of 4 unit from the line $x = 3$ is

a. $y^2 + 6x = 0$ *b.* $y^2 - 6x = 13$
c. $y^2 + 6x = 10$ *d.* $x^2 + 6y = 13$

135. The point $(3, -4)$ lies on both the circles
$$x^2 + y^2 - 2x + 8y + 13 = 0$$
and $x^2 + y^2 - 4x + 6y + 11 = 0$
Then, the angle between the circles is

a. $60°$ *b.* $\tan^{-1}\left(\frac{1}{2}\right)$
c. $\tan^{-1}\left(\frac{3}{5}\right)$ *d.* $135°$

136. The equation of the circle which passes through the origin and cuts orthogonally each of the circles $x^2 + y^2 - 6x + 8 = 0$ and $x^2 + y^2 - 2x - 2y = 7$ is

a. $3x^2 + 3y^2 - 8x - 13y = 0$
b. $3x^2 + 3y^2 - 8x + 29y = 0$
c. $3x^2 + 3y^2 + 8x + 29y = 0$
d. $3x^2 + 3y^2 - 8x - 29y = 0$

137. The number of normals drawn to the parabola $y^2 = 4x$ from the point $(1, 0)$ is

a. 0 *b.* 1 *c.* 2 *d.* 3

138. If the circle $x^2 + y^2 = a^2$ intersects the hyperbola $xy = c^2$ in four points (x_i, y_i), for $i = 1, 2, 3$ and 4, then $y_1 + y_2 + y_3 + y_4$ equals

a. 0 *b.* c *c.* a *d.* c^4

139. The mid point of the chord $4x - 3y = 5$ of the hyperbola $2x^2 - 3y^2 = 12$ is

a. $\left(0, -\frac{5}{3}\right)$ *b.* $(2, 1)$ *c.* $\left(\frac{5}{4}, 0\right)$ *d.* $\left(\frac{11}{4}, 2\right)$

140. If a line in the space makes angle α, β and γ with the coordinate axes, then
$$\cos 2\alpha + \cos 2\beta + \cos 2\gamma + \sin^2 \alpha + \sin^2 \beta + \sin^2 \gamma$$ equals

a. –1 *b.* 0 *c.* 1 *d.* 2

141. The image of the point $(3, 2, 1)$ in the plane $2x - y + 3z = 7$ is

a. $(1, 2, 3)$ *b.* $(2, 3, 1)$
c. $(3, 2, 1)$ *d.* $(2, 1, 3)$

142. $\lim_{x \to \infty} \left(\frac{x+5}{x+2}\right)^{x+3}$ equals

a. e *b.* e^2 *c.* e^3 *d.* e^5

143. If $f : R \to R$ is defined by
$$f(x) = \begin{cases} \frac{2\sin x - \sin 2x}{2x \cos x}, & \text{if } x \neq 0 \\ a, & \text{if } x = 0 \end{cases},$$
then the value of a, so that f is continuous at 0 is

a. 2 *b.* 1 *c.* –1 *d.* 0

144. $x = \frac{1-\sqrt{y}}{1+\sqrt{y}} \Rightarrow \frac{dy}{dx}$ is equal to

a. $\frac{4}{(x+1)^2}$ b. $\frac{4(x-1)}{(1+x)^3}$ c. $\frac{x-1}{(1+x)^3}$ d. $\frac{4}{(x+1)^3}$

145. $\frac{d}{dx}\left[a \tan^{-1} x + b \log\left(\frac{x-1}{x+1}\right)\right] = \frac{1}{x^4-1}$

$\Rightarrow a - 2b$ is equal to

a. 1 b. –1 c. 0 d. 2

146. $y = e^{a\sin^{-1} x} \Rightarrow (1-x^2)\,y_{n+2} - (2n+1)\,xy_{n+1}$ is equal to

a. $-(n^2+a^2)\,y_n$ b. $(n^2-a^2)\,y_n$

c. $(n^2+a^2)\,y_n$ d. $-(n^2-a^2)\,y_n$

147. The function $f(x) = x^3 + ax^2 + bx + c,\ a^2 \le 3b$ has

a. one maximum value

b. one minimum value

c. no extreme value

d. one maximum and one minimum value

148. $\int\left(\frac{2-\sin 2x}{1-\cos 2x}\right)e^x\,dx$ is equal to

a. $-e^x \cot x + c$ b. $e^x \cot x + c$

c. $2e^x \cot x + c$ d. $-2e^x \cot x + c$

149. $\int_0^{\pi} \frac{1}{1+\sin x}\,dx$ is equal to

a. 1 b. 2

c. –1 d. –2

150. The solution of the differential equation $\frac{dy}{dx} = \sin(x+y)\tan(x+y) - 1$ is

a. $\text{cosec}\,(x+y) + \tan(x+y) = x + c$

b. $x + \text{cosec}\,(x+y) = c$

c. $x + \tan(x+y) = c$

d. $x + \sec(x+y) = c$

Mock Test 3

Answers

Physics

1. (d)	2. (c)	3. (a)	4. (a)	5. (a)	6. (c)	7. (b)	8. (*)	9. (d)	10. (d)
11. (c)	12. (a)	13. (c)	14. (a)	15. (b)	16. (c)	17. (a)	18. (b)	19. (b)	20. (d)
21. (a)	22. (c)	23. (d)	24. (b)	25. (d)	26. (a)	27. (c)	28. (c)	29. (c)	30. (c)
31. (c)	32. (c)	33. (b)	34. (d)	35. (d)	36. (c)	37. (c)	38. (a)	39. (b)	40. (a)

Chemistry

41. (c)	42. (a)	43. (b)	44. (a)	45. (a)	46. (a)	47. (c)	48. (a)	49. (a)	50. (b)
51. (a)	52. (c)	53. (c)	54. (d)	55. (c)	56. (c)	57. (b)	58. (c)	59. (a)	60. (d)
61. (b)	62. (c)	63. (c)	64. (c)	65. (b)	66. (c)	67. (d)	68. (c)	69. (a)	70. (b)
71. (a)	72. (d)	73. (b)	74. (b)	75. (c)	76. (c)	77. (c)	78. (b)	79. (b)	80. (b)

English Proficiency

81. (a)	82. (d)	83. (c)	84. (d)	85. (b)	86. (b)	87. (a)	88. (d)	89. (b)	90. (b)
91. (a)	92. (c)	93. (b)	94. (c)	95. (a)					

Logical Reasoning

96. (d)	97. (d)	98. (b)	99. (b)	100. (b)	101. (a)	102. (d)	103. (d)	104. (b)	105. (c)

Mathematics

106. (c)	107. (c)	108. (b)	109. (d)	110. (c)	111. (b)	112. (a)	113. (d)	114. (d)	115. (c)
116. (d)	117. (b)	118. (a)	119. (c)	120. (c)	121. (c)	122. (a)	123. (a)	124. (a)	125. (c)
126. (a)	127. (d)	128. (b)	129. (d)	130. (b)	131. (c)	132. (c)	133. (a)	134. (b)	135. (d)
136. (b)	137. (b)	138. (a)	139. (b)	140. (c)	141. (c)	142. (c)	143. (d)	144. (b)	145. (b)
146. (c)	147. (c)	148. (a)	149. (b)	150. (b)					

Hints & Solutions

Physics

1. ***(d)*** Given, $y = a \sin (bt - cx)$

Comparing the given equation with general wave equation,

$$y = a \sin \left(\frac{2\pi t}{T} - \frac{2\pi x}{\lambda} \right)$$

we get $\quad b = \frac{2\pi}{T}, c = \frac{2\pi}{\lambda}$

(a) Dimensions of $\frac{y}{a} = \frac{\text{metre}}{\text{metre}} = \frac{[\text{L}]}{[\text{L}]}$

$= \text{Dimensionless}$

(b) Dimensions of $bt = \frac{2\pi}{T} \cdot t = \frac{[\text{T}]}{[\text{T}]}$

$= \text{Dimensionless}$

(c) Dimensions of $cx = \frac{2\pi}{\lambda} \cdot x = \frac{[\text{L}]}{[\text{L}]}$

$= \text{Dimensionless}$

(d) Dimensions of $\frac{b}{c} = \frac{2\pi}{T} \Big/ \frac{2\pi}{\lambda}$

$= \lambda / T = [\text{LT}^{-1}]$

Thus, option (d) has dimensions.

2. ***(c)*** Let time taken by the body to fall from point C to B is t'.

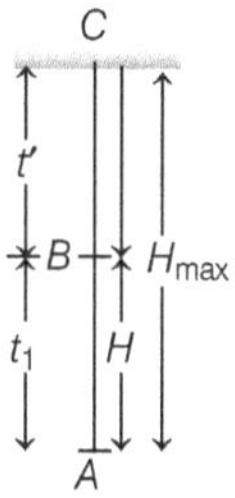

Then, $t_1 + 2t' = t_2$

$$t' = \left(\frac{t_2 - t_1}{2} \right) \quad \ldots\text{(i)}$$

Total time taken to reach point C

$$T = t_1 + t'$$

$$= t_1 + \frac{t_2 - t_1}{2}$$

$$= \frac{2t_1 + t_2 - t_1}{2}$$

$$= \left(\frac{t_1 + t_2}{2} \right)$$

Maximum height attained,

$$H_{\text{max}} = \frac{1}{2} g (T)^2$$

$$= \frac{1}{2} g \left(\frac{t_1 + t_2}{2} \right)^2$$

$$= \frac{1}{2} g \cdot \frac{(t_1 + t_2)^2}{4}$$

$$\Rightarrow \quad H_{\text{max}} = \frac{1}{8} g \cdot (t_1 + t_2)^2$$

3. ***(a)*** Momentum, $p = m \cdot v$

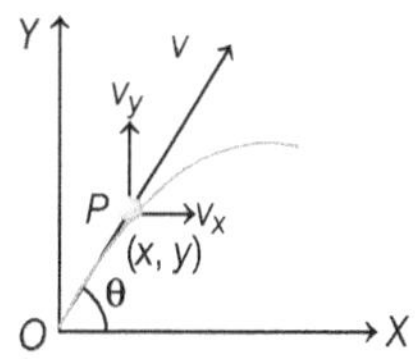

$$\Rightarrow \quad v = \left(\frac{p}{m} \right)$$

Kinetic energy, $\text{KE} = \frac{1}{2} mv^2$

$$= \frac{1}{2} m \left(\frac{p^2}{m^2} \right) = \frac{1}{2m} p^2$$

$$\Rightarrow \quad \text{KE} \propto p^2 \qquad \left(\because \frac{1}{2m} = \text{constant} \right)$$

Hence, the graph between KE and p^2 will be linear as shown below

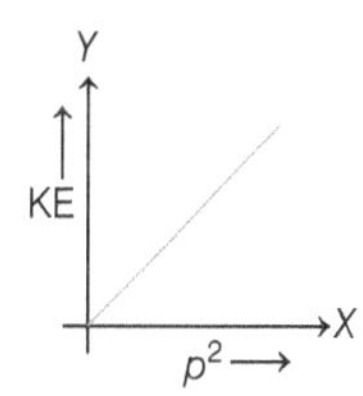

Now, kinetic energy $\text{KE} = \frac{1}{2} mv^2$

The velocity component at point P,

$$v_y = (u \sin \theta - gt)$$

and $\quad v_x = u \cos \theta$

Resultant velocity at point P,

$$\mathbf{v} = v_y \hat{\mathbf{j}} + v_x \hat{\mathbf{i}}$$

$$= (u \sin \theta - gt) \hat{\mathbf{j}} + u \cos \theta \, \hat{\mathbf{i}}$$

$$|\mathbf{v}| = \sqrt{(u \cos \theta)^2 + (u \sin \theta - gt)^2}$$

$$= \sqrt{u^2 \cos^2 \theta + u^2 \sin^2 \theta + g^2 t^2 - 2ugt \sin \theta}$$

$$\therefore \quad = \sqrt{u^2 (\cos^2 \theta + \sin^2 \theta) + g^2 t^2 - 2ugt \sin \theta}$$

$$\text{KE} = \frac{1}{2} m (u^2 + g^2 t^2 - 2ugt \sin \theta)$$

$\Rightarrow \text{KE} \propto t^2$

Hence, graph will be parabolic with intercept on Y-axis.

Hence, the graph between KE and t

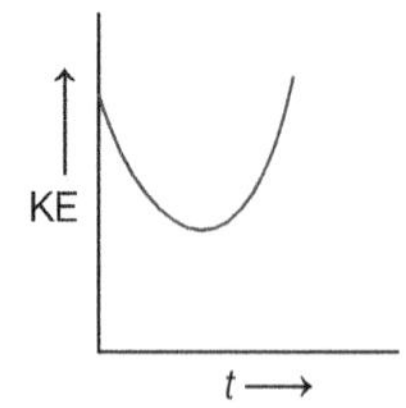

Now, in case of height

$$\text{KE} = \frac{1}{2} m(v^2)$$

and $\quad v^2 = (u^2 - 2gy)$

$\therefore \quad \text{KE} = \frac{1}{2} m(u^2 - 2gy)$

$$\text{KE} = -\, mgy + \frac{1}{2} mu^2$$

Intercept on Y-axis $= \frac{1}{2} mu^2$

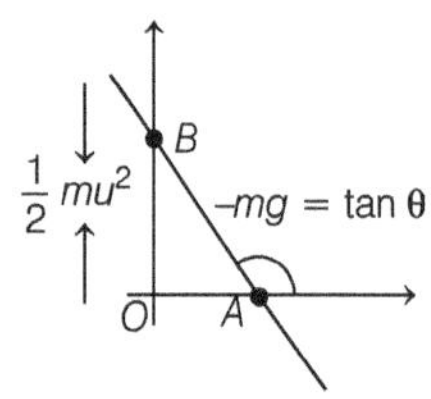

Now, $\quad \text{KE} = \frac{1}{2} mv^2$

$$\text{KE} = \frac{1}{2} m \left(\frac{x}{t}\right)^2$$

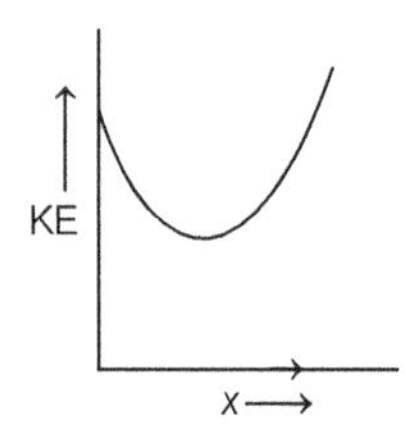

$$\text{KE} \propto x^2$$

Thus graph between KE and x will be parabolic.

4. ***(a)*** Power of motor initially $= P_0$
Let, rate of flow of motor $= x$

Since, power, $P_0 = \frac{\text{work}}{\text{time}} = \frac{mgy}{t}$

$= mg\,(y/t)$,

$\left[\therefore \frac{y}{t} = x = \text{rate of flow of water}\right]$

$= mgx \qquad \ldots\text{(i)}$

If rate of flow of water is increased by n times, *i.e.*, (nx).

Increased power, $P_1 = \frac{mgy'}{t}$

$= mg\left(\frac{y'}{t}\right)$

$= mgn \cdot x$

$= nmgx \qquad \ldots\text{(ii)}$

The ratio of power,

$$\frac{P_1}{P_0} = \frac{nmgx}{mgx}$$

[By using Eqs. (i) and (ii)]

$$\frac{P_1}{P_0} = \frac{n}{1} \Rightarrow P_1 : P_0 = n : 1$$

5. ***(a)*** Mass of the first body $m_1 = 5$ kg, for elastic collision $e = 1$.

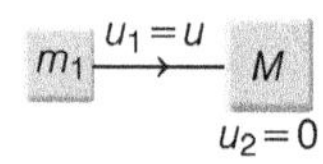

Suppose initially body m_1 moves with velocity v after collision velocity becomes $\left(\frac{u}{10}\right)$.

Let after collision velocity of M block becomes (v_2).
By conservation of momentum,

$$m_1u_1 + m_2u_2 = m_1v_1 + m_2v_2$$

or $\quad 5u + M \times 0 = 5 \times \frac{u}{10} + Mv_2$

or $\quad 5u = \frac{u}{2} + Mv_2 \qquad \ldots\text{(i)}$

Since, $\quad v_1 - v_2 = -\,e(u_1 - u_2)$

$\frac{u}{10} - v_2 = -1(u)$

or $\quad \frac{u}{10} + u = v_2$

$\frac{11u}{10} = v_2 \qquad \ldots\text{(ii)}$

Substituting value of v_2 in Eq. (i) from Eq. (ii), we get

$$5u = \frac{u}{2} + M\left(\frac{11u}{10}\right)$$

or $\quad 5 - \frac{1}{2} = M\left(\frac{11}{10}\right)$

or $\quad M = \frac{9 \times 10}{2 \times 11}$

or $\quad M = \frac{45}{11} = 4.09$ kg

6. ***(c)***

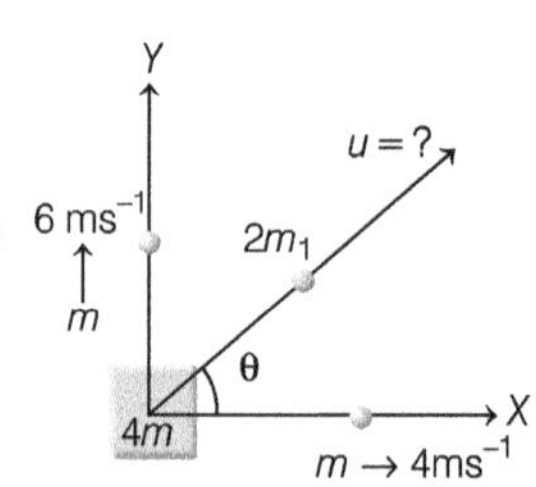

Let, third mass particle $2m$ moves making angle θ with X-axis.

The horizontal component of velocity of $2m$ mass particle $= u \cos\theta$ and vertical component $= u \sin\theta$

From conservation of linear momentum in X-direction,

$$m_1u_1 + m_2u_2 = m_1v_1 + m_2v_2$$

or $$0 = m \times 4 + 2m(u\cos\theta)$$

or $$-4 = 2u\cos\theta$$

or $$-2 = u\cos\theta \quad \text{...(i)}$$

Again, applying law of conservation of linear momentum in Y-direction.

$$0 = m \times 6 + 2m(u\sin\theta)$$

$\Rightarrow$ $$-\frac{6}{2} = u\sin\theta$$

or $$-3 = u\sin\theta \quad \text{...(ii)}$$

Squaring Eqs. (i) and (ii) and adding, we get

$$(4) + (9) = u^2\cos^2\theta + u^2\sin^2\theta$$

$$= u^2(\cos^2\theta + \sin^2\theta)$$

or $$13 = u^2$$

or $$u = \sqrt{13}\ \text{ms}^{-1}$$

7. ***(b)*** Maximum height attained by a projectile

$$h = \frac{v^2R}{2gR - v^2} \quad \text{...(i)}$$

Velocity of body = Half the escape velocity

$$v = \frac{v_e}{2}$$

or $$v = \frac{\sqrt{2gR}}{2} \Rightarrow v^2 = \frac{2gR}{4}$$

or $$v^2 = \left(\frac{gR}{2}\right)$$

Now, putting value of v^2 in Eq. (i), we get

$$h = \frac{\frac{gR}{2}\cdot R}{2gR - \frac{gR}{2}}$$

$$= \frac{gR^2/2}{3gR/2}$$

or $$h = \frac{R}{3}$$

8. (*) The displacement of particle, executing SHM is

$$y = 5\sin\left(4t + \frac{\pi}{3}\right) \quad \text{...(i)}$$

Velocity of particle,

$$\left(\frac{dy}{dt}\right) = \frac{5d}{dt}\sin\left(4t + \frac{\pi}{3}\right)$$

$$= 5\cos\left(4t + \frac{\pi}{3}\right)\cdot 4$$

$$= 20\cos\left(4t + \frac{\pi}{3}\right)$$

Velocity at $t = \left(\frac{T}{4}\right)$

$$\left(\frac{dy}{dt}\right)_{t=\frac{T}{4}} = 20\cos\left(4 \times \frac{T}{4} + \frac{\pi}{3}\right)$$

or $$u = 20\cos\left(T + \frac{\pi}{3}\right) \quad \text{...(ii)}$$

Comparing the given Eq. (i) with standard equation of SHM given by

$$y = a\sin(\omega t + \phi)$$

We get, $$\omega = 4$$

As $$\omega = \frac{2\pi}{T}$$

$\Rightarrow$ $$T = \frac{2\pi}{\omega}$$

or $$T = \frac{2\pi}{4}$$

or $$T = \left(\frac{\pi}{2}\right)$$

Now, putting value of T in Eq. (ii), we get

$$u = 20\cos\left(\frac{\pi}{2} + \frac{\pi}{3}\right)$$

$$= -20\sin\frac{\pi}{6}$$

$$= -20 \times \frac{1}{2}$$

$$= -10$$

The kinetic energy of particle,

$$\text{KE} = \frac{1}{2}mu^2$$

$\because$ $$m = 2\,\text{g} = 2 \times 10^{-3}\ \text{kg}$$

$$= \frac{1}{2} \times 2 \times 10^{-3} \times (-10)^2$$

$$= 10^{-3} \times 100$$

$$= 10^{-1}$$

$$\text{KE} = 0.1\ \text{J}$$

9. ***(d)*** Free body diagram of the two blocks are

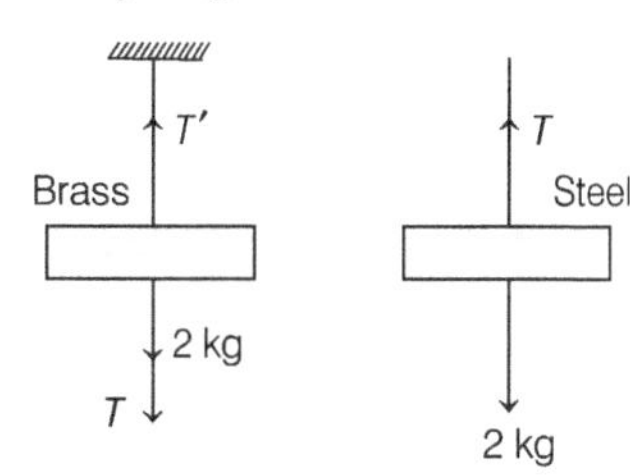

Given, $$\frac{l_1}{l_2} = a, \frac{r_1}{r_2} = b, \frac{Y_1}{Y_2} = c$$

Let Young's modulus of steel is Y_1 and of brass is Y_2.

$\therefore$ $$Y_1 = \frac{F_1 \cdot l_1}{A_1 \cdot \Delta l_1} \quad \text{...(i)}$$

Mock Test 3

and $$Y_2 = \frac{F_2 \cdot l_2}{A_2 \Delta l_2} \quad \ldots(ii)$$

Dividing Eq. (i) by Eq. (ii), we get

$$\frac{Y_1}{Y_2} = \frac{\dfrac{F_1 \cdot l_1}{A_1 \cdot \Delta l_1}}{\dfrac{F_2 \cdot l_2}{A_2 \cdot \Delta l_2}}$$

or $$\frac{Y_1}{Y_2} = \frac{F_1 \cdot A_2 \cdot l_1 \cdot \Delta l_2}{F_2 \cdot A_1 \cdot l_2 \cdot \Delta l_1} \quad \ldots(iii)$$

Force on steel wire from free body diagram,

$$T = F_1 = (2g)\text{ N}$$

Force on brass wire from free body diagram,

$$F_2 = T' = T + 2g = (4g)\text{ N}$$

Now, putting the value of F_1, F_2, in Eq. (iii), we get

$$\frac{Y_1}{Y_2} = \left(\frac{2g}{4g}\right) \cdot \left(\frac{\pi r_2^2}{\pi r_1^2}\right) \cdot \left[\frac{l_1}{l_2}\right] \cdot \left(\frac{\Delta l_2}{\Delta l_1}\right)$$

or $$c = \frac{1}{2}\left(\frac{1}{b^2}\right) \cdot a\left(\frac{\Delta l_2}{\Delta l_1}\right)$$

or $$\frac{\Delta l_1}{\Delta l_2} = \left(\frac{a}{2b^2c}\right)$$

10. ***(d)*** Initially area of soap bubble,

$$A_1 = 4\pi r^2$$

Under isothermal condition radius becomes $2r$,

Then, area, $A_2 = 4\pi(2r)^2$

$$= 4\pi \cdot 4r^2$$

$$= 16\pi r^2$$

Increase in surface area

$$\Delta A = 2(A_2 - A_1)$$

$$= 2(16\pi r^2 - 4\pi r^2)$$

$$= 24\pi r^2$$

Energy spent,

$$W = T \times \Delta A$$

$$= T \cdot 24\pi r^2$$

or $$W = 24\pi T r^2 \text{ J}$$

11. ***(c)*** Let radius of big drop is R.

Then, $$\frac{4}{3}\pi R^3 = \frac{4}{3} \times \pi r^3 \cdot 8$$

$$R = 2r$$

where, r is radius of small drops. Now, terminal velocity of drop in liquid,

$$v_e = \frac{2}{9} \times \frac{r^2}{\eta}(\rho - \sigma) g$$

where, η is coefficient of viscosity, ρ is density of drop and σ is density of liquid.

Terminal speed drop is 6 cm s^{-1}.

$\therefore$ $$6 = \frac{2}{9} \times \frac{r^2}{\eta}(\rho - \sigma) g \quad \ldots(i)$$

Let terminal velocity becomes v' after coalesce, then

$$v' = \frac{2}{9}\frac{R^2}{\eta}(\rho - \sigma) g \quad \ldots(ii)$$

Dividing Eq. (i) by Eq. (ii), we get

$$\frac{6}{v'} = \frac{\dfrac{2}{9}\dfrac{r^2}{\eta}(\rho - \sigma) g}{\dfrac{2}{9}\dfrac{R^2}{\eta}(\rho - \sigma) g}$$

or $$\frac{6}{v'} = \frac{r^2}{(2r)^2}$$

or $$v' = 24 \text{ cm s}^{-1}$$

12. ***(a)*** Time period of oscillation,

$$T = 2\pi\sqrt{\frac{l}{g}}$$

$\Rightarrow$ $$\frac{dT}{T} = \frac{1}{2}\frac{dl}{l}$$

As, thermal expansion,

$$\frac{dl}{l} = \alpha dt$$

$\Rightarrow$ $$\frac{dT}{T} = \frac{1}{2}\alpha dt$$

$$= \frac{1}{2} \times 9 \times 10^{-7} \times (30 - 20)$$

$$= 4.5 \times 10^{-6}$$

$\therefore$ Loss in time $= 4.5 \times 10^{-6} \times 0.5$

$$= 2.25 \times 10^{-6} \text{ s}$$

13. ***(c)*** The volume of the metal at 30°C is

$$V_{30} = \frac{\text{loss of weight}}{\text{specific gravity} \times g}$$

$$= \frac{(45 - 25) g}{1.5 \times g} = 13.33 \text{ cm}^3$$

Similarly, volume of metal at 40°C is

$$V_{40} = \frac{(45 - 27)g}{1.25 \times g}$$

$$= 14.40 \text{ cm}^3$$

Now, $$V_{40} = V_{30}[1 + \gamma(t_2 - t_1)]$$

or $$\gamma = \frac{V_{40} - V_{30}}{V_{30}(t_2 - t_1)}$$

$$= \frac{14.40 - 13.33}{13.33(40 - 30)}$$

$$= 8.03 \times 10^{-3}/°\text{C}$$

$\therefore$ Coefficient of linear expansion of the metal is

$$\alpha = \frac{\gamma}{3} = \frac{8.03 \times 10^{-3}}{3} \approx 2.6 \times 10^{-3} /°C$$

14. *(a)* $A \to B \to C \to D \to A$ is clockwise process. During $A \to B$, pressure is constant and $B \to C$, process follows $p \propto \frac{1}{V}$, it means T is constant. During process $C \to D$, both p and V changes and process $D \to A$ follows $p \propto \frac{1}{V}$ which means T is constant.

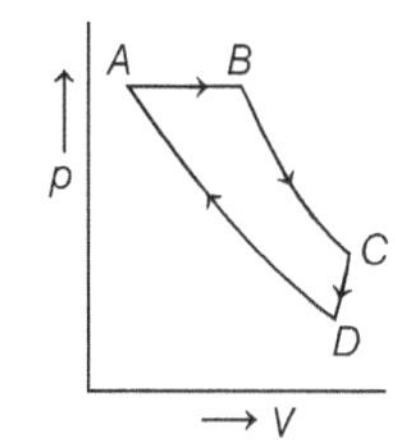

Hence, from above data it is clear that equivalent cyclic process is

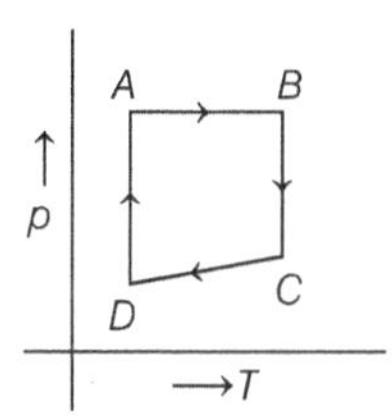

15. *(b)* From first law of thermodynamics

$$Q = \Delta U + W$$

or $$\Delta U = Q - W$$

$\therefore$ $\Delta U_1 = Q_1 - W_1 = 6000 - 2500 = 3500$ J

$\Delta U_2 = Q_2 - W_2 = -5500 + 1000 = -4500$ J

$\Delta U_3 = Q_3 - W_3 = -3000 + 1200 = -1800$ J

$\Delta U_4 = Q_4 - W_4 = 3500 - x$ J

For cyclic process $\Delta U = 0$

$\therefore$ $3500 - 4500 - 1800 + 3500 - x = 0$

or $x = 700$ J

Efficiency, $$\eta = \frac{\text{output}}{\text{input}} \times 100 = \frac{W_1 + W_2 + W_3 + W_4}{Q_1 + Q_4} \times 100$$

$$= \frac{(2500 - 1000 - 1200 + 700)}{6000 + 3500} \times 100$$

$$= \frac{1000}{9500} \times 100$$

$$\eta = 10.5\%$$

16. *(c)* From first law of thermodynamics,

$$Q = \Delta U + W$$

For cylinder A pressure remains constant,

$\therefore$ Work done by a system

$$W = \frac{\mu R}{\gamma - 1}(T_1 - T_2)$$

For monoatomic gases,

$$\mu = 1$$
$$\gamma = \frac{5}{3}$$

$\therefore$ $$W = \frac{1 \times R}{\frac{5}{3} - 1}(442 - 400) = \frac{3}{2} R \times 42$$

or $$W = 63R$$

But $\Delta U = 0$, for cylinder A

$\therefore$ $$Q = 0 + 63R$$
$$Q = 63R$$

For cylinder B volume is constant,

$\therefore$ $$W = 0$$

and $$Q = \mu C_V \Delta T$$

For monoatomic gas,

$$C_V = \frac{3}{2} R$$

$$Q = 1 \times \frac{3}{2} R \Delta T$$

As heat given to both cylinder is same

$\therefore$ $$63R = \frac{3}{2} R \,\Delta T$$

$$\Delta T = 42 \text{ K}$$

17. *(a)* According to the figure

$$H = H_1 + H_2$$

$\Rightarrow$ $$\frac{3KA(100 - T)}{l} = \frac{2KA(T - 50)}{l} + \frac{KA(T - 0)}{l}$$

$$300 - 3T = 2T - 100 + T$$

$\Rightarrow$ $$6T = 400$$

or $$T = \frac{200}{3} °C$$

18. *(b)* Listener go from $A \to B$ with velocity u let the apparent frequency of sound from source A by listener

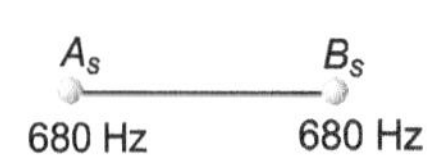

$$n' = n\left(\frac{v - v_o}{v + v_s}\right)$$

or $$n' = 680\left(\frac{340 - u}{340 + 0}\right)$$

The apparent frequency of sound from source B by listener

$$n'' = n\left(\frac{v + v_o}{v - v_s}\right) = 680\left(\frac{340 + u}{340 - 0}\right)$$

But listener hear 10 beats per second.

Mock Test 3

Hence, $n'' - n' = 10$

or $680\left(\frac{340+u}{340}\right) - 680\left(\frac{340-u}{340}\right) = 10$

or $2(340 + u - 340 + u) = 10$

or $u = 2.5\ \text{ms}^{-1}$

19. ***(b)*** Beats per second when both the wires vibrate simultaneously.

$$n_1 \pm n_2 = 6$$

or $\frac{1}{2l}\sqrt{\frac{T}{m}} \pm \frac{1}{2l}\sqrt{\frac{T'}{m}} = 6$

or $\frac{1}{2l}\sqrt{\frac{T'}{m}} - \frac{1}{2l}\sqrt{\frac{T}{m}} = 6$

or $\frac{1}{2l}\sqrt{\frac{T'}{m}} - 600 = 6$

$$\frac{1}{2l}\sqrt{\frac{T'}{m}} = 606 \quad \ldots\text{(i)}$$

Given that, fundamental frequency,

$$\frac{1}{2l}\sqrt{\frac{T}{m}} = 600 \quad \ldots\text{(ii)}$$

Dividing Eq. (i) by Eq. (ii), we get

$$\frac{\frac{1}{2l}\sqrt{\frac{T'}{m}}}{\frac{1}{2l}\sqrt{\frac{T}{m}}} = \frac{606}{600}$$

or $\sqrt{\frac{T'}{T}} = (1.01)$

or $\frac{T'}{T} = (1.02)$

or $T' = T(1.02)$

Increase in tension,

$$\Delta T' = T \times 1.02 - T$$
$$= (0.02T)$$

Hence, $\Delta T' = 0.02$

20. (d)

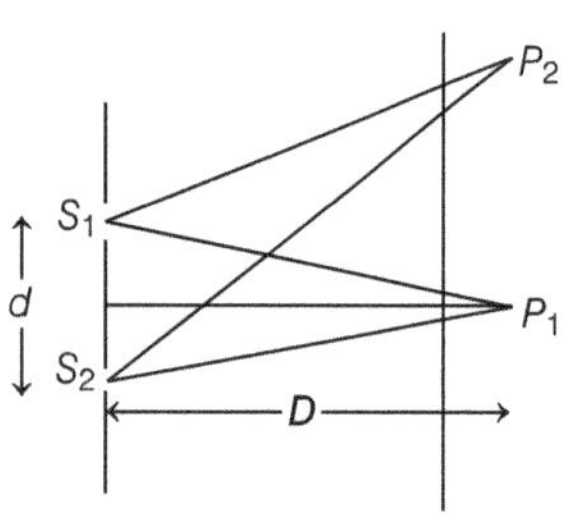

Fringe width, $\beta = \frac{\lambda D}{d}$

Let the amplitude of that place where constructive inference takes place is a.

The position of fringe at p_2 is

$$x = \frac{n\lambda D}{d}$$

Given, $\beta' = \left(\frac{\beta}{4}\right)$

$\therefore$ $\frac{\lambda D}{4d} = \frac{n\lambda D}{d}$

or $n = \frac{1}{4}$

$\therefore$ $\frac{I_1}{I_2} = \frac{a^2}{\left(\frac{a}{4}\right)^2}$

or $I_1 : I_2 = 16 : 1$

21. ***(a)*** Position fringe from central maxima,

$$y_1 = \frac{n\lambda_1 D}{d}$$

Given, $n = 10$

$\therefore$ $y_1 = \frac{10\lambda_1 D}{d} \quad \ldots\text{(i)}$

For second source

$$y_2 = \frac{5\lambda_2 D}{d} \quad \ldots\text{(ii)}$$

From Eqs. (i) and (ii), we get

$\therefore$ $\frac{y_1}{y_2} = \frac{\frac{10\lambda_1 D}{d}}{\frac{5\lambda_2 D}{d}}$

$\Rightarrow$ $\frac{y_1}{y_2} = \frac{2\lambda_1}{\lambda_2}$

22. ***(c)*** Interference phenomenon takes place between two waves which have equal frequency and propagate in same direction.

Hence, $y_1 = a \sin(\omega t + \phi_1)$

$y_3 = a' \sin(\omega t + \phi_2)$

will give rise to interference as the two waves have same frequency ω.

23. ***(d)*** The two lenses of an achromatic doublet should have, sum of the product of their powers and dispersive power equal to zero.

24. ***(b)*** Ratio of magnetic moments of two magnets of equal size when in sum and difference position is

$$\frac{M_A}{M_B} = \frac{T_d^2 + T_s^2}{T_d^2 - T_s^2} = \frac{v_s^2 + v_d^2}{v_s^2 - v_d^2}$$

$$= \frac{\left(\frac{1}{20}\right)^2 + \left(\frac{1}{15}\right)^2}{\left(\frac{1}{15}\right)^2 - \left(\frac{1}{20}\right)^2}$$

$$= \frac{400 + 225}{400 - 225}$$

$$= \frac{625}{175} = \frac{25}{7}$$

$$\Rightarrow \quad M_A : M_B = 25 : 7$$

25. ***(d)*** Length of magnet = 10 cm = 10×10^{-2} m,

$$r = 15 \times 10^{-2} \text{ m}$$

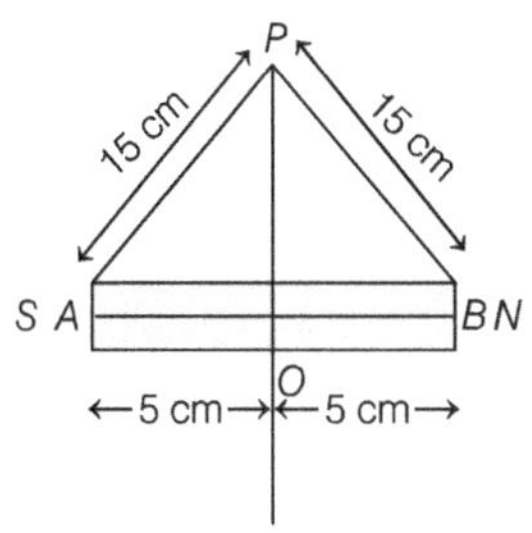

$$OP = \sqrt{225 - 25} = \sqrt{200} \text{ cm}$$

Since, at the neutral point, magnetic field due to the magnet is equal to B_H

$$B_H = \frac{\mu_0}{4\pi} \cdot \frac{M}{(OP^2 + AO^2)^{3/2}}$$

$$0.4 \times 10^{-4} = 10^{-7} \times \frac{M}{(200 \times 10^{-4} + 25 \times 10^{-4})^{3/2}}$$

$$\frac{0.4 \times 10^{-4}}{10^{-7}} \times (225 \times 10^{-4})^{3/2} = M$$

$$0.4 \times 10^{3} \times 10^{-6} (225)^{3/2} = M$$

$$M = 1.35 \text{ A-m}$$

26. ***(a)*** Charge density of long wire

$$\lambda = \frac{1}{3} \text{ C-m}$$

and $\quad r = 18 \times 10^{-2}$ m

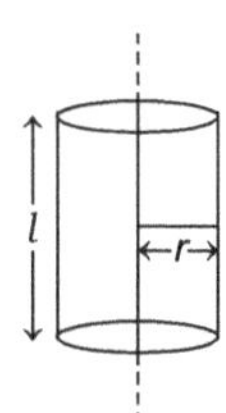

From Gauss's theorem

$$\oint \mathbf{E} \cdot d\mathbf{S} = \frac{q}{\varepsilon_0}$$

$$E \oint dS = \frac{q}{\varepsilon_0}$$

or $\quad E \times 2\pi r l = \dfrac{q}{\varepsilon_0}$

or $\quad E = \dfrac{q}{2\pi\varepsilon_0 r l} = \dfrac{q/l}{2\pi\varepsilon_0 r}$

$$= \frac{\lambda \times 2}{2\pi\varepsilon_0 r \times 2} = \frac{\lambda \times 2}{4\pi\varepsilon_0 r}$$

$$= 9 \times 10^9 \times \frac{1}{3} \times 2 \times \frac{1}{18 \times 10^{-2}}$$

$$= \frac{1}{3} \times 10^{11} = 0.33 \times 10^{11}$$

$$= 0.33 \times 10^{11} \text{ NC}^{-1}$$

27. ***(c)*** Potential at P due to $(+q)$ charge,

$$V_1 = \frac{1}{4\pi\varepsilon_0} \cdot \frac{q}{(z - a)}$$

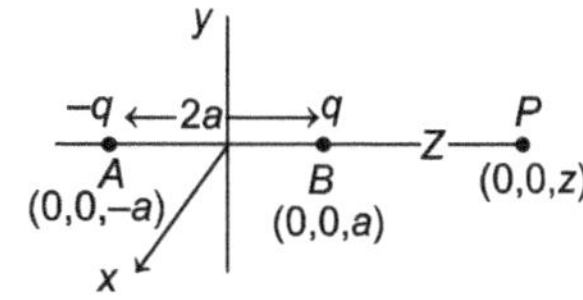

Potential at P due to $(-q)$ charge,

$$V_2 = \frac{1}{4\pi\varepsilon_0} \cdot \frac{-q}{(z + a)}$$

Total potential at P due to (AB) electric dipole

$$V = V_1 + V_2$$

$$= \frac{1}{4\pi\varepsilon_0} \cdot \frac{q}{(z - a)} - \frac{1}{4\pi\varepsilon_0} \frac{q}{(z + a)}$$

$$= \frac{q}{4\pi\varepsilon_0} \frac{(z + a - z + a)}{(z - a)(z + a)}$$

$$\Rightarrow \quad V = \frac{2qa}{4\pi\varepsilon_0 (z^2 - a^2)}$$

28. ***(c)***

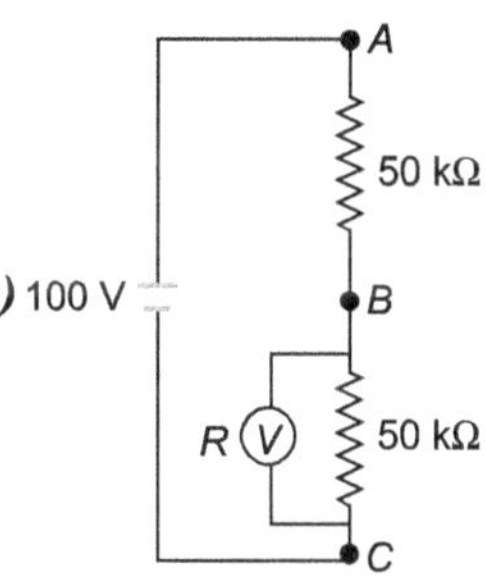

Internal resistance of voltmeter is R.

Therefore effective resistance across B and C, R' is given by

$$\frac{1}{R'} = \frac{1}{R} + \frac{1}{50} = \frac{50 + R}{50R}$$

or $\quad R' = \left(\dfrac{50R}{50 + R}\right)$

According to Ohm's law

$$V' = IR'$$

or $\quad \dfrac{100}{3} = I \cdot \left(\dfrac{50R}{50 + R}\right)$

or $\quad \dfrac{100}{3}\left(\dfrac{50 + R}{50R}\right) = I \quad \ldots\text{(i)}$

Now, total resistance of circuit,

$$R'' = 50 + \frac{50R}{50 + R}$$

or $\quad R'' = \dfrac{(2500 + 100R)}{(50 + R)}$

Now, $V'' = IR''$

$$\Rightarrow \quad 100 = \frac{100}{3}\left(\frac{50+R}{50R}\right)\frac{2500+100R}{(50+R)}$$

[Using Eq. (i)]

or $150R = 2500 + 100R$

or $50R = 2500$

or $R = 50\ \text{k}\Omega$

29. *(c)* Resistance of potentiometer wire,

$$R = \rho \times \frac{l}{A}$$

or $$R = \left(\rho \times \frac{10}{A}\right)$$

The value of 2.5 m length wire,

$$R' = \frac{\rho \times 10}{A \times 10} \times 2.5$$

or $$R' = \left(\frac{2.5\rho}{A \times 10}\right)$$

Potential $V' = I \times R'$

$$= I\left(\frac{2.5\rho}{A \times 10}\right)$$

Now, again the length of potentiometer wire is increased by 1 m, then resistance of null position wire.

$$R'' = \left(\frac{\rho \times l}{11 \times A}\right)$$

$$V'' = IR''$$

and $V = V'$

$$\frac{I \times 2.5\rho}{A \times 10} = \frac{\rho \times l}{11 \times A} \times I$$

or $$\frac{2.5 \times 11}{10} = l = 2.75\ \text{m}$$

30. *(c)*

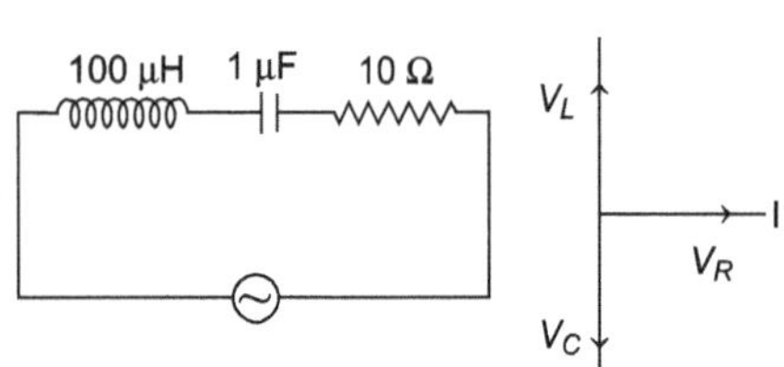

Impedance, $Z = \sqrt{(X_L \sim X_C)^2 + R^2}$

or $$Z = \sqrt{\left(\omega L \sim \frac{1}{\omega C}\right)^2 + R^2}$$

Inductive reactance,

$$X_L = \omega L = 70 \times 10^3 \times 100 \times 10^{-6}$$
$$= 7\ \Omega$$

Capacitance reactance,

$$X_C = \frac{1}{\omega C} = \frac{1}{70 \times 10^3 \times 1 \times 10^{-6}}$$

$$= \frac{1}{7 \times 10^{-2}} = \frac{10^2}{7} = \frac{100}{7}$$

As $X_C > X_L$

Hence, circuit behave like as *R-C* circuit.

31. *(c)* Magnetic field at the centre of the loop

$$B = \frac{\mu_0}{4\pi} \cdot \frac{I \cdot 2\pi R}{R^2} \quad \ldots(i)$$

For the wire which is looped double let radius becomes *r*.

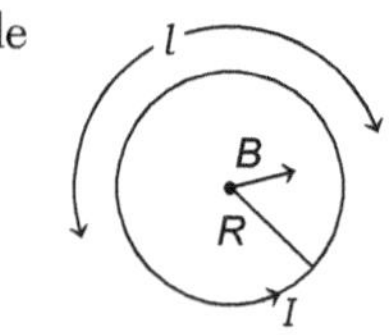

Then, $$\frac{l}{2} = 2\pi r$$

or $$\frac{l}{4\pi} = (r)$$

$\therefore$ $$B' = \frac{\mu_0}{4\pi} \cdot \frac{I \cdot 2\pi r \times 2}{r^2}$$

or $$B' = \frac{\mu_0}{4\pi} \cdot \frac{I \cdot \frac{l}{2} \cdot 2}{\left(\frac{l}{4\pi}\right)^2}$$

or $$B' = \frac{\mu_0}{4\pi} \cdot \frac{Il \times 16\pi^2}{l^2} \quad \ldots(ii)$$

Now, $$B = \frac{\mu_0}{4\pi} \cdot \frac{I \cdot l}{\left(\frac{l}{2\pi}\right)^2}\left[R = \frac{l}{2\pi}\right] \quad \ldots(iii)$$

Dividing Eq. (ii) by Eq. (iii), we get

$$\frac{B'}{B} = \frac{\frac{\mu_0}{4\pi} \cdot \frac{I \cdot l \cdot 16\pi^2}{l^2}}{\frac{\mu_0}{4\pi} \cdot \frac{Il \cdot 4\pi^2}{l^2}}$$

or $$\frac{B'}{B} = 4$$

or $B' = 4B$

32. *(c)*

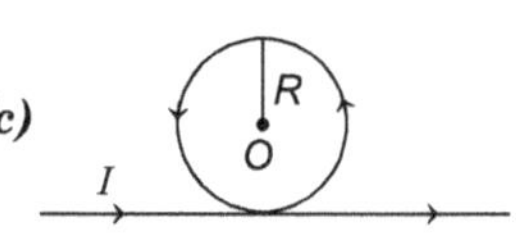

Magnetic field due to long wire at *O* point,

$$B_1 = \frac{\mu_0}{2\pi}\left(\frac{I}{R}\right) \quad \text{(upward direction)}$$

Magnetic field due to loop at *O* point,

$$B_2 = \frac{\mu_0}{4\pi} \cdot \frac{I \cdot 2\pi R}{R^2}$$

$$B_2 = \frac{\mu_0}{2} \cdot \frac{I}{R} \quad \text{(in upward direction)}$$

Hence, resultant magnetic field at centre *O*

$$B = B_1 + B_2$$

$$B = \frac{\mu_0 I}{2\pi \cdot R}(\pi + 1)$$

33. ***(b)*** Work function, $W_0 = 3.31 \times 10^{-19}$ J

Wavelength of incident radiation

$\lambda = 5000 \times 10^{-10}$ m

According to Einstein equation

$$E = W_0 + \text{KE}$$

$$\frac{hc}{\lambda} = 3.31 \times 10^{-19} + \text{KE}$$

$$\text{KE} = -3.31 \times 10^{-19} + \frac{6.62 \times 10^{-34} \times 3 \times 10^8}{5000 \times 10^{-10}}$$

$$= -3.31 \times 10^{-19} + \frac{6.62 \times 3}{5} \times 10^{-19}$$

$$= (-3.31 \times 1.324 \times 3) \times 10^{-19}$$

$$= (3.972 - 3.31) \times 10^{-19} = 0.662 \times 10^{-19} \text{ J}$$

$$\Rightarrow E = \frac{0.662 \times 10^{-19}}{1.6 \times 10^{-19}} = 0.41 \text{ eV}$$

34. ***(d)*** From Einstein's equation,

$$E = W_0 + \frac{1}{2}mv^2$$

$$\sqrt{\frac{2(E - W_0)}{m}} = v \qquad \text{...(i)}$$

A charged particle placed in uniform magnetic field experience a force

$$F = \frac{mv^2}{r}$$

or $$evB = \frac{mv^2}{r}$$

or $$r = \frac{mv}{eB}$$

or $$r = \frac{m\sqrt{\frac{2(E - W_0)}{m}}}{eB}$$

[By using Eq. (i)]

$$\Rightarrow r = \frac{\sqrt{2m(E - W_0)}}{eB}$$

35. ***(d)*** $N_1 = N_0 e^{-10\lambda t}$

and $N_2 = N_0 e^{-\lambda t}$

$$\Rightarrow \frac{N_1}{N_2} = \frac{1}{e} = e^{-1} = e^{(-10\lambda + \lambda)t} = e^{-9\lambda t}$$

$$\Rightarrow t = \frac{1}{9\lambda}$$

36. ***(c)*** In circuit *A*, both (*p-n*) junction diode act as forward biasing. Hence, current flows in circuit *A*. Total resistance *R* is given by

$$\frac{1}{R} = \frac{1}{4} + \frac{1}{4} \text{ or } \frac{1}{R} = \frac{2}{4}$$

or $$R = 2\,\Omega$$

According to Ohm's law,

$$V = I_A R$$

or $$8 = I_A \times 2$$

or $$I_A = 4 \text{ A}$$

In circuit *B*, lower *p-n*-junction diode is reverse biased. Hence, no current will flow is lower diode but upper diode is forward biased so current can flow through it

$$V = I_B R$$

or $$8 = I_B \times 4$$

or $$I_B = 2 \text{ A}$$

37. ***(c)*** After impact the bullet and block move together and comes to rest after covering a distance of 40 m.

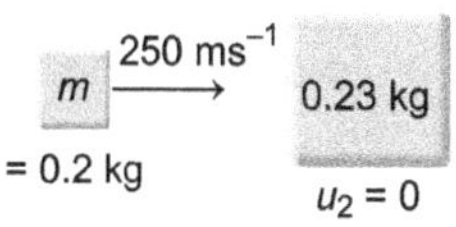

By conservation of momentum,

$$m_1u_1 + m_2u_2 = m_1v_1 + m_2v_2$$

or $$0.02 \times 250 + 0.23 \times 0 = 0.02\,v + 0.23\,v$$

$$5 + 0 = v\ (0.25)$$

$$\frac{500}{25} = v = 20 \text{ ms}^{-1}$$

Now, by conservation of energy,

$$\frac{1}{2}Mv^2 = \mu R \cdot d$$

or $$\frac{1}{2} \times 0.25 \times 400 = \mu \times 0.25 \times 9.8 \times 40$$

$$\Rightarrow \mu = \frac{200}{9.8 \times 40} = 0.51$$

38. ***(a)*** Let after the time t the position of A is $(0, v_At)$ and position of $B = (v_Bt, 10)$. Distance between them

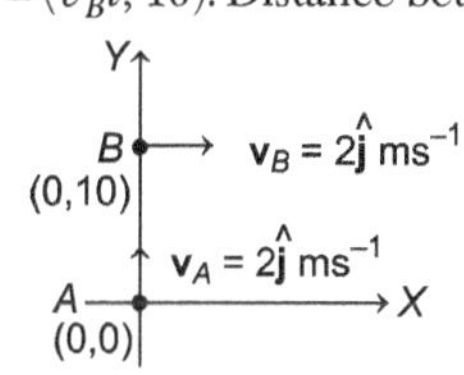

$$y = \sqrt{(0 - v_Bt)^2 + (v_At - 10)^2}$$

or $$y^2 = (2t)^2 + (2t - 10)^2$$

or $$y^2 = l = 4t^2 + 4t^2 + 100 - 40t$$

$$\Rightarrow l = 8t^2 + 100 - 40t$$

Now, $$\frac{dl}{dt} = (16t - 40) = 0$$

$$t = \frac{40}{16} = 2.5 \text{ s}$$

As $$\frac{d^2l}{dt^2} = 16 = (+\text{ve})$$

Hence, l will be minimum.

39. ***(b)*** In this process potential energy of the metre stick will be converted into rotational kinetic energy.

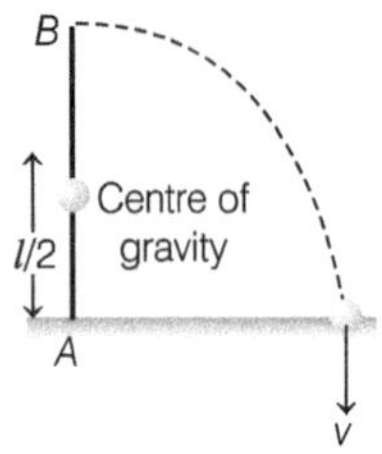

PE of metre stick $= mgl/2$

Because its centre of gravity lies at the middle of the rod.

Rotational kinetic energy $E = \frac{1}{2} I\omega^2$

I = moment of inertia of metre stick about point $A = ml^2/3$.

By the law of conservation of energy

$$mg\left(\frac{l}{2}\right) = \frac{1}{2} I\omega^2 = \frac{1}{2}\frac{ml^2}{3}\left(\frac{v_B}{l}\right)^2$$

By solving, we get

$$v_B = \sqrt{3gl}$$

40. ***(a)*** Given, $r = 0.4$ m,

$\alpha = 8 \text{ rad s}^{-1}$,

$m = 4$ kg, $I = ?$

Torque, $\tau = I\alpha$

$mgr = I \cdot \alpha$

or $4 \times 10 \times 0.4 = I \times 8$

$\Rightarrow I = \frac{16}{8} = 2 \text{ kg-m}^2$

or $I = 2 \text{ kg-m}^2$

Chemistry

41. ***(c)*** Given : $\Delta H_f(\text{H}) = 218$ kJ/mol

i.e., $\frac{1}{2}H_2 \longrightarrow H;\ \Delta H = 218$ kJ/mol

or $H_2 \longrightarrow 2H;\ \Delta H = 436$ kJ/mol

Bond energy of H—H bond $= \frac{436}{4.18} = 104.3$ kcal/mol

Thus, 104.3 kcal/mol energy is absorbed for breaking one mole of H—H bond. Hence, H—H bond energy is 104.3 kcal/mol.

42. ***(a)*** In Wacker process, alkene is oxidised into aldehyde.

$$CH_2{=}CH_2 + \frac{1}{2}O_2 \xrightarrow[H_2O]{PdCl_2 \cdot CuCl_2} \underset{(B)}{CH_3CHO}$$

Since, on ozonolysis, only alkenes produce aldehydes, 'A' must be an alkene. To decide the structure of alkene that undergoes ozonolysis, bring the products together in such a way that O-atoms are face to face and, replace O by double (=) bond. Thus,

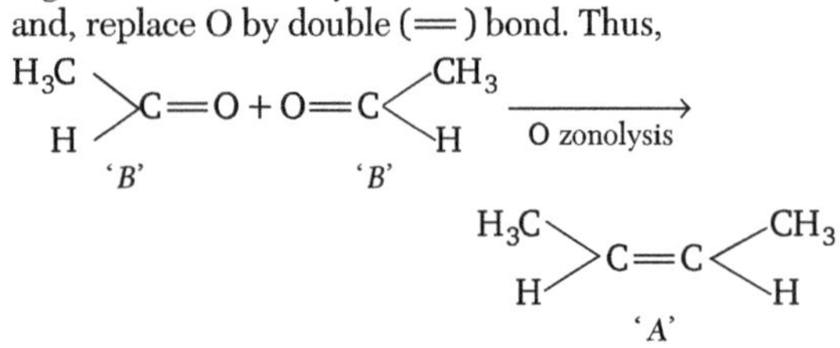

Therefore, alkyne must be but-2-yne.

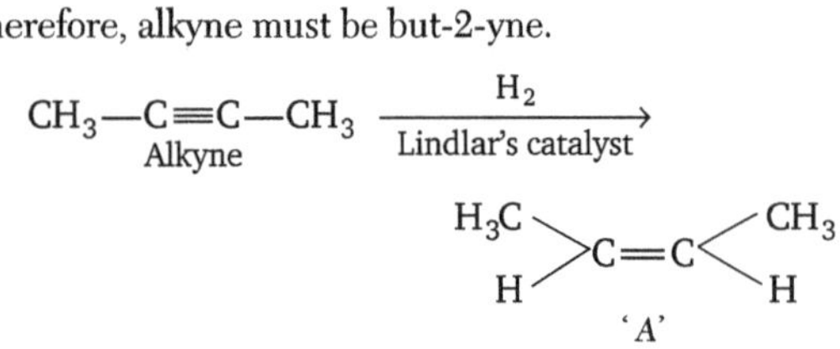

43. ***(b)*** $2F_2 + \underset{\text{dilute}}{2NaOH} \longrightarrow 2NaF + \underset{(A)}{OF_2}\uparrow + H_2O$

The structure of $A\,(OF_2)$ is as

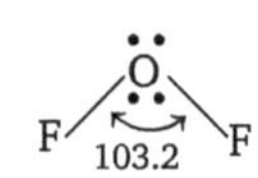

σ-bonds made by O = 2

Lone pairs of electrons on O = 2

$\therefore$ No. of orbitals used by O for hybridisation $= 2 + 2 = 4$

$\therefore$ Hybridisation of O in $OF_2 = sp^3$

Due to repulsion between two lone pairs of electrons, its shape gets distorted. Therefore, the bond angle in the molecule is 103°.

44. ***(a)*** To decide the structure of alkene that undergoes ozonolysis, bring the products together in such a way that O- atoms are face to face, and replace O by double (=) bond. Thus,

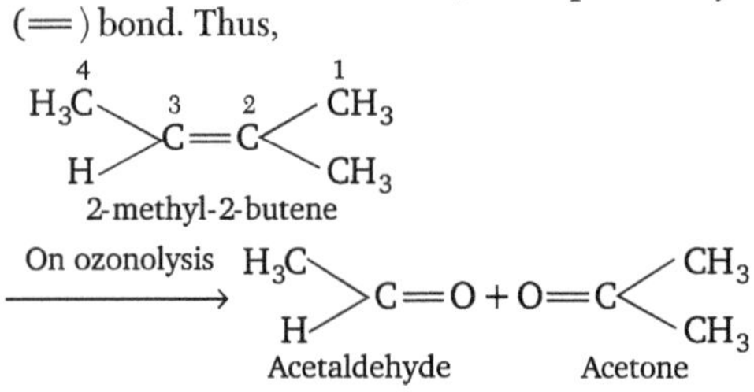

45. ***(a)*** Structure of XeO_3

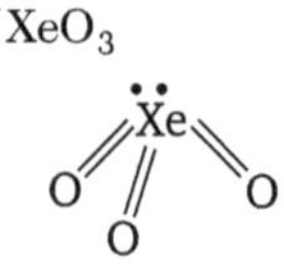

$\Rightarrow$ Three $p\pi$-$d\pi$ bonds are present.

Structure of XeO_4

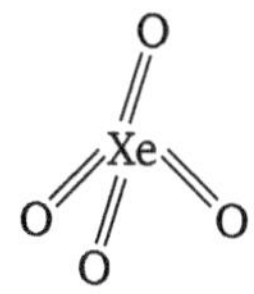

$\Rightarrow$ Four $p\pi$-$d\pi$ bonds are present.

46. ***(a)*** From de-Broglie's equation

$$\lambda = \frac{h}{mv}$$

$$\Rightarrow \quad \lambda^2 = \frac{h^2}{m^2v^2}$$

$$\Rightarrow \quad mv^2 = \frac{h^2}{m\lambda^2} \qquad \left[KE = \frac{1}{2}mv^2\right]$$

$$\therefore \quad KE\,(K) = \frac{1}{2}\frac{h^2}{m\lambda^2}$$

$$\Rightarrow \quad \frac{K_1}{K_2} = \left(\frac{\lambda_2}{\lambda_1}\right)^2 = \left(\frac{5}{3}\right)^2$$

$$\therefore \quad K_1 : K_2 = 25 : 9$$

47. ***(c)*** Paramagnetic property depends upon the number of unpaired electrons. Higher the number of unpaired electrons, higher will be the paramagnetic property will be.

$Cu^{2+} = [Ar]\,3d^9$, no. of unpaired electrons = 1

$V^{2+} = [Ar]\,3d^3$, no. of unpaired electrons = 3

$Cr^{2+} = [Ar]\,3d^4$, no. of unpaired electrons = 4

$Mn^{2+} = [Ar]\,3d^5$, no. of unpaired electrons = 5

Hence, correct order is

$$Cu^{2+} < V^{2+} < Cr^{2+} < Mn^{2+}$$

48. ***(a)*** $\because$ 1 mol = 6.023×10^{23} atoms

KE of 1 mol = 6.023×10^4 J

or KE of 6.023×10^{23} atoms

$$= 6.023 \times 10^4 \text{ J}$$

$$\therefore \text{ KE of 1 atom} = \frac{6.023 \times 10^4}{6.023 \times 10^{23}}$$

$$= 1.0 \times 10^{-19} \text{ J}$$

$$h\nu_{energy} = \frac{hc}{\lambda} = \frac{6.626 \times 10^{-34} \times 3 \times 10^8}{600 \times 10^{-9}}$$

$$= 3.313 \times 10^{-19} \text{ J}$$

Minimum amount of energy required to remove an electron from the metal ion (i.e., Threshold energy)

$$= h\nu - KE$$

$$= 3.313 \times 10^{-19} - 1.0 \times 10^{-19}$$

$$= 2.313 \times 10^{-19} \text{ J}$$

49. ***(a)*** The thermosphere is the fourth layer of the Earth's atmosphere and is located above the mesosphere.

The air is thin in the thermosphere. The Earth's thermosphere also includes the region of the atmosphere, called the ionosphere. The ionosphere is the region of the atmosphere that is filled with charged particles such as O_2^+, O^+, NO^+. The high temperature in the thermosphere can cause molecules to ionise.

50. ***(b)*** Sulphuric anhydride is SO_3 and its structure is as follows :

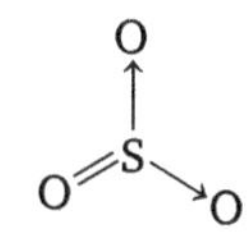

$\Rightarrow$ 3σ, $1p\pi$-$p\pi$ and $2\ p\pi$-$d\pi$ bonds are present.

51. ***(a)*** The Gattermann reaction takes place as follows:

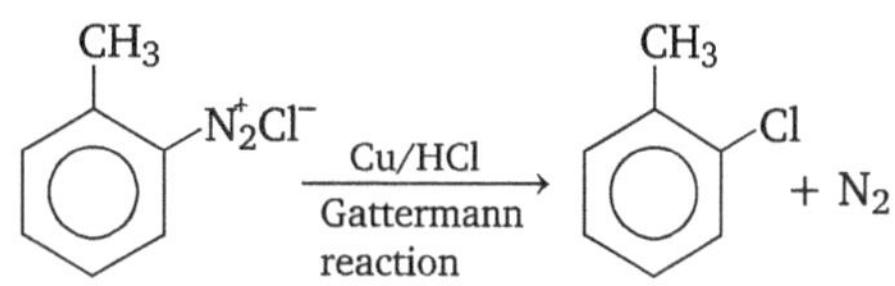

52. ***(c)*** Structures of H_3PO_3 and H_3PO_2 are as follows:

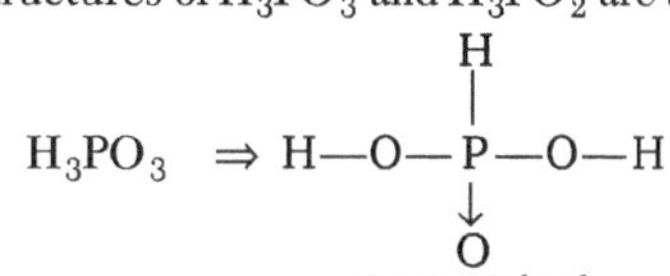

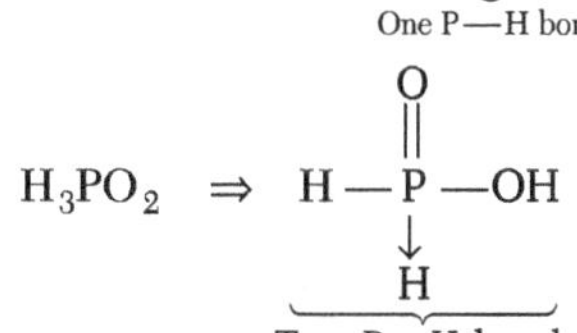

53. ***(c)*** From the definition of dipole moment,

$$\mu = \delta \times d$$

where, δ = magnitude of electric charge

d = distance between particles (i.e. bond length)

$$\therefore \quad \delta = \frac{\mu}{d}$$

$$\text{or,} \quad \frac{\delta_{HCl}}{\delta_{HI}} = \frac{\mu_{HCl}}{d_{HCl}} \times \frac{d_{HI}}{\mu_{HI}}$$

$$= \frac{1.03 \times 1.6}{1.3 \times 0.38} = 3.3 : 1$$

54. ***(d)*** The complete reactions take place as follows :

$$SiCl_4 + 4H_2O \longrightarrow H_4SiO_4 + 4HCl$$

$$H_4SiO_4 \xrightarrow[1000°C]{\Delta} SiO_2 + 2H_2O$$

55. ***(c)*** % of Cd in $CdCl_2 = \frac{0.9}{1.5} \times 100$

$$= 60\%$$

Therefore, % of Cl_2 in $CdCl_2$ = 100 − 60 = 40%

$\because$ 40% part (Cl_2) has atomic weight

$$= 2 \times 35.5 = 71.0$$

$\therefore$ 60% part (Cd) has atomic weight

$$= \frac{71.0 \times 60}{40} = 106.5$$

56. ***(c)*** $2Al + 2NaOH + 2H_2O \longrightarrow \underset{\text{Sodium meta aluminate}}{2NaAlO_2} + 3H_2$

Sodium meta aluminate, is soluble in water and changes into the complex $[Al(H_2O)_2(OH)_4]^-$, in which coordination number of Al is 6.

57. ***(b)*** Average kinetic energy per molecule $= \frac{3}{2} kT$

or $$= \frac{3}{2} \frac{R}{N_0} T$$

$$= \frac{3}{2} \times \frac{8.314}{6.023 \times 10^{23}} \times 300$$

$$= 6.21 \times 10^{-21} \text{ JK}^{-1} \text{ molecule}^{-1}$$

58. ***(c)*** Superoxides are the species having an O—O bond and O in an oxidation state of $-\frac{1}{2}$ (superoxide ion is O_2^-). Usually these are formed by active metals such as KO_2, RbO_2 and CsO_2.

For the salts of larger anions (like O_2^-), lattice energy increases in a group. Since, lattice energy is the driving force for the formation of an ionic compound and its stability, the stability of the superoxides from 'K' to 'Cs' also increases.

$\therefore$ A is true but R is not true.

59. ***(a)*** Perhydrol means 30% solution of H_2O_2.

H_2O_2 decomposes as

$$2H_2O_2 \longrightarrow 2H_2O + O_2$$

Volume strength of 30% H_2O_2 solution is 100 that means 1 mL of this solution on decomposition gives 100 mL oxygen.

$$SO_2 + \frac{1}{2}O_2 \longrightarrow SO_3$$

1 L	$\frac{1}{2}$ L	1 L
2 L	1 L	2 L

Since, 100 mL of oxygen is obtained by

$$= 1 \text{ mL of } H_2O_2$$

$\therefore$ 1000 mL of oxygen will be obtained by

$$= \frac{1}{100} \times 1000 \text{ mL of } H_2O_2$$

$$= 10 \text{ mL of } H_2O_2$$

60. ***(d)*** Buffer capacity, $\beta = \frac{dC_{HA}}{d_{pH}}$,

where, dC_{HA} = number of moles of acid added per litre

d_{pH} = change in pH

$$dC_{HA} = \frac{\text{moles of acetic acid}}{\text{volume}}$$

$$= \frac{0.12/60}{250/1000} = \frac{1}{125}$$

$\therefore$ $$\beta = \frac{1/125}{0.02} = \frac{1}{2.5} = 0.4$$

61. ***(b)*** (A) Feldspar (orthoclase) ($KAlSi_3O_8$).
It is used in the manufacture of porcelain.

(B) Asbestos $\{CaMg_3(SiO_3)_4\}$
It is used for fireproof sheets, clothes etc.

(C) Pyrargyrite (Ruby silver) (Ag_3SbS_3).
It is an ore of silver.

(D) Diaspore ($Al_2O_3 \cdot H_2O$).
It is an ore of aluminium.

62. ***(c)*** First ionisation energy increases in a period. Thus, the first IE of the elements of the second period should be as follows

$$\text{Be} < \text{B} < \text{N} < \text{O}$$

But in practice, the elements do not follow the above order. The first IE of these elements is

$$\text{B} < \text{Be} < \text{O} < \text{N}$$

The lower IE of B than that of Be is because in B ($1s^2$, $2s^2 2p^1$), electron is to be removed from $2p$ which is easy while in Be ($1s^2$, $2s^2$), electron is to be removed from $2s$ which is difficult.

The low IE of O than that of N is because of the half-filled $2p$ orbitals in N ($1s^2$, $2s^2 2p^3$).

63. ***(c)*** The complete reaction as follows:

$$\underset{\text{Ethanol}}{CH_3CH_2OH} + Cl_2 \xrightarrow[-2HCl]{} \underset{\text{Acetaldehyde }(X)}{CH_3CHO} \xrightarrow[-3HCl]{3Cl_2} \underset{\text{Chloral }(Y)}{CCl_3CHO}$$

64. ***(c)*** The reactions take place as follows:

I. $$\begin{matrix} CH_3COO \\ CH_3COO \end{matrix} \Big> Ca \xrightarrow[-CaCO_3]{\Delta} \underset{(A)}{CH_3COCH_3}$$

II. $$CH_3COOH + 6HI \xrightarrow{\text{Red P}} \underset{(B)}{CH_3CH_3} + 3I_2 + 2H_2O$$

III. $$\begin{matrix} CH_3CO\,OH \\ CH_3COO\,H \end{matrix} \xrightarrow{\Delta, P_4O_{10}} \underset{(C)}{\begin{matrix} CH_3CO \\ CH_3CO \end{matrix} \Big> O} + H_2O$$

65. ***(b)*** C = 85.71% $= \frac{85.71}{12} = 7.14$; $\frac{7.14}{7.14} = 1$

H = 14.29% $= \frac{14.29}{1} = 14.29$; $\frac{14.29}{7.14} = 2$

$\therefore$ Empirical formula = CH_2

and, empirical formula weight = 12 + 2 = 14

Again, molecular formula weight

$$= 2 \times \text{vapour density}$$
$$= 2 \times 14 = 28$$

$$\therefore \quad n = \frac{28}{14} = 2$$

$\therefore$ Molecular formula $= (CH_2)_2 = C_2H_4$

$$\underset{(A)}{CH_2{=}CH_2} + HOCl \longrightarrow \underset{(B)}{\underset{\displaystyle OH}{CH_2{-}CH_2{-}Cl}}$$

$$\xrightarrow[KCN]{EtOH} \underset{\displaystyle OH}{CH_2{-}CH_2{-}CN} \xrightarrow{H_3O^+} \underset{(C)}{\underset{\displaystyle OH}{CH_2{-}CH_2{-}COOH}}$$

66. ***(c)*** Tripeptides are amino acids polymers in which three individual amino acid units, called residues, are linked together by amide bonds.

In these, an amine group from one residue forms an amide bond with the carboxyl group of a second residue, the amino group of the second forms an amide bond with the carboxyl group of the third.

Therefore, glycine ($NH_2{-}CH_2{-}COOH$), alanine $\left(\underset{\displaystyle NH_2}{CH_3{-}CH{-}COOH}\right)$ and phenyl alanine $\left(\underset{\displaystyle NH_2}{C_6H_5{-}CH_2{-}CHCOOH}\right)$ can be linked in six different ways.

67. ***(d)*** A codon is a specific sequence of three adjacent bases on a strand of DNA or RNA that provides genetic code information for a particular amino acid. That is to be incorporated into protein.

68. ***(c)*** Dopamine is produced in several areas of the brain. If the amount of dopamine increases in the brain, the patient may be affected with Parkinson's disease. The IUPAC name of dopamine is 2-(3,4-dihydroxyphenyl) ethylamine and its structure is as follows :

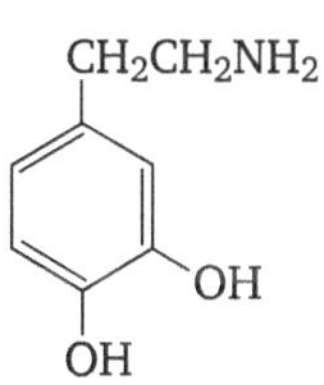

69. ***(a)*** Freezing point of a substance is the temperature at which the solid solvent and the liquid solvent are in equilibrium.

If a non-volatile solute is added to the solvent, there is decrease in vapour pressure of the solution and thus the freezing point of the solution is less than that of pure solvent. It is called depression in freezing point.

70. ***(b)*** The reaction takes place as follows:

$$C_2H_5{-}Cl + AgCN \xrightarrow[H_2O]{EtOH/} \underset{(X)}{C_2H_5{-}NC} + AgCl$$

N-linked to ethyl carbon

$\therefore$ Option (b) is a correct statement.

71. ***(a)*** For the given cell,

$$Ag \mid Ag^+ \mid AgCl \mid Cl^{\ominus} \mid Cl_2, Pt$$

the cell reactions are as follows

At anode

$$Ag(s) \longrightarrow Ag^+ + e^-$$

At cathode

$$AgCl + e^- \longrightarrow Ag(s) + Cl^-$$

Net cell reaction

$$AgCl \longrightarrow Ag^+ + Cl^-$$

$$\therefore \quad \Delta G^\circ_{\text{reaction}} = \Sigma\Delta G^\circ_p - \Sigma\Delta G^\circ_R$$
$$= (78 - 129) - (-109)$$
$$= +58 \text{ kJ/mol}$$
$$\Delta G^\circ = -nFE^\circ$$
$$58 \times 10^3 \text{ J} = -1 \times 96500 \times E^\circ_{\text{cell}}$$
$$E^\circ_{\text{cell}} = \frac{-58 \times 1000}{96500}$$
$$= -0.6 \text{ V}$$

72. ***(d)*** Crotonaldehyde is produced by the aldol condensation of acetaldehyde.

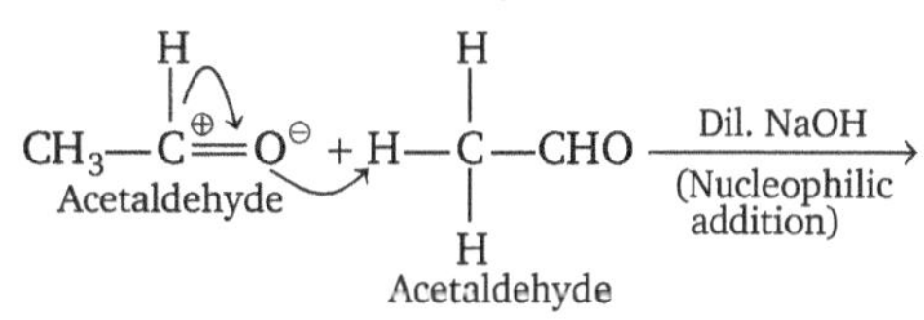

$$\underset{\text{Aldol}}{\underset{\displaystyle OH\ \ H}{\overset{\displaystyle H\ \ H}{CH_3{-}C{-}C{-}CHO}}} \xrightarrow[\substack{-H_2O \\ \text{(Elimination)}}]{\Delta} \underset{\text{Crotonaldehyde}}{CH_3CH{=}CHCHO}$$

73. ***(b)*** $BaCl_2 + 2NaOH \longrightarrow Ba(OH)_2 + 2NaCl$

$$\lambda^\infty_{m\ Ba(OH)_2} = \lambda^\infty_{m\ BaCl_2} + 2\lambda^\infty_{m\ NaOH} - 2\lambda^\infty_{m\ NaCl}$$
$$= 280 \times 10^{-4} + 2 \times 248 \times 10^{-4} - 2 \times 126 \times 10^{-4}$$
$$= (280 + 496 - 252) \times 10^{-4}$$
$$= 524 \times 10^{-4} \text{ Sm}^2 \text{ mol}^{-1}$$

74. *(b)* Density, $d = \frac{MZ}{N_A a^3}$

where, Z = number of atoms in unit cell

$$\therefore \quad Z = \frac{dN_A a^3}{M}$$

$$= \frac{8.92 \times 6.023 \times 10^{23} \times (362 \times 10^{-10})^3}{63.55} = 4.0$$

Thus, metal has face centered unit cell.

75. *(c)* $N_2 + 2O_2 \rightleftharpoons 2NO_2$

$$K_1 = \frac{[NO_2]^2}{[N_2][O_2]^2}$$

or $$100 = \frac{[NO_2]^2}{[N_2][O_2]^2} \quad \ldots(i)$$

Again, $[NO_2] \rightleftharpoons \frac{1}{2}N_2 + O_2$

$$K_2 = \frac{[N_2]^{1/2}[O_2]}{[NO_2]}$$

or $$K_2^2 = \frac{[N_2][O_2]^2}{[NO_2]^2} \quad \ldots(ii)$$

Eqs. (i) × (ii), we get

$$100 \times K_2^2 = 1$$

or $$K_2^2 = \frac{1}{100} \text{ or } K_2 = \frac{1}{10} = 0.1$$

76. *(c)* For a first order reaction,

$$t = \frac{2.303}{\lambda} \log_{10} \frac{a}{a-x}$$

Let initial amount of reactant is 100.

$$\frac{t_1}{t_2} = \frac{\log \frac{100}{100-75}}{\log \frac{100}{100-25}}$$

[$\because$ λ remains constant]

$$= \frac{\log \frac{100}{25}}{\log \frac{100}{75}} = \frac{\log 4}{\log 4/3}$$

$$= \frac{\log 4}{\log 4 - \log 3}$$

$$= \frac{2 \times 0.3010}{2 \times 0.3010 - 0.4771}$$

$$= \frac{0.6020}{0.1249} = 4.81$$

77. *(c)* Specific rotation of compound

$$[\alpha] = \frac{[\alpha]_{observed}}{l \times C}$$

where $[\alpha]_{Observed}$ = Observed rotation

l = length of polarimeter tube

C = concentration of solute

$$= \frac{-1.2}{5 \times \frac{6.15}{1000}} = -39°$$

78. *(b)* Let the concentration of potassium acetate is x.

From Henderson's equation,

$$pH = pK_a + \log \frac{[salt]}{[acid]}$$

$$4.8 = -\log(1.8 \times 10^{-5}) + \log \frac{x \times 50}{20 \times 0.1}$$

$$4.8 = 4.74 + \log 25x$$

or $\log 25x = 0.06$

$$25x = 1.148$$

$$\therefore \quad x = 0.045 \text{ M}$$

79. *(b)* For, $Na_2O + SO_3 \longrightarrow Na_2SO_4$;

$$\Delta H° = -2 \times 146 + \frac{259}{2} - 418$$

or $$\Delta H° = -580.5$$

$$\approx -581 \text{ kJ}$$

80. *(b)* As_2S_3 is a negative sol. It is obvious that cations are effective in coagulating negative sols. According to Hardy Schulze rule, greater the valency of the coagulating ion, greater is its coagulating power. Thus, out of the given, $AlCl_3(Al^{3+})$ is most effective for causing coagulation of As_2S_3 sol.

a. English Proficiency

81. ***(a)*** The most appropriate alternative to fill the blank is 'discrepancies'. Discrepancy means the quality or state of disagreeing or being at variance.

82. ***(d)*** 'Embezzle' should be used in the given blank. Embezzle means steal or misappropriate.

83. ***(c)*** The most appropriate alternative to fill the blank is 'bagged'. Bagged means to winning something or succeed in catching.

84. ***(d)*** The correct sequence is SPRQ.

85. ***(b)*** The correct sequence is QPSR.

86. ***(b)*** The correct sequence is RSQP.

87. ***(a)*** 'Denouement' means the outcome of a situation, when something is decided or made clear. So, 'outcome' is the correct answer.

88. ***(d)*** 'Gauche' means lacking social experience or grace; socially awkward or not tactful. So, 'Tactless' would be its correct similar meaning word.

89. ***(b)*** 'Accolade' means praise, honour and approval. So, 'appreciation' would be its correct similar meaning word as it also means the same.

90. ***(b)*** 'Reprimand' means a formal expression of disapproval. So, 'Appreciate' would be its correct antonym.

91. ***(a)*** 'Impertinent' means not showing proper respect; rude. So, 'polite' would be its correct opposite meaning word.

92. ***(c)*** 'Equivocal' means unclear or usually used to mislead or confuse. So, 'clear' would be its correct opposite meaning word.

93. ***(b)*** A picture or a design made by putting together coloured pieces of glass or stones is called Mosaic.

94. ***(c)*** The doctrine that human soul passes from one body to another at the time of death is called Transmigration.

95. ***(a)*** Pedantic is the correct answer.

b. Logical Reasoning

96. ***(d)*** From problem figure (1) to (2), double figure is converted into single figure and *vice-versa*. Also, figures change place in a set order. Hence, answer figure (d) will replace the sign ?

97. ***(d)*** 'Nurse' receives instructions from 'Doctor' and 'Follower' receives instructions from 'Leader'.

98. ***(b)*** $24 \times 2 + 4 = 52,$

$52 \times 2 + 4 = 108 \neq \boxed{112}$

$108 \times 2 + 4 = 220,$

$220 \times 2 + 4 = 444,$

and so on. Hence, number 112 is wrong and should be replaced by 108.

99. ***(b)*** Only I and II are implicit because in the relief camp the facilities of food, water and shelter are available.

100. ***(b)*** It is clear that answer figure (b) completes the original figure, which looks like as shown in the adjacent figure.

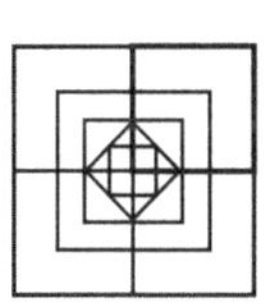

Hence, alternative (b) is the correct answer.

101. ***(a)*** Clearly figure (x) is embedded in alternative figure (a). The portion which figure (x) occupies in the alternative figure has been shown in the adjacent figure.

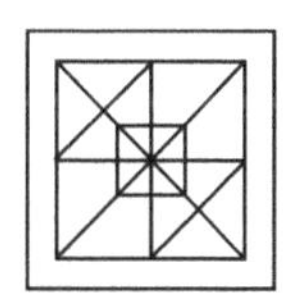

Hence, the correct answer figure is (a).

102. ***(d)*** Symbol appearing on the faces of dice can be shown as given in the figure. We see from the figure that symbol will appear on the opposite face symbol ×.

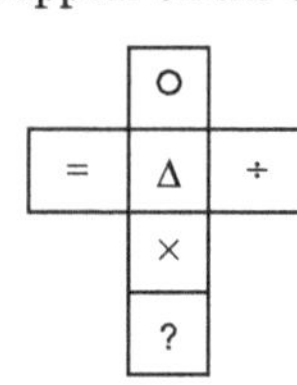

103. ***(d)*** Figure X is the first step in which a circular piece of paper is folded from upper to the lower half along the diameter. In figure Y both the extreme ends of the figure X have been folded to form a triangle and then as given in figure Z, a cut has been marked from the right side. It is clear that this cut will result into two marks, one in the lower half and one in the upper half of the paper, when it will be unfolded.

Answer figure ***(b)*** represents the correct design of the unfolded paper and hence, the correct answer is (d).

104. ***(b)*** As,

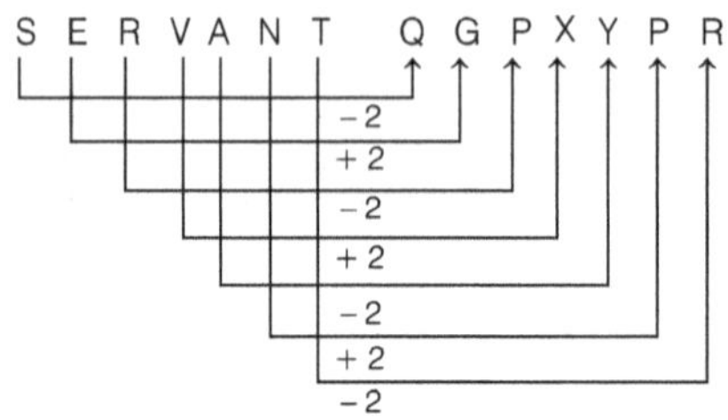

Similarly,

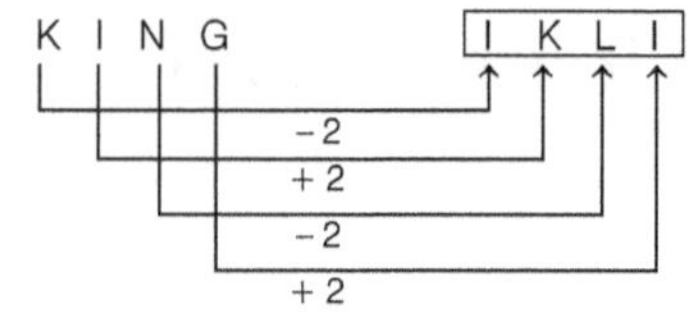

105. ***(c)*** Converting alphabets into mathematical symbols as given above, we get

$$18 \times 12 \div 4 + 5 - 6$$
$$= 18 \times \frac{12}{4} + 5 - 6$$
$$= 18 \times 3 + 5 - 6$$
$$= 54 + 5 - 6$$
$$= 59 - 6 = 53$$

Mathematics

106. ***(c)*** Let $A = \left\{x \in R : \frac{2x-1}{x^3 + 4x^2 + 3x}\right\}$

Now, $x^3 + 4x^2 + 3x = x(x^2 + 4x + 3)$

$$= x(x+3)(x+1)$$

$\therefore \quad A = R - \{0, -1, -3\}$

107. ***(c)*** The total number of subsets of given set is $2^9 = 512$

Even numbers are $\{2, 4, 6, 8\}$.

***Case* I** When selecting only one even number.

$$= {}^4C_1 = 4$$

***Case* II** When selecting only two even numbers

$$= {}^4C_2 = 6$$

***Case* III** When selecting only three even numbers

$$= {}^4C_3 = 4$$

***Case* IV** When selecting only four even numbers

$$= {}^4C_4 = 1$$

$\therefore$ Required number of ways

$$= 512 - (4 + 6 + 4 + 1) - 1$$
$$= 496$$

[Here, we subtract 1 for due to the null set]

108. ***(b)*** Now, $(1 + x^2)^{12}(1 + x^{12} + x^{24} + x^{36})$

$$= [1 + {}^{12}C_1(x^2) + {}^{12}C_2(x^2)^2 + {}^{12}C_3(x^2)^3 + {}^{12}C_4(x^2)^4 + {}^{12}C_5(x^2)^5 + {}^{12}C_6(x^2)^6 + \ldots + {}^{12}C_{12}(x^2)^{12}] \times (1 + x^{12} + x^{24} + x^{36})$$

Coefficient of $x^{24} = {}^{12}C_6 + {}^{12}C_{12} + 1$

$$= {}^{12}C_6 + 2$$

109. ***(d)*** $\frac{1}{(x-1)^2(x-2)} = \frac{1}{-2(1-x)^2\left(1 - \frac{x}{2}\right)}$

$$= -\frac{1}{2}\left[(1-x)^{-2}\left(1 - \frac{x}{2}\right)^{-1}\right]$$

$$= -\frac{1}{2}\left[(1 + 2x + \ldots)\left(1 + \frac{x}{2} + \ldots\right)\right]$$

$\therefore$ Coefficient of constant term is $-\frac{1}{2}$.

110. ***(c)*** Given, $(x-a)(x-a-1) + (x-a-1)(x-a-2) + (x-a)(x-a-2) = 0$

Let $x - a = t$, then

$$t(t-1) + (t-1)(t-2) + t(t-2) = 0$$

$\Rightarrow \quad t^2 - t + t^2 - 3t + 2 + t^2 - 2t = 0$

$\Rightarrow \quad 3t^2 - 6t + 2 = 0$

$\Rightarrow \quad t = \frac{6 \pm \sqrt{36 - 24}}{2(3)} = \frac{6 \pm 2\sqrt{3}}{2(3)}$

$\Rightarrow \quad x - a = \frac{3 \pm \sqrt{3}}{3} \Rightarrow x = a + \frac{3 \pm \sqrt{3}}{3}$

Hence, x is real and distinct.

111. ***(b)*** Given, $f(x) = x^2 + ax + b$ has imaginary roots.

$\therefore$ Discriminant, $D < 0 \Rightarrow a^2 - 4b < 0$

Now, $\quad f'(x) = 2x + a$

$\quad f''(x) = 2$

Also, $\quad f(x) + f'(x) + f''(x) = 0 \quad \ldots(i)$

$\Rightarrow \quad x^2 + ax + b + 2x + a + 2 = 0$

$\Rightarrow \quad x^2 + (a+2)x + b + a + 2 = 0$

$\therefore \quad x = \frac{-(a+2) \pm \sqrt{(a+2)^2 - 4(a+b+2)}}{2}$

$$= \frac{-(a+2) \pm \sqrt{a^2 - 4b - 4}}{2}$$

Mock Test 3

Since, $a^2 - 4b < 0$

$\therefore$ $a^2 - 4b - 4 < 0$

Hence, Eq. (i) has imaginary roots.

112. ***(a)*** Given, $\begin{vmatrix} 3 & 5 & x \\ 7 & x & 7 \\ x & 5 & 3 \end{vmatrix} = 0$

$\Rightarrow$ $3(3x - 35) - 5(21 - 7x) + x(35 - x^2) = 0$

$\Rightarrow$ $9x - 105 - 105 + 35x + 35x - x^3 = 0$

$\Rightarrow$ $x^3 - 79x + 210 = 0$

$\Rightarrow$ $(x + 10)(x - 3)(x - 7) = 0$

$\Rightarrow$ $x = -10, 3, 7$

113. ***(d)*** Let a and R be the first term and common ratio of a GP.

$\therefore$ $T_p = aR^{p-1} = x$

$T_q = aR^{q-1} = y$

and $T_r = aR^{r-1} = z$

$\Rightarrow$ $\log x = \log a + (p - 1) \log R$

$\log y = \log a + (q - 1) \log R$

and $\log z = \log a + (r - 1) \log R$

$$\therefore \begin{vmatrix} \log x & p & 1 \\ \log y & q & 1 \\ \log z & r & 1 \end{vmatrix} = \begin{vmatrix} \log a + (p-1)\log R & p & 1 \\ \log a + (q-1)\log R & q & 1 \\ \log a + (r-1)\log R & r & 1 \end{vmatrix}$$

$$= \begin{vmatrix} \log a & p & 1 \\ \log a & q & 1 \\ \log a & r & 1 \end{vmatrix} + \begin{vmatrix} (p-1)\log R & p & 1 \\ (q-1)\log R & q & 1 \\ (r-1)\log R & r & 1 \end{vmatrix}$$

$$= \log a \begin{vmatrix} 1 & p & 1 \\ 1 & q & 1 \\ 1 & r & 1 \end{vmatrix} + \log R \begin{vmatrix} p-1 & p-1 & 1 \\ q-1 & q-1 & 1 \\ r-1 & r-1 & 1 \end{vmatrix}$$

$(C_2 \to C_2 - C_3)$

$= 0 + 0 = 0$ $\quad$ ($\because$ two columns are identical)

114. ***(d)*** If matrix has no inverse it means the value of determinant should be zero.

$$\therefore \begin{vmatrix} 1 & -1 & x \\ 1 & x & 1 \\ x & -1 & 1 \end{vmatrix} = 0$$

If we put $x = 1$, then column Ist and IIIrd are identical.

Hence, option (d) is correct.

115. ***(c)*** Let $z = x + iy$

Given, $\left| \dfrac{z + 2i}{2z + i} \right| < 1$

$$\Rightarrow \frac{\sqrt{(x)^2 + (y+2)^2}}{\sqrt{(2x)^2 + (2y+1)^2}} < 1$$

$\Rightarrow$ $x^2 + y^2 + 4 + 4y < 4x^2 + 4y^2 + 1 + 4y$

$\Rightarrow$ $3x^2 + 3y^2 > 3$

$\Rightarrow$ $x^2 + y^2 > 1$

116. ***(d)*** Let $f(x) = \sin^4 x + \cos^4 x$

$= (\sin^2 x + \cos^2 x)^2 - 2\sin^2 x \cos^2 x$

$= 1 - \dfrac{1}{4} \cdot 2(\sin 2x)^2$

$= 1 - \dfrac{1}{4}(1 - \cos 4x)$

$= \dfrac{3}{4} + \dfrac{\cos 4x}{4}$

$\therefore$ Period of $f(x) = \dfrac{2\pi}{4} = \dfrac{\pi}{2}$

117. ***(b)*** Now, $\tan(x - y) \tan y$

$$= \frac{\sin(x-y)\sin y}{\cos(x-y)\cos y} \times \frac{2}{2}$$

$$= \frac{\cos(x-2y) - \cos(x)}{\cos(x-2y) + \cos(x)}$$

$$= \frac{1 - \dfrac{\cos x}{\cos(x-2y)}}{1 + \dfrac{\cos(x)}{\cos(x-2y)}}$$

$$= \frac{1-\lambda}{1+\lambda} \qquad \left(\text{Given, } \lambda = \frac{\cos x}{\cos(x-2y)}\right)$$

118. ***(a)*** It is a standard result.

$\cos A \cos 2A \cos 2^2 A \ldots \cos 2^{n-1} A$

$$= \frac{\sin 2^n A}{2^n \sin A}$$

119. ***(c)*** $\sin^2 x - \cos 2x = 2 - \sin 2x$

$\Rightarrow$ $1 - \cos^2 x - (2\cos^2 x - 1) = 2 - 2\sin x \cos x$

$\Rightarrow$ $-3\cos^2 x + 2\sin x \cos x = 0$

$\Rightarrow$ $\cos x (2\sin x - 3\cos x) = 0$

$\Rightarrow$ $\cos x = 0, \quad (\because 2\sin x - 3\cos x \neq 0)$

$\Rightarrow$ $x = 2n\pi \pm \dfrac{\pi}{2}$

$\Rightarrow$ $x = (4n \pm 1)\dfrac{\pi}{2}$

120. ***(c)*** We know that, $2s = a + b + c$

$$\therefore \frac{(a+b+c)(b+c-a)(c+a-b)(a+b-c)}{4b^2c^2}$$

$$= \frac{2s(2s-2a)(2s-2b)(2s-2c)}{4b^2c^2}$$

$$= 4\frac{s(s-a)}{bc} \times \frac{(s-b)(s-c)}{bc}$$

$$= 4\cos^2\frac{A}{2} \times \sin^2\frac{A}{2}$$

$$= \sin^2 A$$

121. *(c)* In ΔAPD,

$$\tan 45° = \frac{a}{AP} \Rightarrow AP = a$$

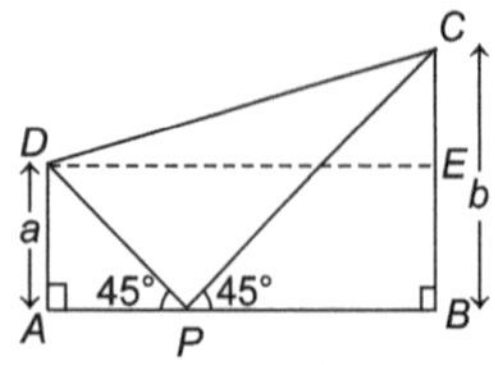

P—H bond

and in ΔBPC, $\tan 45° = \frac{b}{PB} \Rightarrow PB = b$—H bond

$\therefore \quad DE = a + b$ and $CE = b - a$

In ΔDEC, $\quad DC^2 = DE^2 + EC^2$

$$= (a+b)^2 + (b-a)^2$$

$$= 2(a^2 + b^2)$$

122. *(a)* Now, $\mathbf{AB} + 2\mathbf{AD} + \mathbf{BC} - 2\mathbf{DC}$

$$= \mathbf{AC} + 2\mathbf{AD} - 2\mathbf{DC}$$

$$= \mathbf{AC} + 2(\mathbf{AC} + \mathbf{CD}) - 2\mathbf{DC}$$

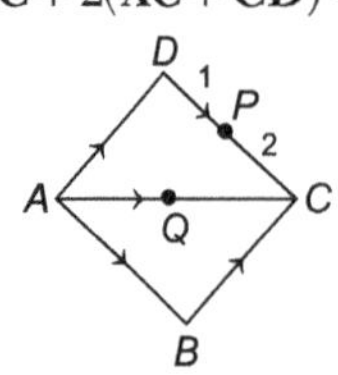

$$= 3\mathbf{AC} - 4\mathbf{DC}$$

$$= 3(2\mathbf{QC}) - 4\left(\frac{3}{2}\mathbf{PC}\right)$$

$$= 6\mathbf{QC} - 6\mathbf{PC} = 6(\mathbf{QC} + \mathbf{CP})$$

$\Rightarrow \quad k\mathbf{PQ} = 6\mathbf{QP} = -6\mathbf{PQ}$ (given)

$\Rightarrow \quad k = -6$

123. *(a)* Given, $m_1 = |\mathbf{a}_1| = \sqrt{2^2 + (-1)^2 + (1)^2} = \sqrt{6}$

$$m_2 = |\mathbf{a}_2| = \sqrt{3^2 + (-4)^2 + (-4)^2} = \sqrt{41}$$

$$m_3 = |\mathbf{a}_3| = \sqrt{1^2 + 1^2 + (-1)^2} = \sqrt{3}$$

and $\quad m_4 = |\mathbf{a}_4| = \sqrt{(-1)^2 + (3)^2 + (1)^2} = \sqrt{11}$

$\therefore \; m_3 < m_1 < m_4 < m_2$

124. *(a)* Let $\mathbf{a} = \hat{\mathbf{i}} + 2\hat{\mathbf{j}} - \hat{\mathbf{k}}$, $\mathbf{b} = \hat{\mathbf{i}} + \hat{\mathbf{j}} + \hat{\mathbf{k}}$

and $\mathbf{c} = \hat{\mathbf{i}} - \hat{\mathbf{j}} + \lambda\hat{\mathbf{k}}$

Since, volume of tetrahedron $= \frac{1}{6}[\mathbf{a}\ \mathbf{b}\ \mathbf{c}]$

$$\Rightarrow \quad \frac{2}{3} = \frac{1}{6}\begin{vmatrix} 1 & 2 & -1 \\ 1 & 1 & 1 \\ 1 & -1 & \lambda \end{vmatrix}$$

$$\Rightarrow \quad \frac{2}{3} = \frac{1}{6}[1(\lambda+1) - 2(\lambda-1) - 1(-1-1)]$$

$$\Rightarrow \quad 4 = [-\lambda + 5]$$

$$\Rightarrow \quad \lambda = 1$$

125. *(c)* Given, $P(\overline{A} \cup \overline{B}) = P(\overline{A \cap B}) = \frac{7}{10}$

Since, $P(A \cap B) + P(\overline{A \cap B}) = 1$

$$\Rightarrow \quad P(A \cap B) = 1 - \frac{7}{10} = \frac{3}{10}$$

Also, $\quad P(A \cup B) = P(A) + P(B) - P(A \cap B)$

$$\Rightarrow \quad \frac{4}{5} = P(A) + \frac{2}{5} - \frac{3}{10}$$

$$\Rightarrow \quad P(A) = \frac{4}{5} - \frac{2}{5} + \frac{3}{10}$$

$$= \frac{2}{5} + \frac{3}{10} = \frac{7}{10}$$

126. *(a)* Here, $n = 6$

According to the question

$${}^6C_2 p^2 q^4 = 4 \cdot {}^6C_4 p^4 q^2$$

$$\Rightarrow \quad q^2 = 4p^2$$

$$\Rightarrow \quad (1-p)^2 = 4p^2$$

$$\Rightarrow \quad 3p^2 + 2p - 1 = 0$$

$$\Rightarrow \quad (p+1)(3p-1) = 0$$

$$\Rightarrow \quad p = \frac{1}{3} \quad (\because p \text{ cannot be negative})$$

127. *(d)* Since, given lines are parallel.

$$\therefore \quad d = \frac{15-5}{\sqrt{4^2+3^2}} = \frac{10}{5}$$

$\Rightarrow \quad d = 2 =$ diameter of the circle

$\therefore \quad$ Radius of circle $= 1$

$\therefore \quad$ Area of circle $= \pi r^2 = \pi$ sq unit.

128. *(b)* Let point (x_1, y_1) be on the line $3x + 4y = 5$.

$$\therefore \quad 3x_1 + 4y_1 = 5 \quad \ldots(\text{i})$$

Also, $\quad (x_1 - 1)^2 + (y_1 - 2)^2 = (x_1 - 3)^2 + (y_1 - 4)^2$

$$\Rightarrow \quad x_1^2 + y_1^2 - 2x_1 - 4y_1 + 5 = x_1^2 + y_1^2 - 6x_1 - 8y_1 + 25$$

$$\Rightarrow \quad 4x_1 + 4y_1 = 20 \quad \ldots(\text{ii})$$

On solving Eqs. (i) and (ii), we get

$$x_1 = 15, y_1 = -10$$

129. *(d)* The point of intersection of lines $x + 3y - 1 = 0$ and $x - 2y + 4 = 0$ is $(-2, 1)$.

Let equation of line perpendicular to the given line is $2x - 3y + \lambda = 0$.

Since, it passes through $(-2, 1)$.

$$\therefore \quad 2(-2) - 3(1) + \lambda = 0$$

$$\Rightarrow \quad \lambda = 7$$

$\therefore \quad$ Required line is $2x - 3y + 7 = 0$

130. *(b)* Given equation is

$$2x^2 - 10xy + 12y^2 + 5x + \lambda y - 3 = 0$$

Here, $a = 2, h = -5, b = 12, g = \frac{5}{2}, f = \frac{\lambda}{2}, c = -3$

For pair of lines $\begin{vmatrix} a & h & g \\ h & b & f \\ g & f & c \end{vmatrix} = 0$

$$\Rightarrow \begin{vmatrix} 2 & -5 & 5/2 \\ -5 & 12 & \lambda/2 \\ 5/2 & \lambda/2 & -3 \end{vmatrix} = 0$$

$$\Rightarrow 2\left(-36 - \frac{\lambda^2}{4}\right) + 5\left(15 - \frac{5\lambda}{4}\right) + \frac{5}{2}\left(\frac{-5\lambda}{2} - 30\right) = 0$$

$$\Rightarrow -72 - \frac{\lambda^2}{2} + 75 - \frac{25\lambda}{4} - \frac{25\lambda}{4} - 75 = 0$$

$$\Rightarrow \lambda^2 + 25\lambda + 144 = 0$$

$$\Rightarrow (\lambda + 9)(\lambda + 16) = 0$$

$$\Rightarrow \lambda = -9 \quad (\because |\lambda| < 16)$$

131. *(c)* Given, $x^2 - 2xy - xy + 2y^2 = 0$

$$\Rightarrow (x - 2y)(x - y) = 0$$

$$\Rightarrow x = 2y, \quad x = y \quad \ldots(i)$$

Also, $x + y + 1 = 0 \quad \ldots(ii)$

On solving Eqs. (i) and (ii), we get

$$A\left(-\frac{2}{3}, -\frac{1}{3}\right), B\left(-\frac{1}{2}, -\frac{1}{2}\right), C(0, 0)$$

$$\therefore \text{ Area of } \Delta ABC = \frac{1}{2}\begin{vmatrix} -\frac{2}{3} & -\frac{1}{3} & 1 \\ -\frac{1}{2} & -\frac{1}{2} & 1 \\ 0 & 0 & 1 \end{vmatrix}$$

$$= \frac{1}{2}\left[\frac{1}{3} - \frac{1}{6}\right] = \frac{1}{2}\left[\frac{1}{6}\right] = \frac{1}{12}$$

132. *(c)* Given pair of lines are $x^2 - 3xy + 2y^2 = 0$

and $x^2 - 3xy + 2y^2 + x - 2 = 0$

$$\therefore (x - 2y)(x - y) = 0$$

and $(x - 2y + 2)(x - y - 1) = 0$

$\Rightarrow x - 2y = 0, \quad x - y = 0$ and $x - 2y + 2 = 0,$

$$x - y - 1 = 0$$

Since, the lines $x - 2y = 0, x - 2y + 2 = 0$ and $x - y = 0, x - y - 1 = 0$ are parallel.

Also, angle between $x - 2y = 0$ and $x - y = 0$ is not $90°$.

$\therefore$ It is a parallelogram.

133. *(a)* In ΔOAC,

$$OC^2 = 2^2 + 4^2 = 20$$

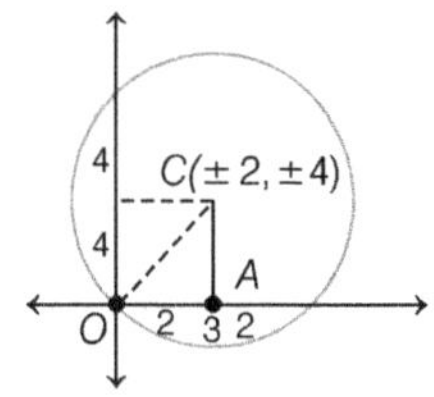

$\therefore$ Required equation of circle is

$$(x \pm 2)^2 + (y \pm 4)^2 = 20$$

$$\Rightarrow x^2 + y^2 \pm 4x \pm 8y = 0$$

134. *(b)* Let centre of circle be $C(-g, -f)$, then equation of circle passing through origin be

$$x^2 + y^2 + 2gx + 2fy = 0$$

$\therefore$ Distance, $d = |-g - 3| = g + 3$

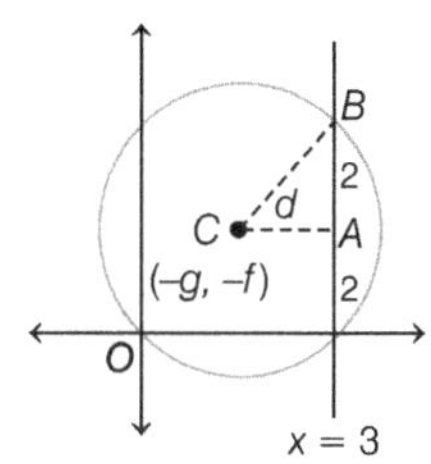

In ΔABC, $(BC)^2 = AC^2 + BA^2$

$$\Rightarrow g^2 + f^2 = (g + 3)^2 + 2^2$$

$$\Rightarrow g^2 + f^2 = g^2 + 6g + 9 + 4$$

$$\Rightarrow f^2 = 6g + 13$$

Hence, required locus is $y^2 - 6x = 13$

135. *(d)* Given circles are $x^2 + y^2 - 2x + 8y + 13 = 0$ and $x^2 + y^2 - 4x + 6y + 11 = 0$.

Here, $C_1 = (1, -4), C_2 = (2, -3),$

$$\Rightarrow r_1 = \sqrt{1 + 16 - 13} = 2$$

and $r_2 = \sqrt{4 + 9 - 11} = \sqrt{2}$

Now, $d = C_1C_2 = \sqrt{(2 - 1)^2 + (-3 + 4)^2} = \sqrt{2}$

$$\therefore \cos\theta = \frac{d^2 - r_1^2 - r_2^2}{2r_1r_2} = \frac{2 - 4 - 2}{2 \times 2 \times \sqrt{2}} = -\frac{1}{\sqrt{2}}$$

$$\Rightarrow \theta = 135°$$

136. *(b)* Let the required equation of circle be $x^2 + y^2 + 2gx + 2fy = 0$. Since, the above circle cuts the given circles orthogonally.

$$\therefore 2(-3g) + 2f(0) = 8$$

$$\Rightarrow 2g = -\frac{8}{3}$$

and $-2g - 2f = -7$

$$\Rightarrow 2f = 7 + \frac{8}{3} = \frac{29}{3}$$

$\therefore$ Required equation of circle is

$$x^2 + y^2 - \frac{8}{3}x + \frac{29}{3}y = 0$$

or $3x^2 + 3y^2 - 8x + 29y = 0$

137. ***(b)*** Given, parabola is $y^2 = 4x$

Here, $a = 1$

$\therefore$ Equation of normal is

$y = mx - 2m - m^3$ which passes through the point (1, 0)

$\therefore \quad 0 = m(1) - 2m - m^3$

$\Rightarrow \quad tm^3 + m = 0$

$\Rightarrow \quad m(m^2 + 1) = 0$

$\Rightarrow \quad m = 0$ is the only solution.

So, only one normal can be drawn from (1, 0).

138. ***(a)*** Given, $x^2y^2 = c^4$

$\Rightarrow \quad y^2(a^2 - y^2) = c^4$

$\Rightarrow \quad y^4 - a^2y^2 + c^4 = 0$

Let y_1, y_2, y_3 and y_4 are the roots.

$\therefore \quad y_1 + y_2 + y_3 + y_4 = 0$

139. ***(b)*** Given, $4x - 3y = 5$ and $2x^2 - 3y^2 = 12$

$\therefore \quad 2\left(\frac{5 + 3y}{4}\right)^2 - 3y^2 = 12$

$\Rightarrow \quad \frac{(25 + 9y^2 + 30y)}{8} - 3y^2 = 12$

$\Rightarrow \quad 15y^2 - 30y + 71 = 0$

$\Rightarrow \quad y = \frac{30 \pm \sqrt{900 - 4260}}{30}$

$$= 1 \pm \frac{\sqrt{-3360}}{30}$$

Also, $2x^2 - 3\left(\frac{4x - 5}{3}\right)^2 = 12$

$\Rightarrow \quad 10x^2 - 40x + 61 = 0$

$\Rightarrow \quad x = \frac{40 \pm \sqrt{1600 - 4 \times 10 \times 61}}{2 \times 10}$

$$= \frac{40 \pm \sqrt{-840}}{20}$$

$$= 2 \pm \frac{\sqrt{-840}}{20}$$

$\therefore$ Points are $A\left(2 + \frac{\sqrt{-840}}{20}, 1 + \frac{\sqrt{-3360}}{30}\right)$ and $B\left(2 - \frac{\sqrt{-840}}{20}, 1 - \frac{\sqrt{-3360}}{30}\right)$.

$\therefore$ Mid-point of AB is $(2, 1)$.

140. ***(c)*** $\cos 2\alpha + \cos 2\beta + \cos 2\gamma + \sin^2 \alpha + \sin^2 \beta + \sin^2 \gamma$

$= (\cos^2 \alpha - \sin^2 \alpha) + (\cos^2 \beta - \sin^2 \beta)$

$\quad + (\cos^2 \gamma - \sin^2 \gamma) + \sin^2 \alpha + \sin^2 \beta + \sin^2 \gamma$

$= \cos^2 \alpha + \cos^2 \beta + \cos^2 \gamma = 1$

141. ***(c)*** We know that image (x, y, z) of a point (x_1, y_1, z_1) in a plane $ax + by + cz + d = 0$ is

$$\frac{x - x_1}{a} = \frac{y - y_1}{b} = \frac{z - z_1}{c} = \frac{-2(ax_1 + by_1 + cz_1 + d)}{a^2 + b^2 + c^2}$$

Here, point is (3, 2, 1) and plane is $2x - y + 3z = 7$.

$\therefore \quad \frac{x - 3}{2} = \frac{y - 2}{-1} = \frac{z - 1}{3} = \frac{-2[2(3) - (2) + 3(1) - 7]}{2^2 + 1^2 + 3^2}$

$\Rightarrow \quad \frac{x - 3}{2} = \frac{y - 2}{-1} = \frac{z - 1}{3} = -2(0)$

$\Rightarrow \quad x = 3, \; y = 2, \; z = 1$

142. ***(c)*** $\lim_{x \to \infty} \left(\frac{x + 5}{x + 2}\right)^{x+3} = \lim_{x \to \infty} \left(1 + \frac{3}{x + 2}\right)^{x+3}$

$$= \lim_{x \to \infty} \left[\left(1 + \frac{3}{x + 2}\right)^{\frac{x+2}{3}}\right]^{\frac{3(x+3)}{x+2}}$$

$$= e^{\lim_{x \to \infty} 3\left(\frac{1 + \frac{3}{x}}{1 + \frac{2}{x}}\right)} = e^3$$

143. ***(d)*** Given, $f(x) = \begin{cases} \dfrac{2\sin x - \sin 2x}{2x \cos x}, & \text{if } x \neq 0 \\ a, & \text{if } x = 0 \end{cases}$

Now, $\lim_{x \to 0} f(x) = \lim_{x \to 0} \frac{2\sin x - \sin 2x}{2x \cos x} \quad \left(\frac{0}{0} \text{ form}\right)$

$$= \lim_{x \to 0} \frac{2\cos x - 2\cos 2x}{2(\cos x - x \sin x)}$$

$$= \lim_{x \to 0} \frac{2 - 2}{2(1 - 0)} = 0$$

Since, $f(x)$ is continuous at $x = 0$

$\therefore \quad f(0) = \lim_{x \to 0} f(x)$

$\Rightarrow \quad a = 0$

144. ***(b)*** Given, $\frac{x}{1} = \frac{1 - \sqrt{y}}{1 + \sqrt{y}}$

Applying componendo and dividendo, we get

$$\frac{1 + x}{1 - x} = \frac{(1 + \sqrt{y}) + (1 - \sqrt{y})}{(1 + \sqrt{y}) - (1 - \sqrt{y})}$$

$$\Rightarrow \quad \frac{1+x}{1-x} = \frac{2}{2\sqrt{y}}$$

$$\Rightarrow \quad y = \left(\frac{1-x}{1+x}\right)^2$$

On differentiating w.r.t. x, we get

$$\frac{dy}{dx} = \frac{-2(1+x)^2(1-x)-(1-x)^2 \cdot 2(1+x)}{(1+x)^4}$$

$$= \frac{(1-x)(1+x)(-2-2x-2+2x)}{(1+x)^4}$$

$$= \frac{4(x-1)}{(x+1)^3}$$

145. ***(b)*** Given, $\frac{d}{dx}\left[a\tan^{-1}x + b\log\left(\frac{x-1}{x+1}\right)\right] = \frac{1}{x^4-1}$

On integrating both sides, we get

$$a\tan^{-1}x + b\log\left(\frac{x-1}{x+1}\right) = \frac{1}{2}\int\left[\frac{1}{x^2-1} - \frac{1}{x^2+1}\right]dx$$

$$\Rightarrow \quad a\tan^{-1}x + b\log\left(\frac{x-1}{x+1}\right) = \frac{1}{4}\log\left(\frac{x-1}{x+1}\right) - \frac{1}{2}\tan^{-1}x$$

$$\Rightarrow \quad a = -\frac{1}{2}, \quad b = \frac{1}{4}$$

$$\therefore \quad a - 2b = -\frac{1}{2} - 2\left(\frac{1}{4}\right) = -1$$

146. ***(c)*** Given, $\quad y = e^{a\sin^{-1}x}$

On differentiating w.r.t. x, we get

$$y_1 = e^{a\sin^{-1}x}\, a \cdot \frac{1}{\sqrt{1-x^2}}$$

$$\Rightarrow \quad y_1\sqrt{1-x^2} = ay$$

$$\Rightarrow \quad (1-x^2)y_1^2 = a^2y^2$$

Again, differentiating w.r.t. x, we get

$$(1-x^2)2y_1y_2 - 2xy_1^2 = a^2 2yy_1$$

$$\Rightarrow \quad (1-x^2)y_2 - xy_1 - a^2y = 0$$

Using Leibnitz's rule,

$$(1-x^2)y_{n+2} + {}^nC_1 y_{n+1}(-2x) + {}^nC_2 y_n(-2) - xy_{n+1} - {}^nC_1 y_n - a^2y_n = 0$$

$$\Rightarrow \quad (1-x^2)y_{n+2} + xy_{n+1}(-2n-1) + y_n[-n(n-1)-n-a^2] = 0$$

$$\Rightarrow (1-x^2)y_{n+2} - (2n+1)xy_{n+1} = (n^2+a^2)y_n$$

147. ***(c)*** Given, $f(x) = x^3 + ax^2 + bx + c,\ a^2 \le 3b.$

On differentiating w.r.t. x, we get

$$f'(x) = 3x^2 + 2ax + b$$

Put $\quad f'(x) = 0$

$$\Rightarrow \quad 3x^2 + 2ax + b = 0$$

$$\Rightarrow \quad x = \frac{-2a \pm \sqrt{4a^2-12b}}{2\times 3} = \frac{-2a \pm 2\sqrt{a^2-3b}}{3}$$

Since, $\quad a^2 \le 3b,$

$\therefore \quad x$ has an imaginary value.

Hence, no extreme value of x exist.

148. ***(a)*** Let $I = \int\left(\frac{2-\sin 2x}{1-\cos 2x}\right)e^x\,dx$

$$= \int\left(\frac{2-2\sin x\cos x}{2\sin^2 x}\right)e^x\,dx$$

$$= \int \underset{\text{II}}{\operatorname{cosec}^2 x}\, e^x\,dx - \int \underset{\text{I}}{\cot x}\, e^x\,dx$$

$$= -\cot x\, e^x - \int(-\cot x)\,e^x\,dx$$

$$= -\cot x\, e^x + c$$

149. ***(b)*** Let $I = \int_0^\pi \frac{1}{1+\sin x}\,dx = \int_0^\pi \frac{1}{1+\frac{2\tan\frac{x}{2}}{1+\tan^2\frac{x}{2}}}\,dx$

$$= \int_0^\pi \frac{\sec^2\frac{x}{2}}{\left(1+\tan\frac{x}{2}\right)^2}\,dx$$

Put $\tan\frac{x}{2} = t \quad \Rightarrow \quad \frac{1}{2}\sec^2\frac{x}{2}\,dx = dt$

$$\therefore \quad I = \int_0^\infty \frac{2\,dt}{(1+t)^2} = \left[-\frac{2}{1+t}\right]_0^\infty = 2$$

150. ***(b)*** Given, $\frac{dy}{dx} = \sin(x+y)\tan(x+y) - 1$

Put $x+y = z \quad \Rightarrow \quad 1 + \frac{dy}{dx} = \frac{dz}{dx}$

$$\therefore \quad \frac{dz}{dx} - 1 = \sin z\tan z - 1$$

$$\Rightarrow \quad \int\frac{\cos z}{\sin^2 z}\,dz = \int dx$$

Put $\quad \sin z = t \Rightarrow \cos z\,dz = dt$

$$\therefore \quad \int\frac{1}{t^2}\,dt = x - c \Rightarrow -\frac{1}{t} = x - c$$

$\Rightarrow -\operatorname{cosec} z = x - c$

$\Rightarrow x + \operatorname{cosec}(x+y) = c.$

BITSAT

Birla Institute of Technology & Science Admission Test

Mock Test 4

Instructions

- There are 150 questions in all. The number of questions in each part is as given below.

	No. of Questions
Part I Physics	1–40
Part II Chemistry	41–80
Part III a. English Proficiency	81–95
b. Logical Reasoning	96–105
Part IV Mathematics	106–150

- All questions are Multiple Choice Questions having four options out of which **only one** is correct.
- Each correct answer fetches 3 marks while incorrect answer has a penalty of 1 mark.
- Time allotted to complete this paper is 3 hrs.

PART I

Physics

1. The velocity of water waves may depend on their wavelength λ, the density of water ρ and the acceleration due to gravity g. The method of dimensional analysis gives the relation between these quantities as

a. $v^2 = k\lambda^{-1}g^{-1}\rho^{-1}$ **b.** $v^2 = k\lambda g$

c. $v^2 = k\lambda\rho g$ **d.** $v^2 = k\lambda^3 g^{-1}\rho^{-1}$

2. A particle located at $x = 0$ and $t = 0$ starts moving along the positive x-direction with a velocity v that varies as $v = \alpha\sqrt{x}$, the displacement of the particle varies with time as

a. $x \propto \sqrt{t}$ **b.** $x \propto t$ **c.** $x \propto t^2$ **d.** $x \propto \frac{1}{t^2}$

3. From an elevated point P, a stone is projected vertically upwards. When the stone reaches a distance h below P, its velocity is double of its velocity at a height h above P. The greatest height attained by the stone from the point of projection P is

a. $\frac{3}{5}h$ **b.** $\frac{5}{3}h$ **c.** $\frac{7}{5}h$ **d.** $\frac{5}{7}h$

4. A train accelerates from rest at a constant rate α for distance x_1 and time t_1. After that it retards to rest at constant rate β for distance x_2 and time t_2. Then, it is found that

a. $\frac{x_1}{x_2} = \frac{\alpha}{\beta} = \frac{t_1}{t_2}$ **b.** $\frac{x_1}{x_2} = \frac{\beta}{\alpha} = \frac{t_1}{t_2}$

c. $\frac{x_1}{x_2} = \frac{\alpha}{\beta} = \frac{t_2}{t_1}$ **d.** $\frac{x_1}{x_2} = \frac{\beta}{\alpha} = \frac{t_2}{t_1}$

5. An ant crawls up a hemispherical surface very slowly as shown in figure. The coefficient of friction between the ant and the surface is 1/3. If the line joining the centre of hemispherical surface to ant makes an angle α with the vertical, the maximum possible value of α is

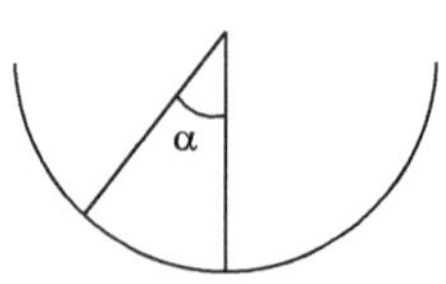

a. $\sec^{-1}(3)$ **b.** $\cot^{-1}(3)$

c. $\tan^{-1}(3)$ **d.** $\mathrm{cosec}^{-1}(3)$

6. A block of weight 5 N is pushed against a vertical wall by a force 12 N. The coefficient of friction between the wall and block is 0.6. The magnitude of the force exerted by wall on the block is

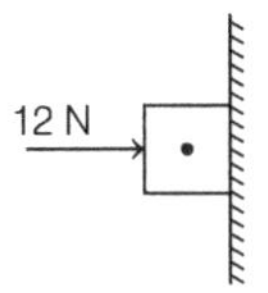

a. 12 N *b.* 5 N *c.* 7.2 N *d.* 13 N

7. A vertical spring with force constant k is fixed on a table. A ball of mass m at a height h above the free upper end of spring falls vertically on the springs, so that the spring is compressed by a distance d. The net work done in this process is

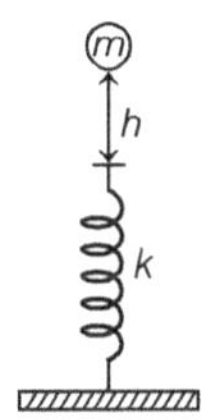

a. $mg(h+d)+\frac{1}{2}kd^2$ *b.* $mg(h+d)-\frac{1}{2}kd^2$

c. $mg(h-d)-\frac{1}{2}kd^2$ *d.* $mg(h-d)+\frac{1}{2}kd^2$

8. A thin uniform rod of length l and mass m is swinging freely about a horizontal axis passing through its end. Its maximum angular velocity is ω. Its centre of mass rises to a maximum height of

a. $\frac{l^2\omega^2}{6g}$ *b.* $\frac{l^2\omega^2}{3g}$ *c.* $\frac{l^2\omega^2}{2g}$ *d.* $\frac{l\omega}{6g}$

9. Two long metallic strips are joined together by two rivets each of radius 0.1cm (see figure). Each rivet can withstand a maximum shearing stress of $4\times 10^8\ \text{Nm}^{-2}$. The maximum tangential force which a strip can exert is

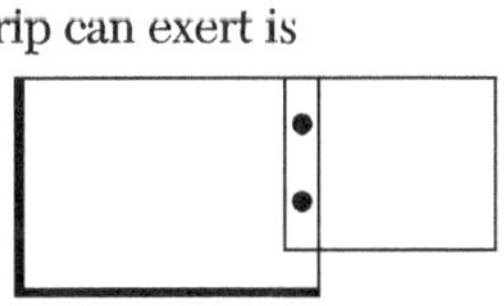

a. 185 N *b.* 1885 N *c.* 855 N *d.* 1555 N

10. Two rods P and Q of same length and same diameter having thermal conductivity ratio 2 : 3 joined end to end. If temperature at one end of P is 100°C and at one end of Q is 0°C, then temperature of interface is

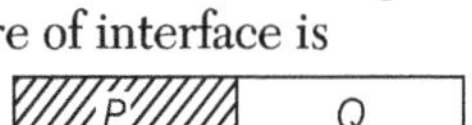

a. 60°C *b.* 70°C
c. 50°C *d.* 40°C

11. A square wire frame of size L is dipped in a liquid. On taking out, a membrane is formed. If the surface tension of liquid is T, force acting on the frame will be

a. $2TL$ *b.* $4TL$ *c.* $8TL$ *d.* $10TL$

12. Assume that a drop of liquid evaporates by decrease in its surface energy, so that its temperature remains unchanged. What should be the minimum radius of the drop for this to be possible ? The surface tension is T, density of liquid is ρ and L is its latent heat of vaporisation.

a. $\sqrt{T/\rho L}$ *b.* $T/\rho L$ *c.* $2T/\rho L$ *d.* $\rho L/T$

13. A liquid is kept in a cylindrical vessel which is rotated along its axis. The liquid rises at the sides (figure). If the radius of the vessel is 0.05 m and the speed of rotation is 2 rad/s, then find the difference in the height of liquid at centre and sides.

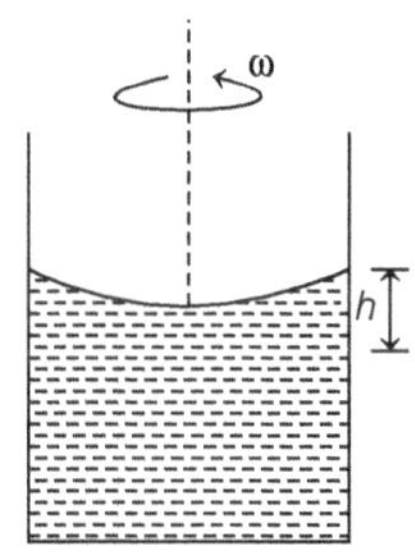

a. 20 cm *b.* 4 cm *c.* 2 cm *d.* 0.2 cm

14. An ideal gas is expanding such that pT^2 = constant. The coefficient of volume expansion of the gas is

a. $\frac{1}{T}$ *b.* $\frac{2}{T}$ *c.* $\frac{3}{T}$ *d.* $\frac{4}{T}$

15. The ratio of the speed of sound in nitrogen gas to that in helium gas at 300 K is

a. $\sqrt{\frac{1}{7}}$ *b.* $\sqrt{\frac{2}{7}}$ *c.* $\frac{\sqrt{3}}{5}$ *d.* $\frac{\sqrt{6}}{5}$

16. A point mass oscillates along X-axis according to the law $x = x_0\cos\left(\omega t-\frac{\pi}{4}\right)$. If the acceleration of the particle is written as $a = A\cos(\omega t+\delta)$, then

a. $A = x_0\omega^2,\ \delta=\pi/4$

b. $A = x_0\omega^2,\ \delta=-\pi/4$

c. $A = x_0\omega^2,\ \delta=3\pi/4$

d. $A = x_0,\ \delta=\pi/4$

17. The fundamental frequency of a closed end pipe is 480 Hz. What is the fundamental frequency when $\left(\frac{1}{4}\right)$th length is filled with water?

a. 120 Hz *b.* 240 Hz *c.* 640 Hz *d.* 960 Hz

18. The electric flux for Gaussian surface A that encloses the charged particles in free space is (Given, $q_1 = -14\text{nC}$, $q_2 = 78.85\text{ nC}, q_3 = -56\text{ nC}$)

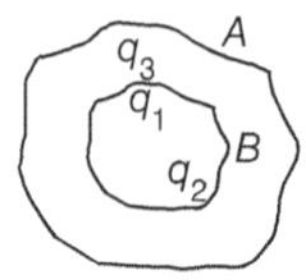

a. $10^3\text{ CN}^{-1}\text{ m}^{-2}$ *b.* $10^3\text{ Nm}^2\text{ C}^{-1}$
c. $6.32\times 10^3\text{ Nm}^2\text{ C}^{-1}$ *d.* $6.32\times 10^3\text{ C N}^{-1}\text{ m}^{-2}$

19. A sphere of radius 4 cm is suspended within a hollow sphere of radius 6 cm. The inner sphere is charged to a potential of 3 esu. When the outer sphere is earthed, then the charge on the inner sphere is

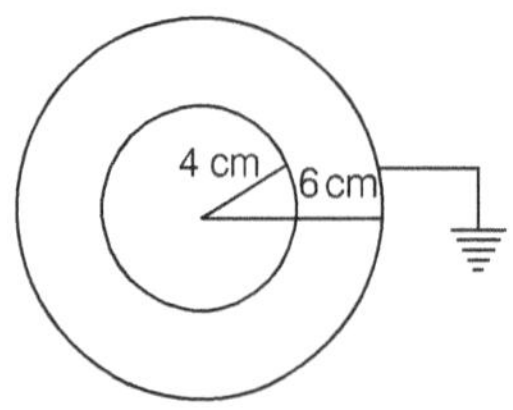

a. 54 esu *b.* 0.25 esu *c.* 30 esu *d.* 36 esu

20. Six identical capacitors joined in parallel and charged to a potential difference of 10 V, are separated and then connected in series, i.e. the positive plate of one is connected to negative plate of other. Then, potential difference between free plates is

a. 10 V *b.* 30 V *c.* 60 V *d.* $\frac{10}{5}$V

21. Kirchhoff's first law $(\Sigma i = 0)$ and second law $(\Sigma iR = \Sigma E)$, where symbols have their usual meanings, are respectively based on
a. conservation of charge, conservation of energy
b. conservation of charge, conservation of momentum
c. conservation of energy, conservation of charge
d. conservation of momentum, conservation of energy

22. In the given figure, find the resistance between A and B. Both the circle and the diameter are made of uniform wire of resistance $1\times 10^{-4}\,\Omega/\text{m}$. The length AB is $2\,\text{m}$.

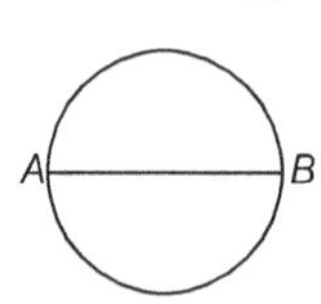

a. $\frac{2}{3}\times 10^{-4}\,\Omega$ *b.* $2\pi\times 10^{-4}\,\Omega$
c. $14.56\times 10^{-4}\,\Omega$ *d.* $0.88\times 10^{-4}\,\Omega$

23. The resistance of series combination of 2 resistors is S. When they are joined in parallel, the total resistance is P. If $S = nP$, then minimum possible value of n is
a. 4 *b.* 3 *c.* 2 *d.* 1

24. Find R, if power dissipated in the circuit is 150 W.

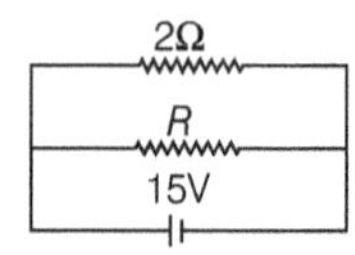

a. 2Ω *b.* 6Ω *c.* 7Ω *d.* 5Ω

25. The oscillating frequency of a cyclotron is 10 MHz. If the radius of its dees is 0.5 m, the kinetic energy of a proton, which is accelerated by the cyclotron is
a. 10.2 MeV *b.* 2.55 MeV
c. 20.4 MeV *d.* 5.1 MeV

26. Two particles A and B of masses m_A and m_B respectively and having the same charge are moving in a plane. A uniform magnetic field exists perpendicular to this plane. The speeds of particles are v_A and v_B respectively and the trajectories are as shown in figure, then

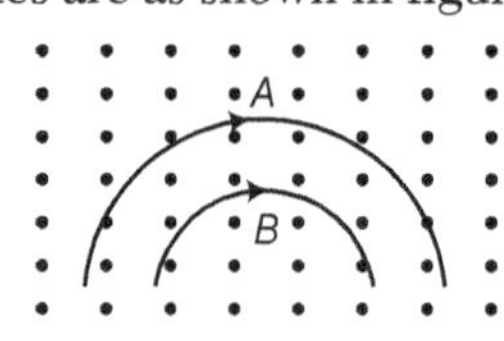

a. $m_A v_A < m_B v_B$
b. $m_A v_A > m_B v_B$
c. $m_A < m_B$ and $v_A < v_B$
d. $m_A = m_B$ and $v_A = v_B$

27. A magnetic needle suspended by a silk thread is vibrating in the Earth's magnetic field. If the temperature of the needle is increased by 500° C, then
a. the time period decreases
b. the time period remains unchanged
c. the time period increases
d. the needle stops vibrating

28. A circular disc of radius 0.2 m is placed in a uniform magnetic field of induction $\left(\frac{1}{\pi}\right)$ Wb/m^2 in such a way that its axis makes an angle of 60° with **B**. The magnetic flux linked with the disc is
a. 0.01 Wb *b.* 0.02 Wb
c. 0.06 Wb *d.* 0.08 Wb

29. A wire of length 50 cm moves with velocity of 300 m/min perpendicular to a magnetic field. If the emf induced in the wire is 2 V, magnitude of field in tesla is

a. 2 *b.* 5 *c.* 0.4 *d.* 0.8

30. In an *L-C-R* series AC circuit, the voltage across each of the components *L*, *C* and *R* is 50 V. The voltage across *LC* combination will be

a. 50 V *b.* $50\sqrt{2}$ V *c.* 100 V *d.* 0

31. The refractive index of air is 1.0003. The thickness of air column which has one more wavelength of yellow light ($\lambda = 600$ nm), then the same thickness of vacuum is

a. 0.02 mm *b.* 0.2 mm *c.* 2 mm *d.* 2 cm

32. In an interference experiment, third bright fringe is obtained at a point on the screen with a light of 700 nm. What should be the wavelength of the light source in order to obtain 5th bright fringe at the same point ?

a. 420 nm *b.* 750 nm *c.* 630 nm *d.* 500 nm

33. Two coherent light beams of intensity *I* and 4*I* are superposed. The maximum and minimum possible intensities in the resulting beam are

a. 5*I* and 3*I* *b.* 9*I* and *I*
c. 9*I* and 3*I* *d.* 5*I* and *I*

34. Two convex lenses of focal lengths 0.3 m and 0.05 m are used to make a telescope. Distance kept between the two in order to obtain image at infinity is

a. 0.25 m *b.* 0.3 m
c. 0.35 m *d.* 0.20 m

35. The number of photons of wavelength 660 nm emitted per second by an electric bulb of 60 W is (Take, 6.6×10^{-34} Js)

a. 2×10^{20} *b.* 2×10^{-20}
c. 3×10^{20} *d.* 1.5×10^{20}

36. When $_{88}\text{Ra}^{236}$ decays in a series by emission of 3α-particles and one β-particle, isotope *X* formed as

a. $_{83}X^{224}$ *b.* $_{84}X^{218}$ *c.* $_{84}X^{220}$ *d.* $_{82}X^{223}$

37. The radioactivity of a sample is A_1 at time t_1 and A_2 at time t_2. If the mean life of the specimen is *T*, then the number of atoms that have disintegrated in the time interval of $(t_2 - t_1)$ is

a. $(A_1 - A_2)$ *b.* $\frac{(A_1 - A_2)}{T}$
c. $(A_1 - A_2)T$ *d.* $(A_1t_1 - A_1t_2)$

38. The radioactive isotope *X* with a half-life of 10^9 year decays to *Y* which is stable. A sample of rocks was found to contain both the elements *X* and *Y* in the ratio 1 : 7. What is the age of the rocks ?

a. 2×10^9 yr *b.* 3×10^9 yr
c. 6×10^9 yr *d.* 7×10^9 yr

39. A common emitter amplifier has a voltage gain of 50, an input impedance of 100Ω and an output impedance of 200Ω. The power gain of the amplifier is

a. 500 *b.* 1000 *c.* 1250 *d.* 100

40. A 600 W carrier is modulated to a depth of 75% by a 400 Hz sine wave. Find the total power of antenna.

a. 769 W *b.* 796 W
c. 679 W *d.* 637.5 W

Mock Test 4

PART II

Chemistry

41. Choose the correct statement from the following reaction.

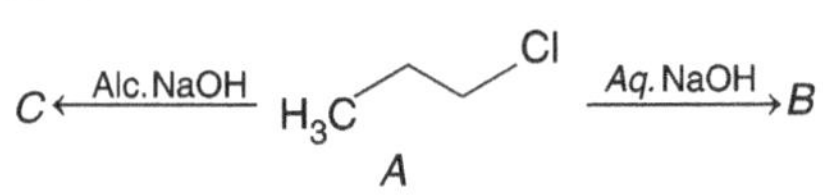

a. *B* is obtained by elimination reaction.

b. *B* is an isomer of ethyl methyl ether, while *C* is the dehydrated compound of *B*.

c. *C* is obtained by substitution reaction.

d. The molecular formula of *B* is C_3H_6 and that of *C* is C_3H_8O.

42. An inorganic halide *P* reacts with water to form two acids *Q* and *R*. *P* also reacts with NaOH to form two salts *S* and *T* which remain in solution. The solution gives white precipitate with both $AgNO_3$ and $BaCl_2$ solutions respectively. *P* is a useful reagent. What is *P* in the reaction?

a. AgCl *b.* NaCl *c.* SO_2Cl_2 *d.* HCl

43. The vapour density of completely dissociated NH_4Cl would be

a. half of that of ammonium chloride

b. double that of ammonium chloride

c. slightly less than half of that of ammonium chloride
d. determined by the amount of solid ammonium chloride used

44. A binary solid (A^+B^-) has a zinc blende structure with B^- ions constituting the lattice and A^+ ions occupying 25% tetrahedral holes, then the formula of solid is

a. AB *b.* A_2B *c.* AB_2 *d.* AB_4

45. Which of the following reactions will not give propane?

a. 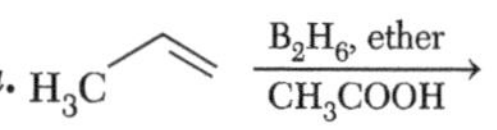

b. 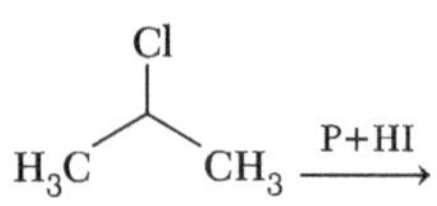

c. 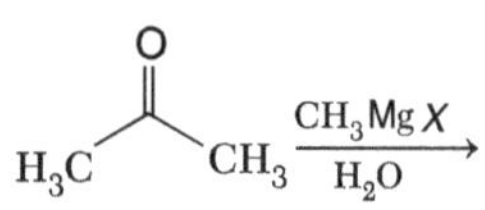

d. 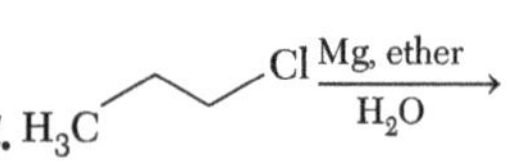

46. The molecular formula of certain compound has X_4O_6 · If 10 g of X_4O_6 has 5.72 g X, then the atomic mass of X is

a. 42 *b.* 30 *c.* 24 *d.* 32

47. 3.15 g oxalic acid $[(COOH)_2 \cdot x\ H_2O]$ was dissolved to make 500 mL solution. On titration, 33.36 mL of this solution were neutralised by 50 mL $N/15$ NaOH. The value of x will be

a. 3 *b.* 2 *c.* 1 *d.* 4

48. Which of the following represent the given mode of hybridisation sp^2-sp^2-sp-sp from right to

a. $H_2C=CH—C\equiv CH$
b. $H_2C=C=C=CH_2$
c. 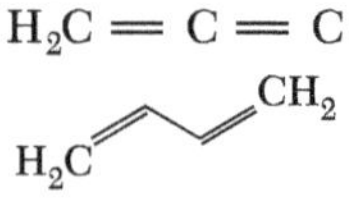

d. $HC\equiv C—CH_2—C\equiv CH$

49. Which of the following statements is incorrect?

a. The degree of dissociation of a weak electrolyte decreases as its concentration decreases.
b. The freezing point of water is depressed by the addition of glucose.
c. If two liquids that form an ideal solution, are mixed, the change in entropy is positive.
d. Energy is released when a substance dissolves in water provided that the hydration energy of the substance is more than its lattice energy.

50. The hardness of water sample which contains 0.001 mole of $MgSO_4$ dissolved per litre of water, is

a. 150 ppm *b.* 100 ppm *c.* 120 ppm *d.* 180 ppm

51. Density of a gas is found to be 5.46 g / dm^3 at 27°C and 2 bar pressure. At STP, its density will be

a. 3 g /cm^3 *b.* 0.3 g / dm^3
c. 3 g / dm^3 *d.* 4 g / dm^3

52. On the basis of intermolecular forces, predict the correct decreasing order of boiling points of the compounds.

a. $CH_3OH > H_2 > CH_4$ *b.* $CH_3OH > CH_4 > H_2$
c. $CH_4 > CH_3OH > H_2$ *d.* $H_2 > CH_4 > CH_3OH$

53. Which of the following options does not have isoelectronic species?

a. CN^-, N_2, C_2^{2-} *b.* SO_3^{2-}, CO_3^{2-}, NO_3^-
c. BO_3^{3-}, CO_3^{2-}, NO_3^- *d.* PO_4^{3-}, SO_4^{2-}, ClO_4^-

54. Actinoids exhibit more number of oxidation states than lanthanoids. The main reason for this is

a. more active nature of the actinoids
b. more energy difference between 5*f* and 6*d*-orbitals than between 4*f* and 5*d*-orbitals
c. lesser energy difference between 5*f* and 6*d*-orbitals than that between 4*f* and 5*d*-orbitals
d. greater metallic character of the lanthanoids than that of the corresponding actinoids

55. Among the elements Li, K, Ca, S and Kr, which one is expected to have the lowest first ionisation enthalpy and which have the highest first ionisation enthalpy? The increasing order of first ionisation enthalpy of the elements Li, K, Ca, S and Kr, is

a. Li < K < Ca < S < Kr
b. K < Li < Ca < S < Kr
c. Kr < Ca < S < Li < K
d. Ca < Li < S < K < Kr

56. Toluene is nitrated and the resulting product is reduced with tin and HCl. The product so obtained is diazotised and then heated with cuprous bromide. The final product is

a. mixture of *o*-and *p*-dibromobenzene
b. mixture of *o*-and *p*-bromotoluene
c. mixture of *o*-and *p*-bromoaniline
d. mixture of *o*-and *m*-bromotoluene

57. In which of the following compounds, the size of cation and anion ratio is minimum?

a. LiI *b.* LiF *c.* CsF *d.* CsI

58. Which of the following is not an antipyretic drug?

a. Paracetamol *b.* Aspirin
c. Phenacetin *d.* Marijuane

59. Which of the following represent the heat of formation of $CO_2(g)$?

a. $CO(g) + \frac{1}{2}O_2(g) \longrightarrow CO_2(g)$

b. $C_6H_6(l) + 7\frac{1}{2}O_2(g) \longrightarrow 6CO_2(g) + 3H_2O(l)$

c. $C\,(\text{graphite}) + O_2(g) \longrightarrow CO_2(g)$

d. $CH_4(g) + 2O_2(g) \longrightarrow CO_2(g) + 2H_2O(l)$

60. Which of the following factors is of no significance for roasting sulphide ores to the oxides and not subjecting the sulphide ores to carbon reduction directly?

a. Metal sulphides are thermodynamically more stable than CS_2
b. Metal sulphides are less stable than the corresponding oxides
c. CO_2 is more volatile than CS_2
d. CO_2 is thermodynamically more stable than CS_2

61. The reaction of acetaldehyde with excess of ethanol in the presence of H_2SO_4 gives the product

a. $CH_3CH(OC_2H_5)_2$
b. $CH_3CH(OH)_2$
c. Ketal
d. $C_2H_5OCH_2CH_2OC_2H_5$

62. When mercuric iodide is added to the aqueous solution of KI, the

a. freezing point is raised
b. freezing point does not change
c. freezing point is lowered
d. boiling point does not change

63. $KMnO_4$ oxidises NO_2^- to NO_3^- in basic medium, the number of moles of NO_2^- oxidised by 1 mole of $KMnO_4$ is

a. $\frac{3}{2}$ *b.* 2
c. $\frac{4}{3}$ *d.* None of these

64. The correct decreasing order of reactivity of the following alkenes, is

I. 2-butene
II. 2-methyl-2-butene
III. 2,3-dimethyl-2-butene IV. ethene

a. III > I > II > IV *b.* IV > I > II > III
c. IV > II > I > III *d.* I > III > IV > II

65. The equilibrium constant for the reaction given below is 2.0×10^{-7} at 300 K.

$$PCl_5(g) \rightleftharpoons PCl_3(g) + Cl_2(g)$$

The standard entropy change, if $\Delta H° = 28.40$ kJ mol^{-1} is

a. 33.6 JK^{-1} *b.* – 33.6 JK^{-1}
c. 41.2 JK^{-1} *d.* – 41.2 JK^{-1}

66. The compound obtained when copper nitrate is strongly heated, is

a. copper *b.* copper nitride
c. copper nitrite *d.* copper oxide

67. If the half-life of a first order reaction is 60 min, then how long will it take to consume 90% of the reactant?

a. 250 min *b.* 200 min *c.* 180 min *d.* 300 min

68. The decreasing order of stability of the following compound is

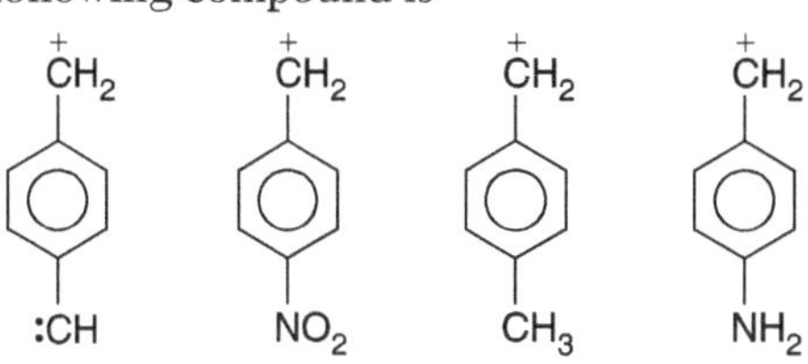

(I) (II) (III) (IV)

a. IV > I > III > II *b.* II > III > I > IV
c. IV > III > I > II *d.* III > IV > I > II

69. Which of the following yield alkanes, alkenes and alkynes?

a. Kolbe's reaction *b.* Wurtz reaction
c. Sandmeyer's reaction *d.* Williamson's synthesis

70. Sodium aluminium silicate (zeolite) is used for softening of hard water. This causes

a. adsorption of Ca^{2+} and Mg^{2+} ions of hard water replacing Al^{3+} ions
b. adsorption of Ca^{2+} and Mg^{2+} ions of hard water replacing Na^+ ions
c. Both (a) and (b) are true
d. None of the above is true

71. The molar conductivity for 0.01M acetic acid is 3.907 S cm^2 mol^{-1} and at infinite dilution, it is 390.7. The pH of the solution is

a. 2 *b.* 3 *c.* 6 *d.* 4

72. The hydrogen electrode is dipped in a solution of pH = 3 at 25°C. The potential of the cell would be $\left(\text{the value of } \frac{2.303\,RT}{F} = 0.059\text{ V}\right)$

a. – 0.732V *b.* 0.059 V
c. – 0.177 V *d.* – 0.059 V

73. During the test of nitrogen by Lassaigne's method, the prussian blue colour is obtained due to the formation of

a. $Fe_2[Fe(CN)_6]$ *b.* $Na_4[Fe(CN)_6]$
c. $Fe_4[Fe(CN)_6]_3$ *d.* $Fe_3[Fe(CN)_6]_4$

74. On the excess addition of NaOH solution to a solution of $ZnCl_2$ produces

a. Na_2ZnO_2 *b.* $Zn(OH)_2$
c. ZnO *d.* None of these

75. Which of the following is a 3° amine?

a. Allyl amine
b. Propan-2-amine
c. N,N-diethyl butan-1-amine
d. N-methyl ethanamine

76. The geometry of XeF_4 is

a. square planar *b.* tetrahedral
c. trigonal *d.* trigonal bipyramidal

77. The vapour density of a mixture containing NO_2 and N_2O_4 is 38.3 at 27° C. In 100 g mixture, the mole of NO_2 is

a. 0.321 *b.* 0.437 *c.* 0.812 *d.* 0.617

78. The chief ores of Sn, Fe, Al and Ag and methods for the purification of their ores respectively are given below. Choose the incorrect match from the following.

a. Sn-Cassiterite-Gravity separation
b. Fe-Haematite-Magnetic separation
c. Al-Bauxite-Leaching
d. Ag-Cinnabar-Thermal decomposition

79. Which of the following reactions is non-spontaneous?

a. $NO_2^- + CH_3COOH \longrightarrow CH_3COO^- + HNO_2$ (in aqueous medium)

b. $HCl + CH_3COOH \longrightarrow CH_3COOH_2^+ + Cl^-$ (in non-aqueous medium)

c. $HNO_2 + CH_3COO^- \longrightarrow CH_3COOH + NO_2^-$ (in aqueous medium)

d. $HF + HCl \longrightarrow H_2^+Cl + F^-$ (in non-aqueous medium)

80. Choose the correct answer from the following.

a. $Ni(PPh_3)_2Cl_2$ has tetrahedral geometry with magnetic moment 2.82 M.
b. $[FeCl_6]^{4-}$ is an inner orbital complex with zero magnetic moment.
c. The magnetic moment of $[Fe(CN)_6]^{4-}$ is zero.
d. Wavelength of absorption in the visible region for $[Ni(H_2O)_6]^{2+}$.

Mock Test 4

PART III

a. English Proficiency

Directions (Q. Nos. 81 to 84) *Spot the error in the following sentences and indicate your answer.*

81. It is a pity (a)/ that even five years old boys (b)/ are engaged in hazardous factories. (c)/ No error (d)

82. I gave him(a)/ two hundred rupees notes (b)/ for depositing. (c)/ No error (d)

83. It is not my business (a)/ to give an advice to those (b)/ who are not sensible enough to deal with their own problems. (c)/ No error (d)

84. I don't think(a)/ it is your house, (b)/ it is someboy's else. (c)/ No error (d)

Directions (Q. Nos. 85 to 88) *Rearrange the jumbled parts of the sentence and then indicate the correct option.*

85. P : it is a pity that
Q : by offering a handsome dowry
R : a number of parents think that
S : they will able to ensure the happiness of their daughters.
The proper sequence should be

a. SQRP *b.* PRSQ *c.* PSRQ *d.* PRQS

86. The common man
P : in nuturing
Q : a more active role
R : communal harmony
S : should play
The proper sequence should be

a. PRSQ *b.* SQPR *c.* SQRP *d.* PRQS

87. The doctor
P : able to find out
Q : what had caused
R : the food poisoning
S : had not been

The proper sequence should be

a. SPRQ *b.* PRQS
c. PRSQ *d.* SPQR

88. P : was suspended
Q : the officer being corrupt
R : before his dismissal
S : from serivce
The proper sequence should be

a. QPSR *b.* QPRS
c. RSQP *d.* RSQP

Directions (Q. Nos. 89 to 92) *Find the synonym of the word written in capital letters in the given sentences.*

89. EXAGGERATION of facts would always lead to confusion

a. Simplification *b.* Negation
c. Emancipation *d.* Amplification

90. After his father's death,he became INSOLVENT.

a. Rich *b.* Poor
c. Bankrupt *d.* Nonchalant

91. He INDUCES human beings to want things they don't want.

a. Influences *b.* Dictates
c. Persuades *d.* Appreciates

92. His information is not AUTHENTIC.

a. Real *b.* Reliable
c. Believable *d.* Genuine

Directions (Q. Nos. 93 to 95) *Find the antonym of the word written in the capital letters.*

93. He finally CONCEDED that he was involved in smuggling.

a. Admitted *b.* Accepted
c. Denied *d.* Concealed

94. The accused emphatically DENIED the charge in the court.

a. Accepted *b.* Agreed
c. Asserted *d.* Affirmed

95. The cloth is COARSE of touch.

a. Delicate *b.* Rough
c. Painful *d.* Harsh

b. Logical Reasoning

96. 'Doctor' is related to 'Patient', in the same way 'Lawyer' is related to

a. Customer
b. Criminal
c. Magistrate
d. Client

97. Find the odd one out.

a. 8-11 *b.* 1-4
c. 7-10 *d.* 3-5

98. Which figure will best represent the relationship amongst the three classes doctor, teacher, women ?

a. 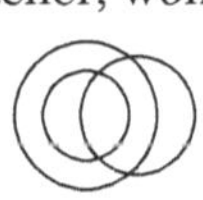 *b.* 

c.  *d.* 

99. Find the missing number.

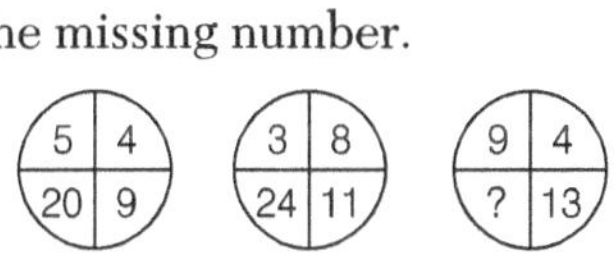

a. 26 *b.* 36
c. 32 *d.* 117

100. Find the odd one out.

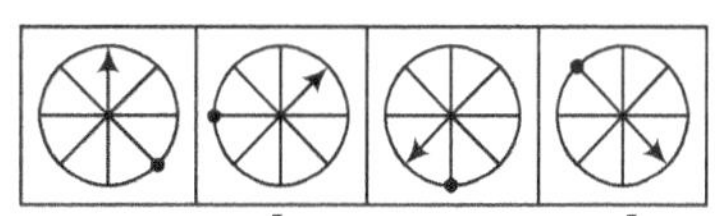

a. *b.* *c.* *d.*

101. Count the number of squares in the following figure.

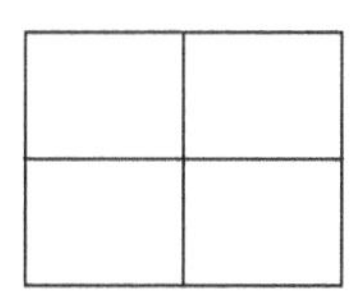

a. 4 *b.* 5
c. 6 *d.* 7

102. In the question below, figure (X) is embedded in one of the following. Then, find out the correct alternative.

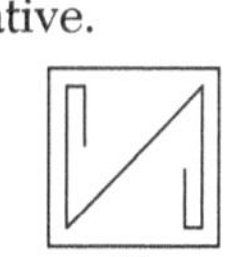

(X)

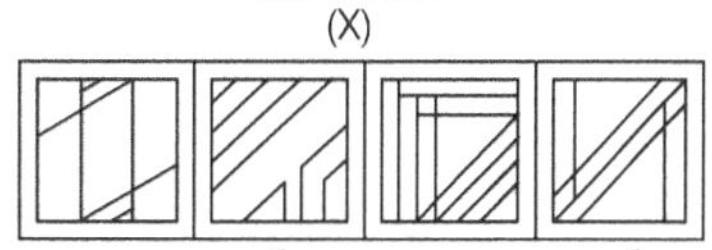

a. *b.* *c.* *d.*

103. If LOFTY is coded as LPFUY, then DWARF will be written as

a. DXASF
b. DXBSG
c. DXATF
d. DWBSG

104. If the day before yesterday was Thursday, when will Sunday be ?

a. Today *b.* Two days after today
c. Tomorrow *d.* Day after Tomorrow

105. Find the next term in the series 5, 16, 51, 158.

a. 1452 *b.* 483 *c.* 481 *d.* 1454

PART IV

Mathematics

106. If the median of 25 observations is 45 and if the observations greater than the median are increased by 4, then the median of the new data is

a. 49 *b.* 41 *c.* 45 *d.* 40

107. The differential equation of rectangular hyperbolas whose axes are asymptotes of the hyperbola $x^2 - y^2 = a^2$, is

a. $y\frac{dy}{dx} = x$ *b.* $x\frac{dy}{dx} = -y$
c. $x\frac{dy}{dx} = y$ *d.* $x\,dy + y\,dx = C$

108. If $f(x) = \max\left\{\sin x, \cos x, \frac{1}{2}\right\}$, then the area of the region bounded by the curves $y = f(x)$, X-axis, Y-axis and $x = \frac{5\pi}{3}$, is

a. $\sqrt{2} - \sqrt{3} + \frac{5\pi}{12}$
b. $\sqrt{2} + \frac{\sqrt{3}}{2} + \frac{5\pi}{12}$
c. $\sqrt{2} + \sqrt{3} + \frac{5\pi}{12}$
d. None of these

109. The value of $\int_0^{[x]} \frac{2^x}{2^{[x]}}\, dx$ is

a. $[x]\log 2$ *b.* $\frac{[x]}{\log 2}$
c. $\frac{1}{2}\cdot\frac{[x]}{\log 2}$ *d.* None of these

110. If $\int \frac{1}{x + x^5}\, dx = f(x) + C$, then the value of $\int \frac{x^4}{x + x^5}\, dx$ is

a. $\log x - f(x) + C$ *b.* $f(x) + \log x + C$
c. $f(x) - \log x + C$ *d.* None of these

111. If $f(x) = a\log|x| + bx^2 + x$ has its extremum values at $x = -1$ and $x = 2$, then

a. $a = 2, b = -1$
b. $a = 2, b = -1/2$
c. $a = -2, b = 1/2$
d. None of the above

112. Let $f(x) = \tan^{-1}\{g(x)\}$, where $g(x)$ is monotonically increasing for $0 < x < \frac{\pi}{2}$. Then, $f(x)$ is

a. increasing on $(0, \pi/2)$
b. decreasing on $(0, \pi/2)$
c. increasing on $(0, \pi/4)$ and decreasing on $\left(\frac{\pi}{4}, \frac{\pi}{2}\right)$
d. None of the above

113. The equation $\sin x + x\cos x = 0$ has atleast one root in the interval

a. $(-\pi/2, 0)$ *b.* $(0, \pi)$
c. $(-\pi/2, \pi/2)$ *d.* None of the above

114. If the tangent at (1, 1) on $y^2 = x(2 - x)^2$ meets the curve again at P, then P is

a. (4, 4) *b.* (−1, 2)
c. (9/4, 3/8) *d.* None of these

115. If $y = |x - x^2|$, then $\frac{dy}{dx}$ at $x = 1$ is

a. −1 *b.* 1
c. Does not exist *d.* None of the above

116. If $f(x) = |x - 2|$ and $g(x) = f\{f(x)\}$, then $g'(x)$ for $x > 20$ is

a. 1 *b.* 2
c. −1 *d.* None of these

117. If α and β are the roots of the equation $ax^2 + bx + c = 0$ then $\lim_{x \to \alpha} (1 + ax^2 + bx + c)^{1/x-\alpha}$ is

a. $e^{a(\alpha - \beta)}$ *b.* $e^{\alpha(\beta - \alpha)}$
c. 1 *d.* None of these

118. If $f(x)=ax^2+bx+c$ satisfies the identity $f(x+1)-f(x)=8x+3$ for all $x \in R$. Then, (a, b) equals

a. $(2, 1)$ **b.** $(4, -1)$ **c.** $(-1, 4)$ **d.** $(-1, 1)$

119. The number of times a die must be tossed to obtain a 6 atleast once with probability exceeding 0.9 is atleast

a. 13 **b.** 19
c. 25 **d.** None of these

120. For all $n \in N$, $3^{3n}-26n-1$ is divisible by

a. 24 **b.** 64 **c.** 17 **d.** 676

121. If $x=5$ and $y=-2$, then $x-2y=9$. The contrapositive of this proposition is

a. if $x-2y \neq 9$, then $x \neq 5$ or $y \neq -2$
b. if $x-2y=9$, then $x \neq 5$ and $y \neq 2$
c. $x-2y=9$ if and only if $x=5$ and $y=-2$
d. None of the above

122. The inverse of the function $f: R \to R$ given by $f(x)=\log_a(x+\sqrt{x^2+1})$; $(a>0, a\neq 1)$, is

a. $\frac{1}{2}(a^x+a^{-x})$ **b.** $\frac{1}{2}(a^x-a^{-x})$
c. $\frac{1}{2}\left(\frac{a^x+a^{-x}}{a^x-a^{-x}}\right)$ **d.** Not defined

123. The relation R defined on the set $A=\{1, 2, 3, 4, 5\}$ by $R=\{(a, b): |a^2-b^2|<16\}$ is given by

a. $\{(1, 1), (2, 1), (3, 1), (4, 1), (2, 3)\}$
b. $\{(2, 2), (3, 2), (4, 2), (2, 4)\}$
c. $\{(3, 3), (4, 3), (5, 4), (3, 4)\}$
d. None of the above

124. If

$$A=\left\{p: p=\frac{(x+2)(2x^5+3x^4+4x^3+5x^2+6}{x^2+2x}; \quad x, p \in Z^+\right\}$$

,

then the number of elements in the set A, is

a. 2 **b.** 3 **c.** 4 **d.** 6

125. If $(21.4)^a=(0.00214)^b=100$, then the value of $\frac{1}{a}-\frac{1}{b}$ is

a. 0 **b.** 1 **c.** 2 **d.** 4

126. If $\sin\theta+\cos\theta=\frac{1}{5}$ and $0 \leq \theta < \pi$, then $\tan\theta$ equals

a. $\frac{-4}{3}$ **b.** $\frac{-3}{4}$ **c.** $\frac{3}{4}$ **d.** $\frac{4}{3}$

127. If a, b, c are the sides of ΔABC, right angled at C, then the minimum value of $\left(\frac{c}{a}+\frac{c}{b}\right)^2$ is

a. 0 **b.** 4 **c.** 6 **d.** 8

128. The equation $\sin^4 x-2\cos^2 x+a^2=0$ can be solved, if

a. $-\sqrt{3} \leq a \leq \sqrt{3}$ **b.** $-\sqrt{2} \leq a \leq \sqrt{2}$
c. $-1 \leq a \leq 1$ **d.** None of these

129. Complete solution set of $[\cot^{-1} x]+2[\tan^{-1} x]=0$, where [ ] denotes the greatest integer function, is equal to

a. $(0, \cot 1)$ **b.** $(0, \tan 1)$
c. $(\tan 1, \infty)$ **d.** $(\cot 1, \tan 1)$

130. In ΔABC, base BC and area of triangle are fixed. The locus of the centroid of ΔABC is a straight line that is

a. parallel to side BC
b. right bisector of side BC
c. right angle of BC
d. inclined at an angle $\sin^{-1}(\sqrt{\Delta}/BC)$ to side BC

131. A triangle is formed by the lines $x+y=0$, $x-y=0$ and $lx+my=1$. If l and m vary subject to the condition $l^2+m^2=1$, then the locus of its circumcentre is

a. $(x^2-y^2)^2=x^2+y^2$ **b.** $(x^2+y^2)^2=x^2-y^2$
c. $x^2+y^2=4x^2y^2$ **d.** $(x^2-y^2)^2=(x^2+y^2)^2$

132. If the pairs of lines $x^2+2xy+ay^2=0$ and $ax^2+2xy+y^2=0$ have exactly one line in common, then the joint equation of the other two lines is given by

a. $3x^2+8xy-3y^2=0$
b. $3x^2+10xy+3y^2=0$
c. $y^2+2xy-3x^2=0$
d. $x^2+2xy-3y^2=0$

133. If the line $x\cos\theta+y\sin\theta=2$ is the equation of a transverse common tangent to the circles $x^2+y^2=4$ and $x^2+y^2-6\sqrt{3}x-6y+20=0$, then the value of θ is

a. $5\pi/6$ **b.** $2\pi/3$
c. $\pi/3$ **d.** $\pi/6$

134. If the parabola $y=ax^2-6x+b$ passes through $(0, 2)$ and has its tangent at $x=3/2$ parallel to X-axis, then

a. $a=2, b=-2$ **b.** $a=2, b=2$
c. $a=-2, b=2$ **d.** $a=-2, b=-3$

135. An ellipse is sliding along the coordinate axes. If the foci of the ellipse are (1, 1) and (3, 3), then the area of the director circle of the ellipse (in sq units) is

a. 2π *b.* 4π *c.* 6π *d.* 8π

136. A normal to the hyperbola $\frac{x^2}{4}-\frac{y^2}{1}=1$ has equal intercepts on the positive X-axis and Y-axis. If this normal touches the ellipse $\frac{x^2}{a^2}+\frac{y^2}{b^2}=1$, then a^2+b^2 is equal to

a. 5 *b.* 25
c. 16 *d.* None of these

137. India plays two matches each with West Indies and Australia. In any match, the probabilities of India getting points 0, 1 and 2 are 0.45, 0.05 and 0.50, respectively. Assuming that the outcomes are independent, the probability of India getting atleast 7 points is

a. 0.8750 *b.* 0.0875 *c.* 0.0625 *d.* 0.0250

138. The number of 3×3 matrices A whose entries are either 0 or 1 and for which the system $A\begin{bmatrix}x\\y\\z\end{bmatrix}=\begin{bmatrix}1\\0\\0\end{bmatrix}$ has exactly two distinct solutions, is

a. 0 *b.* 2^9-1 *c.* 168 *d.* 2

139. A determinant is chosen at random from the set of all determinants of order 2 with elements 0 or 1 only. What is the probability that the value of determinant chosen is positive?

a. 3/16 *b.* 3/6
c. 13/16 *d.* 5/16

140. The coefficient of t^{24} in $(1+t^2)^{12}(1+t^{12})(1+t^{24})$ is

a. ${}^{12}C_6+3$ *b.* ${}^{12}C_6+1$
c. ${}^{12}C_6$ *d.* ${}^{12}C_6+2$

141. How many different nine-digit numbers can be formed from the number 223355888 by rearranging its digits so that the odd digits occupy even positions?

a. 16 *b.* 36 *c.* 60 *d.* 180

142. If a, b, c, d are positive real numbers such that $a+b+c+d=2$, then $M=(a+b)(c+d)$ satisfies the relation

a. $0\le M\le 1$ *b.* $1\le M\le 2$
c. $2\le M\le 3$ *d.* $3\le M\le 4$

143. In the quadratic equation ax^2+bx+c, if $\Delta=b^2-4ac$ and $\alpha+\beta, \alpha^2+\beta^2, \alpha^3+\beta^3$ are in GP, where α, β are the roots of $ax^2+bx+c=0$, then

a. $\Delta\ne 0$ *b.* $b\Delta=0$
c. $c\Delta=0$ *d.* $\Delta=0$

144. Let $z=x+iy$ be a complex number, where x and y are integers. Then, the area of the rectangle whose vertices are the roots of the equation $\bar{z}z^3+z\bar{z}^3=350$, is

a. 48 sq units *b.* 32 sq units
c. 40 sq units *d.* 80 sq units

145. The largest interval for which $x^{12}-x^9+x^4-x+1>0$, is

a. $-4<x\le 0$ *b.* $0<x<1$
c. $-100<x<100$ *d.* $-\infty<x<\infty$

146. The set of all real numbers x for which $x^2-|x+2|+x>0$, is

a. $(-\infty,-2)$ *b.* $(-\infty,-\sqrt{2})\cup(\sqrt{2},\infty)$
c. $(-\infty,-1)\cup(1,\infty)$ *d.* $(\sqrt{2},\infty)$

147. Let **a**, **b**, **c** be three vectors of magnitude $\sqrt{3}$, 1, 2 such that $\mathbf{a}\times(\mathbf{a}\times\mathbf{c})+3\mathbf{b}=\mathbf{0}$. If θ is the angle between **a** and **c**, then $\cos^2\theta$ is equal to

a. $\frac{3}{4}$ *b.* $\frac{1}{2}$
c. $\frac{1}{4}$ *d.* None of these

148. If resolved part of vector **a** along the vector **b** is $\mathbf{a}_1$ and that perpendicular to **b** is $\mathbf{a}_2$, then $\mathbf{a}_1\times\mathbf{a}_2$ is equal to

a. $\frac{(\mathbf{a}\times\mathbf{b})\,\mathbf{b}}{|\mathbf{b}|^2}$ *b.* $\frac{(\mathbf{a}\times\mathbf{b})\,\mathbf{a}}{|\mathbf{a}|^2}$

c. $\frac{(\mathbf{a}\cdot\mathbf{b})(\mathbf{b}\times\mathbf{a})}{|\mathbf{b}|^2}$ *d.* $\frac{(\mathbf{a}\cdot\mathbf{b})(\mathbf{b}\times\mathbf{a})}{|\mathbf{b}\times\mathbf{a}|}$

149. The graph of the equation $x^2+y^2=0$ in the three dimensional space, is

a. X-axis *b.* Y-axis
c. Z-axis *d.* XY-plane

150. If the straight lines $\frac{x-1}{k}=\frac{y-2}{2}=\frac{z-3}{3}$ and $\frac{x-2}{3}=\frac{y-3}{k}=\frac{z-1}{2}$ intersect at a point, then integer k is equal to

a. 2 *b.* -2
c. -5 *d.* 5

Answers

Physics

1. (b)	**2.** (c)	**3.** (b)	**4.** (b)	**5.** (b)	**6.** (d)	**7.** (b)	**8.** (a)	**9.** (b)	**10.** (d)
11. (c)	**12.** (c)	**13.** (c)	**14.** (c)	**15.** (c)	**16.** (c)	**17.** (c)	**18.** (b)	**19.** (d)	**20.** (c)
21. (a)	**22.** (d)	**23.** (a)	**24.** (b)	**25.** (d)	**26.** (b)	**27.** (d)	**28.** (b)	**29.** (d)	**30.** (d)
31. (c)	**32.** (a)	**33.** (b)	**34.** (c)	**35.** (a)	**36.** (a)	**37.** (c)	**38.** (b)	**39.** (c)	**40.** (a)

Chemistry

41. (b)	**42.** (c)	**43.** (a)	**44.** (c)	**45.** (c)	**46.** (d)	**47.** (b)	**48.** (a)	**49.** (a)	**50.** (b)
51. (c)	**52.** (b)	**53.** (b)	**54.** (c)	**55.** (b)	**56.** (b)	**57.** (a)	**58.** (d)	**59.** (c)	**60.** (b)
61. (a)	**62.** (c)	**63.** (d)	**64.** (b)	**65.** (b)	**66.** (d)	**67.** (b)	**68.** (c)	**69.** (a)	**70.** (b)
71. (d)	**72.** (c)	**73.** (c)	**74.** (a)	**75.** (c)	**76.** (a)	**77.** (b)	**78.** (d)	**79.** (a)	**80.** (c)

English Proficiency

81. (b)	**82.** (b)	**83.** (b)	**84.** (c)	**85.** (b)	**86.** (b)	**87.** (d)	**88.** (a)	**89.** (d)	**90.** (c)
91. (c)	**92.** (d)	**93.** (c)	**94.** (a)	**95.** (a)					

Logical Reasoning

96. (d)	**97.** (d)	**98.** (b)	**99.** (b)	**100.** (d)	**101.** (b)	**102.** (d)	**103.** (a)	**104.** (c)	**105.** (c)

Mathematics

106. (c)	**107.** (b)	**108.** (b)	**109.** (b)	**110.** (a)	**111.** (b)	**112.** (a)	**113.** (b)	**114.** (c)	**115.** (c)
116. (a)	**117.** (a)	**118.** (b)	**119.** (a)	**120.** (d)	**121.** (a)	**122.** (b)	**123.** (d)	**124.** (c)	**125.** (c)
126. (a)	**127.** (d)	**128.** (b)	**129.** (d)	**130.** (a)	**131.** (a)	**132.** (b)	**133.** (d)	**134.** (b)	**135.** (d)
136. (d)	**137.** (b)	**138.** (a)	**139.** (a)	**140.** (d)	**141.** (c)	**142.** (a)	**143.** (c)	**144.** (a)	**145.** (d)
146. (b)	**147.** (a)	**148.** (c)	**149.** (c)	**150.** (c)					

Hints & Solutions

Physics

1. ***(b)*** $v \propto \lambda^a \rho^b g^c \Rightarrow \quad v = k\lambda^a \rho^b g^c$

Using dimensions,

$$[LT^{-1}] = [L]^a\,[ML^{-3}]^b[LT^{-2}]^c$$

$$[LT^{-1}] = [L^{a-3b+c}M^bT^{-2c}]$$

Comparing powers on both sides, we get

$$1 = a - 3b + c;\ 0 = b;\ -1 = -2c$$

$$\therefore \quad 1 = a + c, \quad b = 0, \quad c = 1/2$$

$$\Rightarrow \quad 1 = a + 1/2$$

$$a = 1/2$$

$$\therefore \quad v = k\lambda^{1/2}\rho^0 g^{1/2}$$

$$v^2 = k\lambda g$$

2. ***(c)*** $v = \alpha\sqrt{x} \quad v = \dfrac{dx}{dt}$

$$\therefore \quad \frac{dx}{dt} = \alpha\sqrt{x}$$

$$\Rightarrow \quad \frac{dx}{\sqrt{x}} = \alpha dt$$

Integrating both sides, we get

$$2\sqrt{x} = \alpha t$$

$$\Rightarrow \quad x = \frac{\alpha^2 t^2}{4}$$

$$\therefore \quad x \propto t^2$$

3. ***(b)*** For upward motion, velocity at height h

$$v^2 = u^2 - 2gh \quad \ldots\text{(i)}$$

For downward motion from point Q to point R, distance h below P,

$$(2v)^2 = v^2 + 2g \times 2h$$

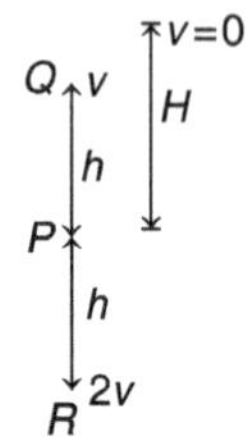

$$4v^2 - v^2 = 4gh$$

$$3v^2 = 4gh \Rightarrow v^2 = \frac{4}{3}gh$$

Putting this value in Eq. (i), we get

$$\frac{4}{3}gh = u^2 - 2gh$$

$$\frac{10}{3}gh = u^2$$

$\therefore$ Maximum height $= \dfrac{u^2}{2g}$

$$\Rightarrow \quad H = \frac{10gh}{3 \times 2g} = \frac{5}{3}h$$

$$\therefore \quad = \frac{5}{3}h$$

4. ***(b)*** Let v be the velocity of the train after time t_1.

Then, $v = \alpha\, t_1 = \beta t_2,\ x_1 = \dfrac{1}{2}\alpha t_1^2$

and $x_2 = \dfrac{1}{2}\beta t_2^2$

$$\therefore \quad \frac{\beta}{\alpha} = \frac{t_1}{t_2} \text{ and } \frac{x_1}{x_2} = \frac{\alpha}{\beta} \times \frac{t_1^2}{t_2^2}$$

$$\frac{x_1}{x_2} = \frac{t_2}{t_1} \times \frac{t_1^2}{t_2^2}$$

$$\Rightarrow \quad \frac{x_1}{x_2} = \frac{t_1}{t_2} \Rightarrow \frac{x_1}{x_2} = \frac{\beta}{\alpha} = \frac{t_1}{t_2}$$

5. ***(b)*** The diagram is pictorial representation of forces acting on (by) ant.

$\therefore$ The ant will remain on the surface, if

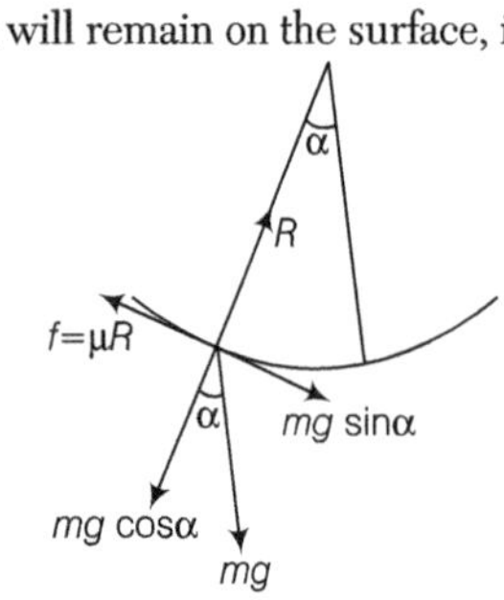

$$R = mg\cos\alpha$$

and $f(=\mu R) = mg\sin\alpha$

$$\therefore \quad \mu = \tan\alpha \text{ or } \cot\alpha = \frac{1}{\mu}$$

$$\therefore \quad \cot\alpha = \frac{1}{1/3} = 3 \Rightarrow \cot\alpha = 3$$

$$\Rightarrow \quad \alpha = \cot^{-1}(3)$$

6. ***(d)*** Wall applies two forces on the block,

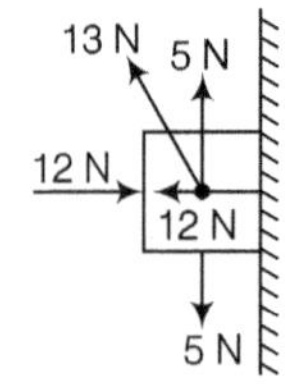

(i) Normal reaction (N)

$$R = 12\,\text{N}$$

(ii) Frictional force, $f_s = mg = 5\,\text{N}$

$\therefore$ Total force exerted by wall on block,

$$F = \sqrt{N^2 + f_s^2}$$

$$= \sqrt{12^2 + 5^2} = \sqrt{144 + 25}$$

$$= 13\,\text{N}$$

7. ***(b)*** Work done is equal to change in energy of body. Situation is shown in figure. When mass m falls vertically on spring, then spring is compressed by distance d.

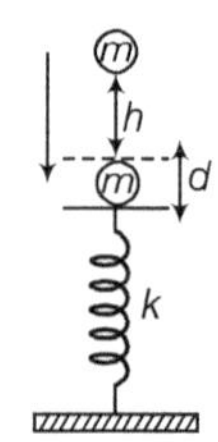

$\therefore$ Net work done,

W = Potential energy stored in spring + Loss in potential energy of mass.

$$= mg(h + d) - \frac{1}{2}kd^2$$

8. ***(a)*** Rotational KE of the rod is converted into the potential energy of the rod at the maximum height.

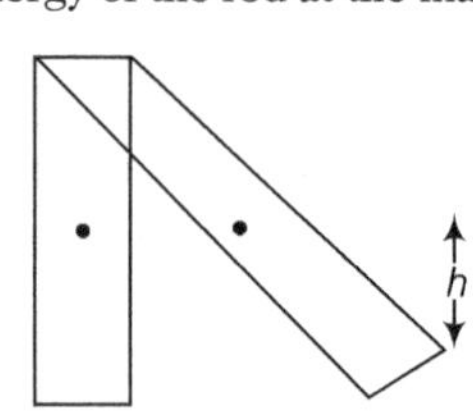

i.e. $$\frac{1}{2}I\omega^2 = mgh$$

As, $$I = \frac{ml^2}{3}$$

$$\therefore \quad \frac{1}{2} \times \frac{ml^2}{3} \times \omega^2 = mgh$$

$$\Rightarrow \quad h = \frac{l^2 \omega^2}{6g}$$

9. ***(b)*** Let F be the tensile force applied. Since, each rivet shares the stretching force equally, so the shearing force on each rivet $= \frac{F}{2}$. If A is the area of each rivet, then shearing stress on each rivet $= \frac{F}{2A}$.

Now, maximum shearing stress on each strip

$$= 4 \times 10^8$$

i.e. $$\frac{F_{max}}{2A} = 3 \times 10^8 \text{ Nm}^{-2}$$

$$\Rightarrow \quad F_{max} = 6 \times 10^8 \times A$$

$$= 6 \times 10^8 \times \pi r^2 \qquad [\because A = \pi r^2]$$

$$= 6 \times 10^8 \times \pi \times (10^{-3})^2$$

$$= 1884.9 \text{ N}$$

$$\Rightarrow \quad \simeq 1885 \text{ N}$$

10. ***(d)*** Temperature of interface connecting two rods of different lengths is given by

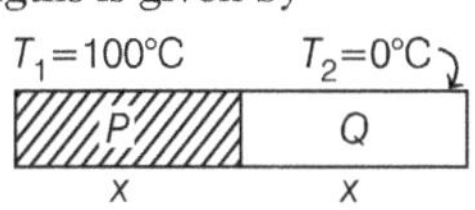

$$T = \frac{K_1 x_2 T_1 + K_2 x_1 T_2}{K_1 x_2 + K_2 x_1}$$

Here $\quad x_1 = x_2 = x$

$$\therefore \quad T = \frac{K_1 T_1 + K_2 T_2}{K_1 + K_2}$$

Dividing above relation by K_2 in numerator and denominator, we get

$$T = \frac{\frac{K_1}{K_2} T_1 + \frac{K_2}{K_2} T_2}{\frac{K_1}{K_2} + \frac{K_2}{K_2}} = \frac{\left(\frac{K_1}{K_2}\right) T_1 + T_2}{\frac{K_1}{K_2} + 1}$$

$$T = \frac{\frac{2}{3} T_1 + T_2}{\frac{2}{3} + 1}$$

$$= \frac{2T_1 + 3T_2}{5} = \frac{2 \times 100 + 0}{5} = 40^\circ\text{C}$$

$$\Rightarrow \quad T = 40^\circ\text{C}$$

11. ***(c)*** $F = T \times$ length of membrane

$F = T \times 2 \times 4L \qquad$ [$\because$ liquid film has two surfaces]

$\therefore \quad F = T\,8L$

$= 8TL$

12. ***(c)*** Let ΔR = decrease in radius of drop due to evaporation. Now, decrease in surface energy = energy needed for evaporation

$$T \times [4\pi r^2 - 4\pi(R - \Delta R)^2] = \Delta mL$$

$$T \times [4\pi R^2 - 4\pi R^2 - 4\pi \Delta R^2 + 8\pi R \Delta R] = \Delta mL$$

$$T \times [-4\pi \Delta R^2 + 8\pi R \Delta R] = \Delta mL$$

As, $\Delta R^2 << R^2$ (Neglected)

$$\therefore \quad T \times (8\pi R\, \Delta R) = \Delta mL$$

As, $$\Delta mL = [(4\pi R^2) \Delta R] \rho \times L$$

$$\therefore \quad T \times (8\pi R \Delta R) = 4\pi R^2\, \Delta R \rho L$$

$$\Rightarrow \quad T(8) = 4R\rho L$$

$$\Rightarrow \quad R = \frac{2T}{\rho L}$$

13. ***(c)*** According to Bernoulli's theorem,

$$p = \frac{1}{2} \rho v^2 = \text{constant}$$

Near the ends, the velocity is higher, so pressure is lower as a result the liquid rises at the sides to compensate the drop of pressure,

$$\therefore \quad \rho gh = \frac{1}{2} \rho v^2 = \frac{1}{2} \rho r^2 \omega^2$$

$$\Rightarrow \quad h = \frac{r^2 \omega^2}{2g}$$

$$\Rightarrow \quad h = \frac{r^2 (2\pi\, v)^2}{2g} = \frac{2\pi^2 r^2 v^2}{g}$$

$$= \frac{2 \times \pi^2 \times (0.05)^2 \times 2^2}{9.8}$$

$$= 2\text{cm}$$

14. ***(c)*** As given, $pT^2 = c$

For ideal gas, $\quad pV = RT$ or $p = \frac{RT}{V}$

$$\therefore \quad \frac{RT}{V} \times T^2 = c$$

or $$V^{-1} T^3 = c$$

Differentiating both sides, we get

$$-V^{-2} \Delta V T^3 + 3T^2 \Delta T V^{-1} = 0$$

$$\Rightarrow \quad 3T^2\, \Delta T V^{-1} = V^{-2} \Delta V T^3$$

$$3\Delta T = TV^{-1} \Delta V$$

$$3\left(\frac{\Delta T}{T}\right) = \left(\frac{\Delta V}{V}\right) \Rightarrow \frac{3}{T} = \left(\frac{\Delta V}{V}\right) \frac{1}{\Delta T} \qquad \ldots\text{(i)}$$

$\therefore$ Coefficient of volume expansion,

$$\gamma = \frac{\Delta V}{V} \times \frac{1}{\Delta T}$$

$$\gamma = \frac{3}{T} \qquad \text{[by Eq. (i)]}$$

15. ***(c)*** Velocity of sound in a gas,

$$v = \sqrt{\frac{\gamma RT}{M}}$$

$$\frac{v_{N_2}}{v_{He}} = \sqrt{\left(\frac{\gamma}{M}\right)_{N_2} \Big/ \left(\frac{\gamma}{M}\right)_{He}}$$

$$\Rightarrow \quad = \sqrt{\frac{\gamma_{N_2}}{\gamma_{He}} \times \frac{M_{He}}{M_{N_2}}}$$

Since, He is a monoatomic gas, ($M_{He} = 4$)

$$\Rightarrow \quad \gamma_{He} = 5/3$$

N_2 is diatomic gas ($M_{N_2} = 28$)

$$\Rightarrow \quad \gamma_{N_2} = \frac{7}{5}$$

$$\therefore \quad \frac{v_{N_2}}{v_{He}} = \sqrt{\frac{7}{5} \times \frac{3}{5} \times \frac{4}{28}}$$

$$\Rightarrow \quad \frac{v_{N_2}}{v_{He}} = \sqrt{\frac{21}{25} \times \frac{4}{28}} = \sqrt{\frac{3}{25}} = \frac{\sqrt{3}}{5}$$

16. ***(c)*** $x = x_0 \cos\left(\omega t - \frac{\pi}{4}\right)$

$$\therefore \quad v = \frac{dx}{dt} = -x_0 \omega \sin\left(\omega t - \frac{\pi}{4}\right)$$

and

$$a = -x_0 \omega^2 \cos\left(\omega t + \frac{\pi}{4}\right)$$

$$a = x_0 \omega^2 \cos\left(\omega t + \frac{3\pi}{4}\right)$$

Comparing it with $a = A\cos(\omega t + \delta)$

Clearly, $A = x_0\omega^2$

$$\delta = \frac{3\pi}{4}$$

17. ***(c)*** Fundamental frequency of closed pipe is given by,

$$\nu = \frac{v}{4L}$$

When $\frac{1}{4}$th length is filled with water, then the length of pipe in which air vibrates

$$= L - \frac{L}{4} = \frac{3L}{4}$$

$$\therefore \quad \nu' = \frac{v}{4 \times \frac{3L}{4}} = \frac{v}{3L}$$

$$\therefore \quad \frac{\nu'}{\nu} = \frac{4}{3}$$

$$\nu' = \nu \times \frac{4}{3}$$

$$\nu' = 480 \times \frac{4}{3}$$

$$= 160 \times 4 \text{Hz}$$

$$\Rightarrow \quad \nu' = 640 \text{Hz}$$

18. ***(b)*** Total charge $= -14 + 78.85 - 56$

$$= +8.85 \text{ nC}$$

Electric flux, $\phi_E = \frac{q}{\varepsilon_0}$

$$\varepsilon_0 = 8.85 \times 10^{-12} \text{ in SI units}$$

$$\therefore \quad \phi = \frac{q}{\varepsilon_0} = \frac{+8.85 \times 10^{-9}}{8.85 \times 10^{-12}} = 10^3$$

$$\therefore \quad \phi_E = 10^3 \text{ Nm}^2\text{C}^{-1}$$

19. ***(d)*** If q is charge on the inner sphere, then induced charge on outer sphere is $-q$.

$\therefore$ Potential difference, $V = \frac{q}{r_1} - \frac{q}{r_2}$

or

$$3 = \frac{q}{4} - \frac{q}{6}$$

$$3 = \frac{2q - 3q}{12}$$

$$\Rightarrow \quad q = -36 \text{ esu}$$

$\therefore$ Inner sphere has a charge of magnitude 36 esu (Negative in nature).

20. ***(c)*** If C is capacity of each capacitor, then charge on each capacitor = 10C.

When connected in series, potential difference between free plates $= \frac{\text{Total charge}}{\text{Total capacity}} = \frac{10C}{C/6} = 60\text{V}$

21. ***(a)*** According to Kirchoff's first law,

$$\Sigma I = 0 \Rightarrow \Sigma\left(\frac{Q}{t}\right) = 0$$

$$\Sigma Q = 0 \qquad \text{(conservation of charge)}$$

According to second law,

$$\Sigma iR = \Sigma E$$

$$\Rightarrow \quad \Sigma E - \Sigma iR = 0$$

$$\Rightarrow \quad \Sigma V = 0$$

$$\Rightarrow \quad \Sigma \frac{W}{Q} = 0$$

$$\Rightarrow \quad \Sigma W = 0$$

$$\Sigma W = \Sigma(\Delta E)$$

$$\therefore \quad \Delta E = 0 \qquad \text{(conservation of energy)}$$

22. ***(d)*** Resistance of each semi-circle $= \pi r \times 1 \times 10^{-4}$ $[\because r = 1\text{m}]$

$$= 3.14 \times 1 \times 10^{-4}$$

$$= 3.14 \times 10^{-4} \, \Omega$$

Resistance of wire $AB = 2 \times 1 \times 10^{-4}$

$$= 2 \times 10^{-4} \, \Omega$$

Two semi-circles and straight wire AB are parallel to each other, therefore

$$\frac{1}{R} = \frac{2}{3.14 \times 10^{-4}} + \frac{1}{2 \times 10^{-4}}$$

$$= \frac{2 \times 10^4}{3.14} + \frac{10^4}{2}$$

$\therefore$ $$R = 0.88 \times 10^{-4}\,\Omega$$

23. ***(a)*** Here, $S = R_1 + R_2$ and $P = \dfrac{R_1 R_2}{R_1 + R_2}$

As, $S = nP$, we get

$$(R_1 + R_2) = n\frac{R_1 R_2}{(R_1 + R_2)}$$

i.e. $$(R_1 + R_2)^2 = n(R_1 R_2)$$

If $R_1 = R_2 = R$ we get

$$4R^2 = nR^2$$

i.e. $$n = 4$$

24. ***(b)*** Total resistance, $R_{eq} = \dfrac{2 \times R}{2 + R}$

$\Rightarrow$ $$R_{eq} = \frac{2R}{2 + R}$$

$$\text{Power, } P = \frac{V^2}{R_{eq}}$$

i.e. $$150 = \frac{15 \times 15\,(2 + R)}{2R}$$

$$20R = 30 + 15R$$

$$5R = 30$$

$\Rightarrow$ $$R = 6\,\Omega$$

25. ***(d)*** Cyclotron frequency,

$$f = \frac{qB}{2\pi m}$$

or $$qB = 2\pi m f$$

$\therefore$ $$\text{KE} = \frac{q^2B^2}{2m}R^2 = \frac{4\pi^2 m^2 f^2 R^2}{2m}$$

$\Rightarrow$ $$\text{KE} = 2\pi^2 m f^2 R^2$$

$\Rightarrow$ $$\text{KE} = 2 \times (3.14)^2 \times 1.67 \times 10^{-27} \times (10 \times 10^6)^2 \times (0.5)^2$$

$$\text{KE} = 8.23 \times 10^{-13}\ \text{J}$$

$$\text{KE} - \frac{8.23 \times 10^{-13}}{1.6 \times 10^{-13}}\ \text{MeV}$$

$\Rightarrow$ $$\text{KE} = 5.1\ \text{MeV}$$

26. ***(b)*** Radius, $R = \dfrac{mv}{qB}$,

If B and q are constant, then

$$R \propto mv$$

As, $$R_A > R_B$$

$\therefore$ $$m_A v_A > m_B v_B$$

27. ***(d)*** When the needle is heated in such a high temperature, then the magnetic needle losses its magnetism. So, the restoring torque no longer acts and the needle stops vibrating.

28. ***(b)*** Magnetic flux, $\phi = BA\cos\theta$

$\Rightarrow$ $$\phi = \frac{1}{\pi} \times \pi r^2 \times \cos 60^\circ$$

$$= r^2 \times \frac{1}{2}$$

$$= (0.2)^2 \times \frac{1}{2} = 0.02\ \text{Wb}$$

29. ***(d)*** Emf indued, $\varepsilon = Blv$

So, $$B = \frac{\varepsilon}{vl}$$

Here, $$l = 50 \times 10^{-2}\,\text{m} = 0.5\ \text{m}$$

$$v = \frac{300}{60}\ \text{m/s} = 5\ \text{m/s}$$

and $$\varepsilon = 2\,\text{V}$$

$\therefore$ $$B = \frac{2}{5 \times 0.5} = \frac{20}{25}$$

$\Rightarrow$ $$= 0.8\ \text{T}$$

30. ***(d)*** In *L-C-R* circuit, the voltage across inductor and capacitor are in opposite phase, so net voltage across the combination becomes zero.

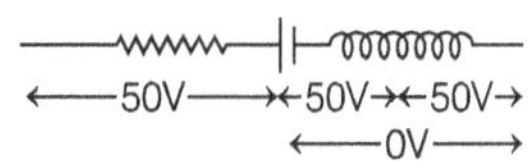

31. ***(c)*** Wavelength of light in air, $\lambda' = \dfrac{\lambda}{n}$

where, n is refractive index.

Let thickness of air column be x, then

$$x = m\lambda$$

For same thickness of vacuum, $x = (m + 1)\lambda'$

or $$m\lambda = (m + 1)\lambda'$$

$\Rightarrow$ $$m\lambda = (m + 1)\frac{\lambda}{n}$$

$\therefore$ $$\frac{1}{m} = n - 1$$

$$\frac{1}{m} = 1.0003 - 1 = 0.0003$$

$\Rightarrow$ $$m = \frac{1}{0.0003} = \frac{1000}{3}$$

Using, $$x = m\lambda$$

$\Rightarrow$ $$x = \frac{1000}{3} \times 600\,\text{nm}$$

$$x = 10000 \times \frac{600}{3} \times 10^{-9}\,\text{m}$$

$$X = 2 \times 10^{-3}\,\text{m}$$

$$X = 2\ \text{mm}$$

32. ***(a)*** Using $n_1\lambda_1 = n_2\lambda_2$, we get

$$3 \times 700 = 5 \times \lambda_2$$

$$\lambda_2 = \frac{2100}{5}\ \text{nm}$$

$$\lambda_2 = 420\ \text{nm}$$

33. ***(b)*** $I_{max} = (\sqrt{I_1} + \sqrt{I_2})^2 = I_1 + I_2 + 2\sqrt{I_1 I_2}$

$$= 5I + 2\sqrt{4I^2} = 5I + 4I = 9I$$

$$I_{min} = I_1 + I_2 - 2\sqrt{I_1 I_2}$$

$$= 5I - 4I = I$$

$\therefore$ The maximum and minimum possible intensities in the beam are $9\,I$ and I, respectively.

34. ***(c)*** As, $d = f_0 + f_e$

$$d = 0.3 + 0.05\,,\ d = 0.35 \text{ m}$$

This set up is for normal adjustment in which image is formed at infinity.

35. ***(a)*** Number of photons emitted per second,

$$\frac{N}{t} = \frac{\text{Power}}{\left(\frac{hc}{\lambda}\right)} = \frac{60 \times 660 \times 10^{-9}}{6.6 \times 10^{-34} \times 3 \times 10^{8}}$$

$$= \frac{60 \times 100 \times 10^{-9}}{3 \times 10^{-26}}$$

$$\frac{N}{t} = 2 \times 10^{20}$$

36. ***(a)*** ${}_{88}\text{Ra}^{236} \xrightarrow{(3\alpha + \beta)} {}_{Z}X^{A}$

$\therefore$ Decrease in mass $= 3 \times 4 = 12$

Decrease in charge $= 3 \times 2 + 1(-1) = 6 - 1 = 5$

$\therefore$ Mass number $= 236 - 12 = 224$

Atomic number $= 88 - 5 = 83$

$\therefore$ The isotope is ${}_{83}X^{224}$.

37. ***(c)*** $A_1 = N_1\lambda,\ A_2 = N_2\lambda$

Mean life, $T = \dfrac{1}{\lambda}$

$$A_1 - A_2 = (N_1 - N_2)\lambda = (N_1 - N_2)\frac{1}{T}$$

So, number of atoms disintegrated in $(t_2 - t_1)$ is

$$= N_1 - N_2$$

$$\therefore \quad = (A_1 - A_2)\,T$$

38. ***(b)*** As, $\dfrac{N}{N_0} = \dfrac{1}{1+7} = \dfrac{1}{8}$ and $\dfrac{N}{N_0} = \left(\dfrac{1}{2}\right)^n$

$$\therefore \quad n = 3$$

So, three half-lives would have been passed

$$\therefore \quad t = nT_{1/2}$$

$$t = 3 \times 10^9 \text{ yr}$$

39. ***(c)*** Power gain $= \dfrac{\Delta V_c}{\Delta V_i} \times \dfrac{\Delta i_c}{\Delta i_b} = A_V \times \beta_{AC}$

where, A_V = voltage gain

β_{AC} = current gain

As, $\quad A_V = \beta_{AC} \times \left(\dfrac{R_0}{R_i}\right)$

$$A_V = 50,\ R_0 = 200\,\Omega,\ R_i = 100\,\Omega$$

As, $\quad A_V = \beta_{AC}\dfrac{R_0}{R_i}$

$$\therefore \quad \beta_{AC} = A_V \times \frac{R_i}{R_0} = 50 \times \frac{100}{200}$$

$$\beta_{AC} = 25$$

Now, AC power gain $= A_C \times \beta_{AC}$

$= 50 \times 25$

$= 1250$

40. ***(a)*** Total power of modulated wave,

$$P_{\text{total}} = P_{\text{carrier}}\left(1 + \frac{m_a^{\ 2}}{2}\right)$$

$$= 600\left(1 + \frac{(0.75)^2}{2}\right)$$

$$= 600\left(1 + \frac{0.5625}{2}\right)$$

$$= 600 \times (1 + 0.28125)$$

$$= 600 \times (1.28125)$$

$$= (768.75 \text{ W}) \text{ approx.}$$

$$\simeq 769 \text{ W}$$

Chemistry

41. ***(b)*** The given reactions take place as follows:

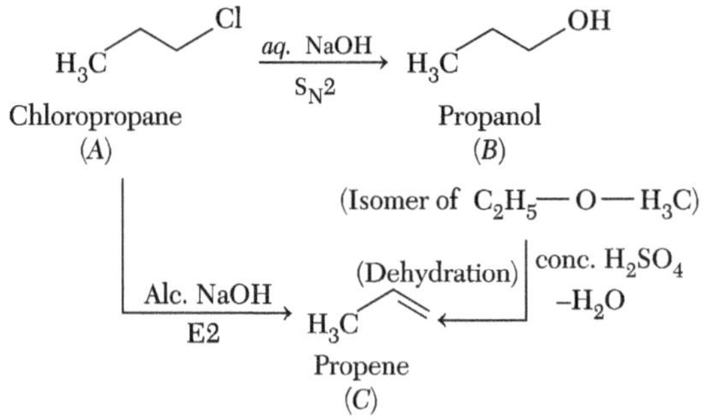

42. ***(c)*** The given reactions take place as follows:

$$\underset{P}{SO_2Cl_2} + 2\,H_2O \longrightarrow \underset{Q}{H_2SO_4} + \underset{R}{2HCl}$$

$$SO_2Cl_2 + 4NaOH \longrightarrow \underset{S}{Na_2SO_4} + \underset{T}{2NaCl} + 2H_2O$$

$$S \xrightarrow{BaCl_2} \underset{\text{White ppt.}}{BaSO_4} \qquad T \xrightarrow{aq\,NO_3} \underset{\text{White ppt.}}{AgCl}$$

43. ***(a)*** For, $NH_4Cl \rightleftharpoons NH_3 + HCl$

$$\alpha = 1$$

$$\therefore \quad \text{Exp. molar mass} = \frac{\text{Normal molar mass}}{1+\alpha} = \frac{\text{Normal molar mass}}{2}$$

$\therefore$ Vapour density of completely dissociated NH_4Cl will be half of that of ammonium chloride.

44. ***(c)*** Number of B^- ions in unit cell $= 8 \times \frac{1}{8} + 6 \times \frac{1}{2} = 4$

Since, A^+ ion occupies 25% of tetrahedral holes

$$= \frac{8 \times 25}{100} = 2$$

$\therefore$ Ratio of B^- to A^+ is 2 : 1.

Hence, formula of solid is AB_2.

45. ***(c)*** The complete given reactions are as follows:

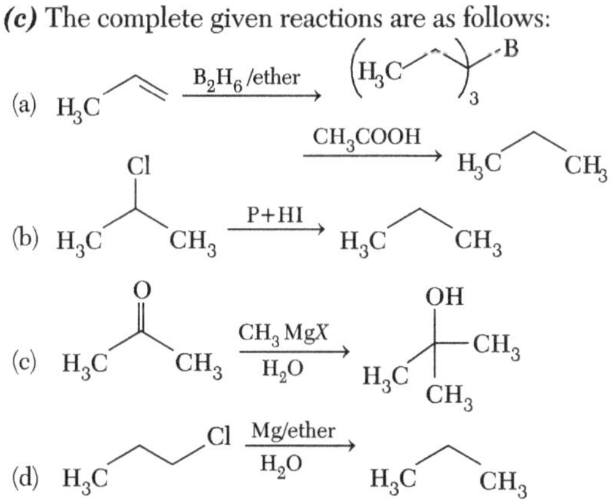

$\therefore$ Reaction (c) will not give propane.

46. ***(d)*** We have,

$$\frac{\text{Weight of } X_4O_6}{\text{Atomic weight of } X_4O_6} = \frac{\text{Weight of } X}{\text{Atomic weight of } 4X}$$

$$\frac{10}{4x + 6 \times 16} = \frac{5.72}{4x} \Rightarrow x = 32$$

47. ***(b)*** Milliequivalent of oxalic acid taken $= \frac{3.15}{E} \times 1000$

Volume of NaOH required for 33.36 mL oxalic acid solution = 50 mL

Volume of NaOH required for 500 mL oxalic acid solution

$$= \frac{50}{33.36} \times 500 \text{ mL}$$

Milliequivalent of NaOH required for complete neutralisation = Milliequivalent of oxalic acid

$$\Rightarrow \quad \frac{50 \times 500}{33.36} \times \frac{1}{15} = \frac{3.15 \times 1000}{E}$$

$$\Rightarrow \quad E = 63$$

For oxalic acid,

$$E = \frac{\text{Molecular mass}}{2} = \frac{[M_{(COOH)_2} + M_{xH_2O}]}{2}$$

$$63 = \frac{90 + 18x}{2}$$

$$\Rightarrow \quad x = 2$$

48. ***(a)*** $\overset{4}{H_2C} = \overset{3}{CH} — \overset{2}{C} \equiv \overset{1}{CH}$

$$\uparrow sp^2 \quad \uparrow sp^2 \quad \uparrow sp \quad \uparrow sp$$

49. ***(a)***

(a) α increases as concentration decreases.

(b) There will be a depression in the freezing point by the addition of glucose in water.

(c) Mixing of solution is always accompanied by an increase in entropy.

(d) $\Delta_{sol} H = \underset{(-\,ve)}{\Delta_{lattice} H} + \underset{(-\,ve)}{\Delta_{hyd} H}$

50. ***(b)*** Mole of $MgSO_4 = 0.001$

$\therefore$ Mole of $CaCO_3 = 0.001$

[$\because$ Mg and Ca both are bivalent]

$\therefore$ Mass of $CaCO_3 = 0.001 \times 100 = 0.1$ g

$\therefore$ 10^3 mL water contains 0.1 g $CaCO_3$

$\therefore$ 10^6 mL water contains $= \frac{0.1 \times 10^6}{10^3} = 100$ ppm

51. ***(c)*** $d_1 = 5.46 \text{ g/dm}^3, T_1 = 300 \text{ K}, p_1 = 2 \text{ bar}$

$d_2 = ?, T_2 = 273 \text{ K}, p_2 = 1 \text{ bar}$

$\because \quad \dfrac{d_1}{d_2} = \dfrac{p_1 m_1}{RT_1} \times \dfrac{RT_2}{p_2 m_2}$

[$\because$ For same gas, $m_1 = m_2$]

$\therefore \quad \dfrac{5.46}{d_2} = \dfrac{2 \times 273}{1 \times 300} \Rightarrow d_2 = 3\ \text{g/dm}^3$

52. ***(b)*** Alcohol involves H-bonding ; also molecular weight of CH_4 > molecular weight of H_2. Greater the molecular weight of covalent compound, higher is its boiling point.

Thus, the correct decreasing order of boiling points of the compounds is $CH_3OH > CH_4 > H_2$.

53. ***(b)*** SO_3^{2-} has 42 electrons, CO_3^{2-} has 32 electrons and NO_3^- has 32 electrons.

54. ***(c)*** Actinoids show a range of oxidation states (+ 3 to + 7) which is due to comparable energies of 5*f*, 6*d* and 7*s*-orbitals.

55. ***(b)*** K has lowest first ionisation energy because of larger size. Kr has highest first ionisation energy because of smallest size and ns^2, np^6 configurations. Ionisation energy in the order K < Li < Ca < S < Kr.

56. ***(b)*** The complete reactions take place as follows:

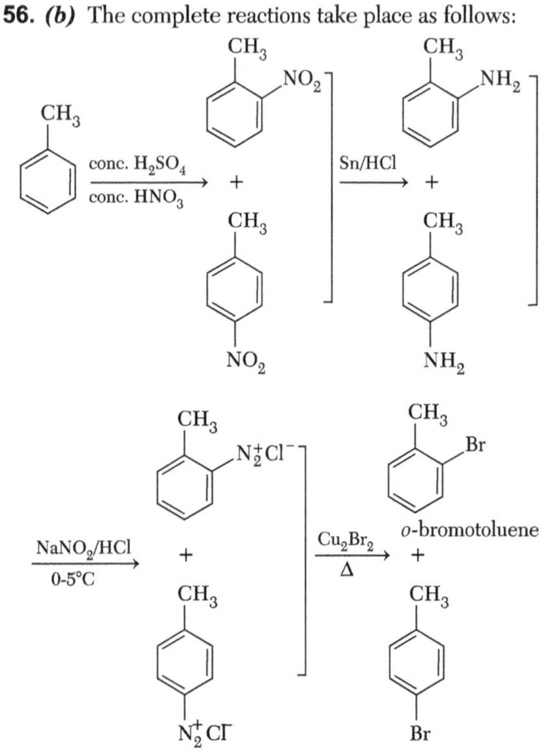

p-bromotoluene

57. ***(a)*** The size of cation is smaller for Li^+ and size of anion is largest for I^- ion.

$\therefore$ The ratio is minimum for LiI.

58. ***(d)*** Antipyretic drugs reduce body temperature in fever. e.g. aspirin, paracetamol, phenacetin.

59. ***(c)*** CO_2 must be obtained from initial components, i.e. carbon and oxygen.

60. ***(b)*** $2MS + C \longrightarrow 2M + CS_2\ ;\ \Delta G_1 =$ positive

$2MO + C \longrightarrow 2M + CO_2\ ;\ \Delta G_2 =$ negative

Therefore, CO_2 is more stable than CS_2, while *MS* are more stable than *MO*.

61. ***(a)*** Acetaldehyde gives ketal on reaction with excess of C_2H_5OH (ethanol).

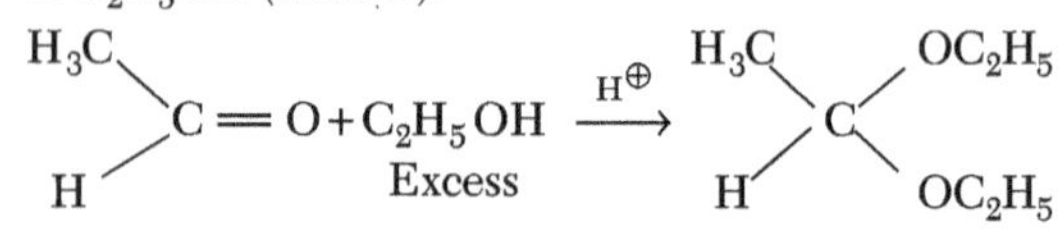

62. ***(c)*** HgI_2 although insoluble in water but shows complex formation with KI and therefore, freezing point decreases.

63. ***(d)*** $[Mn^{7+} + 5e^- \longrightarrow Mn^{2+}] \times 2$

$[N^{3+} \longrightarrow N^{5+} + 2e^-] \times 5$

$2Mn^{7+} + 5\,N^{3+} \longrightarrow 2\,Mn^{2+} + 5\,N^{5+}$

$\therefore$ 2 moles of $KMnO_4 \equiv$ 5 moles of NO_2^-

$\therefore$ 1 mole of $KMnO_4 \equiv \dfrac{5}{2}$ moles of NO_2^-

64. ***(b)*** Less substituted alkene is less stable and therefore more reactive.

Decreasing order of reactivity is

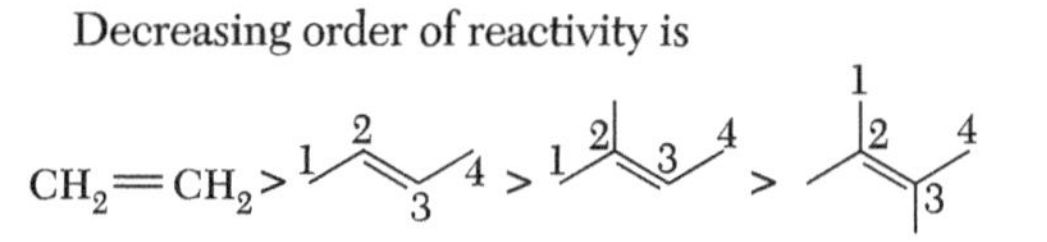

(Ethene) IV; (2-butene) I; (2-methyl-2-butene) II; (2, 3-dimethyl-2-butene) III

65. ***(b)*** $\Delta G^\circ = -\,2.303 \times 8.314 \times 300 \times \log\,(2.0 \times 10^{-7})$

$= +\,38479.8\ \text{J mol}^{-1} = +\,38.48\ \text{kJ mol}^{-1}$

Also, $\Delta G^\circ = \Delta H^\circ - T\Delta S^\circ$

$\therefore \quad \Delta S^\circ = \dfrac{\Delta H^\circ - \Delta G^\circ}{T} = \dfrac{28.40 - 38.48}{300}$

$= -\,0.0336\ \text{kJ K}^{-1}$

$= -\,33.6\ \text{JK}^{-1}$

66. ***(d)*** When $Cu(NO_3)_2$ copper nitrate is strongly heated, copper oxide is obtained.

$2Cu(NO_3)_2 \xrightarrow{\Delta} \underset{\text{Copper oxide}}{2CuO} + 4NO_2 + O_2$

67. ***(b)*** For the first order reaction,

$k = \dfrac{0.693}{t_{1/2}} = \dfrac{0.693}{60} = 11.55 \times 10^{-3}\ \text{min}^{-1}$

Now, according to first order kinetic equation,

$t = \dfrac{2.303}{k} \log \dfrac{a}{(a-x)}$

$= \dfrac{2.303}{11.5 \times 10^{-3}} \log \dfrac{100}{10} = 200\ \text{min}$

68. ***(c)*** EDG stabilises carbocation, whereas EWG destabilises. Electron donating power of $\ddot{N}H_2$ (+ R and − I) > —CH_3 (+ I-effect) but + R > HC : (triplet carbene have the least tendency to release electron) > NO_2 (− I and − R, net electron withdrawing power).

Thus order of stability is IV> III > I > II

69. ***(a)*** **Kolbe's reaction** is used for the preparation of alkanes, alkenes and alkynes.

70. ***(b)*** Hardness of water is due to carbonate or sulphate of Ca^{2+} and Mg^{2+} ions. When sodium aluminium silicate is added, Na^+ replaces Ca^{2+} and Mg^{2+} by adsorption. Therefore, water becomes soft.

71. ***(d)*** $\alpha = \dfrac{\Lambda_m^\circ}{\Lambda_m^\infty} = \dfrac{3.907}{390.7} = 0.01$

$$CH_3COOH \rightleftharpoons CH_3COO^- + H^+$$

	CH_3COOH	CH_3COO^-	H^+
Initial conc.	C	0	0
Final conc.	$C(1-\alpha)$	$C\alpha$	$C\alpha$

$\therefore \quad [H^+] = C\alpha = 0.01 \times 0.01 = 10^{-4}$ M

$pH = -\log[H^+] = -\log(10^{-4}) = 4$

72. ***(c)*** We have, the reduction potential of hydrogen electrode, $E_H = -\dfrac{2.303RT}{F}\log\dfrac{1}{[H^+]}$

$= -0.059\ pH = -0.059 \times 3 = -0.177$ V

73. ***(c)*** During the test of nitrogen by Lassaigne's method, the prussian blue coloured ferri-ferro cyanide $Fe_4[Fe(CN)_6]_3$ is obtained.

$$4Fe^{3+} + 3[Fe(CN)_6]^{4-} \longrightarrow \underset{\text{Ferri-ferrocyanide (Prussian blue)}}{Fe_4[Fe(CN)_6]_3}$$

74. ***(a)*** Excess of addition of $ZnCl_2$ to NaOH give soluble Na_2ZnO_2.

$$ZnCl_2 + 2NaOH \longrightarrow 2NaCl + \underset{\text{Insoluble (white ppt.)}}{Zn(OH)_2}$$

$$Zn(OH)_2 + 2NaOH \longrightarrow \underset{\text{Soluble (sodium zincate)}}{Na_2ZnO_2} + 2H_2O$$

75. ***(c)***

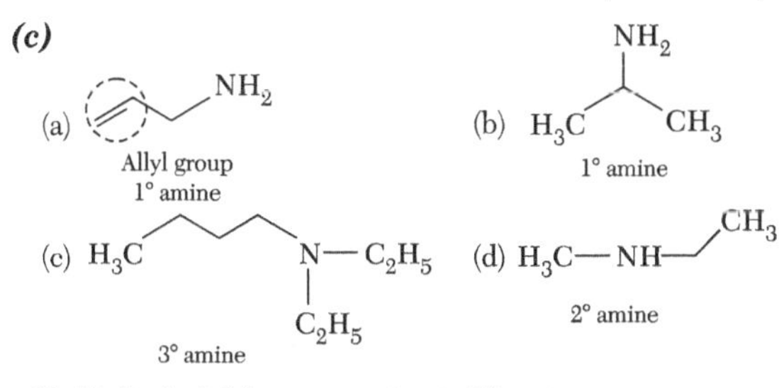

∴N, N-diethyl-1-butane amine is 3° amine.

76. ***(a)*** XeF_4

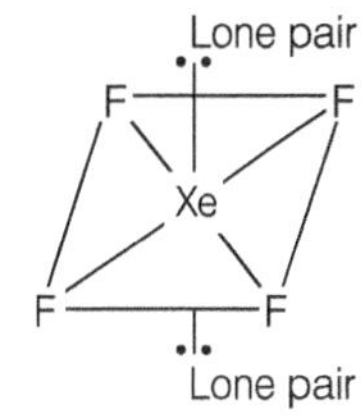

XeF_4 involves sp^3d^2 hybridisation of Xe-atom which gives with two positions occupied by lone pair and thus, XeF_4 is square planar.

77. ***(b)*** Molecular mass of mixture ($NO_2 + N_2O_4$)

$= 2 \times$ vapour density $= 38.3 \times 2$

$= 76.6 \simeq 77$

Total moles/100 g gas

$= \dfrac{100}{77} = 1.3$ moles

Let n is of NO_2 moles, then, $(1.3 - n) = N_2O_4$ mole

Mol. wt. of $NO_2 = 46$, mol. wt. of $N_2O_4 = 92$

$\therefore n \times 46 + (1.3 - n) \times 92 = 100$

$\Rightarrow \quad n = 0.437$

78. ***(d)*** Option (d) is incorrect because the chief ore of Ag is argentite (Ag_2S) and special methods involving use of sodium cyanide is used for their extraction.

79. ***(a)*** The reaction occurs from strong acid to weaker acid direction is spontaneous as HNO_2 is stronger acid than CH_3COOH in aqueous medium. In non-aqueous medium, HF is stronger than HCl and HCl is stronger than CH_3COOH.

Thus, option (a) is non-spontaneous. While reactions (b), (c) and (d) are spontaneous.

80. ***(c)*** Options (a), (b) and (d) are incorrect but (c) is correct.

$[Fe(CN)_6]^{4-}$: Electronic configuration of $Fe^{2+} = 3d^6$

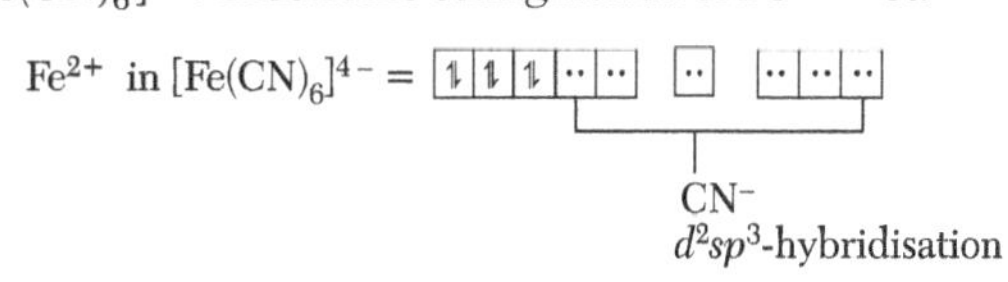

Number of unpaired electrons = 0

Hence, $l_B = \sqrt{n(n+2)}$ BM

$= 0$ BM

a. English Proficiency

81. ***(b)*** Use 'year' in place of 'years'.

82. ***(b)*** Use 'rupee' in place of 'rupees'.

83. ***(b)*** Delete 'an'.

84. ***(c)*** Use 'somebody else's'.

85. ***(b)*** PRSQ

86. ***(b)*** SQPR

87. ***(d)*** SPQR

88. ***(a)*** QPSR

89. ***(d)*** When someone makes something seem larger, more important, better or worse than it really is, then it is called an act of exaggeration. Its synonym is 'amplification'.

90. *(c)* 'Insolvent' means not having enough money to pay debts, buy goods. Its synonym is 'bankrupt'.

91. *(c)* 'Induce' means to persuade someone to do something. So, the answer choice is 'persuades'.

92. *(d)* 'Authentic' means true, real. Among the given options 'genuine' is its obvious synonym.

93. *(c)* The word 'Concede' means 'to admit, often unwillingly that something is true.' Among the given options, 'denied' is its antonym.

94. *(a)* The word 'denied' is the past form of the verb 'deny' that means 'do not admit or accept'. So, 'accepted' is the obvious choice as the answer.

95. *(a)* The word 'coarse', here in this sentence, means 'rough'. So, its antonym would be 'delicate'.

b. Logical Reasoning

96. *(d)* Doctor deals with patient. Similarly, lawyer deals with client.

97. *(d)* As, $8+3=11$

$1+3=4 \Rightarrow 7+3=10$

and $3+3=6 \neq 5$

98. *(b)*

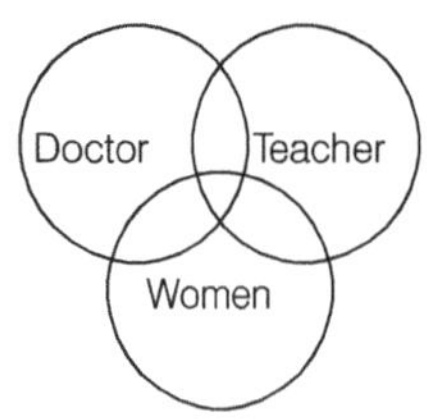

Some women may be doctor and teacher both and some doctors are women and teachers both.

99. *(b)* We have, $(5+4)=9$ and $(5\times4)=20$

and $(3+8)=11$ and $(3\times8)=24$

Similarly $(9+4)=13$ and $(9\times4)=36$

100. *(d)* Only in figure (d), the arrow and the black circle lie at the two ends of the same diameter.

101. *(b)* The figure can be labelled as

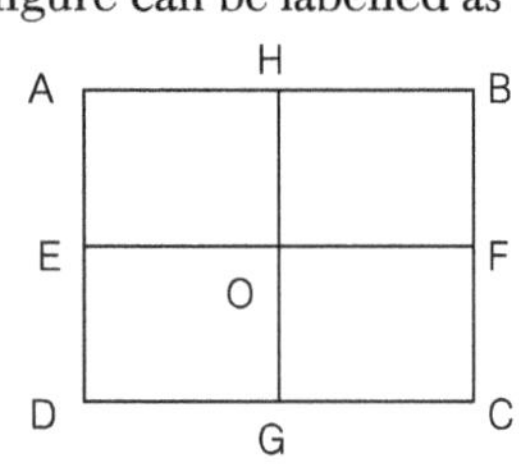

Clearly, there are five squares in the given figure *viz.*, AHOE, HOFB, FOGC, OGDE and ABCD.

Hence, option (b) is correct.

102. *(d)* Figure (X) can be traced out in figure (d) as given below.

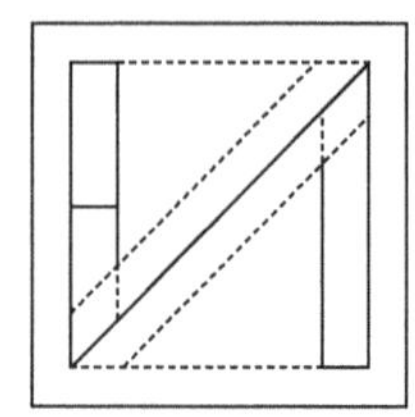

Hence, option (d) is correct.

103. *(a)* Given,

L O F T Y
+0 +1 +0 +1 +0
L P F U Y

Then,

D W A R F
+0 +1 +0 +1 +0
D X A S F

104. *(c)* Today is Thursday +2 = Saturday.

Therefore, tomorrow will be Sunday.

105. *(c)* The pattern of the series is as follows

$$5\times3+1=16$$
$$16\times3+3=51$$
$$51\times3+3=158$$
$$158\times3+3=\boxed{481}$$

Mathematics

106. *(c)* As the median is the positional average and by adding 4 to each observation greater than the median, the positions of the values are unchanged. Hence, the median remains same.

107. *(b)* The equation of the rectangular hyperbola whose axes are asymptotes of the hyperbola $x^2-y^2=a^2$, is

$$xy=c^2, \text{ where } c^2=a^2/2$$

which is one parameter family of curves. On differentiating w.r.t. x, we get

$$x\frac{dy}{dx}+y=0$$

$$\Rightarrow \quad x\frac{dy}{dx}=-y$$

108. *(b)* We have, $f(x) = \max\left\{\sin x, \cos x, \frac{1}{2}\right\}$

$$= \begin{cases} \cos x, & 0 \le x \le \pi/4 \\ \sin x, & \pi/4 \le x \le 5\pi/6 \\ \frac{1}{2}, & \frac{5\pi}{6} \le x \le \frac{5\pi}{3} \end{cases}$$

Let A be the required area. Then,

$$A = \int_0^{5\pi/3} f(x)dx$$

$$= \int_0^{\pi/4} \cos x\,dx + \int_{\pi/4}^{5\pi/6} \sin x\,dx + \int_{5\pi/6}^{5\pi/3} \frac{1}{2}dx$$

$$= [\sin x]_0^{\pi/4} + [-\cos x]_{\pi/4}^{5\pi/6} + \frac{1}{2}[x]_{5\pi/6}^{5\pi/3}$$

$$= \frac{1}{\sqrt{2}} - \left(-\frac{\sqrt{3}}{2} - \frac{1}{\sqrt{2}}\right) + \frac{5\pi}{12} = \frac{5\pi}{12} + \frac{\sqrt{3}}{2} + \sqrt{2}$$

109. *(b)* We have, $\int_0^{[x]} \frac{2^x}{2^{[x]}}dx = \int_0^{[x]} 2^{x-[x]}dx$

Since, $2^{x-[x]}$ is a periodic function with period one unit.

$$\therefore \quad \int_0^{[x]} 2^{x-[x]}dx = [x]\int_0^1 2^{x-[x]}dx = [x]\int_0^1 2^x\,dx$$

$$\Rightarrow \int_0^{[x]} 2^{x-[x]}dx = \frac{[x]}{\log 2}[2^x]_0^1 = \frac{[x]}{\log 2}(2-1) = \frac{[x]}{\log 2}$$

110. *(a)* We have, $I = \int \frac{x^4}{x+x^5}dx = \int \frac{x^4+1-1}{x(1+x^4)}dx$

$$= \int \frac{x^4+1}{x(1+x^4)}dx - \int \frac{1}{x+x^5}dx$$

$$= \int \frac{1}{x}dx - f(x)$$

$$= \log x - f(x) + C$$

111. *(b)* We have, $f(x) = a\log|x| + bx^2 + x$

$$\Rightarrow \quad f'(x) = \frac{a}{x} + 2bx + 1$$

Since, $f(x)$ attains its extremum values at $x = -1, 2$.

$$\therefore \quad f'(-1) = 0 \text{ and } f'(2) = 0$$

$$\Rightarrow \quad -a - 2b + 1 = 0 \text{ and } \frac{a}{2} + 4b + 1 = 0$$

$$\Rightarrow \quad a = 2 \text{ and } b = -1/2$$

112. *(a)* We have, $f(x) = \tan^{-1}\{g(x)\}$

$$\Rightarrow \quad f'(x) = \frac{1}{1+\{g(x)\}^2}\cdot\frac{d}{dx}g(x)$$

For $f(x)$ to be increasing, we must have

$$f'(x) > 0 \Rightarrow \frac{1}{1+g\{(x)\}^2}\cdot g'(x) > 0$$

$$\Rightarrow \quad g'(x) > 0$$

$$\Rightarrow \quad x \in (0, \pi/2)$$

113. *(b)* Consider the function $f(x)$ given by

$$f(x) = \int(\sin x + x\cos x)dx = x\sin x$$

We observe that

$$f(0) = f(\pi) = 0$$

Therefore, 0 and π are the two roots of $f(x) = 0$. and $f'(x)$ exists i.e. $f(x)$ is differentiable in $(0, \pi)$.

Hence, from Rolle's theorem, $f'(x) = 0$ for $x \in (0, \pi)$ i.e. $\sin x + x\cos x = 0$ has atleast one root in $(0, \pi)$.

114. *(c)* We have, $y^2 = x(2-x)^2$ ...(i)

$$\Rightarrow \quad y^2 = x^3 - 4x^2 + 4x$$

On differentiating w.r.t. x, we get

$$2y\frac{dy}{dx} = 3x^2 - 8x + 4$$

$$\Rightarrow \quad \frac{dy}{dx} = \frac{3x^2 - 8x + 4}{2y}$$

$$\Rightarrow \quad \left[\frac{dy}{dx}\right]_{(1,1)} = \frac{3-8+4}{2} = \frac{-1}{2}$$

The equation of the tangent at (1, 1) is

$$y - 1 = \frac{-1}{2}(x-1) \Rightarrow x + 2y - 3 = 0 \quad \text{...(ii)}$$

On solving Eqs. (i) and (ii), we get

$$x = 9/4 \text{ and } y = 3/8$$

Hence, the coordinates of P are (9/4, 3/8).

115. *(c)* We have, $y = |x - x^2| = \begin{cases} x - x^2, & \text{if } 0 \le x \le 1 \\ x^2 - x, & \text{if } x \ge 1 \end{cases}$

Clearly, y is continuous at $x = 1$ but it is not differentiable at $x = 1$, because

$$(\text{LHD at } x = 1) = \left[\frac{d}{dx}(x - x^2)\right]_{\text{at } x=1} = 1 - 2 = -1$$

$$(\text{RHD at } x = 1) = \left[\frac{d}{dx}(x^2 - x)\right]_{\text{at } x=1} = 2 - 1 = 1$$

Hence, $\frac{dy}{dx}$ at $x = 1$ does not exist.

116. *(a)* For $x > 20$, we have

$$f(x) = x - 2 \Rightarrow g(x) = f\{f(x)\}$$

$$= f(x-2) = (x-2) - 2 = x - 4$$

$$\therefore \quad g'(x) = 1, \text{ for } x > 20$$

117. *(a)* We have, $\lim_{x\to\alpha}(1 + ax^2 + bx + c)^{\frac{1}{x-\alpha}}$

$$= e^{\lim_{x\to\alpha}\frac{ax^2+bx+c}{x-\alpha}} = e^{\lim_{x\to\alpha}\frac{a(x-\alpha)(x-\beta)}{(x-\alpha)}} = e^{a(\alpha-\beta)}$$

118. *(b)* We have, $f(x) = ax^2 + bx + c$

$$\therefore \quad f(x+1) - f(x) = 8x + 3, \forall x \in R$$

$$\Rightarrow \{a(x+1)^2 + b(x+1) + c\} - \{ax^2 + bx + c\} = 8x + 3$$

$$\Rightarrow \quad x(2a - 8) + (a + b - 3) = 0$$

$\Rightarrow \quad 2a - 8 = 0$ and $a + b - 3 = 0$

$\Rightarrow \quad a = 4$ and $b = -1$

119. ***(a)*** Let the die be tossed n times and let X denotes the number of sixes obtained in n tosses.

Clearly, p = Probability of getting a six in a toss = $\frac{1}{6}$

Now, $p(X \geq 1) \geq 0.9 \Rightarrow 1 - p(X = 0) \geq 0.9$

$\Rightarrow \quad 1 - (5/6)^n \geq 0.9 \Rightarrow (5/6)^n \leq 0.1$

$\Rightarrow \quad n(\log_{10} 5 - \log_{10} 6) \leq -1$

$$\Rightarrow \quad n \geq \frac{1}{\log_{10} 6 - \log_{10} 5} = \frac{1}{0.7781 - 0.6989} = 12.6$$

$\Rightarrow \quad n = 13, 14, 15, \ldots$

Hence, the least value of n is 13.

120. ***(d)*** We have,

$$3^{3n} - 26^n - 1 = 27^n - 26^n - 1$$

$$\Rightarrow \quad = (1 + 26)^n - 26n - 1$$

$$\Rightarrow \quad = {}^nC_2 \times 26^2 + {}^nC_3 \times 26^3 + \cdots + {}^nC_n \times 26^n$$

Clearly, it is divisible by 26^2, i.e. 676.

121. ***(a)*** Let p, q and r be three propositions given by

$$p : x = 5, \; q : y = -2 \text{ and } r : x - 2y = 9$$

Then, the given statement is $(p \wedge q) \to r$.

Its contrapositive is $\sim r \to \sim (p \wedge q)$

i.e. $\quad \sim r \to \sim p \vee \sim q$

i.e. if $x - 2y \neq 9$, then $x \neq 5$ or $y \neq -2$.

122. ***(b)*** It can be easily obtained for $f : R \to R$ given by $f(x) = \log_a(x + \sqrt{x^2 + 1})$ is a bijection.

Put, $\quad f\{f^{-1}(x)\} = x$

$$\Rightarrow \quad \log_a [f^{-1}(x) + \sqrt{\{f^{-1}(x)\}^2 + 1}] = x$$

$$\Rightarrow \quad f^{-1}(x) + \sqrt{\{f^{-1}(x)\}^2 + 1} = a^x \quad \ldots\text{(i)}$$

$$\Rightarrow \quad \frac{1}{f^{-1}(x) + \sqrt{\{f^{-1}(x)\}^2 + 1}} = a^{-x}$$

After rationalisation of LHS, we have

$$-f^{-1}(x) + \sqrt{f^{-1}(x)\}^2 + 1} = a^{-x} \quad \ldots\text{(ii)}$$

On subtracting Eq. (ii) from Eq. (i), we get

$$2f^{-1}(x) = a^x - a^{-x} \Rightarrow f^{-1}(x) = \frac{1}{2}(a^x - a^{-x})$$

123. ***(d)*** We have, $(a, b) \in R \Leftrightarrow |a^2 - b^2| < 16$

For $a = 1$, $|1 - b^2| < 16$

$$|b^2 - 1| < 16 \Rightarrow -15 < b^2 < 17$$

$\Rightarrow \quad 0 < b^2 < 17 \Rightarrow b = 1, 2, 3, 4$

For $a = 2$ $|4 - b^2| < 16$

$$|b^2 - 4| < 16 \Rightarrow -12 < b^2 < 20$$

$\Rightarrow \quad 0 < b^2 < 20 \Rightarrow b = 1, 2, 3, 4$

For $a = 3 \Rightarrow |9 - b^2| < 16$

$$|b^2 - 9| < 16 \Rightarrow -7 < b^2 < 25$$

$\Rightarrow \quad 0 < b^2 < 25 \Rightarrow b = 1, 2, 3, 4$

For $a = 4$, $|16 - b^2| < 16$

$$|b^2 - 16| < 16 \Rightarrow 0 < b^2 < 32$$

$\Rightarrow \quad b = 1, 2, 3, 4, 5$

and for $a = 5$, $|25 - b^2| < 16$

$$|b^2 - 25| < 16 \Rightarrow 9 < b^2 < 41$$

$\Rightarrow \quad b = 4, 5$

$\therefore \quad R = \{(1, 1), (1, 2), (1, 3), (1, 4), (2, 1), (2, 2), (2, 3), (2, 4), (3, 1), (3, 2), (3, 3), (3, 4), (4, 1), (4, 2), (4, 3), (4, 4), (4, 5), (5, 4), (5, 5)\}$

124. ***(c)*** We have, $p = \dfrac{(x + 2)(2x^5 + 3x^4 + 4x^3 + 5x^2 + 6)}{x^2 + 2x}$

$$\Rightarrow \quad p = 2x^4 + 3x^3 + 4x^2 + 5x + \frac{6}{x}$$

Clearly, $p \in Z^+$ iff $x = 1, 2, 3, 6$

So, A has 4 elements.

125. ***(c)*** $\quad (21.4)^a = 100 \Rightarrow a \log(21.4) = \log 100$

$$\Rightarrow \quad a \log(21.4) = 2 \Rightarrow \log(21.4) = \frac{2}{a}$$

Again, $\quad (0.00214)^b = 100$

$\Rightarrow \quad b \log(0.00214) = 2$

$\Rightarrow \quad b \log(21.4 \times 10^{-4}) = 2$

$$\Rightarrow \quad b = \frac{2}{\log 21.4 - 4} = \frac{2}{\frac{2}{a} - 4} = \frac{1}{\frac{1}{a} - 2}$$

$$\Rightarrow \quad b = \frac{a}{1 - 2a} \Rightarrow \frac{1}{b} = \frac{1 - 2a}{a}$$

$$\Rightarrow \quad \frac{1}{a} - \frac{1}{b} = 2$$

126. ***(a)*** Given, $\cos\theta + \sin\theta = \frac{1}{5}$

$$\Rightarrow \quad 1 + \tan\theta = \frac{1}{5}\sec\theta \qquad [\text{dividing by } \cos\theta]$$

$$\Rightarrow \quad 25 + 50\tan\theta + 25\tan^2\theta = \sec^2\theta$$

$$25 + 50\tan\theta + 25\tan^2\theta = 1 + \tan^2\theta$$

$$\Rightarrow \quad 12\tan^2\theta + 25\tan\theta + 12 = 0$$

$$\Rightarrow \quad 12\tan^2\theta + 16\tan\theta + 9\tan\theta + 12 = 0$$

$$\Rightarrow \quad (4\tan\theta + 3)(3\tan\theta + 4) = 0$$

$$\Rightarrow \quad \tan\theta = \frac{-3}{4} \text{ or } \tan\theta = \frac{-4}{3}$$

Since, $\tan\theta = \frac{-3}{4}$ is rejected as $\tan\theta$ is negative.

So, the angle θ is in second quadrant but it gives $\sec\theta$ as a positive value in second quadrant which is not possible.

Hence, $\tan\theta = \frac{-4}{3}$

127. (d) $\because a = c\sin\theta$

and $b = c\cos\theta$

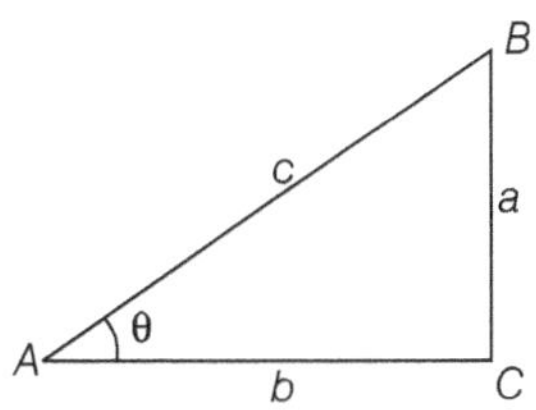

$$\therefore \left(\frac{c}{a}+\frac{c}{b}\right)^2 = \left(\frac{1}{\sin\theta}+\frac{1}{\cos\theta}\right)^2$$

$$= \frac{4(1+\sin 2\theta)}{\sin^2 2\theta}$$

$$= 4\left(\frac{1}{\sin^2 2\theta}+\frac{1}{\sin 2\theta}\right),$$

where $0 < \theta < \frac{\pi}{2}$

$$\Rightarrow \left(\frac{c}{a}+\frac{c}{b}\right)^2_{\min} = 8, \text{ where } 2\theta = 90^\circ$$

128. (b) We have, $\sin^4 x - 2\cos^2 x + a^2 = 0$

$\Rightarrow \quad y^2 - 2(1-y) + a^2 = 0$, where $\sin^2 x = y$

$\Rightarrow \quad y^2 + 2y + a^2 - 2 = 0$

For y to be real, discriminant ≥ 0.

Thus, $4 - 4(a^2 - 2) \geq 0$ or $a^2 \leq 3$

But $\quad \sin^2 x = y$

Therefore, $\quad 0 \leq y \leq 1 \Rightarrow 0 \leq \sqrt{3-a^2} - 1 \leq 1$

$\Rightarrow \quad 1 \leq \sqrt{3-a^2} \leq 2 \Rightarrow 1 \leq 3 - a^2 \leq 4$

$\Rightarrow \quad 2 - a^2 \geq 0 \Rightarrow a^2 \leq 2$

$\Rightarrow \quad -\sqrt{2} \leq a \leq \sqrt{2}$

129. (d) We have, $[\cot^{-1} x] + 2[\tan^{-1} x] = 0$

$\Rightarrow \quad [\cot^{-1} x] = 0, [\tan^{-1} x] = 0$

$\Rightarrow \quad [\cot^{-1} x] = 2, [\tan^{-1} x] = -1$

Now, $\quad [\cot^{-1} x] = 0 \Rightarrow x \in (\cot 1, \infty)$

and $\quad [\tan^{-1} x] = 0 \Rightarrow x \in (0, \tan 1)$

Therefore, $[\cot^{-1} x] = [\tan^{-1} x] = 0$

$\Rightarrow \quad x \in (\cot 1, \tan 1)$

$\quad [\cot^{-1} x] = 2 \Rightarrow x \in (\cot 3, \cot 2]$

and $\quad [\tan^{-1} x] = -1 \Rightarrow x \in [-\tan 1, 0)$

Hence, no such x exists.

Thus, the solution set is $(\cot 1, \tan 1)$.

130. (a) $\Delta = \frac{1}{2}(BC)h$, where h is the distance of vertex A from side BC.

$$\Delta GBC = \frac{\Delta}{3} = \frac{(BC)h}{6}, \text{ where } G \text{ is the centroid.}$$

$$\Rightarrow \quad h = \frac{2\Delta}{BC} = \text{Constant}$$

Thus, distance of vertex A from the side is fixed. Thus, in turn, implies that the distance of centroid from side BC will be fixed. Hence, locus of G will be a line parallel to BC.

131. (a) The coordinates of the circumcentre are

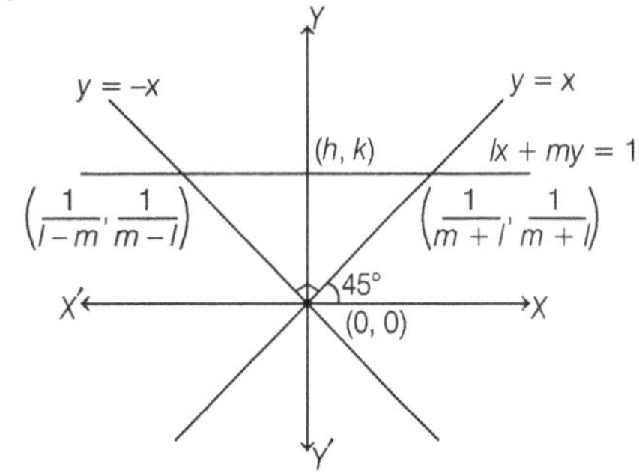

$$\frac{l}{l^2 - m^2}, \frac{m}{m^2 - l^2},$$

Hence, $\quad h = \frac{l}{l^2 - m^2} \quad \ldots\text{(i)}$

$\quad k = \frac{m}{m^2 - l^2} \quad \ldots\text{(ii)}$

On squaring and adding Eqs. (i) and (ii), we get

$$h^2 + k^2 = \frac{l^2 + m^2}{(l^2 - m^2)^2} = \frac{1}{(l^2 - m^2)^2} \quad [\because l^2 + m^2 = 1]$$

$$\therefore \quad \frac{1}{(l^2 - m^2)^2} = (h^2 - k^2)^2$$

$\therefore$ Locus is $\quad x^2 + y^2 = (x^2 - y^2)^2$

132. (b) Let $y = mx$ be a line common to the given pairs of lines.

Then, $am^2 + 2m + 1 = 0$ and $m^2 + 2m + a = 0$

$$\Rightarrow \quad \frac{m^2}{2(1-a)} = \frac{m}{a^2 - 1} = \frac{1}{2(1-a)}$$

$$\Rightarrow \quad m^2 = 1 \text{ and } m = \frac{-a+1}{2}$$

$$\Rightarrow \quad (a+1)^2 = 4 \Rightarrow a = 1, -3$$

But for $a = 1$, the two pairs have both the lines common. So, $a = -3$ and the slope m of the line common to both the pairs, is 1.

Now, $x^2 + 2xy + ay^2 = x^2 + 2xy - 3y^2$

$$= (x - y)(x + 3y)$$

and $ax^2 + 2xy + y^2 = -3x^2 + 2xy + y^2$

$$= -(x - y)(3x + y)$$

So, the equation of the required line is

$$(x + 3y)(3x + y) = 0 \Rightarrow 3x^2 + 10xy + 3y^2 = 0$$

133. ***(d)*** **Circle 1 :** $x^2 + y^2 = 4 \Rightarrow (y - 0)^2 + (x - 8)^2 = (2)^2$

Circle 2 : $x^2 + y^2 - 6\sqrt{3}\,x - 6y + 20 = 0$

$\Rightarrow \quad (x - 3\sqrt{3})^2 + (y - 3)^2 = 4^2$

$C_1 \equiv (0, 0);\ C_2 \equiv (3\sqrt{3}, 3)$ and $r_1 = 2, r_2 = 4$

$C_1C_2 = r_1 + r_2$

The circle touches each other externally.

Equation $C_1C_2 \equiv y - 0 = \dfrac{3}{3\sqrt{3}}(x - c).$

$\Rightarrow \quad xy = \dfrac{x}{\sqrt{3}}$

$\therefore$ Slope of tangent will be $-\sqrt{3}$.

From the point of touch and slope $-\sqrt{3}$, we get.

$$\sqrt{3}x + y - 4 = 0$$

On comparing it with $x\cos\theta + y\sin\theta = 2$,

we get $\quad \theta = \pi/6$

134. ***(b)*** Since, $y = ax^2 - 6x + b$ passes through (0, 2).

$\therefore \quad 2 = a(0)^2 - 6(0) + b$

$\Rightarrow \quad b = 2$

Also, $\quad \dfrac{dy}{dx} = 2ax - 6$

$\therefore \quad \left[\dfrac{dy}{dx}\right]_{x = 3/2} = 2a\left(\dfrac{3}{2}\right) - 6$

$\Rightarrow \quad 3a - 6 = 0$

$\Rightarrow \quad a = 2$

135. ***(d)*** Since, X-axis and Y-axis are perpendicular tangents to the ellipse (0, 0) lies on the director circle and the mid-point of foci (2, 2) is the centre of the circle.

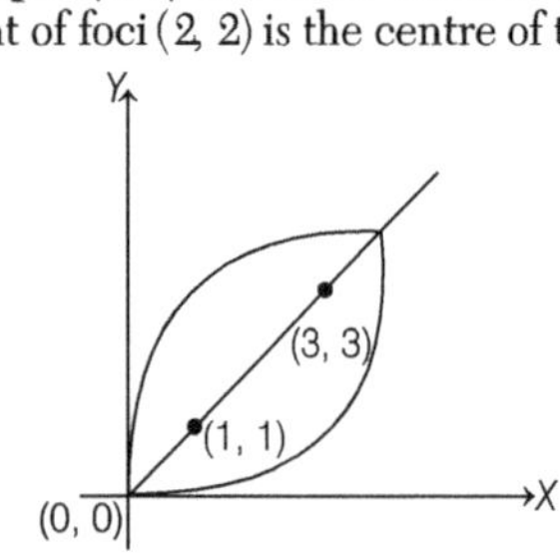

Hence, radius $= 2\sqrt{2}$

Therefore, area $= \pi(r^2) = \pi(2\sqrt{2})^2$

$$= 8\pi \text{ sq units}$$

136. ***(d)*** The equation of the normal to the hyperbola

$\dfrac{x^2}{4} - \dfrac{y^2}{1} = 1$ at $(2\sec\theta, \tan\theta)$ is

$$2x\cos\theta + y\cot\theta = 5$$

The slope of the normal is

$$-2\sin\theta = -1$$

$\Rightarrow \quad \sin\theta = \dfrac{1}{2}$ or $\theta = \dfrac{\pi}{6}$

y-intercept of the normal $= \dfrac{5}{\cot\theta} = \dfrac{5}{\sqrt{3}}$

As it touches the ellipse

$$\frac{x^2}{a^2} + \frac{y^2}{b^2} = 1$$

$\therefore \quad a^2 + b^2 = \dfrac{25}{9}$ [Using $a^2m^2 + b^2 = c^2$]

137. ***(b)*** P (atleast 7 points) = P(7 points) + P(8 points)

[$\because$ atmost 8 points can be scored]

Now, 7 points can be scored by scoring 2 points in 3 matches and 1 point in one match. Similarly, 8 points can be scored by scoring 2 points in each of 4 matches.

$\therefore$ Required probability

$= {}^4C_1 \times [P(2\text{ points})]^3\, P(1\text{ point}) + [P(2\text{ points})]^4$

$= 4(0.5)^3 \times 0.05 + (0.50)^4$

$= 0.0250 + 0.0625 = 0.0875$

138. ***(a)*** Let $A = \begin{bmatrix} a_1 & b_1 & c_1 \\ a_2 & b_2 & c_2 \\ a_3 & b_3 & c_3 \end{bmatrix}$. Now, the system becomes

$a_1x + b_1y + c_1z = 1$, $a_2x + b_2y + c_2z = 0$ and $a_3x + b_3y + c_3z = 0$ representing 3 planes. And three planes meet at (i) exactly one point (ii) no point or (iii) infinite number of points. They cannot meet at exactly two points for any (A).

139. ***(a)*** With 0 and 1 as elements, there are $2 \times 2 \times 2 \times 2 = 16$ determinants of order 2×2 out of

which only $\begin{vmatrix} 1 & 0 \\ 0 & 1 \end{vmatrix}, \begin{vmatrix} 1 & 1 \\ 0 & 1 \end{vmatrix}, \begin{vmatrix} 1 & 0 \\ 1 & 1 \end{vmatrix}$ are the three

determinants whose values are positive. Therefore, the required probability is 3/16.

140. ***(d)*** We have, $(1 + t^2)^{12}(1 + t^{12})(1 + t^{24})$

$$= (1 + t^{12} + t^{24} + t^{36})(1 + t^2)^{12}$$

$\therefore$ Coefficient of $t^{24} = 1 \times$ Coefficient of t^{24} in $(1 + t^2)^{12}$

$+ 1 \times$ Coefficient of t^{12} in $(1 + t^2)^{12}$

$+ 1 \times$ Constant term in $(1 + t^2)^{12}$

$$= {}^{12}C_{12} + {}^{12}C_6 + {}^{12}C_0$$

$$= 1 + {}^{12}C_6 + 1 = {}^{12}C_6 + 2$$

141. ***(c)*** ×_×_×_×_×

The four digits 3, 3, 5, 5 can be arranged at (–) places in $\dfrac{4!}{2!,\ 2!} = 6$ ways.

The five digits 2, 2, 8, 8, 8 can be arranged at (×) place in $\frac{5!}{2!\,3!} = 10$ ways.

$\therefore$ Total number of arrangements is $6 \times 10 = 60$

142. ***(a)*** As AM ≥ GM for positive real numbers, we get

$$\frac{(a+b)+(c+d)}{2} \geq \sqrt{(a+b)(c+d)}$$

$$\Rightarrow \quad M \leq 1$$

Also, $(a+b)(c+d) > 0$ $[\because a, b, c, d > 0]$

$$\therefore \quad 0 \leq M \leq 1$$

143. ***(c)*** Since, $\alpha+\beta, \alpha^2+\beta^2, \alpha^3+\beta^3$ are in GP.

Hence, $(\alpha^2+\beta^2)^2 = (\alpha+\beta)(\alpha^3+\beta^3)$

$$\Rightarrow \quad \alpha^4+\beta^4+2\alpha^2\beta^2 = \alpha^4+\beta^4+\alpha\beta^3+\beta\alpha^3$$

or $\alpha\beta(\alpha^2+\beta^2-2\alpha\beta) = 0$

or $\alpha\beta(\alpha-\beta)^2 = 0$

$$\Rightarrow \quad \alpha.\beta = 0 \text{ or } \alpha-\beta = 0$$

$$\Rightarrow \quad \frac{c}{a} = 0 \text{ or } \alpha = \beta$$

$$\Rightarrow \quad c = 0 \text{ or } \Delta = 0 \text{ [equal roots]}$$

$$\therefore \quad c\Delta = 0$$

144. ***(a)*** $z\bar{z}(\bar{z}^2+z^2) = 350$

Putting $z = x+iy$, we have

$$\Rightarrow \quad (x^2+y^2)(x^2-y^2) = 175$$

$$\Rightarrow \quad (x^2+y^2)(x^2-y^2) = 5 \times 5 \times 7$$

$$x^2+y^2 = 25$$

and $x^2-y^2 = 7$

[as other combinations give non-integral values of x and y]

$\therefore$ $x = \pm 4, y = \pm 3$ $[\because x, y \in I]$

Hence, area is $8 \times 6 = 48$ sq units.

145. ***(d)*** Civen, expression $x^{12}-x^9+x^4-x+1 = f(x)$

For $x < 0$, put $x = -y$, where $y > 0$, we get

$f(x) = y^{12}+y^9+y^4+y+1 > 0$ for $y > 0$

For $0 < x < 1, x^9 < x^4$

$$\Rightarrow \quad -x^9+x^4 > 0$$

Also, $1-x > 0$ and $x^{12} > 0$

$$\Rightarrow \quad x^{12}-x^9+x^4+1-x > 0$$

$$\Rightarrow \quad f(x) > 0$$

For $x > 1$, $f(x) = x(x^3-1)(x^8+1)+1 > 0$

So, $f(x) > 0$ for $-\infty < x < \infty$

146. ***(b)*** We have, $x^2-|x+2|+x > 0$

Case I $x+2 \geq 0$ or $x \geq -2$

$$\therefore \quad x^2-x-2+x > 0 \Rightarrow x^2-2 > 0$$

$$\Rightarrow \quad x \in (-\infty, -\sqrt{2}) \cup (\sqrt{2}, \infty)$$

$$\Rightarrow \quad x \in [-2, -\sqrt{2}) \cup (\sqrt{2}, \infty) \qquad [\because x \geq -2] \quad \ldots\text{(i)}$$

Case II $x+2 < 0 \Rightarrow x < -2$

$$\therefore \quad x^2+x+2+x > 0$$

$\Rightarrow x^2+2x+2 > 0$, which is true for all $x \in R$.

$$\Rightarrow \quad x < -2 \qquad \ldots\text{(ii)}$$

or $x \in (-\infty, -2)$

From Eqs. (i) and (ii),

$$x \in (-\infty, -\sqrt{2}) \cup (\sqrt{2}, \infty)$$

147. ***(a)*** We have, $\mathbf{a} \times (\mathbf{a} \times \mathbf{c}) + 3\mathbf{b} = 0$

$$\Rightarrow \quad (\mathbf{a}\cdot\mathbf{c})\mathbf{a} - (\mathbf{a}\cdot\mathbf{a})\mathbf{c} + 3\mathbf{b} = 0$$

$$\Rightarrow \quad (2\sqrt{3}\cos\theta)\mathbf{a} - 3\mathbf{c} + 3\mathbf{b} = 0$$

$$\Rightarrow \quad (2\cos\theta)\mathbf{a} - \sqrt{3}\mathbf{c} + \sqrt{3}\mathbf{b} = 0$$

$$\Rightarrow \quad |(2\cos\theta)\mathbf{a} - \sqrt{3}\,\mathbf{c}| = |-\sqrt{3}\,\mathbf{b}|$$

$$\Rightarrow \quad |2\cos\theta\,\mathbf{a} - \sqrt{3}\,\mathbf{c}|^2 = |-\sqrt{3}\,\mathbf{b}|^2$$

$$\Rightarrow \quad 4\cos^2\theta\,|\mathbf{a}|^2 + 3|\mathbf{c}|^2 - 4\sqrt{3}\cos\theta(\mathbf{a}\cdot\mathbf{c}) = 3|\mathbf{b}|^2$$

$$\Rightarrow \quad 12\cos^2\theta + 12 - 4\sqrt{3}\cos\theta \times \sqrt{3} \times 2\cos\theta = 3$$

$$\Rightarrow \quad 12\cos^2\theta + 9 - 24\cos^2\theta = 0$$

$$\Rightarrow \quad 12\cos^2\theta = 9$$

$$\Rightarrow \quad \cos^2\theta = \frac{9}{12} = \frac{3}{4}$$

148. ***(c)*** $\mathbf{a}_1 = (\mathbf{a}\cdot\hat{\mathbf{b}})\hat{\mathbf{b}} = \frac{(\mathbf{a}\cdot\mathbf{b})\mathbf{b}}{|\mathbf{b}|^2}$

$$\Rightarrow \quad \mathbf{a}_2 = \mathbf{a} - \mathbf{a}_1 = \mathbf{a} - \frac{(\mathbf{a}\cdot\mathbf{b})\,\mathbf{b}}{|\mathbf{b}|^2}$$

Thus, $\mathbf{a}_1 \times \mathbf{a}_2 = \frac{(\mathbf{a}\cdot\mathbf{b})\mathbf{b}}{|\mathbf{b}|^2} \times \left(\frac{\mathbf{a} - (\mathbf{a}\cdot\mathbf{b})\mathbf{b}}{|\mathbf{b}|^2}\right)$

$$= \frac{(\mathbf{a}\cdot\mathbf{b})(\mathbf{b}\times\mathbf{a})}{|\mathbf{b}|^2}$$

149. ***(c)*** We have, $x^2+y^2 = 0$

$$\Rightarrow \quad x = 0$$

and $y = 0$

which represents the equation of Z-axis.

150. ***(c)*** If the given lines intersect, then they are coplanar.

$$\therefore \quad \begin{vmatrix} 1-2 & 2-3 & 3-1 \\ k & 2 & 3 \\ 3 & k & 2 \end{vmatrix} = 0$$

$$\Rightarrow -(4-3k) + (2k-9) + 2(k^2-6) = 0$$

$$\Rightarrow \quad 2k^2+5k-25 = 0$$

$$\Rightarrow \quad (k+5)(2k-5) = 0$$

$$\Rightarrow \quad k = -5, 5/2$$

BITSAT

Birla Institute of Technology & Science Admission Test

Mock Test 5

Instructions

- There are 150 questions in all. The number of questions in each part is as given below.

	No. of Questions
Part I Physics	1-40
Part II Chemistry	41-80
Part III a. English Proficiency	81-95
b. Logical Reasoning	96-105
Part IV Mathematics	106-150

- All questions are Multiple Choice Questions having four options out of which **only one** is correct.
- Each correct answer fetches 3 marks while incorrect answer has a penalty of 1 mark.
- Time allotted to complete this paper is 3 hrs.

PART I

Physics

1. Suppose the gravitational force varies inversely as the nth power of distance. Then, the time period of a planet in circular orbit of radius R around the sun will be proportional to

a. $R^{(n+1)/2}$ *b.* $R^{(n-1)/2}$

c. R^n *d.* $R^{(n-2)/2}$

2. Two wires are made of the same material and have the same volume. However wire 1 has cross-sectional area A and wire 2 has cross-sectional area 3A. If length of wire 1 increased by Δx on applying force F, how much force is needed to stretch wire 2 by the same amount?

a. $4F$ *b.* $6F$

c. $9F$ *d.* F

3. The satellite of mass m revolving in a circular orbit of radius r around the earth has kinetic energy E. Then, its angular momentum will be

a. $\sqrt{\frac{E}{mr^2}}$ *b.* $\frac{E}{2mr^2}$

c. $\sqrt{2Emr^2}$ *d.* $\sqrt{2Emr}$

4. A galvanometer of resistance 100 Ω gives full scale deflection with 0.01 A current. How much resistance should be connected in parallel to convert it into an ammeter of range 10 A?

a. 0.100 Ω *b.* 1.00 Ω

c. 10.00 Ω *d.* 100.00 Ω

5. A car is moving on a circular road of diameter 50 m with a speed of 5 m/s. It is suddenly accelerated at a rate of 1 m/s^2. If the mass of the car is 500 kg, then the net force acting on the car is

a. 5 N *b.* 1000 N

c. $500\sqrt{2}$ N *d.* $\frac{500}{\sqrt{2}}$ N

6. Hard X-rays for the study of fractures in bones should have a minimum wavelength of 10^{-11} m. The accelerating voltage for electrons in X-ray machine should be

a. <124 kV

b. >124 kV

c. between 60 kV and 70 kV

d. $=100$ kV

7. Natural length of a spring is 60 cm and its spring constant is 4000 N/m. A mass of 20 kg is hung from it. The extension produced in the spring is (Take $g = 9.8\ m/s^2$)

a. 4.9 cm *b.* 0.49 cm *c.* 9.4 cm *d.* 0.94 cm

8. A point source of light is placed 4 m below the surface of water of refractive index 5/3. The minimum diameter of a disc, which should be placed over the source, on the surface of water to cut-off all light coming out of water is

a. infinite *b.* 6 m *c.* 4 m *d.* 3 m

9. What is the maximum acceleration of the particle doing the SHM?

$y = 2\sin\left[\frac{\pi t}{2} + \phi\right]$, where y is in cm

a. $\frac{\pi}{2}$ cm/s² *b.* $\frac{\pi^2}{2}$ cm/s²

c. $\frac{\pi}{4}$ cm/s² *d.* $\frac{\pi^2}{4}$ cm/s²

10. The source of sound generating a frequency of 3 kHz reaches an observer with a speed of 0.5 times the velocity of sound in air. The frequency heard by the observer is

a. 1 kHz *b.* 2 kHz *c.* 4 kHz *d.* 6 kHz

11. Motion of a fluid in a tube is best described by

a. Bernoulli's theorem *b.* Poiseuille's principle

c. Archimedes principle *d.* Stoke's law

12. If a radioactive substance reduces to $\frac{1}{16}$ of its original mass in 40 days, its half-life is

a. 10 days *b.* 20 days

c. 40 days *d.* None of these

13. The velocity of efflux of a liquid through an orifice in the bottom of the tank does not depend upon

a. size of orifice

b. height of liquid

c. acceleration due to gravity

d. density of liquid

14. A neutron with velocity v strikes a stationary deuterium atom, its KE changes by a factor of

a. $\frac{15}{16}$ *b.* $\frac{1}{2}$

c. $\frac{2}{1}$ *d.* None of these

15. The Poisson's ratio of a material is 0.5. If a force is applied to a wire of this material, there is a decrease in the cross-sectional area by 4%. The percentage increase in the length is

a. 1% *b.* 2% *c.* 2.5% *d.* 4%

16. Lenz's law of electromagnetic induction corresponds to the

a. law of conservation of charge

b. law of conservation of energy

c. law of conservation of momentum

d. law of conservation of angular momentum

17. The moment of inertia of the body about an axis is 1.2 kg-m². Initially the body is at rest. In order to produce a rotational kinetic energy of 1500 J, an angular acceleration of 25 rad/s² must be applied about the axis for the duration of

a. 2 s *b.* 4 s *c.* 8 s *d.* 10 s

18. A toroidal solenoid with an air core has an average radius of 15 cm, area of cross-section 12 cm² and 1200 turns. Ignoring the field variation across the cross-section of the toroid the self-inductance of the toroid is

a. 4.6 mH *b.* 6.9 mH

c. 2.3 mH *d.* 9.2 mH

19. The force constant of a spring gun is 50 N/m. If a ball of 20 g when shot by the gun, its spring is compressed by 10 cm, the velocity of the ball is

a. 5 m/s *b.* 15 m/s

c. 25 m/s *d.* 20 m/s

20. 1 g of water (volume 1 cm³) becomes 1671 cm³ of steam when boiled at a pressure of 1 atm. The latent heat of vaporisation is 540 cal/g, then the external work done is
($1\ atm = 1.013 \times 10^5\ N/m^2$)

a. 499.7 J *b.* 40.3 J

c. 169.2 J *d.* 128.57 J

21. A cube has a side of length 1.2×10^{-2} m. Calculate its volume.

a. $1.7 \times 10^{-6} m^3$ *b.* $1.73 \times 10^{-6} m^3$

c. $1.70 \times 10^{-6} m^3$ *d.* $1.732 \times 10^{-6} m^3$

22. A ball is dropped from height h and another from $2h$. The ratio of time taken by the two balls to reach the ground is

a. $1 : \sqrt{2}$ *b.* $\sqrt{2} : 1$ *c.* $2 : 1$ *d.* $1 : 2$

23. The linear momentum p of a body moving in one dimension varies with time t according to the equation $p = a + bt^2$, where a and b are positive constant. The net force acting on the body is

a. a constant

b. proportional to t^2

c. inversely proportional to t

d. proportional to t

24. Which of the following is not an example of perfectly inelastic collision?

a. A bullet fired into a block, if bullet gets embedded into block
b. Capture of an electron by an atom
c. A man jumping onto a moving boat
d. A ball bearing striking another ball bearing

25. If a new planet is discovered rotating around sun with the orbital radius double that of the earth, then what will be its time period? (in earth's days)

a. 1032 *b.* 1023
c. 1024 *d.* 1043

26. If density of earth increases 4 times and its radius becomes half of what it is, our weight will

a. be 4 times its present value
b. be doubled
c. remain same
d. be halved

27. The magnitude of electric field intensity E, such that an electron placed in it would experience an electrical force equal to its weight, is given by

a. mge *b.* $\frac{mg}{e}$ *c.* $\frac{e}{mg}$ *d.* $\frac{e^2}{m^2}g$

28. The work done in placing a charge of 8×10^{-18} C on a capacitor of capacity 100 μF is

a. 32×10^{-32} J *b.* 16×10^{-32} J
c. 3.1×10^{-26} J *d.* 4×10^{-10} J

29. A steady current flows in a metallic conductor of non-uniform cross-section.
The quantity/quantities remaining constant along the whole length of the conductor is/are

a. current, electric field and drift speed
b. drift speed only
c. current and drift speed
d. current only

30. A galvanometer of 50 Ω resistance has 25 divisions. A current of 4×10^{-4} A gives a deflection of one division. To convert this galvanometer into a voltmeter having a range of 25 V, it should be connected with a resistance of

a. 2500 Ω as a shunt
b. 2950 Ω as a shunt
c. 2550 Ω in series
d. 2450 Ω in series

31. The cyclotron frequency of an electron gyrating in a magnetic field of 1 T is approximately

a. 28 MHz *b.* 280 MHz
c. 2.8 GHz *d.* 28 GHz

32. If **M** is magnetic moment and **B** is the magnetic field, then the torque is given by

a. $\mathbf{M} \cdot \mathbf{B}$ *b.* $\frac{|\mathbf{M}|}{|\mathbf{B}|}$
c. $\mathbf{M} \times \mathbf{B}$ *d.* $|\mathbf{M}||\mathbf{B}|$

33. A coil of inductance L is carrying a steady current I. What is the nature of its stored energy?

a. Magnetic *b.* Electrical
c. Both (a) and (b) *d.* Heat

34. Energy conversion in a photoelectric cell takes place from

a. chemical to electrical
b. magnetic to electrical
c. optical to electrical
d. mechanical to electrical

35. If the ionisation potential of helium atom is 24.6 V, the energy required to ionise it will be

a. 24.6 eV *b.* 24.6 V
c. 13.6 eV *d.* 13.6 V

36. Fast neutrons can easily be slowed down by

a. the use of lead shielding
b. passing them through water
c. elastic collision with heavy nuclei
d. applying a strong electric field

37. A film projector magnifies a 100 cm^2 film strip on a screen. If the linear magnification is 4, the area of the magnified film on the screen is

a. 1600 cm^2 *b.* 400 cm^2
c. 800 cm^2 *d.* 6400 cm^2

38. If v_m is the speed of sound in moist air and v_d is the speed of sound in dry air under identical conditions of pressure and temperature, then

a. $v_m > v_d$ *b.* $v_m < v_d$
c. $v_m = v_d$ *d.* $v_m \cdot v_d = 1$

39. A hot and a cold body are kept in vacuum separated from each other. Which of the following cause decreases the temperature of the hot body?

a. Radiation
b. Convection
c. Conduction
d. Temperature remains unchanged

40. An ideal refrigerator has a freezer at a temperature of $-13°$C. The coefficient of performance of the engine is 5. The temperature of the air (to which heat is rejected) will be

a. 325°C *b.* 325 K
c. 39°C *d.* 320°C

PART II

Chemistry

41. The mutual heat of neutralisation of 40 g NaOH and 60 g CH_3COOH will be

a. 57.1 kJ *b.* less than 57.1 kJ
c. more than 57.1 kJ *d.* 13.7 kJ

42. Which has the smallest size?

a. Al^{3+} *b.* Mg^{2+} *c.* P^{5+} *d.* Na^+

43. The treatment of benzene with *iso*-butene in the presence of sulphuric acid gives

a. *iso*-butylbenzene *b.* *tert*-butylbenzene
c. *n*-butylbenzene *d.* no reaction

44. Toluene on reaction with N-bromo-succinimide gives

a. *p*-bromomethylbenzene
b. *o*-bromomethylbenzene
c. phenyl bromomethane
d. *m*-bromomethylbenzene

45. Pinacolone is

a. 2, 3-dimethyl-2, 3-butanediol
b. 3, 3-dimethyl-2-butanone
c. 1-phenyl-2-propanone
d. 1, 1-diphenyl-1, 2-ethandiol

46. A synthetic rubber which is resistant to the action of oils, gasoline and other solvents is

a. buna-S *b.* polyisoprene
c. neoprene *d.* polystyrene

47. Ozone depletion over Antarctica is due to the

a. formation of chlorine nitrate ($ClONO_2$)
b. formation of HCl
c. formation of HOCl and Cl_2 which are converted back into reactive Cl atoms
d. None of the above

48. 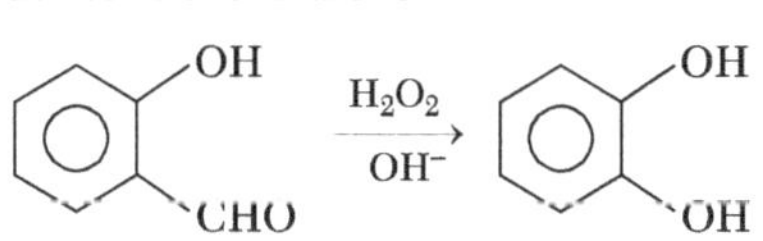

This reaction is called

a. Reimer-Tiemann reaction
b. Liebermann's nitroso reaction
c. Dakin reaction
d. Leader-Manase reaction

49. Which anion is the weakest base?

a. $C_2H_5O^-$ *b.* NO_3^- *c.* F^- *d.* CH_3COO^-

50. K_b for water is 0.52 K/m. Then 0.1 m solution of NaCl will boil approximately at

a. 100.52°C *b.* 100.052°C
c. 101.04°C *d.* 100.104°C

51. One mole of P_2O_5 undergoes hydrolysis as

$$P_2O_5 + H_2O \longrightarrow H_3PO_4$$

The normality of the phosphoric acid formed is (The volume of solution is 1 L.)

a. 2 *b.* 12 *c.* 24 *d.* 4

52. 1 L of a gas is at a pressure of 10^{-6} mm of Hg at 25°C. How many molecules are present in the vessel?

a. 3.2×10^6 *b.* 3.2×10^{13}
c. 3.2×10^{10} *d.* 3×10^4

53. Which of the following has the largest de-Broglie wavelength, given that all have equal velocity?

a. CO_2 molecule *b.* NH_3 molecule
c. Electron *d.* Proton

54. 1 g of U^{235} is converted into UF_6. The radioactivity of UF_6 thus obtained is

a. zero
b. less than that of 1 g of U^{235}
c. more than that of 1 g of U^{235}
d. same as that of 1 g of U^{235}

55. In which of the following molecules S-atom does not assume sp^3 hybridisation?

a. SO_4^{2-} *b.* SF_4 *c.* SF_2 *d.* S_8

56. For the reaction,

$$N_2(g) + 3H_2(g) \rightleftharpoons 2NH_3(g)$$

the unit of *K* in

a. $L\,mol^{-1}$ *b.* $L^2\,mol^{-2}$
c. $mol\,L^{-1}$ *d.* No units

57. A sulphuric acid solution has pH = 3. Its normality is

a. 1/1000 *b.* 1/200 *c.* 1/2000 *d.* 1/100

58. The oxidation number of N and Cl in $NOClO_4$ respectively are

a. +2 and +7 *b.* +3 and +7
c. −3 and +5 *d.* +2 and −7

59. Pyrolusite is a/an

a. oxide ore *b.* sulphide ore
c. carbide ore *d.* Not an ore

60. When potassium ferrocyanide crystals are heated with conc. H_2SO_4, the gas evolved is

a. SO_2 *b.* NH_3
c. CO_2 *d.* CO

61. The product/s of the reaction,

$Na_2CO_3 + CO_2 + H_2O \longrightarrow$ is/are

a. $2NaOH + CO_2$ *b.* $Na_2CO_3 + H_2CO_3$
c. $2NaHCO_3$ *d.* None of these

62. Which among the following is likely to show geometrical isomerism?

a. $CH_3CH{=}NOH$
b. $CH_3CH{=}CH_2$
c. $CH_2{=}CH{-}CH{=}CCl_2$
d. $CH_3C(Cl){=}C(CH_3)_2$

63. A fuel has the same knocking property as a mixture of 70% *iso*-octane (2, 2, 4-trimethyl pentane) and 30% *n*-heptane by volume. The octane number of the fuel is

a. 100 *b.* 70 *c.* 50 *d.* 40

64. Sodium carbonate reacts with SO_2 in aqueous medium to give

a. $NaHSO_3$ *b.* $Na_2S_2O_3$
c. $NaHSO_4$ *d.* Na_2SO_4

65. For a given reaction $t_{1/2} = 1/ka$. The order of this reaction is

a. 0 *b.* 1 *c.* 2 *d.* 3

66. Which of the following compounds will react with two moles of CH_3MgBr?

a. C_2H_5COOH *b.* CH_3COOH
c. $CH_3C{\equiv}CCH_3$ *d.* $HC{\equiv}C{-}CH_2OH$

67. The number of polypeptide chains present in a molecule of haemoglobin is

a. four *b.* one *c.* two *d.* three

68. The pentose sugar in DNA and RNA has the

a. open chain structure *b.* pyranose structure
c. furanose structure *d.* All of these

69. Which of the following is an artificial edible colour?

a. Saffron *b.* Carotene
c. Tetrazine *d.* Melamine

70. The number of unpaired electrons in nickel carbonyl is

a. zero *b.* one
c. four *d.* five

71. The time taken for 90% of a first order reaction to complete is approximately

a. 1.1 times that of half-life
b. 2.2 times that of half-life
c. 3.3 times that of half-life
d. 4.4 times that of half-life

72. The pH of a 0.01 M HCN solution for which pK_a is 4 is

a. 0.47 *b.* 1.2 *c.* 3.0 *d.* 4.0

73. Which of the following does not contain any coordinate bond?

a. H_3O^+ *b.* BF_4^- *c.* HF_2^- *d.* NH_4^+

74. If E is the energy of the combining atomic orbitals, E_1 and E_2 are the energies of the bonding and anti-bonding molecular orbitals formed, then

a. $E - E_1 > E_2 - E$
b. $E - E_1 < E_2 - E$
c. $E - E_1 = E_2 - E$
d. Any one of the above is possible

75. The equilibrium constant (K) for the reaction

$Cu(s) + 2\,Ag^+(aq) \longrightarrow Cu^{2+}(aq) + 2\,Ag(s)$

will be [Given, $E°_{cell} = 0.46$ V]

a. $K_C =$ Antilog 15.6 *b.* $K_C =$ Antilog 2.5
c. $K_C =$ Antilog 1.5 *d.* $K_C =$ Antilog 12.2

76. $E°$ for Fe/Fe^{2+} is +0.44 V and $E°$ for Cu/Cu^{2+} is − 0.32 V. Then, in the cell

a. Cu oxidises Fe^{2+} ion
b. Cu^{2+} oxidises ion
c. Cu reduces Fe^{2+} ion
d. Cu^{2+} ion reduces Fe

77. Which of the following carbon atoms is most electronegative?

$$\overset{\text{III}}{CH_3}{-}\overset{\text{II}}{CH_2}{-}\overset{\text{I}}{C}{\equiv}CH$$

a. I
b. II
c. III
d. All are equally electronegative

78. The reaction/method that does not give an alkane is

a. catalytic hydrogenation of alkenes
b. hydrolysis of alkylmagnesium bromide
c. Kolbe's electrolytic method
d. dehydrohalogenation of an alkyl halide

79. Which of the following will yield a mixture of 2-chlorobutene and 3-chlorobutene on treatment with HCl?

a. $CH_2{=}C{=}CH{-}CH_3$
b. $H_2C{=}\underset{\substack{|\\ CH_3}}{C}{-}CH{=}CH_2$
c. $CH_2{=}CH{-}CH{=}CH_2$
d. $HC{\equiv}C{-}CH{=}CH_2$

80. The well known urinary antiseptic urotropine is formed when formaldehyde reacts with

a. NH_2OH *b.* NH_3
c. $NH_2 \cdot NH_2$ *d.* $C_6H_5NH \cdot NH_2$

PART III

a. English Proficiency

Directions (Q. Nos. 81-83) *Choose the word, which is most nearly the SAME in meaning as the word given in **bold**.*

81. ENLIGHTENMENT

a. Lightness *b.* Insight
c. Twilight *d.* Proficiency

82. SUBLIMITY

a. Despression *b.* Reduction
c. Subversion *d.* Proficiency

83. DEDICATED

a. Devoted *b.* Loyal
c. Submissive *d.* Sincere

Directions (Q. Nos. 84-86) *Choose the word, which is most OPPOSITE in meaning to the word given in bold.*

84. FRAGRANCE

a. Aroma *b.* Perfume
c. Smell *d.* Stink

85. PECULIAR

a. Characteristic *b.* Special
c. Specific *d.* Universal

86. ETERNAL

a. Momentary *b.* Continual
c. Everlasting *d.* Endless

Directions (Q. Nos. 87-90) *In the following questions, a part of the sentence is printed in **bold**. Below are given alternatives to the bold part as* (a), (b) *and* (c) and (d) *which may improve the sentence. Choose the correct alternative. In case no improvement is needed, your answer is* (d).

87. She teaches us grammar, **isn't it**?

a. isn't she? *b.* doesn't she?
c. doesn't it? *d.* No improvement

88. The struggle for independent is gaining **movement** every day.

a. motion *b.* movement
c. momentum *d.* No improvement

89. She is quite without **affection** and has no false pride.

a. admiration *b.* affliction
c. affectation *d.* No improvement

90. I complimented Raju **for** his promotion.

a. with *b.* on
c. about *d.* No improvement

Directions (Q. Nos. 91-95) *In the following passages, the first and the last parts of the sentence are numbered* 1 *and* 6. *The rest of the sentence is split into four parts and named,* P, Q, R *and* S. *These four parts are not given in their proper order. Read the parts and find out which of the four combinations is correct. Then find the correct answer.*

91. 1. Making ourselves
P. our language
Q. part of growing into
R. masters of
S. is an important
6. full manhood or womanhood

a. PSRQ *b.* SQPR
c. RPSQ *d.* PRSQ

92. 1. The very first battle they fought
P. and they had to fall back
Q. cross the border
R. was lost
S. letting the enemy
6. an enter the country

a. RQSP *b.* RPSQ *c.* QRPS *d.* QPRS

93. 1. A nation
P. the material assets it possesses
Q. is not made by
R. and collective determination
S. but by the will
6. of the people

a. PQRS *b.* QPSR *c.* RSPQ *d.* SRPQ

94. 1. When the Governor
P. the bell had rung
Q. justice should be immediately
R. he ordered that
S. found out why
6. done to the horse

a. RSPQ *b.* PQSR *c.* SPRQ *d.* SQRP

95. 1. When you ponder over
P. that the only hope
Q. you will realize
R. of world peace lies
S. the question deeply
6. in the United Nations

a. QRSP *b.* SPQR
c. SQPR *d.* RSPQ

b. Logical Reasoning

96. Which one number is wrong in the given series?

5, 10, 17, 24, 37

a. 10 *b.* 17
c. 24 *d.* 37

97. Find the next two letters in the given series.

E F H K O ?

a. T, Z *b.* Z, T
c. S, Z *d.* T, Y

98. If MONKEY is coded as NNOJFX, what will be the code for TARGET?

a. ZUSFFS *b.* SFFSZU
c. UZSFSF *d.* UZSFFS

99. Among six friends L, M, N, P, Q and S, each having a different height, N is shorter than Q and P but taller than M. S is shorter than only L. Which of the following represents the tallest among six friends?

a. P *b.* Q
c. L *d.* Cannot be determined

100. Manick is fourteenth from the right end in a row of 40 boys. What is his position from the left end?

a. 24th *b.* 25th
c. 26th *d.* 27th

101. The missing number in the given figure is

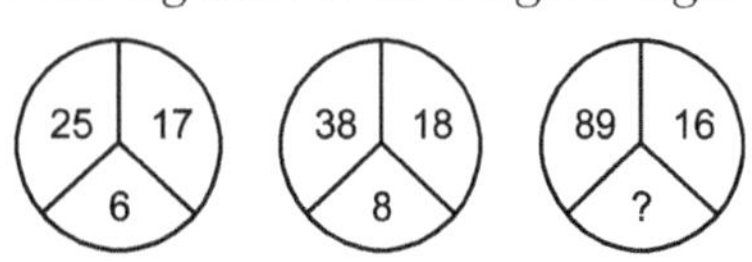

a. 13 *b.* 15
c. 17 *d.* 19

102. Select the combination of numbers so that the letters arranged will form a meaningful word.

H N R C A B
1 2 3 4 5 6

a. 2, 5, 3, 4, 1, 6
b. 3, 5, 6, 4, 1, 2
c. 4, 1, 5, 6, 2, 3
d. 6, 3, 5, 2, 4, 1

103. Which of the given Venn diagrams out of (a), (b), (c) or (d) correctly represents the relationship among the following classes?
Rose, Flower, Lotus

a.

b.

c.
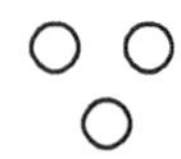
d.

104. A piece of paper is folded and a cut is made as shown below. From the given responses indicate how it will appear when opened?

Question figures

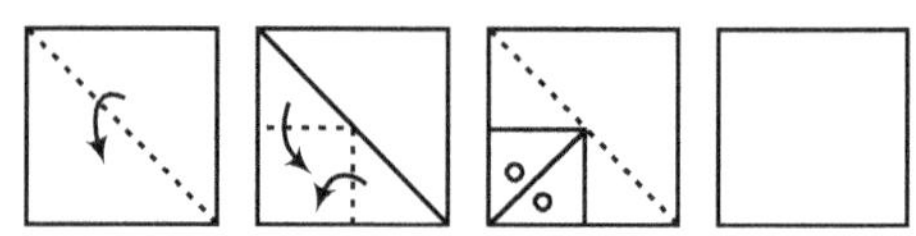

Answer figures

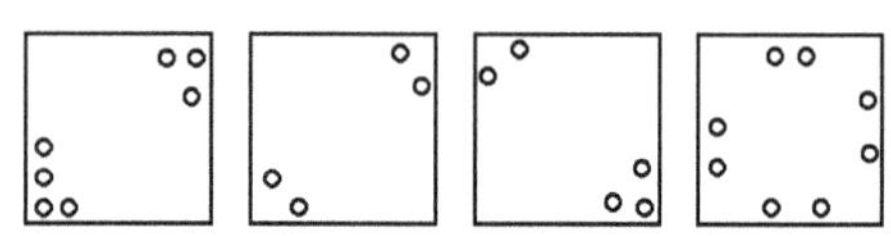

a. *b.* *c.* *d.*

105. Which answer figure will complete the question figure?

Question figure

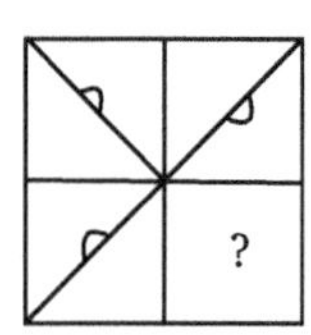

Answer figures

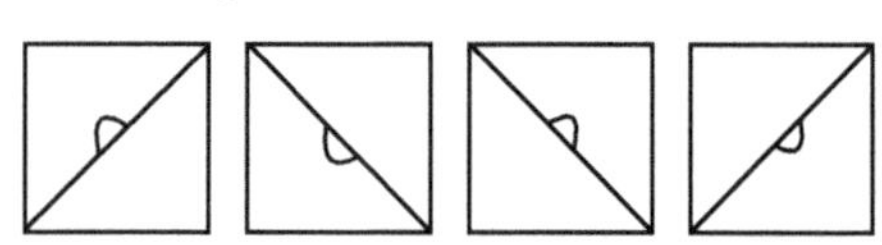

a. *b.* *c.* *d.*

PART IV

Mathematics

106. The equation of the normal to the circle $x^2+y^2=a^2$ at point (x', y') will be

a. $x'y-xy'=0$ b. $xx'-yy'=0$
c. $x'y+xy'=0$ d. $xx'+yy'=0$

107. Equation of the bisector of the acute angle between lines $3x+4y+5=0$ and $12x-5y-7=0$ is

a. $21x+77y+100=0$ b. $99x-27y+30=0$
c. $99x+27y+30=0$ d. $21x-77y-100=0$

108. If $z=\cos\theta+i\sin\theta$, then the value of $z^n+\frac{1}{z^n}$ will be

a. $\sin 2n\theta$ b. $2\sin n\theta$
c. $2\cos n\theta$ d. $\cos 2n\theta$

109. If α and β are the roots of the equation $x^2-2x+4=0$, then the value of $\alpha^n+\beta^n$ will be

a. $i2^{n+1}\sin(n\pi/3)$ b. $2^{n+1}\cos(n\pi/3)$
c. $i2^{n-1}\sin(n\pi/3)$ d. $2^{n-1}\cos(n\pi/3)$

110. If $A=\begin{bmatrix}3 & 1\\ -1 & 2\end{bmatrix}$ and $I=\begin{bmatrix}1 & 0\\ 0 & 1\end{bmatrix}$, then the correct statement is

a. $A^2+5A-7I=O$ b. $-A^2+5A+7I=O$
c. $A^2-5A+7I=O$ d. $A^2+5A+7I=O$

111. The value of the determinant $\begin{vmatrix}a-b-c & 2a & 2a\\ 2b & b-c-a & 2b\\ 2c & 2c & c-a-b\end{vmatrix}$ will be

a. $(a-b-c)(a^2+b^2+c^2)$
b. $(a+b+c)^3$
c. $(a+b+c)(ab+bc+ca)$
d. None of the above

112. If $(1+x)^n=C_0+C_1x+C_2x^2+\ldots+C_nx^n$, then $C_0-C_1+C_2-C_3+\ldots+(-1)^n\cdot C_n$ is equal to

a. 3^n b. 2^n c. 1 d. 0

113. If AM and HM between two numbers are 27 and 12 respectively, then their GM is

a. 9 b. 18 c. 24 d. 36

114. For any two events A and B, if $P(A\cup B)=5/6$, $P(A\cap B)=1/3$, $P(B)=1/2$, then $P(A)$ is

a. 1/2 b. 2/3
c. 1/3 d. None of these

115. A bag contains 3 white and 5 black balls. One ball is drawn at random. Then, the probability that it is white, is

a. $\frac{1}{8}$ b. $\frac{3}{8}$ c. $\frac{5}{8}$ d. $\frac{3}{5}$

116. $\mathbf{a}\cdot(\mathbf{b}\times\mathbf{c})=0$, then the correct statement is

a. out of $\mathbf{a}$, $\mathbf{b}$, $\mathbf{c}$ any two vectors are parallel
b. $\mathbf{a}$, $\mathbf{b}$, $\mathbf{c}$ are coplanar
c. any two are equal $\mathbf{a}$, $\mathbf{b}$, $\mathbf{c}$
d. at least one above statement is correct

117. If $2\hat{i}+\hat{j}-\hat{k}$ and $\hat{i}-4\hat{j}+\lambda\hat{k}$ are perpendicular to each other, then λ is equal to

a. -3 b. -2 c. -1 d. 0

118. If $\frac{d}{dx}(\phi(x))=f(x)$, then $\int_1^2 f(x)\,dx$ is equal to

a. $f(1)-f(2)$ b. $\phi(1)-\phi(2)$
c. $f(2)-f(1)$ d. $\phi(2)-\phi(1)$

119. $\int_0^2 |1-x|\,dx$ is equal to

a. 0 b. 1 c. $\frac{3}{2}$ d. $\frac{1}{2}$

120. $\int \frac{\sin 2x}{\sin^4 x+\cos^4 x}\,dx$ is equal to

a. $2\tan^{-1}(\tan^2 x)+C$
b. $\tan^{-1}(x\tan^2 x)+C$
c. $\tan^{-1}(\tan^2 x)+C$
d. None of the above

121. The function $\sin x+\cos x$ is maximum when x is equal to

a. $\frac{\pi}{6}$ b. $\frac{\pi}{4}$ c. $\frac{\pi}{3}$ d. $\frac{\pi}{2}$

122. $\frac{d}{dx}(x^x)$ is equal to

a. $x^x\log(e/x)$ b. $x^x\log ex$
c. $\log ex$ d. $x^x\log x$

123. $\lim_{x\to 0}\frac{\sin x}{x}$ is equal to

a. 2 b. -1 c. 1 d. 0

124. The set $A=\{x: x\in R, x^2=16 \text{ and } 2x=6\}$ equals

a. ϕ b. $\{14, 3, 4\}$ c. $\{3\}$ d. $\{4\}$

125. In how many ways can 5 prizes be distributed among four students when every student can take one or more prizes?

a. 1024 b. 625 c. 120 d. 600

126. The value of $(\sqrt{5}+1)^5-(\sqrt{5}-1)^5$ is

a. 252 b. 352 c. 452 d. 552

127. The value of
$7\log\left(\frac{16}{15}\right)+5\log\left(\frac{25}{24}\right)+3\log\left(\frac{81}{80}\right)$ is equal to

a. $\log 2$ b. 3
c. 5 d. 7

128. The value of $\frac{2}{1!}+\frac{2+4}{2!}+\frac{2+4+6}{3!}+\ldots\infty$ is

a. e b. $2e$
c. $3e$ d. None of these

129. The sum of the series
$\log_4 2-\log_8 2+\log_{16} 2-\ldots$ is

a. e^2 b. $\log_e 2$
c. $\log_e 3-2$ d. $1-\log_e 2$

130. If the domain of the function $f(x)=x^2-6x+7$ is $(-\infty,\infty)$, then the range of function is

a. $(-\infty,\infty)$ b. $[-2,\infty)$
c. $(-2,3)$ d. $(-\infty,-2)$

131. $\lim_{x\to 0}\frac{\cos(\sin x)-1}{x^2}$ is equal to

a. 1 b. -1 c. $\frac{1}{2}$ d. $-\frac{1}{2}$

132. In order that the function $f(x)=(x+1)^{1/x}$ is continuous at $x=0$, $f(0)$ must be defined as

a. $f(0)=0$ b. $f(0)=e$
c. $f(0)=\frac{1}{e}$ d. $f(0)=1$

133. The function $f(x)=|x|$ at $x=0$ is

a. continuous but non-differentiable
b. discontinuous and differentiable
c. discontinuous and non-differentiable
d. continuous and differentiable

134. The point $(0, 5)$ is closer to the curve $x^2=2y$ at

a. $(2\sqrt{2}, 0)$ b. $(0, 0)$
c. $(2, 2)$ d. None of these

135. The function $f(x)=x^{1/x}$ is

a. increasing in $(1,\infty)$
b. decreasing in $(1,\infty)$
c. increasing in $(1, e)$, decreasing in (e,∞)
d. decreasing in $(1, e)$, increasing in (e,∞)

136. The area bounded by the X-axis and the curve $y=\sin x$ and $x=0$, $x=\pi$ is

a. 1 sq unit b. 2 sq units
c. 0 d. 4 sq units

137. The order and degree of the differential equation $\sqrt{\frac{dy}{dx}}-4\frac{dy}{dx}-7x=0$ are

a. 1 and $\frac{1}{2}$ b. 2 and 1
c. 1 and 1 d. 1 and 2

138. The line $x+y=4$ divides the line joining the points $(-1, 1)$ and $(5, 7)$ in the ratio

a. 2 : 1
b. 1 : 2 internally
c. 1 : 2 externally
d. None of the above

139. The angle between the pair of lines given by equation $x^2+2xy-y^2=0$, is

a. $\frac{\pi}{3}$ b. $\frac{\pi}{6}$ c. $\frac{\pi}{2}$ d. 0

140. The length of tangent from point $(5, 1)$ to the circle $x^2+y^2+6x-4y-3=0$ is

a. 81 b. 29 c. 7 d. 21

141. The length of the latusrectum of the parabola $169\{(x-1)^2+(y-3)^2\}=(5x-12y+17)^2$ is

a. $\frac{14}{13}$ b. $\frac{12}{13}$
c. $\frac{28}{13}$ d. None of these

142. The angle of intersection between the curves $x^2=8y$ and $y^2=8x$ at $(0, 0)$ is

a. $\frac{\pi}{4}$ b. $\frac{\pi}{3}$ c. $\frac{\pi}{6}$ d. $\frac{\pi}{2}$

143. If the centre, one of the foci and semi-major axis of an ellipse be $(0, 0)$, $(0, 3)$ and 5, then its equation is

a. $\frac{x^2}{16}+\frac{y^2}{25}=1$ b. $\frac{x^2}{25}+\frac{y^2}{16}=1$
c. $\frac{x^2}{9}+\frac{y^2}{25}=1$ d. None of these

144. The radius of the director circle to the hyperbola $\frac{x^2}{a^2}-\frac{y^2}{b^2}=1$ is

a. $a-b$ b. $\sqrt{a-b}$
c. $\sqrt{a^2-b^2}$ d. $\sqrt{a^2+b^2}$

145. If projection of any line on coordinate axes 3, 4 and 5, then its length is

a. 12 b. 50
c. $5\sqrt{2}$ d. $3\sqrt{2}$

146. If $\tan\theta = \frac{1}{2}$ and $\tan\phi = \frac{1}{3}$, then the value of $\theta + \phi$ is

a. $\frac{\pi}{6}$ *b.* π *c.* 0 *d.* $\frac{\pi}{4}$

147. If $\sin\theta = \frac{1}{2}$, $\tan\theta = \frac{1}{\sqrt{3}}$, $\forall\, n \in I$, then most general values of θ is

a. $2n\pi + \frac{\pi}{6}, \forall\, n \in I$ *b.* $2n\pi + \frac{\pi}{4}, \forall\, n \in I$

c. $2n\pi + \frac{\pi}{3}, \forall\, n \in I$ *d.* $2n\pi + \frac{\pi}{3}, \forall\, n \in I$

148. The principal value of $\sin^{-1}\left(-\frac{\sqrt{3}}{2}\right)$ is

a. $-\frac{2\pi}{3}$ *b.* $-\frac{\pi}{3}$ *c.* $\frac{4\pi}{3}$ *d.* $\frac{5\pi}{3}$

149. A ladder rests against a wall so that its top touches the roof of the house. If the ladder makes an angle of 60° with the horizontal and height of the house be $6\sqrt{3}$ m, then the length of the ladder is

a. $12\sqrt{3}$ m *b.* 12 m

c. $\frac{12}{\sqrt{3}}$ m *d.* None of these

150. If angles A, B and C are in AP, then $\frac{a+c}{b}$ is equal to

a. $2\sin\left(\frac{A-C}{2}\right)$ *b.* $2\cos\left(\frac{A-C}{2}\right)$

c. $\cos\left(\frac{A-C}{2}\right)$ *d.* $\sin\left(\frac{A-C}{2}\right)$

Answers

Physics

1. (a)	2. (c)	3. (c)	4. (a)	5. (c)	6. (a)	7. (a)	8. (b)	9. (b)	10. (d)
11. (a)	12. (a)	13. (a)	14. (d)	15. (d)	16. (b)	17. (a)	18. (c)	19. (a)	20. (c)
21. (a)	22. (a)	23. (d)	24. (d)	25. (a)	26. (b)	27. (b)	28. (a)	29. (d)	30. (d)
31. (d)	32. (c)	33. (a)	34. (c)	35. (a)	36. (b)	37. (a)	38. (a)	39. (a)	40. (c)

Chemistry

41. (b)	42. (c)	43. (b)	44. (c)	45. (b)	46. (c)	47. (c)	48. (c)	49. (b)	50. (d)
51. (b)	52. (b)	53. (c)	54. (d)	55. (b)	56. (b)	57. (a)	58. (b)	59. (a)	60. (d)
61. (c)	62. (a)	63. (b)	64. (a)	65. (c)	66. (d)	67. (a)	68. (c)	69. (c)	70. (a)
71. (c)	72. (c)	73. (c)	74. (b)	75. (a)	76. (b)	77. (a)	78. (d)	79. (a)	80. (b)

English Proficiency

81. (b)	82. (d)	83. (a)	84. (d)	85. (d)	86. (a)	87. (b)	88. (c)	89. (c)	90. (b)
91. (c)	92. (b)	93. (b)	94. (c)	95. (c)					

Logical Reasoning

96. (c)	97. (a)	98. (d)	99. (c)	100. (d)	101. (b)	102. (d)	103. (a)	104. (d)	105. (b)

Mathematics

106. (a)	107. (c)	108. (c)	109. (b)	110. (c)	111. (b)	112. (d)	113. (b)	114. (b)	115. (b)
116. (b)	117. (b)	118. (d)	119. (b)	120. (d)	121. (b)	122. (b)	123. (c)	124. (a)	125. (a)
126. (b)	127. (a)	128. (c)	129. (d)	130. (b)	131. (d)	132. (b)	133. (a)	134. (d)	135. (c)
136. (b)	137. (d)	138. (b)	139. (c)	140. (c)	141. (c)	142. (d)	143. (a)	144. (c)	145. (c)
146. (d)	147. (a)	148. (b)	149. (b)	150. (b)					

Hints & Solutions

Physics

1. *(a)* The necessary centripetal force required for a planet to move around the sun

= gravitational force exerted on it

$$\frac{mv^2}{R} = \frac{GMm}{R^n}$$

or $$v = \left(\frac{GM}{R^{n-1}}\right)^{1/2}$$

As $$T = \frac{2\pi R}{v} = 2\pi R \times \left(\frac{R^{n-1}}{GM}\right)^{1/2}$$

$$T = 2\pi\left[\frac{R^{\frac{(n+1)}{2}}}{(GM)^{1/2}}\right]$$

$\therefore$ $$T \propto R^{(n+1)/2}$$

2. *(c)* Young's modulus, $Y = \dfrac{FL}{A\Delta L}$

or $$F = \frac{YA\Delta L}{L} = \frac{YA^2\Delta L}{AL}$$

$$= \frac{YA^2\Delta L}{V} = \frac{YA^2\Delta x}{V}$$

where $AL = V$ = Volume of wire, Young's modulus is same as both the wires are made of same material. It is given that both the wires have same volume and same extension in length

$\therefore$ $$\frac{F'}{F} = \frac{A'^2}{A^2} = \frac{(3A)^2}{A^2} = 9$$

$$F' = 9F$$

3. *(c)* KE of a satellite,

$$E = \frac{1}{2}mv^2 \text{ or } mv = \sqrt{2Em}$$

Angular momentum,

$$L = mvr = (\sqrt{2Em}) \times r = \sqrt{2mEr^2}$$

4. *(a)* Shunt is given by

$$S = \frac{I_g \times R_g}{I - I_g} = \frac{0.01 \times 100}{10 - 0.01}$$

$$= \frac{0.01 \times 100}{9.99} = 0.100\ \Omega$$

5. *(c)* Given, $r = \dfrac{50}{2} = 25$ m, $v = 5$ m/s, $m = 500$ kg

$$a_t = 1\text{ m/s}^2,\ a_r = \frac{v^2}{r} = \frac{5 \times 5}{25} = 1\text{ m/s}^2$$

$$\Rightarrow a_{net} = \sqrt{a_t^2 + a_r^2} = \sqrt{1^2 + 1^2} = \sqrt{2}\text{ m/s}^2$$

$\therefore$ $$F = ma_{net} = 500\sqrt{2}\text{ N}$$

6. *(a)* From conservation of energy the electron's kinetic energy equals the maximum photon energy (we neglect the work function ϕ because it is normally so small compared to eV_0)

$$eV_0 = h\nu_{max}$$

$\Rightarrow$ $$eV_0 = \frac{hc}{\lambda_{min}}$$

$$V_0 = \frac{hc}{e\lambda_{min}} = \frac{12400 \times 10^{-10}}{10^{-11}} = 124\text{ kV}$$

Hence, accelerating voltage for electrons in X-ray machine should be less than 124 kV.

7. *(a)* Given, $l = 60$ cm $= 60 \times 10^{-2}$ m

$$m = 20\text{ kg},\quad k = 4000\text{ N/m}$$

The weight hung from the spring

$$= mg = 20 \times 9.8 = 196\text{ N}$$

Suppose x is the extension produced in spring.

Now, force applied by the spring

= downward force on the spring

$\therefore$ $$kx = mg \Rightarrow x = \frac{mg}{k}$$

$$x = \frac{196}{4000} = 0.049\text{ m} = 4.9\text{ cm}$$

8. *(b)* For critical angle case $\sin\theta_C = \dfrac{1}{\mu} = \dfrac{3}{5}$

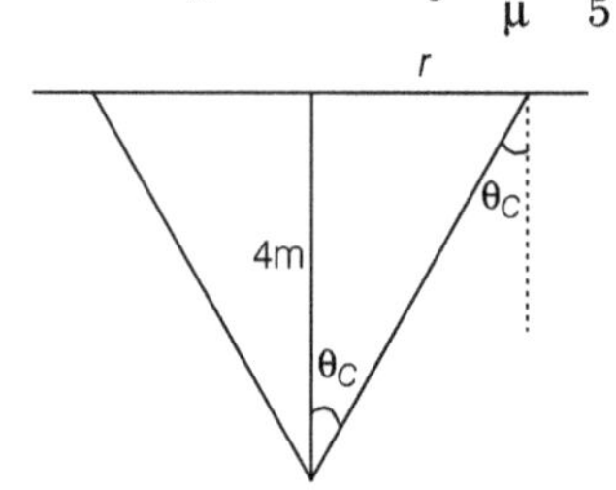

From figure, $\dfrac{r}{4} = \tan\theta_C = \dfrac{3}{4}$

Radius, $r = 3$ m

$\therefore$ Diameter, $d = 6$ m

9. *(b)* $y = 2\sin\left(\dfrac{\pi t}{2} + \phi\right)$

Comparing the equation with the standard equation,

$$y = A\sin(\omega t + \phi)$$

So, we get $\quad A = 2\text{ cm}, \omega = \dfrac{\pi}{2}$

Acceleration of particle is

$$a = \omega^2 x \qquad \text{(numerically)}$$

At $x = +A$, $a = a_{max}$

$$\therefore \quad a_{max} = \omega^2 A$$

$$= \left(\frac{\pi}{2}\right)^2 \times 2 = 2 \times \frac{\pi^2}{4}$$

$$= \frac{\pi^2}{2}\text{ cm/s}^2$$

10. ***(d)*** Apparent frequency heard will be

$$n' = n\left(\frac{v}{v - v_s}\right)$$

where, v = velocity of sound, in air

v_s = velocity of source of sound = $0.5v$

n = frequency = 3 kHz

$$\therefore \quad n' = 3 \times \frac{v}{v - 0.5v}$$

$$= 3 \times \frac{v}{0.5v}$$

$$= 6\text{ kHz}$$

11. ***(a)*** Bernoulli's theorem is applicable only for flow in tube of non-uniform cross-section.

12. ***(a)*** From the formula,

$$N = N_0\left(\frac{1}{2}\right)^n$$

$$\frac{N_0}{16} = N_0\left(\frac{1}{2}\right)^n$$

N_0 = original number of atom

$$\left(\frac{1}{2}\right)^4 = \left(\frac{1}{2}\right)^n$$

$$\Rightarrow \quad n = 4$$

So, 4 half-lives have been passed.

$$\therefore \quad 4T_{1/2} = 40$$

$$T_{1/2} = \frac{40}{4} = 10\text{ days}$$

13. ***(a)*** v = velocity of efflux through an orifice

$$= \sqrt{2gH}$$

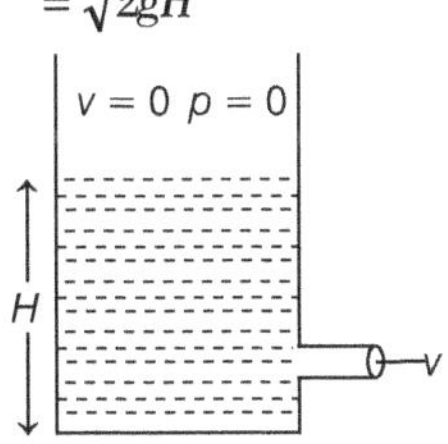

It is independent of the size of orifice.

14. ***(d)*** Neutron velocity = v, mass of neutron, $m_1 = m$

Deuteron contains 1 neutron and 1 proton, so its mass, $m_2 = 2m$

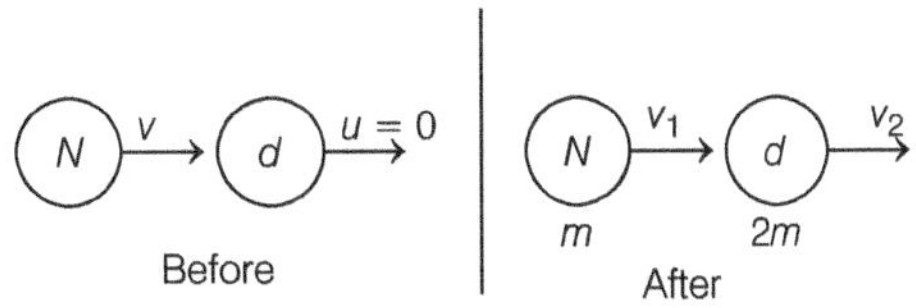

In elastic collision, both momentum and KE are conserved

$$\Rightarrow \quad p_i = p_f$$

$$mv = m_1v_1 + m_2v_2$$

$$mv = mv_1 + 2mv_2 \qquad \ldots\text{(i)}$$

By kinetic energy,

$$\frac{1}{2}mv^2 = \frac{1}{2}mv_1^2 + \frac{1}{2}(2m)v_2^2 \qquad \ldots\text{(ii)}$$

By solving Eqs. (i) and (ii), we get

$$v_1 = \frac{m_1 - m_2}{m_1 + m_2}v + \frac{2m_2}{(m_1 + m_2)}u$$

$$v_1 = \frac{m - 2m}{3m}v$$

$$v_1 = -\frac{v}{3}$$

As, $\quad K_i = \dfrac{1}{2}mv^2$

$$K_f = \frac{1}{2}mv_1^2$$

$$\therefore \quad \frac{K_i - K_f}{K_i} = 1 - \frac{v_1^2}{v^2} = 1 - \frac{1}{9}$$

$$= \frac{8}{9}\text{ fractional change in KE}$$

15. ***(d)*** Poisson's ratio = 0.5

Since, density is constant therefore change in volume is zero, we have

$$V = A \times l = \text{constant}$$

$$\Rightarrow \quad \log V = \log A + \log l$$

or $\quad \dfrac{dA}{A} + \dfrac{dl}{l} = 0 \qquad (\because v = \text{constant})$

$$\Rightarrow \quad \frac{dl}{l} = -\frac{dA}{A} = -(-4\%) = 4\%$$

$\therefore$ Percentage increase in length = 4%

16. ***(b)*** Lenz's law of electromagnetic induction corresponds to the law of conservation of energy.

17. ***(a)*** Rotational kinetic energy = $\dfrac{1}{2}I\omega^2$

According to question,

$$\frac{1}{2}I\omega^2 = 1500$$

$$\frac{1}{2}I(\alpha t)^2 = 1500 \qquad [\because \omega = \alpha t]$$
$$(1.2) \times (25)^2 \times t^2 = 3000$$
$$1.2 \times 625 \times t^2 = 3000$$
$$t^2 = \frac{3000}{1.2 \times 625} = 4$$
$$t = 2\,\text{s}$$

18. ***(c)*** For a solenoid,
$$B = \mu_0 nI$$
$$\because \quad n = \frac{N}{2\pi r}$$
$$\therefore \quad B = \frac{\mu_0 NI}{2\pi r}$$
Flux linked with the solenoid,
$$\phi = NBA$$
$$\phi = \frac{\mu_0 N^2 IA}{2\pi r}$$
$$= \frac{4\pi \times 10^{-7} \times (1200)^2 \times 12 \times 10^{-4}}{2\pi \times 15 \times 10^{-2}}$$
$$L = 2.3 \times 10^{-3}\ \text{H} = 2.3\ \text{mH}$$

19. ***(a)*** By law of conservation of energy,
PE stored in spring = KE of ball
$$\frac{1}{2}kx^2 = \frac{1}{2}mv^2$$
$$\Rightarrow \quad v = \sqrt{\frac{k}{m}}x$$
$$= \sqrt{\frac{50}{20 \times 10^{-3}}}(10 \times 10^{-2})$$
$$= 50 \times 10^{-1} = 5\ \text{m/s}$$

20. ***(c)*** Work done, $W = p\Delta V$
$$= 1.013 \times 10^5 \times (1671 - 1) \times 10^{-6}$$
$$= 1.013 \times 10^5 \times 1670 \times 10^{-6}$$
$$= 169.2\ \text{J}$$

21. ***(a)*** Volume, $V = l^3 = (1.2 \times 10^{-2}\text{m})^3$
$$= 1.728 \times 10^{-6}\text{m}^3$$
Since length (l) has two significant figure, the volume (V) will also have two significant figure.

Therefore, the correct answer is
$$V = 1.7 \times 10^{-6}\text{m}^3$$

22. ***(a)*** Time of fall, $t = \sqrt{\frac{2h}{g}}$ and $t' = \sqrt{\frac{2(2h)}{g}}$
$$\therefore \quad \frac{t}{t'} = \frac{1}{\sqrt{2}}$$

23. ***(d)*** Given, $p = a + bt^2$
$$\frac{dp}{dt} = 2bt$$
$$\because \quad F = \frac{dp}{dt}$$
$$\therefore \quad F = 2bt$$
$$\text{or} \quad F \propto t$$

24. ***(d)*** A ball bearing striking another ball bearing is not an example of perfectly inelastic collision.

25. ***(a)*** By Kepler's third law, $T^2 \propto R^3$
$$\therefore \quad \left(\frac{T_2}{365}\right)^2 = \left(\frac{2R}{R}\right)^3$$
$$T_2 = 365 \times 2\sqrt{2} = 1032\ \text{days}$$

26. ***(b)*** Weight, $w = mg = m\frac{4}{3}\pi GR\rho$
$$\therefore \quad \frac{w'}{w} = \frac{R'\rho'}{R\rho} = \left(\frac{1}{2}\right) \times 4 = 2 \qquad \left[\because \rho' = 4\rho,\ R' = \frac{R}{2}\right]$$
$$\Rightarrow \quad w' = 2w$$

27. ***(b)*** Force on electron,
$$|F| = qE = eE = mg$$
$$\Rightarrow \quad E = \frac{mg}{e}$$

28. ***(a)*** Work done, $W = \frac{q^2}{2C} = \frac{(8 \times 10^{-18})^2}{2 \times 100 \times 10^{-6}}$
$$= 32 \times 10^{-32}\ \text{J}$$

29. ***(d)*** In a metallic conductor of non-uniform cross-section, only the current remains constant along the entire length of the conductor.

30. ***(d)*** Here, $I_g = 25 \times 4 \times 10^{-4}\text{A} = 10^{-2}\ \text{A}$

To convert the galvanometer into a voltmeter we must join a series resistance of
$$R = \frac{V}{I_g} - G = \frac{25}{10^{-2}} - 50$$
$$= 2500 - 50 = 2450\ \Omega$$

31. ***(d)*** Frequency of cyclotron, $\nu = \frac{Bq}{2\pi m}$
$$= \frac{1 \times 1.6 \times 10^{-19}}{2 \times 3.14 \times 9.1 \times 10^{-31}}$$
$$= 2.8 \times 10^{10}\ \text{Hz} = 28\ \text{GHz}$$

32. ***(c)*** Torque, $\tau = \mathbf{M} \times \mathbf{B}$

33. ***(a)*** Energy is stored in an inductor in the form of magnetic potential energy.

34. ***(c)*** In a photoelectric cell, optical energy is being transformed into electrical energy because light photons are being absorbed and photoelectric current is being produced.

Mock Test 5

35. *(a)* Ionisation energy $E = \text{eV}$

$$= 1.6 \times 10^{-19} \times 24.6 \text{ J} = 24.6 \text{ eV}$$

36. *(b)* Fast neutrons can easily be slowed down by passing them through water. Slowing down process is due to collision between neutron and hydrogen nucleus present in water.

37. *(a)* As linear magnification $m = 4$

Hence, a real magnification $m_a = m^2 = (4)^2 = 16$

$\therefore$ Surface area of film image on the screen

$$= 16 \times 100 = 1600 \text{ cm}^2$$

38. *(a)* Under identical pressure and temperature condition, speed of sound in moist air is more than that in dry air, *i.e.*, $v_m > v_d$

39. *(a)* Heat flow through vacuum is possible in radiation mode due to which temperature of hot body falls.

40. *(c)* Given, $T_2 = -13°\text{C} = 260 \text{ K}$ and $\beta = 5$

As, $$\beta = \frac{T_2}{T_1 - T_2}$$

$$5 = \frac{260}{T_1 - 260}$$

$$T_1 = 312 \text{ K} \quad \text{or} \quad T_1 = 312 \text{ K} = 39°\text{C}$$

Chemistry

41. *(b)* 40g of NaOH $= \frac{40}{40} = 1$ mole of NaOH

60g of $CH_3COOH = \frac{60}{60} = 1$ mole of CH_3COOH

Since, acetic acid is a weak acid, some of the heat is utilised to ionise it. So, enthalpy of neutralisation of 1 mole of NaOH by 1 mole of CH_3COOH is less than 57.1 kJ.

Enthalpy of neutralisation of a strong acid by a strong base is always 57.1 kJ.

42. *(c)* Ionic size varies inversely with nuclear charge. Higher the nuclear charge, smaller the radii. Thus, P^{5+} because of the high nuclear charge, has the smallest size.

43. *(b)* The treatment of benzene with *iso*-butene in presence of sulphuric acid gives *t*-butylbenzene as follows

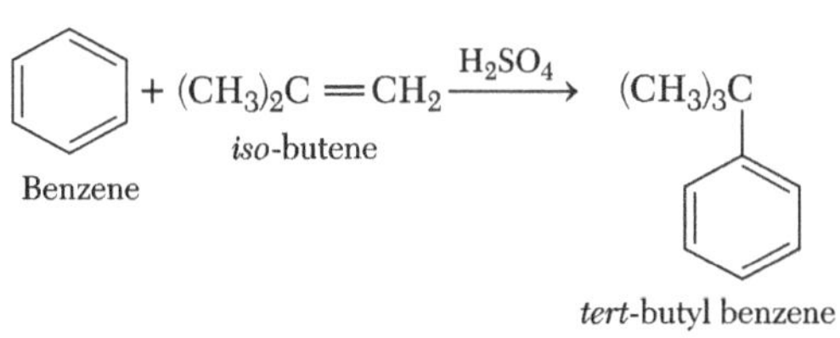

44. *(c)* NBS (N-bromosuccinimide) causes bromination at allylic position.

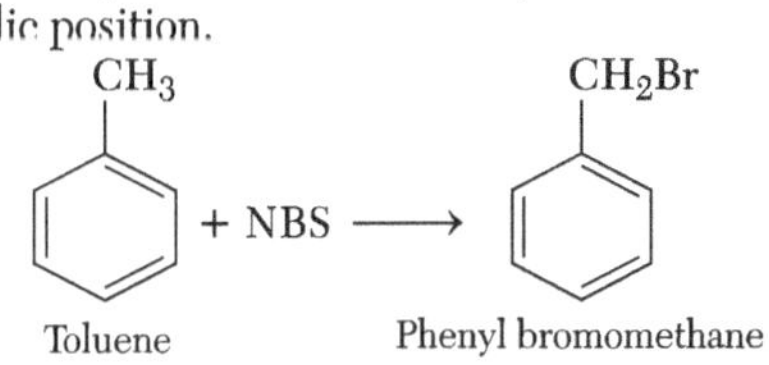

45. *(b)* The structure of pinacolone is

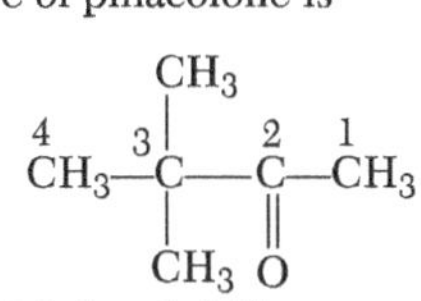

3,3-dimethyl-2-butanone

46. *(c)* Neoprene is synthetic rubber. It is a polymer of chloroprene and is resistant to the action of oils, gasoline and other solvents.

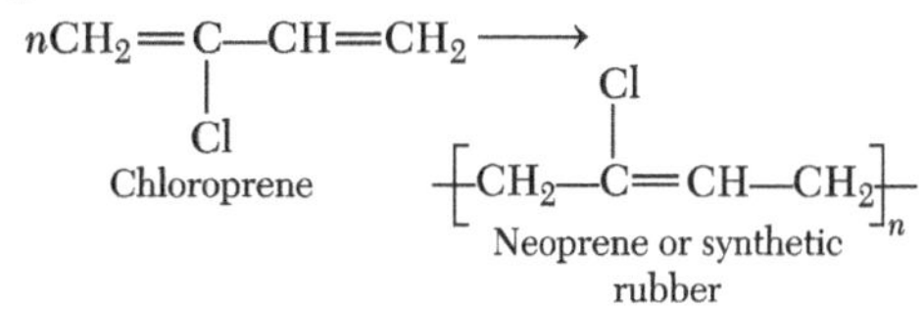

47. *(c)* HOCl and Cl_2 are formed over Antarctica. These are converted back into reactive Cl-atoms which start the chain reaction with O_3 causing its depletion.

48. *(c)* The reduction of —CHO group to —OH by H_2O_2/OH^- is called Dakin reaction.

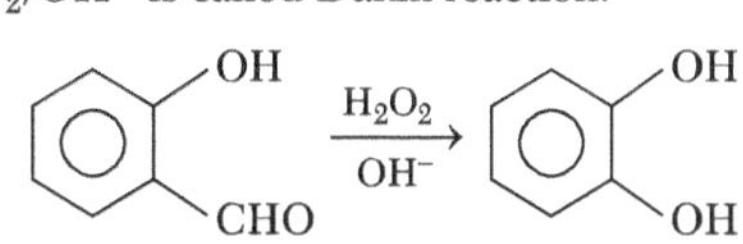

49. *(b)* Conjugated base of a stronger acid is weak.

The corresponding acids of the given conjugated bases are as (in order of acidity)

$$C_2H_5OH < CH_3COOH < HF < HNO_3$$

$\therefore$ The order of basicity is

$$C_2H_5O^- > CH_3COO^- > F^- > NO_3^-$$

Thus, NO_3^- is the weakest base among the given.

50. *(d)* $\Delta T_b = iK_b m$

where, ΔT_b = elevation in b.p. $= T_b^\circ - T_b$

i = van't Hoff factor

m = molality

K_b = b.p elevation constant

$$\Delta T_b = 2 \times 0.52 \times 0.1 = 0.104°\text{C}$$

$$T_b = T^\circ + \Delta T_b = 100 + 0.104°\text{C} = 100.104°\text{C}$$

51. *(b)* $\underset{\text{1 mole}}{2P_2O_5} + 6H_2O \longrightarrow \underset{\text{4 moles}}{4H_3PO_4}$

$\therefore$ 1 L solution contains 4 moles of H_3PO_4.

$\therefore$ Molarity of $H_3PO_4 = 4$ M

Normality = Molarity × Basicity

$= 4 \times 3 = 12$ N

52. ***(b)*** $\frac{p_1V_1}{T_1} = \frac{p_2V_2}{T_2}$

$\Rightarrow \frac{10^{-6} \times 1000}{298} = \frac{760 \times V_2}{273}$

$\Rightarrow V_2 = 1.2 \times 10^{-6}$ cc (at STP)

Number of molecules $= \frac{6.02 \times 10^{23}}{22400} \times 1.2 \times 10^{-6}$

$= 3.2 \times 10^{13}$

53. ***(c)*** $\lambda = h/mv$

For same velocity $\lambda \propto 1/m$

Electron has the least mass, so its wavelength is maximum.

54. ***(d)*** Radioactivity does not depend upon the state of combination, so it remains unaffected.

55. ***(b)*** (a) In $SO_4^{2-} \Rightarrow bp = 4;\ lp = 0\ \therefore\ sp^3$ hybridisation

(b) In $SF_4 \Rightarrow bp = 4;\ lp = 1\ \therefore sp^3d$ hybridisation

(c) In $SF_2 \Rightarrow bp = 2;\ lp = 2\ \therefore\ sp^3$ hybridisation

(d) In $S_8 \Rightarrow bp = 2;\ lp = 2\ \therefore\ sp^3$ hybridisation

Thus, only SF_4 does not have sp^3 hybridisation.

56. ***(b)*** $N_2(g) + 3H_2(g) \rightleftharpoons 2NH_3(g)$

$$K = \frac{[NH_3]^2}{[N_2][H_2]^3} = \frac{\left[\frac{mol}{L}\right]^2}{\left[\frac{mol}{L}\right]\left[\frac{mol}{L}\right]^3}$$

$$= \left[\frac{mol}{L}\right]^{-2} = L^2mol^{-2}$$

57. ***(a)*** For pH = 3,

$[H^+] = 1 \times 10^{-3}$ mol/L

$H_2SO_4 \longrightarrow 2H^+ + SO_4^{2-}$

$[H_2SO_4] = \frac{1 \times 10^{-3}}{2}$

$= \frac{1}{2000}$ M

$N = 2$ M

For H_2SO_4,

Normality $= \frac{2}{2000} = \frac{1}{1000}$ N

58. ***(b)*** $NOClO_4$ is actually $NO^+ClO_4^-$.

Let the oxidation state of N in NO^+ is x.

then, $x + (-2) = +1 \Rightarrow x = +3$

Let the oxidation state of Cl in ClO_4^- is y.

then, $y + (-2) \times 4 = -1$

$y = +7$

59. ***(a)*** Pyrolusite is MnO_2. Thus, it is an oxide ore.

60. ***(d)*** Carbon monoxide gas is evolved when potassium ferrocyanide crystals are heated with conc. H_2SO_4.

$$K_4[Fe(CN)_6] + 6H_2SO_4 + 6H_2O \longrightarrow 2K_2SO_4 + FeSO_4 + 3(NH_4)_2SO_4 + 6CO$$

61. ***(c)*** When sodium carbonate is treated with CO_2 and H_2O, it gets converted into sodium bicarbonate.

$$\underset{\text{Sodium carbonate}}{Na_2CO_3} + CO_2 + H_2O \longrightarrow \underset{\text{Sodium bicarbonate or sodium hydrogen carbonate}}{2NaHCO_3}$$

62. ***(a)*** Among the given, only $CH_3CH{=}NOH$ (oxime) satisfy the conditions essential for exhibiting geometrical isomerism. So, it will exhibit *syn-anti* geometrical isomerism.

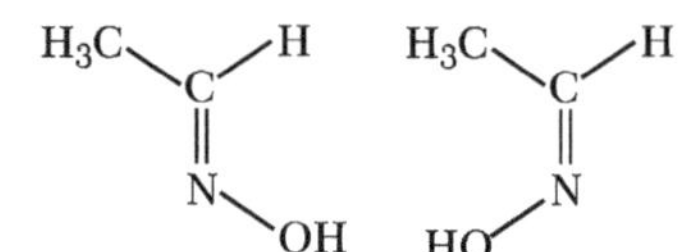

syn *anti*

63. ***(b)*** Octane number is defined as the percentage of *iso*-octane (by volume) in a mixture of *iso*-octane and *n*-heptane, which has the same anti-knocking properties as the fuel under consideration.

Thus, the octane number of the given fuel is 70 as it contains 70% *iso*-octane.

64. ***(a)*** When SO_2 is passed in the sodium carbonate solution, CO_2 gas is evolved and sodium carbonate is converted into $NaHSO_3$ (sodium bisulphite).

$$\underset{\text{Sodium carbonate}}{Na_2CO_3} + 2SO_2 + H_2O \longrightarrow \underset{\text{Sodium bisulphite}}{2NaHSO_3} + CO_2$$

65. ***(c)*** We know that,

$$t_{1/2} \propto \frac{1}{(a)^{n-1}}$$

where, $t_{1/2}$ = half-life

n = order of reactive

a = concentration of reactant

Given, $t_{1/2} \propto \frac{1}{a}$

On comparing, $a^{n-1} = a^1$

$n - 1 = 1$

$n = 1 + 1 = 2$

66. ***(d)*** In prop-2-en-1-ol ($HC{\equiv}C{-}CH_2OH$) two active hydrogen atoms are present, hence it will react with two moles of CH_3MgBr (Grignard reagent).

$$HC{\equiv}C{-}CH_2OH + 2CH_3MgBr \longrightarrow 2CH_4 + BrMgC{\equiv}CCH_2OMgBr$$

67. ***(a)*** Haemoglobin molecule contains four polypeptide chains.

68. ***(c)*** The structure of sugar of DNA (i.e., deoxyribose) and that of RNA (i.e., ribose) is as

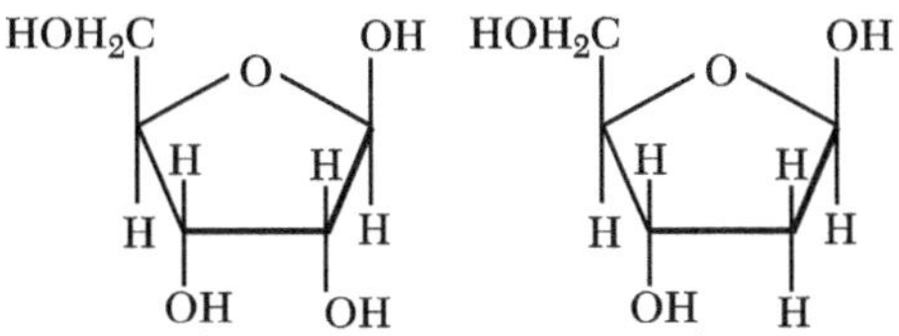

Ribose sugar   Deoxyribose sugar

Thus, it is clear that these have furanose structure.

69. ***(c)*** Tetrazine is an artificial edible colour.

70. ***(a)*** In nickel carbonyl, $Ni(CO)_4$, Ni is present in zero oxidation state.

$Ni = [Ar]\ 3d^8 4s^2$

CO being strong field ligand, shifts electrons from 4*s* to 3*d* orbital.

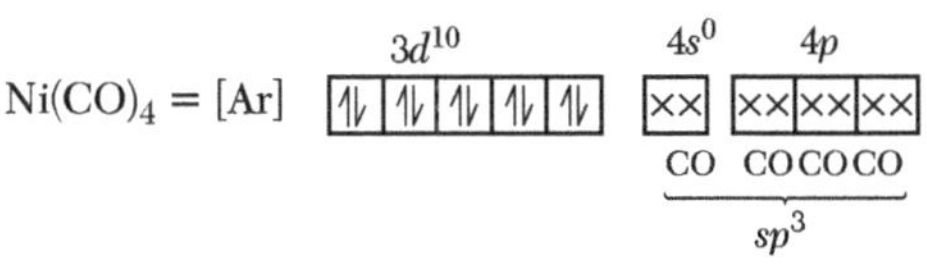

Therefore, number of unpaired electrons in nickel carbonyl is 0.

71. ***(c)*** For a first order reaction,

$$k = \frac{2.303}{t} \log \frac{a}{a-x}$$

$$\Rightarrow \quad t = \frac{2.303 \times t_{1/2}}{0.693} \log \frac{100}{(100-90)} \quad \left[\because k = \frac{t_{1/2}}{0.693}\right]$$

$$= \frac{2.303 \times t_{1/2}}{0.693} \log 10 = 3.3\, t_{1/2}$$

= 3.3 times that of half-life

72. ***(c)*** Given, $pK_a = 4$

$$\therefore \quad K_a - 1 \times 10^{-4}$$

$$[H^+] = \sqrt{K_a \cdot C}$$

$$= \sqrt{1 \times 10^{-4} \times 0.01}$$

$$= \sqrt{10^{-6}} = 10^{-3} M$$

$$pH = -\log[H^+] = -\log 10^{-3}$$

$$= 3$$

73. ***(c)*** Among the given, only HF_2^- has H-bonding $[F—H...F]^-$. Rest all the molecules have coordinate bonds.

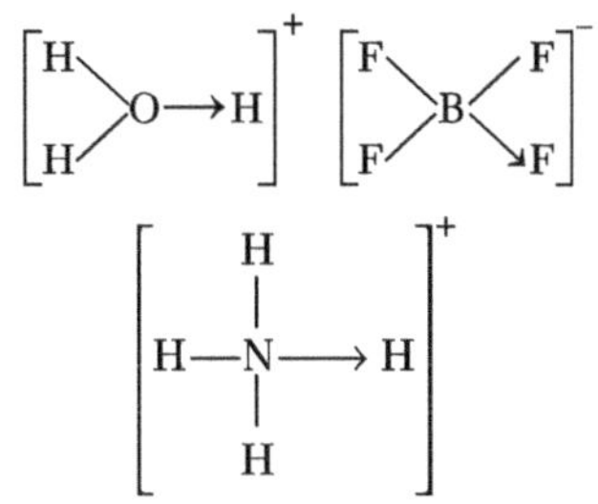

74. ***(b)*** Anti-bonding molecular orbital is raised more in energy $(E_2 - E)$ than the energy by which bonding molecular orbital is lowered $(E - E_1)$.

75. ***(a)***

$$E^\circ = \frac{0.059}{n} \log K_C$$

$$0.46 = \frac{0.059}{2} \log K_C$$

$$\log K_C = \frac{0.46 \times 2}{0.059} = 15.6$$

$$K_C = \text{antilog } 15.6$$

76. ***(b)*** $Fe + Cu^{2+} \longrightarrow Fe^{2+} + Cu$ (Reduction: $Cu^{2+} \to Cu$; Oxidation: $Fe \to Fe^{2+}$)

$$E^\circ_{cell} = E^\circ_{Fe/Fe^{2+}} + E^\circ_{Cu^{2+}/Cu}$$

$$= 0.44 + 0.32 = 0.76\ V$$

77. ***(a)*** $\overset{III}{\underset{sp^3}{CH_3}}—\overset{II}{\underset{sp^3}{CH_2}}—\underset{sp}{C} \equiv \overset{I}{\underset{sp}{CH}}$

Electronegativity ∝ *s*-character.

In *sp* hybrid orbitals, *s*-character = 50%

and in sp^3 hybrid orbitals, *s*-character = 33.3%

Thus, I is the most electronegative.

78. ***(d)*** Dehydrohalogenation of alkyl halide gives alkenes but not alkane.

$$R—CH_2—CH_2X \xrightarrow[-HX]{\text{Alc. KOH}} RCH{=}CH_2$$

79. ***(a)*** $2CH_2{=}C{=}CH—CH_3 + H^+ \longrightarrow$

$$[CH_2{=}\overset{+}{C}—CH_2CH_3 + CH_2{=}CH—\overset{+}{C}HCH_3]$$

$$\xrightarrow{Cl^-} \underset{\text{2-chlorobutene}}{CH_2{=}C(Cl)CH_2CH_3} + \underset{\text{3-chlorobutene}}{CH_2{=}CH—CH(Cl)CH_3}$$

80. ***(b)*** When formaldehyde reacts with ammonia, a well known urinary antiseptic urotropine (also called hexamethylene tetramine) is obtained.

$$\underset{\text{Formaldehyde}}{6HCHO} + \underset{\text{Ammonia}}{4NH_3} \longrightarrow \underset{\text{Urotropine}}{(CH_2)_6N_4} + 4H_2O$$

Mock Test 5

a. English Proficiency

81. ***(b)*** Enlightenment is the action or state of attaining or having attained spiritual knowledge or insight. So, option (b) is its correct synonym.

82. ***(d)*** Sublimity means the state or quality of being sublime, majestic, noble etc. So, option (d) is its correct synonym.

83. ***(a)*** Dedicated means devoted to a task or purpose.

84. ***(d)*** Fragrance means a pleasant sweet smell. So, 'stink' would be its correct antonym, as stink means have a strong unpleasant smell.

85. ***(d)*** Peculiar means belonging to/relating to only particular people or things. So, 'universal' would be its correct anotonym. Universal means relating to or done by all people or things.

86. ***(a)*** Eternal means lasting or existing forever; without end. So 'momentary' (lasting for a very short time) would be its correct opposite meaning word.

87. ***(b)*** 'doesn't she'! should be used to improve the given sentence.

88. ***(c)*** According to the given sentence, 'momentum' should be used in place of given bold word.

89. ***(c)*** Here, 'affectation' should be used in place of 'affection' as affection did not make any sense here.

Affectation means speech or conduct not natural to oneself or an unnatural behaviour, meant especially to 'impress others'.

90. ***(b)*** Here, preposition 'on' should be used in place of 'for'.

91. ***(c)*** The correct sequence is RPSQ.

92. ***(b)*** The correct sequence is RPSQ.

93. ***(b)*** The correct sequence is QPSR.

94. ***(c)*** The correct sequence is SPRQ.

95. ***(c)*** The correct sequence is SQPR.

b. Logical Reasoning

96. ***(c)*** The pattern of the series is

5	10	17	24 [26]	37
↑	↑	↑	↑	↑
$2^2 + 1$	$3^2 + 1$	$4^2 + 1$	$5^2 + 1$	$6^2 + 1$

Hence, 24 is the wrong number.

97. ***(a)*** The pattern of the series is

E → F → H → K → O → [T] → [Z]

+1, +2, +3, +4, +5, +6

98. ***(d)*** As,

			Similarly,			
M	$\xrightarrow{+1}$	N		T	$\xrightarrow{+1}$	U
O	$\xrightarrow{-1}$	N		A	$\xrightarrow{-1}$	Z
N	$\xrightarrow{+1}$	O		R	$\xrightarrow{+1}$	S
K	$\xrightarrow{-1}$	J		G	$\xrightarrow{-1}$	F
E	$\xrightarrow{+1}$	F		E	$\xrightarrow{+1}$	F
Y	$\xrightarrow{-1}$	X		T	$\xrightarrow{-1}$	S

99. ***(c)*** L > S > (P, Q) > N > M

Hence, L is the tallest among six friends.

100. ***(d)*** Position of Manick from left end

$= (40 - 14) + 1 = 26 + 1 = 27$

So, Manick is 27th from the left end.

101. ***(b)*** The sum of the two numbers in the upper part is seven times the number in the lower part.

$(25 + 17) \div 7 = 6$

and $(38 + 18) \div 7 = 8$

So, missing term $= (89 + 16) \div 7$

$= 105 \div 7$

$= 15$

102. ***(d)*** Clearly, the given letters, when arranged in the order of '6, 3, 5, 2, 4, 1' form the word 'BRANCH'.

103. ***(a)*** Rose and Lotus are different type of flowers.

Hence, the diagram will be

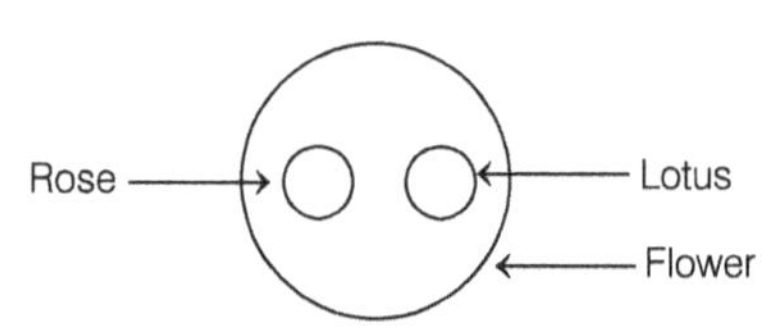

104. ***(d)*** Answer figure (d) will appear when a piece of paper is folded and cut.

105. ***(b)*** Answer figure (b) will complete the question figure.

Mathematics

106. *(a)* Given, $x^2 + y^2 = a^2$

On differentiating w.r.t. x, we get

$$2x + 2y\frac{dy}{dx} = 0$$

$$\Rightarrow \quad \frac{dy}{dx} = -\frac{x}{y}$$

$$\Rightarrow \quad \left(\frac{dy}{dx}\right)_{(x', y')} = -\frac{x'}{y'}$$

$\therefore$ Equation of normal is

$$y - y' = \frac{y'}{x'}(x - x')$$

$$\Rightarrow \quad x'y - y'x' = xy' - y'x'$$

$$\Rightarrow \quad x'y - xy' = 0$$

107. *(c)* Given equations are

$$3x + 4y + 5 = 0$$

and $12x - 5y - 7 = 0$

$$\therefore \quad a_1a_2 + b_1b_2 = 3 \times 12 + 4 \times (-5) = 16 > 0$$

$\therefore$ For acute angle bisector

$$\frac{a_1x + b_1y + c_1}{\sqrt{a_1^2 + b_1^2}} = -\frac{(a_2x + b_2y + c_2)}{\sqrt{a_2^2 + b_2^2}}$$

$$\therefore \quad \frac{3x + 4y + 5}{\sqrt{9 + 16}} = -\frac{(12x - 5y - 7)}{\sqrt{12^2 + (-5)^2}}$$

$$\Rightarrow \quad \frac{3x + 4y + 5}{5} = -\frac{(12x - 5y - 7)}{13}$$

$$\Rightarrow 39x + 52y + 65 = -60x + 25y + 35$$

$$\Rightarrow \quad 99x + 27y + 30 = 0$$

108. *(c)* Given, $z = \cos\theta + i\sin\theta$

$$\therefore \ z^n + \frac{1}{z^n} = (\cos\theta + i\sin\theta)^n + (\cos\theta + i\sin\theta)^{-n}$$

$$= \cos n\theta + i\sin n\theta + \cos n\theta - i\sin n\theta$$

$$= 2\cos n\theta$$

109. *(b)* Since, α and β are the roots of $x^2 - 2x + 4 = 0$

$$\therefore \quad \alpha + \beta = 2 \text{ and } \alpha\beta = 4$$

Now, $(\alpha - \beta) = \sqrt{(\alpha + \beta)^2 - 4\alpha\beta}$

$$= \sqrt{4 - 16} = 2\sqrt{3}\,i$$

On solving, we get

$$2\alpha = 2 + 2\sqrt{3}\,i$$

$$\Rightarrow \quad \alpha = 2\left(\frac{1}{2} + \frac{\sqrt{3}}{2}i\right) = 2\left(\cos\frac{\pi}{3} + i\sin\frac{\pi}{3}\right)$$

and $\beta = \dfrac{2 - 2\sqrt{3}\,i}{2} = 2\left(\cos\dfrac{\pi}{3} - i\sin\dfrac{\pi}{3}\right)$

$$\therefore \ \alpha^n + \beta^n = \left[2\left(\cos\frac{\pi}{3} + i\sin\frac{\pi}{3}\right)\right]^n + \left[2\left(\cos\frac{\pi}{3} - i\sin\frac{\pi}{3}\right)\right]^n$$

$$= 2^n\left[2\cos\frac{n\pi}{3}\right] = 2^{n+1}\cos\frac{n\pi}{3}$$

110. *(c)* Now, $A^2 = \begin{bmatrix} 3 & 1 \\ -1 & 2 \end{bmatrix}\begin{bmatrix} 3 & 1 \\ -1 & 2 \end{bmatrix} = \begin{bmatrix} 8 & 5 \\ -5 & 3 \end{bmatrix}$

$$\therefore \ A^2 - 5A + 7I = \begin{bmatrix} 8 & 5 \\ -5 & 3 \end{bmatrix} - \begin{bmatrix} 15 & 5 \\ -5 & 10 \end{bmatrix} + \begin{bmatrix} 7 & 0 \\ 0 & 7 \end{bmatrix} = \begin{bmatrix} 0 & 0 \\ 0 & 0 \end{bmatrix}$$

111. *(b)* We have, $\begin{vmatrix} a-b-c & 2a & 2a \\ 2b & b-c-a & 2b \\ 2c & 2c & c-a-b \end{vmatrix}$

Applying $R_1 \to R_1 + R_2 + R_3$ and taking $(a+b+c)$ common from R_1, we get

$$\Delta = (a+b+c)\begin{vmatrix} 1 & 1 & 1 \\ 2b & b-c-a & 2b \\ 2c & 2c & c-a-b \end{vmatrix}$$

Applying $C_2 \to C_2 - C_1$ and $C_3 \to C_3 - C_1$,

$$\Delta = (a+b+c)\begin{vmatrix} 1 & 0 & 0 \\ 2b & -(a+b+c) & 0 \\ 2c & 0 & -(a+b+c) \end{vmatrix}$$

$$= (a+b+c)[1(a+b+c)(a+b+c) - 0]$$

$$= (a+b+c)^3$$

112. *(d)* Given, $(1 + x)^n = C_0 + C_1x + C_2x^2 + \ldots + C_nx^n$

Put $x = -1$, we get

$$C_0 - C_1 + C_2 - \ldots + (-1)^n \cdot C_n = 0$$

113. *(b)* Let a and b be two numbers, then

$$\text{AM} = \frac{a+b}{2}$$

$$\Rightarrow \quad 27 = \frac{a+b}{2} \Rightarrow a + b = 54$$

and $\text{HM} = \dfrac{2ab}{a+b} \Rightarrow 12 = \dfrac{2ab}{54}$

$$\Rightarrow \quad ab = 324$$

$$\therefore \quad \text{GM} = \sqrt{ab} = \sqrt{324} = 18$$

114. *(b)* Given, $P(A \cup B) = \frac{5}{6}$, $P(A \cap B) = \frac{1}{3}$, $P(B) = \frac{1}{2}$,

$\because \quad P(A \cup B) = P(A) + P(B) - P(A \cap B)$

$\therefore \quad P(A) = \frac{5}{6} - \frac{1}{2} + \frac{1}{3}$

$= \frac{5-3+2}{6} = \frac{4}{6} = \frac{2}{3}$

115. *(b)* $\therefore$ Required probability $= \frac{{}^3C_1}{{}^8C_1} = \frac{3}{8}$

116. *(b)* Since, scalar triple product is zero, then **a**, **b** and **c** are coplanar.

117. *(b)* Since, given vectors are perpendicular.

$\therefore \quad (2\hat{i} + \hat{j} - \hat{k}) \cdot (\hat{i} - 4\hat{j} + \lambda\hat{k}) = 0$

$\Rightarrow \quad 2 - 4 - \lambda = 0$

$\Rightarrow \quad \lambda = -2$

118. *(d)* $\int_1^2 f(x)dx = \int_1^2 \frac{d}{dx}(\phi(x))dx$

$= [\phi(x)]_1^2 = \phi(2) - \phi(1)$

119. *(b)* $\int_0^2 |1-x|dx = \int_0^1 (1-x)\,dx + \int_1^2 (x-1)\,dx$

$= \left[x - \frac{x^2}{2}\right]_0^1 + \left[\frac{x^2}{2} - x\right]_1^2$

$= 1 - \frac{1}{2} + \left[2 - 2 - \left(\frac{1}{2} - 1\right)\right]$

$= \frac{1}{2} + \frac{1}{2} = 1$

120. *(d)* Let $I = \int \frac{\sin 2x}{\sin^4 x + \cos^4 x} dx$

$= \int \frac{\sin 2x}{(\sin^2 x + \cos^2 x)^2 - 2\sin^2 x \cos^2 x} dx$

$= \int \frac{\sin 2x}{1 - \frac{1}{2}(\sin 2x)^2} dx$

$[\because \sin 2\theta = 2\sin\theta\cos\theta \text{ and } \sin^2\theta + \cos^2\theta = 1]$

$= \int \frac{\sin 2x}{1 - \frac{1}{2}(1 - \cos^2 2x)} dx$

$= \int \frac{\sin 2x}{\frac{1}{2}(1 + \cos^2 2x)} dx$

Put $\cos 2x = t \Rightarrow -2\sin 2x\,dx = dt$

$\therefore \quad I = -\int \frac{dt}{1+t^2} = -\tan^{-1} t + C$

$= -\tan^{-1}(\cos 2x) + C$

121. *(b)* Let $y = \sin x + \cos x = \sqrt{2}\left(\frac{\sin x + \cos x}{\sqrt{2}}\right)$

$= \sqrt{2}\left(\sin\left(\frac{\pi}{4} + x\right)\right)$

$[\because \sin(x+y) = \sin x \cos y + \cos x \sin y]$

Here, y will be maximum when

$\sin\left(\frac{\pi}{4} + x\right) = 1$

But $\quad \sin\frac{\pi}{2} = 1$

So, $\quad \frac{\pi}{4} + x = \frac{\pi}{2}$

Hence, $\quad x = \frac{\pi}{4}$

122. *(b)* Let $y = x^x$

$\Rightarrow \quad \log y = x \log x$

On differentiating w.r.t. x, we get

$\frac{1}{y}\frac{dy}{dx} = \frac{x}{x} + \log x$

$\Rightarrow \quad \frac{dy}{dx} = y(1 + \log x)$

$\Rightarrow \quad = x^x(\log e + \log x)$

$= x^x(\log ex)$

123. *(c)* $\lim_{x \to 0} \frac{\sin x}{x} = \lim_{x \to 0} \frac{\cos x}{1}$

[Using L' Hospital's rule]

$= \cos 0 = 1$

124. *(a)* Since, $x^2 = 16 \Rightarrow x = \pm 4$

and $\quad 2x = 6 \Rightarrow x = 3$

Hence, no value of x is satisfied.

$\therefore \quad A = \phi$

125. *(a)* Required number of ways

$= 4^5 = 2^{10} = 1024$

126. *(b)* Since, $(\sqrt{5}+1)^5 - (\sqrt{5}-1)^5$

$= 2\{{}^5C_1(\sqrt{5})^4 + {}^5C_3(\sqrt{5})^2 + {}^5C_5 \cdot 1\}$

$= 2\{5 \times 25 + 50 + 1\}$

$= 2(176) = 352$

127. *(a)* $\therefore 7\log\left(\frac{16}{15}\right) + 5\log\left(\frac{25}{24}\right) + 3\log\left(\frac{81}{80}\right)$

$[\because \log m^n = n \log m]$

$= \log\left[\left(\frac{16}{15}\right)^7 \cdot \left(\frac{25}{24}\right)^5 \cdot \left(\frac{81}{80}\right)^3\right] = \log 2$

128. *(c)* Here, $T_n = \frac{n(n+1)}{n!} = \frac{n-1+2}{(n-1)!}$

$$= \frac{1}{(n-2)!} + \frac{2}{(n-1)!}$$

$$\therefore \quad S = \sum_{n=1}^{\infty} T_n = \sum_{n=2}^{\infty} \frac{1}{(n-2)!} + 2\sum_{n=1}^{\infty} \frac{1}{(n-1)!}$$

$$= e + 2e = 3e$$

129. *(d)* $\log_4 2 - \log_8 2 + \log_{16} 2 - \ldots$

$$= \frac{\log 2}{\log 4} - \frac{\log 2}{\log 8} + \frac{\log 2}{\log 16} \ldots\ldots \qquad \left[\because \log_a b = \frac{\log b}{\log a}\right]$$

$$= \frac{\log 2}{\log 2^2} - \frac{\log 2}{\log 2^3} + \frac{\log 2}{\log 2^4} \ldots\ldots$$

$$= \frac{\log 2}{2\log 2} - \frac{\log 2}{3\log 3} + \frac{\log 2}{4\log 2} \ldots\ldots \qquad [\because \log m^n = n \log m]$$

$$= \frac{1}{2} - \frac{1}{3} + \frac{1}{4} \ldots\ldots = \left(\frac{1}{2} - \frac{1}{3} + \frac{1}{4} - \ldots\right) - 1 + 1$$

$$= -\left(1 - \frac{1}{2} + \frac{1}{3} - \frac{1}{4} + \ldots\right) + 1$$

$$= 1 - \log(1+1) = 1 - \log_e 2$$

130. *(b)* Now, $x^2 - 6x + 7 = (x-3)^2 - 2$

It is obvious that minimum value is -2 and maximum value is ∞.

131. *(d)* $\lim_{x \to 0} \frac{\cos(\sin x) - 1}{x^2} = \lim_{x \to 0} \frac{-2\sin^2\left(\frac{\sin x}{2}\right)}{x^2}$

$$[\because \cos 2x = 1 - 2\sin^2 x]$$

$$= -2 \lim_{x \to 0} \frac{\sin^2\left(\frac{\sin x}{2}\right) \cdot \left(\frac{\sin x}{2}\right)^2}{\left(\frac{\sin x}{2}\right)^2 \times x^2}$$

$$= -2(1)^2 \frac{1}{4} = -\frac{1}{2}$$

132. *(b)* $\lim_{x \to 0} f(x) = f(0) = \lim_{x \to 0} (1+x)^{1/x}$

$$= e^{\lim_{x \to 0} (1+x-1) \times \frac{1}{x}}$$

$$= e^{\lim_{x \to 0} x \times \frac{1}{x}} = e^{\lim_{x \to 0} 1} = e$$

133. *(a)* Given, $f(x) = |x|$

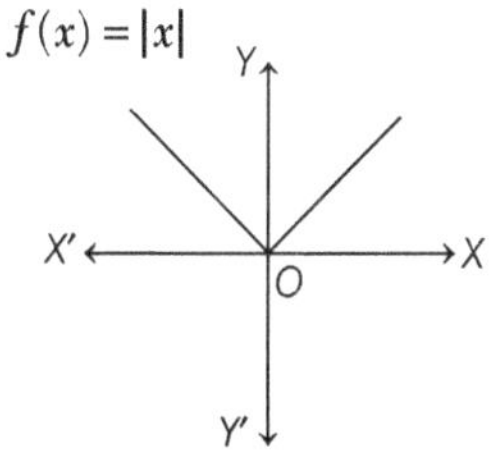

It is clear from the graph, $f(x)$ is continuous but non-differentiable at $x = 0$.

134. *(d)* Let any point on the curve be (h, k).

Then, $\quad h^2 = 2k$

$\therefore$ Distance $D = \sqrt{h^2 + (k-5)^2}$

$$\Rightarrow \quad D = \sqrt{2k + (k-5)^2}$$

$$\Rightarrow \frac{dD}{dk} = \frac{1}{2\sqrt{2k + (k-5)^2}} \times \{2 + 2(k-5)\} = 0$$

$$\therefore \quad k = 4$$

Then, point will be $(\pm 2\sqrt{2}, 4)$.

135. *(c)* Let $y = x^{\frac{1}{x}} \Rightarrow \log y = \frac{1}{x} \log x$

$$\Rightarrow \quad \frac{1}{y}\frac{dy}{dx} = \frac{1}{x^2} - \frac{\log_e x}{x^2}$$

$$\Rightarrow \quad \frac{dy}{dx} = x^{1/x}\left(\frac{1 - \log_e x}{x^2}\right)$$

For $\quad 1 < x < \infty, \; x^{1/x} > 0$

and $\quad \frac{1 - \log_e x}{x^2} > 0$ in $(1, e)$

and $\quad \frac{1 - \log_e x}{x^2} < 0$ in (e, ∞)

Hence, $f(x)$ is increasing in $(1, e)$ and decreasing in (e, ∞).

136. *(b)* $\therefore$ Required area $= \int_0^{\pi} \sin x \, dx$

$$= [-\cos x]_0^{\pi}$$

$$= [+1 + 1]$$

$$= 2 \text{ sq units}$$

137. *(d)* Given, $\sqrt{\frac{dy}{dx}} - 4\frac{dy}{dx} - 7x = 0$

On squaring, we get

$$\frac{dy}{dx} = 16\left(\frac{dy}{dx}\right)^2 + 49x^2 + 56x\frac{dy}{dx}$$

Here, order is 1 and degree is 2.

138. *(b)* The straight line $ax + by + c = 0$ divides the line joining the points (x_1, y_1) and (x_2, y_2) in the ratio

$$-\frac{ax_1 + by_1 + c}{ax_2 + by_2 + c}$$

$\therefore$ Required ratio $= -\frac{(-1+1-4)}{(5+7-4)} = \frac{4}{8} = \frac{1}{2}$

Hence, the lines divides the points in 1:2 internally.

139. *(c)* Given equation is $x^2 + 2xy - y^2 = 0$

As we know lines are perpendicular if coefficient of x^2 and coefficient of y^2 adds to zero.

Here, $a + b = 1 - 1 = 0$

Hence, pair of lines are perpendicular.

140. *(c)* Length of tangent $= \sqrt{S_1}$

$$= \sqrt{x^2 + y^2 + 6x - 4y - 3}$$

$$= \sqrt{5^2 + 1^2 + 30 - 4 - 3} \qquad [\because x = 5,\ y = 1]$$

$$= \sqrt{49} = 7$$

141. *(c)* Given equation can be rewritten as

$$(x-1)^2 + (y-3)^2 = \left(\frac{5x - 12y + 17}{13}\right)^2$$

$$\Rightarrow \qquad SP = PM$$

Here, focus is (1, 3), directrix

$$5x - 12y + 17 = 0$$

$\therefore$ The distance of the focus from the directrix

$$= \left|\frac{5 - 36 + 17}{\sqrt{25 + 144}}\right|$$

$$= \frac{14}{13}$$

$$\therefore \text{Latusrectum} = 2 \times \frac{14}{13} = \frac{28}{13}$$

142. *(d)* It is clear that angle between the curves

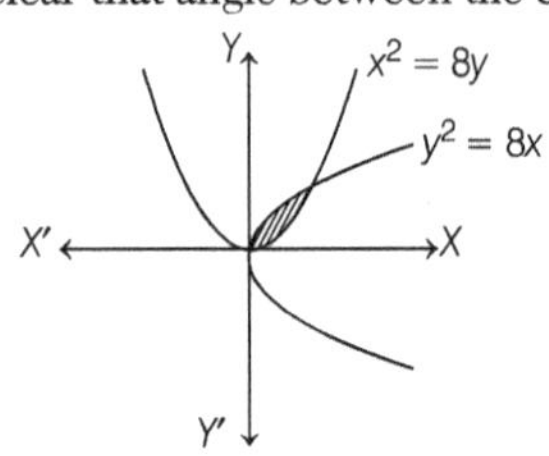

$= \text{angle between the } X\text{-axis and } Y\text{-axis} = \frac{\pi}{2}$.

143. *(a)* Given centre (0, 0), focus (0, 3), $b = 5$

$$\Rightarrow \qquad be = 3 \Rightarrow e = \frac{3}{5}$$

$$\therefore \qquad a = b\sqrt{1 - e^2} = 5\sqrt{1 - \frac{9}{25}} = 4$$

Hence, required equation is

$$\frac{x^2}{16} + \frac{y^2}{25} = 1$$

144. *(c)* The equation of director circle to the hyperbola is

$$x^2 + y^2 = a^2 - b^2$$

$$\therefore \quad \text{Radius} = \sqrt{a^2 - b^2}$$

145. *(c)* Let d be the length of line, then projection on X-axis $= dl = 3$, projection of Y-axis $= dm = 4$ and projection on Z-axis $= dn = 5$

$$\text{Now,} \qquad d^2(l^2 + m^2 + n^2) = 50$$

$$\Rightarrow \qquad d^2 = 50 \Rightarrow d = 5\sqrt{2}$$

146. *(d)* $\therefore \tan(\theta + \phi) = \dfrac{\tan\theta + \tan\phi}{1 - \tan\theta \cdot \tan\phi}$

$$= \left(\frac{1}{2} + \frac{1}{3}\right) \Big/ \left(1 - \frac{1}{2}\cdot\frac{1}{3}\right) = 1$$

$$\left[\because \tan\theta = \frac{1}{2} \text{ and } \tan\phi = \frac{1}{3}\right]$$

$$\Rightarrow \qquad \theta + \phi = \frac{\pi}{4}$$

147. *(a)* $\because \sin\theta = \dfrac{1}{2} = \sin\dfrac{\pi}{6}$

$$\Rightarrow \qquad \theta = \frac{\pi}{6},\ \pi - \frac{\pi}{6}$$

$$\text{and} \qquad \tan\theta = \frac{1}{\sqrt{3}} = \tan\left(\frac{\pi}{6}\right)$$

$$\Rightarrow \qquad \theta = \frac{\pi}{6},\ \pi + \frac{\pi}{6}$$

$\therefore$ Common value of θ is $\dfrac{\pi}{6}$.

$\therefore$ General value of θ is

$$2n\pi + \frac{\pi}{6},\ \forall\, n \in I$$

148. *(b)* $\therefore \ \sin^{-1}\left(-\dfrac{\sqrt{3}}{2}\right) = \sin^{-1}\left(\sin\left(-\dfrac{\pi}{3}\right)\right) = -\dfrac{\pi}{3}$

149. *(b)* $\therefore$ Length of ladder, $AC = \dfrac{6\sqrt{3}}{\sin 60°} = 12$ m

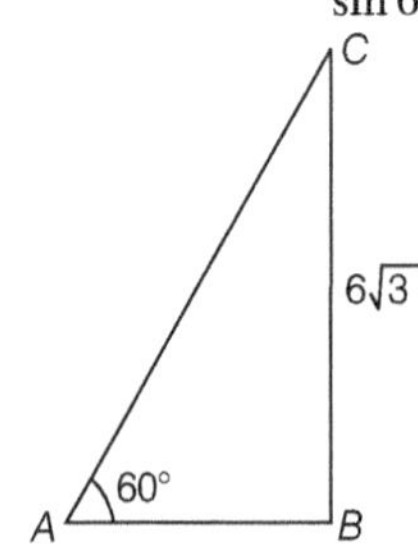

150. *(b)* Since, $A + B + C = \pi$

$$\text{But} \qquad 2B = A + C$$

$$\therefore \qquad 3B = \pi \Rightarrow B = \pi/3$$

$$\therefore \qquad \frac{a + c}{b} = \frac{\sin A + \sin C}{\sin B}$$

$$= \frac{2\sin\left(\frac{A+C}{2}\right)\cos\left(\frac{(A-C)}{2}\right)}{\sin\frac{\pi}{3}}$$

$$= \frac{2\sin\frac{\pi}{3}\cos\left(\frac{A-C}{2}\right)}{\sin\frac{\pi}{3}} = 2\cos\left(\frac{A-C}{2}\right)$$

www.ingramcontent.com/pod-product-compliance
Lightning Source LLC
Chambersburg PA
CBHW080208170726
48282CB00030B/1040
* 9 7 8 9 3 2 5 7 9 9 8 4 4 *